Arithmetic Operations:

$$ab + ac = a(b+c)$$

$$\frac{a}{b} + \frac{c}{d} = \frac{ad+bc}{bd}$$

$$\frac{a+b}{c} = \frac{a}{c} + \frac{b}{c}$$

$$\frac{\left(\frac{a}{b}\right)}{\left(\frac{c}{d}\right)} = \frac{ad}{bc}$$

$$a\left(\frac{b}{c}\right) = \frac{ab}{c}$$

$$\frac{a-b}{c-d} = \frac{b-a}{d-c}$$

$$\frac{ab+ac}{a} = b+c$$

$$\frac{\left(\frac{a}{b}\right)}{c} = \frac{a}{bc}$$

$$\frac{a}{\left(\frac{b}{c}\right)} = \frac{ac}{b}$$

Exponents and Radicals:

$$a^0 = 1 \; (a \neq 0)$$

$$\frac{a^x}{a^y} = a^{x-y}$$

$$\left(\frac{a}{b}\right)^x = \frac{a^x}{b^x}$$

$$\sqrt[n]{a^m} = a^{m/n} = (\sqrt[n]{a})^m$$

$$a^{-x} = \frac{1}{a^x}$$

$$(a^x)^y = a^{xy}$$

$$\sqrt{a} = a^{1/2}$$

$$\sqrt[n]{ab} = \sqrt[n]{a}\,\sqrt[n]{b}$$

$$a^x a^y = a^{x+y}$$

$$(ab)^x = a^x b^x$$

$$\sqrt[n]{a} = a^{1/n}$$

$$\sqrt[n]{\left(\frac{a}{b}\right)} = \frac{\sqrt[n]{a}}{\sqrt[n]{b}}$$

Algebraic Errors to Avoid:

$\dfrac{a}{x+b} \neq \dfrac{a}{x} + \dfrac{a}{b}$ (To see this error, let $a = b = x = 1$.)

$\sqrt{x^2+a^2} \neq x + a$ (To see this error, let $x = 3$ and $a = 4$.)

$a - b(x-1) \neq a - bx - b$ (Remember to distribute negative signs. The equation should be $a - b(x-1) = a - bx + b$.)

$\dfrac{\left(\frac{x}{a}\right)}{b} \neq \dfrac{bx}{a}$ (To divide fractions, invert and multiply. The equation should be

$$\frac{\frac{x}{a}}{b} = \frac{\frac{x}{a}}{\frac{b}{1}} = \left(\frac{x}{a}\right)\left(\frac{1}{b}\right) = \frac{x}{ab}.)$$

$\sqrt{-x^2+a^2} \neq -\sqrt{x^2-a^2}$ (We can't factor a negative sign outside of the square root.)

$\dfrac{\cancel{a}+bx}{\cancel{a}} \neq 1 + bx$ (This is one of many examples of incorrect cancellation. The equation should be

$$\frac{a+bx}{a} = \frac{a}{a} + \frac{bx}{a} = 1 + \frac{bx}{a} \, .)$$

$\dfrac{1}{x^{1/2} - x^{1/3}} \neq x^{-1/2} - x^{-1/3}$ (This error is a sophisticated version of the first error.)

$(x^2)^3 \neq x^5$ (The equation should be $(x^2)^3 = x^2 x^2 x^2 = x^6$.)

Conversion Table:

1 centimeter =	0.394 inches	1 joule =	0.738 foot-pounds	1 mile =	1.609 kilometers
1 meter =	39.370 inches	1 gram =	0.035 ounces	1 gallon =	3.785 liters
=	3.281 feet	1 kilogram =	2.205 pounds	1 pound =	4.448 newtons
1 kilometer =	0.621 miles	1 inch =	2.540 centimeters	1 foot-lb =	1.356 joules
1 liter =	0.264 gallons	1 foot =	30.480 centimeters	1 ounce =	28.350 grams
1 newton =	0.225 pounds	=	0.305 meters	1 pound =	0.454 kilograms

CALCULUS

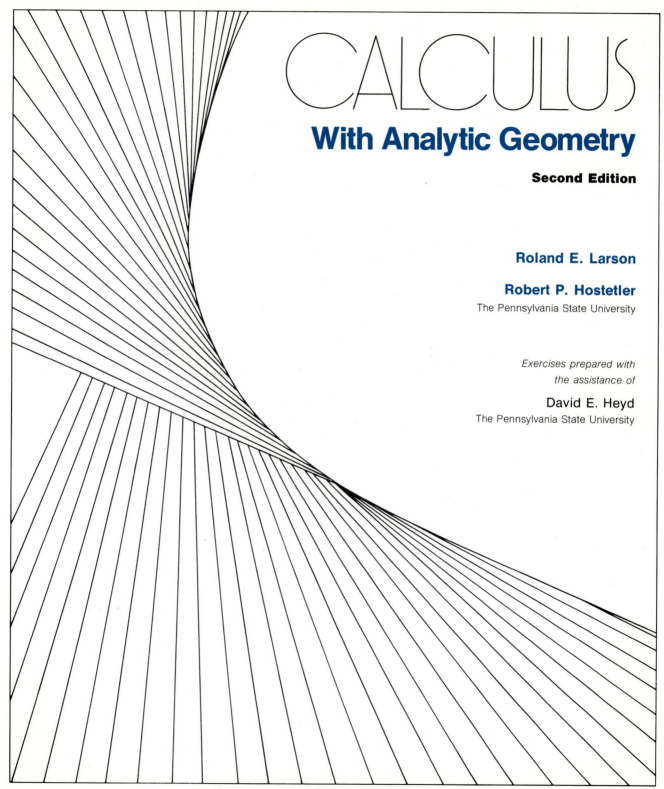

CALCULUS
With Analytic Geometry
Second Edition

Roland E. Larson

Robert P. Hostetler
The Pennsylvania State University

*Exercises prepared with
the assistance of*

David E. Heyd
The Pennsylvania State University

D. C. HEATH AND COMPANY   Lexington, Massachusetts Toronto

Preface

"The last thing one settles in writing a book is what one should put in first." —Blaise Pascal 1623–1662

Calculus, Second Edition, is designed for use in a beginning calculus course for students in mathematics, engineering, the physical sciences, the biological sciences, and economics. In writing this book, we were guided by two primary objectives that have crystalized over many years of teaching calculus. For the student, our objective was to write in a precise, readable manner with the basic concepts and rules of calculus clearly defined and demonstrated. For the instructor, our objective was to design a comprehensive teaching instrument that employs proven pedagogical techniques, thus freeing the instructor to make the most efficient use of classroom time.

Changes in the Second Edition

This edition has benefited greatly from the suggestions given by instructors who used the first edition and by our own students. New material was added, primarily two new chapters: Line Integrals (Chapter 17) and Differential Equations (Chapter 19). The teachability of this edition is improved by substantial changes in exposition in the following areas: inverse functions (Section 1.6), applications of integration (Sections 7.5 and 7.7), exponential and logarithmic functions (Sections 8.1 and 8.2), integration techniques (Sections 10.5, 10.7, and 10.11), and infinite series (Sections 18.2 and 18.3). In addition we made minor exposition changes throughout that we feel will enhance students' understanding of calculus. We added summaries of important rules and concepts at strategic locations.

Some sections were relocated to allow for a more natural flow of topics. Newton's Method was moved from Chapter 4 to Chapter 5 into the more natural setting of differentials. The introduction of line and surface integrals in the first edition as a part of Chapter 16 was completely revised and is now part of the new Chapter 17. In addition, we split the second section of infinite series into two sections and added the Root Test.

Consistent with our emphasis on the value of sketches in problem solving, we revised and refined many of the figures. New figures were added in both the examples and the problem sets. The most noticeable change in the figures is the inclusion of computer-generated figures in

Chapters 14, 15, and 16. This addition greatly enhances the three-dimensional perception of the given examples.

One of the most significant changes in this edition is the inclusion of additional applications problems as both new examples and new exercises. We believe that these new applications will increase the motivation of the students as they experience more variety in the uses of calculus.

Where suggested by our users we added drill exercises to improve skill development. This is most noticeable in the exercise sets for the first derivative test, applications of the derivative, volumes of revolution, review of trigonometric and inverse trigonometric functions, and multiple integrals.

Order of Presentation

In determining the order of topics in the text, we were influenced by three basic considerations: a natural flow from one topic to another, sufficient precalculus material to meet the needs of the average calculus student, and early presentation of differentiation and integration. The nineteen chapters readily adapt to semester or quarter systems.

In each system both differentiation *and* integration can be introduced in the first course of the sequence. There is some flexibility in the order and depth in which the various chapters can be covered. For instance, some users of the first edition introduced trigonometric functions (Sections 9.1 and 9.2) just after Differentiation (Chapter 3). If this is done, appropriate exercises from Chapter 9 can then be assigned as students progress through Chapters 4 and 5. In a similar way, the integrals of trigonometric functions (Section 9.3) can be introduced after Chapter 6. We give a guide to early coverage of the trigonometric functions on p. xii. Integration (Chapter 6) can be studied before Applications of Differentiation (Chapter 5), and Infinite Series (Chapter 18), and Differential Equations (Chapter 19) can be studied any time after Chapter 10. Moreover, a minimal coverage of Conic Sections (Chapter 11) and Polar Coordinates (Chapter 12) will suffice for Chapters 13–16. The new Chapter 17, Line Integrals, naturally follows Chapter 16, Multiple Integrals. However, exclusion of Chapter 17 will not interrupt the flow of material since the remaining chapters on infinite series and differential equations are presented in such a way that they can be studied at any time after Chapter 10.

Instructional Features

During our combined teaching experience of over thirty years and, more specifically, during the nine years of development and classroom testing of this text, we have found the following features to be valuable aids to the teaching and the learning of calculus.

1. *Format* Each section is developed in a systematic manner beginning with a statement of purpose and an introductory paragraph designed to motivate the study of the topics in the section. This is followed by concisely stated definitions and theorems, detailed examples, and summary statements that identify the *working* rules of calculus. The page design is open and uncluttered, with the definitions and theorems highlighted through the use of boxes and a second color.

2. *Definitions and Theorems* Special care has been taken to state the definitions and theorems simply without sacrificing correctness. For example, we chose not to obscure the presentation of L'Hôpital's Rule with two or four separate cases. Instead, we combined the cases into a single statement that emphasizes the basic hypothesis as well as the basic result. (See Section 8.5.)

3. *Proofs* Most instructors agree that the inclusion of proofs in a calculus course is sound pedagogy. A student who understands the proof of a theorem is not apt to use it incorrectly by overlooking its basic hypothesis. The dilemma faced by most calculus instructors is that there is barely enough time to demonstrate the *use* of each theorem, let alone discuss its proof. With this in mind, we have chosen to include only those proofs that we have found to be both instructive and within the grasp of a beginning calculus student. For example, we do not include a detailed proof of the Chain Rule.

4. *Graphics* In calculus, the ability to visualize a problem goes hand in hand with the ability to solve the problem. In writing this text we have attempted to nurture the student's ability to visualize problems by stressing graphics in the exercise sets as well as in the body of the text. The figures have been made *self-descriptive* by labeling them with appropriate captions or equations.

5. *Intuitive Approach* Throughout the text, we guide the student into each new topic with illustrations and examples that appeal to intuition and are consistent with the student's background knowledge. If we have gained anything from the new math era it is the reinforcement of the theory that the teaching process is more productive when students are required to build up a repertoire of skills, examples, and techniques before studying the underlying theory of a mathematical system. The introduction to limits is an important case in point. Chapter 2 begins with an intuitive development of limits followed by an *optional* ϵ–δ approach. The visualization of limits is enhanced by a section on horizontal and vertical asymptotes.

6. *Summaries* Many sections of the book have summaries that identify the core ideas and procedures of the section (see, for example, Sec-

tions 5.1, 8.3, and 18.5). In some instances, entire sections are devoted to summarizing and applying the preceding topics in the chapter. For example, in Section 4.5, the curve-sketching techniques are summarized in a general procedure that incorporates symmetry, asymptotes, continuity, range and domain, relative extrema, and concavity.

7. *Applications*   There are a variety of applications throughout the text. In some instances an application is used to introduce and motivate a concept, while in other instances it supplements a more abstract or geometric presentation. The presentation of practical applications is an age-old problem for calculus instructors. If no applications are given, students are robbed of a great deal of insight, motivation, and reward in their study of calculus. The problem is that many applications of calculus are complex and may involve extensive knowledge of other fields. When such an application is presented, students are frustrated and some lose confidence in their ability to solve problems involving calculus. On the other hand, if an application is too fanciful or simple-minded it makes calculus seem foolish and unnecessary. In our choice of applications we have tried for variety, a minimal knowledge of other fields, and integrity.

8. *Exercises*   Students learn calculus by working problems; there is no other way. Thus, an abundant source of well-chosen exercises is of utmost importance. In the exercise sets for this text, we have incorporated the following features:

Nearly 6000 exercises with a wide variety of applications are presented in the text—more than an ample number of exercises to build skills, without overwhelming the student.

The exercises are graded, progressing from skill-development problems to more challenging problems involving applications and proofs.

The answers to odd-numbered exercises are provided at the end of the text. A great deal of effort has been made to provide answers that are *correct*. Few things are more annoying to students than spending half an hour working a problem correctly and then spending another hour trying to find a nonexistent error. The answers have been checked and rechecked. If an error that we have somehow overlooked is found, we would appreciate your calling it to our attention.

Miscellaneous exercises are included at the end of each chapter.

A set of true-false questions is included at the end of most of the exercise sets under the title *For Review and Class Discussion*. These questions are designed to identify common errors and to emphasize the hypotheses and conclusions of the theorems. All true-false questions are answered at the end of the text.

Many of the exercise sets contain problems identified as calculator problems. Although most of these problems can be solved without a calculator, the solutions can be obtained more easily and accurately with a calculator. Special emphasis is given to the use of hand calculators in sections dealing with limits, Newton's Method, and numerical integration. A hand calculator is not a prerequisite for this text, but we do recommend its use as a valuable tool in the process of learning calculus.

9. *Study Aids*   Since much of calculus requires some knowledge of trigonometry, geometry, and (to a greater extent) algebra, pertinent formulas from these areas are listed inside the front and back covers of the text. As an additional aid to students who need algebraic help, the *Study and Solutions Guide* contains detailed solutions to several representative problems from each exercise set. These solutions are generally given in greater detail than the examples in the text, with special care taken to show the *algebra* of the solutions. In addition, the *Study and Solutions Guide* contains a brief review of algebra. For students wanting briefer solutions to *all* exercises, the *Solutions Guide to Accompany Calculus, Second Edition* in three volumes (I, Chapters 1–7; II, Chapters 7–13; and III, Chapters 13–19) is also available.

10. *Examples*   Together, the text and the *Study and Solutions Guide* contain over 1,600 examples of solved problems. The examples were chosen as illustrative of the basic problem types in each section without being overly repetitive. They are set in an open format that clearly identifies the various solution steps.

Ernest Hemingway made the following observation about writing in a changing world. "A writer's problem does not change. He himself changes and the world he lives in changes but his problem remains the same. It is always how to write truly and, having found what is true, to project it in such a way that it becomes a part of the experience of the person who reads it." Calculus, along with almost everything else, changes from year to year. The applications change. Calculating devices change. Teaching techniques change. Most importantly, students change. *Calculus, Second Edition,* represents our answer to the changing needs of today's calculus classes.

Roland E. Larson
Robert P. Hostetler

The Behrend College
Erie, Pennsylvania

Acknowledgments

We would like to thank the many people who have helped us at various stages of this project during the past five years. Their encouragement, criticisms, and suggestions have been invaluable to us.

Special thanks to:

Our consulting editor
 Frank Kocher, The Pennsylvania State University

Our reviewers for the second edition
 Harry L. Baldwin, Jr., San Diego City College
 Phillip Ferguson, Fresno City College
 Thomas M. Green, Contra Costa College
 Arnold Insel, Illinois State University
 William J. Keane, Boston College
 David C. Lantz, Colgate University
 Richard E. Shermoen, Washburn University
 Thomas W. Shilgalis, Illinois State University
 Florence A. Warfel, University of Pittsburgh

Our reviewers for the first edition
 Paul Davis, Worcester Polytechnic Institute
 Eric R. Immel, Georgia Institute of Technology
 Joseph F. Krebs, Boston College
 Maurice L. Monahan, South Dakota State University
 Robert A. Nowlan, Southern Connecticut State College
 N. James Schoonmaker, University of Vermont
 Bert K. Waits, Ohio State University

Our users, and especially those instructors whose suggestions for improving the first edition helped shape this edition:
 Dennis Albér, James J. Ball, Wilson Banks, Kenneth N. Berk, Thomas G. Burgess, Robert Burghardt, Louis Bush, Rose R. Carroll, Robert A. Close, Arlo Davis, Joyce T. Davis, Tony DiJulio, Charles Vanden Eynden, Lisa Grenier, Joseph Guerriero, Marcia Guza, Francis H. S. Hall, Harry L. Hancock, John J. Hanevy, Marvin C. Henry, Richard S. Hyman, Ann R. Kraus, Joseph F. Krebs, Archille J. Laferriere, Jean Lane, Samuel M. Laposata, Wilmer R. Lehman, Grace Morrissey, John Murray, Christopher Nevison, Jean-Michel Pomarede, Edward A. Race, Wayne Rhea, Young H. Rhie, Joyce Riseberg, Daniel Ross, Jack E. Schlossnagel, Stanley E. Seltzer, John H. Smith, Delbert W. Snyder, Richard Stout, G. G. Taylor, Angel H. Tellez, Morris Tepper, Edward Thomas, Linda L. Tully, Paul J. Welsh, Jr.

Our colleague, who created and solved the exercise sets
   David E. Heyd
Our typist
   Deanna Larson
Our publisher
   D. C. Heath and Company
Our editor
   John Rudolf
Our production editor
   Cathy Cantin
Our designer
   Esther Agonis
Our assistant editor
   Jackie Unch
Our proofreader and colleague
   Dianna Zook
Our student assistants
   Bernard Badger, Joseph DeRiggi, Margaret Fisher, Patricia Juchno, William
   McDonough, Beth Ann Morrison, Terry O'Brien, and Raymond Steinbacher
Our calculus students, who purchased mimeographed preliminary versions of
   the manuscript and whose many valuable suggestions helped shape this text
Our wives
   Deanna Larson and Eloise Hostetler

Guide to Early Trigonometry Coverage

Recommended Coverage Point	Section 9.1 (Trig. Review)		Section 9.2 (Trig. Differentiation)		Section 9.3 (Trig. Integration)		Miscellaneous Exercises
	Examples	Exercises	Examples	Exercises	Examples	Exercises	
Following Chapter 3 (Differentiation)	All	All	1–4	1–26, 32, 33, 36–42			1–3, 5–8, 13, 14, 17, 18, 20–22, 24, 51
Section 4.4 (Relative Extrema)			6, 7, 10, 11	52–59			
Section 5.2 (Related Rates)			12	63–70			54–58
Section 5.5 (Newton's Method)				60–62			49, 53
Following Section 6.4 (Integration)					1–7, 11	1–6, 10–16, 19, 20, 31, 33–35, 41, 42, 47, 48	
Section 7.3 (Volume)						37–40	
Section 8.2 (Log derivatives)				28, 29, 34, 35			
Section 8.3 (Log integrals)					8–10	7–9, 18–22, 25–29, 32, 43–46	32, 34, 39 45, 50
Section 8.4 (Exponentials)			5	27, 30, 31, 43		17, 23, 30	4, 23, 33, 35, 52
Section 8.5 (L'Hôpital's Rule)			8, 9	44–51, 71			

Contents

---

* Chapter 18 may be covered any time after Chapter 10.

What Is Calculus?

We begin to answer this question by saying that calculus is the reformulation of elementary mathematics through the use of a limit process. If limit processes are unfamiliar to you this answer is, at least for now, somewhat less than illuminating. From an elementary point of view, we may think of calculus as a "limit machine" that generates new formulas from old. Actually, the study of calculus involves three distinct stages of mathematics: *precalculus mathematics* (the length of a line segment, the area of a rectangle, and so forth), the *limit process,* and new *calculus* formulations (derivatives, integrals, and so forth). Some students try to learn calculus as if it were simply a collection of new formulas. This is unfortunate. When students reduce calculus to the memorization of differentiation and integration formulas, they miss a great deal of understanding, self confidence, and satisfaction.

On the following two pages we have listed some familiar precalculus concepts coupled with their more powerful calculus versions. Throughout this text, our goal is to show you how precalculus formulas and techniques are used as building blocks to produce the more general calculus formulas and techniques. Don't worry if you are unfamiliar with some of the "old formulas" listed on the following two pages—we will be reviewing all of them.

As you proceed through this text, we suggest that you come back to this discussion repeatedly. Try to keep track of where you are relative to the three stages involved in the study of calculus. For example, the first three chapters break down as follows: precalculus (Chapter 1), the limit process (Chapter 2), and new calculus formulas (Chapter 3). This cycle is repeated many times on a smaller scale throughout the text. We wish you well in your venture into calculus.

WITHOUT CALCULUS	WITH DIFFERENTIAL CALCULUS
value of $f(x)$ when $x = c$	limit of $f(x)$ as x approaches c
slope of a line	slope of a curve
secant line to a curve	tangent line to a curve
average rate of change between $t = a$ and $t = b$	instantaneous rate of change at $t = c$
curvature of a circle	curvature of a curve
height of a curve when $x = c$	maximum height of a curve on an interval
tangent plane to a sphere	tangent plane to a surface
direction of motion along a straight line	direction of motion along a curved line

WITHOUT CALCULUS	WITH INTEGRAL CALCULUS
area of a rectangle	area under a curve
work done by a constant force	work done by a variable force
center of a rectangle	centroid of a region
length of a line segment	length of an arc
surface area of a cylinder	surface area of a solid of revolution
mass of a solid of constant density	mass of a solid of variable density
volume of a rectangular solid	volume of a region under a surface
sum of a finite number of terms $\quad a_1 + a_2 + \cdots + a_n = S$	sum of an infinite number of terms $\quad a_1 + a_2 + a_3 \cdots = S$

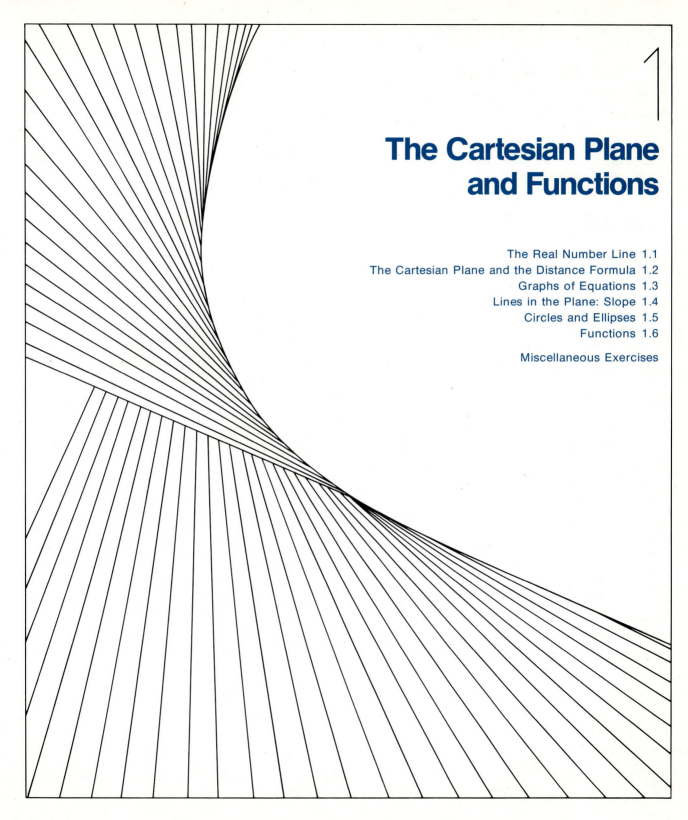

The Cartesian Plane and Functions

1.1 The Real Number Line

Purpose
- To review the real line.
- To discuss the notation for inequalities, absolute value, and intervals on the real line.

One of the main reasons which keeps those who are beginning these studies out of the true road they ought to follow is the notion they get at the start that the good things are inaccessible because they bear the names: great, high, exalted, sublime. That spoils everything. I should like to call them humble, common, familiar. These names suit them better. —*Blaise Pascal (1623–1662)*

This first chapter sets the stage for our study of calculus. The props we require include basic algebra and analytic geometry. A good working knowledge of basic algebra is essential for the study of calculus and we assume that you possess such a knowledge. As for analytic geometry, we assume you have a less familiarity with this topic and we will discuss its basic concepts as the need arises.

The basic goal of analytic geometry is to create visual representations of mathematical concepts through the use of coordinate systems. For example, the coordinate system we use to represent the real numbers is called the **real line** or *x*-axis (Figure 1.1). The **positive direction** (to the right) is denoted by an arrowhead and indicates the direction of increasing values of *x*. The real number corresponding to a particular point on the real line is called the **coordinate** of the point. As shown in Figure 1.1, it is customary to identify those points whose coordinates are **integers.**

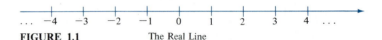

FIGURE 1.1   The Real Line

Real numbers may be categorized as either rational or irrational. Recall that a **rational** number is one that can be expressed as the quotient of two integers. An **irrational** number is one that is not rational; that is, it cannot be expressed as the quotient of two integers. Figure 1.2 identifies the points on the real line corresponding to the rational numbers $-\frac{5}{2}, \frac{3}{4}$, and $\frac{7}{5}$ as well as the points corresponding to the irrational numbers $-\sqrt{2}$, $\sqrt{5}$, and π. Each point on the real line corresponds to one and only one real number, and each real number corresponds to one and only one point on the real line. This type of relationship is referred to as a **one-to-one correspondence.**

FIGURE 1.2

One important property of real numbers is that they are **ordered;** 0 is less than 1, -3 is less than -2.5, π is less than $\frac{22}{7}$, and so on. We can visualize this property on the real line by observing that a is less than b if and only if a lies to the left of b. Symbolically, we denote "a is less than b" by the inequality

$$a < b$$

From Figure 1.2, we can see that $\frac{3}{4} < \frac{7}{5}$ because $\frac{3}{4}$ lies to the left of $\frac{7}{5}$ on the real line.

In practice, we often work with subsets of the real line rather than with the entire real line. For example, water is normally in a liquid state between the temperatures of 32 and 212 degrees Fahrenheit. We can denote this subset of the real numbers by the inequality

$$32 < x < 212$$

or by the **interval** $\qquad$ (32, 212)

Likewise, the **infinite interval**

$$[-273, \infty)$$

which corresponds to the inequality

$$-273 \leq x < \infty$$

can be used to denote the range of temperatures on the Celsius scale. Note that a square bracket is used to denote "less than or equal to" ($\leq$). Furthermore, we use the symbols ∞ and $-\infty$ to refer to positive and negative infinity. These symbols do *not* denote real numbers; they merely enable us to write certain statements more concisely.

An interval of the form (a, b) does not contain its endpoints a and b, and it is called an **open** interval. The interval $[a, b]$ does contain its endpoints and is called a **closed** interval. Intervals of the form $[a, b)$ and $(a, b]$ are called **half-open** intervals. Geometrically, an interval is a line segment; Table 1.1 pictures the nine types of intervals on the real line.

The following four rules are often used when working with inequalities. These same rules hold if $<$ is replaced by $\leq$.

Four Rules of Inequalities	
	1. If $a < b$ and $b < c$, then $a < c$.
	2. If $a < b$ and c is any number, then $a + c < b + c$ and $a - c < b - c$.
	3. If $a < b$ and $c < d$, then $a + c < b + d$.
	4. If $a < b$, then
	$\qquad ac < bc, \qquad$ provided $c > 0$
	while $\qquad ac > bc, \qquad$ provided $c < 0$

Note that Rule 3 does not allow the subtraction of two inequalities. For instance, consider $3 < 5$ and $1 < 6$. We can add these two inequalities and obtain $4 < 11$, but the subtraction of these inequalities would yield

TABLE 1.1 Intervals on the Real Line

Interval Notation	Inequality Notation	Graph
$(-\infty, a]$	$-\infty < x \leq a$	
$(-\infty, a)$	$-\infty < x < a$	
(a, b)	$a < x < b$	
$[a, b]$	$a \leq x \leq b$	
$[a, b)$	$a \leq x < b$	
$(a, b]$	$a < x \leq b$	
(b, ∞)	$b < x < \infty$	
$[b, \infty)$	$b \leq x < \infty$	
$(-\infty, \infty)$	$-\infty < x < \infty$	

the incorrect statement that $2 < -1$. Rule 4 is critical in that multiplying an inequality by a negative number *reverses* the direction of the inequality. For instance, if we multiply the inequality $-2 < 3$ by -5, we reverse the direction of the inequality and obtain $10 > -15$.

Example 1

Compare the following inequality and interval notations.

Inequality	Interval
$-\infty < x < 2$	$(-\infty, 2)$
$0 \leq x \leq 5$	$[0, 5]$
$-4 \leq x < 1$	$[-4, 1)$
x is greater than 2	$(2, \infty)$

Example 2

Write the interval form for the set of *x*-values that satisfy the inequality $2x - 5 < 7$ and graph the solution.

Solution: By Rule 2 for inequalities, we can add 5 to both members; thus

$$2x - 5 + 5 < 7 + 5$$
$$2x < 12$$

By Rule 4, multiplying by $\frac{1}{2}$ gives us

$$\tfrac{1}{2}(2x) < \tfrac{1}{2}(12)$$
$$x < 6$$

The interval form of the solution can be written as $(-\infty, 6)$. (See Figure 1.3.)

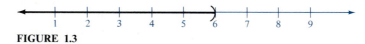

FIGURE 1.3

Example 3

Write the interval form for the set of all *x*-values that satisfy the inequality $-3 \le 2 - 5x \le 12$.

Solution: Although two inequalities are involved in this problem, we can work with both of them simultaneously. Thus subtracting 2 from both inequalities yields

$$-3 - 2 \le 2 - 5x - 2 \le 12 - 2$$
$$-5 \le -5x \le 10$$

Dividing by -5 reverses the direction of both inequalities, and we obtain

$$\frac{-5}{-5} \ge \frac{-5x}{-5} \ge \frac{10}{-5}$$

or

$$1 \ge x \ge -2$$

The corresponding interval form is $[-2, 1]$.

Given two distinct points *a* and *b* on the real line, there are three distances of interest (see Figure 1.4):

1.   the **directed distance from a to b,** denoted by $b - a$
2.   the **directed distance from b to a,** denoted by $a - b$
3.   the **distance between a and b,** denoted by $|a - b|$ or $|b - a|$

The symbol $|a - b|$ is called the **absolute value** of $a - b$ and it denotes the *magnitude* of the difference $a - b$. The precise definition of the absolute value of any real number is as follows:

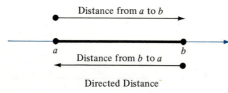

FIGURE 1.4

Definition of Absolute Value	If x is any real number, then the **absolute value** of x, denoted by $|x|$, is determined by $$|x| = \begin{cases} x, & \text{if } x \geq 0 \\ -x, & \text{if } x < 0 \end{cases}$$

At first glance it may appear from this definition that an absolute value can be negative. Such is not the case. For instance, let $x = -4$. Then by the definition we have

$$|-4| = -(-4) \quad \text{since} \quad -4 < 0$$

and therefore, $\qquad\qquad |-4| = 4$

The following alternative definition of absolute value avoids potential confusion with signs and it fits well with the definition of distance between two points. Since the square root symbol $\sqrt{}$ denotes only the *nonnegative* root, we can define absolute value as

$$|x| = \sqrt{x^2}$$

Thus the distance between two points on the real line can be defined as follows:

Definition of Distance Between Two Points on Real Line	The distance d between points x_1 and x_2 on the real line is given by $$d = |x_2 - x_1| = \sqrt{(x_2 - x_1)^2}$$

Note that the order of subtracting x_1 and x_2 does not matter since

$$|x_2 - x_1| = |x_1 - x_2| \quad \text{and} \quad (x_2 - x_1)^2 = (x_1 - x_2)^2$$

Example 4

Determine the distance between -3 and 4 on the real line. What is the distance from -3 to 4? From 4 to -3?

Solution: The distance between -3 and 4 is given by

$$|4 - (-3)| = |7| = 7 \quad \text{or by} \quad |-3 - 4| = |-7| = 7$$

The directed distance from -3 to 4 is $4 - (-3) = 7$. The directed distance from 4 to -3 is $-3 - 4 = -7$. ∎

Example 5

What interval on the real line contains all numbers that lie no more than two units from 3?

Solution: Let x be any point in this interval. Then we wish to find all x such that the distance between x and 3 is less than or equal to 2. Symbolically we write this as

$$|x - 3| \leq 2$$

This relationship is shown geometrically in Figure 1.5.

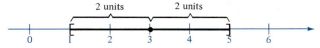

FIGURE 1.5

Requiring the absolute value of $x - 3$ to be less than or equal to 2 means that $x - 3$ must lie between 2 and -2, and hence we write

$$-2 \leq x - 3 \leq 2$$

Solving this pair of inequalities, we have

$$-2 + 3 \leq x - 3 + 3 \leq 2 + 3$$
$$1 \leq x \leq 5$$

Therefore the desired interval is $[1, 5]$, as shown in Figure 1.5.   ■

Example 6

Determine the intervals on the real line that contain all numbers that lie more than three units from -2.

Solution: Refer to Figure 1.6. Since the distance between x and -2 is given by $|x - (-2)|$, we write

$$|x - (-2)| > 3$$
$$|x + 2| > 3$$

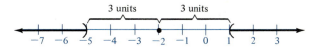

FIGURE 1.6

This means that $x + 2$ must be numerically greater than 3. Hence we seek x-values that satisfy *either one* of the inequalities

$$x + 2 < -3 \qquad \text{or} \qquad x + 2 > 3$$

which implies that $\qquad x < -5 \qquad$ or $\qquad x > 1$

Therefore x can lie in either of the intervals

$$(-\infty, -5) \qquad \text{or} \qquad (1, \infty)$$ ■

Examples 5 and 6 suggest the following general results for absolute value and inequalities:

1. $|x - a| \leq d$ if and only if $-d \leq x - a \leq d$.
2. $|x - a| \geq d$ if and only if $x - a \leq -d$ or $x - a \geq d$.

To find the **midpoint** of an interval with endpoints a and b, we simply find the average value of a and b. That is, the midpoint c of the interval $[a, b]$ is

$$c = \frac{a + b}{2}$$

Example 7

Find the midpoint of each of the following intervals:
(a) $[-5, 7]$ (b) $(-12, -1)$ (c) $(3, 21]$

Solution:

(a) midpoint $= \dfrac{-5 + 7}{2} = \dfrac{2}{2} = 1$

(b) midpoint $= \dfrac{-12 + (-1)}{2} = -\dfrac{13}{2}$

(c) midpoint $= \dfrac{3 + 21}{2} = \dfrac{24}{2} = 12$ ∎

Section Exercises (1.1)

1. Complete the accompanying chart, showing the appropriate interval notation, inequality notation, and graph on the real line.

Interval Notation	Inequality Notation	Graph
		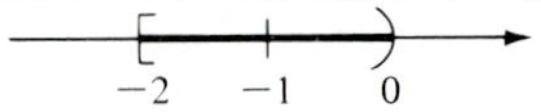
$(-\infty, -4]$		
	$3 \leq x \leq \frac{11}{2}$	
$(-1, 7)$		
		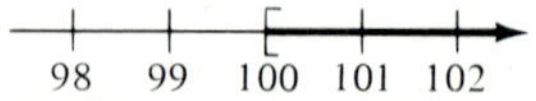
	$10 < x < \infty$	
$(\sqrt{2}, 8]$		
	$\frac{1}{3} < x \leq \frac{22}{7}$	

In Exercises 2–25, solve the inequality and graph the solution on the real number line.

2. $2x > 3$

⟨S⟩ **3.** $x - 5 \geq 7$

4. $2x + 7 < 3$

5. $4x + 1 < 2x$

6. $3x + 1 \geq 2x + 2$

7. $2x - 1 \geq 0$

8. $x - 4 \leq 2x + 1$

9. $4 - 2x < 3x - 1$

10. $0 \leq x + 3 < 5$

⟨S⟩ **11.** $-4 < 2x - 3 < 4$

12. $-1 < -\dfrac{x}{3} < 1$

13. $\dfrac{3}{4}x > x + 1$

14. $x > \dfrac{1}{x}$

15. $\dfrac{x}{2} + \dfrac{x}{3} > 5$

16. $\dfrac{x}{2} - \dfrac{x}{3} > 5$

17. $|x| < 1$

18. $\left|\dfrac{x}{2}\right| > 3$

⟨S⟩ **19.** $\left|\dfrac{x - 3}{2}\right| \geq 5$

20. $|x + 2| < 5$

21. $|x - a| < b$

22. $|3x + 1| \geq 4$

23. $|2x + 1| < 5$

24. $|9 - 2x| < 1$

25. $|1 - \tfrac{2}{3}x| < 1$

In Exercises 26–29, find (a) the directed distance from a to b, (b) the directed distance from b to a, (c) the distance between a and b, and (d) the midpoint of the interval $[a, b]$.

26. $a = -\tfrac{7}{2}$, $b = 5$

27. $a = -10$, $b = \tfrac{1}{3}$

28. $a = 5$, $b = 14$

29. $a = -3$, $b = -\tfrac{3}{2}$

30. The interval $[5, 14]$ is to be trisected. Find the points of trisection.

⟨S⟩ **31.** The interval $[-4, 12]$ is to be trisected. Find the points of trisection.

In Exercises 32–35, prove each property.

⟨S⟩ **32.** $|ab| = |a|\,|b|$

33. $|a - b| = |b - a|$ [Hint: Use Exercise 32 and the fact that $(a - b) = (-1)(b - a)$.]

34. $\left|\dfrac{a}{b}\right| = \dfrac{|a|}{|b|}$

35. $|a + b| \leq |a| + |b|$ (triangle inequality)

⟨S⟩ Detailed solution in *Student Solutions Manual.*

For Review and Class Discussion

True or False

1. ____ If a and b are any two real numbers, then $a < b$ or $a > b$.

2. ____ $\pi = 355/113$.

3. ____ If a and b are any two real numbers, then $|a - b| \leq |a| - |b|$.

4. ____ It is possible for two intervals on the real line to have only one point in common.

5. ____ It is possible for two open intervals on the real line to have only one point in common.

6. ____ If d is the directed distance from a to b, then $-d$ is the directed distance from b to a.

7. ____ The absolute value of every real number is positive.

8. ____ The absolute value of every real number is non-negative.

9. ____ If $x < 0$, $\sqrt{x^2} = -x$.

10. ____ For every x, $\sqrt{(-x)^2} = -x$.

1.2 The Cartesian Plane and the Distance Formula

Purpose
- To review the Cartesian plane.
- To develop the formula for the distance between two points in the plane.
- To describe the Midpoint Rule.

I hope that posterity will judge me kindly, not only as to the things which I have explained, but also as to those which I have intentionally omitted so as to leave to others the pleasure of discovery. —René Descartes (1596–1650)

Just as the real numbers can be represented geometrically by points on the real line, we can represent ordered pairs of real numbers by points in

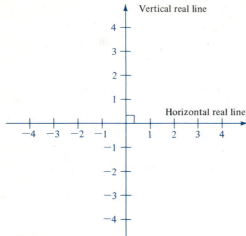

FIGURE 1.7

a plane. The model we develop for representing ordered pairs of real numbers is called the **rectangular coordinate system,** or the **Cartesian plane.** We develop this model by considering two real lines intersecting at right angles (Figure 1.7).

The horizontal real line is traditionally called the **x-axis** and the vertical real line is called the **y-axis.** Their point of intersection is called the **origin,** and the lines divide the plane into four parts called **quadrants** (Figure 1.8).

We identify each point in the plane by an ordered pair (x, y) of real numbers x and y called the **coordinates** of the point. The number x represents the directed distance from the y-axis to the point, and y represents the directed distance from the x-axis to the point (Figure 1.9). For the point (x, y), the first coordinate is referred to as the x-coordinate or **abscissa** and the second or y-coordinate is referred to as the **ordinate.**

Perhaps you are a bit concerned that we have used the notation (x, y) to denote a point in the plane as well as to describe an open interval on the real line. Generally there is no confusion because the nature of a specific problem will show whether we are talking about points in the plane or about intervals on the real line.

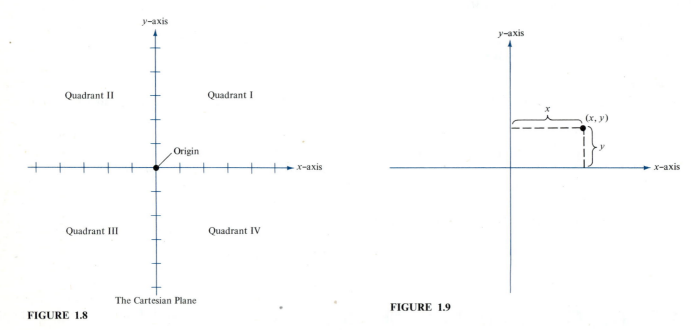

FIGURE 1.8

FIGURE 1.9

Example 1

Locate the points $(-1, 2)$, $(3, 4)$, $(0, 0)$, $(3, 0)$, and $(-2, -3)$ in the Cartesian plane.

Solution: The solution is shown in Figure 1.10.

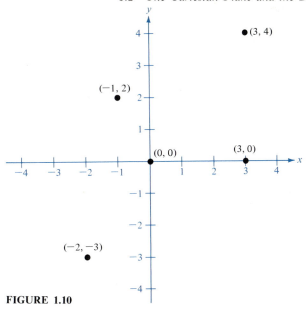

FIGURE 1.10

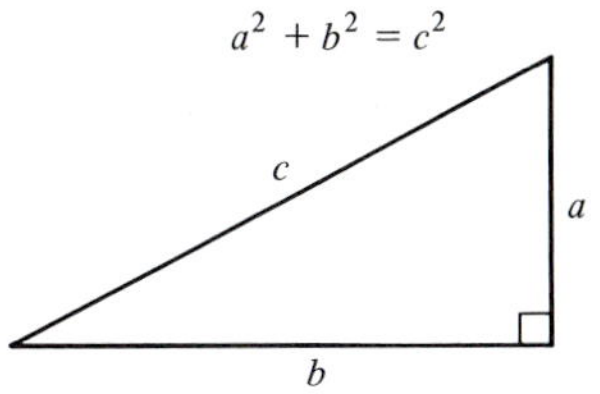

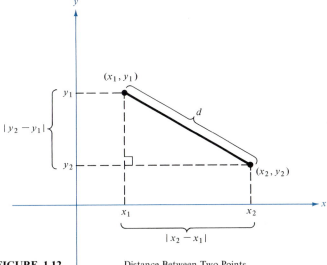

$$a^2 + b^2 = c^2$$

Pythagorean Theorem

FIGURE 1.11

We have seen how to determine the distance between two points x_1 and x_2 on the real line. We now turn our attention to the slightly more difficult problem of finding the distance between two points in the plane. Recall from the Pythagorean Theorem that for a right triangle with hypotenuse c and sides a and b, we have the relationship $a^2 + b^2 = c^2$. Conversely, if $a^2 + b^2 = c^2$, then the triangle is a right triangle (Figure 1.11).

Suppose we wish to determine the distance d between two points (x_1, y_1) and (x_2, y_2) in the plane. Using these two points, a right triangle can be formed, as shown in Figure 1.12.

FIGURE 1.12   Distance Between Two Points

The length of the vertical side of the triangle is simply the distance between y_1 and y_2, or $|y_2 - y_1|$. Similarly, the length of the horizontal side of the triangle is simply $|x_2 - x_1|$. By the Pythagorean Theorem we then have

$$d^2 = |x_2 - x_1|^2 + |y_2 - y_1|^2$$

or

$$d = \sqrt{|x_2 - x_1|^2 + |y_2 - y_1|^2}$$

Replacing $|x_2 - x_1|^2$ and $|y_2 - y_1|^2$ by the equivalent expressions $(x_2 - x_1)^2$ and $(y_2 - y_1)^2$, we can write

$$d = \sqrt{(x_2 - x_1)^2 + (y_2 - y_1)^2}$$

We choose the positive square root for d because the distance *between* two points is not a directed distance. Of course, we can interchange the order of subtraction and write the equivalent form

$$d = \sqrt{(x_1 - x_2)^2 + (y_1 - y_2)^2}$$

We have therefore established the following theorem:

THEOREM 1.1
(*Distance Formula*)

The distance d between two points (x_1, y_1) and (x_2, y_2) in the plane is given by

$$d = \sqrt{(x_2 - x_1)^2 + (y_2 - y_1)^2}$$

Example 2

Find the distance between the points $(-2, 1)$ and $(3, 4)$.

Solution: Applying Theorem 1.1 we have

$$d = \sqrt{[3 - (-2)]^2 + (4 - 1)^2} = \sqrt{(5)^2 + (3)^2}$$
$$= \sqrt{25 + 9} = \sqrt{34} \approx 5.83$$

(Note the use of $\approx$ to represent *approximately equal*.) ∎

Example 3

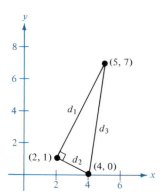

FIGURE 1.13

Use the Distance Formula to show that the points $(2, 1)$, $(4, 0)$, and $(5, 7)$ are the vertices of a right triangle.

Solution: Refer to Figure 1.13. The three sides have lengths

$$d_1 = \sqrt{(5 - 2)^2 + (7 - 1)^2} = \sqrt{9 + 36} = \sqrt{45}$$
$$d_2 = \sqrt{(4 - 2)^2 + (0 - 1)^2} = \sqrt{4 + 1} = \sqrt{5}$$
$$d_3 = \sqrt{(5 - 4)^2 + (7 - 0)^2} = \sqrt{1 + 49} = \sqrt{50}$$

Since

$$d_1^2 + d_2^2 = 45 + 5 = 50 = d_3^2$$

the triangle must be a right triangle. ∎

Note in Example 3 that the figure provided was not really essential to the solution of the problem. *Nevertheless,* we strongly recommend that you get in the habit of including sketches with your problem solutions even if they are not specifically required.

We conclude this section with a rule for finding the coordinates of the midpoint of the line segment joining two points in the plane.

THEOREM 1.2
(*Midpoint Rule*)

The midpoint of the line segment joining points (x_1, y_1) and (x_2, y_2) is

$$\left(\frac{x_1 + x_2}{2}, \frac{y_1 + y_2}{2}\right)$$

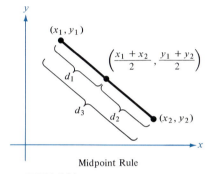

Midpoint Rule

FIGURE 1.14

Proof: The result of this theorem is just what we would expect. To find the midpoint of a line segment, we merely find the "average" value of the respective coordinates of the two endpoints. The formal proof requires that, in Figure 1.14, we show

$$d_1 = d_2 \qquad \text{and} \qquad d_1 + d_2 = d_3$$

Using the Distance Formula we obtain

$$d_1 = \sqrt{\left(\frac{x_1 + x_2}{2} - x_1\right)^2 + \left(\frac{y_1 + y_2}{2} - y_1\right)^2}$$

$$= \frac{1}{2}\sqrt{(x_2 - x_1)^2 + (y_2 - y_1)^2}$$

$$d_2 = \sqrt{\left(x_2 - \frac{x_1 + x_2}{2}\right)^2 + \left(y_2 - \frac{y_1 + y_2}{2}\right)^2}$$

$$= \frac{1}{2}\sqrt{(x_2 - x_1)^2 + (y_2 - y_1)^2}$$

$$d_3 = \sqrt{(x_2 - x_1)^2 + (y_2 - y_1)^2}$$

Thus it follows that

$$d_1 = d_2 \qquad \text{and} \qquad d_1 + d_2 = d_3$$

(You are asked to fill in the algebraic details of this proof in Exercise 27.)

■

Example 4

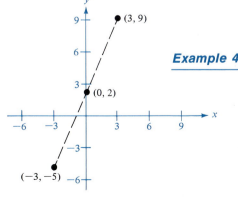

FIGURE 1.15

Find the midpoint of the line segment joining the points $(-3, -5)$ and $(3, 9)$.

Solution: By Theorem 1.2 the midpoint is

$$\left(\frac{-3 + 3}{2}, \frac{-5 + 9}{2}\right) = (0, 2)$$

(See Figure 1.15.)

■

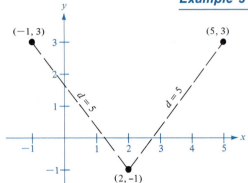

FIGURE 1.16

Example 5

Find x so that the distance between $(x, 3)$ and $(2, -1)$ is 5.

Solution: Using the Distance Formula, we have

$$d = 5 = \sqrt{(x - 2)^2 + (3 + 1)^2}$$
$$25 = (x^2 - 4x + 4) + 16$$
$$0 = x^2 - 4x - 5$$
$$0 = (x - 5)(x + 1)$$

Therefore, $x = 5$ or $x = -1$ and we conclude that both of the points $(5, 3)$ and $(-1, 3)$ lie 5 units from the point $(2, -1)$. (See Figure 1.16.) ■

Section Exercises (1.2)

In Exercises 1–8, plot the points, find the distance between the points, and find the midpoint of the line segment joining the points.

S 1. $(2, 1), (4, 5)$

2. $(-3, 2), (3, -2)$

3. $(\frac{1}{2}, 1), (-\frac{3}{2}, -5)$

4. $(\frac{2}{3}, -\frac{1}{3}), (\frac{5}{6}, 1)$

5. $(2, 2), (4, 14)$

6. $(-3, 7), (1, -1)$

7. $(1, \sqrt{3}), (-1, 1)$

8. $(-2, 0), (0, \sqrt{2})$

S 9. Show that the points $(4, 0), (2, 1), (-1, -5)$ are vertices of a right triangle.

10. Show that the points $(1, -3), (3, 2), (-2, 4)$ are vertices of an isosceles triangle.

11. Show that the points $(0, 0), (1, 2), (2, 1), (3, 3)$ are vertices of a rhombus.

12. Show that the points $(0, 1), (3, 7), (4, 4), (1, -2)$ are vertices of a parallelogram.

13. Use the Distance Formula to determine if the points $(0, -4), (2, 0), (3, 2)$ lie on a straight line.

c 14. Use the Distance Formula to determine if the points $(0, 4), (7, -6), (-5, 11)$ lie on a straight line.

c s 15. Use the Distance Formula to determine if the points $(-2, 1), (-1, 0), (2, -2)$ lie on a straight line.

16. Find y so that the distance from the origin to the point $(3, y)$ is 5.

17. Find x so that the distance from the origin to the point $(x, -4)$ is 5.

18. Find x so that the distance from $(2, -1)$ to the point $(x, 2)$ is 5.

S 19. Find the relationship between x and y so that the point (x, y) is equidistant from $(4, -1)$ and $(-2, 3)$.

20. Find the relationship between x and y so that the point (x, y) is equidistant from $(3, \frac{5}{2})$ and $(-7, -1)$.

S 21. Use the Midpoint Rule successively to find the three points that divide the line segment joining (x_1, y_1) and (x_2, y_2) into four equal parts.

22. Use the results of Exercise 21 to find the points that divide into four equal parts the line segment joining these points:
(a) $(1, -2)$ and $(4, -1)$ (b) $(-2, -3)$ and $(0, 0)$

23. Prove that

$$\left(\frac{2x_1 + x_2}{3}, \frac{2y_1 + y_2}{3} \right)$$

is one of the points of trisection of the line segment joining (x_1, y_1) and (x_2, y_2). Also, find the midpoint of the line segment joining

$$\left(\frac{2x_1 + x_2}{3}, \frac{2y_1 + y_2}{3} \right) \quad \text{and} \quad (x_2, y_2)$$

to find the second point of trisection of the line segment joining (x_1, y_1) and (x_2, y_2).

24. Use the results of Exercise 23 to find the points of trisection of the line segment joining these points:
(a) $(1, -2)$ and $(4, 1)$ (b) $(-2, -3)$ and $(0, 0)$

25. Prove that the line segments joining the midpoints of the opposite sides of any quadrilateral bisect each other.

26. Prove that the midpoint of the hypotenuse of any right triangle is equidistant from each of the three vertices.

27. Prove Theorem 1.2.

S Detailed solution in *Student Solutions Manual.*   **C** Calculator may be helpful.

For Review and Class Discussion

True or False

1. _____ If $ab < 0$, then the point (a, b) lies in either quadrant II or quadrant IV.

2. _____ The points (a, b) and $(-a, b)$ can lie in the same quadrant.

3. _____ If $\sqrt{(x_2 - x_1)^2 + (y_2 - y_1)^2} = |y_2 - y_1|$, then the points (x_1, y_1) and (x_2, y_2) both lie on the same vertical line.

4. _____ If two points lie in the same quadrant, then the midpoint of the line segment joining them must also lie in that quadrant.

5. _____ If (x_0, y_0) is equidistant from (x_1, y_1) and (x_2, y_2), then (x_0, y_0) must be the midpoint of the line segment joining (x_1, y_1) and (x_2, y_2).

6. _____ The distance between the points $(a + b, a)$ and $(a - b, a)$ is $2b$.

7. _____ The distance between $(a + b, a)$ and $(a - b, a)$ is $|2b|$.

8. _____ If the distance between two points is zero, then the two points must coincide.

9. _____ If $ab = 0$, then the point (a, b) lies on the x-axis or the y-axis.

10. _____ The distance between $(a, 0)$ and $(b, 2\sqrt{ab})$ is $|a + b|$.

1.3 Graphs of Equations

Purpose
- To define the graph of an equation.
- To introduce the point-plotting method of graphing equations.
- To investigate the use of symmetry in sketching graphs of equations.

As long as algebra and geometry traveled separate paths, their advance was slow and their applications limited. But when these two sciences joined company, they drew from each other fresh vitality and then forward marched at a rapid pace towards perfection. It is to Descartes that we owe the application of algebra to geometry—an application which has furnished the key to the greatest discoveries in all branches of mathematics. —Joseph Lagrange (1736–1813)

The idea of using a graph to show how two quantities are related to each other is familiar to all of us. Newsmagazines frequently show graphs that compare the rate of inflation, the gross national product, wholesale prices, or the unemployment rate to the time of year. Industrial firms and businesses use graphs to report their monthly production and sales statistics. The value of such graphs is that they provide a simple geometrical picture of the way one quantity changes with respect to another.

Frequently the relationship between two quantities is expressed in the form of an equation. For instance, degrees on the Fahrenheit temperature scale are related to degrees on the Celsius scale by the equation $F = \frac{9}{5}C + 32$. In this section we introduce the basic procedure for determining the geometric picture associated with such an equation.

Consider the equation

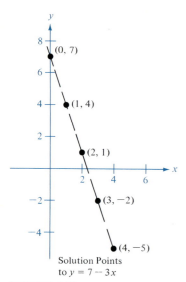

FIGURE 1.17

$$3x + y = 7$$

If $x = 2$ and $y = 1$, the equation is satisfied and we call the point $(2, 1)$ a **solution point** of the equation. Of course, there are other solution points, such as $(1, 4)$ and $(0, 7)$. We can make up a **table of values** for x and y by choosing arbitrary values for x and determining the corresponding values for y. To determine the values for y, it is convenient to write the equation in the form

$$y = 7 - 3x$$

x	0	1	2	3	4
$y = 7 - 3x$	7	4	1	-2	-5

Thus $(0, 7)$, $(1, 4)$, $(2, 1)$, $(3, -2)$, and $(4, -5)$ are all solution points of the equation $3x + y = 7$. We could continue this process indefinitely and obtain infinitely many solution points for the equation $3x + y = 7$. We call the set of all such solution points the **graph** of the equation $3x + y = 7$, as shown in Figure 1.17.

Definition of Graph of Equation in Two Variables	The **graph of an equation** involving two variables x and y is the collection of all points in the plane that are solution points to the equation.

The graph of an equation is sometimes called its **locus**. Thus when we refer to the locus of all points satisfying a particular equation, we are merely referring to its graph.

Example 1

Sketch the graph of the equation $y = x^2 - 2$.

Solution: First, we make a table of values by choosing several convenient values of x and calculating the corresponding values of y.

x	-2	-1	0	1	2	3
$y = x^2 - 2$	2	-1	-2	-1	2	7

Next, we locate these points in the plane, as in Figure 1.18. Finally, we connect these points by a *smooth curve,* as in Figure 1.19.

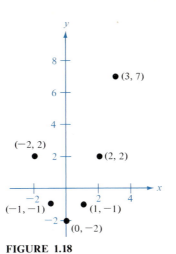

Graph of $y = x^2 - 2$

FIGURE 1.18

FIGURE 1.19 ■

We call this method of sketching a graph the **point-plotting method.** Its basic features are the following:

The Point-Plotting Method of Graphing	1. Make up a table of several solution points of the equation.
	2. Plot these points in the plane.
	3. Connect the points with a smooth curve.

Steps 1 and 2 of the point-plotting method can usually be accomplished with ease. However, step 3 can be the source of some major difficulties. For instance, how would you connect the four points in Figure 1.20? Without additional points or further information about the equation, any one of the three graphs in Figure 1.21 would be reasonable.

Obviously, with too few solution points, we could badly misrepresent the graph of a given equation. It is hard to say just how many points should be plotted. For a straight-line graph two points are sufficient; for more complicated graphs we need many more points. In spite of this difficulty with the point-plotting procedure, it is a good foundation upon which to build the more sophisticated techniques discussed in later chapters. In the meantime we suggest that you plot enough points so as to reveal the essential behavior of the graph, and the more solution points

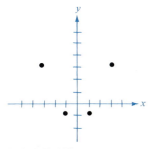

FIGURE 1.20

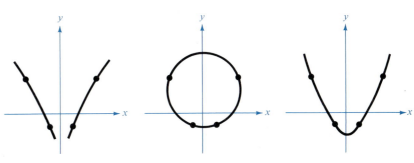

FIGURE 1.21

you plot, the more accurate your graph will be. (A programmable calculator is a very useful device for determining the many solution points needed for an accurate graph.)

In choosing points to plot, we suggest that you start with those that are easiest to calculate. Two points that are usually easy to determine are those having zero as either their *x*- or *y*-coordinate. Such points are called **intercepts,** because they are points at which the graph intersects the *x*- or *y*-axis.

Definition of Intercepts	The point $(a, 0)$ is called an ***x*-intercept** of the graph of an equation if it is a solution point of the equation. The point $(0, b)$ is called a ***y*-intercept** of the graph of an equation if it is a solution point of the equation.

[Note: Some authors denote the *x*-intercept as the *x*-coordinate of the point $(a, 0)$ rather than the point itself. Unless it is necessary to make a distinction, we will use "intercept" to mean either the point or the coordinate.]

Of course, it is possible that a particular graph will have no intercepts, or it may have several. For instance, consider the four graphs in Figure 1.22.

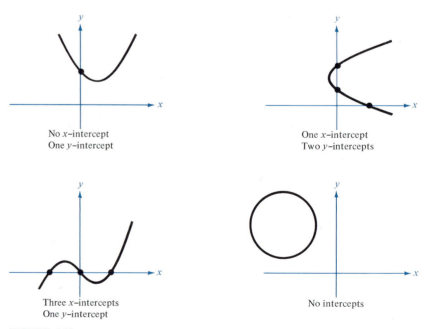

No *x*–intercept
One *y*–intercept

One *x*–intercept
Two *y*–intercepts

Three *x*–intercepts
One *y*–intercept

No intercepts

FIGURE 1.22

Finding Intercepts	To find the x-intercepts, let y be zero and solve the equation for x.
	To find the y-intercepts, let x be zero and solve the equation for y.

Example 2

Find the x- and y-intercepts for the graphs of the following:

(a) $y = x^3 - 4x$ (b) $y^2 - 3 = x$

Solution:

(a) Let $y = 0$; then $0 = x(x^2 - 4)$ has solutions $x = 0$ and $x = \pm 2$.

$$x\text{-intercepts: } (0, 0),\ (2, 0),\ (-2, 0)$$

Let $x = 0$; then $y = 0$.

$$y\text{-intercept: } (0, 0)$$

(b) Let $y = 0$; then $-3 = x$.

$$x\text{-intercept: } (-3, 0)$$

Let $x = 0$; then $y^2 - 3 = 0$ has solutions $y = \pm\sqrt{3}$.

$$y\text{-intercepts: } (0, \sqrt{3}),\ (0, -\sqrt{3})$$ ∎

Example 3

Sketch the graph of the equation $x^2 + 4y^2 = 16$.

Solution: Sometimes it is beneficial to rewrite an equation before calculating solution points. For example, if we rewrite the equation

$$x^2 + 4y^2 = 16$$

as

$$x = \pm\sqrt{16 - 4y^2} = \pm 2\sqrt{4 - y^2}$$

then we can easily determine several solution points by choosing values for y and calculating the corresponding values for x. Note $x = \pm 2\sqrt{4 - y^2}$ is defined only when $|y| \leq 2$, since the square root of a negative number is not real.

$x = \pm 2\sqrt{4 - y^2}$	0	$\pm 2\sqrt{3}$	± 4	$\pm 2\sqrt{3}$	0
y	-2	-1	0	1	2

By plotting these points and connecting them with a smooth curve, we have the graph shown in Figure 1.23.

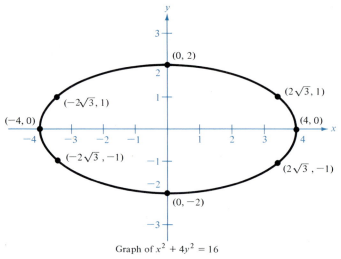

Graph of $x^2 + 4y^2 = 16$

FIGURE 1.23

The graphs shown in Figures 1.19, 1.21, and 1.23 are all said to be "symmetric" with respect to the y-axis. This means that if the Cartesian plane were folded along the y-axis, the portion of the graph to the left of the y-axis would then coincide with the portion to the right of the y-axis. Another way to describe this symmetry is to say that the graphs are reflections of themselves with respect to the y-axis. Symmetry with respect to the x-axis can be described in a similar manner.

Knowing the symmetry of a graph *before* attempting to sketch it is quite beneficial, for then we need only half as many solution points as we otherwise would. We define three basic types of symmetry as follows (see Figure 1.24):

Definition of Symmetry	A graph is said to be **symmetric with respect to the y-axis** if, whenever (x, y) is a point on the graph, $(-x, y)$ is also on the graph.
	A graph is said to be **symmetric with respect to the x-axis** if, whenever (x, y) is on the graph, $(x, -y)$ is also on the graph.
	A graph is said to be **symmetric with respect to the origin** if, whenever (x, y) is on the graph, $(-x, -y)$ is also on the graph.

Suppose we apply this definition of symmetry to the graph of the equation $y = x^2 - 2$. By replacing x by $-x$, we obtain

$$y = (-x)^2 - 2 \qquad \text{or} \qquad y = x^2 - 2$$

Since this substitution does not change the equation, it follows that if (x, y) is a solution point of the equation, then $(-x, y)$ must also be a

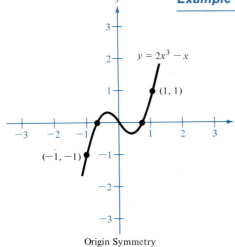

y–axis Symmetry   *x*–axis Symmetry   Origin Symmetry

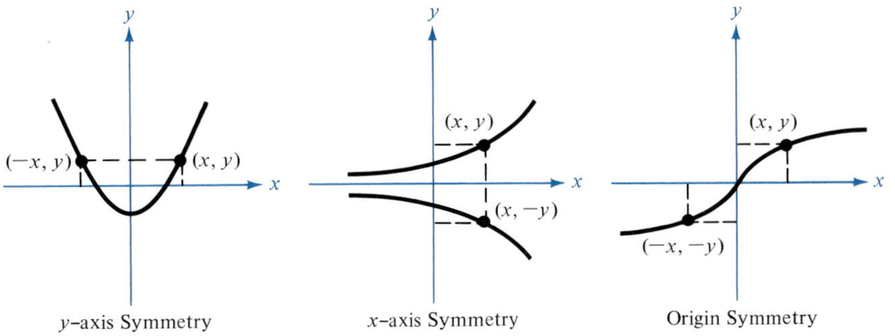

FIGURE 1.24

solution point. Therefore, the graph of $y = x^2 - 2$ is symmetric with respect to the *y*-axis.

A similar test can be made for symmetry with respect to the *x*-axis or to the origin. These three tests are summarized in the following theorem, which we state without proof.

THEOREM 1.3
(Tests for Symmetry)

The graph of an equation is symmetric with respect to:

i.   the *y*-axis if replacing *x* by $-x$ yields an equivalent equation
ii. the *x*-axis if replacing *y* by $-y$ yields an equivalent equation
iii. the origin if replacing *x* by $-x$ *and* *y* by $-y$ yields an equivalent equation

---

Example 4

Show that the graph of $y = 2x^3 - x$ is symmetric with respect to the origin.

Solution: By replacing *x* by $-x$ and *y* by $-y$, we have

$$-y = 2(-x)^3 - (-x)$$
$$-y = -2x^3 + x$$

Now by multiplying both sides of the equation by -1, we have

$$y = 2x^3 - x$$

which is the original equation. Therefore, the graph of $y = 2x^3 - x$ must be symmetric with respect to the origin. See Figure 1.25.   ∎

Origin Symmetry

FIGURE 1.25

Example 5

Sketch the graph of $x - y^2 = 1$.

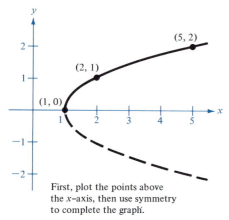

First, plot the points above the x–axis, then use symmetry to complete the graph.

FIGURE 1.26

Solution: The graph is symmetric with respect to the x-axis since replacing y by $-y$ yields

$$x - (-y)^2 = 1$$
$$x - y^2 = 1$$

This means that the graph below the x-axis is a mirror image of the graph above the x-axis. Hence we can first sketch the graph above the x-axis and then reflect it to obtain the entire graph (Figure 1.26).

$x = y^2 + 1$	1	2	5
y	0	1	2

Since each point of a graph is a solution point of its corresponding equation, a **point of intersection** of two graphs is simply a solution point that satisfies both equations. Moreover, the points of intersection of two graphs can be found by solving the given equations simultaneously.

Example 6

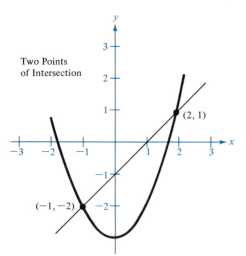

Two Points of Intersection

FIGURE 1.27

Find all points of intersection of the graphs of

$$x - y = 1 \quad \text{and} \quad x^2 - y = 3$$

Solution: Applying the methods of this section, we make a sketch for each equation on the *same* coordinate plane (Figure 1.27).

From Figure 1.27 it appears that the two graphs have two points of intersection. Solving each equation for y, we obtain

$$y = x - 1 \quad \text{and} \quad y = x^2 - 3$$

By equating the two expressions for y, we obtain

$$x^2 - 3 = x - 1$$

or

$$x^2 - x - 2 = 0$$

Factoring yields

$$(x - 2)(x + 1) = 0$$

and thus we have $x = 2$ or $x = -1$. The corresponding values of y are obtained by substituting $x = 2$ and $x = -1$ into either of the original equations. For instance, if we choose the equation $y = x - 1$, then the values of y are 1 and -2, respectively. Therefore, the two points of intersection are

$$(2, 1) \quad \text{and} \quad (-1, -2)$$

Section Exercises (1.3)

In Exercises 1–10, check for symmetry about both axes and the origin.

1. $y = x^2 - 2$
2. $y = x^4 - x^2 + 3$
3. $x^2y - x^2 + 4y = 0$
4. $x^2y - x^2 - 4y = 0$
[S] 5. $y^2 = x^3 - 4x$
6. $xy^2 = -10$
7. $y = x^3 + x$
8. $xy = 1$
[S] 9. $y = \dfrac{x}{x^2 + 1}$
10. $y = x^3 + x - 3$

In Exercises 11–20, find the intercepts.

11. $y = 2x - 3$
12. $y = (x - 1)(x - 3)$
[S] 13. $y = x^2 + x - 2$
14. $y^2 = x^3 - 4x$
15. $y = x^2\sqrt{9 - x^2}$
16. $xy = 4$
17. $y = \dfrac{x - 1}{x - 2}$
18. $y = \dfrac{x^2 + 3x}{(3x + 1)^2}$
[S] 19. $x^2y - x^2 + 4y = 0$
20. $y = 2x - \sqrt{x^2 + 1}$

In Exercises 21–45, use the methods of this section to sketch the graph of each equation. Identify the intercepts and test for symmetry.

21. $y = x$
22. $y = x - 2$
23. $y = x + 3$
24. $y = 2x - 3$
25. $y = -3x + 2$
26. $y = -\frac{1}{2}x + 2$
27. $y = \frac{1}{2}x - 4$
28. $y = x^2 + 3$
29. $y = 1 - x^2$
30. $y = 2x^2 + x$
31. $y = -2x^2 + x + 1$
32. $y = x^3 - 1$
33. $y = x^3 + 2$
34. $y = \sqrt{9 - x^2}$
35. $x^2 + 4y^2 = 4$
36. $x = y^2 - 4$
37. $y = (x + 2)^2$
38. $y = \dfrac{1}{x^2 + 1}$
39. $y = \dfrac{1}{x}$
40. $y = 2x^4$
41. $y = \sqrt{x - 3}$
42. $y = |x| - 2$
[S] 43. $y = |x - 2|$
44. $y = -|x - 2|$
[C] 45. $y = x^3 - 3x$

[C] 46. (a) Sketch the graph of $y = 3x^4 - 4x^3$ by completing the accompanying table and plotting the resulting points.

x	-1	0	1	2
y				

(b) Find additional points satisfying $y = 3x^4 - 4x^3$ by completing the accompanying table. Now refine the graph of part (a).

x	-0.75	-0.50	-0.25	0.25	0.5	0.75	1.33
y							

In Exercises 47–56, find the points of intersection of the graphs of the equations; check your results.

47. $x + y = 2,\ 2x - y = 1$
48. $2x - 3y = 13,\ 5x + 3y = 1$
[S] 49. $x + y = 7,\ 3x - 2y = 11$
50. $x^2 + y^2 = 25,\ 2x + y = 10$
[S] 51. $x^2 + y^2 = 5,\ x - y = 1$
52. $x^2 + y = 4,\ 2x - y = 1$
53. $y = x^3,\ y = x$
54. $y = x^4 - 2x^2 + 1,\ y = 1 - x^2$
[S] 55. $y = x^3 - 2x^2 + x - 1,\ y = -x^2 + 3x - 1$
56. $x = 3 - y^2,\ y = x - 1$

57. Determine whether the points $(1, 2)$, $(1, -1)$, $(4, 5)$ lie on the graph of $2x - y - 3 = 0$.
58. Determine whether the points $(1, -\sqrt{3})$, $(\frac{1}{2}, -1)$, $(\frac{3}{2}, \frac{7}{2})$ lie on the graph of $x^2 + y^2 = 4$.
59. Determine whether the points $(1, \frac{1}{5})$, $(2, \frac{1}{2})$, $(-1, -2)$ lie on the graph of $x^2y - x^2 + 4y = 0$.
60. Determine whether the points $(0, 2)$, $(-2, -\frac{1}{6})$, $(3, -6)$ lie on the graph of $x^2 - xy + 4y = 3$.
[S] 61. For what values of k does the graph of $y = kx^3$ pass through these points?
    (a) $(1, 4)$ (b) $(-2, 1)$
    (c) $(0, 0)$ (d) $(-1, -1)$
62. For what values of k does the graph of $y^2 = 4kx$ pass through these points?
    (a) $(1, 1)$ (b) $(2, 4)$
    (c) $(0, 0)$ (d) $(3, 3)$
63. Prove that if a graph is symmetric with respect to the x-axis and to the y-axis, then it is symmetric with respect to the origin. Give an example to show that the converse is not true.
64. Prove that if a graph is symmetric with respect to one axis and the origin, then it is symmetric with respect to the other axis also.

[S] Detailed solution in *Student Solutions Manual.*   [C] Calculator may be helpful.

For Review and Class Discussion

True or False

1. _____ The graph of an equation may contain an infinite number of solution points.

2. _____ The graph of an equation must contain an infinite number of solution points.

3. _____ The graph of an equation may have no x- or y-intercepts.

4. _____ The graph of $y = x^2 + a$ has no x-intercepts.

5. _____ If the graph of $y = x^2 + a$ has only one x-intercept, then $a = 0$.

6. _____ If a graph is symmetric with respect to the origin, then it is also symmetric with respect to either the x-axis or the y-axis.

7. _____ If $(1, -2)$ is a point on a graph that is symmetric with respect to the x-axis, then $(-1, -2)$ is also a point on the graph.

8. _____ If $(1, -2)$ is a point on a graph that is symmetric with respect to the y-axis, then $(-1, -2)$ is also a point on the graph.

9. _____ If $b^2 - 4ac > 0$ and $a \neq 0$, then the graph of $y = ax^2 + bx + c$ has two x-intercepts.

10. _____ If $b^2 - 4ac = 0$ and $a \neq 0$, then the graph of $y = ax^2 + bx + c$ has only one x-intercept.

1.4 Lines in the Plane; Slope

Purpose
- To define the slope of a line.
- To investigate various forms of the equation of a line.

If any could work with perfect accuracy, he would be the most perfect mechanic of all, for the description of right lines and circles, upon which geometry is founded, belongs to mechanics. Geometry does not teach us to draw these lines, but requires them to be drawn, for it requires that the learner should first be taught to describe these accurately before he enters upon geometry, then it shows how by these operations problems may be solved. —Isaac Newton (1642–1727)

In Section 1.3 we were introduced to one of the major problems of analytic geometry. That is, "Given an equation, what is its graph?" We begin this section by looking at another major problem of analytic geometry: "Given a graph, what is its equation?"

Let us first consider one of the simplest graphs—a straight line—and see if we can find its equation. (Throughout this text we will use the term "line" to mean a *straight* line, unless indicated otherwise.) From Figure 1.28 we can see that a horizontal line can be identified by its constant y-coordinate and that a vertical line can be identified by its constant x-coordinate. Thus $x = a$ is the equation of a *vertical* line through the point (a, b), whereas $y = b$ is the equation of a *horizontal* line through the point (a, b).

It is somewhat more complicated to determine the equation of a line that is neither vertical nor horizontal. To determine the equation of such a line, we introduce the notion of *slope*. By the **slope** m of a line we mean

Horizontal Line

Vertical Line

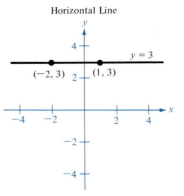

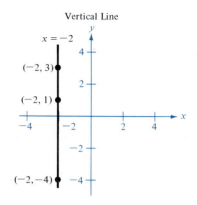

FIGURE 1.28

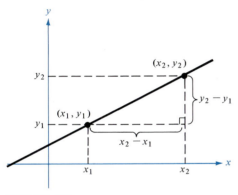

FIGURE 1.29

the number of units a line rises (or falls) vertically for each unit of horizontal change from left to right. For instance, consider the two points (x_1, y_1) and (x_2, y_2) on the line in Figure 1.29. As we move from left to right along this line, a change of $(y_2 - y_1)$ units in the vertical direction corresponds to a change of $(x_2 - x_1)$ units in the horizontal direction. We denote these two changes by the symbols

$$\Delta y = y_2 - y_1 = \text{the change in } y$$

and $$\Delta x = x_2 - x_1 = \text{the change in } x$$

(Δ is the Greek capital letter delta, and the symbols Δy and Δx are read "delta y" and "delta x.") We use the ratio of Δy to Δx to define the slope of a line as follows:

Definition of Slope of a Line	The **slope** m of the line passing through the points (x_1, y_1) and (x_2, y_2) is $$m = \frac{\Delta y}{\Delta x} = \frac{y_2 - y_1}{x_2 - x_1}$$ where $x_1 \neq x_2$.

We make the following observations about the slope,

$$m = \frac{y_2 - y_1}{x_2 - x_1} = \frac{\Delta y}{\Delta x} = \frac{\text{change in } y}{\text{change in } x} = \frac{\text{rise}}{\text{run}}$$

1. If a line is *vertical*, then $x_1 = x_2$ and its slope is *undefined*. Informally, vertical lines may be said to have infinite slope.
2. If a line is *horizontal*, then $y_1 = y_2$ and its slope is *zero*.
3. If a line *rises* to the right, then its slope is *positive*, because both Δy and Δx are positive (see Figure 1.30).

4. If a line *falls* to the right, then its slope is *negative,* because Δy is negative and Δx is positive (see Figure 1.30).

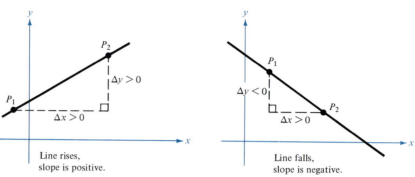

FIGURE 1.30

5. The order in which the two points are chosen does not matter, because

$$\frac{y_2 - y_1}{x_2 - x_1} = \frac{y_1 - y_2}{x_1 - x_2}$$

Example 1

Find the slopes of the lines containing each of the following pairs of points:

$$L_1: (-2, 0) \text{ and } (3, 1)$$
$$L_2: (-1, 2) \text{ and } (2, 2)$$
$$L_3: (0, 4) \text{ and } (1, -1)$$

Solution: For L_1 the slope is

$$m_1 = \frac{1 - 0}{3 - (-2)} = \frac{1}{3 + 2} = \frac{1}{5}$$

For L_2 the slope is

$$m_2 = \frac{2 - 2}{2 - (-1)} = \frac{0}{3} = 0$$

For L_3 the slope is

$$m_3 = \frac{-1 - 4}{1 - 0} = \frac{-5}{1} = -5$$

See Figure 1.31.

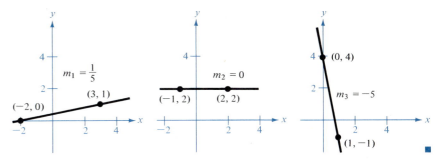

FIGURE 1.31

Example 2

In Exercise 13 of Section 1.2, we used the Distance Formula to show that the points

$$(0, -4), (2, 0), (3, 2)$$

all lie on a straight line. Show that the slope of this line is 2, no matter which pair of points are used in the calculation.

Solution: Using $(0, -4)$ and $(2, 0)$, we have

$$m = \frac{0 - (-4)}{2 - 0} = \frac{4}{2} = 2$$

Using $(2, 0)$ and $(3, 2)$, we have

$$m = \frac{2 - 0}{3 - 2} = \frac{2}{1} = 2$$

Using $(0, -4)$ and $(3, 2)$, we have

$$m = \frac{2 - (-4)}{3 - 0} = \frac{6}{3} = 2$$

Example 2 suggests that any two points on a line can be used to calculate its slope. This can be verified from the similar triangles shown in Figure 1.32. (Recall that the ratios of corresponding sides of similar triangles are equal.)

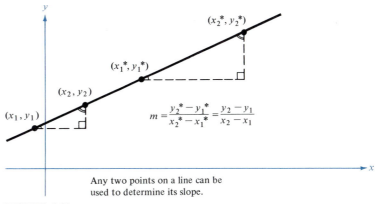

Any two points on a line can be
used to determine its slope.

FIGURE 1.32

The following theorem explicitly identifies the relationship between points on a line and the slope of the line.

THEOREM 1.4	The distinct points (x, y), (x_1, y_1), and (x_2, y_2) lie on the same nonvertical line if and only if $$\frac{y - y_1}{x - x_1} = m = \frac{y_2 - y_1}{x_2 - x_1}$$

Example 3

Show that the point $(0, 5)$ is on the line that passes through $(-2, 4)$ and $(4, 7)$.

Solution: Applying Theorem 1.4 we have

$$\frac{5 - 4}{0 - (-2)} = \frac{1}{2} \quad \text{and} \quad \frac{7 - 4}{4 - (-2)} = \frac{3}{6} = \frac{1}{2}$$

Thus $(0, 5)$ must be on the line containing $(-2, 4)$ and $(4, 7)$.   ∎

We began this section with the objective of determining the equation of a straight line. For a nonvertical line Theorem 1.4 gives us such an equation. For, if (x, y) and (x_1, y_1) are any distinct points that lie on the line of slope m that passes through (x_1, y_1), then

$$\frac{y - y_1}{x - x_1} = m$$

The restriction that (x, y) and (x_1, y_1) be distinct points can be removed if this last equation is replaced with

$$y - y_1 = m(x - x_1)$$

which is commonly referred to as the **point-slope** equation of a line.

Point-Slope Equation of a Line	The equation of the line with slope m passing through the point (x_1, y_1) is given by $$y - y_1 = m(x - x_1)$$

Example 4

Find the equation of the line that has a slope of 3 and passes through the point $(1, -2)$.

Solution: Using the point-slope form,

$$y - y_1 = m(x - x_1)$$

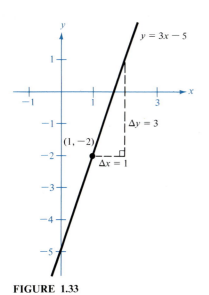

$y = 3x - 5$

$(1, -2)$

$\Delta y = 3$

$\Delta x = 1$

FIGURE 1.33

we have
$$y - (-2) = 3(x - 1)$$
$$y + 2 = 3x - 3$$
or
$$y = 3x - 5$$

See Figure 1.33. ■

In Example 4 note that the line given by
$$y = 3x - 5$$
has a slope of 3 and crosses the y-axis when $y = -5$. In keeping with this observation, we say that the equation
$$y = mx + b$$
is in the **slope-intercept form,** where m is the slope and $(0, b)$ is the y-intercept. The equation of a vertical line cannot be written in the slope-intercept form, since its slope is undefined. However, *every* line has an equation that can be written in the form
$$Ax + By + C = 0$$
where A and B are not both zero. For vertical and horizontal lines the equations $x = a$ and $y = b$ can be written as
$$x - a = 0 \quad \text{and} \quad y - b = 0$$
Furthermore, the slope-intercept equation can be written as
$$-mx + y - b = 0$$
Therefore, the form $\qquad Ax + By + C = 0$

is called the **general equation** of a line.

We now have identified the following five forms of equations of lines:

Equation of Lines	General equation: $Ax + By + C = 0$
	Point-slope equation: $y - y_1 = m(x - x_1)$
	Slope-intercept equation: $y = mx + b$
	Vertical line: $x = a$
	Horizontal line: $y = b$

The slope of a line is a convenient tool for determining when two lines are parallel or perpendicular. This is seen in the following two theorems.

THEOREM 1.5 (*Parallel Lines*)	Two distinct nonvertical lines are parallel if and only if their slopes are equal.

Proof: The phrase "if and only if" is a way of stating two theorems in one. One theorem says that "if two nonvertical lines are parallel, then they must have equal slopes." The other theorem is the converse, which says, "If two distinct nonvertical lines have equal slopes, they must be parallel." We will prove the first of these two theorems and leave the proof of the converse as an exercise (see Exercise 62).

Assume that we have two parallel lines L_1 and L_2 with slopes m_1 and m_2. If these lines are both horizontal, then $m_1 = m_2 = 0$, and the theorem is established. If L_1 and L_2 are not horizontal, then they must intersect the x-axis at points $(x_1, 0)$ and $(x_2, 0)$, as shown in Figure 1.34. Since L_1 and L_2 are parallel, their intersection with the x-axis must produce equal angles α_1 and α_2. (α is the Greek lowercase letter alpha.) Therefore, the two right triangles with vertices

$$(x_1, 0), (x_3, 0), (x_3, y_1)$$

and

$$(x_2, 0), (x_3, 0), (x_3, y_2)$$

must be similar. From this we conclude that the ratios of their corresponding sides must be equal, and thus

$$m_1 = \frac{y_1 - 0}{x_3 - x_1} = \frac{y_2 - 0}{x_3 - x_2} = m_2$$

Hence the lines L_1 and L_2 must have equal slopes. ∎

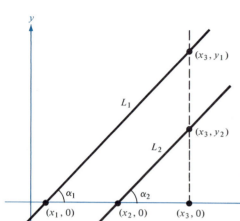

FIGURE 1.34

THEOREM 1.6
(*Perpendicular Lines*)

Two nonvertical lines are perpendicular if and only if their slopes are related by the equation

$$m_1 = \frac{-1}{m_2}$$

Proof: As in the previous theorem, we will prove one direction of the theorem and leave the other as an exercise (see Exercise 63). Let us assume that we are given two nonvertical perpendicular lines L_1 and L_2 with slopes m_1 and m_2. For simplicity's sake let these two lines intersect at the origin, as in Figure 1.35. The vertical line $x = 1$ will intersect L_1 and L_2 at the respective points $(1, m_1)$ and $(1, m_2)$. Since the triangle formed by these two points and the origin is a right triangle, we can apply the Pythagorean Theorem and conclude that

$$\left(\begin{matrix}\text{distance between}\\ (0, 0) \text{ and } (1, m_1)\end{matrix}\right)^2 + \left(\begin{matrix}\text{distance between}\\ (0, 0) \text{ and } (1, m_2)\end{matrix}\right)^2 = \left(\begin{matrix}\text{distance between}\\ (1, m_1) \text{ and } (1, m_2)\end{matrix}\right)^2$$

Using the Distance Formula we have

$$(\sqrt{1 + m_1{}^2})^2 + (\sqrt{1 + m_2{}^2})^2 = (\sqrt{0^2 + (m_1 - m_2)^2})^2$$
$$1 + m_1{}^2 + 1 + m_2{}^2 = (m_1 - m_2)^2$$

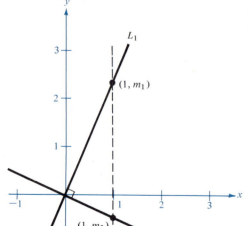

FIGURE 1.35

$$2 + m_1{}^2 + m_2{}^2 = m_1{}^2 - 2m_1m_2 + m_2{}^2$$
$$2 = -2m_1m_2$$
$$-1 = m_1m_2$$

Dividing both sides of this equation by m_2, we have our desired conclusion,

$$\frac{-1}{m_2} = m_1 \qquad \blacksquare$$

Example 5

Find the equation of the line that passes through the point $(2, -1)$ and is

(a) parallel to the line $2x - 3y = 5$
(b) perpendicular to the line $2x - 3y = 5$

Solution: By writing the equation $2x - 3y = 5$ in the slope-intercept form, we have

$$3y = 2x - 5$$

or

$$y = (\tfrac{2}{3})x - \tfrac{5}{3}$$

Therefore, the given line has a slope of $m = \tfrac{2}{3}$.

(a) Any line parallel to the given line $2x - 3y = 5$ must have a slope of $\tfrac{2}{3}$. Thus the line through $(2, -1)$ that is parallel to the line $2x - 3y = 5$ has an equation of the form

$$y - (-1) = \tfrac{2}{3}(x - 2)$$
$$3(y + 1) = 2(x - 2)$$
$$3y + 3 = 2x - 4$$
$$-2x + 3y = -7$$

or

$$2x - 3y = 7$$

(Note the similarity to the original equation $2x - 3y = 5$.) See Figure 1.36.

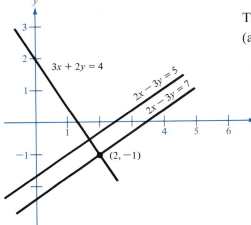

3x + 2y = 4

2x − 3y = 5

2x − 3y = 7

(2, −1)

FIGURE 1.36

(b) Any line perpendicular to the line $2x - 3y = 5$ (whose slope is $\tfrac{2}{3}$) must have a slope of $-\tfrac{3}{2}$. Therefore, the line through $(2, -1)$ that is perpendicular to the line $2x - 3y = 5$ has the equation

$$y - (-1) = -\tfrac{3}{2}(x - 2)$$
$$2(y + 1) = -3(x - 2)$$

or

$$3x + 2y = 4$$

See Figure 1.36. $\qquad \blacksquare$

Section Exercises (1.4)

In Exercises 1–8, plot the points and find the slope of the line passing through the given points.

[S] **1.** $(3, -4)$, $(5, 2)$ **2.** $(-2, 1)$, $(4, -3)$
3. $(\frac{1}{2}, 2)$, $(6, 2)$ **4.** $(-\frac{3}{2}, -5)$, $(\frac{5}{6}, 4)$
5. $(-6, -1)$, $(-6, 4)$ **6.** $(2, 1)$, $(2, 5)$
7. $(1, 2)$, $(-2, -2)$ **8.** $(\frac{7}{8}, \frac{3}{4})$, $(\frac{5}{4}, -\frac{1}{4})$

In Exercises 9–27, write the general form $(Ax + By + C = 0)$ of the equation of the indicated line and sketch its graph.

[S] **9.** through $(2, 1)$ and $(0, -3)$
10. through $(-3, -4)$ and $(1, 4)$
11. through $(0, 0)$ and $(-1, 3)$
12. through $(-3, 6)$ and $(1, 2)$
13. through $(2, 3)$ and $(2, -2)$
14. through $(6, 1)$ and $(10, 1)$
15. through $(1, -2)$ and $(3, -2)$
16. through $(\frac{7}{8}, \frac{3}{4})$ and $(\frac{5}{4}, -\frac{1}{4})$
[S] **17.** through $(0, 3)$, $m = \frac{3}{4}$
18. through $(-1, 2)$, m is undefined
19. through $(0, 0)$, $m = \frac{2}{3}$
20. through $(-1, -4)$, $m = \frac{1}{4}$
21. through $(0, 5)$, $m = -2$
22. through $(-2, 4)$, $m = -\frac{3}{5}$
23. y-intercept 2, $m = 4$
24. y-intercept $-\frac{2}{3}$, $m = \frac{1}{6}$
25. y-intercept $\frac{2}{3}$, $m = \frac{3}{4}$
26. y-intercept 4, $m = 0$
27. vertical line with x-intercept 3

28. Show that an equation of the straight line with x-intercept a and y-intercept b is

$$\frac{x}{a} + \frac{y}{b} = 1, \quad a \neq 0, \quad b \neq 0$$

In Exercises 29–36, use the result of Exercise 28 to write an equation of the line indicated.

[S] **29.** x-intercept 2, y-intercept 3
30. x-intercept -3, y-intercept 4
31. x-intercept $-\frac{1}{6}$, y-intercept $-\frac{2}{3}$
32. x-intercept $-\frac{2}{3}$, y-intercept -2
33. through the point $(1, 2)$ with equal intercepts
34. through the point $(-3, 4)$ with equal intercepts
35. through the point $(\frac{3}{2}, \frac{1}{2})$ with the x-intercept twice the y-intercept

36. through the point $(-3, 1)$ with the intercepts of equal absolute value and opposite signs

In Exercises 37–42, write an equation of the line through the given point (a) parallel and (b) perpendicular to the given line.

[S] **37.** $(2, 1)$; $4x - 2y = 3$ **38.** $(-3, 2)$; $x + y = 7$
39. $(\frac{7}{8}, \frac{3}{4})$; $5x + 3y = 0$ **40.** $(-6, 4)$; $3x + 4y = 1$
41. $(2, 5)$; $x = 4$ **42.** $(-1, 0)$; $y = -3$

In Exercises 43–45, use Theorem 1.4 to determine if the three given points are collinear (lie on the same straight line). (See Section 1.2, Exercises 13–15.)

[S] **43.** $(0, -4)$, $(2, 0)$, $(3, 2)$
44. $(0, 4)$, $(7, -6)$, $(-5, 11)$
45. $(-2, 1)$, $(-1, 0)$, $(2, -2)$

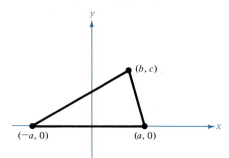

FIGURE 1.37

In Exercises 46–49, refer to the triangle in Figure 1.37.
46. Find the point of intersection of the medians.
47. Find the point of intersection of the perpendicular bisectors of the sides.
48. Find the point of intersection of the altitudes.
49. Show that the points of intersection of Exercises 46, 47, and 48 are collinear.

50. Prove analytically that the diagonals of a rhombus intersect at right angles.
51. Prove that the figure formed by connecting consecutive midpoints of the sides of any quadrilateral is a parallelogram.
52. Find the equation of the line giving the relationship between the temperature in degrees Celsius C and in degrees Fahrenheit F, knowing that water freezes at 0°C, or 32°F, and boils at 100°C, or 212°F.

53. Use the result of Exercise 52 to complete the accompanying table.

C		−10°	10°			177°
F	0°			68°	90°	

In Exercises 54–60, use the result that **the distance from the point** (x_1, y_1) **to the line** $Ax + By + C = 0$ is

$$\frac{|Ax_1 + By_1 + C|}{\sqrt{A^2 + B^2}}$$

54. Find the distance from the origin to the line $4x + 3y = 10$.

S **55.** Find the distance from the point $(2, 3)$ to the line $4x + 3y = 10$.

56. Find the distance from the point $(-2, 1)$ to the line $x - y - 2 = 0$.

57. Find the distance from the point $(6, 2)$ to the line $x = -1$.

58. Find the distance between the parallel lines $x + y = 1$ and $x + y = 5$.

S **59.** Find the distance between the parallel lines $3x - 4y = 1$ and $3x - 4y = 10$.

60. Find the equation of the line that bisects the acute angle formed by the lines $y = \frac{3}{4}x$ and $y = 2$.

61. Find the distance from the origin to the line $4x + 3y = 10$ by first finding the point of intersection of the given line and the line through the origin perpendicular to the given line. Now find the distance between the given point and the point of intersection. Compare the result with that of Exercise 54.

62. Complete the proof of Theorem 1.5.

63. Complete the proof of Theorem 1.6.

For Review and Class Discussion

True or False

1. ____ A horizontal line has a slope of zero.

2. ____ If a line contains points in both the first and third quadrants, then its slope must be positive.

3. ____ The equation of any line can be written in point-slope form.

4. ____ The lines represented by $ax + by = c_1$ and $bx - ay = c_2$ are perpendicular.

5. ____ If the x- and y-intercepts of a line are nonzero and rational, then the slope of the line is rational.

6. ____ The equation of any line can be written in general form.

7. ____ It is possible for two mutually perpendicular lines to both have positive slope.

8. ____ If $0 < a < b$, then the line passing through (a, b) and (a^2, b^2) has a positive slope.

9. ____ If $a < b$, then the line whose intercepts are $(a, 0)$ and $(0, b)$ has a negative slope.

10. ____ If two distinct lines have the same slope, they must be parallel.

1.5 Circles and Ellipses

Purpose

■ To introduce the standard forms of equations of circles and ellipses.

Some problems require only circles and straight lines for their construction, while others require a conic section and still others require more complex curves. —René Descartes (1596–1650)

Once we developed the idea of the slope of a line, it was a simple matter to determine the equation of the line having slope m and passing through the point (x_1, y_1). This line simply consists of all the points (x, y) that

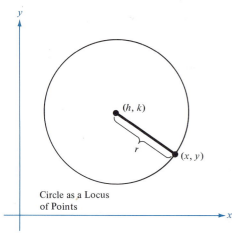

Circle as a Locus
of Points

FIGURE 1.38

satisfy the equation

$$y - y_1 = m(x - x_1)$$

In a similar manner, we can find the equation of any circle in the plane having the point (h, k) as its center and r as its radius. A point (x, y) is on this circle if and only if its distance from the center (h, k) is r (Figure 1.38).

This means that a circle consists of the set of all points (x, y) that are at a given positive distance r from a fixed point (h, k). Expressing this relationship in terms of the Distance Formula, we have

$$\sqrt{(x - h)^2 + (y - k)^2} = r$$

By squaring both sides of this equation, we obtain the *standard form* of the equation of the circle, which is part of the following theorem.

THEOREM 1.7 (*Standard Equation of a Circle*)	The point (x, y) lies on the circle of radius r and center (h, k) if and only if $$(x - h)^2 + (y - k)^2 = r^2$$

It follows from Theorem 1.7 that the equation of a circle with its center at the origin is simply

$$x^2 + y^2 = r^2$$

Example 1

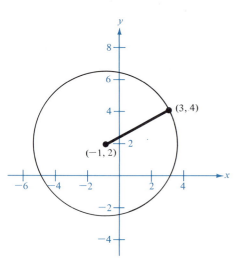

FIGURE 1.39

The point $(3, 4)$ lies on a circle whose center is at $(-1, 2)$ (Figure 1.39). Find an equation for the circle.

Solution: The radius of the circle is the distance between $(-1, 2)$ and $(3, 4)$. Thus

$$r = \sqrt{[3 - (-1)]^2 + (4 - 2)^2} = \sqrt{16 + 4} = \sqrt{20}$$

Therefore, by Theorem 1.7 the standard equation for this circle is

$$[x - (-1)]^2 + (y - 2)^2 = (\sqrt{20})^2$$
$$(x + 1)^2 + (y - 2)^2 = 20$$

∎

If we remove parentheses in the standard equation of Example 1, we obtain

$$(x + 1)^2 + (y - 2)^2 = 20$$
$$x^2 + 2x + 1 + y^2 - 4y + 4 = 20$$
$$x^2 + y^2 + 2x - 4y - 15 = 0$$

where the latter equation is in the **general form of the equation of a circle:**

$$Ax^2 + Ay^2 + Dx + Ey + F = 0, \qquad A \neq 0$$

(Note that a general second-degree equation in x and y has the form

$$Ax^2 + Bxy + Cy^2 + Dx + Ey + F = 0$$

We will discuss this equation in more detail in Section 11.4.) The general form of the equation is less useful than the corresponding standard form. For instance, we know little about the circle with general equation

$$x^2 + y^2 - 6x + 10y + 24 = 0$$

However, from the corresponding standard form of this equation,

$$(x - 3)^2 + (y + 5)^2 = 10$$

we can readily see that the circle is centered at $(3, -5)$ and that its radius is $\sqrt{10}$. This observation suggests that to graph the equation of a circle, it is best to write the equation in standard form. This can be accomplished by using the algebraic process called **completing the square,** which we demonstrate in the following example.

Example 2

Sketch the graph of the circle whose general equation is

$$4x^2 + 4y^2 + 20x - 16y + 37 = 0$$

Solution: To complete the square we will first divide by 4 so that the coefficients of x^2 and y^2 are both 1. Thus we have

$$x^2 + y^2 + 5x - 4y + \tfrac{37}{4} = 0$$

Then we write

$$(x^2 + 5x +) + (y^2 - 4y +) = -\tfrac{37}{4}$$

reserving space to add the square of half the coefficient of x and the square of half the coefficient of y to both sides of the equation. Thus we obtain

$$(x^2 + 5x + \tfrac{25}{4}) + (y^2 - 4y + 4) = -\tfrac{37}{4} + \tfrac{25}{4} + 4$$
$$(x + \tfrac{5}{2})^2 + (y - 2)^2 = 1$$

Therefore, the circle is centered at $(-\tfrac{5}{2}, 2)$ and its radius is 1 (Figure 1.40). ∎

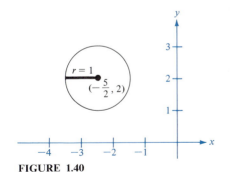

FIGURE 1.40

Example 3

Discuss the graph of

$$3x^2 + 3y^2 + 24x - 6y + 51 = 0$$

Solution: First, dividing by 3 we obtain

$$x^2 + y^2 + 8x - 2y + 17 = 0$$

Then we write

$$(x^2 + 8x + \quad) + (y^2 - 2y + \quad) = -17$$
$$(x^2 + 8x + 16) + (y^2 - 2y + 1) = -17 + 16 + 1$$
$$(x + 4)^2 + (y - 1)^2 = 0$$

The sum on the left side of the last equation can be zero only if both terms are zero. This is true only when $x = -4$ and $y = 1$. Therefore, the graph consists of the single point $(-4, 1)$.   ∎

Example 3 shows that the general equation $Ax^2 + Ay^2 + Dx + Ey + F = 0$ may not always represent a circle. In fact, such an equation may have no solution at all if the procedure of completing the square yields the impossible result

$$(x - h)^2 + (y - k)^2 = \text{(negative number)}$$

The circle is one example of a class of curves called **conic sections.** We will study these in detail in Chapter 11. In the meantime we give an informal introduction to one other type of conic section, the ellipse. We give here the standard equation of the ellipse and postpone a technical development of this form until later.

Standard Equation of Ellipse

The graph of each of the equations

$$\frac{(x - h)^2}{a^2} + \frac{(y - k)^2}{b^2} = 1, \qquad \text{major axis horizontal } (a > b)$$

$$\frac{(x - h)^2}{b^2} + \frac{(y - k)^2}{a^2} = 1, \qquad \text{major axis vertical } (a > b)$$

is an ellipse with center at (h, k), where $2a$ is the length of the *major axis* and $2b$ the length of the *minor axis* of each ellipse (Figure 1.41).

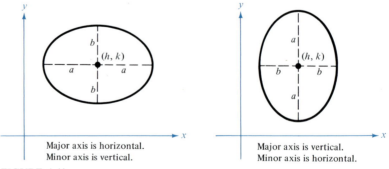

Major axis is horizontal.
Minor axis is vertical.

Major axis is vertical.
Minor axis is horizontal.

FIGURE 1.41

A circle is sometimes referred to as a "degenerate" ellipse whose major and minor axes are equal in length. This is readily seen by letting $a = b$

in the standard equation of the ellipse. We obtain

$$\frac{(x-h)^2}{a^2} + \frac{(y-k)^2}{a^2} = 1$$

which can be written in the form

$$(x-h)^2 + (y-k)^2 = a^2$$

which is the standard equation of a circle with center at (h, k) and radius a.

Example 4

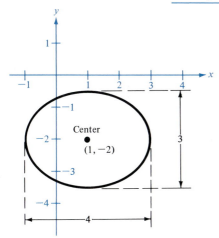

FIGURE 1.42

Find the general equation for the ellipse shown in Figure 1.42.

Solution: Considering the standard form

$$\frac{(x-h)^2}{a^2} + \frac{(y-k)^2}{b^2} = 1$$

we have $h = 1$, $k = -2$, $a = \frac{4}{2} = 2$, and $b = \frac{3}{2}$. Thus the equation is

(standard form) $\qquad \dfrac{(x-1)^2}{2^2} + \dfrac{[y-(-2)]^2}{(\frac{3}{2})^2} = 1$

$$\frac{(x-1)^2}{4} + \frac{4(y+2)^2}{9} = 1$$

$$9(x-1)^2 + 16(y+2)^2 = 36$$

$$9x^2 - 18x + 9 + 16y^2 + 64y + 64 = 36$$

(general form) $\qquad 9x^2 + 16y^2 - 18x + 64y + 37 = 0 \qquad \blacksquare$

Example 5

Sketch the graph of the ellipse whose general equation is

$$x^2 + 4y^2 + 6x - 8y + 9 = 0$$

Solution: As in the case of a circle, we begin by completing the square:

$$(x^2 + 6x + \quad) + 4(y^2 - 2y + \quad) = -9$$

(Note that 4 is factored from the terms involving y.) Then we have

$$(x^2 + 6x + 9) + 4(y^2 - 2y + 1) = -9 + 9 + 4$$

$$(x+3)^2 + 4(y-1)^2 = 4$$

$$\frac{(x+3)^2}{2^2} + \frac{(y-1)^2}{1^2} = 1$$

General Form:
$x^2 + 4y^2 + 6x - 8y + 9 = 0$
Standard Form:
$\dfrac{(x+3)^2}{4} + \dfrac{(y-1)^2}{1} = 1$

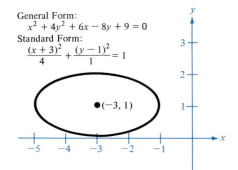

FIGURE 1.43

Thus the ellipse has its center at $(-3, 1)$, a horizontal major axis of length 4, and a vertical minor axis of length 2 (see Figure 1.43). $\qquad \blacksquare$

Section Exercises (1.5)

In Exercises 1–10, write the general equation of the circle with:

1. center at the origin and radius 3
2. center at the origin and radius 5
[S] 3. center at $(2, -1)$ and radius 4
4. center at $(-4, 3)$ and radius $\frac{5}{8}$
5. center at $(-1, 2)$ and passing through the origin
6. center at $(3, -2)$ and passing through $(-1, 1)$
7. points $(2, 5)$ and $(4, -1)$ as ends of a diameter
8. points $(0, 0)$, $(0, 8)$, and $(6, 0)$ on the circle
[S] 9. center at the origin and tangent to $x - y = 10$
10. center at $(-1, -2)$ and tangent to $x - y = 10$

In Exercises 11–18, write each equation in standard form and sketch its graph.

11. $x^2 + y^2 - 2x + 6y + 6 = 0$
12. $x^2 + y^2 - 2x + 6y - 15 = 0$
13. $x^2 + y^2 - 2x + 6y + 10 = 0$
14. $3x^2 + 3y^2 - 6y - 1 = 0$
[S] 15. $2x^2 + 2y^2 - 2x - 2y - 3 = 0$
16. $4x^2 + 4y^2 - 4x + 2y - 1 = 0$
17. $16x^2 + 16y^2 + 16x + 40y - 7 = 0$
18. $x^2 + y^2 - 4x + 2y + 3 = 0$

In Exercises 19–33, sketch the graph of the given equation.

19. $\dfrac{x^2}{25} + \dfrac{y^2}{16} = 1$

20. $\dfrac{x^2}{9} + \dfrac{y^2}{4} = 1$

21. $\dfrac{x^2}{16} + \dfrac{y^2}{25} = 1$

22. $\dfrac{x^2}{4} + \dfrac{y^2}{9} = 1$

23. $16x^2 + 9y^2 = 144$

24. $2x^2 + 3y^2 = 6$

25. $4x^2 + y^2 = 1$

26. $x^2 + 16y^2 = 1$

27. $\dfrac{(x - 1)^2}{25} + \dfrac{(y + 2)^2}{16} = 1$

28. $\dfrac{(x + 5)^2}{4} + \dfrac{(y - 3)^2}{9} = 1$

[S] 29. $4x^2 + 9y^2 - 8x + 9y + 4 = 0$
30. $16x^2 + 9y^2 + 96x + 36y + 36 = 0$
31. $4x^2 + y^2 - 4x + 2y + 1 = 0$
32. $9x^2 + 4y^2 - 36x + 8y + 31 = 0$
33. $3x^2 + 4y^2 - 18x + 8y + 19 = 0$

34. Find the equation of the circle passing through the points $(1, 2)$, $(-1, 2)$, and $(2, 1)$.
[S] 35. Find the equation of the circle passing through the points $(4, 3)$, $(-2, -5)$, and $(5, 2)$.
36. Find the equations of the circles passing through the points $(4, 1)$ and $(6, 3)$ and having radius $\sqrt{10}$.
[S] 37. Write an equation of the ellipse centered at $(4, 5)$ with the major axis vertical and 8 units in length and the minor axis 6 units in length.
38. Write an equation of the ellipse centered at $(-2, 1)$ with the major axis horizontal and 4 units in length and the minor axis 2 units in length.
39. Sketch the set of all points for which $x^2 + y^2 - 4x + 2y + 1 \leq 0$.
40. Sketch the set of all points for which $x^2 + y^2 - 4x + 2y + 1 > 0$.
41. Sketch the set of points for which $(x + 3)^2 + (y - 1)^2 < 9$.
42. Sketch the set of points for which $(x - 1)^2 + (y - \frac{1}{2})^2 > 1$.
[S] 43. Prove that an angle inscribed in a semicircle is a right angle.
44. Prove that the perpendicular bisector of any chord of a circle passes through the center of the circle.

[S] Detailed solution in *Student Solutions Manual.* [C] Calculator may be helpful.

For Review and Class Discussion

True or False

1. _____ It is possible to construct a circle whose radius and circumference are both rational.

2. _____ If $ax^2 + by^2 + cx + dy + e = 0$ represents the equation of a circle, then a and b must be equal.

3. _____ Any three distinct points determine a circle.

4. _____ If $a \neq b$, then the graph of the equation $ax^2 + by^2 + cx + dy + e = 0$ is an ellipse.

5. _____ It is possible for a circle and an ellipse to have one, two, three, or four points of intersection.

1.6 **Functions**

Purpose
- To define the term "function."
- To introduce functional notation.
- To characterize the inverse of a function.

We here denote by a function of a variable quantity, a quantity composed in some way or other of this variable quantity and constants.   —*John Bernoulli (1667–1748)*

In many common relationships between two variables, the value of one of the variables depends on the value of the other. For example, the sales tax on an item depends on its selling price; the distance an object moves in a given time depends on its speed; the pressure of a gas in a closed container depends on its temperature; and the area of a circle depends on its radius.

Consider the relationship between the area of a circle and its radius. This relationship can be expressed by the simple equation

$$A = \pi r^2$$

Considering the radius of a circle to be positive, we have within the set of positive numbers a free choice for the value of r. The value of A then depends on our choice of r. Thus we refer to A as the **dependent variable** and r as the **independent variable.**

The *type* of relationship between two variables is of extreme importance in calculus. Specifically, we are interested in relationships such that to every value of the independent variable there corresponds *one and only one* value of the dependent variable. Mathematically we call this type of correspondence a **function.**

Definition of Function	A **function** is a relationship between two variables such that to each value of the independent variable there corresponds exactly one value of the dependent variable.

A **function** is a relationship between two variables such that to each value of the independent variable there corresponds exactly one value of the dependent variable.

The collection of all values assumed by the independent variable is called the **domain** of the function, and the collection of all values assumed by the dependent variable is called the **range** of the function.

If to each value in the range there corresponds exactly one value in the domain, the function is said to be **one-to-one.**

This definition means that if y is a one-to-one (also denoted by 1–1) function of x, then two different x-values cannot correspond to the same y-value and vice versa.

Although functions can be described by various means, we usually specify functions by formulas, or equations.

Example 1

Which of the following equations define functional relationships between the variables x and y? (See Figure 1.44.)

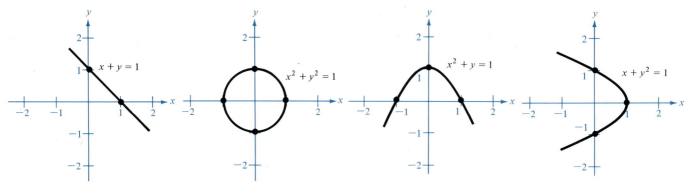

FIGURE 1.44

(a) $x + y = 1$
(b) $x^2 + y^2 = 1$
(c) $x^2 + y = 1$
(d) $x + y^2 = 1$

Solution: The standard procedure in writing an equation for a function is to isolate the dependent variable on the left-hand side. Thus we have Table 1.2.

Notice that those equations that assign two values ($\pm$) to the dependent variable for each assigned value of the independent variable do not define functions. For instance, if $y = 0$, then the equation $x = \pm\sqrt{1 - y^2}$ indicates that $x = +1$ or $x = -1$. ∎

TABLE 1.2

Original equation	x as the dependent variable	Is x a function of y?	y as the dependent variable	Is y a function of x?	Is y a 1–1 function of x?
(a) $x + y = 1$	$x = 1 - y$	yes	$y = 1 - x$	yes	yes
(b) $x^2 + y^2 = 1$	$x = \pm\sqrt{1 - y^2}$	no (two values of x for some y)	$y = \pm\sqrt{1 - x^2}$	no (two values of y for some x)	no
(c) $x^2 + y = 1$	$x = \pm\sqrt{1 - y}$	no	$y = 1 - x^2$	yes	no (two values of x yield some y)
(d) $x + y^2 = 1$	$x = 1 - y^2$	yes	$y = \pm\sqrt{1 - x}$	no	no

Example 2

Determine the domain and range for the function of x defined by $y = \sqrt{x - 1}$.

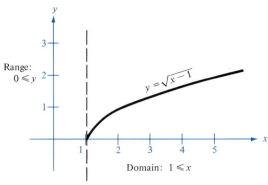

Range:
$0 \leqslant y$

$y = \sqrt{x-1}$

Domain: $1 \leqslant x$

FIGURE 1.45

Solution: Since $\sqrt{x-1}$ is not defined for $x < 1$, we must have $x \geq 1$. Therefore, the interval $[1, \infty)$ is the domain of the function. Since $\sqrt{x-1}$ is always positive or zero on its domain, and since $\sqrt{x-1}$ increases as x increases, the range of the function is the interval $[0, \infty)$. The graph of the function lends further support to our conclusions (Figure 1.45). ∎

On occasion we define a function by using more than one equation.

Example 3

Determine the domain and range for the function of x given by

$$y = \begin{cases} \sqrt{x-1}, & \text{if } x \geq 1 \\ 1-x, & \text{if } x < 1 \end{cases}$$

Solution: Since $x \geq 1$ or $x < 1$, the domain of the function is the entire set of real numbers. On the portion of the domain for which $x \geq 1$, the function behaves as in Example 2. For $x < 1$, $1 - x$ is positive, and, therefore, the range of the function is the interval $[0, \infty)$. Again, a graph of the function helps to verify our conclusions (Figure 1.46).

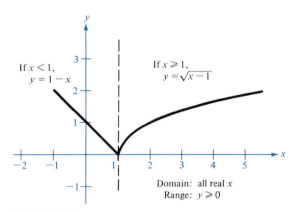

If $x < 1$,
$y = 1 - x$

If $x \geq 1$,
$y = \sqrt{x-1}$

Domain: all real x
Range: $y \geqslant 0$

FIGURE 1.46 ∎

When equations are used to describe functions, sometimes the domain is clearly specified as in Example 3: ($x \geq 1$ or $x < 1$). On other occasions the domain is *implied*, as in Example 2: ($\sqrt{x-1}$ is defined only if $x \geq 1$). On still other occasions the physical nature of the problem may restrict the domain to a certain subset of the real numbers. For instance, the equation for the area of a circle, $A = \pi r^2$, has no specified restrictions; yet physically we always consider the radius of a circle to be positive. Thus we list the domain of this function as all $r > 0$.

When an equation is used to define a function, it is customary to isolate the dependent variable on the left side of the equation. For instance, writing the equation $x + 2y = 1$ in the form

$$y = \frac{1 - x}{2}$$

indicates that y is the dependent variable. In **functional notation** this equation has the form

$$f(x) = \frac{1 - x}{2}$$

This notation has the advantage of clearly identifying the dependent variable as $f(x)$ while at the same time providing a name "f" for the function. [The symbol $f(x)$ is read "f of x."]

To denote the value of the function $f(x) = (1 - x)/2$ when $x = 3$, we use the symbol $f(3)$, and its value is given by

$$f(3) = \frac{1 - (3)}{2} = \frac{-2}{2} = -1$$

Similarly,

$$f(0) = \frac{1 - (0)}{2} = \frac{1}{2}$$

$$f(-2) = \frac{1 - (-2)}{2} = \frac{3}{2}$$

The values $f(3)$, $f(0)$, and $f(-2)$ are called **functional values,** and they lie in the range of f. This means that the values $f(3)$, $f(0)$, and $f(-2)$ are y-values and thus the points $(3, f(3))$, $(0, f(0))$, and $(-2, f(-2))$ lie on the graph of f.

Example 4

If a function f is defined by the equation $f(x) = 2x^2 - 4x + 1$, find the value of f when x is -1, 0, and 2. Is f one-to-one?

Solution: When $x = -1$, the value of f is given by

$$f(-1) = 2(-1)^2 - 4(-1) + 1 = 2 + 4 + 1 = 7$$

When $x = 0$, the value of f is given by

$$f(0) = 2(0)^2 - 4(0) + 1 = 0 - 0 + 1 = 1$$

When $x = 2$, the value of f is given by

$$f(2) = 2(2)^2 - 4(2) + 1 = 8 - 8 + 1 = 1$$

Note that two different values of x may yield the same value for $f(x)$. Thus f is *not* one-to-one.   ∎

Example 4 suggests that the role of the variable x in the equation

$$f(x) = 2x^2 - 4x + 1$$

is simply that of a "placeholder." The same function f can also be described by using parentheses instead of x. Thus f can be defined by the equation

$$f(\ ) = 2(\ )^2 - 4(\ ) + 1$$

Therefore, to evaluate $f(-2)$, we simply place -2 in each parentheses:

$$f(-2) = 2(-2)^2 - 4(-2) + 1 = 2(4) + 8 + 1 = 17$$

Example 5

For the function f defined by $f(x) = x^2 - 4x + 7$, evaluate

$$f(3x), \qquad f(x - 1), \qquad f(x + \Delta x), \qquad \frac{f(x + \Delta x) - f(x)}{\Delta x}$$

Solution: Writing the equation for f in the form

$$f(\ ) = (\ )^2 - 4(\ ) + 7$$

we have

$$\begin{aligned} f(3x) &= (3x)^2 - 4(3x) + 7 \\ &= 9x^2 - 12x + 7 \end{aligned}$$

Similarly,

$$\begin{aligned} f(x - 1) &= (x - 1)^2 - 4(x - 1) + 7 \\ &= x^2 - 2x + 1 - 4x + 4 + 7 \\ &= x^2 - 6x + 12 \end{aligned}$$

and

$$\begin{aligned} f(x + \Delta x) &= (x + \Delta x)^2 - 4(x + \Delta x) + 7 \\ &= x^2 + 2x\,\Delta x + (\Delta x)^2 - 4x - 4\,\Delta x + 7 \end{aligned}$$

Finally,

$$\begin{aligned} \frac{f(x + \Delta x) - f(x)}{\Delta x} &= \frac{[(x + \Delta x)^2 - 4(x + \Delta x) + 7] - [x^2 - 4x + 7]}{\Delta x} \\ &= \frac{x^2 + 2x\,\Delta x + (\Delta x)^2 - 4x - 4\,\Delta x + 7 - x^2 + 4x - 7}{\Delta x} \\ &= \frac{2x\,\Delta x + (\Delta x)^2 - 4\,\Delta x}{\Delta x} \\ &= 2x + \Delta x - 4 \end{aligned}$$

■

Although we generally use f as a convenient function name and x as the independent variable, we can use other symbols. For instance, the following equations all define the same function:

$$\begin{aligned} f(x) &= x^2 - 4x + 7 \\ f(t) &= t^2 - 4t + 7 \\ g(s) &= s^2 - 4s + 7 \end{aligned}$$

Two functions can be combined in various ways to create new functions. For example, if

$$f(x) = 2x - 3 \quad \text{and} \quad g(x) = x^2 + 1$$

we can form the functions

$$f(x) + g(x) = (2x - 3) + (x^2 + 1) = x^2 + 2x - 2 \qquad \text{(sum)}$$
$$f(x) - g(x) = (2x - 3) - (x^2 + 1) = -x^2 + 2x - 4 \qquad \text{(difference)}$$
$$f(x)\, g(x) = (2x - 3)(x^2 + 1) = 2x^3 - 3x^2 + 2x - 3 \qquad \text{(product)}$$
$$\frac{f(x)}{g(x)} = \frac{2x - 3}{x^2 + 1} \qquad \text{(quotient)}$$

We can combine two functions in yet another way, called the **composition** of two functions.

Definition of Composite Function	Let f and g be functions such that the range of g is in the domain of f. Then the function whose values are given by $f(g(x))$ is called the **composite** of f with g.

It is important to realize that the composite of f *with* g may not be equal to the composite of g *with* f. This is illustrated in the following example.

Example 6

Given $f(x) = 2x - 3$ and $g(x) = x^2 + 1$, find $f(g(x))$ and $g(f(x))$.

Solution: Since

$$f(\ \ ) = 2(\ \ ) - 3$$

we have

$$f(g(x)) = 2(g(x)) - 3 = 2(x^2 + 1) - 3 = 2x^2 - 1$$

And since
$$g(\ \ ) = (\ \ )^2 + 1$$

we have

$$g(f(x)) = (f(x))^2 + 1 = (2x - 3)^2 + 1 = 4x^2 - 12x + 10 \qquad \blacksquare$$

Note in Example 6 that $f(g(x)) \neq g(f(x))$, and, in general, this is the case. An important case in which these two composite functions are equal occurs when

$$f(g(x)) = g(f(x)) = x$$

We call such functions **inverses** of each other, as stated in the following definition.

Definition of Inverse Functions	Two functions f and g are **inverses** of each other if

$$f(g(x)) = x \quad \text{for each } x \text{ in the domain of } g$$

and
$$g(f(x)) = x \quad \text{for each } x \text{ in the domain of } f$$

We denote g by f^{-1} (read "f inverse").

Note that for inverse functions f and g, the range of g must be equal to the domain of f and vice versa.

Example 7

Show that the following functions are inverses of each other:

$$f(x) = 2x^3 - 1 \quad \text{and} \quad g(x) = \sqrt[3]{\frac{x+1}{2}}$$

Solution: First note that both composite functions exist since the domain and range of both f and g consist of the set of all real numbers.

The composite of f with g is given by

$$f(g(x)) = 2\left(\sqrt[3]{\frac{x+1}{2}}\right)^3 - 1 = 2\left(\frac{x+1}{2}\right) - 1 = x + 1 - 1 = x$$

The composite of g with f is given by

$$g(f(x)) = \sqrt[3]{\frac{(2x^3 - 1) + 1}{2}} = \sqrt[3]{\frac{2x^3}{2}} = \sqrt[3]{x^3} = x$$

Since $f(g(x)) = g(f(x)) = x$, f and g are inverses of each other.   ■

The following theorem suggests a geometrical interpretation of inverse functions.

THEOREM 1.8 The graph of f contains the point (a, b) if and only if the graph of f^{-1} contains the point (b, a).

Proof: Observe from the definition of inverse functions that
$$f^{-1}(f(x)) = x$$

Now assume that the point (a, b) lies on the graph of f. This implies that $f(a) = b$, which in turn implies that

$$f^{-1}(b) = f^{-1}(f(a)) = a$$

Thus (b, a) lies on the graph of f^{-1}. Similarly, we can show that if (b, a) lies on the graph of f^{-1}, then (a, b) lies on the graph of f.   ■

Theorem 1.8 can be interpreted geometrically to mean that the graph of f^{-1} can be obtained by reflecting the graph of f in the line $y = x$ (Figure 1.47).

Example 8

Find the inverse of the function given by $f(x) = \sqrt{2x - 3}$.

Solution: Substituting y for $f(x)$, we have $y = \sqrt{2x - 3}$. Now to find the inverse function, we simply solve for x in terms of y. Since y is nonnegative, squaring both sides gives an equivalent equation:

$$\sqrt{2x - 3} = y$$
$$2x - 3 = y^2, \qquad y \geq 0$$
$$2x = y^2 + 3$$
$$x = \frac{y^2 + 3}{2}, \qquad y \geq 0$$

Thus the inverse function has the form

$$f^{-1}(\ ) = \frac{(\ )^2 + 3}{2}$$

Using x as the independent variable, we write

$$f^{-1}(x) = \frac{x^2 + 3}{2}, \qquad x \geq 0$$

Note from the graphs of these two functions in Figure 1.48 that the domain of f^{-1} coincides with the range of f.

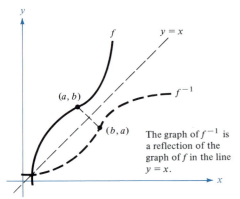

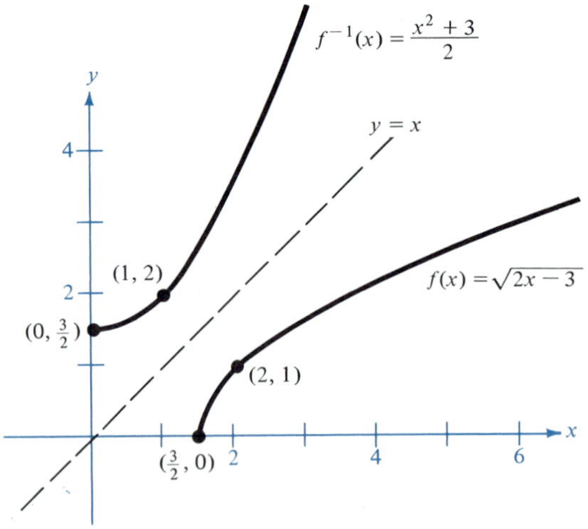

The graph of f^{-1} is a reflection of the graph of f in the line $y = x$.

FIGURE 1.47

$f^{-1}(x) = \frac{x^2 + 3}{2}$

$y = x$

$(1, 2)$

$(0, \frac{3}{2})$

$(2, 1)$

$(\frac{3}{2}, 0)$

$f(x) = \sqrt{2x - 3}$

FIGURE 1.48

Not all functions possess an inverse. In fact, for a function f to have an inverse, it is necessary that f be one-to-one. The next example is a case in point.

Example 9

Find the inverse (if it exists) of the function given by

$$f(x) = x^2 - 1$$

(Assume the domain of f is the set of all real numbers.)

Solution: First we note that

$$f(2) = (2)^2 - 1 = 3$$

and

$$f(-2) = (-2)^2 - 1 = 3$$

Thus f is not one-to-one and it has no inverse. This same conclusion can be obtained by substituting y for $f(x)$ and solving for x as follows:

$$x^2 - 1 = y$$
$$x^2 = y + 1$$
$$x = \pm\sqrt{y + 1}$$

This last equation does not define x as a function of y and thus f has no inverse. ∎

Section Exercises (1.6)

1. Given $f(x) = 2x - 3$, find:
   (a) $f(1)$ (b) $f(0)$
   (c) $f(-3)$ (d) $f(b)$
   (e) $f(x - 1)$ (f) $f(\frac{1}{4})$

2. Given $f(x) = x^2 - 2x + 2$, find:
   (a) $f(\frac{1}{2})$ (b) $f(3)$
   (c) $f(-1)$ (d) $f(c)$
   (e) $f(x + \Delta x)$ (f) $f(2)$

[S] 3. Given $f(x) = \sqrt{x + 3}$, find:
   (a) $f(-3)$ (b) $f(-2)$
   (c) $f(0)$ (d) $f(6)$
   (e) $f(x + \Delta x)$ (f) $f(c)$

4. Given $f(x) = 1/\sqrt{x}$, find:
   (a) $f(1)$ (b) $f(4)$
   (c) $f(2)$ (d) $f(\frac{1}{4})$
   (e) $f(x + \Delta x)$ (f) $f(x + \Delta x) - f(x)$

5. Given $f(x) = |x|/x$, find:
   (a) $f(2)$ (b) $f(-2)$

   (c) $f(-100)$ (d) $f(100)$
   (e) $f(x^2)$ (f) $f(x - 1)$

6. Given $f(x) = |x| + 4$, find:
   (a) $f(2)$ (b) $f(-2)$
   (c) $f(3)$ (d) $f(x^2)$
   (e) $f(x + \Delta x)$ (f) $f(x + \Delta x) - f(x)$

7. Given $f(x) = x^2 - x + 1$, find $\dfrac{f(2 + \Delta x) - f(2)}{\Delta x}$.

8. Given $f(x) = \dfrac{1}{x}$, find $\dfrac{f(1 + \Delta x) - f(1)}{\Delta x}$.

[S] 9. Given $f(x) = x^3$, find $\dfrac{f(x + \Delta x) - f(x)}{\Delta x}$.

10. Given $f(x) = 3x - 1$, find $\dfrac{f(x) - f(1)}{x - 1}$.

11. Given $f(x) = \dfrac{1}{\sqrt{x - 1}}$, find $\dfrac{f(x) - f(2)}{x - 2}$.

12. Given $f(x) = x^3 - x$, find $\dfrac{f(x) - f(1)}{x - 1}$.

13. Given $f(x) = x^2 - x$ and $g(x) = 3x - 1$, find:
(a) $g(f(0))$ (b) $f(g(0))$
(c) $f(g(3))$ (d) $g(f(3))$
(e) $g(f(\frac{1}{2}))$ (f) $f(g(x))$

14. Given $f(x) = \sqrt{x}$ and $g(x) = x^2 - 1$, find:
(a) $f(g(1))$ (b) $g(f(1))$
(c) $g(f(0))$ (d) $f(g(-4))$
(e) $f(g(x))$ (f) $g(f(x))$

[S] **15.** Given $f(x) = 1/x$ and $g(x) = x^2 - 1$, find:
(a) $f(g(2))$ (b) $g(f(2))$
(c) $f(g(1/\sqrt{2}))$ (d) $g(f(1/\sqrt{2}))$
(e) $g(f(x))$ (f) $f(g(x))$

16. Given $f(x) = 1/(x - 2)$ and $g(x) = \sqrt{2x + 3}$, find:
(a) $f(g(-1))$ (b) $g(f(3))$
(c) $g(f(1))$ (d) $f(g(0))$
(e) $g(f(0))$ (f) $g(f(x))$

17. Let $f(x) = x^2 - 1$. Find all real numbers x such that $f(x) = 8$.

18. Let $f(x) = x^3 - x$. Find all real numbers x such that $f(x) = 0$.

[S] **19.** Let
$$f(x) = \frac{3}{x - 1} + \frac{4}{x - 2}$$
Find all real numbers x such that $f(x) = 0$.

20. Let $f(x) = a + (b/x)$. Find all real numbers x such that $f(x) = 0$.

In Exercises 21–30, sketch the graph of each function and give its domain and range.

21. $f(x) = \sqrt{x - 1}$ **22.** $f(x) = \sqrt{1 - x}$
23. $f(x) = x^2$ **24.** $f(x) = 4 - x^2$
[S] **25.** $f(x) = \sqrt{9 - x^2}$ **26.** $f(x) = \sqrt{25 - x^2}$
27. $f(x) = \dfrac{1}{|x|}$ **28.** $f(x) = |x - 2|$
29. $f(x) = \dfrac{|x|}{x}$ **30.** $f(x) = \sqrt{x^2 - 4}$

In Exercises 31–40, identify the equations that determine y as a function of x.

31. $x^2 + y^2 = 4$
32. $x = y^2$
33. $x^2 + y = 4$
34. $x + y^2 = 4$
35. $2x + 3y = 4$
36. $x^2 + y^2 - 2x - 4y + 1 = 0$
37. $y^2 = x^2 - 1$
38. $y = \pm\sqrt{x}$
[S] **39.** $x^2y - x^2 + 4y = 0$
40. $xy - y - x - 2 = 0$

In Exercises 41–48, find the inverse of f, then graph both f and f^{-1}.

41. $f(x) = 2x - 3$ **42.** $f(x) = 3x$
[S] **43.** $f(x) = x^3$ **44.** $f(x) = x^3 + 1$
45. $f(x) = \sqrt{x}$ **46.** $f(x) = x^2, 0 \leq x$
47. $f(x) = \dfrac{1}{x}$ **48.** $f(x) = \sqrt[3]{x}$

In Exercises 49–54, determine if the given function is one-to-one and, if so, find its inverse.

49. $f(x) = ax + b$ **50.** $f(x) = \sqrt{x - 2}$
[S] **51.** $f(x) = x^2$ **52.** $f(x) = x^4$
53. $f(x) = |x - 2|$ **54.** $f(x) = \dfrac{1}{x - 1}$

[S] Detailed solution in *Student Solutions Manual.* [C] Calculator may be helpful.

For Review and Class Discussion

True or False

1. ____ If the domain of a function consists of a single number, then its range must also consist of only one number.

2. ____ If the range of a function consists of a single number, then its domain must also consist of only one number.

3. ____ If $f(a) = f(b)$, then $a = b$.

4. ____ A vertical line can intersect the graph of a function at most once.

5. ____ If $f(x) = f(-x)$ for all x in the domain of f, then the graph of f is symmetric with respect to the y-axis.

6. _____ If f possesses an inverse function and $f(a) = f(b)$, then $a = b$.

7. _____ If f is a function, then $f(x + \Delta x) - f(x) = f(\Delta x)$.

8. _____ If f is a function, then $f(ax) = af(x)$.

9. _____ If $f(f(x)) = x$, then $f(1) = 1$.

10. _____ If f possesses an inverse function, then f^{-1} also possesses an inverse function.

Miscellaneous Exercises (Ch. 1)

In Exercises 1–6, express the described function by a formula and give its domain.

S 1. the value v of a farm at \$850 per acre, with buildings, livestock, and equipment worth \$300,000, as a function of the number of acres a

2. the value v of wheat at \$3.25 per bushel as a function of the number of bushels b

3. the surface area s of a cube as a function of the length of an edge x

4. the surface area s of a sphere as a function of the radius r

5. the distance d traveled by a car at a speed of 45 miles per hour as a function of time t

6. the area a of an equilateral triangle as a function of the length of one of its sides

7. Find the midpoint of the interval $[\frac{7}{8}, \frac{10}{4}]$.

8. Find the points of trisection of the interval $[-2, 6]$.

9. Find the midpoint of the line segment joining the points $(1, -3)$ and $(5, 11)$.

10. Determine if the points $(-1, 3)$, $(2, 9)$, and $(3, 11)$ lie on the same straight line.

In Exercises 11–14, find the slope and y-intercept and sketch each of the lines.

11. $4x - 2y = 6$

12. $0.02x + 0.15y = 0.25$

13. $-\frac{1}{3}x + \frac{5}{6}y = 1$

14. $51x + 17y = 102$

15. Find the point of intersection of the lines $3x - 4y = 8$ and $x + y = 5$.

    Let θ (the Greek lowercase letter theta) be the angle measured counterclockwise from the x-axis to the line L. The angle θ is called the **angle of inclination** and the slope m of L can be defined by $m = \tan \theta$. This definition of slope allows us to use properties of the trigonometric functions to derive properties of the slope. For example, let L_1 and L_2 be two nonvertical lines (Figure 1.49) with inclinations θ_1 and θ_2 and slopes m_1 and m_2, respectively. Then the acute angle ϕ (the Greek lowercase letter phi) between L_1 and L_2 is such that

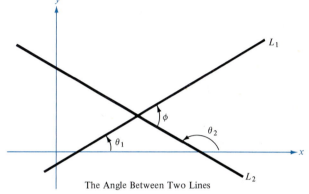

The Angle Between Two Lines

FIGURE 1.49

$$\tan \phi = \left| \frac{\tan \theta_2 - \tan \theta_1}{1 + \tan \theta_1 \tan \theta_2} \right|$$

    Use the results of the preceding paragraph for Exercises 16–19.

16. Show that the definition of slope as the tangent of the angle of inclination is equivalent to the definition in Section 1.4.

S 17. Find the angle of inclination of the line $x - y = 4$.

C 18. Find the angle of inclination of the line $x - 3y = 11$.

C 19. Find the angle of inclination of the line $5x + 3y = 10$.

    If two lines having slopes m_1 and m_2 intersect at an angle ϕ, then

$$\tan \phi = \left| \frac{m_2 - m_1}{1 + m_1 m_2} \right|$$

(See Figure 1.49.) Use this formula to solve Exercises 20–24.

20. Find the angle between the lines $y = x + 2$ and $y = 3$.

21. Find the angle between the lines $2x - 3y = 1$ and $x + 5y = 2$.

[c] **22.** Find the angle between the lines $4x + 3y + 2 = 0$ and $3x + 4y - 7 = 0$.

[c][s] **23.** Find the angle between the lines $2x - y + 7 = 0$ and $x + y + 2 = 0$.

[c] **24.** Use the result of Exercise 23 to write an equation of the line that bisects the angle between $2x - y + 7 = 0$ and $x + y + 2 = 0$.

25. Determine the equation of the line passing through the point $(-2, 4)$ and:
  (a) with slope $\frac{7}{16}$
  (b) parallel to the line $5x - 3y = 3$
  (c) perpendicular to the line $5x - 3y = 3$
  (d) parallel to the x-axis
  (e) parallel to the y-axis
  (f) passing through the origin

26. Determine the radius and center of the circle $x^2 + y^2 + 6x - 2y + 1 = 0$ and sketch its graph.

27. Determine the radius and center of the circle $4x^2 + 4y^2 - 4x + 8y = 11$ and sketch its graph.

28. Determine the radius and center of the circle $x^2 + y^2 + 6x - 2y + 10 = 0$ and sketch its graph.

29. Determine the value of c so that the circle given by $x^2 - 6x + y^2 + 8y = c$ has a radius of 2.

[s] **30.** Determine the equation that the coordinates of (x, y) must satisfy if the distance between this point and $(-2, 0)$ is twice the distance from the point to $(3, 1)$. Sketch the graph of the equation.

31. Sketch the graph of the equation $9x^2 + 16y^2 - 18x - 128y + 121 = 0$.

32. Sketch the graph of the equation $16x^2 + y^2 - 32x + 6y + 9 = 0$.

In Exercises 33–36, sketch the intervals defined by the given inequalities.

33. $|x - 2| \le 3$

34. $|3x - 2| \le 0$

[s] **35.** $4 < (x + 3)^2$

36. $\dfrac{1}{|x|} < 1$

In Exercises 37–40, use the point-plotting method to sketch the graphs of the equations.

37. $y = 1 + \dfrac{1}{x}$

38. $y = \dfrac{-x + 3}{2}$

39. $y = 7 - 6x - x^2$

40. $y = 6x - x^3$

41. Find the vertices of the triangle whose sides have the midpoints $(0, 2)$, $(1, -1)$, $(2, 1)$.

42. Find the vertices of the triangle whose sides have the midpoints $(0, 0)$, $(-\frac{1}{2}, 3)$, $(\frac{1}{2}, 2)$.

43. Find the equation of the circle whose center is $(1, 2)$ and whose radius is 3. Then determine if the following points are inside, outside, or on the circle.
  (a) $(1, 5)$ (b) $(0, 0)$
  (c) $(-2, 1)$ (d) $(0, 4)$
  (e) $(0, 5)$ (f) $(3.5, 4)$
  (g) $(\frac{7}{2}, 0)$ (h) $(2, 5)$

44. If one end of a line segment is $(2, 3)$ and the midpoint is $(-1, 4)$, find the coordinates of the other end of the segment.

In Exercises 45–50, sketch the graph of each equation to observe if the equation determines y as a function of x.

45. $x^2 - y = 0$

46. $x^2 + 4y^2 = 16$

47. $x - y^2 = 0$

[s] **48.** $x^3 - y^2 + 1 = 0$

49. $y = x^2 - 2x$

50. $y = 36 - x^2$

51. If $f(x) = 1 - x^2$ and $g(x) = 2x + 1$, find:
  (a) $f(x) + g(x)$ (b) $f(x) - g(x)$
  (c) $f(x) g(x)$ (d) $f(x)/g(x)$
  (e) $f(g(x))$ (f) $g(f(x))$

---

[s] Detailed solution in *Student Solutions Manual*.   [c] Calculator may be helpful.

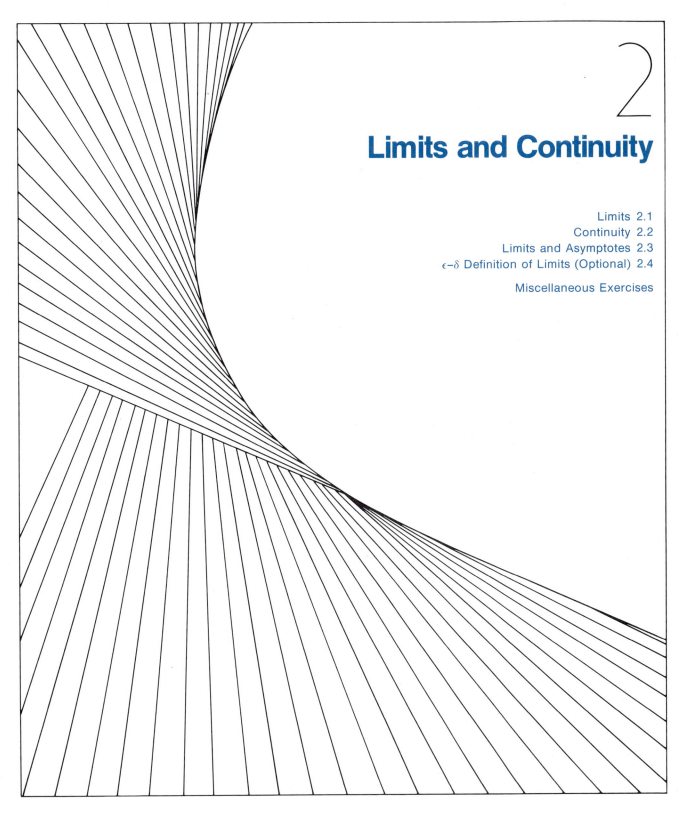

2

Limits and Continuity

2.1 **Limits**

Purpose
- To provide an intuitive discussion of the limit concept.
- To identify some properties of limits.
- To provide some general guidelines for evaluating limits.

If a nonnegative quantity were so small that it is smaller than any given one, then it certainly could not be anything but zero. To those who ask what the infinitely small quantity in mathematics is, we answer that it is actually zero. Hence there are not so many mysteries hidden in this concept as there are usually believed to be. These supposed mysteries have rendered the calculus of the infinitely small quite suspect to many people. Those doubts that remain we shall thoroughly remove in the following pages, where we shall explain this calculus. —Leonhard Euler (1707–1783)

The notion of a limit is fundamental to the study of calculus. As such, it is important for students in calculus to acquire a good working knowledge of limits before moving on to other calculus topics. In this chapter we will discuss limits in two stages. In the first three sections, we build up an intuitive understanding of limits by examining various kinds of limits, ways to evaluate limits, and the role of limits in defining continuity. This background is followed, in Section 2.4, by a more rigorous development of the limit concept. This two-stage approach is quite in keeping with the historical evolution of limits. In fact, limits were not formally defined as in Section 2.4 until well after much of the early work in calculus was completed.

Just what do we mean by the term "limit"? In everyday language we refer to the speed limit, a wrestler's weight limit, the limit of one's endurance, the limits of modern technology, or stretching a spring to its limits. These terms all suggest that a limit is a type of bound, which on some occasions may not be reached while on other occasions may be reached or even exceeded.

Suppose we are given an ideal spring, which is made so that it will break only if a weight of 10 or more pounds is attached to it. Our task is to determine how far the spring will stretch without breaking (see Figure 2.1).

We could carry out our experiment by increasing the weight attached to the spring and measuring the spring length s at each successive weight. As our attached weight nears 10 pounds, we would need to use smaller and smaller increments so as not to reach the 10-pound maximum. By recording the successive spring lengths, we should be able to determine the value L, which s approaches as the weight w approaches 10 pounds. Symbolically we write

$$s \to L \qquad \text{as} \qquad w \to 10, \qquad (w < 10)$$

and we say that L is the **limit** of the length s.

Mathematically, our notion of a limit is much like the limit of a spring. For instance, consider the function f given by

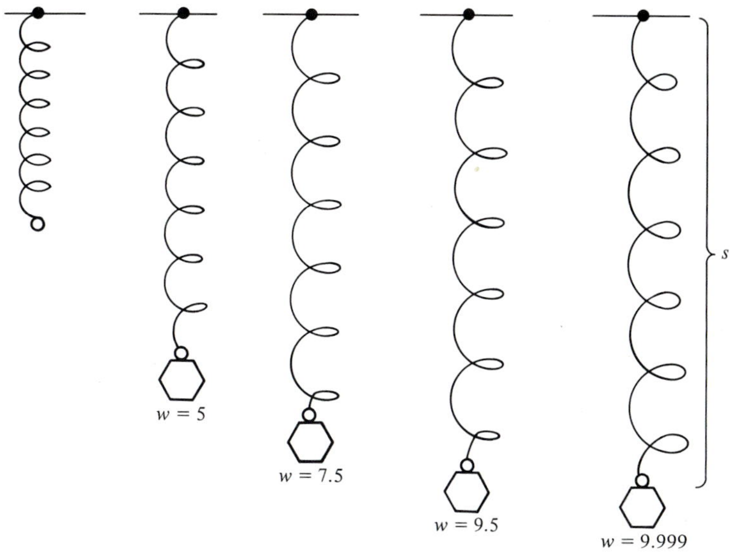

What is the limit of s as w approaches 10 lb?

FIGURE 2.1

$$f(x) = \frac{x^3 - 1}{x - 1}$$

where the domain includes all real numbers other than $x = 1$. Suppose our objective is to determine the limit (if it exists) of $f(x)$ as x approaches 1. Note that we are not interested in determining the value of $f(1)$, because f is not defined at $x = 1$. What we seek, however, is the value that $f(x)$ approaches as x approaches 1.

To perform this experiment we could program the function into a calculator and evaluate $f(x)$ for several values of x near 1. In approaching 1 from the left, we use the values

$$x = 0.5, 0.75, 0.9, 0.99, 0.999$$

and from the right, we use

$$x = 1.5, 1.25, 1.1, 1.01, 1.001$$

Table 2.1 gives the values of $f(x)$ that correspond to these ten different values of x.

TABLE 2.1

x	0.5	0.75	0.9	0.99	0.999 →	1	← 1.001	1.01	1.1	1.25	1.5
$f(x)$	1.750	2.313	2.710	2.970	2.997 →	?	← 3.003	3.030	3.310	3.813	4.750

From Table 2.1 it seems reasonable to conclude that the limit of $f(x)$, as x approaches 1, is 3. We denote this limit by

$$f(x) \to 3 \qquad \text{as} \qquad x \to 1$$

or by the equation

$$\lim_{x \to 1} f(x) = 3$$

Even though $f(x)$ is undefined when $x = 1$, it appears from our table that we can force $f(x)$ to be arbitrarily close to the value 3 by choosing values of x closer and closer to 1. This suggests the following definition:

Informal Definition of Limit

If $f(x)$ becomes arbitrarily close to a single number L as x approaches c from either side, then we write

$$\lim_{x \to c} f(x) = L$$

and say that the limit of $f(x)$, as x approaches c, is L.

The phrase "x approaches c" means that no matter how close x comes to the value c, there is always another value of x (different from c) in the domain of f that is even closer to c.

Inherent in this definition is the assumption that a function cannot approach two different limits at the same time. This means that *if the limit of a function exists, it is unique.*

It is important to realize that the limit L of $f(x)$ as $x \to c$ does *not* depend on the value of $f(x)$ *at* $x = c$. Rather, this limit is determined solely from the values of $f(x)$ when x is *near* c. Since the value of L does not depend on the value of $f(x)$ at $x = c$, any one of the following cases can occur:

1. f is undefined when $x = c$; hence $f(c)$ cannot equal L.
2. $f(c)$ exists, but $f(c) \neq L$.
3. $f(c)$ exists and $f(c) = L$.

The following example illustrates these three cases.

Example 1

Sketch the graphs of

$$f(x) = \frac{x^3 - 1}{x - 1} \qquad g(x) = \begin{cases} \dfrac{x^3 - 1}{x - 1}, & x \neq 1 \\ 0, & x = 1 \end{cases}$$

$$h(x) = x^2 + x + 1$$

Solution: The graphs of these three functions are shown in Figure 2.2.

By factoring $x^3 - 1$ we have

$$\frac{x^3 - 1}{x - 1} = \frac{(x - 1)(x^2 + x + 1)}{(x - 1)} = x^2 + x + 1, \qquad \text{for all } x \neq 1$$

From this equation and the graphs in Figure 2.2, we can see that the three functions f, g, and h are equal for all x other than $x = 1$. Thus the limit of each function, as $x \to 1$, is the same:

$$\lim_{x \to 1} f(x) = \lim_{x \to 1} g(x) = \lim_{x \to 1} h(x) = 3$$

(Keep in mind that the limit as x approaches 1 does not depend on the value of the function *at* $x = 1$.)

Finally, we note the three cases:

1. $f(1)$ is undefined and cannot equal 3.
2. $g(1)$ exists, but $g(1) = 0 \neq 3$.
3. $h(1)$ exists and $h(1) = 3$.   ∎

Of the three functions in Example 1, only the polynomial function h possesses the property that its limit, as $x \to 1$, corresponds to the value of the function at $x = 1$. That is,

$$\lim_{x \to 1} h(x) = h(1) = 3$$

All polynomial functions share this special property, that the limit as $x \to c$ can be determined by substituting c for x in the polynomial. Thus if $p(x)$ is any polynomial, then

$$\lim_{x \to c} p(x) = p(c)$$

We will see later that the limit (as $x \to c$) of some other functions can also be determined by substituting c for x.

In our original example of a spring, we could test for the limit of the length of the spring only by using weights that were less than 10 pounds. However, our definition of the limit,

$$\lim_{x \to c} f(x) = L$$

requires that we let x approach c from the right as well as from the left. To denote these two possible approaches to c, we use the symbols

$$x \to c^-$$

read "x approaches c from the left," and

$$x \to c^+$$

read "x approaches c from the right." Because of the importance of the "limits" obtained from these one-sided approaches, we give each a special name.

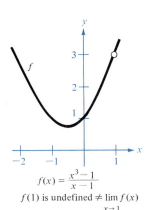

$f(x) = \dfrac{x^3 - 1}{x - 1}$

$f(1)$ is undefined $\neq \lim\limits_{x \to 1} f(x)$

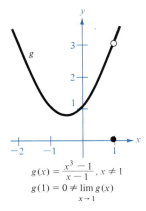

$g(x) = \dfrac{x^3 - 1}{x - 1}, x \neq 1$

$g(1) = 0 \neq \lim\limits_{x \to 1} g(x)$

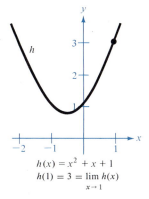

$h(x) = x^2 + x + 1$

$h(1) = 3 = \lim\limits_{x \to 1} h(x)$

FIGURE 2.2

One-Sided Limits

$\lim\limits_{x \to c^-} f(x)$ is called the **limit from the left.**

$\lim\limits_{x \to c^+} f(x)$ is called the **limit from the right.**

One-sided limits provide us with a practical way to determine whether or not $\lim_{x \to c} f(x)$ exists.

THEOREM 2.1

If f is a function and c and L are real numbers, then

$$\lim\limits_{x \to c} f(x) = L$$

if and only if

$$\lim\limits_{x \to c^-} f(x) = L \qquad \text{and} \qquad \lim\limits_{x \to c^+} f(x) = L$$

Example 2

For the function

$$f(x) = \begin{cases} 4 - x, & \text{for } x < 1 \\ 4x - x^2, & \text{for } x \geq 1 \end{cases}$$

find $\lim_{x \to 1} f(x)$, if it exists.

Solution: Remember, we are not concerned about the value of f at $x = 1$ but rather near $x = 1$. Thus for $x < 1$

$$\lim\limits_{x \to 1^-} f(x) = \lim\limits_{x \to 1^-} (4 - x) = 4 - 1 = 3$$

and for $x > 1$

$$\lim\limits_{x \to 1^+} f(x) = \lim\limits_{x \to 1^+} (4x - x^2) = 4 - 1 = 3$$

Since the one-sided limits both exist and are equal to 3, we have

$$\lim\limits_{x \to 1} f(x) = 3$$

A sketch will further illustrate our result (see Figure 2.3). ∎

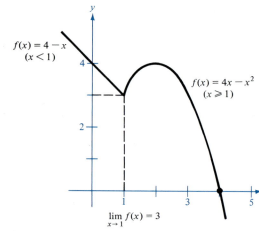

$f(x) = 4 - x$
$(x < 1)$

$f(x) = 4x - x^2$
$(x \geq 1)$

$\lim\limits_{x \to 1} f(x) = 3$

FIGURE 2.3

Another interesting situation involving one-sided limits arises with the "post office" function given in the next example.

Example 3

Suppose the cost of sending first-class mail is 15¢ for the first ounce and 13¢ for each additional ounce. Letting x represent the weight of a letter

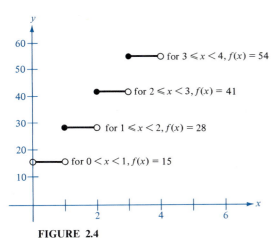

FIGURE 2.4

and $f(x)$ the cost of mailing the letter first class, we have

$$f(x) = 15 + 13n, \qquad n < x \le n + 1 \qquad (n = 0, 1, 2, 3, \ldots)$$

Show that the limit of $f(x)$, as $x \to 2$, does not exist.

Solution: The graph of f is shown in Figure 2.4.
Observe that

$$f(x) = \begin{cases} 28, & 1 \le x < 2 \\ 41, & 2 \le x < 3 \end{cases}$$

Thus

$$\lim_{x \to 2^-} f(x) = 28$$

whereas

$$\lim_{x \to 2^+} f(x) = 41$$

Since these one-sided limits are not equal, the limit of $f(x)$, as $x \to 2$, *does not exist*.   ∎

Example 4

If $f(x) = (x^2 + x - 6)/(x + 3)$, determine $\lim_{x \to -3} f(x)$.

Solution: Direct substitution of -3 for x yields the meaningless result

$$\lim_{x \to -3} f(x) = \frac{9 - 3 - 6}{-3 + 3} = \frac{0}{0}$$

By factoring $f(x)$ we obtain

$$f(x) = \frac{x^2 + x - 6}{x + 3} = \frac{(x + 3)(x - 2)}{x + 3}$$

Now for all $x \ne -3$, and therefore for all x near -3, we can cancel like factors and obtain

$$\frac{x^2 + x - 6}{x + 3} = x - 2$$

It follows that

$$\lim_{x \to -3} \frac{x^2 + x - 6}{x + 3} = \lim_{x \to -3} (x - 2) = -5$$

See Figure 2.5.   ∎

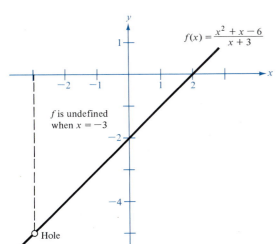

FIGURE 2.5

In our informal development of the limit concept, we have used, without identifying them, several basic properties of limits—properties that, for the most part, are ones we would expect the limit to possess. We provide here a list of basic properties, some of whose proofs will be discussed in Section 2.4.

Properties of Limits

If b, c, n, A, and B are real numbers, with f and g functions such that $\lim_{x \to c} f(x) = A$ and $\lim_{x \to c} g(x) = B$, then:

1. $\lim_{x \to c} b = b$

2. $\lim_{x \to c} x = c$

3. $\lim_{x \to c} b \cdot f(x) = bA$

4. $\lim_{x \to c} [f(x) \pm g(x)] = A \pm B$

5. $\lim_{x \to c} f(x) g(x) = AB$

6. $\lim_{x \to c} \dfrac{f(x)}{g(x)} = \dfrac{A}{B}$, provided $B \neq 0$

7. $\lim_{x \to c} x^n = c^n$

As an illustration let us identify the use of these properties in evaluating the limit $\lim_{x \to 2} (x^2 + 2x - 3)$. By Property 7,

$$\lim_{x \to 2} x^2 = 2^2 = 4$$

By Properties 2 and 3,

$$\lim_{x \to 2} 2x = 2 \lim_{x \to 2} x = 2(2) = 4$$

By Property 1,

$$\lim_{x \to 2} 3 = 3$$

Finally, by Property 4,

$$\lim_{x \to 2} (x^2 + 2x - 3) = \lim_{x \to 2} x^2 + \lim_{x \to 2} 2x - \lim_{x \to 2} 3 = 4 + 4 - 3 = 5$$

Before concluding this section we provide two general guidelines for evaluating the types of limits encountered thus far. As we proceed further into this chapter, we will identify additional suggestions for evaluating limits.

Guidelines for Evaluating $\lim_{x \to c} f(x)$

1. Check to see if f is defined differently for $x \to c^-$ than for $x \to c^+$. If so, evaluate each one-sided limit and draw appropriate conclusions about the limit as $x \to c$.
2. Use direct substitution. If f involves a quotient and $0/0$ is obtained, try to change the fraction algebraically to remove the zero from the denominator. Then take the limit as $x \to c$.

Example 5

Evaluate the following limits, if they exist:

(a) $\lim_{x \to -2} \dfrac{x^2 + x - 2}{x - 2}$

(b) $\lim\limits_{x \to 1} f(x)$, where $f(x) = \begin{cases} 2x - x^3, & x < 1 \\ 2x^2 - 2, & x \geq 1 \end{cases}$

(c) $\lim\limits_{x \to -1} f(x)$, using $f(x)$ in part b

(d) $\lim\limits_{x \to 0} \dfrac{(x+2)^2 - 4}{x}$

Solution:

(a) By Property 6 and direct substitution, we obtain

$$\lim_{x \to -2} \frac{x^2 + x - 2}{x - 2} = \frac{4 - 2 - 2}{-2 - 2} = \frac{0}{-4} = 0$$

(b) Since f is defined differently for $x < 1$ than for $x \geq 1$, we consider the one-sided limits

$$\lim_{x \to 1^-} f(x) = \lim_{x \to 1^-} (2x - x^3) = 2 - 1 = 1$$

and

$$\lim_{x \to 1^+} f(x) = \lim_{x \to 1^+} (2x^2 - 2) = 2 - 2 = 0$$

Because these one-sided limits are not equal, we conclude that $\lim_{x \to 1} f(x)$ does not exist.

(c) By direct substitution and Property 4, we have, for $x < 1$,

$$\lim_{x \to -1} f(x) = \lim_{x \to -1} (2x - x^3) = -2 + 1 = -1$$

(d) By direct substitution and Property 6, we obtain

$$\lim_{x \to 0} \frac{(x+2)^2 - 4}{x} = \frac{4 - 4}{0} = \frac{0}{0}$$

which is meaningless. Factoring and reducing we have

$$\lim_{x \to 0} \frac{x^2 + 4x + 4 - 4}{x} = \lim_{x \to 0} \frac{x(x + 4)}{x} = \lim_{x \to 0} (x + 4) = 4 \qquad \blacksquare$$

Example 6

Determine

$$\lim_{x \to 0} \frac{\sqrt{x + 1} - 1}{x}$$

Solution: By direct substitution we get the meaningless result

$$\lim_{x \to 0} \frac{\sqrt{x + 1} - 1}{x} = \frac{0}{0}$$

In this case we simplify the fraction by rationalizing the numerator:

$$\frac{\sqrt{x+1}-1}{x} = \left(\frac{\sqrt{x+1}-1}{x}\right)\left(\frac{\sqrt{x+1}+1}{\sqrt{x+1}+1}\right) = \frac{(x+1)-1}{x(\sqrt{x+1}+1)}$$

$$= \frac{x}{x(\sqrt{x+1}+1)} = \frac{1}{\sqrt{x+1}+1}$$

Therefore,

$$\lim_{x\to 0}\frac{\sqrt{x+1}-1}{x} = \lim_{x\to 0}\frac{1}{\sqrt{x+1}+1} = \frac{1}{1+1} = \frac{1}{2} \quad\blacksquare$$

Section Exercises (2.1)

In Exercises 1–30, determine if the limit exists, and if it does, determine the limit.

1. $\lim\limits_{x\to 2} x^2$

2. $\lim\limits_{x\to -3} (3x+2)$

3. $\lim\limits_{x\to 0} (5x+4)$

4. $\lim\limits_{x\to 0} \dfrac{x^2-1}{x+1}$

S 5. $\lim\limits_{x\to -1^-} \dfrac{x^2-1}{x+1}$

6. $\lim\limits_{x\to -1} \dfrac{2x^2-x-3}{x+1}$

7. $\lim\limits_{x\to 3} \dfrac{x-3}{x^2-9}$

8. $\lim\limits_{x\to 1} f(x)$, where $f(x) = \begin{cases} x, & x\le 1 \\ 1-x, & x>1 \end{cases}$

S 9. $\lim\limits_{x\to 3} f(x)$, where $f(x) = \begin{cases} \dfrac{x+2}{2}, & x\le 3 \\ \dfrac{12-2x}{3}, & x>3 \end{cases}$

10. $\lim\limits_{x\to 2} f(x)$, where $f(x) = \begin{cases} x^2-4x+6, & x<2 \\ -x^2+4x-2, & x\ge 2 \end{cases}$

S 11. $\lim\limits_{x\to 1} f(x)$, where $f(x) = \begin{cases} x^3+1, & x<1 \\ x+1, & x\ge 1 \end{cases}$

12. $\lim\limits_{x\to -1} \dfrac{x^3+1}{x+1}$

13. $\lim\limits_{x\to -2} \dfrac{x^3+8}{x+2}$

14. $\lim\limits_{x\to 0} \dfrac{|x|}{x}$

S 15. $\lim\limits_{x\to 2} \dfrac{|x-2|}{x-2}$

16. $\lim\limits_{\Delta x\to 0} \dfrac{(x+\Delta x)^2-x^2}{\Delta x}$

17. $\lim\limits_{\Delta x\to 0^+} \dfrac{2(x+\Delta x)-2x}{\Delta x}$

18. $\lim\limits_{\Delta x\to 0} \dfrac{(x+\Delta x)^3-x^3}{\Delta x}$

S 19. $\lim\limits_{\Delta x\to 0} \dfrac{(x+\Delta x)^2-2(x+\Delta x)+1-(x^2-2x+1)}{\Delta x}$

20. $\lim\limits_{\Delta x\to 0} \dfrac{(1+\Delta x)^3-1}{\Delta x}$

21. $\lim\limits_{x\to 5^+} \dfrac{x-5}{x^2-25}$

22. $\lim\limits_{x\to 2^+} \dfrac{2-x}{x^2-4}$

23. $\lim\limits_{x\to 1} \dfrac{x^2+x-2}{x^2-1}$

24. $\lim\limits_{x\to 5} \dfrac{x^2-25}{x+5}$

25. $\lim\limits_{x\to -2^-} \dfrac{x}{\sqrt{x^2-4}}$

26. $\lim\limits_{x\to 0} \dfrac{\sqrt{2+x}-\sqrt{2}}{x}$

S 27. $\lim\limits_{x\to 0} \dfrac{\sqrt{3+x}-\sqrt{3}}{x}$

28. $\lim\limits_{x\to 0} \dfrac{[1/(x+4)]-\frac{1}{4}}{x}$

S 29. $\lim\limits_{x\to 0} \dfrac{[1/(2+x)]-\frac{1}{2}}{x}$

30. $\lim\limits_{x\to 4^-} \dfrac{\sqrt{x}-2}{x-4}$

31. If $\lim\limits_{x\to c} f(x) = 2$ and $\lim\limits_{x\to c} g(x) = 3$, find:

(a) $\lim\limits_{x\to c} [f(x)-g(x)]$　　(b) $\lim\limits_{x\to c} [f(x)\,g(x)]$

(c) $\lim\limits_{x\to c} \dfrac{f(x)}{g(x)}$

32. If $\lim\limits_{x\to c} f(x) = \frac{3}{2}$ and $\lim\limits_{x\to c} g(x) = \frac{1}{2}$, find:

(a) $\lim\limits_{x\to c} [f(x)+g(x)]$　　(b) $\lim\limits_{x\to c} [f(x)\,g(x)]$

(c) $\lim\limits_{x\to c} \dfrac{f(x)}{g(x)}$

In Exercises 33–40, compile a table to evaluate the limit of the given function.

C 33. $\lim\limits_{x\to 2} (5x+4)$

C 34. $\lim\limits_{x\to 2} \dfrac{x-2}{x^2-x-2}$

C 35. $\lim\limits_{x\to 2} \dfrac{x-2}{x^2-4}$

C 36. $\lim\limits_{x\to 0} \dfrac{\sqrt{x+3}-\sqrt{3}}{x}$

C **37.** $\lim_{x \to 0} \dfrac{\sqrt{x + 2} - \sqrt{2}}{x}$ C **38.** $\lim_{x \to 2^-} \dfrac{2 - x}{\sqrt{4 - x^2}}$ C **39.** $\lim_{x \to 0} \dfrac{[1/(2 + x)] - \frac{1}{2}}{x}$ C **40.** $\lim_{x \to 2} \dfrac{x^5 - 32}{x - 2}$

In Exercises 41–46, use the graph to visually determine:

(a) $\lim_{x \to c^+} f(x)$ (b) $\lim_{x \to c^-} f(x)$ (c) $\lim_{x \to c} f(x)$

41.

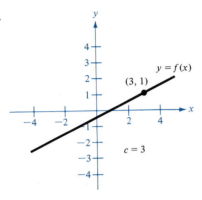

42.

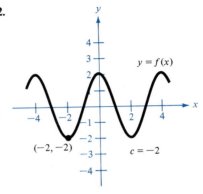

S **43.**

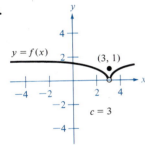

44.

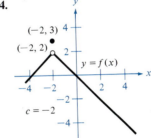

45.

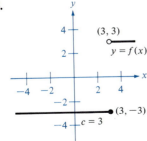

46.

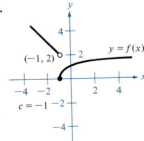

For Review and Class Discussion

True or False

1. _____ If $\lim_{x \to c} f(x) = L$, then $f(c) = L$.

2. _____ If $f(x) = g(x)$ for all real numbers other than $x = 0$, and $\lim_{x \to 0} f(x) = L$, then $\lim_{x \to 0} g(x) = L$.

3. _____ If $f(x) \to 0$ as $x \to 0$, then there must exist a number c such that $f(c) < 0.001$.

4. _____ For polynomial functions the right- and left-hand limits always exist and are equal to each other.

5. _____ If f is undefined for $x = c$, then the limit of $f(x)$ as $x \to c$ does not exist.

2.2 Continuity

Purpose
- To define the term "continuity."
- To identify some properties of continuous functions.
- To introduce the Intermediate and Extreme Value Theorems.

Consider the definition of the limit of a function (if any) and the definition of what is meant by a continuous function. Both of these ideas are somewhat technical, and would hardly demand treatment in a mere introduction to mathematical philosophy but for the fact that, especially through the so-called infinitesimal calculus, wrong views upon our present topics have become so firmly embedded in the minds of professional philosophers that a prolonged and considerable effort is required for their uprooting.

—*Bertrand Russell* (1872–1970)

In mathematics the term "continuous" has much the same meaning as it does in our everyday usage. To say that a function is continuous at $x = c$ means that there is no interruption in the graph of f at c. Its graph is unbroken at c and there are no holes, jumps, or gaps. Roughly speaking, we say that a function is continuous if its graph can be traced without lifting the pencil from the paper. As simple as this concept may seem initially, its precise definition eluded mathematicians for many years. In fact, it was not until the early 1800s that a careful definition was finally developed.

Before looking at this definition, let us consider the function whose graph is shown in Figure 2.6. This figure identifies three values of x at which the graph of f is not continuous. At all other points ($x \neq c_1, c_2, c_3$) of the interval (a, b), the graph of f is uninterrupted and we say it is continuous at such points. Where the graph is discontinuous, we observe the following:

1. At $x = c_1$, $f(c_1)$ is not defined.
2. At $x = c_2$, $\lim_{x \to c_2} f(x)$ does not exist.
3. At $x = c_3$, $f(c_3) \neq \lim_{x \to c_3} f(x)$.

Thus it appears that continuity at $x = c$ is destroyed under one or more of the following conditions:

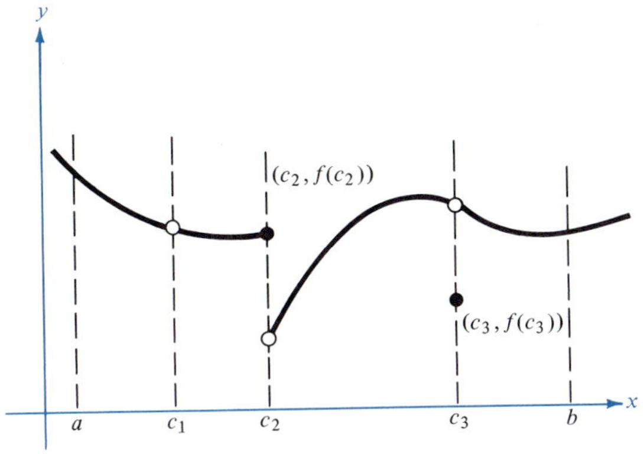

Discontinuities when $x = c_1, c_2, c_3$

FIGURE 2.6

1. if $f(c)$ is not defined
2. if $\lim_{x \to c} f(x)$ does not exist
3. if $f(c) \neq \lim_{x \to c} f(x)$

This brings us to the following definition:

Definition of Continuity	A function is said to be **continuous at** c if the following three conditions are met:

1. $f(c)$ is defined.
2. $\lim_{x \to c} f(x)$ exists.
3. $\lim_{x \to c} f(x) = f(c)$.

A function is said to be **continuous on an interval** (a, b) if it is continuous at each point in the interval.

A function is said to be **discontinuous at** c if it is not continuous at c. Discontinuities fall into two categories: **removable** and **nonremovable**. A discontinuity at $x = c$ is called removable if f can be made continuous by redefining f at $x = c$. For instance, in Figure 2.6 the discontinuity at $x = c_1$ could be removed by defining $f(c_1)$ as $\lim_{x \to c_1} f(x)$. Furthermore, the discontinuity at c_3 can be removed by redefining $f(c_3)$ so as to move the point $(c_3, f(c_3))$ up to plug the hole. Finally, the discontinuity at $x = c_2$ in Figure 2.6 is not removable since we cannot make the graph continuous by merely redefining $f(c_2)$.

For a function to be continuous on a closed interval $[a, b]$, it is suffi-cient that it be continuous on the open interval (a, b) and that

$$\lim_{x \to a^+} f(x) = f(a) \qquad \text{and} \qquad \lim_{x \to b^-} f(x) = f(b)$$

In this situation we say that f is continuous *from the right at* a and contin-uous *from the left at* b.

In Section 2.1 we stated without proof that for any polynomial function $p(x)$, we have

$$\lim_{x \to c} p(x) = p(c)$$

Now we can see that this property is a requirement for continuity, as indicated in the following theorem:

THEOREM 2.2
(Continuity of a Polynomial)

A polynomial function is continuous at every real number.

Proof: Let f be the polynomial

$$f(x) = a_0 x^n + a_1 x^{n-1} + \cdots + a_{n-1} x + a_n \qquad (a_0 \neq 0)$$

where n is a nonnegative integer and $a_0, a_1, \ldots, a_n$ are real numbers. To show that f is continuous at any real number c, we must establish that

$$\lim_{x \to c} f(x) = f(c)$$

By an extension of Property 4 of limits (Section 2.1), we have

$$\lim_{x \to c} f(x) = \lim_{x \to c} (a_0 x^n) + \lim_{x \to c} (a_1 x^{n-1}) + \cdots + \lim_{x \to c} (a_n)$$

By limit Properties 1 and 7, we obtain

$$\lim_{x \to c} f(x) = a_0 c^n + a_1 c^{n-1} + \cdots + a_n = f(c)$$

Therefore, we conclude that f is continuous at every real number.   ∎

The sum, difference, and product of two polynomials are also polynomials. The quotient of two polynomials is called a **rational function.** For those points at which the denominator of a rational function is zero, the function is undefined and hence not continuous; for all other points rational functions are continuous.

THEOREM 2.3
(Continuity of a Rational Function)

A rational function is continuous at every real number in its domain.

Proof: If f is a rational function, it can be defined as the ratio of two polynomials. Thus let

$$f(x) = \frac{g(x)}{h(x)}$$

where g and h are polynomial functions. The domain of f is all x for which $h(x) \neq 0$. Let c be in the domain of f; then by Property 6 of limits

(Section 2.1), we have

$$\lim_{x \to c} f(x) = \frac{\lim\limits_{x \to c} g(x)}{\lim\limits_{x \to c} h(x)}$$

By the continuity of the polynomials g and h, we have

$$\lim_{x \to c} f(x) = \frac{g(c)}{h(c)} = f(c)$$

which implies that f is continuous at every real number c in its domain. ■

Example 1

Discuss the continuity of

$$g(x) = \begin{cases} 5 - x, & -1 \le x \le 2 \\ x^2 - 1, & 2 < x \le 3 \end{cases}$$

Solution: By Theorem 2.2 the polynomial functions $5 - x$ and $x^2 - 1$ are continuous on the intervals $[-1, 2)$ and $(2, 3]$, respectively. Thus to conclude that g is continuous on the entire interval $[-1, 3]$, we need only to check the behavior of g when $x = 2$. By taking the one-sided limits when $x = 2$, we see that

$$\lim_{x \to 2^-} g(x) = \lim_{x \to 2^-} (5 - x) = 3$$

and

$$\lim_{x \to 2^+} g(x) = \lim_{x \to 2^+} (x^2 - 1) = 3$$

Since these two limits are equal, we have

$$\lim_{x \to 2} g(x) = g(2) = 3$$

Thus g is continuous at $x = 2$, and consequently it is continuous on the entire interval $[-1, 3]$. The graph of g is shown in Figure 2.7. ■

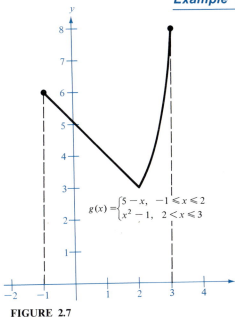

$$g(x) = \begin{cases} 5 - x, & -1 \le x \le 2 \\ x^2 - 1, & 2 < x \le 3 \end{cases}$$

FIGURE 2.7

Example 2

Discuss the continuity of

$$f(x) = \begin{cases} \dfrac{x^2 - 2x - 3}{x - 3}, & x \ne 3 \\ 3, & x = 3 \end{cases}$$

on the entire real line.

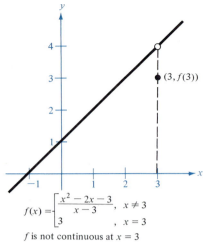

$$f(x) = \begin{cases} \dfrac{x^2 - 2x - 3}{x - 3}, & x \neq 3 \\ 3, & x = 3 \end{cases}$$

f is not continuous at $x = 3$

FIGURE 2.8

Solution: For $x \neq 3$ the rational function $f(x) = (x^2 - 2x - 3)/(x - 3)$ is defined and by Theorem 2.3 it is continuous for all $x \neq 3$. Thus we need only test for continuity at $x = 3$. Note that for $x \neq 3$,

$$\frac{x^2 - 2x - 3}{x - 3} = \frac{(x - 3)(x + 1)}{x - 3} = x + 1$$

and so

$$\lim_{x \to 3} f(x) = \lim_{x \to 3} (x + 1) = 4 \neq f(3)$$

Therefore, we conclude that f is not continuous at $x = 3$ (see Figure 2.8). Note that this is an example of a removable discontinuity. ∎

Example 3

Discuss the continuity of $f(x) = \sqrt{3 - x}$.

Solution: First, f is defined for all $x \leq 3$. Furthermore, f is continuous from the left at $x = 3$, because

$$\lim_{x \to 3^-} f(x) = \lim_{x \to 3^-} \sqrt{3 - x} = 0 = f(3)$$

For all $x \leq 3$ the function f obviously satisfies the three requirements for continuity and we conclude that f is continuous on the interval $(-\infty, 3]$. (See Figure 2.9.) ∎

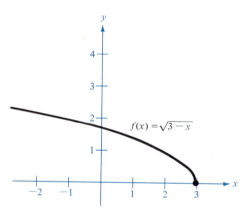

$f(x) = \sqrt{3 - x}$

FIGURE 2.9

In the context of one-sided limits and one-sided continuity, it is interesting to look at the *greatest-integer function*. This function is denoted by

$$\lfloor x \rfloor = (\text{greatest integer} \leq x)$$

For instance,

$$\lfloor -2.1 \rfloor = (\text{greatest integer} \leq -2.1) = -3$$
$$\lfloor -2 \rfloor = (\text{greatest integer} \leq -2) = -2$$
$$\lfloor 1.5 \rfloor = (\text{greatest integer} \leq 1.5) = 1$$

From the graph of the greatest-integer function in Figure 2.10, notice how the function jumps up one unit at each integer.

Example 4

Discuss the continuity of the greatest-integer function $f(x) = \lfloor x \rfloor$.

Solution: Obviously, the function has a jump discontinuity at each integer k, since

$$\lim_{x \to k^-} \lfloor x \rfloor = k - 1, \qquad \text{whereas} \qquad \lim_{x \to k^+} \lfloor x \rfloor = k$$

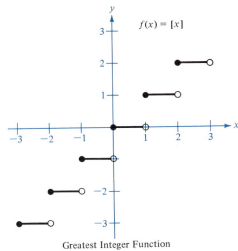

$f(x) = [x]$

Greatest Integer Function

FIGURE 2.10

Furthermore, the greatest-integer function is continuous from the right but discontinuous from the left at each integer k, because

$$\lim_{x \to k^+} [x] = k = f(k) \qquad \text{(continuous from right)}$$

whereas

$$\lim_{x \to k^-} [x] = k - 1 \neq f(k) \qquad \text{(discontinuous from left)}$$

Finally, the greatest-integer function is continuous at all noninteger values because it is constant on each interval $(k, k + 1)$, where k is an integer. ∎

Because continuity is defined in terms of limits, it is not surprising to learn that continuous functions and limits possess many of the same properties. We list some properties here and give an illustration of the type of proof required for these properties.

Properties of Continuous Functions	If the functions f and g are continuous at c, then the following functions are also continuous at c:

1. $f \pm g$
2. af, where a is any constant
3. fg
4. f/g, if $g(c) \neq 0$
5. $f(g(x))$, provided f is continuous at $g(c)$

Proof of 1: Let $F(x) = f(x) + g(x)$. Since f and g are both continuous at c, we have

$$\lim_{x \to c} f(x) = f(c) \qquad \text{and} \qquad \lim_{x \to c} g(x) = g(c)$$

Therefore, by Property 4 of limits (Section 2.1), we write

$$\lim_{x \to c} [f(x) + g(x)] = \lim_{x \to c} f(x) + \lim_{x \to c} g(x) = f(c) + g(c)$$

And it follows that

$$\lim_{x \to c} F(x) = F(c)$$

which implies F is continuous at c.

Similar arguments, incorporating properties of limits, can be used to prove Properties 2, 3, and 4. Property 5 will be discussed again in Section 2.4. ∎

We conclude this section with two important theorems concerning the behavior of continuous functions. By referring to texts on advanced calculus, you will find that most proofs of these two theorems are based on a property of real numbers called "completeness." A discussion of this property would be inappropriate at this point. Hence we will omit the proofs of both theorems. However, the theorems are not difficult to understand and we feel that you will see the reasonableness of these two important results.

The first theorem is called the Intermediate Value Theorem, and it states that for a continuous function f, if x assumes all values between a and b, then $f(x)$ must assume all values between $f(a)$ and $f(b)$. As a simple example of the theorem, consider a person's height. Suppose that a boy is 5 feet tall on his thirteenth birthday and 5 feet 7 inches tall on his fourteenth birthday. Then, for any height h between 5 feet and 5 feet 7 inches, there must have been a time t when his height was exactly h. This seems reasonable since we believe that normal human growth is continuous and a person's height could not abruptly change from one value to another.

The precise statement of the theorem is as follows:

THEOREM 2.4 (*Intermediate Value Theorem*)	If f is continuous on $[a, b]$ and k is any number between $f(a)$ and $f(b)$, then there is at least one number c between a and b such that $f(c) = k$.

This theorem asserts that if the domain of a function is a closed interval and the function is continuous, then there are no holes or gaps in the graph of that function. Discontinuous functions, however, may not possess the intermediate value property, as a comparison of Figures 2.11 and 2.12 indicates.

You may be familiar with one application of the Intermediate Value Theorem in algebra. That is, if we are given a polynomial $f(x)$ and there are two numbers a and b such that $f(a)$ is negative and $f(b)$ is positive, then the polynomial must have a root (it must cross the x-axis at least once) between a and b. The following example illustrates this application of the Intermediate Value Theorem to polynomials.

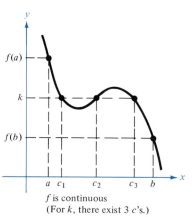

f is continuous
(For k, there exist 3 c's.)

FIGURE 2.11

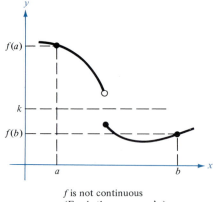

f is not continuous
(For k, there are no c's.)

FIGURE 2.12

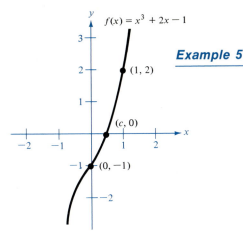

$f(x) = x^3 + 2x - 1$

(1, 2)

(c, 0)

(0, −1)

FIGURE 2.13

Example 5

Use the Intermediate Value Theorem to show that the polynomial $f(x) = x^3 + 2x - 1$ has a root in the interval $[0, 1]$.

Solution: Since $f(0) = -1$ and $f(1) = 2$, we apply the Intermediate Value Theorem to conclude that there must be some c in $[0, 1]$ such that $f(c) = 0$ (see Figure 2.13).

Note that the Intermediate Value Theorem tells us that (at least) one c exists, but it does not give us a method for finding c. Such theorems are called *existence theorems*.   ■

Prior to our final theorem in this section, we define what we mean by the minimum and maximum values of a function on an interval.

Definition of Extrema	Let c be a number in the interval $[a, b]$; then we have the following:

1. $f(c)$ is a **minimum** (value) of f on $[a, b]$ if $f(c) \leq f(x)$ for every x in $[a, b]$.
2. $f(c)$ is a **maximum** (value) of f on $[a, b]$ if $f(c) \geq f(x)$ for every x in $[a, b]$.

The minimum and maximum values of a function on an interval are often called the **extreme values,** or **extrema,** of the function on the interval.

THEOREM 2.5 (*Extreme Value Theorem*)	If f is continuous on $[a, b]$, then f takes on both a minimum value and a maximum value on $[a, b]$.

For a continuous function on a closed interval, the existence of the maximum and minimum are illustrated in Figure 2.14. Notice in Figure 2.15 that on the *open* interval $(-1, 2)$, $f(x) = x^2 + 1$ has a minimum value of 1 but does not take on a maximum value. This is the case because as x approaches 2 (but never takes on the value 2) from the left, $f(x)$ increases toward (but never takes on) the value 5.

If we drop the continuity requirement from the hypothesis of Theorem 2.5, we may destroy the existence of a maximum or a minimum. On the interval $[-1, 2]$ consider the function

$$g(x) = \begin{cases} x^2 + 1, & x \neq 0 \\ 2, & x = 0 \end{cases}$$

This function is discontinuous at $x = 0$ and has a maximum but no minimum on the interval $[-1, 2]$, as shown in Figure 2.16.

In summary, we remind you that Theorems 2.4 and 2.5 guarantee the *existence* of specific points but do not give techniques for finding the

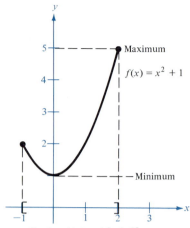

On closed interval $[-1, 2]$,
f has both a maximum and
a mimimum.

FIGURE 2.14

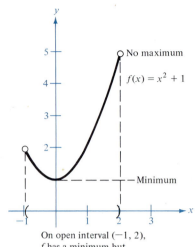

On open interval $(-1, 2)$,
f has a minimum but
no maximum.

FIGURE 2.15

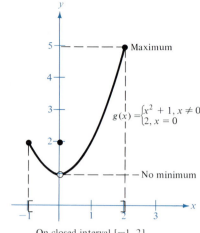

On closed interval $[-1, 2]$,
g has a maximum but
no minimum.

FIGURE 2.16

points. As a matter of fact, the determination of minimums and maximums and intermediate values will be among our major concerns once we have developed the preliminary concepts of differential calculus.

Section Exercises (2.2)

In Exercises 1–20, find the discontinuities (if any) for each function. Which of the discontinuities are removable?

1. $f(x) = x^2 - 2x + 1$

2. $f(x) = \dfrac{1}{x^2 + 1}$

S 3. $f(x) = \dfrac{1}{x - 1}$

4. $f(x) = \dfrac{x}{x^2 - 1}$

5. $f(x) = \dfrac{x}{x^2 + 1}$

6. $f(x) = \dfrac{x - 3}{x^2 - 9}$

S 7. $f(x) = \dfrac{x + 2}{x^2 - 3x - 10}$

8. $f(x) = \dfrac{x - 1}{x^2 + x - 2}$

9. $f(x) = \begin{cases} x, & x \le 1 \\ x^2, & x > 1 \end{cases}$

10. $f(x) = \begin{cases} -2x + 3, & x < 1 \\ x^2, & x \ge 1 \end{cases}$

S 11. $f(x) = \begin{cases} \dfrac{x}{2} + 1, & x \le 2 \\ 3 - x, & x > 2 \end{cases}$

12. $f(x) = \begin{cases} -2x, & x \le 2 \\ x^2 - 4x + 1, & x > 2 \end{cases}$

13. $f(x) = \dfrac{|x + 2|}{x + 2}$

14. $f(x) = \begin{cases} |x - 2| + 3, & x < 0 \\ x + 5, & x \ge 0 \end{cases}$

S 15. $f(x) = \begin{cases} 3 + x, & x \le 2 \\ x^2 + 1, & x > 2 \end{cases}$

16. $f(x) = [x - 1]$ (greatest-integer function)

17. $f(x) = x - [x]$ (greatest-integer function)

18. $f(g(x))$, $f(x) = x^2$, $g(x) = x - 1$

19. $f(g(x))$, $f(x) = \dfrac{1}{\sqrt{x}}$, $g(x) = x - 1$

20. $f(g(x))$, $f(x) = \dfrac{1}{x - 1}$, $g(x) = x^2 + 5$

21. Sketch the graph of $y = (x^2 - 16)/(x - 4)$ to determine any points of discontinuity.

22. Sketch the graph of $y = (x^3 - 8)/(x - 2)$ to determine any points of discontinuity.

23. Sketch the graph of $y = |x| - x$ to determine any points of discontinuity.

24. The function $f(x) = x^2 - 4x + 3$ is negative at $x = 2$ and positive at $x = 4$. What theorem guarantees that the function has at least one zero in the interval $[2, 4]$?
25. The function $f(x) = x^3 + 3x - 2$ is negative at $x = 0$ and positive at $x = 1$. What theorem guarantees that the function has at least one zero in the interval $[0, 1]$?

In Exercises 26–29, verify the applicability of the Intermediate Value Theorem and find the value of c guaranteed by the theorem.

26. $f(x) = x^2 + x - 1$, $k = 11$, $[0, 5]$
27. $f(x) = x^2 - 6x + 8$, $k = 0$, $[0, 3]$
28. $f(x) = x^3 - x^2 + x - 2$, $k = 4$, $[0, 3]$
S 29. $f(x) = (x^2 + x)/(x - 1)$, $k = 6$, $[\frac{5}{2}, 4]$

30. Determine the constant a so that

$$f(x) = \begin{cases} x^3, & x \le 2 \\ ax^2, & x > 2 \end{cases}$$

is continuous.

31. Determine the constants a and b so that the function

$$f(x) = \begin{cases} 2, & x \le -1 \\ ax + b, & -1 < x < 3 \\ -2, & x \ge 3 \end{cases}$$

is continuous.
32. Is the function $f(x) = \sqrt{1 - x^2}$ continuous at $x = 1$? Give the reason for your answer.
33. Does the function $f(x) = x^3$ have a maximum and minimum on the interval $(0, 1)$? Explain your answer.
34. Prove that if f and g are both continuous at c, then their sum is continuous at c.
S 35. Prove that if f and g are both continuous at c, then their product is continuous at c.
36. Prove that if f and g are both continuous at c, and $g(c) \ne 0$, then their quotient is continuous at c.

For Review and Class Discussion

True or False

1. ____ If $\lim_{x \to c} f(x) = L$ and $f(c) = L$, then f is continuous at c.
2. ____ If $f(x) = g(x)$ for $x \ne c$ and $f(c) \ne g(c)$, then either f or g must be discontinuous at c.
3. ____ If f is continuous at c, $f(x) = g(x)$ for $x \ne c$, and $f(c) \ne g(c)$, then g has a removable discontinuity at c.
4. ____ If f is continuous on $(a, b]$, then f must take on both a minimum and a maximum in $(a, b]$.
5. ____ A rational function can have infinitely many discontinuities.

2.3 Limits and Asymptotes

Purpose
- To introduce the notions of infinite limits and limits at infinity.
- To interpret graphically these two new types of limits as vertical and horizontal asymptotes.

The asymptote to any branch is, therefore, found by seeking the tangent to a point in that branch at an infinite distance. The direction of the branch may be found, by determining the position of a straight line parallel to the tangent to a point in the curve infinitely distant; for such a straight line will have the same direction as the infinite branch itself.
—*Isaac Newton (1642–1727)*

Up to this point our interpretation of the equation $\lim_{x \to c} f(x) = L$ requires both c and L to be finite. In this section we expand our interpretation to include "infinite limits" (L is infinite) and "limits at infinity" (c is

TABLE 2.2

x	0	1	1.5	1.9	1.999 $\longrightarrow$	2	$\longleftarrow$ 2.001	2.1	2.5	3	5
$f(x)$	$-\frac{3}{2}$	-3	-6	-30	$-3000 \longrightarrow -\infty$	$\infty \longleftarrow 3000$		30	6	3	1

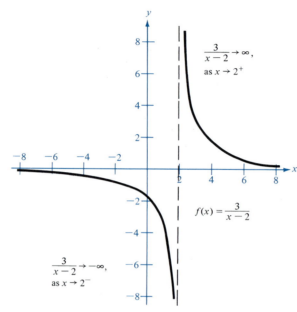

FIGURE 2.17

infinite). For example, let f be a function defined by

$$f(x) = \frac{3}{x - 2}$$

From Figure 2.17 and Table 2.2, we can see that $f(x)$ *decreases without bound* as x gets closer and closer to 2 from the left; $f(x)$ *increases without bound* as x approaches 2 from the right. Symbolically we can write this as

$$\frac{3}{x - 2} \to -\infty \quad \text{as} \quad x \to 2^-$$

and

$$\frac{3}{x - 2} \to \infty \quad \text{as} \quad x \to 2^+$$

We denote the type of behavior in Figure 2.17 by these limit statements:

$$\lim_{x \to 2^-} \frac{3}{x - 2} = -\infty$$

and

$$\lim_{x \to 2^+} \frac{3}{x - 2} = \infty$$

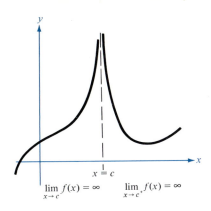

$$\lim_{x \to c^-} f(x) = \infty \qquad \lim_{x \to c^+} f(x) = \infty$$

FIGURE 2.18   Even Asymptotes

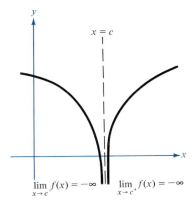

$$\lim_{x \to c^-} f(x) = -\infty \qquad \lim_{x \to c^+} f(x) = -\infty$$

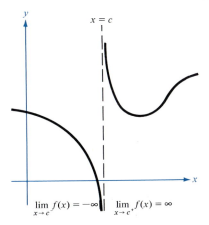

$$\lim_{x \to c^-} f(x) = -\infty \qquad \lim_{x \to c^+} f(x) = \infty$$

FIGURE 2.19   Odd Asymptotes

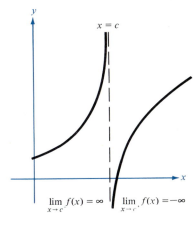

$$\lim_{x \to c^-} f(x) = \infty \qquad \lim_{x \to c^+} f(x) = -\infty$$

[Note: The statement $\lim_{x \to c} f(x) = \infty$ does not mean that the limit exists (has a specific value); on the contrary, it tells us how the limit fails to exist by denoting the unbounded behavior of $f(x)$ as x approaches c.]

In general, the types of limits where $f(x)$ becomes infinite as x approaches c from the left or right are referred to as **infinite limits.**

Infinite Limits

For a function f there are four possible types of one-sided **infinite limits** at c:

1. $\displaystyle\lim_{x \to c^-} f(x) = \infty$ 2. $\displaystyle\lim_{x \to c^-} f(x) = -\infty$

3. $\displaystyle\lim_{x \to c^+} f(x) = \infty$ 4. $\displaystyle\lim_{x \to c^+} f(x) = -\infty$

In each of these cases, the line $x = c$ is called a **vertical asymptote** of f.

If both one-sided limits of f at c are infinite and of like sign, then the line $x = c$ is called an **even** vertical asymptote (see Figure 2.18). If both one-sided limits of f at c are infinite and have unlike signs, then the line $x = c$ is called an **odd** vertical asymptote (see Figure 2.19).

The next two theorems, given without proof, are useful in determining infinite limits and vertical asymptotes.

THEOREM 2.6
(*Infinite Limits*)

If n is a positive integer, then

$$\lim_{x \to 0} \frac{1}{x^n}$$

is infinite. Specifically,

1. $\lim\limits_{x \to 0^+} \dfrac{1}{x^n} = \infty$

2. $\lim\limits_{x \to 0^-} \dfrac{1}{x^n} = \begin{cases} -\infty, & \text{if } n \text{ is odd} \\ \infty, & \text{if } n \text{ is even} \end{cases}$

Theorem 2.6 suggests that the undefined ratio $1/0$ can be considered to be ∞ or $-\infty$, depending on the direction from which zero is approached as a limiting value. In general, a vertical asymptote occurs when a ratio has a zero denominator and a nonzero numerator. This is stated formally in the following theorem:

THEOREM 2.7
(*Vertical Asymptotes*)

If $F(x) = f(x)/g(x)$ is such that $f(c) \neq 0$, $g(c) = 0$, and f and g are continuous in an open interval containing c, then the graph of F has a vertical asymptote at $x = c$.

Example 1

Determine and classify as odd or even any vertical asymptote for the function

$$f(x) = \frac{x + 2}{x^2 - 2x}$$

Solution: By Theorem 2.7 possible vertical asymptotes occur where the denominator of $f(x)$ is zero. Thus we consider the one-sided limits at $x = 2$ and at $x = 0$. We have

$$\lim_{x \to 2^+} \frac{x + 2}{x^2 - 2x} = \frac{4}{0^+} = \infty$$

Note that 0^+ means that $x^2 - 2x$ approaches zero through positive values. We also have

$$\lim_{x \to 2^-} \frac{x + 2}{x^2 - 2x} = \frac{4}{0^-} = -\infty$$

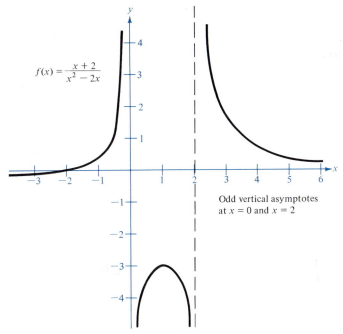

$$f(x) = \frac{x + 2}{x^2 - 2x}$$

Odd vertical asymptotes at $x = 0$ and $x = 2$

FIGURE 2.20

where 0^- indicates that the denominator approaches zero through negative values. Similarly, at $x = 0$ we have

$$\lim_{x \to 0^+} \frac{x + 2}{x^2 - 2x} = \frac{2}{0^-} = -\infty$$

$$\lim_{x \to 0^-} \frac{x + 2}{x^2 - 2x} = \frac{2}{0^+} = \infty$$

Therefore, the lines $x = 2$ and $x = 0$ are both *odd* vertical asymptotes (see Figure 2.20). ∎

[*Note:* In Example 1 the symbol $4/0^+$ does not mean we are dividing by zero. It is simply a convenient way to indicate that the limit is $+\infty$ rather than $-\infty$.]

Example 2

Determine and classify any vertical asymptotes of

$$g(x) = \frac{2x}{(x + 3)^2}$$

Solution: The only possible vertical asymptote occurs when $x = -3$ (denominator is zero). Therefore, the one-sided limits are

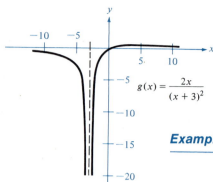

$g(x) = \dfrac{2x}{(x+3)^2}$

Even vertical asymptote
at $x = -3$

FIGURE 2.21

$$\lim_{x \to -3^+} \frac{2x}{(x+3)^2} = \frac{-6}{0^+} = -\infty$$

and

$$\lim_{x \to -3^-} \frac{2x}{(x+3)^2} = \frac{-6}{0^+} = -\infty$$

We conclude that the line $x = -3$ is an *even* vertical asymptote (compare Figure 2.21 with Figure 2.18). ■

Example 3

Determine and classify any vertical asymptotes of

$$f(x) = \frac{x^2 + 2x - 8}{x^2 - 4}$$

Solution: Possible vertical asymptotes occur at $x = 2$ and $x = -2$. However, the statement

$$\lim_{x \to 2^+} \frac{x^2 + 2x - 8}{x^2 - 4} = \frac{0}{0}$$

suggests that we simplify before taking the limit. Thus

$$\frac{x^2 + 2x - 8}{x^2 - 4} = \frac{(x+4)(x-2)}{(x+2)(x-2)} = \frac{x+4}{x+2}, \qquad (x \neq 2)$$

and we can eliminate the line $x = 2$ as a possible vertical asymptote.

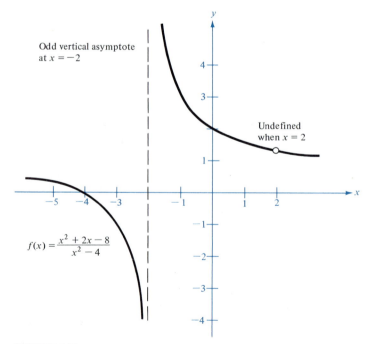

Odd vertical asymptote
at $x = -2$

Undefined
when $x = 2$

$f(x) = \dfrac{x^2 + 2x - 8}{x^2 - 4}$

FIGURE 2.22

TABLE 2.3

x	$-\infty \longleftarrow -100$	-10	-2	0	2	10	$100 \longrightarrow \infty$
$f(x)$	$3 \longleftarrow 2.9997$	2.97	2.4	0	2.4	2.97	$2.9997 \longrightarrow 3$

From

$$\lim_{x \to -2^+} \frac{x+4}{x+2} = \frac{2}{0^+} = \infty$$

and

$$\lim_{x \to -2^-} \frac{x+4}{x+2} = \frac{2}{0^-} = -\infty$$

it follows that the line $x = -2$ is an odd vertical asymptote (see Figure 2.22). ■

Another type of limit to be considered is one for which the values of a function approach some finite number as x increases (or decreases) without bound. For instance, the graph in Figure 2.23 and Table 2.3 for the function

$$f(x) = \frac{3x^2}{x^2 + 1}$$

suggest that the value of $f(x)$ approaches 3 as x increases without bound $(x \to \infty)$. Similarly, $f(x)$ approaches 3 as $x \to -\infty$.

In general, the types of limits where $f(x)$ approaches some finite value as x becomes infinite are called **limits at infinity.**

Limits at Infinity

If f is a function and L is a real number, then the statements:

1. $\displaystyle\lim_{x \to \infty} f(x) = L$

2. $\displaystyle\lim_{x \to -\infty} f(x) = L$

denote **limits at infinity.** In either case the line $y = L$ is called a **horizontal asymptote** of f.

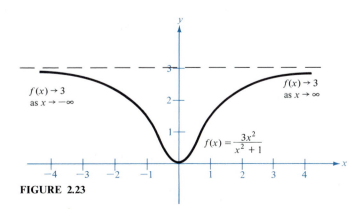

$$f(x) \to 3 \text{ as } x \to -\infty$$

$$f(x) \to 3 \text{ as } x \to \infty$$

$$f(x) = \frac{3x^2}{x^2 + 1}$$

FIGURE 2.23

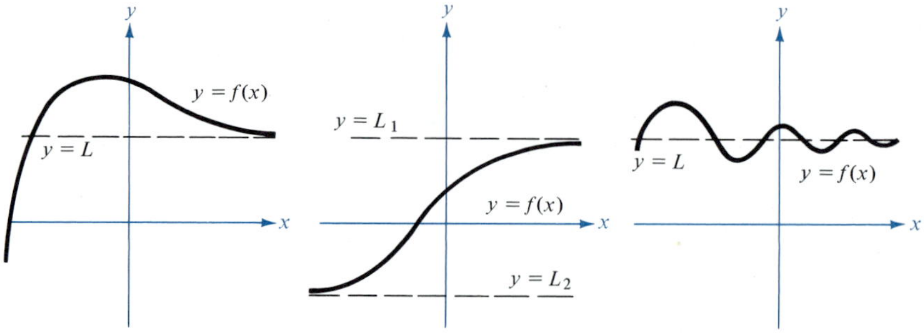

The graph of a function may approach
a horizontal asymptote in various ways.

FIGURE 2.24

Figure 2.24 suggests some ways in which the graph of a function f may approach one or more horizontal asymptotes. Observe in Figure 2.24 that it is possible for a function to cross its horizontal asymptote.

When evaluating limits at infinity, the following theorem is frequently used.

THEOREM 2.8
(*Limits at Infinity*)

If r is positive and c is any real number, then

$$\lim_{x \to \infty} \frac{c}{x^r} = 0$$

Furthermore,

$$\lim_{x \to -\infty} \frac{c}{x^r} = 0$$

if x^r is defined when $x < 0$.

The properties of limits at infinity are the same as those for finite limits discussed in Section 2.1. For example, the properties

$$\lim_{x \to c} b = b \qquad \text{and} \qquad \lim_{x \to c} b \cdot f(x) = b \left(\lim_{x \to c} f(x) \right)$$

can be rewritten as

$$\lim_{x \to \infty} b = b \qquad \text{and} \qquad \lim_{x \to \infty} b \cdot f(x) = b \left(\lim_{x \to \infty} f(x) \right)$$

Example 4

Evaluate

$$\lim_{x \to \infty} \left(5 - \frac{2}{x^2} \right)$$

Solution: Since

$$\lim_{x \to \infty} \left(5 - \frac{2}{x^2} \right) = \lim_{x \to \infty} 5 - \lim_{x \to \infty} \frac{2}{x^2}$$

it follows that $\lim\limits_{x\to\infty}\left(5-\dfrac{2}{x^2}\right)=5-0=5$ ∎

Example 5

Evaluate $\lim\limits_{x\to\infty}\dfrac{2x-1}{x+1}$

Solution: Since $\lim\limits_{x\to\infty}\dfrac{2x-1}{x+1}=\dfrac{\lim\limits_{x\to\infty}(2x-1)}{\lim\limits_{x\to\infty}(x+1)}$

we obtain the meaningless result

$$\lim\limits_{x\to\infty}\dfrac{2x-1}{x+1}=\dfrac{\infty}{\infty}$$

To apply Theorem 2.8 to each term of the fraction, we divide both numerator and denominator by x. Thus

$$\dfrac{2x-1}{x+1}=\dfrac{2-(1/x)}{1+(1/x)}$$

Now we see that

$$\lim\limits_{x\to\infty}\dfrac{2x-1}{x+1}=\lim\limits_{x\to\infty}\dfrac{2-(1/x)}{1+(1/x)}=\dfrac{2-0}{1+0}=2$$

The line $y=2$ is a horizontal asymptote. (See Figure 2.25.) (You may wish to verify this result by substituting large positive values for x.) ∎

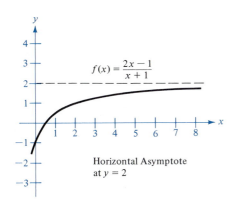

$f(x)=\dfrac{2x-1}{x+1}$

Horizontal Asymptote at $y=2$

FIGURE 2.25

Previously (Section 2.1) we encountered the meaningless expression $0/0$. In Example 5 we obtained another meaningless expression, ∞/∞. In both cases we were able to resolve the difficulty by rewriting the given expression in an equivalent form. Later (in Section 8.5) we will discuss an alternative procedure for dealing with "indeterminate forms" such as $0/0$ and ∞/∞. In the meantime, it is of benefit to know which forms can be evaluated, that is, which are "determinate." For ready reference we list here six determinate forms.

Determinate Forms for $\lim\dfrac{f(x)}{g(x)}$			
1. Limit is zero:	$\dfrac{0}{\pm\infty}$,	$\dfrac{L}{\pm\infty}$,	$\dfrac{0}{L}$
2. Limit is infinite:	$\dfrac{\pm\infty}{0}$,	$\dfrac{\pm\infty}{L}$,	$\dfrac{L}{0}$

Example 6

Evaluate the following limits:

(a) $\lim\limits_{x\to\infty} \dfrac{-2x + 3}{3x^2 + 1}$

(b) $\lim\limits_{x\to\infty} \dfrac{-2x^2 + 3}{3x^2 + 1}$

(c) $\lim\limits_{x\to\infty} \dfrac{-2x^3 + 3}{3x^2 + 1}$

Solution:

(a) To apply Theorem 2.8 to each term of the fraction, divide both numerator and denominator by x^2. Thus it follows that

$$\lim_{x\to\infty} \frac{-2x + 3}{3x^2 + 1} = \lim_{x\to\infty} \frac{(-2/x) + (3/x^2)}{3 + (1/x^2)} = \frac{-0 + 0}{3 + 0} = \frac{0}{3} = 0$$

(b) In this case we also divide by x^2 to apply Theorem 2.8 to each term of the fraction. Thus we obtain

$$\lim_{x\to\infty} \frac{-2x^2 + 3}{3x^2 + 1} = \lim_{x\to\infty} \frac{-2 + (3/x^2)}{3 + (1/x^2)} = \frac{-2 + 0}{3 + 0} = \frac{-2}{3}$$

(c) In this case we divide each term of the fraction by x^3, and by Theorem 2.8 it follows that

$$\lim_{x\to\infty} \frac{-2x^3 + 3}{3x^2 + 1} = \lim_{x\to\infty} \frac{-2 + (3/x^3)}{(3/x) + (1/x^3)} = \frac{-2 + 0}{0 + 0} = \frac{-2}{0^+} = -\infty \quad \blacksquare$$

Observe that in part (a) of Example 6, the degree of the numerator was *less* than the degree of the denominator and the limit of the ratio was zero. In part (b) the degrees of the numerator and denominator were *equal* and the limit was merely the ratio of the coefficients of the highest-powered terms. Finally, in part (c) the degree of the numerator was *greater* than that of the denominator and the limit was infinite. These results suggest an informal alternative way of determining the limits at infinity for functions expressed as the ratio of two polynomials.

Limits at Infinity for Rational Functions

For the rational function $f(x)/g(x)$, where

$$f(x) = a_n x^n + a_{n-1}x^{n-1} + \cdots + a_0$$

and

$$g(x) = b_m x^m + b_{m-1}x^{m-1} + \cdots + b_0$$

we have

$$\lim_{x\to\pm\infty} \frac{f(x)}{g(x)} = \begin{cases} 0, & \text{if } n < m \\ \dfrac{a_n}{b_m}, & \text{if } n = m \\ \pm\infty, & \text{if } n > m \end{cases}$$

This result seems reasonable if we realize that for large values of x the highest-powered term of a polynomial is the most "influential" term. That is, a polynomial tends to behave like its highest-powered term as x becomes sufficiently large in either the positive or negative sense.

There are many examples of asymptotic behavior in the physical sciences. For instance, the following example describes the asymptotic recovery of oxygen in a pond.

Example 7

Suppose that $f(t)$ measures the level of oxygen in a pond, where $f(t) = 1$ is the normal level and the time t is measured in weeks. When $t = 0$, some organic waste is dumped into the pond, and as the waste material oxidizes, the amount of oxygen in the pond is given by

$$f(t) = \frac{t^2 - t + 1}{t^2 + 1}$$

What percentage of the normal level of oxygen exists in the pond after 1 week? After 2 weeks? After 10 weeks? What is the limit as t approaches infinity?

Solution: When $t = 1, 2,$ and 10, the levels of oxygen are as follows:

(1 week) $f(1) = \dfrac{1^2 - 1 + 1}{1^2 + 1} = \dfrac{1}{2} = 50\%$

(2 weeks) $f(2) = \dfrac{2^2 - 2 + 1}{2^2 + 1} = \dfrac{3}{5} = 60\%$

(10 weeks) $f(10) = \dfrac{10^2 - 10 + 1}{10^2 + 1} = \dfrac{91}{101} = 90.1\%$

To take the limit as t approaches infinity, we note that the degrees of the numerator and denominator of this rational function are equal. Thus the limit is

$$\lim_{t \to \infty} \frac{t^2 - t + 1}{t^2 + 1} = 1 = 100\%$$

(See Figure 2.26.)

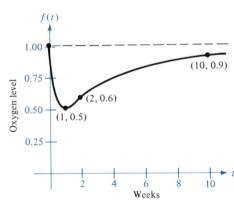

FIGURE 2.26

Example 8

Determine

$$\lim_{x \to \pm\infty} \frac{5x^3 + 2}{(2x - 1)^3}$$

Solution: Expanding the denominator yields a highest-powered term of $(2x)^3$, or $8x^3$. Thus we see that the numerator and denominator are of equal degree and we conclude that

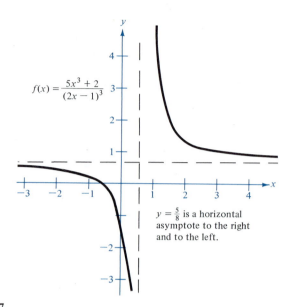

$$f(x) = \frac{5x^3 + 2}{(2x - 1)^3}$$

$y = \frac{5}{8}$ is a horizontal asymptote to the right and to the left.

FIGURE 2.27

$$\lim_{x \to \pm\infty} \frac{5x^3 + 2}{(2x - 1)^3} = \frac{5}{8}$$

See Figure 2.27. (You are encouraged to verify this result by the method of Example 6.) ■

In Figure 2.27 we can see that the rational function $f(x) = (5x^3 + 2)/(2x - 1)^3$ approaches the same horizontal asymptote to the right and left. That is,

$$\lim_{x \to -\infty} f(x) = \frac{5}{8} = \lim_{x \to \infty} f(x)$$

This is always the case with rational functions. However, functions that are not rational may approach different horizontal asymptotes to the right and left, as in the next example.

Example 9

Determine the following:

(a) $\displaystyle\lim_{x \to \infty} \frac{3x - 2}{\sqrt{2x^2 + 1}}$

(b) $\displaystyle\lim_{x \to -\infty} \frac{3x - 2}{\sqrt{2x^2 + 1}}$

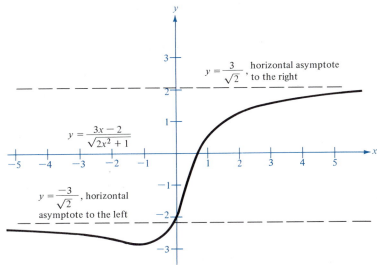

FIGURE 2.28

Solution:

(a)   To apply Theorem 2.8 we divide numerator and denominator by x. Since $\sqrt{x^2} = x$ for $x > 0$, we have

$$\frac{3x - 2}{\sqrt{2x^2 + 1}} = \frac{(3x - 2)/x}{\sqrt{2x^2 + 1}/\sqrt{x^2}} = \frac{3 - (2/x)}{\sqrt{(2x^2 + 1)/x^2}} = \frac{3 - (2/x)}{\sqrt{2 + (1/x^2)}}$$

Therefore,

$$\lim_{x \to \infty} \frac{3x - 2}{\sqrt{2x^2 + 1}} = \lim_{x \to \infty} \frac{3 - (2/x)}{\sqrt{2 + (1/x^2)}} = \frac{3 - 0}{\sqrt{2 + 0}} = \frac{3}{\sqrt{2}} \approx 2.12$$

See Figure 2.28.

(b)   For $x \to -\infty$ we have $x < 0$ and $x = -\sqrt{x^2}$. Thus we write

$$\frac{3x - 2}{\sqrt{2x^2 + 1}} = \frac{(3x - 2)/x}{\sqrt{2x^2 + 1}/(-\sqrt{x^2})} = \frac{3 - (2/x)}{-\sqrt{2 + (1/x^2)}}$$

Therefore,

$$\lim_{x \to -\infty} \frac{3x - 2}{\sqrt{2x^2 + 1}} = \lim_{x \to -\infty} \frac{3 - (2/x)}{-\sqrt{2 + (1/x^2)}} = \frac{3 - 0}{-\sqrt{2}} = -\frac{3}{\sqrt{2}} \approx -2.12$$

See Figure 2.28.   ■

Example 10

Determine

$$\lim_{x \to \infty} (x - \sqrt{x^2 - 2})$$

Solution: Since

$$\lim_{x \to \infty} (x - \sqrt{x^2 - 2}) = \infty - \infty$$

is meaningless, we proceed to rewrite the expression by rationalizing the numerator. Thus

$$(x - \sqrt{x^2 - 2})\left(\frac{x + \sqrt{x^2 - 2}}{x + \sqrt{x^2 - 2}}\right) = \frac{x^2 - (x^2 - 2)}{x + \sqrt{x^2 - 2}} = \frac{2}{x + \sqrt{x^2 - 2}}$$

By Theorem 2.8 $\displaystyle \lim_{x \to \infty} \frac{2}{x + \sqrt{x^2 - 2}} = \frac{2}{\infty} = 0$ ∎

Section Exercises (2.3)

In Exercises 1–10, find the vertical asymptotes for each function and write the corresponding left- and right-hand limits. Then classify each vertical asymptote as odd or even.

1. $f(x) = \dfrac{1}{x^2}$

2. $f(x) = \dfrac{4}{(x - 2)^3}$

3. $f(x) = \dfrac{2 + x}{1 - x}$

4. $f(x) = \dfrac{x^3}{x^2 - 1}$

S 5. $f(x) = 1 - \dfrac{4}{x^2}$

6. $f(x) = \dfrac{-2}{(x - 2)^2}$

7. $f(x) = \dfrac{1}{(x + 3)^4}$

8. $f(x) = \dfrac{-4x}{x^2 + 4}$

S 9. $f(x) = \dfrac{x}{x^2 + x - 2}$

10. $f(x) = \dfrac{x^2 - 1}{x^2 - x - 2}$

In Exercises 11–29, find the indicated limit.

11. $\displaystyle \lim_{x \to \infty} \frac{2x - 1}{3x + 2}$

12. $\displaystyle \lim_{x \to \infty} \frac{5x^3 + 1}{10x^3 - 3x^2 + 7}$

S 13. $\displaystyle \lim_{x \to \infty} \frac{x}{x^2 - 1}$

14. $\displaystyle \lim_{x \to \infty} \frac{2x^{10} - 1}{10x^{11} - 3}$

S 15. $\displaystyle \lim_{x \to -\infty} \frac{5x^2}{x + 3}$

16. $\displaystyle \lim_{x \to \infty} \frac{x^3 - 2x^2 + 3x + 1}{x^2 - 3x + 2}$

17. $\displaystyle \lim_{x \to \infty} \left(2x - \frac{1}{x^2}\right)$

18. $\displaystyle \lim_{x \to \infty} (x + 3)^{-2}$

S 19. $\displaystyle \lim_{x \to -\infty} \left(\frac{2x}{x - 1} + \frac{3x}{x + 1}\right)$

20. $\displaystyle \lim_{x \to \infty} \left(\frac{2x^2}{x - 1} + \frac{3x}{x + 1}\right)$

21. $\displaystyle \lim_{x \to -\infty} (x + \sqrt{x^2 + 3})$

22. $\displaystyle \lim_{x \to \infty} (2x - \sqrt{4x^2 + 1})$

S 23. $\displaystyle \lim_{x \to \infty} (x - \sqrt{x^2 + x})$

24. $\displaystyle \lim_{x \to -\infty} (3x + \sqrt{9x^2 - x})$

S 25. $\displaystyle \lim_{x \to -\infty} \frac{x}{\sqrt{x^2 - x}}$

26. $\displaystyle \lim_{x \to \infty} \frac{x}{\sqrt{x^2 + 1}}$

27. $\displaystyle \lim_{x \to \infty} \frac{2x + 1}{\sqrt{x^2 - x}}$

28. $\displaystyle \lim_{x \to -\infty} \frac{-3x + 1}{\sqrt{x^2 + x}}$

29. $\displaystyle \lim_{x \to \infty} \frac{x^2 - x}{\sqrt{x^4 + x}}$

In Exercises 30–41, sketch the graph of each equation. As a sketching aid examine each equation for intercepts, symmetry, and asymptotes.

30. $y = \dfrac{x - 3}{x - 2}$

S 31. $y = \dfrac{2 + x}{1 - x}$

32. $y = \dfrac{x^2}{x^2 - 9}$

33. $y = \dfrac{x^2}{x^2 + 9}$

34. $x^2 y = 4$

35. $xy^2 = 4$

36. $y = \dfrac{2x}{1 - x^2}$

37. $y = \dfrac{2x}{1 - x}$

38. $y = 1 + \dfrac{1}{x}$

39. $y = 2 - \dfrac{3}{x^2}$

40. $y = \dfrac{x}{\sqrt{x^2 - 4}}$

S 41. $y = \dfrac{x^3}{\sqrt{x^2 - 4}}$

In Exercises 42–44, complete the table for each function and estimate $\lim_{x \to \infty} f(x)$.

C 42. $f(x) = \dfrac{x + 1}{x\sqrt{x}}$

x	1	10	10^2	10^4	10^6
$f(x)$					

[c] **43.** $f(x) = x - \sqrt{x(x-1)}$

x	1	10	10^2	10^4	10^6
$f(x)$					

[c] **44.** $f(x) = x^2 - x\sqrt{x(x-1)}$

x	1	10	10^2	10^4	10^6
$f(x)$					

If $D(x)$ is a polynomial of degree n and $N(x)$ is a polynomial of degree $n + 1$, then, by division, we obtain

$$f(x) = \frac{N(x)}{D(x)} = (mx + b) + \frac{c}{D(x)}$$

Since $\lim_{x \to \infty} [c/D(x)] = 0$, we know that $f(x)$ approaches the line $y = mx + b$ as x approaches ∞. This line is called a **slant asymptote** and it is helpful when sketching the graph of f. In Exercises 45–48, sketch the graph of each equation. Use any of the sketching aids that we have developed, including slant asymptotes.

45. $y = \dfrac{x^2 + 1}{x}$

46. $y = \dfrac{x^3}{x^2 - 1}$

[s] **47.** $y = \dfrac{x^3}{2x^2 - 8}$

48. $y = \dfrac{2x^2 - 5x + 5}{x - 2}$

49. A business has a cost of $C = 0.5x + 500$ for producing x units. The average cost per unit is given by C/x. Evaluate $\lim_{x \to \infty} (C/x)$ and interpret the result.

50. According to Einstein's theory of relativity, the mass m of a particle depends upon its velocity v according to the equation $m = m_0/\sqrt{1 - (v^2/c^2)}$, where m_0 is the mass when at rest and c is the speed of light. Find the limit of the mass as v approaches c.

51. The efficiency of an internal combustion engine is defined to be

$$\text{efficiency } (\%) = 100\left[1 - \frac{1}{(v_1/v_2)^{c-1}}\right]$$

where v_1/v_2 is the compression ratio and c is a constant $(c > 1)$. Theoretically, what happens to the efficiency as the compression ratio increases without bound?

52. In Section 6.3 it will be shown that the area A of the region bounded by $y = x^2$, $y = 0$, and $x = 2$ is given by

$$A = \lim_{n \to \infty}\left[\left(\frac{4}{3n^3}\right)(2n^3 + 3n^2 + n)\right]$$

Find this limit.

53. Using the techniques of Section 7.2, it can be shown that the volume V of a sphere of radius 2 is

$$V = 16\pi \lim_{n \to \infty}\left(1 - \frac{2n^3 + 3n^2 + n}{6n^3}\right)$$

Find this limit.

54. If $f(x) = a_n x^n + a_{n-1}x^{n-1} + \cdots + a_0$ and $g(x) = b_m x^m + b_{m-1}x^{m-1} + \cdots + b_0$, use Theorem 2.8 to prove that

$$\lim_{x \to \pm\infty} \frac{f(x)}{g(x)} = \begin{cases} 0, & \text{if } n < m \\ \dfrac{a_n}{b_m}, & \text{if } n = m \\ \pm\infty, & \text{if } n > m \end{cases}$$

For Review and Class Discussion

True or False

1. ____ A function can have at most one horizontal asymptote.

2. ____ A function can have at most two horizontal asymptotes.

3. ____ If $p(x)$ is a polynomial, then the function given by $f(x) = p(x)/(x - 1)$ has a vertical asymptote at $x = 1$.

4. ____ If $\lim_{x \to c} f(x) = \infty$ and $\lim_{x \to c} g(x) = \infty$, then $\lim_{x \to c} [f(x) - g(x)] = 0$.

5. ____ The graph of f may cross a horizontal asymptote of f.

6. ____ Polynomial functions have no vertical or horizontal asymptotes.

7. ____ If f has a vertical asymptote at $x = 0$, then f is undefined at $x = 0$.

8. ____ If f has a vertical asymptote at $x = 0$ and f is symmetric with respect to the y-axis, then $x = 0$ is an even asymptote.

9. ____ If $p(x)$ is a polynomial and the function given by $f(x) = p(x)/(x - 1)^2$ has an odd vertical asymptote at $x = 1$, then $(x - 1)$ is a factor of $p(x)$.

10. ____ If f is continuous on $(-\infty, \infty)$, then it has no vertical asymptotes.

2.4 ϵ–δ Definition of Limits (Optional)

Purpose
- To give a formal definition of limits.
- To prove some properties of limits by the limit definition.
- To demonstrate the use of the limit definition to verify a limit.

One cannot help concluding that, since calculus yields reliable, simple, and exact methods, the principles on which it depends must also be simple and certain.   —*Jean D'Alembert* (1717–1783)

Our discussion of limits thus far has been strictly at the intuitive level. We have used expressions such as these:

For x near c, $f(x)$ is near L.

$f(x)$ approaches L as x approaches c.

$f(x)$ becomes infinite as x approaches c.

$f(x)$ approaches L as x tends toward infinity.

Each of these expressions is an informal attempt to describe the limiting behavior of a function. In this section we give formal analytic statements of the various types of limits previously encountered in this chapter.

Keep in mind that in this text it is generally sufficient to operate with limits at the intuitive level. However, the formal definitions given in this section may serve to further clarify the meaning of limits.

Intuitively, we said that the limit statement

$$\lim_{x \to c} f(x) = L$$

means "$f(x)$ approaches L as x approaches c." This statement can be interpreted to mean that the difference $|f(x) - L|$ is arbitrarily small provided the difference $|x - c|$ is sufficiently small. In other words, for each small $\epsilon > 0$ (the Greek lowercase letter epsilon), there must exist a small $\delta > 0$ (the Greek lowercase letter delta) such that

$$|f(x) - L| < \epsilon$$

whenever

$$|x - c| < \delta$$

Recall that for limits at c we are not interested in the value of f at $x = c$. Thus we will further require $0 < |x - c|$, which implies that $x \neq c$. In the following definition of the *limit of f at c*, we assume that f is defined on an open interval containing c, except possibly at c itself.

Definition of Limit	The statement

$$\lim_{x \to c} f(x) = L$$

means that for each $\epsilon > 0$ there exists a $\delta > 0$ such that

$$|f(x) - L| < \epsilon \qquad \text{whenever} \qquad 0 < |x - c| < \delta$$

Note that the definition has an *order* to it: "for each $\epsilon > 0$ there exists a $\delta > 0$." This means that if we choose a particular value for ϵ, then for this choice of ϵ there exists a δ that works. We do not require any specific δ to work for more than one choice of ϵ. Furthermore, the number δ is not unique, for if a specific δ works, then any smaller positive number will also work.

To gain a better understanding of our definition of limit and the correspondence between ϵ and δ, let us consider a graphical description, as shown in Figure 2.29.

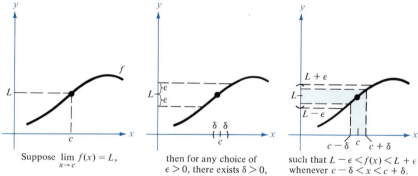

Suppose $\lim_{x \to c} f(x) = L$,   then for any choice of $\epsilon > 0$, there exists $\delta > 0$,   such that $L - \epsilon < f(x) < L + \epsilon$ whenever $c - \delta < x < c + \delta$.

FIGURE 2.29

Figure 2.30 shows how a new value, ϵ_1, can require a new choice for δ. The original value for δ is now too large and we must find a smaller value δ_1 that works for ϵ_1.

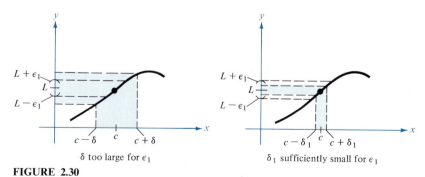

δ too large for ϵ_1 δ_1 sufficiently small for ϵ_1

FIGURE 2.30

The examples that follow illustrate some ways to determine the connection between ϵ and δ for a particular limit. Though they may appear to be rather complicated, studying them may further clarify your understanding of limits.

Example 1

By the ϵ-δ definition prove that

$$\lim_{x \to 3} (2x - 5) = 1$$

Solution: We are required to show that for each $\epsilon > 0$, there exists a $\delta > 0$ such that

$$|(2x - 5) - 1| < \epsilon \qquad \text{whenever} \qquad 0 < |x - 3| < \delta$$

Since we know that our choice for δ is dependent on our choice of ϵ, we must establish a connection between δ and ϵ. This in turn requires that we establish a connection between the absolute values $|(2x - 5) - 1|$ and $|x - 3|$. Simplifying the first absolute value, we get

$$|(2x - 5) - 1| = |2x - 6| = |2(x - 3)| = 2|x - 3|$$

Thus the statement

$$|(2x - 5) - 1| = 2|x - 3| < \epsilon$$

requires

$$|x - 3| < \frac{\epsilon}{2}$$

which means we can choose $\delta \leq \epsilon/2$. This choice works because whenever

$$0 < |x - 3| < \delta \leq \frac{\epsilon}{2}$$

then

$$|(2x - 5) - 1| = 2|x - 3| < 2\left(\frac{\epsilon}{2}\right) = \epsilon$$

∎

Example 2

By the limit definition prove

$$\lim_{x \to 5} \frac{x^2 - 13x + 40}{4x - 20} = -\frac{3}{4}$$

Solution: Since $\dfrac{x^2 - 13x + 40}{4x - 20} = \dfrac{(x - 8)(x - 5)}{4(x - 5)}$

we can cancel $(x - 5)$ from the numerator and denominator to obtain

$$\frac{x - 8}{4} = \frac{x}{4} - 2$$

Now the definition involves the inequalities

$$\left|\left(\frac{x}{4} - 2\right) - \left(-\frac{3}{4}\right)\right| < \epsilon \quad \text{and} \quad 0 < |x - 5| < \delta$$

We wish to establish a connection between the quantities

$$\left|\left(\frac{x}{4} - 2\right) - \left(-\frac{3}{4}\right)\right| \quad \text{and} \quad |x - 5|$$

Simplifying the first we get

$$\left|\left(\frac{x}{4} - 2\right) - \left(-\frac{3}{4}\right)\right| = \left|\frac{x}{4} - \frac{8}{4} + \frac{3}{4}\right| = \frac{1}{4}|x - 8 + 3| = \frac{1}{4}|x - 5|$$

Thus

$$\frac{1}{4}|x - 5| < \epsilon$$

requires

$$|x - 5| < 4\epsilon$$

which means we can choose $\delta \leq 4\epsilon$. This choice works because whenever

$$0 < |x - 5| < \delta \leq 4\epsilon$$

then

$$\left|\left(\frac{x}{4} - 2\right) + \frac{3}{4}\right| = \frac{1}{4}|x - 5| < \frac{1}{4}(4\epsilon) = \epsilon \qquad \blacksquare$$

Example 3

Let

$$f(x) = \begin{cases} 1, & x \leq 0 \\ 2, & x > 0 \end{cases}$$

Use the ε-δ definition of limit to show that $\lim_{x \to 0} f(x)$ *does not* exist.

Solution: Assume that $\lim_{x \to 0} f(x)$ exists and is equal to L. Choose $\epsilon = \frac{1}{2}$, which is less than the jump discontinuity at $x = 0$ (see Figure 2.31). Then there exists a $\delta > 0$ such that

$$|f(x) - L| < \tfrac{1}{2} \quad \text{whenever} \quad 0 < |x - 0| < \delta$$

The limit from the left requires

$$|1 - L| < \tfrac{1}{2} \quad \text{whenever} \quad -\delta < x < 0$$

The limit from the right requires

$$|2 - L| < \tfrac{1}{2} \quad \text{whenever} \quad 0 < x < \delta$$

Together these statements imply that

$$-\tfrac{1}{2} < L - 1 < \tfrac{1}{2} \quad \text{and} \quad -\tfrac{1}{2} < L - 2 < \tfrac{1}{2}$$

or

$$\tfrac{1}{2} < L < \tfrac{3}{2} \quad \text{and} \quad \tfrac{3}{2} < L < \tfrac{5}{2}$$

But no single value L can satisfy both of these inequalities. Thus our assumption is false and we conclude that $\lim_{x \to 0} f(x)$ does not exist. ■

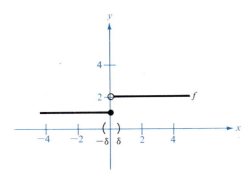

FIGURE 2.31

In Example 3 we made use of one-sided limits. For future reference we provide the formal ϵ–δ definition of one-sided limits.

ϵ–δ Definition of One-Sided Limits	$\lim_{x \to c^-} f(x) = L$ if, for each $\epsilon > 0$, there exists a $\delta > 0$ such that $$\|f(x) - L\| < \epsilon \qquad \text{whenever} \qquad -\delta < x - c < 0$$ $\lim_{x \to c^+} f(x) = L$ if, for each $\epsilon > 0$, there exists a $\delta > 0$ such that $$\|f(x) - L\| < \epsilon \qquad \text{whenever} \qquad 0 < x - c < \delta$$

There are several equivalent forms of the limit statement

$$\lim_{x \to c} f(x) = L$$

In the future we will use whichever form is most convenient for the task at hand. We list three equivalent forms next and suggest that you use the ϵ–δ definition to establish their equivalence:

1. $\lim_{x \to c} f(x) = L$

2. $\lim_{x \to c} [f(x) - L] = 0$

3. $\lim_{h \to 0} f(c + h) = L$

Your appreciation of the ϵ–δ definition of the limit should grow as you see how useful it is in proving properties of limits. We now demonstrate the type of argument needed to establish the properties of limits identified in Section 2.1.

Proofs

Proof of Limit Property 2 (Section 2.1): To show that $\lim_{x \to c} x = c$, we need to show that for each $\epsilon > 0$ there exists a $\delta > 0$ such that

$$\|x - c\| < \epsilon \qquad \text{whenever} \qquad 0 < \|x - c\| < \delta$$

Quite obviously, the connection between ϵ and δ is that we can choose $\delta \leq \epsilon$. This choice works because

whenever $\qquad\qquad 0 < \|x - c\| < \delta$

then $\qquad\qquad \|x - c\| < \delta \leq \epsilon$

Proof of Limit Property 4 (Section 2.1): To show that

$$\lim_{x \to c} [f(x) + g(x)] = \lim_{x \to c} f(x) + \lim_{x \to c} g(x) = A + B$$

we need to establish that for each $\epsilon > 0$ there exists a $\delta > 0$ such that

$$\|[f(x) + g(x)] - (A + B)\| < \epsilon$$

whenever $\qquad\qquad 0 < \|x - c\| < \delta$

Certainly for $\frac{1}{2}\epsilon > 0$ there exists a $\delta_1 > 0$ such that

$$|f(x) - A| < \frac{\epsilon}{2} \qquad \text{whenever} \qquad 0 < |x - c| < \delta_1$$

because $\lim_{x \to c} f(x) = A$. Similarly, $\lim_{x \to c} g(x) = B$ implies that for $\frac{1}{2}\epsilon > 0$ there exists a $\delta_2 > 0$ such that

$$|g(x) - B| < \frac{\epsilon}{2} \qquad \text{whenever} \qquad 0 < |x - c| < \delta_2$$

Therefore, if we let δ be the smaller of δ_1 and δ_2, we have

$$|f(x) - A| < \frac{\epsilon}{2} \qquad \text{whenever} \qquad 0 < |x - c| < \delta$$

and $\qquad |g(x) - B| < \frac{\epsilon}{2} \qquad \text{whenever} \qquad 0 < |x - c| < \delta$

Combining these statements we have

$$\begin{aligned}
|[f(x) + g(x)] - (A + B)| &= |[f(x) - A] + [g(x) - B]| \\
&\leq |f(x) - A| + |g(x) - B| \\
&< \frac{\epsilon}{2} + \frac{\epsilon}{2} = \epsilon
\end{aligned}$$

whenever $0 < |x - c| < \delta$ (see Exercise 35, Section 1.1). We have thereby established that

$$\lim_{x \to c} [f(x) + g(x)] = A + B$$

A proof for the difference $f(x) - g(x)$ is left as an exercise. ■

Proofs of Limit Properties 5 and 6 of Section 2.1 require some rather sophisticated manipulations with the limit definition and will be omitted. These proofs can be found in advanced calculus texts, if you are interested. Limit Property 7 of Section 2.1 is easily established by using Properties 2 and 5. The remaining proofs of the limit properties are left as exercises.

Our next theorem is one that is used quite often in the proofs of other theorems. It concerns the limiting behavior of a function that is trapped between two other functions, each of which has the same limit at a given point. We refer to this theorem as the Squeeze Theorem.

THEOREM 2.9 (*Squeeze Theorem*)	If $h(x) \leq f(x) \leq g(x)$ for all x in an open interval containing c, except possibly at c itself, and if $$\lim_{x \to c} h(x) = L = \lim_{x \to c} g(x)$$ then $\qquad\qquad\qquad \lim_{x \to c} f(x) = L$

Proof: Since $\lim_{x \to c} h(x) = \lim_{x \to c} g(x) = L$, then for each $\epsilon > 0$ there exist $\delta_1 > 0$ and $\delta_2 > 0$ such that

$$|h(x) - L| < \epsilon \qquad \text{whenever} \qquad 0 < |x - c| < \delta_1$$
$$|g(x) - L| < \epsilon \qquad \text{whenever} \qquad 0 < |x - c| < \delta_2$$

Let δ be the smaller of δ_1 and δ_2; then

$$|h(x) - L| < \epsilon \qquad \text{and} \qquad |g(x) - L| < \epsilon$$

whenever $\qquad\qquad 0 < |x - c| < \delta$

This implies that if $0 < |x - c| < \delta$, then

$$-\epsilon < h(x) - L < \epsilon \qquad \text{and} \qquad -\epsilon < g(x) - L < \epsilon$$

and consequently

$$L - \epsilon < h(x) \qquad \text{and} \qquad g(x) < L + \epsilon$$

From these last two inequalities and the fact that $h(x) \leq f(x) \leq g(x)$ for $0 < |x - c| < \delta$, it follows that

$$L - \epsilon < f(x) < L + \epsilon$$

Consequently,

$$|f(x) - L| < \epsilon \qquad \text{whenever} \qquad 0 < |x - c| < \delta$$

and therefore, $\qquad\qquad \lim_{x \to c} f(x) = L$ ∎

In Section 2.2 we described *continuity* in terms of a limit, and in Section 2.3 we presented informal descriptions of *infinite limits* and *limits at infinity*. The remainder of this section is devoted to formal descriptions of each of these terms.

ϵ–δ Definition of Continuity

A function f is **continuous** at c if for each $\epsilon > 0$ there exists a $\delta > 0$ such that

$$|f(x) - f(c)| < \epsilon \qquad \text{whenever} \qquad |x - c| < \delta$$

Note that this definition assumes f is defined on some open interval containing c. Comparing this definition with the ϵ–δ definition of the *limit of f at c*, we observe that for continuity it is not necessary to require $0 < |x - c|$ because we know $f(c)$ is defined and obviously the inequality $|f(x) - f(c)| < \epsilon$ is satisfied at $x = c$.

We use the ϵ–δ definition of continuity to prove the following important theorem concerning the limit of a composite function.

THEOREM 2.10 (*Limit of a Composite Function*)	If $\lim_{x \to c} g(x) = L$ and f is continuous at L, then $$\lim_{x \to c} f(g(x)) = f\left(\lim_{x \to c} g(x) \right) = f(L)$$

Proof: Consider any $\epsilon > 0$. By the continuity of f at L, there exists a $\delta_1 > 0$ such that

$$|f(y) - f(L)| < \epsilon \quad \text{whenever} \quad |y - L| < \delta_1$$

Replacing y by $g(x)$ yields

$$|f(g(x)) - f(L)| < \epsilon \quad \text{whenever} \quad |g(x) - L| < \delta_1$$

Because $\lim_{x \to c} g(x) = L$, then for any $\delta_1 > 0$ there exists a $\delta > 0$ such that

$$|g(x) - L| < \delta_1 \quad \text{whenever} \quad 0 < |x - c| < \delta$$

(Note that δ_1 is used as our epsilon in this case.) Combining these results we have

$$|f(g(x)) - f(L)| < \epsilon \quad \text{whenever} \quad 0 < |x - c| < \delta$$

Therefore,

$$\lim_{x \to c} f(g(x)) = f(L) = f\left(\lim_{x \to c} g(x) \right) \qquad \blacksquare$$

Two useful results follow directly from Theorem 2.10.

1. If g is continuous at c and f is continuous at $g(c)$, then $f(g(x))$ is continuous at c.

2. $\lim_{x \to c} \sqrt[n]{f(x)} = \sqrt[n]{\lim_{x \to c} f(x)}$, provided the nth root exists.

Recall from Section 2.3 our work with infinite limits, such as

$$\lim_{x \to c} f(x) = \pm \infty$$

We described such limits by saying that $f(x)$ increases (decreases) without bound as x nears c. The formal description is as follows:

Definition of Infinite Limits	1. $\lim_{x \to c} f(x) = \infty$ if for each $M > 0$ there exists a $\delta > 0$ such that $$f(x) > M \quad \text{whenever} \quad 0 < |x - c| < \delta$$ 2. $\lim_{x \to c} f(x) = -\infty$ if for each $N < 0$ there exists a $\delta > 0$ such that $$f(x) < N \quad \text{whenever} \quad 0 < |x - c| < \delta$$

Graphic descriptions of the definition of infinite limits are given in Figure 2.32.

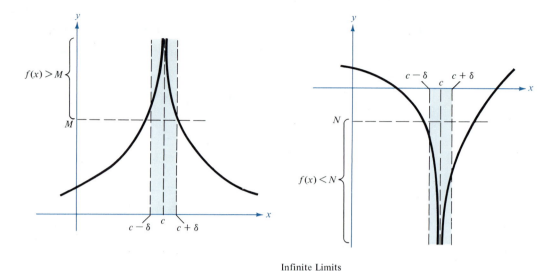

Infinite Limits

FIGURE 2.32

To define the infinite limit from the left,

$$\lim_{x \to c-} f(x) = \infty$$

we merely replace

$$0 < |x - c| < \delta \qquad \text{by} \qquad c - \delta < x < c$$

For the limit from the right, we replace

$$0 < |x - c| < \delta \qquad \text{by} \qquad c < x < c + \delta$$

Our definitions of limits at infinity take the following forms:

Definition of Limits at Infinity	
	1. $\lim_{x \to \infty} f(x) = L$ if for each $\epsilon > 0$ there exists an $M > 0$ such that $$|f(x) - L| < \epsilon \qquad \text{whenever} \qquad x > M$$
	2. $\lim_{x \to -\infty} f(x) = L$ if for each $\epsilon > 0$ there exists an $N < 0$ such that $$|f(x) - L| < \epsilon \qquad \text{whenever} \qquad x < N$$

Figure 2.33 describes this definition graphically.

Example 4

Use the definition of infinite limits to show that

$$\lim_{x \to 2^+} \frac{4}{x - 2} = \infty$$

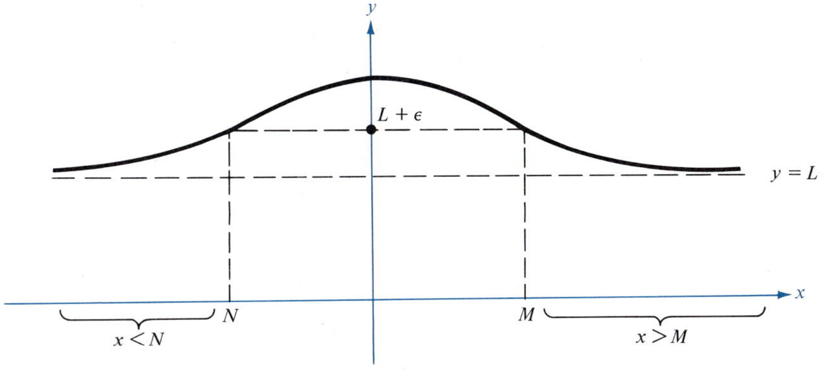

Limits at Infinity

FIGURE 2.33

Solution: For each $M > 0$ we must show that there exists a $\delta > 0$ such that

$$\frac{4}{x-2} > M \qquad \text{whenever} \qquad 2 < x < 2 + \delta$$

To determine the connection between M and δ, we consider

$$\frac{4}{x-2} > \frac{4}{(2+\delta)-2} = \frac{4}{\delta}$$

Thus if we choose $4/\delta > M$ (i.e., $\delta < 4/M$), we will have

$$\frac{4}{x-2} > \frac{4}{\delta} > \frac{4}{(4/M)} = M \qquad \text{whenever} \qquad 2 < x < 2 + \delta \qquad \blacksquare$$

Example 5

Use the definition of infinite limits to show that

$$\lim_{x \to -\infty} \frac{2x}{x-3} = 2$$

Solution: For each $\epsilon > 0$ we must show that there exists an $N < 0$ such that

$$\left| \frac{2x}{x-3} - 2 \right| < \epsilon \qquad \text{whenever} \qquad x < N$$

To determine the connection between ϵ and N, consider

$$\left| \frac{2x}{x-3} - 2 \right| = \left| \frac{2x - 2x + 6}{x-3} \right| = \left| \frac{6}{x-3} \right| < \frac{6}{|N-3|} < \frac{6}{|N|}$$

for $$x < N < 0$$

Therefore, we consider $$\frac{6}{|N|} < \epsilon$$

and choose $6/\epsilon < |N|$ or, equivalently, $N < -6/\epsilon$ for $N < 0$. This choice works, since

$$\left| \frac{2x}{x-3} - 2 \right| = \left| \frac{6}{x-3} \right| < \frac{6}{|N-3|} < \frac{6}{|N|} < \frac{6}{(6/\epsilon)} = \epsilon$$

whenever $\qquad\qquad x < N$ ∎

Section Exercises (2.4)

In Exercises 1–6, use the ϵ–δ definition of limits to find a satisfactory δ if $\epsilon = 0.01$ in each of the indicated limits. Also, sketch a graph of the function, identifying the limit.

1. $\lim_{x \to 2} (3x + 2) = 8$

2. $\lim_{x \to 4} \left(4 - \frac{x}{2} \right) = 2$

3. $\lim_{x \to 2} (x^2 - 3) = 1$

4. $\lim_{x \to 4^+} \sqrt{x - 4} = 0$

S 5. $\lim_{x \to 2} \frac{x^2 - 3x + 2}{x - 2} = 1$

6. $\lim_{x \to -2} \frac{x^2 + 5x + 6}{x + 2} = 1$

In Exercises 7–28, use the definitions of this section to verify each limit.

S 7. $\lim_{x \to 2} (x + 3) = 5$

8. $\lim_{x \to -3} (2x + 5) = -1$

9. $\lim_{x \to 6} 3 = 3$

10. $\lim_{x \to 2} -1 = -1$

S 11. $\lim_{x \to 0} \sqrt[3]{x} = 0$

12. $\lim_{x \to 3} |x - 3| = 0$

13. $\lim_{x \to 2} x^2 = 4$

14. $\lim_{x \to 2} x^3 = 8$

15. $\lim_{x \to 2} \frac{x^2 + x - 6}{x - 2} = 5$

16. $\lim_{x \to -10} \frac{x^2 + 10x}{x + 10} = -10$

17. $\lim_{x \to 2} \frac{1}{x} = \frac{1}{2}$

18. $\lim_{x \to -1} \frac{3}{x + 2} = 3$

S 19. $\lim_{x \to 2} (x^2 - x) = 2$

20. $\lim_{x \to -1} x^2 = 1$

21. $\lim_{x \to 0^+} \sqrt{x} = 0$

22. $\lim_{x \to 3^-} f(x) = 2, f(x) = \begin{cases} 2, x \le 3 \\ 0, x > 3 \end{cases}$

S 23. $\lim_{x \to -1^+} \frac{1}{x + 1} = \infty$

24. $\lim_{x \to -1^-} \frac{1}{x + 1} = -\infty$

25. $\lim_{x \to 2} \frac{1}{(x - 2)^2} = \infty$

26. $\lim_{x \to \infty} \frac{2x}{x + 1} = 2$

S 27. $\lim_{x \to \infty} \frac{x + 1}{x - 1} = 1$

28. $\lim_{x \to \infty} \frac{1}{x} = 0$

S 29. Prove that $f(x) = x^2$ is continuous at $x = 3$.

30. Prove that $f(x) = 4 - 3x$ is continuous at $x = 1$.

31. Prove that $\lim_{x \to c} f(x) = 0$ if $\lim_{x \to c} |f(x)| = 0$.

32. Prove that

$$\lim_{x \to 0^+} \frac{|x|}{x} = 1$$

33. Use the Squeeze Theorem to prove that

$$\lim_{x \to 0} \frac{x}{x^2 + 1} = 0$$

34. Prove that $\lim_{x \to 1} f(x)$ does not exist if

$$f(x) = \begin{cases} x, x \le 1 \\ 1 - x, x > 1 \end{cases}$$

Miscellaneous Exercises (Ch. 2)

In Exercises 1–20, evaluate (if possible) the given limits.

1. $\lim_{x \to 2} (5x - 3)$

2. $\lim_{x \to 2} (3x + 5)$

3. $\lim_{x \to 2} (5x - 3)(3x + 5)$

4. $\lim_{x \to 2} \frac{3x + 5}{5x - 3}$

5. $\lim_{t \to 3} \frac{t^2 + 1}{t}$

6. $\lim_{t \to 3} \frac{t^2 - 9}{t - 3}$

7. $\lim_{t \to -2} \frac{t + 2}{t^2 - 4}$

8. $\lim_{x \to 0} \frac{\sqrt{4 + x} - 2}{x}$

9. $\lim_{x \to 0} \frac{[1/(x + 1)] - 1}{x}$

S 10. $\lim_{s \to 0} \frac{(1/\sqrt{1 + s}) - 1}{s}$

11. $\lim_{x \to -1} \frac{x^3 + 1}{x + 1}$

S 12. $\lim_{x \to -2} \frac{x^2 - 4}{x^3 + 8}$

13. $\lim_{x \to \infty} \left(1 - \frac{1}{x^3} \right)$

14. $\lim_{x \to \infty} \frac{1}{\sqrt[3]{x^2 + 2}}$

15. $\lim_{x \to \infty} \frac{2x^2 + x + 1}{x^2 + 1}$

16. $\lim_{x \to \infty} \frac{2x - 1}{5x - 3}$

17. $\lim\limits_{x\to\infty} \dfrac{x+1}{x^2+3}$

18. $\lim\limits_{x\to\infty} \dfrac{x^2+3}{x+1}$

19. $\lim\limits_{x\to\infty} \dfrac{x^2-2x+1}{x-1}$

20. $\lim\limits_{x\to1} \dfrac{2x^2}{1-x^2}$

c **21.** Complete the following table to estimate $\lim_{x\to\infty} f(x)$, where $f(x) = \sqrt{2x+1} - \sqrt{2x-1}$.

x	1	10^2	10^4	10^6
$f(x)$				

22. Evaluate $\lim_{x\to\infty} (\sqrt{2x+1} - \sqrt{2x-1})$ by first rationalizing the numerator, and then compare your result to the approximation obtained in Exercise 21.

c **23.** Complete the following table to estimate $\lim_{x\to\infty} f(x)$, where $f(x) = \sqrt{x^2+x} - x$.

x	1	10^2	10^4	10^6
$f(x)$				

S **24.** Evaluate $\lim_{x\to\infty} (\sqrt{x^2+x} - x)$ by first rationalizing the numerator, and then compare your result to the approximation obtained in Exercise 23.

In Exercises 25–29, determine if the given limit statement is true or false.

25. $\lim\limits_{x\to0} \dfrac{|x|}{x} = 1$

26. $\lim\limits_{x\to2} f(x) = 3$, where $f(x) = \begin{cases} 3, x \le 2 \\ 0, x > 2 \end{cases}$

27. $\lim\limits_{x\to3} f(x) = 1$, where $f(x) = \begin{cases} x-2, x \le 3 \\ -x^2 + 8x - 14, x > 3 \end{cases}$

28. $\lim\limits_{x\to0} x^3 = 0$

29. $\lim\limits_{x\to0} f(x) = 0$, where $f(x) = \begin{cases} 1, x \le 0 \\ 0, x > 0 \end{cases}$

In Exercises 30–36, determine the points of discontinuity of f.

30. $f(x) = [x + 3]$

31. $f(x) = \dfrac{3x^2 - x - 2}{x-1}$

32. $f(x) = \begin{cases} \dfrac{3x^2 - x - 2}{x-1}, x \ne 1 \\ 0, x = 1 \end{cases}$

33. $f(x) = \begin{cases} 5 - x, x \le 2 \\ 2x - 3, x > 2 \end{cases}$

34. $f(x) = \dfrac{1}{(x-2)^2}$

35. $f(x) = \sqrt{\dfrac{x+2}{x}}$

36. $f(x) = \dfrac{3}{x+1}$

S **37.** Determine the value of c so that the function

$$f(x) = \begin{cases} x + 3, x \le 2 \\ cx + 6, x > 2 \end{cases}$$

is continuous for all x.

38. Determine the values of b and c so that the function

$$f(x) = \begin{cases} x + 1, 1 < x < 3 \\ x^2 + bx + c, x \le 1 \text{ or } 3 \le x \end{cases}$$

is continuous for all x.

S **39.** Sketch the graph of $y = \dfrac{x^2-1}{x^2+1}$.

40. Sketch the graph of $y = \dfrac{x}{x^2-9}$.

S **41.** Sketch the graph of $y^2 = \dfrac{x+2}{x(x-2)}$.

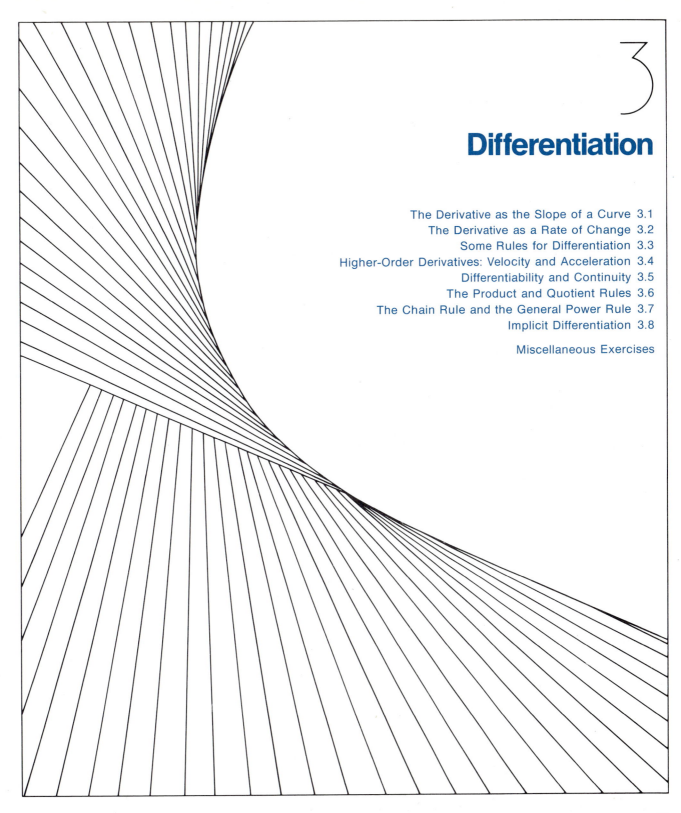

3

Differentiation

3.1　**The Derivative as the Slope of a Curve**

Purpose
- To define the slope of a curve at a point.
- To define the derivative and use it to find the slope of a curve.
- To calculate the derivative by its limit definition.

Differential calculus consists merely in algebraically determining the limit of a ratio. . . . This provides us with the slope of the tangent line we are looking for. This is perhaps the most precise and neatest possible definition of the differential calculus. —Jean D'Alembert (1717–1783)

In Section 1.4 we showed that the slope of a line is given by

$$m = \frac{\Delta y}{\Delta x} = \frac{y_2 - y_1}{x_2 - x_1}$$

where (x_1, y_1) and (x_2, y_2) are *any* two points on the line. Thus the slope of a line is constant (Figure 3.1). This is not the case for other curves. For instance, in Figure 3.2 it appears that the curve is steeper (has a greater slope) at point Q than at point P. This suggests that for a nonlinear curve the slope is not constant but actually varies from point to point. In this section we introduce a procedure for determining the *slope of a curve at a point*.

To determine the slope of a curve at some point, we make use of the *tangent line* to the curve at the point. For instance, in Figure 3.3 the tangent line to the graph of f at P is the line that best approximates the graph of f at that point. Our problem of finding the slope of a curve at a point thus becomes one of finding the slope of the line tangent to the curve at that point.

Recall that the familiar "tangent line to a circle" requires the circle and the line to have only one point in common. Such is not the case in Figure 3.3, as the dotted portion of the tangent line indicates. Informally, when we refer to a tangent line to a curve, we mean a "local" tangent to the curve at a specific point and we do not care if the line and the curve intersect at some other point.

We can identify the tangent line at a point in the following manner: Consider the **secant line** through the two points P and Q on the graph of $y = f(x)$ in Figure 3.4. Suppose point Q moves along the curve toward P, determining secant lines from P to Q, from P to Q_1, and so on. Now as Q

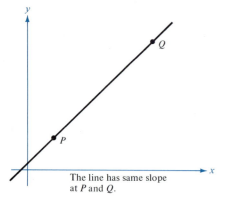

The line has same slope at P and Q.

FIGURE 3.1

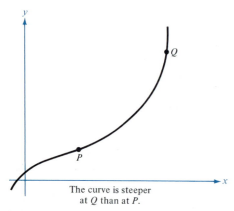

The curve is steeper at Q than at P.

FIGURE 3.2

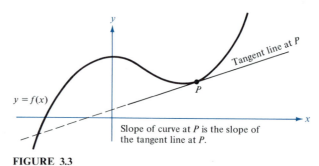

$y = f(x)$

Tangent line at P

Slope of curve at P is the slope of the tangent line at P.

FIGURE 3.3

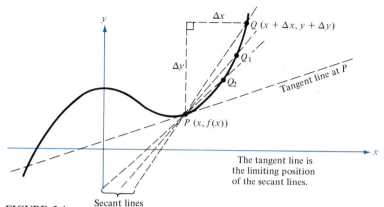

FIGURE 3.4

moves closer and closer to P, observe that these secant lines approach a "limiting position." This limiting position of the secant lines is called the **tangent line to the curve at** P.

We can determine the tangent line at P by determining its slope. Since the tangent line is the limiting position of the secant lines, its slope m_{tan} is the limiting value of the slopes m_{sec} of the secant lines as Q approaches P. Informally, this can be represented by

$$m_{\text{tan}} = \lim_{Q \to P} m_{\text{sec}}$$

From Figure 3.4 we see that the slope of the secant line through P and Q is

$$m_{\text{sec}} = \frac{\Delta y}{\Delta x} = \frac{f(x + \Delta x) - f(x)}{\Delta x}$$

As $Q \to P$, we have $\Delta x \to 0$. Therefore, the slope of the tangent line at P is

$$m_{\text{tan}} = \lim_{\Delta x \to 0} \frac{\Delta y}{\Delta x} = \lim_{\Delta x \to 0} \frac{f(x + \Delta x) - f(x)}{\Delta x}$$

We are now ready to define the slope of a curve at a point on the curve.

Definition of Slope of a Curve

At $(x, f(x))$ the slope m of the graph of $y = f(x)$ is equal to the slope of its tangent line at $(x, f(x))$, and it is determined by the formula

$$m = \lim_{\Delta x \to 0} \frac{\Delta y}{\Delta x} = \lim_{\Delta x \to 0} \frac{f(x + \Delta x) - f(x)}{\Delta x}$$

provided this limit exists.

To calculate the slope of the tangent line to a curve by its limit definition, we follow a general four-step process.

Four-Step Process

Given $y = f(x)$:

1. Determine $f(x + \Delta x)$.
2. Calculate $f(x + \Delta x) - f(x)$.
3. Divide by Δx to obtain

$$\frac{f(x + \Delta x) - f(x)}{\Delta x}$$

4. Let $\Delta x \to 0$ to obtain

$$\lim_{\Delta x \to 0} \frac{f(x + \Delta x) - f(x)}{\Delta x} = m$$

Example 1

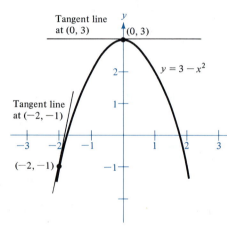

Tangent line at (0, 3)

$y = 3 - x^2$

Tangent line at $(-2, -1)$

$(-2, -1)$

FIGURE 3.5

Determine the slope of the line tangent to $f(x) = 2x - 3$ at any point (x, y).

Solution: By the four-step process:

1. $$f(x + \Delta x) = 2(x + \Delta x) - 3$$

2. $$f(x + \Delta x) - f(x) = (2x + 2\,\Delta x - 3) - (2x - 3) = 2\,\Delta x$$

3. $$\frac{f(x + \Delta x) - f(x)}{\Delta x} = \frac{2\,\Delta x}{\Delta x} = 2$$

4. $$\lim_{\Delta x \to 0} \frac{f(x + \Delta x) - f(x)}{\Delta x} = \lim_{\Delta x \to 0} 2 = 2$$

Therefore, $m = 2$ ∎

Note in Example 1 that the slope of f is constant. This is not surprising since we know from Chapter 1 that the graph of f is a line. Of course, not all graphs have constant slope, as the next example illustrates.

Example 2

Determine the formula for the slope of the graph of $y = 3 - x^2$ (see Figure 3.5). What is its slope at $(0, 3)$? At $(-2, -1)$?

Solution: By the four-step process:

1. $$f(x + \Delta x) = 3 - (x + \Delta x)^2$$

2. $$f(x + \Delta x) - f(x) = 3 - x^2 - 2x(\Delta x) - (\Delta x)^2 - 3 + x^2$$
$$= -2x(\Delta x) - (\Delta x)^2$$

3. $$\frac{f(x + \Delta x) - f(x)}{\Delta x} = -2x - (\Delta x)$$

4. $\displaystyle\lim_{\Delta x \to 0} \frac{f(x + \Delta x) - f(x)}{\Delta x} = \lim_{\Delta x \to 0} (-2x - \Delta x) = -2x$

Therefore, $m = -2x$

Now at $(0, 3)$, $m = -2(0) = 0$

At $(-2, -1)$ $m = -2(-2) = 4$ ∎

Example 3

Write the equation of the line tangent to $f(x) = \sqrt{x}$ at the point $(4, 2)$. (See Figure 3.6.)

Solution:

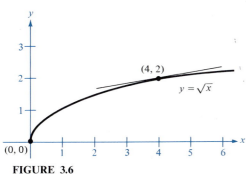

FIGURE 3.6

1. $f(x + \Delta x) = \sqrt{x + \Delta x}$

2. $f(x + \Delta x) - f(x) = \sqrt{x + \Delta x} - \sqrt{x}$

Rationalizing the numerator we obtain

$$f(x + \Delta x) - f(x) = (\sqrt{x + \Delta x} - \sqrt{x})\left(\frac{\sqrt{x + \Delta x} + \sqrt{x}}{\sqrt{x + \Delta x} + \sqrt{x}}\right)$$

$$= \frac{(x + \Delta x) - x}{\sqrt{x + \Delta x} + \sqrt{x}} = \frac{\Delta x}{\sqrt{x + \Delta x} + \sqrt{x}}$$

3. $\displaystyle\frac{f(x + \Delta x) - f(x)}{\Delta x} = \frac{1}{\sqrt{x + \Delta x} + \sqrt{x}}$

4. $\displaystyle\lim_{\Delta x \to 0} \frac{f(x + \Delta x) - f(x)}{\Delta x} = \lim_{\Delta x \to 0} \left(\frac{1}{\sqrt{x + \Delta x} + \sqrt{x}}\right) = \frac{1}{2\sqrt{x}}$

Therefore, at $(4, 2)$ the slope of the tangent line is

$$m = \tfrac{1}{4}$$

and the equation of this tangent line is

$$y - 2 = \tfrac{1}{4}(x - 4)$$
$$y = \tfrac{1}{4}x + 1$$ ∎

In our definition of the slope of a curve, we have the phrase "provided this limit exists." This suggests that some curves may not have a slope at each point. This is the case in Example 3, since the limit

$$\lim_{\Delta x \to 0} \frac{f(x + \Delta x) - f(x)}{\Delta x} = \frac{1}{2\sqrt{x}}$$

does not exist at $(0, 0)$.

We have now arrived at a crucial point in our study of calculus, for the limit

$$\lim_{\Delta x \to 0} \frac{f(x + \Delta x) - f(x)}{\Delta x}$$

is used to define one of the two fundamental quantities of calculus.

Definition of Derivative

The limit
$$\lim_{\Delta x \to 0} \frac{f(x + \Delta x) - f(x)}{\Delta x}$$

is called the **derivative** of f at x (provided the limit exists), and it is denoted by $f'(x)$.

A function is said to be **differentiable** at x if its derivative exists at x, and the process of finding the derivative is called **differentiation.**

In addition to $f'(x)$ (read "f prime of x"), various other notations are used to denote the derivative. The most commonly used notations are

$$\frac{dy}{dx}, \qquad y', \qquad \frac{d}{dx}[f(x)], \qquad D_x(y)$$

We will frequently use dy/dx to denote the derivative. We read dy/dx as the "derivative of y with respect to x," and, using limit notation, we write

$$\frac{dy}{dx} = \lim_{\Delta x \to 0} \frac{\Delta y}{\Delta x}$$

Thus we have

$$\frac{dy}{dx} = \lim_{\Delta x \to 0} \frac{\Delta y}{\Delta x} = \lim_{\Delta x \to 0} \frac{f(x + \Delta x) - f(x)}{\Delta x} = f'(x)$$

Since the derivative $f'(x)$ and the slope of the tangent line to the graph of f are both defined by the limit

$$\lim_{\Delta x \to 0} \frac{f(x + \Delta x) - f(x)}{\Delta x}$$

we can use the four-step process to determine $f'(x)$.

Example 4

Find the derivative of $f(x) = x^3 + 2x$.

Solution:

1. $$f(x + \Delta x) = (x + \Delta x)^3 + 2(x + \Delta x)$$

2. $$f(x + \Delta x) - f(x) = x^3 + 3x^2(\Delta x) + 3x(\Delta x)^2 + (\Delta x)^3$$
$$+ 2x + 2(\Delta x) - (x^3 + 2x)$$
$$= 3x^2(\Delta x) + 3x(\Delta x)^2 + (\Delta x)^3 + 2(\Delta x)$$

3. $$\frac{f(x+\Delta x)-f(x)}{\Delta x} = 3x^2 + 3x(\Delta x) + (\Delta x)^2 + 2$$

4. $$\lim_{\Delta x\to 0} \frac{f(x+\Delta x)-f(x)}{\Delta x} = \lim_{\Delta x\to 0} [3x^2 + 3x(\Delta x) + (\Delta x)^2 + 2]$$
$$= 3x^2 + 2$$

Therefore, $$f'(x) = 3x^2 + 2$$ ■

Example 5

Given $y = 2/t$, determine the derivative of y with respect to t.

Solution: Considering $y = f(t)$ we have

1. $$f(t+\Delta t) = \frac{2}{t+\Delta t}$$

2. $$f(t+\Delta t) - f(t) = \frac{2}{t+\Delta t} - \frac{2}{t} = \frac{2t - 2t - 2(\Delta t)}{t(t+\Delta t)}$$
$$= \frac{-2(\Delta t)}{t(t+\Delta t)}$$

3. $$\frac{f(t+\Delta t)-f(t)}{\Delta t} = \frac{-2}{t(t+\Delta t)}$$

4. $$\lim_{\Delta t\to 0} \frac{f(t+\Delta t)-f(t)}{\Delta t} = \lim_{\Delta t\to 0} \frac{-2}{t(t+\Delta t)} = \frac{-2}{t^2}$$

Therefore, the derivative of y with respect to t is

$$\frac{dy}{dt} = \frac{-2}{t^2}$$ ■

Before concluding this section we point out that a function may have a tangent line at each point on its graph and yet not be differentiable at each point. In particular, at any point where the tangent line is vertical, the slope is infinite; hence the derivative does not exist.

Section Exercises (3.1)

In Exercises 1–10, find the derivative by the four-step process.

1. $f(x) = 3$

2. $f(x) = 3x + 2$

3. $f(x) = -5x$

4. $f(x) = 1 - x^2$

[S] **5.** $f(x) = 2x^2 + x - 1$

6. $f(x) = \sqrt{x-4}$

[S] **7.** $f(x) = \dfrac{1}{x-1}$

8. $f(x) = \dfrac{1}{x^2}$

9. $f(t) = t^3 - 12t$

10. $f(t) = t^3 + t^2$

In Exercises 11–18, use the four-step process to find the derivative of each function. Sketch the graph of each function and find the equation of the tangent line at the given point.

11. $f(x) = x^2 + 1$; $(2, 5)$

12. $f(x) = x^2 + 2x + 1$; $(-3, 4)$

[S] **13.** $f(x) = x^3$; $(2, 8)$

14. $f(x) = x^3$; $(-2, -8)$

[S] **15.** $f(x) = \sqrt{x+1}$; $(3, 2)$

16. $f(x) = \dfrac{1}{\sqrt{x}}$; $(4, \frac{1}{2})$

17. $f(x) = \dfrac{1}{x}$; $(1, 1)$ **18.** $f(x) = \dfrac{1}{x + 1}$; $(0, 1)$

[S] **19.** Find the equation of a line that is tangent to the curve $y = x^3$ and is parallel to the line $3x - y + 1 = 0$.

20. Find the equation of the line that is tangent to the curve $y = 1/\sqrt{x}$ and is parallel to the line $x + 2y - 6 = 0$.

[S] **21.** There are two tangent lines to the curve $y = 4x - x^2$ that pass through the point $(2, 5)$. Find the equations of those two lines and make a sketch to verify your results.

22. Two lines through the point $(1, -3)$ are tangent to the curve $y = x^2$. Find the equations of these two lines and make a sketch to verify your results.

For Review and Class Discussion

True or False

1. _____ It is possible to construct a curve whose slope is 1 at one point and -1 at some other point.

2. _____ The slope of the graph of $y = x^2$ is different at every point on the curve.

3. _____ The slope of the graph of $y = x^3$ is different at every point on the curve.

4. _____ The tangent line to a curve at a point can touch the curve at only one point.

5. _____ The slope of the graph of $y = f(x)$ at the point $(1, f(1))$ is $m = f'(1)$.

6. _____ The equation of the line that is tangent to the graph of $y = x^2$ at the point $(-1, 1)$ is $y - 1 = 2x(x + 1)$.

7. _____ If the derivative of a function is zero at a point, then the tangent line at that point is horizontal.

8. _____ The functions given by $f(x) = x^2$ and $g(x) = x^2 + 2$ have the same derivative.

9. _____ If $f(x) = ax + b$, then $f'(x) = a$.

10. _____ If there is a point on a curve where the slope is undefined, then the curve has no tangent at that point.

3.2 The Derivative as a Rate of Change

Purpose
- To distinguish between average and instantaneous rates of change.
- To use the derivative to calculate instantaneous rates of change.

The calculus was the first achievement of modern mathematics, and it is difficult to overestimate its importance. I think it defines more unequivocally than anything else the inception of modern mathematics, and the system of mathematical analysis, which is its logical development, still constitutes the greatest technical advance in exact thinking.   —*John von Neumann (1903–1957)*

We have already seen how the derivative

$$f'(x) = \lim_{\Delta x \to 0} \frac{f(x + \Delta x) - f(x)}{\Delta x}$$

can be interpreted as the slope of a curve. We now consider another interpretation, namely, as a way of determining the rate of change in one variable with respect to another. There are numerous applications of the notion of rate of change. A few examples are population growth rates, unemployment rates, production rates, and rate of water flow. Although rates of change usually involve change with respect to time (see Section 3.4), we can actually investigate the rate of change with respect to any related variable. It is in this more general context that we begin our discussion.

Suppose a ball is thrown into the air at a certain angle and with a certain velocity so that the equation that represents its path is

$$y = -0.005x^2 + x$$

as shown in Figure 3.7. We will focus our attention on the rate at which the height y of the ball changes as the horizontal distance x increases. For instance, at $x = 20$ the height is $y = 18$, while at $x = 60$ the height of the ball is $y = 42$. We can determine the **average rate of change** of y on the interval from $x = 20$ to $x = 60$ to be

$$\frac{\text{change in } y}{\text{change in } x} = \frac{\Delta y}{\Delta x} = \frac{42 - 18}{60 - 20} = 0.6$$

On the interval from $x = 60$ to $x = 100$, the average rate of change is

$$\frac{\Delta y}{\Delta x} = \frac{50 - 42}{100 - 60} = 0.2$$

In general, we can calculate the average rate of change of y with respect to x over an interval Δx by the following rule:

$$\frac{\text{change in } y}{\text{change in } x} = \frac{\Delta y}{\Delta x} = \frac{\text{average rate of change}}{\text{of } y \text{ with respect to } x}$$

Suppose, however, that we are not interested in the average rate of change but in the *exact* rate of change in y at a particular x-value. For instance, when $x = 20$, at what rate is y increasing for each unit increase in x? This rate of change in y at a specific x-value is called the **instantaneous rate of change** of y with respect to x. This instantaneous rate of change is determined by letting Δx approach zero.

Table 3.1 shows the average rates of change of $y = f(x) = -0.005x^2 + x$ on several intervals from $x = 20$ to $x = 20 + \Delta x$. Ob-

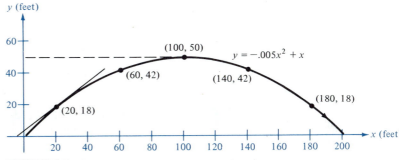

FIGURE 3.7

TABLE 3.1

x	Δx	$x + \Delta x$	$\Delta y = f(x + \Delta x) - f(x)$	$\dfrac{\Delta y}{\Delta x}$ = average rate of change
20	1.0	21	0.795	0.795
20	0.1	20.1	0.07995	0.7995
20	0.01	20.01	0.0079995	0.79995
20	0.001	20.001	0.000799995	0.799995

serve how the average rate of change in y gets closer and closer to 0.8 as the Δx interval approaches zero. This suggests that the instantaneous rate of change in y at $x = 20$ is 0.8. Compare this result with an estimate of the slope of $y = -0.005x^2 + x$ at $x = 20$ [see the tangent line at (20, 18) in Figure 3.7].

The discussion above suggests the following definition:

Definition of Instantaneous Rate of Change

The **instantaneous rate of change** of y at x is given by

$$\frac{dy}{dx} = \lim_{\Delta x \to 0} \frac{\Delta y}{\Delta x} = \lim_{\Delta x \to 0} \frac{f(x + \Delta x) - f(x)}{\Delta x}$$

Note that the limit in the preceding definition is the same as the limit in the definition of the derivative of f at x. This is an important observation, for we have arrived at a second major interpretation of the derivative—as an *instantaneous rate of change in one variable with respect to another.*

We make the following summary statement concerning the derivative and its interpretations:

Interpretations of the Derivative

Let the function given by $y = f(x)$ be differentiable at x; then its derivative

$$\frac{dy}{dx} = f'(x) = \lim_{\Delta x \to 0} \frac{f(x + \Delta x) - f(x)}{\Delta x}$$

denotes both:

1. the *slope* of the graph of f at x
2. the *instantaneous rate of change* in y with respect to x

In future work with the derivative, we will use "rate of change" to mean "instantaneous rate of change." Also, it is important to bear in mind that the three quantities, the derivative, the slope of a graph, and the rate of change, *are all equivalent.*

Example 1

Given $y = -0.005x^2 + x$, determine the rate of change in y with respect to x when $x = 20$. When $x = 100$.

Solution: Since rate of change is determined by the derivative, we consider $y = f(x)$ and apply the four-step process.

1. $$f(x + \Delta x) = -0.005(x + \Delta x)^2 + (x + \Delta x)$$
2. $$f(x + \Delta x) - f(x) = -0.005x^2 - 0.01x(\Delta x) - 0.005(\Delta x)^2$$
$$+ x + \Delta x + 0.005x^2 - x$$
$$= -0.01x(\Delta x) - 0.005(\Delta x)^2 + \Delta x$$
3. $$\frac{f(x + \Delta x) - f(x)}{\Delta x} = -0.01x - 0.005(\Delta x) + 1$$
4. $$\lim_{\Delta x \to 0} \frac{f(x + \Delta x) - f(x)}{\Delta x} = -0.01x + 1$$

Therefore, at $x = 20$ the rate of change in y with respect to x is

$$\frac{dy}{dx} = f'(20) = -0.01(20) + 1 = 0.8$$

which corresponds to the slope of the tangent line at (20, 18). (See Figure 3.7.) At $x = 100$, we have

$$\frac{dy}{dx} = f'(100) = -0.01(100) + 1 = 0$$

Example 2

Given $s = t/(1 + t)$, determine the formula for the rate of change in s with respect to t. Apply this formula when $t = 0$ and when $t = 3$.

Solution:

1. $$f(t + \Delta t) = \frac{t + \Delta t}{1 + (t + \Delta t)}$$

2. $$f(t + \Delta t) - f(t) = \frac{t + \Delta t}{1 + (t + \Delta t)} - \frac{t}{1 + t}$$
$$= \frac{t + \Delta t + t^2 + t \Delta t - t - t^2 - t \Delta t}{[1 + (t + \Delta t)](1 + t)}$$
$$= \frac{\Delta t}{[1 + (t + \Delta t)](1 + t)}$$

3. $$\frac{f(t + \Delta t) - f(t)}{\Delta t} = \frac{1}{[1 + (t + \Delta t)](1 + t)}$$

4. $\lim\limits_{\Delta t \to 0} \dfrac{f(t + \Delta t) - f(t)}{\Delta t} = \dfrac{1}{(1 + t)^2}$

Therefore, the rate of change in s when $t = 0$ is

$$\frac{ds}{dt} = \frac{1}{(1 + 0)^2} = 1$$

When $t = 3$ it is $\qquad \dfrac{ds}{dt} = \dfrac{1}{(1 + 3)^2} = \dfrac{1}{16}$ ∎

Section Exercises (3.2)

In Exercises 1–8, find the average rate of change of each of the functions between the two given points. Compare this average rate of change to the instantaneous rate of change at each point.

1. $f(t) = 2t + 7$; $(1, 9)$, $(2, 11)$
2. $f(x) = 3x - 1$; $(0, -1)$, $(\frac{1}{3}, 0)$
S 3. $f(t) = \dfrac{1}{(t + 1)}$; $(0, 1)$, $(3, \frac{1}{4})$
4. $f(x) = x^2 - 3$; $(2, 1)$, $(2.1, 1.41)$
5. $h(s) = s^2 - 6s - 1$; $(-1, 6)$, $(3, -10)$
6. $g(x) = \sqrt{x^2 - 1}$; $(1, 0)$, $(3, \sqrt{8})$
S 7. $h(t) = \sqrt{t^2 - 16}$; $(4, 0)$, $(5, 3)$
8. $h(x) = x^3 + 4x$; $(0, 0)$, $(0.1, 0.401)$

9. A ball is thrown and follows a path described by $y = x - 0.02x^2$.
   (a) Sketch a graph of the path.
   (b) Find the total horizontal distance the ball was thrown.
   (c) For what x-value does the ball reach its maximum height? (Use the symmetry of the path.)
   (d) Find the equation that gives the instantaneous rate of change in the height of the ball with respect to the horizontal change and evaluate this equation at $x = 0$, 10, 25, 30, 50.
   (e) What is the instantaneous rate of change of the height when the ball reaches its maximum height?

10. The path of a projectile thrown at an angle of $45°$ with level ground is given by

$$y = x - \frac{32}{v_0^2}(x^2)$$

where the initial velocity is v_0 ft/s.
   (a) Sketch the path followed by the projectile.
   (b) Find the x-coordinate of the point where the projectile strikes the ground. Use the symmetry of the path of the projectile to locate the abscissa of the point where the projectile reaches its maximum height.
   (c) What is the instantaneous rate of change of the height when the projectile is at its maximum height?

C S 11. (a) Sketch the graph of $f(x) = x^2 + 2x + 1$.
   (b) Find $\quad m_{\text{sec}} = \dfrac{f(x + \Delta x) - f(x)}{\Delta x}$

   the average rate of change of the function from point $P(x, f(x))$ to $Q(x + \Delta x, f(x + \Delta x))$.
   (c) Assume that P has coordinates $(-3, 4)$. Find the value of m_{sec} for the following values of Δx:

i. $\Delta x = 2$	ii. $\Delta x = 1$
iii. $\Delta x = 0.25$	iv. $\Delta x = 0.01$
v. $\Delta x = -2$	vi. $\Delta x = -1$
vii. $\Delta x = -0.25$	viii. $\Delta x = -0.01$

   (d) Find, at $(-3, 4)$,

$$\lim_{\Delta x \to 0} \frac{f(x + \Delta x) - f(x)}{\Delta x}$$

For Review and Class Discussion

True or False

1. _____ The average rate of change of y with respect to x is given by

$$\frac{\Delta y}{\Delta x} = \frac{f(x + \Delta x) - f(x)}{\Delta x}$$

2. _____ The average rate of change approaches the instantaneous rate of change as Δx approaches zero.

3. _____ The average rate of change is always larger than the instantaneous rate of change.

4. ____ The average rate of change can be equal to the instantaneous rate of change.

5. ____ For $f(x) = x - x^2$, $x = 0$, and $\Delta x = 1$, the average rate of change is zero.

3.3 Some Rules for Differentiation

Purpose

- To prove and demonstrate these differentiation rules: the Constant Rule, the Power Rule, the Scalar Multiple Rule, and the Sum Rule.

Knowing the algorithm of this calculus, which I call differential calculus, we can find maxima and minima as well as tangents without the necessity of removing fractions, irrationals, and other restrictions, as had to be done according to the methods that have been published hitherto. —*Gottfried Leibniz (1646–1716)*

Up to this point we have found derivatives by the limit definition of the derivative. This procedure is rather tedious even for simple functions, but fortunately there are rules that greatly simplify the differentiation process. These rules permit us to calculate the derivative without the direct use of limits. We now derive some of these rules, and in each case we assume that the derivative of the given function exists.

Constant Rule

The derivative of a constant is zero.

$$\frac{d}{dx}[c] = 0, \qquad \text{where } c \text{ is a constant}$$

Proof: Let $f(x) = c$; then by the limit definition

$$\frac{d}{dx}[f(x)] = f'(x) = \lim_{\Delta x \to 0} \frac{f(x + \Delta x) - f(x)}{\Delta x} = \lim_{\Delta x \to 0} \frac{c - c}{\Delta x} = \lim_{\Delta x \to 0} 0 = 0$$

Therefore, $\qquad\qquad \dfrac{d}{dx}[c] = 0$ ∎

Example 1

(a) $\dfrac{d}{dx}[7] = 0.$

(b) If $f(x) = 0$, then $f'(x) = 0$.

(c) If $y = 2$, then $dy/dx = 0$.

(d) If $g(t) = \frac{-3}{2}$, then $g'(t) = 0$. ∎

Before deriving the next rule, we review the procedure for expanding a binomial. Recall that

$$(x + \Delta x)^2 = x^2 + 2x\,\Delta x + (\Delta x)^2$$

and $\qquad (x + \Delta x)^3 = x^3 + 3x^2\,(\Delta x) + 3x\,(\Delta x)^2 + (\Delta x)^3$

and, in general, the binomial expansion is

$$(x + \Delta x)^n = x^n + nx^{n-1}(\Delta x) + \frac{n(n-1)x^{n-2}}{2}(\Delta x)^2$$

$$+ \frac{n(n-1)(n-2)x^{n-3}}{2(3)}(\Delta x)^3 + \cdots + (\Delta x)^n$$

where n is a positive integer. We use this binomial expansion in proving a special case of the following rule:

(Simple) Power Rule
$$\frac{d}{dx}[x^n] = nx^{n-1}, \qquad \text{where } n \text{ is any real number}$$

Proof: Although the Power Rule is true for any real number n, the binomial expansion applies only when n is a positive integer. Thus for the time being, we give a proof for the case when n is a positive integer. Sections 3.6 and 3.8 contain proofs for negative integers and arbitrary rationals, respectively.

$$\frac{d}{dx}[f(x)] = \lim_{\Delta x \to 0} \frac{f(x + \Delta x) - f(x)}{\Delta x} = \lim_{\Delta x \to 0} \frac{(x + \Delta x)^n - x^n}{\Delta x}$$

Now applying the binomial expansion rule, we obtain

$$\frac{d}{dx}[x^n]$$

$$= \lim_{\Delta x \to 0} \frac{x^n + nx^{n-1}(\Delta x) + [n(n-1)x^{n-2}/2](\Delta x)^2 + \cdots + (\Delta x)^n - x^n}{\Delta x}$$

$$= \lim_{\Delta x \to 0} \left[nx^{n-1} + \frac{n(n-1)x^{n-2}}{2}(\Delta x) + \cdots + (\Delta x)^{n-1} \right]$$

$$= nx^{n-1} + 0 + \cdots + 0 = nx^{n-1} \qquad \blacksquare$$

Before showing some examples of the use of the Power Rule, we establish yet a third differentiation rule.

Scalar Multiple Rule
$$\frac{d}{dx}[cf(x)] = cf'(x), \qquad c \text{ is a constant}$$

Proof: Applying the limit definition, we have

$$\frac{d}{dx}[cf(x)] = \lim_{\Delta x \to 0} \frac{cf(x + \Delta x) - cf(x)}{\Delta x}$$

$$= \lim_{\Delta x \to 0} c \left[\frac{f(x + \Delta x) - f(x)}{\Delta x} \right]$$

$$= c \left[\lim_{\Delta x \to 0} \frac{f(x + \Delta x) - f(x)}{\Delta x} \right] = cf'(x)$$ ∎

Informally, the Scalar Multiple Rule states that constants can be factored out of the differentiation process:

$$\frac{d}{dx}[cf(x)] = c \frac{d}{dx}[\bigcirc f(x)] = cf'(x)$$

The power and versatility of this rule is often overlooked, especially when the constant appears in the denominator, as follows:

$$\frac{d}{dx}\left[\frac{f(x)}{c}\right] = \frac{d}{dx}\left[\left(\frac{1}{c}\right)f(x)\right] = \left(\frac{1}{c}\right)\frac{d}{dx}[\bigcirc f(x)] = \left(\frac{1}{c}\right)f'(x)$$

To use the Scalar Multiple Rule to best advantage, be on the lookout for constants that can be factored out *before* differentiating.

Example 2

(a) If $f(x) = x^3$, then by the Power Rule $f'(x) = 3x^2$.

(b) If $y = 2x^{1/2}$, then by the Scalar Multiple Rule and the Power Rule, we have

$$\frac{dy}{dx} = \frac{d}{dx}[2x^{1/2}] = 2\frac{d}{dx}[x^{1/2}] = 2\left(\frac{1}{2}x^{-1/2}\right) = x^{-1/2}$$

(c) If $f(t) = 4t^2/5$, we rewrite $f(t)$ as $f(t) = \frac{4}{5}t^2$. Then by the Scalar Multiple and Power Rules, we have

$$f'(t) = \frac{d}{dt}\left[\frac{4}{5}t^2\right] = \left(\frac{4}{5}\right)\frac{d}{dt}[t^2] = \frac{4}{5}(2t) = \frac{8}{5}t$$

(d) If $y = x^{-2}$, then by the Power Rule

$$\frac{dy}{dx} = \frac{d}{dx}[x^{-2}] = -2x^{-3}$$ ∎

It is of benefit to see that the Scalar Multiple and Power Rules can be combined into one rule. The combination rule is

$$\frac{d}{dx}[cx^n] = cnx^{n-1}, \qquad \text{where } n \text{ is any real number}$$

Two special cases of the Power Rule occur when $n = 0$ and $n = 1$. If $n = 0$, we actually have the Constant Rule. If $n = 1$, we have

$$\frac{d}{dx}[cx^1] = c\frac{d}{dx}[x^1] = c(1)(x^0) = c$$

Example 3

(a) $\dfrac{d}{dx}\left[\dfrac{-3}{2}x\right] = -\dfrac{3}{2}.$

(b) $\dfrac{d}{dx}[3\pi x] = 3\pi.$

(c) If $y = -x/2$, then $y' = -\frac{1}{2}$.

(d) If $f(x) = -5x/6$, then $f'(x) = -\frac{5}{6}$. ∎

The four problems in Example 3 are very simple, yet errors are frequently made in differentiating a constant multiple of the first power of x. Keep in mind that

$$\frac{d}{dx}[cx] = c, \qquad \text{where } c \text{ is any constant}$$

The next rule is one that we certainly would expect to be true and it is often used without thinking about it. For instance, if you were to find the derivative of $y = 3x + 2x^3$, you would probably write

$$y' = 3 + 6x^2$$

without questioning your answer. The validity of differentiating a sum "term by term" is given in the following rule.

Sum Rule

The derivative of the sum of two functions is the sum of their derivatives:

$$\frac{d}{dx}[f(x) + g(x)] = f'(x) + g'(x)$$

Proof: Let $F(x) = f(x) + g(x)$. Then the derivative of F is

$$F'(x) = \lim_{\Delta x \to 0} \frac{F(x + \Delta x) - F(x)}{\Delta x}$$

$$= \lim_{\Delta x \to 0} \frac{f(x + \Delta x) + g(x + \Delta x) - f(x) - g(x)}{\Delta x}$$

$$= \lim_{\Delta x \to 0} \frac{f(x + \Delta x) - f(x) + g(x + \Delta x) - g(x)}{\Delta x}$$

$$= \lim_{\Delta x \to 0} \left[\frac{f(x + \Delta x) - f(x)}{\Delta x} + \frac{g(x + \Delta x) - g(x)}{\Delta x}\right]$$

$$= \lim_{\Delta x \to 0} \frac{f(x + \Delta x) - f(x)}{\Delta x} + \lim_{\Delta x \to 0} \frac{g(x + \Delta x) - g(x)}{\Delta x}$$

$$= f'(x) + g'(x)$$

Thus
$$\frac{d}{dx}[f(x) + g(x)] = f'(x) + g'(x)$$

∎

By a similar procedure it can be shown that

$$\frac{d}{dx}[f(x) - g(x)] = f'(x) - g'(x)$$

Furthermore, either rule can be extended to the derivative of any finite number of functions. For instance, if

$$F(x) = f(x) + g(x) - h(x) - k(x)$$

then $\qquad F'(x) = f'(x) + g'(x) - h'(x) - k'(x)$

Example 4

If $f(x) = x^3 - 4x + 5$, find the value of $f'(2)$.

Solution: $\qquad f'(x) = 3x^2 - 4$

Therefore, $\qquad f'(2) = 3(2)^2 - 4 = 12 - 4 = 8$ ∎

Example 5

If $g(x) = (-x^4/2) + 3x^3 - 2x$, find the value of $g'(-1)$.

Solution:

$$g'(x) = \frac{-4}{2}x^3 + 9x^2 - 2 = -2x^3 + 9x^2 - 2$$
$$g'(-1) = -2(-1)^3 + 9(-1)^2 - 2 = 2 + 9 - 2 = 9$$ ∎

Example 6

If $y = 3x^{-2}$, find the value of dy/dx when $x = 2$.

Solution: By the Power Rule with $n = -2$, we obtain

$$\frac{dy}{dx} = 3(-2)x^{-3} = -6x^{-3} = \frac{-6}{x^3}$$

Therefore, when $x = 2$,

$$\frac{dy}{dx} = \frac{-6}{(2)^3} = \frac{-6}{8} = \frac{-3}{4}$$ ∎

Example 7

If $g(t) = 5 - (1/2t^3)$, find $g'(2)$.

Solution: We can rewrite $g(t)$ as

$$g(t) = 5 - \tfrac{1}{2}t^{-3}$$

then $\qquad g'(t) = 0 - \dfrac{1}{2}(-3)t^{-4} = \dfrac{3}{2t^4}$

Therefore, $$g'(2) = \frac{3}{2(2)^4} = \frac{3}{32}$$

∎

Notice in Example 7 that the term

$$\frac{1}{2t^3} \qquad \text{was rewritten as} \qquad \frac{1}{2}t^{-3}$$

Errors are frequently made in rewriting fractions in negative exponent form. Study the next examples carefully.

Example 8

Differentiate $$y = \frac{5}{2x^3}$$

Solution: First, write

$$y = \tfrac{5}{2}(x^{-3})$$

then $$y' = \frac{5}{2}(-3x^{-4}) = -\frac{15}{2}x^{-4} = -\frac{15}{2x^4}$$

∎

Example 9

Differentiate $$y = \frac{5}{(2x)^3}$$

Solution: First, write

$$y = \frac{5}{2^3 x^3} = \frac{5}{2^3}(x^{-3}) = \frac{5}{8}x^{-3}$$

Then $$y' = \frac{5}{8}(-3x^{-4}) = \frac{-15}{8}x^{-4} = \frac{-15}{8x^4}$$

∎

Notice the effect of the parentheses in the denominator of Example 9. The exponent applies to the factor 2, whereas in Example 8 it does *not* apply.

Example 10

Differentiate $$y = \frac{7}{3x^{-2}}$$

Solution: First, write

$$y = \tfrac{7}{3}x^2$$

then $$y' = \frac{7}{3}(2x) = \frac{14x}{3}$$

∎

Example 11

Differentiate $$y = \frac{7}{(3x)^{-2}}$$

Solution: First, write
$$y = 7(3x)^2 = 7(3^2 x^2) = 63x^2$$
then $$y' = 63(2x) = 126x$$ ∎

Example 12

Differentiate $$y = \sqrt{3x}$$

Solution: Rewriting we have
$$y = (\sqrt{3})(x^{1/2})$$

By the Power and Scalar Multiple Rules, we have

$$y' = \sqrt{3}\left(\frac{1}{2}\right)x^{-1/2} = \sqrt{3}\left(\frac{1}{2\sqrt{x}}\right) = \frac{\sqrt{3}}{2\sqrt{x}}$$ ∎

Note that when differentiating functions involving radical signs, it is best to rewrite the function in terms of rational exponents.

Example 13

Differentiate $$y = \frac{1}{2\sqrt[3]{x^2}}$$

Solution: Rewriting with a rational exponent, we have

$$y = \left(\frac{1}{2}\right)\left(\frac{1}{x^{2/3}}\right) = \left(\frac{1}{2}\right)x^{-2/3}$$

By the Power Rule and Scalar Multiple Rule, we have

$$\frac{dy}{dx} = \left(\frac{1}{2}\right)\left(\frac{-2}{3}\right)x^{-5/3} = -\frac{1}{3x^{5/3}}$$ ∎

Example 14

Suppose a guitar string is plucked and vibrates with a frequency of

$$F = 200\sqrt{T}$$

where F is measured in vibrations per second and the tension T is measured in pounds. As the string is tightened, at what rate is the frequency increasing when the tension is 4 pounds? At 9 pounds?

Solution: The rate of change of the frequency with respect to the tension is given by dF/dT. Thus we differentiate F as follows:

$$F = 200T^{1/2}$$

$$\frac{dF}{dT} = 200\left(\frac{1}{2}\right)T^{-1/2} = \frac{100}{\sqrt{T}}$$

When $T = 4$, we have

$$\frac{dF}{dT} = \frac{100}{\sqrt{4}} = 50$$

and when $T = 9$, we have

$$\frac{dF}{dT} = \frac{100}{\sqrt{9}} = 33\frac{1}{3}$$

■

Section Exercises (3.3)

In Exercises 1–10, differentiate the functions.

1. $y = 3$
2. $f(x) = -2$
3. $f(x) = x + 1$
4. $g(x) = 3x - 1$
5. $g(x) = x^2 + 4$
6. $y = t^2 + 2t - 3$
7. $f(t) = -2t^2 + 3t - 6$
8. $y = x^3 - 9$
S **9.** $s(t) = t^3 - 2t + 4$
10. $f(x) = 2x^3 - x^2 + 3x - 1$

In Exercises 11–18, differentiate the functions and evaluate each derivative at the indicated point.

11. $f(x) = \frac{1}{x}$; $(1, 1)$

12. $f(x) = -\frac{1}{2} + \frac{7}{5}x^3$; $(0, -\frac{1}{2})$

S **13.** $f(t) = 3 - \frac{3t}{5t^2}$; $(\frac{3}{5}, 2)$

14. $y = \frac{1}{(3x)^3}$; $(1, \frac{1}{27})$

15. $y = \frac{1}{3x^3}$; $(1, \frac{1}{3})$

16. $y = 3x\left(x^2 - \frac{2}{x}\right)$; $(2, 18)$

S **17.** $y = (2x + 1)^2$; $(0, 1)$
18. $f(x) = 3(5 - x)^2$; $(5, 0)$

In Exercises 19–36, find $f'(x)$.

19. $f(x) = -2(1 - 4x^2)^2$
20. $f(x) = [(x - 2)(x + 4)]^2$

S **21.** $f(x) = x^2 - \frac{4}{x}$

22. $f(x) = x^2 - 3x - 3x^{-2} + 5x^{-3}$

23. $f(x) = x^3 - 3x - \frac{2}{x^4}$

24. $f(x) = \frac{2x^2 - 3x + 1}{x}$

S **25.** $f(x) = \frac{x^3 - 3x^2 + 4}{x^2}$

26. $f(x) = \frac{2}{3x^2}$

S **27.** $f(x) = \frac{\pi}{(3x)^2}$

28. $f(x) = (x^2 + 2x)(x + 1)$
29. $f(x) = x(x^2 - 1)$

30. $f(x) = x + \frac{1}{x^2}$

31. $f(x) = x^{4/5}$

32. $f(x) = x^{1/3} - 1$

33. $f(x) = \sqrt[3]{x} + \sqrt[5]{x}$

34. $f(x) = \dfrac{1}{\sqrt[3]{x^2}}$

35. $f(x) = \dfrac{1}{x^{1/2}} + \dfrac{1}{x^2} + \dfrac{1}{x^4}$

36. $f(x) = 5x^{3/2} - 3x^{1/2} - x^{-1/2}$

S 37. Find the equation of the line tangent to $y = x^4 - 3x^2 + 2$ at the point $(1, 0)$.

38. Find the equation of the line tangent to $y = x^3 + x$ at the point $(-1, -2)$.

S 39. At what points, if any, does $y = x^4 - 3x^2 + 2$ have horizontal tangents?

40. At what points, if any, does $y = x^3 + x$ have horizontal tangents?

41. At what points, if any, does $y = 1/x^2$ have horizontal tangents?

S 42. Sketch the graphs of the two equations $y = x^2$ and $y = -x^2 + 6x - 5$ and sketch the two lines that are tangent to both graphs. Find equations for these lines.

43. The area of a square with sides of length l is given by $A = l^2$. Find the rate of change of the area with respect to l when $l = 4$.

44. The volume of a cube with sides of length l is given by $V = l^3$. Find the rate of change of the volume with respect to l when $l = 4$.

S 45. Suppose that a certain company finds that by charging p dollars per unit, its monthly revenue R will be

$$R = 12{,}000p - 1{,}000p^2, \qquad 0 \le p \le 12$$

(Note that the revenue is zero when $p = 12$ since no one is willing to pay that much.) Find the rate of change of R with respect to p when:

(a) $p = 1$ (b) $p = 4$

(c) $p = 6$ (d) $p = 10$

46. Suppose that the profit P obtained in selling x units of a certain item each week is given by

$$P = 50\sqrt{x} - 0.5x - 500, \qquad 0 \le x \le 8000$$

Find the rate of change of P with respect to x when:

(a) $x = 900$ (b) $x = 1600$

(c) $x = 2500$ (d) $x = 3600$

(Note that the eventual decline in profit occurs because the only way to sell larger quantities is to decrease the price per unit.)

47. Suppose that the effectiveness E of a painkilling drug t hours after entering the bloodstream is given by

$$E = \frac{1}{27}(9t + 3t^2 - t^3), \qquad 0 \le t \le 4.5$$

Find the rate of change of E with respect to t when:

(a) $t = 1$ (b) $t = 2$

(c) $t = 3$ (d) $t = 4$

48. In a certain chemical reaction, the amount in grams Q of a substance produced in t hours is given by the equation $Q = 16t - 4t^2$, where $0 < t \le 2$. What is the rate in grams per hour at which the substance is being produced at these times?

(a) $t = \frac{1}{2}$ h   (b) $t = 1$ h

(c) $t = 2$ h

For Review and Class Discussion

True or False

1. ____ If the tangent line to the graph of f is horizontal at the point $(a, f(a))$, then $f'(a) = 0$.

2. ____ If $f'(x) = g'(x)$, then $f(x) = g(x)$.

3. ____ If $f(x) = g(x) + c$, then $f'(x) = g'(x)$.

4. ____ If $y = 1/3x$, then $dy/dx = -3x^{-2}$.

5. ____ $\dfrac{d}{dx}[\sqrt{cx}] = c\dfrac{d}{dx}[\sqrt{x}]$

6. ____ If $y = \pi^2$, then $dy/dx = 2\pi$.

7. ____ If $y = x/\pi$, then $dy/dx = 1/\pi$.

8. ____ The derivative of the sum of two functions is the sum of their derivatives.

9. ____ If $f(x) = \sqrt{2x}$, then $f'(x) = 1/\sqrt{2x}$.

10. ____ If $f(x) = \sqrt{ax}$, then $f'(x) = 1/\sqrt{ax}$ for any constant a.

3.4 Higher-order Derivatives: Velocity and Acceleration

Purpose
- To define the second-, third-, and higher-order derivatives of a function.
- To define instantaneous velocity and acceleration in straight-line motion.
- To use derivatives to determine the velocity and acceleration of objects traveling along linear paths.

Galileo, the originator of dynamical theories, discovered the laws of motion of heavy bodies. Within this new science Newton comprised the whole system of the universe. The successors of these philosophers have extended these theories, and given them an admirable perfection: they have taught us that the most diverse phenomena are subject to a small number of fundamental laws which are reproduced in all the acts of nature. —*Joseph Fourier (1768–1830)*

Since the derivative of a function is itself a function, we may consider finding its derivative. If this is done, the result is again a function that may be differentiated. If we continue in this manner, we have what are called **higher-order derivatives.** For instance, the derivative of f' is called the **second derivative of** f and is denoted by f''. Similarly, we define the **third derivative of** f as the derivative of f'' and we denote it by f'''. So long as each successive derivative is differentiable, we can continue in this manner to obtain derivatives of higher orders.

Example 1

If $f(x) = 2x^4 - 3x^2$, then:

(first derivative)	$f'(x) = 8x^3 - 6x$
(second derivative)	$f''(x) = 24x^2 - 6$
(third derivative)	$f'''(x) = 48x$
(fourth derivative)	$f^{(4)}(x) = 48$
(fifth derivative)	$f^{(5)}(x) = 0$
	$\vdots$
(nth derivative)	$f^{(n)}(x) = 0$

∎

Notation for Higher-Order Derivatives

First derivative: $\quad y', \; f'(x), \; \dfrac{dy}{dx}, \dfrac{d}{dx}[f(x)], \; D_x(y)$

Second derivative: $\quad y'', \; f''(x), \; \dfrac{d^2y}{dx^2}, \dfrac{d^2}{dx^2}[f(x)], \; D_x^2(y)$

Third derivative: $\quad y''', \; f'''(x), \; \dfrac{d^3y}{dx^3}, \dfrac{d^3}{dx^3}[f(x)], \; D_x^3(y)$

Fourth derivative: $\quad y^{(4)}, \; f^{(4)}(x), \; \dfrac{d^4y}{dx^4}, \dfrac{d^4}{dx^4}[f(x)], \; D_x^4(y)$

$\vdots$

nth derivative: $\quad y^{(n)}, \; f^{(n)}(x), \; \dfrac{d^ny}{dx^n}, \dfrac{d^n}{dx^n}[f(x)], \; D_x^n(y)$

Notice that we drop the prime notation after the third derivative and use parentheses around the order of the derivative. Observe further that in this particular example the "fourth derivative of $f(x)$," denoted by $f^{(4)}(x)$, is constant and therefore each succeeding derivative is zero by our Constant Rule for differentiation.

Example 2

If $g(t) = -t^4 + 2t^3 + t + 4$, find the value of $g'''(2)$.

Solution:

$$g'(t) = -4t^3 + 6t^2 + 1$$
$$g''(t) = -12t^2 + 12t$$
$$g'''(t) = -24t + 12$$

Therefore,

$$g'''(2) = -24(2) + 12 = -36 \quad \blacksquare$$

Example 3

Find the first four derivatives of $y = 3/x$.

Solution: Since $y = 3x^{-1}$, by the Power Rule we obtain

$$y' = -3x^{-2} = \frac{-3}{x^2}$$

$$y'' = 6x^{-3} = \frac{6}{x^3}$$

$$y''' = -18x^{-4} = \frac{-18}{x^4}$$

$$y^{(4)} = 72x^{-5} = \frac{72}{x^5} \quad \blacksquare$$

What interpretation do we give to the second derivative, or the third, or the fourth? We know that $f'(x)$ represents the *slope* of the graph of f. Similarly, $f''(x)$ represents the slope of the graph of f', $f'''(x)$ represents the slope of the graph of f'', and so on.

We know also that $f'(x)$ represents the *rate of change* of $f(x)$. Hence we can generalize this to mean that $f''(x)$ gives the rate of change of $f'(x)$; $f'''(x)$ gives the rate of change of $f''(x)$, and so forth. A common "rate of change" interpretation of the first and second derivatives arises from the position of an object moving along a linear path.

For objects that move along a linear path, it is customary to use a horizontal line, with a designated origin, to represent the line of motion. Movement to the right is considered to be in the positive direction and to the left, negative. The equation of motion gives the position (relative to the origin) of the object as a function of time. This equation of motion is commonly referred to as the **position equation.**

Suppose the position of an object at time t is given by the equation $s = s(t)$. Then over an interval of time Δt, the object changes its position by the amount

$$\Delta s = s(t + \Delta t) - s(t)$$

Recalling the familiar formula

$$\text{distance} = (\text{rate} \times \text{time})$$

or

$$\frac{\text{distance}}{\text{time}} = \text{rate}$$

we say the **average velocity** of the object on the interval Δt is given by

$$\frac{\Delta s}{\Delta t} = \frac{s(t + \Delta t) - s(t)}{\Delta t} = \text{average velocity}$$

Average velocity indicates the rate of change of s with respect to t over an interval of time. The rate of change of s with respect to t *at* a specific time is referred to as the **instantaneous velocity** of the object. For instance, if an automobile travels 60 miles in $1\frac{1}{2}$ hours, then its average velocity during that interval of time is $60 \div 1.5 = 40$ miles per hour. However, at any specific time during the $1\frac{1}{2}$ hours, the instantaneous velocity of the car corresponds to the speedometer reading at that instant of time.

The instantaneous velocity of an object at time t_1 can be determined by considering the average velocity over short time intervals about t_1. For instance, if the car was traveling 60 miles per hour at $t = 10$ seconds, then its average velocity on the interval from $t = 10$ to $t = 12$ seconds would be a good approximation of its instantaneous velocity (60 miles per hour) because there would be very little time to change the speed of the car. If we use shorter and shorter time intervals (let $\Delta t \to 0$), the average velocities become better approximations of the instantaneous velocity at t. We refer to this instantaneous velocity simply as the *velocity* v at time t and write

$$\text{velocity} = \lim_{\Delta t \to 0} \frac{\Delta s}{\Delta t} = \lim_{\Delta t \to 0} \frac{s(t + \Delta t) - s(t)}{\Delta t}$$

Of course, we recognize this limit to be the definition of the derivative of s with respect to t. Therefore, we have the following definition:

Definition of Velocity	If $s = s(t)$ is the position equation for an object moving along a linear path, then the **velocity** of the object at time t is given by

$$v(t) = \lim_{\Delta t \to 0} \frac{s(t + \Delta t) - s(t)}{\Delta t} = s'(t)$$

Example 4

An object moves along a linear path according to the equation $s = 2t^2 - 12t + 10$, where s is measured in feet and t in seconds. Determine its velocity when $t = 4$ and when $t = 2$. When is the velocity zero?

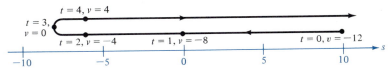

FIGURE 3.8

Note: The graph in Figure 3.8 is used to denote the path of an object moving back and forth along a *straight line*. The rounded portion of the path does not really exist, it merely indicates when the object reverses its direction.

Solution:
$$v(t) = s'(t) = 4t - 12$$

Therefore,

$$v(4) = s'(4) = 4 \text{ ft/s} \qquad \text{and} \qquad v(2) = s'(2) = -4 \text{ ft/s}$$

Furthermore, the velocity is zero when $4t - 12 = 0$, or when $t = 3$ seconds. ∎

Notice in Example 4 that the velocity may sometimes be negative. For horizontal motion we consider velocity to be negative when the object is moving to the left and positive when it is moving to the right. In our example the object moves first in the negative direction, then, at $t = 3$, the velocity is zero as the object reverses its direction. In the case of an object thrown straight up into the air, we usually consider velocity to be positive while it is rising, zero at its maximum height, and negative when it falls.

Sometimes there is confusion between the terms "speed" and "velocity." We define **speed** as the absolute value of velocity and, as such, it is always nonnegative. This simply means that speed indicates only how fast an object is moving, whereas velocity also indicates the direction of its motion.

If velocity is the rate of change of the position, what is the rate of change of the velocity? Suppose we consider the velocity as the first derivative,

$$v(t) = s'(t)$$

Then the rate of change of the velocity with respect to time is simply

$$v'(t) = s''(t)$$

We call this second derivative, $s''(t)$, the **acceleration** of the object at time t, and it indicates how fast the velocity of an object is changing.

Definition of Acceleration	If $s = s(t)$ is the position equation for an object moving along a linear path, then the **acceleration** of the object at time t is given by $$a(t) = v'(t) = s''(t)$$ where $v(t)$ is the velocity at t.

In general, for our velocity and acceleration problems, time will be given in *seconds* and distance in *feet*. Therefore, velocity will be expressed in feet per second (ft/s) and acceleration will be given in feet per second per second (ft/s²).

Example 5

An object is moving along a linear path (see Figure 3.9) according to the equation

$$s(t) = \tfrac{1}{4}t^4 - 2t^3 + 4t^2$$

(a) Determine its position, velocity, and acceleration when $t = 0$ and when $t = 3$ seconds.
(b) When is the velocity zero?
(c) On what interval(s) is the object moving to the right? To the left?

Solution:

(a) Since
$$s(t) = \tfrac{1}{4}t^4 - 2t^3 + 4t^2$$

we have $s(0) = 0$ ft and $s(3) = \tfrac{9}{4}$ ft. Since

$$v(t) = s'(t) = t^3 - 6t^2 + 8t$$

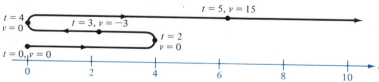

FIGURE 3.9

we have $v(0) = 0$ ft/s and $v(3) = -3$ ft/s. Furthermore, since

$$a(t) = v'(t) = 3t^2 - 12t + 8$$

we have $a(0) = 8$ ft/s^2 and $a(3) = -1$ ft/s^2.

(b)   The velocity is zero when

$$v(t) = t(t^2 - 6t + 8) = t(t - 2)(t - 4) = 0$$

or when $t = 0$, $t = 2$, and $t = 4$ seconds.

(c)   Since the object reverses direction when the velocity is zero, we consider the intervals determined by the times $t = 0$, $t = 2$, and $t = 4$. From Figure 3.9 we conclude that the object is moving to the right when $0 < t < 2$ and again when $4 < t$. Finally, the object is moving to the left when $2 < t < 4$.   ■

Example 6

Suppose the equation $s(t) = -16t^2 + 48t + 160$ represents the position (in feet) above ground of a ball thrown into the air from the top of a cliff.
(a)   What is the initial height s_0 from which the ball is thrown; in other words, what is the height of the cliff?
(b)   What is the initial velocity v_0 at which the ball is thrown?
(c)   What is the acceleration a of the ball?

Solution:

(a)   Since the initial height is attained when $t = 0$, we have
$s_0 = s(0) = 160$ ft.

(b)   Since $\qquad v(t) = s'(t) = -32t + 48$

the initial velocity is

$$v_0 = v(0) = 48 \text{ ft/s}$$

(c)   The acceleration is

$$a(t) = v'(t) = -32 \text{ ft/s}^2 \qquad ■$$

Note in Example 6 that the initial height and the initial velocity correspond to the constant term and to the coefficient of t, respectively, in the position equation of the ball. In general, the position of an object moving freely (neglecting air resistance) under the influence of gravity can be represented by the equation

$$s(t) = \tfrac{1}{2}gt^2 + v_0 t + s_0$$

where s_0 is the *initial height* of the object, v_0 is the *initial velocity* at which the object is thrown, and g is the *acceleration* due to gravity. Considering the acceleration due to the earth's gravity as $g = -32$ ft/s^2, we have the position equation

$$s(t) = -16t^2 + v_0 t + s_0$$

Remember that for objects moving freely under the influence of gravity, we consider the velocity to be positive for upward motion and negative for downward motion.

Example 7

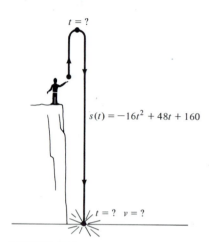

$t = ?$

$s(t) = -16t^2 + 48t + 160$

$t = ?$ $v = ?$

FIGURE 3.10

For the ball in Example 6, find the following (refer to Figure 3.10):
(a) the maximum height reached by the ball
(b) when and with what velocity the ball hits the ground

Solution:

(a) The ball reaches its maximum height at the moment when the velocity changes from positive to negative. This occurs when $v = 0$. Therefore, if $v = s'(t) = 0$, then $-32t + 48 = 0$, and $t = \frac{3}{2}$. The maximum height of the ball is then

$$s(\tfrac{3}{2}) = -16(\tfrac{3}{2})^2 + 48(\tfrac{3}{2}) + 160 = 196 \text{ ft}$$

(b) The object reaches the ground when $s = 0$. Thus

$$s = -16t^2 + 48t + 160 = 0$$
$$t^2 - 3t - 10 = 0$$
$$(t - 5)(t + 2) = 0$$

and $s = 0$ when $t = 5$ or $t = -2$ seconds. Considering only positive time, the ball hits the ground when $t = 5$, and its velocity then is

$$v(5) = s'(5) = -32(5) + 48 = -112 \text{ ft/s}$$

Section Exercises (3.4)

In Exercises 1–14, find the indicated derivative of the given function.

1. $f(x) = x^2 - 2x + 1$; $f''(x)$
2. $f(x) = 3x - 1$; $f''(x)$
[S] 3. $f(x) = 5 - 4x$; $f''(x)$
4. $f(x) = x^5 - 3x^4$; $f'''(x)$
5. $f(x) = x^4 - 2x^3$; $f'''(x)$
6. $f(x) = \sqrt{x}$; $f'''(x)$

[S] 7. $f(x) = x^2 - \dfrac{1}{x} + 1$; $f^{(4)}(x)$

8. $f(t) = 1 - 2t + \dfrac{1}{t^2}$; $f^{(4)}(t)$

[S] 9. $f(t) = t^{-1/3}$; $f'''(t)$

10. $f(t) = \dfrac{3}{4t^2}$; $f''(t)$

11. $f(x) = \dfrac{3}{(4x)^2}$; $f'''(x)$

12. $f(x) = (x - 1)^2$; $f'''(x)$

[S] 13. $f(x) = 2x(x - 1)^2$; $f'''(x)$

14. $f(x) = x\sqrt[3]{x}$; $f''(x)$

In Exercises 15–20, the position equation is given for a particle moving along a horizontal line. For each equation determine (a) the equations for the velocity and acceleration of the particle and (b) on what interval it is moving to the right and when it is moving to the left. (Assume $0 \le t$.)

15. $s = t^2 + 2$ 16. $s = t^2 - 6t + 4$
17. $s = t^3 - 3t^2 + 4$ 18. $s = t^4 - 4t^3 + 4t^2$

S 19. $s = 3t^4 - 28t^3 + 60t^2$ 20. $s = \dfrac{t}{4} + \dfrac{1}{t}$

c 21. Given the position equation $s(t) = t^3 - 3t^2 + 4$.

t	0	0.5	1.0	1.5	2.0	2.5	3.0	3.5	4.0
$s(t)$									
$v(t)$									
$a(t)$									

(a) Complete the accompanying table.
(b) Draw a diagram (as in Figures 3.8 and 3.9) to describe the motion of the particle.
(c) If $v(t)$ and $a(t)$ have the same sign, what observation can be made about the speed of the particle?
(d) If $v(t)$ and $a(t)$ have opposite signs, what observations can be made about the speed of the particle?

22. Repeat Exercise 21, using the position equation $s(t) = t^2 - 6t + 9$.

S 23. A projectile P is shot directly upward from the surface of the earth with an initial velocity $v_0 = 384$ ft/s.
(a) Write the position equation for the motion of the projectile.

(b) Find the average velocity over the interval from $t = 5$ to $t = 10$ seconds.
(c) Find the instantaneous velocity at $t = 5$ and $t = 10$.
(d) Find the maximum height reached by the projectile.

24. Repeat Exercise 23 with $v_0 = 256$ ft/s.

25. A pebble is dropped from a height of 600 ft. Find the instantaneous velocity of the stone when it reaches the ground.

26. A ball is thrown vertically downward from the top of a 220-ft building with an initial velocity of -22 ft/s.
(a) What will its velocity be after 3 s?
(b) What will its velocity be after falling 108 ft?

27. With what initial velocity must an object be thrown vertically upward in order to reach a man on a scaffold 15 ft above?

S 28. A ball is thrown vertically upward with an initial velocity v_0. Show that the maximum height of the ball is $[(v_0^2/64) + s_0]$ ft.

S 29. To estimate the height of a building, a man drops a stone from the top of the building into a pool of water at ground level. How high is the building if the splash is seen 6.8 s after the stone is dropped?

c 30. To estimate the level of the water in a deep well, a woman drops a stone into it. It is impossible to see the stone hit the water, but the splash is heard 4.3 s after the stone is dropped. Estimate the distance to the water, assuming that the speed of sound is 1100 ft/s.

31. A ball is dropped from a height of 100 ft. One second later another ball is dropped from a height of 75 ft. Which one hits the ground first?

For Review and Class Discussion

True or False

1. _____ If $f(x)$ is an nth-degree polynomial, then $f^{(n)}(x) = 0$.

2. _____ If $f(x)$ is an nth-degree polynomial, then $f^{(n+1)}(x) = 0$.

3. _____ The second derivative represents the rate of change of the first derivative.

4. _____ Velocity can be positive, negative, or zero, but speed is always positive.

5. _____ The acceleration of an object can be negative.

6. _____ The speed of an object can be increasing even though the velocity is decreasing.

7. _____ If $f(x) = 1/x$, then $f^{(4)}(x) = 24/x^5$.

8. _____ If $s(t) = -16t^2 + 100t + 400$ represents the position of a moving object, then the initial velocity is 100.

9. _____ A speed of 55 miles per hour is less than a speed of 55 feet per second.

10. _____ If the velocity of an object is constant, then its acceleration is zero.

3.5 **Differentiability and Continuity**

Purpose
- To investigate the relationship between the differentiability and continuity of a function.

It was not till Leibniz and Newton, by the discovery of the differential calculus, had dispelled the ancient darkness which enveloped the conception of the infinite, and had clearly established the conception of the continuous and continuous change, that a full and productive application of the newly found mechanical conceptions made any progress.   —Hermann von Helmholtz *(1821–1894)*

In calculus two of the most useful properties of functions are continuity and differentiability, both of which are defined in terms of a limit. It is our concern in this section to describe the interrelationship of these two concepts. If a function is continuous at a point, is it differentiable there? If a function is differentiable, must it also be continuous? Or perhaps the two properties are equivalent. It will be to our advantage to answer these questions before attempting (in the next section) to derive further rules for differentiation.

We begin by developing an alternative limit form for the derivative. Recall that the derivative of f at c is given by

$$f'(c) = \lim_{\Delta x \to 0} \frac{f(c + \Delta x) - f(c)}{\Delta x}$$

In Figure 3.11 we can see that if $x = c + \Delta x$, then $x \to c$ as $\Delta x \to 0$. Thus if we replace $c + \Delta x$ by x, we can write

$$f'(c) = \lim_{\Delta x \to 0} \frac{f(c + \Delta x) - f(c)}{\Delta x} = \lim_{x \to c} \frac{f(x) - f(c)}{x - c}$$

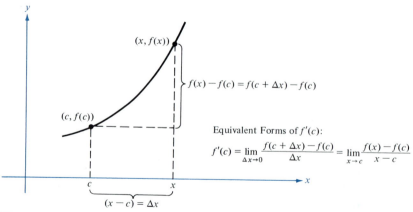

Equivalent Forms of $f'(c)$:

$$f'(c) = \lim_{\Delta x \to 0} \frac{f(c + \Delta x) - f(c)}{\Delta x} = \lim_{x \to c} \frac{f(x) - f(c)}{x - c}$$

FIGURE 3.11

Alternative Form of the Derivative

The **derivative of** f **at** c is given by

$$f'(c) = \lim_{x \to c} \frac{f(x) - f(c)}{x - c}$$

provided this limit exists.

Note that the existence of the limit in this alternative form requires the equality of the one-sided limits

$$\lim_{x \to c^-} \frac{f(x) - f(c)}{x - c} \quad \text{and} \quad \lim_{x \to c^+} \frac{f(x) - f(c)}{x - c}$$

For convenience we refer to these one-sided limits as the *derivatives from the left and the right*, respectively. However, keep in mind that, if these one-sided limits are not equal at c, then the derivative does not exist at c.

To get some idea of the relationship between continuity and differentiability, consider the following functions and their graphs.

The function $f(x) = |x - 2|$ in Figure 3.12 is continuous at $x = 2$. However, the derivative (slope) from the left is

$$\lim_{x \to 2^-} \frac{f(x) - f(2)}{x - 2} = -1$$

and the derivative (slope) from the right is

$$\lim_{x \to 2^+} \frac{f(x) - f(2)}{x - 2} = 1$$

Therefore, the limit $\qquad \lim_{x \to 2} \frac{f(x) - f(2)}{x - 2}$

does not exist and f is not differentiable at $x = 2$. Thus we see that a function may be continuous but not differentiable if its graph possesses a "sharp turn."

A slightly different situation occurs with $f(x) = x^{1/3}$ at $x = 0$ (see Figure 3.13). In this case

$$\lim_{x \to 0} \frac{f(x) - f(0)}{x - 0} = \lim_{x \to 0} \frac{x^{1/3} - 0}{x} = \lim_{x \to 0} \frac{1}{x^{2/3}} = \infty$$

which means that the tangent line is vertical at $x = 0$. In such cases we say that the derivative does not exist. Therefore, we have a continuous function that is not differentiable everywhere because its graph possesses a vertical tangent at some point of continuity.

Now let us drop the continuity requirement and see what happens.

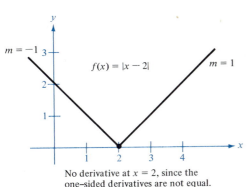

No derivative at $x = 2$, since the one–sided derivatives are not equal.

FIGURE 3.12

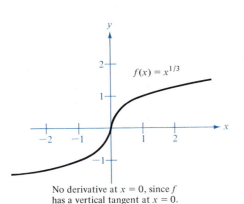

No derivative at $x = 0$, since f has a vertical tangent at $x = 0$.

FIGURE 3.13

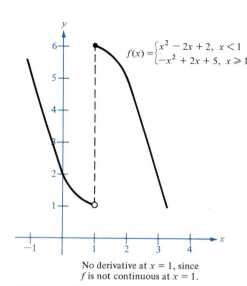

$$f(x) = \begin{cases} x^2 - 2x + 2, & x < 1 \\ -x^2 + 2x + 5, & x \geq 1 \end{cases}$$

No derivative at $x = 1$, since
f is not continuous at $x = 1$.

FIGURE 3.14

Suppose

$$f(x) = \begin{cases} x^2 - 2x + 2, & \text{for } x < 1 \\ -x^2 + 2x + 5, & \text{for } x \geq 1 \end{cases}$$

(See Figure 3.14.) Since

$$\lim_{x \to 1^-} f(x) = 1 \neq f(1)$$

f is not continuous at $x = 1$. The derivative from the left is

$$\lim_{x \to 1^-} \frac{f(x) - f(1)}{x - 1} = \lim_{x \to 1^-} \frac{(x^2 - 2x + 2) - 6}{x - 1}$$

$$= \lim_{x \to 1^-} \frac{x^2 - 2x - 4}{x - 1} = \frac{-5}{0} \text{ (undefined)}$$

Therefore,

$$\lim_{x \to 1} \frac{f(x) - f(1)}{x - 1}$$

does not exist and f is not differentiable at $x = 1$. This example suggests that if a function is not continuous, then it is not differentiable.

From these examples we list some *characteristics that destroy differentiability:*

1. sharp turns in the graph
2. vertical tangents
3. discontinuities

Thus on the one hand, continuity is not sufficient to guarantee differentiability, but on the other hand, discontinuity is sufficient to destroy it. This brings us to the following theorem:

THEOREM 3.1
(*Differentiability Implies Continuity*)

If f is differentiable at $x = c$, then f is continuous there.

Proof: To prove f is continuous at $x = c$, we will show that $\lim_{x \to c} f(x) = f(c)$, or, equivalently, that $\lim_{x \to c} [f(x) - f(c)] = 0$. First, we write

$$\lim_{x \to c} [f(x) - f(c)] = \lim_{x \to c} (x - c) \left(\frac{f(x) - f(c)}{x - c} \right)$$

$$= \left(\lim_{x \to c} (x - c) \right) \left(\lim_{x \to c} \frac{f(x) - f(c)}{x - c} \right)$$

$$= (0) \left(\lim_{x \to c} \frac{f(x) - f(c)}{x - c} \right)$$

Now by the differentiability of f at c, we have

$$\lim_{x \to c} [f(x) - f(c)] = (0) f'(c) = 0$$

or, equivalently, $$\lim_{x \to c} f(x) = f(c)$$

which means that f is continuous at $x = c$. ■

Example 1

Sketch the graph of the continuous function $f(x) = |4 - x^2|$ and indicate where f is not differentiable. Where it is not differentiable, state why.

Solution: The graph is shown in Figure 3.15. We see from this figure that the graph of f has sharp turns at $x = 2$ and $x = -2$, where the slope of the graph is negative from the left and positive from the right. This suggests that f is not differentiable at $x = 2$ and $x = -2$. We verify that f is not differentiable at $x = 2$ and leave the case where $x = -2$ to the reader. For $0 < x < 2$ we have

$$\lim_{x \to 2^-} \frac{f(x) - f(2)}{x - 2} = \lim_{x \to 2^-} \frac{(4 - x^2) - 0}{x - 2} = \lim_{x \to 2^-} \frac{(2 - x)(2 + x)}{x - 2}$$

$$= \lim_{x \to 2^-} (-1)(2 + x) = -4$$

For $x \to 2^+$ we have

$$\lim_{x \to 2^+} \frac{f(x) - f(2)}{x - 2} = \lim_{x \to 2^+} \frac{-(4 - x^2) - 0}{x - 2} = \lim_{x \to 2^+} \frac{(x - 2)(x + 2)}{x - 2}$$

$$= \lim_{x \to 2^+} (1)(x + 2) = +4$$

Therefore, $$\lim_{x \to 2} \frac{f(x) - f(2)}{x - 2}$$

does not exist and f is not differentiable at $x = 2$. ■

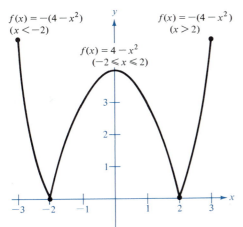

$f(x) = -(4 - x^2)$
$(x < -2)$

$f(x) = -(4 - x^2)$
$(x > 2)$

$f(x) = 4 - x^2$
$(-2 \leq x \leq 2)$

FIGURE 3.15

Example 2

Sketch the graph of the continuous function $y = x^{2/3}$ and indicate where y is not differentiable.

Solution: The graph is shown in Figure 3.16. Differentiating we have

$$\frac{dy}{dx} = \frac{2}{3}x^{-1/3} = \frac{2}{3\sqrt[3]{x}}$$

We see that the only point at which dy/dx does not exist is the origin. ■

The type of sharp turn encountered in Figure 3.16 is called a **cusp,** while those in Figures 3.12 and 3.15 are called **nodes.** Since we will have no reason to distinguish between these two types of sharp turns, we omit their technical definitions.

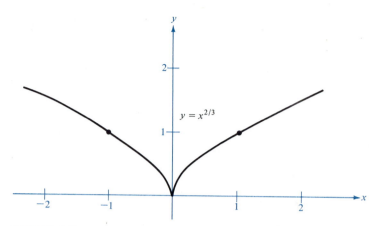

$y = x^{2/3}$

FIGURE 3.16

Section Exercises (3.5)

In Exercises 1–14, (a) indicate where, if ever, the function is not differentiable and (b) state why the function is not differentiable at each point found in part (a).

1. $f(x) = |x + 3|$

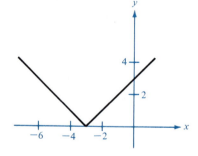

2. $f(x) = |x - 4|$

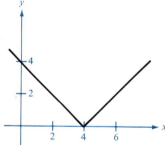

S 3. $f(x) = |x(x + 2)(x - 2)|$

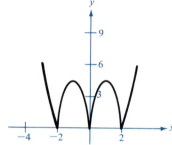

4. $f(x) = |x^2 - 9|$

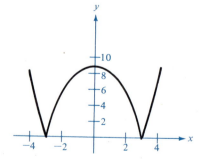

5. $f(x) = \dfrac{1}{x + 1}$

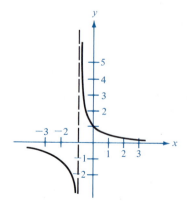

6. $f(x) = \dfrac{2x}{x - 1}$

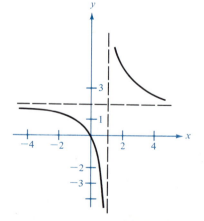

[S] **7.** $f(x) = x^{2/5}$

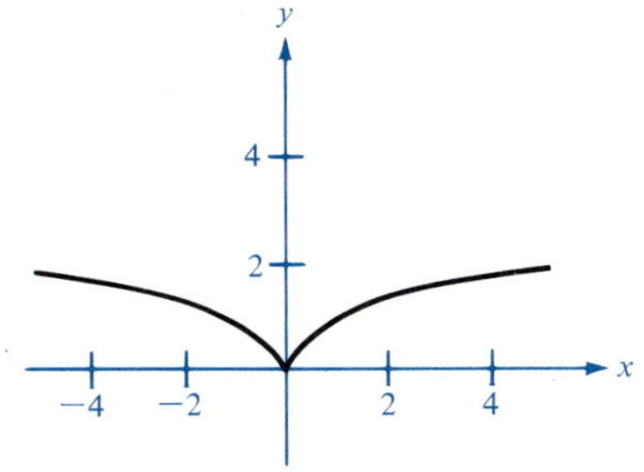

8. $f(x) = (x - 3)^{2/3}$

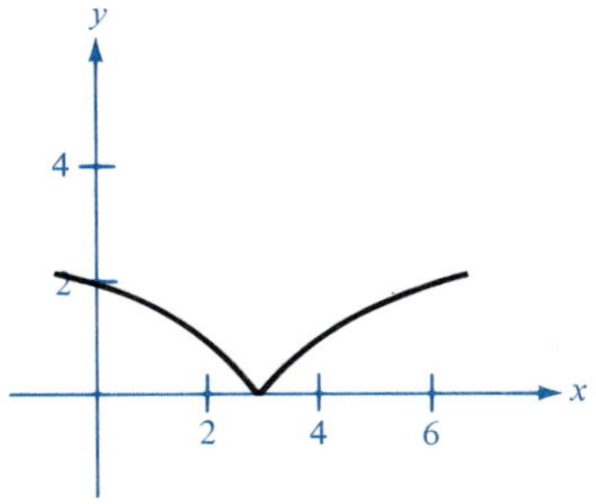

9. $f(x) = \dfrac{x^2}{x^2 + 4}$

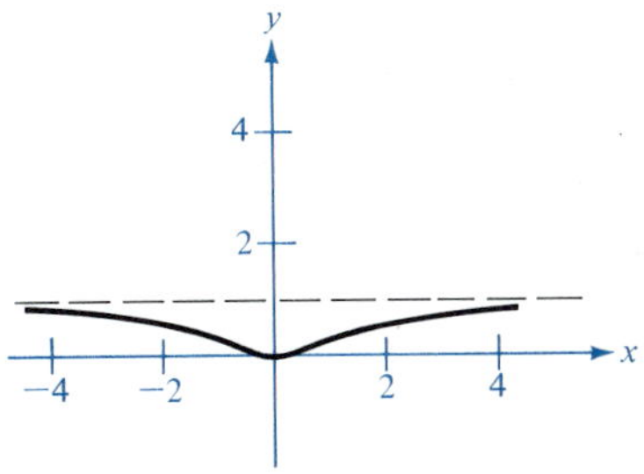

10. $f(x) = \dfrac{x^2}{x^2 - 4}$

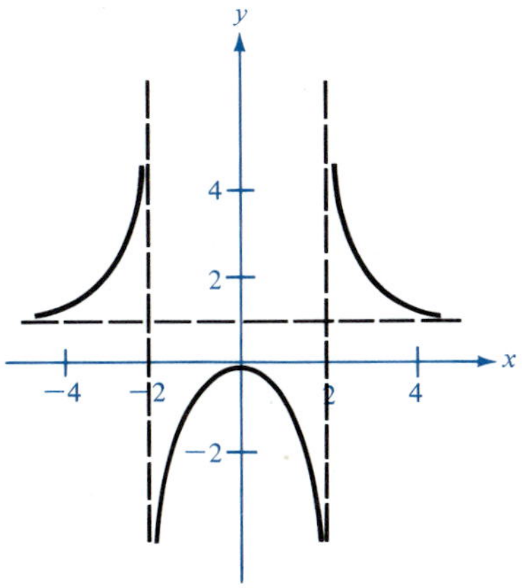

11. $f(x) = \begin{cases} 4 - x^2, & 0 < x \\ x^2 - 4, & x \leq 0 \end{cases}$

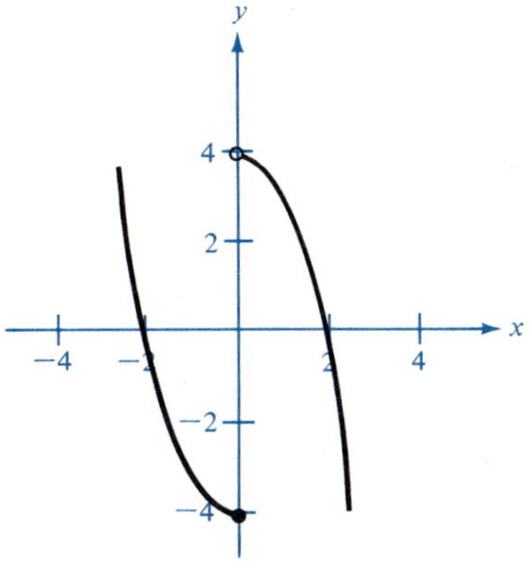

12. $f(x) = \begin{cases} x^2 - 2x, & x > 1 \\ x^3 - 3x^2 + 3x, & x \leq 1 \end{cases}$

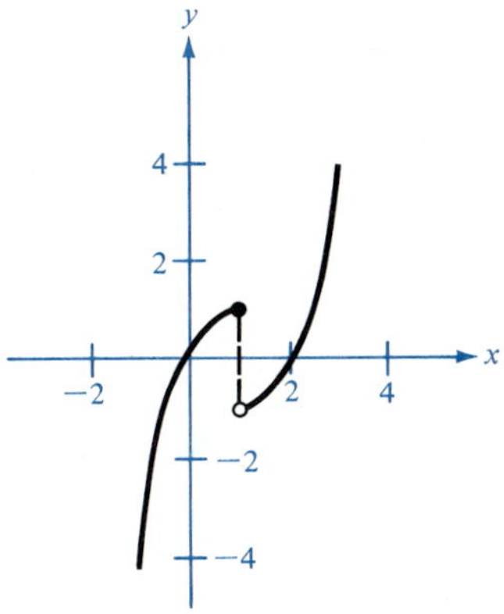

[S] **13.** $f(x) = \sqrt{x - 1}$

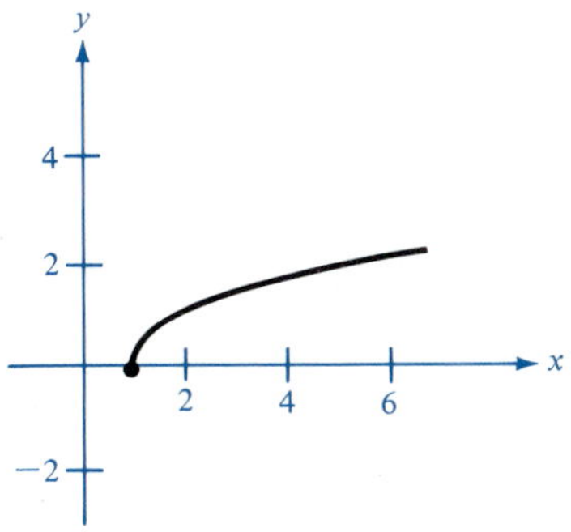

14. $f(x) = \sqrt{3 - x}$

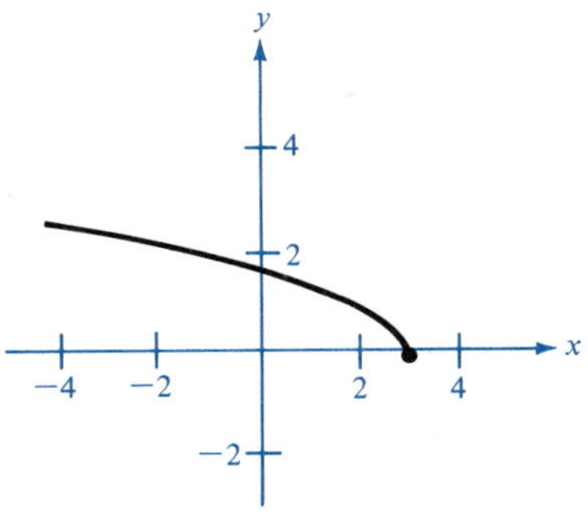

15. Given $f(x) = \sqrt{4 - x^2}$, show that neither the derivative from the right at -2 nor the derivative from the left at 2 exist. Sketch the graph of the function.

16. Given

$$f(x) = \begin{cases} x^3 - 3x^2 + 3x, & \text{if } x \leq 1 \\ x^2 - 2x, & \text{if } x > 1 \end{cases}$$

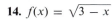

show that the derivative from the right and the derivative from the left at $x = 1$ are not equal.

[S] **17.** Given

$$f(x) = \begin{cases} x, & \text{if } x \leq 0 \\ x^2, & \text{if } x > 0 \end{cases}$$

show that the derivative from the right and the derivative from the left are not equal at $x = 0$.

18. Given

$$f(x) = \begin{cases} x, & \text{if } x \leq 1 \\ x^2, & \text{if } x > 1 \end{cases}$$

show that the derivative from the right and the derivative from the left are not equal at $x = 1$.

For Review and Class Discussion

True or False

1. _____ If a function is continuous, then it is differentiable.

2. _____ If a function is differentiable, then it is continuous.

3. _____ If a function has derivatives from both the right and the left at a point, then it is differentiable at that point.

4. _____ If f is differentiable at $x = c$, then f' is differentiable at $x = c$.

5. _____ If the graph of a function possesses a tangent line at a point, then it is differentiable at that point.

3.6 **The Product and Quotient Rules**

Purpose
- To derive and use rules for the differentiation of products and quotients.
- To point out some pitfalls in the use of these differentiation rules.

The derivative of the product of two quantities is equal to the product of the derivative of the first of those quantities with the second plus the product of the derivative of the second with the first. —*Guillaume L'Hôpital (1661–1701)*

In Section 3.3 we noted that the derivative of a sum or difference of two functions is simply the sum or difference of their derivatives. The rules for the derivative of a product or quotient of two functions are not so simple and you may find the results surprising. As we derive each of the rules, we will assume that the derivatives of the given functions exist. Furthermore, we strongly recommend that you memorize each rule, especially the *verbal* statements of the product and quotient rules.

Product Rule

The derivative of the product of two functions is equal to the first function times the derivative of the second, plus the second times the derivative of the first.

$$\frac{d}{dx}[f(x)\,g(x)] = f(x)\,g'(x) + g(x)\,f'(x)$$

Proof: Let $F(x) = f(x) g(x)$; then

$$F'(x) = \lim_{\Delta x \to 0} \frac{F(x + \Delta x) - F(x)}{\Delta x}$$

$$= \lim_{\Delta x \to 0} \frac{f(x + \Delta x) g(x + \Delta x) - f(x) g(x)}{\Delta x}$$

Now because it is both legal and useful, let us subtract and add $f(x + \Delta x) g(x)$ in the numerator, giving us

$F'(x)$

$$= \lim_{\Delta x \to 0} \frac{f(x + \Delta x) g(x + \Delta x) - f(x + \Delta x) g(x) + f(x + \Delta x) g(x) - f(x) g(x)}{\Delta x}$$

$$= \lim_{\Delta x \to 0} \left[f(x + \Delta x) \left(\frac{g(x + \Delta x) - g(x)}{\Delta x} \right) + g(x) \left(\frac{f(x + \Delta x) - f(x)}{\Delta x} \right) \right]$$

$$= \lim_{\Delta x \to 0} f(x + \Delta x) \left(\frac{g(x + \Delta x) - g(x)}{\Delta x} \right) + \lim_{\Delta x \to 0} g(x) \left(\frac{f(x + \Delta x) - f(x)}{\Delta x} \right)$$

$$= \left[\lim_{\Delta x \to 0} f(x + \Delta x) \right] \left[\lim_{\Delta x \to 0} \frac{g(x + \Delta x) - g(x)}{\Delta x} \right]$$

$$+ \left[\lim_{\Delta x \to 0} g(x) \right] \left[\lim_{\Delta x \to 0} \frac{f(x + \Delta x) - f(x)}{\Delta x} \right]$$

From the differentiability of f and g, we obtain the following limits:

$$\lim_{\Delta x \to 0} f(x + \Delta x) = f(x)$$

$$\lim_{\Delta x \to 0} \frac{g(x + \Delta x) - g(x)}{\Delta x} = g'(x)$$

$$\lim_{\Delta x \to 0} g(x) = g(x)$$

$$\lim_{\Delta x \to 0} \frac{f(x + \Delta x) - f(x)}{\Delta x} = f'(x)$$

Therefore, we have

$$F'(x) = f(x) g'(x) + g(x) f'(x)$$

and

$$\frac{d}{dx}[f(x) g(x)] = f(x) g'(x) + g(x) f'(x)$$ ∎

Example 1

Find the derivative of $F(x) = (3x - 2x^2)(5 + 4x)$.

Solution: By the Product Rule

$$F'(x) = (3x - 2x^2)\frac{d}{dx}[5 + 4x] + (5 + 4x)\frac{d}{dx}[3x - 2x^2]$$

$$= (3x - 2x^2)(4) + (5 + 4x)(3 - 4x)$$

$$= (12x - 8x^2) + (15 - 8x - 16x^2) = 15 + 4x - 24x^2$$

Of course, we could have bypassed the Product Rule by first multiplying $(3x - 2x^2)$ by $(5 + 4x)$ to get $15x + 2x^2 - 8x^3$ and then differentiating this result. Although in this case this alternative procedure would have been just as simple, we cannot always avoid the use of the Product Rule by first multiplying the given factors. ∎

It is helpful to realize that, in general,

$$\begin{pmatrix} \text{the derivative of the} \\ \text{product of two functions} \end{pmatrix} \neq \begin{pmatrix} \text{the product of the derivatives} \\ \text{of the two functions} \end{pmatrix}$$

This is evident in Example 1, since

$$\frac{d}{dx}[(3x - 2x^2)(5 + 4x)] = 15 + 4x - 24x^2$$

whereas

$$\left(\frac{d}{dx}[3x - 2x^2]\right)\left(\frac{d}{dx}[5 + 4x]\right) = (3 - 4x)(4) = 12 - 16x$$

Example 2

Differentiate $f(x) = (1 + x^{-1})(x - 1)$.

Solution: By the Product Rule

$$f'(x) = (1 + x^{-1})(1) + (x - 1)(-x^{-2})$$

$$= 1 + \frac{1}{x} - \frac{x - 1}{x^2} = \frac{x^2 + x - x + 1}{x^2} = \frac{x^2 + 1}{x^2}$$
∎

Quotient Rule

The derivative of the quotient of two functions is equal to the denominator times the derivative of the numerator minus the numerator times the derivative of the denominator, all divided by the square of the denominator:

$$\frac{d}{dx}\left[\frac{f(x)}{g(x)}\right] = \frac{g(x)f'(x) - f(x)g'(x)}{[g(x)]^2}, \qquad \text{where} \qquad g(x) \neq 0$$

Proof: Let $F(x) = f(x)/g(x)$. Then

$$F'(x) = \lim_{\Delta x \to 0} \frac{F(x + \Delta x) - F(x)}{\Delta x} = \lim_{\Delta x \to 0} \frac{\dfrac{f(x + \Delta x)}{g(x + \Delta x)} - \dfrac{f(x)}{g(x)}}{\Delta x}$$

$$= \lim_{\Delta x \to 0} \frac{g(x) f(x + \Delta x) - f(x) g(x + \Delta x)}{\Delta x \, g(x) \, g(x + \Delta x)}$$

Now by adding and subtracting $f(x) g(x)$ in the numerator, we have

$$F'(x) = \lim_{\Delta x \to 0} \frac{g(x) f(x + \Delta x) - f(x) g(x) + f(x) g(x) - f(x) g(x + \Delta x)}{\Delta x \, g(x) \, g(x + \Delta x)}$$

$$= \frac{\displaystyle\lim_{\Delta x \to 0} \frac{g(x)[f(x + \Delta x) - f(x)]}{\Delta x} - \lim_{\Delta x \to 0} \frac{f(x)[g(x + \Delta x) - g(x)]}{\Delta x}}{\displaystyle\lim_{\Delta x \to 0} [g(x) g(x + \Delta x)]}$$

$$= \frac{g(x)\left[\displaystyle\lim_{\Delta x \to 0} \frac{f(x + \Delta x) - f(x)}{\Delta x}\right] - f(x)\left[\displaystyle\lim_{\Delta x \to 0} \frac{g(x + \Delta x) - g(x)}{\Delta x}\right]}{\displaystyle\lim_{\Delta x \to 0} [g(x) g(x + \Delta x)]}$$

$$= \frac{g(x) f'(x) - f(x) g'(x)}{[g(x)]^2} \qquad \blacksquare$$

As suggested previously, you should memorize the verbal statement of the Quotient Rule. The following form may assist you in memorizing the rule:

$$\frac{d}{dx}[\text{quotient}] = \frac{(\text{denom.}) \dfrac{d}{dx}[\text{numerator}] - (\text{numerator}) \dfrac{d}{dx}[\text{denom.}]}{(\text{denom.})^2}$$

This form certainly points out that

$$\left(\begin{array}{c} \text{the derivative of the} \\ \text{quotient of two functions} \end{array}\right) \neq \left(\begin{array}{c} \text{the quotient of the derivatives} \\ \text{of the two functions} \end{array}\right)$$

Example 3

Differentiate $$y = \frac{2x^2 - 4x + 3}{2 - 3x}$$

Solution: By the Quotient Rule

$$y' = \frac{(2 - 3x) \dfrac{d}{dx}[2x^2 - 4x + 3] - (2x^2 - 4x + 3) \dfrac{d}{dx}[2 - 3x]}{(2 - 3x)^2}$$

$$= \frac{(2 - 3x)(4x - 4) - (2x^2 - 4x + 3)(-3)}{(2 - 3x)^2}$$

$$= \frac{-12x^2 + 20x - 8 + (6x^2 - 12x + 9)}{4 - 12x + 9x^2} = \frac{-6x^2 + 8x + 1}{9x^2 - 12x + 4} \quad \blacksquare$$

Example 4

If
$$f(x) = \frac{-7}{3x^2 - 5}$$

find $f'(1)$.

Solution: By the Quotient Rule
$$f'(x) = \frac{(3x^2 - 5)(0) - (-7)(6x)}{(3x^2 - 5)^2} = \frac{42x}{(3x^2 - 5)^2}$$

Therefore,
$$f'(1) = \frac{42}{(-2)^2} = \frac{42}{4} = \frac{21}{2} \qquad \blacksquare$$

You can avoid many algebraic errors when differentiating quotients if you use parentheses liberally. It is a good idea to enclose all factors and derivatives in parentheses and pay special attention to the subtraction required in the numerator of the Quotient Rule. In fact, liberal use of parentheses is a sound guideline in *all* types of differentiation problems.

Example 5

Find the derivative of
$$y = \frac{3 - (1/x)}{x + 5}$$

Solution: First, let us rewrite y as
$$y = \frac{(3x - 1)/x}{x + 5} = \frac{3x - 1}{x(x + 5)} = \frac{3x - 1}{x^2 + 5x}$$

Now by the Quotient Rule we have
$$\frac{dy}{dx} = \frac{(x^2 + 5x)(3) - (3x - 1)(2x + 5)}{(x^2 + 5x)^2}$$

$$= \frac{(3x^2 + 15x) - (6x^2 + 13x - 5)}{(x^2 + 5x)^2} = \frac{-3x^2 + 2x + 5}{(x^2 + 5x)^2}$$

(Notice the parentheses.) $\qquad \blacksquare$

Example 6

Not every quotient needs to be differentiated by the Quotient Rule. The following quotients can be considered as products of a constant times a function of x. In such cases the Scalar Multiple Rule (Section 3.3) is more convenient and is used in place of the Quotient Rule.

Given	Write	Then
$y = \dfrac{x^2 + 3x}{-6}$	$y = \dfrac{-1}{6}(x^2 + 3x)$	$y' = \dfrac{-1}{6}(2x + 3)$
$y = \dfrac{5x^4}{8}$	$y = \dfrac{5}{8}x^4$	$y' = \dfrac{5}{8}(4x^3) = \dfrac{5}{2}x^3$
$y = \dfrac{-3(3x - 2x^2)}{7x}$	$y = \dfrac{-3}{7}(3 - 2x)$	$y' = \dfrac{-3}{7}(-2) = \dfrac{6}{7}$
$y = \dfrac{9}{-5x^2}$	$y = \dfrac{9}{-5}(x^{-2})$	$y' = \dfrac{9}{-5}(-2x^{-3}) = \dfrac{18}{5x^3}$

When we introduced the Power Rule,

$$\frac{d}{dx}[x^n] = nx^{n-1}, \qquad \text{where } n \text{ is any real number}$$

in Section 3.3, we proved only the case where n is a positive integer. We now prove this rule for the case when n is a negative integer.

Proof: Suppose n is a negative integer; then there exists a positive integer k such that $n = -k$. Let

$$y = x^n = x^{-k} = \frac{1}{x^k}$$

By the Quotient Rule

$$\frac{dy}{dx} = \frac{x^k(0) - (1)(kx^{k-1})}{(x^k)^2} = \frac{0 - kx^{k-1}}{x^{2k}} = -kx^{-k-1} = nx^{n-1}$$

Therefore, if n is a negative integer, we also have

$$\frac{d}{dx}[x^n] = nx^{n-1}$$

∎

Example 7 (*Combining Rules*)

Differentiate

$$y = \frac{(1 - 2x)(3x + 2)}{5x - 4}$$

Solution: Here we have a product within a quotient, and we could first multiply the factors in the numerator, then apply the Quotient Rule. However, to show how the Product Rule can be used within the Quotient Rule, we proceed with the equation as written.

$$y' = \frac{(5x - 4)\dfrac{d}{dx}[(1 - 2x)(3x + 2)] - (1 - 2x)(3x + 2)\dfrac{d}{dx}[5x - 4]}{(5x - 4)^2}$$

$$= \frac{(5x - 4)[(1 - 2x)(3) + (3x + 2)(-2)] - (1 - 2x)(3x + 2)(5)}{(5x - 4)^2}$$

$$= \frac{(5x - 4)(-12x - 1) - (1 - 2x)(15x + 10)}{(5x - 4)^2}$$

$$= \frac{(-60x^2 + 43x + 4) - (-30x^2 - 5x + 10)}{(5x - 4)^2}$$

$$= \frac{-30x^2 + 48x - 6}{(5x - 4)^2}$$

∎

Example 8

As blood moves from the heart through the major arteries out to the capillaries and back through the veins, the systolic pressure continuously drops. Suppose that this pressure is given by

$$P = \frac{25t^2 + 125}{t^2 + 1} \qquad (0 \le t \le 10)$$

where P is measured in millimeters of mercury and t is measured in seconds. At what rate is the pressure dropping 5 seconds after leaving the heart?

Solution: By the Quotient Rule we have

$$\frac{dP}{dt} = \frac{(t^2 + 1)(50t) - (25t^2 + 125)(2t)}{(t^2 + 1)^2}$$

$$= \frac{50t^3 + 50t - 50t^3 - 250t}{(t^2 + 1)^2} = -\frac{200t}{(t^2 + 1)^2}$$

Therefore, when $t = 5$ the pressure is dropping at the rate of

$$\frac{dP}{dt} = -\frac{200(5)}{26^2} = 1.48 \text{ mm/s}$$

∎

Note in the examples in this section that much of the work in obtaining the final form of the answer occurs *after* the differentiation is complete. This is often the case; direct application of differentiation rules yields answers that are not in simplest form, as reviewed in Table 3.2.

TABLE 3.2

	$f'(x)$ after differentiating	$f'(x)$ after simplifying
Example 1	$(3x - 2x^2)(4) + (5 + 4x)(3 - 4x)$	$15 + 4x - 24x^2$
Example 2	$(1 + x^{-1})(1) + (x - 1)(-x^{-2})$	$\dfrac{x^2 + 1}{x^2}$
Example 3	$\dfrac{(2 - 3x)(4x - 4) - (2x^2 - 4x + 3)(-3)}{(2 - 3x)^2}$	$\dfrac{-6x^2 + 8x + 1}{(2 - 3x)^2}$
Example 7	$\dfrac{(5x - 4)\,[(1 - 2x)(3) + (3x + 2)(-2)] - (1 - 2x)(3x + 2)(5)}{(5x - 4)^2}$	$\dfrac{-30x^2 + 48x - 6}{(5x - 4)^2}$

As seen in the table, two characteristics of the simplest form of an algebraic expression are (1) the absence of negative exponents and (2) the combining of like terms.

Section Exercises (3.6)

In Exercises 1–10, differentiate and find $f'(x)$ at the given values of x.

1. $f(x) = \frac{1}{3}(2x^3 - 4)$; $x = 0$

2. $f(x) = \dfrac{5 - 6x^2}{7}$; $x = 1$

3. $f(x) = \dfrac{7}{3x^3}$; $x = 1$

4. $f(x) = 5x^{-2}(x + 3)$; $x = 0$

5. $f(x) = (x^2 - 2x + 1)(x^3 - 1)$; $x = 1$

6. $f(x) = (x^3 - 3x)(2x^2 + 3x + 5)$; $x = 0$

⒮ **7.** $f(x) = (x - 1)(x^2 - 3x + 2)$; $x = 0$

8. $f(x) = \left(x^2 - \dfrac{1}{x}\right)(x^2 + 1)$; $x = 1$

9. $f(x) = (x^5 - 3x)\left(\dfrac{1}{x^2} + x\right)$; $x = -1$

10. $f(x) = \dfrac{x + 1}{x - 1}$; $x = 2$

In Exercises 11–28, differentiate each function.

11. $f(x) = \dfrac{3x - 2}{2x - 3}$

12. $f(x) = \dfrac{x^3 + 3x + 2}{x^2 - 1}$

⒮ **13.** $f(x) = \dfrac{3 - 2x - x^2}{x^2 - 1}$

14. $f(x) = \dfrac{9}{3x^2 - 2x}$

15. $f(x) = \dfrac{1}{4 - 3x^2}$

16. $f(x) = x^4\left(1 - \dfrac{2}{x + 1}\right)$

⒮ **17.** $f(x) = \sqrt[3]{x}(\sqrt{x} + 3)$

18. $f(x) = \dfrac{x + 1}{\sqrt{x}}$

19. $h(t) = \dfrac{t + 1}{t^2 + 2t + 2}$

20. $h(x) = (x^2 - 1)^2$

21. $h(s) = (s^3 - 2)^2$

22. $f(x) = \left(\dfrac{x^2 - x - 3}{x^2 + 1}\right)(x^2 + x + 1)$

⒮ **23.** $g(x) = \left(\dfrac{x + 1}{x + 2}\right)(2x - 5)$

24. $f(x) = (x^2 - x)(x^2 + 1)(x^2 + x + 1)$

⒮ **25.** $f(x) = (3x^3 + 4x)(x - 5)(x + 1)$

26. $f(x) = \dfrac{x^2 + c^2}{x^2 - c^2}$ (c is a constant)

27. $f(x) = \dfrac{c^2 - x^2}{c^2 + x^2}$

28. $f(x) = \dfrac{x(x^2 - 1)}{x + 3}$

⒮ **29.** Find an equation of the line tangent to the graph of $f(x) = x/(x - 1)$ at the point $(2, 2)$.

30. Find an equation of the line tangent to the graph of $f(x) = (x - 1)(x^2 - 2)$ at the point $(0, 2)$.

31. Find an equation of the tangent line to $f(x) = (x^3 - 3x + 1)(x + 2)$ at $(1, -3)$.

32. Find an equation of the tangent line to $f(x) = (x - 1)/(x + 1)$ at the point $(2, \frac{1}{3})$.

S 33. At what point(s) does $f(x) = x^2/(x - 1)$ have a horizontal tangent?

34. A certain automobile depreciates according to the formula

$$V = \frac{7500}{1 + 0.4t + 0.1t^2}$$

where $t = 0$ represents the time of purchase (in years). At what rate is the car depreciating:
(a) 1 year after purchase?
(b) 2 years after purchase?

S 35. A population of 500 bacteria is introduced into a culture and grows in number according to the equation

$$P(t) = 500\left(1 + \frac{4t}{50 + t^2}\right)$$

where t is measured in hours. Find the rate at which the population is growing when $t = 2$.

36. (Doppler effect) The frequency F of a fire truck siren heard by a stationary observer is given by

$$F = \frac{132,400}{331 \pm v}$$

where $\pm v$ represents the velocity of the accelerating fire truck. Find the rate of change of F with respect to v when (a) the fire truck is approaching at a velocity of 30 m/s [use $-v$] and (b) the fire truck is moving away at a velocity of 30 m/s [use $+v$].

$$F = \frac{132,400}{331 + v} \qquad F = \frac{132,400}{331 - v}$$

Doppler Effect

37. In Section 2.3, Example 7, the function

$$f(t) = \frac{t^2 - t + 1}{t^2 + 1}$$

measured the percentage of the normal level of oxygen in a pond, where t is the time in weeks after organic waste is dumped into the pond. Find the rate of change of f with respect to t when:
(a) $t = 0.5$ (b) $t = 2$
(c) $t = 8$

38. Suppose that the temperature T of food placed in a freezer drops according to the equation

$$T = \frac{700}{t^2 + 4t + 10}$$

where t is the time in hours. Find the rate of change of T with respect to t when:
(a) $t = 1$ (b) $t = 3$
(c) $t = 5$ (d) $t = 10$

S 39. Suppose that the temperature T of food placed in a refrigerator drops according to the equation

$$T = 10\left(\frac{4t^2 + 16t + 75}{t^2 + 4t + 10}\right)$$

where t is the time in hours. What is the initial temperature of the food? What is the limit of T as t approaches infinity? Find the rate of change of T with respect to t when:
(a) $t = 1$ (b) $t = 3$
(c) $t = 5$ (d) $t = 10$

40. Derive the equations for the velocity and acceleration of a particle that moves according to the law $s(t) = t + [1/(t + 1)]$.

41. Derive the equations for the velocity and acceleration of a particle that moves according to the law $s(t) = 1/(t^2 + 2t + 1)$.

S 42. Show that $dy/dx = f'(x)\, g(x)\, h(x) + f(x)\, g'(x)\, h(x) + f(x)\, g(x)\, h'(x)$ if $y = f(x)\, g(x)\, h(x)$.

For Review and Class Discussion

True or False

1. _____ If $y = f(x)\, g(x)$, then $dy/dx = f'(x)\, g'(x)$.

2. _____ If $(x + 1)^2$ is a factor of $f(x)$, then $(x + 1)$ is a factor of $f'(x)$.

3. _____ If $y = (x + 1)(x + 2)(x + 3)(x + 4)$, then $d^5y/dx^5 = 0$.

4. _____ If $y = f(x)\, g(x)$, then $d^2y/dx^2 = f(x)\, g''(x) + g(x)\, f''(x)$.

5. _____ If $y = [f(x)]^2$, then $dy/dx = 2f'(x)\, f(x)$.

6. _____ The derivative of a rational function is a rational function.

7. ____ If $y = 1/f(x)$, then $dy/dx = 1/f'(x)$.

8. ____ If $f'(c)$ and $g'(c)$ are zero and $h(x) = f(x) g(x)$, then $h'(c) = 0$.

9. ____ If f' and g' both exist at $x = c$, then $\dfrac{d}{dx}[f(x) g(x)]$ exists at $x = c$.

10. ____ If f' and g' both exist at $x = c$, then $\dfrac{d}{dx}\left[\dfrac{f(x)}{g(x)}\right]$ exists at $x = c$.

3.7 The Chain Rule and the General Power Rule

Purpose
- To derive and use the Chain and Power Rules for differentiating.
- To summarize the rules for differentiation and illustrate their use in combination with one another.

Civilization advances by extending the number of operations which we can perform without thinking about them. —*Alfred Whitehead (1861–1947)*

In our development of differentiation rules, we have yet to discuss one of the most important rules in differential calculus, the Chain Rule. We begin this section with an intuitive development of the Chain Rule without giving a formal proof of it. The term "chain" in the Chain Rule refers to functions that are composed as a chain of other functions. We have previously defined such functions as being *composite*. For example, consider the following composite function. Suppose

$$y = f(u) = 3u^{15}$$

where

$$u = g(x) = 2x - 1$$

Though y is originally given as a function of u, we can express y as a composite function of x as follows:

$$y = f(u) = f(g(x)) = 3(2x - 1)^{15}$$

We wish to determine the derivative of y with respect to x.

One way of determining dy/dx would be to expand $(2x - 1)^{15}$ in the equation $y = 3(2x - 1)^{15}$ and then proceed to calculate dy/dx. Of course, we would very much like to avoid such a cumbersome expansion. Hence we seek a rule for finding the derivative of a composite function more easily.

Consider, in general, two differentiable functions

$$y = f(u) \qquad \text{and} \qquad u = g(x)$$

with the composite function $y = f(g(x))$. Since a derivative indicates a "rate of change," we can say that

$$y \text{ changes } \frac{dy}{du} \text{ times as fast as } u$$

$$u \text{ changes } \frac{du}{dx} \text{ times as fast as } x$$

From this it seems reasonable to conclude that

$$y \text{ changes } \frac{dy}{du}\frac{du}{dx} \text{ times as fast as } x$$

which is equivalent to saying

$$\frac{dy}{dx} = \frac{dy}{du}\frac{du}{dx}$$

In functional notation this can be written as

$$\frac{d}{dx}[f(g(x))] = f'(u)\,g'(x)$$

We now give the detailed statement of the Chain Rule.

Chain Rule

If $y = f(u)$ is a differentiable function of u, and $u = g(x)$ is a differentiable function of x, then $y = f(g(x))$ is a differentiable function of x and

$$\frac{dy}{dx} = \frac{dy}{du}\frac{du}{dx}$$

or, equivalently,

$$\frac{d}{dx}[f(g(x))] = f'(u)\,g'(x)$$

As an aid to memorizing the Chain Rule, it is helpful to think of dy/du and du/dx as two "fractions." When the two fractions are multiplied together, we can imagine that the two du's cancel each other to produce

$$\frac{dy}{d\!\!\!/u}\frac{d\!\!\!/u}{dx} = \frac{dy}{dx}$$

Example 1

If $y = 3u^{15}$ and $u = 2x - 1$, find dy/dx.

Solution: By the Chain Rule we have

$$\frac{dy}{dx} = \frac{dy}{du}\frac{du}{dx} = 45(u)^{14}(2) = 90(u)^{14}$$

Substituting for u we get

$$\frac{dy}{dx} = 90(2x - 1)^{14}$$

■

The function in Example 1 is an instance of perhaps the most common type of composite function, that is, functions of the form

$$y = [u(x)]^n$$

The rule for differentiating such "power functions" is called the *General Power Rule,* and it is a special case of the Chain Rule.

(General) **Power Rule**	If $y = [u(x)]^n$, where u is a differentiable function of x and n is a real number, then $$\frac{dy}{dx} = n[u(x)]^{n-1}\frac{du}{dx}$$ or, equivalently, $\dfrac{d}{dx}[u^n] = nu^{n-1}\,u'$

Proof: Let $y = u^n$, where u is a differentiable function of x and n is any real number. Then by the Chain Rule we have

$$\frac{dy}{dx} = \frac{dy}{du}\frac{du}{dx} = \frac{d}{du}[u^n]\frac{du}{dx}$$

But by the simple Power Rule of Section 3.3 (replacing x with u), we have

$$\frac{d}{du}[u^n] = nu^{n-1}$$

Therefore, $\dfrac{dy}{dx} = nu^{n-1}\dfrac{du}{dx}$ ∎

Example 2

Differentiate $f(x) = (3x - 2x^2)^3$

Solution: Since $f(x) = (3x - 2x^2)^3 = u^n$

we have

$$f'(x) = \overbrace{3}^{n}\overbrace{(3x - 2x^2)^2}^{u^{n-1}}\overbrace{\frac{d}{dx}[3x - 2x^2]}^{u'}$$

$$= 3(3x - 2x^2)^2(3 - 4x)$$

$$= (9 - 12x)(3x - 2x^2)^2 \qquad ∎$$

Keep in mind that the Power Rule is applicable to fractional powers of a function.

Example 3

Find the derivative of $y = \sqrt[3]{(x^2 + 2)^2}$.

Solution: Rewrite the equation as $y = (x^2 + 2)^{2/3}$. By the Power Rule

$$y' = \overset{n}{\frac{2}{3}}\overbrace{(x^2 + 2)^{-1/3}}^{u^{n-1}}\overbrace{(2x)}^{u'} = \frac{4x}{3\sqrt[3]{x^2 + 2}}$$ ∎

The derivative of a quotient may sometimes be found more readily by using the Power Rule rather than the Quotient Rule, especially when the numerator is a constant. Our next example is a case in point.

Example 4

Differentiate $$g(t) = \frac{-7}{(2t - 3)^2}$$

Solution: If we rewrite the equation as $g(t) = -7(2t - 3)^{-2}$, then by the Power Rule we obtain

$$g'(t) = (-7)(-2)(2t - 3)^{-3}(2) = 28(2t - 3)^{-3} = \frac{28}{(2t - 3)^3}$$

Alternative Solution: By the Quotient and Power Rules,

$$g'(t) = \frac{(2t - 3)^2(0) - (-7)(2)(2t - 3)(2)}{[(2t - 3)^2]^2} = \frac{28(2t - 3)}{(2t - 3)^4} = \frac{28}{(2t - 3)^3}$$ ∎

Example 5

Differentiate $$f(x) = 3x^2 \sqrt[3]{9 - 4x^2}$$

Solution: Write $f(x) = 3x^2(9 - 4x^2)^{1/3}$. Then by the Product and Power Rules, we obtain

$$f'(x) = 3x^2 \frac{d}{dx}[(9 - 4x^2)^{1/3}] + (9 - 4x^2)^{1/3} \frac{d}{dx}[3x^2]$$

$$= 3x^2[\tfrac{1}{3}(9 - 4x^2)^{-2/3}(-8x)] + (9 - 4x^2)^{1/3}(6x)$$

$$= -8x^3(9 - 4x^2)^{-2/3} + 6x(9 - 4x^2)^{1/3}$$

(Note the use of the Power Rule in the first term of the derivative.) Factoring out the *least* powers of x and $(9 - 4x^2)$, we have

$$f'(x) = x(9 - 4x^2)^{-2/3}[-8x^2(1) + 6(9 - 4x^2)]$$

$$= \frac{x(-8x^2 + 54 - 24x^2)}{(9 - 4x^2)^{2/3}} = \frac{x(54 - 32x^2)}{(9 - 4x^2)^{2/3}}$$

Alternative Method of Simplifying: Rather than factoring, you may prefer to remove negative exponents by rationalizing, as follows:

$$f'(x) = -8x^3(9 - 4x^2)^{-2/3} + 6x(9 - 4x^2)^{1/3}$$

$$= [-8x^3(9 - 4x^2)^{-2/3} + 6x(9 - 4x^2)^{1/3}]\frac{(9 - 4x^2)^{2/3}}{(9 - 4x^2)^{2/3}}$$

$$= \frac{-8x^3(1) + 6x(9 - 4x^2)}{(9 - 4x^2)^{2/3}} = \frac{54x - 32x^3}{(9 - 4x^2)^{2/3}} \quad \blacksquare$$

In Example 5 note that we subtract exponents when factoring. Thus when $(9 - 4x^2)^{-2/3}$ is factored out of $(9 - 4x^2)^{1/3}$, the *remaining* factor has an exponent of $(1/3) - (-2/3) = 1$. The next example further demonstrates this principle.

Example 6

Differentiate $\qquad y = \dfrac{(2x - 3)^3}{\sqrt{4x - 9}}$

Solution: Using the Quotient and Power Rules, we have

$$\frac{dy}{dx} = \frac{(4x - 9)^{1/2}(3)(2x - 3)^2(2) - (2x - 3)^3(\frac{1}{2})(4x - 9)^{-1/2}(4)}{(4x - 9)}$$

Factoring out the least powers of $(4x - 9)$ and $(2x - 3)$, we have

$$\frac{dy}{dx} = \frac{(4x - 9)^{-1/2}(2x - 3)^2[6(4x - 9)^1(2x - 3)^0 - 2(2x - 3)^1(4x - 9)^0]}{(4x - 9)}$$

$$= \frac{(4x - 9)^{-1/2}(2x - 3)^2[6(4x - 9) - 2(2x - 3)]}{(4x - 9)}$$

$$= \frac{(2x - 3)^2(24x - 54 - 4x + 6)}{(4x - 9)^{3/2}} = \frac{(2x - 3)^2(20x - 48)}{(4x - 9)^{3/2}} \quad \blacksquare$$

After studying this chapter, you may be asking the question, "Do I have to simplify my derivatives?" The answer is, "Yes, if you expect to use them." As you will quickly see in the next chapter, most applications of the derivative require the derivative to be written in simplified form. Furthermore, if you are asking the question because you have discovered that your algebraic muscles are weak, we encourage you to strengthen them through exercise.

Example 7

Find the derivative of $\qquad y = \left(\dfrac{3x - 1}{x^2 + 3}\right)^2$

First Solution: By using the Power Rule, we have

$$\frac{dy}{dx} = 2\left(\frac{3x - 1}{x^2 + 3}\right)\frac{d}{dx}\left[\frac{3x - 1}{x^2 + 3}\right]$$

$$= \left[\frac{2(3x-1)}{x^2+3}\right]\left[\frac{(x^2+3)(3)-(3x-1)(2x)}{(x^2+3)^2}\right]$$

$$= \frac{2(3x-1)(3x^2+9-6x^2+2x)}{(x^2+3)^3} = \frac{2(3x-1)(-3x^2+2x+9)}{(x^2+3)^3}$$

Second Solution: By rewriting and using the Quotient Rule, we have

$$y = \frac{(3x-1)^2}{(x^2+3)^2}$$

$$\frac{dy}{dx} = \frac{(x^2+3)^2(2)(3x-1)(3)-(3x-1)^2(2)(x^2+3)(2x)}{(x^2+3)^4}$$

$$= \frac{(x^2+3)(3x-1)[(x^2+3)(2)(3)-(3x-1)(2)(2x)]}{(x^2+3)^4}$$

$$= \frac{(3x-1)[6x^2+18-12x^2+4x]}{(x^2+3)^3}$$

$$= \frac{2(3x-1)(-3x^2+2x+9)}{(x^2+3)^3} \qquad\blacksquare$$

There is one more frequently encountered function for which we have not given a differentiation rule, namely, the absolute value function.

Absolute Value Rule

If $y = |u|$, where u is a differentiable function of x, then

$$\frac{d}{dx}[|u|] = \frac{u}{|u|}\frac{du}{dx}$$

wherever $u(x) \neq 0$.

Proof: This rule is actually just a special case of the Power Rule, since

$$|u| = \sqrt{u^2}$$

Thus $\qquad \dfrac{d}{dx}[|u|] = \dfrac{d}{dx}[\sqrt{u^2}] = \dfrac{d}{dx}[(u^2)^{1/2}]$

$$= \left(\frac{1}{2}\right)(u^2)^{-1/2}(2u)\frac{du}{dx} = \frac{u}{\sqrt{u^2}}\frac{du}{dx} = \frac{u}{|u|}\frac{du}{dx} \qquad\blacksquare$$

Recall that the absolute value of a function created some difficulties with respect to the existence of its derivative. The graph of the absolute value of a function frequently had "sharp turns" (see Figure 3.15) at which the derivative did not exist. However, except at such points, the rule above can be used to determine the derivative of the absolute value of a function. Note that

$$\frac{u}{|u|} = \begin{cases} +1, & \text{for } u > 0 \\ -1, & \text{for } u < 0 \end{cases}$$

Example 8

Differentiate $\qquad\qquad f(x) = |4 - x^2|$

Solution: By the Absolute Value Rule,

$$f'(x) = \frac{4 - x^2}{|4 - x^2|}(-2x)$$

Therefore,

$$f'(x) = (1)(-2x) = -2x, \qquad \text{when} \qquad 4 - x^2 > 0$$

and $\qquad f'(x) = (-1)(-2x) = 2x, \qquad \text{when} \qquad 4 - x^2 < 0$

Furthermore, the derivative of f does not exist at $x = \pm 2$ (see Figure 3.15). ∎

We have now discussed all the rules needed to differentiate any algebraic function. For convenience we list these rules here. We strongly urge you to memorize these rules, as this will greatly increase your efficiency in differentiating.

Differentiation Rules

(u and v are functions of x)

Constant Rule: $\qquad\qquad \dfrac{d}{dx}[c] = 0$

Scalar Multiple Rule: $\qquad \dfrac{d}{dx}[cu] = c\dfrac{du}{dx}$

Sum Rule: $\qquad\qquad \dfrac{d}{dx}[u \pm v] = \dfrac{du}{dx} \pm \dfrac{dv}{dx}$

Product Rule: $\qquad\qquad \dfrac{d}{dx}[uv] = u\dfrac{dv}{dx} + v\dfrac{du}{dx}$

Quotient Rule: $\qquad\qquad \dfrac{d}{dx}\left[\dfrac{u}{v}\right] = \dfrac{v\dfrac{du}{dx} - u\dfrac{dv}{dx}}{v^2}$

Chain Rule (y is a function of u):

$$\dfrac{d}{dx}[y] = \dfrac{dy}{du}\dfrac{du}{dx}$$

Power Rule: $\qquad\qquad \dfrac{d}{dx}[u^n] = nu^{n-1}\dfrac{du}{dx}$

$$\dfrac{d}{dx}[x^n] = nx^{n-1}$$

Absolute Value Rule: $\qquad \dfrac{d}{dx}[|u|] = \dfrac{u}{|u|}\dfrac{du}{dx}$

Section Exercises (3.7)

In Exercises 1–54, find the first derivative.

1. $y = (2x - 7)^3$

2. $y = (3x^2 + 1)^4$

S 3. $f(x) = 2(x^2 - 1)^3$

4. $g(x) = 3(9x - 4)^4$

5. $y = \dfrac{1}{x - 2}$

6. $s(t) = \dfrac{1}{t^2 + 3t - 1}$

S 7. $f(t) = \left(\dfrac{1}{t - 3}\right)^2$

8. $y = \dfrac{-4}{(t + 2)^2}$

9. $f(x) = \dfrac{3}{x^3 - 4}$

10. $f(x) = \dfrac{1}{(x^2 - 3x)^2}$

S 11. $f(x) = x^2(x - 2)^4$

12. $f(x) = x(3x - 9)^3$

13. $y = (x - 2)(x + 3)^3$

14. $g(t) = (3t^2 - 2)^2(t + 1)^3$

15. $f(t) = \sqrt{t + 1}$

16. $g(x) = \sqrt{2x + 3}$

17. $s(t) = \sqrt{t^2 + 2t - 1}$

18. $y = \sqrt[3]{3x^3 + 4x}$

S 19. $y = \sqrt[3]{9x^2 + 4}$

20. $g(x) = \sqrt{x^2 - 2x + 1}$

21. $y = 2\sqrt{x^2 + 4}$

22. $f(x) = -3\sqrt[4]{9x + 2}$

23. $f(x) = (x^2 - 9)^{2/3}$

24. $f(t) = (9t + 2)^{2/3}$

S 25. $y = \dfrac{1}{\sqrt{x + 2}}$

26. $g(t) = \sqrt{\dfrac{1}{t^2 - 2}}$

27. $g(x) = \dfrac{3}{\sqrt[3]{x^3 - 1}}$

28. $s(x) = \dfrac{1}{\sqrt{x^2 - 3x + 4}}$

29. $y = \dfrac{-1}{\sqrt{x + 1}}$

30. $y = \dfrac{1}{2\sqrt{t - 3}}$

31. $y = |x|$

32. $f(x) = |-2x|$

S 33. $y = |x^2 - 1|$

34. $g(t) = |t^2 - 2t - 3|$

35. $y = \dfrac{1}{\sqrt{x} + \sqrt[3]{x}}$

36. $y = \dfrac{1}{\sqrt{t} - \sqrt[3]{t}}$

37. $g(x) = \dfrac{2x}{\sqrt{x + 1}}$

38. $f(x) = \dfrac{-2x^2}{x - 1}$

39. $y = \dfrac{\sqrt{x} + 1}{x^2 + 1}$

40. $f(x) = \dfrac{x + 1}{2x - 3}$

41. $f(t) = \dfrac{3t + 2}{t - 1}$

42. $y = \sqrt{\dfrac{2x}{x + 1}}$

S 43. $g(t) = \dfrac{3t^2}{\sqrt{t^2 + 2t - 1}}$

44. $f(x) = \sqrt{x}(x - 2)^2$

45. $f(t) = \sqrt{t + 1}\,\sqrt[3]{t + 1}$

46. $y = \sqrt{3x}(x + 2)^3$

47. $y = \sqrt{\dfrac{x + 1}{x}}$

48. $y = (t^2 - 9)\sqrt{t + 2}$

49. $s(t) = \dfrac{-2(2 - t)\sqrt{1 + t}}{3}$

50. $g(x) = \sqrt{x - 1} + \sqrt{x + 1}$

S 51. $s(t) = \sqrt{\dfrac{t^2 - 4}{4t}}$

52. $y = \dfrac{\sqrt{x^2 + 1}}{x}$

53. $f(x) = \dfrac{x + 2}{\sqrt{x^2 + x + 1}}$

54. $s(t) = \dfrac{-\sqrt{t^2 + 4t - 7}}{2t}$

In Exercises 55–58, find the equation of the tangent line to the curve $y = f(x)$ at the given point.

55. $f(x) = \sqrt{3x^2 - 2}$; (3, 5)

56. $f(x) = x\sqrt{x^2 + 5}$; (2, 6)

S 57. $f(x) = \left(x - \dfrac{1}{x}\right)^{3/2}$; (1, 0)

58. $f(x) = \dfrac{\sqrt[3]{6x - 4}}{x}$; (2, 1)

In Exercises 59–64, find the second derivative of the function.

S 59. $f(x) = 2(x^2 - 1)^3$

60. $f(x) = \dfrac{1}{x - 2}$

61. $f(x) = |x^2 - 1|$ (Hint: See Exercise 69.)

62. $f(x) = \sqrt{2x}(x + 2)^2$

S 63. $f(t) = \dfrac{\sqrt{t^2 + 1}}{t}$

64. $g(x) = \sqrt[3]{9x^2 + 4}$

65. The emergent velocity v of a liquid from a hole in the bottom of a tank is given by $v = \sqrt{2gh}$, where g is the acceleration due to gravity (32 ft/s²) and h is the depth of the liquid in the tank. Find the rate of change of v with respect to h when:
   (a) $h = 9$ (b) $h = 4$

66. The speed S of blood that is r centimeters from the center of an artery is given by $S = C(R^2 - r^2)$, where C is a constant, R is the radius of the artery, and S is measured in centimeters per second. Suppose a drug is administered and the artery begins dilating at the rate of dR/dt. Find the rate at which S changes with respect to t for $C = 1.76 \times 10^5$, $R = 1.2 \times 10^{-2}$, and $dR/dt = 10^{-5}$.

(Hint: Hold r constant and use

$$\frac{dS}{dt} = \frac{dS}{dR}\frac{dR}{dt}$$

the Chain Rule.)

S **67.** Given $y = u^2 - 2u - 3$ and $u = x + (1/x)$, find dy/dx by (a) using the Chain Rule and (b) forming the composite function and then differentiating. Show that the results are equivalent.

68. Given $y = 1/(x + 3)$ and $x = 64t - 16t^2$, find dy/dt by (a) using the Chain Rule and (b) forming the composite function and then differentiating. Show that the results are equivalent.

69. Prove that

$$\frac{d^2}{dx^2}[|u|] = \frac{u}{|u|}\frac{d^2u}{dx^2}$$

where $u \neq 0$.

For Review and Class Discussion

True or False

1. ___ If $y = (1 - x)^{1/2}$, then $y' = \frac{1}{2}(1 - x)^{-1/2}$.

2. ___ If $y = (x^2 + 1)^{10}$, then y' is an eighteenth-degree polynomial.

3. ___ If y is a differentiable function of u and u is a differentiable function of x, then y is a differentiable function of x.

4. ___ If y is a function of u, u is a function of x, and d^2y/dx^2 exists, then

$$\frac{d^2y}{dx^2} = \frac{dy}{du}\frac{d^2u}{dx^2} + \left(\frac{du}{dx}\right)^2\frac{d^2y}{du^2}$$

5. ___ If y is a differentiable function of u, u is a differentiable function of v, and v is a differentiable function of x, then

$$\frac{dy}{dx} = \frac{dy}{du}\frac{du}{dv}\frac{dv}{dx}$$

3.8 **Implicit Differentiation**

Purpose
- To distinguish between explicit and implicit forms of an equation.
- To demonstrate the technique of implicit differentiation and apply it to equations in implicit form.

The term function was used by the first analysts to denote the powers of a given quantity. Since then the meaning of this term has been extended to any quantity formed in any manner from any other quantity. Leibniz and the Bernoullis were the first to use it in this general sense, which nowadays is the accepted one. —Joseph Lagrange (1736–1813)

Up to this point our equations involving two variables were generally expressed in the **explicit form**

$$y = f(x)$$

That is, one of the two variables was explicitly given in terms of the other. For example, the equations

$$y = 3x - 5, \qquad s = -16t^2 + 20t, \qquad u = 3w - w^2$$

are all written in explicit form, and we say that y, s, and u are functions of x, t, and w, respectively.

However, many relationships are not given explicitly and are only implied by a given equation. For instance, the equations

1. $2x - y = 4$
2. $xy = 1$
3. $x^2 + y^2 = 1$
4. $x^3 + xy + 2y^3 = 0$

are not given in explicit form. We say such equations are given in **implicit form.** Of course, it may be possible to change the form of an equation from implicit to explicit. For example, solving for y in Equation 1 yields

$$y = 2x - 4$$

In Equation 2 we have
$$y = \frac{1}{x}$$

and from Equation 3 we can define the functions
$$y = \sqrt{1 - x^2} \quad \text{or} \quad y = -\sqrt{1 - x^2}$$

Occasionally we encounter equations that are either very difficult or impossible to convert to explicit form. For instance, we would be hard pressed to solve for y in Equation 4.

Fortunately it is not always necessary to rewrite an equation in explicit form in order to differentiate it. We can use a procedure called **implicit differentiation.** In this procedure we simply differentiate separately each term of the given equation. We illustrate this process in the next five examples. We assume that each of the given equations determines at least one differentiable function $y = f(x)$ on some interval about a fixed x value.

Example 1

Find dy/dx given that $x^2 - 2y^3 + 4x = 2$.

Solution: By differentiating each term with respect to x, we have

$$\frac{d}{dx}[x^2 - 2y^3 + 4x] = \frac{d}{dx}[2]$$

$$2x - 6y^2\left(\frac{dy}{dx}\right) + 4 = 0$$

(Note the use of the Chain Rule on the $2y^3$ term.) Now solving for dy/dx we get

$$-6y^2\left(\frac{dy}{dx}\right) = -2x - 4$$

$$\frac{dy}{dx} = \frac{-2(x + 2)}{-6y^2}$$

Therefore,
$$\frac{dy}{dx} = \frac{x+2}{3y^2}$$ ∎

The fact that the derivative in Example 1 involves the two variables x and y is generally of no disadvantage. For instance, in using the derivative to determine the slope of a curve at a point, we are usually given both coordinates, and if not, the remaining one can probably be calculated with a reasonable amount of effort.

Example 2

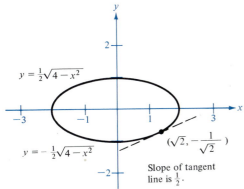

$y = \frac{1}{2}\sqrt{4-x^2}$

$y = -\frac{1}{2}\sqrt{4-x^2}$

$(\sqrt{2}, -\frac{1}{\sqrt{2}})$

Slope of tangent line is $\frac{1}{2}$.

FIGURE 3.17

Determine the slope of the tangent line to the graph of $x^2 + 4y^2 = 4$ at the point $(\sqrt{2}, -1/\sqrt{2})$. (See Figure 3.17.)

Solution: Implicitly differentiating the equation $x^2 + 4y^2 = 4$ with respect to x yields

$$2x + 8y\left(\frac{dy}{dx}\right) = 0$$

and
$$\frac{dy}{dx} = \frac{-2x}{8y} = \frac{-x}{4y}$$

Therefore, at $(\sqrt{2}, -1/\sqrt{2})$ the slope is

$$\frac{dy}{dx} = \frac{-\sqrt{2}}{-4/\sqrt{2}} = \frac{1}{2}$$

Alternative Solution: Explicitly solving for y in terms of x, we obtain the equations

$$y = \tfrac{1}{2}\sqrt{4-x^2} \quad \text{and} \quad y = -\tfrac{1}{2}\sqrt{4-x^2}$$

whose graphs are, respectively, the top and bottom halves of the ellipse in Figure 3.17. The point $(\sqrt{2}, -1/\sqrt{2})$ satisfies the equation

$$y = -\tfrac{1}{2}\sqrt{4-x^2}$$

Therefore, the slope is determined by

$$\frac{dy}{dx} = \frac{-1}{4}(4-x^2)^{-1/2}(-2x) = \frac{x}{2\sqrt{4-x^2}}$$

At the point $(\sqrt{2}, -1/\sqrt{2})$ we have

$$\frac{dy}{dx} = \frac{\sqrt{2}}{2\sqrt{4-2}} = \frac{1}{2}$$

as was obtained by implicit differentiation. ∎

To implicitly differentiate a term involving both x and y, we use the Product or Quotient Rules, as illustrated in the next two examples.

Example 3

Given $x^3 + xy + y^2 = 4 - x$, use implicit differentiation to find dy/dx. Evaluate dy/dx at $(0, 2)$ and when $x = 1$.

Solution: Given $x^3 + xy + y^2 = 4 - x$ and differentiating each term, we get

$$3x^2 + \left[x \left(\frac{dy}{dx} \right) + y \left(\frac{dx}{dx} \right) \right] + 2y \left(\frac{dy}{dx} \right) = 0 - 1$$

(Note the use of the Product Rule for the xy term.) To solve for dy/dx, we use the fact that $dx/dx = 1$ and group terms to obtain

$$x \left(\frac{dy}{dx} \right) + 2y \left(\frac{dy}{dx} \right) = -1 - 3x^2 - y$$

$$(x + 2y) \frac{dy}{dx} = -1 - 3x^2 - y$$

$$\frac{dy}{dx} = \frac{-1 - 3x^2 - y}{x + 2y}$$

At the point $(0, 2)$

$$\frac{dy}{dx} = \frac{-1 - 2}{4} = \frac{-3}{4}$$

When $x = 1$, we determine y by solving the equation

$$1^3 + 1(y) + y^2 = 4 - 1$$
$$y^2 + y - 2 = 0$$
$$(y + 2)(y - 1) = 0$$
$$y = 1 \quad \text{or} \quad y = -2$$

Therefore, we consider the slope at two points (Figure 3.18).

At $(1, 1)$

$$\frac{dy}{dx} = \frac{-1 - 3 - 1}{1 + 2} = \frac{-5}{3}$$

At $(1, -2)$

$$\frac{dy}{dx} = \frac{-1 - 3 + 2}{1 - 4} = \frac{2}{3}$$

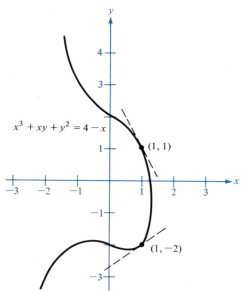

$x^3 + xy + y^2 = 4 - x$

$(1, 1)$

$(1, -2)$

FIGURE 3.18

Example 4

Find the slope of the curve $3x^2 - 2xy + xy^3 = 7$ at $(1, 2)$.

Solution: Differentiating implicitly and using y' for dy/dx, we obtain

$$6x - [2xy' + y(2)] + [x(3y^2)y' + y^3] = 0$$

$$(-2x + 3xy^2)y' = 2y - 6x - y^3$$

$$y' = \frac{2y - 6x - y^3}{3xy^2 - 2x}$$

At the point $(1, 2)$

$$y' = \frac{4 - 6 - 8}{12 - 2} = \frac{-10}{10} = -1$$

■

Example 5

Given $x^2 + y^2 = 25$, find y''.

Solution: Differentiating each term with respect to x, we obtain

$$2x + 2yy' = 0$$
$$2yy' = -2x$$
$$y' = \frac{-2x}{2y} = -\frac{x}{y}$$

Differentiating a second time with respect to x yields

$$y'' = -\frac{(y)(1) - (x)(y')}{y^2}$$

Finally, substituting $-x/y$ for y', we have

$$y'' = -\frac{y - (x)(-x/y)}{y^2} = -\frac{y^2 + x^2}{y^3} = -\frac{25}{y^3}$$

■

In the previous five examples we assumed that the given equation defined at least one differentiable function of x, $y = f(x)$, near a fixed x. In each case we could have assumed that the equation defined at least one differentiable function of y, say $x = g(y)$. We could then have determined dx/dy by implicit differentiation.

To demonstrate differentiation with respect to y rather than x, let us reconsider the equation of Example 1. Given $x^2 - 2y^3 + 4x = 2$, we can differentiate both sides *with respect to* y to obtain

$$2x\frac{dx}{dy} - 6y^2 + 4\frac{dx}{dy} = 0$$

(Note that the derivatives, *with respect to* y, of the terms x^2 and $4x$ each contain the factor dx/dy.) Solving for dx/dy, we have

$$(2x + 4)\frac{dx}{dy} = 6y^2$$

$$\frac{dx}{dy} = \frac{6y^2}{2x + 4} = \frac{3y^2}{x + 2}$$

Note that this result is the reciprocal of that obtained in Example 1. This suggests the general result that

$$\frac{dy}{dx} = \frac{1}{\dfrac{dx}{dy}}$$

provided both derivatives exist.

In Section 3.3 we listed the Power Rule as

$$\frac{d}{dx}[x^n] = nx^{n-1}$$

Although we claimed that this rule holds when n is any real number, we have given proofs only for the cases when n is a positive integer (Section 3.3) or a negative integer (Section 3.6). Using implicit differentiation we can now demonstrate the validity of the Power Rule when n is any rational number.

Proof: Let $y = x^n$, where n is the rational number p/q (p and q are integers). By raising both sides of this equation to the qth power, we have

$$y = x^{p/q}$$
$$y^q = x^p$$

Since p and q are both integers, we implicitly differentiate both sides of the equation with respect to x to obtain

$$qy^{q-1}\frac{dy}{dx} = px^{p-1}$$

Solving for dy/dx yields

$$\frac{dy}{dx} = \frac{px^{p-1}}{qy^{q-1}}$$

Finally, substituting $x^{p/q}$ for y, we have

$$\frac{dy}{dx} = \frac{px^{p-1}}{q(x^{p/q})^{q-1}} = \left(\frac{p}{q}\right)\left[\frac{x^{p-1}}{x^{p(q-1)/q}}\right]$$

$$= \left(\frac{p}{q}\right)x^{(p-1)-[p(q-1)/q]} = \left(\frac{p}{q}\right)x^{(pq-q-pq+p)/q}$$

$$= \left(\frac{p}{q}\right)x^{(p-q)/q} = \left(\frac{p}{q}\right)x^{(p/q)-1}$$

∎

As a final comment we note that we have not shown that the Power Rule is valid for irrational numbers. For a proof of this see a text on advanced calculus.

Section Exercises (3.8)

In Exercises 1–16, find dy/dx by implicit differentiation and evaluate the derivative at the indicated point.

1. $x^2 + y^2 = 16$; $(3, \sqrt{7})$
2. $x^2 - y^2 = 16$; $(4, 0)$
S 3. $xy = 4$; $(-4, -1)$
4. $x^2 - y^3 = 0$; $(-1, 1)$
5. $x^{1/2} + y^{1/2} = 9$; $(16, 25)$
6. $x^3 + y^3 = 8$; $(0, 2)$
S 7. $x^3 - xy + y^2 = 4$; $(0, -2)$
8. $x^2y + y^2x = -2$; $(2, -1)$
9. $y^2 = \dfrac{x^2 - 9}{x^2 + 9}$; $(3, 0)$
10. $(x + y)^3 = x^3 + y^3$; $(-1, 1)$
11. $x^3y^3 - y = x$; $(0, 0)$
12. $\sqrt{xy} = x - 2y$; $(4, 1)$
13. $x^{2/3} + y^{2/3} = 5$; $(8, 1)$
S 14. $(x - y^2)(x + xy) = 4$; $(2, 1)$
15. $x^3 - 2x^2y + 3xy^2 = 38$; $(2, 3)$
16. $x^3 + y^3 = 2xy$; $(1, 1)$

In Exercises 17–20, (a) find two explicit functions defined by the given equation and state their domains; (b) find the derivatives of the functions obtained in part (a); (c) differentiate the given equation implicitly and verify that the result is the same as the result of part (b); (d) sketch the graph of the equation and label the parts of the graph given by the functions of part (a).

17. $x^2 + y^2 = 16$
18. $x^2 + y^2 - 4x + 6y + 9 = 0$
S 19. $9x^2 + 16y^2 = 144$
20. $4y^2 - x^2 = 4$

In Exercises 21–26, find d^2y/dx^2 in terms of x and y.

21. $x^2 + xy = 5$
22. $x^2y^2 - 2x = 3$
S 23. $x^2 - y^2 = 16$
24. $1 - xy = x - y$
25. $y^2 = x^3$
26. $y^2 = 4x$

S 27. Show that the **normal line** (the line perpendicular to the tangent line to a curve) at any point on the circle $x^2 + y^2 = r^2$ passes through the origin.
28. Find the equations of the tangent line and normal line to the circle $x^2 + y^2 = 25$ at these points:
(a) $(4, 3)$ (b) $(-3, 4)$
29. Find equations of the tangent line and normal line to the circle $x^2 + y^2 = 9$ at $(0, 3)$ and at $(2, \sqrt{5})$.
30. Use implicit differentiation to find the vertical and horizontal tangents to the curve whose equation is $25x^2 + 16y^2 + 200x - 160y + 400 = 0$. Sketch the graph of the ellipse.
S 31. Show that the graphs of the equations $2x^2 + y^2 = 6$ and $y^2 = 4x$ are **orthogonal** (the curves intersect at right angles). Sketch the graph of each equation.
32. Show that the graphs of $y^2 = x^3$ and $2x^2 + 3y^2 = 5$ are orthogonal.
33. Show that the graphs of $x^3 = 3(y - 1)$ and $x(3y - 29) = 3$ are orthogonal.
34. Two circles of radius 4 are tangent to the graph of $y^2 = 4x$ at the point $(1, 2)$. Find equations for these two circles.
35. Show that

$$\frac{dy}{dx} = -\left(\frac{x - h}{y - k}\right) \quad \text{and} \quad \frac{d^2y}{dx^2} = \frac{\dfrac{dy}{dx}\left[1 + \left(\dfrac{dy}{dx}\right)^2\right]}{x - h}$$

for the equation of the circle $(x - h)^2 + (y - k)^2 = r^2$.

Miscellaneous Exercises (Ch. 3)

In Exercises 1–18, find the derivative of the given function.

1. $f(x) = x^3 - 3x^2$
2. $f(x) = \dfrac{2x^3 - 1}{x^2}$
3. $f(x) = x^{1/2} - x^{-1/2}$
4. $f(x) = \dfrac{x + 1}{x - 1}$
5. $g(t) = \dfrac{2}{3t^2}$
6. $h(x) = \dfrac{2}{(3x)^2}$
7. $f(x) = \sqrt{x^3 + 1}$
8. $f(x) = \sqrt[3]{x^2 - 1}$
S 9. $f(x) = (3x^2 + 7)(x^2 - 2x + 3)$
10. $f(x) = \left(x^2 + \dfrac{1}{x}\right)^5$
11. $f(s) = (s^2 - 1)^{5/2}(s^3 + 5)^{5/3}$
12. $h(\theta) = \dfrac{\theta}{(1 - \theta)^3}$
13. $g(x) = \sqrt{x}\sqrt{x^2 + 1}$
S 14. $f(x) = \dfrac{6x - 5}{x^2 + 1}$
15. $f(x) = \dfrac{x^2 + x - 1}{x^2 - 1}$
16. $f(t) = t^2(t - 1)^5(t + 2)^3$
17. $f(x) = |x^3|$
18. $f(x) = |x^2 - 2x - 5|$

In Exercises 19–26, find the second derivative of the given function.

S **19.** $f(x) = \sqrt{x^2 + 9}$

20. $h(x) = x\sqrt{x^2 - 1}$

21. $f(t) = \dfrac{t}{(1 - t)^2}$

22. $h(x) = x^2 + \dfrac{3}{x}$

S **23.** $g(x) = \dfrac{6x - 5}{x^2 + 1}$

24. $f(x) = (3x^2 + 7)(x^2 - 2x + 3)$

25. $f(x) = |x^5|$

26. $f(x) = |x^3 - 1|$

In Exercises 27–32, use implicit differentiation to find dy/dx.

S **27.** $x^2 + 3xy + y^3 = 10$

28. $x^2 + 9y^2 - 4x + 3y - 7 = 0$

29. $y\sqrt{x} - x\sqrt{y} = 16$

30. $y^2 + x^2 - 6y - 2x - 5 = 0$

31. $y^2 - x^2 = 25$

32. $y^2 = (x - y)(x^2 + y)$

In Exercises 33–38, find the equation of the tangent line and the normal line to the given curve at the given point.

33. $y = (x + 3)^3$; $(-2, 1)$ **34.** $y = (x - 2)^2$; $(2, 0)$

35. $x^2 + y^2 = 20$; $(2, 4)$ **36.** $x^2 - y^2 = 16$; $(5, 3)$

S **37.** $y = \sqrt[3]{(x - 2)^2}$; $(3, 1)$ **38.** $y = \dfrac{2x}{1 - x^2}$; $(0, 0)$

S **39.** What is the smallest initial velocity that is required to throw a stone to the top of a 49-ft silo?

40. A bomb is dropped from a plane at an altitude of 14,400 ft. How long will it take to reach the ground? (Even though it will not be a vertical drop due to the motion of the plane, the time will be the same as that for a vertical fall.) Suppose the plane is moving at 600 mi/h. How far will the bomb move horizontally after it is released from the plane?

41. Let $f(x) = \frac{1}{3}x^3 + x^2 - x - 1$. Find the points on the graph of f at which the slope is:

(a) -1 (b) 2

(c) 0

In Exercises 42–45, find the derivative of the given function by the four-step process.

42. $f(x) = \dfrac{1}{x^2}$ **43.** $f(x) = \dfrac{x + 1}{x - 1}$

44. $f(x) = \sqrt{x + 2}$ S **45.** $f(x) = \dfrac{1}{\sqrt{x}}$

46. Show that

$$y^{(n)} = \frac{(-1)^n n!}{x^{n+1}} \quad \text{if} \quad y = \frac{1}{x}$$

[Note: $n! = n(n - 1)(n - 2) \cdots 3 \cdot 2 \cdot 1$.]

47. Sketch the graph of $f(x) = 4 - |x - 2|$.

(a) Is f continuous at $x = 2$?

(b) Is f differentiable at $x = 2$? Why or why not?

48. Sketch the graph of

$$f(x) = \begin{cases} x^2 + 4x + 2, & \text{if } x < -2 \\ 1 - 4x - x^2, & \text{if } x \geq -2 \end{cases}$$

(a) Is f continuous at $x = -2$?

(b) Is f differentiable at $x = -2$? Why or why not?

S **49.** The path of a projectile, thrown at an angle of 45° with the ground, is given by $y = x - (32/v_0^2)(x^2)$, where the initial velocity is v_0 feet per second. Show that doubling the initial velocity of the projectile multiplies both the maximum height and the range by a factor of 4.

50. Use the equation given in Exercise 49 to find the maximum height and range of a projectile thrown with an initial velocity of 70 ft/s.

51. The geometric mean of x and $x + n$ is $g = \sqrt{x(x + n)}$ and the arithmetic mean is $a = [x + (x + n)]/2$. Show that

$$\frac{dg}{dx} = \frac{a}{g}$$

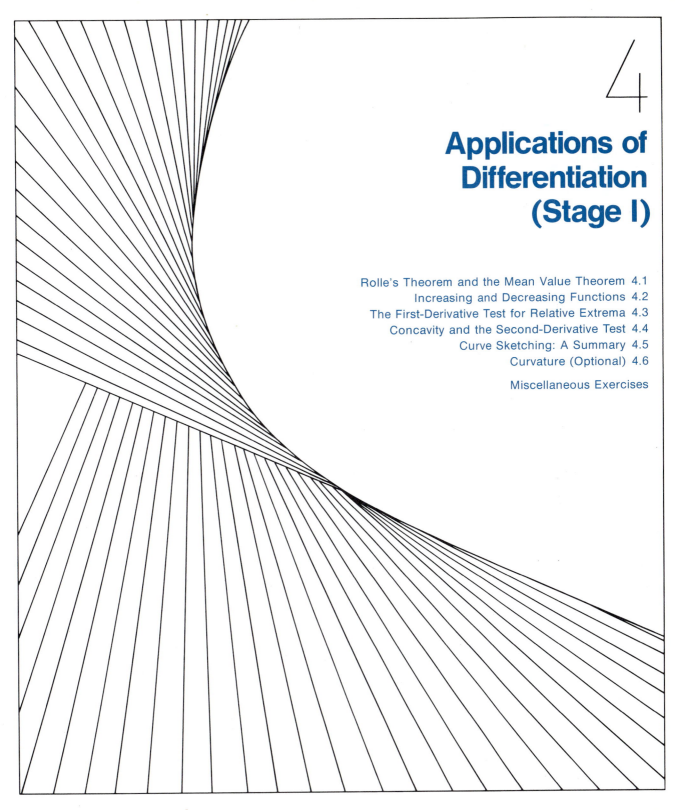

4

Applications of Differentiation (Stage I)

4.1 Rolle's Theorem and the Mean Value Theorem

Purpose
- To introduce Rolle's Theorem.
- To use Rolle's Theorem to prove the Mean Value Theorem.

An abstract analysis which is accepted without any synthetic examination of the question under discussion is likely to surprise rather than enlighten us. —*Daniel Bernoulli (1700–1782)*

In calculus a great deal of effort is devoted to determining the characteristics of a given function defined on a certain interval. Characteristics such as these are examined: Where is the function increasing? Where is it decreasing? Does it have a maximum value, a minimum value? Where is its tangent horizontal? There are many theorems related to the identification of these characteristics, and their proofs frequently depend on two important results—Rolle's Theorem and the Mean Value Theorem.

In this section we state and prove Rolle's Theorem and the Mean Value Theorem as a necessary prelude to the remaining theorems in this chapter. Because of the rather theoretical nature of these two theorems, you may wish to scan Section 4.2 before completing this section to see how these theorems are used in the proofs in that section.

We begin with a discussion leading up to the statement of Rolle's Theorem. Consider the following situation. Suppose the graph of a function crosses the *x*-axis twice; does it necessarily have at least one horizontal tangent? At first glance we might be convinced that it does, especially if the function is continuous, as are those in Figure 4.1. However, further investigation will reveal the continuity is not sufficient to guarantee the existence of a horizontal tangent. For example, Figure 4.2 shows a continuous function whose graph crosses the *x*-axis twice and yet it has no horizontal tangent.

The catch in Figure 4.2 is that, although the function is continuous, it is *not* differentiable at all points in the interval (a, b). By requiring the function to be differentiable, we can guarantee the existence of a horizontal tangent, and we state this result as Rolle's Theorem.

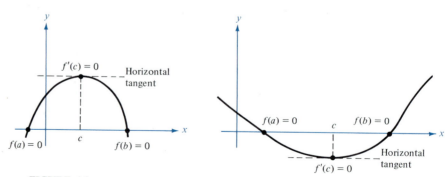

FIGURE 4.1

| **THEOREM 4.1** | If f is continuous on $[a, b]$, differentiable on (a, b), and $f(a) = f(b) = 0$, then there is at least one number c in (a, b) such that $f'(c) = 0$. |
| (*Rolle's Theorem*) | |

Proof:

Case 1: If $f(x) = 0$ for all x in (a, b), the result easily follows, since the function is constant in the interval and $f'(x) = 0$ for *all* x in (a, b).

Case 2: If $f(x) > 0$ for some x in (a, b), then since f is continuous in $[a, b]$, by the Extreme Value Theorem $f(x)$ must achieve a maximum in the interval $[a, b]$. Say $f(c)$ is such a maximum; then c is neither a nor b because $f(c) > 0$ and $f(a) = f(b) = 0$. Thus c lies in the interval (a, b). Now since f is differentiable at c, we know that

$$\lim_{x \to c^-} \frac{f(x) - f(c)}{x - c} = f'(c) = \lim_{x \to c^+} \frac{f(x) - f(c)}{x - c}$$

Because $f(c)$ is a maximum, for x sufficiently close to c, we have $f(x) - f(c) \leq 0$ and it follows that

$$\frac{f(x) - f(c)}{x - c} \geq 0$$

when x approaches c from the left. Similarly, when x approaches c from the right, we have

$$\frac{f(x) - f(c)}{x - c} \leq 0$$

Therefore, by the existence of $f'(c)$, we have

$$0 \leq \lim_{x \to c^-} \frac{f(x) - f(c)}{x - c} = \lim_{x \to c^+} \frac{f(x) - f(c)}{x - c} \leq 0$$

which implies that

$$f'(c) = \lim_{x \to c} \frac{f(x) - f(c)}{x - c} = 0$$

Case 3: If $f(x) < 0$, for some x in (a, b), an argument similar to that used in case 2 will work. ∎

We illustrate Rolle's Theorem in the following two examples.

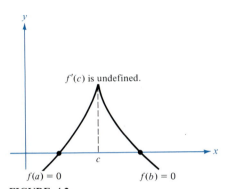

$f'(c)$ is undefined.

$f(a) = 0$ c $f(b) = 0$

FIGURE 4.2

Example 1

Find the two x-intercepts of the function $f(x) = x^2 - 3x + 2$ and show that $f'(x) = 0$ at some point between the two intercepts.

Solution: By setting $f(x)$ equal to zero, we have

$$x^2 - 3x + 2 = 0$$
$$(x - 1)(x - 2) = 0$$

Thus $f(1) = f(2) = 0$, and since f is differentiable everywhere, Rolle's Theorem guarantees the existence of some c in the interval $(1, 2)$ such that $f'(c) = 0$. To find c we solve the equation

$$f'(x) = 2x - 3 = 0$$
$$2x = 3$$
$$x = \tfrac{3}{2}$$

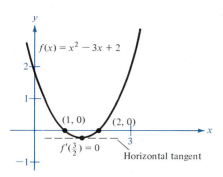

$f(x) = x^2 - 3x + 2$

$(1, 0)$ $(2, 0)$

$f'(\tfrac{3}{2}) = 0$

Horizontal tangent

FIGURE 4.3

Since $\tfrac{3}{2}$ is in the interval $(1, 2)$ and $f'(\tfrac{3}{2}) = 0$, we let $c = \tfrac{3}{2}$. (See Figure 4.3.) ∎

Note that Rolle's Theorem states that there must be *at least* one point between a and b where the derivative is zero. There may, of course, be more than one such point, as we see in the next example.

Example 2

Let $f(x) = x^4 - 2x^2 - 8$. Find all c in the interval $(-2, 2)$ such that $f'(c) = 0$.

Solution: Since $f(-2) = f(2) = 0$, and f is differentiable everywhere, by Rolle's Theorem there is at least one c in $(-2, 2)$ such that $f'(c) = 0$. Solving the equation

$$f'(x) = 4x^3 - 4x = 0$$
$$4x(x^2 - 1) = 0$$

we obtain

$$x = 0, 1, -1$$

Thus in the interval $(-2, 2)$ the values for c are $c = -1$, $c = 0$, or $c = 1$. (See Figure 4.4.) ∎

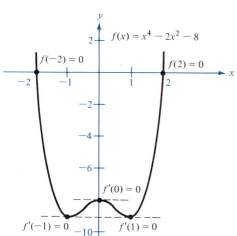

$f(x) = x^4 - 2x^2 - 8$

$f(-2) = 0$ $f(2) = 0$

$f'(0) = 0$

$f'(-1) = 0$ $f'(1) = 0$

FIGURE 4.4

Rolle's Theorem can be used to prove a more widely applicable result called the Mean Value Theorem. For differentiable functions this theorem guarantees the existence of a tangent line that is parallel to the secant line through the points $(a, f(a))$ and $(b, f(b))$, as shown in Figure 4.5.

THEOREM 4.2
(*Mean Value Theorem*)

If f is continuous on $[a, b]$ and differentiable on (a, b), then there exists a number c in (a, b) such that

$$f'(c) = \frac{f(b) - f(a)}{b - a}$$

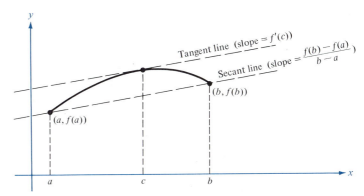

FIGURE 4.5

Proof: Refer to Figure 4.5. The equation for the secant line containing the points $(a, f(a))$ and $(b, f(b))$ is given by

$$y = \left[\frac{f(b) - f(a)}{b - a}\right](x - a) + f(a)$$

Let $g(x)$ be the difference between $f(x)$ and this secant line; then

$$g(x) = f(x) - \left[\frac{f(b) - f(a)}{b - a}\right](x - a) - f(a)$$

By letting $x = a$ and $x = b$ in the equation above, we see that

$$g(a) = 0 = g(b)$$

Furthermore, since f is differentiable, g is also differentiable, and we can apply Rolle's Theorem to the function g. Thus there exists a point c in (a, b) such that $g'(c) = 0$. This means that

$$0 = g'(c) = f'(c) - \frac{f(b) - f(a)}{b - a}$$

Therefore, there exists a point c in (a, b) such that

$$f'(c) = \frac{f(b) - f(a)}{b - a}$$ ∎

Example 3

Given $f(x) = 3 - (6/x)$, find all c in the interval $(2, 6)$ such that $f'(c) = [f(6) - f(2)]/(6 - 2)$.

Solution: First,

$$\frac{f(6) - f(2)}{6 - 2} = \frac{(3 - 1) - (3 - 3)}{4} = \frac{2}{4} = \frac{1}{2}$$

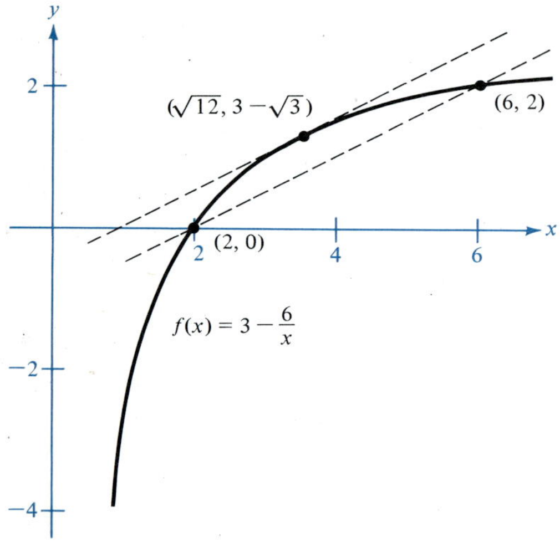

FIGURE 4.6

Since f satisfies the conditions of the Mean Value Theorem, there exists at least one c in (2, 6) such that $f'(c) = \frac{1}{2}$. Solving the equation

$$f'(x) = \frac{6}{x^2} = \frac{1}{2}$$

$$12 = x^2$$

we obtain

$$x = \pm\sqrt{12}$$

Finally, in the interval (2, 6) we choose $c = +\sqrt{12}$. (See Figure 4.6.) ■

Although on some occasions the Mean Value Theorem is used directly in problem solving, it is used primarily in the proofs of other theorems, as we will see in the remaining sections of this chapter. A useful alternative form of the Mean Value Theorem is this: If f is continuous on $[a, b]$ and differentiable on (a, b), then there exists a number c in (a, b) such that

$$f(b) = f(a) + (b - a)f'(c)$$

In working the following exercise set, keep in mind that (1) every polynomial function is differentiable on the entire real line, and (2) every rational function is differentiable at every point in its domain.

Section Exercises (4.1)

In Exercises 1–12, determine any intervals on which Rolle's Theorem can be applied. For each such interval find all values of c in the interval such that $f'(c) = 0$.

1. $f(x) = x^2 - 2x$
2. $f(x) = x^2 - 3x + 2$
[S] 3. $f(x) = x^3 - 6x^2 + 11x - 6$
4. $f(x) = x(x^2 - x - 2)$
5. $f(x) = |x| - 1$
6. $f(x) = 3 - |x - 3|$
[S] 7. $f(x) = x^{2/3} - 1$
8. $f(x) = x - x^{1/3}$
9. $f(x) = \dfrac{x^2 - 2x - 3}{x + 2}$
10. $f(x) = (x - 3)(x + 1)^2$
11. $f(x) = x^2 - 6x + 10$
12. $f(x) = \dfrac{x + 1}{x}$

In Exercises 13–19, apply the Mean Value Theorem to f on the indicated interval. In each case find all values of c in the interval (a, b) such that $f'(c) = [f(b) - f(a)]/(b - a)$.

13. $f(x) = x^2$; $[-2, 1]$
14. $f(x) = x(x^2 - x - 2)$; $[-1, 1]$
[S] 15. $f(x) = x^{2/3}$; $[0, 1]$
16. $f(x) = \dfrac{x + 1}{x}$; $\left[\dfrac{1}{2}, 2\right]$
[S] 17. $f(x) = \dfrac{x}{x + 1}$; $\left[-\dfrac{1}{2}, 2\right]$
18. $f(x) = \sqrt{x - 2}$; $[2, 6]$
19. $f(x) = x^3$; $[0, 1]$
20. If f is continuous on $[a, b]$, differentiable on (a, b), and $f(a) = f(b) = d$, prove that there exists at least one number c in (a, b) such that $f'(c) = 0$.
[S] 21. Show that $x^{2n+1} + ax + b$ cannot have two real roots, where $a > 0$ and n is any positive integer.
22. Let $p(x)$ be a nonconstant polynomial. Show that between any two consecutive roots of the equation $p'(x) = 0$, there is at most one root of the equation $p(x) = 0$.
23. Use the Mean Value Theorem to prove the following:
    (a) If the derivative of a function f is 0 for all x in an interval, then the function is constant on the interval.
    (b) If two functions f and g have the same derivatives on an interval, then they differ only by a constant on that interval. [Hint: Let $h(x) = f(x) - g(x)$ and use part (a).]

For Review and Class Discussion

True or False

1. ____ The Mean Value Theorem can be applied to $f(x) = 1/x$ on the interval $[-1, 1]$.

2. ____ If the graph of a function has three x-intercepts, then it must have at least two points at which its tangent line is horizontal.

3. ____ If the graph of a polynomial function has three x-intercepts, then it must have at least two points at which its tangent line is horizontal.

4. ____ If $0 < a < b < 1$ and f is differentiable on $(0, 1)$, then f is continuous on $[a, b]$.

5. ____ If f is continuous on (a, b) and both $f(a)$ and $f(b)$ exist, then f is continuous on $[a, b]$.

4.2 Increasing and Decreasing Functions

Purpose
- To define increasing and decreasing functions.
- To use the derivative to determine when a function is increasing or decreasing.

Geometry may sometimes appear to take the lead over analysis, but in fact precedes it only as a servant goes before his master to clear the path and light him on his way. —*James Sylvester (1814–1897)*

Given the graph of a continuous function, it is a simple matter to decide where (on what intervals) the function is increasing, is constant, or is decreasing. For instance, in Figure 4.7 we have identified the behavior of a function on the intervals (a, c_1), (c_1, c_2), (c_2, c_3), (c_3, c_4), and (c_4, b). However, without the graph of a function, it is much more difficult to determine when the function is increasing (or decreasing). We can get some help from the following definition:

Definition of Increasing and Decreasing Functions	A function f is said to be **increasing** on an interval if for any two numbers x_1 and x_2 in the interval, $$x_1 < x_2 \quad \text{implies} \quad f(x_1) < f(x_2)$$ A function f is said to be **decreasing** on an interval if for any two numbers x_1 and x_2 in the interval, $$x_1 < x_2 \quad \text{implies} \quad f(x_1) > f(x_2)$$

The derivative is useful in determining if a function is increasing or decreasing on a specified interval. We know that if the derivative of f is positive at $x = c$, then its graph slopes upward there, and the function is increasing at that point. Similarly, a negative derivative (or downward slope) implies the function is decreasing at that point. The following theorem indicates how the first derivative may be used to determine intervals on which a function is increasing or decreasing.

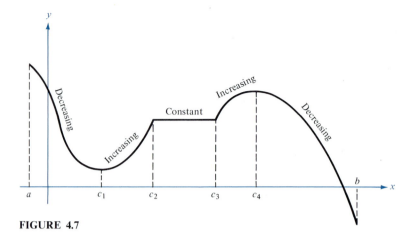

FIGURE 4.7

THEOREM 4.3 (*Test for Increasing or Decreasing Functions*)	Let f be a function that is differentiable on the interval (a, b). i. If $f'(x) > 0$ for all x in (a, b), then f is increasing on (a, b). ii. If $f'(x) < 0$ for all x in (a, b), then f is decreasing on (a, b). iii. If $f'(x) = 0$ for all x in (a, b), then f is constant on (a, b).

Proof: We prove only the first case and leave the other two cases as exercises. Assume $f'(x) > 0$ for all x in the interval (a, b) and let x_1 and x_2 be any two points in the interval such that $x_1 < x_2$. By the Mean Value Theorem, we know there exists a number c such that $x_1 < c < x_2$, and

$$f'(c) = \frac{f(x_2) - f(x_1)}{x_2 - x_1}$$

Since $f'(c) > 0$ and $x_2 - x_1 > 0$, we know that $f(x_2) - f(x_1) > 0$ and so $f(x_1) < f(x_2)$. Thus f is increasing on the interval.   ■

To apply Theorem 4.3 we observe that $f'(x)$ can change signs at points where $f'(x) = 0$, as well as at points where f' is discontinuous. This suggests the following steps for finding intervals on which f is increasing or decreasing.

1. Locate, on the real line, those points where $f'(x) = 0$ and where f' is discontinuous.
2. Test the sign of $f'(x)$ at an arbitrary point in each of the intervals determined by the points in the first step.
3. Use Theorem 4.3 to decide whether f is increasing or decreasing on the intervals in question.

Example 1

Find the intervals on which $f(x) = x^3 - \frac{3}{2}x^2$ is increasing or decreasing.

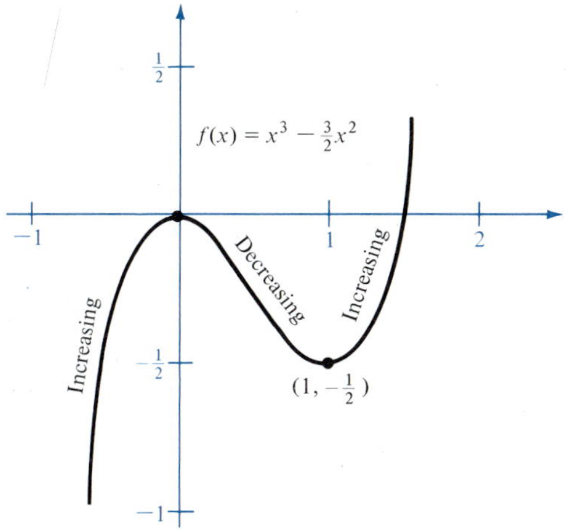

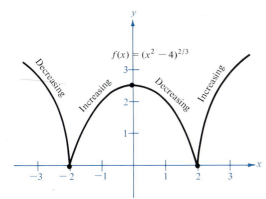

$f(x) = x^3 - \frac{3}{2}x^2$

$(1, -\frac{1}{2})$

Increasing *Decreasing* *Increasing*

FIGURE 4.8

Solution: Since

$$f'(x) = 3x^2 - 3x = 3(x)(x - 1)$$

we know $f'(x)$ is zero for $x = 0$ and $x = 1$. Thus the intervals to be tested are

$$(-\infty, 0), \qquad (0, 1), \qquad (1, \infty)$$

In $(-\infty, 0)$ let $x = -1$; then $f'(-1) = 6 > 0$ implies f is increasing.
In $(0, 1)$ let $x = \frac{1}{2}$; then $f'(\frac{1}{2}) = -\frac{3}{4} < 0$ implies f is decreasing.
In $(1, \infty)$ let $x = 2$; then $f'(2) = 6 > 0$ implies f is increasing. (See Figure 4.8.) ∎

Example 2

Find the intervals on which $f(x) = (x^2 - 4)^{2/3}$ is increasing or decreasing.

Solution: Since

$$f'(x) = \frac{2}{3}(x^2 - 4)^{-1/3}(2x) = \frac{4x}{3(x^2 - 4)^{1/3}}$$

we see that $f'(x)$ is zero at $x = 0$ and, furthermore, f' is undefined (hence discontinuous) at $x = \pm 2$. Thus the intervals to be tested are

$$(-\infty, -2), \qquad (-2, 0), \qquad (0, 2), \qquad (2, \infty)$$

In $(-\infty, -2)$ let $x = -3$; then

$$f'(-3) = \frac{-12}{3\sqrt[3]{5}} < 0$$

implies f is decreasing on $(-\infty, -2)$. Similarly,

$$f'(-1) = \frac{-4}{3\sqrt[3]{-3}} > 0$$

implies f is increasing on $(-2, 0)$, and

$$f'(1) = \frac{4}{3\sqrt[3]{-3}} < 0$$

implies f is decreasing on $(0, 2)$, and

$$f'(3) = \frac{12}{3\sqrt[3]{5}} > 0$$

implies f is increasing on $(2, \infty)$. (See Figure 4.9.) ∎

$f(x) = (x^2 - 4)^{2/3}$

Decreasing *Increasing* *Decreasing* *Increasing*

FIGURE 4.9

Since the zeros and undefined points of f' play an important role in studying the behavior of a function f, we give these values of x a special name.

Definition of a Critical Number   If a number c is in the domain of a function f, then c is called a **critical number** of f provided $f'(c) = 0$ or $f'(c)$ does not exist.

Example 3

Find the critical numbers for the function $f(x) = (x^2 + 3)/(x - 1)$.

Solution: Since

$$f'(x) = \frac{(x - 1)(2x) - (x^2 + 3)(1)}{(x - 1)^2} = \frac{x^2 - 2x - 3}{(x - 1)^2} = \frac{(x - 3)(x + 1)}{(x - 1)^2}$$

then $f'(x)$ is zero at $x = 3$ and $x = -1$ and f' is undefined at $x = 1$. However, $x = 1$ is not in the domain of f and hence the only critical numbers of f are $x = 3$ and $x = -1$.   ∎

We should observe that the converse of Theorem 4.3 is not valid. This means that it is possible for a function to be increasing on an interval even though the derivative is not *positive* at every point in the interval. The next example illustrates such a case.

Example 4

Show that the function $f(x) = x^3 - 3x^2 + 3x$ is increasing on the entire real line.

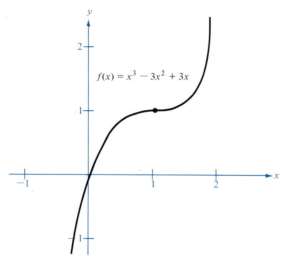

FIGURE 4.10

Solution: Since
$$f'(x) = 3x^2 - 6x + 3 = 0$$
$$3(x^2 - 2x + 1) = 0$$
$$3(x - 1)^2 = 0$$

f' is defined everywhere and has only one critical number, $x = 1$. Thus the intervals to be tested are $(-\infty, 1)$ and $(1, \infty)$.

In $(-\infty, 1)$ let $x = 0$; then $f'(0) = 3 > 0$ implies f is increasing on $(-\infty, 1)$.

In $(1, \infty)$ let $x = 2$; then $f'(2) = 3 > 0$ implies f is increasing on $(1, \infty)$.

Finally, since f is continuous, it is increasing on the entire real line even though $f'(1) = 0$. (See Figure 4.10.) ∎

Section Exercises (4.2)

In Exercises 1–18, find the critical numbers (if any exist) and the intervals on which the function is increasing or decreasing. Sketch the graph of each function.

1. $f(x) = 2x - 3$

2. $f(x) = 5 - 3x$

3. $f(x) = x^2 - 2x$

4. $f(x) = -(x^2 - 2x)$

[S] **5.** $f(x) = x^3 - 3x^2$

6. $f(x) = x^4 - 4x^3 + 15$

[S] **7.** $y = \dfrac{x^2}{x^2 - 4}$

8. $y = \dfrac{1}{x + 1}$

9. $y = \dfrac{x^2}{x^2 + 4}$

10. $f(x) = \sqrt{4 - x^2}$

[S] **11.** $f(x) = x + \sqrt{x^2 - 1}$

12. $f(x) = x\sqrt{x + 1}$

13. $f(x) = x^{2/3}$

14. $f(x) = \sqrt[3]{x - 1}$

15. $f(x) = x + \dfrac{32}{x^2}$

16. $f(x) = \begin{cases} 2x + 1, & x \le -1 \\ x^2 - 2, & x > -1 \end{cases}$

[S] **17.** $f(x) = \begin{cases} 4 - x^2, & x \le 0 \\ -2x + 2, & x > 0 \end{cases}$

18. $f(x) = \begin{cases} -x^3 + 1, & x \le 0 \\ -x^2 + 2x + 1, & x > 0 \end{cases}$

19. Prove case 2 of Theorem 4.3.

20. Prove case 3 of Theorem 4.3.

For Review and Class Discussion

True or False

1. ____ The sum of two increasing functions is increasing.

2. ____ The product of two increasing functions is increasing.

3. ____ The composition of two increasing functions is increasing.

4. ____ If f' is increasing on (a, b), then f is increasing on (a, b).

5. ____ If $f(x) = ax^3 + b$ and f is increasing on $(-1, 1)$, then $a > 0$.

6. ____ Every nth-degree polynomial has $(n - 1)$ critical numbers.

7. ____ An nth-degree polynomial has at most $(n - 1)$ critical numbers.

8. ____ If f is increasing on (a, b), then $f'(x)$ is positive for all x in (a, b).

9. ____ If f is increasing and continuous on (a, b), then f is differentiable on (a, b).

10. ____ If f' is undefined at $x = c$, then c is a critical number of f.

4.3 **The First-Derivative Test for Relative Extrema**

Purpose
- To use the derivative of a function to locate its relative minimum and relative maximum points.

Since differential calculus can be reduced to the problem of tangents, it follows that one can always use tangents to solve various problems of calculus, for instance to find maxima and minima, points of inflection, and cusps.   —*Jean D'Alembert (1717–1783)*

In the preceding section we used the derivative of a function to determine the intervals in which the function was increasing or decreasing. In this section we examine the points at which a function changes from increasing to decreasing or vice versa. It is often the case that at such points the function achieves a local maximum or a local minimum. We use the terms *relative maximum* and *relative minimum* to mean the local extrema of *f*.

Definition of Relative Extrema	Let *f* be a function defined at *c*.

1. $f(c)$ is called a **relative maximum** of *f* if there exists an interval (a, b) containing *c* such that $f(x) \leq f(c)$ for all *x* in (a, b).
2. $f(c)$ is called a **relative minimum** of *f* if there exists an interval (a, b) containing *c* such that $f(x) \geq f(c)$ for all *x* in (a, b).

(The plurals of maximum and minimum are *maxima* and *minima*.)

Recall from Section 2.2 (Extreme Value Theorem) that the maximum and minimum values of a function *f* on an interval were, respectively, the largest and smallest function values on that interval. For instance, in Figure 4.11 the maximum value of *f* on the interval [*a*, *b*] occurs at $x = b$, whereas a relative maximum occurs at $x = c$. The minimum value of *f*, which occurs at $x = d$, also happens to be a relative minimum. The mini-

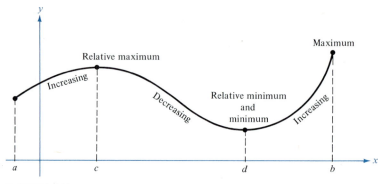

FIGURE 4.11

mum and maximum values of a function f on an interval $[a, b]$ are sometimes referred to as the *absolute minimum* and *absolute maximum* of f on $[a, b]$.

This discussion suggests that an *extremum* (maximum or minimum) of a function on an interval $[a, b]$ can occur:

(a)   at an *interior point* of $[a, b]$ and thus it is also a relative extremum, or

(b)   at an *endpoint* of $[a, b]$, in which case we call it an **endpoint extremum.**

By observing the location of the relative maxima and minima shown in Figure 4.11, we might conclude that they always occur at a critical number of the function. This is indeed the case, as the following theorem indicates.

THEOREM 4.4	If f has a relative minimum or relative maximum when $x = c$, then either (i) $f'(c) = 0$ or (ii) $f'(c)$ is undefined. That is, c is a critical number of f.

Theorem 4.4 states that if a function has a relative minimum or a relative maximum when $x = c$, then c *must* be a critical number of f. Unfortunately, the converse is not true. Indeed, it is possible for a function to have a critical number that does not yield a relative minimum or a relative maximum. This can be seen in the next example.

Example 1

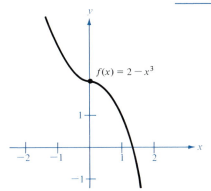

FIGURE 4.12

Show that the function $f(x) = 2 - x^3$ has a critical number but no relative maxima or minima.

Solution: We have

$$f(x) = 2 - x^3 \qquad \text{and} \qquad f'(x) = -3x^2$$

Now $f'(x)$ has only one critical number, $x = 0$. However, since

$$f'(x) < 0 \qquad \text{for} \qquad x < 0$$

and

$$f'(x) < 0 \qquad \text{for} \qquad x > 0$$

we conclude that f is *decreasing* for all x and hence has no relative extrema. (See Figure 4.12.)   ∎

Even though Example 1 shows that a critical number may not yield a relative extremum, all relative extrema *must* occur at a critical number. Thus we must test each critical number to determine if it yields a relative maximum, a relative minimum, or possibly neither. The following theorem provides such a test.

THEOREM 4.5
(First-Derivative Test for Relative Extrema)

Suppose that f is continuous on the interval (a, b) and c is the only critical number of f in the interval. If f is differentiable on the interval (except possibly at c), then $f(c)$ can be classified as shown in Table 4.1.

TABLE 4.1

$f(c)$	Sign of f' in (a, c)	Sign of f' in (c, b)	Graphically
Relative maximum	$+$	$-$	
Relative minimum	$-$	$+$	
Neither	$+$	$+$	
Neither	$-$	$-$	

Proof: The proof is straightforward, using Theorem 4.3 (the test for increasing or decreasing functions), and is left as an exercise. ■

Example 2

Locate all relative extrema for the function

$$f(x) = 2x^3 - 3x^2 - 36x + 14$$

Solution: By setting the derivative of f equal to zero, we have

$$f'(x) = 6x^2 - 6x - 36 = 0$$
$$6(x^2 - x - 6) = 0$$
$$6(x - 3)(x + 2) = 0$$

Since f' is defined for all real numbers, the only critical numbers of f are -2 and 3. By the method used to test for increasing and decreasing

TABLE 4.2

Interval	$(-\infty, -2)$	$(-2, 3)$	$(3, \infty)$
Point in interval	-3	0	4
Sign of $f'(x)$	$+$	$-$	$+$

functions, we test the sign of f' in each of the intervals $(-\infty, -2)$, $(-2, 3)$, and $(3, \infty)$. See Table 4.2.

Using theorem 4.5 we conclude that the critical number -2 yields a relative maximum (f' changes sign from $+$ to $-$) and 3 yields a relative minimum (f' changes sign from $-$ to $+$). (See Figure 4.13.)

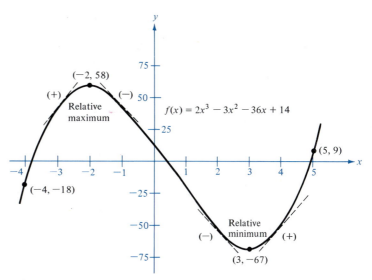

FIGURE 4.13

Example 3

Locate all relative extrema for the function $f(x) = x^4 - x^3$.

Solution:
$$f'(x) = 4x^3 - 3x^2 = 0$$
$$x^2(4x - 3) = 0$$

Since f' is defined everywhere, the only critical numbers of f are 0 and $\frac{3}{4}$. See Table 4.3.

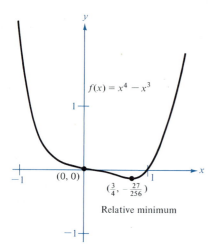

$f(x) = x^4 - x^3$

$(0, 0)$

$(\frac{3}{4}, -\frac{27}{256})$

Relative minimum

FIGURE 4.14

TABLE 4.3

Interval	$(-\infty, 0)$	$(0, \frac{3}{4})$	$(\frac{3}{4}, \infty)$
Point in interval	-1	$\frac{1}{2}$	1
Sign of $f'(x)$	$-$	$-$	$+$

Using Theorem 4.5 we conclude that the critical number $\frac{3}{4}$ yields a relative minimum, whereas 0 yields neither a relative minimum nor a relative maximum (see Figure 4.14). ∎

Example 4

Locate all relative extrema for the function $f(x) = 2x - 3x^{2/3}$.

Solution: Differentiating we have

$$f'(x) = 2 - \frac{2}{x^{1/3}}$$

Since $f'(x) = 0$ if $x = 1$ and f' is undefined when $x = 0$, the critical numbers of f are 0 and 1. See Table 4.4. Therefore, 0 yields a relative maximum and 1 yields a relative minimum, as shown in Figure 4.15.

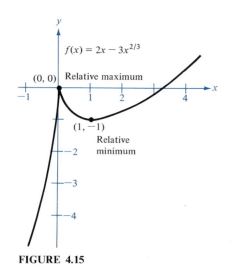

$f(x) = 2x - 3x^{2/3}$

$(0, 0)$ Relative maximum

$(1, -1)$
Relative minimum

FIGURE 4.15

TABLE 4.4

Interval	$(-\infty, 0)$	$(0, 1)$	$(1, \infty)$
Point in interval	-1	$\frac{1}{2}$	2
Sign of $f'(x)$	$+$	$-$	$+$

∎

Example 5

An object is projected vertically upward from ground level so that its height h at time t is given by $h = -16t^2 + 144t$, where h is measured in feet and t is measured in seconds. What is its maximum height?

Solution: By differentiating we have

$$\frac{dh}{dt} = -32t + 144$$

Since $dh/dt = 0$ when $t = 4.5$ s, we know that the maximum height must be

$$h = -16(4.5)^2 + 144(4.5) = 324 \text{ ft}$$ ■

In Example 5 we stated that the maximum height occurred at $t = 4.5$ without actually verifying this maximum by our first-derivative test. It is often the case that the nature of a physical problem clearly indicates whether a critical number yields a maximum or minimum without a formal test. This is further demonstrated by the next example.

Example 6

Recall from Example 7 of Section 2.3 that the oxygen level in a pond polluted by organic waste is given by the equation

$$f(t) = \frac{t^2 - t + 1}{t^2 + 1} \qquad 0 \le t < \infty$$

where t represents the time in weeks.
(a) When is the oxygen level lowest?
(b) When is the oxygen level highest?

Solution:

(a) Since

$$f'(t) = \frac{(t^2 + 1)(2t - 1) - (t^2 - t + 1)(2t)}{(t^2 + 1)^2}$$

$$= \frac{2t^3 - t^2 + 2t - 1 - 2t^3 + 2t^2 - 2t}{(t^2 + 1)^2}$$

$$= \frac{t^2 - 1}{(t^2 + 1)^2}$$

is zero when $t^2 - 1 = 0$, and $t = -1$ is not in the domain, the only critical number is $t = 1$. The nature of the problem indicates that this value ($t = 1$ week) yields the minimum oxygen level. (See Figure 4.16.)

(b) Since $t = 1$ is a horizontal asymptote, it is clear from Figure 4.16 that the maximum level of oxygen occurs when $t = 0$ and is an example of an *endpoint extremum*. ■

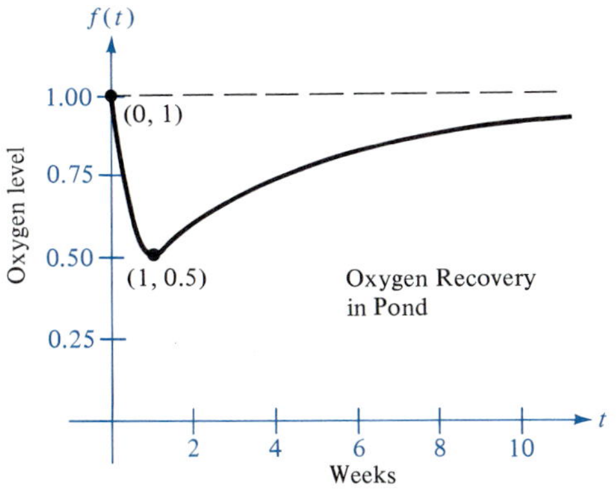

f(t)
1.00
0.75
0.50
0.25
(0, 1)
(1, 0.5)
Oxygen level
Oxygen Recovery in Pond
2
4
6
8
10
t
Weeks

FIGURE 4.16

Section Exercises (4.3)

In Exercises 1–20, locate all relative extrema.

1. $f(x) = -2x^2 + 4x + 3$
2. $f(x) = x^2 + 8x + 10$
3. $f(x) = x^2 - 6x$
4. $f(x) = (x - 1)^2(x + 2)$
[S] **5.** $f(x) = 2x^3 + 3x^2 - 12x$

6. $f(x) = (x - 3)^3$

7. $f(x) = x^3 - 6x^2 + 15$ **8.** $f(x) = x^4 - 1$

9. $f(x) = x^4 - 2x^3$ **10.** $f(x) = (x - 1)^{2/3}$

$\boxed{S}$ **11.** $f(x) = x^{1/3} + 1$ **12.** $f(x) = x^{2/3}(x - 5)$

13. $f(x) = x + \dfrac{1}{x}$ **14.** $f(x) = \dfrac{x}{x + 1}$

15. $f(x) = \dfrac{x^2}{x^2 - 9}$ **16.** $f(x) = \dfrac{x + 3}{x^2}$

17. $f(x) = x^4 - 32x + 4$ **18.** $f(x) = \dfrac{x^5 - 5x}{5}$

$\boxed{S}$ **19.** $f(x) = \dfrac{x^2 - 2x + 1}{x + 1}$ **20.** $f(x) = \dfrac{x^2 - 3x - 4}{x - 2}$

In Exercises 21–30, locate the absolute extrema of the function on the indicated interval.

21. $f(x) = 2(3 - x)$; $[-1, 2]$ **22.** $f(x) = \dfrac{2x + 5}{3}$; $[0, 5]$

$\boxed{S}$ **23.** $f(x) = -x^2 + 4x$; $[0, 3]$

24. $f(x) = x^2 + 2x - 4$; $[-1, 1]$

25. $f(x) = x^3 - 3x^2$; $[-1, 3]$ **26.** $f(x) = x^3 - 12x$; $[0, 4]$

27. $f(x) = 3x^{2/3} - 2x$; $[-1, 1]$

28. $g(t) = \dfrac{t^2}{t^2 + 3}$; $[-1, 1]$

29. $h(s) = \dfrac{1}{s - 2}$; $[0, 1]$ **30.** $h(t) = \dfrac{t}{t - 2}$; $[3, 5]$

$\boxed{S}$ **31.** Find a, b, c, and d so that the function $f(x) = ax^3 + bx^2 + cx + d$ has a relative minimum at $(0, 0)$ and a relative maximum at $(2, 2)$.

32. Find a, b, and c so that the function $f(x) = ax^2 + bx + c$ has a relative maximum at $(5, 20)$ and passes through $(2, 10)$.

In Exercises 33–40, determine from the graph of f if f possesses a relative minimum in the interval (a, b).

33.

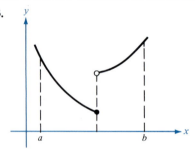

34.

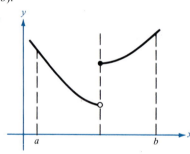

35.

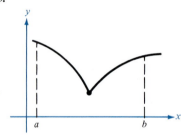

36.

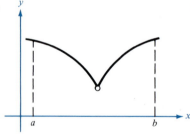

$\boxed{S}$ **37.**

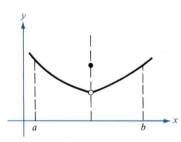

38.

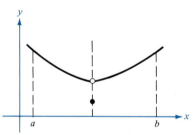

$\boxed{S}$ **39.**

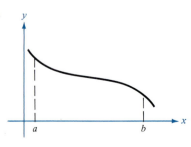

40.

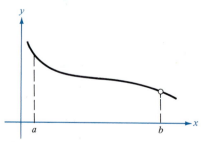

41. The height of a ball at time t is given by the equation $h(t) = 96t - 16t^2$.
    (a) What was the initial velocity of the ball?
    (b) How high did the ball go?
    (c) What direction was the ball moving at time $t = 4$?
    (d) Find the height of the ball at time $t = 1$.
42. Repeat Exercise 41 using $h(t) = 64t - 16t^2$ as the position function.
43. Coughing forces the trachea (windpipe) of a person to contract, which in turn affects the velocity v of the air through the trachea. Suppose the velocity of the air during coughing is

$$v = k(R - r)r^2$$

where k is a constant, R is the normal radius of the trachea, and r is the radius during coughing. What radius will produce the maximum air velocity?
44. The concentration C of a certain chemical in the bloodstream t hours after injection into muscle tissue is given by

$$C = \frac{3t}{27 + t^3}$$

When is the concentration greatest?

[S] 45. The resistance R of a certain type of resistor is given by

$$R = \sqrt{0.001T^4 - 4T + 100}$$

where R is measured in ohms and the temperature T is measured in degrees Celsius. What temperature produces a minimum resistance for this type of resistor?
46. The electric power P in watts in a (direct current) circuit with two resistors of resistance R_1 and R_2 connected in series is

$$P = \frac{vR_1R_2}{(R_1 + R_2)^2}$$

where v is the voltage. If v and R_1 are held constant, what resistance R_2 produces the maximum power?

The error estimate for the Trapezoidal Rule (see Section 10.10) involves the maximum of the absolute value of the second derivative in an interval. In Exercises 47–50, find the maximum value of $|f''(x)|$ in the indicated interval.

47. $f(x) = \dfrac{1}{x^2 + 1}$; $[0, 3]$

48. $f(x) = \dfrac{1}{x^2 + 1}$; $[\frac{1}{2}, 3]$

[S] 49. $f(x) = \sqrt{1 + x^3}$; $[0, 2]$

50. $f(x) = x^3(3x^2 - 10)$; $[0, 1]$

The error estimate for Simpson's Rule (see Section 10.10) involves the maximum of the absolute value of the fourth derivative in an interval. In Exercises 51–54, find the maximum value of $|f^{(4)}(x)|$ in the indicated interval.

51. $f(x) = 15x^4 - \left(\dfrac{2x - 1}{2}\right)^6$; $[0, 1]$

52. $f(x) = x^5 - 5x^4 + 20x^3 + 600$; $[0, \frac{3}{2}]$

[S] 53. $f(x) = (x + 1)^{2/3}$; $[0, 2]$

54. $f(x) = \dfrac{1}{x^2 + 1}$; $[-1, 1]$

55. Assume that f is differentiable for all x and $f'(x) > 0$ on $(-\infty, -4)$, $f'(x) < 0$ on $(-4, 6)$, and $f'(x) > 0$ on $(6, \infty)$. Supply the appropriate inequality in the following:
    (a) If $h(x) = f(x) + 5$, then $h'(0)$____0.
    (b) If $g(x) = -f(x)$, then $g'(-6)$____0 and $g'(0)$____0.
    (c) If $j(x) = f(x - 10)$, then $j'(0)$____0 and $j'(8)$____0.
56. Prove Theorem 4.5.

For Review and Class Discussion

True or False

1. ____ Every second-degree polynomial possesses precisely one relative extremum.

2. ____ If $a < c < b$ and c is a critical number of f, then f has a relative extremum in (a, b).

3. ____ If f has a maximum at $x = c$, then f has a relative maximum at $x = c$.

4. ____ If f is increasing on (a, b), then f has no relative extrema in (a, b).

5. ____ If $a < c < b$ and $f(c)$ is a relative maximum of f in (a, b), then $g(c)$ is a relative maximum of g in (a, b), where $g(x) = \sqrt[3]{f(x)}$.

4.4 Concavity and the Second-Derivative Test

Purpose

- To use the first and second derivatives of a function to determine concavity and points of inflection.
- To introduce the Second-Derivative Test for relative extrema.

When the first derivative vanishes, if at the same time the second derivative is positive, the ordinate is then a minimum, but is a maximum if the second derivative is negative. —*Colin Maclaurin (1698–1746)*

We have already seen that locating the intervals in which a function f increases or decreases is helpful in determining its graph. In this section we extend this idea and show that by locating the intervals in which f' increases or decreases, we can determine where the graph of f is curving upward or curving downward. We define this notion of curving upward or downward as **concavity.**

Definition of Concavity

Let f be differentiable on (a, b). We say that the graph of f is:

1. **concave upward** on (a, b) if f' is increasing on (a, b)
2. **concave downward** on (a, b) if f' is decreasing on (a, b). (See Figure 4.17.)

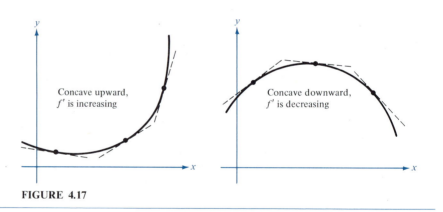

FIGURE 4.17

Notice the following in Figure 4.17:

1. If a curve lies *above* its tangent lines, then it is concave upward.
2. If a curve lies *below* its tangent lines, then it is concave downward.

This visual test for concavity is useful when the graph of a function is given. To determine concavity without seeing a graph, we need an analytic test for finding the intervals on which the derivative is increasing or decreasing. It turns out that we can use the second derivative to determine the intervals in which f' is increasing or decreasing, just as we used the first derivative to determine the intervals in which f is increasing or decreasing. We can see this parallel by comparing Theorem 4.3 to the following theorem.

THEOREM 4.6	Let f be a function whose second derivative exists on the interval (a, b).
(Test for Concavity)	i. If $f''(x) > 0$ for *all* x in (a, b), then the graph of f is concave upward on (a, b).
	ii. If $f''(x) < 0$ for *all* x in (a, b), then the graph of f is concave downward on (a, b).

Proof: The proof consists of a straightforward application of Theorem 4.3 in which f is replaced by f'. ∎

Example 1

Determine the intervals in which the graph of $f(x) = 6/(x^2 + 3)$ is concave upward or downward.

Solution: Since $f(x) = 6(x^2 + 3)^{-1}$, then

$$f'(x) = (-6)(2x)(x^2 + 3)^{-2} = \frac{-12x}{(x^2 + 3)^2}$$

and

$$f''(x) = \frac{(x^2 + 3)^2(-12) - (-12x)(2)(2x)(x^2 + 3)}{(x^2 + 3)^4}$$

TABLE 4.5

Interval	$(-\infty, -1)$	$(-1, 1)$	$(1, \infty)$
Sign of f''	$+$	$-$	$+$
f'	Increasing	Decreasing	Increasing
Graph of f	Concave upward	Concave downward	Concave upward

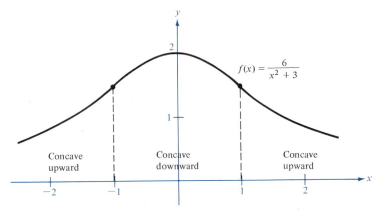

FIGURE 4.18

$$= \frac{-12(x^2 + 3) + (48x^2)}{(x^2 + 3)^3} = \frac{36(x^2 - 1)}{(x^2 + 3)^3}$$

Since $f''(x) = 0$ when $x = \pm 1$, we test f'' in the intervals $(-\infty, -1)$, $(-1, 1)$, and $(1, \infty)$. The results are shown in Table 4.5 and Figure 4.18. ■

Notice in Example 1 that f' is increasing on the interval $(1, \infty)$ even though f is decreasing there, as we see from Figure 4.18. The increasing or decreasing of f' does not necessarily correspond to the respective increasing or decreasing of f.

The graph in Figure 4.18 has two points at which the concavity changes. If at such a point the tangent line to the graph exists, we call the point a **point of inflection.**

Definition of Point of Inflection	If the graph of a continuous function possesses a tangent line at a point where its concavity changes from upward to downward (or vice versa), we call the point a **point of inflection.** (See Figure 4.19.)

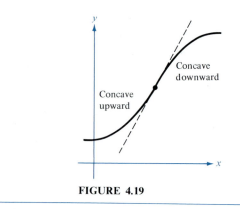
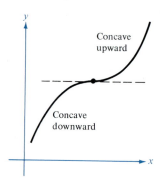
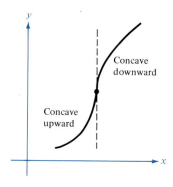

FIGURE 4.19

Note in Figure 4.19 that at each point of inflection, the graph *crosses* its tangent line.

Since a point of inflection occurs where the concavity of a graph changes, it is also true (see Theorem 4.6) that the sign of f'' changes at such points. Thus to locate possible points of inflection, we need only determine the values of x for which $f''(x) = 0$ or for which $f''(x)$ does not exist. This parallels the procedure for locating relative extrema of f by determining the critical numbers of f (see Theorem 4.4). We state, without further derivation, the following property of points of inflection.

Property of Points of Inflection	If $(c, f(c))$ is a point of inflection of f, then either $f''(c) = 0$ or $f''(c)$ does not exist.

Example 2

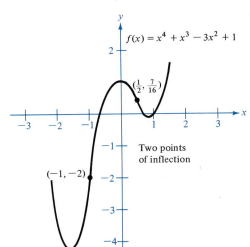

FIGURE 4.20

Determine the points of inflection and discuss the concavity of the graph of $f(x) = x^4 + x^3 - 3x^2 + 1$.

Solution: Differentiating twice we have

$$f'(x) = 4x^3 + 3x^2 - 6x$$
$$f''(x) = 12x^2 + 6x - 6 = 6(2x^2 + x - 1) = 6(2x - 1)(x + 1)$$

Possible points of inflection occur at $x = -1$ and $x = \frac{1}{2}$. By testing the intervals $(-\infty, -1)$, $(-1, \frac{1}{2})$, and $(\frac{1}{2}, \infty)$, we determine that f is concave upward in $(-\infty, -1)$, concave downward in $(-1, \frac{1}{2})$, and concave upward in $(\frac{1}{2}, \infty)$. Thus the numbers -1 and $\frac{1}{2}$ both yield points of inflection. (See Figure 4.20.) ∎

We should point out that it is possible for the second derivative to be zero at a point which is *not* a point of inflection. For example, compare the graphs of $f(x) = x^3$ and $g(x) = x^4$ (see Figure 4.21). Both second derivatives are zero when $x = 0$, but only the graph of f has a point of inflection at $x = 0$. This shows that before concluding that a point of inflection exists at a value of x for which $f''(x) = 0$, we should test to be certain that the concavity actually changes there.

If the second derivative exists, we can often use it as a simple test for relative minima and relative maxima. The test is based on the idea that if $f(c)$ is a relative maximum of a differentiable function f, then its graph is concave downward in some interval containing c. Similarly, if $f(c)$ is a relative minimum, its graph is concave upward in some interval containing c (see Figure 4.22). This test is referred to as the Second-Derivative Test for Relative Extrema.

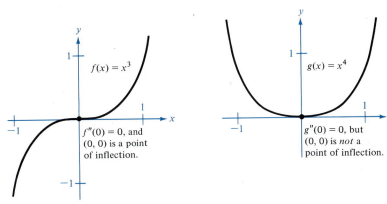

FIGURE 4.21

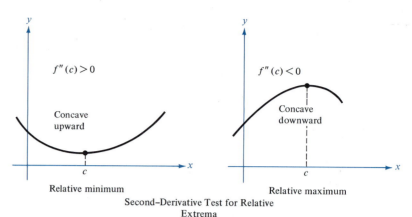

Second–Derivative Test for Relative
Extrema

FIGURE 4.22

THEOREM 4.7 (*Second-Derivative Test for Relative Extrema*)	Let f'' exist on some open interval containing c and $f'(c) = 0$. i. If $f''(c) > 0$, then $f(c)$ is a relative minimum. ii. If $f''(c) < 0$, then $f(c)$ is a relative maximum.

Proof: This theorem follows readily from the Second-Derivative Test for concavity. Since $f'(c) = 0$, the tangent line must be horizontal at $(c, f(c))$. If $f''(c) > 0$, then f is concave upward in some interval containing c, and thus the graph of f cannot extend below the horizontal tangent line in that interval. It follows that $f(c)$ is a relative minimum. Similarly, if $f''(c) < 0$, then $f(c)$ is a relative maximum.   ■

Note that if $f''(c) = 0$, the Second-Derivative Test does not apply. In such cases we can use the First-Derivative Test.

Example 3

Find the relative minimum and relative maximum points for the function $f(x) = -3x^5 + 5x^3$.

Solution: Differentiating twice we have

$$f'(x) = -15x^4 + 15x^2 = 15(-x^4 + x^2) = 15x^2(1 - x^2)$$
$$f''(x) = 15(-4x^3 + 2x)$$

Now $f'(x) = 0$ when x is -1, 0, or 1. By the Second-Derivative Test, $f''(-1) > 0$ implies $(-1, -2)$ is a relative minimum, and $f''(1) < 0$ implies $(1, 2)$ is a relative maximum. Since $f''(0) = 0$, the Second-Derivative Test does not apply. In this case the First-Derivative Test shows that $(0, 0)$ is neither a relative maximum nor a relative minimum. [A test of concavity would show that $(0, 0)$ is a point of inflection.] (See Figure 4.23.)

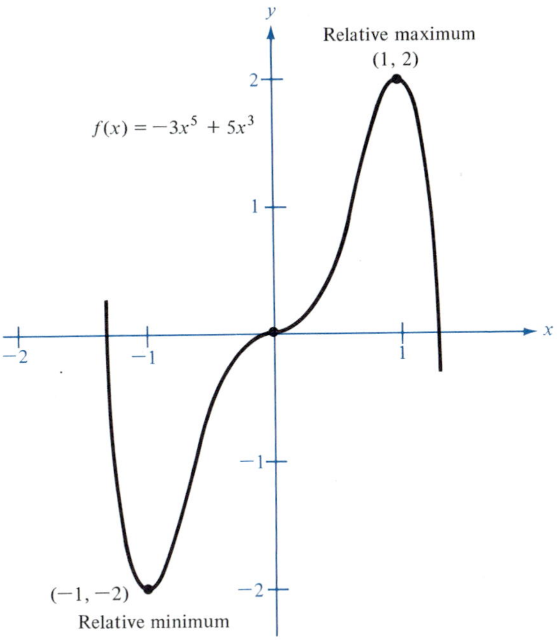

Relative maximum
(1, 2)

$f(x) = -3x^5 + 5x^3$

$(-1, -2)$
Relative minimum

FIGURE 4.23

Section Exercises (4.4)

In Exercises 1–12, identify all relative extrema. Use the Second-Derivative Test when applicable.

1. $f(x) = 6x - x^2$ **2.** $f(x) = x^2 + 3x - 8$

3. $f(x) = (x - 5)^2$ **4.** $f(x) = -(x - 5)^2$

Ⓢ **5.** $f(x) = x^3 - 3x^2 + 3$ **6.** $f(x) = 5 + 3x^2 - x^3$

7. $f(x) = x^4 - 4x^3 + 2$

8. $f(x) = x^3 - 9x^2 + 27x - 26$

9. $f(x) = x^{2/3} - 3$ **10.** $f(x) = \sqrt{x^2 + 1}$

Ⓢ **11.** $f(x) = x + \dfrac{4}{x}$ **12.** $f(x) = \dfrac{x}{x - 1}$

In Exercises 13–25, identify all relative extrema and points of inflection. Sketch the graphs.

13. $f(x) = x^3 - 6x^2 + 12x - 8$

14. $f(x) = x^3 + 1$ **15.** $f(x) = x^3 - 12x$

16. $f(x) = 2x^3 - 3x^2 - 12x + 8$

Ⓢ **17.** $f(x) = \dfrac{x^4}{4} - 2x^2$ **18.** $f(x) = 2x^4 - 8x + 3$

19. $f(x) = x^2 + \dfrac{1}{x^2}$ **20.** $f(x) = \dfrac{x^2}{x^2 - 1}$

Ⓢ **21.** $f(x) = x\sqrt{x + 3}$ **22.** $f(x) = x\sqrt{x + 1}$

23. $f(x) = \dfrac{x}{x^2 - 4}$ **24.** $f(x) = \dfrac{1}{x^2 - x - 2}$

25. $f(x) = \dfrac{x - 2}{x^2 - 4x + 3}$

26. Sketch the graph of a continuous function f having the following characteristics:

  i. $f(2) = f(4) = 0$
  ii. $f'(x) > 0$ if $x < 3$; $f'(x) < 0$ if $x > 3$; $f'(3)$ is undefined.
  iii. $f''(x) > 0$ for all x in the domain of f.

Ⓢ **27.** Find a, b, c, and d so that the function $f(x) = ax^3 + bx^2 + cx + d$ has a relative maximum at $(3, 3)$, a relative minimum at $(5, 1)$, and a point of inflection at $(4, 2)$.

28. Prove that the midpoint of the line segment joining the relative maximum and relative minimum of the cubic $f(x) = ax^3 + bx^2 + cx + d$ is a point of inflection.

29. Use the result of Exercise 28 to find the point of inflection of $f(x) = x(6 - x)^2$.

Ⓢ **30.** Assume that the zeros of the function $f(x) = ax^3 + bx^2 + cx + d$ are all real. Prove that the abscissa of the point of inflection is the average of these zeros. [Hint: Express the cubic in the form $f(x) = a(x - r_1)(x - r_2)(x - r_3)$.]

31. The equation $E = T/(x^2 + a^2)^{3/2}$ gives the electric field intensity on the axis of a uniformly charged ring, where T is the total charge on the ring and a is the radius of the ring. At what value of x is E maximum?

32. A manufacturer has determined that the total cost C of operating a certain facility is given by $C = 0.5x^2 + 15x + 5000$, where x is the number of units produced. At what level of production will the average cost per unit be minimum? (The average cost per unit is given by C/x.)

[S] **33.** Find the optimal order size if the total cost C for ordering and storing x units is $C = 2x + (300{,}000/x)$.

34. The deflection D of a particular type of beam of length L is given by $D = 2x^4 - 5Lx^3 + 3L^2x^2$, where x is the distance (in feet) from one end of the beam. Find the value of x that yields the maximum deflection.

For Review and Class Discussion

True or False

1. ____ If $f''(c) = 0$, then the graph of f has a point of inflection at $(c, f(c))$.

2. ____ The graph of $f(x) = ax + b$ is neither concave upward nor concave downward.

3. ____ The graph of every cubic polynomial has precisely one point of inflection.

4. ____ The graph of a fourth-degree polynomial can have zero or two points of inflection, but not one.

5. ____ The graph of $f(x) = 1/x$ is concave downward for $x < 0$ and concave upward for $x > 0$, and thus it has a point of inflection when $x = 0$.

4.5 Curve Sketching: A Summary

Purpose

- To summarize curve-sketching techniques presented in this and the previous chapters.

Here we have an opportunity of expounding more clearly what has already been said. —*René Descartes (1596–1650)*

The cliché "a picture is worth a thousand words" properly identifies the importance of curve sketching in mathematics. Descartes' introduction of this fruitful concept not only preceded but to a great extent was responsible for the rapid advances in mathematics during the last half of the seventeenth century. Today government, science, industry, business, education, and the social and health sciences all make widespread use of graphs to describe and predict relationships between variables within their domain of interest.

Although we can gain some idea of the relationship between two variables directly from the mathematical equation relating the variables, the true relationship is often best seen from a graph of the equation. In many instances the graph of a functional relationship is easily made by merely plotting a collection of points. Yet we have seen how some functions behave rather oddly and to sketch their graphs requires considerable ingenuity.

We have previously discussed several concepts that are useful in sketching the graph of a function. For instance, the following properties of a function f and its graph have been discussed at some length: domain and range of f; continuity; symmetry; x- and y-intercepts; horizontal and vertical asymptotes; relative minima and maxima; concavity and points of inflection. In this section we incorporate these concepts into an effective procedure for sketching curves.

Suggestions for Sketching the Graph of a Function	1. Make a rough preliminary sketch, using any easily determined intercepts, symmetry, vertical asymptotes, or other points of discontinuity. 2. Locate the x-values where $f'(x)$ and $f''(x)$ are either zero or undefined. 3. Test the behavior of f at each of these x-values as well as *within* each interval determined by them. (See Table 4.6 in Example 1.) 4. Sharpen the accuracy of the final sketch by plotting the relative extrema points, the points of inflection, and a few points between.

Example 1

Sketch a graph of
$$f(x) = \frac{x^2 - 2x + 4}{x - 2}$$

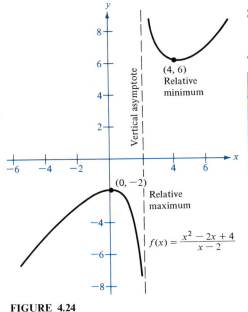

FIGURE 4.24

TABLE 4.6

	$f(x)$	$f'(x)$	$f''(x)$	Shape of graph
x in $(-\infty, 0)$		$+$	$-$	increasing, concave down
$x = 0$	-2	0	$-$	relative maximum
x in $(0, 2)$		$-$	$-$	decreasing, concave down
$x = 2$	does not exist	does not exist	does not exist	odd vertical asymptote
x in $(2, 4)$		$-$	$+$	decreasing, concave up
$x = 4$	6	0	$+$	relative minimum
x in $(4, \infty)$		$+$	$+$	increasing, concave up

Solution: f has a vertical asymptote at $x = 2$ and a y-intercept at $(0, -2)$.

$$f'(x) = \frac{x(x-4)}{(x-2)^2} \quad \text{and} \quad f''(x) = \frac{8}{(x-2)^3}$$

The critical numbers are $x = 0$ and $x = 4$. There are no possible points of inflection since $f''(x)$ is never zero. Testing, we have the results shown in Table 4.6. The graph is shown in Figure 4.24. ∎

Example 2

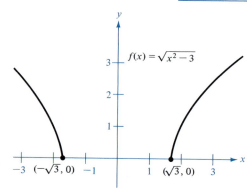

FIGURE 4.25

Sketch a graph of $f(x) = \sqrt{x^2 - 3}$.

Solution: The function is not defined for $-\sqrt{3} < x < \sqrt{3}$; hence there is no y-intercept. The graph is symmetric with respect to the y-axis.

$$f'(x) = \frac{x}{\sqrt{x^2 - 3}} \quad \text{and} \quad f''(x) = \frac{-3}{(x^2 - 3)^{3/2}}$$

The critical numbers are $x = \pm\sqrt{3}$, and there are no possible points of inflection. The test results are given in Table 4.7; the graph is shown in Figure 4.25.

TABLE 4.7

	$f(x)$	$f'(x)$	$f''(x)$	Shape of graph
x in $(-\infty, -\sqrt{3})$		$-$	$-$	decreasing, concave down
$x = -\sqrt{3}$	0	does not exist	does not exist	minimum, vertical tangent
x in $(-\sqrt{3}, \sqrt{3})$	does not exist	does not exist	does not exist	undefined
$x = \sqrt{3}$	0	does not exist	does not exist	minimum, vertical tangent
x in $(\sqrt{3}, \infty)$		$+$	$-$	increasing, concave down

∎

Example 3

Sketch a graph of $f(x) = |x^2 - 2x - 3|$.

TABLE 4.8

	$f(x)$	$f'(x)$	$f''(x)$	Shape of graph
x in $(-\infty, -1)$		$-$	$+$	decreasing, concave up
$x = -1$	0	does not exist	does not exist	minimum
x in $(-1, 1)$		$+$	$-$	increasing, concave down
$x = 1$	4	0	$-$	relative maximum
x in $(1, 3)$		$-$	$-$	decreasing, concave down
$x = 3$	0	does not exist	does not exist	minimum
x in $(3, \infty)$		$+$	$+$	increasing, concave up

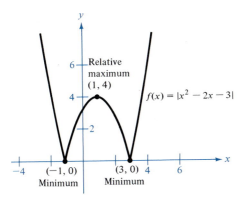

FIGURE 4.26

Solution: There are no vertical asymptotes, and the y-intercept is $(0, 3)$.

$$f'(x) = \frac{x^2 - 2x - 3}{|x^2 - 2x - 3|}(2x - 2) = \frac{(x - 3)(x + 1)(2x - 2)}{|(x - 3)(x + 1)|}$$

$$f''(x) = \frac{x^2 - 2x - 3}{|x^2 - 2x - 3|}(2)$$

The critical numbers are $x = -1$, $x = 1$, and $x = 3$. Testing, we have the results given in Table 4.8. The graph is shown in Figure 4.26. ∎

Example 4

Sketch a graph of $f(x) = 2x^{5/3} - 5x^{4/3}$.

Solution: Since $f(x) = x^{4/3}(2x^{1/3} - 5)$

the intercepts are $(0, 0)$ and $(125/8, 0)$. The first and second derivatives are

$$f'(x) = \frac{10x^{1/3}(x^{1/3} - 2)}{3} \quad \text{and} \quad f''(x) = \frac{20(x^{1/3} - 1)}{9x^{2/3}}$$

Testing, we have the results given in Table 4.9. The graph is shown in Figure 4.27.

TABLE 4.9

	$f(x)$	$f'(x)$	$f''(x)$	Shape of graph
x in $(-\infty, 0)$		+	−	increasing, concave down
$x = 0$	0	0	does not exist	relative maximum
x in $(0, 1)$		−	−	decreasing, concave down
$x = 1$	−3	−	0	point of inflection
x in $(1, 8)$		−	+	decreasing, concave up
$x = 8$	−16	0	+	relative minimum
x in $(8, \infty)$		+	+	increasing, concave up

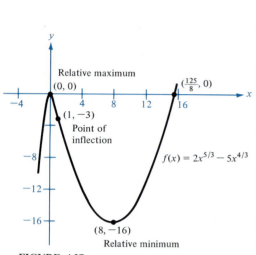

FIGURE 4.27

Example 5

Sketch a graph of $\qquad f(x) = \dfrac{x - \sqrt{x^2 + 1}}{x}$

Solution:

$$f'(x) = \frac{1}{x^2 \sqrt{x^2 + 1}} \qquad \text{and} \qquad f''(x) = \frac{-(3x^2 + 2)}{x^3(x^2 + 1)^{3/2}}$$

The test results are given in Table 4.10. The graph is shown in Figure 4.28.

TABLE 4.10

	$f(x)$	$f'(x)$	$f''(x)$	Shape of graph
x in $(-\infty, 0)$		+	+	increasing, concave up, horizontal asymptote when $y = 2$
$x = 0$	does not exist	does not exist	does not exist	odd vertical asymptote
x in $(0, \infty)$		+	−	increasing, concave down, horizontal asymptote when $y = 0$

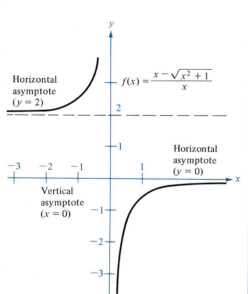

FIGURE 4.28

Section Exercises (4.5)

In Exercises 1–22, sketch the graph of each function, choosing a scale that allows all relative extrema and points of inflection to be identified on the sketch.

1. $y = x^3 - 3x^2 + 3$

2. $3y = -x^3 + 3x - 2$

3. $y = 2 - x - x^3$

4. $y = x^3 + 3x^2 + 3x + 2$

5. $y = 3x^3 - 9x + 1$

6. $y = (x + 1)(x - 2)(x - 5)$

7. $y = -x^3 + 3x^2 + 9x - 2$

8. $3y = (x - 1)^3 + 6$

[S] **9.** $y = 3x^4 + 4x^3$

10. $y = 3x^4 - 6x^2$

11. $y = x^4 - 4x^3 + 16x$

12. $y = x^4 - 8x^3 + 18x^2 - 16x + 5$

13. $y = x^4 - 4x^3 + 16x - 16$

14. $y = x^5 + 1$

15. $y = x^5 - 5x$

16. $y = |2x - 3|$

[S] **17.** $y = |x^2 - 6x + 5|$

18. $y = \dfrac{x}{x^2 + 1}$

19. $y = \dfrac{x^2}{x^2 + 3}$

20. $y = 3x^{2/3} - x^2$

[S] **21.** $y = 3x^{2/3} - 2x$

22. $y = \dfrac{x}{\sqrt{x^2 + 7}}$

In Exercises 23–30, sketch the graph of each function. In each case label the intercepts, relative extrema, points of inflection, and the domain.

[S] **23.** $y = \dfrac{1}{x - 2} - 3$

24. $y = \dfrac{x^2 + 1}{x^2 - 2}$

25. $y = \dfrac{2x}{x^2 - 1}$

26. $y = \dfrac{x^2 - 6x + 12}{x - 4}$

[S] **27.** $y = x\sqrt{4 - x}$

28. $y = x\sqrt{4 - x^2}$

29. $y = \dfrac{x + 2}{x}$

30. $y = x + \dfrac{32}{x^2}$

In Exercises 31–36, determine conditions on the coefficients a, b, and c such that the graph of $f(x) = ax^3 + bx^2 + cx + d$ will resemble the given graph.

[S] **31.**

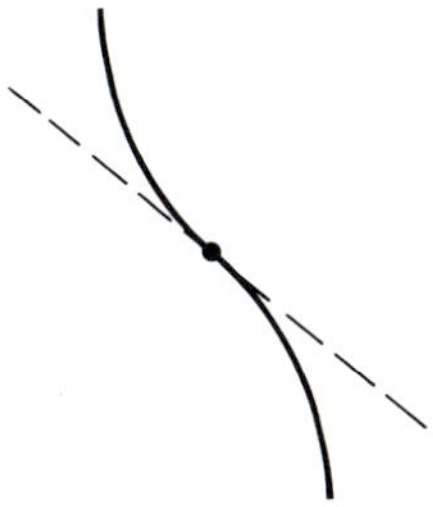

32.

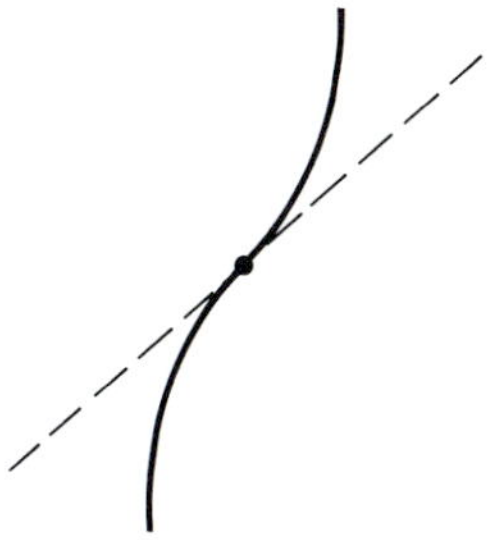

33.

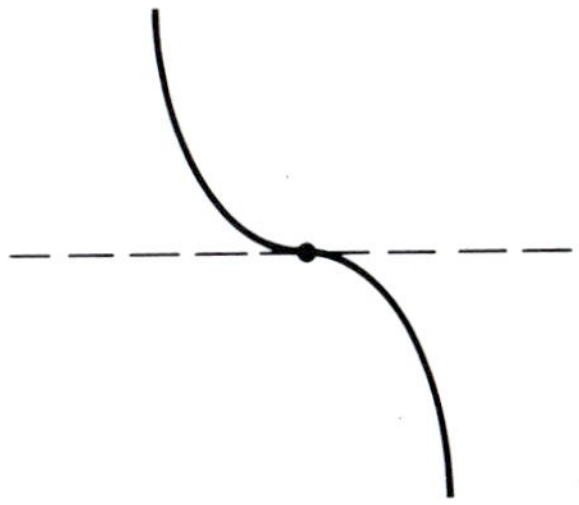

34.

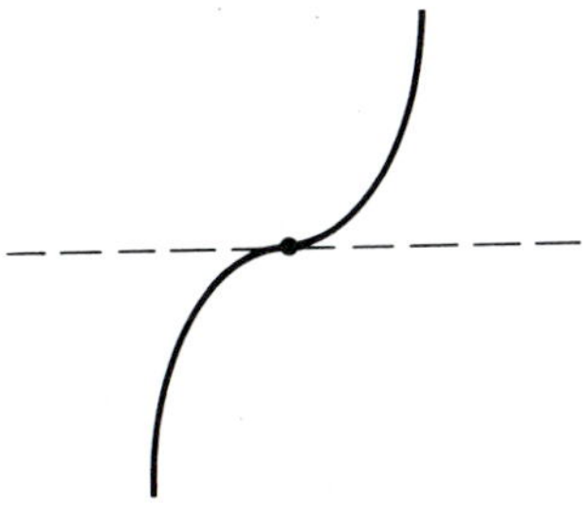

35.

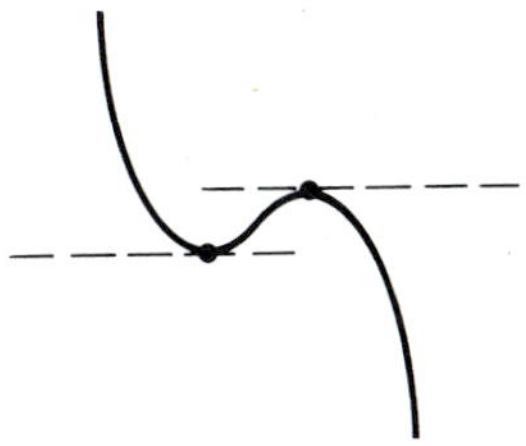

36.

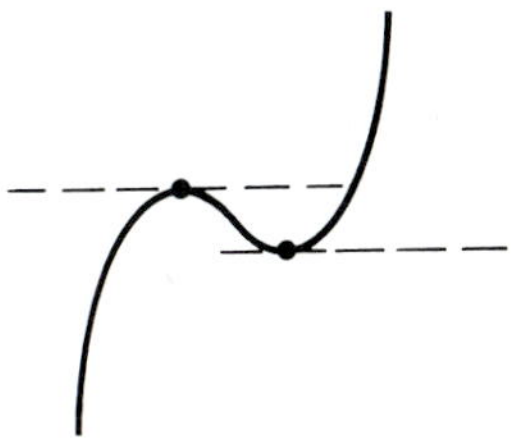

4.6 Curvature (Optional)

Purpose

- To introduce the concept of the curvature of a function and show how it can be calculated by using the first and second derivatives.

The (centripetal) force in this (noncircular) orbit at any point is the same as in a circle of the same curvature. —*Isaac Newton (1642–1727)*

In Section 4.4 we discussed concavity and discovered that the concavity of a graph can be determined from the sign of the second derivative. In this section we will show how the second derivative is used to measure the "amount of curve" a graph has at any one of its points.

We can get some idea of the meaning of the term "amount of curve" by examining the graph shown in Figure 4.29 at points P and Q. Informally we say that the graph of f is "more curved" or has greater **curvature** at P than at Q. Observe further that the curvature of the graph in Figure 4.29 changes from point to point. Such is not the case for all curves. For instance, in Figure 4.30 the smaller circle has the *same* curvature at all of its points. The same is true for the larger circle; however, because the larger circle bends less sharply, its curvature is *less* than that of the smaller circle.

From this comparison it appears that the curvature of a circle is constant, and the larger its radius, the smaller is its curvature. Thus we have the following definition:

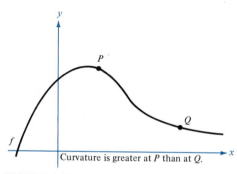

Curvature is greater at P than at Q.

FIGURE 4.29

Definition of Curvature of a Circle	A circle of radius *r* has a **curvature** of $$K = \frac{1}{r}$$

We extend this notion of curvature to graphs other than circles by means of circular approximations to the graphs. For example, in Figure 4.31, to measure the curvature of the graph at the indicated point, we try

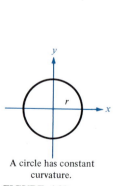

A circle has constant curvature.

FIGURE 4.30

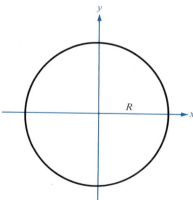

The larger the radius, the smaller the curvature.

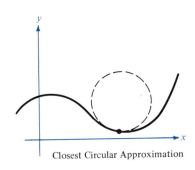

Tangent Circles   Closest Circular Approximation

FIGURE 4.31

to find the circle that most "snugly fits" the curve near the desired point. We call this circle the **closest circular approximation** to the graph at the point in question.

Many circles can be tangent to a curve at the same point. Of all possible circles tangent to the graph of a function f at a point P, we say the closest circular approximation is the circle whose second derivative agrees with the second derivative of f at P.

Definition of Closest Circular Approximation	The circle $(x - h)^2 + (y - k)^2 = r^2$ is called the **closest circular approximation** to the graph of f at the point P if at P:

1.   the circle is tangent to the graph of f

2. $\dfrac{d^2y}{dx^2} = f''(x)$

Example 1

Show that $x^2 + (y - 2)^2 = 4$ is the closest circular approximation to the graph of $f(x) = x^2/4$ at the origin.

Solution: The circle $x^2 + (y - 2)^2 = 4$ passes through the origin, and by implicit differentiation we find that

$$2x + 2(y - 2)\frac{dy}{dx} = 0$$

$$\frac{dy}{dx} = \frac{-x}{y - 2}$$

which is zero at the point $(0, 0)$. Since $f'(0) = 0$, we conclude that the graph of f and the circle are tangent at the origin (see Figure 4.32).

Calculating d^2y/dx^2 we have

$$\frac{d^2y}{dx^2} = \frac{(y - 2)(-1) - (-x)(dy/dx)}{(y - 2)^2}$$

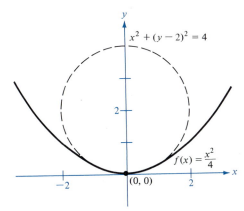

FIGURE 4.32

At the origin,

$$\frac{d^2y}{dx^2} = \frac{(0-2)(-1)-(0)(0)}{(0-2)^2} = \frac{2}{4} = \frac{1}{2}$$

which agrees with $f''(0) = \frac{1}{2}$

Thus by the definition, $x^2 + (y-2)^2 = 4$ is the closest circular approximation to $f(x) = x^2/4$ at $(0,0)$.   ∎

Definition of Curvature

The graph of a function f has **curvature K** at a point if K is the curvature of the closest circular approximation to the graph of f at the point.

Note that this definition of curvature defines only *nonzero* curvature. As a special case we say that the graph of f has zero curvature at a point of inflection or at any point for which $f''(x) = 0$. For instance, the graphs of $f(x) = x^3$ and $g(x) = x^4$ both have zero curvature at the origin and straight lines have zero curvature everywhere.

THEOREM 4.8

If f' and f'' exist when $x = a$, then the curvature of f at the point (a, b) is given by

$$K = \left| \frac{f''(a)}{(1 + [f'(a)]^2)^{3/2}} \right|$$

Proof: Since $(x-h)^2 + (y-k)^2 = r^2$ is the equation of the closest circular approximation, and $K = 1/r$, we have

$$K = \frac{1}{\sqrt{(x-h)^2 + (y-k)^2}}$$

The remainder of this proof consists in replacing $(x-h)$ and $(y-k)$ in the equation above with expressions involving $f'(a)$ and $f''(a)$.

To find these expressions we differentiate $(x-h)^2 + (y-k)^2 = r^2$ implicitly, obtaining

$$\frac{dy}{dx} = \frac{-(x-h)}{(y-k)} \quad \text{and} \quad \frac{d^2y}{dx^2} = \frac{-(y-k) + (x-h)(dy/dx)}{(y-k)^2}$$

The first of these gives

$$x - h = -(y-k)\frac{dy}{dx}$$

Substituting this value of $x - h$ in the second expression gives

$$\frac{d^2y}{dx^2} = -\frac{1 + (dy/dx)^2}{y-k}$$

which can be solved for $y - k$, giving

$$y - k = -\frac{1 + (dy/dx)^2}{d^2y/dx^2}$$

In the equation for K we now replace $x - h$ with $-(y - k)(dy/dx)$, getting

$$K = \frac{1}{\sqrt{(y - k)^2(dy/dx)^2 + (y - k)^2}} = \frac{1}{|y - k|\sqrt{1 + (dy/dx)^2}}$$

Replacing $y - k$ with

$$-\frac{1 + (dy/dx)^2}{(d^2y/dx^2)}$$

we obtain

$$K = \frac{|d^2y/dx^2|}{[1 + (dy/dx)^2]^{3/2}} = \left|\frac{d^2y/dx^2}{[1 + (dy/dx)^2]^{3/2}}\right|$$

Finally, since dy/dx at (a, b) is the same as $f'(a)$, and d^2y/dx^2 at (a, b) is the same as $f''(a)$, we have

$$K = \left|\frac{f''(a)}{(1 + [f'(a)]^2)^{3/2}}\right|$$ ∎

Example 2

Find the equation of the closest circular approximation to the graph of $f(x) = x^2 - 3x + 1$ when $x = 2$.

Solution: $f'(x) = 2x - 3$ and $f''(x) = 2$

Using Theorem 4.8 we have

$$K = \frac{f''(2)}{(1 + [f'(2)]^2)^{3/2}} = \frac{2}{(2)^{3/2}} = \frac{1}{\sqrt{2}} \approx 0.707$$

Thus the radius of the closest circular approximation is $1/K$, or $\sqrt{2}$ (see Figure 4.33).

Since the closest circular approximation is tangent to the graph of f at $(2, -1)$, its center will lie on the normal line through $(2, -1)$. This normal line has a slope of $m = -1/f'(2) = -1$, and hence the equation of the normal line is $y + 1 = -1(x - 2)$, or $y = 1 - x$. Since $r = \sqrt{2}$ is the radius of the closest circular approximation, its center will lie on the normal line at a distance of $\sqrt{2}$ units from $(2, -1)$. Using the distance formula and the fact that $y = 1 - x$, we have

$$\sqrt{2} = \sqrt{(x - 2)^2 + (y + 1)^2} = \sqrt{(x - 2)^2 + (2 - x)^2}$$

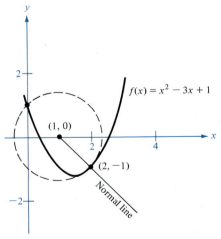

$f(x) = x^2 - 3x + 1$

$(1, 0)$

$(2, -1)$

Normal line

FIGURE 4.33

$$2 = 2x^2 - 8x + 8$$
$$0 = x^2 - 4x + 3 = (x - 3)(x - 1)$$

Therefore, the center of the closest circular approximation is $(1, 0)$, its radius is $\sqrt{2}$, and its equation is

$$(x - 1)^2 + y^2 = 2$$

∎

Section Exercises (4.6)

In Exercises 1–6, find the curvature and the radius of curvature at the given point.

1. $y = 3x - 2$; $x = a$

2. $y = mx + b$; $x = a$

$\boxed{S}$ **3.** $y = 2x^2 + 3$; $x = -1$

4. $y = x + \dfrac{1}{x}$; $x = 1$

5. $y = \sqrt{a^2 - x^2}$; $x = 0$

6. $y = \frac{3}{4}\sqrt{16 - x^2}$; $x = 0$

$\boxed{S}$ **7.** Given the function $y = (x - 1)^2 + 3$.
  (a) Sketch the graph of the function.
  (b) Find an expression for the curvature K of the function.
  (c) Find the coordinates of the point where K is maximum.
  (d) What does K approach as $x \to \pm\infty$?

8. Repeat Exercise 7, using $y = x^3$.

$\boxed{S}$ **9.** The function $f(x) = x + (1/x)$ has a relative minimum at $(1, 2)$. Find the equation of the circle that most closely approximates the graph of f at that point.

$\boxed{C}$ **10.** Find the equation of the circle that most closely approximates the graph of $f(x) = x + (1/x)$ at the point $(2, \frac{5}{2})$.

$\boxed{S}$ **11.** Verify for the ellipse $x^2 + 4y^2 = 4$ that the curvature is greatest at the endpoints of the major axis and least at the endpoints of the minor axis.

12. Find all points at which the curvature is zero for the graph of $y = (x - 1)^3 + 3$.

$\boxed{S}$ **13.** The two curves $y_1 = ax(b - x)$ and $y_2 = x/(x + 2)$ intersect at only one point P; they have a common tangent line at P and the same curvature at P.
  (a) Find all a and b that satisfy these conditions.
  (b) Make sketches of the curves for each possible choice of a and b.

$\boxed{C}$ **14.** The smaller the curvature in a bend of a road, the faster a car can travel. Assume that the maximum speed around a turn is inversely proportional to the square root of the curvature. If a car, moving on the path $y = \frac{1}{3}x^3$ (x and y measured in miles), can safely go 30 miles per hour at $(1, \frac{1}{3})$, how fast can it go at $(\frac{3}{2}, \frac{9}{8})$?

15. An engineer lays out a highway as indicated in Figure 4.34. BC is an arc of a circle. AB and CD are straight lines tangent to the circular arc. Criticize the design.

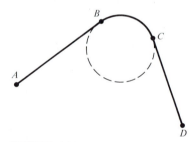

FIGURE 4.34

16. Given the function $f(x) = x^{2/3}$.
  (a) Sketch the graph of the function.
  (b) Find an expression for the curvature.
  (c) What is the curvature at the origin?
  (d) What does the curvature approach as x approaches infinity?

For Review and Class Discussion

True or False

1. _____ A circle of radius 10 has a greater curvature than a circle of radius 5.

2. _____ The graph of a second-degree polynomial function has its greatest curvature at the point where its tangent line is horizontal.

3. _____ The graph of a function has zero curvature at a point of inflection.

4. ____ The graph of every polynomial function has a point at which the curvature is greater than 1.

5. ____ The graph of every polynomial function has a point at which the curvature is less than 1.

Miscellaneous Exercises (Ch. 4)

In Exercises 1–24, make use of domain, range, symmetry, asymptotes, intercepts, relative extrema, or points of inflection to obtain an accurate graph of each function.

1. $f(x) = 4x - x^2$

2. $f(x) = 4x^3 - x^4$

3. $f(x) = x\sqrt{16 - x^2}$

4. $f(x) = x + \dfrac{4}{x^2}$

$\boxed{\text{S}}$ **5.** $f(x) = \dfrac{x + 1}{x - 1}$

6. $f(x) = x^2 + \dfrac{1}{x}$

7. $f(x) = x^3 + x + \dfrac{4}{x}$

8. $f(x) = x^3(x + 1)$

9. $f(x) = (x - 1)^3(x - 3)^2$

10. $f(x) = (x - 3)(x + 2)^3$

11. $f(x) = (5 - x)^3$

12. $f(x) = (x^2 - 4)^2$

$\boxed{\text{S}}$ **13.** $f(x) = x^{1/3}(x + 3)^{2/3}$

14. $f(x) = (x - 2)^{1/3}(x + 1)^{2/3}$

15. $f(x) = x^3 + \dfrac{243}{x}$

16. $f(x) = \dfrac{2x}{1 + x^2}$

17. $f(x) = \dfrac{4}{1 + x^2}$

18. $f(x) = \dfrac{x^2}{1 + x^4}$

19. $f(x) = |x^2 - 9|$

20. $f(x) = |9 - x^2|$

21. $f(x) = |x^3 - 3x^2 + 2x|$

$\boxed{\text{S}}$ **22.** $f(x) = |x - 1| + |x - 3|$

23. $f(x) = \dfrac{1}{|x - 1|}$

24. $f(x) = \dfrac{x - 1}{1 + 3x^2}$

25. Find any maximum and minimum points that exist on the graph of $x^2 + 4y^2 - 2x - 16y + 13 = 0$:
(a) without using calculus
(b) using calculus

$\boxed{\text{S}}$ **26.** Consider the function $f(x) = x^n$ for positive integer values of n.
(a) For what values of n does the function have a relative minimum at the origin?
(b) For what values of n does the function have a point of inflection at the origin? Explain.

In Exercises 27–31, verify the Mean Value Theorem for each function on the given interval. Sketch the graph of each function and identify the point(s) guaranteed by the theorem.

27. $f(x) = \dfrac{2x + 3}{3x + 2}$; [1, 5]

28. $f(x) = \dfrac{1}{x}$; [1, 4]

29. $f(x) = x^{2/3}$; [1, 8]

30. $f(x) = |x^2 - 9|$; [0, 2]

31. $f(x) = \sqrt{x} - 2x$; [0, 4]

32. Can the Mean Value Theorem be applied to the function $f(x) = 1/x^2$ on $[-2, 1]$? Explain.

33. If $f(x) = 3 - |x - 4|$, verify that $f(1) = f(7) = 0$ and yet f' does not equal 0 for any x in [1, 7]. Does this contradict Rolle's Theorem? Explain.

$\boxed{\text{S}}$ **34.** For the function $f(x) = Ax^2 + Bx + C$, determine the value of c guaranteed by the Mean Value Theorem on the interval $[x_1, x_2]$.

35. Demonstrate the result of Exercise 34 for $f(x) = 2x^2 - 3x + 1$ on the interval [0, 4].

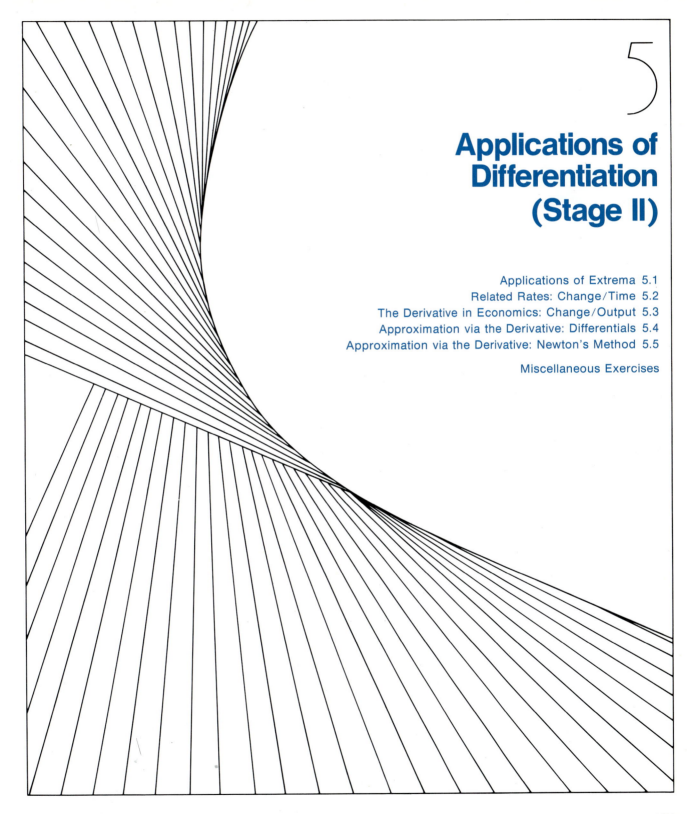

Applications of Differentiation (Stage II)

5

5.1   **Applications of Extrema**

Purpose
- To apply extrema of functions to the solution of problems from physics, engineering, and economics.
- To outline a general approach to solving minimum-maximum problems.

And this is only the beginning of much more sublime Geometry, pertaining to even the most difficult and most beautiful problems of applied mathematics, which without our differential calculus or something similar no one could attack with any such ease.   —Gottfried Leibniz (1646–1716)

Up to this point in our study of calculus most of the problems have been given in the form of mathematical equations. Our primary task has been to choose and then apply an appropriate calculus procedure to arrive at a solution. In this chapter we focus on application problems that are not originally stated in terms of mathematical equations; hence an additional *problem formulation* stage is required in the solution process. In this and the next two sections, we provide some guidelines to assist you in constructing mathematical formulations for several common applications of calculus.

One of the most common applications of calculus in fields other than mathematics involves the determination of minimum or maximum values. Consider how frequently we hear or read terms like greatest profit, least cost, cheapest product, least time, greatest voltage, optimum size, least area, greatest strength, or greatest distance. Before outlining a general method of solution for such problems, we present an example.

Example 1

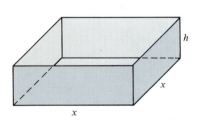

Open Box with Square Base

FIGURE 5.1

An open box having a square base is to be constructed from 108 in.² of material. What should be the dimensions of the box to obtain a maximum volume?

Solution: Since the box has a square base, its volume is given by (see Figure 5.1)

$$V = x^2h$$

Furthermore, since the box is open at the top, its surface area is given by

$$S = 108 = x^2 + 4xh$$

Now since V is to be maximized, it is helpful to express V as a function of just one variable. To do this we solve for h in terms of x to obtain

$$4xh = 108 - x^2$$
$$h = \frac{108 - x^2}{4x}, \qquad 0 < x \leq \sqrt{108}$$

Substituting for h we get

$$V = x^2\left(\frac{108 - x^2}{4x}\right) = 27x - \frac{x^3}{4}$$

On the interval $0 < x \le \sqrt{108}$, the critical numbers for V are the solutions to

$$\frac{dV}{dx} = 27 - \frac{3x^2}{4} = 0$$

$$3x^2 = 108$$

$$x = 6$$

The Second-Derivative Test will verify that V has a relative maximum when $x = 6$. Furthermore, this relative maximum is the maximum of V on $(0, \sqrt{108}]$ and we conclude that the maximum volume of the box occurs when $x = 6$ in. and $h = 3$ in. ∎

There are several obvious stages in the solution of Example 1. We first made a sketch and assigned symbols to all *known* quantities and quantities *to be determined*. Second, we identified an equation for the quantity to be maximized. Then we reduced this equation to obtain a function of one independent variable. And, finally, we applied the techniques of calculus to find the value of x that yielded the desired maximum. This pattern of solution is quite helpful for most minimum-maximum problems (often called "min-max problems"). We suggest the following four-step outline for solving maximum and minimum problems:

Solving for Minimum or Maximum Values

1. Assign symbols to all given quantities and quantitites to be determined. When feasible, make a sketch.
2. Write a "primary" equation for the quantity to be maximized or minimized.
3. Reduce this "primary" equation to one having a single independent variable—this may involve the use of "secondary" equations (restrictions) relating the independent variables of the "primary" equation.
4. Determine the desired maximum or minimum value by the techniques of calculus.

Note about step 4: Recall that to determine the maximum or minimum value of a continuous function f on a closed interval, we compare the values of f at its relative extrema to the values of f at the endpoints of the interval. The largest (smallest) of these values is the desired maximum (minimum).

Example 2

Find two positive numbers that minimize the sum of twice the first number plus the second if the product of the two numbers is 288.

Solution:

1. Let x be the first number, y the second, and S the sum to be minimized.
2. Since we wish to minimize S, the *primary* equation is

$$S = 2x + y$$

3. Since the product of the two numbers is 288, we have the *secondary* equation

$$xy = 288 \qquad \text{or} \qquad y = \frac{288}{x}$$

Thus we can rewrite the primary equation in terms of x alone as

$$S = 2x + \frac{288}{x}, \qquad 0 < x$$

4. Differentiating to find the critical values yields

$$\frac{dS}{dx} = 2 - \frac{288}{x^2} = 0$$

$$x^2 = 144$$

$$x = \pm 12$$

Choosing the positive x-value, we find that the two numbers are

$$x = 12 \qquad \text{and} \qquad y = \frac{288}{12} = 24$$

The Second-Derivative Test readily verifies that $x = 12$ yields a minimum for S. ∎

Example 3

Find the points on the graph of $y = 4 - x^2$ that are closest to the point $(0, 2)$.

Solution:

1. Figure 5.2 indicates that there are two points at a minimum distance from $(0, 2)$.
2. We are asked to minimize the distance d; hence we use the Distance Formula to obtain the primary equation

$$d = \sqrt{(x - 0)^2 + (y - 2)^2}$$

3. Using the secondary equation $y = 4 - x^2$, we can rewrite the primary equation as

$$d = \sqrt{(x)^2 + (4 - x^2 - 2)^2} = \sqrt{x^2 + (2 - x^2)^2}$$

$$= \sqrt{x^4 - 3x^2 + 4}$$

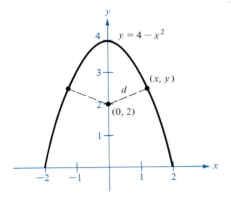

$y = 4 - x^2$

(x, y)

d

$(0, 2)$

FIGURE 5.2

Since d is smallest when the expression under the radical is smallest, we need only find the critical values of

$$f(x) = x^4 - 3x^2 + 4$$

4. Differentiation yields

$$f'(x) = 4x^3 - 6x = 0$$
$$2x(2x^2 - 3) = 0$$
$$x = 0, \ \sqrt{\tfrac{3}{2}}, \ -\sqrt{\tfrac{3}{2}}$$

The Second-Derivative Test verifies that $x = 0$ yields a relative maximum, while both $x = \sqrt{\tfrac{3}{2}}$ and $x = -\sqrt{\tfrac{3}{2}}$ yield a minimum distance; hence on the graph of $y = 4 - x^2$, the closest points to $(0, 2)$ are $(\sqrt{\tfrac{3}{2}}, \tfrac{5}{2})$ and $(-\sqrt{\tfrac{3}{2}}, \tfrac{5}{2})$. ■

Example 4

A rectangular page is to contain 24 in.^{2} of print. The margins at the top and bottom of the page are each $1\frac{1}{2}$ in. wide. The margins on each side are 1 in. What should the dimensions of the page be so that the least amount of paper is used?

Solution:

1. See Figure 5.3.
2. Letting A be the area to be minimized, our primary equation is $A = xy$.
3. The printed area inside the margins is given by $24 = (x - 3)(y - 2)$. Solving this equation for y, we have

$$y = \frac{24}{x - 3} + 2 = \frac{2x + 18}{x - 3}$$

Thus the primary equation becomes

$$A = (x)\left(\frac{2x + 18}{x - 3}\right) = 2\left(\frac{x^2 + 9x}{x - 3}\right)$$

Since the margins at the top and the bottom of the page must add up to 3, we are only interested in values of A when $x > 3$.

4. Now to find the minimum of

$$A = 2\left(\frac{x^2 + 9x}{x - 3}\right), \qquad 3 < x < \infty$$

we differentiate with respect to x and obtain

$$\frac{dA}{dx} = 2\left[\frac{(x - 3)(2x + 9) - (x^2 + 9x)}{(x - 3)^2}\right] = 2\left[\frac{x^2 - 6x - 27}{(x - 3)^2}\right]$$

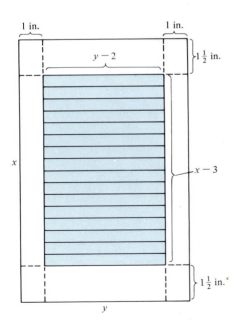

1 in.   1 in.

$y - 2$

$1\frac{1}{2}$ in.

x

$x - 3$

$1\frac{1}{2}$ in.

y

FIGURE 5.3

Setting $dA/dx = 0$ we have

$$x^2 - 6x - 27 = 0$$
$$(x - 9)(x + 3) = 0$$
$$x = 9 \quad \text{or} \quad x = -3$$

Since x must be greater than 3, we choose $x = 9$, and the First-Derivative Test will confirm that A is a relative minimum at this point. Therefore, the dimensions of the page should be $x = 9$ in. and $y = [24/(9 - 3)] + 2 = 6$ in. ∎

In Example 4 we eliminated the negative critical number because it was not a feasible solution. In many physical applications it is not necessary to test all the critical numbers and the endpoints of the given interval to see if they yield the desired maximum or minimum because common sense often permits us to eliminate all but one of them.

Example 5

Two posts, one 20 ft high and the other 28 ft high, stand 30 ft apart. They are to be stayed by wires attached to a single stake, running from ground level to the tops of the posts. Where should the stake be placed to use the least wire?

Solution:

1. See Figure 5.4.
2. Let W be the length of the wire (to be minimized); then our primary equation is $W = y + z$.
3. In this problem, rather than solving for y in terms of z (or vice versa), we solve for both y and z in terms of a third variable x (see Figure 5.4).

$$x^2 + 20^2 = y^2, \quad y = \sqrt{x^2 + 400}$$
$$(30 - x)^2 + 28^2 = z^2, \quad z = \sqrt{x^2 - 60x + 1684}$$

Thus we have

$$W = y + z = \sqrt{x^2 + 400} + \sqrt{x^2 - 60x + 1684}, \quad 0 \leq x \leq 30$$

4. Differentiating W yields

$$\frac{dW}{dx} = \frac{x}{\sqrt{x^2 + 400}} + \frac{x - 30}{\sqrt{x^2 - 60x + 1684}}$$

From $dW/dx = 0$ we obtain

$$x\sqrt{x^2 - 60x + 1,684} = (30 - x)\sqrt{x^2 + 400}$$
$$x^4 - 60x^3 + 1,684x^2 = x^4 - 60x^3 + 1,300x^2$$
$$- 24,000x + 360,000$$

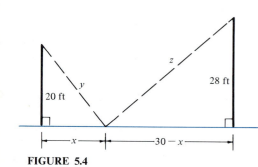

20 ft

z

28 ft

y

x — $30 - x$

FIGURE 5.4

$$384x^2 + 24{,}000x - 360{,}000 = 0$$
$$2x^2 + 125x - 1{,}875 = 0$$
$$(x + 75)(2x - 25) = 0$$
$$x = -75 \quad \text{or} \quad x = 12.5$$

Since $x = -75$ is not in the interval $[0, 30]$ and since both $x = 0$ and $x = 30$ are not feasible solutions (see Figure 5.4), we conclude that the wire should be stayed at $x = 12.5$ ft from the 20-ft pole. ■

Section Exercises (5.1)

1. A rectangle has a perimeter of 100 ft. What length and width should it have so that its area is maximum?
2. What positive number x minimizes the sum of x and its reciprocal?

[S] 3. The sum of one number and two times a second number is 24. What numbers should be selected so that their product is as large as possible?

4. The difference of two numbers is 50. Select the two numbers so that their product is as small as possible.
5. Find two positive numbers whose sum is 110 and whose product is a maximum.
6. Find two positive numbers such that the sum of the first and twice the second is 100 and whose product is a maximum.
7. Find two positive numbers whose product is 192 and whose sum is a minimum.

[S] 8. The product of two positive numbers is 192. What numbers should be chosen so that the sum of the first plus three times the second is a minimum?

9. A rancher has 200 ft of fencing to enclose two adjacent rectangular corrals. (See Figure 5.5.) What dimensions should be used so that the enclosed area will be a maximum?
10. A dairy farmer plans to fence in a rectangular pasture adjacent to a river. He figures that the pasture must contain 180,000 m^2 in order to provide enough grass for his herd. What dimensions would require the least amount of fencing if no fencing is needed along the river?

[S] 11. Find the coordinates of the point on the curve $y = \sqrt{x}$ closest to the point $(4, 0)$.

12. Find the coordinates of the point on the curve $y = x^2$ that is closest to the point $(2, \frac{1}{2})$.
13. An open box is to be made from a square piece of material, 12 in. on a side, by cutting equal squares from each corner and turning up the sides. Find the volume of the largest box that can be made in this manner. (See Figure 5.6.)

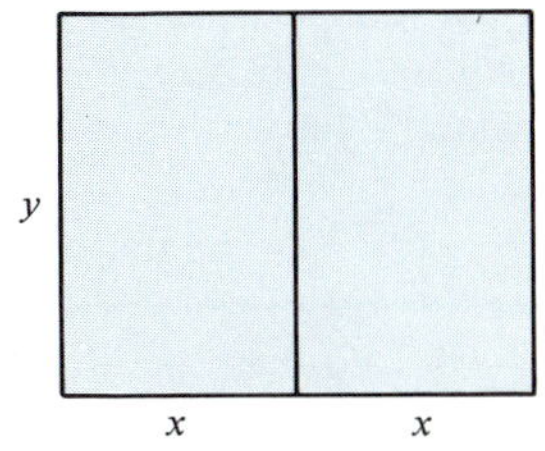

FIGURE 5.5

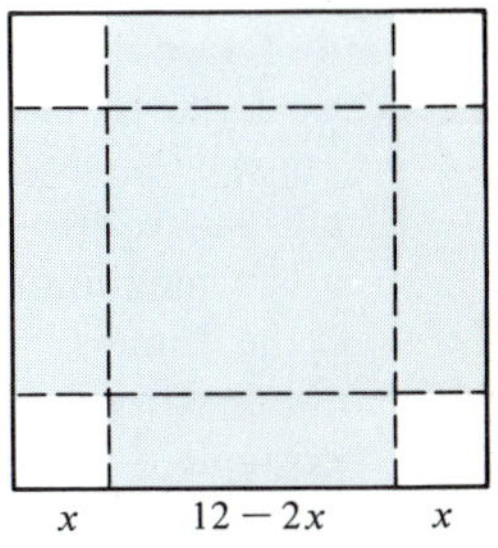

FIGURE 5.6

[S] 14. (a) Solve Exercise 13 if the square piece of material is s inches on a side.
(b) If the dimensions of the square piece of material are doubled, how does the volume change?

[C] 15. An open box is to be made from a rectangular piece of material by cutting equal squares from each corner and turning up the sides. Find the dimensions of the box of maximum volume if the material has dimensions 2 ft by 3 ft.

16. A net enclosure for golf practice is open at one end, as shown in Figure 5.7. Find the dimensions that require the least amount of netting if the volume of the enclosure is to be $83\frac{1}{3}$ m^3.
17. An indoor physical fitness room consists of a rectangular region with a semicircle on each end. If the perimeter of the room is to be a 200-m running track, find the dimensions that will make the area of the *rectangular* region as large as possible.
18. A page is to contain 30 in.^{2} of print. The margins at the top and the bottom of the page are each 2 in. wide. The margins on each side are only 1 in. wide. Find the dimensions of the page so that the least paper is used.

[C][S] 19. A right circular cylinder is to be designed to hold 12 fluid ounces of a soft drink and to use the minimal amount of material in its construction. Find the required dimen-

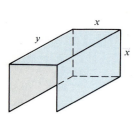

FIGURE 5.7

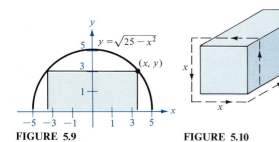

FIGURE 5.8

sions, assuming that 1 fluid ounce requires 1.80469 cubic inches.

20. Work Exercise 19 if the right circular cylinder has a volume of V_0 cubic inches.

21. A rectangle is bounded by the x- and y-axes and the graph of $y = (6 - x)/2$, as shown in Figure 5.8. What length and width should the rectangle have so that its area is a maximum?

S **22.** Find the dimensions of the largest isosceles triangle that can be inscribed in a circle of radius r.

23. A rectangle is bounded by the x-axis and the semicircle $y = \sqrt{25 - x^2}$, as shown in Figure 5.9. What length and width should the rectangle have so that its area is a maximum?

24. Find the dimensions of the largest rectangle that can be inscribed in a semicircle of radius r. (See Exercise 23.)

25. A rectangular package to be sent by a postal service can have a maximum combined length and girth (perimeter of a cross section) of 100 in. Find the dimensions of the package of maximum volume that can be sent. (See Figure 5.10.) (Assume the cross section is square.)

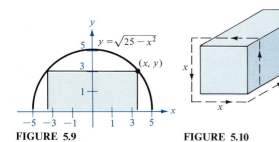

FIGURE 5.9   **FIGURE 5.10**

26. The combined perimeter of an equilateral triangle and a square is 10. Find the dimensions of the triangle and square that produce a minimum total area.

27. The combined perimeter of a circle and a square is 16. Find the dimensions of the circle and square that produce a minimum total area.

28. A figure is formed by adjoining two hemispheres to each end of a right circular cylinder. The total volume of the figure is 12 in.³. Find the radius of the cylinder that produces the minimum surface area.

S **29.** A man is in a boat 2 mi from the nearest point on the coast. He is to go to a point Q, 3 mi down the coast and 1 mi inland. If he can row at 2 mi/h and walk at 4 mi/h, toward what point on the coast should he row in order to reach point Q in the least time? (See Figure 5.11.)

30. The conditions are the same as in Exercise 29 except that the man can row at 4 mi/h.

S **31.** A right triangle is formed in the first quadrant by the x- and y-axes and a line through the point $(2, 3)$. Find the vertices of the triangle so that its area is minimum.

C **32.** A right triangle is formed in the first quadrant by the x- and y-axes and a line through the point $(1, 2)$. Find the vertices of the triangle so that the length of the hypotenuse is minimum. (See Figure 5.12.)

S **33.** A wooden beam has a rectangular cross section of height h and width w, as shown in Figure 5.13. The strength S of the beam is directly proportional to the width and the square of the height. What are the dimensions of the strongest beam that can be cut from a round log of diameter 24 in.? (Hint: $S = kh^2w$, where k is the proportionality constant.)

34. The illumination from a light source is directly proportional to the strength of the source and inversely proportional to the square of the distance from the source. Two light sources of intensities I_1 and I_2, respectively, are d units apart. At what point on the line segment joining the two sources is the illumination least?

35. Show that among all positive numbers x and y with $x^2 + y^2 = r^2$, the sum $x + y$ is largest when $x = y$.

36. Find the volume of the largest right circular cone that can be inscribed in a sphere of radius r.

S **37.** Find the volume of the largest right circular cylinder that can be inscribed in a sphere of radius r.

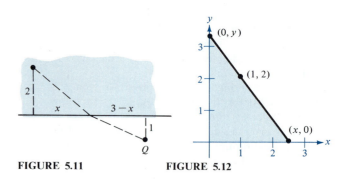

FIGURE 5.11   **FIGURE 5.12**

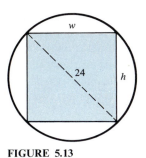

FIGURE 5.13

38. Find the dimensions of the trapezoid of greatest area that can be inscribed in a semicircle of radius r.

39. The formula for the output P of a battery is given by $P = VI - RI^2$, where V is the voltage, R is the resistance, and I is the current. Find the current (measured in amperes, A) that corresponds to a maximum value for P in a battery where $V = 12$ volts (V) and $R = 0.5$ ohms (Ω).

5.2 **Related Rates: Change/Time**

Purpose
- To develop a method for calculating the rates of change of two or more related variables, each of which changes with respect to time.

Before setting out to attack any definite problem it behooves us first, without making any selection, to assemble those truths that are obvious as they present themselves to us and afterwards, proceeding step by step, to inquire whether any others can be deduced from these. —René Descartes (1596–1650)

In this section our primary interest lies in the relationship between the *rates of change* of two or more related variables. For example, if water is flowing into a bucket at a specific number of gallons per minute, how fast is the depth of the water increasing? Or, if a person walks away from a lamp post at x feet per second, how fast does the end of his or her shadow move away from the lamp post? Problems such as these, which involve variables whose rates of change with respect to time are related in some manner, are called **related-rate** problems.

If two or more time-dependent variables are related by a given equation, then the derivative of this equation with respect to time shows how the individual rates of change of the variables are related. This is illustrated in the next example.

Example 1

A pebble is dropped into a calm pool of water, causing ripples in the form of concentric circles. If each ripple moves out from the center at a rate of 1 ft/s, at what rate is the total area of disturbed water increasing at the end of 4 s?

Solution: First, we make a sketch (see Figure 5.14).

Let r be the radius of the largest ripple. Then the area inside that circle is given by $A = \pi r^2$. By differentiating *with respect to t,* we have the related-rate equation

$$\frac{dA}{dt} = 2\pi r \frac{dr}{dt}$$

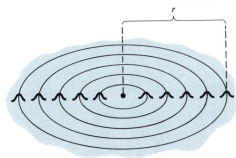

Movement of Circular Ripples

FIGURE 5.14

The fact that each of the variables is a function of t gives us the right to differentiate implicitly with respect to t even though t does not appear in the equation. The result of this differentiation gives us an equation that describes the relationship between the rates of change of each of the variables. Since we know that the radius r of a ripple increases 1 ft/s ($dr/dt = 1$), it follows that $r = 4$ when $t = 4$, and by the end of 4 s, we have

$$\frac{dA}{dt} = 2(\pi)(4 \text{ ft})(1 \text{ ft/s}) = 8\pi \text{ ft}^2/\text{s}$$

Therefore, the total area of the disturbed water is increasing at the rate of 8π ft^2/s at the end of 4 s. ∎

As with maximum and minimum problems, we recommend the following four-step procedure for solving related-rate problems.

Suggested Procedure for Solving Related-Rate Problems	1. Assign symbols to all *given* quantities and quantities *to be determined.* Make a labeled sketch when feasible.
	2. Write an equation involving the variables whose rates of change either are given or are to be determined.
	3. By the Chain Rule, implicitly differentiate both sides of this equation with *respect to time t.*
	4. Into the resulting equation substitute all known values for the variables and their rates of change, and then solve for the required rate of change.

Example 2

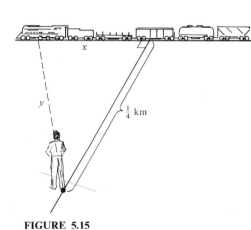

FIGURE 5.15

Suppose a person is standing along a straight road $\frac{1}{4}$ kilometer (km) from a railroad that crosses the road at right angles. If a train is moving at a constant speed of 80 kilometers per hour (km/h), at what rate is the distance between the observer and the engine changing when the engine is $\frac{1}{2}$ km past the crossing?

Solution:

1. See Figure 5.15. We know $dx/dt = 80$ km/h and we are to find dy/dt when $x = \frac{1}{2}$ km.
2. The variables x and y are related by the equation

$$y^2 = x^2 + (\tfrac{1}{4})^2$$

3. Implicitly differentiating this equation with respect to t, we obtain

$$2y\frac{dy}{dt} = 2x\frac{dx}{dt} + 0$$

or
$$\frac{dy}{dt} = \frac{x}{y}\frac{dx}{dt}$$

4.   When $x = \frac{1}{2}$, $y = \sqrt{(\frac{1}{2})^2 + (\frac{1}{4})^2} = \sqrt{5}/4$, and thus
$$\frac{dy}{dt} = \frac{1/2}{\sqrt{5}/4}(80) = \frac{160}{\sqrt{5}} \approx 71.6 \text{ km/h}$$

Note that because the distance was given in kilometers and the time in hours, the velocity has the units kilometers per hour. ∎

Example 3

A windlass is used to tow a boat to the dock. The rope is attached to the boat at a point 15 ft below the level of the windlass. If the windlass pulls in the rope at a rate of 30 ft/min, at what rate is the boat approaching the dock when there is 75 ft of rope out? When there is 25 ft of rope out?

Solution:

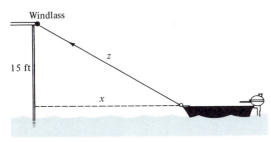

FIGURE 5.16

1.   See Figure 5.16.
2.   The variables x and z are related by the equation $x^2 + 15^2 = z^2$.
3.   By implicitly differentiating this equation with respect to time t, we obtain the equation
$$2x\frac{dx}{dt} + 0 = 2z\frac{dz}{dt}$$
4.   When $z = 75$, $x = \sqrt{75^2 - 15^2} = 30\sqrt{6}$. Thus it follows that
$$2(30\sqrt{6})\frac{dx}{dt} = 2(75)(-30) = -4500$$

or
$$\frac{dx}{dt} = -\frac{75}{\sqrt{6}} \approx -30.6 \text{ ft/min}$$

When $z = 25$, $x = 20$ and
$$2(20)\frac{dx}{dt} = 2(25)(-30)$$

or
$$\frac{dx}{dt} = -37.5 \text{ ft/min}$$
∎

Example 4

Water is flowing into a trough at a rate of 100 cubic centimeters per second (cm³/s). The trough has a length of 3 m and cross sections in the form of a trapezoid, whose height is 50 cm, whose lower base is 25 cm, and whose upper base is 1 m. At what rate is the water level rising when the depth of the water is 25 cm?

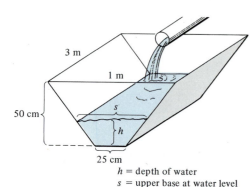

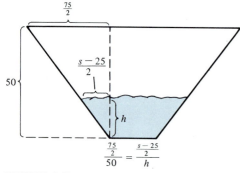

h = depth of water
s = upper base at water level

FIGURE 5.17

FIGURE 5.18

Solution:

1. See Figure 5.17. We know $dV/dt = 100$ cm³/s, and we are to find dh/dt when $h = 25$ cm.

2. By converting to centimeters and looking at a cross section of the trough (Figure 5.17), we calculate the volume of water in terms of h and s to be

$$V = (\text{length})(\text{area of trapezoidal cross section})$$
$$= (300)(h)\left(\frac{25 + s}{2}\right)$$

Since no specific information is given about s or its derivative ds/dt, we must write s in terms of the depth h. From similar triangles (see Figure 5.18) we have

$$\frac{75/2}{50} = \frac{(s - 25)/2}{h}$$

and it follows that $\qquad s = \frac{3}{2}h + 25$

Thus the volume in terms of h is given by

$$V = (300h)\left[\frac{(3h/2) + 50}{2}\right] = 75(3h^2 + 100h)$$

3. By differentiating with respect to t, we obtain

$$\frac{dV}{dt} = 75(6h + 100)\frac{dh}{dt}$$

4. When $h = 25$, it follows that

$$100 = 75(250)\frac{dh}{dt}$$

or $\qquad \dfrac{dh}{dt} = \dfrac{100}{75(250)} = \dfrac{2}{375}$ cm/s   ∎

Example 5

The combined electrical resistance R resulting from two resistances R_1 and R_2 connected in parallel is given by

$$\frac{1}{R} = \frac{1}{R_1} + \frac{1}{R_2}$$

where R, R_1, and R_2 are measured in ohms (Ω). Assume that R_1 and R_2 are increasing at rates of 1 and 1.5 Ω/s, respectively. At what rate is R increasing when $R_1 = 50\ \Omega$ and $R_2 = 75\ \Omega$?

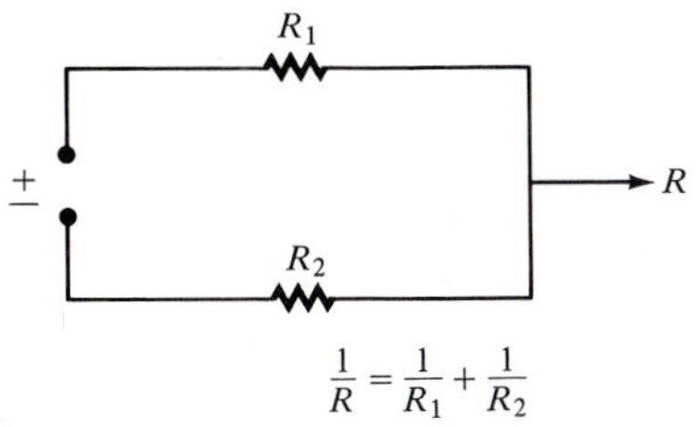

FIGURE 5.19

Solution:

1. See Figure 5.19.
2. The equation relating the variables is the given equation.
3. By differentiating the equation

$$\frac{1}{R} = \frac{1}{R_1} + \frac{1}{R_2}$$

with respect to t, we have

$$\left(\frac{-1}{R^2}\right)\left(\frac{dR}{dt}\right) = \left(\frac{-1}{R_1^{\,2}}\right)\left(\frac{dR_1}{dt}\right) + \left(\frac{-1}{R_2^{\,2}}\right)\left(\frac{dR_2}{dt}\right)$$

4. When $R_1 = 50$ and $R_2 = 75$, we have

$$R = \frac{1}{(1/50) + (1/75)} = \frac{1}{(3 + 2)/150} = 30$$

Therefore, since we know $dR_1/dt = 1$ and $dR_2/dt = 1.5$, it follows that

$$\left(\frac{-1}{30^2}\right)\frac{dR}{dt} = \left(\frac{-1}{50^2}\right)(1) + \left(\frac{-1}{75^2}\right)(1.5)$$

$$\frac{dR}{dt} = \frac{900}{2500} + \frac{900}{3750} = \frac{9}{25} + \frac{6}{25} = \frac{3}{5} = 0.6\ \Omega/\text{s}$$

■

Section Exercises (5.2)

[S] **1.** A ladder 25 ft long is leaning against a house. If the base of the ladder is pulled away from the house wall at a rate of 2 ft/s, how fast is the top moving down the wall when the base of the ladder is:
(a) 7 ft from the wall? (b) 15 ft from the wall?
(c) 24 ft from the wall?

2. A boat is pulled in by means of a winch on the dock 12 ft above the deck of the boat. If the winch pulls in rope at the rate of 4 ft/s, determine the speed of the boat when there is 13 ft of rope out. What happens to the speed of the boat as it gets closer to the dock?

3. An air traffic controller spots two planes at the same altitude converging on a point as they fly at right angles to one another. (See Figure 5.20.) One plane is 150 mi from the point and is moving at 450 mi/h. The other plane is 200 mi from the point and has a speed of 600 mi/h.
(a) At what rate is the distance between the planes decreasing?
(b) How much time does the traffic controller have to get one of the planes on a different flight path?

4. A point $P(0, y)$ moves along the y-axis at a constant rate of R feet per second, while a point $Q(x, 0)$ moves along the x-axis at a constant rate of r feet per second. Find an expression for the rate of change of the distance between P and Q.

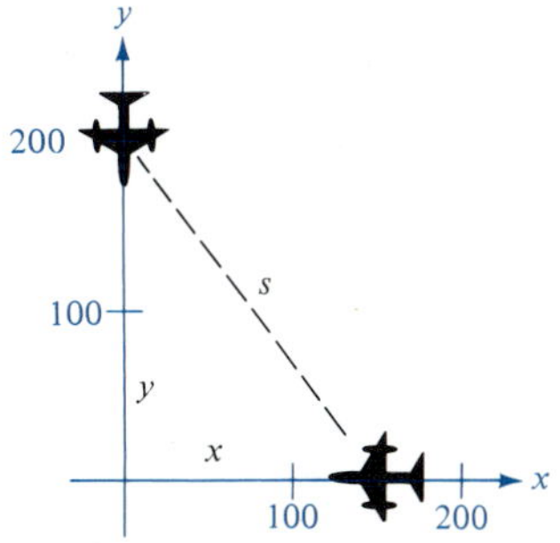

FIGURE 5.20

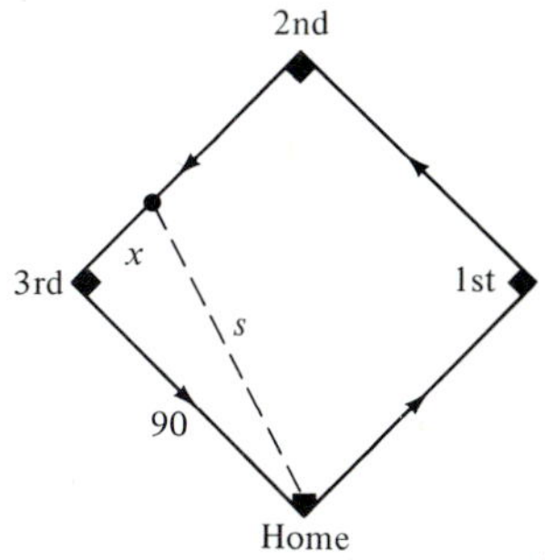

FIGURE 5.21

5. A baseball diamond has the shape of a square with sides 90 ft long. (See Figure 5.21, p. 209.) If a player is running from second to third base at a speed of 28 ft/s, at what rate is his distance from home plate changing when he is 30 ft from third?

6. For the baseball diamond in Exercise 5, suppose the player is running from first to second at a speed of 28 ft/s. Find the rate at which the distance from home plate is changing when the player is 30 ft from second.

7. Let A be the area of a circle of radius r. Find dA/dt. If dr/dt is constant, is dA/dt constant? Explain why or why not.

8. Let V be the volume of a sphere of radius r. Find the relationship between dV/dt and dr/dt. If dr/dt is constant, is dV/dt constant? Explain why or why not.

9. Use the result of Exercise 8 to find the rate at which the volume of a sphere is changing when the radius is 6 in., if the radius is changing at a rate of $\frac{1}{2}$ in./min.

10. A spherical balloon is inflated with gas at the rate of 20 ft³/min. How fast is the radius of the balloon increasing at the instant the radius is:
    (a) 1 ft?   (b) 2 ft?

S 11. At a sand and gravel plant, sand is falling off a conveyer and onto a conical pile at the rate of 10 ft³/min. The diameter of the base of the cone is approximately three times the altitude. At what rate is the height of the pile changing when it is 15 ft high?

12. A conical tank (with vertex down) is 10 ft across the top and 12 ft deep. If water is flowing into the tank at the rate of 10 ft³/min, find the rate of change of the depth of the water the instant it is 8 ft deep.

S 13. A swimming pool is 40 ft long, 10 ft wide, 4 ft deep at the shallow end, and 9 ft deep at the deep end, the bottom being an inclined plane. (See Figure 5.22.) Assume that water is being pumped into the pool at 10 ft³/min and that there is 4 ft of water at the deep end.
    (a) What percentage of the pool is filled?
    (b) At what rate is the water level rising?

14. A trough is 12 ft long and 3 ft across the top, and its ends are isosceles triangles with an altitude of 3 ft. If water is being pumped into the trough at 2 ft³/min, how fast is the water level rising when it is 1 ft deep?

S 15. A man 6 ft tall walks at a rate of 5 ft/s away from a light that is 15 ft above the ground. (See Figure 5.23.) When the man is 10 ft from the base of the light:
    (a) at what rate is the tip of his shadow moving?
    (b) at what rate is the length of his shadow changing?

16. As a spherical raindrop falls, it reaches a layer of dryer air at the lower levels of the atmosphere and begins to evaporate. If this evaporation occurs at a rate proportional to the surface area ($s = 4\pi r^2$) of the droplet, show that the radius shrinks at a constant rate.

17. The edge of a cube is expanding at the rate of 3 cm/s. How fast is the volume changing when each edge is:
    (a) 1 cm?   (b) 10 cm?

18. The conditions are the same as in Exercise 17. Now measure how fast the surface area is changing when each edge is:
    (a) 1 cm   (b) 10 cm

19. A point is moving along the curve $y = x^2$ so that dx/dt is 2 cm/min. Find dy/dt when:
    (a) $x = 0$ (b) $x = 3$

20. The conditions are the same as in Exercise 19 but now measure the rate of change of the distance between the point and the origin.

21. A point is moving along the curve $y = 1/(1 + x^2)$ so that $dx/dt = 2$ cm/min. Find dy/dt when:
    (a) $x = -2$ (b) $x = 0$
    (c) $x = 2$ (d) $x = 10$

22. Under the same conditons as in Exercise 21, what does dy/dt approach as x approaches infinity? Sketch the graph of the function.

S 23. An airplane is flying at 31,680 ft above ground level and the flight path passes directly over a radar antenna. The radar picks up the plane and determines that the distance s from the unit to the plane is changing at the rate of 4 mi/min when that distance is 10 mi. Compute the speed of the plane in miles per hour. (See Figure 5.24.)

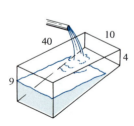

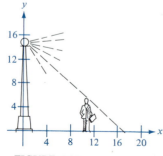

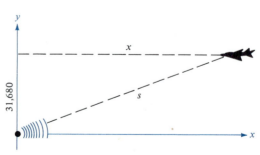

FIGURE 5.22   **FIGURE 5.23**   **FIGURE 5.24**

5.3 The Derivative in Economics: Change/Output

Purpose
- To introduce two common uses of the derivative in economics, (1) to compute marginals and (2) to compute elasticities.

The economic world is a misty region. The first explorers used unaided vision. Mathematics is the lantern by which what before was dimly visible now looms up in firm, bold outlines. The old illusions disappear. We see better. We also see further. —*Irving Fisher (1867–1947)*

In the previous section we mentioned that one of the most common ways to measure change is with respect to time. In this section we will look at a few important rates of change in the field of economics, rates of change that are not measured with respect to time. For example, economists refer to **marginal profit, marginal revenue,** and **marginal cost** as the rates of change of the profit, revenue, and cost with respect to the number of units produced or sold. An equation that relates these three quantities is

$$P = R - C$$

where

$$P = \text{total profit}, \qquad R = \text{total revenue}, \qquad C = \text{total cost}$$

Differentiating each of these gives the *marginals,* a term used in economics to denote derivatives:

$$\frac{dP}{dx} = \text{marginal profit}$$

$$\frac{dR}{dx} = \text{marginal revenue}$$

$$\frac{dC}{dx} = \text{marginal cost}$$

It is worthwhile noting that the problems in this section are primarily minimum and maximum problems; hence the four-step procedure used in Section 5.1 is an appropriate model to follow.

Example 1

Suppose a company has determined that its total revenue R for a certain product is given by $R = -x^3 + 450x^2 + 52{,}500x$, where R is measured in dollars and x is the number of units produced. What production level will yield a maximum revenue?

Solution:

1. A sketch is given in Figure 5.25.
2. The *primary* equation is the given revenue equation:

$$R = -x^3 + 450x^2 + 52{,}500x$$

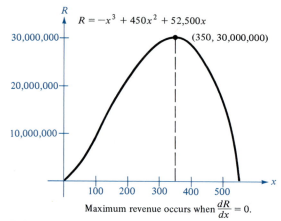

Maximum revenue occurs when $\dfrac{dR}{dx} = 0$.

FIGURE 5.25

3. Since R is already given as a function of one variable, we do not need a *secondary* equation.

4. Differentiating and setting the derivative equal to zero yields

$$\frac{dR}{dx} = -3x^2 + 900x + 52,000 = 0$$

$$-x^2 + 300x + 17,500 = 0$$

$$(-x + 350)(x + 50) = 0$$

The critical values are $x = 350$ and $x = -50$. Choosing the positive value of x, we conclude that the maximum revenue is obtained when 350 units are produced. ∎

To study the effect of production levels on cost, economists use the **average cost function** $\overline{C}$ defined as

$$\overline{C} = \frac{C}{x}$$

where $C = f(x)$ is the total cost function.

Example 2

A company estimates that the cost (in dollars) of producing x units of a certain product is given by

$$C = 800 + 0.04x + 0.0002x^2$$

Find the production level that minimizes the average cost per unit. Compare this minimal average cost to the average cost when 400 units are produced.

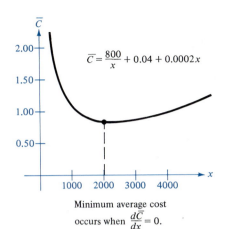

$$\overline{C} = \frac{800}{x} + 0.04 + 0.0002x$$

Minimum average cost occurs when $\frac{d\overline{C}}{dx} = 0$.

FIGURE 5.26

Solution:

1. We let $\overline{C}$ be the average cost and x be the number of units produced. (See Figure 5.26.)

2. The primary equation representing the quantity to be minimized is

$$\overline{C} = \frac{C}{x}$$

3. Substituting from the given equation for C, we have

$$\overline{C} = \frac{800 + 0.04x + 0.0002x^2}{x} = \frac{800}{x} + 0.04 + 0.0002x$$

4. Setting the derivative equal to zero yields

$$\frac{d\overline{C}}{dx} = -\frac{800}{x^2} + 0.0002 = 0$$

$$x^2 = \frac{800}{0.0002} = 4{,}000{,}000$$

$$x = 2{,}000$$

For 2000 units the average cost is

$$\overline{C} = \frac{800}{2000} + 0.04 + (0.0002)(2000) = \$0.84$$

For 400 units the average cost is

$$\overline{C} = \frac{800}{400} + 0.04 + (0.0002)(400) = \$2.12$$

■

Example 3

A certain business sells 2000 items per month at a price of $10 each. If it can sell 250 more items per month for each $0.25 reduction in price, what price per item will maximize its monthly revenue?

Solution:

1. Let

$x = $ number of items sold per month

$p = $ price of each item (in dollars)

$R = $ monthly revenue (in dollars)

2. Since the monthly revenue is to be maximized, the primary equation is

revenue = (number of units sold)(price per unit)

$$R = xp$$

3. The number of items x is related to p in that x increases 250 units every time p drops $0.25 from the original cost of $10. This is described by the equation

$$x = 2{,}000 + 250 \left(\frac{10 - p}{0.25} \right) = 12{,}000 - 1{,}000p$$

or
$$p = 12 - \frac{x}{1{,}000}$$

Substituting this result into the revenue equation, we have

$$R = x \left(12 - \frac{x}{1000} \right) = 12x - \frac{x^2}{1000}$$

4. Setting the derivative equal to zero, we obtain

$$\frac{dR}{dx} = 12 - \frac{x}{500} = 0$$

$$x = 12(500) = 6000$$

Therefore, the critical number is $x = 6000$ and we conclude that the price that maximizes the revenue is

$$p = 12 - \frac{6000}{1000} = \$6.00$$

∎

In Example 3 the equation

$$x = 2000 + 250 \left(\frac{10 - p}{0.25} \right) = 2000 + 1000(10 - p) = 1000(12 - p)$$

determines the number of units that can be sold—that is, the *demand*—at a given price. We call the function given by $x = f(p) = 1000(12 - p)$ the **demand function.** Note that as the price increases, the demand decreases, and vice versa.

Example 4

A business, in marketing a certain item, has discovered that the demand for the item is represented by the equation

$$x = \frac{2500}{p^2}$$

Assuming that the total revenue R is given by $R = xp$ and the cost for producing x items is given by $C = 0.5x + 500$, find the price per unit that yields a maximum profit.

Solution:

1. Let P represent the profit.

2. Since we are seeking a maximum profit, the primary equation is

$$P = R - C$$

3. Using the given equations for R and C, we have

$$P = R - C = xp - (0.5x + 500)$$

Solving for p in the equation $x = 2500/p^2$, we have

$$p = \frac{50}{\sqrt{x}}$$

Thus the profit equation is

$$P = x\left(\frac{50}{\sqrt{x}}\right) - 0.5x - 500 = 50\sqrt{x} - 0.5x - 500$$

4. Setting the derivative equal to zero, we obtain

$$\frac{dP}{dx} = \frac{25}{\sqrt{x}} - 0.5 = 0$$

$$\sqrt{x} = \frac{25}{0.5} = 50$$

$$x = 2500$$

Finally, we conclude that the maximum profit occurs when the price is

$$p = \frac{50}{\sqrt{2500}} = \frac{50}{50} = \$1.00$$

■

Notice that to find the maximum profit, we differentiate the equation $P = R - C$ and set dP/dx equal to zero. From the equation

$$\frac{dP}{dx} = \frac{dR}{dx} - \frac{dC}{dx} = 0$$

it follows that the maximum profit occurs when the marginal revenue is equal to the marginal cost.

One way economists describe the behavior of a demand function is by a term called the **price elasticity of demand.** It describes the relative responsiveness of consumers to a change in the price of an item. If $x = f(p)$ is a differentiable demand function, then the price elasticity of demand is given by

$$\eta = \frac{p}{x}\frac{dx}{dp}$$

(where η is the Greek lowercase letter eta). For a given price, if $|\eta| < 1$, the demand is said to be **inelastic;** if $|\eta| > 1$, the demand is said to be **elastic.**

Example 5

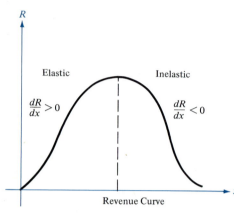

$\frac{dR}{dx} > 0$ Elastic   Inelastic $\frac{dR}{dx} < 0$

Revenue Curve

FIGURE 5.27

Show that the demand function in Example 4, $x = 2500/p^2$, is elastic.

Solution:

$$\eta = \left(\frac{p}{x}\right)\left(\frac{dx}{dp}\right) = \left(\frac{p}{2500/p^2}\right)\left[\frac{(-2)(2500)}{p^3}\right] = \frac{-5000p^3}{2500p^3} = -2$$

Since $|\eta| = 2$, we conclude that the demand is elastic for any price. ∎

Elasticity has an interesting relationship to the total revenue:

1. If the demand is *elastic,* then a decrease in the price per unit is accompanied by a sufficient increase in unit sales to cause the total revenue to increase.
2. If the price is *inelastic,* then the increase in unit sales accompanying a decrease in price per unit is *not* sufficient to increase revenue and the total revenue decreases. (See Figure 5.27.)

The next example verifies this relationship between elasticity and total revenue.

Example 6

Show that for a differentiable demand function, the marginal revenue is positive when the demand is elastic and negative when the demand is inelastic. (Assume that the quantity demanded increases as the price decreases and thus dx/dp is negative.)

Solution: Since the revenue is given by $R = xp$, we calculate the marginal revenue to be

$$\frac{dR}{dx} = x\left(\frac{dp}{dx}\right) + p = p\left[\left(\frac{x}{p}\right)\left(\frac{dp}{dx}\right) + 1\right]$$

$$= p\left[\frac{1}{\left(\frac{p}{x}\right)\left(\frac{dx}{dp}\right)} + 1\right] = p\left(\frac{1}{\eta} + 1\right)$$

If the demand is inelastic, then $|\eta| < 1$ implies that $1/\eta < -1$, since dx/dp is negative. (We assume that x and p are positive.) Therefore,

$$\frac{dR}{dx} = p\left(\frac{1}{\eta} + 1\right)$$

is negative.

Similarly, if the demand is elastic, then $|\eta| > 1$ and $-1 < 1/\eta$, which implies that dR/dx is positive. ∎

Section Exercises (5.3)

In Exercises 1–4, find the number of units x that produce a maximum revenue R.

1. $R = 900x - 0.1x^2$

2. $R = 600x^2 - 0.02x^3$

S **3.** $R = 1{,}000{,}000x/(0.02x^2 + 1{,}800)$

4. $R = 30x^{2/3} - 2x$

In Exercises 5–8, find the number of units x that produce the minimum average cost per unit $\overline{C}$ ($\overline{C} = C/x$).

5. $C = 0.125x^2 + 20x + 5000$

S **6.** $C = 0.001x^3 - 5x + 250$

7. $C = 3000x - x^2\sqrt{300 - x}$

8. $C = (2x^3 - x^2 + 5000x)/(x^2 + 2500)$

In Exercises 9–12, find the price per unit p that produces the maximum profit P ($P = R - C$).

S **9.** $C = 100 + 30x;\ x = 90 - p$

10. $C = 2400x + 5200;\ x = \sqrt{15{,}000 - 2.5p}$

11. $C = 4000 - 40x + 0.02x^2;\ x = 5000 - 100p$

12. $C = 35x + 2\sqrt{x - 1};\ x = (40 - p)^2 + 1$

13. A manufacturer of lighting fixtures has daily production costs of $C = 800 - 10x + \frac{1}{4}x^2$. How many fixtures x should be produced each day to minimize costs?

14. Let x be the amount (in hundreds of dollars) a company spends on advertising and let P be the profit. If $P = 230 + 20x - \frac{1}{2}x^2$, what amount of advertising gives the maximum profit?

S **15.** A manufacturer of radios charges $90 per unit when the average production cost per unit is $60. However, to encourage large orders from distributors, the manufacturer will reduce the charge by $0.10 per unit for each unit ordered in excess of 100 (for example, there would be a charge of $88 per radio for an order size of 120). Find the largest order size the manufacturer should allow so as to realize maximum profit.

16. A real estate office handles 50 apartment units. When the rent is $180 per month, all units are occupied. However, on the average, for each $10 increase in rent, one unit becomes vacant. Each occupied unit requires an average of $12 per month for service and repairs. What rent should be charged to realize the most profit?

17. A power station is on one side of a river that is $\frac{1}{2}$ mi wide, and a factory is 6 mi downstream on the other side. It costs $6 per foot to run power lines overland and $8 per foot to run them underwater. Find the most economical path for the transmission line from the power station to the factory.

18. An offshore oil well is 1 mi off the coast. The refinery is 2 mi down the coast. If the cost of laying pipe in the ocean is twice as expensive as on land, in what path should the pipe be constructed in order to minimize the cost?

S **19.** Assume that the amount of money deposited in a bank is proportional to the square of the interest rate the bank pays on this money. Furthermore, the bank can reinvest this money at 12%. Find the interest rate the bank should pay to maximize profit. (Use the simple interest formula.)

20. Prove that the average cost is minimum at the value of x where the average cost equals the marginal cost.

21. Given the cost function $C = 2x^2 + 5x + 18$.
(a) Find the value of x where the average cost is minimum.
(b) For the value of x found in part (a), show that the marginal cost and average cost are equal (see Exercise 20).

22. Given the cost function $C = x^3 - 6x^2 + 13x$.
(a) Find the value of x where the average cost function is minimum.
(b) For the value of x found in part (a), show that the marginal cost and average cost are equal.

S **23.** The demand function for a certain product is given by $x = 20 - 2p^2$.
(a) Consider the point $(2, 12)$. If the price decreases by 5%, determine the corresponding percentage increase in quantity demanded.
(b) **Average elasticity of demand** is defined to be the percentage change in quantity by the percentage change in price. Use the percentage of part (a) to find the average elasticity at $(2, 12)$.
(c) Find the exact elasticity at $(2, 12)$ by using the formula in this section. Compare the result with that of part (b).
(d) Find an expression for total revenue ($R = xp$), and find the values of x and p that maximize R.
(e) For the value of x found in part (d), show that $|\eta| = 1$.

24. Assume that the demand equation is $p^3 + x^3 = 9$.
(a) Find η when $x = 2$.
(b) Find the values of x and p that maximize the total revenue ($R = xp$).
(c) Show that $|\eta| = 1$ for the value of x found in part (b).

25. The demand function for a particular commodity is

given by $p = (16 - x)^{1/2}$, $0 \leq x \leq 16$. Determine the price and quantity for which revenue is maximum.

26. The demand function is given by the equation $x = a/p^m$, where $m > 1$. Show that $\eta = -m$ (i.e., in terms of approximate price changes, a 1% increase in price results in a $m\%$ decrease in quantity demanded).

27. A given commodity has a demand function given by $p = 100 - \frac{1}{2}x^2$ and the total cost function of $C = 40x + 375$.
    (a) What price gives the maximum profit?

(b) What is the average cost per unit if production is set to give maximum profit?

28. Rework Exercise 27, using the cost function $C = 50x + 375$.

29. When a wholesaler sold a certain product at $25 per unit, sales were 800 units each week. However, after a price raise of $5, the average number of units sold dropped to 775 per week. Assume that the demand function is linear and find the price that will maximize the total revenue.

5.4 Approximation via the Derivative: Differentials

Purpose
- To define the differentials dy and dx and demonstrate their use in approximating Δy or Δx.

The differential of a given expression is the difference of the values which this expression receives on the given curve and on the curve that results from it by an infinitely small change —Leonhard Euler (1707–1783)

We previously defined the derivative as the limit (as $\Delta x \to 0$) of the ratio $\Delta y/\Delta x$, and it seemed natural to retain the ratio symbolism for the limit itself. Thus we denoted the derivative by

$$\frac{dy}{dx} = \lim_{\Delta x \to 0} \frac{\Delta y}{\Delta x}$$

even though we did not think of dy/dx as the quotient of the two separate quantities dy and dx. In this section we give separate meanings to dy and dx in such a way that their quotient, when $dx \neq 0$, is equal to the derivative of y with respect to x. We do this in the following manner:

Definition of Differentials, dx and dy	Let $y = f(x)$, where $f'(x)$ exists. The **differential of x** (denoted by dx) is any real number. The **differential of y** (denoted by dy) is given by $$dy = f'(x)\,dx$$

Note that in this definition dx can have any value; however, when using dy to approximate Δy, it is best to choose dx small, and we denote this choice by $dx = \Delta x$.

Example 1

Find dy when $x = 1$ and $dx = 0.01$, where $y = 3x^2 - 8x + 2$.

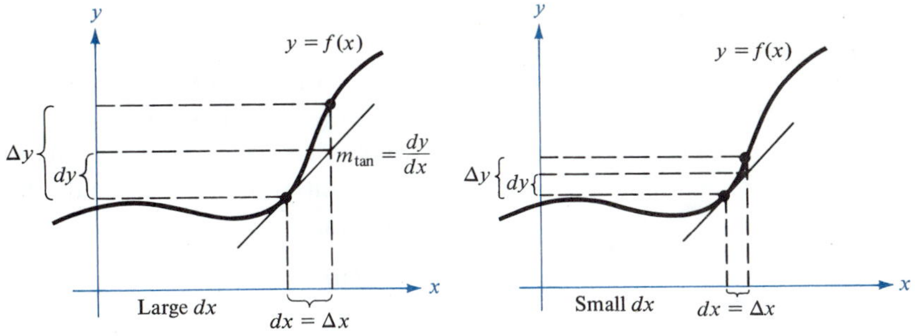

FIGURE 5.28

dy as an approximation of Δy

Solution: Consider

$$y = f(x) = 3x^2 - 8x + 2$$

then

$$dy = f'(x)\,dx = (6x - 8)\,dx = [(6)(1) - 8](0.01)$$
$$= (-2)(0.01) = -0.02$$

∎

Geometrically, from Figure 5.28 we see that if $dx = \Delta x$ is small, then dy can be used as an approximation for Δy, because the difference $\Delta y - dy$ becomes smaller as Δx gets smaller.

Example 2

Compare the values of dy and Δy as x changes from 2 to 2.01, where $y = x^3 - x^2 + 3x - 2$.

Solution: As x changes from 2 to 2.01, y changes from 8 to 8.110501. Thus $\Delta y = 0.110501$.

Calculating the differential of y, we have

$$dy = (3x^2 - 2x + 3)\,dx$$

When $x = 2$ and $\Delta x = dx = 0.01$,

$$dy = (12 - 4 + 3)(0.01) = 0.11$$

Thus Δy and dy differ by only

$$\Delta y - dy = 0.000501$$

∎

The calculation of Δy often involves considerably more effort than the calculation of dy. For instance, in Example 2, to calculate Δy we had to evaluate the cubic $x^3 - x^2 + 3x - 2$ at both $x = 2$ and $x = 2.01$, whereas the calculation of dy merely required the evaluation of $(3x^2 - 2x + 3)(0.01)$ at $x = 2$. Because dy is often simpler to calculate than Δy, it is frequently used to estimate errors that arise from physical measurements. This use is demonstrated in the next example.

Example 3

The circumference of a solid sphere was measured to be 30 in. However, the measuring technique may have introduced an error of ±0.1 in. If the volume of the sphere is calculated by using 30 in. as the circumference, use differentials to estimate the maximum possible error in the calculated volume.

Solution: If x is the circumference of the sphere, then $x = 2\pi r$ and $x/2\pi = r$. Thus the volume is given by

$$V = \left(\frac{4\pi}{3}\right)\left(\frac{x}{2\pi}\right)^3 = \left(\frac{4\pi}{3}\right)\left(\frac{1}{2\pi}\right)^3(x^3)$$

We estimate the maximum error ΔV by dV and obtain

$$dV = \left(\frac{4\pi}{3}\right)\left(\frac{1}{2\pi}\right)^3(3x^2)\,dx = \left(\frac{x^2}{2\pi^2}\right)dx$$

For $x = 30$ and $dx = \Delta x = \pm0.1$,

$$dV = \left[\frac{(30)^2}{2\pi^2}\right](\pm0.1) \approx \pm4.56 \text{ in.}^3 \qquad\blacksquare$$

The estimated error in the calculated volume in Example 3 may appear to be rather large. However, when we compare this error with the total volume of the sphere, we obtain the *relative* error

$$\frac{dV}{V} = \frac{4.56}{456} \approx 0.01$$

which corresponds to a *percentage* error of

$$\frac{dV}{V}(100) = 1\%$$

The relative and percentage errors usually give a clearer picture of the effect of an error in measurement. In general,

$$\text{relative error in } y = \left(\frac{\text{error in } y}{y}\right) \approx \frac{dy}{y}$$

$$\text{percentage error in } y = \left(\frac{\text{error in } y}{y}\right)(100) \approx \frac{dy}{y}(100)$$

Example 4

Use differentials to approximate $\sqrt[3]{126}$.

Solution: To estimate $\sqrt[3]{126}$, we use the cube root function $y = \sqrt[3]{x}$. A convenient choice for x is $x = 125$. Thus $dx = \Delta x = 1$, and $y = \sqrt[3]{125} = 5$.

Considering

$$\sqrt[3]{126} = y + \Delta y = 5 + \Delta y$$

and estimating Δy by

$$dy = y'\,dx = \tfrac{1}{3}x^{-2/3}\,dx$$

it follows that, for $x = 125$,

$$dy = \left(\frac{1}{3}\right)\left(\frac{1}{125^{2/3}}\right)(1) = \frac{1}{75}$$

Therefore,
$$\sqrt[3]{126} \approx 5 + \frac{1}{75} \approx 5.0133$$

This result lies within 0.0001 of the actual value of $\sqrt[3]{126}$. ■

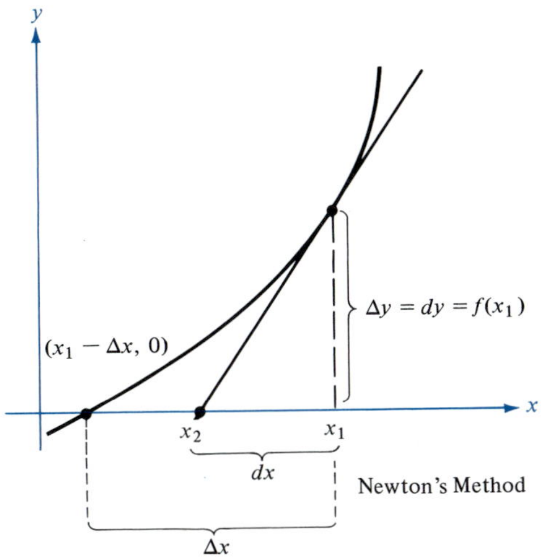

FIGURE 5.29

We have indicated that it is often simpler to approximate Δy by dy than it is to perform the more complicated calculations necessary to determine the value of Δy. However, with today's ready accessibility to electronic calculators, the need to avoid messy calculations has lessened. Ironically, it is this same accessibility to calculators that has generated increased interest in another type of approximation by differentials, the *approximation of* Δx *by* dx, *where* $dy = \Delta y$. In fact, this approximation of Δx by dx is precisely the basis for Newton's Method discussed in the next section.

As we will see in Section 5.5, Newton's Method is an iterative process for approximating an x-intercept of a function based upon an initial estimate x_1. In Figure 5.29 if x_1 is the initial estimate, then $x_2 = x_1 - dx$ is a second and better estimate of the actual x-intercept, $x_1 - \Delta x$.

Section Exercises (5.4)

In Exercises 1–6, find the differential dy of each function.

1. $y = 3x^2 - 4$

2. $y = 2x^{3/2}$

3. $y = \dfrac{x+1}{2x-1}$

4. $y = \sqrt{x^2 - 4}$

[S] **5.** $y = x\sqrt{1 - x^2}$

6. $y = \sqrt{x} + \dfrac{1}{\sqrt{x}}$

In Exercises 7–10, compute dy and Δy for the given values of x and dx.

[S] **7.** $y = x^2$
(a) $x = 2,\ dx = \frac{1}{4}$
(b) $x = -1,\ dx = \frac{1}{4}$
(c) $x = -1,\ dx = \frac{1}{2}$

8. $y = x^3$
(a) $x = 1,\ dx = \frac{1}{8}$
(b) $x = 1,\ dx = \frac{1}{2}$
(c) $x = -2,\ dx = -\frac{1}{4}$

9. $y = \dfrac{1}{x^2},\ x = 2,\ dx = \frac{1}{4}$

[C] **10.** $y = \dfrac{1}{\sqrt[3]{x}},\ x = 2,\ dx = 0.01$

11. The area of a square of side x is given by $A(x) = x^2$.
(a) Compute dA and ΔA in terms of x and Δx.
(b) In Figure 5.30 shade the region whose area is dA.
(c) Using a different colored pencil, shade the region whose area is $\Delta A - dA$.

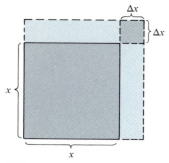

FIGURE 5.30

In Exercises 12–16, approximate the given quantity by means of differentials.

12. $\sqrt[3]{63}$

S 13. $\sqrt[4]{83}$

14. $\dfrac{1}{\sqrt{101}}$

15. $\dfrac{1}{\sqrt[3]{25}}$

16. $\sqrt{50}$

C 17. For each of Exercises 12–16, find the actual value of the quantity, correct to five decimal places.

18. The measurement of the edge of a cube is found to be 12 in., with a possible error of 0.03 in. Using differentials, approximate the maximum possible error in computing the following:
(a) the volume of the cube
(b) the surface area of the cube

S 19. The radius of a sphere is claimed to be 6 in., with a possible error of 0.02 in. Using differentials, approximate the maximum possible error in calculating the following:
(a) the volume of the sphere
(b) the surface area of the sphere
(c) What is the relative error in parts (a) and (b)?

20. The profit P for a company is given by $P =$ $(500x - x^2) - (\frac{1}{2}x^2 - 77x + 3000)$. Approximate the change in profit as production changes from $x = 115$ to $x = 120$ units. Approximate the percentage change in profit when x changes from $x = 115$ to $x = 120$ units.

S 21. The period of a pendulum is given by $T = 2\pi\sqrt{L/g}$, where L is the length of the pendulum in feet, g is the acceleration due to gravity, and T is time in seconds. Suppose that the pendulum has been subjected to an increase in temperature so that the length increases by $\frac{1}{2}\%$.
(a) What is the approximate percentage change in the period?
(b) Using the result of part (a), find the approximate error in this pendulum clock in one day.

22. The acceleration of gravity varies somewhat from point to point on the earth's surface. Suppose a pendulum is to be used to find the acceleration of gravity at a given point on the earth's surface. Use the formula $T = 2\pi\sqrt{L/g}$ to approximate the percentage error in the value of g if you can identify the period T to within $\frac{1}{10}\%$ of its true value.

23. Each of the rules for finding derivatives can be stated in differential form. For example, let u and v be functions of x; then

$$\frac{d}{dx}[u^n] = nu^{n-1}\frac{du}{dx}$$

has the differential form $d(u^n) = nu^{n-1}\,du$, and

$$\frac{d}{dx}[u + v] = \frac{du}{dx} + \frac{dv}{dx}$$

has the differential form $d(u + v) = du + dv$. Write the product and quotient rules in differential form by following the pattern above.

For Review and Class Discussion

True or False

1. ____ If $y = x + c$, then $dy = dx$.

2. ____ If $y = x^3$, then for $x \neq 0$, it follows that $0 < dy$.

3. ____ If $y = ax + b$, then $\Delta y/\Delta x = dy/dx$.

4. ____ If y is differentiable, then $\lim_{\Delta x \to 0}(\Delta y - dy) = 0$.

5. ____ If $y = f(x)$, f is increasing and differentiable, and $\Delta x > 0$, then $\Delta y \geq dy$.

5.5 **Approximation via the Derivative: Newton's Method**

Purpose
- To develop Newton's Method for approximating zeros of functions.

And by repeating the computation the solution may be found continually to greater and greater accuracy. —*Isaac Newton (1642–1727)*

In Chapters 4 and 5 we frequently needed to find the zeros of a function. [The zeros of f are those values of x for which $f(x) = 0$.] Until now our functions have been carefully chosen so that elementary algebraic techniques suffice for finding their zeros. For example, the zeros of

$$f(x) = x^2 - 6x + 8, \qquad g(x) = 2x^2 - 3x - 7,$$
$$h(x) = x^3 - 2x^2 - x + 2$$

can all be found by factoring or by the quadratic formula. However, in practice we frequently encounter functions whose zeros are more difficult to find. For example, the zeros of a function as simple as

$$f(x) = x^3 - x + 1$$

cannot be found by elementary algebraic methods. In such cases the tangent line may serve as a convenient tool for approximating the zeros.

To illustrate this, suppose we wish to find a zero for a function f, which is differentiable on the interval (a, b). If $f(a)$ and $f(b)$ differ in sign, then by the Intermediate Value Theorem, f must possess at least one zero $(x = c)$ in (a, b). Suppose we estimate this zero to occur when $x = x_1$ (see Figure 5.31).

Newton's Method for approximating zeros is based on the assumption that f and the tangent to f at $(x_1, f(x_1))$ both cross the x-axis at *about* the same point. Since we can easily calculate the x-intercept for this tangent line, we use it as our second (and hopefully better) estimate for the zero of f. The tangent line passes through the point $(x_1, f(x_1))$ with a slope of $f'(x_1)$. In point-slope form the equation of the tangent line is therefore

$$y - f(x_1) = f'(x_1)(x - x_1)$$

or

$$y = f'(x_1)(x - x_1) + f(x_1)$$

By setting $y = 0$ and solving for x, we have

$$x = x_1 - \frac{f(x_1)}{f'(x_1)}$$

Thus from our initial guess we arrive at a new estimate

$$x_2 = x_1 - \frac{f(x_1)}{f'(x_1)}$$

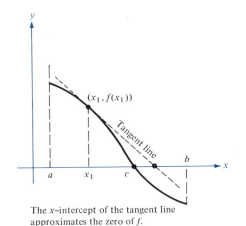

The x-intercept of the tangent line approximates the zero of f.

FIGURE 5.31

We may improve upon x_2 and calculate yet a third estimate:

$$x_3 = x_2 - \frac{f(x_2)}{f'(x_2)}$$

Repeated application of this process is called **Newton's Method.**

Newton's Method for Approximating the Zeros of a Function	Let f be differentiable on (a, b) and $f(c) = 0$, where c is in (a, b). Then to approximate c we do the following: 1. Make an initial estimate x_1 "close" to c. 2. Determine a new approximation by $$x_{n+1} = x_n - \frac{f(x_n)}{f'(x_n)}$$ 3. If $\|x_n - x_{n+1}\|$ is less than the desired accuracy, let x_{n+1} serve as our final approximation. Otherwise, return to step 2 and calculate a new approximation. The process of calculating each successive approximation is called an **iteration.**

Example 1

Calculate three iterations of Newton's Method to approximate a zero of $f(x) = x^2 - 2$. Use $x_1 = 1$ as the initial guess. The calculations are shown in Table 5.1.

TABLE 5.1

n	x_n	$f(x_n)$	$f'(x_n)$	$\dfrac{f(x_n)}{f'(x_n)}$	$x_n - \dfrac{f(x_n)}{f'(x_n)}$
1	1.000000	-1.000000	2.000000	-0.500000	1.500000
2	1.500000	0.250000	3.000000	0.083333	1.416667
3	1.416667	0.006945	2.833334	0.002451	1.414216
4	1.414216				

Solution: Of course, in this example we know that the two zeros of the function are $\pm\sqrt{2}$. To six decimal places, $\sqrt{2} = 1.414214$. Thus after only three iterations of Newton's Method, we have obtained an approximation (1.414216) that is within 2 millionths of the actual root.   ■

As you might expect, Newton's Method has become more popular with the emergence of electronic calculators.

Example 2

Use Newton's Method to approximate the zeros of $f(x) = 2x^3 + x^2 - x + 1$. Continue the iterations until two successive approximations differ by less than 0.0001.

Solution: We begin by sketching a graph of f and observing that it has only one zero, which occurs near $x = -1.2$ (see Figure 5.32). The calculations are shown in Table 5.2.

TABLE 5.2

n	x_n	$f(x_n)$	$f'(x_n)$	$\dfrac{f(x_n)}{f'(x_n)}$	$x_n - \dfrac{f(x_n)}{f'(x_n)}$
1	-1.20000	0.18400	5.24000	0.03511	-1.23511
2	-1.23511	-0.00771	5.68276	-0.00136	-1.23375
3	-1.23375	0.00001	5.66533	0.00000	-1.23375
4	-1.23375				

Thus we estimate the zero of f to be -1.23375, since two successive approximations differ by less than the required 0.0001. ∎

When, as in Examples 1 and 2, the approximations approach a zero of the function, we say that the method **converges.** It is important to realize that Newton's Method does not always converge. Two ways in which this may happen are as follows:

1. if $f'(x_n) = 0$ for some n (see Figure 5.33)
2. if $\lim_{n \to \infty} x_n$ does not exist (see Figure 5.34)

The type of problem encountered in Figure 5.33 can usually be overcome with a better choice for x_1. However, the problem illustrated in Figure 5.34 is usually more serious and may be independent of the choice

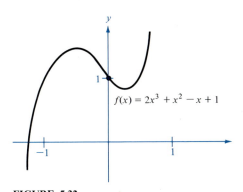

$f(x) = 2x^3 + x^2 - x + 1$

FIGURE 5.32

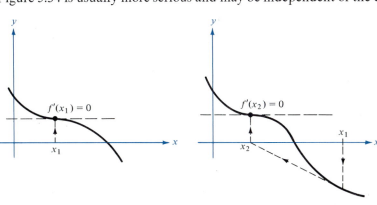

FIGURE 5.33

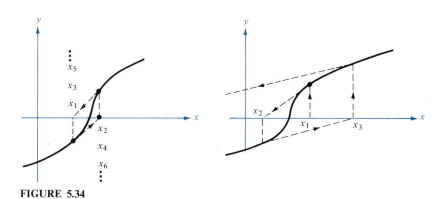

FIGURE 5.34

of x_1. For example, Newton's Method does not converge for any choice of x_1 (other than the actual zero) for the function $f(x) = x^{1/3}$.

Example 3

Refer to Figure 5.35. Using $x_1 = 0.1$, show that $\lim_{n \to \infty} x_n$ does not exist for the function $f(x) = x^{1/3}$.

Solution: The calculations are given in Table 5.3.
    Table 5.3 and Figure 5.35 indicate that x_n increases in magnitude as $n \to \infty$ and thus $\lim_{n \to \infty} x_n$ does not exist.

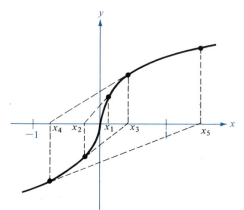

FIGURE 5.35

TABLE 5.3

n	x_n	$f(x_n)$	$f'(x_n)$	$\dfrac{f(x_n)}{f'(x_n)}$	$x_n - \dfrac{f(x_n)}{f'(x_n)}$
1	0.10000	0.46416	1.54720	0.30000	−0.20000
2	−0.20000	−0.58480	0.97467	−0.60000	0.40000
3	0.40000	0.73681	0.61401	1.20000	−0.80000
4	−0.80000	−0.92832	0.38680	−2.40000	1.60000

Section Exercises (5.5)

[c] **1.** Do one iteration of Newton's Method on the function $f(x) = x^2 - 3$, using $x_1 = 1.7$ as the initial guess.

[c] **2.** Do one iteration of Newton's Method on the function $f(x) = 3x^3 - 2$, using $x_1 = 1$ as the initial guess.

In Exercises 3–8, approximate the zero of the function in the given interval. Use Newton's Method and continue the process until you are correct to three decimal places.

[c][s] **3.** $f(x) = x^3 + x - 1$; $[0, 1]$

[c] **4.** $f(x) = x^5 + x - 1$; $[0, 1]$

[c][s] **5.** $f(x) = 3\sqrt{x - 1} - x$; $[1, 2]$

[c] **6.** $f(x) = x^3 - 3.9x^2 + 4.79x - 1.881$; $[0, 1]$

[c] **7.** $f(x) = x^4 - 10x^2 - 11$; $[3, 4]$

[c] **8.** $f(x) = x^3 + 3$; $[-2, -1]$

[c] **9.** Approximate to two decimal places the abscissa of the point of intersection of the graphs of $f(x) = 2x + 1$ and $g(x) = \sqrt{x + 4}$.

In Exercises 10–12, apply Newton's Method, using the indicated initial guess, and explain why the method fails.

10. $y = 2x^3 - 6x^2 + 6x - 1$; $x_1 = 1$

$\boxed{\text{s}}$ **11.** $y = x^4 + x^2 - 6x + 2$; $x_1 = 1$

12. $y = x^4 - x + 1$; $x_1 = 1$

$\boxed{\text{s}}$ **13.** Use Newton's Method to obtain a general formula for

approximating $\sqrt[n]{a}$. [Hint: Apply Newton's Method to the function $f(x) = x^n - a$.]

$\boxed{\text{c}}$ **14.** Use the result of Exercise 13 to aproximate $\sqrt[3]{7}$ correct to three decimal places.

$\boxed{\text{c}}$ **15.** Use the result of Exercise 13 to approximate $\sqrt[4]{6}$ correct to three decimal places.

For Review and Class Discussion

True or False

1. ____ If $b^2 < 3ac$, then the function $f(x) = ax^3 + bx^2 + cx + d$ has precisely one zero.

2. ____ The zeros of $f(x) = p(x)/q(x)$ coincide with the zeros of $p(x)$.

3. ____ If the coefficients of a polynomial function are all positive, then the polynomial has no positive zeros.

4. ____ If $f(x)$ is a cubic polynomial such that $f'(x)$ is never zero, then any initial guess will force Newton's Method to converge to the zero of f.

5. ____ The roots of $\sqrt{f(x)} = 0$ coincide with the roots of $f(x) = 0$.

Miscellaneous Exercises (Ch. 5)

$\boxed{\text{c}}$ **1.** At noon ship A was 100 mi due east of ship B. Ship A is sailing west at 12 mi/h and ship B is sailing south at 10 mi/h. At what time will the ships be nearest to one another and what will this distance be?

2. The total cost C of producing x units per day is $C = \frac{1}{4}x^2 + 62x + 125$, and the price p per unit at which they are sold is $p = 75 - \frac{1}{3}x$.
  (a) What should be the daily output to obtain maximum profit?
  (b) What should be the daily output to obtain minimum average cost?
  (c) Find the elasticity of demand.

$\boxed{\text{s}}$ **3.** Find the maximum profit if the demand equation is $p = 36 - 4x$ and the total cost is $C = 2x^2 + 6$.

4. Find the dimensions of the rectangle of maximum area, with sides parallel to the coordinate axes, that can be inscribed in the ellipse $(x^2/144) + (y^2/16) = 1$.

$\boxed{\text{c}}$ **5.** A right triangle, in the first quadrant, has the coordinate axes as sides and the hypotenuse passes through the point $(1, 8)$. Find the vertices of the triangle so that the length of the hypotenuse is minimum.

$\boxed{\text{c}}$ **6.** The side wall of a building is to be braced by a beam that must pass over a parallel wall 5 ft high and 4 ft from the building. Find the length of the shortest beam that can be used.

$\boxed{\text{s}}$ **7.** Find the length of the longest pipe that can be carried level around a right-angle corner if the two intersecting corridors are of width 4 ft and 6 ft.

8. A college class has asked a bus company for the rates for a certain tour. The bus company agrees to run buses if at least 80 people will go. The fare is to be $8 if 80 go and will decrease by $0.05 for everybody for each person over 80 that goes. What number of passengers will give the bus company maximum revenue?

9. Show that the greatest area of any rectangle inscribed in a triangle is one-half that of the triangle.

10. Three sides of a trapezoid have the same lengths. Of all such possible trapezoids, show that the one of maximum area has its fourth side of length $2s$.

$\boxed{\text{c}}$ **11.** The cost of fuel in running a locomotive is proportional to the $\frac{3}{2}$ power of the speed and is $50 per hour for a speed of 25 mi/h. Other fixed costs amount to an average of $100 per hour. Find the speed that will minimize the cost per mile.

12. Do Exercise 7 if the corridors are of width a feet and b feet.

13. A point moves along the curve $y = \sqrt{x}$ in such a way that the abscissa is increasing at the rate of 2 units per second. At what rate is the ordinate changing for the following values of x?

(a) $x = \frac{1}{2}$ (b) $x = 1$
(c) $x = 4$

14. The same conditions exist as in Exercise 13. Find the rate the distance from the origin is increasing for the following:

(a) $x = \frac{1}{2}$ (b) $x = 1$
(c) $x = 4$

S **15.** The cross section of a 5-ft trough is an isosceles trapezoid with its lower base 2 ft, upper base 3 ft, and altitude 2·ft. If water is running into the trough at the rate of 1 ft³/min, how fast is the water level rising when the water is 1 ft deep?

16. The pressure p and the volume V of an ideal gas are related by the adiabatic compression formula $pV^k = C$. The constant C is determined by an observation that $p = p_1$ when $V = V_1$, and k is a constant determined by the heat capacity of the gas. If V is changing at the rate of 1 cm³/s, at what rate is p changing?

17. The kinetic energy of a particle is one-half the product of the mass of the particle and the square of its speed, or $E_k = \frac{1}{2}mV^2$. Use differentials to approximate the change in kinetic energy in a 3000-pound (lb) automobile if the driver increases the speed from 30 to 35 mi/h. (Hint: $m = 3000$ lb/32 ft/s².)

C **18.** Find $f(10.5)$ to five-decimal-place accuracy, given that $f(x) = \sqrt[3]{x^2 + 25}$.

19. Estimate, using differentials, $f(10.5)$ if $f(x) = \sqrt[3]{x^2 + 25}$.

20. The diameter of a sphere is 18 in., with a maximum possible error of 0.05 in. Use the differential to discuss possible error in the surface area and the volume of the sphere.

21. A company finds that the demand for x units of its commodity is related to the price p by the equation $p = 75 - \frac{1}{4}x$. If x changes from 7 to 8, find the corresponding change in p. Find Δp and dp. Do the values of Δp and dp agree?

22. Repeat Exercise 21, using $p = 75 - \frac{1}{4}\sqrt{x}$ as the demand equation.

23. If a 1% error is made in measuring the edge of a cube, approximately what percentage error will be made in computing the following?
(a) the surface area of the cube
(b) the volume of the cube

C **24.** Approximate, to three decimal places, the zero of $f(x) = x^3 - 3x - 1$ which is in the interval $[-1, 0]$.

C **25.** Approximate, to three decimal places, the abscissa of the points of intersection of the equations $y = x^4$ and $y = x + 3$.

C **26.** Approximate, to three decimal places, the real root of $x^3 + 2x + 1 = 0$.

S **27.** The demand and cost equations for a certain product are

$$p = 600 - 3x \qquad \text{and} \qquad C = 0.3x^2 + 6x + 600$$

where p is the price per unit, x is the number of units, and C is the cost of producing x units. If t is the excise tax per unit, the profit for producing x units is

$$P = xp - C - xt$$

Find the maximum profit for the following:
(a) $t = 5$ (b) $t = 10$
(c) $t = 20$

28. The cost of inventory depends on ordering cost and storage cost. In the following inventory model, we assume that sales occur at a steady rate, Q is the number of units sold per year, r is the cost of storing one unit for one year, s is the cost of placing an order, and x is the number of units per order.

$$C = \left(\frac{Q}{x}\right)s + \left(\frac{x}{2}\right)r$$

Determine the order size that will minimize the cost.

Integration

6.1 **The Indefinite Integral**

Purpose
- To define integration as the inverse operation of differentiation.
- To derive some rules for evaluating indefinite integrals.

The basic problem of differentiation is: given the path of a moving point, to calculate its velocity, or given a curve, to calculate its slope. The basic problem of integration is the inverse: given the velocity of a moving point at every instant, to calculate its path, or given the slope of a curve at each of its points, to calculate the curve. —*Hans Hahn (1879–1934)*

Up to this point in our study of calculus, we have been concerned primarily with this problem:

Given a function, find its derivative.

Many important applications of calculus involve the inverse problem:

Given the derivative of a function, find the original function.

For example, suppose we are given the following derivatives:

$$f'(x) = 2, \qquad g'(x) = 3x^2, \qquad s'(t) = 4t$$

Our problem is to determine functions f, g, and s that have these respective derivatives. If we make some educated guesses, we might come up with the following functions:

$$f(x) = 2x \qquad \text{because} \qquad \frac{d}{dx}[2x] = 2$$

$$g(x) = x^3 \qquad \text{because} \qquad \frac{d}{dx}[x^3] = 3x^2$$

$$s(t) = 2t^2 \qquad \text{because} \qquad \frac{d}{dt}[2t^2] = 4t$$

This operation of determining the original function from its derivative is the inverse operation of differentiation and we call it **antidifferentiation.**

Definition of Antiderivative	A function F is called an **antiderivative** of a function f if $F'(x) = f(x)$ for every x in the domain of f.

We will use the phrase "$F(x)$ is an antiderivative of $f(x)$" synonymously with "F is an antiderivative of f."

It should be emphasized that if $F(x)$ is an antiderivative of $f(x)$, then $F(x) + C$ (where C is any constant) is also an antiderivative of $f(x)$. For if

$$\frac{d}{dx}[F(x)] = f(x)$$

then it is also true that

$$\frac{d}{dx}[F(x) + C] = \frac{d}{dx}[F(x)] + \frac{dC}{dx} = f(x) + 0 = f(x)$$

For instance,

$$F(x) = x^3, \qquad G(x) = x^3 - 5, \qquad H(x) = x^3 + 0.3$$

are all antiderivatives of $3x^2$, because

$$\frac{d}{dx}[x^3] = \frac{d}{dx}[x^3 - 5] = \frac{d}{dx}[x^3 + 0.3] = 3x^2$$

As it turns out, *all* the antiderivatives of $3x^2$ are of the form $x^3 + C$. This is verified in the following theorem.

THEOREM 6.1

If F and G possess identical derivatives on an open interval, then $F(x) - G(x) = C$ on that interval. That is, any two antiderivatives of a function differ at most by a *constant*.

Proof: Let $H(x) = F(x) - G(x)$, and assume $H(x)$ is *not* constant on the given interval. Then in this interval there exist a and b $(a < b)$ such that $H(a) \neq H(b)$. Since F and G are differentiable, their difference satisfies the Mean Value Theorem $[a, b]$. Therefore, there exists some c in (a, b) such that

$$H'(c) = \frac{H(b) - H(a)}{b - a}$$

Now since $H(b) \neq H(a)$, it follows that $H'(c) \neq 0$. However, since F' and G' are identical, we know that

$$H'(c) = F'(c) - G'(c) = 0$$

and we have arrived at a contradiction. Consequently, our assumption that $H(x)$ is not constant must be false and we conclude that $H(x) = F(x) - G(x) = C$.   ■

It is now clear that the process of antidifferentiation does not determine a unique function but rather a *family* of functions, each differing from the other by a constant. The antidifferentiation process is commonly referred to as **integration** and is denoted by the symbol

$$\int$$

called an **integral sign.** The symbol

$$\int f(x)\, dx$$

is called the **indefinite integral** of $f(x)$ and it denotes the antiderivatives of $f(x)$. More specifically, if $F'(x) = f(x)$ for all x, then

$$\int f(x)\, dx = F(x) + C$$

where $f(x)$ is called the **integrand** and C the **constant of integration.**

For the present we simply note that the differential *dx* in the indefinite integral identifies the variable of integration. That is, the symbol $\int f(x)\,dx$ denotes the "antiderivative of *f with respect to x*" just as the symbol *dy/dx* denotes the "derivative of *y with respect to x*." Additional uses of this notation will become apparent later.

The inverse nature of the operations of integration and differentiation can be shown symbolically as follows:

Differentiation is the inverse of integration:

$$\frac{d}{dx}\left[\int f(x)\,dx\right] = f(x)$$

Integration is the inverse of differentiation:

$$\int f'(x)\,dx = f(x) + C$$

Having defined the integration process, we have not yet provided any rules for determining antiderivatives. Fortunately, since integration (anti-differentiation) is the inverse operation of differentiation, we can readily obtain integration rules from differentiation rules. The following rules are easily verified by differentiation.

Basic Integration Rules

1. $\int dx = x + C$

2. $\int kf(x)\,dx = k \int f(x)\,dx$

3. $\int [f(x) \pm g(x)]\,dx = \int f(x)\,dx \pm \int g(x)\,dx$

4. $\int x^n\,dx = \dfrac{x^{n+1}}{n+1} + C$, if $n \neq -1$

Applications of these rules are demonstrated in the following examples.

Example 1

Evaluate the indefinite integral $\int (3x - 7)\,dx$.

Solution:

$$\int (3x - 7)\,dx = \int 3x\,dx - \int 7\,dx \qquad \text{(Rule 3)}$$

$$= 3 \int x \, dx - 7 \int dx \qquad\qquad \text{(Rule 2)}$$

$$= 3 \left(\frac{x^2}{2} + C_1 \right) - 7(x + C_2) \qquad\qquad \text{(Rules 4 and 1)}$$

$$= \tfrac{3}{2}x^2 - 7x + 3C_1 - 7C_2$$

Because $3C_1 - 7C_2$ is just another constant, we write our answer as

$$\int (3x - 7) \, dx = \tfrac{3}{2}x^2 - 7x + C \qquad\qquad \blacksquare$$

We frequently use the properties of integrals without specifically identifying them, and many times we use more than one rule in a given step so as to reduce the total number of steps in solving a problem.

Example 2

Evaluate the indefinite integral $\int \sqrt[3]{y} \, dy$.

Solution:

$$\int \sqrt[3]{y} \, dy = \int y^{1/3} \, dy = \frac{y^{\frac{1}{3}+1}}{\frac{1}{3}+1} + C = \frac{y^{4/3}}{\frac{4}{3}} + C = \tfrac{3}{4}y^{4/3} + C \qquad\qquad \blacksquare$$

Example 3

Find

$$\int \left(\frac{3}{x^2} - \frac{1}{\sqrt{x^3}} \right) dx$$

Solution:

$$\int \left(\frac{3}{x^2} - \frac{1}{\sqrt{x^3}} \right) dx = \int (3x^{-2} - x^{-3/2}) \, dx$$

$$= 3 \int x^{-2} \, dx - \int x^{-3/2} \, dx$$

$$= 3 \left(\frac{x^{-1}}{-1} \right) - \frac{x^{-1/2}}{-1/2} + C = \frac{-3}{x} + \frac{2}{\sqrt{x}} + C \qquad\qquad \blacksquare$$

So far we have been finding a family of functions $F(x) + C$, each having the derivative $F'(x)$. Sometimes additional information is given that allows us to determine the unique member of the family we want.

Example 4

Given $f'(x) = 6 - x^{1/2}$ and $f(1) = \tfrac{4}{3}$. Find $f(x)$.

Solution:

$$f(x) = \int (6 - x^{1/2})\, dx = 6x - \frac{x^{3/2}}{\frac{3}{2}} + C = 6x - \frac{2}{3}x^{3/2} + C$$

But
$$f(1) = \tfrac{4}{3} = 6(1) - \tfrac{2}{3}(1)^{3/2} + C$$

implies
$$\tfrac{4}{3} = 6 - \tfrac{2}{3} + C$$
$$\tfrac{6}{3} - 6 = C$$
$$-4 = C$$

Therefore, the particular antiderivative we want is $f(x) = 6x - \frac{2}{3}x^{3/2} - 4$. ■

Example 5

Suppose the marginal cost for producing x units of a commodity is given by

$$\frac{dC}{dx} = 32 - 0.04x$$

If it costs \$50 to make 1 unit, find the total cost of making 200 units.

Solution: Since dC/dx is the derivative of the total cost function, we have

$$C = \int (32 - 0.04x)\, dx = 32x - 0.04\left(\frac{x^2}{2}\right) + K = 32x - 0.02x^2 + K$$

When $x = 1$, $C = 50$; thus

$$50 = 32(1) - 0.02(1)^2 + K$$
$$18.02 = K$$

Therefore, the total cost function is

$$C = 32x - 0.02x^2 + 18.02$$

The cost of making 200 units is, therefore,

$$C = 32(200) - 0.02(40{,}000) + 18.02 = \$5{,}618.02$$ ■

Example 6

A ball is thrown upward with an initial velocity of 64 ft/s from a height of 80 ft. Use antidifferentiation to verify that the equation for the position function is

$$s(t) = -16t^2 + v_0 t + s_0 = -16t^2 + 64t + 80$$

Solution: Symbolically the conditions in this problem are

$$s''(t) = -32, \qquad s'(0) = 64, \qquad s(0) = 80$$

By antidifferentiation it follows that

$$s'(t) = \int s''(t)\, dt = \int -32\, dt = -32t + C_1$$

where $C_1 = s'(0) = 64$. Similarly,

$$s(t) = \int s'(t)\, dt = \int (-32t + 64)\, dt = -16t^2 + 64t + C_2$$

where $C_2 = s(0) = 80$. Therefore, we have

$$s(t) = -16t^2 + 64t + 80$$ ■

Before completing this section we identify and prove the General Power Rule for integration.

General Power Rule for Integration

If u is a differentiable function of x, then

$$\int u^n u'\, dx = \frac{u^{n+1}}{n+1} + C$$

where $n \neq -1$ and C is a constant.

Proof: By the General Power Rule for differentiation, we have

$$\frac{d}{dx}\left[\frac{u^{n+1}}{n+1} + C\right] = \left(\frac{n+1}{n+1}\right)u^n u' + 0$$

which in differential form is

$$d\left[\frac{u^{n+1}}{n+1} + C\right] = u^n u'\, dx$$

By the inverse property of integration, we have

$$\int d\left[\frac{u^{n+1}}{n+1} + C\right] = \int u^n u'\, dx$$

or

$$\frac{u^{n+1}}{n+1} + C = \int u^n u'\, dx$$

where $n \neq -1$. ■

An important consideration that is often overlooked when using this rule is the existence of u' as a factor of the integrand. We must first determine u by identifying, within the integrand, a function u that is raised to some power. Then, secondly, we must show that its derivative u' is also a factor of the integrand. Simply stated, the clues for using the Power Rule for integration are as follows:

1.   Identify which function u is raised to a power.
2.   Check to see if u' is also a factor of the integrand.

We demonstrate these considerations in the next examples.

Example 7

Evaluate the indefinite integral $\int 2(1 + 2x)^3 \, dx$.

Solution: Obviously, the function raised to a power is

$$u = 1 + 2x$$

thus

$$u' = 2$$

and, indeed, u' is a factor of the integrand. So we write

$$\int 2(1 + 2x)^3 \, dx = \int \overbrace{(1 + 2x)^3}^{u^3} \overbrace{(2)}^{u'} \, dx$$

and we apply the Power Rule to obtain

$$\int 2(1 + 2x)^3 \, dx = \frac{(1 + 2x)^4}{4} + C \qquad \blacksquare$$

Recall from our definition of differentials that if u is a differentiable function of x, then

$$du = u' \, dx$$

Thus the General Power Rule can be written in the form

$$\int u^n \, du = \frac{u^{n+1}}{n + 1} + C$$

Although the two forms of the General Power Rule are equivalent, in practice, this second form involves a separate step in which the change of variables from x to u is *written out* before integrating. For instance, the solution to Example 7 would proceed as follows:

To evaluate $\qquad \int 2(1 + 2x)^3 \, dx$

let $u = 1 + 2x$; then $du = 2 \, dx$, and we have

$$dx = \frac{du}{2}$$

Thus

$$\int 2(1 + 2x)^3 \, dx = \int 2u^3 \, \frac{du}{2} = \int u^3 \, du = \frac{u^4}{4} + C$$

Resubstituting for u, we have

$$\int 2(1 + 2x)^3 \, dx = \frac{(1 + 2x)^4}{4} + C$$

If you prefer this *du* form of the integration rules, just keep in mind that throughout this text we generally use the *u′ dx* form and that $du = u′\, dx$.

Many times, part of *u′* is missing from the integrand, and in *some* such cases we can make the necessary adjustments in order to apply the Power Rule. In other instances we cannot adjust appropriately and, therefore, cannot apply the Power Rule.

Example 8

Evaluate $\int x(3 - 4x^2)^2\, dx$.

Solution: Let $u = 3 - 4x^2$; then $u′ = -8x$. Now we see that the factor -8 is not a part of the integrand. However, we can adjust the integrand in the following way by multiplying by -8 and its reciprocal:

$$\int x(3 - 4x^2)^2\, dx = \int \frac{1}{-8}(3 - 4x^2)^2(-8x)\, dx$$

Now, because $-\frac{1}{8}$ is a constant, the right side of this equation becomes

$$-\frac{1}{8} \int \overbrace{(3 - 4x^2)^2}^{u^2}\overbrace{(-8x)}^{u′}\, dx$$

to which we apply the Power Rule to get

$$\left(-\frac{1}{8}\right)\left[\frac{(3 - 4x^2)^3}{3}\right] + C = \frac{-(3 - 4x^2)^3}{24} + C \qquad ■$$

Example 9

Evaluate $\int -8(3 - 4x^2)^2\, dx$.

Solution: If we let $u = 3 - 4x^2$, then $u′ = -8x$, and again part of *u′* is missing from the integrand. Since the missing part of *u′* is a *variable* rather than a constant, the adjustment would require that we move the variable quantity $1/x$ outside the integral sign. However, this is not possible. That is,

$$\int -8(3 - 4x^2)^2\, dx = \int \left(\frac{1}{x}\right)(3 - 4x^2)^2(-8x)\, dx$$

$$\neq \frac{1}{x}\int (3 - 4x^2)^2(-8x)\, dx$$

(Note: If we were permitted to move variable quantities outside the integral sign, why not move the entire integrand out and eliminate the problem entirely?) In this example we cannot apply the Power Rule since we cannot make the necessary adjustments for *u′*. However, we can (in this particular case) expand the integrand and write

$$\int -8(3 - 4x^2)^2 \, dx = \int -8(9 - 24x^2 + 16x^4) \, dx$$

$$= \int (-72 + 192x^2 - 128x^4) \, dx$$

$$= -72x + 64x^3 - \frac{128}{5}x^5 + C$$

$\blacksquare$

Sometimes an integrand contains an extra constant factor that is *not* needed as part of u'. In such cases we simply move this factor outside the integral sign, insert the necessary factor, and adjust accordingly. The next example illustrates this situation.

Example 10

Evaluate $\qquad\qquad \displaystyle\int \frac{7x^2 \, dx}{\sqrt{4x^3 - 5}}$

Solution: Write

$$\int \frac{7x^2 \, dx}{\sqrt{4x^3 - 5}} = \int 7x^2(4x^3 - 5)^{-1/2} \, dx$$

and let $u = 4x^3 - 5$; then $u' = 12x^2$. We need the factor 12, rather than 7, so we write

$$\int 7x^2(4x^3 - 5)^{-1/2} \, dx = 7 \int \frac{1}{12}(4x^3 - 5)^{-1/2}(12x^2) \, dx$$

$$= \frac{7}{12} \int (4x^3 - 5)^{-1/2}(12x^2) \, dx$$

$$= \left(\frac{7}{12}\right)\left[\frac{(4x^3 - 5)^{1/2}}{\frac{1}{2}}\right] + C = \frac{7}{6}\sqrt{4x^3 - 5} + C$$

$\blacksquare$

Make sure that you see the distinction between the two types of constant factors moved outside the integral sign in Example 10. In the equation

$$\int 7x^2(4x^3 - 5)^{-1/2} \, dx = (7)\left(\frac{1}{12}\right)\int (4x^3 - 5)^{-1/2}(12x^2) \, dx$$

the 7 is an unnecessary factor that is moved out *as is,* whereas the $\frac{1}{12}$ is the *reciprocal* adjustment for the factor 12 used to create $u' = 12x^2$.

Section Exercises (6.1)

In Exercises 1–50, evaluate the indefinite integrals and check your results by differentiation.

1. $\int (x^3 + 2)\, dx$

2. $\int (x^2 - 2x + 3)\, dx$

S 3. $\int (x^{3/2} + 2x + 1)\, dx$

4. $\int \left(\sqrt{x} + \dfrac{1}{2\sqrt{x}} \right) dx$

5. $\int \sqrt[3]{x^2}\, dx$

6. $\int (\sqrt[4]{x^3} + 1)\, dx$

7. $\int \dfrac{1}{x^3}\, dx$

8. $\int \dfrac{1}{x^2}\, dx$

9. $\int \dfrac{1}{4x^2}\, dx$

10. $\int (2x + x^{-1/2})\, dx$

11. $\int \dfrac{x^2 + x + 1}{\sqrt{x}}\, dx$

12. $\int \dfrac{x^2 + 1}{x^2}\, dx$

13. $\int (x + 1)(3x - 2)\, dx$

14. $\int (2t^2 - 1)^2\, dt$

S 15. $\int \dfrac{t^2 + 2}{t^2}\, dt$

16. $\int (1 - 2y + 3y^2)\, dy$

17. $\int y^2 \sqrt{y}\, dy$

18. $\int (1 + 3t)t^2\, dt$

19. $\int dx$

20. $\int 3\, dt$

21. $\int (1 + 2x)^4 2\, dx$

22. $\int (x^2 - 1)^3 2x\, dx$

S 23. $\int x^2(x^3 - 1)^4\, dx$

24. $\int x(1 - 2x^2)^3\, dx$

25. $\int x(x^2 - 1)^7\, dx$

26. $\int \dfrac{x^2}{(x^3 - 1)^2}\, dx$

27. $\int \dfrac{4x}{\sqrt{1 + x^2}}\, dx$

28. $\int \dfrac{6x}{(1 + x^2)^3}\, dx$

S 29. $\int 5x \sqrt[3]{1 + x^2}\, dx$

30. $\int 3(x - 3)^{5/2}\, dx$

31. $\int \dfrac{-3}{\sqrt{2x + 3}}\, dx$

32. $\int \dfrac{4x + 6}{(x^2 + 3x + 7)^3}\, dx$

33. $\int \dfrac{x + 1}{(x^2 + 2x - 3)^2}\, dx$

34. $\int u^3 \sqrt{u^4 + 2}\, du$

S 35. $\int \dfrac{1}{\sqrt{x}(1 + \sqrt{x})^2}\, dx$

S 36. $\int \left(1 + \dfrac{1}{t}\right)^3 \left(\dfrac{1}{t^2}\right) dt$

37. $\int \dfrac{x^2}{(1 + x^3)^2}\, dx$

38. $\int \dfrac{x^2}{\sqrt{1 + x^3}}\, dx$

39. $\int \dfrac{x^3}{\sqrt{1 + x^4}}\, dx$

40. $\int \dfrac{t + 2t^2}{\sqrt{t}}\, dt$

41. $\int \dfrac{1}{2\sqrt{x}}\, dx$

42. $\int \dfrac{1}{(3x)^2}\, dx$

S 43. $\int \dfrac{1}{\sqrt{2x}}\, dx$

44. $\int \dfrac{1}{3x^2}\, dx$

45. $\int \dfrac{-6x}{\sqrt{36 + x^2}}\, dx$

46. $\int (ax^2 + bx + c)\, dx$

S 47. $\int t^2 \left(t - \dfrac{2}{t} \right)^2 dt$

48. $\int \left(\dfrac{t^3}{3} + \dfrac{1}{4t^2} \right) dt$

49. $\int (9 - y)\sqrt{y}\, dy$

50. $\int 2\pi y(8 - y^{3/2})\, dy$

51. Find $\int (2x - 1)^2\, dx$ in two ways. Explain the difference in the appearance of the two answers.

52. Find $\int x(x^2 - 1)^2\, dx$ in two ways. Explain the difference in the appearance of the two answers.

In Exercises 53–56, find the equation of the curve, given the derivative and one point on the curve.

53. $\dfrac{dy}{dx} = 2x - 1$; $(1, 1)$

54. $\dfrac{dy}{dx} = 2(x - 1)$; $(3, 2)$

S 55. $\dfrac{dy}{dx} = x\sqrt{1 - x^2}$; $(0, \frac{4}{3})$

56. $\dfrac{dy}{dx} = 1 - \dfrac{1}{x^2}$; $(1, 3)$

In Exercises 57–60, find $y = f(x)$ satisfying the given conditions.

57. $f''(x) = 2$; $f'(2) = 5$; $f(2) = 10$

58. $f''(x) = x^2$; $f'(0) = 6$; $f(0) = 3$

S 59. $f''(x) = x^{-3/2}$; $f'(4) = 2$; $f(0) = 0$

60. $f''(x) = x^{-3/2}$; $f'(1) = 2$; $f(9) = -4$

In Exercises 61–65, use $a(t) = -32$ ft/s^2 as the acceleration due to gravity. (Neglect air resistance.)

61. An object is dropped from a balloon, which is stationary at 1600 ft. Express its height above the ground as a function of t. How long does it take the object to reach the ground?

62. A ball is thrown vertically upward with an initial velocity of 60 ft/s. How high will the ball go?

S 63. With what initial velocity must an object be thrown upward from the ground to reach a maximum height of 550 ft? (Approximate height of the Washington Monument.)

64. If an object is thrown upward from a point s_0 feet above the ground with an initial velocity of v_0 feet per second, show that its height above the ground is given by the function $f(t) = -16t^2 + v_0t + s_0$.

ⓒ **65.** A balloon, rising vertically with a velocity of 16 ft/s, releases a sandbag at an instant when the balloon is 64 ft above the ground.
   (a) How many seconds after its release will the bag strike the ground?
   (b) With what velocity will it reach the ground?

66. Assume that as a fully loaded plane starts from rest, it has a constant acceleration while moving down the runway. Find this acceleration if the plane requires, on the average 0.7 mi of runway and a speed of 160 mi/h before lifting off.

ⓒ **67.** The makers of a certain automobile advertise that it will accelerate from 15 to 50 mi/h in high gear in 13 s. Assuming constant acceleration, compute the following:
   (a) the acceleration in feet per second per second
   (b) the distance the car travels in the given time

ⓒ **68.** In a car traveling at 45 mi/h, the brakes were applied and the car was brought to a stop in 132 ft. How far had the car moved by the time its speed was reduced to (a) 30 mi/h? (b) 15 mi/h? Draw the real number line from 0 to 132 and plot the points found in parts (a) and (b). What conclusions can you draw?

ⓢ **69.** At the instant the traffic light turns green, an automobile that has been waiting at an intersection starts ahead with a constant acceleration of 6 ft/s². At the same instant a truck traveling with a constant velocity of 30 ft/s overtakes and passes the car.
   (a) How far beyond its starting point will the automobile overtake the truck?
   (b) How fast will it be traveling?

70. A ball is released from rest and rolls down an inclined plane, requiring 4 s to cover a distance of 100 cm. What was its acceleration, in centimeters per second per second?

71. Galileo Galilei (1564–1642) stated the following proposition: "The time in which any space is traversed by a body starting from rest and uniformly accelerated is equal to the time in which that same space would be traversed by the same body moving at a uniform speed whose value is the mean of the highest speed and the speed just before acceleration began." Use the techniques of this section to verify the statement.

72. If marginal cost is constant (i.e., for each unit increase in output, the increase in cost is always the same), show that the cost function is a straight line.

ⓢ **73.** The marginal cost for production is $dC/dx = 2x - 12$. Find the total cost function and the average cost function if fixed costs are $50.

74. If marginal revenue is $dR/dx = 100 - 5x$, find the revenue and demand functions.

75. If marginal revenue is $dR/dx = 10 - 6x - 2x^2$, determine the revenue and demand functions.

For Review and Class Discussion

True or False

1. ____ $\int 7x \, dx = \frac{1}{7} \int x \, dx = (x^2/14) + C$.

2. ____ $\int (2x + 1)^2 \, dx = [(2x + 1)^3/3] + C$.

3. ____ $\int x(x^2 + 1) \, dx = (x^2/2)[(x^3/3) + x] + C$.

4. ____ $\int (1/x) \, dx = -(1/x^2) + C$.

5. ____ Each antiderivative of an nth-degree polynomial function is an $(n + 1)$st-degree polynomial function.

6. ____ $\int (x^2 + 1)^2 \, dx = (1/2x) \int (x^2 + 1)^2(2x) \, dx = (1/2x)[(x^2 + 1)^3/3] + C$.

7. ____ If $F(x)$ and $G(x)$ are antiderivatives of $f(x)$, then $F(x) = G(x) + C$.

8. ____ If f and g have the same antiderivative, then $f(x) = g(x)$.

9. ____ If $f'(x) = g(x)$, then $\int g(x) \, dx = f(x) + C$.

10. ____ If $p(x)$ is a polynomial function, then p has exactly one antiderivative whose graph contains the origin.

6.2 Sigma Notation

Purpose
- To introduce sigma notation for expressing sums of many terms.
- To develop some properties of sigma notation and rules for using it.

Obviousness is always the enemy to correctness. Hence we invent some new and difficult symbolism, in which nothing seems obvious. Then we set up certain rules for operating on the symbols, and the whole thing becomes mechanical. —Bertrand Russell (1872–1970)

We temporarily delay further discussion of integration in order to examine a shortcut for denoting sums involving many terms. This shortcut will be used later to provide a geometric interpretation of integration. The notation that facilitates the writing of these sums is called **sigma notation,** because it uses the Greek capital letter sigma, written as Σ. Some illustrations of the use of sigma notation are given in the following example.

Example 1

Sum	Σ **Notation**
(a) $1 + 2 + 3 + 4 + 5 + 6$	$\sum_{i=1}^{6} i$
(b) $3^2 + 4^2 + 5^2 + 6^2 + 7^2$	$\sum_{j=3}^{7} j^2$
(c) $1 + 2 + 2^2 + 2^3 + \cdots + 2^n$	$\sum_{k=0}^{n} 2^k$
(d) $A_1 + A_2 + A_3 + \cdots + A_n$	$\sum_{i=1}^{n} A_i$
(e) $f(x_1)\,\Delta x + f(x_2)\,\Delta x + f(x_3)\,\Delta x$ $+ \cdots + f(x_n)\,\Delta x$	$\sum_{i=1}^{n} f(x_i)\,\Delta x$

In general, the terms of a sum may be expressed as a function f of a "dummy" variable, say i. We write

$$\sum_{i=m}^{n} f(i) = f(m) + f(m + 1) + f(m + 2) + \cdots + f(n)$$

where i, m, and n are integers and

i is the **index of summation** (any dummy variable may be used) *m* is the **lower limit** of the summation

n is the **upper limit** of the summation

and there are $(n - m + 1)$ terms in the sum.

Example 2

Write the sigma notation for the sum

$$\frac{1}{n}(1^2 + 1) + \frac{1}{n}(2^2 + 1) + \frac{1}{n}(3^2 + 1) + \cdots + \frac{1}{n}(n^2 + 1)$$

Solution: We begin by noting that the *n* terms in this sum are each of the form

$$f(i) = \frac{1}{n}(i^2 + 1)$$

Furthermore, we observe that in the first term, $i = 1$; in the second term, $i = 2$; and so on until we reach the *n*th term. Thus our index *i* runs from 1 to *n* and the sigma notation for the given sum is

$$\sum_{i=1}^{n} f(i) = \sum_{i=1}^{n} \frac{1}{n}(i^2 + 1)$$ ■

Example 3

Evaluate the sum

$$\sum_{i=m}^{n} f(i) = \sum_{i=1}^{n} 3$$

Solution: Since the index for this sum runs from 1 to *n*, there are $n - m + 1 = n - 1 + 1 = n$ terms, and we have

$$\sum_{i=1}^{n} 3 = \underbrace{3 + 3 + 3 + \cdots + 3}_{n \text{ terms}} = 3n$$ ■

Although we commonly use 1 as the lower limit of summation, this is not necessary. For instance, the sums in Examples 2 and 3 could have been written as

$$\sum_{i=1}^{n} \frac{1}{n}(i^2 + 1) = \sum_{i=0}^{n-1} \frac{1}{n}[(i + 1)^2 + 1]$$

and

$$\sum_{i=1}^{n} 3 = \sum_{i=0}^{n-1} 3$$

The following properties of sigma notation are useful. To verify these two properties, we suggest that you write them in expanded form and apply the associative, commutative, and distributive properties of arithmetic.

Properties of Sigma Notation

1. $\displaystyle\sum_{i=1}^{n} cf(i) = c \sum_{i=1}^{n} f(i),$ c is any constant

2. $\displaystyle\sum_{i=1}^{n} [f(i) \pm g(i)] = \sum_{i=1}^{n} f(i) \pm \sum_{i=1}^{n} g(i)$

Property 1 implies that any factor that is not dependent on (not a function of) the index variable can be factored out from "under" the Σ symbol. For instance, since $2/n$ is not dependent on the index variable i, we can write

$$\sum_{i=1}^{n} \frac{2i}{n} = \frac{2}{n} \sum_{i=1}^{n} i$$

In Example 3 we saw that there are instances in which a sum can be written as a simple function of the upper limit n,

$$\sum_{i=1}^{n} 3 = 3n$$

This example represents only one of several summation formulas we will find useful.

Summation Formulas

If n is a positive integer and c is a constant, then:

1. $\displaystyle\sum_{i=1}^{n} c = cn$ 2. $\displaystyle\sum_{i=1}^{n} i = \frac{n(n+1)}{2}$

3. $\displaystyle\sum_{i=1}^{n} i^2 = \frac{n(n+1)(2n+1)}{6}$ 4. $\displaystyle\sum_{i=1}^{n} i^3 = \frac{n^2(n+1)^2}{4}$

5. $\displaystyle\sum_{i=1}^{n} i^4 = \frac{n(n+1)(6n^3 + 9n^2 + n - 1)}{30}$

All these formulas can be proven by mathematical induction. An alternative derivation of formula 2 is given here.

Proof of Formula 2:   Write the equivalent forms

$$\sum_{i=1}^{n} i = 1 + \quad 2 \quad + \quad 3 \quad + \cdots + (n-1) + n$$

$$\sum_{i=1}^{n} i = n + (n-1) + (n-2) + \cdots + \quad 2 \quad + 1$$

Adding these two equations term by term yields the sum

$$2 \sum_{i=1}^{n} i = (n+1) + (n+1) + (n+1) + \cdots + (n+1) + (n+1)$$

which has *n* terms. Therefore,

$$2 \sum_{i=1}^{n} i = n(n+1)$$

and

$$\sum_{i=1}^{n} i = \frac{n(n+1)}{2}$$
■

Example 4

Evaluate $\sum_{k=1}^{n} k(2 - 3k^2)$.

Solution:

$$\sum_{k=1}^{n} k(2 - 3k^2) = \sum_{k=1}^{n} (2k - 3k^3) = \sum_{k=1}^{n} 2k - \sum_{k=1}^{n} 3k^3$$

$$= 2 \sum_{k=1}^{n} k - 3 \sum_{k=1}^{n} k^3$$

$$= (2)\left[\frac{n(n+1)}{2}\right] - (3)\left[\frac{n^2(n+1)^2}{4}\right]$$

$$= n^2 + n - \tfrac{3}{4}(n^4 + 2n^3 + n^2)$$

$$= \frac{4n^2 + 4n - 3n^4 - 6n^3 - 3n^2}{4}$$

$$= \frac{4n + n^2 - 6n^3 - 3n^4}{4}$$
■

Example 5

Evaluate $\sum_{i=0}^{25} (3i - 4)$.

Solution: Note that the lower limit is zero rather than one. Therefore, we write

$$\sum_{i=0}^{25} (3i - 4) = -4 + \sum_{i=1}^{25} (3i - 4) = -4 + 3\sum_{i=1}^{25} i - \sum_{i=1}^{25} 4$$

$$= -4 + 3\left[\frac{25(26)}{2}\right] - 4(25) = 871 \qquad \blacksquare$$

Our next example illustrates the use of summation formulas in evaluating limits of the form

$$\lim_{n \to \infty} \sum_{i=1}^{n} f(i)$$

Such limits are used extensively in the next section.

Example 6

Evaluate
$$\sum_{i=1}^{n} \left(\frac{2}{n}\right)\left(\frac{i+1}{n}\right)^2$$

when $n = 30$; when $n \to \infty$.

Solution:
$$\sum_{i=1}^{n} \left(\frac{2}{n}\right)\left(\frac{i+1}{n}\right)^2 = \sum_{i=1}^{n} \left(\frac{2}{n}\right)\left(\frac{i^2 + 2i + 1}{n^2}\right)$$

Since $2/n^3$ is constant relative to index i, we have

$$\sum_{i=1}^{n} \left(\frac{2}{n}\right)\left(\frac{i+1}{n}\right)^2 = \frac{2}{n^3} \sum_{i=1}^{n} (i^2 + 2i + 1)$$

$$= \frac{2}{n^3}\left\{\frac{n(n+1)(2n+1)}{6} + 2\left[\frac{n(n+1)}{2}\right] + n\right\}$$

$$= \frac{2}{n^3}\left(\frac{2n^3 + 3n^2 + n}{6} + n^2 + n + n\right)$$

$$= \frac{2}{n^3}\left(\frac{2n^3 + 3n^2 + n + 6n^2 + 12n}{6}\right)$$

$$= \frac{1}{3n^3}(2n^3 + 9n^2 + 13n) = \frac{2}{3} + \frac{3}{n} + \frac{13}{3n^2}$$

When $n = 30$,

$$\sum_{i=1}^{n} \left(\frac{2}{n}\right)\left(\frac{i+1}{n}\right)^2 = \frac{2}{3} + \frac{3}{30} + \frac{13}{2700} = \frac{2083}{2700}$$

As $n \to \infty$ we have

$$\lim_{n \to \infty} \sum_{i=1}^{n} \left(\frac{2}{n}\right)\left(\frac{i+1}{n}\right)^2 = \lim_{n \to \infty} \left(\frac{2}{3} + \frac{3}{n} + \frac{13}{3n^2}\right) = \frac{2}{3}$$

■

Section Exercises (6.2)

In Exercises 1–8, find the given sum.

1. $\displaystyle\sum_{i=1}^{5} (2i + 1)$
2. $\displaystyle\sum_{i=1}^{6} 2i$

S **3.** $\displaystyle\sum_{k=0}^{4} \frac{1}{1 + k^2}$
4. $\displaystyle\sum_{j=3}^{5} \frac{1}{j}$

5. $\displaystyle\sum_{k=1}^{4} c$
6. $\displaystyle\sum_{n=1}^{4} \frac{c}{n + 1}$

7. $\displaystyle\sum_{i=1}^{4} [(i - 1)^2 + (i + 1)^3]$
8. $\displaystyle\sum_{k=2}^{5} (k + 1)(k - 3)$

In Exercises 9–18, write the given sum in sigma notation.

9. $\dfrac{1}{3(1)} + \dfrac{1}{3(2)} + \dfrac{1}{3(3)} + \cdots + \dfrac{1}{3(9)}$

10. $\dfrac{5}{1 + 1} + \dfrac{5}{1 + 2} + \dfrac{5}{1 + 3} + \cdots + \dfrac{5}{1 + 15}$

11. $[2(\frac{1}{8}) + 3] + [2(\frac{2}{8}) + 3] + \cdots + [2(\frac{8}{8}) + 3]$

12. $[1 - (\frac{1}{4})^2] + [1 - (\frac{2}{4})^2] + \cdots + [1 - (\frac{4}{4})^2]$

13. $[(\frac{1}{6})^2 + 2](\frac{1}{6}) + [(\frac{2}{6})^2 + 2](\frac{1}{6}) + \cdots + [(\frac{6}{6})^2 + 2](\frac{1}{6})$

14. $\left[\left(\frac{1}{n}\right)^2 + 2\right]\left(\frac{1}{n}\right) + \left[\left(\frac{2}{n}\right)^2 + 2\right]\left(\frac{1}{n}\right) + \cdots$
$$+ \left[\left(\frac{n}{n}\right)^2 + 2\right]\left(\frac{1}{n}\right)$$

S **15.** $\left[\left(\frac{2}{n}\right)^3 - \frac{2}{n}\right]\left(\frac{2}{n}\right) + \left[\left(\frac{4}{n}\right)^3 - \frac{4}{n}\right]\left(\frac{2}{n}\right) + \cdots$
$$+ \left[\left(\frac{2n}{n}\right)^3 - \frac{2n}{n}\right]\left(\frac{2}{n}\right)$$

16. $\left[1 - \left(\frac{2}{n} - 1\right)^2\right]\left(\frac{2}{n}\right) + \left[1 - \left(\frac{4}{n} - 1\right)^2\right]\left(\frac{2}{n}\right) + \cdots$
$$+ \left[1 - \left(\frac{2n}{n} - 1\right)^2\right]\left(\frac{2}{n}\right)$$

17. $\left[2\left(1 + \frac{3}{n}\right)^2\right]\left(\frac{3}{n}\right) + \left[2\left(1 + \frac{6}{n}\right)^2\right]\left(\frac{3}{n}\right) + \cdots$
$$+ \left[2\left(1 + \frac{3n}{n}\right)^2\right]\left(\frac{3}{n}\right)$$

18. $\left(\frac{1}{n}\right)\sqrt{1 - \left(\frac{0}{n}\right)^2} + \left(\frac{1}{n}\right)\sqrt{1 - \left(\frac{1}{n}\right)^2} + \cdots$
$$+ \left(\frac{1}{n}\right)\sqrt{1 - \left(\frac{n-1}{n}\right)^2}$$

Use the properties of sigma notation and the expression for the sums of powers of the first n positive integers to evaluate the sums in Exercises 19–23.

19. $\displaystyle\sum_{i=1}^{20} 2i$
20. $\displaystyle\sum_{i=1}^{10} i(i^2 + 1)$

21. $\displaystyle\sum_{i=1}^{20} (i - 1)^2$
22. $\displaystyle\sum_{i=1}^{15} (2i - 3)$

S **23.** $\displaystyle\sum_{i=1}^{15} \frac{1}{n^3}(i - 1)^2$

In Exercises 24–28, find the limits as $n \to \infty$.

24. $\displaystyle\lim_{n \to \infty} \sum_{i=1}^{n} \frac{1}{n^3}(i - 1)^2$
S **25.** $\displaystyle\lim_{n \to \infty} \sum_{i=1}^{n} \left(1 + \frac{2i}{n}\right)^2\left(\frac{2}{n}\right)$

26. $\displaystyle\lim_{n \to \infty} \sum_{i=1}^{n} \frac{16i}{n^2}$
27. $\displaystyle\lim_{n \to \infty} \sum_{i=1}^{n} \left(\frac{2i}{n}\right)^3\left(\frac{2}{n}\right)$

28. $\displaystyle\lim_{n \to \infty} \sum_{i=1}^{n} \left(1 + \frac{2i}{n}\right)^3\left(\frac{2}{n}\right)$

In Exercises 29–34, determine if the two sums are equal.

29. $\displaystyle\sum_{i=1}^{n} i;\ \sum_{i=0}^{n-1} (i + 1)$

30. $\sum_{i=1}^{n} i$; $1 + \sum_{i=1}^{n-1} i$

31. $\sum_{k=2}^{n} (2k - 1)$; $\sum_{k=1}^{n-1} (2k + 1)$

32. $\sum_{k=2}^{n} (3k - 1)$; $\sum_{k=1}^{n-1} (3k + 2)$

S **33.** $\sum_{k=2}^{n} (3k - 1)$; $\sum_{k=1}^{n-1} (3k + 1)$

34. $\sum_{j=0}^{n} (j^2 - j)$; $\sum_{j=1}^{n-1} (j^2 - j)$

For Exercises 35 and 36, use the following definitions of the arithmetic mean (average) $\bar{x}$ and standard deviation s. Given a set of n measurements $x_1, x_2, x_3, \ldots, x_n$,

$$\bar{x} = \frac{\sum_{i=1}^{n} x_i}{n} \quad \text{and} \quad s = \sqrt{\frac{\sum_{i=1}^{n} (x_i - \bar{x})^2}{n - 1}}$$

35. Prove that $\sum_{i=1}^{n} (x_i - \bar{x}) = 0$.

S **36.** Prove that

$$\sum_{i=1}^{n} (x_i - \bar{x})^2 = \sum_{i=1}^{n} x_i^2 - \frac{\left(\sum_{i=1}^{n} x_i\right)^2}{n}$$

6.3 **Area and the Definite Integral**

Purpose

- To develop the limit procedure for calculating the area under a curve.
- To define the definite integral.

If in any figure terminated by right lines and a curve, there be inscribed and circumscribed any number of rectangles and if the breadth of these rectangles be supposed to be diminished, and their number to be augmented in infinitum, I say, that the ultimate ratios which the inscribed figure and the circumscribed figure and the curvilinear figure will have to one another, are ratios of equality. —Isaac Newton (1642–1727)

Area is a concept familiar to all of us through our study of various geometric figures such as the rectangle, square, triangle, and circle. We generally think of area as a number that in some way suggests the size of a bounded region. Of course, for simple geometric figures we have specific formulas for calculating their areas.

Our problem here is to develop a way to calculate the area of any plane region Q, bounded by the x-axis, the lines $x = a$ and $x = b$, and the graph of a nonnegative continuous function f. (See the shaded region in Figure 6.1.)

Initially we will calculate the areas of such regions by finding the *limit of a sum of areas of rectangles*. It is interesting to know that the use of limits to determine areas is not unique to calculus. In fact, early Greek mathematicians discovered the formula for the area of a circle from the limit (as $n \to \infty$) of the areas of inscribed regular n-sided polygons.

Let us begin our development by subdividing the interval $[a, b]$ into n subintervals, $[x_{i-1}, x_i]$, where

$$a = x_0 < x_1 < x_2 < \cdots < x_n = b$$

We call this subdivision a **partition** of the interval $[a, b]$ and we denote it

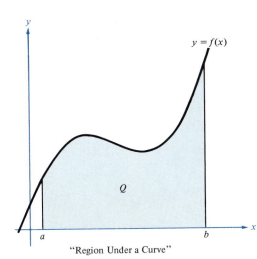

$y = f(x)$

Q

"Region Under a Curve"

FIGURE 6.1

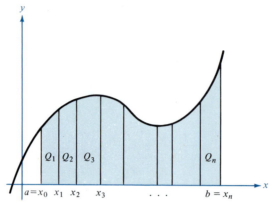

Region Q Partitioned into n Subregions

FIGURE 6.2

by the symbol Δ. This in turn divides region Q into n subregions $Q_1, Q_2, Q_3, \ldots, Q_n$. (See Figure 6.2.)

We can estimate the area of Q by estimating the area of each subregion Q_i and summing the results. Since f is continuous, the Extreme Value Theorem guarantees that $f(x)$ has both a minimum and a maximum on each subinterval. Thus we can estimate the area of each Q_i by the area of an *inscribed* or a *circumscribed* rectangle formed by using the respective minimum or maximum value of $f(x)$ as its height. (See Figure 6.3.) Suppose we let r_i denote the ith inscribed rectangle and R_i the ith circumscribed rectangle. Then since the height of r_i is the minimum value of $f(x)$ on $[x_{i-1}, x_i]$ and the height of R_i is the maximum, we have the following inequality:

$$\text{area of } r_i \leq \text{area of } Q_i \leq \text{area of } R_i$$

To determine the sum of the areas of the inscribed and of the circumscribed rectangles, let

$$\Delta x_i = \text{length of } i\text{th subinterval, } [x_{i-1}, x_i]$$
$$f(m_i) = \text{minimum value of } f \text{ on } [x_{i-1}, x_i]$$
$$f(M_i) = \text{maximum value of } f \text{ on } [x_{i-1}, x_i]$$

Then

$$\text{area of } r_i = f(m_i)(\Delta x_i) = \text{area of } i\text{th inscribed rectangle}$$
$$\text{area of } R_i = f(M_i)(\Delta x_i) = \text{area of } i\text{th circumscribed rectangle}$$

Summing these areas we have, for the partition Δ,

$$\text{lower sum} = s(\Delta) = \sum_{i=1}^{n} f(m_i)\, \Delta x_i \quad \text{(area of inscribed rectangles)}$$

$$\text{upper sum} = S(\Delta) = \sum_{i=1}^{n} f(M_i)\, \Delta x_i \quad \text{(area of circumscribed rectangles)}$$

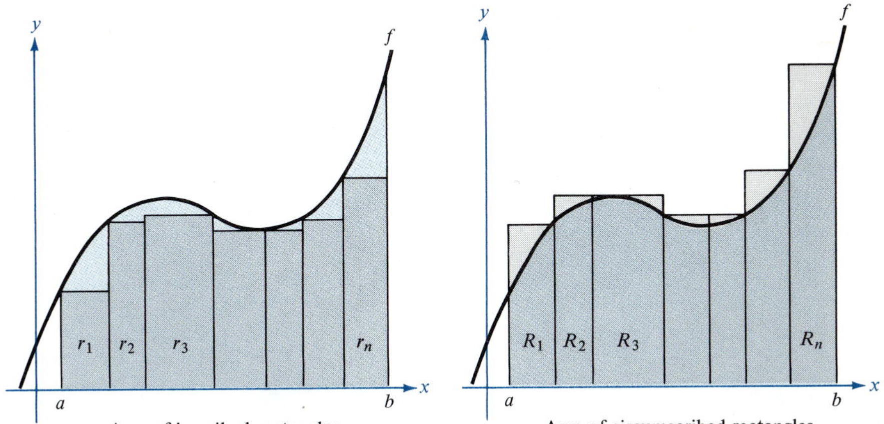

Area of inscribed rectangles
is *less* than area of region Q.

Area of circumscribed rectangles
is *greater* than area of region Q.

FIGURE 6.3

From Figure 6.3 we can see that the lower sum $s(\Delta)$ is *smaller* and the upper sum $S(\Delta)$ is *larger* than the actual area A of the region Q. Thus we have the relationship

$$s(\Delta) \leq A \leq S(\Delta)$$

Just how closely $s(\Delta)$ and $S(\Delta)$ approximate the actual area of Q depends upon the partition Δ. In practice, we improve our approximations by further subdividing the partition Δ in such a way that the width of the largest subinterval approaches zero. We call the width of the largest subinterval of Δ the **norm** of Δ and denote it by $\|\Delta\|$. For example, suppose we divide in half each subinterval in our original partition of $[a, b]$. Then by comparing Figure 6.3 with Figure 6.4, we can see that we have increased the area of the inscribed rectangles and decreased the area of the circumscribed rectangles while still maintaining the relationship $s(\Delta) \leq A \leq S(\Delta)$. This suggests that by subdividing Δ in such a way that $\|\Delta\| \to 0$, we are forcing n to increase ($n \to \infty$), and as a result we have

$$\lim_{\|\Delta\| \to 0} s(\Delta) = A = \lim_{\|\Delta\| \to 0} S(\Delta)$$

or, equivalently,

$$\lim_{n \to \infty} \sum_{i=1}^{n} f(m_i)\, \Delta x_i = A = \lim_{n \to \infty} \sum_{i=1}^{n} f(M_i)\, \Delta x_i$$

In the preceding developing we allowed the partition Δ to have subintervals of different widths. However, in practice, the computation of the limits

$$\lim_{\|\Delta\| \to 0} s(\Delta) \qquad \text{and} \qquad \lim_{\|\Delta\| \to 0} S(\Delta)$$

is usually simplified by partitioning $[a, b]$ into subintervals of equal widths. Our first example demonstrates this procedure.

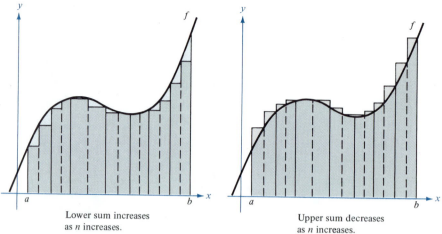

Lower sum increases
as *n* increases.

Upper sum decreases
as *n* increases.

FIGURE 6.4

Example 1

Use both upper and lower sums to find the area of the region between the graph of $f(x) = x^2$ and the x-axis from $x = 0$ to $x = 2$.

Solution: To simplify our calculations we partition the interval [0, 2] into n subintervals of equal length:

$$\Delta x = \frac{b - a}{n} = \frac{2 - 0}{n} = \frac{2}{n}$$

Figure 6.5 shows the points of subdivision and several inscribed and circumscribed rectangles.

Since $f(x) = x^2$ is increasing on the interval [0, 2], the minimum value of f on each subinterval occurs at the *left* endpoints of the subintervals, whereas the maximum value of f occurs at the *right* endpoints. Therefore,

$$m_1 = x_0 = 0 \qquad\qquad M_1 = x_1 = \frac{2}{n}$$

$$m_2 = x_1 = \frac{2}{n} \qquad\qquad M_2 = x_2 = \frac{4}{n}$$

$$m_3 = x_2 = \frac{4}{n} \qquad\qquad M_3 = x_3 = \frac{6}{n}$$

$$\vdots \qquad\qquad\qquad \vdots$$

$$m_i = x_{i-1} = \frac{2(i - 1)}{n} \qquad M_i = x_i = \frac{2i}{n}$$

Thus the lower sum is

$$s(\Delta) = \sum_{i=1}^{n} f(m_i)\,\Delta x = \sum_{i=1}^{n} f\left[\frac{2(i - 1)}{n}\right]\!\left(\frac{2}{n}\right) = \sum_{i=1}^{n} \left[\frac{2(i - 1)}{n}\right]^2\!\left(\frac{2}{n}\right)$$

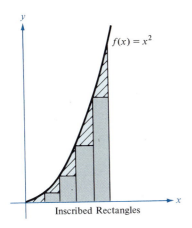

Inscribed Rectangles

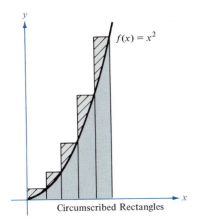

Circumscribed Rectangles

FIGURE 6.5

$$= \sum_{i=1}^{n} \left(\frac{8}{n^3}\right)(i^2 - 2i + 1) = \frac{8}{n^3}\left[\sum_{i=1}^{n} i^2 - 2\sum_{i=1}^{n} i + \sum_{i=1}^{n} 1\right]$$

$$= \frac{8}{n^3}\left\{\frac{n(n + 1)(2n + 1)}{6} - 2\left[\frac{n(n + 1)}{2}\right] + n\right\}$$

$$= \frac{8}{n^3}\left[\frac{2n^3 + 3n^2 + n}{6} - n^2 - n + n\right]$$

$$= \frac{8}{n^3}\left[\frac{2n^3 + 3n^2 + n - 6n^2}{6}\right]$$

$$= \frac{4}{3n^3}(2n^3 - 3n^2 + n) = \frac{8}{3} - \frac{4}{n} + \frac{4}{3n^2}$$

(Note the use of the summation formulas from Section 6.2.) The upper sum is

$$S(\Delta) = \sum_{i=1}^{n} f(M_i)\,\Delta x = \sum_{i=1}^{n} f\left(\frac{2i}{n}\right)\left(\frac{2}{n}\right) = \sum_{i=1}^{n} \left(\frac{2^2 i^2}{n^2}\right)\left(\frac{2}{n}\right) = \frac{8}{n^3}\sum_{i=1}^{n} i^2$$

$$= \frac{8}{n^3}\left[\frac{n(n + 1)(2n + 1)}{6}\right] = \frac{8}{n^3}\left[\frac{2n^3 + 3n^2 + n}{6}\right]$$

$$= \frac{4}{3n^3}(2n^3 + 3n^2 + n) = \frac{8}{3} + \frac{4}{n} + \frac{4}{3n^2}$$

Now since *each* subinterval Δx_i has length $2/n$, we are guaranteed that $n \to \infty$ as $\|\Delta\| \to 0$. Therefore,

$$\lim_{\|\Delta\| \to 0} s(\Delta) = \lim_{n \to \infty}\left(\frac{8}{3} - \frac{4}{n} + \frac{4}{3n^2}\right) = \frac{8}{3}$$

Furthermore,

$$\lim_{\|\Delta\| \to 0} S(\Delta) = \lim_{n \to \infty}\left(\frac{8}{3} + \frac{4}{n} + \frac{4}{3n^2}\right) = \frac{8}{3}$$

Therefore, the area under $f(x) = x^2$ from $x = 0$ to $x = 2$ is $A = \frac{8}{3}$ square units.   ∎

In finding the upper and lower sums, we chose m_i and M_i in $[x_{i-1}, x_i]$ so that $f(m_i)$ and $f(M_i)$ were the minimum and maximum values of $f(x)$ on the ith subinterval. However, since these choices ultimately led to the same limit, it appears that the choice is actually arbitrary and, in fact, we can choose *any* c_i in $[x_{i-1}, x_i]$ and obtain the same limit. Therefore, to save time in subsequent problems, we will not only choose Δ to have n subintervals of equal widths, but we will also choose c_i to be the *right* endpoint of the ith interval. We denote the sum formed with these choices by $S(n)$ and we use the limit

$$\lim_{n \to \infty} S(n)$$

in place of either

$$\lim_{\|\Delta\|\to 0} s(\Delta) \quad \text{or} \quad \lim_{\|\Delta\|\to 0} S(\Delta)$$

Example 2

Find the area of the region between the graph of $f(x) = x^3$ and the x-axis from $x = 1$ to $x = 4$.

Solution: For our subdivision let

$$\Delta x = \frac{4-1}{n} = \frac{3}{n}$$

and choosing the c_i's as right endpoints, we have

$$c_1 = x_1 = 1 + \frac{3}{n}$$

$$c_2 = x_2 = 1 + 2\left(\frac{3}{n}\right)$$

$$c_3 = x_3 = 1 + 3\left(\frac{3}{n}\right)$$

$$\vdots$$

$$c_i = x_i = 1 + i\left(\frac{3}{n}\right)$$

(See Figure 6.6.) Therefore,

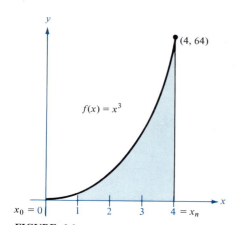

FIGURE 6.6

$$S(n) = \sum_{i=1}^{n} f\left(1 + \frac{3i}{n}\right)\left(\frac{3}{n}\right) = \sum_{i=1}^{n}\left(1 + \frac{3i}{n}\right)^3\left(\frac{3}{n}\right)$$

$$= \frac{3}{n}\sum_{i=1}^{n}\left[1 + 3\left(\frac{3i}{n}\right) + 3\left(\frac{3i}{n}\right)^2 + \left(\frac{3i}{n}\right)^3\right]$$

$$= \frac{3}{n}\sum_{i=1}^{n} 1 + \frac{3}{n}\sum_{i=1}^{n}\frac{9i}{n} + \frac{3}{n}\sum_{i=1}^{n}\frac{27i^2}{n^2} + \frac{3}{n}\sum_{i=1}^{n}\frac{27i^3}{n^3}$$

$$= \frac{3}{n}(n) + \frac{27}{n^2}\sum_{i=1}^{n} i + \frac{81}{n^3}\sum_{i=1}^{n} i^2 + \frac{81}{n^4}\sum_{i=1}^{n} i^3$$

$$= 3 + \frac{27}{n^2}\left[\frac{n(n+1)}{2}\right] + \frac{81}{n^3}\left[\frac{n(n+1)(2n+1)}{6}\right] + \frac{81}{n^4}\left[\frac{n^2(n+1)^2}{4}\right]$$

In the figure: labels y, $(4, 64)$, $f(x) = x^3$, $x_0 = 0$, axis marks 1, 2, 3, $4 = x_n$, and x.

$$= 3 + \frac{27n^2 + 27n}{2n^2} + \frac{162n^3 + 243n^2 + 81n}{6n^3} + \frac{81n^4 + 162n^3 + 81n^2}{4n^4}$$

Hence $\qquad$ area $= \lim_{n\to\infty} S(n) = 3 + \frac{27}{2} + \frac{162}{6} + \frac{81}{4} = 63\frac{3}{4}$ ∎

Example 3

Find the area of the region between the graph of $f(x) = (x - 2)^2$ and the x-axis from $x = 0$ to $x = 4$.

Solution: Let $\qquad\qquad \Delta x = \frac{4 - 0}{n} = \frac{4}{n}$

and choosing right endpoints, we have

$$c_1 = x_1 = 1\left(\frac{4}{n}\right)$$

$$c_2 = x_2 = 2\left(\frac{4}{n}\right)$$

$$c_3 = x_3 = 3\left(\frac{4}{n}\right)$$

$$\vdots$$

$$c_i = x_i = i\left(\frac{4}{n}\right)$$

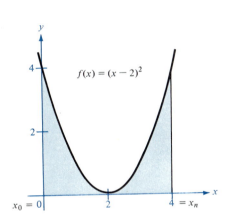

$f(x) = (x - 2)^2$

$x_0 = 0$ $\qquad$ 2 $\qquad$ 4 $= x_n$

FIGURE 6.7

(See Figure 6.7.) Therefore,

$$S(n) = \sum_{i=1}^{n} f\left(\frac{4i}{n}\right)\left(\frac{4}{n}\right) = \sum_{i=1}^{n} \left(\frac{4i}{n} - 2\right)^2\left(\frac{4}{n}\right) = \frac{4}{n}\sum_{i=1}^{n}\left(\frac{16i^2}{n^2} - \frac{16i}{n} + 4\right)$$

$$= \frac{64}{n^3}\sum_{i=1}^{n} i^2 - \frac{64}{n^2}\sum_{i=1}^{n} i + \frac{16}{n}\sum_{i=1}^{n} 1$$

$$= \frac{64}{n^3}\left[\frac{n(n + 1)(2n + 1)}{6}\right] - \frac{64}{n^2}\left[\frac{n(n + 1)}{2}\right] + \left(\frac{16}{n}\right)n$$

$$= \frac{128n^3 + 192n^2 + 64n}{6n^3} - \frac{32n^2 + 32n}{n^2} + 16$$

Hence, $\qquad$ area $= \lim_{n\to\infty} S(n) = \frac{128}{6} - 32 + 16 = \frac{16}{3}$ ∎

Example 4

Using 4 subintervals, estimate the area of the region between the graph of $f(x) = \sqrt{x}$ and the x-axis from $x = 1$ to $x = 4$. (See Figure 6.8.)

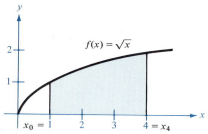

FIGURE 6.8

Solution: Given

$$\Delta x = \frac{4-1}{4} = \frac{3}{4}$$

and using right endpoints, we have

$$c_1 = x_1 = 1 + \tfrac{3}{4} = \tfrac{7}{4}$$
$$c_2 = x_2 = 1 + 2(\tfrac{3}{4}) = \tfrac{10}{4}$$
$$c_3 = x_3 = 1 + 3(\tfrac{3}{4}) = \tfrac{13}{4}$$
$$c_4 = x_4 = 1 + 4(\tfrac{3}{4}) = 4$$

Therefore,

$$A \approx S(4) = \sum_{i=1}^{4} f(c_i)\left(\frac{3}{4}\right) = \left(\sqrt{\frac{7}{4}} + \sqrt{\frac{10}{4}} + \sqrt{\frac{13}{4}} + \sqrt{\frac{16}{4}}\right)\left(\frac{3}{4}\right)$$

$$= \left(\frac{\sqrt{7} + \sqrt{10} + \sqrt{13} + 4}{2}\right)\left(\frac{3}{4}\right)$$

$$\approx \left(\frac{13.41358}{2}\right)\left(\frac{3}{4}\right) \approx 5.03$$

(The actual area is $4.66\overline{6}$.) ∎

Although we have developed the "limit of a sum" process as a means for calculating the area of a plane region, applications of this summation technique extend far beyond the finding of areas of plane regions. They include the finding of volumes of solids, fluid pressure, work, centers of mass, lengths of arcs, areas of surfaces of revolution, and many other geometric and physical quantities.

Because of the wide application of this "limit of a sum" process, we give it the following special name and symbol:

Definition of Definite Integral

Let f be defined on the closed interval $[a, b]$, and suppose c_i lies in the subinterval $[x_{i-1}, x_i]$ of width Δx_i. Then if

$$\lim_{\|\Delta\| \to 0} \sum_{i=1}^{n} f(c_i)\,\Delta x_i$$

exists, we denote this limit by

$$\int_a^b f(x)\,dx$$

and call it the **definite integral of f from a to b**.

On the interval $[a, b]$, if

$$\lim_{\|\Delta\| \to 0} \sum_{i=1}^{n} f(c_i)\,\Delta x_i$$

exists, we say f is **integrable** on $[a, b]$, and we write

$$\int_a^b f(x)\,dx = \lim_{\|\Delta\| \to 0} \sum_{i=1}^{n} f(c_i)\,\Delta x_i$$

We call a the **lower limit** of integration and b the **upper limit.**

Our definition of the definite integral does not require that f be continuous on $[a, b]$. In fact, some discontinuous functions are integrable. For example, it can be readily shown that the greatest-integer function, $f(x) = [x]$, is integrable on any closed interval.

Even though continuity is not necessary for a function to be integrable, it is a sufficient condition for integrability and we state this, without proof, in the following theorem:

THEOREM 6.2 *(Continuity Implies Integrability)*	If f is continuous on $[a, b]$, then it is integrable on $[a, b]$.

Observe that if f is continuous and nonnegative on $[a, b]$, then from our previous limit description of the area of a region, we can see that the definite integral $\int_a^b f(x)\,dx$ denotes the area of the region between the graph of f and the x-axis from $x = a$ to $x = b$.

Example 5

Evaluate the definite integral $\int_{-1}^{2} (2y - 3)\,dy$.

Solution: We obtain

$$\Delta y = \frac{2 - (-1)}{n} = \frac{3}{n}$$

Then we let

$$y_0 = -1$$

$$c_1 = y_1 = -1 + \frac{3}{n}$$

$$c_2 = y_2 = -1 + 2\left(\frac{3}{n}\right)$$

$$c_3 = y_3 = -1 + 3\left(\frac{3}{n}\right)$$

$$\vdots$$

$$c_i = y_i = -1 + i\left(\frac{3}{n}\right)$$

and it follows that

$$S(n) = \sum_{i=1}^{n} f(c_i)(\Delta y_i) = \sum_{i=1}^{n} \left[2\left(-1 + \frac{3i}{n} \right) - 3 \right]\left(\frac{3}{n} \right)$$

$$= \sum_{i=1}^{n} \left(-5 + \frac{6i}{n} \right)\left(\frac{3}{n} \right) = \sum_{i=1}^{n} \frac{-15}{n} + \sum_{i=1}^{n} \frac{18i}{n^2}$$

$$= \frac{-15}{n}(n) + \frac{18}{n^2}\frac{n(n+1)}{2} = -15 + \frac{9n^2 + 9n}{n^2}$$

Therefore,

$$\int_{-1}^{2} (2y - 3)\,dy = \lim_{n \to \infty} \sum_{i=1}^{n} f(c_i)(\Delta y_i)$$

$$= \lim_{n \to \infty}\left[-15 + \frac{(9n^2 + 9n)}{n^2} \right] = -15 + 9 = -6 \quad \blacksquare$$

Note in Example 5 that we were merely asked to *evaluate* the definite integral $\int_{-1}^{2} (2y - 3)\,dy$ and not to interpret the result. Since a definite integral can be applied to more than the calculation of areas, we cannot necessarily interpret a definite integral as area.

One final clarifying distinction needs to be made regarding notation involving the integral sign. The *indefinite integral* $\int f(x)\,dx$ denotes a family of *functions of x*, namely, the antiderivatives of $f(x)$. However, the *definite integral* $\int_{a}^{b} f(x)\,dx$ is a *number* that is the limit of a sum. Although definite and indefinite integrals are defined in entirely different ways, the notation used for each suggests a close connection. This relationship will be explained in the next section.

Section Exercises (6.3)

In Exercises 1–6, use the upper and lower sums to approximate the area of the region between the graph of each function and the x-axis over the given interval. In each case use the number of subdivisions indicated by n.

[c] **1.** $y = \sqrt{x}$; [0, 1]; $n = 4$

[c] **2.** $y = \sqrt{x} + 1$; [0, 2]; $n = 8$

3. $y = \dfrac{1}{x}$; [1, 2]; $n = 5$

4. $y = \dfrac{1}{x - 2}$; [4, 6]; $n = 4$

[c][s] **5.** $y = \sqrt{1 - x^2}$; [0, 1]; $n = 5$

[c] **6.** $y = \sqrt{x} + 1$; [0, 1]; $n = 4$

In Exercises 7–16, use the "limit of a sum" process to find the area of the region between the graph of each function and the x-axis over the given interval. Sketch each region.

[s] **7.** $y = -2x + 3$; [0, 1]

8. $y = 1 - x^2$; [−1, 1]

9. $y = x^2 + 2$; [0, 1]

10. $y = 3x - 4$; [2, 5]

[s] **11.** $y = 2x^2$; [1, 3]

12. $y = 2x - x^3$; [0, 1]

13. $y = 1 - x^3$; [0, 1]

14. $y = x^2 - x^3$; [0, 1]

[s] **15.** $y = x^2 - x^3$; [−1, 0]

16. $y = 2x^2 - x + 1$; [0, 2]

In Exercises 17–22, evaluate the definite integrals.

17. $\displaystyle\int_{4}^{10} 6\,dx$

18. $\displaystyle\int_{-2}^{3} x\,dx$

19. $\boxed{\text{S}}$ $\int_{-1}^{1} x^3 \, dx$

20. $\int_{0}^{1} x^3 \, dx$

21. $\int_{1}^{2} (x^2 + 1) \, dx$

22. $\int_{1}^{2} 4x^2 \, dx$

25. $\boxed{\text{S}}$ $\int_{0}^{2} (2x + 5) \, dx$

26. $\int_{-1}^{1} (1 - |x|) \, dx$

27. $\int_{-r}^{r} \sqrt{r^2 - x^2} \, dx$

28. $\int_{-4}^{4} \sqrt{1 - \frac{x^2}{16}} \, dx$

In Exercises 23–28, sketch the region whose area is indicated by the given definite integrals. Then in each case use a geometric formula to evaluate the integral.

23. $\int_{0}^{3} 4 \, dx$

24. $\int_{0}^{4} x \, dx$

For Review and Class Discussion

True or False

1. ____ If f is increasing on $[a, b]$, then the minimum value of $f(x)$ on $[a, b]$ is $f(a)$.

2. ____ The area of the region bounded by the graph of $y = x^2 + 2$, the x-axis, and the lines $x = 1$ and $x = 5$ is given by $\int_{1}^{5} (x^2 + 2) \, dx$.

3. ____ If the norm of a partition approaches zero, then the number of subintervals approaches infinity.

4. ____ If Δ is a partition of $[1, 4]$ where $\Delta x = 3/n$, then the right endpoints of the subintervals are $c_1 = 3/n$, $c_2 = 2(3/n)$, . . . , $c_n = n(3/n)$.

5. ____ The value of $\int_{a}^{b} f(x) \, dx$ must be positive.

6. ____ If f is continuous on $[a, b]$, then f is integrable on $[a, b]$.

7. ____ $\int_{-1}^{1} x \, dx = \lim_{n \to \infty} \sum_{i=1}^{n} \left(-1 + \frac{2i}{n}\right)\left(\frac{2}{n}\right) = 0.$

8. ____ $\int_{0}^{1} x^2 \, dx = \lim_{n \to \infty} \sum_{i=1}^{n} \frac{i}{n^3}.$

9. ____ If $b_k \neq 0$, then

$$\lim_{n \to \infty} \frac{a_k n^k + a_{k-1} n^{k-1} + \cdots + a_0}{b_k n^k + b_{k-1} n^{k-1} + \cdots + b_0} = \frac{a_k}{b_k}.$$

10. ____ $\int_{0}^{1} x^2 \, dx < \int_{1}^{2} x^2 \, dx.$

6.4   The Fundamental Theorem of Calculus

Purpose

- To provide an informal development of the Fundamental Theorem of Calculus.
- To use the Fundamental Theorem to calculate areas under a curve.
- To derive some properties of the definite integral.

The further a mathematical theory is developed, the more harmoniously and uniformly does its construction proceed, and unsuspected relations are disclosed between hitherto separated branches of the science.   —David Hilbert *(1862–1943)*

Perhaps by now you are thinking, "If the definite integral has such wide application, why isn't there a simpler way to evaluate it than by the limit of a sum?" Well, in many cases there is an easier way. In fact, we are already familiar with the concepts involved in this "easier" way. This

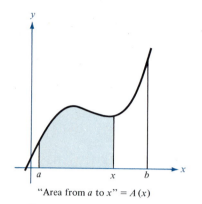

"Area from *a* to *x*" = $A(x)$

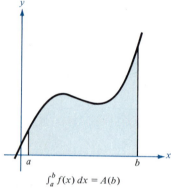

$\int_a^b f(x)\, dx = A(b)$

FIGURE 6.9

section includes an informal development of a theorem that describes a simple procedure for evaluating a definite integral without using the limit of a sum. This theorem is referred to as the **Fundamental Theorem of Calculus,** for it shows the relationship between the two basic operations of calculus—differentiation and integration.

The objective of the following discussion is to show the feasibility of the Fundamental Theorem and not to prove it. As in the previous section, we use the area of a region as the setting for our discussion.

Suppose *f* is continuous and nonnegative on [*a*, *b*]. We denote the area of the region under the graph of *f* from *a* to *x* by the area function $A(x)$. (See Figure 6.9.)

Now, if we let *x* increase by an amount Δx, then the area of the region under the graph of *f* increases by ΔA. Furthermore, if $f(m)$ and $f(M)$ denote the minimum and maximum values of *f* on the interval [*x*, *x* + Δx], then we have the relationship

$$f(m)\, \Delta x \leq \Delta A \leq f(M)\, \Delta x$$

(See Figure 6.10.) Dividing each term in

$$f(m)\, \Delta x \leq \Delta A \leq f(M)\, \Delta x$$

by Δx, we have

$$f(m) \leq \frac{\Delta A}{\Delta x} \leq f(M)$$

Since both $f(m)$ and $f(M)$ approach $f(x)$ as Δx approaches 0, and since

$$\lim_{\Delta x \to 0} \frac{\Delta A}{\Delta x} = A'(x)$$

it follows that

$$f(x) \leq A'(x) \leq f(x)$$

which means that

$$f(x) = A'(x)$$

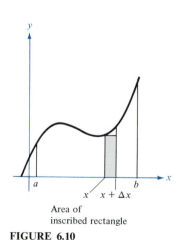

Area of inscribed rectangle

$\leqslant$

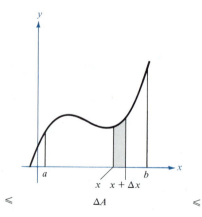

ΔA

$\leqslant$

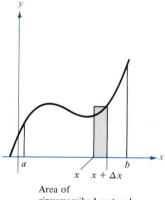

Area of circumscribed rectangle

FIGURE 6.10

Thus we have established that the area function $A(x)$ is an antiderivative of f and, consequently, must be of the following form:

$$A(x) = F(x) + C$$

where $F(x)$ is any antiderivative of f (Section 6.1). To solve for C we note that $A(a) = 0$ and thus $C = -F(a)$. Furthermore, by evaluating $A(b)$ we have

$$A(b) = F(b) + C = F(b) - F(a)$$

Finally, by replacing $A(b)$ by its integral form (see Figure 6.9), we have

$$\int_a^b f(x)\, dx = F(b) - F(a)$$

This equation tells us that *if we can find an antiderivative for f,* then we can evaluate the definite integral $\int_a^b f(x)\, dx$ without having to use the limit of a sum. We summarize these results in the following theorem:

THEOREM 6.3 (*Fundamental Theorem of Calculus*)	If a function f is continuous on the interval $[a, b]$, then $$\int_a^b f(x)\, dx = F(b) - F(a)$$ where F is any function such that $F'(x) = f(x)$ for all x in $[a, b]$.

Several comments regarding the Fundamental Theorem of Calculus are in order. First, this theorem describes a means for *evaluating* a definite integral, not a procedure for finding antiderivatives. Secondly, when applying this theorem it is helpful to use the formulation

$$\int_a^b f(x)\, dx = F(x)\bigg]_a^b = F(b) - F(a)$$

For instance, we write

$$\int_1^3 x^3\, dx = \frac{x^4}{4}\bigg]_1^3 = \frac{(3)^4}{4} - \frac{(1)^4}{4} = \frac{81}{4} - \frac{1}{4} = 20$$

Finally, we observe that the constant of integration C can be dropped from the antiderivative, because

$$\int_a^b f(x)\, dx = \left[F(x) + C\right]_a^b = [F(b) + C] - [F(a) + C]$$

$$= F(b) - F(a) + C - C = F(b) - F(a)$$

The Fundamental Theorem is useful for proving certain properties of definite integrals of continuous functions. For instance, if f is continuous on $[a, b]$, $F'(x) = f(x)$ for all x in $[a, b]$, and $a < c < b$, then

$$\int_a^c f(x)\,dx + \int_c^b f(x)\,dx = [F(c) - F(a)] + [F(b) - F(c)]$$

$$= F(b) - F(a)$$

and it follows that

$$\int_a^c f(x)\,dx + \int_c^b f(x)\,dx = \int_a^b f(x)\,dx$$

When using the interval $[a, b]$, it is assumed that $a < b$. As a result, our limit definition of the definite integral, $\int_a^b f(x)\,dx$, does not take into account the cases where $a \geq b$. Thus we define the following special cases:

1. If $a > b$, then

$$\int_a^b f(x)\,dx = -\int_b^a f(x)\,dx$$

provided the latter integral exists.

2. If $f(a)$ exists, then

$$\int_a^a f(x)\,dx = 0$$

We now list some useful properties of the definite integral. In each case we assume the integrability of f and g on $[a, b]$.

Properties of Definite Integrals

1. $\displaystyle\int_a^b kf(x)\,dx = k\int_a^b f(x)\,dx$, where k is a constant

2. $\displaystyle\int_a^b f(x)\,dx = \int_a^c f(x)\,dx + \int_c^b f(x)\,dx$, where $a < c < b$

3. $\displaystyle\int_a^b [f(x) \pm g(x)]\,dx = \int_a^b f(x)\,dx \pm \int_a^b g(x)\,dx$

Example 1

Use the Fundamental Theorem to evaluate $\int_1^2 (x^2 - 3)\,dx$.

Solution:

$$\int_1^2 (x^2 - 3)\,dx = \left[\frac{x^3}{3} - 3x\right]_1^2 = \left(\frac{8}{3} - 6\right) - \left(\frac{1}{3} - 3\right) = \frac{-2}{3} \quad \blacksquare$$

Example 2

Evaluate the definite integral $\int_0^1 (4t + 1)^2\,dt$.

Solution:

$$\int_0^1 (4t + 1)^2 \, dt = \frac{1}{4} \int_0^1 (4t + 1)^2 (4) \, dt$$

$$= \frac{1}{4} \left[\frac{(4t + 1)^3}{3} \right]_0^1 = \frac{1}{4} \left[\frac{125}{3} - \frac{1}{3} \right] = \frac{31}{3}$$

(Note the use of Property 1 in the first line of the solution.)   ■

Example 3

Evaluate $\int_1^4 3\sqrt{x} \, dx$.

Solution:

$$\int_1^4 3\sqrt{x} \, dx = 3 \int_1^4 x^{1/2} \, dx = 3 \left[\frac{x^{3/2}}{\frac{3}{2}} \right]_1^4 = 2(4)^{3/2} - 2(1)^{3/2} = 14$$ ■

Example 4

Evaluate

$$\int_{-8}^{-1} \frac{x + 2x^2}{\sqrt[3]{x}} \, dx$$

Solution:

$$\int_{-8}^{-1} \frac{x + 2x^2}{\sqrt[3]{x}} \, dx = \int_{-8}^{-1} \left(\frac{x}{x^{1/3}} + \frac{2x^2}{x^{1/3}} \right) dx$$

$$= \int_{-8}^{-1} (x^{2/3} + 2x^{5/3}) \, dx = \left[\frac{x^{5/3}}{\frac{5}{3}} + \frac{2x^{8/3}}{\frac{8}{3}} \right]_{-8}^{-1}$$

$$= \left(\frac{-3}{5} + \frac{3}{4} \right) - \left(\frac{-96}{5} + 192 \right)$$

$$= -0.6 + 0.75 + 19.2 - 192 = -172.65$$ ■

Example 5 (*Integrating Absolute Value*)

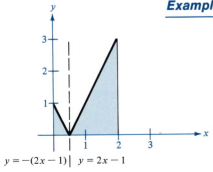

$y = -(2x - 1)|$ $y = 2x - 1$

FIGURE 6.11 $y = |2x - 1|$

Evaluate $\int_0^2 |2x - 1| \, dx$.

Solution: From Figure 6.11 and the definition of absolute value, we note that

$$|2x - 1| = \begin{cases} -(2x - 1), & \text{for } x < \frac{1}{2} \\ (2x - 1), & \text{for } x \geq \frac{1}{2} \end{cases}$$

Hence we rewrite the integral in two parts as

$$\int_0^2 |2x - 1| \, dx = \int_0^{1/2} -(2x - 1) \, dx + \int_{1/2}^2 (2x - 1) \, dx$$

$$= \left[-x^2 + x \right]_0^{1/2} + \left[x^2 - x \right]_{1/2}^2$$

$$= \left(-\frac{1}{4} + \frac{1}{2} \right) - (0 + 0) + (4 - 2) - \left(\frac{1}{4} - \frac{1}{2} \right) = \frac{5}{2}$$

∎

Example 6

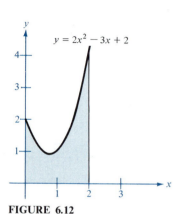

$y = 2x^2 - 3x + 2$

FIGURE 6.12

Use the Fundamental Theorem to find the area under the graph of $y = 2x^2 - 3x + 2$ between $x = 0$ and $x = 2$. (See Figure 6.12.)

Solution:

$$\text{area} = \int_0^2 (2x^2 - 3x + 2)\, dx = \left[\frac{2x^3}{3} - \frac{3x^2}{2} + 2x \right]_0^2$$

$$= \frac{16}{3} - 6 + 4 = \frac{10}{3}$$

∎

Up to this point our discussion of properties of the definite integral has focused on "working properties," that is, properties that aid us in the evaluation of definite integrals. To end this section we take a look at a more theoretical property of the definite integral called the Mean Value Theorem for Integrals.

Recall from our discussion in the previous section that we used rectangles to approximate the area of a region under a curve. In that discussion we observed that the actual area of the region was greater than the area of an inscribed rectangle and less than the area of a circumscribed rectangle. The Mean Value Theorem for Integrals states that somewhere "between" the inscribed and circumscribed rectangles there is a rectangle whose area is precisely equal to the area of the region under the curve. (See Figure 6.13.)

THEOREM 6.4
(*Mean Value Theorem for Integrals*)

If f is continuous on $[a, b]$, then there exists a number c in (a, b) such that

$$\int_a^b f(x)\, dx = f(c)(b - a)$$

Proof:

Case 1: If f is constant over the interval $[a, b]$, the result is trivial since c can be any point in (a, b).

Case 2: If f is not constant on $[a, b]$, then by the Extreme Value Theorem, choose $f(m)$ and $f(M)$ to be the minimum and maximum values of f on $[a, b]$. Since

$$f(m) \le f(x) \le f(M)$$

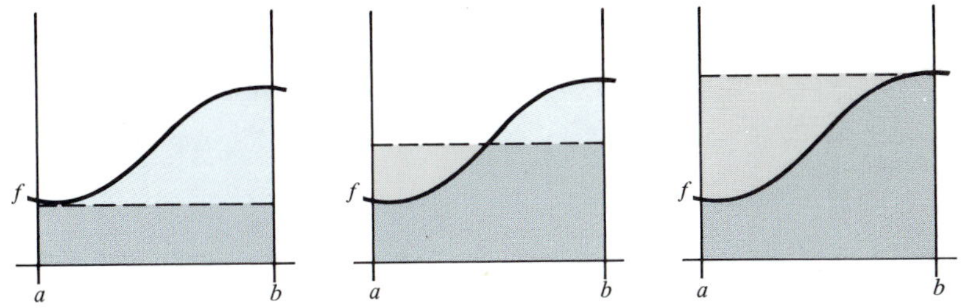

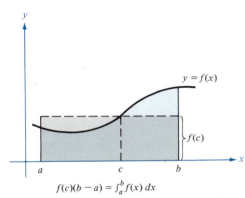

f
a
b
Inscribed Rectangle
(less than actual area)
f
a
b
Mean Value Rectangle
(equal to actual area)
f
a
b
Circumscribed Rectangle
(greater than actual area)

FIGURE 6.13

for all x in $[a, b]$, we conclude (see Exercise 57 in this section) that

$$\int_a^b f(m)\,dx \le \int_a^b f(x)\,dx \le \int_a^b f(M)\,dx$$

Since $f(m)$ and $f(M)$ are both constant, we have

$$\int_a^b f(m)\,dx = f(m)(b-a) \qquad \text{and} \qquad \int_a^b f(M)\,dx = f(M)(b-a)$$

Thus we have

$$f(m)(b-a) \le \int_a^b f(x)\,dx \le f(M)(b-a)$$

Dividing by $b - a$ we get

$$f(m) \le \frac{1}{b-a}\int_a^b f(x)\,dx \le f(M)$$

Applying the Intermediate Value Theorem, we conclude that there exists some c in (a, b) such that

$$f(c) = \frac{1}{b-a}\int_a^b f(x)\,dx$$

y
y = f(x)
f(c)
x
a
c
b

$f(c)(b-a) = \int_a^b f(x)\,dx$

FIGURE 6.14

Thus for this c (Figure 6.14), we have

$$f(c)(b-a) = \int_a^b f(x)\,dx$$

■

Note that the Mean Value Theorem does not specify how to determine c; it merely guarantees the existence of the number c. However, the value of $f(c)$ can be found by the formula

$$f(c) = \frac{\int_a^b f(x)\,dx}{b-a}$$

This value is referred to as the **average value** of f on interval $[a, b]$.

Example 7

Determine the value of c such that

$$\int_1^4 (3x^2 - 2x)\,dx = f(c)(4 - 1)$$

What is the average value of $f(x) = 3x^2 - 2x$ on $[1, 4]$?

Solution:

$$\int_1^4 (3x^2 - 2x)\,dx = \left[\frac{3x^3}{3} - \frac{2x^2}{2}\right]_1^4 = 64 - 16 - (1 - 1) = 48$$

We wish to find c such that $f(c)(4 - 1) = 48$. Thus

$$(3c^2 - 2c)(3) = 48$$
$$3c^2 - 2c = 16$$
$$3c^2 - 2c - 16 = 0$$
$$(3c - 8)(c + 2) = 0$$

Since $c = -2$ is not in the interval $[1, 4]$, we choose $c = \frac{8}{3}$ and obtain

$$\int_1^4 (3x^2 - 2x)\,dx = f(\tfrac{8}{3})(4 - 1)$$

The average value of $f(x) = 3x^2 - 2x$ on $[1, 4]$ is

$$f(c) = \frac{\displaystyle\int_1^4 (3x^2 - 2x)\,dx}{4 - 1} = \frac{48}{3} = 16 \qquad\blacksquare$$

Section Exercises (6.4)

In Exercises 1–30, evaluate the definite integrals.

1. $\displaystyle\int_0^1 2x\,dx$

2. $\displaystyle\int_2^7 3\,dv$

3. $\displaystyle\int_{-1}^0 (x - 2)\,dx$

4. $\displaystyle\int_2^5 (-3v + 4)\,dv$

5. $\displaystyle\int_{-1}^1 (t^2 - 2)\,dt$

6. $\displaystyle\int_0^3 (3x^2 + x - 2)\,dx$

S 7. $\displaystyle\int_0^1 (2t - 1)^2\,dt$

8. $\displaystyle\int_{-1}^1 (t^3 - 9t)\,dt$

9. $\displaystyle\int_1^2 \left(\frac{3}{x^2} - 1\right) dx$

10. $\displaystyle\int_0^1 (3x^3 - 9x + 7)\,dx$

11. $\displaystyle\int_1^2 (5x^4 + 5)\,dx$

12. $\displaystyle\int_{-3}^3 v^{1/3}\,dv$

13. $\displaystyle\int_{-1}^1 (\sqrt[3]{t} - 2)\,dt$

S 14. $\displaystyle\int_{-2}^{-1} \sqrt{\frac{-2}{x}}\,dx$

S 15. $\displaystyle\int_1^4 \frac{u - 2}{\sqrt{u}}\,du$

16. $\displaystyle\int_{-2}^{-1} \left(\frac{-1}{u^2} + u\right) du$

17. $\displaystyle\int_0^1 \frac{x - \sqrt{x}}{3}\,dx$

18. $\displaystyle\int_0^2 (2 - t)\sqrt{t}\,dt$

19. $\displaystyle\int_{-1}^0 (t^{1/3} - t^{2/3})\,dt$

20. $\displaystyle\int_{-8}^{-1} \frac{x - x^2}{2\sqrt[3]{x}}\,dx$

21. $\displaystyle\int_0^4 \frac{1}{\sqrt{2x + 1}}\,dx$

22. $\displaystyle\int_0^1 x\sqrt{1 - x^2}\,dx$

S 23. $\displaystyle\int_{-1}^1 x(x^2 + 1)^3\,dx$

24. $\displaystyle\int_0^2 \frac{x}{\sqrt{1 + 2x^2}}\,dx$

25. $\displaystyle\int_0^2 x\sqrt[3]{4 + x^2}\, dx$

26. $\displaystyle\int_1^9 \frac{1}{\sqrt{x}(1 + \sqrt{x})^2}\, dx$

27. $\displaystyle\int_{-1}^1 |x|\, dx$

28. $\displaystyle\int_0^3 |2x - 3|\, dx$

[S] **29.** $\displaystyle\int_0^4 |x^2 - 4x + 3|\, dx$

30. $\displaystyle\int_{-1}^1 |x^3|\, dx$

In Exercises 31–36, evaluate the definite integrals and make a sketch of the region whose area is given by the integral.

31. $\displaystyle\int_1^3 (2x - 1)\, dx$

32. $\displaystyle\int_0^2 (x + 4)\, dx$

33. $\displaystyle\int_3^4 (x^2 - 9)\, dx$

34. $\displaystyle\int_{-1}^2 (-x^2 + x + 2)\, dx$

[S] **35.** $\displaystyle\int_0^1 (x - x^3)\, dx$

36. $\displaystyle\int_0^1 \sqrt{x}(1 - x)\, dx$

In Exercises 37–46, determine the area of each region having the given boundaries.

37. $y = 3x^2 + 1;\ x = 0,\ x = 2,\ y = 0$
38. $y = 1 + \sqrt{x};\ x = 0,\ x = 4,\ y = 0$
39. $y = \sqrt[3]{2x};\ x = 4,\ y = 0$
40. $y = x - x^2;\ y = 0$
[S] **41.** $y = -x^2 + 2x + 3;\ y = 0$
42. $y = x^3 + x;\ x = 2,\ y = 0$
43. $y = 1 - x^4;\ y = 0$

44. $y = \dfrac{1}{x^2};\ x = 1,\ x = 2,\ y = 0$

[S] **45.** $y = (3 - x)\sqrt{x};\ y = 0$
46. $y = -x^2 + 3x;\ y = 0$

In Exercises 47–52, sketch the graph of each function over the given interval. Find the average value of each function over the given interval and all values of x where the function equals its average.

47. $f(x) = 4 - x^2;\ [-2, 2]$
48. $f(x) = x^2 - 2x + 1;\ [0, 1]$
[C][S] **49.** $f(x) = x\sqrt{4 - x^2};\ [0, 2]$

50. $f(x) = \dfrac{x^2 + 1}{x^2};\ [\tfrac{1}{2}, 2]$

[C] **51.** $f(x) = x - 2\sqrt{x};\ [0, 4]$

52. $f(x) = \dfrac{1}{(x - 3)^2};\ [0, 2]$

53. Show that $\int_{-a}^a f(x)\, dx = 0$ if f is continuous on $[-a, a]$ and symmetric with respect to the origin.

54. Show that $\int_{-a}^a f(x)\, dx = 2\int_0^a f(x)\, dx$ if f is continuous on $[-a, a]$ and symmetric with respect to the y-axis.

55. Knowing that $\int_0^2 x^2\, dx = \tfrac{8}{3}$, find the value of the following definite integrals *without* using the Fundamental Theorem of Calculus.

(a) $\displaystyle\int_{-2}^0 x^2\, dx$

(b) $\displaystyle\int_{-2}^2 x^2\, dx$

(c) $\displaystyle\int_0^2 -x^2\, dx$

(d) $\displaystyle\int_0^2 (x^2 + 1)\, dx$

(e) $\displaystyle\int_{-2}^0 3x^2\, dx$

56. Prove that if f is integrable on $[a, b]$ and if $f(x) \geq 0$ for all x in $[a, b]$, then $\int_a^b f(x)\, dx \geq 0$.

57. Use the result of Exercise 56 to prove that if f and g are integrable on $[a, b]$ and $f(x) \leq g(x)$ for all x in $[a, b]$, then $\int_a^b f(x)\, dx \leq \int_a^b g(x)\, dx$.

For Review and Class Discussion

True or False

1. ____ If $F'(x) = G'(x)$ on the interval $[a, b]$, then $F(b) - F(a) = G(b) - G(a)$.

2. ____ If $f(x) = -f(-x)$ on the interval $[-a, a]$, then $\int_{-a}^a f(x)\, dx = 0$.

3. ____ $\int_1^2 (2x - 3)\, dx = [x^2 - 3x]_1^2$
$\qquad = 2^2 - 3(2) - 1^2 - 3(1) = -6.$

4. ____ If $F'(x) = f(x)$ on the interval $[0, b]$, then $\int_0^b f(x)\, dx = F(b)$.

5. ____ $\int_{-10}^{10} (ax^3 + bx^2 + cx + d)\, dx = 2\int_0^{10} (bx^2 + d)\, dx.$

Miscellaneous Exercises (Ch. 6)

In Exercises 1–10, find the indefinite integral.

1. $\int \frac{2}{3\sqrt[3]{x}} dx$

2. $\int \frac{2}{\sqrt[3]{3x}} dx$

3. $\int (2x^2 + x - 1) dx$

4. $\int \frac{x^3 - 2x^2 + 1}{x^2} dx$

[S] **5.** $\int \frac{(1 + x)^2}{\sqrt{x}} dx$

6. $\int x^2 \sqrt{x^3 + 3} dx$

[S] **7.** $\int \frac{x^2}{\sqrt{x^3 + 3}} dx$

[S] **8.** $\int \frac{x^2 + 2x}{(x + 1)^2} dx$

9. $\int (x^2 + 1)^3 dx$

10. $\int \sqrt{2 - 5x} dx$

11. A family of curves has the derivative $y' = -2x$. Find the equation of the curve that passes through the point $(-1, 1)$.

[S] **12.** A function $y = f(x)$ has a second derivative of $f''(x) = 6(x - 1)$. Find the function if its graph passes through the point $(2, 1)$ and at that point is tangent to the line $3x - y - 5 = 0$.

13. An airplane taking off from a runway travels 3600 ft before lifting off. If it starts from rest, moves with constant acceleration, and makes the run in 30 s, with what velocity does it take off?

[S] **14.** The speed of a car, traveling in a straight line, is reduced from 45 to 30 mi/h in a distance of 264 ft. Find the distance in which the car can be brought to rest from 30 mi/h, assuming the same constant acceleration.

15. A ball is thrown vertically upward from the ground with a velocity of 96 ft/s.
   (a) How long will it take it to rise to its maximum height?
   (b) What is the maximum height?
   (c) When is the velocity of the ball one-half the initial velocity?
   (d) What is the height of the ball when its velocity is one-half the initial velocity?

In Exercises 16–20, evaluate the sum if $x_1 = 2$, $x_2 = -1$, $x_3 = 5$, $x_4 = 3$, and $x_5 = 7$.

16. $\dfrac{\left(\sum\limits_{i=1}^{5} x_i \right)}{5}$

17. $\sum\limits_{i=1}^{5} \dfrac{1}{x_i}$

18. $\sum\limits_{i=1}^{5} (2x_i - x_i^2)$

19. $\sum\limits_{i=2}^{5} (x_i - x_{i-1})$

20. $\sum\limits_{i=1}^{5} x_i^2 - \dfrac{\left(\sum\limits_{i=1}^{5} x_i \right)^2}{5}$

21. Write in sigma notation the sum of the following:
   (a) the first ten positive odd integers
   (b) the cubes of the first n positive integers
   (c) $6 + 10 + 14 + 18 + \cdots + 42$

22. Prove that $\sum_{k=1}^{n} (ax_k + by_k) = a \sum_{k=1}^{n} x_k + b \sum_{k=1}^{n} y_k$.

23. Given the region bounded by $y = mx$, $y = 0$, $x = 0$, and $x = b$, find the following:
   (a) The upper and lower sum to approximate the area of the region when $\Delta x = b/4$.
   (b) The upper and lower sum to approximate the area of the region when $\Delta x = b/n$.
   (c) The area of the region by letting n approach infinity in both sums of part (b). Show that in each case you obtain the formula for the area of a triangle.
   (d) The area of the region by using the Fundamental Theorem of Calculus.

24. (a) Find the area of the region bounded by $y = x^3$, $y = 0$, $x = 1$, and $x = 3$ by the "limit of the sum" technique.
   (b) Find the area of the given region by using the Fundamental Theorem of Calculus.

In Exercises 25–32, use the Fundamental Theorem of Calculus to evaluate the definite integral.

25. $\int_0^4 (2 + x) dx$

26. $\int_{-1}^1 (t^2 + 2)t \, dt$

[S] **27.** $\int_{-1}^1 (4t^3 - 2t) \, dt$

28. $\int_3^6 \dfrac{x}{3\sqrt{x^2 - 8}} dx$

29. $\int_0^3 \dfrac{1}{\sqrt{1 + x}} dx$

30. $\int_0^1 x^2(x^3 + 1)^3 dx$

31. $\int_4^9 x \sqrt{x} \, dx$

32. $\int_1^2 \left(\dfrac{1}{x^2} - \dfrac{1}{x^3} \right) dx$

[C][S] **33.** Approximate the definite integral $\int_0^1 [1/(1 + x^2)] \, dx$ by taking the average of the upper and lower sums when $n = 5$. (The value of the integral to three decimal places is 0.785.)

34. Is the following true or false? Explain.

$$\int_{-1}^1 \frac{1}{x^2} dx = -\frac{1}{x} \Big]_{-1}^1 = (-1) - (1) = -2$$

In Exercises 35–38, find the average value of the function over the given interval. Find the values of x where the function assumes its mean value and sketch the graph of the function.

S **35.** $f(x) = \dfrac{1}{\sqrt{x-1}}$; [5, 10]   **36.** $f(x) = x^3$; [0, 2]

37. $f(x) = x$; [0, 4]   **38.** $f(x) = x^2 - \dfrac{1}{x^2}$; [1, 2)

39. Suppose that gasoline is increasing in price according to the equation

$$p = 1 + 0.1t + 0.02t^2$$

where p is the dollar price per gallon and $t = 0$ represents the year 1978. If a certain automobile is driven 15,000 mi a year and gets M miles per gallon, then the annual fuel cost is

$$C = \frac{15{,}000}{M} \int_t^{t+1} p \, dt$$

Find the annual fuel cost for the following years:
(a) 1980   (b) 1985

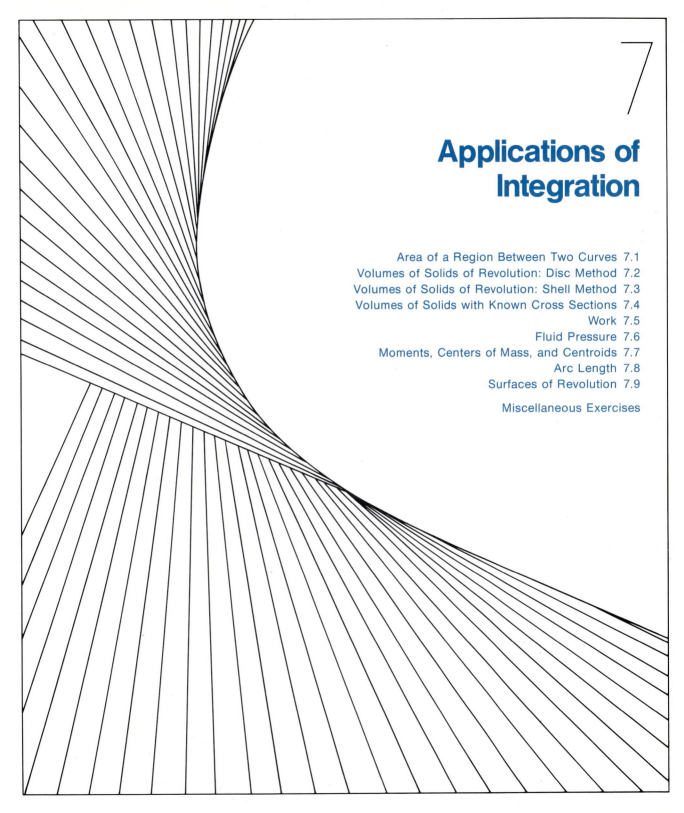

7

Applications of Integration

7.1 Area of a Region Between Two Curves

Purpose
- To demonstrate a procedure for determining the area of a region between two curves.

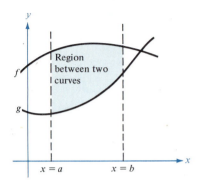

FIGURE 7.1

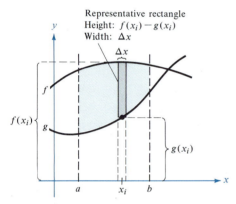

Representative rectangle
Height: $f(x_i) - g(x_i)$
Width: Δx

FIGURE 7.2

The method of determining area is with me only a special case, and indeed an easier one, of the far greater problem which I call the inverse method of tangents. —Gottfried Leibniz (1646–1716)

With a few modifications we can extend the application of definite integrals from the area of a region *under* a curve to the area of a region *between* two curves. Let us consider the region bounded by $y = f(x)$, $y = g(x)$, $x = a$, and $x = b$. (See Figure 7.1.) Assume that both f and g are continuous in $[a, b]$ and that $g(x) \leq f(x)$ for all x in $[a, b]$. We partition $[a, b]$ into n subintervals, each of width Δx and sketch a "representative rectangle" of width Δx and height $f(x_i) - g(x_i)$, where x_i is in the ith subinterval. (See Figure 7.2.) The area of a representative rectangle is

$$\Delta A = (\text{height})(\text{width}) = [f(x_i) - g(x_i)]\Delta x$$

The sum of the areas of these n rectangles is

$$\sum_{i=1}^{n} [f(x_i) - g(x_i)]\Delta x$$

Taking the limit as $\|\Delta\| \to 0$ $(n \to \infty)$, we have

$$\lim_{n \to \infty} \sum_{i=1}^{n} [f(x_i) - g(x_i)]\Delta x$$

Since f and g are continuous on $[a, b]$, $f - g$ is also continuous on this interval and the limit exists. Therefore, the area A of the given region is

$$A = \lim_{n \to \infty} \sum_{i=1}^{n} [f(x_i) - g(x_i)]\Delta x = \int_a^b [f(x) - g(x)]\,dx$$

We summarize this result as follows:

Area Between Two Curves

If f and g are continuous on $[a, b]$ and $g(x) \leq f(x)$ for all x in $[a, b]$, then the area of the region bounded by $y = f(x), y = g(x), x = a$, and $x = b$ is given by

$$A = \int_a^b [f(x) - g(x)]\,dx$$

If both f and g lie above the x-axis, we can geometrically interpret the area of the region between f and g as simply the area of the region under g subtracted from the area of the region under f. (See Figure 7.3.)

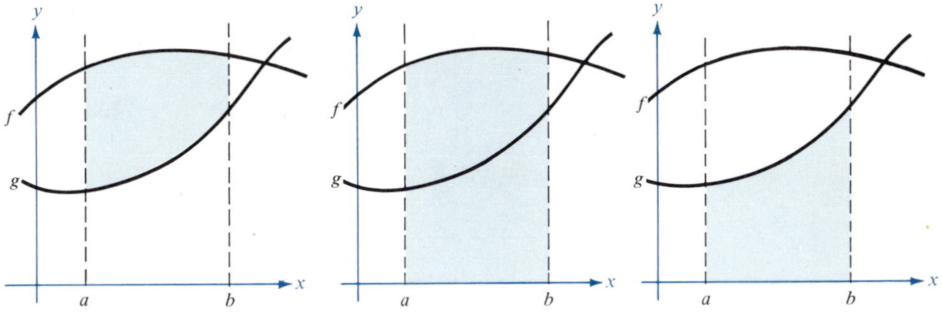

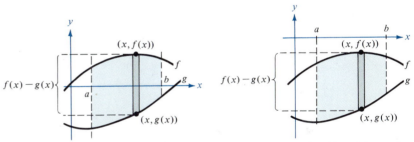

(Area of region between f and g) $=$ (Area of region under f) $-$ (Area of region under g)

$$\int_a^b [f(x) - g(x)]\, dx \quad = \quad \int_a^b f(x)\, dx \quad - \quad \int_a^b g(x)\, dx$$

FIGURE 7.3

It is important to realize that the development of our area formula, $A = \int_a^b [f(x) - g(x)]\, dx$, depended *only* on the continuity of f and g and the assumption that $g \leq f$, not on their position with respect to the x-axis. Try to convince yourself of this by referring to Figure 7.4.

Height of representative rectangle is $f(x) - g(x)$ regardless of the relative position of the x-axis.

FIGURE 7.4

Example 1

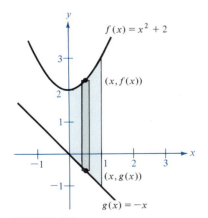

FIGURE 7.5

Find the area of the region bounded by $y = x^2 + 2$, $y = -x$, $x = 0$, and $x = 1$.

Solution: If we let $g(x) = -x$ and $f(x) = x^2 + 2$, then $g(x) \leq f(x)$ for all x in $[0, 1]$ (see Figure 7.5). Thus we determine the area of the representative rectangle to be

$$\Delta A = [f(x) - g(x)]\, \Delta x = [(x^2 + 2) - (-x)]\, \Delta x$$

Summing these areas we obtain a total area of

$$A = \int_0^1 [(x^2 + 2) - (-x)]\, dx = \int_0^1 (x^2 + x + 2)\, dx$$

$$= \left[\frac{x^3}{3} + \frac{x^2}{2} + 2x \right]_0^1 = \frac{1}{3} + \frac{1}{2} + 2 = \frac{17}{6}$$ ∎

In Example 1 the two curves $f(x) = x^2 + 2$ and $g(x) = -x$ do not intersect, and the values of a and b were explicitly given. A more common type of problem involves the area of the region bounded by two *intersecting* curves. In this type of problem, the values of a and b must be calculated.

Example 2

Find the area of the region bounded by the curve $g(x) = 2 - x^2$ and the line $f(x) = x$.

Solution: In this case a and b are determined by the points of intersection of $f(x) = x$ and $g(x) = 2 - x^2$. In order to find these points, we set these two functions equal to each other and solve for x.

$$x = 2 - x^2$$
$$x^2 + x - 2 = 0$$
$$(x + 2)(x - 1) = 0$$

The zeros are $x = -2$ and $x = 1$. Since $f(x) \leq g(x)$ on the interval $[-2, 1]$, we have

$$\Delta A = [g(x) - f(x)]\Delta x = [(2 - x^2) - x]\Delta x$$

Hence the desired area is

$$A = \int_{-2}^{1} [(2 - x^2) - (x)]\,dx = \int_{-2}^{1} (2 - x^2 - x)\,dx$$

$$= \left[2x - \frac{x^3}{3} - \frac{x^2}{2}\right]_{-2}^{1}$$

$$= \left(2 - \frac{1}{3} - \frac{1}{2}\right) - \left(-4 + \frac{8}{3} - 2\right) = \frac{9}{2}$$

(See Figure 7.6.) ∎

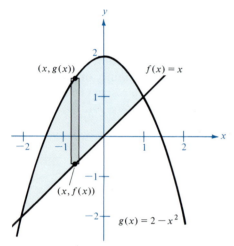

FIGURE 7.6

Example 3

Find the area of the region bounded by $y = x^2 - 3x - 4$ and the x-axis.

Solution: Refer to Figure 7.7. Since $x^2 - 3x - 4$ intersects the x-axis when $x = -1$ and $x = 4$, and since $x^2 - 3x - 4 \leq 0$ for all x in $[-1, 4]$, then

$$\Delta A = [0 - f(x)]\Delta x = [0 - (x^2 - 3x - 4)]\Delta x$$

Hence the desired area is

$$A = \int_{-1}^{4} [0 - (x^2 - 3x - 4)]\,dx = \int_{-1}^{4} (-x^2 + 3x + 4)\,dx$$

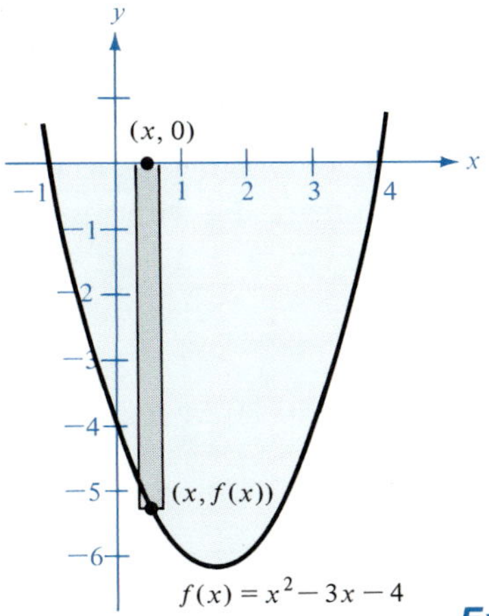

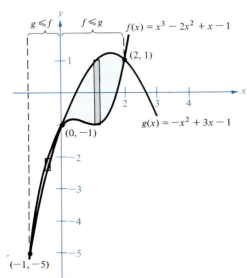

$f(x) = x^2 - 3x - 4$

FIGURE 7.7

$$= \left[\frac{-x^3}{3} + \frac{3x^2}{2} + 4x \right]_{-1}^{4}$$

$$= \left(\frac{-64}{3} + 24 + 16 \right) - \left(\frac{1}{3} + \frac{3}{2} - 4 \right) = \frac{125}{6} \qquad \blacksquare$$

If two curves intersect in *more* than two points, then to find the area of the region between these curves, we must find *all* points of intersection and check to see which curve is above the other in each interval determined by these points.

Example 4

Find the area of the region between the curves $f(x) = x^3 - 2x^2 + x - 1$ and $g(x) = -x^2 + 3x - 1$.

Solution: Solving for x in the equation $f(x) = g(x)$, we have

$$f(x) - g(x) = x^3 - x^2 - 2x = 0$$
$$x(x^2 - x - 2) = 0$$
$$x(x - 2)(x + 1) = 0$$

Thus the zeros are $x = -1$, $x = 0$, and $x = 2$.

By graphing these two functions, we see that $g(x) \leq f(x)$ on $[-1, 0]$, but the two curves switch at the point $(0, -1)$, and $f(x) \leq g(x)$ on $[0, 2]$. (See Figure 7.8.) Hence we need to use two integrals to determine the area of the region between f and g, one for the interval $[-1, 0]$ and one for $[0, 2]$.

$$A = \int_{-1}^{0} [f(x) - g(x)]\, dx + \int_{0}^{2} [g(x) - f(x)]\, dx$$

$$= \int_{-1}^{0} (x^3 - x^2 - 2x)\, dx + \int_{0}^{2} (-x^3 + x^2 + 2x)\, dx$$

$$= \left[\frac{x^4}{4} - \frac{x^3}{3} - x^2 \right]_{-1}^{0} + \left[\frac{-x^4}{4} + \frac{x^3}{3} + x^2 \right]_{0}^{2}$$

$$= -\left(\frac{1}{4} + \frac{1}{3} - 1 \right) + \left(-4 + \frac{8}{3} + 4 \right) = \frac{37}{12} \qquad \blacksquare$$

Up to this point we have dealt only with areas of regions bounded by graphs of functions of x. If the graph of a function of y is a boundary of a region, it is often convenient to use *horizontal* rectangles and find the area by integrating with respect to y. This procedure is shown in the next example.

FIGURE 7.8

Example 5 (*Horizontal Rectangles*)

Find the area of the region bounded by $x = 3 - y^2$ and $y = x - 1$.

Solution: Note from Figure 7.9 that we would need two integrals to find the area by integrating with respect to x, but by integrating with respect to y, we need only one integral. We consider $g(y) = 3 - y^2$ and $f(y) = y + 1$. Since these two curves intersect when $y = -2$ and $y = 1$, and since $y + 1 \leq 3 - y^2$ on this interval, then

$$\Delta A = [g(y) - f(y)]\,\Delta y = [(3 - y^2) - (y + 1)]\,\Delta y$$

Hence the area is

$$A = \int_{-2}^{1} [(3 - y^2) - (y + 1)]\,dy = \int_{-2}^{1} (-y^2 - y + 2)\,dy$$

$$= \left[\frac{-y^3}{3} - \frac{y^2}{2} + 2y \right]_{-2}^{1}$$

$$= \left(\frac{-1}{3} - \frac{1}{2} + 2 \right) - \left(\frac{8}{3} - 2 - 4 \right) = \frac{9}{2} \qquad \blacksquare$$

Although we will continue to work primarily with equations that represent y as a function of x, we will later encounter additional applications of the definite integral that involve interchanging the role of x and y as we have done in Example 5. In general, to determine the area between two curves, we have, for vertical rectangles,

$$A = \int_{x_1}^{x_2} \underbrace{(\text{top curve}) - (\text{bottom curve})}_{\text{in variable } x}\,dx$$

and for horizontal rectangles,

$$A = \int_{y_1}^{y_2} \underbrace{(\text{right curve}) - (\text{left curve})}_{\text{in variable } y}\,dy$$

where (x_1, y_1) and (x_2, y_2) are either adjacent points of intersection of the two curves involved or are points on specified boundary lines.

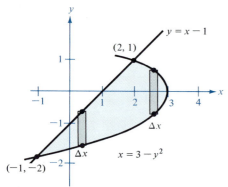

Vertical Rectangles
(Integration with respect to x)

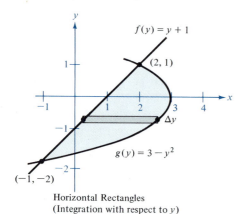

Horizontal Rectangles
(Integration with respect to y)

FIGURE 7.9

Example 6

Based upon U.S. Department of Energy statistics, the total consumption of gasoline in the United States from 1950 to 1974 followed a growth pattern described by the equation

$$f(t) = 2.158 + 0.082t + 0.001t^2$$

where $f(t)$ is measured in billions of gallons and t in years, with $t = 0$

representing the year 1970. With the onset of dramatic increases in crude oil prices in 1974, the growth pattern for gasoline consumption changed and began following the pattern described by the equation

$$g(t) = 1.81 + 0.174t - 0.009t^2$$

Find the total amount of gasoline saved from 1976 to 1982 when gasoline was consumed at the post-1974 rate rather than the pre-1974 rate.

Solution: Since the pre-1974 curve *f* lies above the post-1974 curve *g* on the interval [6, 12], the gasoline saved is given by the integral

$$\int_6^{12} [f(t) - g(t)]\, dt = \int_6^{12} [(2.158 + 0.082t + 0.001t^2)$$

$$- (1.81 + 0.174t - 0.009t^2)]\, dt$$

$$= \int_6^{12} (0.348 - 0.092t + 0.01t^2)\, dt$$

$$= \left[0.348t - 0.046t^2 + (0.01)\frac{t^3}{3} \right]_6^{12}$$

$$= 3.312 - 1.152 = 2.16$$

Therefore, 2.16 billion barrels of gasoline were saved. (See Figure 7.10.) ∎

y

4

3

Billions of Barrels

2

1

(1970) 2 4 6 8 10 12 14 → t

Pre-1974 rate f(t)

Gasoline saved

g(t)

Post-1974 rate

U. S. Gasoline Consumption

FIGURE 7.10

In this section we developed our integration formula for area between two curves by using a rectangle as the *representative element*. For each new application in the remaining sections of this chapter, we will construct an appropriate representative element using precalculus formulas you already know. Each integration formula will then be obtained by summing these representative elements.

Known Precalculus Formula	increment form →	Representative Element	summation process →	New Integration Formula

For example, in this section we have

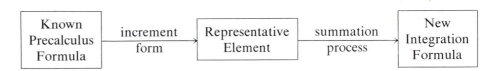

Area = (Height)(Width) ⟼ $\Delta A = [f(x) - g(x)]\Delta x$ ⟼ $A = \int_a^b [f(x) - g(x)]\, dx$

We suggest that you learn to set up the form of each representative element in this chapter, because this will help you to see how integration works as a summation process. Furthermore, setting up representative elements minimizes the need to memorize integral forms.

Section Exercises (7.1)

In Exercises 1–20, sketch the region bounded by the graphs of the given equations and find the area of each region by means of definite integrals.

1. $f(x) = x^2 - 6x$, $g(x) = 0$
2. $f(x) = 3 - 2x - x^2$, $g(x) = 0$
[S] 3. $f(x) = x^2 + 2x + 1$, $g(x) = 3x + 3$
4. $f(x) = -x^2 + 4x + 2$, $g(x) = x + 2$
5. $f(x) = x^2 - 4x + 3$, $g(x) = -x^2 + 2x + 3$
6. $f(x) = x^2 + 5x - 6$, $g(x) = 6x - 6$
[S] 7. $f(x) = 2x^2 - 4x + 1$, $g(x) = x^2 - 4x + 3$
8. $f(x) = 3x^2 + 2x$, $g(x) = 8$
9. $f(x) = x^3$, $g(x) = x^2$
10. $f(x) = x(x^2 - 3x + 3)$, $g(x) = x^2$
11. $f(x) = 3(x^3 - x)$, $g(x) = 0$
12. $f(x) = x^3 - 2x + 1$, $g(x) = -2x$, $x = 1$
[C][S] 13. $f(x) = \dfrac{4}{x^2}$, $g(x) = x^2 - 6x + 9$
14. $f(x) = \sqrt[3]{x}$, $g(x) = x$
15. $f(x) = \sqrt{3x} + 1$, $g(x) = x + 1$
16. $f(x) = x^4 - x^3 + x - 1$, $g(x) = -x^3 + x$, $x = 2$
17. $f(y) = y^2$, $g(y) = y + 2$
18. $f(y) = y(2 - y)$, $g(y) = -y$
[S] 19. $f(y) = y^2 + 1$, $g(y) = 0$, $y = -1$, $y = 2$
20. $f(y) = \dfrac{y}{\sqrt{16 - y^2}}$, $g(y) = 0$, $y = 3$

In Exercises 21–24, use integration to find the areas of the triangles having the given vertices.

21. $(0, 0)$, $(4, 0)$, $(4, 4)$
22. $(0, 0)$, $(4, 0)$, $(6, 4)$
[S] 23. $(0, 0)$, $(a, 0)$, (b, c)
24. $(2, -3)$, $(4, 6)$, $(6, 1)$

25. The graphs of $y = x^4 - 2x^2 + 1$ and $y = 1 - x^2$ intersect at three points; however, the area between the curves *can* be found by a single integral. Explain why this is so and write an integral for this area.

26. The graphs of $y = x^3$ and $y = x$ intersect at the three points $(-1, -1)$, $(0, 0)$, and $(1, 1)$, and the area between the curves *cannot* be found by the single integral $\int_{-1}^{1} (x^3 - x)\, dx$. Explain why this is so. Making use of symmetry, write a single integral that does represent the area.

27. Based upon U.S. Department of Agriculture statistics, the total consumption of beef in the United States from 1950 to 1970 followed a growth pattern approximated by

$$f(t) = 23.703 + 1.002t + 0.015t^2$$

where $f(t)$ is measured in billions of pounds and t is measured in years, with $t = 0$ representing the year 1970. From 1970 to 1980 the growth pattern was more closely approximated by

$$g(t) = 22.93 + 0.678t - 0.037t^2$$

Estimate the total reduction of consumption of beef from 1970 to 1980 due to the change in consumption rate.

28. In economics the equilibrium price p_0 for a product is defined to be the price at which the supply and demand curves intersect. The *consumers' surplus* is defined to be the area between the demand curve and the lines $p = p_0$ and $x = 0$. The *producers' surplus* is defined similarly, using the supply curve. Find the consumers' and producers' surpluses for the following:

(supply curve) $q(x) = \sqrt{0.1x + 9} - 2$
(demand curve) $p(x) = \sqrt{25 - 0.1x}$

For Review and Class Discussion

True or False

1. ___ If f and g are both continuous on the interval $[a, b]$, then $f - g$ is integrable on $[a, b]$.

2. ___ If the area of the region bounded by the graphs of f and g is 1, then the area of the region bounded by the graphs of $h(x) = f(x) + C$ and $k(x) = g(x) + C$ is also 1.

3. ___ $\int_0^1 (x^2 - x)\, dx$ represents the area of the region bounded by the graph of $y = x^2 - x$ and the x-axis.

4. ___ If

$$\int_a^b [f(x) - g(x)]\, dx = A$$

then

$$\int_a^b [g(x) - f(x)]\, dx = -A$$

5. ___ $\int_a^c [f(x) - g(x)]\, dx + \int_c^b [g(x) - f(x)]\, dx$

$$= \int_a^b [f(x) - g(x)]\, dx$$

7.2 Volumes of Solids of Revolution: Disc Method

Purpose
- To define a solid of revolution.
- To introduce the Disc Method for using a definite integral to find the volume of a solid of revolution.

I consider mathematical quantities not as consisting of very small parts; but as described by continued motion. Lines are described, and thereby generated not by the apposition of parts, but by the continued motion of points; surfaces by the motion of lines; solids by the motion of surfaces; angles by the rotation of the sides; portions of time by a continued flux: and so in other quantities. —*Isaac Newton (1642–1727)*

As mentioned in Section 6.3 of Chapter 6, finding area is only *one* of the applications of the definite integral. Another important application is its use in finding the volume of three-dimensional figures. In this section we consider a particular type of three-dimensional figure called a *solid of revolution*. Some common examples of solids of revolution are rockets, funnels, pills, inner tubes, and bottles.

| **Definition of Solid of Revolution** | If a region in the plane is revolved about a line, the resulting solid is called a **solid of revolution** and the line is called the **axis of revolution**. (See Figure 7.11.) |

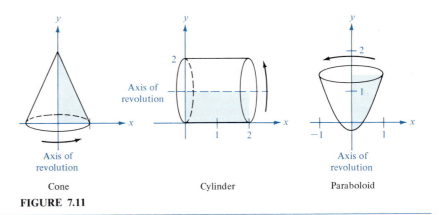

Cone   Cylinder   Paraboloid

FIGURE 7.11

The three solids depicted in Figure 7.11 result from revolving a triangle, a rectangle, and a parabolic region about a line. With some simple formulas from geometry, we can find the volume of the first two solids shown in Figure 7.11. However, we have not yet encountered a method for calculating the volume of the third solid of revolution.

We begin the development of one useful method for finding volumes of solids of revolution by deriving a formula for the volume of a disc. Consider the rectangle in Figure 7.12, where w, R, and r are defined by

w = width of rectangle
R = outside distance from the rectangle to the axis of revolution
r = inside distance from the rectangle to the axis of revolution

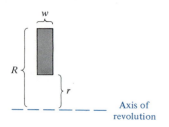

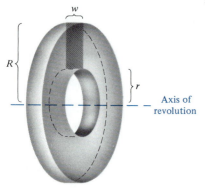

FIGURE 7.12

When the rectangle is revolved about its axis of revolution, it forms a disc whose volume is the difference $\pi R^2 w - \pi r^2 w$. Note that we subtract $\pi r^2 w$ since it represents the volume of the hole in the center of the disc. Therefore, we have

$$\text{volume of disc} = \pi(R^2 - r^2)w$$

Now suppose a solid of revolution is formed by revolving the plane region in Figure 7.13 about the indicated axis. To determine the volume of this solid, we first draw a rectangle of width Δx in the plane region, as shown in Figure 7.13. (We consider this rectangle to be "perpendicular" to the axis of revolution.) Now as the plane region is revolved about its axis of revolution, the rectangle generates a representative disc whose volume is

$$\Delta V = \pi[R(x)^2 - r(x)^2]\,\Delta x$$

If we approximate the volume of the solid by n such discs of width Δx, as shown in Figure 7.14, we have

$$\text{volume of solid} \approx \sum_{i=1}^{n} \pi[R(x_i)^2 - r(x_i)^2]\,\Delta x$$

$$= \pi \sum_{i=1}^{n} [R(x_i)^2 - r(x_i)^2]\,\Delta x$$

By taking the limit as $\|\Delta\| \to 0$ ($n \to \infty$), we have

$$\text{volume of solid} = \lim_{n \to \infty} \pi \sum_{i=1}^{n} [R(x_i)^2 - r(x_i)^2]\,\Delta x$$

$$= \pi \int_a^b [R(x)^2 - r(x)^2]\,dx$$

A similar formula can be derived if the axis of revolution is vertical. To

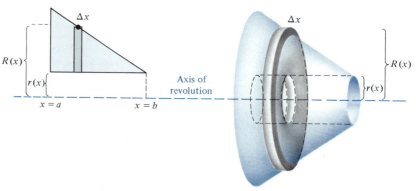

FIGURE 7.13

FIGURE 7.14

simplify our notation, we will generally use R and r in place of the functional notation $R(x)$ [or $R(y)$] and $r(x)$ [or $r(y)$]. We summarize the results in the following description of the **Disc Method** for determining volumes of solids of revolution.

Disc Method (Perpendicular Rectangles)

If the axis of revolution is horizontal, then

$$\text{volume of solid of revolution} = \pi \int_a^b (R^2 - r^2)\, dx$$

If the axis of revolution is vertical, then

$$\text{volume of solid of revolution} = \pi \int_c^d (R^2 - r^2)\, dy$$

See Figure 7.15.

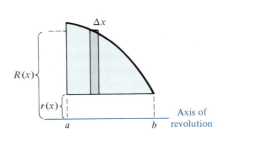

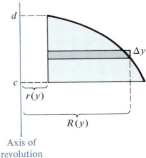

FIGURE 7.15

When using the Disc Method, it often happens that there is *no hole* in the disc [that is, $r(x) = 0$ or $r(y) = 0$] and the formulas simplify to

$$\pi \int_a^b R(x)^2\, dx \qquad \text{or} \qquad \pi \int_c^d R(y)^2\, dy$$

Also note that the variable of integration is x when Δx is the width of the rectangle and y when Δy is the width.

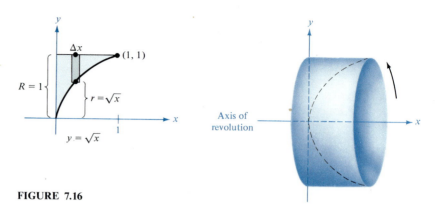

FIGURE 7.16

Example 1

Find the volume of the solid of revolution formed by revolving the region bounded by $y = \sqrt{x}$, $y = 1$, and $x = 0$ about the x-axis. (See Figure 7.16.)

Solution: Using the Disc Method we must place our rectangle perpendicular to the axis of revolution. (See Figure 7.16.) The corresponding representative disc has width Δx, $r = \sqrt{x}$, $R = 1$, and hence its volume is

$$\Delta V = \pi(R^2 - r^2)\,\Delta x = \pi[(1)^2 - (\sqrt{x})^2]\,\Delta x = \pi(1 - x)\,\Delta x$$

Now since x ranges from 0 to 1, the volume of the solid is

$$V = \pi \int_0^1 (1 - x)\,dx = \pi \left[x - \frac{x^2}{2} \right]_0^1 = \frac{\pi}{2}$$

∎

Notice in Example 1 that the entire problem was worked *without* any reference to the three-dimensional sketch given in Figure 7.16. In general, when you set up the integral for calculating the volume of a solid of revolution, a sketch of the plane region is more useful than a sketch of the solid, since R and r are more readily visualized in the plane region.

Example 2

Find the volume of the solid formed by revolving the region bounded by $y = -x^2 + x$ and $y = 0$ about the x-axis. (See Figure 7.17.)

Solution: We begin by placing a rectangle perpendicular to the axis of revolution. In this case the disc has no hole in it and thus $r = 0$. Since $R = -x^2 + x$, it follows that the volume of the representative disc is

$$\Delta V = \pi R^2 \,\Delta x = \pi(-x^2 + x)^2 \,\Delta x = \pi(x^4 - 2x^3 + x^2)\,\Delta x$$

Now since x ranges from 0 to 1, the volume of the solid is

$$V = \pi \int_0^1 (x^4 - 2x^3 + x^2)\,dx = \pi \left[\frac{x^5}{5} - \frac{x^4}{2} + \frac{x^3}{3} \right]_0^1 = \frac{\pi}{30} \approx 0.105$$

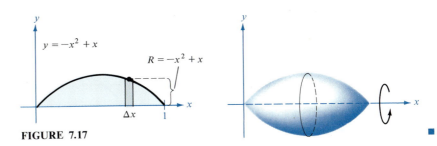

FIGURE 7.17

Example 3 (*Nonadjacent Axis of Revolution*)

Suppose the plane region in Example 2 is revolved about the line $y = 1$. Determine the volume of the resulting solid of revolution.

Solution: From Figure 7.18 we can see that $r = 1 - (-x^2 + x) = x^2 - x + 1$, and $R = 1$. Therefore, the volume is given by

$$V = \pi \int_0^1 (R^2 - r^2)\, dx = \pi \int_0^1 [1^2 - (x^2 - x + 1)^2]\, dx$$

$$= \pi \int_0^1 (-x^4 + 2x^3 - 3x^2 + 2x)\, dx$$

$$= \pi \left[-\frac{x^5}{5} + \frac{x^4}{2} - x^3 + x^2 \right]_0^1$$

$$= \pi \left(-\frac{1}{5} + \frac{1}{2} - 1 + 1 \right) = \frac{3\pi}{10} \approx 0.942$$

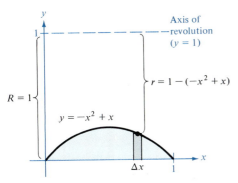

FIGURE 7.18

Note that although the same region was used to form the solids of revolution in both Example 2 and Example 3, the volume in Example 3 is larger because the region is further from its axis of revolution.

Example 4 (*Two-Integral Case*)

Find the volume of the solid formed by revolving the region bounded by $y = x^2 + 1$, $y = 0$, $x = 0$, and $x = 1$ about the y-axis. (See Figure 7.19.)

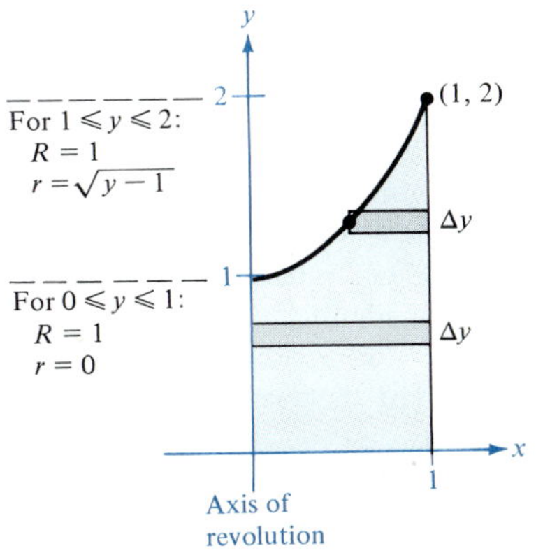

Solution: In this problem we use two integrals since we do not have a single formula to represent r. When $0 \leq y \leq 1$, then $r = 0$, but for $1 \leq y \leq 2$, r is determined by the equation $y = x^2 + 1$, which implies that $r = \sqrt{y - 1}$. Since $R = 1$, we form the two integrals

$$V = \pi \int_0^1 (1^2 - 0^2)\, dy + \pi \int_1^2 [1^2 - (\sqrt{y - 1})^2]\, dy$$

$$= \pi \int_0^1 1\, dy + \pi \int_1^2 (2 - y)\, dy = \pi y \Big]_0^1 + \pi \left[2y - \frac{y^2}{2} \right]_1^2$$

$$= \pi \left(1 + 4 - 2 - 2 + \frac{1}{2} \right) = \frac{3}{2}\pi$$

FIGURE 7.19

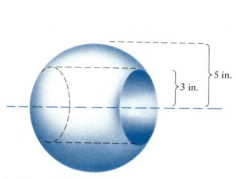

FIGURE 7.20

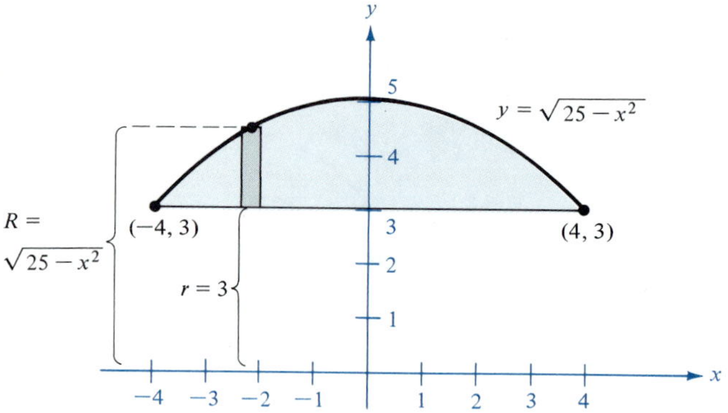

FIGURE 7.21

Example 5

A manufacturer plans to drill a hole through the center of a metal sphere of radius 5 in. (See Figure 7.20.) If the hole has a radius of 3 in., what is the volume of the resulting ring?

Solution: We imagine the ring to be generated by a segment of the circle $x^2 + y^2 = 25$, as shown in Figure 7.21.

Since the radius of the hole is $y = 3$ in., its intersection with the circle $x^2 + y^2 = 25$ occurs at $x = -4$ and $x = 4$. Thus we have

$$r = 3 \quad \text{and} \quad R = \sqrt{25 - x^2}$$

and the volume is given by

$$V = \pi \int_{-4}^{4} [(\sqrt{25 - x^2})^2 - (3)^2] \, dx$$

$$= \pi \int_{-4}^{4} (16 - x^2) \, dx = \pi \left[16x - \frac{x^3}{3} \right]_{-4}^{4} = \frac{256\pi}{3} \text{ in.}^3 \qquad \blacksquare$$

Section Exercises (7.2)

1. Find the volume of the solid generated by revolving the region (see Figure 7.22) bounded by $y = 6$, $y = 0$, $x = 0$, and $x = 4$ about:
   (a) the x-axis   (b) the y-axis
2. Find the volume of the solid generated by revolving the region (see Figure 7.23) bounded by $y = x$, $y = 0$, and $x = 2$ about:
   (a) the x-axis   (b) the y-axis
3. Find the volume of the solid generated by revolving the region (see Figure 7.24) bounded by $y = \sqrt{1 - x^2}$ and $y = 0$ about the x-axis.

[S] 4. Find the volume of the solid generated by revolving the region (see Figure 7.25) bounded by $y = x^2$ and $y = 4$ about the x-axis.

5. Find the volume of the solid generated by revolving the region (see Figure 7.26) bounded by $y = 1 - x$, $y = 0$, and $x = 0$ about:
   (a) the x-axis   (b) the y-axis
   (c) the line $y = -1$
6. Find the volume of the solid generated by revolving the region (see Figure 7.27) bounded by $y = x$, $y = 0$, and $2 \le x \le 4$ about:

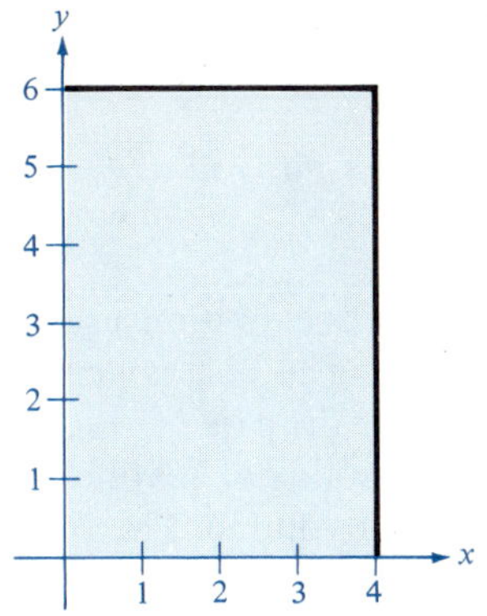

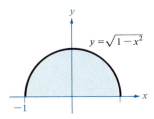

FIGURE 7.22

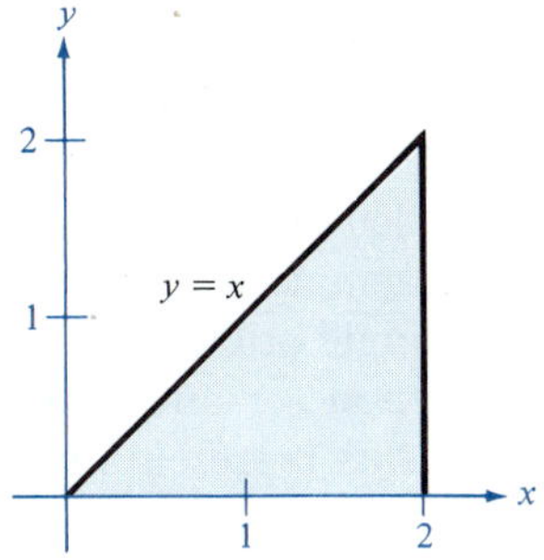

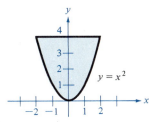

FIGURE 7.23

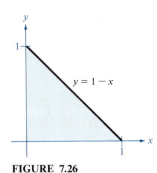

FIGURE 7.24

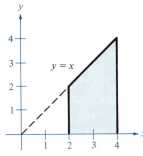

FIGURE 7.25

FIGURE 7.26

FIGURE 7.27

(a) the x-axis (b) the y-axis
(c) the line $x = 4$

S **7.** Find the volume of the solid generated by revolving the region bounded by $y = \sqrt{x}$, $y = 0$, and $x = 4$ about:
(a) the x-axis (b) the y-axis
(c) the line $x = 4$ (d) the line $x = 6$

8. Find the volume of the solid generated by revolving the region bounded by $y = 2x^2$, $y = 0$, and $x = 2$ about:
(a) the y-axis (b) the x-axis
(c) the line $y = 8$

S **9.** Find the volume of the solid formed by revolving the region bounded by $y = x^2$ and $y = 4x - x^2$ about:
(a) the x-axis (b) the line $y = 6$

10. Find the volume of the solid formed by revolving the region bounded by $y = 6 - 2x - x^2$ and $y = x + 6$ about:
(a) the x-axis (b) the line $y = 3$

11. The region bounded by the parabola $y = 4x - x^2$ and the x-axis is revolved about the x-axis. Find the volume of the solid.

12. If the equation of the parabola in Exercise 11 were changed to $y = 4 - x^2$, would the volume of the solid generated be different? Why or why not?

13. If the portion of the line $y = \frac{1}{2}x$ lying in the first quadrant is revolved about the x-axis, a cone is generated. Find the volume of the cone extending from $x = 0$ to $x = 6$.

14. Use the Disc Method to verify that the volume of a right circular cone is $\frac{1}{3}\pi r^2 h$, where r is the radius of the base and h is the height.

S **15.** Use the Disc Method to verify that the volume of a sphere of radius r is $\frac{4}{3}\pi r^3$.

16. The upper half of the ellipse $9x^2 + 25y^2 = 225$ is revolved about the x-axis to form a prolate spheroid (shaped like a football). Find the volume of the spheroid.

17. The right half of the ellipse $9x^2 + 25y^2 = 225$ is revolved about the y-axis to form an oblate spheroid. Find the volume of this spheroid.

18. The region bounded by $y = \sqrt{x}$, $y = 0$, $x = 0$, and $x = 4$ is revolved about the x-axis. Find the value of x in the interval $[0, 4]$ that divides the solid into two parts of equal volume.

S **19.** The tank on a water tower is a sphere of radius 50 ft. Determine the depth of the water when the tank is filled to 21.6% of its total capacity.

C **20.** Repeat Exercise 19, determining the depth of the water when the tank is filled to $\frac{2}{3}$ its total capacity.

21. Match the following integrals with the solid whose volume it represents and give the dimensions of each solid.

____ $\pi \int_0^h \left(\dfrac{rx}{h}\right)^2 dx$ (a) right circular cylinder

____ $\pi \int_{-r}^r (\sqrt{r^2 - x^2})^2\, dx$ (b) ellipsoid

____ $\pi \int_0^h r^2\, dx$ (c) sphere

____ $\pi \int_{-r}^r [(R + \sqrt{r^2 - x^2})^2 - (R - \sqrt{r^2 - x^2})^2]\, dx$ (d) right circular cone

____ $\pi \int_{-b}^b \left(a\sqrt{1 - \dfrac{x^2}{b^2}}\right)^2 dx$ (e) torus

7.3 Volumes of Solids of Revolution: Shell Method

Purpose
- To introduce the Shell Method for finding the volume of a solid of revolution.
- To compare the Shell and Disc Methods.

Moreover, this method is valid for all cylindric bodies or parts of apples or figs. —*Johann Kepler* (1571–1630)

The Disc Method for calculating volumes of solids of revolution uses a circular disc as its representative element of volume. The disc is generated by revolving a rectangle oriented *perpendicular* to the axis of revolution. An alternative method for finding volumes of solids of revolution uses a cylindrical shell as its representative element of volume. The shell is formed by revolving a rectangle placed *parallel* to the axis of revolution (see Figure 7.28).

We develop the formula for the volume of a cylindrical shell by considering the rectangle in Figure 7.28, where

w = width of rectangle

h = length of rectangle

p = distance between the axis of revolution and the center of rectangle

When the rectangle is revolved about its axis of revolution, it forms a cylindrical shell, whose volume is the difference

$$\pi\left(p + \frac{w}{2}\right)^2 h - \pi\left(p - \frac{w}{2}\right)^2 h = \pi\left[\left(p + \frac{w}{2}\right)^2 - \left(p - \frac{w}{2}\right)^2\right]h$$

$$\text{volume of shell} = \pi\left(p^2 + pw + \frac{w^2}{4} - p^2 + pw - \frac{w^2}{4}\right)h$$

$$= \pi(2pw)h = 2\pi phw = 2\pi(\text{radius})(\text{height})(\text{thickness})$$

We use this formula to calculate the volume of a solid of revolution as follows. Assume that the plane region in Figure 7.29 is revolved about a line to form the indicated solid. If we place a rectangle of width Δy *parallel* to the axis of revolution, then as the plane region is revolved about its axis of revolution, the rectangle generates a representative shell (Figure 7.29), whose volume is

$$\Delta V = 2\pi p(y)\, h(y)\, \Delta y$$

If we approximate the volume of the solid by n such shells, we have

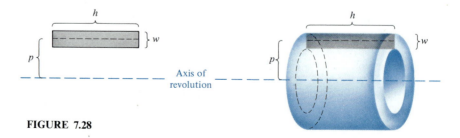

FIGURE 7.28

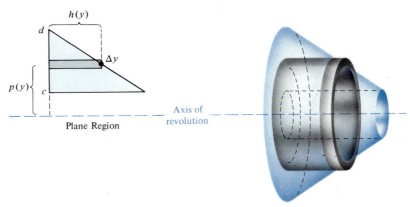

FIGURE 7.29

$$\text{volume of solid} \approx \sum_{i=1}^{n} 2\pi p(y_i)\, h(y_i)\, \Delta y$$

By taking the limit as $\|\Delta\| \to 0$ $(n \to \infty)$, we have

$$\text{volume of solid} = \lim_{n \to \infty}\left[\sum_{i=1}^{n} 2\pi p(y_i)\, h(y_i)\, \Delta y\right] = 2\pi \int_{c}^{d} p(y)\, h(y)\, dy$$

A similar result is obtained for a vertical axis of revolution. We summarize the results as follows:

Shell Method (Parallel Rectangles)

If the axis of revolution is horizontal, then

$$\text{volume of solid of revolution} = 2\pi \int_{c}^{d} ph\, dy$$

If the axis of revolution is vertical, then

$$\text{volume of solid of revolution} = 2\pi \int_{a}^{b} ph\, dx$$

See Figure 7.30.

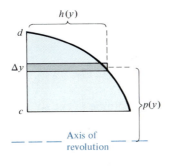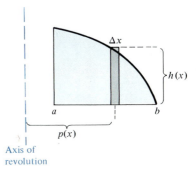

FIGURE 7.30

Example 1

Use the Shell Method to find the volume of the solid of revolution formed by revolving the region bounded by $y = x - x^3$ and $0 \le x \le 1$ about the y-axis.

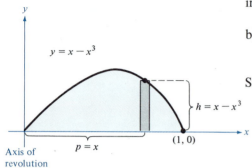

$y = x - x^3$

$h = x - x^3$

$p = x$

$(1, 0)$

Axis of revolution

FIGURE 7.31

Solution: Since we are to use the Shell Method, we begin by drawing a rectangle parallel to the axis of revolution (see Figure 7.31). The width Δx indicates that x is the variable of integration.

Since the radius of the shell is $p = x$ and the height of the shell is given by $h = x - x^3$, the volume of the representative shell is

$$\Delta V = 2\pi p h \, \Delta x = 2\pi (x)(x - x^3) \, \Delta x$$

Since x ranges from 0 to 1, the volume of the solid is

$$V = 2\pi \int_0^1 x(x - x^3) \, dx = 2\pi \int_0^1 (-x^4 + x^2) \, dx$$

$$= 2\pi \left[\frac{-x^5}{5} + \frac{x^3}{3} \right]_0^1 = \frac{4\pi}{15}$$

Note that in presenting the Disc and Shell Methods for finding the volume of a solid of revolution, we have stressed the placement of the rectangle as either perpendicular or parallel to the axis of revolution. We summarize this as follows:

Comparison of Disc and Shell Methods

1. If we place a rectangle *perpendicular* to the axis of revolution, we have chosen the *Disc Method,* and the variable of integration is denoted by the width of the rectangle.

2. If we place a rectangle *parallel* to the axis of revolution, then we have chosen the *Shell Method,* where the variable of integration is again denoted by the width of the rectangle.

Sometimes one method is more convenient to use than the other. The following example illustrates a case in which the Shell Method is preferable.

Example 2 (Shell Method Preferable)

Find the volume of the solid formed by revolving the region bounded by $y = x^2 + 1$, $y = 0$, $x = 0$, and $x = 1$ about the y-axis.

Solution: Refer to Figure 7.32.

For this problem the Shell Method is preferable because the Disc Method involves *two* distinct integrals. (See Example 4 of the previous

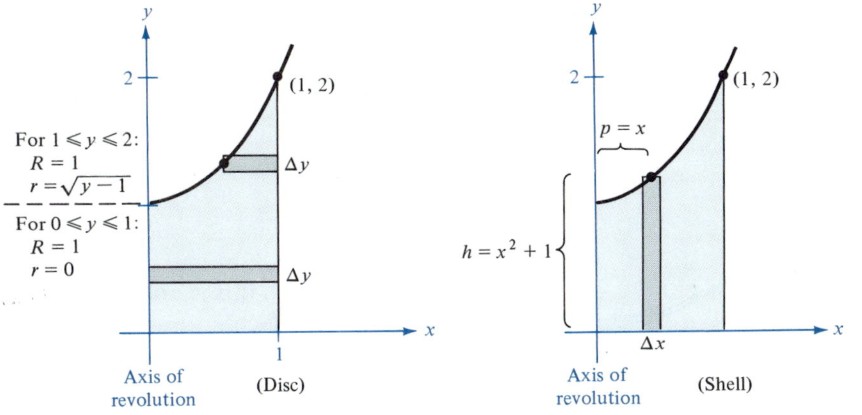

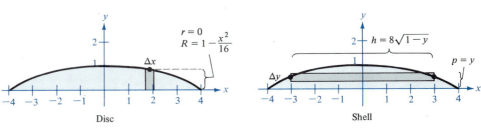

FIGURE 7.32

section.) By the Disc Method

$$V = \pi \int_0^1 (1^2 - 0^2)\, dy + \pi \int_1^2 [1^2 - (\sqrt{y-1})^2]\, dy$$

By the Shell Method

$$\Delta V = 2\pi p h\, \Delta x = 2\pi x(x^2 + 1)\, \Delta x$$

Hence the volume is

$$V = 2\pi \int_0^1 x(x^2 + 1)\, dx = 2\pi \left[\frac{x^4}{4} + \frac{x^2}{2} \right]_0^1 = 2\pi \left(\frac{3}{4} \right) = \frac{3\pi}{2} \qquad \blacksquare$$

Example 3 (*Disc Method Preferable*)

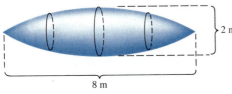

FIGURE 7.33

A fuel tank is to be made in the shape shown in Figure 7.33. The tank was designed by rotating the segment of the parabola $y = 1 - (x^2/16)$, $-4 \le x \le 4$, about the x-axis, where x and y are measured in meters. How many cubic meters of fuel will the tank hold?

Solution: Again, let us compare the Disc and Shell Methods (Figure 7.34).

Using the Shell Method we see that Δy is the width of the rectangle,

FIGURE 7.34

$p = y$, and $h = 2x$. Solving for x in the equation $y = 1 - (x^2/16)$, we have

$$x = 4\sqrt{1 - y}$$

and thus

$$h = 2x = 8\sqrt{1 - y}$$

Hence the Shell Method results in the integral

$$2\pi \int_0^1 (y)(8\sqrt{1 - y})\, dy$$

Although this is not a particularly difficult integral, we have not yet studied the techniques necessary for evaluating it.

On the other hand, the Disc Method yields the relatively simple integral

$$V = \int_{-4}^4 \pi \left(1 - \frac{x^2}{16}\right)^2 dx = \pi \int_{-4}^4 \left(1 - \frac{x^2}{8} + \frac{x^4}{256}\right) dx$$

$$= \pi \left[x - \frac{x^3}{24} + \frac{x^5}{1280}\right]_{-4}^4 = \frac{64\pi}{15} \approx 13.4 \text{ m}^3 \qquad \blacksquare$$

Note that for the Shell Method in Example 3, we had to solve for x in terms of y in the equation $y = 1 - (x^2/16)$. Sometimes it may be very difficult (or even impossible) to express x in terms of y when the equation is originally given in the form $y = f(x)$. In such cases we must use a (vertical) rectangle of width Δx, thus making x the variable of integration. The position (horizontal or vertical) of the axis of revolution then determines the method to be used. This is illustrated in the following example.

Example 4 (*Shell Method Necessary*)

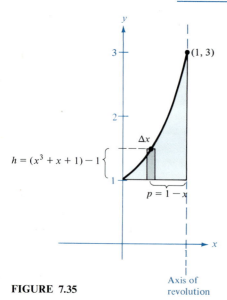

FIGURE 7.35

Find the volume of the solid formed by revolving, about the line $x = 1$, the region bounded by $y = x^3 + x + 1$, $y = 1$, and $x = 1$.

Solution: In the equation $y = x^3 + x + 1$, we cannot solve for x in terms of y, and, therefore, our variable of integration must be x. This means that our rectangle must be vertical. (See Figure 7.35.)

Now since the rectangle is parallel to the axis of revolution, we must use the Shell Method, and the volume is

$$V = 2\pi \int_0^1 ph\, dx = 2\pi \int_0^1 (1 - x)(x^3 + x + 1 - 1)\, dx$$

$$= 2\pi \int_0^1 (-x^4 + x^3 - x^2 + x)\, dx$$

$$= 2\pi \left[-\frac{x^5}{5} + \frac{x^4}{4} - \frac{x^3}{3} + \frac{x^2}{2}\right]_0^1$$

$$= 2\pi \left(-\frac{1}{5} + \frac{1}{4} - \frac{1}{3} + \frac{1}{2}\right) = \frac{13\pi}{30} \qquad \blacksquare$$

Section Exercises (7.3)

In Exercises 1–18, use the Shell Method to find the volume generated by revolving the given plane region about the given line.

1. $y = x$, $y = 0$, $x = 2$, about the y-axis
2. $y = 1 - x$, $y = 0$, $x = 0$, about the y-axis
S 3. $y = x$, $y = 0$, $x = 2$, about the x-axis
4. $y = 2 - x$, $y = 0$, $x = 0$, about the x-axis
5. $y = \sqrt{x}$, $y = 0$, $x = 4$, about the y-axis
6. $y = x^2 + 4$, $y = 8$, $x \geq 0$, about the y-axis
7. $y = x^2$, $y = 0$, $x = 2$, about the y-axis
8. $y = x^2$, $y = 0$, $x = 4$, about the y-axis
S 9. $y = x^2$, $y = 4x - x^2$, about the y-axis
10. $y = x^2$, $y = 4x - x^2$, about the line $x = 2$
S 11. $y = x^2$, $y = 4x - x^2$, about the line $x = 4$
12. $y = 1/x$, $x = 1$, $x = 2$, $y = 0$, about the x-axis
13. $y = 4x - x^2$, $y = 0$, about the line $x = 5$
S 14. $x + y^2 = 9$, $x = 0$, about the x-axis
15. $y = 4x - x^2$, $x = 0$, $y = 4$, about the y-axis
16. $y = 4 - x^2$, $y = 0$, about the y-axis
17. $y = \sqrt{x}$, $y = 0$, $x = 4$, about the line $x = 6$
18. $y = 2x$, $y = 4$, $x = 0$, about the y-axis

S 19. Use the Disc or Shell Method to find the volume of the solid generated by revolving the region bounded by $y = x^3$, $y = 0$, and $x = 2$ about:
    (a) the x-axis   (b) the y-axis
    (c) the line $x = 4$ (d) the line $y = 8$
20. Find the volume of the solid generated by revolving the region bounded by $y = -2x^2 + 8x - 6$ and $y = 0$ about the y-axis.
21. If possible, find the volume of the solid generated by revolving the plane region bounded by $x^{1/2} + y^{1/2} = a^{1/2}$, $x = 0$, and $y = 0$ about:
    (a) the x-axis   (b) the y-axis
    (c) the line $x = a$

22. Find the volume of the solid generated by revolving the region bounded by $x^{2/3} + y^{2/3} = a^{2/3}$ (hypocycloid of four cusps) about the x-axis.
S 23. A solid is generated by revolving the region bounded by $y = \frac{1}{2}x^2$ and $y = 2$ about the y-axis. A hole, centered along the axis of revolution, is drilled through this solid so that $\frac{1}{4}$ the volume is removed. Find the diameter of the hole.
C 24. A solid is generated by revolving the region bounded by $y = \sqrt{9 - x^2}$ and $y = 0$ about the y-axis. A hole, centered along the axis of revolution, is drilled through this solid so that $\frac{1}{3}$ the volume is removed. Find the diameter of the hole.
25. Let a sphere of radius r be cut by a plane, thus forming a segment of height h. Show that the volume of this segment is $\frac{1}{3}\pi h^2(3r - h)$.
26. Match the following integrals with the solid whose volume it represents and give the dimensions of each solid.

$$\underline{\qquad}\quad 2\pi \int_0^r hx \, dx$$ \qquad (a) right circular cone

$$\underline{\qquad}\quad 2\pi \int_0^r hx\left(1 - \frac{x}{r}\right) dx$$ \qquad (b) torus

$$\underline{\qquad}\quad 2\pi \int_0^r 2x\sqrt{r^2 - x^2} \, dx$$ \qquad (c) sphere

$$\underline{\qquad}\quad 2\pi \int_0^b 2ax\sqrt{1 - \frac{x^2}{b^2}} \, dx$$ \qquad (d) right circular cylinder

$$\underline{\qquad}\quad 2\pi \int_{-r}^r (R - x)(2\sqrt{r^2 - x^2}) \, dx$$ \qquad (e) ellipsoid

7.4 Volumes of Solids with Known Cross Sections

Purpose
- To extend the class of solids whose volumes we can determine by integration to include solids with known cross sections.

The mathematical facts worthy of being studied are those which, by their analogy with other facts, are capable of leading us to the knowledge of a mathematical law just as experimental facts lead us to the knowledge of a physical law. They are those which reveal to us unsuspected kinship between other facts, long known, but wrongly believed to be strangers to one another. —Henri Poincaré (1854–1912)

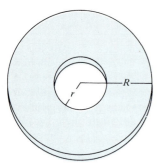

FIGURE 7.36

In our discussion of the Disc Method, we developed the formula for the volume of a solid of revolution:

$$\text{volume} = \pi \int_a^b (R^2 - r^2)\, dx$$

where $\pi(R^2 - r^2)$ is the *area* of a circular cross section and Δx is considered to be the thickness of the disc. (See Figure 7.36.)

We can generalize this method so that it applies to solids of any shape as long as we know a formula for the area of an arbitrary cross section. For instance, if $A(x)$ denotes the area of a cross section at x, taken perpendicular to the x-axis (see Figure 7.37), then the volume of the solid can be found by integrating $A(x)$ with respect to x. If the cross sections are taken perpendicular to the y-axis (see Figure 7.37), then we find the volume by integrating the area of the cross section with respect to y.

Volumes of Solids with Known Cross Sections

If the cross sections are perpendicular to the x-axis, then the volume is given by

$$\text{volume} = \int_a^b A(x)\, dx$$

where $A(x)$ is the area of a cross section at x.

If the cross sections are perpendicular to the y-axis, then the volume is given by

$$\text{volume} = \int_c^d A(y)\, dy$$

where $A(y)$ is the area of a cross section at y.

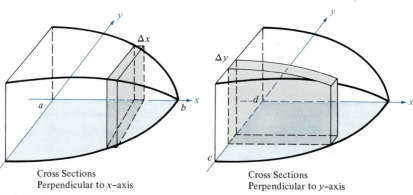

Cross Sections
Perpendicular to x–axis

Cross Sections
Perpendicular to y–axis

FIGURE 7.37

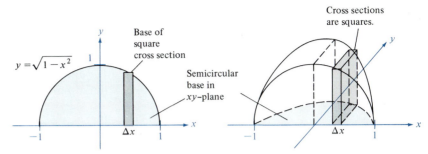

FIGURE 7.38

Example 1

Find the volume of the solid whose base is the semicircle $y = \sqrt{1 - x^2}$ and whose cross sections perpendicular to the x-axis are squares. (See Figure 7.38.)

Solution: The length of the base of each square cross section is $\sqrt{1 - x^2}$. Therefore, the area of any cross section is given by

$$A(x) = (\sqrt{1 - x^2})^2$$

and its thickness by Δx. Hence

$$V = \int_{-1}^{1} (1 - x^2)\, dx = \left[x - \frac{x^3}{3} \right]_{-1}^{1} = 1 - \frac{1}{3} + 1 - \frac{1}{3} = \frac{4}{3} \quad \blacksquare$$

Example 2

Find the volume of the solid whose base is the area bounded by $f(x) = 1 - (x/2)$, $g(x) = -1 + (x/2)$, and $x = 0$ and whose cross sections perpendicular to the x-axis are equilateral triangles. (See Figure 7.39.)

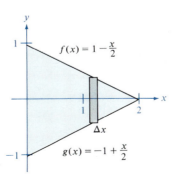

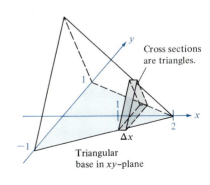

FIGURE 7.39

Solution: The base of each triangular cross section has a length of

$$\left(1 - \frac{x}{2}\right) - \left(-1 + \frac{x}{2}\right) = 2 - x$$

The area of an equilateral triangle with side b is $\sqrt{3}b^2/4$. Since x ranges from 0 to 2, the volume of the solid in Figure 7.39 is

$$V = \int_0^2 \frac{\sqrt{3}}{4}(2 - x)^2 \, dx = \frac{-\sqrt{3}}{4}\left[\frac{(2 - x)^3}{3}\right]_0^2 = \frac{2\sqrt{3}}{3} \qquad \blacksquare$$

Example 3

Find the volume of the solid whose base is the area bounded by $y = x^2 - 1$ and $y = 0$ and whose cross sections taken perpendicular to the x-axis are semicircles. (See Figure 7.40.)

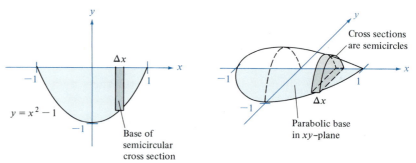

FIGURE 7.40

Solution: The area $A(x)$ of each semicircular cross section is

$$A(x) = \frac{1}{2}\pi \left(\frac{y}{2}\right)^2 = \frac{\pi}{8}(x^2 - 1)^2$$

Therefore, since x ranges from -1 to 1, the volume is

$$V = \frac{\pi}{8}\int_{-1}^1 (x^2 - 1)^2 \, dx = \frac{\pi}{8}\int_{-1}^1 (x^4 - 2x^2 + 1) \, dx$$

$$= \frac{\pi}{8}\left[\frac{x^5}{5} - \frac{2x^3}{3} + x\right]_{-1}^1 = \frac{\pi}{8}\left(\frac{1}{5} - \frac{2}{3} + 1 + \frac{1}{5} - \frac{2}{3} + 1\right)$$

$$= \frac{\pi}{8}\left(\frac{16}{15}\right) = \frac{2}{15}\pi \qquad \blacksquare$$

Example 4

Justify the formula for the volume of a cone,

$$\text{volume} = \tfrac{1}{3}(\text{area of the base})(\text{altitude})$$

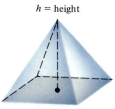
h = height

Square base
of area A

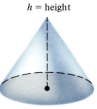
h = height

Circular base
of area A

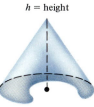
h = height

Irregular base
of area A

FIGURE 7.41

Solution: Note that the volume of a cone depends only on the *area* of the base (not its shape) and the height of the vertex (even if the vertex is not centered over the base). Thus the three "cones" in Figure 7.41 have the same volume.

Assume that the cone has a base of area A and height of h. Let B_1 be the base of the cone and B_2 be a cross section taken at height y. (See Figure 7.42.) From geometry we have the following ratio:

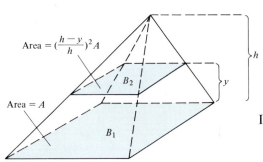

Area $= (\frac{h-y}{h})^2 A$

Area $= A$

B_2

B_1

h

y

FIGURE 7.42

$$\frac{\text{area of } B_2}{\text{area of } B_1} = \left(\frac{\text{distance between } B_2 \text{ and vertex}}{\text{distance between } B_1 \text{ and vertex}}\right)^2$$

$$\frac{\text{area of } B_2}{A} = \left(\frac{h-y}{h}\right)^2$$

$$\text{area of } B_2 = \left(\frac{h-y}{h}\right)^2 A$$

Integrating between 0 and h, we have

$$\text{volume} = \int_0^h \left(\frac{h-y}{h}\right)^2 (A)\, dy = \frac{A}{h^2} \int_0^h (h^2 - 2hy + y^2)\, dy$$

$$= \frac{A}{h^2}\left[h^2 y - hy^2 + \frac{y^3}{3} \right]_0^h = \frac{A}{h^2}\left(h^3 - h^3 + \frac{h^3}{3} \right) = \frac{Ah}{3} \quad \blacksquare$$

Section Exercises (7.4)

1. Find the volume of the solid if the base is bounded by the circle $x^2 + y^2 = 4$ and the cross sections perpendicular to the x-axis are:
   (a) squares
   (b) equilateral triangles
   (c) semicircles
   (d) isosceles right triangles whose hypotenuses lie on the base of the solid

2. The base of the solid is bounded by $y = x + 1$ and $y = x^2 - 1$. Find the volume of the solid if the cross sections perpendicular to the x-axis are:

   (a) squares
   (b) rectangles of height 1
   (c) equilateral triangles
   (d) semiellipses of height 2 (area of an ellipse is πab)
   (e) isosceles right triangles whose hypotenuses lie on the base of the solid

⬚ 3. The base of the solid is bounded by $y = x^3$, $y = 0$, and $x = 1$. Find the volume of the solid if the cross sections perpendicular to the y-axis are:
   (a) semicircles
   (b) squares

(c) equilateral triangles

(d) trapezoids for which $h = b_1 = \frac{1}{2}(b_2)$, where b_1 and b_2 are the lengths of the upper and lower bases, respectively

(e) semiellipses whose heights are twice the lengths of their bases

⟨S⟩ 4. Find the volume of the solid of intersection (the solid common to both) of two right circular cylinders of radius r whose axes meet at right angles. (See Figure 7.43.)

⟨S⟩ 5. A wedge is cut from a right circular cylinder of radius r inches by a plane through the diameter of the base,

which makes a 45° angle with the plane of the base. Find the volume of the wedge cut out. (See Figure 7.44.)

6. A barn is 100 ft long and 40 ft wide at the base. A cross section of the barn forms the parabola $y = 40 - \frac{1}{10}x^2$. Find the amount of storage space in the barn. (See Figure 7.45.)

7. Determine the volume (in gallons, gal) of a bulk milk tank if its top is a rectangle 5 ft by 6 ft and a cross section forms the curve $y = -\sqrt{9 - x^2}$. (Note: 1 ft³ holds 7.48 gal.) (See Figure 7.46.)

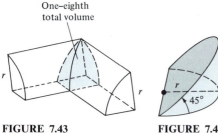

FIGURE 7.43

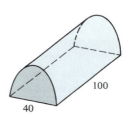

FIGURE 7.44

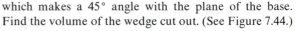

FIGURE 7.45

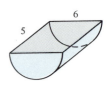

FIGURE 7.46

7.5 Work

Purpose

- To demonstrate the use of the definite integral to determine the work done by a variable force.

Mathematical analysis is as extensive as nature itself; it defines all perceptible relations, measures, times, spaces, forces, temperatures; this science is formed slowly, but it preserves every principle which it has once acquired. —*Joseph Fourier (1768–1830)*

The concept of *work* is important to scientists and engineers in determining the energy needed to perform various physical tasks. For instance, it is useful to know the amount of work to be done when a crane lifts a steel girder, when a spring is compressed, when a rocket is shot into the air, or when a truck pulls a load along a highway.

In general, we say that work is done when an applied force moves an object through some distance. If the applied force is *constant*, we have the following simple definition of work:

Work Done by a Constant Force   If an object is moved a distance D in the direction of an applied constant force F, then the **work** W done by the force is defined as $W = FD$.

Example 1

Determine the work done in lifting a 150-lb object a distance of 4 ft.

Solution: We consider the magnitude of the required force F to be the weight of the object. Thus the work done in lifting the object 4 ft is given by

$$W = FD = 150(4) = 600 \text{ foot pounds} \qquad \blacksquare$$

Work may also be expressed in units other than foot pounds (ft · lb), for example, inch pounds, foot tons, or any other combination of units of length and force. In the metric system the basic units of work are the dyne centimeter (erg) and the newton meter (joule, J), where $1 \text{ J} = 10^7$ ergs.

In Example 1 the force involved is *constant*. If a *variable* force is applied to an object, then the methods of calculus are needed to determine the work done because the amount of force required to move the object changes as the object changes position. For instance, the force required to compress a spring increases as the spring is compressed.

In the following development suppose that an object is moved along a straight line from $x = a$ to $x = b$ by a continuously varying force $F(x)$. Let Δ be a partition that divides the interval $[a, b]$ into n subintervals determined by

$$a = x_0 < x_1 < x_2 < \cdots < x_n = b$$

and let $\Delta x_i = x_i - x_{i-1}$. For each i choose c_i such that $x_{i-1} \leq c_i \leq x_i$. Then at c_i the force is given by $F(c_i)$. Since F is continuous, and assuming that the Δx_i's are small, we conclude that the force is nearly constant on each subinterval. Therefore, the work done in moving the object through the ith subinterval can be approximated by the increment

$$\Delta W_i = F(c_i) \, \Delta x_i$$

By adding the work done in each subinterval, we can approximate the total work done as the object moves from a to b by

$$W \approx \sum_{i=1}^{n} \Delta W_i = \sum_{i=1}^{n} F(c_i) \, \Delta x_i$$

Moreover, the smaller we make Δx_i, the better this approximation becomes. By taking the limit of this sum as $\|\Delta\| \to 0$, $(n \to \infty)$, we have

$$W = \lim_{n \to \infty} \sum_{i=1}^{n} F(c_i) \, \Delta x_i = \int_{a}^{b} F(x) \, dx$$

We therefore define work as follows:

Work Done by a Variable Force   If an object is moved along a straight line by a continuously varying force $F(x)$, then the **work** W done by the force as the object is moved from $x = a$ to $x = b$ is given by

$$W = \lim_{n \to \infty} \sum_{i=1}^{n} \Delta W_i = \int_a^b F(x)\, dx$$

In the next several examples the importance of setting up the increment of work ΔW should become evident.

Example 2

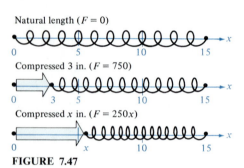

Natural length ($F = 0$)

Compressed 3 in. ($F = 750$)

Compressed x in. ($F = 250x$)

FIGURE 7.47

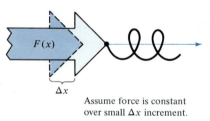

Assume force is constant over small Δx increment.

FIGURE 7.48

A force of 750 lb is required to compress a spring 3 in. from its natural length of 15 in. Find the work done in compressing the spring an additional 3 in.

Solution: According to Hooke's Law, the force $F(x)$ required to compress (or stretch) a spring x units from its natural length is given by $F(x) = kx$, where k is a constant whose value depends on the particular nature of the spring. Using the given data, we have

$$F(3) = 750 = (k)(3)$$

so that $k = 250$ and $F(x) = 250x$. (See Figure 7.47.) To find the increment of work, we assume that the force required to compress the spring over the small increment Δx is constant. (See Figure 7.48.) Thus the increment of work is

$$\Delta W = (\text{force})(\text{distance increment}) = (250x)\, \Delta x$$

Now since the spring is compressed from $x = 3$ to $x = 6$ inches less than its natural length, the work required is

$$W = \int_3^6 250x\, dx = 125x^2 \Big]_3^6 = 4500 - 1125 = 3375 \text{ in.} \cdot \text{lb}$$

Note that we do not integrate from $x = 0$ to $x = 6$ because we were asked to determine the work done in compressing the spring an *additional* 3 in., not including the first 3 in.   ■

Example 3

If a space module weighs 15 tons on the surface of the earth, then, neglecting air resistance, how much work is done in propelling the module to a height of 800 mi above the earth? (See Figure 7.49.)

Solution: Since the weight of a body varies inversely as the square of its distance from the center of the earth, we can write the force $F(x)$ exerted by gravity as

$$F(x) = \frac{k}{x^2}$$

Since the module weighs 15 tons on the ground and the radius of the earth is approximately 4,000 mi, we get the equation

$$15 = \frac{k}{(4,000)^2} \quad \text{or} \quad k = 240,000,000$$

Thus the increment of work is

$$\Delta W = \text{(force)(distance increment)} = \frac{240,000,000}{x^2} \Delta x$$

Since the module is propelled from $x = 4,000$ to $x = 4,800$ miles, the total work done is

$$W = \int_{4,000}^{4,800} \frac{240,000,000}{x^2} \, dx = \frac{-240,000,000}{x} \Big]_{4,000}^{4,800}$$

$$= -50,000 - (-60,000) = 10,000 \text{ mile tons} \quad \blacksquare$$

The solutions to Examples 2 and 3 conform to our formulation of work as the summation of increments in the form

$$\Delta W = \text{(force)(distance increment)} = F \Delta x$$

An equally useful way to formulate the increment of work is

$$\Delta W = \text{(force increment)(distance)} = \Delta F x$$

This second interpretation of ΔW is useful in work problems involving the moving of nonrigid substances such as fluids or chains. Our next two examples illustrate this alternative interpretation of the increment of work.

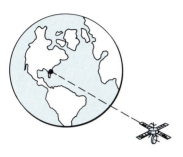

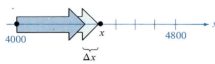

Space module is moved 800 miles above the earth.

FIGURE 7.49

Example 4

A tank has a height of 8 ft and a radius of 2 ft at its top, as shown in Figure 7.50. If the tank is filled to a depth of 6 ft with oil weighing 50 lb/ft³, find the work required to pump all the oil over the edge of the tank.

Solution: If we consider the oil to be subdivided into "layers" of width Δy, then we determine the work required to pump each layer over the edge by first describing the weight of each layer as the force increment

$$\Delta F = \text{(weight)} = \text{(density)(volume of layer)} = 50(\pi r^2 h)$$

From Figure 7.50 we can see that the layer of oil has a radius $r = x$ and a

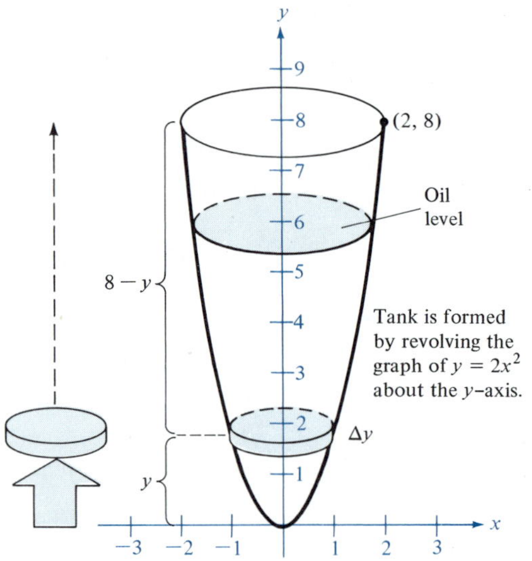

FIGURE 7.50 Disc of height Δy is moved $8 - y$ feet.

height $h = \Delta y$. Since $y = 2x^2$, we have $x^2 = y/2$ and we can write the force increment as

$$\Delta F = 50(\pi x^2 \, \Delta y) = 50(\pi)\left(\frac{y}{2}\right)\Delta y = 25\pi y \, \Delta y$$

Figure 7.50 also shows that each layer is moved a distance of $8 - y$ feet, and, therefore, the increment of work is

$$\Delta W = (\text{force increment})(\text{distance}) = (25\pi y \, \Delta y)(8 - y) = 25\pi(8y - y^2) \, \Delta y$$

Finally, since the heights of the various layers range from $y = 0$ to $y = 6$, the work required to empty the tank is

$$W = \int_0^6 25\pi(8y - y^2) \, dy = 25\pi\left[4y^2 - \frac{y^3}{3}\right]_0^6 = 25\pi(72) = 1800\pi \ \text{ft} \cdot \text{lb}$$

∎

Example 5

A 20-ft chain, weighing 5 lb/ft, is lying coiled on the ground. How much work is required to raise one end of the chain to a height of 20 ft so that it is fully extended? (See Figure 7.51.)

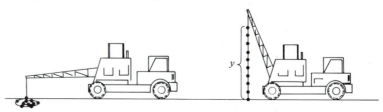

FIGURE 7.51

Solution: Imagine that the chain is divided into small sections, each of length Δy. Then the weight of each section is the increment of force

$$\Delta F = (\text{weight}) = (\text{density})(\text{length}) = 5\,\Delta y$$

Since a typical section (initially on the ground) is raised to a height y (see Figure 7.51), we conclude that the increment of work is

$$\Delta W = (\text{force increment})(\text{distance}) = (5\,\Delta y)y = 5y\,\Delta y$$

Finally, since y ranges from 0 to 20, the total work is

$$W = \int_0^{20} 5y\,dy = \frac{5y^2}{2}\bigg]_0^{20} = \frac{5(400)}{2} = 1000 \text{ ft} \cdot \text{lb}$$

Section Exercises (7.5)

1. A force of 5 lb compresses a 15-in. spring a total of 4 in. How much work is done in compressing the spring 7 in.?
2. How much work is done in compressing the spring in Exercise 1 from a length of 10 in. to a length of 6 in.?

[S] 3. A force of 60 lb stretches a spring 1 ft. How much work is done in stretching the spring from 9 in. to 15 in.?

4. A cylindrical water tank 12 ft high with a radius of 8 ft is buried so that the top of the tank is 3 ft below ground level. How much work is done in pumping a full tank of water up to ground level? (Density of water is 62.4 lb/ft^3.) (See Figure 7.52.)
5. Suppose the tank in Exercise 4 is located on a tower so that the bottom of the tank is 20 ft aboveground. How much work is done in filling the tank half full of water through a hole in the bottom, using a water source at ground level? (See Figure 7.53.)
6. How much work is done in filling the tank in Exercise 5 half full if the water is pumped in through the top?

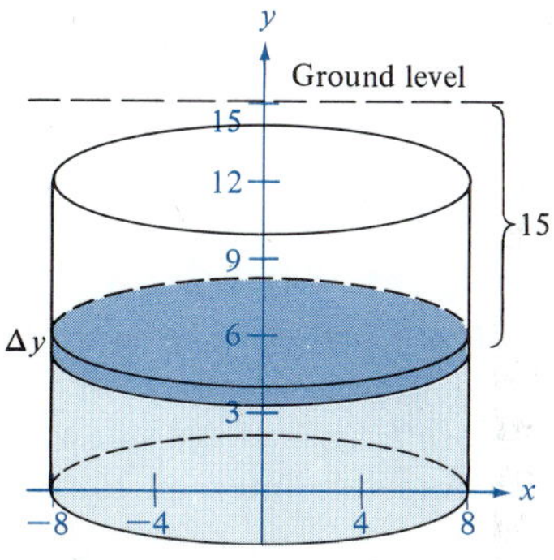

FIGURE 7.52

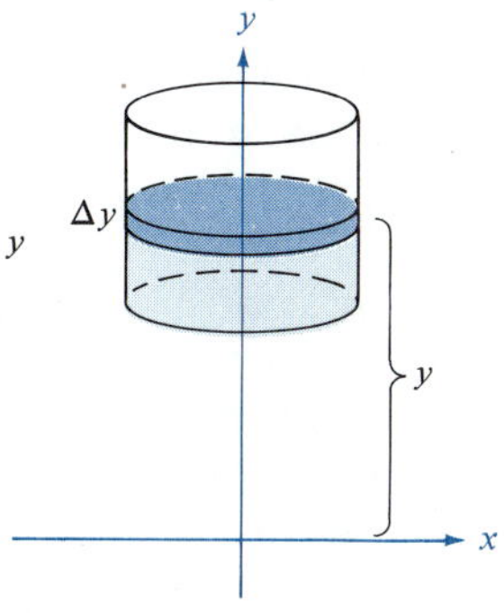

FIGURE 7.53

[S] 7. A hemispherical tank of radius 6 ft is positioned so that its base is circular. The water source is at the base.
   (a) How much work is required to fill the tank with water through a hole in the top?
   (b) Through a hole in the base?

8. Suppose the tank in Exercise 7 is inverted. How much work is required to pump the top 2 ft of water out of a hole in the top?

[S] 9. An open tank has the shape of a right circular cone. If the tank is 8 ft across the top and has a height of 6 ft, how much work is required to empty the tank of water by pumping the water over the edge of the tank? (See Figure 7.54.)

10. If water is pumped in through the bottom, how much work is required to fill the tank in Exercise 9:
    (a) to a depth of 2 ft?
    (b) from a depth of 4 ft to a depth of 6 ft?

[C] 11. If a lunar module weighs 12 tons on the surface of the earth, how much work is done in propelling the module to a height of 50 mi above the surface of the moon? (Hint: Consider the radius of the moon to be 1100 mi and its force of gravity to be one-sixth that of the earth's.)

12. Two electrons repel each other with a force that varies inversely as the square of the distance between them. If one electron is fixed at the point (2, 4), find the work done in moving a second electron from $(-2, 4)$ to $(1, 4)$.
13. A chain 15 ft long and weighing 3 lb/ft is suspended vertically from a height of 15 ft. How much work is required to raise the entire chain to the 15-ft level?
14. For the chain in Exercise 13, how much work is required to raise the chain vertically so that one-third of the chain

is raised to a level of 15 ft? Compare this to the answer to Exercise 13.

S **15.** For the chain in Exercise 13, how much work is required to take the bottom of the chain and raise it to the 15-ft level, leaving the chain doubled but still hanging vertically? (See Figure 7.55.)

16. For the chain in Exercise 13, how much work is required to raise the chain vertically so that the bottom of the chain is at the 10-ft level?

17. A demolition crane has a 500-lb ball suspended from a 40-ft cable weighing 0.7 lb/ft. How much work is required to wind up 15 ft of this apparatus?

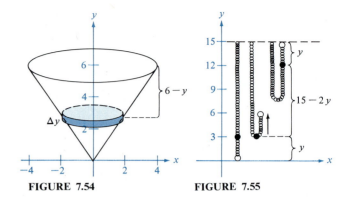

FIGURE 7.54   **FIGURE 7.55**

7.6 Fluid Pressure

Purpose

- To use the definite integral to calculate the force exerted by a fluid against a vertical wall.

This device of mechanics for multiplying forces, when once understood, gives the reason why liquids weigh in accordance with their height and not in accordance with their expanse. —*Blaise Pascal (1623–1662)*

Certainly swimmers are aware of the fact that the deeper an object is submerged into a fluid, the greater is the pressure upon the object. (**Pressure** is referred to as the force per unit of area.) This concept is described explicitly by the formula

$$p = \rho h$$

(ρ is the Greek lowercase letter rho), where

p = pressure of the fluid

h = depth below the surface

ρ = density (weight per unit volume) of the fluid

For example, the weight of water is 62.4 pounds per cubic foot, and so at a depth of 6 feet, the pressure upon a submerged body is

$$p = (62.4 \text{ lb/ft}^3)(6 \text{ ft}) = 374.4 \text{ lb/ft}^2$$

This amount of pressure corresponds to the weight of the 6-foot-high "column of water" above each square foot of area. Furthermore, according to Pascal's Principle, the pressure exerted by a fluid at a specific depth is *equal in all directions*. Thus the pressure at depth h against the sidewall of a container is equal to the pressure on an object submerged to this same depth.

It may be surprising to realize that the formula for the pressure exerted by a fluid,

$$\text{pressure} = (\text{density})(\text{depth below fluid surface})$$

is independent of the size of the container. This means that at 3 feet below the surface, the pressure (force per unit of area) against the side-wall of a backyard swimming pool is just as great as the pressure against the face of a dam, assuming the density of the water is the same. The pressure is determined by depth alone; all other dimensions of the container are of no consequence.

Our main interest in fluid pressure is to determine the total force exerted by a fluid upon the walls of its container. For a container having vertical sides, it is a simple matter to calculate the total force on the *bottom* of the container. The pressure at the bottom is constant; the total force is simply the product of the pressure times the area of the bottom. In general, for a horizontally submerged plane region, we have

$$\text{total force on plane region} = (\text{pressure})(\text{area of region})$$
$$F = (\text{density})(\text{depth})(\text{area}) = \rho h A$$

We face a more difficult problem in determining the total force against the vertical *sides* of a container, because the pressure is not constant at each point; it increases as the depth increases. Thus our simple formula for total force is not directly applicable because we do not know what depth to use. Fortunately, in the case of a continuously varying depth, we can apply the techniques of calculus to determine the force exerted by the fluid.

Suppose a vertical plane region is submerged in a fluid of density ρ, as shown in Figure 7.56. We wish to determine the total force against this region from depth $h - a$ to depth $h - b$.

First, we subdivide interval $[a, b]$ into n subintervals each of length Δy. Now consider the representative rectangle of width Δy and length L_i, where y_i is in the ith subinterval. The force against this representative rectangle is

$$\Delta F_i = \rho(\text{depth})(\text{area}) = \rho(h - y_i) L_i \Delta y$$

The force against n such rectangles is

$$\sum_{i=1}^{n} \Delta F_i = \rho \sum_{i=1}^{n} (h - y_i) L_i \Delta y$$

Note that the density ρ is considered to be constant and is factored out of the summation. Therefore, taking the limit as $\|\Delta\| \to 0$ $(n \to \infty)$, we find the total force against the region to be

$$F = \rho \lim_{n \to \infty} \sum_{i=1}^{n} (h - y_i) L_i \Delta y$$

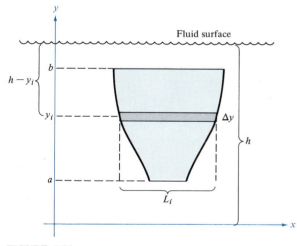

FIGURE 7.56

Force Exerted by a Fluid

The force F exerted by a fluid of constant density ρ against a vertically submerged plane region from $y = a$ to $y = b$ is given by

$$F = \rho \int_a^b (h - y) L \, dy$$

where h is the total depth of the fluid and L is the horizontal length of the region at y.

Example 1

A vertical gate in a dam has the shape of an isosceles trapezoid 8 ft across the top and 6 ft across the bottom, with a height of 5 ft. What is the total force against the gate if the top of the gate is 4 ft below the surface of the water?

Solution: Our problem can be simplified a bit by locating the y-axis so as to divide the trapezoid in half. (See Figure 7.57.)

The equation of the line through $(3, 0)$ and $(4, 5)$ is

$$y - 0 = 5(x - 3)$$

or

$$\frac{y + 15}{5} = x$$

Thus the representative rectangle has width Δy and length

$$L = 2x = 2\left(\frac{y + 15}{5}\right) = \frac{2}{5}(y + 15)$$

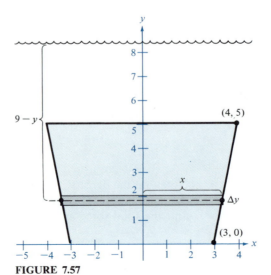

FIGURE 7.57

and is at a depth of $9 - y$. Since the density of water is 62.4 lb/ft³, the force against the rectangle is

$$\Delta F = \text{(density)(depth)(area)} = (62.4)(9 - y)[\tfrac{2}{5}(y + 15)\,\Delta y]$$

Since y ranges from 0 to 5, the total force is

$$F = \int_0^5 (62.4)(9 - y)(\tfrac{2}{5})(y + 15)\,dy$$

$$= 62.4(0.4)\int_0^5 (9 - y)(y + 15)\,dy = 24.96\int_0^5 (135 - 6y - y^2)\,dy$$

$$= 24.96\left[135y - 3y^2 - \frac{y^3}{3}\right]_0^5 = 24.96\left(\frac{1{,}675}{3}\right) = 13{,}936 \text{ lb} \quad\blacksquare$$

Example 2

The vertical dam shown in Figure 7.58 is 120 ft across the top and 40 ft high. Determine the force against the dam if the water level is 15 ft below the top of the dam.

Solution: Solving for x in Figure 7.58, we get $x = \pm\sqrt{90y}$, and the force against a representative rectangle of length $2x$ is

$$\Delta F = \text{(density)(depth)(area)} = (62.4)(25 - y)(2x\,\Delta y)$$

$$= 62.4(25 - y)(2\sqrt{90y})\,\Delta y$$

Since y ranges from 0 to 25, the total force against the dam is

$$F = \int_0^{25} 62.4(25 - y)2\sqrt{90y}\,dy = 124.8\sqrt{90}\int_0^{25}(25y^{1/2} - y^{3/2})\,dy$$

$$= 374.4\sqrt{10}\left[\frac{50}{3}y^{3/2} - \frac{2}{5}y^{5/2}\right]_0^{25} = 374.4\sqrt{10}\left(\frac{6{,}250}{3} - 1{,}250\right)$$

$$\approx 986{,}630.6 \text{ lb} \approx 493.3 \text{ tons}$$

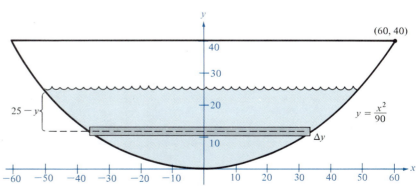

FIGURE 7.58 $\quad\blacksquare$

Example 3

The bottom of a swimming pool is an inclined plane. The pool is 2 ft deep at one end and 10 ft deep at the other. If the pool is 40 ft long and 30 ft wide with vertical sides, find the total force against one 40-ft side.

Solution:
By placing the x- and y-axes as in Figure 7.59, we see that the equation representing the bottom of the pool is

$$y = mx + b = \tfrac{1}{5}x \qquad \text{or} \qquad x = 5y$$

However, we must note that the equation $x = 5y$ is only valid for $0 \le y \le 8$. When y is between 8 and 10, x is a constant 40. Thus the force against a representative rectangle of length x is

$$\Delta F = 62.4(10 - y)(x)(\Delta y)$$
$$= \begin{cases} 62.4(10 - y)(5y)\,\Delta y, & \text{for } 0 \le y \le 8 \\ 62.4(10 - y)(40)\,\Delta y, & \text{for } 8 \le y \le 10 \end{cases}$$

Therefore, we determine the total force F to be

$$F = \int_0^8 62.4(10 - y)(5y)\,dy + \int_8^{10} 62.4(10 - y)(40)\,dy$$

$$= 312 \int_0^8 (10y - y^2)\,dy + 2{,}496 \int_8^{10} (10 - y)\,dy$$

$$= 312 \left[5y^2 - \frac{y^3}{3} \right]_0^8 + 2{,}496 \left[10y - \frac{y^2}{2} \right]_8^{10}$$

$$= 312 \left(\frac{448}{3} \right) + 2{,}496(2)$$

$$= 51{,}584 \text{ lb}$$

Note that the 30-ft width of the pool is a "red herring" in the sense that the force against the 40-ft sides is not dependent on the width of the pool.

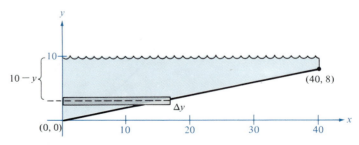

FIGURE 7.59

Section Exercises (7.6)

In Exercises 1–6, find the force on a vertical side of a tank if the tank is full of water and the side has the given shape.

1. rectangle

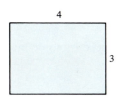

2. triangle

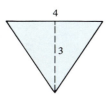

[S] **3.** trapezoid

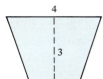

4. semicircle

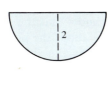

5. parabolic ($y = x^2$)

6. semiellipse

In Exercises 7–10, find the total force on the given vertical plates submerged in water.

7. square

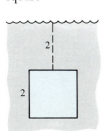

8. diamond

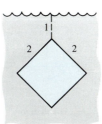

[S] **9.** triangle

10. rectangle

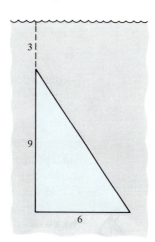

[S] **11.** A cylindrical gasoline tank is placed so that the axis of the cylinder is horizontal. If the tank is half full, find the force on a circular end of the tank, assuming that the diameter is 3 ft and gasoline has a density of 50 lb/ft³.

12. A rectangular plate of height h feet and base b feet is submerged vertically in a tank of water. The base is kept k feet below the surface of the water, where $h \le k$. Show that the total pressure on the surface of the rectangle is given by

$$\text{pressure} = (62.4) \binom{\text{area of}}{\text{rectangle}} \binom{\text{depth of center}}{\text{of rectangle}}$$

13. A porthole on a vertical side of a ship is 1 ft square. Find the total pressure on the porthole, assuming that the top of the square is 15 ft below the surface.

[S] **14.** A swimming pool is 20 ft wide, 40 ft long, 4 ft deep at one end, and 8 ft deep at the other. The bottom is an inclined plane. Find the total force on each of the vertical walls of the pool.

7.7 Moments, Centers of Mass, and Centroids

Purpose

- To provide a general description of the concepts of moment, center of mass, and centroid.
- To develop formulas for calculating centroids of plane regions.

This method never fails and could be extended to a number of beautiful problems; with its aid, we have found the centers of gravity of figures bounded by straight lines and curves, as well as those of solids.

—*Pierre de Fermat (1601–1665)*

Loosely speaking, the **moment** about a point P produced by some mass is simply

$$\text{moment} = (\text{mass})(\text{moment arm})$$

where the moment arm is the distance of the mass from point P. This concept of moment can be simply demonstrated by a seesaw. Suppose a child weighing 50 pounds sits 6 feet to the left of the center point P, while a child weighing 60 pounds sits 6 feet to the right of P, as shown in Figure 7.60. From experience we know that the seesaw would begin to rotate clockwise in a vertical plane about the point P. This rotation occurs because the moment produced by the child on the left,

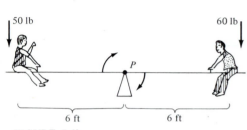

50 lb
60 lb
P
6 ft
6 ft

FIGURE 7.60

$$\text{left moment} = (50)(6) = 300 \text{ lb} \cdot \text{ft}$$

is less than the moment produced by the child on the right,

$$\text{right moment} = (60)(6) = 360 \text{ lb} \cdot \text{ft}$$

For the seesaw to balance, the two moments must be equal. Thus if the 60-pound child sits 5 feet from point P, then the seesaw would balance, because the respective moments are now equal:

$$\text{left moment} = (50)(6) = 300 \text{ lb} \cdot \text{ft}$$
$$\text{right moment} = (60)(5) = 300 \text{ lb} \cdot \text{ft}$$

To be a bit more precise, if we let P be the origin, and introduce coordinates $x_1 = -6$ and $x_2 = 5$, then the seesaw balances because the resultant moment about the origin is zero; that is,

$$\text{moment} = m_1x_1 + m_2x_2 = (50)(-6) + (60)(5) = 0$$

To generalize our development, suppose several masses $m_1, m_2, \ldots, m_n$ are located along the x-axis at the respective points $x_1, x_2, \ldots, x_n$ (Figure 7.61). Then the measure of the tendency of this system to rotate about the

m_1
m_2
m_3
m_n
0
x
x_1
x_2
x_3
x_n

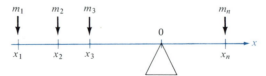

If the moment, $x_1m_1 + x_2m_2 + \cdots + x_nm_n$, equals zero then the system is in equilibrium.

FIGURE 7.61

origin 0 is called the **moment** of the system about the origin, and it is denoted by

$$M_0 = m_1 x_1 + m_2 x_2 + \cdots + m_n x_n = \sum_{i=1}^{n} m_i x_i$$

If the moment is zero, then the system is said to be in **equilibrium.**

Consider a system that is not in equilibrium, and suppose we move the pivot point to some point $x = \overline{x}$ so that equilibrium is attained. Then for this system, with pivot point $\overline{x}$, we have

$$\sum_{i=1}^{n} m_i(x_i - \overline{x}) = m_1(x_1 - \overline{x}) + m_2(x_2 - \overline{x}) + \cdots + m_n(x_n - \overline{x}) = 0$$

or

$$\sum_{i=1}^{n} m_i x_i - \sum_{i=1}^{n} m_i \overline{x} = 0$$

and solving for $\overline{x}$, we have

$$\overline{x} = \frac{\displaystyle\sum_{i=1}^{n} m_i x_i}{\displaystyle\sum_{i=1}^{n} m_i} = \frac{\text{moment of system about origin}}{\text{total mass of system}}$$

This balancing point, $\overline{x}$, is called the **center of mass** of the system.

Definition of the Moment of a Linear System	The **moment about the origin** of a system of masses $m_1, m_2, \ldots, m_n$ located along the x-axis at the points $x_1, x_2, \ldots, x_n$ is $$M_0 = m_1 x_1 + m_2 x_2 + \cdots + m_n x_n$$ If m is the total mass of the system, then the **center of mass** $\overline{x}$ is given by $$\overline{x} = \frac{M_0}{m}$$

Example 1

Find the center of mass of the following linear system:

$$m_1 = 10, \quad x_1 = -5; \qquad m_2 = 15, \quad x_2 = 0$$
$$m_3 = 5, \quad x_3 = 4; \qquad m_4 = 10, \quad x_4 = 7$$

Solution: The moment about the origin is given by

$$M_0 = m_1 x_1 + m_2 x_2 + m_3 x_3 + m_4 x_4$$
$$= 10(-5) + 15(0) + 5(4) + 10(7) = -50 + 0 + 20 + 70 = 40$$

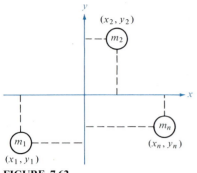

FIGURE 7.62

Since the total mass of the system is $m = 10 + 15 + 5 + 10 = 40$, the center of mass is

$$\bar{x} = \frac{M_0}{m} = \frac{40}{40} = 1$$

We can extend these concepts to two dimensions by considering the masses to be located in the xy-plane at points $(x_1, y_1), (x_2, y_2), \ldots, (x_n, y_n)$, respectively (Figure 7.62). Then we define their moments with respect to the x-axis and with respect to the y-axis as follows:

Definition of the Moments of a Two-Dimensional System

For a system of masses $m_1, m_2, \ldots, m_n$ located in the xy-plane at the points $(x_1, y_1), (x_2, y_2), \ldots, (x_n, y_n)$, the **moment about the y-axis M_y** is

$$M_y = m_1 x_1 + m_2 x_2 + \cdots + m_n x_n$$

and the **moment about the x-axis M_x** is

$$M_x = m_1 y_1 + m_2 y_2 + \cdots + m_n y_n$$

If m is the total mass of the system, then the **center of mass** $(\bar{x}, \bar{y})$ is given by

$$\bar{x} = \frac{M_y}{m} \quad \text{and} \quad \bar{y} = \frac{M_x}{m}$$

The numbers $m\bar{x}$ and $m\bar{y}$ may be regarded as the moments about the y- and x-axes, respectively, of a mass m located at the point $(\bar{x}, \bar{y})$. That is, the center of mass of a system of particles is the point where the *total mass* could be concentrated and still give the same total moments M_y and M_x.

Example 2

Find the center of mass of a system of masses, $m_1 = 6$, $m_2 = 3$, $m_3 = 2$, and $m_4 = 9$, located at the points $(3, -2)$, $(0, 0)$, $(-5, 3)$, and $(4, 2)$, respectively.

Solution:
$$m = 6 + 3 + 2 + 9 = 20$$
$$M_x = (6)(-2) + (3)(0) + (2)(3) + (9)(2) = 12$$
$$M_y = (6)(3) + (3)(0) + (2)(-5) + (9)(4) = 44$$

Therefore,
$$\bar{x} = \frac{M_y}{m} = \frac{44}{20} = \frac{11}{5}$$

$$\bar{y} = \frac{M_x}{m} = \frac{12}{20} = \frac{3}{5}$$

and the center of mass of this system of particles is $(\frac{11}{5}, \frac{3}{5})$.

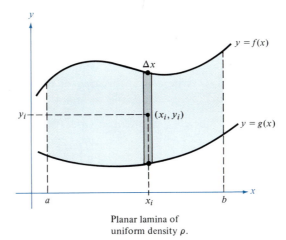

Planar lamina of
uniform density ρ.

FIGURE 7.63

In the preceding discussion we have assumed the total mass of a system to be distributed at discrete points along a line or in the plane. Let us now consider a thin flat plate of material (called a **lamina**) whose total mass is uniformly distributed throughout the plate; that is, its density does not vary from point to point. (Even though the plate is three-dimensional, we treat it as though it were two-dimensional.) We wish to define the center of mass $(\bar{x}, \bar{y})$ of a lamina, so that this point is the "balance point" in the same sense as in a system of discrete particles. For example, the center of mass of a circular lamina is located at the center of the circle, and the center of mass of a rectangular lamina is located at the center of the rectangle.

To find the center of mass of an irregularly shaped lamina, we proceed as follows. Consider the lamina of constant density ρ shown in Figure 7.63. A typical rectangle has been obtained by subdividing the interval $[a, b]$ into n subintervals of widths Δx. If we denote the center of mass of the ith rectangle by (x_i, y_i), then by the Midpoint Rule, we have

$$y_i = \frac{f(x_i) + g(x_i)}{2}$$

With this information we can readily develop our formulas for the center of mass of our region.

First, the mass of the ith rectangle is

$$\text{mass} = (\text{density})(\text{area}) = \rho(\Delta A_i) = \rho[f(x_i) - g(x_i)](\Delta x)$$

Then the total mass of the region is approximated by

$$m \approx \sum_{i=1}^{n} \rho[f(x_i) - g(x_i)](\Delta x)$$

Taking the limit as $\|\Delta\| \to 0$ $(n \to \infty)$, we have the mass defined as

$$m = \rho \int_a^b [f(x) - g(x)]\,dx = \rho A$$

where A is the area of the region.

Second, the moment about the x-axis of the ith rectangle is

$$\text{moment} = (\text{mass})(\text{moment arm}) = (\rho\,\Delta A_i)(y_i) = \rho(y_i)(\Delta A_i)$$

$$= \rho\left[\frac{f(x_i) + g(x_i)}{2}\right][f(x_i) - g(x_i)]\,\Delta x$$

$$= \frac{\rho}{2}[f(x_i)^2 - g(x_i)^2]\,\Delta x$$

Summing all such moments and taking the limit as $n \to \infty$, we have the moment about the x-axis defined as

$$M_x = \frac{\rho}{2}\int_a^b [f(x)^2 - g(x)^2]\,dx$$

In a similar manner, the moment about the y-axis is determined to be

$$M_y = \rho \int_a^b x[f(x) - g(x)]\,dx$$

Moments of a Planar Lamina

Let $g \le f$ be continuous functions on $[a, b]$. Then for the planar lamina of uniform density ρ bounded by $y = g(x)$, $y = f(x)$, $x = a$, and $x = b$, the **moments about the x- and y-axes** are given by

$$M_x = \frac{\rho}{2}\int_a^b [f(x)^2 - g(x)^2]\,dx \qquad \text{and} \qquad M_y = \rho \int_a^b x[f(x) - g(x)]\,dx$$

The **mass** m of the lamina is given by

$$m = \rho \int_a^b [f(x) - g(x)]\,dx$$

and the **center of mass** $(\bar{x}, \bar{y})$ is

$$\bar{x} = \frac{M_y}{m} \qquad \text{and} \qquad \bar{y} = \frac{M_x}{m}$$

Example 3

Find the center of mass of the lamina of uniform density ρ bounded by $y = 4 - x^2$ and the x-axis.

Solution: First, since the center of mass lies on the axis of symmetry, we know that $\bar{x} = 0$. (See Figure 7.64.) Integrating to find the mass, we have

$$m = \rho \int_{-2}^{2} y\,dx = \rho \int_{-2}^{2} (4 - x^2)\,dx = \rho \left[4x - \frac{x^3}{3} \right]_{-2}^{2} = \frac{32\rho}{3}$$

The moment about the x-axis is

$$M_x = \frac{\rho}{2} \int_{a}^{b} [f(x)^2 - g(x)^2]\,dx = \frac{\rho}{2} \int_{-2}^{2} [(4 - x^2)^2 - (0)^2]\,dx$$

$$= \frac{\rho}{2} \int_{-2}^{2} (16 - 8x^2 + x^4)\,dx = \frac{\rho}{2} \left[16x - \frac{8x^3}{3} + \frac{x^5}{5} \right]_{-2}^{2} = \frac{256\rho}{15}$$

Finally, $\bar{y}$ is given by

$$\bar{y} = \frac{M_x}{m} = \frac{256\rho/15}{32\rho/3} = \frac{8}{5}$$

Thus the center of mass of the lamina is $(0, \frac{8}{5})$. ∎

Note that the density ρ in Example 3 is a common factor of the moments and the mass and as such may be canceled in the quotients for the coordinates of the center of mass. In other words, the center of mass of a lamina of uniform density depends only on the shape of the lamina and not on its density. This observation implies that we can generalize the formula for the center of mass of a lamina to find the center of a "massless" region in the plane. When we do this, we call the point $(\bar{x}, \bar{y})$ the **centroid** of the region.

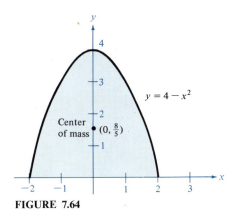

y = 4 − x²

Center of mass $(0, \frac{8}{5})$

FIGURE 7.64

Centroid of a Plane Region

Let $g \le f$ be continuous functions on $[a, b]$. The **centroid** $(\bar{x}, \bar{y})$ of the region bounded by $y = g(x)$, $y = f(x)$, $x = a$, and $x = b$ is given by

$$\bar{x} = \frac{\displaystyle\int_{a}^{b} x[f(x) - g(x)]\,dx}{A}$$

and

$$\bar{y} = \frac{\displaystyle\frac{1}{2} \int_{a}^{b} [f(x)^2 - g(x)^2]\,dx}{A}$$

where A is the area of the region.

Example 4

Find the centroid of the region bounded by the graphs of $f(x) = 4 - x^2$ and $g(x) = x + 2$.

Solution: The two curves intersect at the points $(-2, 0)$ and $(1, 3)$. (See Figure 7.65.) In this case the area is given by

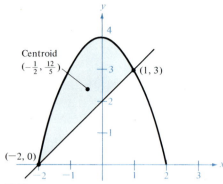

Centroid
$(-\frac{1}{2}, \frac{12}{5})$

$(1, 3)$

$(-2, 0)$

FIGURE 7.65

$$A = \int_{-2}^{1} [f(x) - g(x)]\, dx = \int_{-2}^{1} [(4 - x^2) - (x + 2)]\, dx$$

$$= \int_{-2}^{1} (2 - x - x^2)\, dx = \frac{9}{2}$$

The centroid $(\overline{x}, \overline{y})$ of the rectangle has coordinates

$$\overline{x} = \frac{\int_{-2}^{1} x[(4 - x^2) - (x + 2)]\, dx}{\frac{9}{2}}$$

and

$$\overline{y} = \frac{\frac{1}{2}\int_{-2}^{1} [(4 - x^2)^2 - (x + 2)^2]\, dx}{\frac{9}{2}}$$

Evaluating these integrals we have

$$\overline{x} = \frac{2}{9}\int_{-2}^{1} (-x^3 - x^2 + 2x)\, dx = \frac{2}{9}\left[\frac{-x^4}{4} - \frac{x^3}{3} + x^2\right]_{-2}^{1} = \frac{-1}{2}$$

$$\overline{y} = \frac{1}{9}\int_{-2}^{1} (x^4 - 9x^2 - 4x + 12)\, dx$$

$$= \frac{1}{9}\left[\frac{x^5}{5} - 3x^3 - 2x^2 + 12x\right]_{-2}^{1} = \frac{12}{5}$$

Thus the centroid of the region is $(\frac{-1}{2}, \frac{12}{5})$.   ∎

For simple plane regions we may be able to find the centroid without resorting to integration. Example 5 presents such a case.

Example 5

Find the centroid of the region described in Figure 7.66.

Solution: By superimposing a coordinate system on the region, as indicated in Figure 7.67, we locate the centroids of the three rectangles as

$$(\tfrac{1}{2}, \tfrac{3}{2}), (\tfrac{5}{2}, \tfrac{1}{2}), (5, 1)$$

Suppose we consider the area of each rectangle to be its mass and the centroid of each rectangle to be its center of mass. Then the centroid of the region can be calculated in a manner similar to that given in Example 2. That is,

$$m = \text{area of region} = 3 + 3 + 4 = 10$$

and

$$M_x = (\tfrac{3}{2})(3) + (\tfrac{1}{2})(3) + (1)(4) = 10$$

$$M_y = (\tfrac{1}{2})(3) + (\tfrac{5}{2})(3) + (5)(4) = 29$$

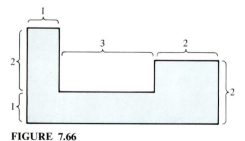

FIGURE 7.66

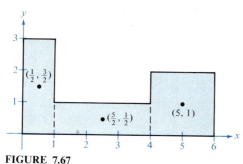

FIGURE 7.67

Then

$$\bar{x} = \frac{M_y}{m} = \frac{29}{10} = 2.9$$

$$\bar{y} = \frac{M_x}{m} = \frac{10}{10} = 1$$

Thus the centroid of the region is (2.9, 1).

∎

Section Exercises (7.7)

In Exercises 1–4, the masses m_i are located at the points x_i on the x-axis. Find the center of mass.

1. $m_1 = 6$, $x_1 = -5$; $m_2 = 3$, $x_2 = 1$; $m_3 = 5$, $x_3 = 3$

S 2. $m_1 = 8$, $x_1 = 6$; $m_2 = 4$, $x_2 = -2$; $m_3 = 7$, $x_3 = -3$; $m_4 = 3$, $x_4 = 5$

3. $m_1 = m_2 = m_3 = m_4 = m_5 = 1$; $x_1 = 18$, $x_2 = 15$; $x_3 = 8$; $x_4 = 12$; $x_5 = 7$

4. $m_1 = 3$, $x_1 = 0$; $m_2 = 12$, $x_2 = -3$; $m_3 = 11$, $x_3 = 4$; $m_4 = 6$, $x_4 = -1$; $m_5 = 1$, $x_5 = -2$

5. Notice that $\bar{x}$ in Exercise 3 is the arithmetic mean of the x-coordinates of the points.
   (a) In Exercise 3, translate each point 5 units to the right. Now calculate $\bar{x}$ and compare the result with the result in Exercise 3.
   (b) Now translate each point h units and show that the center of mass will also be translated h units.

In Exercises 6–9, the masses m_i are located in the xy-plane at the points P_i. Find the center of mass.

6. $m_1 = 5$, $P_1 = (2, 2)$; $m_2 = 1$, $P_2 = (-3, 1)$; $m_3 = 3$, $P_3 = (1, -4)$

7. $m_1 = 10$, $P_1 = (1, -1)$; $m_2 = 2$, $P_2 = (5, 5)$; $m_3 = 5$, $P_3 = (-4, 0)$

8. $m_1 = 3$, $P_1 = (-2, -3)$; $m_2 = 4$, $P_2 = (-1, 0)$; $m_3 = 2$, $P_3 = (7, 1)$; $m_4 = 1$, $P_4 = (0, 0)$; $m_5 = 6$, $P_5 = (-3, 0)$

S 9. $m_1 = 4$, $P_1 = (2, 3)$; $m_2 = 2$, $P_2 = (-1, 5)$; $m_3 = 2.5$, $P_3 = (6, 8)$; $m_4 = 5$, $P_4 = (2, -2)$

In Exercises 10–11, introduce an appropriate rectangular coordinate system and find the coordinates of the center of mass. In each figure assume that the plate has uniform density.

10.

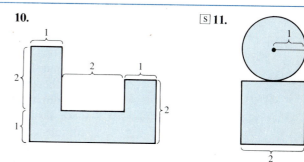

S 11.

12. Suppose that the circle in Exercise 11 has twice the density of the square. Now find the center of mass.

In Exercises 13–16, find M_x, M_y, and $(\bar{x}, \bar{y})$ for the laminas of uniform density ρ bounded by the given curves.

13. $y = \sqrt{x}$, $y = 0$, $x = 4$

14. $y = x^2$, $y = 0$, $x = 4$

S 15. $x = 4 - y^2$, $x = 0$

16. $x = 2y - y^2$, $x = 0$

In Exercises 17–19, find the centroids of the regions bounded by the given curves.

17. $y = \sqrt{a^2 - x^2}$, $y = 0$

18. $y = x^3$, $y = x$, $0 \le x \le 1$

S 19. $y = x^2$, $y = x$

S 20. Find the centroid of the triangle with vertices $(-a, 0)$, $(a, 0)$, (b, c). Show that it is at the point of intersection of its medians. (Assume $-a < b < a$.)

7.8 Arc Length

Purpose
- To use integration to find the length of a curve over a finite interval.

Grant that a curve may be considered as the assemblage of an infinite number of infinitely small line segments. —Guillaume L'Hôpital (1661–1701)

In Chapter 1 we found that the length of the line segment joining the points (x_1, y_1) and (x_2, y_2) is given by

$$d = \sqrt{(x_2 - x_1)^2 + (y_2 - y_1)^2}$$

In this section we use this formula to develop a method for calculating the length of the graph of a function.

Although the notion of the length of a curve should seem reasonable to you, your experience in calculating such lengths is probably limited to line segments and arcs of circles. In practice, calculating the lengths of curves other than circles can be quite difficult. In fact, it may be impossible, since the "length" of a continuous curve between two points on that curve might be infinite.

When a curve has a finite length between two points on that curve, we say that the curve is **rectifiable** between those two points. You might think that the graph of a function which is continuous on a closed interval of its domain would be rectifiable on that interval. Surprisingly, this is not necessarily true. However, if a function possesses a derivative that is continuous on a closed interval of the domain, then its graph is rectifiable on that interval.

Assume that $y = f(x)$ is a function whose first derivative is continuous over the interval $[a, b]$, and let s denote the length of its graph on this interval (Figure 7.68). We approximate the graph of f by n line segments, as in Figure 7.69, where Δ is the corresponding partition of $[a, b]$ such that

$$a = x_0 < x_1 < x_2 < \cdots < x_n = b$$

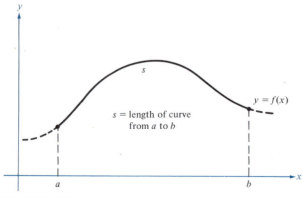

FIGURE 7.68

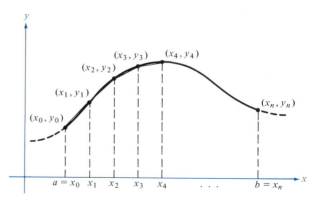

FIGURE 7.69

The total length of these line segments is

$$\sqrt{(x_1 - x_0)^2 + (y_1 - y_0)^2} + \sqrt{(x_2 - x_1)^2 + (y_2 - y_1)^2} + \cdots$$
$$+ \sqrt{(x_n - x_{n-1})^2 + (y_n - y_{n-1})^2}$$

By letting $\Delta x_i = x_i - x_{i-1}$ and $\Delta y_i = y_i - y_{i-1}$, we approximate the total length of the graph of f between a and b by

$$s \approx \sum_{i=1}^{n} \sqrt{(\Delta x_i)^2 + (\Delta y_i)^2}$$

By taking the limit of the right-hand side as $\|\Delta\| \to 0$ ($n \to \infty$), we have the length of the graph of f between a and b to be

$$s = \lim_{n \to \infty} \sum_{i=1}^{n} \sqrt{(\Delta x_i)^2 + (\Delta y_i)^2} = \lim_{n \to \infty} \sum_{i=1}^{n} \sqrt{1 + \left(\frac{\Delta y_i}{\Delta x_i}\right)^2} (\Delta x_i)$$

Since f' is continuous on each $[x_{i-1}, x_i]$, we can apply the Mean Value Theorem to f. Thus there exists c_i in $[x_{i-1}, x_i]$ such that

$$f(x_i) - f(x_{i-1}) = f'(c_i)(x_i - x_{i-1})$$

or, equivalently,

$$\frac{\Delta y_i}{\Delta x_i} = f'(c_i)$$

Thus we write

$$s = \lim_{n \to \infty} \sum_{i=1}^{n} \sqrt{1 + [f'(c_i)]^2} (\Delta x_i)$$

or

$$s = \int_a^b \sqrt{1 + [f'(x)]^2} \, dx$$

We call the value s the **arc length** of f between a and b.

Arc Length

If $y = f(x)$ has a continuous derivative f' on the interval $[a, b]$, then the **arc length** of f between a and b is given by

$$s = \int_a^b \sqrt{1 + [f'(x)]^2} \, dx$$

Similarly, for a curve $x = g(y)$, the arc length of g between c and d is given by

$$s = \int_c^d \sqrt{1 + [g'(y)]^2} \, dy$$

The definite integrals obtained for the arc length of a function are often very difficult to evaluate. For this reason in this section we present only a few simple examples and return to arc length problems in succeeding chapters as our integration skills improve.

Example 1

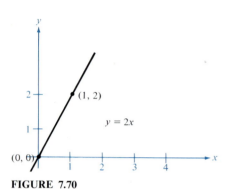

FIGURE 7.70

Find the arc length of the line $y = 2x$ over the interval $[0, 1]$. Show that the answer agrees with the formula for the distance between two points. (See Figure 7.70.)

Solution: Since $y = 2x$, then $dy/dx = 2$ and the arc length is

$$s = \int_0^1 \sqrt{1 + (2)^2}\, dx = \int_0^1 \sqrt{5}\, dx = \sqrt{5}\, x\Big]_0^1 = \sqrt{5}$$

Using the Distance Formula to calculate the distance between $(0, 0)$ and $(1, 2)$, we have

$$\sqrt{(1 - 0)^2 + (2 - 0)^2} = \sqrt{5} \qquad \blacksquare$$

Example 2

Find the arc length of the graph of $y = (x^3/6) + (1/2x)$ over the interval $[\frac{1}{2}, 2]$. (See Figure 7.71.)

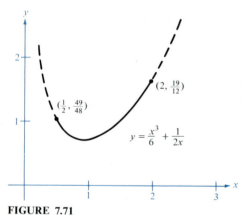

FIGURE 7.71

Solution: Since

$$y = \frac{x^3}{6} + \frac{1}{2x}$$

$$\frac{dy}{dx} = \frac{3x^2}{6} - \frac{1}{2x^2} = \frac{1}{2}\left(x^2 - \frac{1}{x^2}\right)$$

Thus the arc length is

$$s = \int_{1/2}^2 \sqrt{1 + \left[\frac{1}{2}\left(x^2 - \frac{1}{x^2}\right)\right]^2}\, dx = \int_{1/2}^2 \sqrt{1 + \frac{1}{4}\left(x^4 - 2 + \frac{1}{x^4}\right)}\, dx$$

$$= \int_{1/2}^2 \sqrt{\frac{1}{4}\left(x^4 + 2 + \frac{1}{x^4}\right)}\, dx = \int_{1/2}^2 \frac{1}{2}\left(x^2 + \frac{1}{x^2}\right) dx$$

$$= \frac{1}{2}\left[\frac{x^3}{3} - \frac{1}{x}\right]_{1/2}^2 = \frac{1}{2}\left(\frac{13}{6} + \frac{47}{24}\right) = \frac{99}{48} = \frac{33}{16} \qquad \blacksquare$$

Example 3

Find the length of the graph of $(y - 1)^3 = x^2$ over the interval $[0, 8]$. (See Figure 7.72.)

Solution: Since $(y - 1)^3 = x^2$, we can solve for either y or x. Solving for x we have

$$x = \pm(y - 1)^{3/2}$$

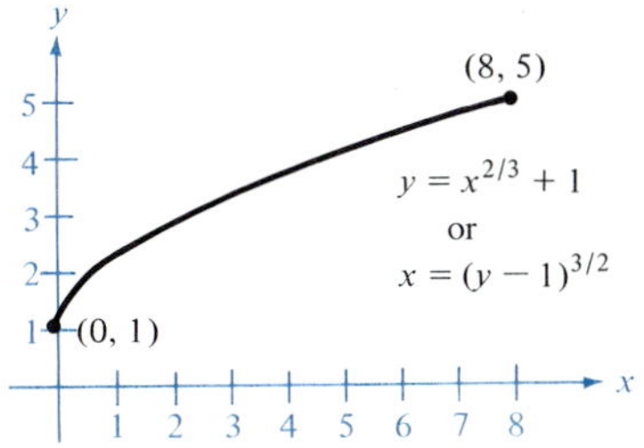

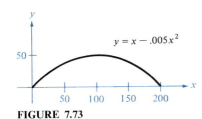

FIGURE 7.72

By choosing positive values for x, we have

$$\frac{dx}{dy} = \frac{3}{2}(y - 1)^{1/2}$$

Therefore, the arc length is

$$s = \int_1^5 \sqrt{1 + \left[\frac{3}{2}(y - 1)^{1/2}\right]^2}\, dy = \int_1^5 \sqrt{\frac{9}{4}y - \frac{5}{4}}\, dy$$

$$= \frac{1}{2} \int_1^5 \sqrt{9y - 5}\, dy = \frac{1}{18}\left[\frac{(9y - 5)^{3/2}}{\frac{3}{2}}\right]_1^5$$

$$= \frac{1}{27}(40^{3/2} - 4^{3/2}) = \frac{8}{27}(10^{3/2} - 1) \approx 9.0734 \qquad \blacksquare$$

Example 4

Consider the problem of the ball discussed in Section 3.2, where the path of the ball was given by $y = x - 0.005x^2$, with x and y measured in feet. *Set up* the integral that represents the total distance traveled by the ball.

Solution: Refer to Figure 7.73. Since

$$\frac{dy}{dx} = 1 - 0.01x = 1 - \frac{x}{100}$$

the arc length is given by the integral

$$s = \int_0^{200} \sqrt{1 + \left(1 - \frac{x}{100}\right)^2}\, dx = \int_0^{200} \sqrt{2 - \frac{x}{50} + \frac{x^2}{10,000}}\, dx$$

We have not yet discussed the techniques needed to evaluate this integral. Thus we delay its evaluation until Chapter 10, where we will show its value is 229.56 ft. $\qquad \blacksquare$

FIGURE 7.73

Section Exercises (7.8)

In Exercises 1–5, find the arc length of the curves.

1. $y = \frac{2}{3}x^{3/2} + 1$ between $x = 0$ and $x = 1$

2. $y = x^{3/2} - 1$ between $x = 0$ and $x = 4$

[S] **3.** $y = \frac{x^4}{8} + \frac{1}{4x^2}$ between $x = 1$ and $x = 2$

[S] **4.** $y = \frac{3}{2}x^{2/3}$ between $x = 1$ and $x = 8$

5. $y = \frac{x^5}{10} + \frac{1}{6x^3}$ between $x = 1$ and $x = 2$

In Exercises 6–10, find a definite integral that represents the arc length of the curves. (Do not evaluate the integral.)

6. $y = x^2$, $x = 0$, $x = 1$

[S] **7.** $y = \frac{1}{x}$, $x = 1$, $x = 3$

8. $y = \frac{1}{x + 1}$, $x = 0$, $x = 1$

9. $y = x^2 + x - 2$, $x = -2$, $x = 1$

10. $y = \frac{x}{2x - 1}$, $x = -1$, $x = 0$

S 11. In a certain pursuit problem, a fleeing object leaves the origin and moves up the *y*-axis. At the same time a pursuer leaves the point $(1, 0)$ and moves always toward the fleeing object. If the pursuer's speed is twice that of the fleeing object, the equation of the path is

$$y = \tfrac{1}{3}[x^{3/2} - 3x^{1/2} + 2]$$

How far has the fleeing object traveled when it is caught? Show that the pursuer traveled twice as far.

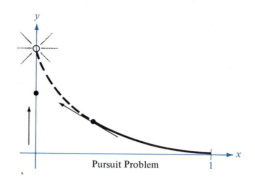

Pursuit Problem

7.9 **Surfaces of Revolution**

Purpose
- To use the definite integral to calculate areas of surfaces of revolution.

The surfaces of revolution are an important class of surfaces, characterized by the property that they can be generated by rotating a plane curve about an axis lying in the plane of the curve. We meet them in the course of everyday living in the guise of drinking glasses, bottles, etc.

—David Hilbert (1862–1943)

We have previously used the definite integral to calculate the volumes of solids of revolution (Sections 7.2 and 7.3). Our interest in this section lies in determining a method for calculating the area of a *surface of revolution*. Again, we will make use of the definite integral. We provide first a general definition of a surface of revolution.

Definition of Surface of Revolution	If the graph of a continuous function is revolved about a line, the resulting surface is called a **surface of revolution.**

In our development of a method for finding the area of a surface of revolution, we make use of the lateral surface area of the frustum of a right circular cone. Consider the line segment in Figure 7.74, where

L = length of line segment

r = distance from one end of line segment to axis of revolution

R = distance from other end of line segment to axis of revolution

When the line segment is revolved about its axis of revolution, it forms a frustum of a right circular cone, whose lateral surface area is

$$\text{lateral surface area} = \left[2\pi \left(\frac{R + r}{2}\right)\right] L = \pi(R + r)L$$

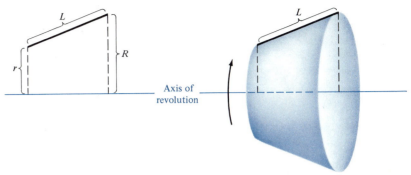

FIGURE 7.74

Suppose the graph of a function f, having a continuous derivative on the interval $[a, b]$, is revolved about the x-axis to form a surface of revolution as shown in Figure 7.75. Let Δ be a partition of $[a, b]$, with subintervals of width Δx_i. Then the line segment of length

$$\Delta L_i = \sqrt{\Delta x_i^2 + \Delta y_i^2}$$

generates a frustum of a cone, whose lateral surface area, ΔS_i, is given by

$$\Delta S_i = \pi(R_i + r_i)\,\Delta L_i = \pi[f(x_{i-1}) + f(x_i)]\,\Delta L_i$$
$$= \pi[f(x_{i-1}) + f(x_i)]\sqrt{\Delta x_i^2 + \Delta y_i^2}$$
$$= \pi[f(x_{i-1}) + f(x_i)]\sqrt{1 + \left(\frac{\Delta y_i}{\Delta x_i}\right)^2}\,\Delta x_i$$

Applying both the Mean Value Theorem and the Intermediate Value Theorem, we may conclude that there exist c_i and d_i in (x_{i-1}, x_i) such that

$$f'(c_i) = \frac{f(x_i) - f(x_{i-1})}{x_i - x_{i-1}} = \frac{\Delta y_i}{\Delta x_i}$$

and
$$f(d_i) = \frac{f(x_{i-1}) + f(x_i)}{2}$$

Therefore,
$$\Delta S_i = 2\pi f(d_i)\sqrt{1 + [f'(c_i)]^2}\,\Delta x_i$$

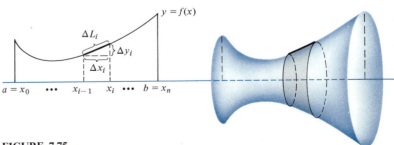

FIGURE 7.75

and the total surface area can be approximated by

$$S \approx 2\pi \sum_{i=1}^{n} f(d_i) \sqrt{1 + [f'(c_i)]^2} \, \Delta x_i$$

By taking the limit as $\|\Delta\| \to 0$ ($n \to \infty$), we have

$$S = 2\pi \lim_{n \to \infty} \left[\sum_{i=1}^{n} f(d_i) \sqrt{1 + [f'(c_i)]^2} \, \Delta x_i \right]$$

$$= 2\pi \int_{a}^{b} f(x) \sqrt{1 + [f'(x)]^2} \, dx$$

In a similar manner, it follows that if the graph of f is revolved about the y-axis, then S is given by

$$S = 2\pi \int_{a}^{b} x \sqrt{1 + [f'(x)]^2} \, dx$$

In both formulas for S, we can regard the products $2\pi f(x)$ and $2\pi x$ as the circumferences of the circles traced by a point (x, y) on the graph of f as it is revolved about the respective x- and y-axes.

Area of Surface of Revolution

If $y = f(x)$ has a continuous derivative on the interval $[a, b]$, then the area S of the surface of revolution formed by revolving the graph of f on $[a, b]$ is:

1. about the x-axis

$$S = 2\pi \int_{a}^{b} f(x) \sqrt{1 + [f'(x)]^2} \, dx$$

2. about the y-axis

$$S = 2\pi \int_{a}^{b} x \sqrt{1 + [f'(x)]^2} \, dx$$

Example 1

Find the area of the surface formed by revolving the graph of $f(x) = x^3$ on the interval $[0, 1]$ about the x-axis. (See Figure 7.76.)

Solution: Since $f'(x) = 3x^2$ and the x-axis is the axis of revolution, the surface area is

$$S = 2\pi \int_{0}^{1} f(x) \sqrt{1 + [f'(x)]^2} \, dx = 2\pi \int_{0}^{1} x^3 \sqrt{1 + (3x^2)^2} \, dx$$

$$= 2\pi \int_{0}^{1} x^3 \sqrt{1 + 9x^4} \, dx = \frac{2\pi}{36} \int_{0}^{1} (36x^3)(1 + 9x^4)^{1/2} \, dx$$

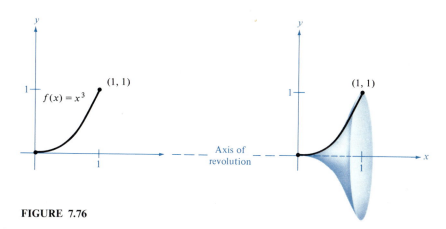

FIGURE 7.76

$$= \frac{\pi}{18} \left[\frac{(1 + 9x^4)^{3/2}}{\frac{3}{2}} \right]_0^1 = \frac{\pi}{27}(10^{3/2} - 1) \approx 3.563$$

Example 2

Use the integral for surface area to verify the equation $S = \pi r \sqrt{r^2 + h^2}$, where S is the lateral surface area of a right circular cone of height h and radius r (Figure 7.77).

Solution: If we revolve the line segment given by $y = (-r/h)x + r$ on the interval $[h, 0]$ about the x-axis, we obtain a cone of radius r and height h. The surface area is given by

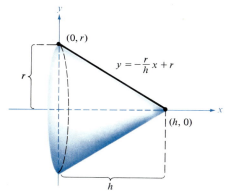

FIGURE 7.77

$$S = 2\pi \int_0^h y\sqrt{1 + (y')^2}\, dx = 2\pi \int_0^h \left(\frac{-rx}{h} + r\right)\sqrt{1 + \left(\frac{-r}{h}\right)^2}\, dx$$

$$= 2\pi \int_0^h r\left(1 - \frac{x}{h}\right)\sqrt{\frac{h^2 + r^2}{h^2}}\, dx = 2\pi \int_0^h \left(\frac{r\sqrt{h^2 + r^2}}{h}\right)\left(1 - \frac{x}{h}\right) dx$$

$$= 2\pi \left(\frac{r\sqrt{h^2 + r^2}}{h}\right)\left[x - \frac{x^2}{2h}\right]_0^h = 2\pi \left(\frac{r\sqrt{h^2 + r^2}}{h}\right)\left(h - \frac{h}{2}\right)$$

$$= \pi r \sqrt{h^2 + r^2}$$

Example 3

Find the area of the surface formed by revolving the graph of $y = x^2$ on the interval $[0, \sqrt{2}]$ about the y-axis. (See Figure 7.78.)

Solution: Since $y' = 2x$ and the y-axis is the axis of revolution, the surface area is

$$S = 2\pi \int_a^b x\sqrt{1 + (y')^2}\, dx = 2\pi \int_0^{\sqrt{2}} x\sqrt{1 + (2x)^2}\, dx$$

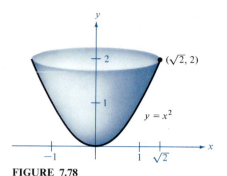

$$= \frac{2\pi}{8} \int_0^{\sqrt{2}} (1 + 4x^2)^{1/2}(8x)\, dx = \frac{2\pi}{8}\left[\frac{(1 + 4x^2)^{3/2}}{\frac{3}{2}}\right]_0^{\sqrt{2}}$$

$$= \frac{\pi}{6}[(1 + 8)^{3/2} - (1)^{3/2}] = \frac{13\pi}{3}$$

■

FIGURE 7.78

Section Exercises (7.9)

In Exercises 1–5, find the area of the indicated surfaces of revolution.

1. $y = 4 - x^2$, $0 \le x \le 2$, revolved about the y-axis
2. $y = x^3/3$, $0 \le x \le 3$, revolved about the x-axis.

$\boxed{\text{S}}$ **3.** $y = \frac{x^3}{6} + \frac{1}{2x}$, $1 \le x \le 2$, revolved about the x-axis

4. $y = \sqrt{x}$, $1 \le x \le 4$, revolved about the x-axis

$\boxed{\text{S}}$ **5.** $y = \sqrt[3]{x} + 2$, $1 \le x \le 8$, revolved about the y-axis.

$\boxed{\text{S}}$ **6.** Find the area of the zone of a sphere formed by revolving the graph of $y = \sqrt{9 - x^2}$, $0 \le x \le 2$, about the y-axis.

7. Find the area of the zone of a sphere formed by revolving the graph of $y = \sqrt{r^2 - x^2}$, $0 \le x \le a$, about the y-axis. Assume that $a < r$.

Miscellaneous Exercises (Ch. 7)

In Exercises 1–14, sketch and find the area of the region bounded by the given equations.

1. $y = \frac{1}{x^2}$, $y = 0$, $x = 1$, $x = 5$

2. $y = \frac{1}{x^2}$, $y = 4$, $x = 5$

3. $y = x$, $y = 2 - x$, $y = 0$

4. $y = 1 - \frac{x}{2}$, $y = x - 2$, $y = 1$

$\boxed{\text{S}}$ **5.** $x = y^2 - 2y$, $x = 0$
6. $x = y^2 - 2y$, $x = -1$, $y = 0$
7. $y = x$, $y = x^3$
8. $x = y^2 + 1$, $x = y + 3$
9. $y = x^2 - 4x + 3$, $y = 3 + 4x - x^2$
$\boxed{\text{C}}\boxed{\text{S}}$ **10.** $y = x^2 - 4x + 3$, $y = x^3$, $x = 0$
11. $y = \sqrt{x - 1}$, $y = 2$, $y = 0$, $x = 0$

12. $y = \sqrt{x - 1}$, $y = \frac{x - 1}{2}$

13. $\sqrt{x} + \sqrt{y} = 1$, $y = 0$, $x = 0$
14. $y = x^4 - 2x^2$, $y = 2x^2$

15. Find the volume of the solid generated by revolving the region bounded by $y = x$, $y = 0$, and $x = 4$ about:
(a) the x-axis   (b) the y-axis
(c) the line $x = 4$ (d) the line $x = 6$

16. Find the volume of the solid generated by revolving the region bounded by $y = \sqrt{x}$, $y = 2$, and $x = 0$ about:
(a) the x-axis   (b) the line $y = 2$
(c) the y-axis   (d) the line $x = -1$

17. A swimming pool is 5 ft deep at one end and 10 ft at the other, and the bottom is an inclined plane. The length and width of the pool are 40 ft and 20 ft, respectively. If the pool is full of water, what is the total force on each of its vertical walls?

$\boxed{\text{S}}$ **18.** Find the volume of the solid generated by revolving the ellipse $(x^2/a^2) + (y^2/b^2) = 1$ about:
(a) its minor axis (oblate spheroid)
(b) its major axis (prolate spheroid)

19. Find the volume of the solid generated by revolving the ellipse $(x^2/16) + (y^2/9) = 1$ about:
(a) its minor axis   (b) its major axis

$\boxed{\text{C}}\boxed{\text{S}}$ **20.** A gas tank is an oblate spheroid generated by revolving the region bounded by $(x^2/16) + (y^2/9) = 1$ about its minor axis (units are in feet).

(a) How many gallons (gal) are contained in the tank? (1 ft^3 contains 7.481 gal.)

(b) Find the depth of gasoline in the tank when it is filled to $\frac{1}{4}$ of its total capacity.

21. Find the work done in stretching a spring from its natural length of 10 in. to a length of 15 in. if a force of 4 lb is needed to stretch it 1 in. from its natural position.

22. Find the amount of work done in stretching a spring from its natural length of 9 in. to double that length if a force of 50 lb is needed to hold the spring at double its natural length.

[S] **23.** A water well has an 8-in. casing (diameter) and is 175 ft deep. If the water is 25 ft from the top of the well, determine the amount of work done in pumping it dry, assuming that no water enters the well while it is being pumped.

[C] **24.** Rework Exercise 23, assuming that water enters the well at the rate of 4 gal/min and the pump works at a rate of 12 gal/min. How many gallons are pumped in this case?

25. A chain 10 ft long and weighing 5 lb/ft is suspended from a platform 20 ft above the ground. How much work is required to raise the entire chain to the 20-ft level?

26. A windlass, situated 200 ft above ground level on the top of a building, uses cable weighing 4 lb/ft. Find the amount of work done in winding up the cable if its end is at ground level.

27. Rework Exercise 26, assuming that a 300-lb load is attached to the end of the cable.

28. Find the length of the graph of $y = (x^3/6) + (1/2x)$ from $x = 1$ to $x = 3$.

[S] **29.** Find the lateral surface area of a right circular cone of height h and radius r.

30. Find the centroid of the region bounded by $y = x^2$ and $y = 2x + 3$.

31. Find the centroid of the region bounded by $\sqrt{x} + \sqrt{y} = \sqrt{a}$, $x = 0$, and $y = 0$.

32. Find the centroid of the region bounded by $y = x^{2/3}$ and $y = \frac{1}{2}x$.

[S] **33.** Show that the force against any vertical region in a liquid is the product of the density of the liquid, the area of the region, and the depth of the centroid of the region.

34. Using the result of Exercise 33, find the total force on one side of a vertical circular plate of radius 4 ft that is submerged in water so that its center is 5 ft below the surface. How much must the water level be raised so that the total force on one side of the plate is doubled?

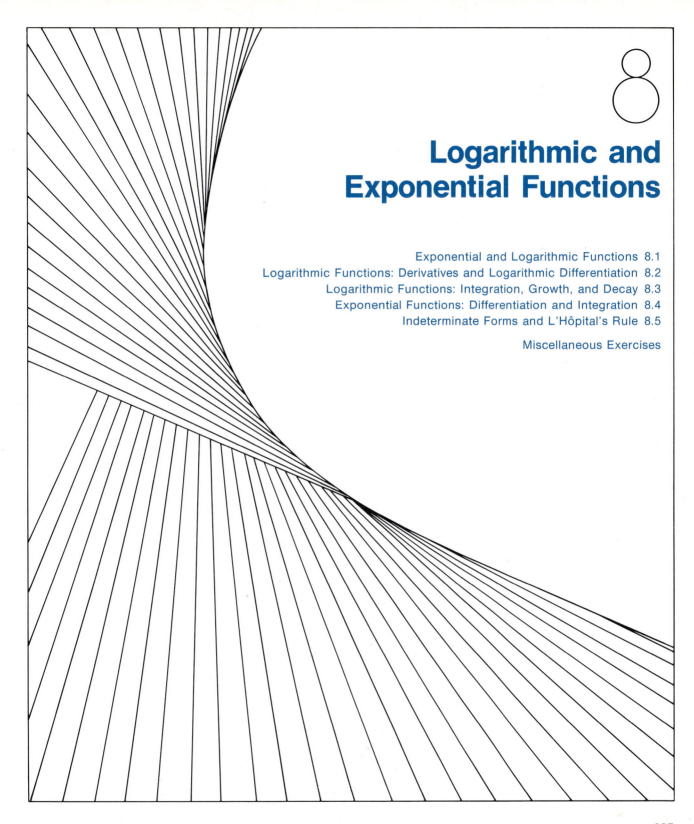

8

Logarithmic and Exponential Functions

8.1 Exponential and Logarithmic Functions

Purpose

- To review the properties of exponents and logarithms.
- To emphasize the inverse relationship between exponential and logarithmic functions.
- To introduce the number e as the base for the natural logarithmic function.

The simplest idea which we can form of the theory of logarithms, as they are found in the ordinary tables, is that of conceiving all numbers as powers of 10; the exponents of these powers then, will be the logarithms of the numbers. —Joseph Lagrange (1736–1813)

We are quite familiar with the behavior of functions like

$$f(x) = x^2, \qquad g(x) = \sqrt{x}, \qquad h(x) = x^{-1}$$

which involve a variable raised to a constant power. By interchanging roles and raising a constant to a variable power, we obtain an important class of functions called **exponential functions.** Some simple examples are

$$f(x) = 2^x, \qquad g(x) = \left(\frac{1}{10}\right)^x = \frac{1}{10^x}, \qquad h(x) = 3^{2x} = 9^x$$

In general, we can use any positive base $a \neq 1$ for exponential functions.

Definition of Exponential Function	If $a > 0$ and $a \neq 1$, then we call the function $y = a^x$ the **exponential function** with base a.

Before discussing the behavior of exponential functions, we list some familiar properties of exponents.

Properties of Exponents $(a, b > 0)$	1. $a^0 = 1$	2. $a^x a^y = a^{x+y}$
	3. $\dfrac{a^x}{a^y} = a^{x-y}$	4. $(a^x)^y = a^{xy}$
	5. $(ab)^x = a^x b^x$	6. $\left(\dfrac{a}{b}\right)^x = \dfrac{a^x}{b^x}$

Although we have generally used these six properties with integral and rational values for x and y, it is important to realize that the properties hold for *any* real values for x and y. With a calculator we can readily obtain approximate values for a^x when x is nonintegral or irrational.

Example 1

Sketch the graphs of the exponential functions

$$f(x) = 2^x, \qquad g(x) = (\tfrac{1}{2})^x = 2^{-x}, \qquad h(x) = 3^x$$

Solution: Table 8.1 shows some values for these functions, and their graphs are shown in Figure 8.1.

TABLE 8.1

x	-3	-2	-1	0	1	2	3	4
2^x	$\frac{1}{8}$	$\frac{1}{4}$	$\frac{1}{2}$	1	2	4	8	16
2^{-x}	8	4	2	1	$\frac{1}{2}$	$\frac{1}{4}$	$\frac{1}{8}$	$\frac{1}{16}$
3^x	$\frac{1}{27}$	$\frac{1}{9}$	$\frac{1}{3}$	1	3	9	27	81

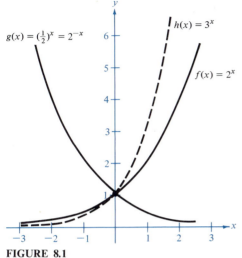

$g(x) = (\frac{1}{2})^x = 2^{-x}$

$h(x) = 3^x$

$f(x) = 2^x$

FIGURE 8.1

The graphs in Figure 8.1 are typical of the behavior of the exponential functions a^x and a^{-x} $(a > 1)$. We summarize these characteristics in Figure 8.2.

We have introduced the exponential function by using an unspecified base a. In calculus it turns out that the natural (or convenient) choice for a base is the irrational number e, whose decimal approximation is

$$e \approx 2.71828\ldots$$

This choice may seem anything but natural; however, the convenience of this particular choice will become apparent as we develop the rules for differentiating and integrating exponential and logarithmic functions. In the limit development of these differentiation rules, we encounter the limit used in the following definition of e.

Definition of e

$$e = \lim_{x \to 0} (1 + x)^{1/x}$$

(To twelve significant digits, $e \approx 2.71828182846$.)

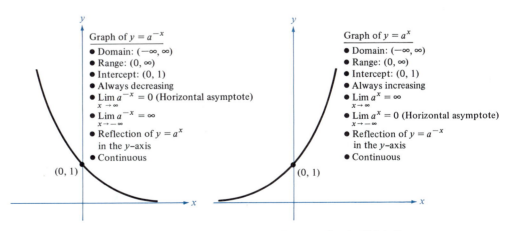

Graph of $y = a^{-x}$
- Domain: $(-\infty, \infty)$
- Range: $(0, \infty)$
- Intercept: $(0, 1)$
- Always decreasing
- $\lim_{x \to \infty} a^{-x} = 0$ (Horizontal asymptote)
- $\lim_{x \to -\infty} a^{-x} = \infty$
- Reflection of $y = a^x$ in the y–axis
- Continuous

$(0, 1)$

Graph of $y = a^x$
- Domain: $(-\infty, \infty)$
- Range: $(0, \infty)$
- Intercept: $(0, 1)$
- Always increasing
- $\lim_{x \to \infty} a^x = \infty$
- $\lim_{x \to -\infty} a^x = 0$ (Horizontal asymptote)
- Reflection of $y = a^{-x}$ in the y–axis
- Continuous

$(0, 1)$

Characteristics of the Exponential Functions a^x and a^{-x} $(a > 1)$

FIGURE 8.2

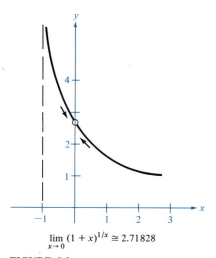

$$\lim_{x \to 0} (1 + x)^{1/x} \cong 2.71828$$

FIGURE 8.3

TABLE 8.2

x	$(1 + x)^{1/x}$
-0.5	4.0000
-0.1	2.8680
-0.01	2.7320
-0.001	2.7196
-0.0001	2.7184
$\downarrow$	$\downarrow$
0	$e \approx 2.71828$

x	$(1 + x)^{1/x}$
1.0	2.0000
0.5	2.2500
0.1	2.5937
0.01	2.7048
0.001	2.7169
0.0001	2.7181
$\downarrow$	$\downarrow$
0	$e \approx 2.71828$

Table 8.2 lists some values of the function $f(x) = (1 + x)^{1/x}$ for x near zero, and Figure 8.3 shows how the graph of f approaches e as x approaches zero.

Example 2

Sketch the graph of $f(x) = e^x$.

Solution: Figure 8.4 shows the graph of $f(x) = e^x$ determined from the values in Table 8.3.

TABLE 8.3

x	-2	-1	0	1	2
e^x	$\dfrac{1}{e^2} \approx 0.135$	$\dfrac{1}{e} \approx 0.368$	1	$e \approx 2.718$	$e^2 \approx 7.389$

■

One of the theorems of advanced calculus states that if a continuous function is always increasing (or always decreasing), then it possesses an inverse. From the properties of the exponential functions identified in Figure 8.2, we see that the function $f(x) = a^x$ has the necessary characteristics for possessing an inverse. We call this inverse of the exponential function the **logarithmic function,** and we define it as follows:

Definition of $\log_a x$

If $a > 0$ and $a \neq 1$, then

$$\log_a x = b \quad \text{if and only if} \quad a^b = x$$

($\log_a x = b$ is read "the log of x, to the base a, is b.")

This definition suggests that logarithmic equations can be written in an equivalent exponential form and vice versa. For example,

	Logarithmic Form	Exponential Form
	$\log_2 8 = 3$	$2^3 = 8$
	$\log_a 1 = 0$	$a^0 = 1$
	$\log_{10} 0.1 = -1$	$10^{-1} = 0.1$
	$\log_{10} 1000 = 3$	$10^3 = 1000$

This comparison clearly shows that *a logarithm is an exponent,* a fact you should keep constantly in mind when working with logarithms!

When e is used as the base for a logarithmic function f, we call f the **natural logarithmic function,** and we use the notation given in the following definition.

Definition of Natural Logarithm	The logarithmic function whose base is e is called the **natural logarithmic function** and is denoted by $$\log_e x = \ln x$$

Since the functions $f(x) = \log_a x$ and $g(x) = a^x = f^{-1}(x)$ are defined to be inverses of each other, their graphs should be reflections of each other in the line $y = x$. (See Section 1.6.) This reflective property is illustrated in the next example.

Example 3

Sketch the graphs of $f(x) = \log_2 x$ and $h(x) = \ln x$.

Solution: By definition the logarithmic function $f(x) = \log_2 x$ is the inverse of the exponential function $g(x) = 2^x$, whose range is the set of

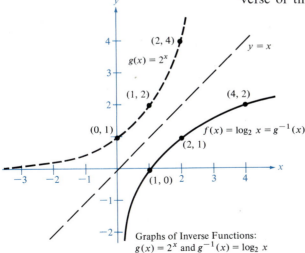

Graphs of Inverse Functions:
$g(x) = 2^x$ and $g^{-1}(x) = \log_2 x$

FIGURE 8.5

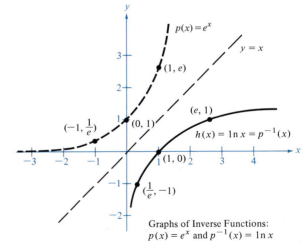

Graphs of Inverse Functions:
$p(x) = e^x$ and $p^{-1}(x) = \ln x$

FIGURE 8.6

TABLE 8.4

x	$f(x) = \log_2 x$
$\frac{1}{4}$	-2
$\frac{1}{2}$	-1
1	0
2	1
4	2

positive real numbers (see Figure 8.1). Therefore, from Section 1.6 we know that the domain of f must lie within the set of positive real numbers and that the graph of f is the reflection, in the line $y = x$, of the graph of g. This reflection is shown in Figure 8.5 (p. 329). Plotting the points listed in Table 8.4 verifies this reflection to be the graph of $f(x) = \log_2 x$. Similarly, $h(x) = \ln x$ is the inverse of $p(x) = e^x$. (See Figure 8.6, p. 329.) ∎

In Figure 8.7 we summarize the basic characteristics of the logarithmic function $y = \log_a x$ $(a > 1)$.

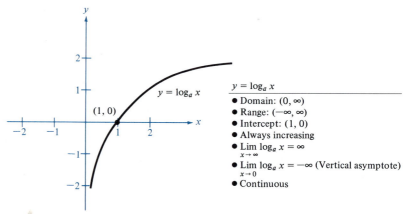

$y = \log_a x$
- Domain: $(0, \infty)$
- Range: $(-\infty, \infty)$
- Intercept: $(1, 0)$
- Always increasing
- $\lim\limits_{x \to \infty} \log_a x = \infty$
- $\lim\limits_{x \to 0} \log_a x = -\infty$ (Vertical asymptote)
- Continuous

FIGURE 8.7

Recall from Section 1.6 that inverse functions possess the property that

$$f(f^{-1}(x)) = x \qquad \text{and} \qquad f^{-1}(f(x)) = x$$

From this we conclude that for $f(x) = \log_a x$ and $f^{-1}(x) = a^x$, we have

$$\log_a (a^x) = f(a^x) = f(f^{-1}(x)) = x$$

and

$$a^{\log_a x} = f^{-1}(f(x)) = x$$

Inverse Properties of Exponential and Logarithmic Functions

If $a > 0$ and $a \neq 1$, then

$$\log_a (a^u) = u \qquad \text{and} \qquad a^{\log_a u} = u$$

If $a = e$, then

$$\ln (e^u) = u \qquad \text{and} \qquad e^{\ln u} = u$$

Example 4

Simplify the following expressions:
(a) $\log_2 (2^{x^2})$ (b) $\ln (e^{\sqrt{2}})$
(c) $\log_a a$ (d) $e^{\ln 3x}$
(e) $10^{\log_{10} 2}$ (f) $b^{\log_b \sqrt{x}}$

Solution: Applying the property $\log_a (a^u) = u$, we have:

(a) $\log_2 (2^{x^2}) = x^2$

(b) $\ln (e^{\sqrt{2}}) = \sqrt{2}$

(c) $\log_a (a^1) = 1$

Applying the property $a^{\log_a u} = u$, we have:

(d) $e^{\ln 3x} = 3x$

(e) $10^{\log_{10} 2} = 2$

(f) $b^{\log_b \sqrt{x}} = \sqrt{x}$ ∎

Since a logarithm is an exponent, we would expect the properties of logarithms to correspond closely to those of exponents. For instance, when multiplying, we add exponents:

$$a^n a^m = a^{m+n}$$

Correspondingly, "the log of the product of two numbers is equal to the sum of the logs of the numbers." That is,

$$\log_a (xy) = \log_a x + \log_a y$$

To prove this we make use of the inverse property of logarithms and exponents, as follows:

$$a^{\log_a (xy)} = xy = a^{\log_a x} a^{\log_a y} = a^{(\log_a x + \log_a y)}$$

Now by equating exponents we conclude that

$$\log_a (xy) = \log_a x + \log_a y$$

Properties of Logarithms
$(a, b > 1)$

1. $\log_a 1 = 0$

2. $\log_a a = 1$

3. $\log_a (uv) = \log_a u + \log_a v$

4. $\log_a \left(\dfrac{u}{v}\right) = \log_a u - \log_a v$

5. $\log_a (u^n) = n \log_a u$

6. $\log_a u = \dfrac{\log_b u}{\log_b a}$

7. $\log_a b = \dfrac{1}{\log_b a}$

(These same properties hold if $\log_a u$ is replaced by $\log_e u = \ln u$.)

Proof: We have already established Properties 2 and 3. Properties 1, 4, and 5 can be proved in a similar manner and are left as exercises. To prove Property 6 consider the following argument. Since

$$a^{\log_a u} = u$$

then

$$\log_b a^{\log_a u} = \log_b u$$

Using Property 5 we have

$$(\log_a u)(\log_b a) = \log_b u$$

$$\log_a u = \frac{\log_b u}{\log_b a}$$

To prove Property 7 we use Properties 6 and 2 and write

$$\log_a b = \frac{\log_b b}{\log_b a} = \frac{1}{\log_b a}$$ ■

Example 5 *(Expanding Logarithmic Expressions)*

Use the properties of logarithms to write each of the following expressions as a sum, difference, or multiple of logarithms:

(a) $\log_5 \frac{10}{9}$ (b) $\ln \sqrt{3x + 2}$

(c) $\log_2 \dfrac{xy}{5}$ (d) $\ln \dfrac{(x - 3)^2}{x \sqrt[3]{x - 1}}$

Solution:

(a)   By Property 4,

$$\log_5 \tfrac{10}{9} = \log_5 10 - \log_5 9$$

(b)   By Property 5,

$$\ln \sqrt{3x + 2} = \ln (3x + 2)^{1/2} = \tfrac{1}{2} \ln (3x + 2)$$

(c)   By Properties 3 and 4,

$$\log_2 \frac{xy}{5} = \log_2 (xy) - \log_2 5 = \log_2 x + \log_2 y - \log_2 5$$

(d)   By Properties 3, 4, and 5,

$$\ln \frac{(x - 3)^2}{x \sqrt[3]{x - 1}} = \ln (x - 3)^2 - \ln (x \sqrt[3]{x - 1})$$
$$= 2 \ln (x - 3) - [\ln x + \ln (x - 1)^{1/3}]$$
$$= 2 \ln (x - 3) - \ln x - \ln (x - 1)^{1/3}$$
$$= 2 \ln (x - 3) - \ln x - \tfrac{1}{3} \ln (x - 1)$$ ■

Example 6 *(Condensing Logarithmic Expressions)*

Use the properties of logarithms to rewrite the following expressions as the logarithm of a single quantity:
(a) $\ln x + 2 \ln y$
(b) $\log_{10} (x + 1) - \tfrac{1}{2} \log_{10} x - \log_{10} (x^2 - 1)$

Solution:

(a) $\ln x + 2 \ln y = \ln x + \ln y^2 = \ln xy^2$

(b) $\log_{10}(x+1) - \frac{1}{2}\log_{10} x - \log_{10}(x^2-1)$

$$= \log_{10}(x+1) - \log_{10} x^{1/2} - \log_{10}(x^2-1)$$
$$= \log_{10}(x+1) - [\log_{10}\sqrt{x} + \log_{10}(x^2-1)]$$
$$= \log_{10}(x+1) - \log_{10}[\sqrt{x}(x^2-1)]$$
$$= \log_{10}\left[\frac{x+1}{\sqrt{x}(x^2-1)}\right] = \log_{10}\left[\frac{1}{\sqrt{x}(x-1)}\right] \quad \blacksquare$$

Example 7

Solve for x in each of the following equations:
(a) $\log_2 x - \log_2(x-8) = 3$
(b) $y = e^{2x-5}$

Solution:

(a)
$$\log_2 x - \log_2(x-8) = 3$$
$$\log_2\left(\frac{x}{x-8}\right) = 3$$

The equivalent exponential form is

$$\frac{x}{x-8} = 2^3 = 8$$

Thus
$$x = (x-8)(8)$$
$$-7x = -64$$
$$x = \tfrac{64}{7}$$

(b) By writing $y = e^{2x-5}$ in logarithmic form, we have

$$\ln y = 2x - 5$$
$$5 + \ln y = 2x$$
$$\frac{5 + \ln y}{2} = x \quad \blacksquare$$

To give you further insight into the usefulness of the natural number e, we conclude this section with a business problem in which the limit definition of e plays an important role.

Suppose P dollars are deposited in a savings account at an annual interest rate r. If accumulated interest is deposited into the account, what is the balance in the account at the end of one year? Of course, the answer depends on the number of times the interest is compounded, as indicated in Table 8.5. For instance, the results for a deposit of $1000 at 8% interest compounded n times a year are as shown in Table 8.6.

As n increases the balance A approaches the limit

$$\lim_{n\to\infty} P\left(1 + \frac{r}{n}\right)^n$$

TABLE 8.5

Number of times compounded	Balance after one year
1 (annually)	$A = P(1 + r)$
2 (semiannually)	$A = P\left(1 + \dfrac{r}{2}\right)^2$
4 (quarterly)	$A = P\left(1 + \dfrac{r}{4}\right)^4$
$\vdots$	$\vdots$
n	$A = P\left(1 + \dfrac{r}{n}\right)^n$

TABLE 8.6

n	Balance
1 (annually)	$1080.00
2 (semiannually)	1081.60
4 (quarterly)	1082.43
12 (monthly)	1083.00
365 (daily)	1083.27

We determine this limit in the following manner:

$$\lim_{n \to \infty} P\left(1 + \frac{r}{n}\right)^n = P \lim_{n \to \infty} \left[\left(1 + \frac{r}{n}\right)^{n/r}\right]^r$$

If we let $x = r/n$, then $x \to 0$ as $n \to \infty$. Thus we have

$$P\left[\lim_{n \to \infty}\left(1 + \frac{r}{n}\right)^{n/r}\right]^r = P\left[\lim_{x \to 0}(1 + x)^{1/x}\right]^r = P[e]^r$$

The equation

$$A = \lim_{n \to \infty} P\left(1 + \frac{r}{n}\right)^n = Pe^r$$

denotes the balance at the end of one year, during which time the interest is said to be compounded *continuously*. For a deposit of $1000 at 8% interest compounded continuously, the balance at the end of one year would be

$$A = 1000e^{0.08} \approx \$1083.29$$

In general, we have the following two formulas:

1. compounded n times per year,

$$A = P\left(1 + \frac{r}{n}\right)^{nt}$$

2. compounded continuously,

$$A = Pe^{rt}$$

where $P =$ amount of deposit

$r =$ interest rate

$n =$ number of times compounded per year

$t =$ number of years

$A =$ balance after t years

Example 8

If P dollars are deposited at 8% interest compounded continuously, how long will it take to double the original deposit?

Solution: To double the deposit, we consider

$$Pe^{0.08t} = 2P$$

Thus
$$e^{0.08t} = 2$$
$$0.08t = \ln 2$$
$$t = \left(\frac{1}{0.08}\right)\ln 2 \approx 8.66$$

Therefore, the balance will double by the end of 8 years and 8 months. ∎

The exercises in this section do not involve calculus; instead, they emphasize the basic properties of the logarithmic and exponential functions. A good working knowledge of these properties is *crucial* to your subsequent understanding of the calculus of the logarithmic and exponential functions. Therefore, make sure that you have mastered the techniques covered in this exercise set before proceeding with the remainder of the chapter.

Section Exercises (8.1)

In Exercises 1–12, write each logarithmic equation as an exponential equation and vice versa.

1. $2^3 = 8$
2. $3^{-1} = \frac{1}{3}$
3. $27^{2/3} = 9$
4. $16^{3/4} = 8$
5. $\log_{10} 0.01 = -2$
6. $\log_{0.5} 8 = -3$
7. $e^0 = 1$
8. $e^2 = 7.389\ldots$
9. $\ln 2 = .6931\ldots$
10. $\ln 8.4 = 2.128\ldots$
11. $\ln 0.5 = -.6931\ldots$
12. $49^{1/2} = 7$

In Exercises 13–30, find the value of the unknown.

13. $\log_{10} 1000 = x$
14. $\log_{10} 0.1 = x$
15. $\log_4 \frac{1}{64} = x$
16. $\log_5 25 = x$
17. $\log_3 x = -1$
18. $\log_2 x = -4$
19. $\log_b 27 = 3$
20. $\log_b 125 = 3$
21. $\log_{27} x = -\frac{2}{3}$
22. $\ln e^x = 3$
23. $e^{\ln x} = 4$
24. $\ln x = 2$
25. $\log_3 x + \log_3 (x - 2) = 1$
26. $\log_{10} (x + 3) - \log_{10} x = 1$
27. $x^2 - x = \log_5 25$
28. $3x + 5 = \log_2 64$
29. $x - 3 = \log_2 32$
30. $x - x^2 = \log_4 \frac{1}{16}$

In Exercises 31–40, sketch the graph of each equation.

31. $y = 3^x$
32. $y = 3^{x-1}$
33. $y = (\frac{1}{3})^x$
34. $y = 2^{x^2}$
35. $y = e^{-x^2}$
36. $y = \log_3 x$
37. $y = \log_3 (x - 5)$
38. $y = 1 + \log_2 x$
39. $y = \ln |x|$
40. $y = \ln 2x$

In Exercises 41–45, show that the given functions are inverses of each other by sketching their graphs on the same coordinate system.

41. $f(x) = 4^x$; $g(x) = \log_4 x$
42. $f(x) = 3^x$; $g(x) = \log_3 x$
43. $f(x) = e^{2x}$; $g(x) = \ln \sqrt{x}$
44. $f(x) = e^x - 1$; $g(x) = \ln (x + 1)$
45. $f(x) = e^{x-1}$; $g(x) = 1 + \ln x$

In Exercises 46–55, use the properties of logarithms to write each as a sum, difference, or multiple of logarithms.

46. $\log_5 \frac{2}{3}$
47. $\log_2 xyz$

48. $\ln \dfrac{xy}{z}$

49. $\ln \sqrt{a-1}$

50. $\log_2 \sqrt{2^3}$

51. $\log_2 \frac{1}{5}$

52. $\ln \left(\dfrac{x^2-1}{x^3}\right)^3$

53. $\ln 3e^2$ [S]

54. $\ln z(z-1)^2$

55. $\ln \dfrac{1}{e}$

In Exercises 56–60, express each as a single logarithm.

56. $\log_3(x-2) - \log_3(x+2)$

57. $3\ln x + 2\ln y - 4\ln z$

58. $\frac{1}{3}[2\ln(x+3) + \ln x - \ln(x^2-1)]$

59. $2[\ln x - \ln(x+1) - \ln(x-1)]$ [S]

60. $2\ln 3 - \frac{1}{2}\ln(x^2+1)$

[C] **61.** Express each of the following as a power of e. (For example, $2 = e^{\ln 2} \approx e^{0.6931}$.)
(a) 4
(b) π
(c) $\sqrt{5}$
(d) $\frac{1}{3}$
(e) $2e$

[C] **62.** Complete the accompanying table to demonstrate that e can also be defined as

$$\lim_{x\to\infty}\left(1+\frac{1}{x}\right)^x$$

x	1	10	10^2	10^4	10^6
$\left(1+\dfrac{1}{x}\right)^x$					

[S] **63.** Using properties of logarithms and the fact that $\ln 2 \approx 0.6931$ and $\ln 3 \approx 1.0986$, find the following:
(a) $\ln 6$
(b) $\ln \frac{2}{3}$
(c) $\ln 81$
(d) $\ln \sqrt{3}$
(e) $\ln 0.25$
(f) $\ln 24$

[C] **64.** Demonstrate Property 6 of logarithms by showing that $\log_{10} 2 = \ln 2/\ln 10$.

[C] **65.** Demonstrate Property 6 of logarithms by showing that $\ln 5 = \log 5/\log e$.

66. For how many real numbers is $5^x = x^2$?

[C] **67.** Compare $271{,}801/99{,}990$ to e.

[C] **68.** Compare $299/110$ to e.

[C] **69.** Compare $1 + 1 + \frac{1}{2} + \frac{1}{6} + \frac{1}{24} + \frac{1}{120} + \frac{1}{720} + \frac{1}{5040}$ to e.

[C] **70.** Find the amount of time necessary for P dollars to double if it is compounded continuously at $7\frac{1}{2}\%$ interest. Find the time necessary for it to triple.

[C][S] **71.** Complete the accompanying table to find the time t necessary for P dollars to double if compounded continuously at the rate r.

r	2%	4%	6%	8%	10%	12%
t						

[C] **72.** If \$1000 is invested at 5% interest, find the amount after 10 years if it is compounded:
(a) annually
(b) semiannually
(c) quarterly
(d) monthly
(e) daily
(f) continuously

[C] **73.** If \$1000 is invested at $7\frac{1}{2}\%$ interest, find the amount after 10 years if it is compounded:
(a) annually
(b) semiannually
(c) quarterly
(d) monthly
(e) daily
(f) continuously

74. Prove Properties 1, 4, and 5 of logarithms as given in this section.

75. In living organic material the ratio of radioactive carbon isotopes to the total number of carbon atoms is about 1 in 10^{12}. When organic material dies, its radioactive carbon isotopes decay with a half-life rate of 5700 years, and the ratio of radioactive carbon isotopes to carbon atoms is given by

$$R = 10^{-12}2^{-t/5700}$$

where t measures the number of years since the material died. If the ratio in a certain fossil is 10^{-33}, how long ago was the fossil living?

76. Use the continuity of $f(x) = a^x$ to prove the continuity of $g(x) = \log_a x$.

For Review and Class Discussion

True or False

1. _____ If $f(x) = 2^{x^2}$, then $f(3) = 64$.

2. _____ If $f(x) = 2x^3$, then $f(1) = 8$.

3. _____ $\log_a e = 1/\ln a$.

4. _____ $\ln(x+25) = \ln x + \ln 25$.

5. ____ The range of the natural logarithmic function is the set of all real numbers.

6. ____ $e = 271,801/99,990$.

7. ____ If $f(x) = \ln x$, then $f(e^{n+1}) - f(e^n) = 1$ for any value of n.

8. ____ The functions given by $f(x) = 2 + e^x$ and $g(x) = \ln (x - 2)$ are inverses of each other.

9. ____ $a^x = e^{x(\ln a)}$.

10. ____ $(\ln x)^{1/2} = \frac{1}{2}(\ln x)$.

8.2 Logarithmic Functions: Derivatives and Logarithmic Differentiation

Purpose
- To develop differentiation formulas for logarithmic functions.
- To apply logarithmic differentiation to functions that are not themselves logarithmic.

It is clear that our method also covers transcendental curves, those that cannot be reduced by algebraic computation, or have no particular degree. —Gottfried Leibniz (1646–1716)

Prior to this chapter we have been concerned exclusively with functions that can be generated by algebraic operations (addition, subtraction, multiplication, division, and raising to constant powers). We call such functions **algebraic.** For example, the functions given by

$$f(x) = 2x^3 - x^2 + x + 1, \qquad g(x) = \frac{2x + 3}{x^2 + 1},$$

$$h(x) = x + \sqrt{x - 1}$$

are algebraic. Functions that are not algebraic are called **transcendental.** The common types of algebraic and transcendental functions are the following:

Algebraic Functions	Transcendental Functions
polynomial functions	exponential functions
rational functions	logarithmic functions
functions involving	trigonometric functions
radicals	inverse trigonometric functions

In this chapter and the following chapter, we will discuss the basic transcendental functions, together with their derivatives and a few examples of their use. Let us begin by considering the derivative of the natural logarithmic function. This derivative will point out one of the peculiarities of transcendental functions. That is, even though the derivative of an algebraic function must be algebraic, the derivative of a transcendental function need not be transcendental.

THEOREM 8.1
(*Derivative of Natural Logarithmic Function*)

$$\frac{d}{dx}[\ln x] = \frac{1}{x}$$

Proof: Recall that the derivative of f is defined to be

$$f'(x) = \lim_{\Delta x \to 0} \frac{f(x + \Delta x) - f(x)}{\Delta x}$$

Applying this limit to the natural logarithmic function, we have

$$\frac{d}{dx}[\ln x] = \lim_{\Delta x \to 0} \frac{\ln (x + \Delta x) - \ln x}{\Delta x} = \lim_{\Delta x \to 0} \frac{\ln [(x + \Delta x)/x]}{\Delta x}$$

$$= \lim_{\Delta x \to 0} \left[\frac{1}{\Delta x} \ln \left(1 + \frac{\Delta x}{x} \right) \right] = \lim_{\Delta x \to 0} \left[\left(\frac{1}{x} \right) \left(\frac{x}{\Delta x} \right) \ln \left(1 + \frac{\Delta x}{x} \right) \right]$$

$$= \lim_{\Delta x \to 0} \left[\frac{1}{x} \ln \left(1 + \frac{\Delta x}{x} \right)^{x/\Delta x} \right]$$

$$= \left(\frac{1}{x} \right) \left[\lim_{\Delta x \to 0} \ln \left(1 + \frac{\Delta x}{x} \right)^{x/\Delta x} \right]$$

Recall that we defined the natural number e as the limit

$$e = \lim_{t \to 0} (1 + t)^{1/t}$$

To make use of this limit, we let $t = \Delta x/x$, then $\Delta x \to 0$ as $t \to 0$, and by the continuity of the natural logarithmic function, we have

$$\lim_{\Delta x \to 0} \left[\ln \left(1 + \frac{\Delta x}{x} \right)^{x/\Delta x} \right] = \lim_{t \to 0} [\ln (1 + t)^{1/t}]$$

$$= \ln \left[\lim_{t \to 0} (1 + t)^{1/t} \right] = \ln e = 1$$

Thus

$$\frac{d}{dx}[\ln x] = \frac{1}{x}$$

∎

At this point you might do well to pause and contemplate the rather startling nature of Theorem 8.1. So far in our development of the natural logarithmic function, it would have been difficult to predict its intimate relationship to the rational function $1/x$. Hidden relationships such as this not only illustrate the joy of mathematical discovery, they also give us logical alternatives in constructing a mathematical system. An alternative that Theorem 8.1 provides is that we could have developed the natural logarithmic function as the *antiderivative* of $1/x$, rather than as the inverse of e^x. If you are interested in pursuing this alternative development of ln x, we suggest that you consult other calculus texts in your school's library.

The next theorem follows directly from Theorem 8.1 and the Chain Rule.

THEOREM 8.2

If u is a differentiable function of x, then

$$\frac{d}{dx}[\ln u] = \frac{1}{u} \frac{du}{dx}$$

Example 1

Find the derivative of $f(x) = \ln(2x^2 + 4)$.

Solution: By letting $u = 2x^2 + 4$, we have

$$f'(x) = \frac{1}{u}\frac{du}{dx} = \left(\frac{1}{2x^2 + 4}\right)(4x) = \frac{2x}{x^2 + 2}$$

$\blacksquare$

Example 2

Differentiate $f(x) = \ln \sqrt{x + 1}$.

Solution:

$$f'(x) = \left(\frac{1}{\sqrt{x+1}}\right)\frac{d}{dx}[(x+1)^{1/2}] = \left(\frac{1}{\sqrt{x+1}}\right)\left(\frac{1}{2}\right)(x+1)^{-1/2}$$

$$= \frac{1}{2\sqrt{x+1}\sqrt{x+1}} = \frac{1}{2(x+1)}$$

$\blacksquare$

We know from our previous work with differentiation that *it is often helpful to rewrite a function before differentiating*. For instance, in Example 2 if we rewrite $\ln \sqrt{x+1}$ as $(\frac{1}{2}) \ln(x+1)$, we can simplify the differentiation process:

$$f'(x) = \frac{d}{dx}[\ln \sqrt{x+1}] = \frac{d}{dx}[\tfrac{1}{2}\ln(x+1)] = \frac{1}{2}\left(\frac{1}{x+1}\right)$$

Our next example shows an even more dramatic illustration of the benefit of rewriting a function before differentiating.

Example 3

Differentiate $\qquad f(x) = \ln\left[\dfrac{x(x^2 + 1)^2}{\sqrt{2x^3 - 1}}\right]$

Solution: Without rewriting the expression for $f(x)$, we have

$$f'(x) = \frac{1}{[x(x^2+1)^2/\sqrt{2x^3-1}]}\frac{d}{dx}\left[\frac{x(x^2+1)^2}{\sqrt{2x^3-1}}\right]$$

Of course, we like to avoid differentiating expressions like

$$\frac{x(x^2+1)^2}{\sqrt{2x^3-1}}$$

if at all possible. In this case we can do just that by using the properties of logarithms to write

$$f(x) = \ln\left[\frac{x(x^2+1)^2}{\sqrt{2x^3-1}}\right]$$

as

$$f(x) = (\ln x) + 2\ln(x^2+1) - \tfrac{1}{2}\ln(2x^3-1)$$

In this form our derivative is simply

$$f'(x) = \frac{1}{x} + \frac{4x}{x^2+1} - \frac{3x^2}{2x^3-1}$$

■

In Section 8.1 we promised to justify the selection of e as the natural choice for the base of our logarithmic functions. One reason for this seemingly unusual choice can be seen in the following theorem.

THEOREM 8.3

$$\frac{d}{dx}[\log_a x] = (\log_a e)\frac{1}{x}$$

Proof: Using logarithm Properties 5 and 6 in the previous section, we have

$$\log_a x = \frac{\log_e x}{\log_e a} = \left(\frac{1}{\log_e a}\right)(\log_e x) = (\log_a e)(\ln x)$$

Therefore,

$$\frac{d}{dx}[\log_a x] = \frac{d}{dx}[(\log_a e)(\ln x)] = (\log_a e)\frac{1}{x}$$

■

By Theorem 8.3 and the Chain Rule, we obtain the more general result

$$\frac{d}{dx}[\log_a u] = (\log_a e)\left(\frac{1}{u}\right)\frac{du}{dx}$$

The coefficient $\log_a e$ is a rather *cumbersome constant* to include as a factor each time we differentiate a logarithmic function. So the *natural* thing to do is to eliminate this constant by choosing e as our base. Thus if $a = e$, we have

$$\log_a e = \log_e e = 1$$

and Theorem 8.3 is reduced to Theorem 8.1.

We summarize the four rules for differentiating logarithmic functions.

Derivatives of Logarithmic Functions

$$\frac{d}{dx}[\ln x] = \frac{1}{x} \qquad \frac{d}{dx}[\log_a x] = (\log_a e)\frac{1}{x}$$

$$\frac{d}{dx}[\ln u] = \frac{u'}{u} \qquad \frac{d}{dx}[\log_a u] = (\log_a e)\frac{u'}{u}$$

Example 4

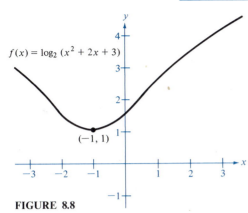

$f(x) = \log_2 (x^2 + 2x + 3)$

(−1, 1)

FIGURE 8.8

Find the minimum value of the function $f(x) = \log_2 (x^2 + 2x + 3)$.

Solution: Differentiating $f(x)$ we obtain

$$f'(x) = (\log_2 e)\left(\frac{1}{x^2 + 2x + 3}\right)(2x + 2)$$

Now $f'(x) = 0$ for $x = -1$, and by the First-Derivative Test, we determine that the point $(-1, f(-1))$ is a relative minimum point. Since f has no other relative extrema and since $\log_2 (x^2 + 2x + 3)$ increases as $x \to \pm\infty$, we conclude that the minimum value of $f(x)$ is

$$f(-1) = \log_2 (2) = 1$$

(See Figure 8.8.) ∎

Occasionally it is convenient to use logarithms as an aid in differentiating nonlogarithmic functions. We call this procedure **logarithmic differentiation,** and we illustrate its use in the next two examples.

Example 5 (*Logarithmic Differentiation*)

Find the derivative of

$$y = \frac{(x - 2)^2(x^3 - 4)}{\sqrt[3]{x^2 + 1}}$$

Solution: Taking the natural logarithm of both sides, we obtain

$$\ln y = \ln\left[\frac{(x - 2)^2(x^3 - 4)}{\sqrt[3]{x^2 + 1}}\right]$$

$$= 2\ln(x - 2) + \ln(x^3 - 4) - \tfrac{1}{3}\ln(x^2 + 1)$$

Differentiating both sides with respect to x, we have

$$\left(\frac{1}{y}\right)\frac{dy}{dx} = 2\left(\frac{1}{x - 2}\right) + \frac{3x^2}{x^3 - 4} - \frac{1}{3}\left(\frac{2x}{x^2 + 1}\right)$$

$$= \frac{2}{x - 2} + \frac{3x^2}{x^3 - 4} - \frac{2x}{3x^2 + 3}$$

Solving for dy/dx we have

$$\frac{dy}{dx} = y\left(\frac{2}{x - 2} + \frac{3x^2}{x^3 - 4} - \frac{2x}{3x^2 + 3}\right)$$

and substituting for y, we conclude that

$$\frac{dy}{dx} = \frac{(x - 2)^2(x^3 - 4)}{\sqrt[3]{x^2 + 1}}\left(\frac{2}{x - 2} + \frac{3x^2}{x^3 - 4} - \frac{2x}{3x^2 + 3}\right)$$ ∎

The complexity of the differentiation procedure used in the solution to Example 5 should be compared with that involved in a solution using the Product, Power, and Quotient Rules. If

$$y = \frac{(x-2)^2(x^3-4)}{\sqrt[3]{x^2+1}}$$

then by a combination of these rules, we obtain

$$\frac{dy}{dx} = \frac{(x^2+1)^{1/3}[(x-2)^2(3x^2) + (x^3-4)(2)(x-2)] - (x-2)^2(x^3-4)(\frac{1}{3})(x^2+1)^{-2/3}(2x)}{(x^2+1)^{2/3}}$$

A considerable amount of algebraic effort is still necessary to obtain a usable form of this result. Of course, the two answers for dy/dx are equivalent, but the one obtained by logarithmic differentiation involves less algebraic manipulation.

Logarithmic differentiation is also convenient to use when differentiating functions having both a variable base and a variable exponent—functions of the form

$$y = [f(x)]^{g(x)}$$

Example 6 (*Logarithmic Differentiation*)

Find the derivative of $y = x^{2x}$, $x > 0$.

Solution: Taking the natural logarithm of both sides, we obtain

$$\ln y = \ln (x^{2x}) = (2x) \ln x$$

Differentiating both sides we have

$$\left(\frac{1}{y}\right)\frac{dy}{dx} = 2x\left(\frac{1}{x}\right) + (2) \ln x = 2 (1 + \ln x)$$

Thus $$\frac{dy}{dx} = 2y(1 + \ln x) = 2x^{2x}(1 + \ln x)$$

In general, we suggest the use of logarithmic differentiation under the following circumstances:

1. to differentiate a function involving many factors
2. to differentiate a function having both a variable base and a variable exponent

Section Exercises (8.2)

In Exercises 1–25, find dy/dx.

1. $y = \ln(x^2)$

2. $y = \ln(x^2 + 3)$

3. $y = \ln \sqrt{x^4 - 4x}$

4. $y = \ln(1 - x)^{3/2}$

5. $y = (\ln x)^4$

6. $y = x(\ln x)$

$\boxed{S}$ **7.** $y = \ln(x\sqrt{x^2 - 1})$

8. $y = \ln\left(\dfrac{x}{x + 1}\right)$

9. $y = \ln\left(\dfrac{x}{x^2 + 1}\right)$

10. $y = \dfrac{\ln x}{x}$

$\boxed{S}$ **11.** $y = \dfrac{\ln x}{x^2}$

12. $y = \ln(\ln x)$

13. $y = \ln(\ln x^2)$

14. $y = \ln\sqrt{\dfrac{x - 1}{x + 1}}$

15. $y = \ln\sqrt{\dfrac{x + 1}{x - 1}}$

16. $y = \ln\sqrt{x^2 - 4}$

17. $y = \ln\left(\dfrac{\sqrt{4 + x^2}}{x}\right)$

18. $y = \ln(x + \sqrt{4 + x^2})$

$\boxed{S}$ **19.** $y = \dfrac{-\sqrt{x^2 + 1}}{x} + \ln(x + \sqrt{x^2 + 1})$

20. $y = \dfrac{-\sqrt{x^2 + 4}}{2x^2} - \left(\dfrac{1}{4}\right)\ln\left(\dfrac{2 + \sqrt{x^2 + 4}}{x}\right)$

21. $y = \ln e^x$

22. $y = \log_3 x$

$\boxed{S}$ **23.** $y = \log_3\left(\dfrac{x\sqrt{x - 1}}{2}\right)$

24. $y = \log_5 \sqrt{x^2 - 1}$

25. $y = \log_{10}\left(\dfrac{x^2 - 1}{x}\right)$

In Exercises 26–35, find dy/dx by using logarithmic differentiation.

26. $y = x\sqrt{x^2 - 1}$

27. $y = \sqrt{(x - 1)(x - 2)(x - 3)}$

28. $y = \dfrac{x^2\sqrt{3x - 2}}{(x - 1)^2}$

$\boxed{S}$ **29.** $y = \sqrt[3]{\dfrac{x^2 + 1}{x^2 - 1}}$

30. $y = \dfrac{x(x - 1)^{3/2}}{\sqrt{x + 1}}$

31. $y = \dfrac{(x + 1)(x + 2)}{(x - 1)(x - 2)}$

32. $y = x^{x-1}$

$\boxed{S}$ **33.** $y = x^{2/x}$

34. $y = (x - 2)^{x+1}$

35. $y = (1 + x)^{1/x}$

36. Show that $y = x(\ln x) - 4x$ is a solution to the differential equation $(x + y) - x(y') = 0$.

$\boxed{S}$ **37.** Show that $y = 2(\ln x) + 3$ is a solution to the differential equation $x(y'') + y' = 0$.

In Exercises 38–42, find any relative extrema and inflection points, and sketch the graph of the function.

$\boxed{C}$ **38.** $y = x - \ln x$

$\boxed{C}\boxed{S}$ **39.** $y = \dfrac{x^2}{2} - \ln x$

$\boxed{C}$ **40.** $y = \dfrac{\ln x}{x}$

$\boxed{C}$ **41.** $y = x(\ln x)$

$\boxed{C}$ **42.** $y = x^2(\ln x)$

$\boxed{C}\boxed{S}$ **43.** Use Newton's Method to approximate, to three decimal places, the value of x satisfying the equation $\ln x = -x$.

$\boxed{C}$ **44.** Use Newton's Method to approximate, to three decimal places, the x-coordinate of the point of intersection of the graphs of $y = 3 - x$ and $y = \ln x$.

45. Show that $f(x) = (\ln x^n)/x$ is a decreasing function for $x > e$ and $n > 0$.

46. A man walking along a dock drags a boat by a 10-ft rope. The boat travels along a path known as a tractrix (see Figure 8.9). The equation of this path is

$$y = 10\ln\left(\dfrac{10 + \sqrt{100 - x^2}}{x}\right) - \sqrt{100 - x^2}$$

What is the slope of this path when:

(a) $x = 10$?   (b) $x = 5$?

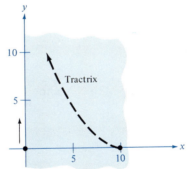

FIGURE 8.9

$\boxed{C}$ **47.** There are 25 prime numbers less than 100. The **Prime Number Theorem** states that the number of primes less than x approaches

$$p(x) \approx \dfrac{x}{\ln x}$$

Use this approximation to estimate the rate (in primes per 100 integers) at which the prime numbers occur when:

(a) $x = 1{,}000$ (b) $x = 1{,}000{,}000$

(c) $x = 1{,}000{,}000{,}000$

For Review and Class Discussion

True or False

1. ____ The derivative of $f(x) = \ln x$ is positive for all x in the domain of f.

2. ____ If $f(x) = \ln u$, where u is a function of x, then $f'(x) = 0$ only if $du/dx = 0$.

3. ____ The derivative of every transcendental function is transcendental.

4. ____ If $f(x) = \ln (ax)$ and $g(x) = \ln (bx)$, then $f'(x) = g'(x)$.

5. ____ If $y = \ln \pi$, then $y' = 1/\pi$.

8.3 Logarithmic Functions: Integration, Growth, and Decay

Purpose
- To evaluate integrals of the form $\int (u'/u)\, dx$.
- To apply this type of integral to problems involving exponential growth and decay.

The right-hand side of the formula $\int x^n\, dx = \dfrac{x^{n+1}}{n+1} + C$ *is infinite if* $n = -1$ *and hence, when* $n = -1$, *we use the formula*

$$\int x^{-1}\, dx = \int \frac{1}{x}\, dx = \ln x + C$$

—Augustin-Louis Cauchy (1789–1857)

The differentiation formulas

$$\frac{d}{dx}[\ln x] = \frac{1}{x} \qquad \text{and} \qquad \frac{d}{dx}[\ln u] = \frac{u'}{u}$$

allow us to patch up the hole in our General Power Rule for integration. Recall from Section 6.1 that

$$\int u^n u'\, dx = \frac{u^{n+1}}{n+1} + C$$

provided $n \neq -1$. Having the derivative formulas for logarithmic functions, we are now in a position to evaluate $\int u^n u'\, dx$ for $n = -1$, as stated in the following theorem:

THEOREM 8.4
(The Log Rule for Integration)

If u is a differentiable function of x, then

$$\int \frac{u'}{u}\, dx = \ln |u| + C$$

In particular,

$$\int \frac{1}{x}\, dx = \ln |x| + C$$

Proof: Differentiating $\ln |u|$ we obtain

$$\frac{d}{dx}[\ln|u|] = \left(\frac{1}{|u|}\right)\frac{d}{dx}[|u|] = \left(\frac{1}{|u|}\right)\left(\frac{|u|}{u}\right)\frac{du}{dx} = \left(\frac{1}{u}\right)\frac{du}{dx} = \frac{u'}{u}$$

Now since
$$\frac{d}{dx}[\ln|u|] = \frac{u'}{u}$$

we know that $\ln|u|$ is an antiderivative of u'/u, and thus we can write

$$\int \frac{u'}{u}dx = \ln|u| + C$$
∎

Example 1

Evaluate the integral $\qquad \int \dfrac{dx}{2x - 1}$

Solution: By letting $u = 2x - 1$, we have $u' = 2$. Thus to apply the Log Rule, we multiply and divide by 2 and write

$$\int \frac{dx}{2x - 1} = \frac{1}{2}\int \frac{2}{2x - 1}dx = \frac{1}{2}\int \frac{u'}{u}dx$$

$$= \frac{1}{2}\ln|u| + C = \frac{1}{2}\ln|2x - 1| + C$$
∎

Integrals to which the Log Rule can be applied are often given in disguised form. For instance, if a rational function has a numerator of degree greater than or equal to the degree of the denominator, division may reveal a form to which we can apply the Log Rule. Our next example is a case in point.

Example 2

Find the area of the region bounded by $y = (x^2 + x + 1)/(x^2 + 1)$, the coordinate axes, and $x = 3$.

Solution: By first dividing we obtain
$$y = \frac{x^2 + x + 1}{x^2 + 1} = 1 + \frac{x}{x^2 + 1}$$

From Figure 8.10 we see that the area of the specified region is

$$A = \int_0^3 \frac{x^2 + x + 1}{x^2 + 1}dx = \int_0^3 \left[1 + \frac{x}{x^2 + 1}\right]dx$$

$$= \int_0^3 (1)dx + \frac{1}{2}\int_0^3 \frac{2x}{x^2 + 1}dx = x\Big]_0^3 + \left[\frac{1}{2}\ln(x^2 + 1)\right]_0^3$$

$$= 3 + \frac{1}{2}[\ln(10) - \ln(1)] = 3 + \frac{1}{2}\ln(10) \approx 4.15$$

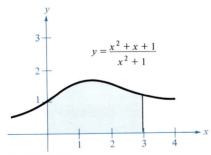

$$y = \frac{x^2 + x + 1}{x^2 + 1}$$

FIGURE 8.10

(Notice that an absolute value sign is unnecessary since $x^2 + 1 > 0$ on the interval under consideration.)   ∎

Example 3

Evaluate

$$\int \frac{2x}{(x + 1)^2} dx$$

Solution: By letting $u = (x + 1)^2$, we have $u' = 2(x + 1) = 2x + 2$. Now to obtain the form u'/u, we add and subtract 2 in the numerator. Thus we have

$$\int \frac{2x}{(x + 1)^2} dx = \int \frac{2x + 2 - 2}{(x + 1)^2} dx = \int \frac{2x + 2}{(x + 1)^2} dx - \int \frac{2}{(x + 1)^2} dx$$

$$= 2 \int \left(\frac{1}{x + 1} \right) dx - 2 \int (x + 1)^{-2} dx$$

$$= 2 \ln |x + 1| - \frac{2(x + 1)^{-1}}{-1} + C$$

$$= 2 \ln |x + 1| + \frac{2}{x + 1} + C$$

Alternative Solution: If we consider $du = u' dx$, then the Log Rule has the form

$$\int \frac{du}{u} = \ln |u| + C$$

Thus, in this case, if we let $u = x + 1$, then $du = (1) dx = dx$, and $x = u - 1$. Then by substituting we have

$$\int \frac{2x}{(x + 1)^2} dx = \int \frac{2(u - 1)}{u^2} du = 2 \int \left[\frac{u}{u^2} - \frac{1}{u^2} \right] du$$

$$= 2 \int \frac{du}{u} - 2 \int u^{-2} du = 2 \ln |u| - 2 \left(\frac{u^{-1}}{-1} \right) + C$$

$$= 2 \ln |u| + \frac{2}{u} + C$$

Resubstituting for u we have

$$\int \frac{2x}{(x + 1)^2} dx = 2 \ln |x + 1| + \frac{2}{x + 1} + C$$ ∎

As you study the two different methods of solving Example 3 you should be aware that both methods involve rewriting a disguised integrand into a form that fits one or more of our integration formulas. In this and the next two chapters, we will be devoting a great deal of time to integration techniques. To master these techniques you must recognize the "form fitting" nature of integration. In this sense integration is not nearly as straightforward as differentiation. In comparing the two operations, we can liken differentiation to "Here is the question. What is the answer?" while integration is more like "Here is the answer. What is the question?" We suggest the following general approach to integration:

Guidelines for Integration

1. Memorize a basic list of integration formulas. (At this point our list consists of two formulas: the Power Rule and the Log Rule. By the beginning of Chapter 10, we will have expanded this list to include 19 basic formulas.)

2. Find an integration formula that resembles all or part of the integrand and try to visualize (mentally or on paper) which choice of u will make the integrand conform to the formula. (Integration involves trial and error, so don't be discouraged by choices that don't pan out.)

3. If step 2 fails to solve the integral, try altering the integrand by means of division or some other algebraic operation. Take advantage of your failures in step 2. If a particular choice of u almost works, try to alter the integrand accordingly.

Example 4

Find
$$\int \frac{1}{x \ln x}\, dx$$

Solution: Since the integrand is a quotient whose denominator is raised to the first power, we try to use the Log Rule. The choices $u = x \ln x$ and $u = x$ both fail to fit the form u'/u. However, the choice $u = \ln x$ gives us $u' = 1/x$, and by rewriting the integral as

$$\int \frac{1}{x \ln x}\, dx = \int \frac{(1/x)}{\ln x}\, dx$$

we obtain the form $\int (u'/u)\, dx$. Thus

$$\int \frac{(1/x)}{\ln x}\, dx = \ln |\ln x| + C \qquad \blacksquare$$

In the next example we consider a piston of radius r in a cylindrical casing. (See Figure 8.11.) As the gas in the cylinder expands, the piston moves and work is done. If p represents the gas pressure (in pounds per square foot) against the piston head and V represents the volume of the gas (in cubic feet), then the work increment in moving the piston Δx feet is

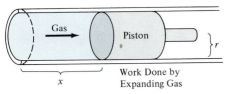

Gas → Piston

Work Done by Expanding Gas

FIGURE 8.11

$$\Delta W = \text{(force)(distance increment)} = F \Delta x = p(\pi r^2) \Delta x = p \Delta V$$

Thus as the volume of the gas expands from V_0 to V_1, the work done in moving the piston is

$$W = \int_{V_0}^{V_1} p \, dV$$

Assuming the pressure of the gas to be inversely proportional to its volume, we have $p = k/V$ and the integral for work becomes

$$W = \int_{V_0}^{V_1} \frac{k}{V} dV$$

Example 5

A quantity of gas with an initial volume of 1 ft³ and pressure of 500 lb/ft² expands to a volume of 2 ft³. Find the work done by the gas. (Assume the pressure is inversely proportional to the volume.)

Solution: Since $p = k/V$ and $p = 500$ when $V = 1$, we have $k = 500$, and the work is

$$W = \int_{V_0}^{V_1} \frac{k}{V} dV = \int_1^2 \frac{500}{V} dV = 500 \ln |V| \Big]_1^2$$

$$= 500 \ln (2) \approx 346.6 \text{ ft} \cdot \text{lb} \qquad \blacksquare$$

One of the more common applications of integrals of the form $\int (u'/u) \, dx$ arises in situations involving exponential growth or decay. In this type of application, we deal with a substance whose rate of growth or decay at a particular time is proportional to the amount of the substance present at that time. For example, the rate of decomposition of a radioactive substance is proportional to the amount of radioactive substance at a given instant. Or, similarly, the rate of growth of a bacteria culture may be proportional to the number of bacteria in the culture at time t. In its simplest form this relationship is described by the differential equation

$$\frac{dy}{dt} = ky$$

where k is a constant and y is a function of t. [A **differential equation** is an equation involving derivatives (of first or higher order) of one variable with respect to another. We discuss such equations in more detail in Chapter 19.]

An obvious solution to the differential equation $dy/dt = ky$ is $y = 0$. To find the other solutions, we assume $y \neq 0$ and divide both sides of the equation by y to obtain

$$\left(\frac{1}{y}\right)\frac{dy}{dt} = k$$

Integrating both sides with respect to t, we have

$$\int\left(\frac{1}{y}\right)\frac{dy}{dt}dt = \int k\, dt$$

$$\ln|y| = kt + C_1$$

(Note that we need to include only one constant of integration.) Now solving for y we have

$$|y| = e^{kt+C_1} = e^{C_1}e^{kt}$$

$$y = \pm e^{C_1}e^{kt}$$

Finally, since $y = \pm e^{C_1}e^{kt}$ represents all nonzero solutions, and $y = 0$ is already known to be a solution, we can write the general form of the solution as

$$y = Ce^{kt}$$

where C is any real number. This equation is often referred to as the **law of exponential growth (or decay)**, and it has wide applications, as we shall see in the following two examples.

Example 6 (*Exponential Decay*)

Let y represent the mass of a particular radioactive element whose *half-life* is 25 years. [In other words, if we began with 1 gram (g) of the element, only $\frac{1}{2}$ g would remain after 25 years, $\frac{1}{4}$ g after 50 years, etc.] How much of the 1 g would remain after 15 years? (See Figure 8.12.)

Solution: Assuming that the rate of decay is proportional to y, we then have

$$\frac{dy}{dt} = ky \qquad \text{or} \qquad y = Ce^{kt}$$

Since it was stated that $y = 1$ when $t = 0$, we find that $C = 1$. Furthermore, since $y = \frac{1}{2}$ when $t = 25$, we have

$$\frac{1}{2} = e^{25k} \qquad \text{and so} \qquad \frac{-\ln 2}{25} = k$$

Therefore, the equation for y as a function of time must be

$$y = e^{-(\ln 2)t/25} = (e^{\ln 2})^{-t/25} = 2^{-t/25}$$

At $t = 15$, $\qquad y = 2^{-3/5} \approx 0.660$ g $\qquad\blacksquare$

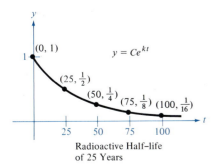

FIGURE 8.12

Radioactive Half–life of 25 Years

Example 7 (*Exponential Growth*)

In a certain research experiment, a population of fruit flies increases according to the law of exponential growth. If there were 180 flies after the second day of the experiment and 300 flies after the fourth day, how many flies were in the original population?

Solution: Let y be the number of flies at time t. Since the population increases according to the law of exponential growth, we know y is of the form $y = Ce^{kt}$, where $y = Ce^{k(0)} = C$ is the number of flies in the original population. Since $y = 180$ when $t = 2$ and $y = 300$ when $t = 4$, we have the equations

$$180 = Ce^{2k} \quad \text{and} \quad 300 = Ce^{4k} = C(e^{2k})^2$$

Substituting $e^{2k} = 180/C$ into the second equation, we have

$$300 = C\left(\frac{180}{C}\right)^2$$

or

$$C = \frac{(180)^2}{300} = 108 \text{ flies}$$

∎

Example 8 (*Newton's Law of Cooling*)

Let y represent the temperature (in degrees) of an object in a room. **Newton's Law of Cooling** states that the rate of change in the temperature of the object is proportional to the difference between its temperature and the temperature of the room. Suppose that in a room kept at a constant temperature of 60°, the object is cooled from 100° to 90° in 10 min. How much longer will it take for the temperature of the object to decrease to 80°?

Solution: According to Newton's Law of Cooling, we have

$$\frac{dy}{dt} = k(y - 60), \quad 80 \leq y \leq 100$$

Therefore,

$$\left(\frac{1}{y - 60}\right)\frac{dy}{dt} = k$$

$$\int\left(\frac{1}{y - 60}\right)\frac{dy}{dt}\, dt = \int k\, dt$$

$$\ln(y - 60) = kt + C$$

[Note that since $(y - 60)$ is positive, we can omit the absolute value signs.] Now since $y = 100$ when $t = 0$, we have

$$C = \ln 40$$

and $$kt = \ln(y - 60) - \ln 40 = \ln\left(\frac{y - 60}{40}\right)$$

Furthermore, $y = 90$ when $t = 10$ implies that

$$k = \frac{1}{10} \ln \frac{3}{4}$$

and thus $$t = \left(\frac{10}{\ln \frac{3}{4}}\right) \ln\left(\frac{y - 60}{40}\right)$$

Finally, when $y = 80$,

$$t = \frac{10 \ln \frac{1}{2}}{\ln \frac{3}{4}} \approx 24.09 \text{ min}$$

Therefore, in approximately 14.09 *more* minutes, the object will cool to a temperature of 80°. ■

Section Exercises (8.3)

In Exercises 1–23, evaluate each integral.

1. $\int \frac{1}{x + 1}\, dx$

2. $\int \frac{1}{x - 5}\, dx$

3. $\int \frac{1}{3 - 2x}\, dx$

4. $\int \frac{1}{6x + 1}\, dx$

[S] **5.** $\int \frac{x}{x^2 + 1}\, dx$

6. $\int \frac{x^2}{3 - x^3}\, dx$

[S] **7.** $\int_{-2}^{-1} \frac{x^2 - 4}{x}\, dx$

8. $\int_{-2}^{-1} \frac{x + 5}{x}\, dx$

9. $\int_1^e \frac{\ln x}{2x}\, dx$

10. $\int_e^{e^2} \frac{1}{x(\ln x)}\, dx$

[S] **11.** $\int_1^e \frac{(1 + \ln x)^2}{x}\, dx$

12. $\int_0^1 \frac{x - 1}{x + 1}\, dx$

13. $\int_0^2 \frac{x^2 - 2}{x + 1}\, dx$

14. $\int \frac{1}{(x + 1)^2}\, dx$

[S] **15.** $\int \frac{1}{\sqrt{x} + 1}\, dx$

16. $\int \frac{x + 3}{x^2 + 6x + 7}\, dx$

17. $\int \frac{x^2 + 2x + 3}{x^3 + 3x^2 + 9x + 1}\, dx$

18. $\int \frac{(\ln x)^2}{x}\, dx$

19. $\int \frac{1}{x^{2/3}(1 + x^{1/3})}\, dx$

20. $\int \frac{1}{x \ln(x^2)}\, dx$

[S] **21.** $\int \frac{\sqrt{x}}{1 - x\sqrt{x}}\, dx$

22. $\int \frac{2x}{(x - 1)^2}\, dx$

[S] **23.** $\int \frac{x(x - 2)}{(x - 1)^3}\, dx$

24. Evaluate the definite integral $\int_0^{1/2} 1/(6x + 1)\, dx$.

25. Evaluate the definite integral

$$\int_0^3 \frac{x - 1}{x + 1}\, dx$$

26. Find the area of the region bounded by $y = (x^2 + 4)/x$, $y = 0$, $x = 1$, and $x = 4$.

[S] **27.** Find the area of the region bounded by $y = (x + 5)/x$, $y = 0$, $x = 1$, and $x = 5$.

[C] **28.** Find the area of the region bounded by $y = 1/x$ and $y = -x^2 + 4x - 2$.

29. Find the volume of the solid generated by revolving the region bounded by $y = 1/x^2$, $y = 0$, $x = 1$, and $x = 4$ about:

(a) the x-axis (b) the y-axis

(c) $x = 4$

30. Find the volume of the solid generated by revolving the region bounded by $y = 1/\sqrt{x + 1}$, $y = 0$, $x = 0$, and $x = 3$ about the x-axis.

31. A certain type of bacteria increases continuously at a rate proportional to the number present. If there are 100 present at a given time and 300 present 5 h later, how many will there be in 10 h after the initial given time?

32. Given the conditions of Exercise 31, how long does it take for the number of bacteria to double?

⌐S⌐ **33.** In 1960 the population of a town was 2500 and in 1970 it was 3350. Assuming the population increases continuously at a constant rate proportional to the existing population, estimate the population in 1990.

⌐C⌐ **34.** Given the conditions of Exercise 33, how many years are necessary for the population to double?

35. Radioactive radium has a half-life of approximately 1600 years. What percentage of a present amount remains after 100 years?

⌐C⌐⌐S⌐ **36.** If radioactive material decays continuously at a rate proportional to the amount present, find the half-life of the material if after 1 year 99.57% of an initial amount still remains.

37. Using Newton's Law of Cooling (see Example 8), determine the reading on a thermometer 5 min after it is taken from a room at 72° Fahrenheit to the outdoors

where the temperature is 20°, if the reading dropped to 48° after 1 min.

⌐C⌐ **38.** A body in a room at 70° cools from 350° to 150° in 45 min. Using Newton's Law of Cooling, find the time necessary for the body to cool to 80°.

⌐S⌐ **39.** Using Newton's Law of Cooling, determine the outdoor temperature if a thermometer is taken from a room where the temperature is 68° to the outdoors, where after $\frac{1}{2}$ min and 1 min, the thermometer reads 53° and 42°, respectively.

40. Determine the work done in compressing an initial volume of 9 ft³ and pressure of 15 lb/in.² to a volume of 3 ft³, assuming the pressure-volume relation is $p = k/V$ (k is a constant).

⌐S⌐ **41.** A quantity of gas with an initial volume of 24 ft³ and pressure of 120 lb/in.² expands until the pressure is 60 lb/in.². Determine the final volume and work done by the gas if the pressure-volume relation is $p = k/V$.

For Review and Class Discussion

True or False

1. ____ $\int \ln x \, dx = (1/x) + C$.

2. ____ $\int \frac{1}{x} \, dx = \ln |cx|$, for $c \neq 0$.

3. ____ If u is a differentiable function of x such that $u \neq 0$, then

$$\int \left(\frac{1}{u}\right) \frac{du}{dx} \, dx = \int \frac{1}{u} \, du$$

4. ____ $\int_{-1}^{e} \frac{1}{x} \, dx = \ln |x| \Big]_{-1}^{e} = \ln e - \ln 1 = 1$.

5. ____ If $p(x) = nx^n + (n-1)x^{n-1} + (n-2)x^{n-2} + \cdots + 2x^2 + x + 1$, then

$$\int \frac{p(x)}{x} \, dx = x^n + x^{n-1} + x^{n-2} + \cdots$$
$$+ x^2 + x + \ln |x| + C$$

8.4 Exponential Functions: Differentiation and Integration

Purpose

- To develop differentiation formulas for exponential functions.
- To develop integration formulas for exponential functions.

Thus, when we wish to demonstrate a general theorem, we must give the rule as applied to a particular case; but if we wish to demonstrate a particular case, we must begin with the general rule. For we always find the thing obscure which we wish to prove and that clear which we use for the proof. —*Blaise Pascal (1623–1662)*

In Section 8.3 we saw that the solution to the differential equation $dy/dt = ky$ is $y = Ce^{kt}$. In other words,

$$\frac{d}{dt}[Ce^{kt}] = kCe^{kt}$$

By letting $C = k = 1$, we obtain

$$\frac{d}{dt}[e^t] = e^t$$

which means that the exponential function e^t has the interesting property that it is its own derivative.

THEOREM 8.5 $$\frac{d}{dx}[e^x] = e^x$$

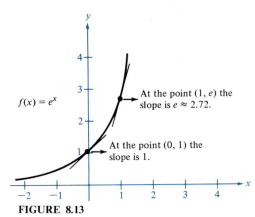

$f(x) = e^x$

At the point $(1, e)$ the slope is $e \approx 2.72$.

At the point $(0, 1)$ the slope is 1.

FIGURE 8.13

Geometrically this means that the slope of the graph of $f(x) = e^x$ at any point (x, e^x) is numerically equal to the y-coordinate of the point (Figure 8.13).

Logarithmic differentiation may be used to find the derivative of the exponential function with base a. If $y = a^x$, then

$$\ln y = \ln (a^x) = x \ln a$$

Differentiating both sides with respect to x, we obtain

$$\left(\frac{1}{y}\right)\frac{dy}{dx} = \ln a$$

$$\frac{dy}{dx} = (\ln a)(y) = (\ln a)\, a^x$$

By applying the Chain Rule to the two formulas

$$\frac{d}{dx}[e^x] = e^x \qquad \text{and} \qquad \frac{d}{dx}[a^x] = (\ln a)\, a^x$$

we can obtain differentiation formulas for e^u and a^u, where u is a differentiable function of x.

Derivatives of Exponential Functions

$$\frac{d}{dx}[e^x] = e^x \qquad \frac{d}{dx}[a^x] = (\ln a)\, a^x$$

$$\frac{d}{dx}[e^u] = e^u u' \qquad \frac{d}{dx}[a^u] = (\ln a)\, a^u u'$$

Once again, as with logarithmic functions, we can see that the choice of e as a base leads to more convenient differentiation formulas.

Example 1

Differentiate the following:
(a) $y = e^{2x-1}$
(b) $y = e^{-3/x}$

Solution:
(a) Considering $u = 2x - 1$, we have $u' = 2$, and thus

$$\frac{dy}{dx} = e^u u' = e^{2x-1}(2) = 2e^{2x-1}$$

(b) Considering $u = -3/x$, we have $u' = 3/x^2$, and thus

$$\frac{dy}{dx} = e^u u' = e^{-3/x}\left(\frac{3}{x^2}\right) = \frac{3e^{-3/x}}{x^2}$$

∎

Example 2

Differentiate $\qquad f(x) = e^2 + \dfrac{e^{-2x}}{x^3}$

Solution: Note that e^2 is a constant; thus we have

$$f'(x) = 0 + \frac{d}{dx}\left[\frac{e^{-2x}}{x^3}\right]$$

Now by the Quotient Rule, we obtain

$$f'(x) = \frac{x^3[e^{-2x}(-2)] - e^{-2x}[3x^2]}{(x^3)^2}$$

$$= \frac{-2x^3 e^{-2x} - 3x^2 e^{-2x}}{x^6} = \frac{-x^2 e^{-2x}(2x + 3)}{x^6}$$

$$= \frac{-e^{-2x}(2x + 3)}{x^4}$$

∎

Example 3

Show that the **normal probability density function**

$$f(x) = \left(\frac{1}{\sqrt{2\pi}}\right)e^{-x^2/2}$$

has points of inflection when $x = \pm 1$.

Solution: First, $\qquad f'(x) = \left(\dfrac{1}{\sqrt{2\pi}}\right)(-x)e^{-x^2/2}$

and by the Product Rule, we have

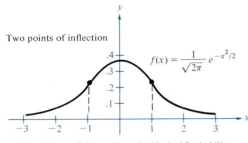

Two points of inflection

$$f(x) = \frac{1}{\sqrt{2\pi}} e^{-x^2/2}$$

"Bell–Shaped" Curve Given By Normal Probability Density Function

FIGURE 8.14

$$f''(x) = \left(\frac{1}{\sqrt{2\pi}}\right)[(-x)(-x)e^{-x^2/2} + (-1)e^{-x^2/2}]$$

$$= \left(\frac{1}{\sqrt{2\pi}}\right)(e^{-x^2/2})(x^2 - 1)$$

Therefore, $f''(x) = 0$ when $x = \pm1$, and we can apply the techniques in Section 4.4 to conclude that these values of x yield the two points of inflection. (See Figure 8.14.) ∎

The general form of a normal probability density function is given by

$$f(x) = \left(\frac{1}{\sigma\sqrt{2\pi}}\right)e^{-x^2/2\sigma^2}$$

where σ is the standard deviation from the mean (σ is the Greek lower-case letter sigma). By following the procedure of Example 3, it can be shown that the "bell-shaped" graph of f has points of inflection when $x = \pm\sigma$.

Example 4

The curvature K of a function $y = f(x)$ is given by

$$K = \frac{|y''|}{[1 + (y')^2]^{3/2}}$$

Show that the curvature of $y = e^x$ is greatest when $x = -\ln\sqrt{2}$.

Solution: Since $y = y' = y''$, the formula for the curvature of $y = e^x$ is

$$K = \frac{e^x}{[1 + (e^x)^2]^{3/2}} = \frac{e^x}{[1 + e^{2x}]^{3/2}}$$

Thus by the Quotient Rule,

$$\frac{dK}{dx} = \frac{(1 + e^{2x})^{3/2}(e^x) - e^x(\frac{3}{2})(1 + e^{2x})^{1/2}(2e^{2x})}{[(1 + e^{2x})^{3/2}]^2}$$

$$= \frac{e^x(1 + e^{2x})^{1/2}[(1 + e^{2x}) - 3e^{2x}]}{(1 + e^{2x})^3} = \frac{e^x(1 - 2e^{2x})}{(1 + e^{2x})^{5/2}}$$

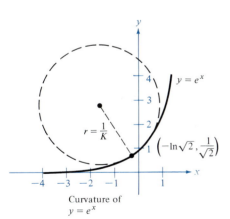

Curvature of $y = e^x$

FIGURE 8.15

By setting $dK/dx = 0$ we have

$$1 - 2e^{2x} = 0$$

Solving for x we obtain the critical value

$$x = -\ln\sqrt{2} \approx -0.347$$

Since $dK/dx > 0$ at $x = -1$ and $dK/dx < 0$ at $x = 0$, we conclude by the First-Derivative Test that $y = e^x$ has maximum curvature when $x = -\ln\sqrt{2}$. (See Figure 8.15.) ∎

Example 5

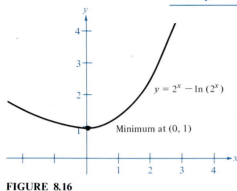

$y = 2^x - \ln (2^x)$

Minimum at (0, 1)

FIGURE 8.16

Find the minimum of $y = 2^x - \ln 2^x$.

Solution: First, we write $y = 2^x - x(\ln 2)$; then

$$y' = (\ln 2)(2^x) - \ln 2 = (\ln 2)(2^x - 1)$$

Thus $y' = 0$ when $2^x - 1 = 0$, and the critical number is $x = 0$. Now since $y'' = (\ln 2)^2 2^x$ is positive when $x = 0$, by the Second-Derivative Test we conclude that the point $(0, 1)$ is a minimum (Figure 8.16).   ■

Each of the four formulas for differentiating exponential functions has its corresponding integration formula, as shown next.

Integrals of Exponential Functions

$$\int e^x \, dx = e^x + C \qquad \int a^x \, dx = \left(\frac{1}{\ln a}\right) a^x + C$$

$$\int e^u u' \, dx = e^u + C \qquad \int a^u u' \, dx = \left(\frac{1}{\ln a}\right) a^u + C$$

Since

$$\frac{1}{\ln a} = \frac{1}{\log_e a} = \log_a e$$

the integration formulas involving base a have the alternative forms

$$\int a^x \, dx = (\log_a e) a^x + C$$

and

$$\int a^u u' \, dx = (\log_a e) a^u + C$$

Example 6

Evaluate $\int e^{3x+1} \, dx$.

Solution: Considering $u = 3x + 1$, we have $u' = 3$. Introducing the missing factor 3 in the integrand and then multiplying the integral by the reciprocal factor $\frac{1}{3}$, we write

$$\int e^{3x+1} \, dx = \frac{1}{3} \int e^{3x+1}(3) \, dx = \frac{1}{3} \int e^u u' \, dx = \frac{1}{3} e^u + C = \frac{e^{3x+1}}{3} + C$$

■

Example 7

Evaluate $\int 5xe^{-x^2} \, dx$.

Solution: If we let $u = -x^2$, then $u' = -2x$. Now we adjust our integrand by moving the unneeded factor 5 outside the integral sign. Then we introduce the needed factor -2 and multiply by $\frac{-1}{2}$. Thus we write

$$\int 5xe^{-x^2}\,dx = 5\int e^{-x^2}(x)\,dx = 5\left(\frac{-1}{2}\right)\int e^{-x^2}(-2x)\,dx$$

$$= \frac{-5}{2}\int e^u u'\,dx = \frac{-5}{2}e^{-x^2} + C \qquad\blacksquare$$

Keep in mind that we cannot introduce a missing *variable* factor in the integrand. For instance, we cannot evaluate $\int 5e^{x^2}\,dx$ by introducing the factor $2x$ and then multiplying the integral by $1/2x$. That is,

$$\int 5e^{x^2}\,dx \neq \frac{5}{2x}\int e^{x^2}(2x)\,dx$$

Example 8

Evaluate

$$\int_0^2 \frac{e^x}{1 + e^x}\,dx$$

Solution: By letting $u = 1 + e^x$, we have $u' = e^x$, and we can apply the formula

$$\int \frac{u'}{u}\,dx = \ln|u| + C$$

We then have

$$\int_0^2 \frac{e^x}{1 + e^x}\,dx = \ln|1 + e^x|\Big]_0^2 = \ln(1 + e^2) - \ln 2$$

$$= \ln\left(\frac{1 + e^2}{2}\right) \approx 1.434 \qquad\blacksquare$$

Example 9

Evaluate

$$\int \frac{2^{1/x}}{x^2}\,dx$$

Solution: By choosing $u = 1/x$, we have

$$u' = \frac{-1}{x^2}$$

Thus $\quad \displaystyle\int \frac{2^{1/x}}{x^2}\,dx = -\int 2^{1/x}\left(\frac{-1}{x^2}\right)dx = \left(-\frac{1}{\ln 2}\right)2^{1/x} + C \qquad\blacksquare$

Section Exercises (8.4)

In Exercises 1–20, find dy/dx.

1. $y = e^{2x}$

2. $y = e^{1-x}$

3. $y = e^{1-2x+x^2}$

4. $y = e^{-x^2}$

$\boxed{\text{S}}$ **5.** $y = e^{\sqrt{x}}$

6. $y = x^2 e^{-x}$

7. $y = (e^{-x} + e^x)^3$

8. $y = e^{-1/x^2}$

$\boxed{\text{S}}$ **9.** $y = \ln(e^{x^2})$

10. $y = \ln\left(\dfrac{1 + e^x}{1 - e^x}\right)$

11. $y = \ln(1 + e^{2x})$

12. $y = \dfrac{2}{e^x + e^{-x}}$

13. $y = \ln\left(\dfrac{e^x + e^{-x}}{2}\right)$

14. $y = xe^x - e^x$

15. $y = x^2 e^x - 2xe^x + 2e^x$

16. $y = \dfrac{e^x - e^{-x}}{2}$

17. $y = 5^{x-2}$

18. $y = x(7^{-3x})$

$\boxed{\text{S}}$ **19.** $y = e^{-x} \ln x$

20. $y = 2^{x^2} 3^{-x}$

In Exercises 21–40, evaluate each integral.

21. $\displaystyle\int_0^1 e^{-2x}\, dx$

22. $\displaystyle\int_1^2 e^{1-x}\, dx$

$\boxed{\text{S}}$ **23.** $\displaystyle\int_0^2 (x^2 - 1)e^{x^3 - 3x + 1}\, dx$

24. $\displaystyle\int x^2 e^{x^3}\, dx$

25. $\displaystyle\int \dfrac{e^{-x}}{1 + e^{-x}}\, dx$

26. $\displaystyle\int \dfrac{e^{2x}}{1 + e^{2x}}\, dx$

27. $\displaystyle\int xe^{ax^2}\, dx$

28. $\displaystyle\int_0^{\sqrt{2}} xe^{-(x^2/2)}\, dx$

$\boxed{\text{S}}$ **29.** $\displaystyle\int_1^3 \dfrac{e^{3/x}}{x^2}\, dx$

30. $\displaystyle\int (e^x - e^{-x})^2\, dx$

31. $\displaystyle\int_{-1}^2 2^x\, dx$

32. $\displaystyle\int x5^{-x^2}\, dx$

$\boxed{\text{S}}$ **33.** $\displaystyle\int e^x \sqrt{1 - e^x}\, dx$

34. $\displaystyle\int \dfrac{e^x - e^{-x}}{e^x + e^{-x}}\, dx$

$\boxed{\text{S}}$ **35.** $\displaystyle\int \dfrac{e^x + e^{-x}}{e^x - e^{-x}}\, dx$

36. $\displaystyle\int \dfrac{2e^x - 2e^{-x}}{(e^x + e^{-x})^2}\, dx$

37. $\displaystyle\int \dfrac{5 - e^x}{e^{2x}}\, dx$

38. $\displaystyle\int (3 - x)7^{(3-x)^2}\, dx$

39. $\displaystyle\int_{-2}^0 (3^3 - 5^2)\, dx$

40. $\displaystyle\int \dfrac{e^{2x} + 2e^x + 1}{e^x}\, dx$

In Exercises 41–45, find (if any exist) the extrema and the points of inflection, and sketch the graph of each function.

41. $f(x) = \left(\dfrac{1}{\sqrt{2\pi}}\right)e^{-(x^2/2)}$

42. $f(x) = \dfrac{e^x - e^{-x}}{2}$

43. $f(x) = \dfrac{e^x + e^{-x}}{2}$

44. $f(x) = xe^{-x}$

$\boxed{\text{S}}$ **45.** $f(x) = x^2 e^{-x}$

46. Find the area of the largest rectangle that can be inscribed under the curve $y = e^{-x^2}$ in the first and second quadrants.

47. Find the equation of the line normal to the graph of $y = e^{-x}$ at the point $(0, 1)$.

$\boxed{\text{C}\,\text{S}}$ **48.** Find, to three decimal places, the value of x so that $e^{-x} = x$. [Use Newton's Method to find the zero of $f(x) = x - e^{-x}$.]

$\boxed{\text{C}}$ **49.** Find any extrema of the graph of the function $f(x) = -2 + e^{3x}(4 - 2x)$. Also find all the zeros of the function and sketch its graph.

$\boxed{\text{C}}$ **50.** Find the point on the graph of $y = e^{-x}$ where the normal line to the curve at that point will pass through the origin.

In Exercises 51–53, find the area of the indicated region.

51. bounded by $y = e^x$, $y = 0$, $x = 0$, $x = 5$

52. bounded by $y = e^x$, $y = 0$, $x = a$, $x = b$

53. bounded by $y = 3^x$, $y = 2x + 1$

54. Find the volume of the solid generated by revolving the region bounded by $y = e^{-x}$, $x = 0$, $x = 1$, and $y = 0$ about the x-axis.

$\boxed{\text{S}}$ **55.** Find the volume of the solid generated by revolving the region bounded by $y = (1/\sqrt{2\pi})e^{-x^2/2}$, $y = 0$, $x = 0$, and $x = 1$ about the y-axis.

56. Find the length of the arc of $y = \frac{1}{2}(e^x + e^{-x})$ from $x = 0$ to $x = 2$.

57. From the graphs of $y = e^x$ and $y = 1$, it is clear that $e^x \geq 1$ for $x \geq 0$. Therefore, we conclude that $\int_0^x e^t\, dt \geq \int_0^x 1\, dt$. Perform this integration to derive the inequality $e^x \geq 1 + x$ for $x \geq 0$.

58. Integrate each term of the following inequalities in a manner similar to that of Exercise 57 to obtain each succeeding inequality, valid for $x \geq 0$.

(a) $e^x \geq 1 + x$

(b) $e^x \geq 1 + x + \dfrac{x^2}{2}$

(c) $e^x \geq 1 + x + \dfrac{x^2}{2} + \dfrac{x^3}{6}$

(d) $e^x \geq 1 + x + \dfrac{x^2}{2} + \dfrac{x^3}{6} + \dfrac{x^4}{24}$

© **59.** Evaluate both sides of the inequality

$$e^x \geq 1 + x + \frac{x^2}{2} + \frac{x^3}{6} + \frac{x^4}{24}$$

at $x = 2$, $x = 1$, $x = \frac{1}{2}$, $x = \frac{1}{10}$, and $x = 0$.

60. If a is constant, find the derivative of each of the following:
   (a) $y = x^a$ (b) $y = a^x$
   (c) $y = x^x$ (d) $y = a^a$

61. A certain lake is stocked with 500 fish and their population increases according to the **logistics curve**

$$p(t) = \frac{10,000}{1 + 19e^{-t/5}}$$

where t is measured in months. At what rate is the fish population of the lake increasing at the end of 1 month? At the end of 10 months? After how many months is the population increasing most rapidly?

62. The Ebbinghaus Model for human memory is

$$p(t) = (100 - a)e^{-bt} + a$$

where $p(t)$ is the percentage retained after t weeks. (The constants a and b vary from one person to another.) If $a = 20$ and $b = 0.5$, at what rate is information being forgotten after 1 week? After 3 weeks?

For Review and Class Discussion

True or False

1. ____ The exponential function $y = Ce^x$ is a solution of the differential equation $d^n y/dx^n = y$, $n = 1, 2, 3, \ldots$.

2. ____ If $y = a^x$, where $0 < a$, and $y' = a^x$, then $a = e$.

3. ____ If $f(x) = x^e$, then $f'(x) = x^e$.

4. ____ The graphs of $f(x) = e^x$ and $g(x) = e^{-x}$ meet at right angles.

5. ____ If u and v are both functions of x, then the only zeros of $f(x) = ve^u$ are the zeros of v.

8.5 Indeterminate Forms and L'Hôpital's Rule

Purpose
- To introduce L'Hôpital's Rule as an aid in determining limits.

If the differential of the numerator be found, and that be divided by the differential of the denominator, we shall have the value we sought.

—*Guillaume L'Hôpital (1661–1701)*

In our introduction to limits in Chapter 2, we encountered the indeterminate forms $0/0$, ∞/∞, and $\infty - \infty$. We described these forms as "indeterminate" because they did not guarantee that a limit existed, nor did they indicate what the limit was, if one existed. When we encountered one of these indeterminate forms, we attempted to rewrite the expression by using various algebraic techniques, as illustrated by the examples in Table 8.7.

Occasionally we can extend these algebraic techniques to find limits of functions that are not algebraic. For instance,

$$\lim_{x \to 0} \frac{e^{2x} - 1}{e^x - 1} = \frac{0}{0}$$

TABLE 8.7 Indeterminate Forms

$\dfrac{0}{0}$	$\dfrac{\infty}{\infty}$	$\infty - \infty$
$\displaystyle\lim_{x \to -1} \frac{2x^2 - 2}{x + 1}$	$\displaystyle\lim_{x \to \infty} \frac{3x^2 - 1}{2x^2 + 1}$	$\displaystyle\lim_{x \to \infty} (x - \sqrt{x^2 + x})$
$= \displaystyle\lim_{x \to -1} 2(x - 1)$	$= \displaystyle\lim_{x \to \infty} \frac{3 - (1/x^2)}{2 + (1/x^2)}$	$= \displaystyle\lim_{x \to \infty} \frac{-x}{x + \sqrt{x^2 + x}}$
$= -4$	$= \dfrac{3}{2}$	$= \displaystyle\lim_{x \to \infty} \frac{-1}{1 + \sqrt{1 + (1/x)}}$
		$= \dfrac{-1}{2}$

Algebraic Technique for Rewriting Limit		
Divide numerator and denominator by $(x + 1)$.	Divide numerator and denominator by x^2.	Rationalize; then divide numerator and denominator by x.

By factoring $(e^{2x} - 1)$ and then dividing, we have

$$\lim_{x \to 0} \frac{e^{2x} - 1}{e^x - 1} = \lim_{x \to 0} \frac{(e^x + 1)(e^x - 1)}{(e^x - 1)} = \lim_{x \to 0} (e^x + 1) = 2$$

However, not all indeterminate forms can be evaluated by algebraic manipulation. This is particularly true when both algebraic and transcendental functions are involved. For instance,

$$\lim_{x \to 0} \frac{e^{2x} - 1}{x} = \frac{0}{0}$$

Dividing both numerator and denominator by x merely produces another indeterminate form:

$$\lim_{x \to 0} \frac{e^{2x}}{x} - \frac{1}{x} = \infty - \infty$$

In this case other algebraic changes will also lead to indeterminate forms.
 Of course, we could use a calculator to estimate this limit by evaluating $(e^{2x} - 1)/x$ near zero. As shown in Table 8.8, the limit appears to be 2. (The exact limit will be determined in Example 1.)

TABLE 8.8

x	-1	-0.5	-0.1	-0.01	0	0.01	0.1	0.5	1
$\dfrac{e^{2x} - 1}{x}$	0.865	1.264	1.813	1.980	?	2.028	2.214	3.437	6.389

Fortunately there is a procedure for determining the exact value of limits such as

$$\lim_{x \to 0} \frac{e^{2x} - 1}{x}$$

We call this procedure **L'Hôpital's Rule.** This rule states that under certain conditions the limit of a quotient $f(x)/g(x)$ is determined by the limit of $f'(x)/g'(x)$. An informal statement of the rule is as follows:

L'Hôpital's Rule	If $\lim f(x)/g(x)$ results in the indeterminate forms $0/0$ or ∞/∞, then $$\lim \frac{f(x)}{g(x)} = \lim \frac{f'(x)}{g'(x)}$$ provided the latter limit exists.

Note: We use the symbol "lim" to represent any one of the following types of limits:

$$\lim_{x \to a}, \quad \lim_{x \to a^+}, \quad \lim_{x \to a^-}, \quad \lim_{x \to \infty}, \quad \lim_{x \to -\infty}$$

Of the several possible cases in which L'Hôpital's Rule can be applied, we present a formal listing and proof of just one of them.

Let f and g be continuous on $[a, b]$ and differentiable on (a, b). If $f(a) = 0$, $g(a) = 0$, and $g'(x) \neq 0$ for all x in (a, b), and if

$$\lim_{x \to a^+} \frac{f'(x)}{g'(x)} = L$$

then

$$\lim_{x \to a^+} \frac{f(x)}{g(x)} = L$$

Proof: To prove this statement we consider the function

$$h(x) = g(b) f(x) - f(b) g(x)$$

and its derivative

$$h'(x) = g(b) f'(x) - f(b) g'(x)$$

Now h is continuous on $[a, b]$ and differentiable on (a, b) because f and g possess these properties. Thus by Rolle's Theorem there exists a number c such that $a < c < b$ and $h'(c) = 0$. Therefore, at $x = c$ we have

$$0 = g(b) f'(c) - f(b) g'(c)$$

or

$$\frac{f(b)}{g(b)} = \frac{f'(c)}{g'(c)}$$

Since $a < c < b$, then as $b \to a^+$, we must have $c \to a^+$, and thus we obtain

$$\lim_{b \to a^+} \frac{f(b)}{g(b)} = \lim_{c \to a^+} \frac{f'(c)}{g'(c)} = L$$

∎

Example 1 $\left(\frac{0}{0}\right)$

Use L'Hôpital's Rule to evaluate

$$\lim_{x \to 0} \frac{e^{2x} - 1}{x}$$

Solution: Since direct substitution results in the indeterminate form $0/0$, we apply L'Hôpital's Rule to obtain

$$\lim_{x \to 0} \frac{e^{2x} - 1}{x} = \lim_{x \to 0} \frac{\dfrac{d}{dx}[e^{2x} - 1]}{\dfrac{d}{dx}[x]} = \lim_{x \to 0} \frac{2e^{2x}}{1} = 2$$

∎

Occasionally it is necessary to apply L'Hôpital's Rule more than once to remove an indeterminate form.

Example 2 $\left(\frac{\infty}{\infty}\right)$

Use L'Hôpital's Rule to find

$$\lim_{x \to -\infty} \frac{x^2}{e^{-x}}$$

Solution: Since direct substitution results in the indeterminate form ∞/∞, we apply L'Hôpital's Rule to obtain

$$\lim_{x \to -\infty} \frac{x^2}{e^{-x}} = \lim_{x \to -\infty} \frac{2x}{-e^{-x}} = \frac{-\infty}{-\infty}$$

Applying L'Hôpital's Rule again, we have

$$\lim_{x \to -\infty} \frac{2x}{-e^{-x}} = \lim_{x \to -\infty} \frac{2}{e^{-x}} = \frac{2}{\infty} = 0$$

∎

In addition to the forms $0/0$, ∞/∞, or $\infty - \infty$, there are other indeterminate forms, such as $0 \cdot \infty$, 1^∞, ∞^0, and 0^0. The following examples indicate methods by which these forms can be evaluated. Basically we attempt to convert all forms to those for which L'Hôpital's Rule is applicable.

Example 3 $(0 \cdot \infty)$

Determine $\lim_{x \to \infty} e^{-x} \sqrt{x}$.

Solution: Since

$$\lim_{x \to \infty} e^{-x} \sqrt{x} = 0 \cdot \infty$$

we rewrite the limit as

$$\lim_{x \to \infty} e^{-x} \sqrt{x} = \lim_{x \to \infty} \frac{\sqrt{x}}{e^x} = \frac{\infty}{\infty}$$

Now by L'Hôpital's Rule we have

$$\lim_{x \to \infty} \frac{\sqrt{x}}{e^x} = \lim_{x \to \infty} \frac{1/(2\sqrt{x})}{e^x} = \frac{0}{\infty} = 0 \qquad \blacksquare$$

You may have questions about how to rewrite a function in order to apply L'Hôpital's Rule, and rightfully so, for there is a great deal of trial and error involved. If rewriting the limit in one of the forms $0/0$ or ∞/∞ does not seem to work, try the other form. For instance, we could have written the limit in Example 3 as

$$\lim_{x \to \infty} e^{-x} \sqrt{x} = \lim_{x \to \infty} \frac{e^{-x}}{x^{-1/2}} = \frac{0}{0}$$

Then by L'Hôpital's Rule we would get

$$\lim_{x \to \infty} \frac{e^{-x}}{x^{-1/2}} = \lim_{x \to \infty} \frac{-e^{-x}}{-1/(2x^{3/2})} = \frac{0}{0}$$

which is still indeterminate. Since the quotient seems to be getting more complicated, we abandon this approach and try the ∞/∞ form, as was shown in Example 3.

The indeterminate forms 1^∞, ∞^0, and 0^0 arise from limits of functions that have a variable base and a variable exponent. When we encountered this type of function in Section 8.2, we used logarithmic differentiation to find the derivative. We use a similar procedure when taking limits. Consider

$$\lim_{x \to \infty} \left(1 + \frac{1}{x}\right)^x$$

We already know from Section 8.1, Exercise 62, that this limit is e. We now show how to find this limit by L'Hôpital's Rule. Since direct substitution yields the indeterminate form 1^∞, we proceed in the following manner. Let

$$y = \lim_{x \to \infty} \left(1 + \frac{1}{x}\right)^x$$

Taking the natural logarithm of both sides, we obtain

$$\ln y = \ln \lim_{x \to \infty} \left(1 + \frac{1}{x}\right)^x$$

Since the natural logarithmic function is continuous, we can rewrite this equation as

$$\ln y = \lim_{x \to \infty} \left[\ln \left(1 + \frac{1}{x}\right)^x\right]$$

$$= \lim_{x \to \infty} \left[x \ln \left(1 + \frac{1}{x}\right)\right]$$

$$= \lim_{x \to \infty} \left\{\frac{\ln [1 + (1/x)]}{1/x}\right\} = \frac{0}{0}$$

By L'Hôpital's Rule we have

$$\lim_{x \to \infty} \left\{\frac{\ln [1 + (1/x)]}{1/x}\right\} = \lim_{x \to \infty} \left\{\frac{(-1/x^2)(1/[1 + (1/x)])}{-1/x^2}\right\} = \lim_{x \to \infty} \frac{1}{1 + (1/x)}$$

and thus

$$\ln y = \lim_{x \to \infty} \frac{1}{1 + (1/x)} = 1$$

$$\lim_{x \to \infty} \left(1 + \frac{1}{x}\right)^x = e$$

Example 4 (0^0)

Determine

$$\lim_{x \to 0^-} (1 - e^x)^x$$

Solution: Since

$$y = \lim_{x \to 0^-} (1 - e^x)^x = 0^0$$

we write

$$\ln y = \ln \left[\lim_{x \to 0^-} (1 - e^x)^x\right] = \lim_{x \to 0^-} \ln [(1 - e^x)^x]$$

$$= \lim_{x \to 0^-} [x \ln (1 - e^x)] = \lim_{x \to 0^-} \frac{\ln (1 - e^x)}{1/x} = \frac{-\infty}{-\infty}$$

By L'Hôpital's Rule

$$\ln y = \lim_{x \to 0^-} \frac{-e^x/(1 - e^x)}{-1/x^2} = \lim_{x \to 0^-} \frac{x^2 e^x}{1 - e^x} = \frac{0}{0}$$

A second application of the rule yields

$$\ln y = \lim_{x \to 0^-} \frac{x^2 e^x + 2x e^x}{-e^x} = \lim_{x \to 0^-} (-x^2 - 2x) = 0$$

Thus $\ln y = 0,$ $y = 1,$ $\displaystyle\lim_{x \to 0^-} (1 - e^x)^x = 1$

Example 5 ($\infty - \infty$)

Evaluate

$$\lim_{x \to 1^+} \left(\frac{1}{\ln x} - \frac{1}{x - 1} \right)$$

Solution: By direct substitution we obtain the indeterminate form

$$\lim_{x \to 1^+} \left(\frac{1}{\ln x} - \frac{1}{x - 1} \right) = \infty - \infty$$

Rewriting the limit in quotient form, we have

$$\lim_{x \to 1^+} \left(\frac{1}{\ln x} - \frac{1}{x - 1} \right) = \lim_{x \to 1^+} \left[\frac{x - 1 - \ln x}{(x - 1) \ln x} \right] = \frac{0}{0}$$

Now by L'Hôpital's Rule we obtain

$$\lim_{x \to 1^+} \left(\frac{1}{\ln x} - \frac{1}{x - 1} \right) = \lim_{x \to 1^+} \left[\frac{1 - (1/x)}{(x - 1)(1/x) + \ln x} \right]$$

$$= \lim_{x \to 1^+} \left[\frac{x - 1}{(x - 1) + x(\ln x)} \right] = \frac{0}{0}$$

Applying L'Hôpital's Rule again, we get

$$\lim_{x \to 1^+} \left(\frac{1}{\ln x} - \frac{1}{x - 1} \right) = \lim_{x \to 1^+} \left[\frac{1}{1 + x(1/x) + \ln x} \right] = \frac{1}{2} \qquad \blacksquare$$

We have identified the forms

$$\frac{0}{0}, \qquad \frac{\infty}{\infty}, \qquad \infty - \infty, \qquad 0 \cdot \infty, \qquad 0^0, \qquad 1^\infty, \qquad \infty^0$$

as *indeterminate*. There are similar-looking forms that you should recognize as *determinate*, such as

$$\infty + \infty = \infty, \qquad -\infty - \infty = -\infty, \qquad 0^\infty = 0, \qquad 0^{-\infty} = \infty$$

Section Exercises (8.5)

In Exercises 1–25, evaluate each limit, using L'Hôpital's Rule where necessary.

1. $\lim\limits_{x \to 2} \dfrac{x^2 - x - 2}{x - 2}$

2. $\lim\limits_{x \to -1} \dfrac{x^2 - x - 2}{x + 1}$

S 3. $\lim\limits_{x \to 0} \dfrac{\sqrt{4 - x^2} - 2}{x}$

4. $\lim\limits_{x \to 2^-} \dfrac{\sqrt{4 - x^2}}{x - 2}$

5. $\lim\limits_{x \to 0} \dfrac{e^x - (1 - x)}{x}$

6. $\lim\limits_{x \to 0^+} \dfrac{e^x - (1 + x)}{x^3}$

S 7. $\lim\limits_{x \to 0^+} \dfrac{e^x - (1 + x)}{x^n}, n = 1, 2, 3, \ldots$

8. $\lim\limits_{x \to 1} \dfrac{\ln x}{x^2 - 1}$

S 9. $\lim\limits_{x \to \infty} \dfrac{\ln x}{x}$

S 10. $\lim\limits_{x \to \infty} \dfrac{e^x}{x}$

11. $\lim\limits_{x \to \infty} \dfrac{3x^2 - 2x + 1}{2x^2 + 3}$

12. $\lim\limits_{x \to \infty} \dfrac{x - 1}{x^2 + 2x + 3}$

13. $\lim\limits_{x\to\infty} \dfrac{x^2 + 2x + 3}{x - 1}$

14. $\lim\limits_{x\to\infty} \dfrac{x^2}{e^x}$

S 15. $\lim\limits_{x\to 0^+} x^2 \ln x$

16. $\lim\limits_{x\to 0} \left(\dfrac{1}{x} - \dfrac{1}{x^2} \right)$

S 17. $\lim\limits_{x\to 2} \left(\dfrac{8}{x^2 - 4} - \dfrac{x}{x - 2} \right)$

18. $\lim\limits_{x\to 2} \left(\dfrac{1}{x^2 - 4} - \dfrac{\sqrt{x - 1}}{x^2 - 4} \right)$

19. $\lim\limits_{x\to\infty} \dfrac{x}{\sqrt{x^2 + 1}}$

20. $\lim\limits_{x\to 1^+} \left(\dfrac{3}{\ln x} - \dfrac{2}{x - 1} \right)$

S 21. $\lim\limits_{x\to 0^+} x^{1/x}$

22. $\lim\limits_{x\to 0^+} (e^x + x)^{1/x}$

23. $\lim\limits_{x\to\infty} x^{1/x}$

24. $\lim\limits_{x\to\infty} \left(1 + \dfrac{a}{x} \right)^x$

25. $\lim\limits_{x\to\infty} (1 + x)^{1/x}$

L'Hôpital's Rule is useful in determining the comparative rates of increase of the functions

$$f(x) = x^m, \qquad g(x) = e^{nx}, \qquad h(x) = (\ln x)^n$$

where $n > 0$, $m > 0$, and $x \to \infty$. The solutions to the next seven exercises suggest that $(\ln x)^n$ tends toward infinity more slowly than x^m, which, in turn, tends toward infinity more slowly than e^{nx}. In Exercises 26–30, use LHôpital's Rule to evaluate each limit.

26. $\lim\limits_{x\to\infty} \dfrac{x^3}{e^{2x}}$

27. $\lim\limits_{x\to\infty} \dfrac{x^2}{e^{5x}}$

28. $\lim\limits_{x\to\infty} \dfrac{(\ln x)^2}{x^3}$

S 29. $\lim\limits_{x\to\infty} \dfrac{(\ln x)^n}{x^m}$, where $0 < n, m$

30. $\lim\limits_{x\to\infty} \dfrac{x^m}{e^{nx}}$, where $0 < n, m$

C 31. Complete the accompanying table to show that x eventually "overpowers" $(\ln x)^4$.

x	10	10^2	10^4	10^6	10^8	10^{10}
$\dfrac{(\ln x)^4}{x}$						

C 32. Complete the following table to show that e^x eventually "overpowers" x^5.

x	1	5	10	20	30	40	50	100
$\dfrac{e^x}{x^5}$								

33. If $f(x) \geq 0$, $\lim_{x\to a} f(x) = 0$, and $\lim_{x\to a} g(x) = \infty$, show that $\lim_{x\to a} f(x)^{g(x)} = 0^\infty = 0$. (Hint: Use logarithmic differentiation.)

34. If $f(x) \geq 0$, $\lim_{x\to a} f(x) = 0$, and $\lim_{x\to a} g(x) = -\infty$, show that $\lim_{x\to a} f(x)^{g(x)} = 0^{-\infty} = \infty$.

35. The Prime Number Theorem states that as x increases, the number of primes less than x approaches $x/\ln x$. Use this approximation together with L'Hôpital's Rule to show that there are infinitely many prime numbers.

36. The Gamma Function $\Gamma(n)$ is defined in terms of the integral of the function

$$f(x) = x^{n-1}e^{-x}, \qquad n > 0$$

Show that for any fixed value of n, the limit of $f(x)$ as x approaches infinity is zero.

S 37. The velocity of an object falling in a resisting medium such as air or water is given by

$$v = \frac{32}{k}\left(1 - e^{-kt} + \frac{v_0 k e^{-kt}}{32} \right)$$

where v_0 is the initial velocity, t is the time, and k is the resistance constant of the medium. Use L'Hôpital's Rule to find the formula for the velocity of a falling body in a vacuum by fixing v_0 and t and letting k approach zero.

38. The formula for the amount A in a savings account compounded n times a year for t years at an interest rate of r and an initial deposit of P is

$$A = P\left(1 + \frac{r}{n} \right)^{nt}$$

Use L'Hôpital's Rule to show that the limiting formula as the number of compoundings per year becomes infinite is

$$A = Pe^{rt}$$

(Note that this limiting formula is used for continuous compounding of interest.)

For Review and Class Discussion

True or False

1. ____ $\lim_{x \to 0} \left[\frac{x^2 + x + 1}{x} \right] = \lim_{x \to 0} \left[\frac{2x + 1}{1} \right] = 1.$

2. ____ If $y = e^x / x^2$, then $y' = e^x / 2x$.

3. ____ If $p(x)$ is a polynomial, then $\lim_{x \to \infty} [p(x)/e^x] = 0$.

4. ____ If
$$\lim_{x \to \infty} \frac{f(x)}{g(x)} = 1$$
then
$$\lim_{x \to \infty} [f(x) - g(x)] = 0$$

5. ____ If both f and g are differentiable and
$$\lim_{x \to 0} \frac{f(x)}{g(x)} = L$$
then
$$\lim_{x \to 0} \frac{f'(x)}{g'(x)} = L$$

Miscellaneous Exercises (Ch. 8)

In Exercises 1–5, find the indicated logarithms using the values $\log_{10} 3 \approx 0.4771$ and $\log_{10} 4 \approx 0.6021$.

1. $\log_{10} 2$
2. $\log_{10} 160$
3. $\log_{10} 1.2$
4. $\log_{10} \frac{1}{12}$
5. $\log_{10} 27$

In Exercises 6–10, solve the given equations for x.

6. $\ln e^x = 8$
7. $e^{\ln x} = 3$
8. $\ln x + \ln (x - 3) = 0$
9. $\log_3 x + \log_3 (x - 1) - \log_3 (x - 2) = 2$
10. $\log_x 125 = 3$

In Exercises 11–30, find dy/dx.

11. $y = \ln \sqrt{x}$

12. $y = \ln \frac{x(x - 1)}{x - 2}$

13. $y = \ln (e^{-x^2})$

14. $y = \ln [x(x^2 - 2)^{2/3}]$

15. $y = x \sqrt{\ln x}$

16. $y = \ln \left(\frac{e^x}{1 + e^x} \right)$

17. $y \ln x + y^2 = 0$

18. $\ln (x + y) = x$

19. $y = \frac{1}{b^2} \left[\ln (a + bx) + \frac{a}{a + bx} \right]$

20. $y = \frac{1}{b^2}[a + bx - a \ln (a + bx)]$

21. $y = \frac{-1}{a} \ln \left(\frac{a + bx}{x} \right)$

22. $y = \frac{-1}{ax} + \frac{b}{a^2} \ln \left(\frac{a + bx}{x} \right)$

23. $y = x^2 e^x$
24. $y = e^{-x^2/2}$
25. $y = \sqrt{e^{2x} + e^{-2x}}$
26. $y = x^{2x+1}$

27. $y = 3^{x-1}$
28. $y = (4e)^x$
29. $ye^x + xe^y = xy$
30. $y = \frac{x^2}{e^x}$

In Exercises 31–50, evaluate the given integral.

31. $\int x e^{-3x^2} \, dx$

32. $\int \frac{e^{1/x}}{x^2} \, dx$

33. $\int \frac{1}{7x - 2} \, dx$

34. $\int \frac{x}{x^2 - 1} \, dx$

35. $\int \frac{1}{x \ln (3x)} \, dx$

36. $\int_1^e \frac{\ln \sqrt{x}}{x} \, dx$

37. $\int \frac{x^2 + 3}{x} \, dx$

38. $\int \frac{x^3 + 1}{x^2} \, dx$

39. $\int_0^1 \frac{e^{4x} - e^{2x} + 1}{e^x} \, dx$

40. $\int \frac{e^{2x} - e^{-2x}}{e^{2x} + e^{-2x}} \, dx$

41. $\int \frac{e^x}{e^x - 1} \, dx$

42. $\int x^2 e^{x^3 + 1} \, dx$

43. $\int_0^1 x e^{-x^2/2} \, dx$

44. $\int_1^2 \frac{2x + 3}{x^2 + 3x - 1} \, dx$

45. $\int \frac{x^2}{x^3 - 1} \, dx$

46. $\int \frac{2x}{(x - 3)^2} \, dx$

47. $\int_0^1 \frac{x}{(x + 3)^2} \, dx$

48. $\int \left(x + \frac{1}{x} \right)^2 \, dx$

49. $\int \frac{1}{x \sqrt{\ln x}} \, dx$

50. $\int \frac{x + 2}{2x + 3} \, dx$

The function $P(t) = (1/\mu)e^{-t/\mu}$ is called the **exponential density function,** where t represents time and μ is the average time between successive occurrences of some event. The definite integral

$$\int_{t_1}^{t_2} \left(\frac{1}{\mu}\right) e^{-t/\mu} \, dt$$

gives the probability that between t_1 and t_2 units of time must elapse before the next occurrence of the event. In Exercises 51 and 52, use this integral to find the required probabilities, assuming that the exponential density function is the appropriate model.

S **51.** Trucks arrive at a terminal at an average rate of 3 per hour (thus $\mu = 20$ min is the average time between arrivals). If a truck has just arrived, find the probability that the next arrival will be:
(a) within the next 10 min
(b) within the next 30 min
(c) between 15 min and 30 min from now
(d) within the next hour

C **52.** The average time between incoming calls at a switchboard is 3 min. Complete the accompanying table to show the probabilities of time elapsed between incoming calls.

Time elapsed between calls	0–2 min	2–4 min	4–6 min	6–8 min	8–10 min
Probability					

53. If $500 is earning interest at the rate of 5% compounded continuously, find its value after:
(a) 1 year   (b) 10 years
(c) 100 years

54. If P dollars bear interest at the rate r compounded continuously and double in value in 10 years, find r.

S **55.** Mr. Jones invests P dollars at 7% interest compounded continuously. Find P if there is to be $10,000 available when his daughter enters college in 15 years.

56. If we assume that a population is growing continuously at the rate of $2\frac{1}{2}$% per year, find the time necessary for the population to (a) double and (b) triple in size.

57. A tank contains 50 gal of salt water in which 25 lb of salt has been dissolved. Fresh water runs into the tank at the rate of 2 gal/min. The mixture is stirred continuously so that it is kept uniform and runs out at the rate of 2 gal/min.
(a) How many pounds of salt remain in the tank after $\frac{1}{2}$ h?
(b) When the brine weakens to the point of having only 15 lb of salt per 50 gal of water, then a 10-lb bag of salt is added to bring the mixture back to 25 lb of

salt per 50 gal of water. How often must this addition of salt be made?

58. Under ideal conditions air pressure decreases continuously with height above sea level at a rate proportional to the pressure at that height. If the barometer reads 30 in. at sea level and 15 in. at 18,000 ft, find the barometric pressure at 35,000 ft.

59. Suppose an object is projected horizontally with an initial velocity of v_0 feet per second and that the resistance to its motion (e.g., air resistance) is proportional to its velocity. Show that the distance traveled from its starting point in t seconds is given by $s = (v_0/K)(1 - e^{-Kt})$.

60. Consider the motion of an object falling from rest and suppose that the resistance due to air is proportional to its velocity. We will measure the distance from which the object started to fall and thus take the downward direction as positive. The change in velocity due to gravity will be 32 ft/s², but the air resistance will cause a force in the opposite direction. Thus the net rate of change in velocity (acceleration) is given by the equation $dv/dt = 32 - Kv$. Integrate this equation to show that the velocity of the object is given by $v(t) = (32/K)(1 - e^{-Kt})$.

61. Assuming that the object of Exercise 60 was projected downward with an initial velocity v_0, find an expression for $v(t)$.

62. Use the result of Exercise 60 to observe the behavior of the velocity of the object as t increases. Does the velocity increase indefinitely? Explain.

S **63.** Integrate the velocity equation of Exercise 60 to find an expression for $s(t)$, the distance the object has fallen.

64. Two numbers are chosen at random between 0 and 10. The probability that their product is less than n $(0 < n < 100)$ is given by

$$P = \frac{1}{100} \left(n + \int_{n/10}^{10} \frac{n}{x} \, dx \right)$$

(a) What is the probability that the product is less than 25?
(b) What is the probability that the product is less than 50?

S **65.** A certain automobile gets 28 mi/gal of gasoline for speeds up to 50 mi/h. Over 50 mi/h the miles per gallon drop at the rate of 12% for each 10 mi/h.
(a) If s is the speed and y is the miles per gallon, find y as a function of s by solving the differential equation

$$\frac{dy}{ds} = -0.012y \qquad (s > 50)$$

(b) Use the function in part (a) to complete the following table.

Speed	50	55	60	65	70
Miles per gallon					

66. A solution of a certain drug contained 500 units per milliliter (mL) when prepared. It was analyzed after 40 days and found to contain 300 units/mL. Assuming that the rate of decomposition is proportional to the amount present, find an equation giving the amount A after t days.

In Exercises 67–74, use L'Hôpital's Rule to evaluate the given limit.

67. $\lim\limits_{x \to 1} \dfrac{(\ln x)^2}{x - 1}$

68. $\lim\limits_{x \to k} \left(\dfrac{\sqrt[3]{x} - \sqrt[3]{k}}{x - k} \right)$

69. $\lim\limits_{x \to \infty} \dfrac{e^{2x}}{x^2}$

70. $\lim\limits_{x \to 1^+} \left(\dfrac{2}{\ln x} - \dfrac{2}{x - 1} \right)$

71. $\lim\limits_{x \to \infty} (\ln x)^{2/x}$

72. $\lim\limits_{x \to 1} (x - 1)^{\ln x}$

73. $\lim\limits_{n \to \infty} 1000 \left(1 + \dfrac{0.09}{n} \right)^n$

74. $\lim\limits_{x \to \infty} x e^{-x^2}$

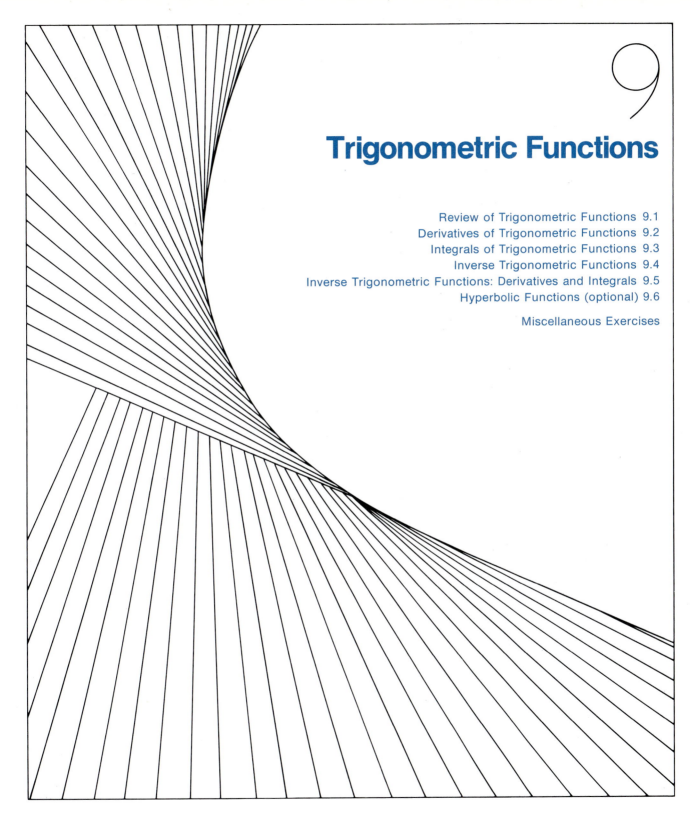

Trigonometric Functions

9

9.1 **Review of Trigonometric Functions**

Purpose
- To review the definition and graphs of the six trigonometric functions.
- To introduce two important limits involving trigonometric functions.

Trigonometry contains the science of continually undulating magnitude: meaning magnitude which becomes alternately greater and less, without any termination to succession of increase and decrease. —*Augustus DeMorgan* (*1806–1871*)

There are two common approaches to the study of trigonometry. In one case the trigonometric functions are defined as ratios of two sides of a right triangle. In the other case these functions are defined in terms of a point on the terminal side of an arbitrary angle. The first approach is the one generally used in surveying, navigation, and astronomy, where a typical problem involves a fixed triangle having three of its six parts (sides and angles) known and three to be determined. The second approach is the one normally used in physics, electronics, and biology, where the periodic nature of the trigonometric functions is emphasized. In the following definition we define the six trigonometric functions from both viewpoints.

Definition of the Six Trigonometric Functions

Refer to Figure 9.1:

$$\sin \theta = \frac{\text{opp.}}{\text{hyp.}} \qquad \csc \theta = \frac{\text{hyp.}}{\text{opp.}}$$

$$\cos \theta = \frac{\text{adj.}}{\text{hyp.}} \qquad \sec \theta = \frac{\text{hyp.}}{\text{adj.}}$$

$$\tan \theta = \frac{\text{opp.}}{\text{adj.}} \qquad \cot \theta = \frac{\text{adj.}}{\text{opp.}}$$

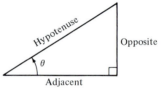

FIGURE 9.1

Refer to Figure 9.2:

$$\sin \theta = \frac{y}{r} \qquad \csc \theta = \frac{r}{y}$$

$$\cos \theta = \frac{x}{r} \qquad \sec \theta = \frac{r}{x}$$

$$\tan \theta = \frac{y}{x} \qquad \cot \theta = \frac{x}{y}$$

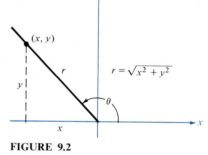

$$r = \sqrt{x^2 + y^2}$$

FIGURE 9.2

In Figure 9.2 the angle θ (measured counterclockwise) is in **standard position;** that is, its vertex is at the origin and its initial side is on the positive x-axis. If two angles in standard position have the same terminal side, they are called **coterminal angles.** For instance, the angles 45°, −315°, and 405° are coterminal.

Note also from Figure 9.2 that r is always positive and thus the quad-

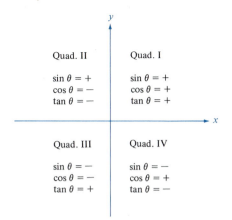

Quad. II   Quad. I

$\sin \theta = +$ $\sin \theta = +$
$\cos \theta = -$ $\cos \theta = +$
$\tan \theta = -$ $\tan \theta = +$

Quad. III   Quad. IV

$\sin \theta = -$ $\sin \theta = -$
$\cos \theta = -$ $\cos \theta = +$
$\tan \theta = +$ $\tan \theta = -$

FIGURE 9.3

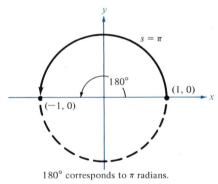

180° corresponds to π radians.

FIGURE 9.4

rant signs of x and y determine the quadrant signs of the various trigonometric functions, as shown in Figure 9.3.

The following formulas are direct consequences of the definitions:

$$\csc \theta = \frac{1}{\sin \theta} \qquad \cot \theta = \frac{1}{\tan \theta} \qquad \cot \theta = \frac{\cos \theta}{\sin \theta}$$

$$\sec \theta = \frac{1}{\cos \theta} \qquad \tan \theta = \frac{\sin \theta}{\cos \theta}$$

Furthermore, since

$$\sin^2 \theta + \cos^2 \theta = \left(\frac{x}{r}\right)^2 + \left(\frac{y}{r}\right)^2 = \frac{x^2 + y^2}{r^2} = \frac{r^2}{r^2} = 1$$

we can readily obtain the Pythagorean Identities [note: $\sin^2 \theta = (\sin \theta)^2$]:

$$\sin^2 \theta + \cos^2 \theta = 1$$
$$\tan^2 \theta + 1 = \sec^2 \theta$$
$$1 + \cot^2 \theta = \csc^2 \theta$$

Additional trigonometric identities are listed on the inside cover of the text.

Although an angle θ may be measured in either degrees or radians (rad), in calculus radian measure is preferred. One reason for this preference will be seen in the proof of Theorem 9.1. In any case, we assume that all angles in this text are measured in radians, unless otherwise stated. Thus when we write sin 3, we mean the sine of 3 radians, not 3 degrees.

To assign a radian measure to θ, we consider θ to be a central angle of a unit circle and measure the length s of the subtended arc. For $\theta = 180°$ in the unit circle shown in Figure 9.4, we have the subtended arc

$$s = \tfrac{1}{2} \text{ circumference} = \pi$$

Thus we have the relationship

$$180° = \pi \text{ radians}$$

Equivalent formulas are

$$1° = \frac{\pi}{180} \text{ radians}$$

$$\left(\frac{180}{\pi}\right)° = 1 \text{ radian} \approx 57.296°$$

The degree and radian measure of several common angles are given in Table 9.1 along with the corresponding values of the sine, cosine, and tangent.

To extend the use of Table 9.1 to angles in the *third* and *fourth* quadrants, we use the reduction formulas

$$\sin \theta = -\sin (\theta - \pi), \qquad \cos \theta = -\cos (\theta - \pi), \qquad \tan \theta = \tan (\theta - \pi)$$

TABLE 9.1

Radians	0	$\dfrac{\pi}{6}$	$\dfrac{\pi}{4}$	$\dfrac{\pi}{3}$	$\dfrac{\pi}{2}$	$\dfrac{2\pi}{3}$	$\dfrac{3\pi}{4}$	$\dfrac{5\pi}{6}$	π	$\dfrac{3\pi}{2}$	2π
Degrees	$0°$	$30°$	$45°$	$60°$	$90°$	$120°$	$135°$	$150°$	$180°$	$270°$	$360°$
$\sin\theta$	0	$\dfrac{1}{2}$	$\dfrac{\sqrt{2}}{2}$	$\dfrac{\sqrt{3}}{2}$	1	$\dfrac{\sqrt{3}}{2}$	$\dfrac{\sqrt{2}}{2}$	$\dfrac{1}{2}$	0	-1	0
$\cos\theta$	1	$\dfrac{\sqrt{3}}{2}$	$\dfrac{\sqrt{2}}{2}$	$\dfrac{1}{2}$	0	$-\dfrac{1}{2}$	$-\dfrac{\sqrt{2}}{2}$	$-\dfrac{\sqrt{3}}{2}$	-1	0	1
$\tan\theta$	0	$\dfrac{\sqrt{3}}{3}$	1	$\sqrt{3}$	undefined	$-\sqrt{3}$	-1	$-\dfrac{\sqrt{3}}{3}$	0	undefined	0

For example, to find the sine of $7\pi/6$, we subtract π from $7\pi/6$ and obtain

$$\sin\left(\frac{7\pi}{6}\right) = -\sin\left(\frac{7\pi}{6} - \pi\right) = -\sin\left(\frac{\pi}{6}\right) = -\frac{1}{2}$$

Example 1

Solve the following equation for θ:

$$\cos 2\theta = 2 - 3\sin\theta \qquad 0 \le \theta \le 2\pi$$

Solution: Using the double-angle identity

$$\cos 2\theta = 1 - 2\sin^2\theta$$

we have

$$1 - 2\sin^2\theta = 2 - 3\sin\theta$$
$$0 = 2\sin^2\theta - 3\sin\theta + 1$$
$$0 = (2\sin\theta - 1)(\sin\theta - 1)$$

If $2\sin\theta - 1 = 0$, we have $\sin\theta = \frac{1}{2}$ and

$$\theta = \frac{\pi}{6} \qquad \text{or} \qquad \frac{5\pi}{6}$$

If $\sin\theta - 1 = 0$, we have $\sin\theta = 1$ and

$$\theta = \frac{\pi}{2}$$

Thus for $0 \le \theta \le 2\pi$, there are three solutions to the given equation:

$$\theta = \frac{\pi}{6}, \qquad \frac{\pi}{2}, \qquad \text{or} \qquad \frac{5\pi}{6}$$

∎

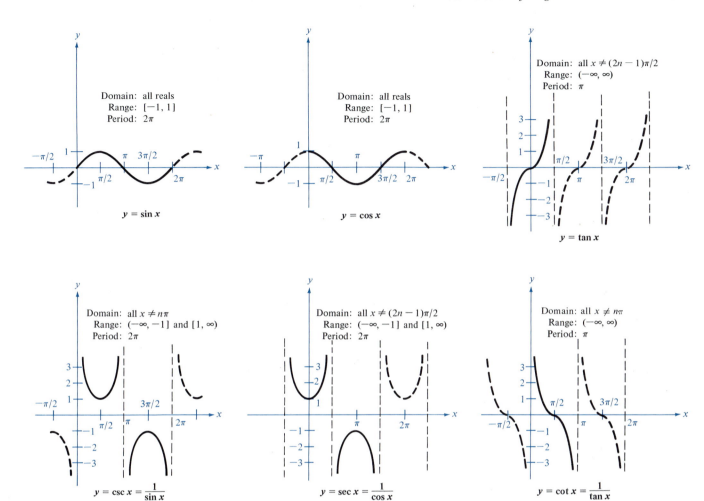

FIGURE 9.5

For ready reference we show, in Figure 9.5, the graphs of one cycle of each of the six basic trigonometric functions. Observe that each function is periodic; its graph repeats after a certain interval called the **period** of the function. The **amplitude** (of the sine and cosine functions) is half the difference between the maximum and minimum values of each function. For instance, the amplitude of $y = \sin x$ is 1, while the amplitude of $y = 2 \sin x$ is 2.

Familiarity with the graphs of the six basic trigonometric functions enables us to readily sketch graphs of more general functions, such as

$$y = a \sin (bx + c) \quad \text{or} \quad y = a \cos (bx + c)$$

as indicated in Table 9.2. Note that by adding a constant value c to the angle, the graph is shifted (left or right) $|c/b|$ units from its normal position.

TABLE 9.2

$y = \sin x$ or $y = \cos x$		$y = a \sin (bx + c)$ or $y = a \cos (bx + c)$	
Amplitude:	1	Amplitude:	$\lvert a \rvert$
Period:	2π	Period:	$\left\lvert \dfrac{2\pi}{b} \right\rvert$
Horizontal Shift: none		If $\dfrac{c}{b} > 0$, Left Shift:	$\dfrac{c}{b}$
		If $\dfrac{c}{b} < 0$, Right Shift:	$\left\lvert \dfrac{c}{b} \right\rvert$

In the next three examples, we demonstrate the procedure for graphing trigonometric functions.

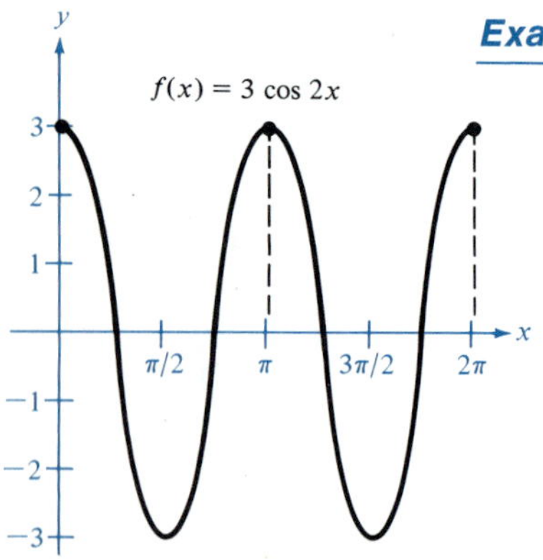

Example 2

Sketch the graph of $f(x) = 3 \cos 2x$.

Solution: The graph of $f(x) = 3 \cos 2x$ has the following characteristics. (Refer to Table 9.2.)

Amplitude: 3
Period: $2\pi/2 = \pi$
No horizontal shift

Two cycles of the graph are shown in Figure 9.6, starting with the maximum point $(0, 3)$. ■

FIGURE 9.6

Example 3

Sketch the graph of $f(x) = 2 \sin [3x - (\pi/2)]$.

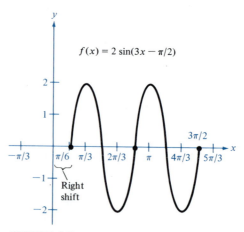

FIGURE 9.7

Solution: The graph of f has the following characteristics. (Refer to Table 9.2.)

> Amplitude: 2
> Period: $2\pi/3$
> Right shift: $|(-\pi/2)/3| = \pi/6$

Since the period is $2\pi/3$, and if we start one cycle at $x = \pi/6$, then this cycle ends at $x = (\pi/6) + (2\pi/3) = (5\pi/6)$. Two complete cycles are shown in Figure 9.7. ■

Example 4 *(Addition of Ordinates)*

Sketch the graph of $f(x) = x + \cos x$ for $0 \leq x \leq 2$.

Solution: This graph can be obtained by a procedure called "addition of ordinates." We first make an accurate sketch of the graphs of $y = x$ and $y = \cos x$ on the same coordinate plane and then geometrically add the ordinates (y-values) for each x. This addition is aided by the use of a compass to measure the displacement of one graph from the x-axis and then to mark off an equal displacement from the other graph. (See Figure 9.8.)

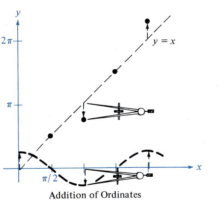 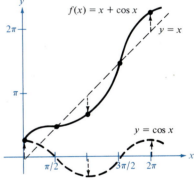

Addition of Ordinates

FIGURE 9.8 ■

In the next section we will see how the derivatives of the trigonometric functions can assist us in sketching more complicated graphs. However, before we discuss the derivatives of the trigonometric functions, we need to evaluate two important limits:

$$\lim_{x \to 0} \frac{\sin x}{x} \quad \text{and} \quad \lim_{x \to 0} \frac{1 - \cos x}{x}$$

In the first instance note that $(\sin x)/x$ is defined and continuous for $x \neq 0$. Furthermore, since

$$\frac{\sin(-x)}{-x} = \frac{-\sin x}{-x} = \frac{\sin x}{x}$$

we need only investigate the limit as $x \to 0^+$ in order to evaluate $\lim_{x \to 0} (\sin x)/x$. We consider this information in the proof of the following theorem:

THEOREM 9.1

$$\lim_{x \to 0} \frac{\sin x}{x} = 1$$

Proof: By direct substitution we obtain

$$\lim_{x \to 0^+} \frac{\sin x}{x} = \frac{0}{0}$$

which is an indeterminate form. Previously under such circumstances we did some algebraic manipulating so as to cancel a factor, which would then allow us to evaluate the limit. Such manipulation is not possible here, nor can we apply L'Hôpital's Rule without knowing the derivative of $\sin x$. Thus we pursue a different course of action.

For $0 < x < \pi/2$, consider the situation shown in Figure 9.9. From this figure it is obvious that

$$\text{area } \triangle BOP \le \text{area sector } BOP \le \text{area } \triangle BOQ$$

Since the radius of the circular arc is 1, we have

$$(\text{area of } \triangle BOP) = \frac{1}{2}rh = \frac{h}{2}$$

$$(\text{area of sector } BOP) = \frac{1}{2}r^2x = \frac{x}{2} \qquad (x \text{ in radians})$$

$$(\text{area of } \triangle BOQ) = \frac{1}{2}rH = \frac{H}{2}$$

Furthermore, we know that

$$\sin x = \frac{h}{r} = h \qquad \text{and} \qquad \tan x = \frac{H}{r} = H$$

Therefore, the inequality

$$\text{area } \triangle BOP \le \text{area sector } BOP \le \text{area } \triangle BOQ$$

can be written as

$$\frac{\sin x}{2} \le \frac{x}{2} \le \frac{\tan x}{2}$$

Multiplication by the positive quantity $2/(\sin x)$ gives us

$$1 \le \frac{x}{\sin x} \le \frac{1}{\cos x}$$

or

$$1 \ge \frac{\sin x}{x} \ge \cos x$$

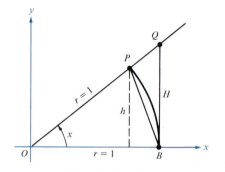

H = height of $\triangle BOQ$
h = height of $\triangle BOP$
r = radius of circular arc

FIGURE 9.9

By the Squeeze Theorem of Section 2.4, it follows that

$$\lim_{x \to 0^+} 1 = \lim_{x \to 0^+} \frac{\sin x}{x} = \lim_{x \to 0^+} \cos x = 1 \quad \blacksquare$$

It is of interest to note that the convenient limit of 1 in Theorem 9.1 is dependent on the radian measure of x. If x were measured in degrees rather than radians, then we would have the cumbersome limit

$$\lim_{x \to 0} \frac{\sin (x^\circ)}{x} = \frac{\pi}{180} \approx 0.0175$$

THEOREM 9.2
$$\lim_{x \to 0} \frac{1 - \cos x}{x} = 0$$

Proof: First, we write

$$\frac{1 - \cos x}{x} = \left(\frac{1 - \cos x}{x} \right)\left(\frac{1 + \cos x}{1 + \cos x} \right) = \frac{1 - \cos^2 x}{x(1 + \cos x)} = \frac{\sin^2 x}{x(1 + \cos x)}$$

Hence

$$\lim_{x \to 0} \frac{1 - \cos x}{x} = \lim_{x \to 0} \left(\frac{\sin x}{x} \right)\left(\frac{\sin x}{1 + \cos x} \right)$$

$$= \left(\lim_{x \to 0} \frac{\sin x}{x} \right)\left(\lim_{x \to 0} \frac{\sin x}{1 + \cos x} \right) = 1 \left(\frac{0}{1 + 1} \right) = 0 \quad \blacksquare$$

Example 5

Use Theorem 9.1 to find the slope of $f(x) = \sin x$ when $x = 0$.

Solution: Recalling the limit formula for the derivative (slope) of a function, we have

$$f'(x) = \lim_{\Delta x \to 0} \frac{f(x + \Delta x) - f(x)}{\Delta x}$$

Therefore, the slope of $f(x) = \sin x$ when $x = 0$ is

$$f'(0) = \lim_{\Delta x \to 0} \frac{f(0 + \Delta x) - f(0)}{\Delta x} = \lim_{\Delta x \to 0} \frac{\sin (\Delta x) - \sin (0)}{\Delta x}$$

$$= \lim_{\Delta x \to 0} \frac{\sin \Delta x}{\Delta x} = 1 \quad \text{(by Theorem 9.1)}$$

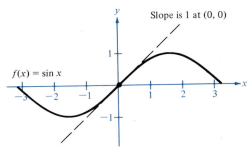

FIGURE 9.10

Figure 9.10 reinforces our conclusion. $\quad \blacksquare$

Section Exercises (9.1)

1. Sketch each angle in standard position and determine a coterminal angle θ such that $0° \leq \theta \leq 360°$.
(a) $390°$ (b) $-45°$ (c) $725°$
(d) $415°30'$ (e) $-300°15'$

2. Sketch each angle in standard position and determine a coterminal angle θ such that $0 \leq \theta \leq 2\pi$.
(a) $8\pi/3$ (b) $15\pi/6$ (c) $-2\pi/3$
(d) $9\pi/4$ (e) $-\pi/12$

S 3. Express each angle in radian measure as a multiple of π.
(a) $90°$ (b) $120°$ (c) $-315°$
(d) $240°$ (e) $20°$

C 4. Express each angle in radian measure (to the nearest hundredth).
(a) $-725°$ (b) $34°12'$ (c) $225°50'$
(d) $36°24'$ (e) $108°$

C 5. Express each angle in degrees (to the nearest hundredth).
(a) $7\pi/6$ (b) $5\pi/4$ (c) $2\pi/5$
(d) $\pi/9$ (e) $2\pi/45$ (f) 2.75
(g) 0.486 (h) 3.0 (i) 7.25
(j) 1.0

6. The given points are on the terminal side of an angle θ in standard position. Determine the values of $\sin\theta$, $\cos\theta$, and $\tan\theta$.
(a) $(3, 4)$ (b) $(8, -15)$ (c) $(-12, -5)$
(d) $(1, -1)$ (e) $(-\sqrt{3}, 1)$

S 7. Without using tables or a calculator, find the indicated value(s) from the given value.
(a) $\sin\theta = \frac{1}{2}$, find $\tan\theta$.   (b) $\sin\theta = \frac{1}{2}$, find $\csc\theta$.
(c) $\cos\theta = \frac{4}{5}$, find $\cot\theta$.   (d) $\sec\theta = \frac{13}{5}$, find $\cot\theta$.
(e) $\tan\theta = \frac{15}{8}$, find $\sec\theta$.

8. Determine the quadrant in which θ lies.
(a) $\sin\theta > 0$ and $\cos\theta < 0$
(b) $\tan\theta > 0$ and $\cos\theta < 0$
(c) $\csc\theta < 0$ and $\cos\theta < 0$
(d) $\sec\theta > 0$ and $\sin\theta > 0$
(e) $\tan\theta < 0$ and $\sin\theta > 0$

C 9. Evaluate to four decimal places.
(a) $\sin(-10°)$ (b) $\sec(225°)$
(c) $\cos(\pi/4)$ (d) $\tan(0.85)$
(e) $\csc(-0.23)$

10. Using the table in Appendix C, find two values of θ, $0° < \theta < 360°$, that satisfy each of the given equations.
(a) $\tan\theta = 0.1763$ (b) $\sin\theta = 0.8746$
(c) $\cos\theta = -0.7234$ (d) $\cot\theta = -0.7133$
(e) $\tan\theta = 2.282$ (f) $\sin\theta = -0.6583$

In Exercises 11–20, solve the given equation for x, $0 \leq x < 2\pi$.

11. $2\sin^2 x = 1$ **12.** $\tan^2 x = 3$

S 13. $\tan^2 x = \tan x$ **14.** $2\cos(x - \pi) = -\sqrt{3}$

15. $2\cos^2 x - \cos x = 1$ **16.** $\sin^2 x = 3\cos^2 x$

S 17. $\sin 2x + \sqrt{2}\sin x = 0$ **18.** $\cos x = \cos(x/2)$

19. $\cos 4x - 7\cos 2x = 8$ **20.** $\sec x \csc x = 2\csc x$

In Exercises 21–28, graph each function through two periods.

21. $f(x) = 2\sin\left(\dfrac{2x}{3}\right)$ **22.** $f(x) = \dfrac{1}{2}\cos\left(\dfrac{x}{3}\right)$

S 23. $f(x) = \cos\left(2x - \dfrac{\pi}{3}\right)$ **24.** $f(x) = -\sin\left(2x + \dfrac{\pi}{2}\right)$

25. $f(x) = \tan\left(\dfrac{x}{2}\right)$ **26.** $f(x) = \csc 2x$

27. $f(x) = \sec\left(x - \dfrac{\pi}{4}\right)$ **28.** $f(x) = \cot 3x$

In Exercises 29–32, use addition of ordinates to sketch the graph of the given equation.

29. $y = x + \cos x$ **30.** $y = x + 2\sin x$

S 31. $f(x) = \sin x + \sin 2x$ **32.** $f(x) = \sin x + \cos 2x$

In Exercises 33–40, solve for x, y, or r, as indicated.

S 33.

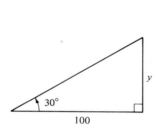

34.

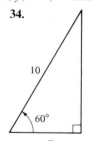

35.

36.

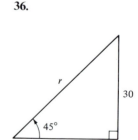

37.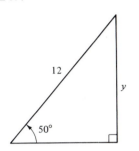

38.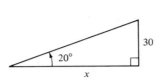

39.

40.

In Exercises 41–45, use the limits

$$\lim_{\Delta x \to 0} \frac{\sin \Delta x}{\Delta x} = 1 \quad \text{and} \quad \lim_{\Delta x \to 0} \frac{1 - \cos \Delta x}{\Delta x} = 0$$

to find $f'(0)$.

41. $f(x) = \tan x$ **42.** $f(x) = \sec x$

43. $f(x) = \sin x + 1$ **44.** $f(x) = 1 - \cos x$

45. $f(x) = \sin 2x = 2 \sin x \cos x$

46. Demonstrate that $\lim_{x \to 0} (\sin x)/x = 1$ by completing the accompanying table.

x	1	10^{-1}	10^{-2}	10^{-3}
$\dfrac{\sin x}{x}$				

47. Make a table of values as in Exercise 46 to evaluate $\lim_{x \to 0} (\sin 3x)/(\sin 4x)$.

For Review and Class Discussion

True or False

1. _____ $\sin (x + y) = \sin x + \sin y$.

2. _____ $\cos (ax) = a \cos x$.

3. _____ The sine and cosine functions are continuous on the entire real line.

4. _____ If $0 < x < \pi/2$, then $\sin x < x$.

5. _____ If $f(x) = \sin^2 x$, then $f(\pi/4) = \frac{1}{2}$.

9.2 Derivatives of Trigonometric Functions

Purpose
- To derive formulas for the derivatives of the trigonometric functions.
- To demonstrate the use of the derivative to evaluate limits and to determine critical points on the graphs of trigonometric functions.

For transcendental problems, wherever dimensions and tangents occur that have to be found by computation, there can hardly be found a calculus more useful, shorter, and more universal than my differential calculus.
             —Gottfried Leibniz (1646–1716)

To differentiate $f(x) = \sin x$, we use the limit definition of the derivative and the four-step process.

1. $f(x + \Delta x) = \sin (x + \Delta x) = \sin x \cos \Delta x + \cos x \sin \Delta x$

2. $f(x + \Delta x) - f(x) = \sin x \cos \Delta x + \cos x \sin \Delta x - \sin x$
$$= \cos x \sin \Delta x - \sin x(1 - \cos \Delta x)$$

3. $$\frac{f(x + \Delta x) - f(x)}{\Delta x} = \cos x \left(\frac{\sin \Delta x}{\Delta x}\right) - \sin x \left(\frac{1 - \cos \Delta x}{\Delta x}\right)$$

4. $$\lim_{\Delta x \to 0} \frac{f(x + \Delta x) - f(x)}{\Delta x}$$

$$= \lim_{\Delta x \to 0} \left[\cos x \left(\frac{\sin \Delta x}{\Delta x}\right) - \sin x \left(\frac{1 - \cos \Delta x}{\Delta x}\right)\right]$$

$$= (\cos x)(1) - (\sin x)(0)$$

(Note the use of the limits derived in Section 9.1.) Therefore,

$$f'(x) = \lim_{\Delta x \to 0} \frac{f(x + \Delta x) - f(x)}{\Delta x} = \cos x$$

and we write

$$\frac{d}{dx}[\sin x] = \cos x$$

This relationship is shown in Figure 9.11. Note that the *slope* of the sine curve determines the function *value* of the cosine curve. We can generalize this formula with the Chain Rule by considering $y = \sin u$, where u is a differentiable function of x. Then

$$\frac{dy}{dx} = \frac{dy}{du}\frac{du}{dx} = \cos u \frac{du}{dx}$$

Therefore we have

$$\frac{d}{dx}[\sin u] = \cos u \frac{du}{dx}$$

The derivatives of the other trigonometric functions could be obtained by using the definition of the derivative, but it is much simpler to use trigonometric identities and the formula just derived. For instance, since $\cos u = \sin[(\pi/2) - u]$, we have

$$\frac{d}{dx}[\cos u] = \frac{d}{dx}\left[\sin\left(\frac{\pi}{2} - u\right)\right] = \cos\left(\frac{\pi}{2} - u\right)\left(\frac{-du}{dx}\right) = -\sin u \frac{du}{dx}$$

Considering

$$\tan u = \frac{\sin u}{\cos u}$$

and applying the Quotient Rule, we obtain

$$\frac{d}{dx}[\tan u] = \frac{(\cos u)(\cos u)u' - (\sin u)(-\sin u)u'}{\cos^2 u}$$

$$= \frac{(\cos^2 u + \sin^2 u)u'}{\cos^2 u} = \left(\frac{1}{\cos^2 u}\right)u' = \sec^2 u \frac{du}{dx}$$

The derivatives of $\sec u$, $\csc u$, and $\cot u$ can also be obtained from the derivatives of $\sin u$ and $\cos u$, and these derivations are left as an exercise. We summarize the results next.

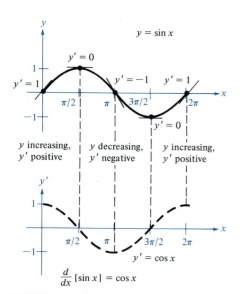

$y = \sin x$

$y' = 0$

$y' = 1$ $y' = -1$ $y' = 1$

$\pi/2$ π $3\pi/2$ 2π

$y' = 0$

y increasing, *y'* positive | *y* decreasing, *y'* negative | *y* increasing, *y'* positive

$\pi/2$ π $3\pi/2$ 2π

$y' = \cos x$

$\frac{d}{dx}[\sin x] = \cos x$

FIGURE 9.11

Derivatives of Trigonometric Functions

$$\frac{d}{dx}[\sin u] = \cos u \, \frac{du}{dx} \qquad \frac{d}{dx}[\cos u] = -\sin u \, \frac{du}{dx}$$

$$\frac{d}{dx}[\tan u] = \sec^2 u \, \frac{du}{dx} \qquad \frac{d}{dx}[\cot u] = -\csc^2 u \, \frac{du}{dx}$$

$$\frac{d}{dx}[\sec u] = \sec u \tan u \, \frac{du}{dx} \qquad \frac{d}{dx}[\csc u] = -\csc u \cot u \, \frac{du}{dx}$$

As an aid to memorization, note that the cofunctions (cosine, cotangent, and cosecant) require a negative sign as part of their derivatives.

Example 1

Differentiate $y = \cos 3x^2$.

Solution: We consider $u = 3x^2$; then

$$y' = -\sin u \, \frac{du}{dx} = -\sin 3x^2 \frac{d}{dx}[3x^2]$$

$$= -(\sin 3x^2)(6x) = -6x \sin 3x^2 \qquad \blacksquare$$

Example 2

Differentiate $y = \tan^4 3x$.

Solution: By the Power Rule, we have

$$y' = 4 \tan^3 3x \frac{d}{dx}[\tan 3x] = 4(\tan^3 3x)(\sec^2 3x)(3)$$

(Don't forget the u' factor 3 in the derivative of $\tan 3x$.) Thus

$$y' = 12 \tan^3 3x \sec^2 3x \qquad \blacksquare$$

Example 3

Differentiate $y = \csc(x/2)$.

Solution: $y' = -\csc \dfrac{x}{2} \cot \dfrac{x}{2} \dfrac{d}{dx}\left[\dfrac{x}{2}\right] = -\dfrac{1}{2} \csc \dfrac{x}{2} \cot \dfrac{x}{2} \qquad \blacksquare$

Example 4

Find the derivative of $f(t) = \sqrt{\sin 4t}$.

Solution: By the Power Rule

$$f'(t) = \frac{1}{2}(\sin 4t)^{-1/2}(\cos 4t)(4) = \frac{2 \cos 4t}{\sqrt{\sin 4t}} \qquad \blacksquare$$

Example 5

Find the derivative of $g(x) = e^{-2x} \cot x$.

Solution: By the Product Rule

$$g'(x) = e^{-2x}(-\csc^2 x) + (\cot x)(e^{-2x})(-2)$$
$$= -e^{-2x}(\csc^2 x + 2 \cot x) \qquad \blacksquare$$

Example 6

Determine the relative extrema of the graph of

$$y = \frac{1 + \cos x}{1 + \sin x}$$

on the interval $(-\pi/2, 3\pi/2)$.

Solution: First,

$$y' = \frac{(1 + \sin x)(-\sin x) - (1 + \cos x)(\cos x)}{(1 + \sin x)^2}$$
$$= \frac{-\sin x - \cos x - (\sin^2 x + \cos^2 x)}{(1 + \sin x)^2} = \frac{-1 - \sin x - \cos x}{(1 + \sin x)^2}$$

On the interval $(-\pi/2, 3\pi/2)$ the equation

$$0 = -1 - \sin x - \cos x$$
$$-1 = \sin x + \cos x$$

has only one solution, $x = \pi$. By the First-Derivative Test, the point $(\pi, 0)$ is a minimum (Figure 9.12). $\blacksquare$

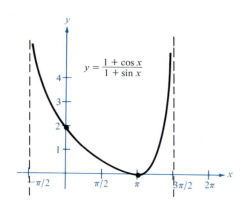

FIGURE 9.12

Example 7

Find the relative extrema and sketch the graph of $f(x) = 2 \sin x - \cos 2x$ on the interval $[0, 2\pi]$.

Solution: By setting $f'(x)$ equal to zero, we have

$$f'(x) = 2 \cos x + 2 \sin 2x = 0$$

Since $\sin 2x = 2 \cos x \sin x$, we have

$$2 \cos x + 4 \cos x \sin x = 0$$
$$2(\cos x)(1 + 2 \sin x) = 0$$

This equation has solutions when $\cos x = 0$ or when $\sin x = -\frac{1}{2}$. Thus $f'(x) = 0$ if

$$x = \frac{\pi}{2}, \frac{3\pi}{2} \qquad \text{or} \qquad x = \frac{7\pi}{6}, \frac{11\pi}{6}$$

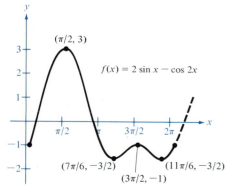

FIGURE 9.13

By the Second-Derivative Test, we can determine that $(\pi/2, 3)$ and $(3\pi/2, -1)$ are relative maxima, and $(7\pi/6, -3/2)$ and $(11\pi/6, -3/2)$ are relative minima (Figure 9.13). ∎

Example 8

Evaluate

$$\lim_{x \to 0} \frac{\tan 2x}{x}$$

Solution: By direct substitution we obtain the indeterminate form 0/0. Applying L'Hôpital's Rule we obtain

$$\lim_{x \to 0} \frac{\tan 2x}{x} = \lim_{x \to 0} \frac{2 \sec^2 2x}{1} = 2 \qquad ∎$$

Example 9

Evaluate

$$\lim_{x \to 0^+} \frac{\cot x}{\ln x}$$

Solution: Since

$$\lim_{x \to 0^+} \frac{\cot x}{\ln x} = \frac{\infty}{-\infty}$$

we apply L'Hôpital's Rule to obtain

$$\lim_{x \to 0^+} \frac{\cot x}{\ln x} = \lim_{x \to 0^+} \frac{-\csc^2 x}{1/x} = \lim_{x \to 0^+} \frac{-x}{\sin^2 x} = \frac{0}{0}$$

Applying L'Hôpital's Rule again we get

$$\lim_{x \to 0^+} \frac{\cot x}{\ln x} = \lim_{x \to 0^+} \frac{-x}{\sin^2 x} = \lim_{x \to 0^+} \frac{-1}{2 \sin x \cos x} = \frac{-1}{0^+} = -\infty$$

and we conclude that the limit does not exist. ∎

Example 10 (*Simple Harmonic Motion*)

The height of an object attached to a spring is given by the *simple harmonic motion* equation

$$y = \tfrac{1}{6} \cos 2t + \tfrac{3}{8} \sin 2t$$

where y is measured in inches and t is measured in seconds. Calculate the height and velocity of the object when $t = 0, 0.25, 0.5, 0.75,$ and 1 second.

Solution: The velocity of the object is given by

$$v = \frac{dy}{dt} = -\frac{1}{3} \sin 2t + \frac{3}{4} \cos 2t$$

The desired values of y and v are shown in Table 9.3, and the motion of the spring is indicated in Figure 9.14.

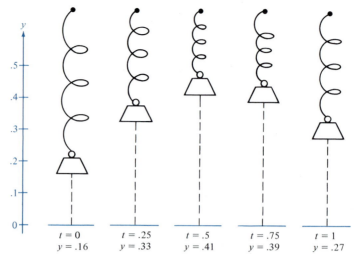

FIGURE 9.14

TABLE 9.3

t (s)	0	0.25	0.5	0.75	1
y (in.)	0.16	0.33	0.41	0.39	0.27
v (in./s)	0.75	0.50	0.13	-0.28	-0.62

Example 11

Find the maximum height of the object in Example 10.

Solution: To find the maximum value of y, we set the derivative equal to zero:

$$\frac{dy}{dt} = -\frac{1}{3} \sin 2t + \frac{3}{4} \cos 2t = 0$$

$$\frac{1}{3} \sin 2t = \frac{3}{4} \cos 2t$$

$$\tan 2t = \frac{9}{4}$$

Now to find the value of y when $\tan 2t = \frac{9}{4}$, we can use the triangle in Figure 9.15. Thus

$$\sin 2t = \frac{9}{\sqrt{97}} \quad \text{and} \quad \cos 2t = \frac{4}{\sqrt{97}}$$

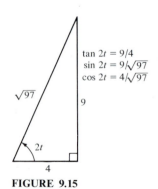

$\tan 2t = 9/4$
$\sin 2t = 9/\sqrt{97}$
$\cos 2t = 4/\sqrt{97}$

FIGURE 9.15

and we have

$$y = \frac{1}{6}\cos 2t + \frac{3}{8}\sin 2t$$

$$= \frac{1}{6}\left(\frac{4}{\sqrt{97}}\right) + \frac{3}{8}\left(\frac{9}{\sqrt{97}}\right) = \frac{\sqrt{97}}{24} \approx 0.4104 \text{ in.}$$

From Figure 9.14 we can see that this maximum occurs between $t = 0.5$ second and $t = 0.75$ second. ∎

Example 12

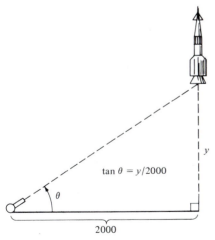

$\tan \theta = y/2000$

2000

FIGURE 9.16

A television camera at ground level is filming the lift-off of a rocket that is rising vertically with a constant acceleration of 100 ft/s². If the camera is 2000 ft from the launch pad, find the rate of change in the angle of elevation of the camera 10 s after lift-off.

Solution: We let θ be the angle of elevation of the camera and y be the height of the rocket. Since the acceleration of the rocket is 100 ft/s², we have

$$\frac{d^2y}{dt^2} = 100$$

$$\frac{dy}{dt} = 100t + v_0$$

$$y = 50t^2 + v_0 t + h_0$$

Since both the initial velocity and the initial height are zero, we have $y = 50t^2$. Now from Figure 9.16, we see that θ and y are related by the equation

$$\tan \theta = \frac{y}{2000} = \frac{50t^2}{2000}$$

Differentiating with respect to t, we have

$$\sec^2 \theta \left(\frac{d\theta}{dt}\right) = \frac{t}{20}$$

or

$$\frac{d\theta}{dt} = \cos^2 \theta \left(\frac{t}{20}\right)$$

Finally, when $t = 10$, then $y = 5000$, $\cos \theta = 2000/\sqrt{2000^2 + 5000^2} = 2/\sqrt{29}$, and, therefore,

$$\frac{d\theta}{dt} = \left(\frac{2}{\sqrt{29}}\right)^2 \left(\frac{10}{20}\right) = \frac{2}{29} \text{ rad/s}$$

∎

Section Exercises (9.2)

In Exercises 1–40, find dy/dx.

1. $y = \sin 2x$

2. $y = \cos 3x$

[S] **3.** $y = 2 \cos \left(\dfrac{x}{2} \right)$

4. $y = 3 \tan 4x$

5. $y = \sec (x^2)$

6. $y = \dfrac{1}{\sec x}$

7. $y = x \sin x$

8. $y = \sin \pi x$

9. $y = \cos \pi x$

[S] **10.** $y = \frac{1}{4} \sin^2 2x$

11. $y = \sin x \cos x$

12. $y = \dfrac{\cos x}{\sin x}$

13. $y = \sqrt{\sin x}$

14. $y = \csc^2 x$

15. $y = \tan \left(\pi x - \dfrac{\pi}{2} \right)$

16. $y = \cot \left(\dfrac{x}{2} + \dfrac{\pi}{4} \right)$

17. $y = x \sin \left(\dfrac{1}{x} \right)$

18. $y = x^2 \sin \left(\dfrac{1}{x} \right)$

[S] **19.** $y = \sec^3 (2x)$

20. $y = \dfrac{\cos x + 1}{x}$

21. $y = \frac{1}{2} \csc 2x$

22. $y = \dfrac{1 + \sin x}{1 - \sin x}$

23. $y = \tan x - x$

24. $y = \frac{2}{5} \sin^{3/2} x - \frac{2}{7} \sin^{7/2} x$

25. $y = \dfrac{x}{2} - \dfrac{\sin 2x}{4}$

26. $y = \dfrac{\sec^7 x}{7} - \dfrac{\sec^5 x}{5}$

27. $y = e^x (\sin x + \cos x)$

28. $y = \ln (\sec x + \tan x)$

[S] **29.** $y = \ln (\csc x - \cot x)$

30. $y = e^{\tan x}$

31. $y = \tan^2 (e^x)$

32. $y = \dfrac{x}{2} + \dfrac{\sin 2x}{4}$

[S] **33.** $y = \sin 3x \cos^2 3x$

34. $y = \ln (\tan x)$

35. $y = \ln (\cot x)$

36. $y = \frac{1}{2}(x \tan x - \sec x)$

37. $y = \sin (\cos x)$

38. $x = \sin y$

[S] **39.** $\tan (x + y) = x$

40. $\cot y = x - y$

41. Show that $y = 2 \sin x + 3 \cos x$ satisfies the differential equation $y'' + y = 0$.

42. Show that $y = (10 - \cos x)/x$ satisfies the differential equation $xy' + y = \sin x$.

[S] **43.** Show that $y = e^x (\cos \sqrt{2}x + \sin \sqrt{2}x)$ satisfies the differential equation $y'' - 2y' + 3y = 0$.

In Exercises 44–51, evaluate each limit, using L'Hôpital's Rule when necessary.

44. $\lim\limits_{x \to \pi} \dfrac{\sin x}{x - \pi}$

45. $\lim\limits_{x \to 0} \dfrac{\sin 2x}{\sin 3x}$

46. $\lim\limits_{\theta \to 0} \dfrac{1 - \cos \theta}{\theta}$

[S] **47.** $\lim\limits_{x \to 0} \dfrac{x - \tan x}{x - \sin x}$

48. $\lim\limits_{x \to \pi/4} (\tan 2x - \sec 2x)$

49. $\lim\limits_{\theta \to 0} \dfrac{1 - \cos 2\theta}{4\theta^2}$

50. $\lim\limits_{x \to 0} \dfrac{1 - e^x}{\sin x}$

[S] **51.** $\lim\limits_{x \to \infty} \left[x \sin \left(\dfrac{1}{x} \right) \right]$

In Exercises 52–59, sketch the graph of each function on the indicated interval, making use of relative extrema and points of inflection.

52. $f(x) = \sin \left(\dfrac{x}{2} \right)$ on $[0, 4\pi]$

53. $f(x) = \sec \left(x - \dfrac{\pi}{2} \right)$ on $[0, 4\pi]$

54. $f(x) = 2 \csc 2x$ on $[0, 2\pi]$

55. $f(x) = \dfrac{2}{3} \cos \left(\dfrac{3x}{2} \right)$ on $[0, 2\pi]$

[S] **56.** $f(x) = \cos x - x$ on $[0, 4\pi]$

57. $f(x) = 2 \sin x + \sin 2x$ on $[0, 2\pi]$

58. $f(x) = 2 \sin x + \cos 2x$ on $[0, 2\pi]$

59. $f(x) = x - \sin x$ on $[0, 4\pi]$

[c] **60.** Show that we can find the critical number of $f(x) = x \cos x$ on $[0, \pi]$ by solving the equation $\cot x - x = 0$. Use Newton's Method to approximate this critical number to three decimal places.

[c] **61.** Show that we can find the critical number of $f(x) = x \sin x$ on $[0, \pi]$ by solving the equation $\tan x + x = 0$. Use Newton's Method to approximate this critical number to three decimal places.

[c] **62.** Use the result of Exercise 61 to sketch the graph of $f(x) = x \sin x$ on $[-\pi, \pi]$.

63. A balloon rises vertically at the rate of 10 ft/s from a point on the ground 100 ft from an observer. Find the rate of change of the angle of elevation of the balloon from the observer when the balloon is 100 ft above the ground.

64. A boy reels in a fish at the rate of 1 ft/s from a bridge 15 ft above the water. At what rate is the angle between the line and the water changing when 25 ft of line is out? (See Figure 9.17.)

[c][s] **65.** An airplane, flying at 600 mi/h and at an altitude of 30,000 ft, flies toward a point directly over an observer on the ground. (See Figure 9.18.) Find the rate at which the angle of elevation of the line of sight is changing

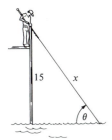

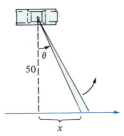

FIGURE 9.17

FIGURE 9.18

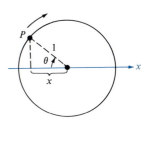

FIGURE 9.19   **FIGURE 9.20**

when the angle is:

(a) 30°   (b) 60°   (c) 75°

66. The height of a weight attached to a spring is given by the harmonic equation $y = \frac{1}{3}\cos(12t) - \frac{1}{4}\sin(12t)$, where y is measured in inches and t is measured in seconds.

(a) Calculate the height and velocity of the weight when $t = \pi/8$ s.

(b) Show that the maximum displacement of the weight is $\frac{5}{12}$ in.

(c) Find the period P of y. Find the frequency f (number of oscillations per second) if $f = 1/P$

[S] 67. The general equation giving the height of an oscillating weight attached to a spring is

$$y = A \sin\left(\sqrt{\frac{k}{m}}\,t\right) + B \cos\left(\sqrt{\frac{k}{m}}\,t\right)$$

where k is the spring constant and m is the mass of the weight.

(a) Show that the maximum displacement of the weight is $\sqrt{A^2 + B^2}$.

(b) Show that the frequency (number of oscillations per second) is $(1/2\pi)\sqrt{k/m}$. How is the frequency changed if the stiffness k of the spring is increased? How is the frequency changed if the mass m of the weight is increased?

68. A police car is parked 50 ft from a long warehouse. If the light on the car revolves at a rate of 30 revolutions per minute (r/min), how fast is the light beam moving along the wall when the beam makes the following angles with the wall? (See Figure 9.19.)

(a) 30°   (b) 60°   (c) 70°

69. A wheel of radius 1 ft revolves at a rate of 10 revolutions per second (r/s). A dot is painted at a point P on the circumference of the wheel and θ is the angle between the horizontal and a line from the center of the wheel to P. (See Figure 9.20.) Find the rate of horizontal movement of P for the following angles:

(a) $\theta = 0°$ (b) $\theta = 30°$ (c) $\theta = 60°$

70. If a connecting rod is fastened between a piston and the point P on the wheel of Exercise 69, show that the piston attains its maximum speed when $\theta = 90°$ and $\theta = 270°$. (See Figure 9.21.)

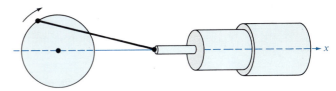

FIGURE 9.21

[C][S] 71. The graph of $y = x \sin(1/x)$ is shown in Figure 9.22.

(a) Find the extreme right-hand minimum.

(b) Show that the graph is concave downward to the right of $x = 1/\pi$.

(c) Use L'Hôpital's Rule to show that the line $y = 1$ is a horizontal asymptote to the right. (See Exercise 51.)

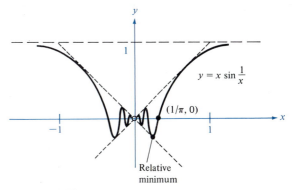

FIGURE 9.22

For Review and Class Discussion

True or False

1. _____ The maximum slope of the graph of $y = \sin(bx)$ is b.

2. _____ The graphs of $f(x) = \sin x$ and $g(x) = \cos x$ intersect at right angles.

3. _____ $y = e^x$, $y = e^{-x}$, $y = \sin x$, and $y = \cos x$ all satisfy the differential equation $d^4y/dx^4 = y$.

4. _____ The maximum value of $y = 3\sin x + 2\cos x$ is 5.

5. _____ If $f(x) = \sin^2(2x)$, then $f'(x) = 2(\sin 2x)(\cos 2x)$.

9.3   Integrals of Trigonometric Functions

Purpose

- To derive formulas for integrating trigonometric functions.

We have seen how to find the derivatives of quantities. Now we shall show inversely how the integrals of the derivatives are found, that is, those quantities from which the derivatives originate.   —*John Bernoulli (1667–1748)*

Corresponding to each formula for differentiating a trigonometric function is an integration formula. For instance, the formula

$$\frac{d}{dx}[\cos u] = -(\sin u)u'$$

corresponds to the formula

$$\int (\sin u)u'\, dx = -\cos u + C$$

We list next the integration formulas corresponding to the derivatives of the six basic trigonometric functions.

Trigonometric Functions	*Integrals*	*Derivatives*
	$\int (\cos u)u'\, dx = \sin u + C$	$\dfrac{d}{dx}[\sin u] = (\cos u)u'$
	$\int (\sin u)u'\, dx = -\cos u + C$	$\dfrac{d}{dx}[\cos u] = (-\sin u)u'$
	$\int (\sec^2 u)u'\, dx = \tan u + C$	$\dfrac{d}{dx}[\tan u] = (\sec^2 u)u'$
	$\int (\sec u \tan u)u'\, dx = \sec u + C$	$\dfrac{d}{dx}[\sec u] = (\sec u \tan u)u'$
	$\int (\csc^2 u)u'\, dx = -\cot u + C$	$\dfrac{d}{dx}[\cot u] = (-\csc^2 u)u'$
	$\int (\csc u \cot u)u'\, dx = -\csc u + C$	$\dfrac{d}{dx}[\csc u] = (-\csc u \cot u)u'$

Example 1

Find $$\int 2 \cos x \, dx$$

Solution: Let $u = x$; then $u' = 1$. Since 2 is not a necessary part of the integrand, we have

$$\int 2 \cos x \, dx = 2 \int \cos x \, dx = 2 \int (\cos u) u' \, dx$$
$$= 2 \sin u + C = 2 \sin x + C$$ ■

Example 2

Find $\int 3x^2 \sin x^3 \, dx$.

Solution: We let $u = x^3$, and thus $u' = 3x^2$. Therefore, we write

$$\int 3x^2 \sin x^3 \, dx = \int (\sin x^3)(3x^2) \, dx = \int (\sin u)(u') \, dx$$
$$= -\cos u + C = -\cos x^3 + C$$ ■

Example 3

Evaluate $\int \sec (3x + 1) \tan (3x + 1) \, dx$.

Solution: Let $u = 3x + 1$; then $u' = 3$, and we write

$$\int \sec (3x + 1) \tan (3x + 1) \, dx = \tfrac{1}{3} \int \sec (3x + 1) \tan (3x + 1)(3) \, dx$$
$$= \tfrac{1}{3} \sec (3x + 1) + C$$ ■

Example 4

Evaluate $$\int \frac{\csc^2 \sqrt{x}}{\sqrt{x}} \, dx$$

Solution: Consider $u = \sqrt{x}$; then $u' = 1/(2\sqrt{x})$. Thus we write

$$\int \frac{\csc^2 \sqrt{x}}{\sqrt{x}} \, dx = \int \left(\csc^2 \sqrt{x} \right) \left(\frac{1}{\sqrt{x}} \right) dx$$
$$= 2 \int \left(\csc^2 \sqrt{x} \right) \left(\frac{1}{2\sqrt{x}} \right) dx$$
$$= 2(-\cot \sqrt{x}) + C = -2 \cot \sqrt{x} + C$$ ■

Example 5

Find $\int_0^{\pi/8} (\sec 2x + \cos 2x)(\sec 2x) \, dx$.

Solution: Since the integrand does not fit any of the given integration formulas, we multiply and obtain

$$\int_0^{\pi/8} (\sec 2x + \cos 2x)(\sec 2x)\, dx = \int_0^{\pi/8} (\sec^2 2x + 1)\, dx$$

$$= \int_0^{\pi/8} \sec^2 2x\, dx + \int_0^{\pi/8} 1\, dx$$

Now considering $u = 2x$, we write

$$\int_0^{\pi/8} (\sec 2x + \cos 2x)(\sec 2x)\, dx = \tfrac{1}{2} \int_0^{\pi/8} (\sec^2 2x)(2)\, dx + \int_0^{\pi/8} 1\, dx$$

$$= [\tfrac{1}{2} \tan 2x + x]_0^{\pi/8}$$

$$= \frac{1}{2} + \frac{\pi}{8} \approx 0.8927 \qquad \blacksquare$$

We frequently use the Power or Log Rules to evaluate integrals containing trigonometric functions. The key to their use lies in identifying, as part of the integrand, the derivative of one of the functions involved. The next three examples show how these rules are used.

Example 6 (*Power Rule*)

Evaluate $\int \sin^2 4x \cos 4x\, dx$.

Solution: The integrand does not fit any of the integration formulas of this section. If we have memorized our derivative formulas, we will readily see that $\cos 4x$ is the essential part of

$$\frac{d}{dx}[\sin 4x] = 4 \cos 4x$$

Thus we consider the Power Rule, with $u = \sin 4x$, and write

$$\int \sin^2 4x \cos 4x\, dx = \frac{1}{4} \int (\sin 4x)^2 (4 \cos 4x)\, dx$$

$$= \frac{1}{4} \int u^2 u'\, dx = \left(\frac{1}{4}\right)\left(\frac{u^3}{3}\right) + C$$

$$= \left(\frac{1}{4}\right)\left(\frac{\sin^3 4x}{3}\right) + C = \frac{1}{12} \sin^3 4x + C \qquad \blacksquare$$

Example 7 (*Power Rule*)

Find

$$\int \frac{\sec^2 x\, dx}{\sqrt{\tan x}}$$

Solution: Since

$$\frac{d}{dx}[\tan x] = \sec^2 x$$

we consider the Power Rule with $u = \tan x$. Thus we write

$$\int \frac{\sec^2 x}{\sqrt{\tan x}} dx = \int (\tan x)^{-1/2}(\sec^2 x) \, dx$$

$$= \int u^{-1/2} u' \, dx = \frac{u^{1/2}}{\frac{1}{2}} + C$$

$$= \frac{(\tan x)^{1/2}}{\frac{1}{2}} + C = 2\sqrt{\tan x} + C$$ ∎

Example 8 (*Log Rule*)

Find

$$\int \frac{\sin x}{\cos x} dx$$

Solution: Knowing that $\frac{d}{dx}[\cos x] = -\sin x$

we consider the Log Rule with $u = \cos x$. Thus we write

$$\int \frac{\sin x}{\cos x} dx = -\int \frac{(-\sin x)}{\cos x} dx = -\ln|\cos x| + C$$

Now since $\tan x = \sin x/\cos x$, we have the following formula:

$$\int \tan x \, dx = -\ln|\cos x| + C$$ ∎

With the result of Example 8, we now have integration formulas for $\sin x$, $\cos x$, and $\tan x$. In the next example we derive the integration formula for $\sec x$, and we leave the derivations for $\csc x$ and $\cot x$ as exercises.

Example 9

Find $\int \sec x \, dx$.

Solution: Consider the following procedure:

$$\int \sec x \, dx = \int \sec x \left(\frac{\sec x + \tan x}{\sec x + \tan x}\right) dx = \int \frac{\sec^2 x + \sec x \tan x}{\sec x + \tan x} dx$$

Now we recognize that if $u = \sec x + \tan x$, then $u' = \sec x \tan x + \sec^2 x$. Therefore, we have

$$\int \sec x \, dx = \int \frac{\sec x \tan x + \sec^2 x}{\sec x + \tan x} dx = \int \frac{u'}{u} dx$$

$$= \ln|\sec x + \tan x| + C$$ ∎

We provide next the integration formulas for each of the six basic trigonometric functions.

Integrals of the Six Basic Trigonometric Functions

$\int (\sin u)u'\, dx = -\cos u + C$ $\qquad$ $\int (\cos u)u'\, dx = \sin u + C$

$\int (\tan u)u'\, dx = -\ln |\cos u| + C$ $\qquad$ $\int (\cot u)u'\, dx = \ln |\sin u| + C$

$\int (\sec u)u'\, dx = \ln |\sec u + \tan u| + C$ $\qquad$ $\int (\csc u)u'\, dx = \ln |\csc u - \cot u| + C$

As you memorize these formulas, note that the three formulas on the right follow the pattern set by the three on the left.

Example 10

Find the arc length of the graph of $y = \ln (\cos x)$ from $x = 0$ to $x = \pi/4$.

Solution: Since $y' = \dfrac{-\sin x}{\cos x} = -\tan x$

the arc length s is given by

$$s = \int_0^{\pi/4} \sqrt{1 + \tan^2 x}\, dx = \int_0^{\pi/4} \sqrt{\sec^2 x}\, dx$$

$$= \int_0^{\pi/4} \sec x\, dx = \ln |\sec x + \tan x| \Big]_0^{\pi/4}$$

Therefore, $s = \ln (\sqrt{2} + 1) - \ln (1) \approx 0.8814$ ∎

Example 11

Suppose that the *electromotive force E* of a particular electrical circuit is given by $E = 3 \sin 2t$, where E is measured in volts and t is measured in seconds. Find the average value of E as t ranges from 0 to 0.5 second.

Solution: The average value (see Section 6.4) of E on this interval is given by

$$\frac{1}{\frac{1}{2} - 0} \int_0^{1/2} E\, dt = 2 \int_0^{1/2} 3 \sin 2t\, dt = 3(-\cos 2t) \Big]_0^{1/2}$$

$$= 3[-\cos (1) + 1] \approx 1.379 \text{ V}$$ ∎

Sometimes a trigonometric integrand that does not seem to fit any integration formula can be made to conform by a judicious use of trigonometric identities (see Example 10). Identities that are especially useful in this context are the Pythagorean and Half-Angle Identities.

Section Exercises (9.3)

In Exercises 1–35, evaluate the given integral.

1. $\int \sin 2x \, dx$

2. $\int x \sin x^2 \, dx$

S 3. $\int x \cos x^2 \, dx$

4. $\int \cos 6x \, dx$

5. $\int_{\pi/2}^{2\pi/3} \sec^2 \left(\frac{x}{2}\right) dx$

6. $\int_{\pi/3}^{\pi/2} \csc^2 \left(\frac{x}{2}\right) dx$

S 7. $\int_0^{\pi/9} \tan 3x \, dx$

8. $\int \tan \left(\frac{\pi}{4} - 2x\right) dx$

9. $\int \cot 2x \, dx$

S 10. $\int_0^{\pi/8} \sin 2x \cos 2x \, dx$

11. $\int_0^1 \sec (1 - x) \tan (1 - x) \, dx$

12. $\int_{\pi/12}^{\pi/4} \csc 2x \cot 2x \, dx$

13. $\int \frac{\csc^2 x}{\cot^3 x} \, dx$

14. $\int \frac{\sin x}{\cos^2 x} \, dx$

S 15. $\int \cot^2 x \, dx$

16. $\int (\sin 2x + \cos 2x)^2 \, dx$

17. $\int e^x \cos e^x \, dx$

18. $\int \sec \left(\frac{3x}{2}\right) dx$

19. $\int \tan^4 x \sec^2 x \, dx$

20. $\int \sqrt{\cot x} \csc^2 x \, dx$

21. $\int_0^{\pi/4} \sec^4 x \tan x \, dx$

22. $\int_0^{\pi/2} (x + \cos x) \, dx$

23. $\int_0^{\pi/2} e^{\sin x} \cos x \, dx$

24. $\int \frac{\sin \sqrt{x}}{\sqrt{x}} \, dx$

S 25. $\int \frac{\sec x \tan x}{\sec x - 1} \, dx$

26. $\int \frac{\csc^2 x}{\sqrt{\cot x - 1}} \, dx$

27. $\int \frac{\cos^2 x}{\sin x} \, dx$

28. $\int_0^{\pi/6} \frac{\tan^2 2x}{\sec 2x} \, dx$

29. $\int \frac{\sin x}{1 + \cos x} \, dx$

30. $\int e^{\tan x} \sec^2 x \, dx$

31. $\int \sin \theta \sqrt{1 + \cos \theta} \, d\theta$

32. $\int \frac{1 - \cos \theta}{\theta - \sin \theta} \, d\theta$

33. $\int_0^{\pi/4} \frac{1 - \sin^2 \theta}{\cos^2 \theta} \, d\theta$

34. $\int \cot^2 \left(x - \frac{\pi}{4}\right) dx$

S 35. $\int (\csc 2\theta - \cot 2\theta)^2 \, d\theta$

36. Show that

$$\int_0^x \sin (a + bt) \, dt = -(1/b)[\cos (a + bx) - \cos a]$$

37. Find the area of the region under one arch of the sine curve $f(x) = \sin x$.

38. Find the volume of the solid generated by rotating the region bounded by $y = \sec x$, $y = 0$, $x = 0$, and $x = \pi/4$ around the x-axis.

S 39. Find the volume of the solid generated by revolving the region bounded by $y = \sqrt{\sin x}$, $y = 0$, $0 \le x \le \pi$ about the x-axis.

40. Find the volume of the solid generated by revolving the region bounded by $y = \sqrt{\cos x}$, $y = 0$, $0 \le x \le \pi/2$ about the x-axis.

41. Evaluate $\int \sin x \cos x \, dx$ two ways. First, make the substitution $u = \sin x$, and second, integrate by letting $u = \cos x$. Explain the difference in the results.

42. The graphs of the sine and cosine are symmetric with respect to the origin and the y-axis, respectively. Use this fact together with the results of Exercises 53 and 54 of Section 6.4 to aid in the evaluation of the following definite integrals:

(a) $\int_{-\pi/4}^{\pi/4} \sin x \, dx$

(b) $\int_{-\pi/4}^{\pi/4} \cos x \, dx$

(c) $\int_{-\pi/2}^{\pi/2} \cos x \, dx$

C (d) $\int_{-1.32}^{1.32} \sin 2x \, dx$

(e) $\int_{-\pi/2}^{\pi/2} \sin x \cos x \, dx$

In Exercises 43–46, show the equivalence of each pair of formulas.

43. $\int \tan x \, dx = -\ln |\cos x| + C$

$\int \tan x \, dx = \ln |\sec x| + C$

44. $\int \cot x \, dx = \ln |\sin x| + C$

$\int \cot x \, dx = -\ln |\csc x| + C$

45. $\int \sec x \, dx = \ln |\sec x + \tan x| + C$

$\int \sec x \, dx = -\ln |\sec x - \tan x| + C$

46. $\int \csc x \, dx = -\ln |\csc x + \cot x| + C$

$\int \csc x \, dx = \ln |\csc x - \cot x| + C$

47. A horizontal plane is ruled with parallel lines 2 in. apart. If a 2-in. needle is randomly tossed onto the plane, it can be shown that the probability of the needle touching a line is given by

$$P = \frac{2}{\pi} \int_0^{\pi/2} \sin \theta \, d\theta$$

where θ is the angle between the needle and any one of the parallel lines. Find this probability.

48. The oscillating current in a certain electric circuit is given by

$$I = 2 \sin (60\pi t) + \cos (120\pi t)$$

where I is measured in amperes and t is measured in seconds. Find the average current for the following time intervals:

(a) $0 < t < \frac{1}{60}$ (b) $0 < t < \frac{1}{240}$

(c) $0 < t < \frac{1}{30}$

For Review and Class Discussion

True or False

1. _____ $\int_a^b \sin x \, dx = \int_a^{b+2\pi} \sin x \, dx$.

2. _____ $4 \int \sin x \cos x \, dx = -\cos 2x + C$.

3. _____ The average value of the sine function over an interval of length 2π is 0.

4. _____ $\int \tan x \, dx = \sec^2 x + C$.

5. _____ $\int \sin^2 2x \cos 2x \, dx = (\sin^3 2x)/3 + C$.

9.4 Inverse Trigonometric Functions

Purpose
- To review the inverse trigonometric functions.

No other language has the capacity for the elegance that arises from a long sequence of expressions linked one to the other and all stemming from one fundamental idea. —Pierre-Simon Laplace (1749–1827)

From the graph of the sine function (Figure 9.23), we can see that to each value of y in the interval $[-1, 1]$, there correspond many values of x. For instance, $x = \pi/6$, $5\pi/6$, $13\pi/6$, and $-7\pi/6$ all correspond to the same value $y = \frac{1}{2}$. To construct an inverse of the sine function, we must restrict the values of x to some finite interval so that to each value of y in $[-1, 1]$, there corresponds one and only one value of x. A convenient set of x-values is the interval $[-\pi/2, \pi/2]$. On this interval (see Figure 9.24), the sine function is continuous and increasing, and thus it possesses an inverse. On this basis we define the inverse sine function to be

$$y = \text{Arcsin } x \qquad \text{if and only if} \qquad \sin y = x$$

where $-1 \leq x \leq 1$ and $-\pi/2 \leq \text{Arcsin } x \leq \pi/2$.

The expression "Arcsin x" is read as "the inverse sine of x," or as "the angle whose sine is x." Some texts denote the inverse sine function by

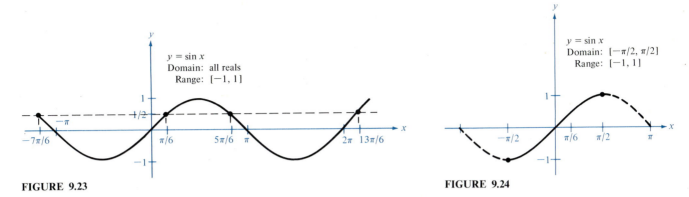

FIGURE 9.23   **FIGURE 9.24**

$\sin^{-1} x$. We will not use this notation because of possible confusion with the function $1/\sin x$.

If their domains are properly restricted, each of the six trigonometric functions possesses an inverse. We define the six inverse trigonometric functions as follows:

Definitions of Inverse Trigonometric Functions

	Definition		Domain	Range
$y = \mathbf{Arcsin}\ x$	if and only if	$\sin y = x$	$[-1, 1]$	$[-\pi/2, \pi/2]$
$y = \mathbf{Arccos}\ x$	if and only if	$\cos y = x$	$[-1, 1]$	$[0, \pi]$
$y = \mathbf{Arctan}\ x$	if and only if	$\tan y = x$	$(-\infty, \infty)$	$(-\pi/2, \pi/2)$
$y = \mathbf{Arccot}\ x$	if and only if	$\cot y = x$	$(-\infty, \infty)$	$(0, \pi)$
$y = \mathbf{Arcsec}\ x$	if and only if	$\sec y = x$	$(-\infty, -1]$ and $[1, \infty)$	$[0, \pi/2)$ and $(\pi/2, \pi]$
$y = \mathbf{Arccsc}\ x$	if and only if	$\csc y = x$	$(-\infty, -1]$ and $[1, \infty)$	$[-\pi/2, 0)$ and $(0, \pi/2]$

The graphs of the six inverse trigonometric functions are shown in Figure 9.25 (p. 398). We know from Section 1.6 that the graph of the inverse of a function is the reflection in the line $y = x$ of the graph of the function itself. As you study the graphs in Figure 9.25, compare them to the graphs of the six trigonometric functions in Figure 9.5.

When evaluating inverse trigonometric functions, it is helpful to keep in mind that Arcsin x, Arccos x, Arctan x, Arccot x, Arcsec x, and Arccsc x all denote *angles in radian measure.*

Example 1

Evaluate the following:
(a) Arcsin $\left(-\frac{1}{2}\right)$ (b) Arccos 0
(c) Arctan $\sqrt{3}$ (d) Arcsec $\left(-\sqrt{2}\right)$

Domain: $[-1, 1]$
Range: $[-\pi/2, \pi/2]$

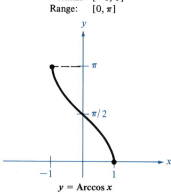

$y = \text{Arcsin } x$

Domain: $(-\infty, -1]$ and $[1, \infty)$
Range: $[-\pi/2, 0)$ and $(0, \pi/2]$

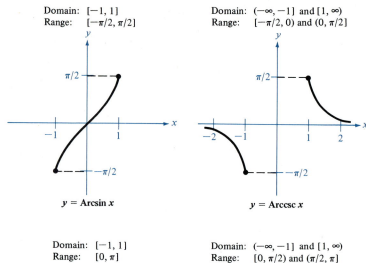

$y = \text{Arccsc } x$

Domain: $(-\infty, \infty)$
Range: $(-\pi/2, \pi/2)$

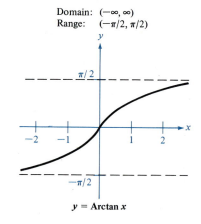

$y = \text{Arctan } x$

Domain: $[-1, 1]$
Range: $[0, \pi]$

$y = \text{Arccos } x$

Domain: $(-\infty, -1]$ and $[1, \infty)$
Range: $[0, \pi/2)$ and $(\pi/2, \pi]$

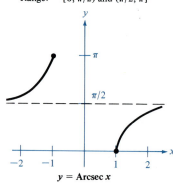

$y = \text{Arcsec } x$

Domain: $(-\infty, \infty)$
Range: $(0, \pi)$

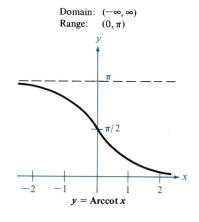

$y = \text{Arccot } x$

FIGURE 9.25

Solution:

(a)   Since $y = \text{Arcsin}\left(-\frac{1}{2}\right)$ means $\sin y = -\frac{1}{2}$, we seek the angle whose sine is $-\frac{1}{2}$. In the interval $[-\pi/2, \pi/2]$ we choose $y = -\pi/6$. Thus $\text{Arcsin}\left(-\frac{1}{2}\right) = -\pi/6$.

(b) $y = \text{Arccos } 0$ means $\cos y = 0$. In the interval $[0, \pi]$ we choose $y = \pi/2$, and thus $\text{Arccos } 0 = \pi/2$.

(c) $y = \text{Arctan } \sqrt{3}$ means $\tan y = \sqrt{3}$. In the interval $(-\pi/2, \pi/2)$ we choose $y = \pi/3$, and thus $\text{Arctan } \sqrt{3} = \pi/3$.

(d) $y = \text{Arcsec}(-\sqrt{2})$ means $\sec y = -\sqrt{2}$. In the intervals $[0, \pi/2)$ and $(\pi/2, \pi]$, we choose $y = 3\pi/4$, and thus $\text{Arcsec}(-\sqrt{2}) = 3\pi/4$. ∎

Example 2

Evaluate $\text{Arcsin } 2x$ when $x = 0.15$.

Solution: For $x = 0.15$, we have $y = \text{Arcsin}(0.30)$, which means $\sin y = 0.30$. From tables for the sine function, we choose $y \approx 17.46° \approx 0.3047$ radians. Therefore, $\text{Arcsin}(0.30) \approx 0.3047$. ∎

Example 3

Solve for x in the equation $\text{Arctan}(2x - 3) = \pi/4$.

Solution: The equation $\dfrac{\pi}{4} = \text{Arctan}(2x - 3)$

means that $$\tan\left(\frac{\pi}{4}\right) = 2x - 3$$

Thus
$$1 = 2x - 3$$
$$4 = 2x$$
$$2 = x$$ ∎

Without using tables it is possible to evaluate trigonometric functions of inverse trigonometric functions. To do so we make use of right triangles, as demonstrated in the next two examples.

Example 4

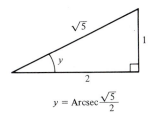

$y = \text{Arcsec}\dfrac{\sqrt{5}}{2}$

FIGURE 9.26

Given $y = \text{Arcsec}(\sqrt{5}/2)$, find $\sin y$ and $\tan y$.

Solution: Since "y is the angle whose secant is $\sqrt{5}/2$," we can sketch this angle as part of a right triangle (Figure 9.26). Therefore,

$$\sin y = \sin\left(\text{Arcsec}\,\frac{\sqrt{5}}{2}\right) = \frac{1}{\sqrt{5}} \approx 0.4472$$

Similarly,
$$\tan y = \tan\left(\text{Arcsec}\,\frac{\sqrt{5}}{2}\right) = \frac{1}{2}$$ ∎

Example 5

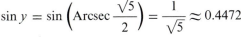

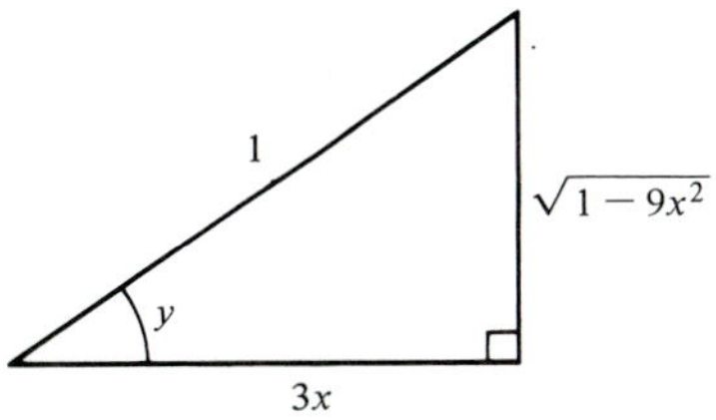

$y = \text{Arccos}\,3x$

FIGURE 9.27

Given $y = \text{Arccos}\,3x$, find $\cot y$ and $\sin y$.

Solution: Since "y is the angle whose cosine is $3x$," we form a right triangle having the acute angle $y = \text{Arccos}\,3x$ (Figure 9.27). Now we have

$$\cot y = \cot(\text{Arccos}\,3x) = \frac{3x}{\sqrt{1 - 9x^2}}$$

and
$$\sin y = \sin(\text{Arccos}\,3x) = \sqrt{1 - 9x^2}$$ ∎

Example 6

In Section 9.2, Example 11, we found that a maximum value of $y = \frac{1}{6}\cos 2t + \frac{3}{8}\sin 2t$ occurred when $\tan 2t = \frac{9}{4}$, $0.5 < t < 0.75$. Find the time t at which the maximum value of y is reached.

Solution: Using the inverse tangent we obtain

$$\tan 2t = \tfrac{9}{4}$$
$$\text{Arctan}\,(\tan 2t) = \text{Arctan}\,(\tfrac{9}{4})$$
$$2t = \text{Arctan}\,(\tfrac{9}{4})$$
$$t = \tfrac{1}{2}\text{Arctan}\,(\tfrac{9}{4}) \approx 0.576 \text{ s} \qquad \blacksquare$$

Section Exercises (9.4)

In Exercises 1–14, evaluate the given expression.
1. Arcsin $(\frac{1}{2})$
2. Arcsin (0)
S 3. Arccos $(\frac{1}{2})$
4. Arccos (0)

5. Arctan $\left(\dfrac{\sqrt{3}}{3}\right)$
S 6. Arccot (-1)

7. Arccsc $(\sqrt{2})$
C 8. Arcsin (-0.39)
C 9. Arcsec (1.269)
C 10. Arctan (-3)
11. sin $[\text{Arcsin}\,(\frac{1}{2})]$
12. cos $[2\,\text{Arcsin}\,(1)]$

S 13. tan $\left[\text{Arccos}\left(\dfrac{\sqrt{2}}{2}\right)\right]$
14. cot $[\text{Arcsin}\,(-0.5)]$

In Exercises 15–20, write an algebraic equation that is equivalent to the given equation.
15. $y = \tan\,(\text{Arctan}\,x)$
S 16. $y = \sin\,(\text{Arccos}\,x)$

17. $y = \cos\,[\text{Arcsin}\,(2x)]$
18. $y = \sec\,[\text{Arctan}\,(3x)]$
S 19. $y = \sin\,(\text{Arcsec}\,x)$
20. $y = \cot\,(\text{Arccot}\,x)$

In Exercises 21–24, solve the given equation for x.
21. Arcsin $(3x - \pi) = \frac{1}{2}$
22. Arctan $2x = -1$
S 23. Arcsin $\sqrt{2x} = \text{Arccos}\,\sqrt{x}$
24. Arccos $x = \text{Arcsec}\,x$

In Exercises 25–28, use the graphs in Figure 9.25 to evaluate the given limit.
25. $\lim\limits_{x \to \infty} \text{Arctan}\,x$
26. $\lim\limits_{x \to -\infty} \text{Arctan}\,x$
27. $\lim\limits_{x \to \infty} \text{Arccot}\,x$
28. $\lim\limits_{x \to -\infty} \text{Arccot}\,x$

9.5 Inverse Trigonometric Functions: Derivatives and Integrals

Purpose
- To derive formulas for the derivatives of the inverse trigonometric functions.
- To identify functions whose antiderivatives are inverse trigonometric functions.

The analytic calculus is thus extended to those curves that hitherto have been excluded for no better reason than that they were thought to be unsuited to it.   —Gottfried Leibniz (1646–1716)

In Section 8.2 we saw that the derivative of the transcendental function $\ln x$ is the algebraic function $1/x$. In this section we will show that the derivatives of the inverse trigonometric functions are also algebraic, even though the inverse trigonometric functions are themselves transcendental. For instance, to find the derivative of $y = \text{Arcsin}\,x$, we write

$$\sin y = x$$

Implicit differentiation yields

$$(\cos y)\frac{dy}{dx} = 1$$

and thus

$$\frac{dy}{dx} = \frac{1}{\cos y}$$

Now from the identity $\sin^2 y + \cos^2 y = 1$, we have

$$\cos y = +\sqrt{1 - \sin^2 y}$$

on the interval $-\pi/2 < y < \pi/2$. Therefore, we have

$$\frac{dy}{dx} = \frac{1}{\cos y} = \frac{1}{\sqrt{1 - \sin^2 y}} = \frac{1}{\sqrt{1 - x^2}}$$

provided $-1 < x < 1$. From the graph (Figure 9.25) of the inverse sine function, we observe that vertical tangents occur at $(-1, -\pi/2)$ and $(1, \pi/2)$, and thus its derivative does not exist at $x = \pm 1$.

If $y = \text{Arcsin } u$, where u is a differentiable function of x, then application of the Chain Rule gives us

$$\frac{d}{dx}[\text{Arcsin } u] = \frac{1}{\sqrt{1 - u^2}}\frac{du}{dx}$$

In a similar manner, the derivative of Arccos u can be shown to be

$$\frac{d}{dx}[\text{Arccos } u] = \frac{-1}{\sqrt{1 - u^2}}\frac{du}{dx}$$

To differentiate Arctan u, let $y = \text{Arctan } u$; then

$$\tan y = u$$

where $-\pi/2 \le y \le \pi/2$. Thus by the Chain Rule,

$$\sec^2 y \frac{dy}{dx} = \frac{du}{dx}$$

$$\frac{dy}{dx} = \frac{1}{\sec^2 y}\frac{du}{dx} = \frac{1}{1 + \tan^2 y}\frac{du}{dx}$$

Since $u = \tan y$, we have

$$\frac{d}{dx}[\text{Arctan } u] = \frac{1}{1 + u^2}\frac{du}{dx}$$

We list the derivatives of the six inverse trigonometric functions next. In each case u denotes a differentiable function of x. Note that the derivatives of Arccos u, Arccot u, and Arccsc u are the negatives of the derivatives of Arcsin u, Arctan u, and Arcsec u.

Derivatives of the Inverse Trigonometric Functions	$\dfrac{d}{dx}[\text{Arcsin } u] = \dfrac{1}{\sqrt{1-u^2}}\dfrac{du}{dx}$	$\dfrac{d}{dx}[\text{Arccos } u] = \dfrac{-1}{\sqrt{1-u^2}}\dfrac{du}{dx}$
	$\dfrac{d}{dx}[\text{Arctan } u] = \dfrac{1}{1+u^2}\dfrac{du}{dx}$	$\dfrac{d}{dx}[\text{Arccot } u] = \dfrac{-1}{1+u^2}\dfrac{du}{dx}$
	$\dfrac{d}{dx}[\text{Arcsec } u] = \dfrac{1}{|u|\sqrt{u^2-1}}\dfrac{du}{dx}$	$\dfrac{d}{dx}[\text{Arccsc } u] = \dfrac{-1}{|u|\sqrt{u^2-1}}\dfrac{du}{dx}$

There is no common agreement on the definition of Arcsec x (or Arccsc x) for negative values of x. When we defined the range of Arcsec x in the previous section, we chose to preserve the identity Arcsec $x = $ Arccos $(1/x)$. One of the consequences of this choice is that the slope of the graph of the inverse secant function is always positive (see Figure 9.25), which accounts for the absolute value sign in the formula for the derivative of Arcsec x.

Example 1

Differentiate $y = \text{Arctan } 3x$.

Solution: Considering $u = 3x$ we have

$$\frac{dy}{dx} = \left[\frac{1}{1+(3x)^2}\right](3) = \frac{3}{1+9x^2}$$ ∎

Example 2

Differentiate $y = \text{Arcsin }\sqrt{x}$.

Solution: Considering $u = \sqrt{x}$ we have

$$\frac{dy}{dx} = \left(\frac{1}{\sqrt{1-x}}\right)\left(\frac{1}{2}x^{-1/2}\right) = \frac{1}{2\sqrt{x}\sqrt{1-x}} = \frac{1}{2\sqrt{x-x^2}}$$ ∎

Example 3

Differentiate $y = \text{Arcsec } e^{2x}$.

Solution: Letting $u = e^{2x}$ we have

$$y' = \left(\frac{1}{e^{2x}\sqrt{(e^{2x})^2-1}}\right)2e^{2x} = \frac{2e^{2x}}{e^{2x}\sqrt{e^{4x}-1}} = \frac{2}{\sqrt{e^{4x}-1}}$$

Note that we omit the absolute value sign since $e^{2x} > 0$. ∎

Example 4

Differentiate $y = 2 \text{ Arccos } 2x + \sqrt{1 - 4x^2}$.

Solution: $y' = \left(\dfrac{-2}{\sqrt{1 - 4x^2}} \right)(2) + \dfrac{1}{2}(1 - 4x^2)^{-1/2}(-8x)$

$$= \frac{-4}{\sqrt{1 - 4x^2}} - \frac{4x}{\sqrt{1 - 4x^2}} = \frac{-4(1 + x)}{\sqrt{1 - 4x^2}}$$ ∎

Example 5

In the rocket problem of Example 12, Section 9.2, the angle of elevation θ was such that

$$\tan \theta = \frac{50t^2}{2000}$$

As an alternative method of solution, use the inverse tangent function to find $d\theta/dt$ when $t = 10$ s.

Solution: Since $\quad \tan \theta = \dfrac{50t^2}{2000} = \dfrac{t^2}{40}$

we have $\quad\quad\quad \theta = \text{Arctan} \left(\dfrac{t^2}{40} \right)$

and it follows that

$$\frac{d\theta}{dt} = \frac{(2t/40)}{[1 + (t^4/1600)]} = \frac{(t/20)}{(1600 + t^4)/1600} = \frac{80t}{1600 + t^4}$$

When $t = 10$, $\quad\quad \dfrac{d\theta}{dt} = \dfrac{800}{11{,}600} = \dfrac{2}{29}$ rad/s ∎

The derivatives of the six inverse trigonometric functions occur in three pairs. In each pair the derivative of one function is the negative of the other. For example,

$$\frac{d}{dx}[\text{Arcsin } x] = \frac{1}{\sqrt{1 - x^2}} \quad \text{and} \quad \frac{d}{dx}[\text{Arccos } x] = \frac{-1}{\sqrt{1 - x^2}}$$

When listing the *antiderivatives* that correspond to each of the inverse trigonometric functions, we need use only one member from each pair. For example, we choose to write

$$\int \frac{1}{\sqrt{1 - x^2}} \, dx = \text{Arcsin } x + C$$

rather than $\quad\quad \displaystyle\int \frac{1}{\sqrt{1 - x^2}} \, dx = -\text{Arccos } x + C$

The next list gives one antiderivative formula for each of the three pairs. We verify the first formula and leave the verification of the remaining two as exercises. In each case u is a differentiable function of x and a is a constant.

Integrals Involving Inverse Trigonometric Functions

$$\int \frac{u'}{\sqrt{a^2 - u^2}}\, dx = \text{Arcsin} \frac{u}{a} + C$$

$$\int \frac{u'}{a^2 + u^2}\, dx = \frac{1}{a} \text{Actan} \frac{u}{a} + C$$

$$\int \frac{u'}{u\sqrt{u^2 - a^2}}\, dx = \frac{1}{a} \text{Arcsec} \frac{|u|}{a} + C$$

Proof: Let
$$y = \text{Arcsin} \frac{u}{a}$$

Then
$$y' = \frac{1}{\sqrt{1 - (u/a)^2}}\left(\frac{u'}{a}\right) = \frac{u'}{a\sqrt{(a^2 - u^2)/a^2}} = \frac{u'}{\sqrt{a^2 - u^2}}$$

Consequently, by the definition of an antiderivative, we have

$$\int \frac{u'}{\sqrt{a^2 - u^2}}\, dx = \text{Arcsin} \frac{u}{a} + C$$ ∎

Example 6

Evaluate
$$\int \frac{dx}{9 + 4x^2}$$

Solution: If we let $u = 2x$, then $u' = 2$, and it follows that

$$\int \frac{dx}{9 + 4x^2} = \int \frac{dx}{3^2 + (2x)^2} = \frac{1}{2}\int \frac{2\, dx}{3^2 + (2x)^2} = \frac{1}{2}\int \frac{u'\, dx}{3^2 + u^2}$$

$$= \frac{1}{2}\left(\frac{1}{3}\right)\text{Arctan} \frac{2x}{3} + C = \frac{1}{6}\text{Arctan} \frac{2x}{3} + C$$ ∎

Integrands involving quotients often require the use of inverse trigonometric functions in conjunction with the Power Rule or the Log Rule. Study the next two examples carefully to see how integration rules can be combined to find antiderivatives.

Example 7

Evaluate
$$\int \frac{x+2}{\sqrt{4-x^2}}\,dx$$

Solution: First, we write

$$\int \frac{x+2}{\sqrt{4-x^2}}\,dx = \int \frac{x\,dx}{\sqrt{4-x^2}} + \int \frac{2\,dx}{\sqrt{4-x^2}}$$

Now the first integral on the right can be evaluated by the Power Rule and the second integral will yield an inverse sine function. Therefore, we write

$$\int \frac{x+2}{\sqrt{4-x^2}}\,dx = -\frac{1}{2}\int (4-x^2)^{-1/2}(-2x)\,dx + 2\int \frac{dx}{\sqrt{4-x^2}}$$

$$= -\frac{1}{2}\left[\frac{(4-x^2)^{1/2}}{\frac{1}{2}}\right] + 2\,\text{Arcsin}\,\frac{x}{2} + C$$

$$= -\sqrt{4-x^2} + 2\,\text{Arcsin}\,\frac{x}{2} + C \qquad \blacksquare$$

Example 8

Evaluate
$$\int \frac{3x^3-2}{x^2+4}\,dx$$

Solution: Since the degree of the numerator is greater than the degree of the denominator, we divide and obtain

$$\int \frac{3x^3-2}{x^2+4}\,dx = \int \left(3x - \frac{12x+2}{x^2+4}\right)dx$$

$$= \int 3x\,dx - 6\int \frac{2x}{x^2+4}\,dx - 2\int \frac{dx}{x^2+4}$$

$$= \frac{3x^2}{2} - 6\ln(x^2+4) - \text{Arctan}\,\frac{x}{2} + C \qquad \blacksquare$$

Example 9

Evaluate
$$\int \frac{dx}{x\sqrt{x^4-1}}$$

Solution: This integral would appear to yield an inverse secant function. Hence if we let $u = x^2$, then $u' = 2x$, and we write

$$\int \frac{dx}{x\sqrt{x^4-1}} = \int \frac{x\,dx}{(x^2)\sqrt{(x^2)^2-1}}$$

Notice that the integrand has been multiplied and divided by x so as to obtain the factor $u = x^2$ in the denominator. Thus, with $a = 1$, we have

$$\int \frac{dx}{x\sqrt{x^4 - 1}} = \frac{1}{2}\int \frac{2x\,dx}{x^2\sqrt{(x^2)^2 - 1}} = \frac{1}{2}\int \frac{u'\,dx}{u\sqrt{u^2 - 1}}$$

$$= \frac{1}{2}\operatorname{Arcsec} x^2 + C$$

■

Example 10

Evaluate

$$\int_0^{1/4} \frac{\operatorname{Arcsin} 2x}{\sqrt{1 - 4x^2}}\,dx$$

Solution: If we let $u = \operatorname{Arcsin} 2x$, then $u' = 2/\sqrt{1 - 4x^2}$. Hence we write

$$\int \frac{\operatorname{Arcsin} 2x}{\sqrt{1 - 4x^2}}\,dx = \int (\operatorname{Arcsin} 2x)\left(\frac{1}{\sqrt{1 - 4x^2}}\right)dx$$

$$= \frac{1}{2}\int (\operatorname{Arcsin} 2x)^1\left(\frac{2}{\sqrt{1 - 4x^2}}\right)dx$$

Therefore, by the Power Rule with $n = 1$, we have

$$\int_0^{1/4} \frac{\operatorname{Arcsin} 2x}{\sqrt{1 - 4x^2}}\,dx = \frac{(\operatorname{Arcsin} 2x)^2}{4}\Big]_0^{1/4} = \frac{\pi^2}{144} \approx 0.069$$

■

Take note from Example 10 that the use of the Power Rule with $n = 1$ is somewhat disguised. Be on the lookout for such cases.

Section Exercises (9.5)

In Exercises 1–20, find the derivative of the given function.

1. $f(x) = \operatorname{Arcsin} 2x$
2. $f(x) = \operatorname{Arcsin}(x^2)$
S **3.** $f(x) = 2\operatorname{Arcsin}(x - 1)$
4. $f(x) = \operatorname{Arccos}\sqrt{x}$
5. $f(x) = 3\operatorname{Arccos}\dfrac{x}{2}$
S **6.** $f(x) = \operatorname{Arctan}\sqrt{x}$

7. $f(x) = \operatorname{Arctan} 5x$
8. $f(x) = x\operatorname{Arctan} x$
S **9.** $f(x) = \operatorname{Arccos}\dfrac{1}{x}$
10. $f(x) = \operatorname{Arcsec}(2x)$

11. $f(x) = \operatorname{Arcsin} x + \operatorname{Arccos} x$
12. $f(x) = \operatorname{Arcsec} x + \operatorname{Arccsc} x$
S **13.** $h(t) = \sin(\operatorname{Arccos} t)$
14. $g(t) = \tan(\operatorname{Arcsin} t)$
15. $f(t) = \dfrac{1}{\sqrt{6}}\operatorname{Arctan}\dfrac{\sqrt{6}t}{2}$

16. $f(x) = \dfrac{1}{2}\left[\dfrac{1}{2}\ln\left(\dfrac{x + 1}{x - 1}\right) - \operatorname{Arctan} x\right]$

S **17.** $f(x) = \dfrac{1}{2}\left[\dfrac{1}{2}\ln\left(\dfrac{1 + x}{1 - x}\right) + \operatorname{Arctan} x\right]$

18. $f(x) = \frac{1}{2}[x\sqrt{1 - x^2} + \operatorname{Arcsin} x]$
S **19.** $f(x) = x\operatorname{Arcsin} x + \sqrt{1 - x^2}$
20. $f(x) = x\operatorname{Arctan} 2x - \frac{1}{4}\ln(1 + 4x^2)$

In Exercises 21–45, evaluate the given integrals.

S **21.** $\displaystyle\int_0^{1/6} \frac{1}{\sqrt{1 - 9x^2}}\,dx$
22. $\displaystyle\int_0^1 \frac{1}{\sqrt{4 - x^2}}\,dx$

23. $\displaystyle\int_0^{\sqrt{3}/2} \frac{1}{1 + 4x^2}\,dx$
24. $\displaystyle\int_{\sqrt{3}}^3 \frac{1}{9 + x^2}\,dx$

25. $\int \dfrac{1}{x\sqrt{4x^2-1}}\,dx$

26. $\int \dfrac{1}{4+(x-1)^2}\,dx$

S 27. $\int \dfrac{x^3}{x^2+1}\,dx$

28. $\int \dfrac{x^4-1}{x^2+1}\,dx$

29. $\int \dfrac{1}{\sqrt{1-(x+1)^2}}\,dx$

30. $\int \dfrac{t}{t^4+16}\,dt$

31. $\int \dfrac{t}{\sqrt{1-t^4}}\,dt$

S 32. $\int \dfrac{1}{x\sqrt{x^4-4}}\,dx$

S 33. $\int \dfrac{1}{(x-1)\sqrt{(x-1)^2-4}}\,dx$

34. $\int \dfrac{\text{Arctan }x}{1+x^2}\,dx$

S 35. $\int_0^{1/\sqrt{2}} \dfrac{\text{Arcsin }x}{\sqrt{1-x^2}}\,dx$

36. $\int_0^{1/\sqrt{2}} \dfrac{\text{Arccos }x}{\sqrt{1-x^2}}\,dx$

37. $\int_{-1/2}^{0} \dfrac{x}{\sqrt{1-x^2}}\,dx$

38. $\int_{-\sqrt{3}}^{0} \dfrac{x}{1+x^2}\,dx$

39. $\int \dfrac{e^x}{\sqrt{1-e^{2x}}}\,dx$

40. $\int \dfrac{\cos x}{\sqrt{4-\sin^2 x}}\,dx$

41. $\int \dfrac{1}{9+(x-3)^2}\,dx$

42. $\int \dfrac{x+1}{x^2+1}\,dx$

S 43. $\int \dfrac{1}{\sqrt{x}(1+x)}\,dx$

44. $\int_1^2 \dfrac{1}{3+(x-2)^2}\,dx$

45. $\int_{\pi/2}^{\pi} \dfrac{\sin x}{1+\cos^2 x}\,dx$

46. Find the point of inflection of $f(x) = \text{Arcsin }x$.

47. Find the point of inflection of $f(x) = \text{Arccot }2x$.

48. Find any relative extrema to the function $f(x) = \text{Arcsec }x - x$.

49. Find any relative extrema to the function $f(x) = \text{Arcsin }x - 2x$.

50. Evaluate

$$\lim_{x \to 0} \frac{\text{Arcsin }x}{x}$$

51. Evaluate

$$\lim_{x \to 1} \frac{\text{Arctan }x - (\pi/4)}{x-1}$$

52. Find the area of the region bounded by $y = 1/(1+x^2)$, $y = 0$, $x = 0$, and $x = 1$.

53. Find the area of the region bounded by $y = 1/\sqrt{4-x^2}$, $y = 0$, $x = 0$, and $x = 1$.

54. Use integration to show that the arc length of the circle $x^2 + y^2 = 4$, from $(-\sqrt{3}, 1)$ clockwise to $(\sqrt{3}, 1)$, is one-third the circumference of the circle.

55. Find the volume of the solid generated by revolving the region bounded by $y = 1/(x^4 + 1)$, $y = 0$, $x = 0$, and $x = 1$ about the y-axis.

C 56. Find the point of intersection of the graphs of $y = \text{Arccos }x$ and $y = \text{Arctan }x$.

S 57. Find the point of intersection of the graphs of $y = \text{Arcsin }x$ and $y = \text{Arccos }x$.

58. A small boat is being pulled toward a dock that is 10 ft above the level of the water. If the rope is being pulled in at a rate of 1.5 ft/s, find the rate at which the angle the rope makes with the horizontal is changing when there is 20 ft of rope out.

S 59. An observer is standing 300 ft from the point at which a balloon is released. If the balloon rises at a rate of 5 ft/s, how fast is the angle of elevation of the observer's line of sight increasing when the balloon is 100 ft high?

60. For a weight of mass m attached to a spring that is oscillating with simple harmonic motion (see Figure 9.28), Hooke's Law gives us

$$my'' = -ky$$

where k is a constant and y is the displacement of the spring. By multiplying this equation by y' and integrating with respect to the time t, we can obtain

$$m(y')^2 = k(A^2 - y^2)$$

where A is the maximum displacement. In differential form this equation becomes

$$\frac{dy}{\sqrt{A^2-y^2}} = \sqrt{\frac{k}{m}}\,dt$$

Integrate this equation and solve for y as a function of t given that $y = 0$ when $t = 0$.

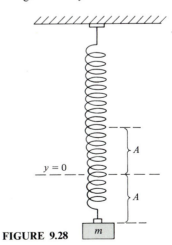

FIGURE 9.28

61. A bullet is projected upward with an initial velocity of 500 ft/s. If the air resistance is proportional to the square of the velocity, we obtain the equation

$$\frac{dv}{dt} = -(32 + kv^2)$$

where 32 ft/s^2 is the acceleration due to gravity and k is a constant. Find the velocity as a function of time by solving the equation

$$\int \frac{dv}{32 + kv^2} = -\int dt$$

62. Verify the following rules:

(a) $\dfrac{d}{dx}[\text{Arccos } u] = \dfrac{-1}{\sqrt{1 - u^2}} \dfrac{du}{dx}$

(b) $\dfrac{d}{dx}[\text{Arcsec } u] = \dfrac{1}{|u|\sqrt{u^2 - 1}} \dfrac{du}{dx}$

63. Verify the following rules by differentiating:

(a) $\displaystyle\int \frac{u'\, dx}{a^2 + u^2} = \frac{1}{a}\text{Arctan }\frac{u}{a} + C$

(b) $\displaystyle\int \frac{u'}{u\sqrt{u^2 - a^2}}\, dx = \frac{1}{a}\text{Arcsec }\frac{|u|}{a} + C$

For Review and Class Discussion

True or False

1. _____ The slope of the graph of the inverse tangent is positive for all x.

2. _____ $\dfrac{d}{dx}[\text{Arctan}(\tan x)] = 1$ for all x.

3. _____ $\displaystyle\int \frac{x}{\sqrt{1 - x^2}}\, dx = \text{Arcsin } x + C$.

4. _____ If $n > 0$ and n is even, then

$$\int \frac{x^n}{1 + x^2}\, dx = \frac{x^{n-1}}{n - 1} - \frac{x^{n-3}}{n - 3} + \frac{x^{n-5}}{n - 5} - \cdots$$
$$\pm x \mp \text{Arctan } x + C$$

but if n is odd, then

$$\int \frac{x^n}{1 + x^2}\, dx = \frac{x^{n-1}}{n - 1} - \frac{x^{n-3}}{n - 3} + \frac{x^{n-5}}{n - 5} - \cdots$$
$$\pm \frac{x^2}{2} \mp \frac{1}{2}\ln(1 + x^2) + C$$

5. _____ If $y = \text{Arcsin } x$, then $dy/dx = 1/(dx/dy)$ for all x in $[-1, 1]$.

9.6 Hyperbolic Functions (optional)

Purpose

- To introduce the hyperbolic functions.
- To derive formulas for the derivatives and integrals of hyperbolic functions.
- To identify some similarities between the six hyperbolic functions and the six trigonometric functions.

The parabola indeed serves in the construction of the catenary, but the two curves are so different that one is algebraic, the other is transcendental.

—John Bernoulli (1667–1748)

A certain class of exponential functions involving combinations of e^x and e^{-x} occur often enough in both pure and applied mathematics to merit special names. We refer to these special exponential functions as the **hyperbolic functions.** Although their names may be unfamiliar, we emphasize that they are simply exponential functions and, as such, they can be integrated and differentiated by our existing rules.

The standard notation for the hyperbolic functions is given in the following definitions:

Definitions of the Hyperbolic Functions

$$\sinh x = \frac{e^x - e^{-x}}{2} \qquad \operatorname{csch} x = \frac{1}{\sinh x}$$

$$\cosh x = \frac{e^x + e^{-x}}{2} \qquad \operatorname{sech} x = \frac{1}{\cosh x}$$

$$\tanh x = \frac{\sinh x}{\cosh x} \qquad \coth x = \frac{1}{\tanh x}$$

($\sinh x$ is read, "the hyperbolic sine of x," $\cosh x$ is read, "the hyperbolic cosine of x," etc.)

The graphs of the six basic hyperbolic functions are shown in Figure 9.29, along with the domain and range of each function. Note that the graphs of $\sinh x$ and $\cosh x$ are readily obtained by *addition of ordinates,* using the exponential functions $f(x) = e^x/2$ and $g(x) = e^{-x}/2$.

Domain: $(-\infty, \infty)$
Range: $(-\infty, \infty)$

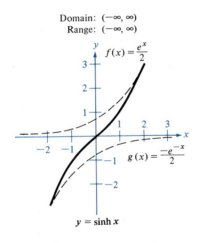

$y = \sinh x$

Domain: $(-\infty, \infty)$
Range: $(-1, 1)$

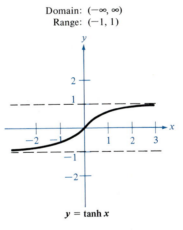

$y = \tanh x$

Domain: $(-\infty, \infty)$
Range: $[1, \infty)$

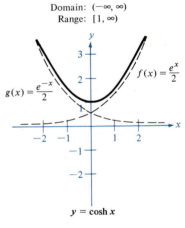

$y = \cosh x$

Domain: $(-\infty, 0)$ and $(0, \infty)$
Range: $(-\infty, 0)$ and $(0, \infty)$

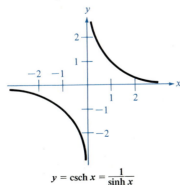

$y = \operatorname{csch} x = \dfrac{1}{\sinh x}$

Domain: $(-\infty, 0)$ and $(0, \infty)$
Range: $(-\infty, -1)$ and $(1, \infty)$

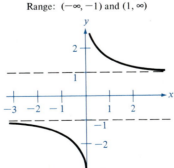

$y = \coth x = \dfrac{1}{\tanh x}$

Domain: $(-\infty, \infty)$
Range: $(0, 1]$

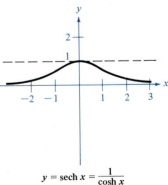
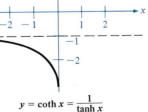

$y = \operatorname{sech} x = \dfrac{1}{\cosh x}$

FIGURE 9.29

The similarity in the names of the hyperbolic functions to those of the trigonometric functions is due primarily to the relationships that exist between the hyperbolic functions; they closely parallel the trigonometric identities. For instance,

$$\cosh^2 x - \sinh^2 x = \left(\frac{e^x + e^{-x}}{2}\right)^2 - \left(\frac{e^x - e^{-x}}{2}\right)^2$$

$$= \frac{e^{2x} + 2 + e^{-2x}}{4} - \frac{e^{2x} - 2 + e^{-2x}}{4} = \frac{4}{4} = 1$$

Or

$$2 \sinh x \cosh x = 2 \left(\frac{e^x - e^{-x}}{2}\right)\left(\frac{e^x + e^{-x}}{2}\right)$$

$$= \frac{e^{2x} - e^{-2x}}{2} = \sinh 2x$$

We list several of the commonly used identities next. Notice the similarity to the trigonometric identities.

Hyperbolic Identities		
$\cosh^2 x - \sinh^2 x = 1$		$\sinh (x + y) = \sinh x \cosh y + \cosh x \sinh y$
$\tanh^2 x + \operatorname{sech}^2 x = 1$		$\sinh (x - y) = \sinh x \cosh y - \cosh x \sinh y$
$\coth^2 x - \operatorname{csch}^2 x = 1$		$\cosh (x + y) = \cosh x \cosh y + \sinh x \sinh y$
		$\cosh (x - y) = \cosh x \cosh y - \sinh x \sinh y$
$\sinh^2 x = \dfrac{-1 + \cosh 2x}{2}$		
		$\sinh 2x = 2 \sinh x \cosh x$
$\cosh^2 x = \dfrac{1 + \cosh 2x}{2}$		$\cosh 2x = \cosh^2 x + \sinh^2 x$

Since the hyperbolic functions are written in terms of the exponential functions e^x and e^{-x}, we can readily derive formulas for their derivatives. For example,

$$\frac{d}{dx}[\sinh x] = \frac{d}{dx}\left[\frac{e^x - e^{-x}}{2}\right] = \frac{e^x + e^{-x}}{2} = \cosh x$$

In a similar manner, we can establish that

$$\frac{d}{dx}[\cosh x] = \sinh x$$

By the Quotient Rule we obtain

$$\frac{d}{dx}[\tanh x] = \frac{d}{dx}\left[\frac{\sinh x}{\cosh x}\right]$$

$$= \frac{\cosh x \, (\cosh x) - \sinh x \, (\sinh x)}{\cosh^2 x}$$

$$= \frac{1}{\cosh^2 x} = \text{sech}^2 x$$

In the following list we give six differentiation formulas for hyperbolic functions, together with their corresponding integration formulas.

Hyperbolic Functions: Derivatives and Integrals	$\frac{d}{dx}[\sinh u] = (\cosh u)\frac{du}{dx}$	$\int (\cosh u)u' \, dx = \sinh u + C$
	$\frac{d}{dx}[\cosh u] = (\sinh u)\frac{du}{dx}$	$\int (\sinh u)u' \, dx = \cosh u + C$
	$\frac{d}{dx}[\tanh u] = (\text{sech}^2 u)\frac{du}{dx}$	$\int (\text{sech}^2 u)u' \, dx = \tanh u + C$
	$\frac{d}{dx}[\coth u] = -(\text{csch}^2 u)\frac{du}{dx}$	$\int (\text{csch}^2 u)u' \, dx = -\coth u + C$
	$\frac{d}{dx}[\text{sech } u] = -(\text{sech } u \tanh u)\frac{du}{dx}$	$\int (\text{sech } u \tanh u)u' \, dx = -\text{sech } u + C$
	$\frac{d}{dx}[\text{csch } u] = -(\text{csch } u \coth u)\frac{du}{dx}$	$\int (\text{csch } u \coth u)u' \, dx = -\text{csch } u + C$

Example 1

Find the derivatives of the following:
(a) $f(x) = \sinh (x^2 - 3)$
(b) $g(x) = \ln (\cosh x)$
(c) $h(x) = x \sinh x - \cosh x$

Solution:

(a) $\qquad f(x) = \sinh (x^2 - 3)$
$\qquad\qquad f'(x) = 2x \cosh (x^2 - 3)$

(b) $\qquad g(x) = \ln (\cosh x)$

$$g'(x) = \left(\frac{1}{\cosh x}\right)(\sinh x) = \frac{\sinh x}{\cosh x} = \tanh x$$

(c) $\qquad h(x) = x \sinh x - \cosh x$
$\qquad\qquad h'(x) = x \cosh x + \sinh x - \sinh x = x \cosh x$ ∎

Example 2

Find the critical points of $f(x) = (x - 1) \cosh x - \sinh x$.

Solution: If we set $f'(x)$ equal to zero, we have

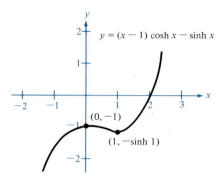

$$y = (x - 1) \cosh x - \sinh x$$

$(0, -1)$

$(1, -\sinh 1)$

FIGURE 9.30

$$f'(x) = (x - 1) \sinh x + \cosh x - \cosh x = 0$$

which implies that $\qquad 0 = (x - 1) \sinh x$

Thus $f'(x) = 0$ if $x = 1$ or if $x = 0$. By the Second-Derivative Test, we can verify that the point $(0, -1)$ is a relative maximum, while $(1, -\sinh 1)$ is a relative minimum. (See Figure 9.30.) ∎

Example 3

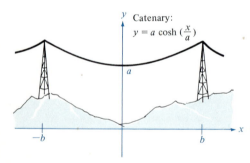

Catenary:
$$y = a \cosh \left(\frac{x}{a}\right)$$

FIGURE 9.31

Find $\qquad\qquad \lim_{x \to 0} \dfrac{\sinh x}{x}$

Solution: Since direct substitution results in the indeterminate form $0/0$, we apply L'Hôpital's Rule to get

$$\lim_{x \to 0} \frac{\sinh x}{x} = \lim_{x \to 0} \frac{\cosh x}{1} = 1$$

∎

When a uniform flexible cable, such as a telephone wire, is suspended from two points, it takes the shape of a **catenary.** (See Figure 9.31.) The equation for such a curve is given in Example 4.

Example 4

Show that the length s of the catenary described by the equation $f(x) = a \cosh (x/a)$, between $x = 0$ and $x = b$, is given by $s = a \sinh (b/a)$.

Solution: Recall from Section 7.8 that the formula for the length s of an arc is given by

$$s = \int_0^b \sqrt{1 + [f'(x)]^2} \, dx$$

Since $\qquad f'(x) = (a) \left(\dfrac{1}{a}\right) \sinh \left(\dfrac{x}{a}\right) = \sinh \left(\dfrac{x}{a}\right)$

we have $\quad s = \int_0^b \sqrt{1 + \sinh^2\left(\dfrac{x}{a}\right)}\, dx = \int_0^b \sqrt{\cosh^2\left(\dfrac{x}{a}\right)}\, dx$

$$= \int_0^b \cosh\left(\frac{x}{a}\right) dx = a \int_0^b \left[\cosh\left(\frac{x}{a}\right)\right]\left(\frac{1}{a}\right) dx$$

$$= a \left[\sinh\left(\frac{x}{a}\right)\right]_0^b = a \sinh\left(\frac{b}{a}\right) \qquad \blacksquare$$

Example 5

Evaluate $\int \cosh 2x \sinh^2 2x\, dx$.

Solution: If we let $u = \sinh 2x$, then by the Power Rule it follows that

$$\int \cosh 2x \sinh^2 2x\, dx = \frac{1}{2} \int (\sinh 2x)^2 (2 \cosh 2x)\, dx$$

$$= \frac{1}{2}\left[\frac{(\sinh 2x)^3}{3}\right] + C = \frac{\sinh^3 2x}{6} + C \qquad \blacksquare$$

Since the hyperbolic functions are defined in terms of exponential functions, it is not surprising to find that the inverse hyperbolic functions can be written in terms of logarithmic functions.

Inverse Hyperbolic Functions *Domain*

Inverse hyperbolic sine: $\sinh^{-1} x = \ln\left(x + \sqrt{x^2 + 1}\right)$ $\qquad (-\infty, \infty)$

Inverse hyperbolic cosine: $\cosh^{-1} x = \ln\left(x + \sqrt{x^2 - 1}\right)$ $\qquad [1, \infty)$

Inverse hyperbolic tangent: $\tanh^{-1} x = \dfrac{1}{2}\ln\left(\dfrac{1 + x}{1 - x}\right)$ $\qquad (-1, 1)$

Inverse hyperbolic cotangent: $\coth^{-1} x = \dfrac{1}{2}\ln\left(\dfrac{x + 1}{x - 1}\right)$ $\qquad (-\infty, -1)$ and $(1, \infty)$

Inverse hyperbolic secant: $\operatorname{sech}^{-1} x = \ln\left(\dfrac{1}{x} + \sqrt{\dfrac{1 - x^2}{x^2}}\right)$ $\qquad (0, 1]$

Inverse hyperbolic cosecant: $\operatorname{csch}^{-1} x = \ln\left(\dfrac{1}{x} + \sqrt{\dfrac{1 + x^2}{x^2}}\right)$ $\qquad (-\infty, 0)$ and $(0, \infty)$

To verify the formulas for these inverses, we can use the definitions of the hyperbolic functions. For example, to show that

$$f(x) = \sinh x = \frac{e^x - e^{-x}}{2}$$

and $\qquad\qquad g(x) = \sinh^{-1} x = \ln\left(x + \sqrt{x^2 + 1}\right)$

are inverses, we show that $f(g(x)) = g(f(x)) = x$. First,

$$f(g(x)) = \frac{e^{\ln(x+\sqrt{x^2+1})} - e^{-\ln(x+\sqrt{x^2+1})}}{2}$$

$$= \frac{(x + \sqrt{x^2 + 1}) - [1/(x + \sqrt{x^2 + 1})]}{2}$$

$$= \frac{x^2 + 2x\sqrt{x^2 + 1} + x^2 + 1 - 1}{2(x + \sqrt{x^2 + 1})}$$

$$= \frac{2x(x + \sqrt{x^2 + 1})}{2(x + \sqrt{x^2 + 1})} = x$$

Conversely,

$$g(f(x)) = \ln\left(\frac{e^x - e^{-x}}{2} + \sqrt{\frac{e^{2x} - 2 + e^{-2x}}{4} + 1}\right)$$

$$= \ln\left(\frac{e^x - e^{-x}}{2} + \sqrt{\frac{e^{2x} + 2 + e^{-2x}}{4}}\right)$$

$$= \ln\left(\frac{e^x - e^{-x}}{2} + \frac{e^x + e^{-x}}{2}\right) = \ln(e^x) = x$$

The graphs of the inverse hyperbolic functions are shown in Figure 9.32.

Inverse Hyperbolic Functions: Derivatives and Integrals

$$\frac{d}{dx}[\sinh^{-1} u] = \frac{1}{\sqrt{u^2 + 1}}\frac{du}{dx}$$

$$\frac{d}{dx}[\cosh^{-1} u] = \frac{1}{\sqrt{u^2 - 1}}\frac{du}{dx}$$

$$\int \frac{u'}{\sqrt{u^2 \pm a^2}}dx = \ln(u + \sqrt{u^2 \pm a^2}) + C$$

$$\frac{d}{dx}[\tanh^{-1} u] = \frac{1}{1 - u^2}\frac{du}{dx}$$

$$\frac{d}{dx}[\coth^{-1} u] = \frac{1}{1 - u^2}\frac{du}{dx}$$

$$\int \frac{u'}{a^2 - u^2}dx = \frac{1}{2a}\ln\left|\frac{a + u}{a - u}\right| + C$$

$$\frac{d}{dx}[\text{sech}^{-1} u] = \frac{-1}{u\sqrt{1 - u^2}}\frac{du}{dx}$$

$$\frac{d}{dx}[\text{csch}^{-1} u] = \frac{-1}{|u|\sqrt{1 + u^2}}\frac{du}{dx}$$

$$\int \frac{u'}{u\sqrt{a^2 \pm u^2}}dx = -\frac{1}{a}\ln\left(\frac{a + \sqrt{a^2 \pm u^2}}{|u|}\right) + C$$

Note that for convenience we listed the integration formulas involving inverse hyperbolic functions in logarithmic form.

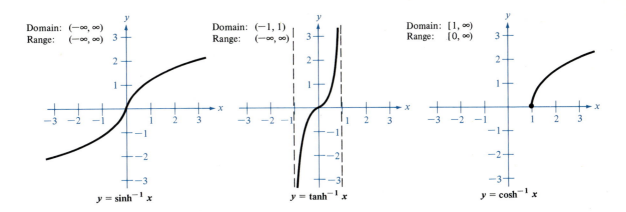

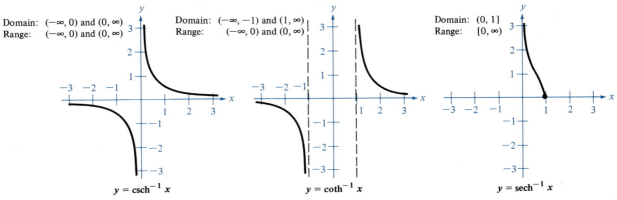

FIGURE 9.32

Example 6

Evaluate the integral $\displaystyle\int \frac{1}{x\sqrt{4-9x^2}}dx$

Solution: Since the denominator contains the form $\sqrt{a^2-u^2}$, we let $u = 3x$ and $a = 2$. Thus $u' = 3$, and we have

$$\int \frac{1}{x\sqrt{4-9x^2}}dx = \int \frac{3}{(3x)\sqrt{4-(3x)^2}}dx$$

$$= -\frac{1}{2}\ln\left(\frac{2+\sqrt{4-9x^2}}{|3x|}\right) + C \qquad \blacksquare$$

Example 7

Evaluate the integral $\displaystyle\int_0^1 \frac{1}{5 - 4x^2}\,dx$

Solution: Since the denominator has the form $a^2 - u^2$, we let $u = 2x$ and $a = \sqrt{5}$. Then $u' = 2$ and we have

$$\int_0^1 \frac{1}{5 - 4x^2}\,dx = \frac{1}{2}\int_0^1 \frac{2}{(\sqrt{5})^2 - (2x)^2}\,dx$$

$$= \frac{1}{2}\left[\frac{1}{2\sqrt{5}}\ln\left(\frac{\sqrt{5} + 2x}{\sqrt{5} - 2x}\right)\right]_0^1$$

$$= \left(\frac{1}{4\sqrt{5}}\right)\ln\left(\frac{\sqrt{5} + 2}{\sqrt{5} - 2}\right)$$

$$= \left(\frac{\sqrt{5}}{20}\right)\ln(9 + 4\sqrt{5}) \approx 0.3228$$ ∎

Section Exercises (9.6)

In Exercises 1–14, find y' and simplify.

1. $y = \sinh(1 - x^2)$ **2.** $y = \coth 3x$

3. $y = \ln(\sinh x)$ **4.** $y = \ln(\cosh x)$

S 5. $y = \ln\left[\tanh\left(\dfrac{x}{2}\right)\right]$

6. $y = x(\sinh x) - \cosh x$

7. $y = \dfrac{1}{4}\sinh(2x) - \dfrac{x}{2}$ **8.** $y = x - \coth x$

9. $y = \text{Arctan}(\sinh x)$ **10.** $y = e^{\sinh x}$

S 11. $y = x^{\cosh x}$ **12.** $y = \text{sech}^2\, 3x$

13. $y = (\cosh x - \sinh x)^2$ **14.** $y = \text{sech}(x + 1)$

In Exercises 15–24, evaluate the given integral.

15. $\displaystyle\int \sinh(1 - 2x)\,dx$ **16.** $\displaystyle\int \frac{\cosh\sqrt{x}}{\sqrt{x}}\,dx$

17. $\displaystyle\int \cosh^2(x - 1)\sinh(x - 1)\,dx$

S 18. $\displaystyle\int \frac{\sinh x}{1 + \sinh^2 x}\,dx$

19. $\displaystyle\int \frac{\cosh x}{\sinh x}\,dx$ **20.** $\displaystyle\int \text{sech}^2(2x - 1)\,dx$

21. $\displaystyle\int x\,\text{csch}^2\left(\frac{x^2}{2}\right)dx$ **22.** $\displaystyle\int \text{sech}^3 x\tanh x\,dx$

23. $\displaystyle\int \frac{\text{csch}(1/x)\coth(1/x)}{x^2}\,dx$

24. $\displaystyle\int \sinh^2 x\,dx$

In Exercises 25–32, find y' and simplify.

25. $y = \cosh^{-1} 3x$ **26.** $y = \tanh^{-1}\dfrac{x}{2}$

27. $y = \sin^{-1}(\tan x)$ **S 28.** $y = \text{sech}^{-1}(\cos 2x)$
                                              $(0 < x < \pi/4)$

29. $y = \coth^{-1}(\sin 2x)$ **30.** $y = (\text{csch}^{-1} x)^2$

S 31. $y = 2x\sinh^{-1} 2x - \sqrt{1 + 4x^2}$

32. $y = x\tanh^{-1} x + \ln\sqrt{1 - x^2}$

In Exercises 33–42, evaluate each integral.

33. $\displaystyle\int_0^4 \frac{1}{25 - x^2}\,dx$ **34.** $\displaystyle\int_0^4 \frac{1}{\sqrt{25 - x^2}}\,dx$

35. $\displaystyle\int_0^{\sqrt{2}/4} \frac{2}{\sqrt{1 - 4x^2}}\,dx$ **36.** $\displaystyle\int \frac{2}{x\sqrt{1 + 4x^2}}\,dx$

37. $\displaystyle\int \frac{x}{x^4 + 1}\,dx$ **38.** $\displaystyle\int \frac{\cot x}{\sqrt{9 - \sin^2 x}}\,dx$

S 39. $\displaystyle\int \frac{1}{\sqrt{1 + e^{2x}}}\,dx$ **40.** $\displaystyle\int \frac{e^x}{1 - e^{2x}}\,dx$

S 41. $\int \dfrac{1}{\sqrt{x}\,\sqrt{1+x}}\,dx$ 42. $\int \dfrac{\sqrt{x}}{\sqrt{1+x^3}}\,dx$

C 43. Find the points of inflection of $y = \operatorname{sech} x$.

C 44. Find any relative extrema and points of inflection of the graph of $y = x(\cosh x) - \sinh x$. Sketch the graph.

C 45. Find any relative extrema and points of inflection of the graph of $y = x - \tanh x$. Sketch the graph.

C 46. A farmer builds a barn that has length 100 ft and width 40 ft. A cross section of the roof is the inverted catenary $y = 31 - 20\cosh(x/20)$ (see Figure 9.33).

    (a) Find the number of cubic feet of feed that can be stored within the roof structure of the barn.

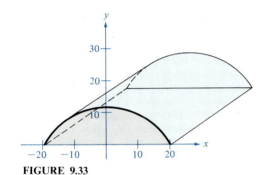

FIGURE 9.33

(b) Find the number of square feet of roofing on the barn.

S 47. The equation of a *tractrix* is given by $y = a\operatorname{sech}^{-1}(x/a) - \sqrt{a^2 - x^2}$. Find dy/dx. Assume $0 < a$.

48. Sketch the graph of the tractrix given in Exercise 47 for $0 < x \le 1$ and $a = 1$.

49. Let L be the tangent line at the point P to the tractrix given in Exercise 47. If L intersects the y-axis at the point Q, show that the distance between P and Q is a.

In Exercises 50–55, prove each statement.

50. $\dfrac{d}{dx}[\tanh x] = \operatorname{sech}^2 x$

51. $\dfrac{d}{dx}[\operatorname{sech} x] = -\operatorname{sech} x \tanh x$

52. $\dfrac{d}{dx}[\cosh^{-1} x] = \dfrac{1}{\sqrt{x^2 - 1}}$

S 53. $\dfrac{d}{dx}[\tanh^{-1} x] = \dfrac{1}{1 - x^2}$

54. $\dfrac{d}{dx}[\operatorname{sech}^{-1} x] = \dfrac{-1}{x\sqrt{1 - x^2}}$

55. $\dfrac{d}{dx}[\sinh^{-1} x] = \dfrac{1}{\sqrt{x^2 + 1}}$

Miscellaneous Exercises (Ch. 9)

In Exercises 1–25, find dy/dx and simplify.

1. $y = 3\cos(3x + 1)$
2. $y = 2\csc^3 \sqrt{x}$
3. $y = \tan\sqrt{1 - x}$
4. $y = \tfrac{1}{2}e^{\sin 2x}$
5. $y = x\sin x^2$
6. $y = 1 - \cos 2x + 2\cos^2 x$
7. $y = \dfrac{\sin x}{x^2}$
8. $y = \csc 3x + \cot 3x$
9. $y = \tan(\operatorname{Arcsin} x)$
10. $y = \operatorname{Arctan}(x^2 - 1)$

S 11. $y = \ln(\sin^2 x)$
12. $y = \csc(1 - 3x)$

13. $y = x\tan x$
14. $y = \dfrac{\cos(x - 1)}{x - 1}$

15. $y = x\operatorname{Arcsec} x$
16. $y = \tfrac{1}{2}\operatorname{Arctan} e^{2x}$

17. $y = \dfrac{\sin 4x}{4} + x$
18. $y = x\cos x - \sin x$

S 19. $y = \left(\dfrac{x^2 + 1}{2}\right)\operatorname{Arctan} x$
20. $x\sin y = y\cos x$

21. $x = 2 + \sin y$
22. $\sin(x + y) = x$

S 23. $\cos x^2 = xe^y$ 24. $\cos(x + y) = x$

25. $y = x(\operatorname{Arcsin} x)^2 - 2x + 2\sqrt{1 - x^2}\operatorname{Arcsin} x$

In Exercises 26–45, evaluate the indefinite integrals.

26. $\int \sin^3 x \cos x\, dx$

S 27. $\int \dfrac{\cos x}{1 + \sin^2 x}\,dx$

28. $\int \dfrac{\cos x}{\sqrt{\sin x}}\,dx$

29. $\int \dfrac{\operatorname{Arctan} 2x}{1 + 4x^2}\,dx$

30. $\int \dfrac{\sin\theta}{\sqrt{1 - \cos\theta}}\,d\theta$

31. $\int \tan^n x \sec^2 x\, dx,\ n \ne -1$

32. $\int \dfrac{x - 1}{3x^2 - 6x - 1}\,dx$

33. $\int \dfrac{e^{-2x}}{1 + e^{-2x}}\,dx$

S 34. $\int \dfrac{\tan(1/x)}{x^2}\,dx$

S 35. $\int \dfrac{1}{e^{2x} + e^{-2x}}\,dx$

36. $\int \dfrac{1}{3 + 25x^2}\, dx$

37. $\int \dfrac{x}{\sqrt{1 - x^4}}\, dx$

38. $\int \dfrac{1}{16 + x^2}\, dx$

39. $\int \dfrac{x}{16 + x^2}\, dx$

40. $\int x \sin 3x^2\, dx$

41. $\int \sec 2x \tan 2x\, dx$

42. $\int \dfrac{\operatorname{Arcsin} x}{\sqrt{1 - x^2}}\, dx$

43. $\int \dfrac{\operatorname{Arctan}(x/2)}{4 + x^2}\, dx$

S **44.** $\int \dfrac{4 - x}{\sqrt{4 - x^2}}\, dx$

45. $\int x \tan x^2\, dx$

46. The curvature of the graph of $y = f(x)$ is given by $|f''(x)| / \{1 + [f'(x)]^2\}^{3/2}$. Find the curvature of the graph of $y = \sin x$ at the point $(\pi/2, 1)$.

47. Find a formula for the curvature of the graph of $y = \tan x$. What is the limit of the curvature as x approaches $\pi/2$ from the left?

48. Find the equation of the tangent line and the equation of the line normal to the graph of $y = \operatorname{Arctan} x$ at the point $(1, \pi/4)$.

C **49.** Find the area of the largest rectangle that can be inscribed between the graph of $y = \cos x$ and the x-axis.

50. Show that $y = [a + \ln(\cos x)]\cos x + (b + x)\sin x$ satisfies the differential equation $y'' + y = \sec x$.

51. Show that $y = a \cos 2x + b \sin 2x + 2x^2 - 1$ satisfies the differential equation $y'' + 4y = 8x^2$.

52. Show that $y = e^x(a \cos 3x + b \sin 3x)$ satisfies the differential equation $y'' - 2y' + 10y = 0$.

C **53.** Approximate, to three decimal places, a value of x such that $2 \sin x = x$ and x lies in the interval $(\pi/2, \pi)$.

54. A rotating beacon is located 1 mi off a straight shoreline. If the beacon rotates at the rate of 3 r/min, how fast (in miles per hour) does the beam of light appear to be moving to a viewer who is $\frac{1}{2}$ mi down the shoreline?

55. Let θ be the angle of displacement (from the vertical) of a pendulum that is L feet long. If $\theta = A \sin \sqrt{32/L}\, t + B \cos \sqrt{32/L}\, t$, where t is measured in seconds, find the maximum rate of change of θ.

S **56.** A hallway of width 6 ft meets a hallway of width 9 ft at right angles. Find the length of longest pipe that can be carried horizontally around this corner. (Hint: If L is the length of the pipe, show that

$$L = 6 \csc \theta + 9 \csc[(\pi/2) - \theta]$$

where θ is the angle between the pipe and wall of the narrower hallway.)

57. Rework Exercise 56 if one of the hallways is of width a and the other is of width b.

58. A certain lawn sprinkler is constructed in such a way that $d\theta/dt$ is constant, where θ ranges between 45° and 135° as shown in the accompanying figure. The distance the water travels horizontally is given by

$$x = \dfrac{v^2 \sin 2\theta}{32}$$

where v is the speed of the water. Find dx/dt and explain why this model of lawn sprinkler does not water evenly. Which portion of the lawn receives the most water?

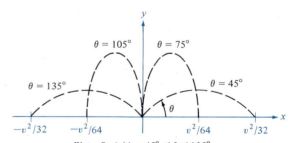

Water Sprinkler: 45° ⩽ θ ⩽ 135°

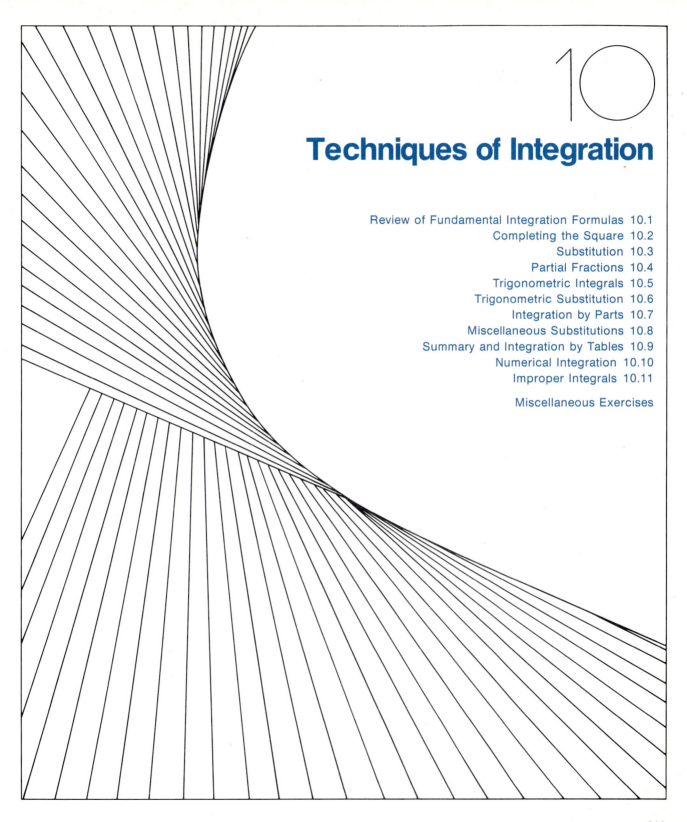

10

Techniques of Integration

10.1   Review of Fundamental Integration Formulas

Purpose

- To review the fundamental rules for integrating algebraic and transcendental functions.

Common integration is only the memory of differentiation. The different devices by which integration is accomplished are changes, not from the known to the unknown, but from forms in which memory will not serve us to those in which it will.   —*Augustus DeMorgan (1806–1871)*

In Chapters 3, 6, 7, 8, and 9 we studied a number of rules for differentiating and integrating algebraic and transcendental functions. In general, the rules for integrating were derived from corresponding differentiation formulas. It may surprise you to learn that although we now have all the necessary tools for *differentiating* elementary functions, our set of tools for *integrating* these functions is by no means complete.

The primary objective of this chapter is to develop several techniques that greatly expand the set of integrals to which the basic integration formulas can be applied. To operate efficiently with the new techniques studied in this chapter, you should memorize the fundamental integration formulas listed in this section.

Even though Formulas 15, 17, and 19 were introduced in the optional Section 9.6, it is not necessary to have studied that section to apply these three formulas. They can easily be verified by differentiation.

We need not work very long with integration problems before we realize that integration is not nearly as straightforward as differentiation. A major part of any integration problem is the recognition of which basic integration formula to use to solve the problem. Skill in recognizing what formula to use requires memorization of the basic formulas and lots of practice in using them.

Example 1

Evaluate $\int (2x + 1)^3 \, dx$.

Solution: We consider Formula 1 and let $u = 2x + 1$. Then $u' = 2$ and we have

$$\int (2x + 1)^3 \, dx = \frac{1}{2} \int (2x + 1)^3 (2) \, dx$$

$$= \frac{1}{2} \int u^3 u' \, dx = \left(\frac{1}{2}\right)\left(\frac{u^4}{4}\right) + C$$

$$= \frac{1}{2}\left[\frac{(2x + 1)^4}{4}\right] + C$$

$$= \frac{(2x + 1)^4}{8} + C$$

Fundamental Integration Formulas

1. $\displaystyle\int u^n u'\, dx = \frac{u^{n+1}}{n+1} + C,\; n \neq -1$ $\displaystyle\left(\int x^n\, dx = \frac{x^{n+1}}{n+1} + C,\; n \neq -1\right)$

2. $\displaystyle\int e^u u'\, dx = e^u + C$ $\displaystyle\left(\int a^u u'\, dx = \frac{1}{\ln a}a^u + C\right)$

3. $\displaystyle\int \frac{u'}{u}\, dx = \ln |u| + C$ $\displaystyle\left(\int \frac{1}{x}\, dx = \ln |x| + C\right)$

4. $\displaystyle\int (\sin u)u'\, dx = -\cos u + C$ 5. $\displaystyle\int (\cos u)u'\, dx = \sin u + C$

6. $\displaystyle\int (\sec^2 u)u'\, dx = \tan u + C$ 7. $\displaystyle\int (\csc^2 u)u'\, dx = -\cot u + C$

8. $\displaystyle\int (\sec u \tan u)u'\, dx = \sec u + C$ 9. $\displaystyle\int (\csc u \cot u)u'\, dx = -\csc u + C$

10. $\displaystyle\int (\tan u)u'\, dx = -\ln |\cos u| + C$ 11. $\displaystyle\int (\cot u)u'\, dx = \ln |\sin u| + C$

12. $\displaystyle\int (\sec u)u'\, dx = \ln |\sec u + \tan u| + C$ 13. $\displaystyle\int (\csc u)u'\, dx = \ln |\csc u - \cot u| + C$

14. $\displaystyle\int \frac{u'}{\sqrt{a^2 - u^2}}\, dx = \operatorname{Arcsin}\frac{u}{a} + C$ 15. $\displaystyle\int \frac{u'}{\sqrt{u^2 \pm a^2}}\, dx = \ln |u + \sqrt{u^2 \pm a^2}| + C$

16. $\displaystyle\int \frac{u'}{a^2 + u^2}\, dx = \frac{1}{a}\operatorname{Arctan}\frac{u}{a} + C$ 17. $\displaystyle\int \frac{u'}{u^2 - a^2}\, dx = \frac{1}{2a}\ln \left|\frac{u-a}{u+a}\right| + C$

18. $\displaystyle\int \frac{u'}{u\sqrt{u^2 - a^2}}\, dx = \frac{1}{a}\operatorname{Arcsec}\frac{|u|}{a} + C$ 19. $\displaystyle\int \frac{u'}{u\sqrt{a^2 \pm u^2}}\, dx = -\frac{1}{a}\ln \left(\frac{a + \sqrt{a^2 \pm u^2}}{|u|}\right) + C$

Example 2

Evaluate $\int 4 \sec^2 3x\, dx$.

Solution: We consider Formula 6 and let $u = 3x$. Then $u' = 3$ and we have

$$\int 4 \sec^2 3x\, dx = \frac{4}{3}\int (\sec^2 3x)(3)\, dx = \frac{4}{3}\int (\sec^2 u)u'\, dx$$

$$= \frac{4}{3}\tan u + C = \frac{4}{3}\tan 3x + C \qquad \blacksquare$$

Example 3

Evaluate the following indefinite integrals:

(a) $\int \dfrac{4}{x^2 + 9}\, dx$ (b) $\int \dfrac{4x}{x^2 + 9}\, dx$ (c) $\int \dfrac{4x^2}{x^2 + 9}\, dx$

Solution:

(a) Considering Formula 16 we let $u = x$ and $a = 3$. Then

$$\int \frac{4}{x^2 + 9}\, dx = 4 \int \frac{1}{x^2 + 3^2}\, dx = 4\left[\frac{1}{3}\, \text{Arctan}\, \frac{x}{3}\right] + C = \frac{4}{3}\, \text{Arctan}\, \frac{x}{3} + C$$

(b) Here Formula 16 does not apply because of the x in the numerator. Considering Formula 3 we let $u = x^2 + 9$. Then $u' = 2x$ and we have

$$\int \frac{4x}{x^2 + 9}\, dx = 2 \int \frac{2x}{x^2 + 9}\, dx = 2 \ln (x^2 + 9) + C$$

(c) Since the degree of the numerator equals the degree of the denominator, we first divide to obtain

$$\frac{4x^2}{x^2 + 9} = 4 - \frac{36}{x^2 + 9}$$

Thus

$$\int \frac{4x^2}{x^2 + 9}\, dx = \int \left(4 - \frac{36}{x^2 + 9}\right) dx = \int 4\, dx - 36 \int \frac{1}{x^2 + 9}\, dx$$

$$= 4x - 36\left[\frac{1}{3}\, \text{Arctan}\, \frac{x}{3}\right] + C = 4x - 12\, \text{Arctan}\, \frac{x}{3} + C$$

by Formulas 1 and 16. ■

Notice in part (c) of Example 3 that some preliminary algebra was required before applying the rules for integrating and subsequently more than one formula was needed to evaluate the resulting integral. This is also the case in the next example.

Example 4

Find
$$\int \frac{x - 3}{\sqrt{9x^2 + 4}}\, dx$$

Solution: By rewriting the integral as

$$\int \frac{x - 3}{\sqrt{9x^2 + 4}}\, dx = \int \frac{x\, dx}{\sqrt{9x^2 + 4}} - \int \frac{3\, dx}{\sqrt{9x^2 + 4}}$$

we can apply Formulas 1 and 15 to obtain

$$\int \frac{x-3}{\sqrt{9x^2+4}}\,dx = \frac{1}{18}\int (9x^2+4)^{-1/2}(18x)\,dx - \int \frac{3\,dx}{\sqrt{(3x)^2+2^2}}$$

$$= \frac{1}{18}\left[\frac{(9x^2+4)^{1/2}}{\frac{1}{2}}\right] - \ln|3x+\sqrt{9x^2+4}| + C$$

$$= \frac{1}{9}\sqrt{9x^2+4} - \ln(3x+\sqrt{9x^2+4}) + C \qquad \blacksquare$$

Example 5

Find $\int_0^{\pi/4}(\sin 2x)e^{-\cos 2x}\,dx$.

Solution: Considering Formula 2 we let $u = -\cos 2x$. Then $u' = 2\sin 2x$ and it follows that

$$\int_0^{\pi/4}\sin 2x\,e^{-\cos 2x}\,dx = \frac{1}{2}\int_0^{\pi/4}e^{-\cos 2x}(2\sin 2x)\,dx$$

$$= \frac{1}{2}e^{-\cos 2x}\Big]_0^{\pi/4} = \frac{1}{2}(e^0 - e^{-1}) \approx 0.3161 \qquad \blacksquare$$

Example 6

Evaluate

$$\int \frac{1+\cos e^{-2x}}{e^{2x}}\,dx$$

Solution: Quite often with a sum or difference in the numerator, we can separate the integrand into two or more parts. In this case we have

$$\int \frac{1+\cos e^{-2x}}{e^{2x}}\,dx = \int \frac{dx}{e^{2x}} + \int \frac{\cos e^{-2x}}{e^{2x}}\,dx$$

$$= \int e^{-2x}\,dx + \int (\cos e^{-2x})(e^{-2x})\,dx$$

$$= -\frac{1}{2}\int e^{-2x}(-2)\,dx - \frac{1}{2}\int (\cos e^{-2x})(-2e^{-2x})\,dx$$

By Formulas 2 and 5 we have

$$\int \frac{1+\cos e^{-2x}}{e^{2x}}\,dx = -\frac{1}{2}e^{-2x} - \frac{1}{2}\sin e^{-2x} + C \qquad \blacksquare$$

Example 7

Evaluate

$$\int \frac{x^2}{\sqrt{16-x^6}}\,dx$$

Solution: The radical in the denominator has the form $\sqrt{a^2-u^2} = \sqrt{4^2-(x^3)^2}$. Formula 14 or 19 seems to be a reasonable choice. We eliminate Formula 19 because of a missing u factor in the denominator. Thus by Formula 14, letting $u = x^3$ and $a = 4$, we have

$$\int \frac{x^2}{\sqrt{16 - x^6}}\, dx = \frac{1}{3}\int \frac{3x^2}{\sqrt{16 - (x^3)^2}}\, dx = \frac{1}{3} \operatorname{Arcsin} \frac{x^3}{4} + C$$ ∎

Example 8

Find the area of the plane region bounded by $y = 1/(x^2 - 2)$, $y = 0$, $x = 0$, and $x = 1$. (See Figure 10.1.)

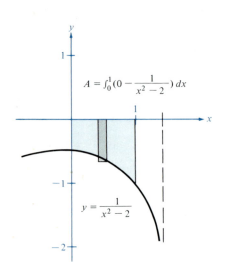

Solution: The area A of the designated region is given by

$$A = \int_0^1 \left(0 - \frac{1}{x^2 - 2} \right) dx$$

By Formula 17 we have

$$-\int_0^1 \frac{dx}{x^2 - 2} = -\int_0^1 \frac{dx}{x^2 - (\sqrt{2})^2} = \frac{-1}{2\sqrt{2}} \left[\ln \left| \frac{x - \sqrt{2}}{x + \sqrt{2}} \right| \right]_0^1$$

$$= -\frac{1}{2\sqrt{2}} \left[\ln \left| \frac{1 - \sqrt{2}}{1 + \sqrt{2}} \right| - \ln 1 \right] \approx 0.6232$$ ∎

The figure shows: $A = \int_0^1 (0 - \frac{1}{x^2 - 2})\, dx$ and $y = \frac{1}{x^2 - 2}$

FIGURE 10.1

The Power and Log Rules (Formulas 1 and 3) are two of the most frequently used integration rules. In many cases their use is somewhat disguised. Therefore, it is a good idea to keep these rules in mind for those integrals that do not seem to fit any of the other fundamental formulas. The next two examples illustrate some of the more subtle applications of the Power and Log Rules.

Example 9

Evaluate $\int \cot x (\ln \sin x)\, dx$.

Solution: This integral does not appear to fit any of our fundamental formulas. However, considering the two primary choices for u,

$$u = \cot x \quad \text{or} \quad u = \ln \sin x$$

we quickly see that the second choice is the appropriate one because

$$\frac{d}{dx}[\ln \sin x] = \frac{\cos x}{\sin x} = \cot x$$

is precisely the factor needed to use the Power Rule on $u = \ln \sin x$. Thus we have

$$\int \cot x (\ln \sin x)\, dx = \int (\ln \sin x)(\cot x)\, dx$$

$$= \int (u^1)u' \, dx = \frac{u^2}{2} + C = \frac{(\ln \sin x)^2}{2} + C \qquad \blacksquare$$

Example 10

Evaluate

$$\int \frac{1}{1 + e^x} \, dx$$

Solution: Again, this integral does not appear to fit any of our fundamental formulas. The Log Rule should be given some consideration since the denominator $(1 + e^x)$ is raised to the first power; however, if we let $u = 1 + e^x$, the numerator lacks the necessary factor $u' = e^x$. By adding and subtracting this factor to the numerator, we have

$$\int \frac{1}{1 + e^x} \, dx = \int \frac{1 + e^x - e^x}{1 + e^x} \, dx = \int \left(\frac{1 + e^x}{1 + e^x} - \frac{e^x}{1 + e^x} \right) dx$$

$$= \int \left(1 - \frac{e^x}{1 + e^x} \right) dx = x - \ln (1 + e^x) + C \qquad \blacksquare$$

An alternative way to solve the integral in Example 10 would be to multiply the numerator and denominator by e^{-x}. See if you can get the same answer by this procedure.

We conclude with a summary of some common procedures for fitting integrands to basic formulas.

TABLE 10.1

Technique	Example	Formulas
Separate numerator	$\dfrac{1 + x}{x^2 + 1} = \dfrac{1}{x^2 + 1} + \dfrac{x}{x^2 + 1}$	16, 3
Divide if rational function is improper	$\dfrac{x^2}{x^2 + 1} = 1 - \dfrac{1}{x^2 + 1}$	1, 16
Add and subtract terms in numerator	$\dfrac{2x}{x^2 + 2x + 1} = \dfrac{2x + 2 - 2}{x^2 + 2x + 1} = \dfrac{2x + 2}{x^2 + 2x + 1} - \dfrac{2}{(x + 1)^2}$	3, 1
Use trigonometric identities	$\tan^2 x = \sec^2 x - 1$	6, 1
Multiply and divide by Pythagorean conjugate	$\dfrac{1}{1 + \sin x} = \left(\dfrac{1}{1 + \sin x} \right)\left(\dfrac{1 - \sin x}{1 - \sin x} \right) = \dfrac{1 - \sin x}{1 - \sin^2 x}$ $= \dfrac{1 - \sin x}{\cos^2 x} = \sec^2 x - \dfrac{\sin x}{\cos^2 x}$	6, 1
Do not separate denominators!	$\dfrac{1}{x^2 + 1} \neq \dfrac{1}{x^2} + \dfrac{1}{1}$	

Section Exercises (10.1)

In Exercises 1–40, evaluate the indefinite integral.

1. $\int (3x - 2)^4 \, dx$

2. $\int \frac{2}{(t - 9)^2} \, dt$

S 3. $\int (-2x + 5)^{3/2} \, dx$

4. $\int x \sqrt{4 - 2x^2} \, dx$

5. $\int \left[v + \frac{1}{(3v - 1)^3} \right] dv$

6. $\int \frac{2t - 1}{t^2 - t + 2} \, dt$

S 7. $\int \frac{t^2 - 3}{-t^3 + 9t + 1} \, dt$

8. $\int \frac{2x}{x - 4} \, dx$

9. $\int \frac{x^2}{x - 1} \, dx$

10. $\int \frac{x + 1}{\sqrt{x^2 + 2x - 4}} \, dx$

11. $\int \left(\frac{1}{3x - 1} - \frac{1}{3x + 1} \right) dx$

12. $\int e^{5x} \, dx$

S 13. $\int t \sin t^2 \, dt$

14. $\int \sec 4u \, du$

15. $\int \cos x \, e^{\sin x} \, dx$

16. $\int \frac{e^x}{1 + e^x} \, dx$

17. $\int \frac{(1 + e^t)^2}{e^t} \, dt$

18. $\int \frac{1 + \sin x}{\cos x} \, dx$

19. $\int \sec 3x \tan 3x \, dx$

20. $\int \frac{1}{\sqrt{x}(1 - 2\sqrt{x})} \, dx$

21. $\int \frac{2}{e^{-x} + 1} \, dx$

22. $\int \frac{1}{2e^x - 3} \, dx$

S 23. $\int \frac{1}{1 - \cos x} \, dx$

24. $\int \frac{1}{\sec x - 1} \, dx$

S 25. $\int \frac{2t - 1}{t^2 + 4} \, dt$

26. $\int \frac{2}{(2t - 1)^2 + 4} \, dt$

27. $\int \frac{3}{\sqrt{t^2 - 1}} \, dt$

28. $\int \frac{3}{\sqrt{1 - t^2}} \, dt$

29. $\int \frac{1}{x \sqrt{x^2 - 4}} \, dx$

30. $\int \frac{-2x}{\sqrt{x^2 - 4}} \, dx$

S 31. $\int \frac{-1}{\sqrt{1 - (2t - 1)^2}} \, dt$

32. $\int \frac{1}{4 + 3x^2} \, dx$

S 33. $\int \frac{1}{16 - 9x^2} \, dx$

34. $\int \frac{1}{x \sqrt{4x^2 - 1}} \, dx$

35. $\int \frac{t}{\sqrt{1 - t^4}} \, dt$

S 36. $\int \frac{x}{(x^2 - 1)^2 - 1} \, dx$

37. $\int \frac{\sec^2 x}{4 + \tan^2 x} \, dx$

38. $\int \tan^2 2x \, dx$

39. $\int \frac{\tan (2/t)}{t^2} \, dt$

40. $\int \frac{e^{1/t}}{t^2} \, dt$

In Exercises 41–50, evaluate the definite integrals.

41. $\int_0^1 x e^{-x^2} \, dx$

42. $\int_0^\pi \sin^2 t \cos t \, dt$

S 43. $\int_1^e \frac{1 - \ln x}{x} \, dx$

44. $\int_1^2 \frac{x - 2}{x} \, dx$

45. $\int_0^4 \frac{2x}{\sqrt{x^2 + 9}} \, dx$

46. $\int_0^{\pi/4} \cos 2x \, dx$

47. $\int_{\pi/4}^{\pi/2} \cot x \, dx$

48. $\int_0^4 \frac{1}{\sqrt{25 - x^2}} \, dx$

S 49. $\int_1^2 \frac{1}{2x \sqrt{4x^2 - 1}} \, dx$

50. $\int_0^{2/\sqrt{3}} \frac{1}{4 + 9x^2} \, dx$

51. Find the area of the region bounded by the graph of $y^2 = x^2(1 - x^2)$.

52. Find the area of the region bounded by $y = \sin 2x$, $y = 0$, $0 \le x \le \pi/2$.

53. The graphs of $f(x) = x$ and $g(x) = ax^2$ intersect at the points $(0, 0)$ and $(1/a, 1/a)$. Find a so that the area of the region bounded by the graphs of these two functions is $\frac{2}{3}$.

54. Find the volume of the solid generated by revolving the region bounded by $y = e^{-x^2}$, $y = 0$, $x = 0$, and $x = 1$ about the y-axis.

C 55. The region bounded by $y = e^{-x^2}$, $y = 0$, $x = 0$, and $x = b$ is revolved around the y-axis. Find b so that the volume of the generated solid is $\frac{4}{3}$ cubic units.

56. Compute the average value of each of the functions over the indicated interval.
(a) $f(x) = \sin nx$, $0 \le x \le \pi/n$, n is a positive integer
(b) $f(x) = 1/(1 + x^2)$, $-3 \le x \le 3$

For Review and Class Discussion

True or False

1. ____ $\int \frac{dx}{x \sqrt{1 - x^3}}$ can be expressed in the form $\int \frac{u'}{u \sqrt{a^2 \pm u^2}} \, dx$, $u = x^{3/2}$.

2. ——— $\int \dfrac{x^3 \, dx}{(x-1)(x^3+x^2+x+1)}$ can be expressed in

the form $\int \dfrac{u'}{u} \, dx$, $u = x^4 - 1$.

3. ——— $\int \dfrac{dx}{1 + \sin^2 x}$ can be expressed in the form

$\int \dfrac{u'}{a^2 + u^2} \, dx$, $u = \sin x$.

4. ——— $\int \sec^2 \theta \sec^2 (\tan \theta) \, d\theta$ can be expressed in the

form $\int (\sec^2 u) \, u' \, dx$, $u = \tan \theta$.

5. ——— $\int \dfrac{dx}{\sqrt{x} \, \sqrt{x-1}}$ can be expressed in the form

$\int \dfrac{u'}{u \sqrt{u^2 - a^2}} \, dx$, $u = \sqrt{x}$.

6. ——— $\int \dfrac{dx}{\sqrt{e^{2x} - 1}}$ can be expressed in the form

$\int \dfrac{u'}{u \sqrt{u^2 - a^2}} \, dx$, $u = e^x$.

7. ——— $\int \sec(\sqrt{x}) \, dx$ can be expressed in the form

$\int (\sec u) \, u' \, dx$, $u = \sqrt{x}$.

8. ——— $\int \dfrac{\ln x^2}{x} \, dx$ can be expressed in the form

$\int u^n u' \, dx$, $u = \ln x^2$.

9. ——— $\int (e^x)^2 \, dx$ can be expressed in the form $\int e^u u' \, dx$,

$u = 2x$.

10. ——— $\int \dfrac{dx}{\sqrt{x} - 1}$ can be expressed in the form $\int \dfrac{u'}{u} \, dx$,

$u = \sqrt{x} - 1$.

10.2 Completing the Square

Purpose
- To expand the application of several fundamental integration formulas by the technique of completing the square.

It is quite important that the student should be able to discern clearly the progress which appears veiled as well as the spontaneous development of the science. —*Pierre de Fermat (1601–1665)*

Six of the 19 fundamental integration formulas listed in Section 10.1 involve the sum or difference of two squares. We can *extend* the application of these formulas by the use of an algebraic technique called "completing the square." This technique provides a means for writing any quadratic polynomial as the sum or difference of two squares. For instance, the polynomial $x^2 + bx + c$ can be written as

$$x^2 + bx + c = x^2 + bx + \left(\frac{b}{2}\right)^2 - \left(\frac{b}{2}\right)^2 + c$$

$$= \left(x + \frac{b}{2}\right)^2 + \left[c - \left(\frac{b}{2}\right)^2\right]$$

Thus we have written $x^2 + bx + c$ in the form $u^2 \pm a^2$, where $u = x + (b/2)$.

Example 1

Evaluate $\qquad\qquad \int \dfrac{dx}{x^2 - 4x + 7}$

Solution: By completing the square we obtain

$$x^2 - 4x + 7 = (x^2 - 4x + 4) - 4 + 7 = (x - 2)^2 + 3$$

Therefore,

$$\int \frac{dx}{x^2 - 4x + 7} = \int \frac{dx}{(x - 2)^2 + 3}$$

Considering $u = x - 2$ and $a = \sqrt{3}$, we have

$$\int \frac{dx}{x^2 - 4x + 7} = \frac{1}{\sqrt{3}} \text{Arctan} \frac{x - 2}{\sqrt{3}} + C$$ ∎

Our procedure for completing the square requires the coefficient of x^2 to be unity. In the next example we expand this procedure to work with any coefficient of x^2.

Example 2

Evaluate

$$\int \frac{dx}{2x^2 - x - 3}$$

Solution: To complete the square we write

$$2x^2 - x - 3 = 2\left(x^2 - \frac{x}{2} - \frac{3}{2}\right) = 2\left[x^2 - \frac{x}{2} + \left(\frac{1}{4}\right)^2 - \left(\frac{1}{4}\right)^2 - \frac{3}{2}\right]$$

$$= 2\left[\left(x - \frac{1}{4}\right)^2 - \frac{25}{16}\right]$$

Therefore,

$$\int \frac{dx}{2x^2 - x - 3} = \frac{1}{2}\int \frac{dx}{(x - \frac{1}{4})^2 - \frac{25}{16}}$$

Letting $u = x - \frac{1}{4}$ and $a = \frac{5}{4}$, then by Formula 17 we have

$$\int \frac{dx}{2x^2 - x - 3} = \left(\frac{1}{2}\right)\left[\frac{1}{2(\frac{5}{4})}\right]\ln\left|\frac{x - \frac{1}{4} - \frac{5}{4}}{x - \frac{1}{4} + \frac{5}{4}}\right| + C$$

$$= \frac{1}{5}\ln\left|\frac{x - \frac{3}{2}}{x + 1}\right| + C = \frac{1}{5}\ln\left|\frac{2x - 3}{2x + 2}\right| + C$$ ∎

In the next example notice how we complete the square when the coefficient of x^2 is negative.

Example 3

Evaluate

$$\int_0^{3/2} \frac{dx}{\sqrt{3x - x^2}}$$

Solution: By completing the square we obtain

$$3x - x^2 = -(x^2 - 3x) = -\left[x^2 - 3x + \left(\frac{3}{2}\right)^2 - \left(\frac{3}{2}\right)^2\right]$$

$$= \frac{9}{4} - \left(x - \frac{3}{2}\right)^2$$

Therefore,
$$\int_0^{3/2} \frac{dx}{\sqrt{3x - x^2}} = \int_0^{3/2} \frac{dx}{\sqrt{\frac{9}{4} - (x - \frac{3}{2})^2}}$$

and considering $u = x - \frac{3}{2}$ and $a = \frac{3}{2}$, we can use Formula 14 to obtain

$$\int_0^{3/2} \frac{dx}{\sqrt{\frac{9}{4} - (x - \frac{3}{2})^2}} = \text{Arcsin}\left(\frac{x - \frac{3}{2}}{\frac{3}{2}}\right)\Bigg]_0^{3/2}$$

$$= \text{Arcsin}\,(0) - \text{Arcsin}\,(-1) = 0 - \left(-\frac{\pi}{2}\right) = \frac{\pi}{2} \quad \blacksquare$$

Example 4

Evaluate
$$\int \frac{dx}{(x - 1)\sqrt{x^2 - 2x - 3}}$$

Solution: Since $x^2 - 2x - 3 = x^2 - 2x + (1)^2 - (1)^2 - 3 = (x - 1)^2 - 4$, we let $u = x - 1$ and $a = 2$ and write

$$\int \frac{dx}{(x - 1)\sqrt{x^2 - 2x - 3}} = \int \frac{dx}{(x - 1)\sqrt{(x - 1)^2 - 2^2}}$$

$$= \frac{1}{2}\,\text{Arcsec}\,\frac{|x - 1|}{2} + C \quad \blacksquare$$

Example 5

Evaluate
$$\int \frac{2x + 1}{\sqrt{x^2 + 2x}}\,dx$$

Solution: Before completing the square in the denominator, we note that if $u = x^2 + 2x$, then $u' = 2x + 2$, which is almost equal to the given numerator. We can create the proper numerator by adding and subtracting 1.

$$\int \frac{2x + 1}{\sqrt{x^2 + 2x}}\,dx = \int \frac{(2x + 1 + 1) - 1}{\sqrt{x^2 + 2x}}\,dx$$

$$= \int \frac{2x + 2}{\sqrt{x^2 + 2x}}\,dx - \int \frac{dx}{\sqrt{x^2 + 2x}}$$

The first integral on the right fits the Power Rule and the other integral can be evaluated by first completing the square. Therefore, we write

$$\int \frac{2x + 1}{\sqrt{x^2 + 2x}} \, dx = \int (x^2 + 2x)^{-1/2}(2x + 2) \, dx - \int \frac{dx}{\sqrt{(x^2 + 2x + 1) - 1}}$$

$$= \int (x^2 + 2x)^{-1/2}(2x + 2) \, dx - \int \frac{dx}{\sqrt{(x + 1)^2 - 1}}$$

$$= \frac{(x^2 + 2x)^{1/2}}{\frac{1}{2}} - \ln |(x + 1) + \sqrt{(x + 1)^2 - 1}| + C$$

$$= 2\sqrt{x^2 + 2x} - \ln |(x + 1) + \sqrt{x^2 + 2x}| + C \qquad ■$$

Example 6

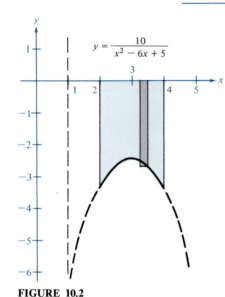

FIGURE 10.2

Determine the area A of the region bounded by $y = 10/(x^2 - 6x + 5)$, $y = 0$, $x = 2$, and $x = 4$.

Solution: A sketch of the region is shown in Figure 10.2. From the sketch we can see that

$$A = \int_2^4 \left(0 - \frac{10}{x^2 - 6x + 5} \right) dx$$

By completing the square we obtain

$$A = -\int_2^4 \frac{10 \, dx}{x^2 - 6x + 5} = -10 \int_2^4 \frac{dx}{(x - 3)^2 - (2)^2}$$

$$= -10 \left[\frac{1}{4} \ln \left| \frac{x - 3 - 2}{x - 3 + 2} \right| \right]_2^4 = -\frac{5}{2} \left[\ln \left| \frac{x - 5}{x - 1} \right| \right]_2^4$$

$$= -\frac{5}{2} \left[\ln \frac{1}{3} - \ln 3 \right] = \frac{5}{2} \ln 9 = 5 \ln 3$$

or

$$A \approx 5.4931 \qquad ■$$

Section Exercises (10.2)

1. Express each polynomial as the sum or difference of squares.
(a) $x^2 + 6x$ \qquad (b) $x^2 - 11x + 1$
(c) $x^4 + 3x^2 - 2$

2. Express each polynomial as the sum or difference of squares.
(a) $2x^2 + 12x + 4$ \qquad (b) $3x^2 - 9x - 6$
(c) $5x^2 - 3x$

In Exercises 3–27, evaluate each integral.

3. $\displaystyle\int_0^2 \frac{1}{x^2 - 2x + 2} \, dx$ \qquad **4.** $\displaystyle\int_{-3}^{-1} \frac{1}{x^2 + 6x + 13} \, dx$

$\boxed{\text{S}}$ **5.** $\displaystyle\int \frac{2x}{x^2 + 6x + 13} \, dx$ \qquad **6.** $\displaystyle\int \frac{2x - 5}{x^2 + 2x + 2} \, dx$

7. $\displaystyle\int_1^2 \frac{1}{2x^2 - 4x - 6} \, dx$ \qquad **8.** $\displaystyle\int_2^4 \frac{1}{\sqrt{5 + 4x - x^2}} \, dx$

$\boxed{\text{S}}$ **9.** $\displaystyle\int \frac{1}{\sqrt{-x^2 - 4x}} \, dx$ \qquad **10.** $\displaystyle\int \frac{x + 2}{\sqrt{-x^2 - 4x}} \, dx$

$\boxed{\text{S}}$ **11.** $\displaystyle\int \frac{1}{(x - 1)\sqrt{x^2 - 2x + 2}} \, dx$

12. $\displaystyle\int \frac{1}{(x - 1)\sqrt{x^2 - 2x}} \, dx$

13. $\int \dfrac{1}{\sqrt{-x^2 + 2x}}\,dx$

14. $\int \dfrac{x-1}{\sqrt{x^2 - 2x}}\,dx$

[S] **15.** $\int_2^3 \dfrac{2x-3}{\sqrt{4x - x^2}}\,dx$

16. $\int_2^3 \dfrac{-1}{4x - x^2}\,dx$

[S] **17.** $\int \dfrac{1}{1 - 4x - 2x^2}\,dx$

18. $\int \dfrac{1}{(x+1)\sqrt{2x^2 + 4x + 8}}\,dx$

19. $\int \dfrac{x}{x^4 + 2x^2 + 2}\,dx$

20. $\int \dfrac{x}{\sqrt{9 + 8x^2 - x^4}}\,dx$

21. $\int \dfrac{1}{\sqrt{-16x^2 + 16x - 3}}\,dx$

22. $\int \dfrac{1}{(x-1)\sqrt{9x^2 - 18x + 5}}\,dx$

23. $\int \dfrac{1}{\sqrt{80 + 8x - 16x^2}}\,dx$

24. $\int \dfrac{1}{(x-1)\sqrt{-4x^2 + 8x - 1}}\,dx$

[S] **25.** $\int \dfrac{x^3 - 21x}{5 + 4x - x^2}\,dx$

26. $\int \dfrac{1 - 2x}{4x - x^2}\,dx$

27. $\int \dfrac{1 - 2x}{\sqrt{9 - 8x - x^2}}\,dx$

28. Find the area of the region bounded by $y = 1/(x^2 - 2x + 5)$, $y = 0$, $x = 1$, and $x = 3$.

[S] **29.** Find the area of the region bounded by $y = 1/\sqrt{3 + 2x - x^2}$, $y = 0$, $x = 0$, and $x = 2$.

30. Find the volume of the solid generated by revolving the region defined in Exercise 29 about the x-axis.

31. Find the volume of the solid generated by revolving the region defined in Exercise 29 about the y-axis.

32. (a) Complete the square to find the extremum of $y = ax^2 + bx + c$.
   (b) Use differential calculus to find the extremum of $y = ax^2 + bx + c$ and compare it with the result of part (a).

33. Suppose that two chemicals A and B combine in a 3-to-1 ratio to form a compound. The amount of compound x being produced at any time t is proportional to the unchanged amounts of A and B remaining in the solution. Thus if 3 kilograms (kg) of A are mixed with 2 kg of B, we have

$$\frac{dx}{dt} = k\left(3 - \frac{3x}{4}\right)\left(2 - \frac{x}{4}\right) = \frac{3k}{16}(x^2 - 12x + 32)$$

If 1 kg of the compound is formed after 10 min, find the amount formed after 20 min by solving the integral

$$\int \frac{3k}{16}\,dt = \int \frac{dx}{x^2 - 12x + 32}$$

For Review and Class Discussion

True or False

1. ____ $-\operatorname{Arctan}(1-x) + C = \int \dfrac{dx}{(1-x)^2 + 1}$

$\qquad = \int \dfrac{dx}{(x-1)^2 + 1} = \operatorname{Arctan}(x-1) + C$

2. ____ If $a \neq 0$, then $\int \dfrac{dx}{ax^2 + bx + c}$ can be evaluated with the formula $\int \dfrac{u'}{a^2 + u^2}\,dx = \dfrac{1}{a}\operatorname{Arctan}\dfrac{u}{a} + C$

or $\int \dfrac{u'}{u^2 - a^2}\,dx = \dfrac{1}{2a}\ln\left|\dfrac{u-a}{u+a}\right| + C.$

3. ____ Any rational function whose denominator is a polynomial of degree one or two can be integrated by means of one of the 19 fundamental formulas.

4. ____ $\int \dfrac{dx}{\sqrt{2x - x^2}} = -\int \dfrac{dx}{\sqrt{x^2 - 2x}}$

5. ____ $\int \dfrac{2\,dx}{(2x-1)\sqrt{x^2 - x}} = \ln\left|2x - 1 + 2\sqrt{x^2 - x}\right| + C$

10.3 **Substitution**

Purpose
- To expand the application of the fundamental integration formulas by substitution.

Note that sometimes quantities occur whose integrals at first sight cannot be found. But they are easily found after a certain change, as in the following cases.   —John Bernoulli (1667–1748)

In this section we demonstrate another technique that extends our use of the fundamental integration formulas. This technique of **substitution** involves a change of variable, which permits us to rewrite an integrand in a form to which we can apply a basic integration rule.

The mechanics of this technique are easy enough to master. However, the question as to what substitution changes an unmanageable integral into a manageable one is not so easily answered. First, let's consider an example that demonstrates the substitution procedure.

Example 1

Evaluate $\int x \sqrt{x - 1}\, dx$.

Solution: First, we must check to see that, as it stands, this integral *does not* fit any of our basic integration formulas. Now consider the substitution $u = \sqrt{x - 1}$. Then

$$u^2 = x - 1 \qquad \text{and} \qquad u^2 + 1 = x$$

To convert the integral to the variable u, we must also convert the differential dx to its equivalent u-variable form. From $u^2 + 1 = x$ we have

$$d[u^2 + 1] = dx$$
$$2u\, du = dx$$

We now substitute for x, $\sqrt{x - 1}$, and dx as follows:

$$\int x \sqrt{x - 1}\, dx = \int (u^2 + 1)(u)2u\, du$$

Therefore,

$$\int x \sqrt{x - 1}\, dx = \int (2u^4 + 2u^2)\, du = \frac{2u^5}{5} + \frac{2u^3}{3} + C$$

Substituting back again to variable x, we have

$$\frac{2u^5}{5} + \frac{2u^3}{3} + C = \frac{2}{5}(\sqrt{x - 1})^5 + \frac{2}{3}(\sqrt{x - 1})^3 + C$$

$$= \frac{6}{15}(x - 1)^{5/2} + \frac{10}{15}(x - 1)^{3/2} + C$$

$$= \frac{2}{15}(x - 1)^{3/2}[3(x - 1) + 5] + C$$

$$= \frac{2}{15}(x - 1)^{3/2}(3x + 2) + C \qquad \blacksquare$$

Although we will encounter several other types of substitution in this chapter, Example 1 shows a very common type, which involves the root of a first-degree polynomial. We outline the steps for this type of substitution as follows:

1. Let $u = \sqrt[n]{ax + b}$.
2. Solve for dx in terms of u and du.
3. Convert the entire integral to u-variable form and integrate.
4. After integrating, rewrite the antiderivative as a function of x.

Example 2

Evaluate

$$\int \frac{1}{\sqrt{x} + 1} dx$$

Solution: Since none of our fundamental formulas work, we let $u = \sqrt{x}$. Then

$$u^2 = x \qquad \text{and} \qquad 2u \, du = dx$$

Thus

$$\int \frac{1}{\sqrt{x} + 1} dx = \int \frac{1}{u + 1}(2u \, du) = \int \frac{2u \, du}{u + 1}$$

Since the degree of the numerator is equal to the degree of the denominator, we divide $2u$ by $u + 1$ and obtain

$$\int \frac{1}{\sqrt{x} + 1} dx = \int \left(2 - \frac{2}{u + 1}\right) du = 2 \int du - 2 \int \frac{du}{u + 1}$$

$$= 2u - 2 \ln |u + 1| + C$$

Since $u = \sqrt{x}$, it follows that

$$\int \frac{1}{\sqrt{x} + 1} dx = 2\sqrt{x} - 2 \ln (\sqrt{x} + 1) + C \qquad \blacksquare$$

The fourth step of our outline suggests that we convert back to variable x. However, for *definite* integrals it is often more convenient to determine the limits of integration for variable u than it is to convert back to variable x and evaluate the antiderivative at the original limits. The next example illustrates the procedure.

Example 3

Find the area A of the region bounded by $y = x/\sqrt{2x - 1}$ and the x-axis, from $x = 1$ to $x = 5$.

Solution: From Figure 10.3 we can see that the area of the region is given by

$$A = \int_1^5 \frac{x}{\sqrt{2x - 1}}\,dx$$

To evaluate this integral we let $u = \sqrt{2x - 1}$; then

$$u^2 = 2x - 1 \qquad \text{or} \qquad \frac{u^2 + 1}{2} = x$$

and

$$u\,du = dx$$

Furthermore,

when $\qquad x = 5, \qquad u = \sqrt{10 - 1} = 3$

and when $\qquad x = 1, \qquad u = \sqrt{2 - 1} = 1$

Thus it follows that

$$\int_1^5 \frac{x}{\sqrt{2x - 1}}\,dx = \int_1^3 \frac{1}{2}\left(\frac{u^2 + 1}{u}\right)u\,du = \frac{1}{2}\int_1^3 (u^2 + 1)\,du$$

$$= \frac{1}{2}\left[\frac{u^3}{3} + u\right]_1^3 = \frac{1}{2}\left(9 + 3 - \frac{1}{3} - 1\right) = \frac{16}{3}$$

Geometrically we can interpret the equation

$$\int_1^5 \frac{x}{\sqrt{2x - 1}}\,dx = \int_1^3 \frac{u^2 + 1}{2}\,du$$

to mean that the two *different* regions shown in Figures 10.3 and 10.4 have the *same* area. ∎

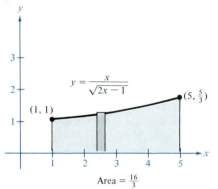

FIGURE 10.3

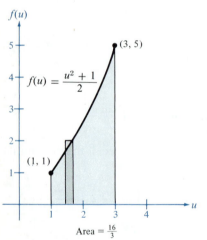

FIGURE 10.4

Whenever a new integration technique is introduced, there is the temptation to overuse it. In the case of substitution, you should be aware that some integrals containing radicals of the form $\sqrt[n]{ax + b}$ can be solved without changing variables. For example,

$$\int \sqrt{2x + 3}\,dx = \frac{1}{2}\int 2(2x + 3)^{1/2}\,dx = \frac{1}{2}\int u'u^{1/2}\,dx$$

$$= \frac{1}{2}\left[\frac{(2x + 3)^{3/2}}{\frac{3}{2}}\right] + C = \frac{1}{3}(2x + 3)^{3/2} + C$$

And

$$\int \frac{3x - 5}{\sqrt[3]{x}}\,dx = \int \frac{3x}{x^{1/3}}\,dx - \int \frac{5}{x^{1/3}}\,dx$$

$$= 3 \int x^{2/3} \, dx - 5 \int x^{-1/3} \, dx$$

$$= \frac{3x^{5/3}}{\frac{5}{3}} - \frac{5x^{2/3}}{\frac{2}{3}} + C = \frac{9x^{5/3}}{5} - \frac{15x^{2/3}}{2} + C$$

$$= \frac{3x^{2/3}}{10}(6x - 25) + C$$

The next example points out yet another situation in which substitution is unnecessary even though the integral contains radicals of the form $\sqrt[n]{ax + b}$.

Example 4

Evaluate
$$\int \frac{dx}{\sqrt{x + 2} - \sqrt{x}}$$

Solution: In this case we can remove the radicals from the denominator by rationalizing it. Thus we write

$$\int \frac{dx}{\sqrt{x + 2} - \sqrt{x}} = \int \left(\frac{1}{\sqrt{x + 2} - \sqrt{x}} \right) \left(\frac{\sqrt{x + 2} + \sqrt{x}}{\sqrt{x + 2} + \sqrt{x}} \right) dx$$

$$= \int \frac{\sqrt{x + 2} + \sqrt{x}}{(x + 2) - (x)} dx = \int \frac{\sqrt{x + 2} + \sqrt{x}}{2} dx$$

Therefore, we have

$$\int \frac{dx}{\sqrt{x + 2} - \sqrt{x}} = \frac{1}{2} \int (\sqrt{x + 2} + \sqrt{x}) \, dx$$

$$= \frac{1}{2} \left[\frac{(x + 2)^{3/2}}{\frac{3}{2}} + \frac{x^{3/2}}{\frac{3}{2}} \right] + C$$

$$= \frac{1}{3}[(x + 2)^{3/2} + x^{3/2}] + C \qquad \blacksquare$$

Example 5

Evaluate $\int_1^{3/2} x \sqrt[3]{3 - 2x} \, dx$.

Solution: Let $u = \sqrt[3]{3 - 2x}$, then

$$u^3 = 3 - 2x$$

$$x = \frac{3 - u^3}{2}$$

$$dx = \frac{-3u^2}{2} \, du$$

Furthermore, when $x = \frac{3}{2}$, $u = 0$; and when $x = 1$, $u = 1$. Yes, in this case, the lower limit for u is larger than the upper limit. Therefore

$$\int_{1}^{3/2} x \sqrt[3]{3 - 2x}\, dx = \int_{1}^{0} \left(\frac{3 - u^3}{2}\right)(u)\left(-\frac{3}{2}u^2\right) du$$

$$= -\frac{3}{4}\int_{1}^{0} (3u^3 - u^6)\, du = -\frac{3}{4}\left[\frac{3u^4}{4} - \frac{u^7}{7}\right]_{1}^{0}$$

$$= -\frac{3}{4}\left(-\frac{3}{4} + \frac{1}{7}\right) = \frac{51}{112}$$

(We suggest that you compare this evaluation with an evaluation using variable x and the limits 1 and $\frac{3}{2}$.) ∎

Section Exercises (10.3)

In Exercises 1–20, find the indefinite integral.

1. $\int x \sqrt{x - 3}\, dx$

2. $\int x \sqrt{2x + 1}\, dx$

[S] 3. $\int x^2 \sqrt{1 - x}\, dx$

4. $\int \frac{2x - 1}{\sqrt{x + 3}}\, dx$

[S] 5. $\int \frac{x^2 - 1}{\sqrt{2x - 1}}\, dx$

6. $\int x^3 \sqrt{x + 2}\, dx$

7. $\int \frac{\sqrt{x - 1}}{x}\, dx$

8. $\int t \sqrt[3]{t + 1}\, dt$

9. $\int \frac{2\sqrt{t} - \sqrt{3t}}{t}\, dt$

10. $\int \frac{\sqrt{x - 2}}{x + 1}\, dx$

[S] 11. $\int \frac{1}{1 + \sqrt{x}}\, dx$

12. $\int \frac{1 - \sqrt{x}}{1 + \sqrt{x}}\, dx$

13. $\int \frac{-x}{x + 1 - \sqrt{x + 1}}\, dx$

14. $\int \frac{1}{t \sqrt{t} + \sqrt{t}}\, dt$

15. $\int \frac{1}{\sqrt{x} - \sqrt{x + 1}}\, dx$

[S] 16. $\int \frac{1}{\sqrt{2x} - \sqrt{2x + 1}}\, dx$

[S] 17. $\int \frac{\sqrt{2x}}{6x + \sqrt{2x}}\, dx$

18. $\int \frac{1}{(\sqrt[6]{x})^3 - (\sqrt[6]{x})^2}\, dx$

19. $\int \frac{1}{\sqrt{x} + \sqrt{2x}}\, dx$

[S] 20. $\int \sqrt{e^t - 3}\, dt$

In Exercises 21–30, evaluate the definite integral.

21. $\int_{1}^{2} (x - 1) \sqrt{2 - x}\, dx$

22. $\int_{2}^{6} \frac{\sqrt{x - 2}}{x}\, dx$

23. $\int_{3}^{7} x \sqrt{x - 3}\, dx$

24. $\int_{0}^{4} \frac{x}{\sqrt{2x + 1}}\, dx$

[S] 25. $\int_{0}^{7} x \sqrt[3]{x + 1}\, dx$

26. $\int_{1}^{4} \frac{\sqrt{x}}{\sqrt{x} - 3}\, dx$

27. $\int_{1}^{5} x^2 \sqrt{x - 1}\, dx$

28. $\int_{0}^{1} \frac{1}{\sqrt{x} + \sqrt{x + 1}}\, dx$

[S] 29. $\int_{0}^{2} \frac{1}{1 + \sqrt{2x}}\, dx$

30. $\int_{-2}^{6} x^2 \sqrt[3]{x + 2}\, dx$

31. Find the area of the region bounded by $y = x \sqrt{x + 1}$ and $y = 0$.

32. The region defined in Exercise 31 is revolved around the x-axis. Find the volume of the solid generated.

33. Find the volume of the solid generated by revolving the region defined in Exercise 31 about the y-axis.

34. The region bounded by $y = 2\sqrt{x}$, $y = 0$, and $x = 3$ is revolved around the x-axis. Find the surface area of the solid generated.

[S] 35. Find the arc length of the graph of $f(x) = \frac{4}{5}x^{5/4}$ from $x = 0$ to $x = 4$.

36. Find the volume of the solid generated by revolving the region bounded by $y = 1/(1 + \sqrt{x - 2})$, $y = 0$, $x = 2$, and $x = 6$ about the y-axis.

The function

$$f(x) = kx^n(1 - x)^m, \qquad 0 \le x \le 1$$

where $n, m > 0$ and k is a constant, can be used to represent various percentage distributions. If k is chosen so that

$$\int_{0}^{1} f(x)\, dx = 1$$

the probability that x will fall between a and b is given by

$$p_{a,b} = \int_a^b f(x)\, dx$$

$$p_{a,b} = \int_a^b \frac{1155}{32} x^3 (1-x)^{3/2}\, dx$$

c 37. The probability of recall in a certain experiment is found to be

$$p_{a,b} = \int_a^b \frac{15}{4} x \sqrt{1-x}\, dx$$

where x represents the percentage of recall. (See Figure 10.5.)

(a) For a randomly chosen individual, what is the probability that he or she will recall between 50% and 75% of the material?

(b) What is the median percentage recall? That is, for what value of b is it true that the probability from 0 to b is 0.5?

c 38. The probability of finding between a and b percentage of iron in ore samples taken from a certain region is given by

(See Figure 10.6.) What is the probability that a sample will contain between:

(a) 0% and 25%? (b) 50% and 100%?

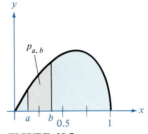

FIGURE 10.5

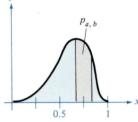

FIGURE 10.6

For Review and Class Discussion

True or False

1. _____ If $u = \sqrt{x-1}$, then $\int_1^5 x \sqrt{x-1}\, dx = \int_0^2 (u^2+1)u\, du$.

2. _____ If $u = \sqrt[3]{3x+1}$, then $dx = u^2\, du$.

3. _____ If $a < b$, $u = \sqrt{4-x}$, and $\int_a^b x \sqrt{4-x}\, dx = 2 \int_c^d (u^4 - 4u^2)\, du$, then $c < d$.

4. _____ If $u = \sqrt{e^x - 1}$, then $dx = 2u\, du/(u^2+1)$.

5. _____ $\int_{-1}^1 \sqrt{x^2 - x^3}\, dx = \int_{-1}^1 x \sqrt{1-x}\, dx$.

10.4 Partial Fractions

Purpose

■ To introduce the method of partial fractions as a means of integrating rational functions.

We have here a demonstration of the proposition, which is usually presupposed in integral calculus, that every polynomial can be resolved into real factors, whether simple ones of the form $x + p$, or double ones of the form $x^2 + px + q$. —Leonhard Euler (*1707–1783*)

In Sections 10.2, and 10.3, we introduced two techniques that allowed us to apply our fundamental integration rules to certain types of algebraic functions. In this section we discuss yet a *third* technique that permits us to rewrite algebraic functions (rational functions, in particular) in a form to which we can apply our fundamental integration rules. This technique involves the decomposition of a rational function into the sum of two or more "simpler" rational functions. We call this procedure the method of **partial fractions.** The need for this procedure can be seen in the discussion of the following problem.

How can we evaluate

$$\int \frac{x+7}{x^2-x-6}dx$$

Using the techniques in Section 10.2, we could solve the problem as follows:

$$\int \frac{x+7}{x^2-x-6}dx = \frac{1}{2}\int \frac{2x+14}{x^2-x-6}dx = \frac{1}{2}\int \frac{2x-1+15}{x^2-x-6}dx$$

$$= \frac{1}{2}\left[\int \frac{2x-1}{x^2-x-6}dx + 15\int \frac{dx}{(x-\frac{1}{2})^2-(\frac{5}{2})^2}\right]$$

$$= \frac{1}{2}\left[\ln|x^2-x-6| + 3\ln\left|\frac{x-3}{x+2}\right|\right] + C$$

$$= \frac{1}{2}\left[\ln|(x-3)(x+2)| + 3\ln\left|\frac{x-3}{x+2}\right|\right] + C$$

$$= \frac{1}{2}[4\ln|x-3| - 2\ln|x+2|] + C$$

$$= 2\ln|x-3| - \ln|x+2| + C$$

However, suppose we knew that

$$\frac{x+7}{x^2-x-6} = \frac{2}{x-3} - \frac{1}{x+2}$$

Then we could write

$$\int \frac{x+7}{x^2-x-6}dx = \int\left(\frac{2}{x-3} - \frac{1}{x+2}\right)dx = 2\int \frac{dx}{x-3} - \int \frac{dx}{x+2}$$

$$= 2\ln|x-3| - \ln|x+2| + C$$

This second method is clearly preferable to the first. However, its use depends on our ability to factor the denominator, (x^2-x-6), and to find the *partial fractions* $2/(x-3)$ and $-1/(x+2)$.

You may recall that one of the basic concerns of algebra is finding the (linear and irreducible quadratic) factors of a polynomial. For instance, the polynomial x^5+x^4-3x+1 can be written as

$$x^5+x^4-3x+1 = (x-1)(x+1)^2(x^2+1)$$

where
$(x-1)$ is a linear factor
$(x+1)^2$ is a repeated linear factor
(x^2+1) is an irreducible quadratic factor

For example, if $N(x)$ is a polynomial of degree less than 5, then the partial fraction decomposition of $N(x)/(x^5+x^4-3x+1)$ has the form

$$\frac{N(x)}{x^5 + x^4 - 3x + 1} = \frac{N(x)}{(x - 1)(x + 1)^2(x^2 + 1)}$$

$$= \frac{A}{x - 1} + \frac{B}{x + 1} + \frac{C}{(x + 1)^2} + \frac{Dx + E}{x^2 + 1}$$

We summarize the steps for partial fraction decomposition as follows:

Decomposition of $\dfrac{N(x)}{D(x)}$ **into**

Partial Fractions

Rule 1. If $N(x)/D(x)$ is not a proper rational function [that is, if the degree of $D(x)$ is not greater than the degree of $N(x)$], then divide $N(x)$ by $D(x)$ to obtain

$$\frac{N(x)}{D(x)} = (\text{a polynomial}) + \frac{N_1(x)}{D(x)}$$

and apply Rules 2, 3, and 4 to the proper rational function $N_1(x)/D(x)$.

Rule 2. Completely factor $D(x)$ into factors of the form

$$(px + q)^m \qquad \text{and} \qquad (ax^2 + bx + c)^n$$

where $ax^2 + bx + c$ is irreducible.

Rule 3. For *each* factor of the form $(px + q)^m$, the partial fraction decomposition must include the following sum of m fractions:

$$\frac{A_1}{(px + q)} + \frac{A_2}{(px + q)^2} + \cdots + \frac{A_m}{(px + q)^m}$$

Rule 4. For *each* factor of the form $(ax^2 + bx + c)^n$, the partial fraction decomposition must include the following sum of n fractions:

$$\frac{B_1x + C_1}{ax^2 + bx + c} + \frac{B_2x + C_2}{(ax^2 + bx + c)^2} + \cdots + \frac{B_nx + C_n}{(ax^2 + bx + c)^n}$$

An algebraic technique for determining the value of the constants in the numerators is demonstrated in the examples that follow.

Example 1 (*Distinct Linear Factors*)

Write the partial fraction decomposition for the rational function

$$\frac{x + 7}{x^2 - x - 6}$$

Solution: Since $x^2 - x - 6 = (x - 3)(x + 2)$

by Rule 3 we include one partial fraction for each factor and write

$$\frac{x + 7}{x^2 - x - 6} = \frac{A}{x - 3} + \frac{B}{x + 2}$$

Multiplying this equation by the lowest common denominator (LCD), $(x - 3)(x + 2)$, leads to the **basic equation**

$$x + 7 = A(x + 2) + B(x - 3)$$

Since this equation is to be true for all x, we can substitute *convenient* values for x to obtain equations in A and B, which we then solve. If $x = -2$, then

$$-2 + 7 = A(0) + B(-5)$$
$$5 = -5B$$
$$-1 = B$$

If $x = 3$, then

$$3 + 7 = A(5) + B(0)$$
$$10 = 5A$$
$$2 = A$$

The decomposition is, therefore,

$$\frac{x + 7}{x^2 - x - 6} = \frac{2}{x - 3} - \frac{1}{x + 2}$$

as indicated at the beginning of this section. ∎

The substitutions for x in Example 1 were chosen for their convenience in determining values for A and B. We chose $x = -2$ so as to eliminate the term $A(x + 2)$, while $x = 3$ was chosen to eliminate the term $B(x - 3)$. The goal is to make *convenient* substitutions, whenever possible.

Example 2 (*Repeated Linear Factors*)

Evaluate

$$\int \frac{5x^2 + 20x + 6}{x^3 + 2x^2 + x} dx$$

Solution: Since

$$x^3 + 2x^2 + x = x(x^2 + 2x + 1) = x(x + 1)^2$$

by Rule 3 we include one fraction for each power of x and $(x + 1)$ and write

$$\frac{5x^2 + 20x + 6}{x(x + 1)^2} = \frac{A}{x} + \frac{B}{x + 1} + \frac{C}{(x + 1)^2}$$

Multiplying by the LCD, $x(x + 1)^2$, leads to the *basic* equation

$$5x^2 + 20x + 6 = A(x + 1)^2 + Bx(x + 1) + Cx$$

Substituting $x = -1$ eliminates the A and B terms and yields

$$5 - 20 + 6 = 0 + 0 - C$$
$$C = 9$$

If $x = 0$, then
$$6 = A(1) + 0 + 0$$
$$6 = A$$

At this point we have exhausted the most convenient choices for x and have yet to find the value of B. Under such circumstances we use *any other value* for x along with the calculated values of A and C. Thus for $x = 1$, $A = 6$, and $C = 9$, we have

$$5 + 20 + 6 = A(4) + B(2) + C$$
$$31 = 6(4) + 2B + 9$$
$$-2 = 2B$$
$$-1 = B$$

Therefore, $\dfrac{5x^2 + 20x + 6}{x(x + 1)^2} = \dfrac{6}{x} - \dfrac{1}{x + 1} + \dfrac{9}{(x + 1)^2}$

and it follows that

$$\int \frac{5x^2 + 20x + 6}{x^3 + 2x^2 + x} dx = \int \frac{6}{x} dx - \int \frac{dx}{x + 1} + \int 9(x + 1)^{-2} dx$$

$$= 6 \ln |x| - \ln |x + 1| + 9 \frac{(x + 1)^{-1}}{-1} + C$$

$$= \ln \left| \frac{x^6}{x + 1} \right| - \frac{9}{x + 1} + C \qquad \blacksquare$$

Note that it is necessary to make as many substitutions for x as there are unknowns (A, B, C, . . .) to be determined.

Example 3 (*Distinct Linear and Quadratic Factors*)

Evaluate $\displaystyle\int \frac{2x^3 - 4x - 8}{(x^2 - x)(x^2 + 4)} dx$

Solution: Since

$$(x^2 - x)(x^2 + 4) = x(x - 1)(x^2 + 4)$$

by Rules 3 and 4 we include one partial fraction for each factor and write

$$\frac{2x^3 - 4x - 8}{x(x - 1)(x^2 + 4)} = \frac{A}{x} + \frac{B}{x - 1} + \frac{Cx + D}{x^2 + 4}$$

Multiplying by the LCD, $x(x - 1)(x^2 + 4)$, yields the *basic* equation

$$2x^3 - 4x - 8 = A(x - 1)(x^2 + 4) + Bx(x^2 + 4) + (Cx + D)(x)(x - 1)$$

If $x = 1$, then
$$-10 = 0 + B(5) + 0$$
$$-2 = B$$

If $x = 0$, then
$$-8 = A(-1)(4) + 0 + 0$$
$$2 = A$$

If $x = -1$, then since $A = 2$ and $B = -2$, we get

$$-6 = (2)(-2)(5) + (-2)(-1)(5) + (-C + D)(-1)(-2)$$
$$2 = -C + D$$

If $x = 2$, then we have

$$0 = (2)(1)(8) + (-2)(2)(8) + (2C + D)(2)(1)$$
$$8 = 2C + D$$

Subtracting the two equations involving C and D, we have

$$\begin{array}{r} -C + D = 2 \\ (-)\quad \underline{2C + D = 8} \\ -3C \quad\quad = -6 \end{array}$$

which yields $C = 2$ and, consequently, $D = 4$. Finally, it follows that

$$\int \frac{2x^3 - 4x - 8}{(x^2 - x)(x^2 + 4)}\,dx = \int \left(\frac{2}{x} - \frac{2}{x-1} + \frac{2x}{x^2+4} + \frac{4}{x^2+4} \right) dx$$

$$= 2\ln|x| - 2\ln|x-1| + \ln(x^2+4)$$

$$+ 2\,\text{Arctan}\,\frac{x}{2} + C$$

$$= \ln\left[\frac{x^2(x^2+4)}{(x-1)^2} \right] + 2\,\text{Arctan}\,\frac{x}{2} + C \qquad \blacksquare$$

In each of the first three examples we began the solution of the basic equation by substituting values of x that make the linear factors become zero. This method works especially well when the partial fraction decomposition involves distinct linear factors. For quadratic factors an alternative method of solving the basic equation is often more convenient. Both methods are outlined in the following summary.

Guidelines for Solving the Basic Equation

Linear Factors

1. Substitute the *roots* of the distinct linear factors into the basic equation.
2. For repeated linear factors use the coefficients determined in part 1 to rewrite the basic equation. Then substitute *other* convenient values of x and solve for the remaining coefficients.

Quadratic Factors

1. Expand the basic equation.
2. Collect terms according to powers of x.
3. Equate the coefficients of like powers to obtain a system of linear equations.
4. Solve the system of linear equations.

The second procedure for solving the basic equation is demonstrated in the next two examples.

Example 4 (*Repeated Quadratic Factors*)

Evaluate

$$\int \frac{2x^5 - 5x}{(x^2 + 2)^2} dx$$

Solution: Since the numerator is of greater degree than the denominator, we first divide to obtain

$$\frac{2x^5 - 5x}{(x^2 + 2)^2} = 2x - \frac{8x^3 + 13x}{(x^2 + 2)^2}$$

By Rule 4 we include one partial fraction for each power of $(x^2 + 2)$, and we write

$$\frac{8x^3 + 13x}{(x^2 + 2)^2} = \frac{Ax + B}{x^2 + 2} + \frac{Cx + D}{(x^2 + 2)^2}$$

Multiplying by the LCD, $(x^2 + 2)^2$, yields the *basic* equation

$$8x^3 + 13x = (Ax + B)(x^2 + 2) + Cx + D$$

By expanding our *basic* equation and collecting like terms, we have

$$8x^3 + 13x = Ax^3 + 2Ax + Bx^2 + 2B + Cx + D$$
$$8x^3 + 13x = Ax^3 + Bx^2 + (2A + C)x + (2B + D)$$

Now we can equate the coefficients of like terms on opposite sides of the equation.

$$8x^3 + 0x^2 + 13x + 0 = Ax^3 + Bx^2 + (2A + C)x + (2B + D)$$

Thus we have

$$8 = A, \quad 0 = B, \quad 13 = 2A + C, \quad 0 = 2B + D$$

From these equations we conclude that

$$A = 8, \quad B = 0, \quad C = -3, \quad D = 0$$

Therefore, we have

$$\frac{8x^3 + 13x}{(x^2 + 2)^2} = \frac{8x}{x^2 + 2} + \frac{-3x}{(x^2 + 2)^2}$$

and it follows that

$$\int \frac{2x^5 - 5x}{(x^2 + 2)^2} dx = \int 2x \, dx - \int \frac{8x}{x^2 + 2} dx + \int \frac{3x \, dx}{(x^2 + 2)^2}$$

$$= x^2 - 4 \ln (x^2 + 2) + \frac{3}{2} \frac{(x^2 + 2)^{-1}}{-1} + C$$

$$= x^2 - 4 \ln (x^2 + 2) - \frac{3}{2(x^2 + 2)} + C \qquad \blacksquare$$

Example 5 *(Repeated Quadratic Factors)*

Evaluate
$$\int \frac{x^2}{(x^2 + 1)^2} \, dx$$

Solution: By Rule 4 we write

$$\frac{x^2}{(x^2 + 1)^2} = \frac{Ax + B}{x^2 + 1} + \frac{Cx + D}{(x^2 + 1)^2}$$

Multiplying by $(x^2 + 1)^2$, we have

$$x^2 = (Ax + B)(x^2 + 1) + Cx + D$$
$$= Ax^3 + Bx^2 + (A + C)x + (B + D)$$

Therefore, by equating coefficients we obtain

$$A = 0, \qquad B = 1, \qquad C = 0, \qquad D = -1$$

and we have

$$\int \frac{x^2}{(x^2 + 1)^2} \, dx = \int \left[\frac{1}{x^2 + 1} - \frac{1}{(x^2 + 1)^2} \right] dx$$

$$= \int \frac{dx}{x^2 + 1} - \int \frac{dx}{(x^2 + 1)^2} = \text{Arctan } x + (?)$$

The second integral is one we cannot solve at this stage and we postpone the completion of this example until we have introduced a method called trigonometric substitution. (See Section 10.6, Example 7.) $\blacksquare$

Our next example shows how to combine effectively the two methods for solving a basic equation.

Example 6

Evaluate
$$\int_0^1 \frac{3x + 4}{x^3 - 2x - 4} \, dx$$

Solution: Since

$$x^3 - 2x - 4 = (x - 2)(x^2 + 2x + 2)$$

we write
$$\frac{3x + 4}{x^3 - 2x - 4} = \frac{A}{x - 2} + \frac{Bx + C}{x^2 + 2x + 2}$$

Multiplying by the LCD yields the basic equation

$$3x + 4 = A(x^2 + 2x + 2) + (Bx + C)(x - 2)$$

which by expanding and collecting terms becomes

$$3x + 4 = (A + B)x^2 + (2A - 2B + C)x + (2A - 2C)$$

If $x = 2$, then $10 = A(10) + 0$ and $A = 1$

Using $A = 1$ and equating coefficients gives us

$$0 = A + B = 1 + B   \text{or}   B = -1$$

and $4 = 2A - 2C = 2 - 2C   \text{or}   C = -1$

Therefore,

$$\int_0^1 \frac{3x + 4}{x^3 - 2x - 4} dx = \int_0^1 \left[\frac{A}{x - 2} + \frac{Bx + C}{x^2 + 2x + 2} \right] dx$$

$$= \int_0^1 \left[\frac{1}{x - 2} + \frac{-x - 1}{x^2 + 2x + 2} \right] dx$$

$$= \left[\ln |x - 2| - \frac{1}{2} \ln (x^2 + 2x + 2) \right]_0^1$$

$$= 0 - \tfrac{1}{2} \ln 5 - \ln 2 + \tfrac{1}{2} \ln 2$$

$$= -\tfrac{1}{2} \ln 10 \approx -1.151 \qquad\blacksquare$$

Before concluding this section a few comments are in order. First, it is not necessary to use the partial fractions technique on all integrals of the form $N(x)/D(x)$. For instance, even though

$$\int \frac{x^2 + 1}{x^3 + 3x - 4} dx$$

can be evaluated by partial fractions as

$$\int \frac{x^2 + 1}{x^3 + 3x - 4} dx = \int \frac{x^2 + 1}{(x - 1)(x^2 + x + 4)} dx$$

$$= \int \left(\frac{A}{x - 1} + \frac{Bx + C}{x^2 + x + 4} \right) dx$$

it is more easily evaluated by using the Log Rule and writing

$$\int \frac{x^2 + 1}{x^3 + 3x - 4} dx = \frac{1}{3} \int \frac{3x^2 + 3}{x^3 + 3x - 4} dx = \frac{1}{3} \ln |x^3 + 3x - 4| + C$$

Second, if $N(x)/D(x)$ is not in reduced form, reducing it may eliminate the need to use partial fractions. For instance,

$$\int \frac{x^2 - x - 2}{x^3 - 2x - 4} dx = \int \frac{(x + 1)(x - 2)}{(x - 2)(x^2 + 2x + 2)} dx = \int \frac{x + 1}{x^2 + 2x + 2} dx$$

and the latter integral can be solved directly by using the Log Rule.

Finally, the partial fractions technique can be used with some quotients involving transcendental functions. For instance, the substitution $u = \sin x$ allows us to write

$$\int \frac{\cos x}{\sin x(\sin x - 1)}\,dx = \int \frac{du}{u(u-1)} = \int \left(\frac{A}{u} + \frac{B}{u-1}\right)du$$

Section Exercises (10.4)

In Exercises 1–25, find the indefinite integrals.

1. $\displaystyle\int \frac{1}{x^2 - 1}\,dx$

2. $\displaystyle\int \frac{1}{4x^2 - 9}\,dx$

S 3. $\displaystyle\int \frac{3}{x^2 + x - 2}\,dx$

4. $\displaystyle\int \frac{x+1}{x^2 + 4x + 3}\,dx$

5. $\displaystyle\int \frac{5-x}{2x^2 + x - 1}\,dx$

6. $\displaystyle\int \frac{3x^2 - 7x - 2}{x^3 - x}\,dx$

S 7. $\displaystyle\int \frac{x^2 + 12x + 12}{x^3 - 4x}\,dx$

8. $\displaystyle\int \frac{x^3 - x + 3}{x^2 + x - 2}\,dx$

9. $\displaystyle\int \frac{2x^3 - 4x^2 - 15x + 5}{x^2 - 2x - 8}\,dx$

10. $\displaystyle\int \frac{x+2}{x^2 - 4x}\,dx$

11. $\displaystyle\int \frac{4x^2 + 2x - 1}{x^3 + x^2}\,dx$

12. $\displaystyle\int \frac{2x-3}{(x-1)^2}\,dx$

S 13. $\displaystyle\int \frac{x^4}{(x-1)^3}\,dx$

14. $\displaystyle\int \frac{4x^2 - 1}{(2x)(x^2 + 2x + 1)}\,dx$

15. $\displaystyle\int \frac{3x}{x^2 - 6x + 9}\,dx$

16. $\displaystyle\int \frac{6x^2 + 1}{x^2(x-1)^3}\,dx$

S 17. $\displaystyle\int \frac{x^2 - 1}{x^3 + x}\,dx$

18. $\displaystyle\int \frac{x}{x^3 - 1}\,dx$

19. $\displaystyle\int \frac{x^2}{x^4 - 2x^2 - 8}\,dx$

20. $\displaystyle\int \frac{2x^2 + x + 8}{(x^2 + 4)^2}\,dx$

21. $\displaystyle\int \frac{x}{16x^4 - 1}\,dx$

22. $\displaystyle\int \frac{x^2 - 4x + 7}{x^3 - x^2 + x + 3}\,dx$

23. $\displaystyle\int \frac{x^2 + x + 2}{(x^2 + 2)^2}\,dx$

24. $\displaystyle\int \frac{x^3}{(x^2 - 4)^2}\,dx$

S 25. $\displaystyle\int \frac{x^2 + 5}{x^3 - x^2 + x + 3}\,dx$

In Exercises 26–30, evaluate the definite integral.

26. $\displaystyle\int_3^4 \frac{1}{x^2 - 4}\,dx$

27. $\displaystyle\int_0^1 \frac{3}{2x^2 + 5x + 2}\,dx$

28. $\displaystyle\int_1^5 \frac{x-1}{x^2(x+1)}\,dx$

S 29. $\displaystyle\int_1^2 \frac{x+1}{x(x^2 + 1)}\,dx$

30. $\displaystyle\int_0^1 \frac{x^2 - x}{x^2 + x + 1}\,dx$

In Exercises 31–35, solve the indefinite integral by using the indicated substitution.

31. $\displaystyle\int \frac{\sin x}{\cos x(\cos x - 1)}\,dx$; let $u = \cos x$.

32. $\displaystyle\int \frac{\sin x}{\cos x + \cos^2 x}\,dx$; let $u = \cos x$.

S 33. $\displaystyle\int \frac{e^x}{(e^x - 1)(e^x + 4)}\,dx$; let $u = e^x$.

34. $\displaystyle\int \frac{e^x}{(e^{2x} + 1)(e^x - 1)}\,dx$; let $u = e^x$.

35. $\displaystyle\int \frac{3\cos x}{\sin^2 x + \sin x - 2}\,dx$; let $u = \sin x$.

36. (a) Express $1/(a^2 - x^2)$ as a sum of partial fractions.
(b) Use the result of part (a) to show that

$$\int \frac{1}{a^2 - x^2}\,dx = \frac{1}{2a}\ln\left|\frac{a+x}{a-x}\right| + C$$

37. Find the area of the region bounded by $y = 7/(16 - x^2)$ and $y = 1$.

38. In Section 8.3 the exponential growth equation was derived under the assumption that the rate of growth was proportional to the existing quantity. In practice, there often exists some upper limit L past which growth cannot occur. (Population growth is limited by the food supply; growth in sales of some new product will stop when the market becomes saturated; etc.) We therefore assume that the rate of growth is proportional not only to the existing quantity but *also* to the difference between the existing quantity y and the upper limit L; that is, $dy/dt = ky(L - y)$. In integral form we can express this relationship as

$$\int \frac{dy}{y(L-y)} = \int k\,dt$$

(a) Use partial fractions to evaluate the integral on the left, and then solve for y as a function of t, where y_0 is the initial quantity.

(b) The graph of the function y is called a *logistic curve* (see Figure 10.7). Show that the rate of growth is a maximum at the point of inflection, and that this occurs when $y = L/2$.

[S] **39.** A single infected individual enters a community of n individuals susceptible to the disease. Let x be the number of newly infected individuals after time t. The common *Epidemic Model* assumes that the disease spreads at a rate proportional to the product of the total number infected and the number of susceptible not yet infected. Thus $dx/dt = k(x + 1)(n - x)$, and we obtain

$$\int \frac{1}{(x + 1)(n - x)}\,dx = \int k\,dt$$

Solve for x as a function of t.

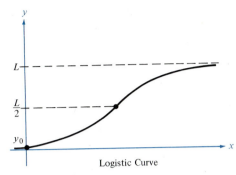

Logistic Curve

FIGURE 10.7

For Review and Class Discussion

True or False

1. _____ It is possible to find constants A and B such that

$$\frac{x^2 - 1}{(x - 2)(x - 3)} = \frac{A}{x - 2} + \frac{B}{x - 3}$$

2. _____ It is possible to find constants A, B, and C such that

$$\frac{1}{(x - 1)(2x - 2)(x - 3)}$$

$$= \frac{A}{x - 1} + \frac{B}{2x - 2} + \frac{C}{x - 3}$$

3. _____ If $|a| \neq 1$, then $x^3 - x^2 - a^2x + a^2$ has three distinct factors.

4. _____ If $p(x) = A(x - a)^3 + B(x - a)^2 + C(x - a) + D$ for all x, then $D = p(a)$, $C = p'(a)$, $B = p''(a)/2$, and $A = p'''(a)/6$.

5. _____ If $a \neq 0$ and

$$\frac{a}{x^2(x - 1)^2} = \frac{A}{x} + \frac{B}{x^2} + \frac{C}{x - 1} + \frac{D}{(x - 1)^2}$$

then A, B, C, and D are each nonzero.

10.5 Trigonometric Integrals

Purpose
- To introduce procedures for evaluating integrals involving powers and products of trigonometric functions.

$\sin^2 \theta$ is odious to me, even though Laplace made use of it; should it be feared that $\sin \theta^2$ might be ambiguous, which would perhaps never occur, or at most very rarely when speaking of $\sin(\theta^2)$, well then, let us write $(\sin \theta)^2$, but not $\sin^2 \theta$, which by analogy should signify $\sin(\sin \theta)$* —Carl Gauss (1777–1855)*

Up to this point in Chapter 10, our discussion of integration techniques has been directed primarily toward the integration of algebraic functions.

---

* Despite Gauss's objection we will continue to follow convention and write $\sin^2 \theta$ for $(\sin \theta)^2$.

In this section we interrupt this discussion to focus on integrals of trigonometric functions. The reason for this digression will become evident in Section 10.6.

We begin our discussion with several trigonometric integrals to which we can apply the Power Rule. For instance, to evaluate

$$\int \sin^5 x \cos x \, dx$$

we consider $u = \sin x$, and then $u' = \cos x$. Therefore, by the Power Rule it follows that

$$\int \sin^5 x (\cos x) \, dx = \int u^n u' \, dx = \frac{u^{n+1}}{n+1} + C = \frac{\sin^6 x}{6} + C$$

For convenience we list next the Power Rules for the sine, cosine, secant, and tangent functions.

Power Rules for Trigonometric Functions

$$\int \sin^n x (\cos x) \, dx = \frac{\sin^{n+1} x}{n+1} + C, \qquad n \neq -1$$

$$\int \cos^n x (-\sin x) \, dx = \frac{\cos^{n+1} x}{n+1} + C, \qquad n \neq -1$$

$$\int \sec^n x (\sec x \tan x) \, dx = \frac{\sec^{n+1} x}{n+1} + C, \qquad n \neq -1$$

$$\int \tan^n x (\sec^2 x) \, dx = \frac{\tan^{n+1} x}{n+1} + C, \qquad n \neq -1$$

Power Rules for $\cot^n x$ and $\csc^n x$ are similar to those for $\tan^n x$ and $\sec^n x$, respectively.

By use of the identities

$$\sin^2 x + \cos^2 x = 1$$

and

$$\tan^2 x + 1 = \sec^2 x$$

we can expand the application of these power rules to integrals of the forms

$$\int \sin^m x \cos^n x \, dx$$

and

$$\int \sec^m x \tan^n x \, dx$$

Example 1

Use the identity $\sin^2 x = 1 - \cos^2 x$ to evaluate

$$\int \sin^3 x \cos^4 x \, dx$$

Solution: With the expectation of using the Power Rule with $u = \cos x$, we *save one sine factor* to serve as u' and convert the remaining sine factors to cosines, as follows:

$$\int \sin^3 x \cos^4 x \, dx = \int \sin^2 x \cos^4 x \overbrace{(\sin x)}^{\text{save for } u'} dx$$

Now since $\sin^2 x = 1 - \cos^2 x$, we have

$$\int \sin^3 x \cos^4 x \, dx = \int (1 - \cos^2 x) \cos^4 x (\sin x) \, dx$$

$$= \int (\cos^4 x - \cos^6 x)(\sin x) \, dx$$

$$= \int \cos^4 x(\sin x) \, dx - \int \cos^6 x(\sin x) \, dx$$

$$= -\int \cos^4 x(-\sin x) \, dx + \int \cos^6 x(-\sin x) \, dx$$

Therefore, by the Power Rule for the cosine function, it follows that

$$\int \sin^3 x \cos^4 x \, dx = -\frac{\cos^5 x}{5} + \frac{\cos^7 x}{7} + C \qquad \blacksquare$$

Integrals Involving Sine and Cosine

1. If the power of the sine is odd and positive, save one sine factor and convert the remaining factors to cosine; then expand and integrate.

$$\int \sin^{\overbrace{2k+1}^{\text{odd}}} x \cos^n x \, dx = \int \sin^{2k} x \cos^n x \overbrace{(\sin x)}^{\text{save for } u'} dx = \int \overbrace{(\sin^2 x)^k}^{\text{convert to cosine}} \cos^n x(\sin x) \, dx$$

$$= \int (1 - \cos^2 x)^k \cos^n x(\sin x) \, dx$$

2. If the power of the cosine is odd and positive, save one cosine factor and convert the remaining factors to sine; then expand and integrate.

$$\int \sin^m x \cos^{\overbrace{2k+1}^{\text{odd}}} x \, dx = \int \sin^m x \cos^{2k} x \overbrace{(\cos x)}^{\text{save for } u'} dx = \int \sin^m x \overbrace{(\cos^2 x)^k}^{\text{convert to sine}}(\cos x) \, dx$$

$$= \int \sin^m x(1 - \sin^2 x)^k(\cos x) \, dx$$

3. If the powers of both sine and cosine are even and positive, make repeated use of the identities

$$\sin^2 x = \frac{1 - \cos 2x}{2} \qquad \text{and} \qquad \cos^2 x = \frac{1 + \cos 2x}{2}$$

to convert the integrand to odd powers of the cosine; then proceed as in case 2.

The key to the solution of Example 1 lies in the fact that the given sine factor has an *odd* power. In the preceding guide to solving integrals of the form

$$\int \sin^m x \cos^n x \, dx$$

note the importance of either beginning with (cases 1 and 2) or obtaining (case 3) an *odd* power of the sine or the cosine.

Example 2

Evaluate $\int \sin^2 x \cos^5 x \, dx$.

Solution: Since the power of the cosine is odd, we have

$$\int \sin^2 x \cos^5 x \, dx = \int \sin^2 x \cos^4 x (\cos x) \, dx$$

$$= \int \sin^2 x (\cos^2 x)^2 (\cos x) \, dx$$

$$= \int \sin^2 x (1 - \sin^2 x)^2 (\cos x) \, dx$$

$$= \int \sin^2 x (1 - 2 \sin^2 x + \sin^4 x)(\cos x) \, dx$$

$$= \int [\sin^2 x - 2 \sin^4 x + \sin^6 x](\cos x) \, dx$$

Applying the Power Rule for $\sin x$ to each term, it follows that

$$\int \sin^2 x \cos^5 x \, dx = \frac{\sin^3 x}{3} - \frac{2 \sin^5 x}{5} + \frac{\sin^7 x}{7} + C \qquad \blacksquare$$

Example 3

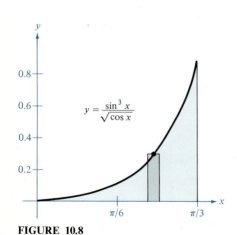

FIGURE 10.8

Evaluate

$$\int_0^{\pi/3} \frac{\sin^3 x}{\sqrt{\cos x}} \, dx$$

Solution: Since the power of the sine is odd, we write

$$\int_0^{\pi/3} \frac{\sin^3 x}{\sqrt{\cos x}} \, dx = \int_0^{\pi/3} \frac{(1 - \cos^2 x)(\sin x)}{\sqrt{\cos x}} \, dx$$

$$= \int_0^{\pi/3} [(\cos x)^{-1/2} \sin x - (\cos x)^{3/2} \sin x] \, dx$$

$$= \int_0^{\pi/3} [-(\cos x)^{-1/2}(-\sin x) + (\cos x)^{3/2}(-\sin x)] \, dx$$

$$= \left[-\frac{(\cos x)^{1/2}}{\frac{1}{2}} + \frac{(\cos x)^{5/2}}{\frac{5}{2}} \right]_0^{\pi/3}$$

$$= \left(-\frac{2}{\sqrt{2}} + \frac{2}{5(2^{5/2})} + 2 - \frac{2}{5}\right) = \frac{36 - 19\sqrt{2}}{20} \approx 0.456$$

∎

Example 4

Evaluate $\int \cos^4 x \, dx$.

Solution: Since m and n are both even ($m = 0$), we replace $\cos^4 x$ by $[(1 + \cos 2x)/2]^2$. Then we write

$$\int \cos^4 x \, dx = \int \left(\frac{1 + \cos 2x}{2}\right)^2 dx$$

$$= \int \left[\frac{1}{4} + \frac{\cos 2x}{2} + \frac{\cos^2 2x}{4}\right] dx$$

$$= \int \left[\frac{1}{4} + \frac{\cos 2x}{2} + \frac{1}{4}\left(\frac{1 + \cos 4x}{2}\right)\right] dx$$

$$= \frac{3}{8}\int dx + \frac{1}{4}\int 2 \cos 2x \, dx + \frac{1}{32}\int 4 \cos 4x \, dx$$

$$= \frac{3x}{8} + \frac{\sin 2x}{4} + \frac{\sin 4x}{32} + C$$

∎

To evaluate integrals of the form

$$\int \sec^m x \tan^n x \, dx$$

we suggest the following guidelines:

Integrals Involving Secant and Tangent

1. If the power of the secant is even and positive, save a secant squared factor and convert the remaining factors to tangents; then expand and integrate.

even   save for u' convert to tangents

$$\int \sec^{2k} x \tan^n x \, dx = \int \sec^{2k-2} x \tan^n x (\sec^2 x) \, dx = \int (\sec^2 x)^{k-1} \tan^n x (\sec^2 x) \, dx$$

$$= \int (1 + \tan^2 x)^{k-1} \tan^n x (\sec^2 x) \, dx$$

2. If the power of the tangent is odd and positive, save a secant-tangent factor and convert the remaining factors to secants; then expand and integrate.

odd   save for u' convert to secants

$$\int \sec^m x \tan^{2k+1} x \, dx = \int \sec^{m-1} x \tan^{2k} x (\sec x \tan x) \, dx = \int \sec^{m-1} x (\tan^2 x)^k (\sec x \tan x) \, dx$$

$$= \int \sec^{m-1} x (\sec^2 x - 1)^k (\sec x \tan x) \, dx$$

3. If there are no secant factors and the power of the tangent is even and positive, convert a tangent squared factor to secants; then expand and repeat if necessary.

convert to secants

$$\int \tan^n x \, dx = \int \tan^{n-2} x \overbrace{(\tan^2 x)} \, dx = \int \tan^{n-2} x(\sec^2 x - 1) \, dx$$

$$= \int \tan^{n-2} x(\sec^2 x) \, dx - \int \tan^{n-2} x \, dx$$

4. If none of the first three cases apply, try rewriting the integrand in terms of sines and cosines.

For integrals involving powers of the cotangent and cosecant, we follow a similar strategy by making use of the identity $\csc^2 u = 1 + \cot^2 u$.

Example 5

Evaluate
$$\int \frac{\tan^3 x}{\sqrt{\sec x}} \, dx$$

Solution: Since the power of the tangent is odd, we write
$$\int (\sec x)^{-1/2} \tan^3 x \, dx = \int (\sec x)^{-3/2}(\tan^2 x)(\sec x \tan x) \, dx$$

$$= \int (\sec x)^{-3/2}(\sec^2 x - 1)(\sec x \tan x) \, dx$$

$$= \int [(\sec x)^{1/2} - (\sec x)^{-3/2}](\sec x \tan x) \, dx$$

$$= \tfrac{2}{3}(\sec x)^{3/2} + 2(\sec x)^{-1/2} + C \qquad \blacksquare$$

Example 6

Evaluate $\int \sec^4 3x \tan^3 3x \, dx$.

Solution: Since the power of the secant is even, we write
$$\int \sec^4 3x \tan^3 3x \, dx = \int \sec^2 3x \tan^3 3x(\sec^2 3x) \, dx$$

$$= \int (1 + \tan^2 3x)\tan^3 3x(\sec^2 3x) \, dx$$

In this case the angle is $u = 3x$; hence $u' = 3$ and we adjust with $(\tfrac{1}{3})(3)$.

$$\int \sec^4 3x \tan^3 3x \, dx = \frac{1}{3} \int [\tan^3 3x + \tan^5 3x](3 \sec^2 3x) \, dx$$

By applying the Power Rule for the tangent function, we obtain

$$\int \sec^4 3x \tan^3 3x \, dx = \frac{1}{3} \left[\frac{\tan^4 3x}{4} + \frac{\tan^6 3x}{6} \right] + C$$

Note that since the power of the tangent is odd, we could have applied the procedure described in case 2. In Exercise 41 of this section you are asked to show that these two results differ by only a constant. ∎

Example 7

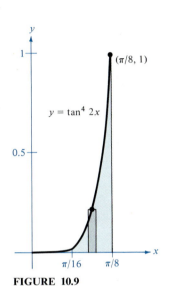

$y = \tan^4 2x$

FIGURE 10.9

Evaluate $\int_0^{\pi/8} \tan^4 2x \, dx$.

Solution: Since there are no secant factors, we write

$$\int_0^{\pi/8} \tan^4 2x \, dx = \int_0^{\pi/8} \tan^2 2x (\tan^2 2x) \, dx$$

$$= \int_0^{\pi/8} \tan^2 2x (\sec^2 2x - 1) \, dx$$

$$= \int_0^{\pi/8} \tan^2 2x (\sec^2 2x) \, dx - \int_0^{\pi/8} \tan^2 2x \, dx$$

Repeating this procedure on the second integral, we obtain

$$\int_0^{\pi/8} \tan^4 2x \, dx = \int_0^{\pi/8} \tan^2 2x (\sec^2 2x) \, dx - \int_0^{\pi/8} (\sec^2 2x - 1) \, dx$$

Adjusting for the angle $u = 2x$, we have

$$\int_0^{\pi/8} \tan^4 2x \, dx = \frac{1}{2} \int_0^{\pi/8} \tan^2 2x (2 \sec^2 2x) \, dx - \frac{1}{2} \int_0^{\pi/8} \sec^2 2x (2) \, dx$$
$$+ \int_0^{\pi/8} dx$$

$$= \left[\frac{1}{2} \frac{\tan^3 2x}{3} - \frac{1}{2} \tan 2x + x \right]_0^{\pi/8}$$

$$= \frac{1}{6} \tan^3 \frac{\pi}{4} - \frac{1}{2} \tan \frac{\pi}{4} + \frac{\pi}{8} = \frac{1}{6} - \frac{1}{2} + \frac{\pi}{8} \approx 0.0594 \quad ∎$$

Example 8

Evaluate $\int \csc^4 x \cot^4 x \, dx$.

Solution: Since the power of the cosecant is even, we write

$$\int \csc^4 x \cot^4 x \, dx = \int \csc^2 x \cot^4 x (\csc^2 x) \, dx$$

$$= \int (1 + \cot^2 x) \cot^4 x (\csc^2 x) \, dx$$

$$= -\int (\cot^4 x + \cot^6 x)(-\csc^2 x) \, dx$$

By the Power Rule for the cotangent function, it follows that

$$\int \csc^4 x \cot^4 x \, dx = -\left(\frac{\cot^5 x}{5} + \frac{\cot^7 x}{7}\right) + C$$

$$= -\frac{\cot^5 x}{5} - \frac{\cot^7 x}{7} + C \qquad \blacksquare$$

Example 9

Evaluate

$$\int \frac{\sec x}{\tan^2 x} \, dx$$

Solution: Since the first three cases do not apply, we can convert the integrand to sines and cosines and write

$$\int \frac{\sec x}{\tan^2 x} \, dx = \int \left(\frac{1}{\cos x}\right)\left(\frac{\cos x}{\sin x}\right)^2 dx = \int \sin^{-2} x (\cos x) \, dx$$

$$= -\sin^{-1} x + C = -\csc x + C \qquad \blacksquare$$

Occasionally we encounter integrals involving the product of sines and cosines of two *different* angles. In such instances we use the product identities

$$\sin mx \sin nx = \tfrac{1}{2}\{\cos[(m-n)x] - \cos[(m+n)x]\}$$
$$\sin mx \cos nx = \tfrac{1}{2}\{\sin[(m+n)x] + \sin[(m-n)x]\}$$
$$\cos mx \cos nx = \tfrac{1}{2}\{\cos[(m+n)x] + \cos[(m-n)x]\}$$

Example 10

Evaluate $\int \sin 5x \cos 4x \, dx$.

Solution: Considering the appropriate product identity, we write

$$\int \sin 5x \cos 4x \, dx = \frac{1}{2} \int (\sin 9x + \sin x) \, dx$$

$$= \frac{1}{2}\left[-\frac{\cos 9x}{9} - \cos x\right] + C$$

$$= -\frac{\cos 9x}{18} - \frac{\cos x}{2} + C \qquad \blacksquare$$

As you review this section, we suggest that you concentrate on the *general pattern* followed throughout the section. Study the examples and guidelines and then try to work the exercises by reasoning each one out rather than looking back to the boxed-in rules.

Section Exercises (10.5)

In Exercises 1–30, find the indefinite integrals.

1. $\int \cos^3 x \sin x \, dx$

2. $\int \cos^3 x \sin^2 x \, dx$

3. $\int \sin^5 2x \cos 2x \, dx$

4. $\int \sin^3 x \, dx$

S 5. $\int \sin^5 x \cos^2 x \, dx$

S 6. $\int \cos^3 \left(\frac{x}{3} \right) dx$

7. $\int \cos^2 x \, dx$

8. $\int \sin^4 2x \, dx$

9. $\int \sin^2 x \cos^4 x \, dx$

10. $\int \frac{\sin^3 \theta}{\sqrt{\cos \theta}} \, d\theta$

S 11. $\int \sin^2 \left(\frac{x}{2} \right) \cos^2 \left(\frac{x}{2} \right) dx$

12. $\int \sin^4 x \cos^2 x \, dx$

13. $\int \sec^2 x \tan x \, dx$

14. $\int \csc^2 3x \cot 3x \, dx$

15. $\int \sec^3 x \tan x \, dx$

16. $\int \tan^3 (3x) \, dx$

S 17. $\int \tan^3 3x \sec 3x \, dx$

18. $\int \sqrt{\tan x} \sec^4 x \, dx$

19. $\int \cot^3 2x \, dx$

20. $\int \tan^4 \left(\frac{x}{2} \right) \sec^4 \left(\frac{x}{2} \right) dx$

21. $\int \csc^4 \theta \, d\theta$

22. $\int \tan^3 t \sec^3 t \, dt$

23. $\int \frac{\cot^2 t}{\csc t} \, dt$

24. $\int \frac{\cot^3 t}{\csc t} \, dt$

S 25. $\int \sin 3x \cos 2x \, dx$

26. $\int \cos 3\theta \cos (-2\theta) \, d\theta$

27. $\int \sin \theta \sin 3\theta \, d\theta$

28. $\int x \tan^2 x^2 \, dx$

29. $\int \frac{1}{\sec x \tan x} \, dx$

30. $\int \frac{\sin^2 x - \cos^2 x}{\cos x} \, dx$

In Exercises 31–40, evaluate each definite integral.

31. $\int_{-\pi}^{\pi} \sin^2 x \, dx$

32. $\int_0^{\pi/4} \tan^2 x \, dx$

33. $\int_0^{\pi/4} \tan^3 x \, dx$

34. $\int_0^{\pi/4} \sec^2 t \sqrt{\tan t} \, dt$

S 35. $\int_0^{\pi/2} \frac{\cos t}{1 + \sin t} \, dt$

36. $\int_{-\pi}^{\pi} \sin 3\theta \cos \theta \, d\theta$

37. $\int_0^{\pi/4} \sin 2\theta \sin 3\theta \, d\theta$

38. $\int_0^{\pi/2} (1 - \cos \theta)^2 \, d\theta$

39. $\int_{-\pi/2}^{\pi/2} \cos^3 x \, dx$

40. $\int_{-\pi/2}^{\pi/2} (\sin^2 x + 1) \, dx$

S 41. In Example 6, $\int \sec^4 3x \tan^3 3x \, dx$ was integrated by procedure 1. Use the second procedure and show that the result differs only by a constant.

42. Integrate $\int \sec^2 x \tan x \, dx$ in two different ways and show that the result differs only by a constant.

43. The *inner product* of two functions f_1 and f_2 on $[a, b]$ is the number (f_1, f_2) defined by the equation $(f_1, f_2) = \int_a^b f_1(x) f_2(x) \, dx$. Two distinct functions are said to be *orthogonal* if $(f_1, f_2) = 0$.

(a) Show that the functions $f_n(x) = \cos nx$ $(n = 0, 1, 2, \dots)$ are orthogonal on the interval $[0, \pi]$.

(b) Show that the functions $\sin x$, $\sin 2x$, $\sin 3x$, $\dots$, 1, $\cos x$, $\cos 2x$, $\cos 3x$, $\dots$ are orthogonal on the interval $[-\pi, \pi]$.

44. Find the area of the region bounded by $y = \tan^2 x$ and $y = 4x/\pi$, where $-\pi/4 \le x \le \pi/4$.

S 45. Find the volume of the solid generated by revolving one arch of the sine curve about the *x*-axis.

In Exercises 46–50, evaluate each definite integral. (Assume *m* and *n* are distinct positive integers.)

46. $\int_0^{2\pi} \sin^2 nx \, dx$

S 47. $\int_0^{2\pi} \cos^2 nx \, dx$

48. $\int_0^{2\pi} \sin (mx) \sin (nx) \, dx$

49. $\int_0^{2\pi} \sin (mx) \cos (nx) \, dx$

50. $\int_0^{2\pi} \cos (mx) \cos (nx) \, dx$

51. If *n* is odd and greater than 1, then

$$\int_0^{\pi/2} \cos^n x \, dx = \left(\frac{2}{3} \right) \left(\frac{4}{5} \right) \left(\frac{6}{7} \right) \cdots \left(\frac{n-1}{n} \right)$$

Verify this version of Wallis's Formula for $n = 3, 5, 7$.

(a) $\int_0^{\pi/2} \cos^3 x \, dx = \frac{2}{3}$

(b) $\int_0^{\pi/2} \cos^5 x \, dx = \frac{8}{15}$

(c) $\int_0^{\pi/2} \cos^7 x \, dx = \frac{16}{35}$

52. If n is even and positive, then

$$\int_0^{\pi/2} \cos^n x \, dx = \left(\frac{1}{2}\right)\left(\frac{3}{4}\right)\left(\frac{5}{6}\right) \cdots \left(\frac{n-1}{n}\right)\left(\frac{\pi}{2}\right)$$

(a) $\displaystyle\int_0^{\pi/2} \cos^2 x \, dx = \frac{\pi}{4}$ (b) $\displaystyle\int_0^{\pi/2} \cos^4 x \, dx = \frac{3\pi}{16}$

(c) $\displaystyle\int_0^{\pi/2} \cos^6 x \, dx = \frac{5\pi}{32}$

Verify this version of Wallis's Formula for $n = 2, 4, 6$.

10.6 Trigonometric Substitution

Purpose

- To introduce the technique of trigonometric substitution as a means of evaluating integrals involving the radical forms $\sqrt{a^2 - u^2}$, $\sqrt{a^2 + u^2}$, and $\sqrt{u^2 - a^2}$.

It is as difficult to find the integral of any differential as it is easy to find the differential of any quantity. Sometimes it is not even certain whether or not we can find the integral of a given quantity. I dare say at any rate that any quantity, integral or rational, that is multiplied or divided by $x^p \sqrt{a^2 - x^2}$, $x^p \sqrt{x^2 - a^2}$, $x^p \sqrt{a^2 + x^2}$ can be integrated. —John Bernoulli (1667–1748)

Now that we can evaluate integrals involving powers of the trigonometric functions, we can use the method of **trigonometric substitution** to evaluate integrals involving the radicals

$$\sqrt{a^2 - u^2}, \qquad \sqrt{a^2 + u^2}, \qquad \sqrt{u^2 - a^2}$$

Just as with algebraic substitution, our objective with trigonometric substitution is to eliminate the radicals in the integrand. The three substitutions that accomplish this objective are as follows:

Trigonometric Substitution

1. For integrals involving $\sqrt{a^2 - u^2}$:

    let $u = a \sin \theta$;   then $\sqrt{a^2 - u^2} = a \cos \theta$

2. For integrals involving $\sqrt{a^2 + u^2}$:

    let $u = a \tan \theta$;   then $\sqrt{a^2 + u^2} = a \sec \theta$

3. For integrals involving $\sqrt{u^2 - a^2}$:

    let $u = a \sec \theta$;   then $\sqrt{u^2 - a^2} = a \tan \theta$

In each substitution we assume that θ is in the range of the corresponding inverse trigonometric function, and in those instances in which $\sqrt{a^2 - u^2}$ or $\sqrt{u^2 - a^2}$ occur as denominators, we assume the further restriction of $u \neq a$.

To show that the radical is eliminated as indicated in each of the three cases, we use the identities

$$\cos^2 \theta = 1 - \sin^2 \theta \qquad \text{and} \qquad \sec^2 \theta = 1 + \tan^2 \theta$$

For example, if $u = a \sin \theta$, then

$$\sqrt{a^2 - u^2} = \sqrt{a^2 - a^2 \sin^2 \theta} = \sqrt{a^2(1 - \sin^2 \theta)}$$
$$= \sqrt{a^2(\cos^2 \theta)} = a \cos \theta$$

Example 1

Evaluate

$$\int \frac{dx}{x^2 \sqrt{9 - x^2}}$$

Solution: First, note that none of the fundamental integration formulas of Section 10.1 apply. Since $\sqrt{9 - x^2}$ has the form $\sqrt{a^2 - u^2}$, we make the substitution

$$x = a \sin \theta = 3 \sin \theta$$

Then

$$dx = 3 \cos \theta \, d\theta, \qquad \sqrt{9 - x^2} = 3 \cos \theta, \qquad x^2 = 9 \sin^2 \theta$$

Therefore,

$$\int \frac{dx}{x^2 \sqrt{9 - x^2}} = \int \frac{3 \cos \theta \, d\theta}{(9 \sin^2 \theta)(3 \cos \theta)} = \frac{1}{9} \int \frac{d\theta}{\sin^2 \theta}$$

$$= \frac{1}{9} \int \csc^2 \theta \, d\theta = -\frac{1}{9} \cot \theta + C$$

Since $x = 3 \sin \theta$, we have $\theta = \text{Arcsin} \dfrac{x}{3}$ and, therefore,

$$\int \frac{dx}{x^2 \sqrt{9 - x^2}} = -\frac{1}{9} \cot \left(\text{Arcsin} \frac{x}{3} \right) + C$$

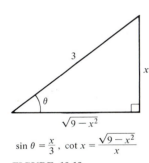

$$\sin \theta = \frac{x}{3}, \quad \cot x = \frac{\sqrt{9 - x^2}}{x}$$

FIGURE 10.10

However, this is *not* considered to be a desirable form for the solution and it is best to simplify the expression cot [Arcsin $(x/3)$]. Following the procedure of Section 9.4 (Examples 4 and 5), we construct the right triangle shown in Figure 10.10. From this triangle we see that

$$\cot \left(\text{Arcsin} \frac{x}{3} \right) = \cot \theta = \frac{\sqrt{9 - x^2}}{x}$$

and it follows that

$$\int \frac{dx}{x^2 \sqrt{9 - x^2}} = -\frac{1}{9} \cot \theta + C = -\frac{\sqrt{9 - x^2}}{9x} + C \qquad \blacksquare$$

Each of the integration formulas 14 through 19 in Section 10.1 can be developed by trigonometric substitution. So if you forget one of these formulas, you have a method to fall back on. To illustrate this, the next example uses trigonometric substitution in place of Formula 15.

Example 2

Evaluate
$$\int \frac{1}{\sqrt{4x^2 + 1}}\, dx$$

Solution: We consider $u = 2x$, $a = 1$, and let $2x = \tan \theta$. Then
$$dx = \tfrac{1}{2} \sec^2 \theta\, d\theta \qquad \text{and} \qquad \sqrt{4x^2 + 1} = \sec \theta$$

Therefore,
$$\int \frac{1}{\sqrt{4x^2 + 1}}\, dx = \frac{1}{2} \int \frac{\sec^2 \theta\, d\theta}{\sec \theta} = \frac{1}{2} \int \sec \theta\, d\theta$$
$$= \tfrac{1}{2} \ln |\sec \theta + \tan \theta| + C = \tfrac{1}{2} \ln |\sqrt{4x^2 + 1} + 2x| + C$$

which is what we would obtain by using Formula 15 with $u = 2x$ and $u' = 2$.   ∎

Example 3

Evaluate
$$\int \frac{dx}{(x^2 + 1)^{3/2}}$$

Solution: If we write $(x^2 + 1)^{3/2}$ as $(\sqrt{x^2 + 1})^3$, then $\sqrt{x^2 + 1}$ has the form $\sqrt{u^2 + a^2}$. Thus consider $u = x$, $a = 1$, and let $x = \tan \theta$. Then
$$dx = \sec^2 \theta\, d\theta \qquad \text{and} \qquad \sqrt{x^2 + 1} = \sec \theta$$

Therefore,
$$\int \frac{dx}{(x^2 + 1)^{3/2}} = \int \frac{dx}{(\sqrt{x^2 + 1})^3} = \int \frac{\sec^2 \theta\, d\theta}{\sec^3 \theta}$$
$$= \int \frac{d\theta}{\sec \theta} = \int \cos \theta\, d\theta = \sin \theta + C$$

From Figure 10.11 we conclude that
$$\int \frac{dx}{(x^2 + 1)^{3/2}} = \sin \theta + C = \frac{x}{\sqrt{x^2 + 1}} + C$$ ∎

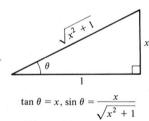

$\tan \theta = x,\ \sin \theta = \dfrac{x}{\sqrt{x^2 + 1}}$

FIGURE 10.11

Example 4   (*Changing Limits of Integration*)

Evaluate
$$\int_{\sqrt{3}}^{2} \frac{\sqrt{x^2 - 3}}{x}\, dx$$

Solution: Since $\sqrt{x^2 - 3}$ has the form $\sqrt{u^2 - a^2}$, we consider $u = x$, $a = \sqrt{3}$, and let $x = \sqrt{3} \sec \theta$. Then
$$dx = \sqrt{3} \sec \theta \tan \theta\, d\theta \qquad \text{and} \qquad \sqrt{x^2 - 3} = \sqrt{3} \tan \theta$$

For definite integrals, it is often more convenient to determine the upper and lower limits of integration for the variable θ and thus avoid the necessity of converting back to the variable x. Using $x = \sqrt{3} \sec \theta$:

when $x = 2$, then $\sec \theta = \dfrac{2}{\sqrt{3}}$ and $\theta = \dfrac{\pi}{6}$

when $x = \sqrt{3}$, then $\sec \theta = 1$ and $\theta = 0$

Therefore, we have

$$\int_{\sqrt{3}}^{2} \frac{\sqrt{x^2 - 3}}{x}\, dx = \int_0^{\pi/6} \frac{(\sqrt{3}\tan\theta)(\sqrt{3}\sec\theta\tan\theta)\, d\theta}{\sqrt{3}\sec\theta}$$

$$= \int_0^{\pi/6} \sqrt{3}\tan^2\theta\, d\theta = \sqrt{3}\int_0^{\pi/6}(\sec^2\theta - 1)\, d\theta$$

$$= \sqrt{3}\Big[\tan\theta - \theta\Big]_0^{\pi/6} = \sqrt{3}\left(\frac{1}{\sqrt{3}} - \frac{\pi}{6}\right)$$

$$= 1 - \frac{\sqrt{3}\pi}{6} \approx 0.0931$$

We leave as an exercise the alternative procedure of converting back to variable x and evaluating the antiderivative at the original limits of integration (see Exercise 33). ∎

We can further expand the range of problems to which trigonometric substitution applies by employing the technique of completing the square. Study the next example carefully.

Example 5 (*Completing the Square*)

Evaluate $$\int \frac{dx}{(x^2 - 4x)^{3/2}}$$

Solution: By completing the square we obtain

$$(x^2 - 4x)^{3/2} = (x^2 - 4x + 4 - 4)^{3/2} = [(x-2)^2 - 4]^{3/2}$$

which fits the form $(\sqrt{u^2 - a^2})^3$ with $u = x - 2$ and $a = 2$. Thus we let $x - 2 = 2\sec\theta$. Then

$$dx = 2\sec\theta\tan\theta\, d\theta \quad \text{and} \quad \sqrt{(x-2)^2 - 4} = 2\tan\theta$$

Therefore,

$$\int \frac{dx}{(x^2 - 4x)^{3/2}} = \int \frac{2\sec\theta\tan\theta\, d\theta}{(2\tan\theta)^3} = \frac{2}{8}\int \frac{\sec\theta}{\tan^2\theta}\, d\theta$$

$$= \frac{1}{4}\int (\sin\theta)^{-2}(\cos\theta)\, d\theta = -\frac{1}{4}\csc\theta + C$$

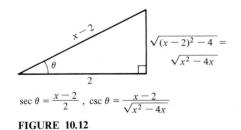

$$\sqrt{(x-2)^2 - 4} = \sqrt{x^2 - 4x}$$

$$\sec\theta = \frac{x-2}{2}, \quad \csc\theta = \frac{x-2}{\sqrt{x^2 - 4x}}$$

FIGURE 10.12

From Figure 10.12 we conclude that

$$\int \frac{dx}{(x^2 - 4x)^{3/2}} = -\frac{1}{4}\csc\theta + C = -\frac{x-2}{4\sqrt{x^2 - 4x}} + C \qquad \blacksquare$$

Example 6

Evaluate $\int \sqrt{a^2 - u^2}\, u'\, dx$.

Solution: Recall that if $u = u(x)$, then $du = u'\, dx$. Thus we can write the given integral in the form $\int \sqrt{a^2 - u^2}\, du$. Now if we let $u = a\sin\theta$, then

$$du = a\cos\theta\, d\theta \qquad \text{and} \qquad \sqrt{a^2 - u^2} = a\cos\theta$$

Therefore,

$$\int \sqrt{a^2 - u^2}\, u'\, dx = \int (a\cos\theta)(a\cos\theta)\, d\theta = a^2 \int \cos^2\theta\, d\theta$$

$$= a^2 \int \frac{1 + \cos 2\theta}{2}\, d\theta = \frac{a^2}{2}\left(\theta + \frac{1}{2}\sin 2\theta\right) + C$$

$$= \frac{a^2}{2}(\theta + \sin\theta\cos\theta) + C$$

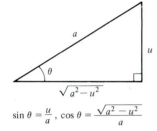

$\sin\theta = \dfrac{u}{a}$, $\cos\theta = \dfrac{\sqrt{a^2 - u^2}}{a}$

FIGURE 10.13

From Figure 10.13 we have

$$\int \sqrt{a^2 - u^2}\, u'\, dx = \frac{a^2}{2}\left[\operatorname{Arcsin}\frac{u}{a} + \left(\frac{u}{a}\right)\left(\frac{\sqrt{a^2 - u^2}}{a}\right)\right] + C$$

$$= \frac{a^2}{2}\left[\frac{a^2 \operatorname{Arcsin}(u/a) + u\sqrt{a^2 - u^2}}{a^2}\right] + C$$

$$= \frac{1}{2}\left(a^2 \operatorname{Arcsin}\frac{u}{a} + u\sqrt{a^2 - u^2}\right) + C \qquad \blacksquare$$

Throughout the remainder of the text, we will occasionally encounter the integral $\int \sqrt{a^2 - u^2}\, u'\, dx$, so for reference purposes, keep the result of Example 6 in mind.

$$\int \sqrt{a^2 - u^2}\, u'\, dx = \frac{1}{2}\left(a^2 \operatorname{Arcsin}\frac{u}{a} + u\sqrt{a^2 - u^2}\right) + C$$

We postpone a discussion of the integrals $\int \sqrt{u^2 + a^2}\, u'\, dx$ and $\int \sqrt{u^2 - a^2}\, u'\, dx$ until we have covered integration by parts in the next section.

Example 7

Evaluate

$$\int \frac{x^2}{(x^2 + 1)^2}\, dx$$

Solution: In Section 10.4, Example 5, we used partial fractions to determine that

$$\int \frac{x^2}{(x^2+1)^2}\,dx = \int \frac{dx}{x^2+1} - \int \frac{dx}{(x^2+1)^2}$$

At that point we were unable to evaluate the integral

$$\int \frac{dx}{(x^2+1)^2}$$

By using trigonometric substitution we can now solve this integral as follows. Let $x = \tan\theta$; then $x^2 + 1 = \sec^2\theta$, and $dx = \sec^2\theta\,d\theta$. Then

$$\int \frac{dx}{(x^2+1)^2} = \int \frac{\sec^2\theta\,d\theta}{\sec^4\theta} = \int \cos^2\theta\,d\theta$$

$$= \frac{1}{2}\int (1 + \cos 2\theta)\,d\theta = \frac{\theta}{2} + \frac{\sin 2\theta}{4} + C$$

$$= \frac{\theta}{2} + \frac{1}{2}\sin\theta\cos\theta + C$$

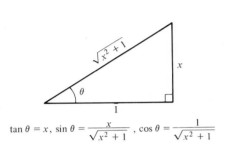

$$\tan\theta = x, \ \sin\theta = \frac{x}{\sqrt{x^2+1}}, \ \cos\theta = \frac{1}{\sqrt{x^2+1}}$$

FIGURE 10.14

From Figure 10.14 it follows that

$$\int \frac{dx}{(x^2+1)^2} = \frac{\text{Arctan } x}{2} + \frac{1}{2}\left(\frac{x}{\sqrt{x^2+1}}\right)\left(\frac{1}{\sqrt{x^2+1}}\right) + C$$

$$= \frac{1}{2}\left(\text{Arctan } x + \frac{x}{x^2+1}\right) + C$$

Therefore, we finally obtain

$$\int \frac{x^2\,dx}{(x^2+1)^2} = \int \frac{dx}{x^2+1} - \int \frac{dx}{(x^2+1)^2}$$

$$= \text{Arctan } x - \frac{1}{2}\left(\text{Arctan } x + \frac{x}{x^2+1}\right) + C$$

$$= \frac{1}{2}\left(\text{Arctan } x - \frac{x}{x^2+1}\right) + C \qquad \blacksquare$$

Example 8

Prove that the area of an ellipse is given by πab, where $2a$ and $2b$ are the lengths of the major and minor axes, respectively. (See Figure 10.15.)

Solution: Considering one-quarter of the ellipse, we will find the area of the region bounded by

$$\left(\frac{x}{a}\right)^2 + \left(\frac{y}{b}\right)^2 = 1$$

in the first quadrant and multiply by 4 to obtain the total area. Thus

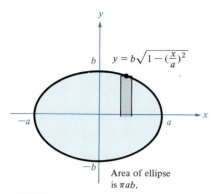

$$A = 4 \int_0^a b \sqrt{1 - \left(\frac{x}{a}\right)^2}\, dx = \frac{4b}{a} \int_0^a \sqrt{a^2 - x^2}\, dx$$

$$= \left(\frac{4b}{a}\right)\left(\frac{1}{2}\right)\left[a^2 \, \text{Arcsin} \, \frac{x}{a} + x \sqrt{a^2 - x^2}\right]_0^a$$

$$= \left(\frac{4b}{a}\right)\left(\frac{1}{2}\right)\left[a^2\left(\frac{\pi}{2}\right) + 0 - 0 - 0\right] = \pi ab \qquad \blacksquare$$

Having studied the examples in this section, you should now be aware that trigonometric substitution is a widely applicable method for evaluating integrands involving rational powers of $a^2 \pm u^2$ and $u^2 - a^2$. You should work enough exercises to gain confidence in the use of this powerful integration technique.

FIGURE 10.15

Area of ellipse is πab.

Section Exercises (10.6)

In Exercises 1–32, evaluate the integral.

1. $\int \dfrac{1}{(25 - x^2)^{3/2}}\, dx$

2. $\int \dfrac{1}{x^2 \sqrt{25 - x^2}}\, dx$

S 3. $\int \dfrac{\sqrt{25 - x^2}}{x}\, dx$

4. $\int \dfrac{1}{\sqrt{25 - x^2}}\, dx$

5. $\int \dfrac{x}{\sqrt{x^2 + 9}}\, dx$

6. $\int \dfrac{1}{x \sqrt{4x^2 + 16}}\, dx$

S 7. $\int_0^2 \sqrt{16 - 4x^2}\, dx$

8. $\int_0^2 x \sqrt{16 - 4x^2}\, dx$

9. $\int \dfrac{1}{\sqrt{x^2 - 9}}\, dx$

10. $\int \dfrac{t}{(1 - t^2)^{3/2}}\, dt$

11. $\int_0^{\sqrt{3}/2} \dfrac{t^2}{(1 - t^2)^{3/2}}\, dt$

12. $\int \dfrac{1}{(1 - t^2)^{5/2}}\, dt$

13. $\int \dfrac{\sqrt{1 - x^2}}{x^4}\, dx$

S 14. $\int \dfrac{\sqrt{4x^2 + 9}}{x^4}\, dx$

S 15. $\int \dfrac{1}{x \sqrt{4x^2 + 9}}\, dx$

16. $\int \dfrac{1}{(x^2 + 3)^{3/2}}\, dx$

17. $\int \dfrac{x}{(x^2 + 3)^{3/2}}\, dx$

18. $\int \dfrac{x^3}{\sqrt{x^2 - 4}}\, dx$

S 19. $\int x^3 \sqrt{x^2 - 4}\, dx$

20. $\int \dfrac{\sqrt{x^2 - 4}}{x}\, dx$

21. $\int e^{2x} \sqrt{1 + e^{2x}}\, dx$

22. $\int_{-1}^1 \dfrac{1}{(1 + x^2)^3}\, dx$

23. $\int_0^1 \dfrac{x^2}{(4 + x^2)^2}\, dx$

24. $\int \dfrac{1}{\sqrt{4x - x^2}}\, dx$

25. $\int (x + 1) \sqrt{x^2 + 2x + 2}\, dx$

S 26. $\int \dfrac{x^2}{\sqrt{2x - x^2}}\, dx$

27. $\int e^x \sqrt{1 - e^{2x}}\, dx$

28. $\int \dfrac{\sqrt{1 - x}}{\sqrt{x}}\, dx$

S 29. $\int \dfrac{1}{4 + 4x^2 + x^4}\, dx$

30. $\int \dfrac{x^3 + x + 1}{x^4 + 2x^2 + 1}\, dx$

31. $\int \dfrac{3x^3 + 3x + 1}{(x^2 + 1)^2}\, dx$

32. $\int \dfrac{x^4 - 1}{x(x^2 + 2)^2}\, dx$

33. Find the value of the definite integral of Example 4 by converting back to the variable x and evaluating the antiderivative at the original limits of integration.

34. Evaluate

$$\int_0^3 \dfrac{x^3}{\sqrt{x^2 + 9}}\, dx$$

in two ways using the given limits and the limits obtained from trigonometric substitution.

35. Evaluate

$$\int_0^{5/3} \sqrt{25 - 9x^2}\, dx$$

in two ways using the given limits and the limits obtained from trigonometric substitution.

36. The field strength H of a magnet of length $2l$ on a particle r units from the center of the magnet is given by

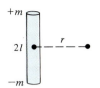

FIGURE 10.16

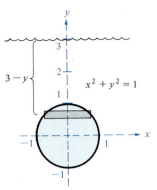

FIGURE 10.17

$$H = \frac{2ml}{(r^2 + l^2)^{3/2}}$$

where $\pm m$ are the poles of the magnet (see Figure 10.16). Find the average field strength as the particle moves from 0 to R units away from the center by evaluating the integral

$$\frac{1}{R} \int_0^R \frac{2ml}{(r^2 + l^2)^{3/2}} \, dr$$

37. A fish hatchery has a large water-filled tank with a circular glass observation port in its side. Find the force on the glass if it has a radius of 1 ft and its center is 3 ft below the surface of the water. (See Figure 10.17.)

38. Find the pressure on the glass of Exercise 37 if its center is d feet below the surface of the water ($d > 1$).

[S] **39.** The region bounded by the circle $(x - 3)^2 + y^2 = 1$ is revolved about the y-axis. The resulting doughnut-shaped solid is called a *torus*. Find the volume of this solid. (See Figure 10.18.)

40. The region bounded by the circle $(x - h)^2 + y^2 = r^2$ ($h > r$) is revolved about the y-axis. Find the volume of the resulting solid.

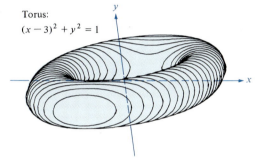

Torus:
$(x - 3)^2 + y^2 = 1$

FIGURE 10.18

For Review and Class Discussion

True or False

1. _____ If $x = \sin \theta$, then $\int \frac{dx}{\sqrt{1 - x^2}} = \int d\theta$.

2. _____ If $x = \sec \theta$, then $\int \frac{\sqrt{x^2 - 1}}{x} \, dx = $

$\int \sec \theta \tan \theta \, d\theta$.

3. _____ If $x = \tan \theta$, then $\int_0^{\sqrt{3}} \frac{dx}{(1 + x^2)^{3/2}} = \int_0^{4\pi/3} \cos \theta \, d\theta$.

4. _____ If $x = \sin \theta$, then $\int_{-1}^1 x^2 \sqrt{1 - x^2} \, dx = $

$2 \int_0^{\pi/2} \sin^2 \theta \cos^2 \theta \, d\theta$.

5. _____ If $x = \sin \theta$, then $\int \frac{\sqrt{1 - x^2}}{x^2} \, dx = $

$\int (\csc \theta - \sin \theta) \, d\theta$.

10.7 **Integration by Parts**

Purpose
- To introduce the technique of integration by parts.

The equation $\int u\, dv = uv - \int v\, du$ reduces the search for the integral $\int u\, dv$ to the integral $\int v\, du$ which, in certain cases, is more easily determined. —*Augustin-Louis Cauchy (1789–1857)*

We have spent the last five sections (10.2–10.6) discussing methods for converting integrands into forms to which we can apply the fundamental integration formulas listed in Section 10.1. The conversion techniques involved two basic approaches:

1. Rewrite the integrand by algebraic or trigonometric manipulations.
2. Make a change of variable by a u substitution or by a trigonometric substitution.

In this section we introduce the third basic method, known as **integration by parts.** This method of integration applies to a wide variety of functions and is particularly useful for integrands involving products of algebraic or transcendental functions. For instance, integration by parts works well with products like

$$x \ln x, \qquad e^x \sin x, \qquad x^2 e^x, \qquad x \sin x$$

Integration by parts is based on the formula for the derivative of a product:

$$\frac{d}{dx}[uv] = u\frac{dv}{dx} + v\frac{du}{dx}$$

where both u and v are differentiable functions of x. In differential form this equation is

$$d[uv] = uv'\, dx + vu'\, dx$$

or

$$uv'\, dx = d[uv] - vu'\, dx$$

Integrating both sides of this equation with respect to x, we obtain the following formula, which should be memorized:

Integration by Parts

$$\int uv'\, dx = uv - \int vu'\, dx$$

Note that the integration by parts formula expresses the original integral in terms of another integral. Depending on the choices for u and v', it may be easier to evaluate the second integral than the original one. Since the choices for u and v' are critical in the integration by parts process, we provide the following general guidelines.

Guidelines for Using Integration by Parts: $\int uv'\,dx$	1. Let v' be the most complicated portion of the integrand that can be "easily" integrated.
	2. Let u be that portion of the integrand whose derivative u' is a "simpler" function than u itself.

These are only suggested guidelines and they should not be followed blindly or without some thought. Furthermore, it is usually best to consider the guidelines above in the stated order, giving greater consideration to the first one.

Example 1

Evaluate $\int xe^x\,dx$.

Solution: Since e^x is easily integrated and the derivative of x is "simpler" than x itself, we let

$$v' = e^x \qquad \text{and} \qquad u = x$$

Then we have $v = e^x + C_1$ and $u' = 1$

Therefore,

$$\int xe^x\,dx = uv - \int vu'\,dx$$

$$= x(e^x + C_1) - \int (e^x + C_1)(1)\,dx$$

$$= xe^x + C_1 x - e^x - C_1 x + C = xe^x - e^x + C \qquad ■$$

Note in Example 1 that the first constant of integration C_1 does not appear in the final result. This will always happen, as we can see by replacing v by $v + C_1$ in the general formula:

$$\int uv'\,dx = u(v + C_1) - \int (v + C_1)u'\,dx$$

$$= uv + C_1 u - \int C_1 u'\,dx - \int vu'\,dx$$

$$= uv + C_1 u - C_1 u - \int vu'\,dx = uv - \int vu'\,dx$$

Thus we may drop the first constant of integration when integrating by parts.

Example 2

Evaluate $\int x^2 \ln x\,dx$.

Solution: In this case x^2 is more easily integrated than $\ln x$ and, further-more, the derivative of $\ln x$ is simpler than $\ln x$. Therefore, we choose $v' = x^2$ and

$$v' = x^2 \dashrightarrow u = \ln x$$
$$v = \frac{x^3}{3} \dashleftarrow u' = \frac{1}{x}$$

(By following this *arrow diagram,* you can generate the terms of the inte-gration by parts rule.)

$$\int v'u \, dx = uv - \int vu' \, dx$$

$$\int x^2 \ln x \, dx = \frac{x^3}{3} \ln x - \int \left(\frac{x^3}{3}\right)\left(\frac{1}{x}\right) dx = \frac{x^3}{3} \ln x - \frac{1}{3} \int x^2 \, dx$$

$$= \frac{x^3}{3} \ln x - \frac{x^3}{9} + C \qquad \blacksquare$$

Example 3

Evaluate $\int_0^1 \text{Arcsin } x \, dx$.

Solution: At first it may appear that integration by parts does not apply. However, if we let $v' = 1$, we have

$$v' = 1 \dashrightarrow u = \text{Arcsin } x$$
$$v = x \dashleftarrow u' = \frac{1}{\sqrt{1 - x^2}}$$

Therefore, we have

$$\int_0^1 \text{Arcsin } x \, dx = \left[x \text{ Arcsin } x \right]_0^1 - \int_0^1 \frac{x}{\sqrt{1 - x^2}} dx$$

$$= \left[x \text{ Arcsin } x \right]_0^1 + \frac{1}{2} \int_0^1 (1 - x^2)^{-1/2}(-2x) \, dx$$

$$= \left[x \text{ Arcsin } x + \sqrt{1 - x^2} \right]_0^1 = \frac{\pi}{2} - 1$$

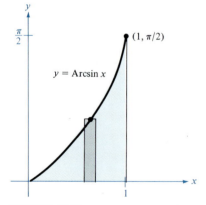

FIGURE 10.19

(See Figure 10.19.) $\blacksquare$

It may happen that a particular integral requires repeated application of integration by parts. This is demonstrated in the next example.

Example 4

Evaluate $\int x^2 \sin x \, dx$.

Solution: We may consider x^2 and $\sin x$ to be equally easy to integrate; however, the derivative of x^2 becomes simpler while the derivative of the sin x does not. Therefore, we let $u = x^2$ and

$$v' = \sin x \dashrightarrow u = x^2$$
$$v = -\cos x \dashleftarrow u' = 2x$$

and it follows that

$$\int x^2 \sin x \, dx = -x^2 \cos x + \int 2x \cos x \, dx$$

Now we apply integration by parts to the new integral and, with the same considerations in mind, we have

$$v' = \cos x \dashrightarrow u = 2x$$
$$v = \sin x \dashleftarrow u' = 2$$

and it follows that

$$\int x^2 \sin x \, dx = -x^2 \cos x + 2x \sin x - \int 2 \sin x \, dx$$

$$= -x^2 \cos x + 2x \sin x + 2 \cos x + C \qquad \blacksquare$$

When making repeated application of integration by parts, we need to be careful not to interchange the substitutions in successive applications. For instance, in Example 4 our first substitutions were

$$v' = \sin x \qquad \text{and} \qquad u = x^2$$

and for the second application of integration by parts, we let

$$v' = \cos x \qquad \text{and} \qquad u = 2x$$

If, in the second application, we had switched our substitutions to

$$v' = 2x \dashrightarrow u = \cos x$$
$$v = x^2 \dashleftarrow u' = -\sin x$$

we would have obtained

$$\int x^2 \sin x \, dx = -x^2 \cos x + \int 2x \cos x \, dx$$

$$= -x^2 \cos x + x^2 \cos x + \int x^2 \sin x \, dx$$

$$= \int x^2 \sin x \, dx$$

which tells us nothing. By switching substitutions we undid the previous integration and therefore returned to our original integral.

There are two things to keep in mind when making repeated application of integration by parts.

Guidelines for Repeated Use of Integration by Parts	1. Be careful *not to switch* the choices for v' and u in successive applications.
	2. After each application watch for the appearance of a *constant multiple* of the original integral.

The latter situation is illustrated in the following example.

Example 5

Evaluate $\int e^x \cos x \, dx$.

Solution: Both e^x and $\cos x$ are easily integrated and neither of their derivatives is simpler than the function itself. Thus our choice for v' is arbitrary and we let $v' = e^x$ and obtain

$$v' = e^x \dashrightarrow u = \cos x$$
$$v = e^x \longleftarrow u' = -\sin x$$

and it follows that

$$\int e^x \cos x \, dx = e^x \cos x + \int e^x \sin x \, dx$$

Making the "same" choices for the next application of integration by parts, we have

$$v' = e^x \dashrightarrow u = \sin x$$
$$v = e^x \longrightarrow u' = \cos x$$

and it follows that

$$\int e^x \cos x \, dx = e^x \cos x + e^x \sin x - \int e^x \cos x \, dx$$

Notice that the negative of our original integral has appeared on the right. By adding this integral to both sides, we obtain

$$2 \int e^x \cos x \, dx = e^x \cos x + e^x \sin x$$

and then dividing by 2 yields the result

$$\int e^x \cos x \, dx = \frac{e^x}{2} \cos x + \frac{e^x}{2} \sin x + C \qquad \blacksquare$$

In Section 10.5 we were unable to evaluate the integral

$$\int \sec^m x \, \tan^n x \, dx$$

where m is odd and n is even. In the next example we demonstrate the integration by parts procedure for evaluating such integrals.

Example 6

Evaluate $\int \sec^3 x \, dx$.

Solution: The most complicated portion of $\sec^3 x$ that can be easily integrated is $\sec^2 x$. Thus we let $v' = \sec^2 x$ and obtain

$$v' = \sec^2 x \dashrightarrow u = \sec x$$
$$v = \tan x \dashleftarrow u' = \sec x \tan x$$

and we obtain

$$\int \sec^3 x \, dx = \sec x \tan x - \int \sec x (\tan^2 x) \, dx$$

$$= \sec x \tan x - \int \sec x (\sec^2 x - 1) \, dx$$

$$= \sec x \tan x - \int \sec^3 x \, dx + \int \sec x \, dx$$

Thus

$$2 \int \sec^3 x \, dx = \sec x \tan x + \int \sec x \, dx$$

and

$$\int \sec^3 x \, dx = \frac{1}{2} \sec x \tan x + \frac{1}{2} \ln |\sec x + \tan x| + C$$

In Section 10.6 we used trigonometric substitution to evaluate certain integrals involving the radicals $\sqrt{u^2 \pm a^2}$, and $\sqrt{a^2 - u^2}$. We also used trigonometric substitution to derive a formula for evaluating $\int \sqrt{a^2 - u^2} \, u' \, dx$. However, we avoided the derivation of formulas for evaluating

$$\int \sqrt{u^2 + a^2} \, u' \, dx \qquad \text{and} \qquad \int \sqrt{u^2 - a^2} \, u' \, dx$$

because they led to integrals involving $\sec^3 \theta$, which we could not then evaluate. With the necessary tools now available, in the next example we show how to derive the integration formulas for two common integrals involving $\sqrt{u^2 + a^2}$ or $\sqrt{u^2 - a^2}$.

Example 7

Evaluate $\int \sqrt{u^2 + a^2} \, u' \, dx$.

Solution: If we let $u = a \tan \theta$, then $du = a \sec^2 \theta \, d\theta$ and $\sqrt{u^2 + a^2} = a \sec \theta$. Thus we have

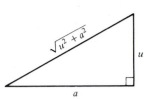

FIGURE 10.20

$$\int \sqrt{u^2 + a^2}\, u'\, dx = \int a^2 \sec^3 \theta\, d\theta$$

which, by the result of Example 6, can be written as

$$\int \sqrt{u^2 + a^2}\, u'\, dx = \frac{a^2}{2} \sec \theta \tan \theta + \frac{a^2}{2} \ln |\sec \theta + \tan \theta| + C_1$$

Figure 10.20 gives us the result

$$\int \sqrt{u^2 + a^2}\, u'\, dx = \left(\frac{a^2}{2}\right)\left(\frac{\sqrt{u^2 + a^2}}{a}\right)\left(\frac{u}{a}\right) + \frac{a^2}{2} \ln \left|\frac{\sqrt{u^2 + a^2}}{a} + \frac{u}{a}\right| + C_1$$

$$= \tfrac{1}{2}[u \sqrt{u^2 + a^2} + a^2 \ln |u + \sqrt{u^2 + a^2}|] + C$$

where $C = C_1 - (a^2/2) \ln |a|$. A similar procedure with $u = a \sec \theta$ can be used to evaluate $\int \sqrt{u^2 - a^2}\, u'\, dx$. ∎

The integrals involved in Examples 6 and 7 will be encountered on occasion in subsequent sections of this text, so for reference purposes we list their integration formulas here.

$$\int \sec^3 u\, u'\, dx = \frac{1}{2}(\sec u \tan u + \ln |\sec u + \tan u|) + C$$

$$\int \sqrt{u^2 \pm a^2}\, u'\, dx = \frac{1}{2}(u \sqrt{u^2 \pm a^2} \pm a^2 \ln |u + \sqrt{u^2 \pm a^2}|) + C$$

The next example requires the application of one of these rules.

Example 8

Find the length of an arc of the graph of $f(x) = 4x - x^2$ from $x = 0$ to $x = 2$. (See Figure 10.21.)

Solution: By the integration formula for arc length s, we have

$$s = \int_0^2 \sqrt{1 + [f'(x)]^2}\, dx = \int_0^2 \sqrt{1 + (4 - 2x)^2}\, dx$$

Considering $u = 4 - 2x$, then $u' = -2$ and we obtain

$$s = \int_0^2 \sqrt{a^2 + u^2}\, u'\, dx = -\frac{1}{2}\int_0^2 \sqrt{1 + (4 - 2x)^2}(-2)\, dx$$

$$= -\frac{1}{4}\left[(4 - 2x)\sqrt{1 + (4 - 2x)^2} + \ln |(4 - 2x) + \sqrt{1 + (4 - 2x)^2}|\right]\Big|_0^2$$

$$= -\frac{1}{4}[0 + \ln 1 - 4\sqrt{17} - \ln (4 + \sqrt{17})]$$

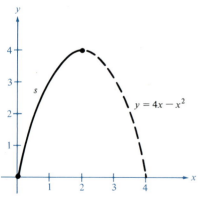

FIGURE 10.21

$$= \sqrt{17} + \frac{1}{4} \ln (4 + \sqrt{17}) \approx 4.65$$

∎

The versatility of integration by parts is further demonstrated in the next example, which involves a product of algebraic functions.

Example 9

Evaluate $\int_0^{\sqrt{3}} x^3 \sqrt{4 - x^2} \, dx$.

Solution: Although we could evaluate this integral by trigonometric substitution, integration by parts is more efficient in this case. The most complicated portion of the integrand that can be easily integrated is $x\sqrt{4 - x^2}$, which fits the Power Rule. Thus we have

$$v' = x\sqrt{4 - x^2} \dashrightarrow u = x^2$$
$$v = -\tfrac{1}{3}(4 - x^2)^{3/2} \dashleftarrow u' = 2x$$

Therefore, we have

$$\int_0^{\sqrt{3}} x^3 \sqrt{4 - x^2} \, dx = \left[\frac{-x^2}{3}(4 - x^2)^{3/2} \right]_0^{\sqrt{3}} - \frac{1}{3} \int_0^{\sqrt{3}} (4 - x^2)^{3/2}(-2x) \, dx$$

The new integral can be evaluated by the Power Rule, and thus

$$\int_0^{\sqrt{3}} x^3 \sqrt{4 - x^2} \, dx = \left[\frac{-x^2}{3}(4 - x^2)^{3/2} - \left(\frac{1}{3}\right)\left(\frac{2}{5}\right)(4 - x^2)^{5/2} \right]_0^{\sqrt{3}}$$

$$= -1 - \frac{2}{15} + 0 + \frac{2}{15}(4)^{5/2} = -\frac{17}{15} + \frac{64}{15} = \frac{47}{15}$$

∎

Example 10

Evaluate $\int \sin x (\ln \cos x) \, dx$.

Solution:
$$v' = \sin x \dashrightarrow u = \ln (\cos x)$$
$$v = -\cos x \dashleftarrow u' = \frac{-\sin x}{\cos x} = -\tan x$$

Therefore,

$$\int \sin x (\ln \cos x) \, dx = -\cos x (\ln \cos x) - \int \sin x \, dx$$

$$= -\cos x (\ln \cos x) + \cos x + C$$

$$= \cos x (1 - \ln \cos x) + C$$

∎

Section Exercises (10.7)

In Exercises 1–29, find the indefinite integral.

1. $\int x \cos x \, dx$

2. $\int x^2 \cos x \, dx$

3. $\int x e^{-x} \, dx$

4. $\int x^3 e^x \, dx$

S 5. $\int \ln x \, dx$

6. $\int x^3 \ln x \, dx$

7. $\int x \sec^2 x \, dx$

8. $\int \theta \sec \theta \tan \theta \, d\theta$

S 9. $\int \text{Arcsec } 2x \, dx$

10. $\int \text{Arccos } x \, dx$

11. $\int \text{Arctan } x \, dx$

S 12. $\int x \text{ Arcsin } x \, dx$

13. $\int \frac{1}{x(\ln x)^3} \, dx$

14. $\int t \ln (t + 1) \, dt$

15. $\int x \sqrt{x - 1} \, dx$

16. $\int x^2 \sqrt{x - 1} \, dx$

17. $\int \frac{x}{\sqrt{2 + 3x}} \, dx$

18. $\int (\ln x)^2 \, dx$

S 19. $\int e^{2x} \sin x \, dx$

20. $\int e^x \cos 2x \, dx$

21. $\int \sec^5 x \, dx$

22. $\int x \sin 2x \, dx$

S 23. $\int x \sin x \cos x \, dx$

24. $\int x^2 \sin^2 x \, dx$

25. $\int x \sin^2 x \, dx$

26. $\int \sqrt{1 + x^2} \, dx$

27. $\int \sqrt{4 + 9x^2} \, dx$

28. $\int x^2 \sqrt{x^2 - 4} \, dx$

S 29. $\int \frac{x^2}{\sqrt{x^2 - 1}} \, dx$

In Exercises 30–34, evaluate the definite integral.

S 30. $\int_0^1 x \text{ Arcsin } x^2 \, dx$

S 31. $\int_0^1 e^x \sin x \, dx$

32. $\int_0^1 x^2 e^x \, dx$

33. $\int_0^\pi x \sin x \, dx$

34. $\int_0^1 \ln (1 + x^2) \, dx$

In Exercises 35–42, use integration by parts to verify each formula.

35. $\int \sin^n x \, dx = -\frac{\sin^{n-1} x \cos x}{n} + \frac{n-1}{n} \int \sin^{n-2} x \, dx$

36. $\int \cos^n x \, dx = \frac{\cos^{n-1} x \sin x}{n} + \frac{n-1}{n} \int \cos^{n-2} x \, dx$

S 37. $\int \cos^m x \sin^n x \, dx = -\frac{\cos^{m+1} x \sin^{n-1} x}{m + n}$
$$+ \frac{n-1}{m + n} \int \cos^m x \sin^{n-2} x \, dx$$

38. $\int x^n \sin x \, dx = -x^n \cos x + n \int x^{n-1} \cos x \, dx$

39. $\int x^n \cos x \, dx = x^n \sin x - n \int x^{n-1} \sin x \, dx$

40. $\int x^n e^{ax} \, dx = \frac{x^n e^{ax}}{a} - \frac{n}{a} \int x^{n-1} e^{ax} \, dx$

S 41. $\int e^{ax} \sin bx \, dx = \frac{e^{ax}(a \sin bx - b \cos bx)}{a^2 + b^2} + C$

42. $\int e^{ax} \cos bx \, dx = \frac{e^{ax}(a \cos bx + b \sin bx)}{a^2 + b^2} + C$

43. Find the length of the arc on the graph of $y = \ln x$ from $x = 1$ to $x = 5$.

44. Find the length of the arc on the graph of $y = x^2$ from $x = 0$ to $x = 3$.

S 45. Find the surface area of the solid generated by revolving the region bounded by $y = x^2$, $y = 0$, $x = 0$, and $x = \sqrt{2}$ about the x-axis.

46. Find the area of the region bounded by $y = \sqrt{4 + x^2}$, $y = 0$, $x = 0$, and $x = 2$.

47. In Example 4 of Section 7.8, the definite integral

$$\int_0^{200} \sqrt{1 + \left(1 - \frac{x}{100}\right)^2} \, dx$$

was set up to give the total distance traveled by a ball that followed the path $y = x - 0.005x^2$. Find that distance.

48. Find the centroid of the region under $y = \sin x$ from $x = 0$ to $x = \pi$.

49. Find the volume of the solid generated by revolving the region bounded by $y = e^x$, $y = 0$, $x = 0$, and $x = 1$ about the y-axis.

50. Find the area of the region bounded by $y = x \sin x$, $y = 0$, $0 \le x \le \pi$.

S 51. A projectile follows the path $y = -(x^2/72) + x$, where

x and y are measured in meters. Find the distance traveled from $x = 0$ to $x = 72$.

52. A damping force effects the vibration of a spring so that the displacement of the spring is given by $y = e^{-4t}(\cos 2t + 5 \sin 2t)$. Find the average value of y on the interval from $t = 0$ to $t = \pi$.

53. Find the fallacy in the following solution:

$$v' = 1 \dashrightarrow u = \frac{1}{x}$$
$$v = x \dashleftarrow u' = -\frac{1}{x^2}$$

$$\int \frac{dx}{x} = \left(\frac{1}{x}\right)(x) - \int \left(-\frac{1}{x^2}\right)(x)\, dx = 1 + \int \frac{dx}{x}$$

Hence $0 = 1$.

54. Is there a fallacy in the following solution of

$$\int \ln(x + 5)\, dx?$$

$$v' = 1 \dashrightarrow u = \ln(x + 5)$$
$$v = x + 5 \dashleftarrow u' = \frac{1}{(x + 5)}$$

$$\int \ln(x + 5)\, dx = (x + 5)\ln(x + 5) - \int dx$$
$$= (x + 5)\ln(x + 5) - x + C$$

55. A string stretched between two points $(0, 0)$ and $(0, 2)$ is plucked by displacing the string h units at its midpoint. The motion of the string is modeled by a *Fourier Sine Series* whose coefficients are given by

$$b_n = h \int_0^1 x \sin \frac{n\pi x}{2}\, dx + h \int_1^2 (-x + 2) \sin \frac{n\pi x}{2}\, dx$$

Evaluate the integral to find b_n.

10.8 Miscellaneous Substitutions

Purpose

- To introduce several types of substitutions that convert radicals, and rational functions of sine and cosine, into rational functions of the variable u.

It seems to be expected of every pilgrim up the slopes of the mathematical Parnassus, that he will at some point or other of his journey sit down and invent a definite integral or two towards the increase of the common stock. —James Sylvester (1814–1897)

In the examples that follow, we demonstrate several special types of substitutions and provide some suggestions for their use.

Example 1

Evaluate

$$\int \frac{dx}{\sqrt{x} - \sqrt[3]{x}}$$

Solution: In this case it is not clear what substitution to make since neither of the substitutions

$$u = \sqrt{x} \quad \text{or} \quad u = \sqrt[3]{x}$$

will eliminate all radicals in the integrand. However, if we consider the "common" root $u = \sqrt[6]{x}$, then

$$\sqrt{x} = x^{1/2} = x^{3/6} = (\sqrt[6]{x})^3 = u^3$$

and

$$\sqrt[3]{x} = x^{1/3} = x^{2/6} = (\sqrt[6]{x})^2 = u^2$$

Now all radicals will be eliminated, and we have

$$u = \sqrt[6]{x}, \qquad u^6 = x, \qquad 6u^5 \, du = dx$$

Therefore,

$$\int \frac{dx}{\sqrt{x} - \sqrt[3]{x}} = \int \frac{6u^5 \, du}{u^3 - u^2} = \int \frac{6u^3 \, du}{u - 1}$$

$$= 6 \int \left(u^2 + u + 1 + \frac{1}{u - 1} \right) du$$

$$= 6 \left(\frac{u^3}{3} + \frac{u^2}{2} + u + \ln|u - 1| \right) + C$$

$$= 2x^{1/2} + 3x^{1/3} + 6x^{1/6} + 6 \ln|x^{1/6} - 1| + C \qquad \blacksquare$$

As shown in Example 1, for integrands containing more than one root of the same quantity, we let u be the root that is the lowest common multiple of the roots involved.

Example 2

Evaluate

$$\int \frac{x^{3/4}}{2(\sqrt{x} + 1)} \, dx$$

Solution: For the two roots $\sqrt[4]{x^3}$ and $\sqrt{x}$, we choose $u = \sqrt[4]{x}$, and hence

$$u^4 = x, \qquad 4u^3 \, du = dx, \qquad u^3 = x^{3/4}, \qquad u^2 = x^{1/2}$$

Therefore,

$$\int \frac{x^{3/4} \, dx}{2(\sqrt{x} + 1)} = \int \frac{u^3 4u^3 \, du}{2(u^2 + 1)} = 2 \int \frac{u^6 \, du}{u^2 + 1}$$

$$= 2 \int \left(u^4 - u^2 + 1 - \frac{1}{u^2 + 1} \right) du$$

$$= 2 \left(\frac{u^5}{5} - \frac{u^3}{3} + u - \text{Arctan } u \right) + C$$

$$= \tfrac{2}{5} x^{5/4} - \tfrac{2}{3} x^{3/4} + 2x^{1/4} - 2 \text{ Arctan } \sqrt[4]{x} + C \qquad \blacksquare$$

Occasionally we encounter rational expressions of $\sin x$ and $\cos x$ that do not fit any of our procedures for integrating trigonometric functions. In such cases there is a substitution that converts the integrand into a rational expression in the variable u. This substitution is

$$u = \frac{\sin x}{1 + \cos x} = \tan \frac{x}{2}$$

Using this substitution it follows that

$$u^2 = \frac{\sin^2 x}{(1 + \cos x)^2} = \frac{1 - \cos^2 x}{(1 + \cos x)^2} = \frac{1 - \cos x}{1 + \cos x}$$

Solving for $\cos x$ in this equation, we have

$$\cos x = \frac{1 - u^2}{1 + u^2}$$

To find $\sin x$ we write

$$u = \frac{\sin x}{1 + \cos x}$$

$$\sin x = u(1 + \cos x) = u\left(1 + \frac{1 - u^2}{1 + u^2}\right) = u\left(\frac{1 + u^2 + 1 - u^2}{1 + u^2}\right)$$

$$= \frac{2u}{1 + u^2}$$

Finally, to find dx we consider $u = \tan (x/2)$. Then $(x/2) = \text{Arctan } u$ and thus

$$dx = \frac{2\ du}{1 + u^2}$$

Substitution for Rational Functions of Sine and Cosine

For integrals involving rational functions of the sine and cosine, use the substitution

$$u = \frac{\sin x}{1 + \cos x} = \tan \frac{x}{2}$$

which implies that

$$\cos x = \frac{1 - u^2}{1 + u^2}, \qquad \sin x = \frac{2u}{1 + u^2}, \qquad dx = \frac{2\ du}{1 + u^2}$$

Example 3

Evaluate

$$\int \frac{dx}{1 + \sin x - \cos x}$$

Solution: Let

$$u = \frac{\sin x}{1 + \cos x}$$

Then

$$\int \frac{dx}{1 + \sin x - \cos x} = \int \frac{2\ du/(1 + u^2)}{1 + [2u/(1 + u^2)] - [(1 - u^2)/(1 + u^2)]}$$

$$= \int \frac{2\ du}{(1 + u^2) + 2u - (1 - u^2)} = \int \frac{2\ du}{2u + 2u^2}$$

$$= \int \frac{du}{u(1 + u)}$$

Using partial fractions we get

$$\int \frac{dx}{1 + \sin x - \cos x} = \int \frac{du}{u(1 + u)} = \int \frac{1}{u} du - \int \frac{1}{1 + u} du$$

$$= \ln |u| - \ln |1 + u| + C = \ln \left| \frac{u}{1 + u} \right| + C$$

$$= \ln \left| \frac{\sin x/(1 + \cos x)}{1 + [\sin x/(1 + \cos x)]} \right| + C$$

$$= \ln \left| \frac{\sin x}{1 + \sin x + \cos x} \right| + C \qquad ■$$

It is important to note that since the six trigonometric functions can be written as rational combinations of the sine and cosine, then the substitution $u = \tan (x/2)$ will work for rational integrands involving any of the six trigonometric functions.

Example 4

Evaluate

$$\int \frac{dx}{\sin x - \tan x}$$

Solution: Since

$$\int \frac{dx}{\sin x - \tan x} = \int \frac{dx}{\sin x - (\sin x/\cos x)} = \int \frac{\cos x \, dx}{\sin x \cos x - \sin x}$$

we use the substitution

$$u = \frac{\sin x}{1 + \cos x}$$

and write

$$\int \frac{dx}{\sin x - \tan x}$$

$$= \int \frac{[(1 - u^2)/(1 + u^2)][2/(1 + u^2)]}{[2u/(1 + u^2)][(1 - u^2)/(1 + u^2)] - [2u/(1 + u^2)]} du$$

$$= \int \frac{2(1 - u^2) \, du}{2u(1 - u^2) - 2u(1 + u^2)} = \int \frac{1 - u^2}{-2u^3} du$$

$$= \frac{-1}{2} \int \left(\frac{1}{u^3} - \frac{1}{u} \right) du = \frac{-1}{2} \left(\frac{-1}{2u^2} - \ln |u| \right) + C$$

$$= \frac{1}{4} \left(\frac{1 + \cos x}{\sin x} \right)^2 + \frac{1}{2} \ln \left| \frac{\sin x}{1 + \cos x} \right| + C \qquad ■$$

The remaining three examples demonstrate additional types of substitutions that are of occasional use.

Example 5

Evaluate

$$\int_1^{2\sqrt{2}} \frac{x^3\,dx}{\sqrt[3]{9-x^2}}$$

Solution: If we let $u = \sqrt[3]{9-x^2} = (9-x^2)^{1/3}$

then $u^3 = 9 - x^2,$ $x^2 = 9 - u^3,$ $2x\,dx = -3u^2\,du$

Furthermore, when $x = 1$, $u = 2$, and when $x = 2\sqrt{2}$, $u = 1$. Therefore,

$$\int_1^{2\sqrt{2}} \frac{x^3\,dx}{\sqrt[3]{9-x^2}} = \int_1^{2\sqrt{2}} \frac{x^2(x\,dx)}{\sqrt[3]{9-x^2}} = \int_2^1 \frac{(9-u^3)(-\tfrac{3}{2}u^2\,du)}{u}$$

$$= -\frac{3}{2}\int_2^1 (9u - u^4)\,du = -\frac{3}{2}\left[\frac{9u^2}{2} - \frac{u^5}{5}\right]_2^1$$

$$= -\frac{3}{2}\left[\frac{9}{2} - \frac{1}{5} - \left(18 - \frac{32}{5}\right)\right]$$

$$= -\frac{3}{2}\left(-\frac{27}{2} + \frac{31}{5}\right) = \frac{219}{20}$$ ∎

Example 6

Evaluate

$$\int \frac{dx}{x^2\sqrt{x^2+2x}}$$

Solution: First, let $u = \dfrac{1}{x}$

Then $x = \dfrac{1}{u},$ $dx = -\dfrac{1}{u^2}\,du,$ $x^2 = \dfrac{1}{u^2}$

Therefore,

$$\int \frac{dx}{x^2\sqrt{x^2+2x}} = \int \frac{(-du/u^2)}{(1/u^2)\sqrt{(1/u^2)+(2/u)}}$$

$$= -\int \frac{du}{\sqrt{(1+2u)/u^2}} = -\int \frac{u\,du}{\sqrt{1+2u}}$$

Now for the square root of a linear expression, we let $z = \sqrt{1+2u}$, and then

$$z^2 = 1 + 2u,\qquad \frac{z^2-1}{2} = u,\qquad z\,dz = du$$

Therefore,

$$\int \frac{dx}{x^2\sqrt{x^2+2x}} = -\int \frac{u\,du}{\sqrt{1+2u}} = -\int \frac{[(z^2-1)/2]z\,dz}{z}$$

$$= \frac{1}{2}\int(1-z^2)\,dz = \frac{1}{2}z - \frac{1}{6}z^3 + C$$

Resubstituting for z gives us

$$\int \frac{dx}{x^2\sqrt{x^2+2x}} = \frac{1}{2}\sqrt{1+2u} - \frac{1}{6}(\sqrt{1+2u})^3 + C$$
$$= \frac{1}{6}\sqrt{1+2u}[3-(1+2u)] + C$$
$$= \frac{1}{3}\sqrt{1+2u}(1-u) + C$$

And finally, resubstituting for u we get

$$\int \frac{dx}{x^2\sqrt{x^2+2x}} = \frac{1}{3}\sqrt{1+\frac{2}{x}}\left(1-\frac{1}{x}\right) + C$$

$$= \frac{x-1}{3x}\sqrt{\frac{x+2}{x}} + C = \frac{x-1}{3x^2}\sqrt{x^2+2x} + C \qquad ■$$

Example 7

Evaluate

$$\int \frac{dx}{\sqrt{1-e^x}}$$

Solution: Let $u = \sqrt{1-e^x}$, and then

$$u^2 = 1 - e^x, \qquad u^2 - 1 = -e^x, \qquad 2u\,du = -e^x\,dx$$

Therefore,

$$dx = \frac{2u\,du}{-e^x} = \frac{2u\,du}{u^2-1}$$

and

$$\int \frac{dx}{\sqrt{1-e^x}} = \int \frac{[2u/(u^2-1)]\,du}{u} = \int \frac{2\,du}{u^2-1}$$

By Formula 17 of Section 10.1, it follows that

$$\int \frac{dx}{\sqrt{1-e^x}} = 2\left(\frac{1}{2}\right)\ln\left|\frac{u-1}{u+1}\right| + C = \ln\left|\frac{\sqrt{1-e^x}-1}{\sqrt{1-e^x}+1}\right| + C \qquad ■$$

Section Exercises (10.8)

In Exercises 1–28, evaluate the given integral.

1. $\displaystyle\int \frac{1}{x\sqrt{1-x}}\,dx$

2. $\displaystyle\int_0^7 \frac{x}{\sqrt[3]{x+1}}\,dx$

3. $\displaystyle\int_2^4 \frac{1}{x\sqrt{x-1}}\,dx$

4. $\displaystyle\int \frac{\sqrt{x+1}}{x}\,dx$

S 5. $\int (\sqrt{x} + \sqrt[3]{x})^{-1} \, dx$

6. $\int \dfrac{1}{(\sqrt{x} + \sqrt[3]{x})^2} \, dx$

7. $\int \dfrac{\sqrt{x^2 - 4}}{x} \, dx$

8. $\int_{4}^{8} \dfrac{\sqrt{x - 4}}{x} \, dx$

9. $\int_{-1}^{5/2} \dfrac{x}{(2x + 3)^{2/3}} \, dx$

10. $\int \dfrac{1}{\sqrt{2x}(1 + \sqrt{2x})} \, dx$

S 11. $\int \dfrac{1}{1 + \sqrt{x - 1}} \, dx$

12. $\int \dfrac{(2x - 1)^{2/3} - (2x - 1)^{1/3} + 1}{x} \, dx$

13. $\int \dfrac{1}{x\sqrt{x^2 + 4x}} \, dx$

14. $\int \dfrac{\sqrt{x - 1} - 1}{\sqrt{x - 1} + 1} \, dx$

S 15. $\int (x^{1/2} + x^{3/4})^{-2} \, dx$

S 16. $\int x^3 \sqrt[3]{x^2 + 1} \, dx$

S 17. $\int \dfrac{1}{x\sqrt{x^2 + x + 1}} \, dx$

18. $\int \dfrac{\sqrt{4x - x^2}}{x^3} \, dx$

19. $\int \dfrac{\sqrt{4 - x^2}}{x^3} \, dx$

20. $\int \dfrac{x - 3}{\sqrt{2x - 1}} \, dx$

S 21. $\int_{0}^{\pi/2} \dfrac{1}{1 + \sin\theta + \cos\theta} \, d\theta$

22. $\int_{0}^{\pi/2} \dfrac{1}{3 - 2\cos\theta} \, d\theta$

23. $\int \dfrac{\sin\theta}{3 - 2\cos\theta} \, d\theta$

24. $\int \dfrac{\sin\theta}{1 + \sin\theta} \, d\theta$

25. $\int \dfrac{\cos\sqrt{\theta}}{\sqrt{\theta}} \, d\theta$

26. $\int \dfrac{\sin\theta}{(\cos\theta)(1 + \sin\theta)} \, d\theta$

27. $\int \dfrac{1}{\sin\theta \tan\theta} \, d\theta$

28. $\int \dfrac{1}{\sec\theta - \tan\theta} \, d\theta$

29. Verify that the following two antiderivatives are equivalent:

$$\int \sec\theta \, d\theta = \ln|\sec\theta + \tan\theta| + C$$

$$\int \sec\theta \, d\theta = \ln\left|\tan\left(\frac{\pi}{4} + \frac{\theta}{2}\right)\right| + C$$

30. Find the area of the region enclosed by $y^2 = x^4(1 + x)$.

31. For the region bounded by $y = x/\sqrt[3]{x + 1}$, $y = 0$, $x = 0$, and $x = 7$, find the following:
(a) its area
(b) the volume of the solid generated by revolving the region about the *x*-axis
(c) the volume of the solid generated by revolving the region about the *y*-axis
(d) its centroid

10.9 Summary and Integration by Tables

Purpose
- To identify some basic forms to which each of the various integrating techniques apply.
- To introduce the technique of integration by tables.
- To derive and demonstrate the use of reduction formulas.

These are the elementary principles of the differential and summatory calculus, by means of which one can deal with highly complicated formulas.
—*Gottfried Leibniz (1646–1716)*

So far in this chapter, we have discussed a number of integrating techniques to use along with the fundamental integration formulas. Certainly we have not considered every possible method for finding an antiderivative, but we have considered the most important ones.

By now you probably have discovered that merely knowing *how* to use the various integrating techniques is not enough; you also need to know *when* to use them. Integration is first and foremost a problem of recognition—recognizing which formula or technique to apply to obtain an antiderivative. Frequently the slightest alteration of an integrand will require the use of a different integration technique. For example, consider the integrals

$$\int x \ln x \, dx, \qquad \int \frac{\ln x}{x} \, dx, \qquad \int \frac{dx}{x \ln x}$$

whose antiderivatives are obtained in the manner shown in Table 10.2.

TABLE 10.2

Integral	Technique	Antiderivative
$\int x \ln x \, dx$	Integration by parts	$\dfrac{x^2}{2} \ln x - \dfrac{x^2}{4} + C$
$\int \dfrac{\ln x}{x} \, dx$	$\int u^n u' \, dx$	$\dfrac{(\ln x)^2}{2} + C$
$\int \dfrac{1}{x \ln x} \, dx$	$\int \dfrac{u'}{u} \, dx$	$\ln \lvert \ln x \rvert + C$

To assist you in identifying which technique to use, we include in this section a list of 91 common types of integrals, together with their antiderivatives. Each integral is classified according to the following basic forms of the integrands:

Forms involving u^n, $a + bu$, $a + bu + cu^2$, $\sqrt{a + bu}$, $(a^2 \pm u^2)$, $\sqrt{u^2 \pm a^2}$, $\sqrt{a^2 - u^2}$.

Forms involving $\sin u$ and/or $\cos u$, $\tan u$, $\cot u$, $\sec u$, $\csc u$, or the inverse trigonometric functions.

Forms involving e^u and $\ln u$.

Each of the integration formulas is further identified by the integration technique used for its derivation. We use the code shown in Table 10.3 for identifying the integration technique used in each case. For example, Formula 4 is listed in the integration table as

$$(\text{PF}) \quad 4. \quad \int \frac{uu'}{(a + bu)^2} \, dx = \frac{1}{b^2} \left(\frac{a}{a + bu} + \ln \lvert a + bu \rvert \right) + C$$

TABLE 10.3

Code	Technique	Section
F1, F2, ..., F19	The 19 Fundamental Formulas	10.1
CS	Completing the Square	10.2
SR	Substitution to Remove Radicals	10.3
PF	Partial Fractions	10.4
TI	Trigonometric Integrals	10.5
TS	Trigonometric Substitution	10.6
IP	Integration by Parts	10.7
MS	Miscellaneous Substitutions	10.8

Separating the integrand into its partial fractions gives us

$$\frac{u}{(a + bu)^2} = \frac{1}{b(a + bu)} - \frac{a}{b(a + bu)^2}$$

Therefore,

$$\int \frac{uu'}{(a + bu)^2}\,dx = \int \frac{u'}{b(a + bu)}\,dx - \int \frac{au'}{b(a + bu)^2}\,dx$$

$$= \frac{1}{b^2}\int \frac{bu'}{a + bu}\,dx - \frac{a}{b^2}\int \frac{bu'}{(a + bu)^2}\,dx$$

$$= \left(\frac{1}{b^2}\right)\ln|a + bu| - \left(\frac{a}{b^2}\right)\left(\frac{-1}{a + bu}\right)$$

and it follows that

$$\int \frac{uu'}{(a + bu)^2}\,dx = \frac{1}{b^2}\left(\frac{a}{a + bu} + \ln|a + bu|\right) + C$$

We encourage you to derive some of the remaining formulas. In so doing, you will discover that some rather tricky substitutions are necessary to apply the indicated techniques. Formula 19 is a case in point.

(IP)   19. $\displaystyle \int \frac{\sqrt{a + bu}}{u}\,u'\,dx = 2\sqrt{a + bu} + a\int \frac{u'}{u\sqrt{a + bu}}\,dx$

To develop this formula using integration by parts, we first write

$$\int \frac{\sqrt{a + bu}}{u}\,u'\,dx = \int \frac{u\sqrt{a + bu}}{u^2}\,u'\,dx$$

Then we let $w = u\sqrt{a + bu}$ and $v' = 1/u^2$. After finding w' and v, and then applying integration by parts to wv', we obtain the form

$$\int \frac{\sqrt{a + bu}}{u}\,u'\,dx = u\sqrt{a + bu}\left(-\frac{1}{u}\right)$$

$$- \int \left(\frac{3bu}{2\sqrt{a + bu}} + \frac{a}{\sqrt{a + bu}}\right)\left(-\frac{1}{u}\right)u'\,dx$$

$$= -\sqrt{a + bu} + \frac{3}{2}\int \frac{bu'\,dx}{\sqrt{a + bu}} + a\int \frac{u'}{u\sqrt{a + bu}}\,dx$$

$$= -\sqrt{a + bu} + 3\sqrt{a + bu} + a\int \frac{u'}{u\sqrt{a + bu}}\,dx$$

Therefore,

$$\int \frac{\sqrt{a + bu}}{u}\,u'\,dx = 2\sqrt{a + bu} + a\int \frac{u'}{u\sqrt{a + bu}}\,dx$$

Note that the integral on the right can be evaluated directly by Formula 17.

INTEGRATION TABLES

Forms Involving u^n

(F1)   1. $\displaystyle\int u^n u' \, dx = \frac{u^{n+1}}{n+1} + C, \; n \neq -1$

(F3)   2. $\displaystyle\int \frac{u'}{u} \, dx = \ln |u| + C$

Forms Involving $a + bu$

(PF)   3. $\displaystyle\int \frac{uu'}{a+bu} \, dx = \frac{1}{b^2}(bu - a \ln |a + bu|) + C$

(PF)   4. $\displaystyle\int \frac{uu'}{(a+bu)^2} \, dx = \frac{1}{b^2}\left(\frac{a}{a+bu} + \ln |a + bu|\right) + C$

(PF)   5. $\displaystyle\int \frac{uu'}{(a+bu)^n} \, dx = \frac{1}{b^2}\left[\frac{-1}{(n-2)(a+bu)^{n-2}} + \frac{a}{(n-1)(a+bu)^{n-1}}\right] + C, \; n \neq 1, 2$

(PF)   6. $\displaystyle\int \frac{u^2 u'}{a+bu} \, dx = \frac{1}{b^3}\left(-\frac{bu}{2}(2a - bu) + a^2 \ln |a + bu|\right) + C$

(PF)   7. $\displaystyle\int \frac{u^2 u'}{(a+bu)^2} \, dx = \frac{1}{b^3}\left(bu - \frac{a^2}{a+bu} - 2a \ln |a + bu|\right) + C$

(PF)   8. $\displaystyle\int \frac{u^2 u'}{(a+bu)^3} \, dx = \frac{1}{b^3}\left[\frac{2a}{a+bu} - \frac{a^2}{2(a+bu)^2} + \ln |a + bu|\right] + C$

(PF)   9. $\displaystyle\int \frac{u^2 u'}{(a+bu)^n} \, dx = \frac{1}{b^3}\left[\frac{-1}{(n-3)(a+bu)^{n-3}} + \frac{2a}{(n-2)(a+bu)^{n-2}} - \frac{a^2}{(n-1)(a+bu)^{n-1}}\right] + C, \; n \neq 1, 2, 3$

(PF)   10. $\displaystyle\int \frac{u'}{u(a+bu)} \, dx = \frac{1}{a} \ln \left|\frac{u}{a+bu}\right| + C$

(PF)   11. $\displaystyle\int \frac{u'}{u(a+bu)^2} \, dx = \frac{1}{a}\left(\frac{1}{a+bu} + \frac{1}{a} \ln \left|\frac{u}{a+bu}\right|\right) + C$

(PF)   12. $\displaystyle\int \frac{u'}{u^2(a+bu)} \, dx = -\frac{1}{a}\left(\frac{1}{u} + \frac{b}{a} \ln \left|\frac{u}{a+bu}\right|\right) + C$

(PF)   13. $\displaystyle\int \frac{u'}{u^2(a+bu)^2} \, dx = -\frac{1}{a^2}\left[\frac{a+2bu}{u(a+bu)} + \frac{2b}{a} \ln \left|\frac{u}{a+bu}\right|\right] + C$

Forms Involving $a + bu + cu^2, \; b^2 \neq 4ac$

(CS)   14. $\displaystyle\int \frac{u'}{a+bu+cu^2} \, dx = \frac{2}{\sqrt{4ac-b^2}} \, \text{Arctan} \, \frac{2cu+b}{\sqrt{4ac-b^2}} + C, \; b^2 < 4ac$

$\displaystyle\qquad = \frac{1}{\sqrt{b^2-4ac}} \ln \left|\frac{2cu+b-\sqrt{b^2-4ac}}{2cu+b+\sqrt{b^2-4ac}}\right| + C, \; b^2 > 4ac$

(CS)   15. $\displaystyle\int \frac{uu'}{a+bu+cu^2} \, dx = \frac{1}{2c}\left(\ln |a+bu+cu^2| - b \int \frac{u'}{a+bu+cu^2} \, dx\right)$

INTEGRATION TABLES (cont.)

Forms Involving $\sqrt{a + bu}$

(IP) 16. $\displaystyle \int u^n \sqrt{a + bu}\, u'\, dx = \frac{2}{b(2n + 3)} \left[u^n(a + bu)^{3/2} - na \int u^{n-1} \sqrt{a + bu}\, u'\, dx \right]$

(SR) 17. $\displaystyle \int \frac{u'}{u \sqrt{a + bu}}\, dx = \frac{1}{\sqrt{a}} \ln \left| \frac{\sqrt{a + bu} - \sqrt{a}}{\sqrt{a + bu} + \sqrt{a}} \right| + C,\ 0 < a$

$\displaystyle \qquad\qquad = \frac{2}{\sqrt{-a}} \operatorname{Arctan} \sqrt{\frac{a + bu}{-a}} + C,\ a < 0$

(IP) 18. $\displaystyle \int \frac{u'}{u^n \sqrt{a + bu}}\, dx = \frac{-1}{a(n - 1)} \left[\frac{\sqrt{a + bu}}{u^{n-1}} + \frac{(2n - 3)b}{2} \int \frac{u'}{u^{n-1} \sqrt{a + bu}}\, dx \right],\ n \neq 1$

(IP) 19. $\displaystyle \int \frac{\sqrt{a + bu}}{u}\, u'\, dx = 2\sqrt{a + bu} + a \int \frac{u'}{u \sqrt{a + bu}}\, dx$

(IP) 20. $\displaystyle \int \frac{\sqrt{a + bu}}{u^n}\, u'\, dx = \frac{-1}{a(n - 1)} \left[\frac{(a + bu)^{3/2}}{u^{n-1}} + \frac{(2n - 5)b}{2} \int \frac{\sqrt{a + bu}}{u^{n-1}}\, u'\, dx \right],\ n \neq 1$

(SR) 21. $\displaystyle \int \frac{uu'}{\sqrt{a + bu}}\, dx = \frac{-2(2a - bu)}{3b^2} \sqrt{a + bu} + C$

(IP) 22. $\displaystyle \int \frac{u^n u'}{\sqrt{a + bu}}\, dx = \frac{2}{(2n + 1)b} \left(u^n \sqrt{a + bu} - na \int \frac{u^{n-1}}{\sqrt{a + bu}}\, u'\, dx \right)$

Forms Involving $a^2 \pm u^2,\ 0 < a$

(F16) 23. $\displaystyle \int \frac{u'}{a^2 + u^2}\, dx = \frac{1}{a} \operatorname{Arctan} \frac{u}{a} + C$

(F17) 24. $\displaystyle \int \frac{u'}{u^2 - a^2}\, dx = -\int \frac{u'}{a^2 - u^2}\, dx = \frac{1}{2a} \ln \left| \frac{u - a}{u + a} \right| + C$

(IP) 25. $\displaystyle \int \frac{u'}{(a^2 \pm u^2)^n}\, dx = \frac{1}{2a^2(n - 1)} \left[\frac{u}{(a^2 \pm u^2)^{n-1}} + (2n - 3) \int \frac{u'}{(a^2 \pm u^2)^{n-1}}\, dx \right],\ n \neq 1$

Forms Involving $\sqrt{u^2 \pm a^2},\ 0 < a$

(TS) 26. $\displaystyle \int \sqrt{u^2 \pm a^2}\, u'\, dx = \frac{1}{2} \left(u \sqrt{u^2 \pm a^2} \pm a^2 \ln |u + \sqrt{u^2 \pm a^2}| \right) + C$

(TS) 27. $\displaystyle \int u^2 \sqrt{u^2 \pm a^2}\, u'\, dx = \frac{1}{8} \left[u(2u^2 \pm a^2) \sqrt{u^2 \pm a^2} - a^4 \ln |u + \sqrt{u^2 \pm a^2}| \right] + C$

(TS) 28. $\displaystyle \int \frac{\sqrt{u^2 + a^2}}{u}\, u'\, dx = \sqrt{u^2 + a^2} - a \ln \left| \frac{a + \sqrt{u^2 + a^2}}{u} \right| + C$

(TS) 29. $\displaystyle \int \frac{\sqrt{u^2 - a^2}}{u}\, u'\, dx = \sqrt{u^2 - a^2} - a \operatorname{Arcsec} \frac{|u|}{a} + C$

(TS) 30. $\displaystyle \int \frac{\sqrt{u^2 \pm a^2}}{u^2}\, u'\, dx = \frac{-\sqrt{u^2 \pm a^2}}{u} + \ln |u + \sqrt{u^2 \pm a^2}| + C$

INTEGRATION TABLES (cont.)

Forms Involving $\sqrt{u^2 \pm a^2}$, $0 < a$

(F15) 31. $\displaystyle\int \frac{u'}{\sqrt{u^2 \pm a^2}}\,dx = \ln|u + \sqrt{u^2 \pm a^2}| + C$

(F19) 32. $\displaystyle\int \frac{u'}{u\sqrt{u^2 + a^2}}\,dx = \frac{-1}{a}\ln\left|\frac{a + \sqrt{u^2 + a^2}}{u}\right| + C$

(F18) 33. $\displaystyle\int \frac{u'}{u\sqrt{u^2 - a^2}}\,dx = \frac{1}{a}\operatorname{Arcsec}\frac{|u|}{a} + C$

(TS) 34. $\displaystyle\int \frac{u^2 u'}{\sqrt{u^2 \pm a^2}}\,dx = \frac{1}{2}\left(u\sqrt{u^2 \pm a^2} \mp a^2\ln|u + \sqrt{u^2 \pm a^2}|\right) + C$

(TS) 35. $\displaystyle\int \frac{u'}{u^2\sqrt{u^2 \pm a^2}}\,dx = \mp\frac{\sqrt{u^2 \pm a^2}}{a^2 u} + C$

(TS) 36. $\displaystyle\int \frac{u'}{(u^2 \pm a^2)^{3/2}}\,dx = \frac{\pm u}{a^2\sqrt{u^2 \pm a^2}} + C$

Forms Involving $\sqrt{a^2 - u^2}$, $0 < a$

(TS) 37. $\displaystyle\int \sqrt{a^2 - u^2}\,u'\,dx = \frac{1}{2}\left(u\sqrt{a^2 - u^2} + a^2\operatorname{Arcsin}\frac{u}{a}\right) + C$

(TS) 38. $\displaystyle\int u^2\sqrt{a^2 - u^2}\,u'\,dx = \frac{1}{8}\left[u(2u^2 - a^2)\sqrt{a^2 - u^2} + a^4\operatorname{Arcsin}\frac{u}{a}\right] + C$

(TS) 39. $\displaystyle\int \frac{\sqrt{a^2 - u^2}}{u}\,u'\,dx = \sqrt{a^2 - u^2} - a\ln\left|\frac{a + \sqrt{a^2 - u^2}}{u}\right| + C$

(TS) 40. $\displaystyle\int \frac{\sqrt{a^2 - u^2}}{u^2}\,u'\,dx = \frac{-\sqrt{a^2 - u^2}}{u} - \operatorname{Arcsin}\frac{u}{a} + C$

(F14) 41. $\displaystyle\int \frac{u'}{\sqrt{a^2 - u^2}}\,dx = \operatorname{Arcsin}\frac{u}{a} + C$

(F19) 42. $\displaystyle\int \frac{u'}{u\sqrt{a^2 - u^2}}\,dx = \frac{-1}{a}\ln\left|\frac{a + \sqrt{a^2 - u^2}}{u}\right| + C$

(TS) 43. $\displaystyle\int \frac{u^2 u'}{\sqrt{a^2 - u^2}}\,dx = \frac{1}{2}\left(-u\sqrt{a^2 - u^2} + a^2\operatorname{Arcsin}\frac{u}{a}\right) + C$

(TS) 44. $\displaystyle\int \frac{u'}{u^2\sqrt{a^2 - u^2}}\,dx = \frac{-\sqrt{a^2 - u^2}}{a^2 u} + C$

(TS) 45. $\displaystyle\int \frac{u'}{(a^2 - u^2)^{3/2}}\,dx = \frac{u}{a^2\sqrt{a^2 - u^2}} + C$

INTEGRATION TABLES (cont.)

Forms Involving sin *u* or cos *u*

(F4)   46. $\int (\sin u) u' \, dx = -\cos u + C$

(F5)   47. $\int (\cos u) u' \, dx = \sin u + C$

(TI)   48. $\int (\sin^2 u) u' \, dx = \frac{1}{2}(u - \sin u \cos u) + C$

(TI)   49. $\int (\cos^2 u) u' \, dx = \frac{1}{2}(u + \sin u \cos u) + C$

(IP)   50. $\int (\sin^n u) u' \, dx = -\frac{\sin^{n-1} u \cos u}{n} + \frac{n-1}{n} \int (\sin^{n-2} u) u' \, dx$

(IP)   51. $\int (\cos^n u) u' \, dx = \frac{\cos^{n-1} u \sin u}{n} + \frac{n-1}{n} \int (\cos^{n-2} u) u' \, dx$

(IP)   52. $\int u(\sin u) u' \, dx = \sin u - u \cos u + C$

(IP)   53. $\int u(\cos u) u' \, dx = \cos u + u \sin u + C$

(IP)   54. $\int u^n(\sin u) u' \, dx = -u^n \cos u + n \int u^{n-1}(\cos u) u' \, dx$

(IP)   55. $\int u^n(\cos u) u' \, dx = u^n \sin u - n \int u^{n-1}(\sin u) u' \, dx$

(TI)   56. $\int \frac{u'}{1 \pm \sin u} \, dx = \tan u \mp \sec u + C$

(TI)   57. $\int \frac{u'}{1 \pm \cos u} \, dx = -\cot u \pm \csc u + C$

(TI)   58. $\int \frac{u'}{\sin u \cos u} \, dx = \ln |\tan u| + C$

Forms Involving tan *u*, cot *u*, sec *u*, csc *u*

(F10)   59. $\int (\tan u) u' \, dx = -\ln |\cos u| + C$

(F11)   60. $\int (\cot u) u' \, dx = \ln |\sin u| + C$

(F12)   61. $\int (\sec u) u' \, dx = \ln |\sec u + \tan u| + C$

(F13)   62. $\int (\csc u) u' \, dx = \ln |\csc u - \cot u| + C$

INTEGRATION TABLES (cont.)

Forms Involving tan *u*, cot *u*, sec *u*, csc *u*

(TI)   63. $\displaystyle\int (\tan^2 u) u'\, dx = -u + \tan u + C$

(TI)   64. $\displaystyle\int (\cot^2 u) u'\, dx = -u - \cot u + C$

(F6)   65. $\displaystyle\int (\sec^2 u) u'\, dx = \tan u + C$

(F7)   66. $\displaystyle\int (\csc^2 u) u'\, dx = -\cot u + C$

(TI)   67. $\displaystyle\int (\tan^n u) u'\, dx = \frac{\tan^{n-1} u}{n-1} - \int (\tan^{n-2} u) u'\, dx, \; n \neq 1$

(TI)   68. $\displaystyle\int (\cot^n u) u'\, dx = -\frac{\cot^{n-1} u}{n-1} - \int (\cot^{n-2} u) u'\, dx, \; n \neq 1$

(IP)   69. $\displaystyle\int (\sec^n u) u'\, dx = \frac{\sec^{n-2} u \tan u}{n-1} + \frac{n-2}{n-1} \int (\sec^{n-2} u) u'\, dx, \; n \neq 1$

(IP)   70. $\displaystyle\int (\csc^n u) u'\, dx = -\frac{\csc^{n-2} u \cot u}{n-1} + \frac{n-2}{n-1} \int (\csc^{n-2} u) u'\, dx, \; n \neq 1$

(TI)   71. $\displaystyle\int \frac{u'}{1 \pm \tan u}\, dx = \frac{1}{2}(u \pm \ln |\cos u \pm \sin u|) + C$

(TI)   72. $\displaystyle\int \frac{u'}{1 \pm \cot u}\, dx = \frac{1}{2}(u \mp \ln |\sin u \pm \cos u|) + C$

(MS)   73. $\displaystyle\int \frac{u'}{1 \pm \sec u}\, dx = u + \cot u \mp \csc u + C$

(MS)   74. $\displaystyle\int \frac{u'}{1 \pm \csc u}\, dx = u - \tan u \pm \sec u + C$

Forms Involving Inverse Trigonometric Functions

(IP)   75. $\displaystyle\int (\text{Arcsin } u) u'\, dx = u \text{ Arcsin } u + \sqrt{1 - u^2} + C$

(IP)   76. $\displaystyle\int (\text{Arccos } u) u'\, dx = u \text{ Arccos } u - \sqrt{1 - u^2} + C$

(IP)   77. $\displaystyle\int (\text{Arctan } u) u'\, dx = u \text{ Arctan } u - \ln \sqrt{1 + u^2} + C$

(IP)   78. $\displaystyle\int (\text{Arccot } u) u'\, dx = u \text{ Arccot } u + \ln \sqrt{1 + u^2} + C$

(IP)   79. $\displaystyle\int (\text{Arcsec } u) u'\, dx = u \text{ Arcsec } u - \ln |u + \sqrt{u^2 - 1}| + C$

(IP)   80. $\displaystyle\int (\text{Arccsc } u) u'\, dx = u \text{ Arccsc } u + \ln |u + \sqrt{u^2 - 1}| + C$

INTEGRATION TABLES (cont.)

Forms Involving e^u

(F2)   81. $\displaystyle\int e^u u' \, dx = e^u + C$

(IP)   82. $\displaystyle\int u e^u u' \, dx = (u - 1)e^u + C$

(IP)   83. $\displaystyle\int u^n e^u u' \, dx = u^n e^u - n \int u^{n-1} e^u u' \, dx$

(MS)   84. $\displaystyle\int \frac{u'}{1 + e^u} \, dx = u - \ln(1 + e^u) + C$

(IP)   85. $\displaystyle\int e^{au}(\sin bu)u' \, dx = \frac{e^{au}}{a^2 + b^2}(a \sin bu - b \cos bu) + C$

(IP)   86. $\displaystyle\int e^{au}(\cos bu)u' \, dx = \frac{e^{au}}{a^2 + b^2}(a \cos bu + b \sin bu) + C$

Forms Involving $\ln u$

(IP)   87. $\displaystyle\int (\ln u)u' \, dx = u(-1 + \ln u) + C$

(IP)   88. $\displaystyle\int u(\ln u)u' \, dx = \frac{u^2}{4}(-1 + 2 \ln u) + C$

(IP)   89. $\displaystyle\int u^n(\ln u)u' \, dx = \frac{u^{n+1}}{(n + 1)^2}[-1 + (n + 1)\ln u] + C, \; n \neq -1$

(IP)   90. $\displaystyle\int (\ln u)^2 u' \, dx = u[2 - 2 \ln u + (\ln u)^2] + C$

(IP)   91. $\displaystyle\int (\ln u)^n u' \, dx = u(\ln u)^n - n \int (\ln u)^{n-1} u' \, dx$

We have now expanded our list of 19 basic formulas to 91. As we add new integration formulas to our basic list, two things occur. On the one hand, it becomes increasingly more difficult to memorize, or even become familiar with, the entire list of formulas. On the other hand, with a longer list we will likely need fewer techniques for fitting an integral to one of the formulas on the list. We call the procedure of integrating by means of a long unmemorized list of formulas **integration by tables.**

Integration by tables is not to be considered a trivial task. It requires considerable thought and insight, and it often involves a substitution procedure. Many persons find a table of integrals to be a valuable supplement to the integration techniques discussed in previous sections of this chapter. As you continue to improve in the use of the various integrating techniques, we encourage you to gain competence in the use of a table of integrals as well. In doing so you should find that a combination of techniques and tables is the most versatile approach to integration.

In the remaining examples of this section, we demonstrate the use of our table of integrals to find antiderivatives.

Example 1

Use the table of integrals to evaluate

$$\int \frac{dx}{x\sqrt{x-1}}$$

Solution: Since the expression under the radical is linear, we consider forms involving $\sqrt{a+bu}$. In this case $a = -1, b = 1$, and $u = x$, and we choose Formula 17:

$$\int \frac{u'}{u\sqrt{a+bu}} dx = \frac{2}{\sqrt{-a}} \text{Arctan} \sqrt{\frac{a+bu}{-a}} + C$$

which yields

$$\int \frac{dx}{x\sqrt{x-1}} = 2 \text{Arctan} \sqrt{x-1} + C$$
∎

Example 2

Use the table of integrals to evaluate $\int x\sqrt{x^4 - 9}\, dx$.

Solution: Since the radical has the form $\sqrt{u^2 - a^2}$ with $u^2 = x^4$ and $a^2 = 9$, we consider Formula 26:

$$\int \sqrt{u^2 - a^2}\, u'\, dx = \frac{1}{2}(u\sqrt{u^2 - a^2} - a^2 \ln|u + \sqrt{u^2 - a^2}|) + C$$

Then we have $u = x^2$, $u' = 2x$, $a = 3$, and we write

$$\int x\sqrt{x^4 - 9}\, dx = \frac{1}{2} \int \sqrt{(x^2)^2 - (3)^2}(2)x\, dx$$

$$= \tfrac{1}{4}(x^2\sqrt{x^4 - 9} - 9\ln|x^2 + \sqrt{x^4 - 9}|) + C$$
∎

Example 3

Evaluate

$$\int \frac{\sin 2x}{2 + \cos x}\, dx$$

Solution: Substituting $2\sin x \cos x$ for $\sin 2x$, we have

$$\int \frac{\sin 2x}{2 + \cos x}\, dx = 2 \int \frac{\sin x \cos x}{2 + \cos x}\, dx$$

A quick check of the forms involving $\sin u$ or $\cos u$ shows that none of those listed apply. Therefore, we consider a form involving $a + bu$, and let $a + bu = 2 + \cos x$. Then we have $u = \cos x$, $a = 2$, $b = 1$, and

$u' = -\sin x$. Now by Formula 3,

$$\int \frac{uu'}{a + bu} \, dx = \frac{1}{b^2}(bu - a \ln |a + bu|) + C$$

it follows that

$$2 \int \frac{\sin x \cos x}{2 + \cos x} \, dx = -2 \int \frac{\cos x(-\sin x)}{2 + \cos x} \, dx$$

$$= -2(\cos x - 2 \ln |2 + \cos x|) + C$$

$$= -2 \cos x + 4 \ln |2 + \cos x| + C \qquad \blacksquare$$

Example 4

Using the table of integrals, evaluate

$$\int \frac{x}{1 + e^{-x^2}} \, dx$$

Solution: Of the forms involving e^u, we consider Formula 84,

$$\int \frac{u'}{1 + e^u} \, dx = u - \ln (1 + e^u) + C$$

with $u = -x^2$ and $u' = -2x$. Thus we obtain

$$\int \frac{x}{1 + e^{-x^2}} \, dx = -\frac{1}{2} \int \frac{-2x}{1 + e^{-x^2}} \, dx$$

$$= -\frac{1}{2}[-x^2 - \ln (1 + e^{-x^2})] + C$$

$$= \frac{1}{2}[x^2 + \ln (1 + e^{-x^2})] + C \qquad \blacksquare$$

Perhaps you have noticed that a number of the formulas in our table of integrals have the form

$$\int f(x) \, dx = g(x) + \int h(x) \, dx$$

where the right-hand member of the formula contains another integral. Such integration formulas are referred to as **reduction formulas,** since they reduce a given integral to the sum of a function and a simpler integral. We demonstrate the use of reduction formulas in the next two examples.

Example 5

Use tables to evaluate $\int x^3 \sin x \, dx$.

Solution: Using Formula 54 and then Formula 55, we obtain

$$\int x^3 \sin dx = -x^3 \cos x + 3 \int x^2 \cos x \, dx$$

$$= -x^3 \cos x + 3\left(x^2 \sin x - 2 \int x \sin x \, dx\right)$$

Now by formula 52 we have

$$\int x^3 \sin x \, dx = -x^3 \cos x + 3x^2 \sin x + 6x \cos x - 6 \sin x + C \quad \blacksquare$$

Example 6

Use tables to evaluate

$$\int \frac{\sqrt{3 - 5x}}{2x} \, dx$$

Solution: By Formula 19,

$$\int \frac{\sqrt{a + bu}}{u} u' \, dx = 2\sqrt{a + bu} + a \int \frac{u'}{u\sqrt{a + bu}} \, dx$$

with $a = 3$, $b = -5$, and $u = x$, it follows that

$$\frac{1}{2} \int \frac{\sqrt{3 - 5x}}{x} \, dx = \frac{1}{2}\left(2\sqrt{3 - 5x} + 3 \int \frac{dx}{x\sqrt{3 - 5x}}\right)$$

$$= \sqrt{3 - 5x} + \frac{3}{2} \int \frac{dx}{x\sqrt{3 - 5x}}$$

Now by Formula 17, with $u = x$, $a = 3$, and $b = -5$, it follows that

$$\int \frac{\sqrt{3 - 5x}}{2x} \, dx = \sqrt{3 - 5x} + \frac{3}{2}\left(\frac{1}{\sqrt{3}} \ln \left|\frac{\sqrt{3 - 5x} - \sqrt{3}}{\sqrt{3 - 5x} + \sqrt{3}}\right|\right) + C$$

$$= \sqrt{3 - 5x} + \frac{\sqrt{3}}{2} \ln \left|\frac{\sqrt{3 - 5x} - \sqrt{3}}{\sqrt{3 - 5x} + \sqrt{3}}\right| + C \quad \blacksquare$$

Section Exercises (10.9)

In Exercises 1–52, use the integration table in this section to find the indefinite integral.

1. $\int \dfrac{x^2}{1 + x} \, dx$

2. $\int \dfrac{x}{\sqrt{1 + x}} \, dx$

3. $\int \dfrac{1}{x^2\sqrt{1 - x^2}} \, dx$

4. $\int x \sin x \, dx$

5. $\int x^2 \ln x \, dx$

6. $\int \text{Arcsec } 2x \, dx$

S 7. $\int \dfrac{1}{x^2\sqrt{x^2 - 4}} \, dx$

8. $\int \dfrac{\sqrt{x^2 - 4}}{x} \, dx$

9. $\int x e^{x^2} \, dx$

10. $\int \dfrac{x}{\sqrt{9 - x^4}} \, dx$

S 11. $\displaystyle\int \frac{2x}{(1-3x)^2}\,dx$

12. $\displaystyle\int \frac{1}{x^2+2x+2}\,dx$

13. $\displaystyle\int e^x \,\text{Arccos}\, e^x \,dx$

14. $\displaystyle\int \frac{\theta^2}{1-\sin\theta^3}\,d\theta$

15. $\displaystyle\int x^3 \ln x \,dx$

16. $\displaystyle\int \cot^3\theta \,d\theta$

17. $\displaystyle\int \frac{x^2}{(3x-5)^2}\,dx$

18. $\displaystyle\int \frac{1}{2x^2(2x-1)^2}\,dx$

19. $\displaystyle\int \frac{x}{1-\sec x^2}\,dx$

20. $\displaystyle\int \frac{e^x}{1-\tan e^x}\,dx$

S 21. $\displaystyle\int \frac{\cos x}{1+\sin^2 x}\,dx$

22. $\displaystyle\int \frac{1}{t[1+(\ln t)^2]}\,dt$

S 23. $\displaystyle\int \frac{1}{1+e^{2x}}\,dx$

24. $\displaystyle\int \frac{1}{\sqrt{x}(1+2\sqrt{x})}\,dx$

25. $\displaystyle\int \frac{\cos\theta}{3+2\sin\theta+\sin^2\theta}\,d\theta$

26. $\displaystyle\int x^2\sqrt{2+9x^2}\,dx$

27. $\displaystyle\int \frac{1}{x^2\sqrt{2+9x^2}}\,dx$

28. $\displaystyle\int \sqrt{3+x^2}\,dx$

29. $\displaystyle\int e^x\sqrt{1+e^{2x}}\,dx$

30. $\displaystyle\int \frac{1}{\sqrt{x}(x-4)^{3/2}}\,dx$

31. $\displaystyle\int \sin^4 2x \,dx$

32. $\displaystyle\int \frac{\cos^3\sqrt{x}}{\sqrt{x}}\,dx$

S 33. $\displaystyle\int \frac{1}{\sqrt{x}(1-\cos\sqrt{x})}\,dx$

34. $\displaystyle\int \frac{1}{1-\tan 5x}\,dx$

35. $\displaystyle\int t^4\cos t \,dt$

36. $\displaystyle\int \sqrt{x}\,\text{Arctan}\, x^{3/2}\,dx$

37. $\displaystyle\int x\,\text{Arcsec}\,(x^2+1)\,dx$

38. $\displaystyle\int (\ln x)^3\,dx$

S 39. $\displaystyle\int \frac{\ln x}{x(3+2\ln x)}\,dx$

40. $\displaystyle\int \frac{e^x}{(1-e^{2x})^{3/2}}\,dx$

41. $\displaystyle\int \frac{\sqrt{2-2x-x^2}}{x+1}\,dx$

42. $\displaystyle\int \frac{1}{(x^2-6x+10)^2}\,dx$

43. $\displaystyle\int \frac{x}{x^4-6x^2+10}\,dx$

44. $\displaystyle\int (2x-3)^2\sqrt{(2x-3)^2+4}\,dx$

45. $\displaystyle\int \frac{x}{\sqrt{x^4-6x^2+5}}\,dx$

46. $\displaystyle\int \frac{\cos x}{\sqrt{\sin^2 x+1}}\,dx$

S 47. $\displaystyle\int \frac{x^3}{\sqrt{4-x^2}}\,dx$

48. $\displaystyle\int \sqrt{\frac{3-x}{3+x}}\,dx$

49. $\displaystyle\int \frac{1}{x^{3/2}\sqrt{1-x}}\,dx$

50. $\displaystyle\int x\sqrt{x^2+2x}\,dx$

51. $\displaystyle\int \frac{e^{3x}}{(1+e^x)^3}\,dx$

52. $\displaystyle\int \sec^5\theta \,d\theta$

In Exercises 53–57, verify the given formula by the indicated method (IP, TS, etc.).

S 53. (PF)   7. $\displaystyle\int \frac{u^2 u'}{(a+bu)^2}\,dx$

$$= \frac{1}{b^3}\left(bu - \frac{a^2}{a+bu} - 2a\ln|a+bu|\right)+C$$

54. (IP)   22. $\displaystyle\int \frac{u^n u'}{\sqrt{a+bu}}\,dx$

$$= \frac{2}{(2n+1)b}\left(u^n\sqrt{a+bu} - na\int \frac{u^{n-1}u'}{\sqrt{a+bu}}\,dx\right)$$

55. (TS)   36. $\displaystyle\int \frac{u'}{(u^2\pm a^2)^{3/2}}\,dx = \frac{\pm u}{a^2\sqrt{u^2\pm a^2}}+C$

56. (IP)   55. $\displaystyle\int u^n(\cos u)\,u'\,dx$

$$= u^n\sin u - n\int u^{n-1}(\sin u)\,u'\,dx$$

57. (IP)   77. $\displaystyle\int (\text{Arctan}\,u)\,u'\,dx$

$$= u\,\text{Arctan}\,u - \ln\sqrt{1+u^2}+C$$

10.10 **Numerical Integration**

Purpose
- To introduce two numerical methods for approximating definite integrals: the Trapezoidal Rule and Simpson's Rule.

The student should not lose any opportunity of exercising himself in numerical calculation. His power of applying mathematics to questions of practical utility is in direct proportion to the facility which he possesses in computation. —*Augustus DeMorgan (1806–1871)*

When we began our discussion of integration techniques, we mentioned that the search for an antiderivative is not nearly as straightforward a process as is differentiation. Occasionally we encounter functions for which we cannot find antiderivatives. Of course, inability to find an antiderivative for a particular function may be due to a lack of cleverness on our part. On the other hand, some elementary functions simply do not possess antiderivatives that are elementary functions. For example, there are no elementary functions that have any one of the following functions as their derivative:

$$\sqrt[3]{x}\,\sqrt{1-x} \qquad \sqrt{x}\,\cos x \qquad e^{-x^2}$$

$$\sqrt{1-x^3} \qquad \frac{\cos x}{x} \qquad \sin x^2$$

$$\sqrt{1+\sin^2 x} \qquad \frac{1}{\ln x} \qquad \frac{1}{\sqrt{x}\,e^x}$$

Up to this point we have been evaluating definite integrals by means of antiderivatives and the Fundamental Theorem of Calculus. However, if we wish to evaluate a definite integral involving a function whose antiderivative we cannot find, or which does not exist, the Fundamental Theorem cannot be applied and we must resort to some other technique. We describe two such techniques in this section.

In Chapter 6 we discussed the approximation of the area of a region by the use of rectangles. However, our primary purpose for that particular approximation was theoretical rather than computational. For computational purposes we can obtain a better approximation to the area of a region under a curve using figures other than rectangles. For example, a comparison of Figures 10.22 and 10.23 indicates that the use of a finite number of *trapezoids* may yield a better approximation to the area of the region under the curve than does the use of the same number of rectangles.

In our development of a method for approximating the value of the definite integral $\int_a^b f(x)\,dx$ using trapezoids, we assume that f is continuous and positive on the interval $[a, b]$ and that this definite integral represents the area of a region bounded by f and the x-axis, from $x = a$ to $x = b$.

First, we partition the interval $[a, b]$ into n equal subintervals, each of width $\Delta x = (b - a)/n$, such that

$$a = x_0 < x_1 < x_2 < \cdots < x_{n-1} < x_n = b$$

We then form trapezoids for each subinterval, as shown in Figure 10.23. The areas of these trapezoids are given by

$$\text{area of first trapezoid} = \left[\frac{f(x_0) + f(x_1)}{2}\right]\left(\frac{b - a}{n}\right)$$

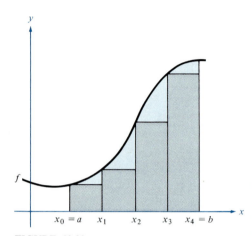

FIGURE 10.22

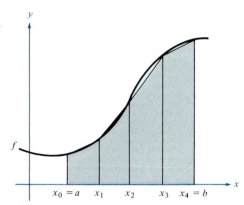

FIGURE 10.23

$$\text{area of second trapezoid} = \left[\frac{f(x_1) + f(x_2)}{2}\right]\left(\frac{b-a}{n}\right)$$

$$\vdots$$

$$\text{area of } n\text{th trapezoid} = \left[\frac{f(x_{n-1}) + f(x_n)}{2}\right]\left(\frac{b-a}{n}\right)$$

Finally, the sum of the areas of the n trapezoids is

$$\left(\frac{b-a}{n}\right)\left[\frac{f(x_0) + f(x_1)}{2} + \frac{f(x_1) + f(x_2)}{2} + \cdots + \frac{f(x_{n-1}) + f(x_n)}{2}\right]$$

$$= \left(\frac{b-a}{2n}\right)[f(x_0) + 2f(x_1) + 2f(x_2) + \cdots + 2f(x_{n-1}) + f(x_n)]$$

Thus we arrive at the following approximation, which we call the Trapezoidal Rule.

Trapezoidal Rule

$$\int_a^b f(x)\,dx \approx \frac{b-a}{2n}[f(x_0) + 2f(x_1) + 2f(x_2) + \cdots + 2f(x_{n-1}) + f(x_n)]$$

Example 1

Use the Trapezoidal Rule to approximate the definite integral $\int_0^\pi \sin x \, dx$. Compare the results for $n = 4$ and $n = 8$. (See Figure 10.24.)

Solution: When $n = 4$, we have $\Delta x = \pi/4$ and

$$x_0 = 0, \qquad x_1 = \frac{\pi}{4}, \qquad x_2 = \frac{\pi}{2}, \qquad x_3 = \frac{3\pi}{4}, \qquad x_4 = \pi$$

Therefore, by the Trapezoidal Rule we have

$$\int_0^\pi \sin x \, dx$$

$$\approx \frac{\pi}{8}\left[\sin(0) + 2\sin\left(\frac{\pi}{4}\right) + 2\sin\left(\frac{\pi}{2}\right) + 2\sin\left(\frac{3\pi}{4}\right) + \sin(\pi)\right]$$

$$= \frac{\pi}{8}[0 + \sqrt{2} + 2 + \sqrt{2} + 0] = \frac{\pi(1 + \sqrt{2})}{4} \approx 1.896$$

When $n = 8$, we have $\Delta x = \pi/8$ and

$$x_0 = 0, \qquad x_1 = \frac{\pi}{8}, \qquad x_2 = \frac{\pi}{4}, \qquad x_3 = \frac{3\pi}{8}, \qquad x_4 = \frac{\pi}{2},$$

$$x_5 = \frac{5\pi}{8}, \qquad x_6 = \frac{3\pi}{4}, \qquad x_7 = \frac{7\pi}{8}, \qquad x_8 = \pi$$

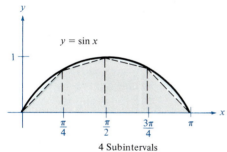

$y = \sin x$

4 Subintervals

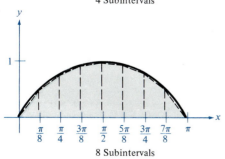

8 Subintervals

FIGURE 10.24

and it follows that

$$\int_0^\pi \sin x \, dx \approx \frac{\pi}{16}\left[\sin(0) + 2\sin\left(\frac{\pi}{8}\right) + 2\sin\left(\frac{\pi}{4}\right) + 2\sin\left(\frac{3\pi}{8}\right)\right.$$

$$+ 2\sin\left(\frac{\pi}{2}\right) + 2\sin\left(\frac{5\pi}{8}\right) + 2\sin\left(\frac{3\pi}{4}\right)$$

$$\left. + 2\sin\left(\frac{7\pi}{8}\right) + \sin(\pi)\right]$$

$$= \frac{\pi}{16}\left[2 + 2\sqrt{2} + 4\sin\left(\frac{\pi}{8}\right) + 4\sin\left(\frac{3\pi}{8}\right)\right] \approx 1.974$$

Of course, for this particular example we could have found an antiderivative and determined that the exact area of the region is 2. ∎

Although you may not yet be excited about using a rather lengthy approximation method to evaluate the integral $\int_0^\pi \sin x \, dx$, two important points must be brought to our attention. First, this approximation method, using trapezoids, becomes more accurate as n increases. (For $n = 16$ in Example 1, the Trapezoidal Rule yields an approximation of 1.994.) Secondly, though we could have used the Fundamental Theorem to evaluate the integral in Example 1, this theorem cannot be used to evaluate an integral so simple looking as $\int_0^\pi \sin x^2 \, dx$, because $\sin x^2$ has no elementary antiderivative. Yet the Trapezoidal Rule can be readily applied to this integral.

One way to view an approximation of the definite integral $\int_a^b f(x) \, dx$ by rectangles is to say that on each of the n subintervals, we approximate f by a horizontal line, which is a polynomial of degree *zero*. With the Trapezoidal Rule we approximate f on each subinterval by a *first*-degree polynomial. And in Simpson's Rule, which follows, we carry this procedure one step further and approximate f by a *second*-degree polynomial.

Before presenting Simpson's Rule, we give the following theorem for evaluating integrals of second-degree polynomials:

THEOREM 10.1

If $p(x) = Ax^2 + Bx + C$, then

$$\int_a^b p(x) \, dx = \left(\frac{b-a}{6}\right)\left[p(a) + 4p\left(\frac{a+b}{2}\right) + p(b)\right]$$

Proof: By integrating we have

$$\int_a^b (Ax^2 + Bx + C) \, dx = \left[\frac{Ax^3}{3} + \frac{Bx^2}{2} + Cx\right]_a^b$$

$$= \frac{A(b^3 - a^3)}{3} + \frac{B(b^2 - a^2)}{2} + C(b - a)$$

$$= \left(\frac{b-a}{6}\right)[2A(a^2 + ab + b^2) + 3B(b + a) + 6C]$$

By regrouping, we obtain

$$\int_a^b (Ax^2 + Bx + C)\,dx$$

$$= \frac{b-a}{6}\left[(Aa^2 + Ba + C) + 4\left\{A\left(\frac{b+a}{2}\right)^2 + B\left(\frac{b+a}{2}\right) + C\right\}\right.$$

$$\left. + (Ab^2 + Bb + C)\right]$$

$$= \frac{b-a}{6}\left[p(a) + 4p\left(\frac{a+b}{2}\right) + p(b)\right]$$

Therefore, we have

$$\int_a^b p(x)\,dx = \frac{b-a}{6}\left[p(a) + 4p\left(\frac{a+b}{2}\right) + p(b)\right] \qquad \blacksquare$$

To develop Simpson's Rule for approximating the value of the definite integral $\int_a^b f(x)\,dx$, we again partition the interval $[a, b]$ into n equal parts, each of width $(b - a)/n$. However, this time we require n to be even and then group the subintervals into pairs such that

$$a = x_0 \underbrace{< x_1 < x_2}_{[x_0, x_2]} \underbrace{< x_3 < x_4}_{[x_2, x_4]} < \cdots \underbrace{< x_{n-2} < x_{n-1} < x_n}_{[x_{n-2}, x_n]} = b$$

Then on each subinterval $[x_{i-2}, x_i]$, we approximate f by a polynomial p of degree less than or equal to 2 that passes through the points

$$(x_{i-2}, f(x_{i-2})), \qquad (x_{i-1}, f(x_{i-1})), \qquad (x_i, f(x_i))$$

Under these conditions, with $i = 2$, it follows that

$$\int_{x_0}^{x_2} p(x)\,dx \approx \int_{x_0}^{x_2} f(x)\,dx$$

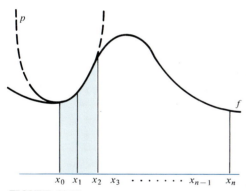

FIGURE 10.25

where p passes through the points $(x_0, f(x_0))$, $(x_1, f(x_1))$, and $(x_2, f(x_2))$. (See Figure 10.25.)

Now by Theorem 10.1 we know that

$$\int_{x_0}^{x_2} p(x)\,dx = \frac{x_2 - x_0}{6}\left[p(x_0) + 4p\left(\frac{x_2 + x_0}{2}\right) + p(x_2)\right]$$

$$= \frac{2[(b - a)/n]}{6}[p(x_0) + 4p(x_1) + p(x_2)]$$

And since $f(x) = p(x)$ for $x = x_0$, $x = x_1$, and $x = x_2$, we have

$$\int_{x_0}^{x_2} p(x)\,dx = \frac{b-a}{3n}[f(x_0) + 4f(x_1) + f(x_2)]$$

Therefore, it follows that

$$\int_{x_0}^{x_2} f(x)\, dx \approx \frac{b-a}{3n}[f(x_0) + 4f(x_1) + f(x_2)]$$

Repeating this procedure on each subinterval $[x_{n-2}, x_n]$ of interval $[a, b]$ results in the formula

$$\int_a^b f(x)\, dx \approx \frac{b-a}{3n}\{[f(x_0) + 4f(x_1) + f(x_2)]$$

$$+ [f(x_2) + 4f(x_3) + f(x_4)] + \cdots$$

$$+ [f(x_{n-2}) + 4f(x_{n-1}) + f(x_n)]\}$$

By grouping like terms we obtain the following approximation formula, known as *Simpson's Rule:*

Simpson's Rule (*n* is even)

$$\int_a^b f(x)\, dx \approx \frac{b-a}{3n}[f(x_0) + 4f(x_1) + 2f(x_2) + 4f(x_3)$$

$$+ \cdots + 4f(x_{n-1}) + f(x_n)]$$

In Example 1 we used the Trapezoidal Rule to estimate $\int_0^\pi \sin x\, dx$. In the next example we see how well Simpson's Rule does for the same integral.

Example 2

Use Simpson's Rule to approximate the integral $\int_0^\pi \sin x\, dx$. Compare the results for $n = 4$ and $n = 8$.

Solution: When $n = 4$, we have

$$\int_0^\pi \sin x\, dx \approx \frac{\pi}{12}\left[\sin(0) + 4\sin\left(\frac{\pi}{4}\right) + 2\sin\left(\frac{\pi}{2}\right) + 4\sin\left(\frac{3\pi}{4}\right) + \sin(\pi)\right]$$

$$= \frac{\pi}{12}(4\sqrt{2} + 2) = \pi\left(\frac{2\sqrt{2}+1}{6}\right) \approx 2.005$$

When $n = 8$, we have

$$\int_0^\pi \sin x\, dx \approx \frac{\pi}{24}\left[\sin(0) + 4\sin\left(\frac{\pi}{8}\right) + 2\sin\left(\frac{\pi}{4}\right)\right.$$

$$+ 4\sin\left(\frac{3\pi}{8}\right) + 2\sin\left(\frac{\pi}{2}\right) + 4\sin\left(\frac{5\pi}{8}\right)$$

$$\left. + 2\sin\left(\frac{3\pi}{4}\right) + 4\sin\left(\frac{7\pi}{8}\right) + \sin(\pi)\right]$$

$$= \frac{\pi}{24}\left[2 + 2\sqrt{2} + 8\sin\left(\frac{\pi}{8}\right) + 8\sin\left(\frac{3\pi}{8}\right)\right] \approx 2.0003 \qquad \blacksquare$$

In Examples 1 and 2 we were able to calculate the exact value of the integral and compare that to our approximations to see how good they were. Of course, in practice we would not bother with an approximation if it were possible to evaluate the integral exactly. However, if it is necessary to use an approximation technique to determine the value of a definite integral, then it is important to know how good we can expect our approximation to be. The following theorem, which we list without proof, gives the formulas for estimating the error involved in the use of Simpson's and the Trapezoidal Rules.

THEOREM 10.2
(*Error in Trapezoidal and Simpson's Rules*)

If f has derivatives up through order four on the interval $[a, b]$, then the error E in approximating $\int_a^b f(x)\,dx$ is:

for the Trapezoidal Rule,

$$E \leq \frac{(b-a)^3}{12n^2}[\max|f''(x)|], \qquad a \leq x \leq b$$

for Simpson's Rule,

$$E \leq \frac{(b-a)^5}{180n^4}[\max|f^{(4)}(x)|], \qquad a \leq x \leq b$$

Theorem 10.2 indicates that the errors generated by the Trapezoidal and Simpson's Rules have upper bounds dependent upon the extreme values of $f''(x)$ and $f^{(4)}(x)$, respectively, in the interval $[a, b]$. Furthermore, it is evident from Theorem 10.2 that the bounds for these errors can be made arbitrarily small by *increasing n*, provided, of course, that f'' and $f^{(4)}$ exist and are bounded in $[a, b]$. The next example shows how to determine a value of n that will bound the error within a predetermined tolerance interval.

Example 3

Using the Trapezoidal Rule, estimate the value of $\int_0^1 e^{-x^2}\,dx$. Determine n so that the approximation error is less than 0.01.

Solution: If $f(x) = e^{-x^2}$, then

$$f'(x) = -2xe^{-x^2}$$
$$f''(x) = 4x^2e^{-x^2} - 2e^{-x^2} = 2e^{-x^2}(2x^2 - 1)$$

By the methods of differential calculus, we can determine that f'' has only one critical value ($x = 0$) in the interval $[0, 1]$ and the maximum value of

$|f''(x)|$ on this interval is $|f''(0)| = 2$. That is,

$$|f''(x)| \le 2, \qquad 0 \le x \le 1$$

Now by Theorem 10.2 we have

$$E \le \frac{(b-a)^3}{12n^2} |f''(x)| \le \frac{1}{12n^2}(2) = \frac{1}{6n^2}$$

To ensure that our approximation has an error less than 0.01, we must choose n so that

$$\frac{1}{6n^2} \le 0.01 = \frac{1}{100}$$

$$100 \le 6n^2$$

$$\frac{50}{3} \le n^2$$

$$4.08 \approx \sqrt{\frac{50}{3}} = n$$

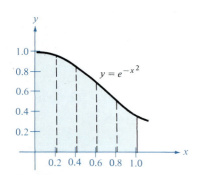

FIGURE 10.26

Therefore, we choose $n = 5$ (since n must be greater than or equal to 4.08) and apply the Trapezoidal Rule (see Figure 10.26) to obtain

$$\int_0^1 e^{-x^2}\, dx \approx \frac{1}{10}\left(\frac{1}{e^0} + \frac{2}{e^{0.04}} + \frac{2}{e^{0.16}} + \frac{2}{e^{0.36}} + \frac{2}{e^{0.64}} + \frac{1}{e^1}\right) \approx 0.744$$

Finally, with an error no larger than 0.01, we know that

$$0.734 \le \int_0^1 e^{-x^2}\, dx \le 0.754 \qquad \blacksquare$$

Example 4

Use Simpson's Rule to estimate the area represented by $\int_0^1 \cos x^2\, dx$. Determine n so that the approximation error is less than 0.001.

Solution: According to Theorem 10.2, the error in Simpson's Rule involves the fourth derivative. Hence by successive differentiation we have

$$f(x) = \cos x^2$$
$$f'(x) = -2x \sin x^2$$
$$f''(x) = -2(2x^2 \cos x^2 + \sin x^2)$$
$$f'''(x) = 4(2x^3 \sin x^2 - 3x \cos x^2)$$
$$f^{(4)}(x) = 4(4x^4 \cos x^2 + 12x^2 \sin x^2 - 3 \cos x^2)$$
$$= 4[(4x^4 - 3)(\cos x^2) + 12x^2 \sin x^2]$$

Since, in the interval $[0, 1]$, $|f^{(4)}(x)|$ is greatest when $x = 1$, we know that

$$|f^{(4)}(x)| \le |f^{(4)}(1)| = 4[\cos(1) + 12 \sin(1)] \approx 42.6 < 45$$

Hence we have

$$E \le \frac{(b-a)^5}{180n^4} |f^{(4)}(x)| = \frac{1}{180n^4} |f^{(4)}(1)| < \frac{45}{180n^4} = \frac{1}{4n^4}$$

Now by choosing n so that

$$\frac{1}{4n^4} < 0.001$$

we can obtain the desired accuracy. Solving for n we have

$$\frac{1000}{4} < n^4$$

$$250 < n^4$$

$$3.97 < n$$

Therefore, we choose $n = 4$ (see Figure 10.27) and obtain

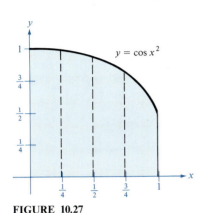

$y = \cos x^2$

FIGURE 10.27

$$\int_0^1 \cos x^2 \, dx \approx \tfrac{1}{12}[\cos(0) + 4\cos\left(\tfrac{1}{16}\right) + 2\cos\left(\tfrac{1}{4}\right) + 4\cos\left(\tfrac{9}{16}\right) + \cos(1)]$$

$$\approx 0.9045$$

and we conclude that

$$0.9035 \le \int_0^1 \cos x^2 \, dx \le 0.9055 \qquad \blacksquare$$

Looking back at the examples in this section, you may wonder why we introduced the Trapezoidal Rule, since for a fixed n, Simpson's Rule usually gives a better approximation. The main reason for including the Trapezoidal Rule is that its error can be more easily estimated than the error involved in Simpson's Rule. Certainly an approximation method is of little benefit if we have no idea of the potential error in the approximation. For instance, if $f(x) = \sqrt{x} \sin(x + 1)$, then to estimate the error in Simpson's Rule, we would need to determine the fourth derivative of f; a monumental task! Therefore, since the estimation of the error in Simpson's Rule involves the fourth derivative, we sometimes prefer to use the Trapezoidal Rule, even though we may have to use a larger n to obtain the desired accuracy.

Section Exercises (10.10)

In Exercises 1–10, use the Trapezoidal Rule and Simpson's Rule to approximate the value of each definite integral. Compare these results with the exact value of the definite integral. Round your answers to four decimal places.

C **1.** $\int_0^2 x^2 \, dx$, $n = 4$

C **2.** $\int_0^1 \left(\frac{x^2}{2} + 1\right) dx$, $n = 4$

C **3.** $\int_0^2 x^3 \, dx$, $n = 4$

C **4.** $\int_1^2 \frac{1}{x} \, dx$, $n = 4$

[c][s] **5.** $\int_0^2 x^3 \, dx, \; n = 8$ [c] **6.** $\int_1^2 \frac{1}{x} \, dx, \; n = 8$

[c] **7.** $\int_1^2 \frac{1}{x^2} \, dx, \; n = 4$ [c] **8.** $\int_0^4 \sqrt{x} \, dx, \; n = 8$

[c] **9.** $\int_0^1 \frac{1}{1 + x^2} \, dx, \; n = 4$ [c] **10.** $\int_0^2 x\sqrt{x^2 + 1} \, dx, \; n = 4$

In Exercises 11–20, approximate each integral using (a) the Trapezoidal Rule and (b) Simpson's Rule.

[c][s] **11.** $\int_0^{\sqrt{\pi/2}} \cos(x^2) \, dx, \; n = 4$

[c] **12.** $\int_0^{\sqrt{\pi/4}} \tan(x^2) \, dx, \; n = 4$

[c] **13.** $\int_0^2 \sqrt{1 + x^3} \, dx, \; n = 2$

[c] **14.** $\int_0^2 \frac{1}{\sqrt{1 + x^3}} \, dx, \; n = 4$

[c][s] **15.** $\int_0^1 \sqrt{x}\sqrt{1 - x} \, dx, \; n = 4$

[c] **16.** $\int_0^{\pi} f(x) \, dx, \; f(x) = \begin{cases} \dfrac{\sin x}{x}, & x > 0 \\ 1, & x = 0 \end{cases}, \; n = 4$

[c] **17.** $\int_0^1 \sin(x^2) \, dx, \; n = 2$

[c] **18.** $\int_0^{\pi} \sqrt{x} \sin x \, dx, \; n = 4$

[c] **19.** $\int_0^{\pi/4} x \tan x \, dx, \; n = 4$

[c] **20.** $\int_0^{\pi/2} \sqrt{1 + \cos^2 x} \, dx, \; n = 2$

In Exercises 21–25, find the maximum possible error if each integral is approximated by (a) the Trapezoidal Rule and (b) Simpson's Rule.

[c] **21.** $\int_0^2 x^3 \, dx, \; n = 4$ [c] **22.** $\int_0^1 \frac{1}{x + 1} \, dx, \; n = 4$

[c][s] **23.** $\int_0^1 e^{x^3} \, dx, \; n = 4$ [c] **24.** $\int_0^1 e^{-x^2} \, dx, \; n = 4$

[c] **25.** $\int_0^1 \sin(x^2) \, dx, \; n = 2$

[c] **26.** Find n so that Simpson's Rule will have an error less than 0.00001 in the approximation of $\int_0^1 e^{-x^2} \, dx$.

[c][s] **27.** Find n so that the Trapezoidal Rule will have an error less than 0.00001 in the approximation of $\int_0^1 e^{-x^2} \, dx$.

[c] **28.** From previous work it is known that

$$\int \frac{1}{1 + x^2} \, dx = \text{Arctan } x + C$$

and, therefore,

$$\int_0^1 \frac{1}{1 + x^2} \, dx = \frac{\pi}{4}$$

Use Simpson's Rule to approximate this definite integral and obtain an approximation of π to five decimal places.

[c][s] **29.** The standard normal probability density function is

$$f(z) = \frac{1}{\sqrt{2\pi}} e^{-z^2/2}$$

The probability that z is in the interval $[a, b]$ is the area of the region defined by $y = f(z)$, $y = 0$, $z = a$, and $z = b$ and is denoted by $\Pr(a \le z \le b)$. Estimate the following probabilities. (Choose n so that the error is less than 0.0001.)
(a) $\Pr(0 \le z \le 1)$ (b) $\Pr(0 \le z \le 2)$

[c] **30.** A farmer has an odd-shaped plot of land bounded by a stream and two relatively straight roads that meet at right angles. At distances of x meters, he measures the distances (y meters) from the one road to the stream. Approximate the number of acres (1 acre $\approx$ 4,047 m²) in the field by the use of Simpson's Rule if the measurements are as follows:

x	0	100	200	300	400	500	600	700	800	900	1000
y	125	125	120	112	90	90	95	88	75	35	0

31. Prove that Simpson's Rule is exact when estimating the integral $\int_{x_1}^{x_2} (a_0 + a_1 x + a_2 x^2 + a_3 x^3) \, dx$.

32. Demonstrate the result of Exercise 31 by using Simpson's Rule to evaluate the integral $\int_0^1 x^3 \, dx$ with $n = 2$.

10.11 **Improper Integrals**

Purpose
- To introduce the concept of an improper integral.
- To use limits to determine the convergence or divergence of improper integrals.

But let us remember that we are dealing with infinities and indivisibles, both of which transcend our finite understanding, the former on account of their magnitude, the latter because of their smallness. In spite of this, men cannot refrain from discussing them, even though it must be done in a roundabout way.   —*Galileo Galilei (1564–1642)*

Our definition (Section 6.3) of the definite integral $\int_a^b f(x)\,dx$ included the requirements that the interval $[a, b]$ is finite and that f is bounded on $[a, b]$. Furthermore, the Fundamental Theorem of Calculus (Section 6.4), by which we have been evaluating definite integrals, requires f to be continuous on $[a, b]$. In this section we discuss a limit procedure for evaluating integrals that do not satisfy these requirements because either:

1. one or both of the limits of integration are infinite, or
2. $f(x)$ has a finite number of infinite discontinuities on the interval $[a, b]$

Integrals possessing either of properties 1 and 2 are called **improper integrals.**
   For instance, the integrals

$$\int_1^\infty \frac{dx}{x} \quad \text{and} \quad \int_{-\infty}^\infty \frac{dx}{x^2 + 1}$$

are improper because one or both of their limits of integration are infinite. On the other hand, the integrals

$$\int_1^5 \frac{dx}{\sqrt{x - 1}} \quad \text{and} \quad \int_{-2}^2 \frac{dx}{(x + 1)^2}$$

are improper because their integrands are infinite somewhere in the interval of integration [$1/\sqrt{x - 1}$ is infinite at $x = 1$, while $1/(x + 1)^2$ is infinite at $x = -1$].
   To get an idea how we might evaluate an improper integral, consider the integral

$$\int_1^b \frac{1}{x^2}\,dx$$

which we can interpret as the area of the shaded region shown in Figure 10.28. Evaluating the integral we have

$$\int_1^b \frac{1}{x^2}\,dx = -\frac{1}{x}\Big]_1^b = -\frac{1}{b} + 1 = 1 - \frac{1}{b}$$

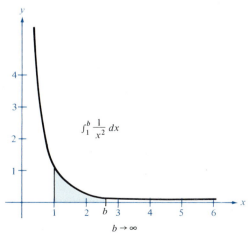

$\int_1^b \frac{1}{x^2}\,dx$

$b \to \infty$

FIGURE 10.28

Now if we let $b \to \infty$, then we have

$$\lim_{b \to \infty} \int_1^b \frac{1}{x^2} dx = \lim_{b \to \infty} \left(\frac{-1}{x} \right]_1^b \right) = \lim_{b \to \infty} \left(1 - \frac{1}{b} \right) = 1$$

We denote this limit by the improper integral

$$\int_1^\infty \frac{1}{x^2} dx$$

and we can interpret it as the area of the unbounded region under $y = 1/x^2$, above the x-axis, and to the right of $x = 1$.

More generally, we describe improper integrals having infinite limits of integration as follows:

Improper Integrals (Infinite Limits of Integration)

1. If f is continuous on the interval $[a, \infty)$, then

$$\int_a^\infty f(x) \, dx = \lim_{b \to \infty} \int_a^b f(x) \, dx$$

2. If f is continuous on the interval $(-\infty, b]$, then

$$\int_{-\infty}^b f(x) \, dx = \lim_{a \to -\infty} \int_a^b f(x) \, dx$$

3. If f is continuous on the interval $(-\infty, \infty)$, then

$$\int_{-\infty}^\infty f(x) \, dx = \int_{-\infty}^c f(x) \, dx + \int_c^\infty f(x) \, dx$$

where c is any real number.

In each case above, if the limit exists, then the improper integral is said to **converge;** otherwise the improper integral **diverges.** This means that in the third case the integral will diverge if either one of the integrals on the right diverges.

Example 1

Determine the convergence or divergence of the integral

$$\int_1^\infty \frac{1}{x} dx$$

Solution:

$$\int_1^\infty \frac{1}{x} dx = \lim_{b \to \infty} \int_1^b \frac{1}{x} dx = \lim_{b \to \infty} \left(\ln x \right]_1^b \right) = \lim_{b \to \infty} (\ln b - 0) = \infty$$

and thus the integral diverges.   ∎

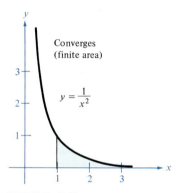

A comparison of the improper integrals

$$\int_1^\infty \frac{1}{x^2}\,dx$$

which converges to 1, and

$$\int_1^\infty \frac{1}{x}\,dx$$

which diverges, suggests the somewhat unpredictable nature of improper integrals. The functions $y = 1/x^2$ and $y = 1/x$ have similar-looking graphs, as shown in Figure 10.29, yet the shaded region under $y = 1/x^2$ to the right of $x = 1$ has *finite* area, whereas the corresponding region under $y = 1/x$ has *infinite* area (see Example 1).

In Example 1 we were careful to use limit notation to determine the divergence of the improper integral $\int_1^\infty (1/x)\,dx$. However, in practice we can use the same notation that we use for definite integrals. For instance, the solution to Example 1 could be written as

$$\int_1^\infty \frac{1}{x}\,dx = \ln x \Big]_1^\infty = \ln(\infty) - \ln(1) = \infty - 0 = \infty$$

Keep in mind that this definite integral notation is a shorthand symbol for an implied limit. Furthermore, the nonexistence of the limit at either endpoint is sufficient to imply divergence.

FIGURE 10.29

Example 2

Evaluate the improper integral

$$\int_{-\infty}^0 \frac{dx}{(1-2x)^{3/2}}$$

Solution:

$$\int_{-\infty}^0 \frac{dx}{(1-2x)^{3/2}} = \frac{1}{\sqrt{1-2x}}\Big]_{-\infty}^0 = \frac{1}{\sqrt{1}} - \frac{1}{\sqrt{\infty}} = 1 - 0 = 1$$

Consequently, this improper integral converges. (See Figure 10.30.) ■

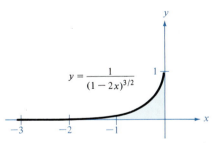

FIGURE 10.30

Example 3

Evaluate the improper integral

$$\int_{-\infty}^\infty \frac{e^x}{1+e^{2x}}\,dx$$

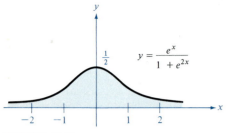

FIGURE 10.31

Solution:

$$\int_{-\infty}^{\infty} \frac{e^x}{1 + e^{2x}}\,dx = \text{Arctan } e^x \Big]_{-\infty}^{\infty}$$

$$= \text{Arctan } \infty - \text{Arctan } 0 = \frac{\pi}{2} - 0 = \frac{\pi}{2}$$

and the integral converges. (See Figure 10.31.) ∎

Example 4

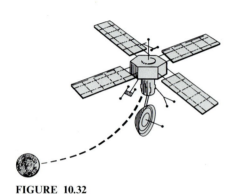

FIGURE 10.32

In Example 3 of Section 7.5, we determined that it would take 10,000 mile tons of work to propel a 15-ton space module to a height 800 miles above the earth. How much work is required to propel this same module an unlimited distance away from the earth's surface? (See Figure 10.32.)

Solution: At first we might think that an infinite amount of work would be required. If so, then it would be impossible to send rockets into outer space. Since this has already been done, the work required must be finite and we can determine the work in the following manner. Using the integral of Example 3, Section 7.5, we replace the upper bound of 4800 miles by ∞ and write

$$W = \int_{4000}^{\infty} \frac{240,000,000}{x^2}\,dx$$

$$= \frac{-240,000,000}{x}\Big]_{4000}^{\infty} = 0 + 60,000 = 60,000 \text{ mile tons}$$ ∎

A second general type of improper integral includes those whose integrands are infinite *at* or *between* the limits of integration. Such integrals are described in the following definition:

Improper Integrals (Infinite Integrands)

1. If f is continuous on the interval $[a, b)$ and becomes infinite at b, then

$$\int_a^b f(x)\,dx = \lim_{c \to b^-} \int_a^c f(x)\,dx$$

2. If f is continuous on the interval $(a, b]$ and becomes infinite at a, then

$$\int_a^b f(x)\,dx = \lim_{c \to a^+} \int_c^b f(x)\,dx$$

3. If f is continuous on the interval $[a, b]$, except for some c in (a, b) at which f is infinite, then

$$\int_a^b f(x)\,dx = \int_a^c f(x)\,dx + \int_c^b f(x)\,dx$$

Example 5

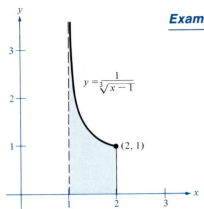

FIGURE 10.33

Evaluate the improper integral

$$\int_1^2 \frac{dx}{\sqrt[3]{x-1}}$$

Solution: Note that the integrand becomes infinite at $x = 1$. In our shorthand form we have

$$\int_1^2 \frac{dx}{\sqrt[3]{x-1}} = \frac{3}{2}(x-1)^{2/3}\Big]_1^2 = \frac{3}{2}[(1)^{2/3} - (0)^{2/3}] = \frac{3}{2}$$

and the integral converges. (See Figure 10.33.)   ■

Example 6

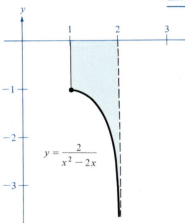

FIGURE 10.34

Evaluate the improper integral

$$\int_1^2 \frac{2\,dx}{x^2 - 2x}$$

Solution: By separating the integrand into its partial fractions, we obtain

$$\int_1^2 \frac{2\,dx}{x^2 - 2x} = \int_1^2 \left(\frac{1}{x-2} - \frac{1}{x}\right) dx = \Big[\ln|x-2| - \ln|x|\Big]_1^2$$

The integral diverges since

$$\lim_{x \to 2^-} (\ln|x-2| - \ln|x|) = -\infty - \ln 2 = -\infty$$

(See Figure 10.34.)   ■

Example 7

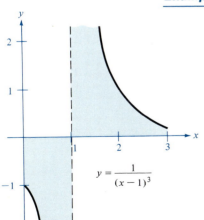

FIGURE 10.35

Evaluate the integral $\int_0^3 (x-1)^{-3}\,dx$.

Solution: This integral is improper because the integrand is infinite at $x = 1$, which lies between the limits of integration. Thus we must write

$$\int_0^3 (x-1)^{-3}\,dx = \int_0^1 (x-1)^{-3}\,dx + \int_1^3 (x-1)^{-3}\,dx$$

$$= \left[\frac{-1}{2(x-1)^2}\right]_0^1 + \left[\frac{-1}{2(x-1)^2}\right]_1^3$$

The integral diverges since

$$\lim_{x \to 1^-} \frac{-1}{2(x-1)^2} = -\infty$$

Remember that the nonexistence of the limit at *any* endpoint is sufficient to imply divergence. (See Figure 10.35.)   ■

Had we not recognized that the integral in Example 7 was improper, we would have obtained the incorrect result

$$\int_0^3 (x-1)^{-3} = \frac{-1}{2(x-1)^2}\Bigg]_0^3 = -\frac{1}{8} + \frac{1}{2} = \frac{3}{8}$$

Improper integrals in which the integrand is infinite at some point *between* the limits of integration are quite often overlooked, so keep alert for such possibilities.

Example 8

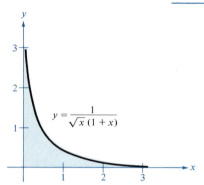

FIGURE 10.36

Evaluate the improper integral

$$\int_0^\infty \frac{dx}{\sqrt{x}(1+x)}$$

Solution: The integral is improper because its upper limit of integration is infinite *and* because the integrand is infinite at the lower limit of integration. Nevertheless we write

$$\int_0^\infty \frac{dx}{\sqrt{x}(1+x)} = 2\,\mathrm{Arctan}\,\sqrt{x}\,\Bigg]_0^\infty = 2\left(\frac{\pi}{2} - 0\right) = \pi$$

and the integral converges. (See Figure 10.36.) ∎

Example 9

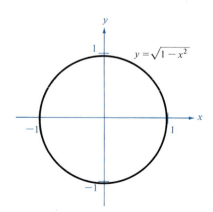

FIGURE 10.37

Use the integral formula for arc length to show that the circumference of the circle $x^2 + y^2 = 1$ is 2π.

Solution: To simplify our work we consider the quarter circle $y = \sqrt{1-x^2}$, where $0 \le x \le 1$. Thus we have

$$y' = \frac{-x}{\sqrt{1-x^2}}$$

Note that y' is continuous on $[0, 1)$ but is infinite when $x = 1$. Thus our formula for arc length results in the improper integral

$$s = \int_a^b \sqrt{1 + (y')^2}\,dx = \int_0^1 \sqrt{1 + \left(\frac{-x}{\sqrt{1-x^2}}\right)^2}\,dx$$

$$= \int_0^1 \frac{dx}{\sqrt{1-x^2}} = \mathrm{Arcsin}\,x\,\Bigg]_0^1 = \frac{\pi}{2}$$

Since the arc length of this quarter circle is $\pi/2$, the circumference of the circle is $4s = 2\pi$. (See Figure 10.37.) ∎

Example 10 (*L'Hôpital's Rule*)

Evaluate the improper integral

$$\int_0^1 \ln x\, dx$$

Solution: Using integration by parts we have

$$\int_0^1 \ln x\, dx = \Big[x \ln x - x\Big]_0^1$$

By L'Hôpital's Rule we have

$$\lim_{x\to 0^+} x \ln x = \lim_{x\to 0^+} \frac{\ln x}{1/x} = \lim_{x\to 0^+} \frac{1/x}{-1/x^2} = \lim_{x\to 0^+} (-x) = 0$$

Thus we have

$$\int_0^1 \ln x\, dx = \Big[x \ln x - x\Big]_0^1 = (0-1) - (0-0) = -1$$

(See Figure 10.38.) ■

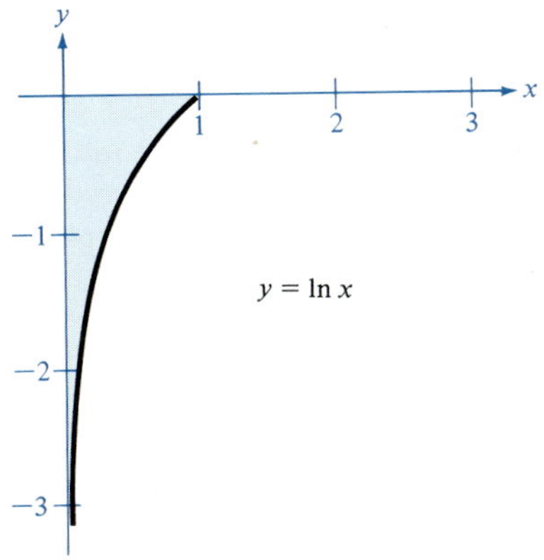

FIGURE 10.38

Section Exercises (10.11)

In Exercises 1–26, determine the divergence or convergence of each improper integral and evaluate those that converge.

1. $\int_0^4 \frac{1}{\sqrt{x}}\, dx$
2. $\int_3^4 \frac{1}{\sqrt{x-3}}\, dx$
[S] 3. $\int_0^8 \frac{1}{\sqrt[3]{8-x}}\, dx$
4. $\int_0^1 \frac{1}{x}\, dx$
5. $\int_0^1 \frac{1}{x^2}\, dx$
6. $\int_0^e \ln x\, dx$
7. $\int_0^1 x \ln x\, dx$
8. $\int_0^{\pi/2} \sec\theta\, d\theta$
9. $\int_0^{\pi/2} \tan\theta\, d\theta$
10. $\int_0^2 \frac{1}{\sqrt{4-x^2}}\, dx$
[S] 11. $\int_2^4 \frac{1}{\sqrt{x^2-4}}\, dx$
12. $\int_0^2 \frac{1}{4-x^2}\, dx$
[S] 13. $\int_0^2 \frac{1}{(x-1)^2}\, dx$
14. $\int_0^2 \frac{1}{(x-1)^{2/3}}\, dx$
15. $\int_0^2 \frac{1}{\sqrt[3]{x-1}}\, dx$
16. $\int_0^2 \frac{1}{(x-1)^{4/3}}\, dx$
17. $\int_{-\infty}^0 xe^{-2x}\, dx$
18. $\int_0^\infty xe^{-x}\, dx$
19. $\int_1^\infty \frac{1}{x^2}\, dx$
20. $\int_1^\infty \frac{1}{\sqrt{x}}\, dx$
[S] 21. $\int_0^\infty e^{-x} \cos x\, dx$
22. $\int_0^\infty e^{-ax} \sin bx\, dx,\ a > 0$
23. $\int_{-\infty}^\infty \frac{1}{1+x^2}\, dx$
24. $\int_0^\infty \frac{x^3}{(x^2+1)^2}\, dx$
25. $\int_0^\infty \frac{1}{e^x+e^{-x}}\, dx$
26. $\int_0^\infty \sin x\, dx$
27. Determine all values of p so that the improper integral $\int_1^\infty (1/x^p)\, dx$ converges.
28. Given the region bounded by $y = 1/x, y = 0$, and $x \geq 1$, find the following (if possible):
    (a) its area
    (b) the volume of the solid generated by revolving the region about the x-axis
    (c) the volume of the solid generated by revolving the region about the y-axis
[S] 29. Sketch the graph of the hypocycloid of four cusps, $x^{2/3} + y^{2/3} = 1$, and find its perimeter.

30. Given the region bounded by $y = e^{-x}, y = 0$, and $x \geq 0$, find the following (if possible):
    (a) its area
    (b) the volume of the solid generated by revolving the region about the y-axis

31. Given two nonnegative continuous functions, f and g, where $f(x) \leq g(x)$ on the interval (a, ∞). Prove the following:
    (a) If $\int_a^\infty g(x)\, dx$ converges, then $\int_a^\infty f(x)\, dx$ will also.
    (b) If $\int_a^\infty f(x)\, dx$ diverges, then $\int_a^\infty g(x)\, dx$ will also.

32. Using the results of Exercises 27 and 31, determine whether or not the following improper integrals converge:
    (a) $\int_1^\infty \dfrac{1}{x^2 + 5}\, dx$
    (b) $\int_2^\infty \dfrac{1}{\sqrt{x-1}}\, dx$

(c) $\int_1^\infty \dfrac{1}{\sqrt{x}(x+1)}\, dx$
(d) $\int_2^\infty \dfrac{1}{\sqrt[3]{x}(x-1)}\, dx$

[S] 33. Use the result of Exercise 31 to determine if $\int_0^\infty e^{-x^2}\, dx$ converges.

34. Use the result of Exercise 31 to determine if $\int_2^\infty 1/(\sqrt{x} \ln x)\, dx$ converges.

35. The region bounded by $(x-2)^2 + y^2 = 1$ is revolved about the y-axis to form a torus. Find the surface area of this torus.

36. A 5-ton rocket is fired from the surface of the earth into outer space. How much work is required to overcome the earth's gravitational force? How far has the rocket traveled when half the total work has occurred?

For Review and Class Discussion

True or False

1. ____ The integral $\int_0^\infty (1/x^p)\, dx$ diverges for all $p > 0$.

2. ____ The integral $\int_1^\infty (1/x^p)\, dx$ converges if and only if $p > 1$.

3. ____ If f is continuous on $[0, \infty)$ and $\lim_{x \to \infty} f(x) = 0$, then $\int_0^\infty f(x)\, dx$ converges.

4. ____ If f' is continuous on $[0, \infty)$ and $\lim_{x \to \infty} f(x) = 0$, then $\int_0^\infty f'(x)\, dx = -f(0)$.

5. ____ If the graph of f is symmetric with respect to the origin or the y-axis, then $\int_0^\infty f(x)\, dx$ converges if and only if $\int_{-\infty}^\infty f(x)\, dx$ converges.

Miscellaneous Exercises (Ch. 10)

In Exercises 1–40, find the indefinite integral.

1. $\int \dfrac{x^2}{x^2 + 2x - 15}\, dx$

2. $\int \dfrac{\sqrt{x^2 - 9}}{x}\, dx$

[S] 3. $\int \dfrac{1}{1 - \sin \theta}\, d\theta$

4. $\int x^2 \sin 2x\, dx$

5. $\int e^{2x} \sin 3x\, dx$

6. $\int (x^2 - 1)e^x\, dx$

[S] 7. $\int \dfrac{\ln (2x)}{x^2}\, dx$

8. $\int 2x \sqrt{2x - 3}\, dx$

9. $\int \sqrt{4 - x^2}\, dx$

10. $\int \dfrac{\sqrt{4 - x^2}}{2x}\, dx$

11. $\int \dfrac{-12}{x^2 \sqrt{4 - x^2}}\, dx$

12. $\int \tan \theta \sec^4 \theta\, d\theta$

[S] 13. $\int \sec^4 \dfrac{x}{2}\, dx$

14. $\int \sec \theta \cos 2\theta\, d\theta$

15. $\int \dfrac{9}{x^2 - 9}\, dx$

16. $\int \dfrac{\sec^2 \theta}{\tan \theta (\tan \theta - 1)}\, d\theta$

[S] 17. $\int \dfrac{x^2 + 2x}{x^3 - x^2 + x - 1}\, dx$

18. $\int \dfrac{4x - 2}{3(x - 1)^2}\, dx$

19. $\int \dfrac{3x^3 + 4x}{(x^2 + 1)^2}\, dx$

20. $\int \sqrt{\dfrac{x - 2}{x + 2}}\, dx$

21. $\int \dfrac{16}{\sqrt{16 - x^2}} dx$

22. $\int \dfrac{\sin \theta}{1 + 2 \cos^2 \theta} d\theta$

23. $\int \dfrac{e^x}{4 + e^{2x}} dx$

24. $\int \dfrac{x}{x^2 - 4x + 8} dx$

25. $\int \dfrac{x}{x^2 + 4x + 8} dx$

26. $\int \dfrac{3}{2x \sqrt{9x^2 - 1}} dx$

27. $\int \theta \sin \theta \cos \theta \, d\theta$

28. $\int \dfrac{\csc \sqrt{2x}}{\sqrt{x}} dx$

29. $\int (\sin \theta + \cos \theta)^2 \, d\theta$

30. $\int \cos 2\theta (\sin \theta + \cos \theta)^2 \, d\theta$

S **31.** $\int \dfrac{x^{1/4}}{1 + x^{1/2}} dx$

32. $\int \sqrt{1 - \cos x} \, dx$

33. $\int \sqrt{1 + \cos x} \, dx$

34. $\int \ln \sqrt{x^2 - 1} \, dx$

35. $\int \ln (x^2 + x) \, dx$

36. $\int x \, \text{Arcsin} \, 2x \, dx$

S **37.** $\int \cos x \ln (\sin x) \, dx$

38. $\int e^x \, \text{Arctan} \, e^x \, dx$

39. $\int \dfrac{x^4 + 2x^2 + x + 1}{(x^2 + 1)^2} dx$

40. $\int \sqrt{1 + \sqrt{x}} \, dx$

In Exercises 41–45, evaluate each integral to two decimal places.

c **41.** $\int_0^{\pi/2} \dfrac{1}{2 - \cos \theta} d\theta$

c **42.** $\int_0^1 \dfrac{x^{3/2}}{2 - x^2} dx$

c S **43.** $\int_0^2 \dfrac{1}{\sqrt{1 + x^3}} dx$

c **44.** $\int_0^{\pi/4} \theta \tan \theta \, d\theta$

c **45.** $\int_1^2 \dfrac{1}{1 + \ln x} dx$

46. Integrate $\int \dfrac{1}{x^2 \sqrt{4 + x^2}} dx$

by letting:
(a) $x = 2 \tan \theta$
(b) $x = 2/u$

47. Integrate $\int \dfrac{1}{x \sqrt{4 + x^2}} dx$

by letting:
(a) $x = 2 \tan \theta$
(b) $u^2 = 4 + x^2$

48. Integrate $\int \dfrac{x^3}{\sqrt{4 + x^2}} dx$

by the following methods:
(a) integration by parts with $v' = x/\sqrt{4 + x^2}$
(b) letting $x = 2 \tan \theta$
(c) letting $u^2 = 4 + x^2$

S **49.** Integrate $\int x \sqrt{4 + x} \, dx$ by the following methods:
(a) integration by parts, letting $v' = \sqrt{4 + x}$
(b) letting $x = 4 \tan^2 \theta$
(c) letting $u = \sqrt{4 + x}$
(d) letting $u = 4 + x$

c **50.** Evaluate to two decimal places:

(a) $\int_0^1 e^x \, dx$

(b) $\int_0^1 xe^x \, dx$

(c) $\int_0^1 xe^{x^2} \, dx$

(d) $\int_0^1 e^{x^2} \, dx$

c **51.** Evaluate to two decimal places:

(a) $\int_0^{\pi/2} \cos x \, dx$

(b) $\int_0^{\pi/2} \cos^2 x \, dx$

(c) $\int_0^{\pi/2} \cos x^2 \, dx$

(d) $\int_0^{\pi/2} \cos \sqrt{x} \, dx$

52. A solid has a base in the form of an ellipse whose equation is given by $(x^2/25) + (y^2/16) = 1$. If the cross sections taken perpendicular to the major axis are isosceles triangles of height 6, find the volume of the solid.

53. The *gamma function* $\Gamma(n)$ is defined to be $\Gamma(n) = \int_0^\infty x^{n-1} e^{-x} \, dx, n > 0$.
(a) Find $\Gamma(1)$.
(b) Find $\Gamma(2)$.
(c) Find $\Gamma(3)$.
(d) Use integration by parts to show that $\Gamma(n + 1) = n\Gamma(n)$.

54. Let $I_n = \int_0^\infty \dfrac{x^{2n-1}}{(x^2 + 1)^{n+3}} dx$

for $n \geq 1$. Prove that $I_n = [(n - 1)/(n + 2)] I_{n-1}$ and then evaluate the following:

(a) $\int_0^\infty \dfrac{x^3}{(x^2 + 1)^5} dx$

(b) $\int_0^\infty \dfrac{x^5}{(x^2 + 1)^6} dx$

(c) $\int_0^\infty \dfrac{x^7}{(x^2 + 1)^7} dx$

c **55.** For $x > 2$ approximate $\int_2^\infty 1/(x^5 - 1) \, dx$ by using the inequality

$$\dfrac{1}{x^5} + \dfrac{1}{x^{10}} + \dfrac{1}{x^{15}} < \dfrac{1}{x^5 - 1} < \dfrac{1}{x^5} + \dfrac{1}{x^{10}} + \dfrac{2}{x^{15}}$$

56. Find the centroid of the region bounded by the semicircle $y = \sqrt{1 - x^2}$ and $y = 0$.

57. Find the centroid of the region bounded by the two circles $(x - 1)^2 + y^2 = 1$ and $(x - 4)^2 + y^2 = 4$.

$\boxed{c}$ **58.** Approximate, to two decimal places, the arc length of $y = \sin x$ between $x = 0$ and $x = \pi$.

$\boxed{c}$ **59.** Approximate, to two decimal places, the arc length of $y = \sin^2 x$ between $x = 0$ and $x = \pi$.

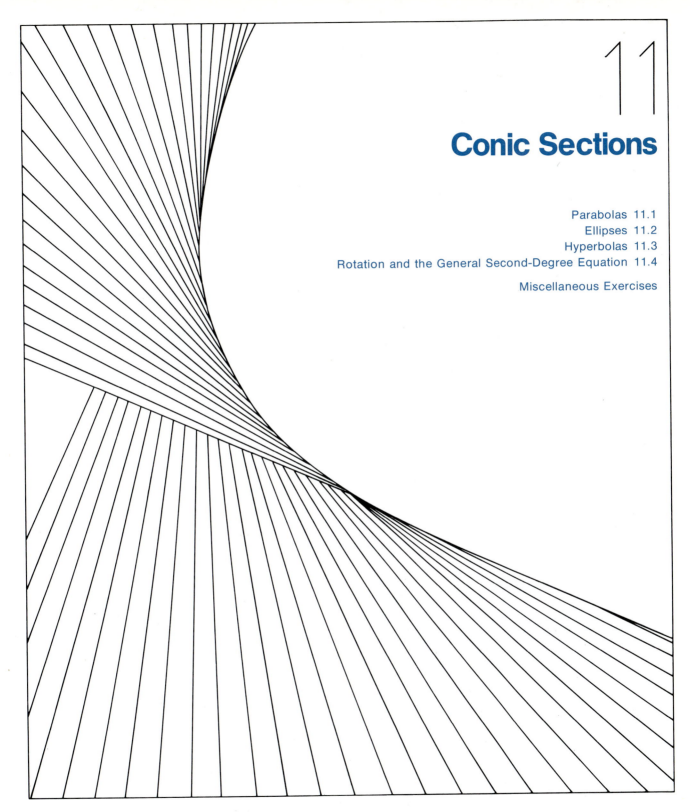

Conic Sections

11.1 **Parabolas**

Purpose
- To identify the four basic conics as the intersection of a cone and a plane.
- To define a parabola and use this definition to investigate some properties of parabolas.

Conic sections were studied for eighteen hundred years merely as an abstract science, and then at the end of this long period of abstract study, they were found to be the necessary key with which to attain the knowledge of the most important laws of nature. —*Alfred Whitehead* (1861–1947)

Conic sections have a rich historical background, going back to early Greek mathematics. Initially, interest in conics centered around construction problems. With the advent of scientific discovery in the seventeenth century, the broad applicability of conics became apparent, and they played a prominent role in the early history of calculus.

The name **conic section,** or simply "conic," refers to the description of a conic as the intersection of a double-napped cone and a plane. Notice from Figure 11.1 that in the formation of the four basic conics, the intersecting plane does not pass through the vertex of the cone. When the plane does pass through the vertex, we call the resulting figure a **degenerate conic,** as shown in Figure 11.2.

There are several ways to approach a study of the conics. We could begin by defining the conics in terms of the intersections of planes and cones, as the Greeks did. Or we could define them algebraically in terms of the general second-degree equation

$$Ax^2 + Bxy + Cy^2 + Dx + Ey + F = 0$$

a procedure to be discussed in Section 11.4. However, a third approach, in which each of the conics is defined as a "locus" (collection) of points satisfying a certain geometric property, suits our needs best. For example, in Section 1.5 we saw how the definition of a circle as "the collection of all points (x, y) that are equidistant from a fixed point (h, k)" led easily to the standard equation of a circle,

$$(x - h)^2 + (y - k)^2 = r^2$$

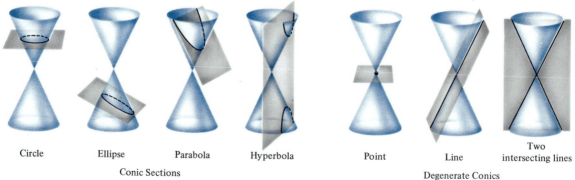

| Circle | Ellipse | Parabola | Hyperbola | Point | Line | Two intersecting lines |

Conic Sections   Degenerate Conics

FIGURE 11.1   **FIGURE 11.2**

In this and the following two sections, we give similar definitions to the other three conics and then discuss their standard equations and properties.

Definition of Parabola

A **parabola** is the set of all points (x, y) that are equidistant from a fixed line (**directrix**) and a fixed point (**focus**) not on the line. (See Figure 11.3.)

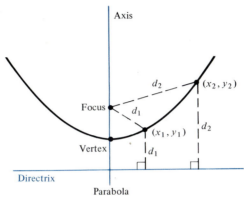

FIGURE 11.3

The midpoint between the focus and the directrix is called the **vertex,** and the line passing through the focus and the vertex is called the **axis** of the parabola. Note that a parabola is symmetric with respect to its axis.

Using this definition of a parabola, we derive the following theorem, which gives the standard form of the equation of a parabola whose directrix is parallel to the x-axis or to the y-axis.

THEOREM 11.1
(*Standard Equation of a Parabola*)

The standard form of the equation of a parabola with vertex at (h, k) and directrix $y = k - p$ is

$$(x - h)^2 = 4p(y - k) \qquad \text{(vertical axis)}$$

For directrix $x = h - p$ the equation is

$$(y - k)^2 = 4p(x - h) \qquad \text{(horizontal axis)}$$

The focus lies on the axis p units (*directed distance*) from the vertex.

Proof: Since the two cases are similar, we give a proof for the first case only. Suppose the directrix ($y = k - p$) is parallel to the x-axis. In Figure 11.4 we assume $p > 0$, and since p is the directed distance from the vertex to the focus, it follows that the focus lies *above* the vertex. Since by definition the point (x, y) is equidistant from $(h, k + p)$ and $y = k - p$, we have the equation

$$\sqrt{(x - h)^2 + [y - (k + p)]^2} = y - (k - p)$$

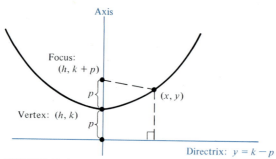

FIGURE 11.4

Squaring both sides of this equation yields

$$(x - h)^2 + [y - (k + p)]^2 = [y - (k - p)]^2$$
$$(x - h)^2 + y^2 - 2y(k + p) + (k + p)^2 = y^2 - 2y(k - p) + (k - p)^2$$
$$(x - h)^2 - 2py + 2pk = 2py - 2pk$$
$$(x - h)^2 = 4p(y - k)$$

■

Example 1

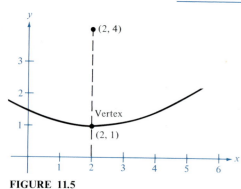

FIGURE 11.5

Find the standard form of the equation of the parabola with vertex $(2, 1)$ and focus $(2, 4)$. (See Figure 11.5.)

Solution: Since the axis of the parabola is vertical, we consider the equation

$$(x - h)^2 = 4p(y - k)$$

where $h = 2$, $k = 1$, and $p = 3$. Thus the standard equation is

$$(x - 2)^2 = 12(y - 1)$$

By expanding this equation we come up with the more common quadratic form

$$y = (\tfrac{1}{12})(x^2 - 4x + 16)$$

■

Example 2

Find the focus of the parabola given by $y = \tfrac{1}{2}(1 - 2x - x^2)$.

Solution: To find the focus we convert the equation to standard form by completing the square. Thus

$$y = \tfrac{1}{2}(1 - 2x - x^2)$$
$$2y = 1 - (x^2 + 2x + \quad)$$
$$2y = 2 - (x^2 + 2x + 1)$$
$$(x + 1)^2 = -2(y - 1)$$

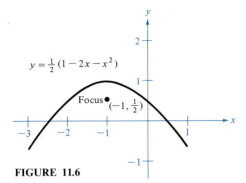

$y = \frac{1}{2}(1 - 2x - x^2)$

Focus $(-1, \frac{1}{2})$

FIGURE 11.6

Comparing this equation to the standard form $(x - h)^2 = 4p(y - k)$, we conclude that

$$h = -1, \qquad k = 1, \qquad p = -\tfrac{1}{2}$$

Since p is negative, the parabola opens downward (see Figure 11.6), and thus the focus of the parabola is

$$(h, k + p) = (-1, \tfrac{1}{2})$$

■

Example 3

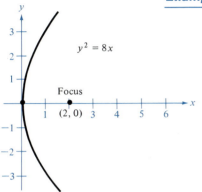

$y^2 = 8x$

Focus $(2, 0)$

FIGURE 11.7

Write the standard equation of the parabola with its vertex at the origin and its focus at $(2, 0)$.

Solution: The axis of the parabola is horizontal, passing through $(0, 0)$ and $(2, 0)$. (See Figure 11.7.) Thus we consider the form

$$(y - k)^2 = 4p(x - h)$$

where $h = k = 0$ and $p = 2$. Therefore, our equation is

$$y^2 = 8x$$

■

Example 4

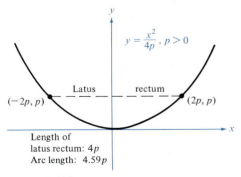

$y = \frac{x^2}{4p}, \, p > 0$

Latus rectum

$(-2p, p)$ $(2p, p)$

Length of
latus rectum: $4p$
Arc length: $4.59p$

FIGURE 11.8

The **latus rectum** of a parabola is the chord through the focus, perpendicular to the axis. For the parabola $y = x^2/4p$, the latus rectum is the line segment joining the points $(-2p, p)$ and $(2p, p)$. (See Figure 11.8.) Find the arc length of the parabola between $(-2p, p)$ and $(2p, p)$.

Solution: The arc length is given by

$$s = \int_{-2p}^{2p} \sqrt{1 + (y')^2}\, dx$$

$$= 2 \int_0^{2p} \sqrt{1 + \left(\frac{x}{2p}\right)^2}\, dx = \frac{1}{p} \int_0^{2p} \sqrt{4p^2 + x^2}\, dx$$

$$= \frac{1}{2p}\left[x\sqrt{4p^2 + x^2} + 4p^2 \ln |x + \sqrt{4p^2 + x^2}| \right]_0^{2p}$$

$$= \frac{1}{2p}[2p\sqrt{8p^2} + 4p^2 \ln(2p + \sqrt{8p^2}) - 4p^2 \ln(2p)]$$

$$= 2p[\sqrt{2} + \ln(1 + \sqrt{2})] \approx 4.59p$$

■

One of the interesting properties of parabolas is illustrated in the following example.

Example 5

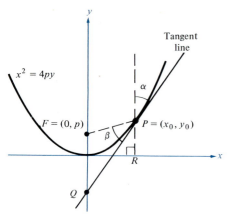

FIGURE 11.9

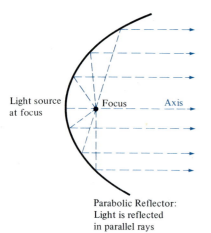

Parabolic Reflector:
Light is reflected
in parallel rays

FIGURE 11.10

Consider the parabola $x^2 = 4py$ and any point $P = (x_0, y_0)$ on the parabola. Using Figure 11.9 show that $\alpha = \beta$.

Solution: If we consider the line through R and P to be parallel to the y-axis, then $\alpha = \angle FQP$ and we can show that $\alpha = \beta$ by showing that $\triangle FQP$ is an isosceles triangle. By differentiating the equation $x^2 = 4py$, we have

$$\frac{dy}{dx} = \frac{x}{2p}$$

and thus the equation for the tangent line at $P = (x_0, y_0)$ is given by

$$y - y_0 = \frac{x_0}{2p}(x - x_0)$$

By letting $x = 0$ we see that

$$Q = \left(0, y_0 - \frac{x_0^2}{2p}\right) = \left(0, \frac{x_0^2}{4p} - \frac{x_0^2}{2p}\right) = \left(0, -\frac{x_0^2}{4p}\right)$$

Thus the distance between F and Q is given by

$$p + \frac{x_0^2}{4p}$$

while the distance between F and P is given by

$$\sqrt{x_0^2 + (y_0 - p)^2} = \sqrt{x_0^2 + \left(\frac{x_0^2}{4p} - p\right)^2} = \sqrt{\left(\frac{x_0^2}{4p} + p\right)^2} = \frac{x_0^2}{4p} + p$$

Therefore, $\triangle FQP$ is isosceles and we conclude that $\alpha = \beta$. ∎

The result of Example 5 is used in the construction of parabolic reflectors. In such a reflector, beams from a light source at the focus are reflected in a direction parallel to the axis of the parabola (Figure 11.10). Conversely, the parabolic mirror of a reflecting telescope reflects through the focus all beams of light coming in parallel to the axis.

Section Exercises (11.1)

In Exercises 1–20, find the vertex, focus, and directrix of the parabola.

1. $y = 4x^2$
2. $y = 2x^2$
S 3. $y^2 = -6x$
4. $y^2 = 3x$
5. $x^2 + 8y = 0$
6. $x + y^2 = 0$
7. $(x - 1)^2 + 8(y + 2) = 0$

8. $(x + 3) + (y - 2)^2 = 0$
9. $(y + \frac{1}{2})^2 = 2(x - 5)$
10. $(x + \frac{1}{2})^2 - 4(y - 3) = 0$
S 11. $y = \frac{1}{4}(x^2 - 2x + 5)$
12. $y = -\frac{1}{6}(x^2 + 4x - 2)$
13. $4x - y^2 - 2y - 33 = 0$

14. $y^2 + x + y = 0$

15. $y^2 + 6y + 8x + 25 = 0$

16. $x^2 - 2x + 8y + 9 = 0$

[S] **17.** $y^2 - 4y - 4x = 0$

18. $y^2 - 4x - 4 = 0$

19. $x^2 + 4x + 4y - 4 = 0$

20. $y^2 + 4y + 8x - 12 = 0$

In Exercises 21–32, find an equation of the specified parabola.

21. vertex: $(0, 0)$; focus: $(0, -\frac{3}{2})$

22. vertex: $(0, 0)$; focus: $(2, 0)$

[S] **23.** vertex: $(3, 2)$; focus: $(1, 2)$

24. vertex: $(-1, 2)$; focus: $(-1, 0)$

25. vertex: $(0, 4)$; directrix: $y = 2$

26. vertex: $(-2, 1)$; directrix: $x = 1$

27. focus: $(0, 0)$; directrix: $y = 4$

28. focus: $(2, 2)$; directrix: $x = -2$

[S] **29.** axis: parallel to y-axis; passes through points $(0, 3)$, $(3, 4)$, $(4, 11)$

30. axis: parallel to x-axis; passes through points $(4, -2)$, $(0, 0)$, $(3, -3)$

[S] **31.**

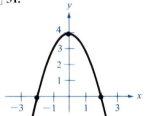

32.

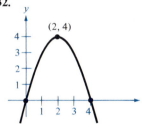

33. Find the equations of the parabolas with a common directrix $y = 1$ and a latus rectum of length 8. (See Example 4.)

34. Find an equation of the parabola with directrix $y = -2$ and latus rectum joining the points $(0, 2)$ and $(8, 2)$. (See Example 4.)

[C][S] **35.** Each cable of a particular suspension bridge is suspended (in the shape of a parabola) between two towers

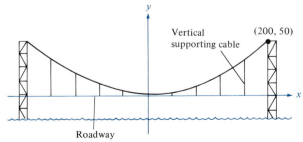

FIGURE 11.11

that are 400 ft apart and 50 ft above the roadway. (See Figure 11.11.) If the cables touch the roadway midway between the towers, find the following:

(a) an equation for the parabolic shape of each cable

(b) the length of one of the suspension cables

36. An earth satellite in a 100-mi-high circular orbit around the earth has a velocity of approximately 17,500 mi/h. If this velocity is multiplied by $\sqrt{2}$, then the satellite will have the minimum velocity necessary to escape the earth's gravity and it will follow a parabolic path with the center of the earth as the focus. (See Figure 11.12.)

(a) Find the escape velocity of the satellite.

(b) Find an equation of its path (assume that the radius of the earth is 4,000 mi).

[S] **37.** Water is flowing from a horizontal pipe 48 ft above the ground at a rate of 10 ft/s. The falling stream of water has the shape of a parabola whose vertex, $(0, 48)$, is at the end of the pipe. (See Figure 11.13.)

(a) Where does the water hit the ground?

(b) Find the equation of the parabola.

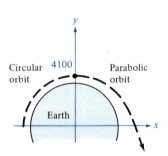

FIGURE 11.12

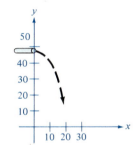

FIGURE 11.13

38. Find the equation of the tangent line to the parabola $y = ax^2$ at $x = x_0$. Prove that the x-intercept of this tangent line is $(x_0/2, 0)$.

39. Prove that any two distinct tangent lines to a parabola always intersect.

40. Prove that if any two tangent lines to a parabola intersect at right angles, then their point of intersection must lie on the directrix.

41. Prove that two tangent lines to a parabola intersect at right angles if and only if the focus of the parabola lies on the line segment connecting the two points of tangency.

42. Demonstrate the results of Exercises 39 and 40 for the parabola $y = \frac{1}{4}(x^2 - 4x + 8)$.

43. Prove that the area enclosed by a parabola and a line parallel to its directrix is $\frac{2}{3}$ the area of the circumscribed rectangle.

For Review and Class Discussion

True or False

1. _____ It is possible for a parabola to intersect its directrix.

2. _____ It is possible to find a parabola passing through the points $(-2, 0)$, $(-1, 3)$, $(0, 4)$, $(1, 3)$, and $(2, 0)$.

3. _____ It is possible to find a parabola passing through the points $(-2, 0)$, $(-1, 2)$, $(0, 3)$, $(1, 2)$, and $(2, 0)$.

4. _____ The graph of $y = x^4$ is a parabola.

5. _____ The point on a parabola closest to its focus is its vertex.

11.2 Ellipses

Purpose

- To define an ellipse and use the definition to investigate some properties of ellipses.

But for the discovery of conic sections, which was probably regarded in Plato's time and long after him, as the unprofitable amusement of a speculative brain, the whole course of practical philosophy of the present day, of the science of astronomy, of the theory of projectiles, of the art of navigation, might have run in a different channel. —James Sylvester (1814–1897)

In this text we have already assumed that you have some familiarity with ellipses, and in Section 1.5 we pointed out that graphs of equations of the form

$$\frac{(x - h)^2}{a^2} + \frac{(y - k)^2}{b^2} = 1$$

are ellipses. In this section we give a precise definition of an ellipse and verify that the standard form of the equation of an ellipse is precisely the form introduced in Section 1.5.

Definition of Ellipse

An **ellipse** is the set of all points (x, y) the sum of whose distances from two distinct fixed points **(foci)** is constant. (See Figure 11.14.)

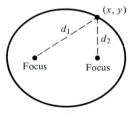

$d_1 + d_2 = \text{constant}$

FIGURE 11.14

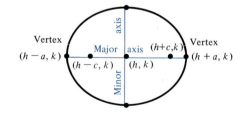

One way to visualize this definition of an ellipse is to consider two thumbtacks placed at the foci (Figure 11.15). If we fasten the ends of a

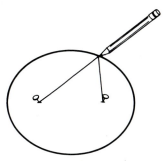

FIGURE 11.15

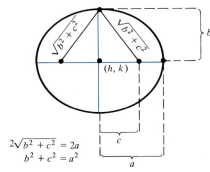

$2\sqrt{b^2 + c^2} = 2a$
$b^2 + c^2 = a^2$

FIGURE 11.16

fixed length of string to the thumbtacks and draw the string taut with a pencil, then the traceable path of the pencil will be an ellipse.

Referring to Figure 11.14, the line through the foci intersects the ellipse at two points called the **vertices.** The chord joining the vertices is called the **major axis** and its midpoint is called the **center** of the ellipse. The chord perpendicular to the major axis at the center is called the **minor axis** of the ellipse.

To derive the standard form of the equation of an ellipse, consider the ellipse in Figure 11.16 with the following points:

$$\text{center: } (h, k); \quad \text{vertices: } (h \pm a, k); \quad \text{foci: } (h \pm c, k)$$

From the definition of an ellipse, we know that the sum of the distances from any point on the ellipse to the two foci is constant. By considering the position of a taut string when the pencil is at one of the *vertices,* we can see that the length of the string would be

$$(a + c) + (a - c) = 2a$$

or simply the length of the major axis.

Now if we let (x, y) be any point on the ellipse, then the sum of the distances between this point and the two foci must also be $2a$. That is,

$$\sqrt{[x - (h - c)]^2 + (y - k)^2} + \sqrt{[x - (h + c)]^2 + (y - k)^2} = 2a$$

which reduces to

$$(a^2 - c^2)(x - h)^2 + a^2(y - k)^2 = a^2(a^2 - c^2)$$

However, from Figure 11.16 we can see that $b^2 = a^2 - c^2$, and therefore the equation of the ellipse is

$$b^2(x - h)^2 + a^2(y - k)^2 = a^2 b^2$$

$$\frac{(x - h)^2}{a^2} + \frac{(y - k)^2}{b^2} = 1$$

Had we chosen the major axis to be parallel to the y-axis, we would have obtained a similar equation. Both results are summarized in the following theorem:

THEOREM 11.2
(*Standard Equation of an Ellipse*)

The standard form of the equation of an ellipse, with center (h, k) and major and minor axes of lengths $2a$ and $2b$, respectively, is

$$\frac{(x - h)^2}{a^2} + \frac{(y - k)^2}{b^2} = 1 \qquad \text{(major axis parallel to the } x\text{-axis)}$$

or

$$\frac{(x - h)^2}{b^2} + \frac{(y - k)^2}{a^2} = 1 \qquad \text{(major axis parallel to the } y\text{-axis)}$$

The foci lie on the major axis, c units from the center, with $c^2 = a^2 - b^2$.

Example 1

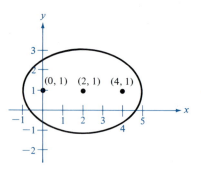

FIGURE 11.17

Find the standard form of the equation of the ellipse having foci at (0, 1) and (4, 1) and with a major axis of length 6. (See Figure 11.17.)

Solution: Since the foci occur at (0, 1) and (4, 1), the center of the ellipse is (2, 1), and $c = 2$ is the distance from the center to one of the foci. Furthermore, $2a = 6$ or $a = 3$, and since $c^2 = a^2 - b^2$, it follows that

$$b = \sqrt{a^2 - c^2} = \sqrt{9 - 4} = \sqrt{5}$$

Therefore, the standard equation of this ellipse, whose major axis is parallel to the *x*-axis, is

$$\frac{(x - 2)^2}{9} + \frac{(y - 1)^2}{5} = 1$$

∎

Example 2

Find the center, vertices, and foci of the ellipse given by the equation $4x^2 - 8x + y^2 + 4y - 8 = 0$.

Solution: By completing the square we can write the equation $4x^2 - 8x + y^2 + 4y - 8 = 0$ in the form

$$4(x^2 - 2x + 1) + (y^2 + 4y + 4) = 8 + 4 + 4$$
$$4(x - 1)^2 + (y + 2)^2 = 16$$
$$\frac{(x - 1)^2}{4} + \frac{(y + 2)^2}{16} = 1$$

This is the equation of an ellipse with its major axis parallel to the *y*-axis, where $h = 1$, $k = -2$, $a = 4$, $b = 2$, and $c = \sqrt{16 - 4} = 2\sqrt{3}$. Therefore, we have

$$\text{center: } (1, -2); \quad \text{vertices: } (1, -6), (1, 2)$$
$$\text{foci: } (1, -2 - 2\sqrt{3}), (1, -2 + 2\sqrt{3})$$

∎

Various physical phenomena involve ellipses. Ellipses are used in acoustical design (see the reflective property of ellipses described in Example 4). Supporting arches and machine gears sometimes have elliptical shapes, and the orbits of satellites and planets are ellipses. In the next example we investigate the elliptical orbit of the moon about the earth.

Example 3

The moon travels about the earth in an elliptical orbit with the earth at one focus. (See Figure 11.18.) If the major and minor axes of the orbit have lengths 774,000 km and 773,000 km, respectively, what are the

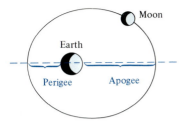

FIGURE 11.18

greatest and least distances (the apogee and perigee) from the earth's center to the moon's center?

Solution: Since

$$2a = 774{,}000 \quad \text{and} \quad 2b = 773{,}000$$

we have

$$a = 387{,}000, \quad b = 386{,}500, \quad c = \sqrt{a^2 - b^2} \approx 19{,}500$$

Therefore, the greatest distance between the centers of the earth and the moon is

$$a + c \approx 406{,}500 \text{ km}$$

and the least distance is

$$a - c \approx 367{,}000 \text{ km} \qquad \blacksquare$$

Example 4

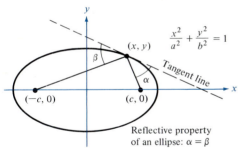

FIGURE 11.19

If (x, y) is any point on the ellipse

$$\frac{x^2}{a^2} + \frac{y^2}{b^2} = 1$$

show that the angles α and β, shown in Figure 11.19, are equal.

Solution: To simplify the proof we assume that $0 < x < c, 0 < y$, and θ is the acute angle between the tangent line and the horizontal axis, as shown in Figure 11.20. Since

$$\tan \theta = \left| \frac{dy}{dx} \right| = \left| \frac{-b^2 x}{a^2 y} \right| = \frac{b^2 x}{a^2 y}$$

and

$$\tan (\theta + \alpha) = \left| \frac{y}{x - c} \right| = \frac{y}{c - x}$$

we can apply the trigonometric identity

$$\tan (\theta + \alpha) = \frac{\tan \theta + \tan \alpha}{1 - \tan \theta \tan \alpha}$$

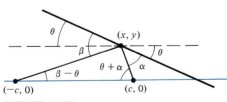

FIGURE 11.20

to obtain

$$\frac{y}{c - x} = \frac{(b^2 x / a^2 y) + \tan \alpha}{1 - (b^2 x / a^2 y)(\tan \alpha)}$$

Finally, we can use the fact that $b^2 x^2 + a^2 y^2 = a^2 b^2$ and $a^2 - b^2 = c^2$ to solve this equation for $\tan \alpha$ and conclude that

$$\tan \alpha = \frac{b^2}{cy}$$

In a similar manner, using angle $\beta - \theta$ in Figure 11.20, we have

$$\tan(\beta - \theta) = \frac{y}{x+c} = \frac{\tan\beta - \tan\theta}{1 + \tan\beta\tan\theta} = \frac{\tan\beta - (b^2x/a^2y)}{1 + (\tan\beta)(b^2x/a^2y)}$$

Solving this equation for $\tan\beta$, we have

$$\tan\beta = \frac{b^2}{cy}$$

and we conclude that $\alpha = \beta$.   ■

In every ellipse the foci are located along the major axis between the vertices and the center. This means that $0 < c < a$, and therefore as $c \to 0$, the foci approach the center and the ellipse becomes more circular. As $c \to a$, the foci approach the vertices and the ellipse flattens out (see Figure 11.21). Thus it appears that the shape of an ellipse can be

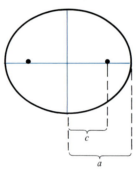

Ellipse becomes circular as $\frac{c}{a} \to 0$

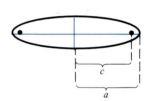

Ellipse flattens as $\frac{c}{a} \to 1$

FIGURE 11.21

described in terms of the relative sizes of a and c. This common measure of the shape of an ellipse is called its **eccentricity**, which is defined as follows:

Definition of Eccentricity

The **eccentricity** e of an ellipse is given by the ratio

$$e = \frac{c}{a}$$

Note that $0 < e < 1$ for every ellipse.

Example 5

Show that the circumference of the ellipse

$$\frac{x^2}{a^2} + \frac{y^2}{b^2} = 1$$

is
$$4a \int_0^{\pi/2} \sqrt{1 - e^2 \sin^2 \theta}\, d\theta$$

where $e = c/a$.

Solution: Since the given ellipse is symmetric with respect to both the x- and y-axis, we know that its circumference C is four times the arc length of $y = (b/a)\sqrt{a^2 - x^2}$ in the first quadrant. Thus

$$C = 4 \int_0^a \sqrt{1 + (y')^2}\, dx$$

$$= 4 \int_0^a \sqrt{1 + \frac{b^2 x^2}{a^2(a^2 - x^2)}}\, dx$$

Using the trigonometric substitution $x = a \sin \theta$, we have

$$C = 4 \int_0^{\pi/2} \sqrt{1 + \frac{b^2 \sin^2 \theta}{a^2 \cos^2 \theta}}(a \cos \theta)\, d\theta$$

Since
$$e^2 = \frac{c^2}{a^2} = \frac{a^2 - b^2}{a^2}$$

we can rewrite this integral (see Exercise 44) in the form

$$4a \int_0^{\pi/2} \sqrt{1 - e^2 \sin^2 \theta}\, d\theta$$

Integrals of this type are called *elliptic integrals*.   ∎

Elliptic integrals do not, in general, have elementary antiderivatives. Thus to find the circumference of an ellipse, we must use an approximation technique, as shown in the following example.

Example 6

Use the elliptic integral in Example 5 to approximate the circumference of the ellipse

$$\frac{x^2}{25} + \frac{y^2}{16} = 1$$

Solution: Since
$$e^2 = \frac{a^2 - b^2}{a^2} = \frac{9}{25}$$

we have

$$\text{circumference} = (4)(5) \int_0^{\pi/2} \sqrt{1 - \frac{9 \sin^2 \theta}{25}}\, d\theta$$

Applying Simpson's Rule ($n = 4$) we have

$$20 \int_0^{\pi/2} \sqrt{1 - \frac{9 \sin^2 \theta}{25}} \, d\theta$$

$$\approx 20 \left(\frac{\pi}{6}\right)\left(\frac{1}{4}\right)[1 + 4(0.9733) + 2(0.9055) + 4(0.8323) + 0.8]$$

$$\approx 28.36 \qquad \blacksquare$$

Section Exercises (11.2)

In Exercises 1–20, find the center, foci, vertices, and eccentricity of each ellipse.

1. $\dfrac{x^2}{25} + \dfrac{y^2}{16} = 1$

2. $\dfrac{x^2}{144} + \dfrac{y^2}{169} = 1$

3. $\dfrac{x^2}{16} + \dfrac{y^2}{25} = 1$

4. $\dfrac{x^2}{169} + \dfrac{y^2}{144} = 1$

5. $\dfrac{x^2}{9} + \dfrac{y^2}{5} = 1$

6. $\dfrac{x^2}{28} + \dfrac{y^2}{64} = 1$

[S] **7.** $x^2 + 4y^2 = 4$

8. $5x^2 + 3y^2 = 15$

9. $3x^2 + 2y^2 = 6$

10. $5x^2 + 7y^2 = 70$

11. $4x^2 + y^2 = 1$

12. $16x^2 + 25y^2 = 1$

13. $\dfrac{(x-1)^2}{9} + \dfrac{(y-5)^2}{25} = 1$

14. $(x+2)^2 + 4(y+4)^2 = 1$

[S] **15.** $9x^2 + 4y^2 + 36x - 24y + 36 = 0$

16. $9x^2 + 4y^2 - 36x + 8y + 31 = 0$

17. $16x^2 + 25y^2 - 32x + 50y + 31 = 0$

18. $9x^2 + 25y^2 - 36x - 50y + 61 = 0$

[S] **19.** $12x^2 + 20y^2 - 12x + 40y - 37 = 0$

20. $36x^2 + 9y^2 + 48x - 36y + 43 = 0$

In Exercises 21–30, find an equation for the specified ellipse.

21. center $(0, 0)$; focus $(2, 0)$; vertex $(3, 0)$

22. center $(0, 0)$; vertex $(2, 0)$; minor axis of length 3

[S] **23.** vertices $(5, 0)$, $(-5, 0)$; eccentricity $\frac{3}{5}$

24. vertices $(0, 8)$, $(0, -8)$; eccentricity $\frac{1}{2}$

[S] **25.** vertices $(0, 2)$, $(4, 2)$; minor axis of length 2

26. foci $(-2, 0)$, $(2, 0)$; major axis of length 8

27. vertices $(3, 1)$, $(3, 9)$; minor axis of length 6

28. center $(0, 0)$; major axis horizontal; curve passes through points $(3, 1)$ and $(4, 0)$

29. foci $(0, 5)$, $(0, -5)$; sum of distances from the foci to any point on the ellipse is 14

[S] **30.** center $(1, 2)$; major axis parallel to the y-axis; curve passes through points $(1, 6)$ and $(3, 2)$

31. A *latus rectum* of an ellipse is a chord passing through a

focus and perpendicular to the major axis. Show that the length of a latus rectum is $2b^2/a$.

32. With the result of Exercise 31, the endpoints of each latus rectum are easily plotted. Sketch the graph of each of the following, making use of the endpoints of the latera recta (plural of latus rectum).

(a) $\dfrac{x^2}{4} + \dfrac{y^2}{1} = 1$ (b) $6x^2 + 4y^2 = 3$

(c) $5x^2 + 3y^2 = 15$

33. Find the equations of the tangent lines to the ellipse $(x^2/16) + (y^2/25) = 1$ when $x = 3$.

34. Use the points where the derivative is zero or undefined to find the endpoints of the major and minor axes of the graph of $9x^2 + 4y^2 + 36x - 24y + 36 = 0$.

[S] **35.** A particle is traveling clockwise on the elliptical orbit given by

$$\frac{x^2}{10^2} + \frac{y^2}{5^2} = 1$$

The particle leaves the orbit at the point $(-8, 3)$ and travels in a straight line tangent to the ellipse. At which point will the particle cross the y-axis?

36. Show that the equation of an ellipse with its center at the origin can be written

$$\frac{x^2}{a^2} + \frac{y^2}{a^2(1 - e^2)} = 1$$

Show that as e approaches zero, with a remaining fixed, the ellipse approaches a circle.

[C] **37.** Given the region bounded by the ellipse $(x^2/4) + (y^2/1) = 1$, find the following:

(a) its area

(b) the volume of the solid generated by revolving the region about its major axis (prolate spheroid)

(c) the surface area of the prolate spheroid of part (b)

(d) the volume of the solid generated by revolving the region about its minor axis (oblate spheroid)

(e) the surface area of the oblate spheroid of part (d)

(f) its circumference

38. Show that the eccentricity of the ellipse $(x^2/a^2) + (y^2/b^2) = 1$ is identical to the eccentricity of

$$\frac{(tx)^2}{a^2} + \frac{(ty)^2}{b^2} = 1$$

for any real t. Give a geometrical explanation of this result.

39. A line segment, 9 in. long, moves so that one endpoint is always on the y-axis and the other always on the x-axis. Find the equation of the curve traced by a point (on the line segment) 6 in. from the endpoint on the y-axis.

40. The earth moves in an elliptical orbit with the sun at one of the foci. If the length of half the major axis is 93 million miles and the eccentricity is 0.017, find the least and greatest distances of the earth from the sun.

41. If the distances to the apogee and the perigee of an elliptical orbit of an earth satellite are measured from the center of the earth, show that the eccentricity of the orbit is given by $e = (A - P)/(A + P)$, where A and P are the apogee and perigee distances, respectively.

42. *Sputnik I*, orbited by the Russians in October 1957, had 583 mi and 132 mi above the earth's surface as the highest and lowest points of its elliptical orbit. What is the eccentricity of this orbit?

43. On 26 November 1963 the United States launched *Explorer 18*. Its low point over the surface of the earth was 119 mi and its high point was 122,000 mi from the surface of the earth.

(a) Find the eccentricity of its elliptical orbit.

(b) Find an equation that describes its orbit.

44. Complete Example 5 by showing that

$$\int_0^{\pi/2} \sqrt{1 + \frac{b^2 \sin^2 \theta}{a^2 \cos^2 \theta}}\,(a \cos \theta)\,d\theta$$
$$= a \int_0^{\pi/2} \sqrt{1 - e^2 \sin^2 \theta}\,d\theta$$

For Review and Class Discussion

True or False

1. _____ If C is the circumference of the ellipse $(x^2/a^2) + (y^2/b^2) = 1$, then $2\pi b \leq C \leq 2\pi a$.

2. _____ The maximum curvature of the ellipse $(x^2/b^2) + (y^2/a^2) = 1$ is $K = a/b^2$.

3. _____ If $0 < p < q$, then the eccentricity of the ellipse $(x^2/a^2) + (y^2/b^2) = q^2$ is greater than the eccentricity of the ellipse $(x^2/a^2) + (y^2/b^2) = p^2$.

4. _____ For every value of θ, the point $(a \sin \theta, b \cos \theta)$ lies on the ellipse $(x^2/a^2) + (y^2/b^2) = 1$.

5. _____ The graph of $(x^2/4) + y^4 = 1$ is an ellipse.

11.3 Hyperbolas

Purpose

- To define a hyperbola and use the definition to investigate some properties of hyperbolas.

In particular, if (in the equation $y^2 = a + bx + cx^2$) the term cx^2 is zero, the conic section is a parabola; if it is positive, it is a hyperbola; and finally, if it is negative, it is an ellipse.   —René Descartes (1596–1650)

The last conic we will consider is the hyperbola. The definition of a hyperbola parallels that of an ellipse. The distinction is that for an ellipse the *sum* of the distances between the foci and a point on the ellipse is fixed, while for a hyperbola the *difference* of these distances is fixed.

Definition of Hyperbola

A **hyperbola** is the set of all points (x, y) the difference of whose distances from two distinct fixed points (foci) is constant. (See Figure 11.22.)

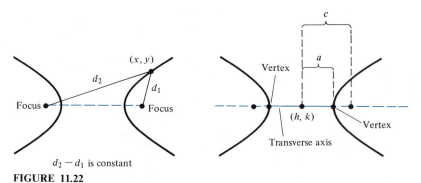

$d_2 - d_1$ is constant

FIGURE 11.22

One distinguishing feature of hyperbolas is that their graphs have two separate branches. Referring to Figure 11.22, the line through the two foci intersects the hyperbola at two points called the **vertices.** The line segment connecting the vertices is called the **transverse axis,** and the midpoint of the transverse axis is called the **center** of the hyperbola.

The development of the standard form of the equation of a hyperbola is similar to that for an ellipse and we list the following theorem without proof.

THEOREM 11.3
(*Standard Equation of a Hyperbola*)

The standard form of the equation of a hyperbola with center at (h, k) is

$$\frac{(x - h)^2}{a^2} - \frac{(y - k)^2}{b^2} = 1 \qquad \text{(transverse axis is } \textit{horizontal}\text{)}$$

$$\frac{(y - k)^2}{a^2} - \frac{(x - h)^2}{b^2} = 1 \qquad \text{(transverse axis is } \textit{vertical}\text{)}$$

where the vertices and foci are, respectively, a and c units from the center and $b^2 = c^2 - a^2$.

Note in Figure 11.22 that the fixed difference is $d_1 - d_2 = 2a$.

Example 1

Find the standard form of the equation of the hyperbola with foci at $(-1, 2)$ and $(5, 2)$ and vertices at $(0, 2)$ and $(4, 2)$.

Solution: By the Midpoint Formula the center of the hyperbola occurs at the point $(2, 2)$. Furthermore, $c = 3$ and $a = 2$, and it follows that

$$b^2 = 3^2 - 2^2 = 9 - 4 = 5$$

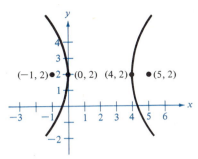

FIGURE 11.23

Thus the equation of this hyperbola is

$$\frac{(x-2)^2}{4} - \frac{(y-2)^2}{5} = 1$$

(See Figure 11.23.)   ∎

An important aid in sketching the graph of a hyperbola is the determination of its **asymptotes** (see Figure 11.24). Each hyperbola has two asymptotes: two straight lines that intersect at the center of the hyperbola. Furthermore, the asymptotes pass through the vertices of a rectangle, of dimension $2a$ by $2b$, with its center at (h, k). The line segment of length $2b$, joining $(h, k + b)$ and $(h, k - b)$, is referred to as the **conjugate axis** of the hyperbola.

The following theorem identifies the equation for the asymptotes of a hyperbola.

THEOREM 11.4
(Asymptotes of a Hyperbola)

If a hyperbola has a *horizontal* transverse axis, then its asymptotes are the lines

$$y = k + \frac{b}{a}(x - h) \quad \text{and} \quad y = k - \frac{b}{a}(x - h)$$

If a hyperbola has a *vertical* transverse axis, then its asymptotes are the lines

$$y = k + \frac{a}{b}(x - h) \quad \text{and} \quad y = k - \frac{a}{b}(x - h)$$

Proof: Solving for y in the equation

$$\frac{(x-h)^2}{a^2} - \frac{(y-k)^2}{b^2} = 1$$

gives us

$$y = k \pm \frac{b}{a}\sqrt{(x-h)^2 - a^2}$$

If the line $y = k + (b/a)(x - h)$ is to be an asymptote to the hyperbola, then the difference of the y-values of a point on the hyperbola and a point on the asymptote must approach zero as x becomes infinite. To prove this, observe that

$$\left[k + \frac{b}{a}\sqrt{(x-h)^2 - a^2} \right] - \left[k + \frac{b}{a}(x - h) \right]$$

$$= \frac{b}{a}[\sqrt{(x-h)^2 - a^2} - (x - h)]$$

$$= \frac{b}{a}\left\{ \frac{[(x-h)^2 - a^2] - (x - h)^2}{\sqrt{(x-h)^2 - a^2} + (x - h)} \right\}$$

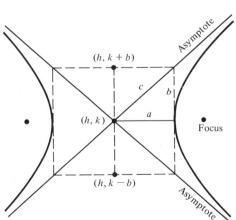

FIGURE 11.24

$$= \frac{-ab}{\sqrt{(x-h)^2 - a^2} + (x-h)}$$

Now by taking the limit as $x \to \infty$, we have

$$\lim_{x \to \infty} \frac{-ab}{\sqrt{(x-h)^2 - a^2} + (x-h)} = 0$$

Therefore, that portion of the hyperbola given by

$$y = k + \frac{b}{a}\sqrt{(x-h)^2 - a^2}$$

has the line

$$y = k + \frac{b}{a}(x-h)$$

as an asymptote. The asymptotic behavior of the other three portions of the hyperbola may be shown in a similar manner, thus establishing our theorem.   ∎

Note from Figure 11.24 that geometrically the asymptotes coincide with the diagonals of the rectangle centered at (h, k), with dimensions $2a$ and $2b$. This provides us with a quick means of sketching the asymptotes, which in turn aids in sketching the hyperbola.

Example 2

Sketch the graph of the equation $4x^2 + 8x - 3y^2 + 16 = 0$.

Solution: Rewriting this equation in standard form, we have

$$4x^2 + 8x - 3y^2 + 16 = 0$$
$$4(x^2 + 2x) - 3y^2 = -16$$
$$-4(x^2 + 2x + 1) + 3y^2 = 16 - 4$$
$$-4(x+1)^2 + 3y^2 = 12$$
$$\frac{y^2}{4} - \frac{(x+1)^2}{3} = 1$$

From this equation we conclude that the hyperbola is centered at $(-1, 0)$, has vertices at $(-1, 2)$ and $(-1, -2)$, and is asymptotic to the lines

$$y = \frac{2}{\sqrt{3}}(x+1) \qquad \text{and} \qquad y = -\frac{2}{\sqrt{3}}(x+1)$$

Once the asymptotes and vertices are sketched, as shown in Figure 11.25, it is relatively easy to complete the sketch of the hyperbola.   ∎

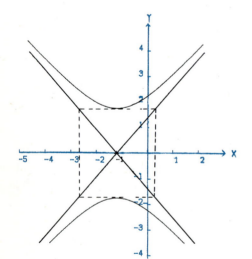

FIGURE 11.25

Example 3

Find the standard form of the equation of the hyperbola having vertices at $(3, -5)$ and $(3, 1)$ and with asymptotes $y = 2x - 8$ and $y = -2x + 4$.

Solution: By the Midpoint Formula the center of the hyperbola is at $(3, -2)$. Furthermore, the hyperbola has a vertical transverse axis with $a = 3$. By Theorem 11.4 the asymptotes have equations whose slopes are, respectively,

$$m_1 = \frac{a}{b} \quad \text{and} \quad m_2 = -\frac{a}{b}$$

From the given equations of the asymptotes, we know that

$$\frac{a}{b} = 2 \quad \text{and} \quad -\frac{a}{b} = -2$$

and since $a = 3$, we conclude that $b = \frac{3}{2}$. Therefore, the standard equation of the hyperbola is

$$\frac{(y + 2)^2}{9} - \frac{(x - 3)^2}{\frac{9}{4}} = 1$$

or we write

$$\frac{(y + 2)^2}{9} - \frac{4(x - 3)^2}{9} = 1 \qquad \blacksquare$$

Example 4

Two microphones, 1 mi apart, record an explosion. Microphone A received the sound 2 seconds before microphone B. Where did the explosion come from?

Solution: Since sound travels at 1100 ft/s, we know that the explosion took place 2200 ft further from B than from A. (See Figure 11.26.) The locus of all points that are 2200 ft closer to A than to B is, by definition, one branch of the hyperbola

$$\frac{x^2}{a^2} - \frac{y^2}{b^2} = 1$$

where

$$c = \frac{5280}{2} = 2640$$

$$a = \frac{2200}{2} = 1100$$

$$b^2 = c^2 - a^2 = 5{,}759{,}600$$

Thus the explosion occurred somewhere on the right branch of the hyperbola

$$\frac{x^2}{1{,}210{,}000} - \frac{y^2}{5{,}759{,}600} = 1 \qquad \blacksquare$$

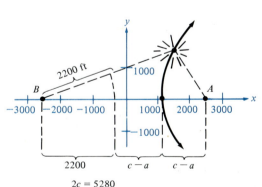

$2c = 5280$
$2200 + 2(c - a) = 5280$

FIGURE 11.26

In Example 4 we were able to determine the hyperbola on which the explosion occurred but not the exact location of the explosion. If, however, we had received the sound from a third position C, two other hyper-

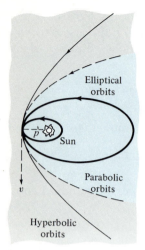

FIGURE 11.27

bolas would be determined. The exact location of the explosion would be the point where these three hyperbolas intersect.

Another interesting application of conic sections involves the orbits of comets in our solar system. Of the 566 comets identified prior to 1960, 211 have elliptical orbits, 290 have parabolic orbits, and 65 have hyperbolic orbits. The center of the sun is a focus point of each of these orbits and each orbit has a vertex at the point where the comet is closest to the sun (Figure 11.27).

If p is the distance between the vertex and the focus, and v is the velocity of the comet at the vertex, then the orbit is:

1.   an ellipse if $v < \sqrt{\dfrac{2GM}{p}}$

2.   a parabola if $v = \sqrt{\dfrac{2GM}{p}}$

3.   a hyperbola if $v > \sqrt{\dfrac{2GM}{p}}$

where M is the mass of the sun and $G \approx 6.67(10^{-8})$ centimeters^3 per gram second^2.

Example 5

For Figure 11.27 prove that the curvature (at the vertex) of each of the elliptical orbits is greater than the curvature of the parabolic orbit, which in turn is greater than the curvature of the hyperbolic orbits.

Solution: For each orbit we place the origin at the vertex and the (major or transverse) axis on the y-axis. Then since $y' = 0$ at the vertex of each orbit, the curvature of each orbit at the vertex is

$$K = \frac{|y''|}{[1 + (y')^2]^{3/2}} = |y''|$$

For the parabolic orbit we have

$$y = \frac{x^2}{4p}, \qquad y' = \frac{x}{2p}, \qquad y'' = \frac{1}{2p}$$

Thus for the parabolic orbit,

$$K = \frac{1}{2p}$$

For the elliptical or hyperbolic orbits, we have

$$\frac{(y \mp a)^2}{a^2} \pm \frac{x^2}{b^2} = 1$$

$$y' = \mp \frac{a^2}{b^2}\left(\frac{x}{y \mp a}\right)$$

$$y'' = \mp \frac{a^2}{b^2}\left[\frac{(y \mp a) - xy'}{(y \mp a)^2}\right]$$

$$= \frac{a}{b^2} \text{ at } (0,0)$$

For the elliptical orbits either $p = a - c$ or $p = a + c$, depending on which focus occurs at the center of the sun, and in either case $b^2 = a^2 - c^2 = (a - c)(a + c) = p(2a - p)$. Thus for the elliptical orbits the curvature at $(0,0)$ is

$$K = \frac{a}{p(2a - p)}$$

For the hyperbolic orbits we have $p = c - a$ and $b^2 = c^2 - a^2 = (c - a)(c + a) = p(2a + p)$. Thus for the hyperbolic orbits the curvature at $(0,0)$ is

$$K = \frac{a}{p(2a + p)}$$

Finally, since $a > p/2$ for the elliptical orbits and $a > 0$ for the hyperbolic ones, we can summarize the various curvatures ($K > 0$) as shown in Table 11.1.

TABLE 11.1

	Curvature at $(0,0)$	Domain of a	Range of K
Elliptical Orbits	$K = \dfrac{a}{p(2a - p)}$	$\dfrac{p}{2} < a$	$K > \dfrac{1}{2p}$
Parabolic Orbit	$K = \dfrac{1}{2p}$	not applicable	$K = \dfrac{1}{2p}$
Hyperbolic Orbits	$K = \dfrac{a}{p(2a + p)}$	$0 < a$	$K < \dfrac{1}{2p}$

■

Section Exercises (11.3)

In Exercises 1–20, find the center, vertices, and foci of the hyperbola and sketch its graph, using asymptotes as an aid.

1. $x^2 - y^2 = 1$

2. $\dfrac{x^2}{9} - \dfrac{y^2}{16} = 1$

[S] **3.** $y^2 - \dfrac{x^2}{4} = 1$

4. $\dfrac{y^2}{9} - x^2 = 1$

5. $\dfrac{y^2}{25} - \dfrac{x^2}{144} = 1$

6. $\dfrac{x^2}{36} - \dfrac{y^2}{4} = 1$

7. $2x^2 - 3y^2 = 6$
8. $3y^2 = 5x^2 + 15$
9. $5y^2 = 4x^2 + 20$
10. $7x^2 - 3y^2 = 21$

$\boxed{S}$ 11. $\dfrac{(x-1)^2}{4} - \dfrac{(y+2)^2}{1} = 1$

12. $\dfrac{(x+1)^2}{144} - \dfrac{(y-4)^2}{25} = 1$

13. $(y+6)^2 - (x-2)^2 = 1$

14. $\dfrac{(y-1)^2}{\frac{1}{4}} - \dfrac{(x+3)^2}{\frac{1}{9}} = 1$

$\boxed{S}$ 15. $9x^2 - y^2 - 36x - 6y + 18 = 0$
16. $x^2 - 9y^2 + 36y - 72 = 0$
$\boxed{S}$ 17. $9y^2 - x^2 + 2x + 54y + 62 = 0$
18. $16y^2 - x^2 + 2x + 64y + 63 = 0$
19. $x^2 - 9y^2 + 2x - 54y - 80 = 0$
20. $9x^2 - y^2 + 54x + 10y + 55 = 0$

In Exercises 21–30, find an equation for the specified hyperbola.

21. center $(0, 0)$; one vertex $(0, 2)$; one focus $(0, 4)$
22. center $(0, 0)$; one vertex $(3, 0)$; one focus $(5, 0)$
$\boxed{S}$ 23. vertices $(-1, 0)$, $(1, 0)$; asymptotes $y = \pm 3x$
24. vertices $(0, -3)$, $(0, 3)$; asymptotes $y = \pm 3x$
25. vertices $(0, 2)$, $(6, 2)$; asymptotes $y = \frac{2}{3}x$ and $y = 4 - \frac{2}{3}x$
26. vertices $(2, 3)$, $(2, -3)$; foci $(2, 5)$, $(2, -5)$
27. vertices $(2, 3)$, $(2, -3)$; passing through point $(0, 5)$
$\boxed{S}$ 28. asymptotes: $y = \pm\frac{3}{4}x$, focus: $(10, 0)$
29. For any point on the hyperbola, the difference of its distances from $(2, 2)$ and $(10, 2)$ is 6.
30. For any point on the hyperbola, the difference of its distances from the points $(-3, 0)$ and $(-3, 3)$ is 2.

31. Find the equations for the tangent lines to the hyperbola $(x^2/9) - y^2 = 1$ when $x = 6$.
32. Find equations for the tangent lines to the hyperbola $(y^2/4) - (x^2/2) = 1$ when $x = 4$.
$\boxed{S}$ 33. Find equations for the normal lines to the hyperbola $(y^2/4) - (x^2/2) = 1$ when $x = 4$.
34. Show that

$$\frac{x_0 x}{a^2} - \frac{y_0 y}{b^2} = 1$$

is the equation of the tangent line to $(x^2/a^2) - (y^2/b^2) = 1$ at the point (x_0, y_0).

35. Given the region bounded by $(x^2/16) - (y^2/9) = 1$, $y = 0$, and $x = 5$, find the following:
(a) the area of the region
(b) the volume of the solid generated when the region is revolved around the x-axis
(c) the volume of the solid when the region is revolved around the y-axis

36. The region bounded by $x^2 - y^2 = 1$, $y = 0$, and $x = 2$ is revolved about the x-axis. Find the surface area of the solid generated.

37. A rifle, positioned at point $(-c, 0)$, is fired at a target positioned at point $(c, 0)$. Show that the positions where the sound of the rifle and the sound of the bullet hitting the target are simultaneous are on one branch of the hyperbola

$$\frac{x^2}{c^2 v_s^2 / v_m^2} - \frac{y^2}{c^2(v_m^2 - v_s^2)/v_m^2} = 1$$

The muzzle velocity of the rifle is v_m and the speed of sound is $v_s = 1100$ ft/s.

$\boxed{S}$ 38. Three listening stations located at $(4400, 0)$, $(4400, 1100)$, and $(-4400, 0)$ hear an explosion. If the latter two stations heard the sound 1 s and 5 s after the first, respectively, where did the explosion occur? Assume that the coordinate system is measured in feet and that sound travels at the rate of 1100 ft/s.

39. Prove that the ellipse

$$\frac{x^2}{a^2} + \frac{y^2}{b^2} = 1$$

and the hyperbola

$$\frac{x^2}{a^2 - b^2} - \frac{y^2}{b^2} = 1$$

intersect at right angles.

In Exercises 40–47, classify the graph of each equation as a circle, a parabola, an ellipse, or a hyperbola.

40. $x^2 + y^2 - 6x + 4y + 9 = 0$
41. $x^2 + 4y^2 - 6x + 16y + 21 = 0$
42. $4x^2 - y^2 - 4x - 3 = 0$
43. $y^2 - 4y - 4x = 0$
44. $4x^2 + 3y^2 + 8x - 24y + 51 = 0$
45. $4y^2 - 2x^2 - 4y - 8x - 15 = 0$
46. $25x^2 - 10x - 200y - 119 = 0$
47. $4x^2 + 4y^2 - 16y + 15 = 0$

For Review and Class Discussion

True or False

1. _____ The hyperbolas

$$\frac{x^2}{a^2} - \frac{y^2}{b^2} = 1 \quad \text{and} \quad \frac{y^2}{b^2} - \frac{x^2}{a^2} = 1$$

have the same asymptotes.

2. _____ If $D \neq 0$ or $E \neq 0$, then the graph of $y^2 - x^2 + Dx + Ey = 0$ is a hyperbola.

3. _____ If the asymptotes of the hyperbola $(x^2/a^2) - (y^2/b^2) = 1$ intersect at right angles, then $a = b$.

4. _____ The graph of $y^2 = (x^2 + 1)^2$ is a hyperbola.

5. _____ Every tangent line to a hyperbola intersects the hyperbola only at the point of tangency.

11.4 Rotation and the General Second-Degree Equation

Purpose

- To investigate the relationship between conic sections and equations of the form $Ax^2 + Bxy + Cy^2 + Dx + Ey + F = 0$.
- To introduce the concept of rotation of axes as an aid to sketching conics whose axes are not parallel to the x- or y-axes.

Geometric curves are best divided into orders, according to the dimensions of the equation expressing the relation between abscissa and ordinate. A curve of the first order will be a straight line and those of the second or quadratic order will be conic sections or circles.   —Isaac Newton (1642–1727)

In previous sections we have shown that the equation of a conic with axes parallel to one of the coordinate axes has one of the following standard forms:

Circle: $(x - h)^2 + (y - k)^2 = r^2$

Parabola: $(x - h)^2 = 4p(y - k)$

$(y - k)^2 = 4p(x - h)$

Ellipse: $\dfrac{(x - h)^2}{a^2} + \dfrac{(y - k)^2}{b^2} = 1$

$\dfrac{(y - k)^2}{a^2} + \dfrac{(x - h)^2}{b^2} = 1$

Hyperbola: $\dfrac{(x - h)^2}{a^2} - \dfrac{(y - k)^2}{b^2} = 1$

$\dfrac{(y - k)^2}{a^2} - \dfrac{(x - h)^2}{b^2} = 1$

Each of these standard forms can be written in the *general* form

$$Ax^2 + Cy^2 + Dx + Ey + F = 0$$

by algebraic manipulations.

In this section we investigate the equations of conics whose axes are not parallel to either the x-axis or the y-axis. Under these circumstances we will see that the general equation for such conics contains an xy term.

Definition (General Second-Degree Equation)	The equation

The equation

$$Ax^2 + Bxy + Cy^2 + Dx + Ey + F = 0$$

where A, B, and C are not all zero, is called the **general second-degree equation.**

Every conic in the xy-plane possesses an equation that fits this general second-degree form. Furthermore, every second-degree equation that has solution points has a graph that is one of the four basic conics (circle, parabola, ellipse, or hyperbola) or one of the three degenerate forms (point, line, or two intersecting lines).

We already know how to sketch the graph of a second-degree equation that contains no xy term. We first write the equation in standard form by completing the square and then sketch the graph based on the information available from the standard form. However, for a general-second degree equation containing an xy term, it is not possible to obtain a standard form by completing the square. We can overcome this problem by a **rotation of axes,** which eliminates the xy term.

In this procedure we introduce a new coordinate system by rotating the x- and y-axes counterclockwise to a new position denoted by the x'- and y'-axes (see Figure 11.28). Therefore, if we are given a general equation of a conic,

$$Ax^2 + Bxy + Cy^2 + Dx + Ey + F = 0$$

our objective is to rotate the x- and y-axes until they are parallel to the axes of the conic. Having accomplished this, then the equation of the conic in the new $x'y'$ system will have the form

$$A'(x')^2 + C'(y')^2 + D'x' + E'y' + F' = 0$$

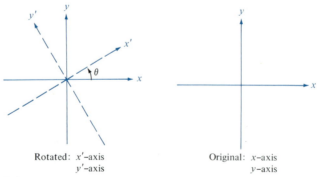

Rotated: x'–axis
y'–axis

Original: x–axis
y–axis

FIGURE 11.28

From this form, with no $x'y'$ term, we can obtain a standard form by completing the square. A sketch of the conic can then be readily made in the $x'y'$ system.

The following theorem identifies how much to rotate the axes to eliminate an xy term and also the equations for determining the new coefficients A', C', D', E', and F'.

THEOREM 11.5

(Rotation of Axes to Eliminate an xy Term)

The equation

$$Ax^2 + Bxy + Cy^2 + Dx + Ey + F = 0$$

can be rewritten as

$$A'(x')^2 + C'(y')^2 + D'x' + E'y' + F' = 0$$

by rotating the coordinate axes through an angle θ, where

$$\cot 2\theta = \frac{A - C}{B}$$

The coefficients of the new equation are given by

$$A' = A\cos^2\theta + B\cos\theta\sin\theta + C\sin^2\theta \qquad E' = -D\sin\theta + E\cos\theta$$
$$C' = A\sin^2\theta - B\cos\theta\sin\theta + C\cos^2\theta \qquad F' = F$$
$$D' = D\cos\theta + E\sin\theta$$

Proof: Using Figure 11.29 we choose a point (x, y) in the original system and attempt to find its coordinates (x', y') in the rotated system. In either system the distance r between the point P and the origin is the same; thus the equations for x, y, x', and y' are those given in Figure 11.29.

Using the formulas for the sine and cosine of the difference of two angles, we have

$$x' = r\cos(\alpha - \theta) = r(\cos\alpha\cos\theta + \sin\alpha\sin\theta)$$
$$= r\cos\alpha\cos\theta + r\sin\alpha\sin\theta = x\cos\theta + y\sin\theta$$
$$y' = r\sin(\alpha - \theta) = r(\sin\alpha\cos\theta - \cos\alpha\sin\theta)$$
$$= r\sin\alpha\cos\theta - r\cos\alpha\sin\theta = y\cos\theta - x\sin\theta$$

Solving the system

$$x' = x\cos\theta + y\sin\theta$$
$$y' = -x\sin\theta + y\cos\theta$$

for x and y yields

$$x = x'\cos\theta - y'\sin\theta$$
$$y = x'\sin\theta + y'\cos\theta$$

Finally, by substituting these values for x and y into the equation

$$Ax^2 + Bxy + Cy^2 + Dx + Ey + F = 0$$

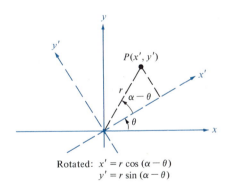

Rotated: $x' = r\cos(\alpha - \theta)$
$y' = r\sin(\alpha - \theta)$

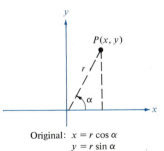

Original: $x = r\cos\alpha$
$y = r\sin\alpha$

FIGURE 11.29

and collecting terms, we obtain

$$[A \cos^2 \theta + B \sin \theta \cos \theta + C \sin^2 \theta](x')^2$$
$$+ [2(C - A) \sin \theta \cos \theta + B(\cos^2 \theta - \sin^2 \theta)](x'y')$$
$$+ [A \sin^2 \theta - B \sin \theta \cos \theta + C \cos^2 \theta](y')^2$$
$$+ [D \cos \theta + E \sin \theta](x')$$
$$+ [E \cos \theta - D \sin \theta](y') + F = 0$$

which is of the form

$$A'(x')^2 + B'x'y' + C'(y')^2 + D'x' + E'y' + F' = 0$$

Note that in order to eliminate the $x'y'$ term, we must select θ so that

$$B' = 2(C - A) \sin \theta \cos \theta + B(\cos^2 \theta - \sin^2 \theta) = 0$$

Since

$$B' = (C - A) \sin 2\theta + B \cos 2\theta = B(\sin 2\theta)\left(\frac{C - A}{B} + \cot 2\theta\right)$$

then B' will be zero if we choose θ so that

$$\cot 2\theta = \frac{A - C}{B}, \qquad B \neq 0$$

Thus we have established the desired results. ∎

Example 1

Write the equation $xy - 1 = 0$ in standard form.

Solution: Since $A = 0$, $B = 1$, $C = D = E = 0$, and $F = -1$, we have

$$\cot 2\theta = \frac{A - C}{B} = 0$$

Therefore, $2\theta = \pi/2$ and $\theta = \pi/4$. By Theorem 11.5 we then have

$$A' = A \cos^2 \theta + B \cos \theta \sin \theta + C \sin^2 \theta$$
$$= 0 + \left(\frac{\sqrt{2}}{2}\right)\left(\frac{\sqrt{2}}{2}\right) + 0 = \frac{1}{2}$$
$$C' = A \sin^2 \theta - B \cos \theta \sin \theta + C \cos^2 \theta$$
$$= 0 - \left(\frac{\sqrt{2}}{2}\right)\left(\frac{\sqrt{2}}{2}\right) + 0 = -\frac{1}{2}$$
$$D' = D \cos \theta + E \sin \theta = 0$$
$$E' = -D \sin \theta + E \cos \theta = 0$$
$$F' = F = -1$$

Hence the new equation is

$$\frac{(x')^2}{2} - \frac{(y')^2}{2} - 1 = 0$$

whose standard form is

$$\frac{(x')^2}{(\sqrt{2})^2} - \frac{(y')^2}{(\sqrt{2})^2} = 1$$

This is the equation of a hyperbola centered at the origin with vertices at $(\pm\sqrt{2}, 0)$ in the $x'y'$ system (see Figure 11.30). From the equations

$$x = x'\cos\theta - y'\sin\theta$$
$$y = x'\sin\theta + y'\cos\theta$$

we can determine the vertices to be $(1, 1)$ and $(-1, -1)$ in the xy system. Note also that the asymptotes of the hyperbola have equations $y' = \pm x'$, which correspond to the original x- and y-axes. ∎

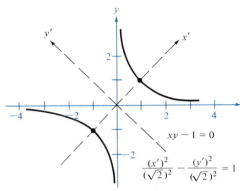

$xy - 1 = 0$

$$\frac{(x')^2}{(\sqrt{2})^2} - \frac{(y')^2}{(\sqrt{2})^2} = 1$$

Vertices: $(\sqrt{2}, 0), (-\sqrt{2}, 0)$ in $x'y'$–system
$(1, 1), (-1, -1)$ in xy–system

FIGURE 11.30

Example 2

Sketch the graph of the equation $7x^2 - 6\sqrt{3}\,xy + 13y^2 - 16 = 0$.

Solution: Applying Theorem 11.5 we have

$$\cot 2\theta = \frac{A - C}{B} = \frac{7 - 13}{-6\sqrt{3}} = \frac{1}{\sqrt{3}}$$

Thus $2\theta = \pi/3$ and $\theta = \pi/6$. Solving for A', C', D', E', and F' yields

$$A' = A\cos^2\theta + B\cos\theta\sin\theta + C\sin^2\theta$$

$$= 7\left(\frac{\sqrt{3}}{2}\right)^2 - 6\sqrt{3}\left(\frac{\sqrt{3}}{2}\right)\left(\frac{1}{2}\right) + 13\left(\frac{1}{2}\right)^2 = \frac{21 - 18 + 13}{4} = 4$$

$$C' = A\sin^2\theta - B\cos\theta\sin\theta + C\cos^2\theta$$

$$= 7\left(\frac{1}{2}\right)^2 + 6\sqrt{3}\left(\frac{\sqrt{3}}{2}\right)\left(\frac{1}{2}\right) + 13\left(\frac{\sqrt{3}}{2}\right)^2 = \frac{7 + 18 + 39}{4} = 16$$

$$D' = D\cos\theta + E\sin\theta = 0$$
$$E' = -D\sin\theta + E\cos\theta = 0$$
$$F' = -16$$

Therefore, the new equation is

$$4(x')^2 + 16(y')^2 - 16 = 0$$

whose standard form is

$$\frac{(x')^2}{4} + \frac{(y')^2}{1} = 1$$

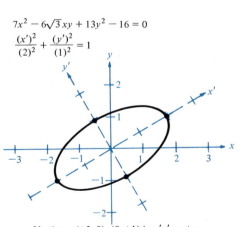

$7x^2 - 6\sqrt{3}\,xy + 13y^2 - 16 = 0$
$$\frac{(x')^2}{(2)^2} + \frac{(y')^2}{(1)^2} = 1$$

Vertices: $(\pm 2, 0), (0, \pm 1)$ in $x'y'$–system
$(\pm\sqrt{3}, \pm 1), (\pm\frac{1}{2}, \mp\frac{\sqrt{3}}{2})$ in xy–system

FIGURE 11.31

This is the equation of an ellipse centered at the origin with vertices at $(\pm 2, 0)$ in the $x'y'$ system (Figure 11.31). ■

In Examples 1 and 2 we carefully chose the equations so that θ would turn out to be one of the common angles 30°, 45°, and so forth. In general, a second-degree equation will not yield such common solutions to the equation $\cot 2\theta = (A - C)/B$. Example 3 illustrates such a case.

Example 3

Sketch the graph of $x^2 - 4xy + 4y^2 + 5\sqrt{5}y + 1 = 0$.

Solution: Since

$$\cot 2\theta = \frac{A - C}{B} \quad \text{and} \quad \cot 2\theta = \frac{\cot^2 \theta - 1}{2 \cot \theta}$$

we have

$$\cot 2\theta = \frac{1 - 4}{-4} = \frac{3}{4} = \frac{\cot^2 \theta - 1}{2 \cot \theta}$$

from which we obtain the equation

$$6 \cot \theta = 4 \cot^2 \theta - 4$$
$$0 = 4 \cot^2 \theta - 6 \cot \theta - 4$$
$$0 = (2 \cot \theta - 4)(2 \cot \theta + 1)$$

Considering $0 < \theta < \pi/2$ we have

$$2 \cot \theta = 4$$
$$\cot \theta = 2$$
$$\theta \approx 26.6°$$

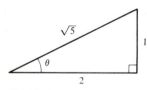

FIGURE 11.32

From Figure 11.32 we obtain $\sin \theta = 1/\sqrt{5}$ and $\cos \theta = 2/\sqrt{5}$. Consequently,

$$A' = A \cos^2 \theta + B \cos \theta \sin \theta + C \sin^2 \theta$$

$$= \left(\frac{2}{\sqrt{5}}\right)^2 - 4\left(\frac{2}{\sqrt{5}}\right)\left(\frac{1}{\sqrt{5}}\right) + 4\left(\frac{1}{\sqrt{5}}\right)^2 = \frac{4 - 8 + 4}{5} = 0$$

$$C' = A \sin^2 \theta - B \cos \theta \sin \theta + C \cos^2 \theta$$

$$= \left(\frac{1}{\sqrt{5}}\right)^2 + 4\left(\frac{2}{\sqrt{5}}\right)\left(\frac{1}{\sqrt{5}}\right) + 4\left(\frac{2}{\sqrt{5}}\right)^2 = \frac{1 + 8 + 16}{5} = 5$$

$$D' = D \cos \theta + E \sin \theta = 0 + 5\sqrt{5}\left(\frac{1}{\sqrt{5}}\right) = 5$$

$$E' = -D \sin \theta + E \cos \theta = 0 + 5\sqrt{5}\left(\frac{2}{\sqrt{5}}\right) = 10$$

$$F' = F = 1$$

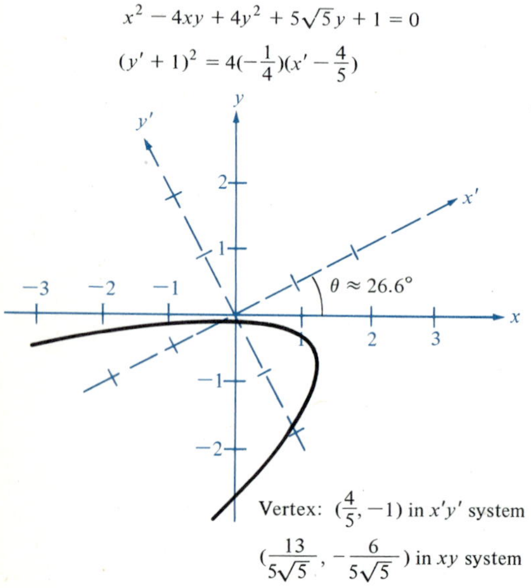

$x^2 - 4xy + 4y^2 + 5\sqrt{5}y + 1 = 0$

$(y' + 1)^2 = 4(-\frac{1}{4})(x' - \frac{4}{5})$

$\theta \approx 26.6°$

Vertex: $(\frac{4}{5}, -1)$ in $x'y'$ system

$(\frac{13}{5\sqrt{5}}, -\frac{6}{5\sqrt{5}})$ in xy system

FIGURE 11.33

The new equation is

$$5(y')^2 + 5x' + 10y' + 1 = 0$$

By completing the square,

$$5(y' + 1)^2 = -5x' + 4$$

we obtain the standard form

$$(y' + 1)^2 = (-1)\left(x' - \frac{4}{5}\right)$$

We conclude that the graph of the equation is a parabola with its vertex at $(\frac{4}{5}, -1)$ and its axis parallel to the x'-axis in the $x'y'$ system (Figure 11.33). ∎

Section Exercises (11.4)

In Exercises 1–16, rotate the axes to eliminate the xy term. Sketch the graph of the resulting equation, showing both sets of axes.

1. $xy + 1 = 0$
2. $xy - 4 = 0$
c 3. $9x^2 + 24xy + 16y^2 + 90x - 130y = 0$
s 4. $9x^2 + 24xy + 16y^2 + 80x - 60y = 0$
s 5. $x^2 - 10xy + y^2 + 1 = 0$
6. $xy + x - 2y + 3 = 0$
7. $xy - 2y - 4x = 0$
8. $2x^2 - 3xy - 2y^2 + 10 = 0$
9. $5x^2 - 2xy + 5y^2 - 12 = 0$
10. $13x^2 + 6\sqrt{3}xy + 7y^2 - 16 = 0$
11. $3x^2 - 2\sqrt{3}xy + y^2 + 2x + 2\sqrt{3}y = 0$
12. $16x^2 - 24xy + 9y^2 - 60x - 80y + 100 = 0$
13. $17x^2 + 32xy - 7y^2 = 75$
s 14. $40x^2 + 36xy + 25y^2 = 52$
c 15. $32x^2 + 50xy + 7y^2 = 52$
16. $4x^2 - 12xy + 9y^2 + (4\sqrt{13} - 12)x - (6\sqrt{13} + 8)y = 91$

The discriminant of $Ax^2 + Bxy + Cy^2 + Dx + Ey + F = 0$ is given by $B^2 - 4AC$. The discriminant of a second-degree equation can be used as follows:

i. If the equation graphs as a parabola, then $B^2 - 4AC = 0$.
ii. If the equation graphs as an ellipse, then $B^2 - 4AC < 0$.
iii. If the equation graphs as a hyperbola, then $B^2 - 4AC > 0$.

17. Each of the following three equations graphs as a (nondegenerate) conic. Use the discriminant of each equation to determine if the graph is a parabola, ellipse, or hyperbola.
    (a) $16x^2 - 24xy + 9y^2 - 30x - 40y = 0$
    (b) $x^2 - 4xy - 2y^2 - 6 = 0$
    (c) $13x^2 - 8xy + 7y^2 - 45 = 0$
s 18. Show that the discriminant of the general second-degree equation is invariant (does not change) under rotation of axes.
19. Show that the equation $x^2 + y^2 = r^2$ is invariant under rotation of axes.

Miscellaneous Exercises (Ch. 11)

In Exercises 1–10, analyze each equation and sketch its graph.

1. $16x^2 + 16y^2 - 16x + 24y - 3 = 0$
2. $y^2 - 12y - 8x + 20 = 0$
3. $3x^2 - 2y^2 + 24x + 12y + 24 = 0$
4. $4x^2 + y^2 - 16x + 15 = 0$
s 5. $3x^2 + 2y^2 - 12x + 12y + 29 = 0$
6. $4x^2 - 4y^2 - 4x + 8y - 11 = 0$

7. $x^2 - 6x + 2y + 9 = 0$

8. $x^2 + y^2 - 2x - 4y + 5 = 0$

9. $x^2 + y^2 + 2xy + 2\sqrt{2}x - 2\sqrt{2}y + 2 = 0$

10. $9x^2 + 6y^2 + 4xy - 20 = 0$

In Exercises 11–20, find an equation of the specified conic.

11. hyperbola; vertices $(0, \pm 1)$; foci $(0, \pm 3)$

12. hyperbola; vertices $(2, 2)$, $(-2, 2)$; foci $(4, 2)$, $(-4, 2)$

13. the ellipse such that the sum of the distances from any of the points on its graph to the points $(0, 0)$ and $(4, 0)$ is 10

14. parabola; vertex $(4, 2)$; focus $(4, 0)$

[S] **15.** parabola; vertex $(0, 0)$; focus $(1, 1)$

16. ellipse; vertices $(2, 0)$, $(2, 4)$; foci $(2, 1)$, $(2, 3)$

17. ellipse passing through points $(1, 2)$ and $(2, 0)$ with center at $(0, 0)$

18. hyperbola; foci $(\pm 4, 0)$; asymptotes $y = \pm 2x$

19. the hyperbola such that the absolute value of difference of the distances from any point on its graph to the points $(4, 0)$ and $(-4, 0)$ is 4

20. parabola; vertex $(0, 2)$; passes through point $(-1, 0)$; vertical axis

21. Find the equation of the line tangent to the parabola $y = x^2 - 2x + 2$ and perpendicular to the line $y = x - 2$.

[S] **22.** Find the equations of the lines tangent to the ellipse $x^2 + 4y^2 - 4x - 8y - 24 = 0$ and parallel to the line $x - 2y - 12 = 0$.

23. Find a so that the hyperbola $(x^2/a^2) - (y^2/4) = 1$ is tangent to the line $2x - y - 4 = 0$.

24. A large parabolic antenna is described as the surface formed by revolving the parabola $y = (1/200)x^2$ on the interval $[0, 100]$ about the y-axis. The receiving and transmitting equipment is positioned at the focus.
   (a) Find the coordinates of the focus.
   (b) Find the surface area of the antenna.

   (c) Find the length of the arc on the parabola $y = (1/200)\, x^2$ from $x = 0$ to $x = 100$.

[S] **25.** The ellipse $(x^2/a^2) + (y^2/b^2) = 1$ is revolved about its minor axis to form an oblate spheroid. Show the following:
   (a) Its volume is $\frac{4}{3}\pi a^2 b$.
   (b) Its surface area is

$$2\pi a^2 + \pi(b^2/e)\ln\left[(1 + e)/(1 - e)\right]$$

   where e is the eccentricity, c/a.

26. The ellipse $(x^2/a^2) + (y^2/b^2) = 1$ is revolved about its major axis to form a prolate spheroid. Show the following:
   (a) Its volume is $\frac{4}{3}\pi ab^2$.
   (b) Its surface area is $2\pi b^2 + 2\pi(ab/e)\,\text{Arcsin } e$, where e is the eccentricity, c/a.

27. Use the results of Exercises 25 and 26 to find the volumes and the surface areas of the prolate and oblate spheroids generated by $(x^2/9) + (y^2/4) = 1$.

[C] **28.** Approximate, to two decimal places, the perimeter of the ellipse $(x^2/9) + (y^2/4) = 1$.

29. Show that the area of the region bounded by the ellipse $(x^2/a^2) + (y^2/b^2) = 1$ is πab.

30. Use the result of Exercise 29 to find the area of the region bounded by the ellipse $(x^2/16) + (y^2/9) = 1$.

31. A milk truck carries a tank 16 ft long and the vertical cross sections of the tank are ellipses described by the equation $(x^2/16) + (y^2/9) = 1$. Find the volume of the tank.

[C] **32.** Approximate, to two-decimal-place accuracy, the surface area of the tank described in Exercise 31.

[C][S] **33.** Find the height of the milk in the tank of Exercise 31 if the truck is on level ground and is carrying $\frac{3}{4}$ of its total capacity.

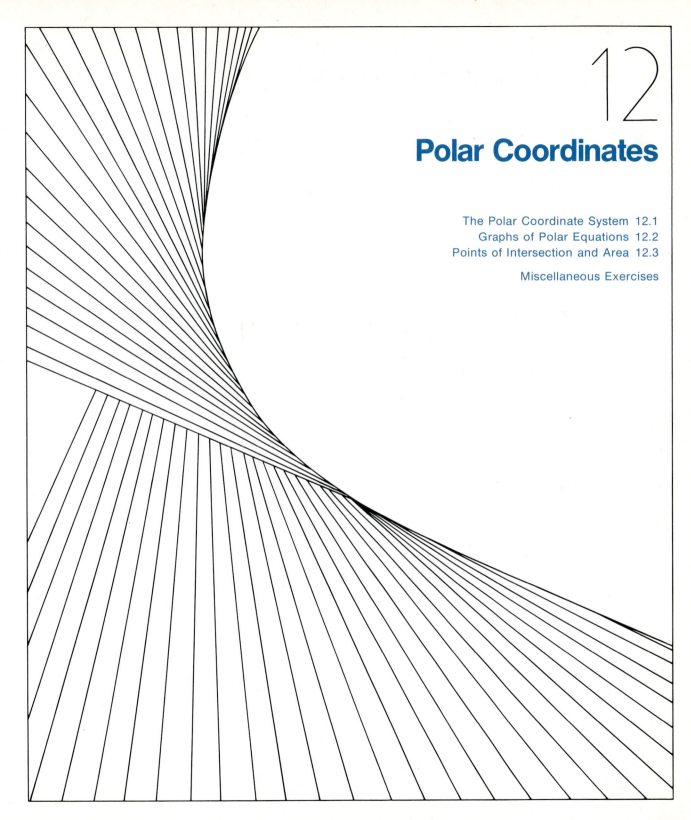

Polar Coordinates

12.1 The Polar Coordinate System

Purpose
- To introduce the polar coordinate system.
- To develop techniques for converting from polar to rectangular coordinates and vice versa.

Is the Euclidean Geometry true? The question is nonsense. One might as well ask whether the metric system is true and the old measures false; whether Cartesian coordinates are true and polar coordinates false.

—Henri Poincaré (1854–1912)

We have been representing a point in the plane by means of an ordered pair (x, y) of rectangular coordinates, which measure the distance of the point from the y- and x-axes, respectively. An alternative way of representing points in the plane is by **polar coordinates.** In this system a point (r, θ) is identified by its distance r from the origin of the system and an angle θ. We make this more specific in the following definition:

Definition of Polar Coordinate System	The **polar coordinate system** is a system of coordinates in which a point in the plane is identified by its distance r along a ray from a fixed point (the **pole**) and by the angle θ between a fixed line (the **polar axis**) and the ray. (See Figure 12.1.)

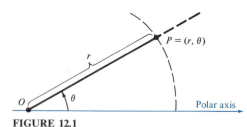

FIGURE 12.1

Every point P, other than the pole, can be represented by an ordered pair of real numbers (r, θ), where

$$r = \text{directed distance from } O \text{ to } P$$
$$\theta = \text{directed angle from polar axis to } OP$$

For an angle θ the positive direction is counterclockwise and the negative direction is clockwise.

Example 1

Plot the points whose polar coordinates are $(2, \pi/4)$, $(3, -\pi/6)$, $(2, 9\pi/4)$, and $(-2, \pi/6)$.

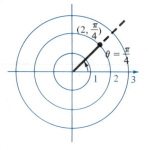

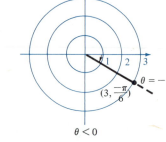

$\theta < 0$

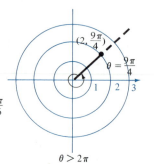

$\theta > 2\pi$

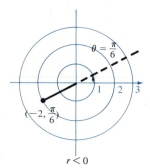

$r < 0$

FIGURE 12.2

Solution: Even though there is only one axis in the polar coordinate system, we usually include a second reference axis corresponding to $\theta = \pi/2$, as shown in Figure 12.2.

Note that the point $(-2, \pi/6)$ is in the third quadrant, because r is negative. *In general, a point (r, θ) lies in the same quadrant as θ if r is positive and in the quadrant on the opposite side of the pole if r is negative.* ∎

In the rectangular coordinate system, each point in the plane has a unique representation, but in the polar coordinate system, each point in the plane has many representations. For example, the polar coordinates $(2, \pi/4)$ and $(2, 9\pi/4)$ represent the same point (Figure 12.2). As a matter of fact, the polar coordinates

$$\left(2, \frac{\pi}{4} + 2n\pi\right) \qquad \text{and} \qquad \left(-2, \frac{5\pi}{4} + 2n\pi\right)$$

all represent the same point for any integer value of n. We summarize this situation as follows:

Multiple Representation of Points in Polar Coordinates	The point (r, θ) can also be represented by $$(r, \theta + 2n\pi) \qquad \text{or by} \qquad (-r, \theta + (2n + 1)\pi)$$ where n is an integer. Furthermore, the pole can be represented by $(0, \theta)$ for any angle θ.

Example 2

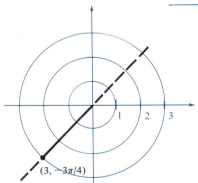

(3, −3π/4)

FIGURE 12.3

Plot the point $(3, -3\pi/4)$ and find three additional polar coordinate representations of this point, using $-2\pi < \theta < 2\pi$.

Solution: The point is plotted in Figure 12.3.
Three other representations for $(3, -3\pi/4)$ are

$$\left(3, \frac{-3\pi}{4} + 2\pi\right) = \left(3, \frac{5\pi}{4}\right)$$

$$\left(-3, \frac{-3\pi}{4} - \pi\right) = \left(-3, \frac{-7\pi}{4}\right)$$

$$\left(-3, \frac{-3\pi}{4} + \pi\right) = \left(-3, \frac{\pi}{4}\right)$$

∎

To establish the relationship between polar and rectangular coordinates, we let the polar axis coincide with the x-axis and the pole with the origin. From Figure 12.4 we see that the following relationships hold:

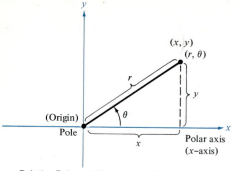

Relating Polar and Rectangular Coordinates

FIGURE 12.4

$$\tan \theta = \frac{y}{x}$$

$$r^2 = x^2 + y^2$$

$$x = r \cos \theta$$

$$y = r \sin \theta$$

These relationships allow us to convert points or equations from one system to the other. Since the representation of a point in rectangular coordinates is unique, the conversion from polar to rectangular coordinates is straightforward and is indicated in the following rule:

Polar-to-Rectangular Conversion　　　To convert a point (r, θ) to the rectangular form (x, y), use the equations

$$x = r \cos \theta \qquad \text{and} \qquad y = r \sin \theta$$

Example 3 (*Polar to Rectangular*)

Find the rectangular coordinates of the points whose polar coordinates are $(-2, 5\pi/6)$ and $(3, 4\pi/3)$.

Solution: For $(r, \theta) = (-2, 5\pi/6)$, we have

$$x = r \cos \theta = -2 \cos \frac{5\pi}{6} = -2\left(-\frac{\sqrt{3}}{2}\right) = \sqrt{3}$$

$$y = r \sin \theta = -2 \sin \frac{5\pi}{6} = -2\left(\frac{1}{2}\right) = -1$$

Therefore, the rectangular coordinates are $(\sqrt{3}, -1)$ (see Figure 12.5).
For $(r, \theta) = (3, 4\pi/3)$, we have

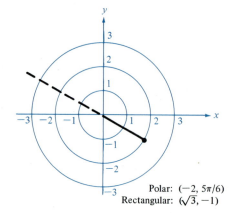

Polar: $(-2, 5\pi/6)$
Rectangular: $(\sqrt{3}, -1)$

FIGURE 12.5

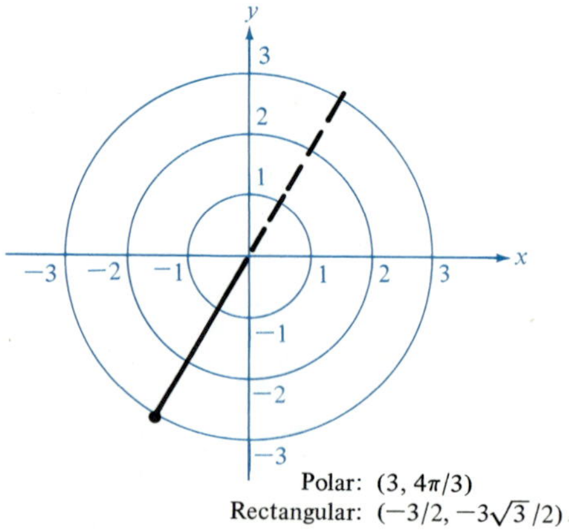

Polar: $(3, 4\pi/3)$
Rectangular: $(-3/2, -3\sqrt{3}/2)$

FIGURE 12.6

$$x = r \cos \theta = 3 \cos \frac{4\pi}{3} = 3\left(-\frac{1}{2}\right) = -\frac{3}{2}$$

$$y = r \sin \theta = 3 \sin \frac{4\pi}{3} = 3\left(-\frac{\sqrt{3}}{2}\right) = -\frac{3\sqrt{3}}{2}$$

and the rectangular coordinates are $(-\frac{3}{2}, -\frac{3}{2}\sqrt{3})$ (see Figure 12.6). ■

Conversion from rectangular to polar coordinates is complicated somewhat by the fact that points in the polar coordinate system have multiple representations.

Rectangular-to-Polar Conversion

To convert points from (x, y) to (r, θ) form, use the equations

$$\tan \theta = \frac{y}{x} \quad \text{and} \quad r = \pm\sqrt{x^2 + y^2}$$

For values of θ in the same quadrant as (x, y), choose r to be positive. For values of θ in the opposite quadrant from (x, y), choose r to be negative.

Example 4 (*Rectangular to Polar*)

Convert the points $(1, -1)$ and $(-\frac{1}{2}, \frac{1}{2}\sqrt{3})$ from rectangular to polar form.

Solution: For $(x, y) = (1, -1)$, we have

$$r = \pm\sqrt{x^2 + y^2} = \pm\sqrt{(1)^2 + (-1)^2} = \pm\sqrt{2}$$

$$\tan \theta = \frac{y}{x} = \frac{-1}{1} = -1$$

and thus

$$\theta = \dots, \frac{-\pi}{4}, \frac{3\pi}{4}, \frac{7\pi}{4}, \dots$$

Two possible polar coordinate representations are

$$\left(\sqrt{2}, \frac{-\pi}{4}\right) \quad \text{and} \quad \left(-\sqrt{2}, \frac{3\pi}{4}\right)$$

For $(x, y) = (-\frac{1}{2}, \frac{1}{2}\sqrt{3})$, we have

$$r = \pm\sqrt{x^2 + y^2} = \pm\sqrt{\frac{1}{4} + \frac{3}{4}} = \pm 1$$

$$\tan \theta = \frac{\sqrt{3}/2}{-\frac{1}{2}} = -\sqrt{3}$$

and hence

$$\theta = \dots, \frac{2\pi}{3}, \frac{5\pi}{3}, \dots$$

Two possible polar representations are

$$\left(1, \frac{2\pi}{3}\right) \quad \text{and} \quad \left(-1, \frac{5\pi}{3}\right) \qquad \blacksquare$$

In the remaining portion of this chapter we will assume, unless otherwise specified, that whenever the coordinates of a point are given, these numbers are *polar* coordinates of the point. Furthermore, we refer to an equation involving the variables r and θ as a **polar equation.**

The relationships between polar and rectangular coordinates also allow us to convert *equations* from one form to another. The next two examples illustrate the procedure.

Example 5 *(Polar to Rectangular)*

Find a rectangular equation that has the same graph as the polar equation $r = 2 \cos \theta$.

Solution: With the conversion equations in mind, we note that it is convenient to multiply both sides of the equation by r. Then we have

$$r^2 = 2(r \cos \theta)$$

Since $\qquad x^2 + y^2 = r^2 \qquad$ and $\qquad x = r \cos \theta$

we have $\qquad x^2 + y^2 = 2x$

Note that in the polar equation $r = 2 \cos \theta$, r is a *function* of θ, but in the rectangular equation $x^2 + y^2 = 2x$, neither variable is a function of the other. $\qquad \blacksquare$

Example 6 *(Rectangular to Polar)*

Find a polar equation that has the same graph as the rectangular equation $y = x^3$.

Solution: Substituting $x = r \cos \theta$ and $y = r \sin \theta$, we have

$$y = x^3$$
$$r \sin \theta = (r \cos \theta)^3$$

Although this equation is acceptable as it stands, it can be simplified by rewriting it as

$$r(\sin \theta - r^2 \cos^3 \theta) = 0$$

This implies that

$$r = 0 \qquad \text{or} \qquad \sin \theta - r^2 \cos^3 \theta = 0$$

Since the pole is the only point satisfying the equation $r = 0$, and since the coordinates $(0, 0)$ also satisfy the second equation, we can drop $r = 0$, leaving us with the polar equation

$$\sin \theta - r^2 \cos^3 \theta = 0 \qquad \blacksquare$$

At this point you might well be asking, "Why another coordinate system?" One answer is that the equations of many curves in the plane are simpler in polar form than in rectangular form. For example, the polar equation of the circle $x^2 + y^2 = a^2$ is simply $r = a$. As another case in point, let us consider the equation of an ellipse whose major axis lies on the x-axis and whose focus lies at the origin:

$$\frac{(x - c)^2}{a^2} + \frac{y^2}{b^2} = 1 \qquad (a^2 - b^2 = c^2)$$

To write this equation in polar coordinates, we proceed as follows:

$$\frac{(x - c)^2}{a^2} + \frac{y^2}{b^2} = 1$$

$$b^2 x^2 - 2cb^2 x + b^2 c^2 + a^2 y^2 = a^2 b^2$$
$$(a^2 - c^2)x^2 + a^2 y^2 = (b^2 + c^2)b^2 + 2cb^2 x - b^2 c^2$$
$$a^2(x^2 + y^2) = b^4 + 2cb^2 x + c^2 x^2$$
$$a^2(x^2 + y^2) = (b^2 + cx)^2$$
$$a^2 r^2 = [b^2 + (cr) \cos \theta]^2$$
$$ar = \pm b^2 \pm (cr) \cos \theta$$
$$r[a \mp (c) \cos \theta] = \pm b^2$$

Thus

$$r = \frac{\pm b^2 / a}{1 \mp e \cos \theta}$$

where $e = c/a$.

By a similar procedure we can determine that the polar equation of any ellipse, hyperbola, or parabola whose focus is at the pole and whose axes lie on the x- or y-axes has one of the following forms:

Polar Equations for Conics

Polar equations of the forms

$$r = \frac{A}{1 \pm e \cos \theta} \qquad \text{(horizontal axis; vertices at } \theta = 0 \text{ and/or } \theta = \pi)$$

$$r = \frac{A}{1 \pm e \sin \theta} \qquad \text{(vertical axis; vertices at } \theta = \pi/2 \text{ and/or } \theta = 3\pi/2)$$

represent conics with a focus at the pole and axis on one of the coordinates axes. In particular, the conic is a parabola if $e = 1$, an ellipse if $0 < e < 1$, and a hyperbola if $e > 1$.

Example 7

Sketch the graph of the polar equation

$$r = \frac{32}{3 + 5 \cos \theta}$$

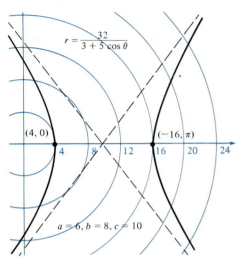

$$r = \frac{32}{3 + 5\cos\theta}$$

(4, 0)

4   8   12   16   20   24

(−16, π)

$a = 6, b = 8, c = 10$

FIGURE 12.7

Solution: Dividing each term by 3, we have

$$r = \frac{\frac{32}{3}}{1 + \left(\frac{5}{3}\right)\cos\theta}$$

Since $e = \frac{5}{3} > 1$, the graph of this equation is a hyperbola for which the pole is a focus. The transverse axis of the hyperbola lies on the *x*-axis and we can determine the vertices by

$$\theta = 0, \qquad r = \frac{\frac{32}{3}}{1 + \frac{5}{3}} = \frac{\frac{32}{3}}{\frac{8}{3}} = 4$$

$$\theta = \pi, \qquad r = \frac{\frac{32}{3}}{1 - \frac{5}{3}} = \frac{\frac{32}{3}}{-\frac{2}{3}} = -16$$

From Figure 12.7 we see that the distance $2a$ between the vertices is 12 and thus $a = 6$. Furthermore, since $e = \frac{5}{3} = c/a$, we know that $c = 10$; and finally, $b = \sqrt{c^2 - a^2} = \sqrt{64} = 8$. Therefore, we have the sketch shown in Figure 12.7. ∎

Section Exercises (12.1)

In Exercises 1–8, the polar coordinates of the points are given. Plot the points and find the rectangular coordinates for the same points.

1. $(4, 3\pi/6)$
2. $(4, 3\pi/2)$
3. $(-1, 5\pi/4)$
4. $(0, -\pi)$
5. $(4, -\pi/3)$
6. $(-1, -3\pi/4)$
C S **7.** $(\sqrt{2}, 2.36)$
8. $(-3, -1.57)$

In Exercises 9–16, the rectangular coordinates of a point are given. In each case find two sets of polar coordinates for the point, using $0 \le \theta < 2\pi$.

9. $(1, 1)$
10. $(0, -5)$
C S **11.** $(-3, 4)$
C **12.** $(3, -1)$
13. $(-\sqrt{3}, -\sqrt{3})$
14. $(-2, 0)$
C **15.** $(4, 6)$
C **16.** $(5, 12)$

In Exercises 17–23, find a polar equation of the graph having the given rectangular equation.

17. $x^2 + y^2 = a^2$
18. $x^2 + y^2 - 2ay = 0$
S **19.** $(x^2 + y^2)^2 - 9(x^2 - y^2) = 0$
20. $4x + 7y - 2 = 0$
21. $x^2 + y^2 - 2ax = 0$
22. $y^2 - 8x - 16 = 0$
S **23.** $x^2 - 4ay - 4a^2 = 0$

In Exercises 24–31, find a rectangular equation of the graph having the given polar equation.

24. $r = 4 \cos \theta$
25. $r = 4 \sin \theta$
26. $r = \dfrac{1}{1 - \cos \theta}$
27. $r = 1 - 2 \sin \theta$
28. $r^2 = \sin 2\theta$
S **29.** $r = 2 \sin 3\theta$
30. $r = \dfrac{6}{2 - 3 \sin \theta}$
31. $r = \dfrac{6}{2 \cos \theta - 3 \sin \theta}$

32. Show that the distance between (r_1, θ_1) and (r_2, θ_2) is $\sqrt{r_1^2 + r_2^2 - 2r_1 r_2 \cos(\theta_1 - \theta_2)}$.

33. Show that (r_1, θ_1), (r_2, θ_2), and (r_3, θ_3) are collinear if and only if $r_2 r_3 \sin(\theta_3 - \theta_2) + r_3 r_1 \sin(\theta_1 - \theta_3) + r_1 r_2 \sin(\theta_2 - \theta_1) = 0$.

S **34.** (a) Find a polar equation of the vertical line whose rectangular equation is $x = k$.
(b) Find a polar equation of the horizontal line whose rectangular equation is $y = k$.

S **35.** Show that the corresponding polar equation of the ellipse $(x^2/a^2) + (y^2/b^2) = 1$ is

$$r^2 = \frac{b^2}{1 - e^2 \cos^2 \theta}$$

36. Show that the corresponding polar equation of the hyperbola $(x^2/a^2) - (y^2/b^2) = 1$ is

$$r^2 = \frac{-b^2}{1 - e^2 \cos^2 \theta}$$

In Exercises 37–42, find a polar equation of the specified conic section.

37. parabola; focus $(0, 0)$; vertex $(1, -\pi/2)$
38. ellipse; focus $(0, 0)$; vertices $(2, 0)$, $(8, \pi)$
S 39. ellipse; one focus $(4, 0)$; vertices $(5, 0)$, $(5, \pi)$
40. parabola; focus $(0, 0)$; vertex $(4, 0)$
S 41. hyperbola; focus $(0, 0)$; vertices $(1, 3\pi/2)$, $(9, 3\pi/2)$
42. hyperbola; one focus $(5, \pi/2)$; vertices $(4, \pi/2)$, $(4, -\pi/2)$

In Exercises 43–47, classify the graph of each equation as a parabola, ellipse, or hyperbola. In each case write the standard (rectangular) form of the equation of the conic.

43. $r = \dfrac{2}{1 - \cos \theta}$

44. $r(1 - 2 \cos \theta) = 4$

S 45. $r = \dfrac{2}{2 - \cos \theta}$

46. $r(3 + 2 \sin \theta) = 3$

47. $r = \dfrac{3}{2 - 6 \sin \theta}$

C 48. On 26 November 1963 the United States launched *Explorer 18*. Its low point in its elliptical orbit over the surface of the earth was 119 mi and its high point was 122,000 mi from the surface of the earth. (The center of the earth is the focus of the orbit.) Find the polar equation that describes its orbit. (See Exercises 41–43 of Section 11.2.)

C S 49. Use the equation of Exercise 48 to find the distance from the earth's surface to the satellite when the angle between the line from the center of the earth to the satellite and the major axis of the ellipse is 60°.

50. Find the polar equation for the parabola described in Exercise 36 of Section 11.1.

For Review and Class Discussion

True or False

1. _____ If (r_1, θ_1) and (r_2, θ_2) represent the same point in the polar coordinate system, then $|r_1| = |r_2|$.

2. _____ If (r, θ_1), and (r, θ_2) represent the same point in the polar coordinate system, then $\theta_1 = \theta_2 + 2\pi n$, for some integer n.

3. _____ If $x > 0$, then the point (x, y) in the rectangular coordinate system can be represented by (r, θ) in the polar coordinate system, where $r = \sqrt{x^2 + y^2}$ and $\theta = \text{Arctan}\,(y/x)$.

4. _____ In the polar coordinate system, the point $(1, 2)$ lies in the first quadrant.

5. _____ The rectangular equation $y = ax$ has the same graph as the polar equation $\theta = \text{Arctan}\, a$.

12.2 Graphs of Polar Equations

Purpose
- To outline a procedure for sketching the graph of the polar equation $r = f(\theta)$, using the period of f, symmetry, the relative extrema of r, and tangents at the pole.

Even so simple a figure as a circle has given rise to so many and such profound investigations that they could constitute a course all by themselves. —*David Hilbert (1862–1943)*

In the early chapters of this text we devoted a considerable amount of time to curve sketching in rectangular coordinates. We began, in Section 1.3, with the basic point-plotting method to which we added a number of techniques as aids to curve sketching. We finally summarized the various aids to curve sketching in Section 4.5. We approach curve sketching in polar coordinates in a similar manner.

A useful aid in sketching the graph of many common polar equations is their periodic nature. For example, if $r = f(\theta)$ has a period of 2π, then we can obtain its *entire* graph using only the interval $[0, 2\pi]$, since the graph merely retraces itself for θ outside this interval.

Example 1

Use the point-plotting method to sketch the graph of the polar equation $r = 2 + \sin\theta$.

Solution: Since $r = 2 + \sin\theta$ has a period of 2π, we can restrict θ in our table of coordinates to the interval $[0, 2\pi]$. See Table 12.1. After plotting and connecting these points, we obtain the graph shown in Figure 12.8.

TABLE 12.1

θ	0	$\dfrac{\pi}{6}$	$\dfrac{\pi}{3}$	$\dfrac{\pi}{2}$	$\dfrac{2\pi}{3}$	$\dfrac{5\pi}{6}$	π
r	2	2.5	2.87	3	2.87	2.5	2

θ	π	$\dfrac{7\pi}{6}$	$\dfrac{4\pi}{3}$	$\dfrac{3\pi}{2}$	$\dfrac{5\pi}{3}$	$\dfrac{11\pi}{6}$	2π
r	2	1.5	1.13	1	1.13	1.5	2

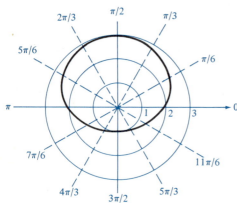

$r = 2 + \sin\theta$ Limaçon

FIGURE 12.8

It appears that the graph in Figure 12.8 is symmetric with respect to the vertical axis, $\theta = \pi/2$. Had we known about this symmetry beforehand, our sketch could have been obtained from half as many points. As with rectangular coordinates, there are some simple tests for symmetry in polar coordinates. Figure 12.9 gives a graphic description of the three types of symmetry we will consider: symmetry with respect to the line

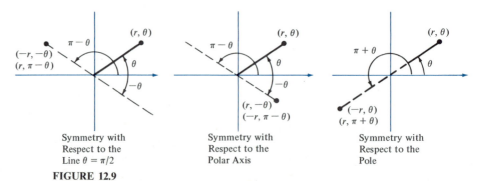

Symmetry with Respect to the Line $\theta = \pi/2$

Symmetry with Respect to the Polar Axis

Symmetry with Respect to the Pole

FIGURE 12.9

$\theta = \pi/2$, the polar axis, and the pole. The following theorem lists two tests for each type of symmetry:

THEOREM 12.1

(*Tests for Symmetry in Polar Coordinates*)

The graph of a polar equation is symmetric with respect to:

i.   *the line* $\theta = \pi/2$ if replacing (r, θ) by either $(r, \pi - \theta)$ or $(-r, -\theta)$ yields an equivalent equation

ii. *the polar axis* if replacing (r, θ) by either $(r, -\theta)$ or $(-r, \pi - \theta)$ yields an equivalent equation

iii. *the pole* if replacing (r, θ) by either $(r, \pi + \theta)$ or $(-r, \theta)$ yields an equivalent equation

---

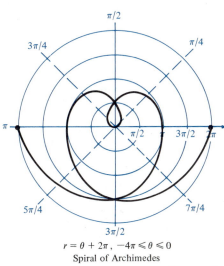

$r = \theta + 2\pi$, $-4\pi \leqslant \theta \leqslant 0$
Spiral of Archimedes

FIGURE 12.10

When applying Theorem 12.1, it is important to remember that we are using only two of the many possible polar representations of the points under consideration. Consequently, the conditions of Theorem 12.1 are *sufficient* to guarantee the various types of symmetry, but they are not necessary. This means that if a test indicates symmetry, then such symmetry will exist. While, on the other hand, if a test fails (suggesting no symmetry), the graph may still have that type of symmetry. Negative results in Theorem 12.1 *cannot* be interpreted as a lack of symmetry. For instance, Figure 12.10 shows the graph of $r = \theta + 2\pi$ to be symmetric with respect to the line $\theta = \pi/2$. Yet the symmetry test fails to indicate symmetry because the equation $r = \theta + 2\pi$ changes when (r, θ) is replaced by either $(-r, -\theta)$ or by $(r, \pi - \theta)$.

Example 2

Use Theorem 12.1 to show the following:
(a) The graph of $r = 2 + \sin\theta$ is symmetric with respect to the line $\theta = \pi/2$.
(b) The graph of $r = 1 - 2\cos\theta$ is symmetric with respect to the polar axis.
(c) The graph of $r = 3\sin 2\theta$ is symmetric with respect to the line $\theta = \pi/2$, the polar axis, and the pole.

Solution:

(a) If we replace (r, θ) by $(r, \pi - \theta)$, we obtain the equation

$$r = 2 + \sin(\pi - \theta) = 2 + \sin\theta$$

which is the same as the original equation, and, therefore, the graph is symmetric with respect to the line $\theta = \pi/2$.

(b) If we replace (r, θ) by $(r, -\theta)$, we obtain the equation

$$r = 1 - 2 \cos(-\theta) = 1 - 2 \cos \theta$$

which is the same as the original equation, and, therefore, the graph is symmetric with respect to the polar axis.

(c)   If we replace (r, θ) by $(r, \pi - \theta)$, we obtain

$$r = 3 \sin(2\pi - 2\theta) = -3 \sin 2\theta$$

which is different from the original equation. Using the other substitution, we replace (r, θ) by $(-r, -\theta)$ and obtain

$$-r = 3 \sin(-2\theta) = -3 \sin 2\theta$$

which is equivalent to the original equation, and hence the graph is symmetric with respect to the line $\theta = \pi/2$. For polar axis symmetry, the second substitution yields an equivalent equation. Finally, since the graph is symmetric with respect to both the line $\theta = \pi/2$ and the polar axis, it must also be symmetric with respect to the pole.   ■

As a final comment about symmetry in polar coordinates, we point out the following quick test for symmetry:

Symmetry for $r = f(\sin \theta)$ and $r = g(\cos \theta)$	1. If r is a function of $\sin \theta$, the graph of r will be symmetric with respect to the line $\theta = \pi/2$.
	2. If r is a function of $\cos \theta$, the graph of r will be symmetric with respect to the polar axis.

For example, the graphs of $r = 2 \sin \theta + \sin^2 \theta$, $r = 1/(3 + \sin \theta)$, and $r = \sqrt{\sin \theta + 1}$ all have vertical axis symmetry, and the graphs of $r = 2 \cos \theta + \cos^2 \theta$, $r = 1/(3 + \cos \theta)$, and $r = \sqrt{\cos \theta + 1}$ all have polar axis symmetry.

Example 3

Sketch the graph of $r = 1 - 2 \cos \theta$.

Solution: Using polar axis symmetry together with the fact that $r = 1 - 2 \cos \theta$ has a period of 2π, we consider values of θ in the interval $[0, \pi]$, as shown in Table 12.2. After plotting the points in the table and using polar axis symmetry, we obtain the graph shown in Figure 12.11.

We have seen that for curve sketching in polar coordinates, it is helpful to consider the periodic nature and the symmetry of the graph. Two additional curve-sketching aids involve the use of calculus. In particular, it is helpful to locate the *relative extrema* of r and the *tangent lines at the pole*.

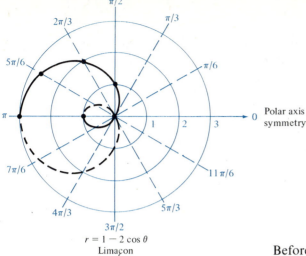

$r = 1 - 2\cos\theta$
Limaçon

FIGURE 12.11

TABLE 12.2

θ	0	$\dfrac{\pi}{6}$	$\dfrac{\pi}{4}$	$\dfrac{\pi}{3}$	$\dfrac{\pi}{2}$
r	-1	-0.73	-0.41	0	1

θ	$\dfrac{\pi}{2}$	$\dfrac{2\pi}{3}$	$\dfrac{3\pi}{4}$	$\dfrac{5\pi}{6}$	π
r	1	2	2.41	2.73	3

Before describing procedures for determining these curve-sketching aids, we consider the two derivatives dy/dx and $dr/d\theta$. The geometrical interpretation of these two derivatives is fundamentally different, as can be seen in Figures 12.12 and 12.13. Since the graph of $r = 1 + 2\cos\theta$ has vertical tangents at the points $(-1, \pi)$ and $(3, 0)$, we know that dy/dx must be undefined at these two points, and yet $dr/d\theta = -2\sin\theta$ is zero at both points. Thus $dr/d\theta$ does not (necessarily) represent the slope of a polar graph. However, from Figure 12.13 we can see that the two points at which $dr/d\theta = 0$ are of graphical interest since they occur at the relative maximum distances of r from the pole. We call such points the **relative extrema of r**.

Having established the fact that $dr/d\theta$ cannot be interpreted as the slope of a tangent line to a polar graph, we now want to show how to find the tangent line to a polar graph, using dy/dx.

To begin, let us assume that we are given a differentiable polar function defined by $r = f(\theta)$. Imagine then that a rectangular coordinate sys-

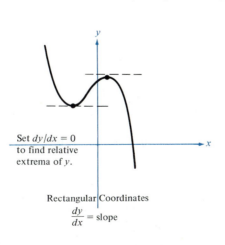

Set $dy/dx = 0$
to find relative
extrema of y.

Rectangular Coordinates

$\dfrac{dy}{dx} = $ slope

FIGURE 12.12

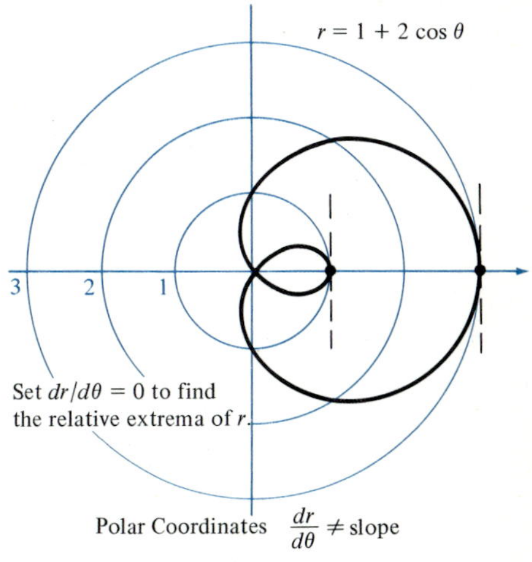

$r = 1 + 2\cos\theta$

Set $dr/d\theta = 0$ to find
the relative extrema of r.

Polar Coordinates $\dfrac{dr}{d\theta} \neq$ slope

FIGURE 12.13

tem is superimposed over the polar system. Now to find the slope of the tangent line at (r, θ), we need to determine the value of dy/dx at (r, θ). Since

$$y = r \sin \theta = f(\theta) \sin \theta$$

and

$$x = r \cos \theta = f(\theta) \cos \theta$$

we have

$$\frac{dy}{dx} = \frac{dy/d\theta}{dx/d\theta} = \frac{f'(\theta) \sin \theta + f(\theta) \cos \theta}{f'(\theta) \cos \theta - f(\theta) \sin \theta}$$

This establishes the following theorem:

THEOREM 12.2 (*Slope in Polar Form*)	If f is a differentiable function of θ, then the slope of the tangent line to $r = f(\theta)$ at the point (r, θ) is given by $$\frac{dy}{dx} = \frac{f'(\theta) \sin \theta + f(\theta) \cos \theta}{f'(\theta) \cos \theta - f(\theta) \sin \theta}$$

One of the consequences of Theorem 12.2 is that when the graph of $r = f(\theta)$ crosses the pole, the formula for dy/dx simplifies. To show this, we assume that for some angle α, $f(\alpha) = 0$ and $f'(\alpha) \neq 0$. Then

$$\frac{dy}{dx} = \frac{f'(\alpha) \sin \alpha + f(\alpha) \cos \alpha}{f'(\alpha) \cos \alpha - f(\alpha) \sin \alpha} = \frac{f'(\alpha) \sin \alpha}{f'(\alpha) \cos \alpha} = \frac{\sin \alpha}{\cos \alpha} = \tan \alpha$$

This means that the slope of the graph of $r = f(\theta)$ at the point $(0, \alpha)$ is equal to $\tan \alpha$. But we know that the slope of the line $\theta = \alpha$ is also equal to $\tan \alpha$. Therefore, we conclude that the line $\theta = \alpha$ is tangent at the pole.

THEOREM 12.3 (*Tangent Lines at Pole*)	If $f(\alpha) = 0$ and $f'(\alpha) \neq 0$, then the line $\theta = \alpha$ is tangent at the pole to the graph of $r = f(\theta)$.

As a consequence of Theorem 12.3, we can find the tangent lines at the pole by finding the zeros of $r = f(\theta)$. Of course, since a polar graph may cross a point more than once, there may be more than one tangent line at the pole. For example, both of the lines $\theta = \pi/3$ and $\theta = 5\pi/3$ are tangent at the pole in Figure 12.11.

Study the next three examples carefully to see how the various aids to curve sketching can be combined.

Example 4

Consider the polar equation $r = 2 \sin 3\theta$.
(a)   Sketch its graph.
(b)   Find the slope of the graph at $\theta = \pi/6$ and $\theta = \pi/4$.

Solution:

(a)   Period: $2\pi/3$

Symmetry: with respect to the line $\theta = \pi/2$
Tangents at the pole: $r = 0$ when
$$\theta = 0, \theta = \pi/3, \text{ and } \theta = 2\pi/3$$
Relative extrema of r: $dr/d\theta = 0$ at
$$(2, \pi/6), (-2, \pi/2), \text{ and } (2, 5\pi/6)$$

(b) Since $dr/d\theta = f'(\theta) = 6 \cos 3\theta$, we have

$$\frac{dy}{dx} = \frac{f'(\theta) \sin \theta + f(\theta) \cos \theta}{f'(\theta) \cos \theta - f(\theta) \sin \theta} = \frac{6 \cos 3\theta \sin \theta + 2 \sin 3\theta \cos \theta}{6 \cos 3\theta \cos \theta - 2 \sin 3\theta \sin \theta}$$

Then at $\theta = \pi/6$ the slope is

$$\frac{dy}{dx} = \frac{0 + 2(1)(\sqrt{3}/2)}{0 - 2(1)(\frac{1}{2})} = -\sqrt{3}$$

At $\theta = \pi/4$, the slope is

$$\frac{dy}{dx} = \frac{6(-\sqrt{2}/2)(\sqrt{2}/2) + 2(\sqrt{2}/2)(\sqrt{2}/2)}{6(-\sqrt{2}/2)(\sqrt{2}/2) - 2(\sqrt{2}/2)(\sqrt{2}/2)} = \frac{-3 + 1}{-3 - 1} = \frac{1}{2}$$

The plotting points are given in Table 12.3 and the graph is shown in Figure 12.14.

TABLE 12.3

θ	0	$\dfrac{\pi}{12}$	$\dfrac{\pi}{6}$	$\dfrac{\pi}{3}$	$\dfrac{5\pi}{12}$	$\dfrac{\pi}{2}$
r	0	1.41	2	0	−1.41	−2

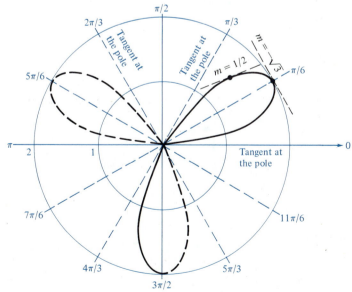

FIGURE 12.14

$r = 2 \sin 3\theta$
Rose Curve

Each of the curves discussed in Examples 1 through 4 can be classified as either a limaçon or a rose curve. Some of the characteristics of these two types of curves are outlined in the following summary.

Special Polar Curves

1. The graphs of equations of the form

$$r = a \pm b \cos \theta \qquad \text{or} \qquad r = a \pm b \sin \theta$$

are called **limaçons.** Furthermore,

  i. if $|a| < |b|$, the limaçon has two loops (Figure 12.11)
  ii. if $|a| = |b|$, the limaçon has one heart-shaped loop and is called a **cardioid** (Figure 12.25)
  iii. if $|a| > |b|$, the limaçon has one flattened loop (Figure 12.8)

2. The graphs of equations of the form

$$r = a \cos n\theta \qquad \text{or} \qquad r = a \sin n\theta \qquad (n \geq 2)$$

are called **rose curves.** Furthermore,

  i. if n is odd, the curve has n petals (Figure 12.14)
  ii. if n is even, the curve has $2n$ petals (Figure 12.15)

If r is replaced by r^2 in the equation for a rose curve, the number of petals will change to $2n$ if n is odd and to n if n is even. Two-petal curves resulting from this substitution are called **lemniscates.**

Example 5

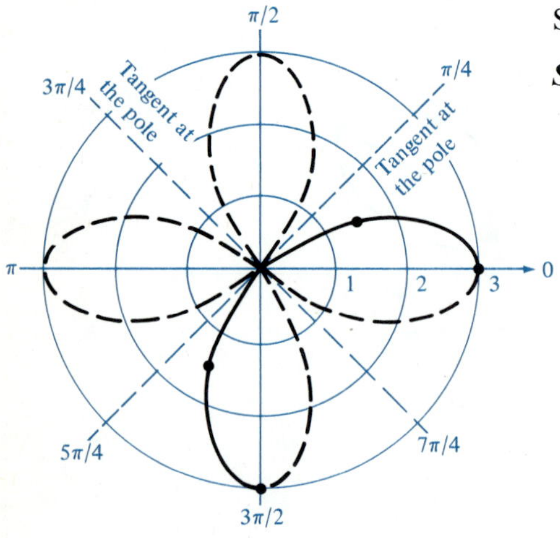

$r = 3 \cos 2\theta$
Rose Curve

FIGURE 12.15

Sketch the graph of $r = 3 \cos 2\theta$.

Solution: Type of curve: rose curve with $2n = 4$ petals
Period: π
Symmetry: with respect to the polar axis and the line $\theta = \pi/2$
Tangents at the pole: $r = 0$ when
$$\theta = \pi/4 \text{ and } \theta = 3\pi/4$$
Relative extrema of r: $dr/d\theta = 0$ at
$$(3, 0), \cdot (-3, \pi/2), (3, \pi), \text{ and } (-3, 3\pi/2)$$
The plotting points are given in Table 12.4. The graph is shown in Figure 12.15.

TABLE 12.4

θ	0	$\dfrac{\pi}{6}$	$\dfrac{\pi}{4}$	$\dfrac{\pi}{3}$	$\dfrac{\pi}{2}$
r	3	1.5	0	-1.5	-3

Example 6

Consider the equation $r = 2 - \csc\theta$, $0 < \theta < \pi$.
(a)   Sketch its graph.
(b)   Show that $y = -1$ is a horizontal asymptote to the graph.

Solution:

(a)   Symmetry: with respect to the line $\theta = \pi/2$
Tangents at the pole: $\theta = \pi/6$ and $\theta = 5\pi/6$
Relative extrema of r: $(1, \pi/2)$

The plotting points are given in Table 12.5. The graph is shown in Figure 12.16.

TABLE 12.5

θ	0	$\dfrac{\pi}{8}$	$\dfrac{\pi}{6}$	$\dfrac{\pi}{4}$	$\dfrac{\pi}{2}$
r	undefined	-0.61	0	0.59	1

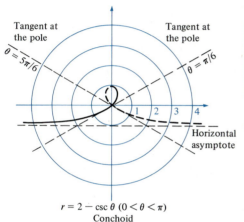

Tangent at the pole

Tangent at the pole

$\theta = 5\pi/6$

$\theta = \pi/6$

Horizontal asymptote

$r = 2 - \csc\theta \ (0 < \theta < \pi)$
Conchoid

FIGURE 12.16

(b)   Since $y = r\sin\theta$, it follows that

$$\frac{r}{y} = \csc\theta$$

and thus from the equation

$$r = 2 - \csc\theta = 2 - \frac{r}{y}$$

we obtain

$$\frac{r}{y} = 2 - r$$

$$y = \frac{r}{2 - r} = -1 + \frac{2}{2 - r}$$

Therefore,

$$\lim_{r \to \pm\infty} y = \lim_{r \to \pm\infty}\left(-1 + \frac{2}{2 - r}\right) = -1$$

and we conclude that $y = -1$ is a horizontal asymptote to the graph of $r = 2 - \csc\theta$.   ■

If a polar equation can be easily converted to rectangular form, it may be worthwhile to do so to see if its graph can be recognized directly from the rectangular equation. For example, the polar equation

$$r = \frac{\sin\theta}{\cos^2\theta}$$

can be converted to rectangular form by multiplying by $r \cos^2 \theta$ to obtain

$$r^2 \cos^2 \theta = r \sin \theta$$
$$(r \cos \theta)^2 = r \sin \theta$$
$$x^2 = y$$

Now in rectangular form we can recognize the graph to be a parabola whose vertex is at the pole.

Section Exercises (12.2)

In Exercises 1–24, sketch the graph of each equation and identify all limaçons, cardioids, rose curves, and conics.

1. $r = 3(1 - \cos \theta)$
2. $r = 2(1 - \sin \theta)$
3. $r = 2 + 3 \sin \theta$ [S]
4. $r = 4 + 5 \cos \theta$
5. $r = 3 - 2 \cos \theta$
6. $r = 5 - 4 \sin \theta$
7. $r = 5$
8. $r = -2$
9. $r = \theta$
10. $r = 3 \cos \theta$
11. $r = 3 \sin \theta$
12. $r = 4 \sin 2\theta$
13. $r = 2 \cos 3\theta$ [S]
14. $r = -\sin 5\theta$
15. $r = 3 \cos 2\theta$
16. $r^2 = 4 \sin \theta$
17. $r^2 = 4 \cos 2\theta$
18. $r^2 = \cos 3\theta$
19. $r^2 = 9 \sin 3\theta$ [S]
20. $r = \dfrac{6}{2 \sin \theta - 3 \cos \theta}$
21. $r = 2 \sec \theta$
22. $r = \dfrac{6}{1 - \cos \theta}$
23. $r = \dfrac{2}{2 - \cos \theta}$
24. $r = \dfrac{4}{6 + \sin \theta}$

In Exercises 25–30, find the slope of the graph at the indicated point.

25. $r = 3(1 - \cos \theta)$ at $\theta = \pi/2$ (see Exercise 1) [S]
26. $r = 2(1 - \sin \theta)$ at $\theta = \pi/6$ (see Exercise 2)
27. $r = 2 + 3 \sin \theta$ at $\theta = \pi/2$ (see Exercise 3)
28. $r = 3 - 2 \cos \theta$ at $\theta = 0$ (see Exercise 5)
29. $r = \theta$ at $\theta = \pi$ (see Exercise 9) [S]
30. $r = 3 \cos \theta$ at $\theta = \pi/4$ (see Exercise 10)

31. Sketch the graph of $r = 2 \cos (3\theta/2)$. [S]

32. Sketch the graph of $r = 3 \sin (5\theta/2)$.
33. Sketch the graph of $r = 2 - \sec \theta$ and show that $x = -1$ is a vertical asymptote of the graph. [S]
34. Sketch the graph of $r = 2 + \csc \theta$, $0 < \theta < \pi$, and show that $y = 1$ is a horizontal asymptote of the graph.
35. Prove that $\tan \psi = |f(\theta)/f'(\theta)|$, where ψ (Greek lowercase letter psi) is the angle between the extended radius vector and the tangent line at the point (r, θ) (see Figure 12.17).

In Exercises 36–41, use the result of Exercise 35 to find the angle ψ for each curve at the given value of θ.

36. $r = 3(1 - \cos \theta)$; $\theta = 3\pi/4$
37. $r = 2(1 - \cos \theta)$; $\theta = \pi$
38. $r = 5$; $\theta = \pi/6$
39. $r = 2 \cos 3\theta$; $\theta = \pi/6$
40. $r = 4 \sin 2\theta$; $\theta = \pi/6$ [C]
41. $r = \dfrac{6}{1 - \cos \theta}$; $\theta = 2\pi/3$ [S]

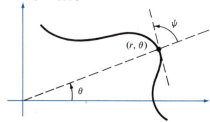

FIGURE 12.17

For Review and Class Discussion

True or False

1. _____ The polar equations $r = \sin 2\theta$ and $r = -\sin 2\theta$ have the same graphs.

2. _____ If the graph of $r = f(\theta)$ has a vertical tangent at (r_1, θ_1), then $f'(\theta_1)$ is undefined.

3. _____ The graph of $r = a \sin 2\theta + b \cos 2\theta$ is symmetric with respect to the polar axis and the line $\theta = \pi/2$.

4. _____ The graph of $r = \sin \theta$ is a circle of radius 1.

5. _____ The slope of the graph of $r = 2 - \sin \theta$ is 1 at the point $(2, \pi)$.

12.3 Points of Intersection and Area

Purpose
- To demonstrate a procedure for finding the points of intersection of two polar graphs.
- To use integration to find the area of a region bounded by polar graphs.

We may have three main objects in the study of truth: first, to find it when we are seeking it; second, to demonstrate it after we have found it; third, to distinguish it from error by examining it. —*Blaise Pascal (1623–1662)*

Two polar equations can have the same polar graph even though they differ algebraically. Perhaps the simplest example of this phenomenon is seen in the two polar equations $r = 1$ and $r = -1$. These two equations are not equivalent and yet the polar graph of each equation is the unit circle with its center at the pole.

This possibility of representing a graph by more than one polar equation leads to complications when we try to find the points of intersection of two polar graphs. As a case in point, let us consider the points of intersection of the graphs of

$$r = 1 - 2 \cos \theta \qquad \text{and} \qquad r = 1$$

as shown in Figure 12.18.

If, as with rectangular equations, we attempt to find the points of intersection by solving the two equations simultaneously, we have

$$1 = 1 - 2 \cos \theta$$
$$2 \cos \theta = 0$$
$$\cos \theta = 0$$
$$\theta = \frac{\pi}{2}, \frac{3\pi}{2} \qquad (0 \le \theta < 2\pi)$$

from which we obtain the two points

$$\left(1, \frac{\pi}{2}\right) \qquad \text{and} \qquad \left(1, \frac{3\pi}{2}\right)$$

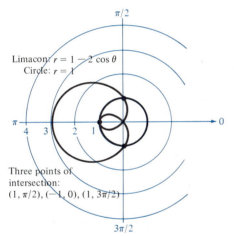

Limacon: $r = 1 - 2 \cos \theta$
Circle: $r = 1$

Three points of intersection:
$(1, \pi/2), (-1, 0), (1, 3\pi/2)$

FIGURE 12.18

To try to find the third point, we can choose an *alternative* polar equation for one of the graphs. For example, if we replace the polar equation $r = 1$ by $r = -1$ and again solve simultaneously, we have

$$-1 = 1 - 2 \cos \theta$$
$$2 \cos \theta = 2$$
$$\cos \theta = 1$$
$$\theta = 0 \qquad (0 \le \theta < 2\pi)$$

which gives us the third point of intersection, $(-1, 0)$.

We can compare the problem of finding the points of intersection of two polar graphs with two satellites in orbit about the earth. (See Figure 12.19.) It is entirely possible that the paths of the two satellites intersect without resulting in a collision as long as the satellites reach the points of

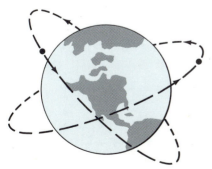

FIGURE 12.19

intersection at different times. A collision will occur only if the satellites reach a particular point of intersection at the *same* time. Similarly, when we solve two polar equations simultaneously, we find only those points of intersection that have the same θ-values. With this in mind, we suggest the following procedure for finding the points of intersection of two polar graphs:

Finding Points of Intersection in Polar Coordinates

1. Solve the given polar equations simultaneously and determine the resulting points of intersection.
2. Replace r by $-r$ and θ by $\theta + \pi$ in *one* equation and solve simultaneously with the *other* equation to determine additional points of intersection.
3. Test to see if the pole is a point of intersection.
4. Make a rough sketch of both graphs to decide if any points of intersection have been missed.

Example 1

Find the points of intersection of the graphs of

$$r = 1 + 3 \cos \theta \qquad \text{and} \qquad r = \frac{2}{2 - \cos \theta}$$

Solution: Solving simultaneously we have

$$1 + 3 \cos \theta = \frac{2}{2 - \cos \theta}$$

$$-3 \cos^2 \theta + 5 \cos \theta + 2 = 2$$

$$\cos \theta (-3 \cos \theta + 5) = 0$$

The second factor is never zero and the first factor is zero when $\theta = \pi/2$ or $\theta = 3\pi/2$. Hence we have the two points of intersection

$$\left(1, \frac{\pi}{2}\right) \qquad \text{and} \qquad \left(1, \frac{3\pi}{2}\right)$$

To test for other points of intersection, we replace r by $-r$ and θ by $\theta + \pi$ in the equation $r = 1 + 3 \cos \theta$. Thus we have

$$-r = 1 + 3 \cos (\theta + \pi) \qquad \text{or} \qquad r = -1 + 3 \cos \theta$$

Now solving the equations

$$r = -1 + 3 \cos \theta \qquad \text{and} \qquad r = \frac{2}{2 - \cos \theta}$$

simultaneously, we have

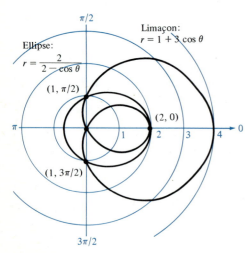

Ellipse:
$$r = \frac{2}{2 - \cos \theta}$$
$(1, \pi/2)$

Limaçon:
$r = 1 + 3 \cos \theta$

$(2, 0)$

$(1, 3\pi/2)$

FIGURE 12.20

$$-1 + 3\cos\theta = \frac{2}{2 - \cos\theta}$$

$$-3\cos^2\theta + 7\cos\theta - 2 = 2$$

$$-3\cos^2\theta + 7\cos\theta - 4 = 0$$

$$(\cos\theta - 1)(-3\cos\theta + 4) = 0$$

Again, the second factor is never zero, but this time the first factor is zero when $\theta = 0$. Hence $(2, 0)$ is another point of intersection. [Note that $(2, 0)$ coincides with $(-2, \pi)$.] Finally, from Figure 12.20 we can see that there are only three points of intersection: $(1, \pi/2)$, $(1, 3\pi/2)$, and $(2, 0)$. ∎

Example 2

Find the points of intersection of the graphs given by

$$r = \sin\theta \qquad \text{and} \qquad r = \cos\left(\theta + \frac{\pi}{6}\right)$$

Solution: Solving simultaneously we have

$$\sin\theta = \cos\left(\theta + \frac{\pi}{6}\right)$$

$$\sin\theta = \cos\theta\cos\frac{\pi}{6} - \sin\theta\sin\frac{\pi}{6}$$

$$\sin\theta = \frac{\sqrt{3}}{2}\cos\theta - \frac{1}{2}\sin\theta$$

$$\frac{3}{2}\sin\theta = \frac{\sqrt{3}}{2}\cos\theta$$

$$\tan\theta = \frac{\sqrt{3}}{3}$$

$$\theta = \frac{\pi}{6}, \frac{7\pi}{6} \qquad (0 \le \theta < 2\pi)$$

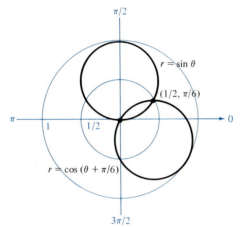

$r = \sin\theta$

$(1/2, \pi/6)$

$r = \cos(\theta + \pi/6)$

FIGURE 12.21

As it turns out, both of these values of θ yield the same point of intersection, $(\frac{1}{2}, \pi/6)$.

Now when we try to construct alternative equations of $r = \sin\theta$ or of $r = \cos[\theta + (\pi/6)]$, we find that the alternative equations are the same as the original ones. Nevertheless, from a sketch of the graphs of these two equations, we can see that the pole is a second point of intersection (Figure 12.21). ∎

Now that we are able to find the points of intersection of polar graphs, we are ready to focus on a procedure for finding the area of a polar region. The development of this procedure parallels that given in Chap-

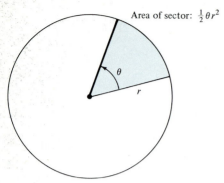

Area of sector: $\frac{1}{2}\theta r^2$

FIGURE 12.22

ter 6 for finding area in the rectangular coordinate system. The difference in the two developments lies in the replacement of rectangles as our basic element of area by sectors of a circle. Recall that the area of a sector of a circle is given by

$$\text{area} = \tfrac{1}{2}\theta r^2$$

where r is the radius of the sector and θ is the radian measure of its central angle (Figure 12.22).

To begin, let us assume that f is a continuous, nonnegative function of θ on the interval $[\alpha, \beta]$, where $0 < \beta - \alpha \le 2\pi$. We can determine the area A of the region bounded by $r = f(\theta)$, $\theta = \alpha$, and $\theta = \beta$ as follows.

First, we partition the interval $[\alpha, \beta]$ into n equal subintervals,

$$\alpha = \theta_0 < \theta_1 < \theta_2 < \cdots < \theta_{n-1} < \theta_n = \beta$$

as indicated in Figure 12.23. Then we approximate the area of the region by the sum of the areas of the n sectors, where

$$\text{radius of } i\text{th sector} = f(\theta_i)$$

$$\text{central angle of } i\text{th sector} = \frac{\beta - \alpha}{n} = \Delta\theta$$

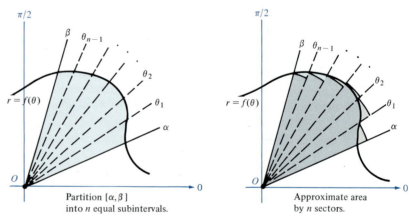

FIGURE 12.23

Thus we have

$$A \approx \sum_{i=1}^{n} \left(\frac{1}{2}\right) \Delta\theta \, f(\theta_i)^2$$

Finally, by taking the limit as $n \to \infty$, we have

$$A = \lim_{n\to\infty} \frac{1}{2} \sum_{i=1}^{n} f(\theta_i)^2 \, \Delta\theta = \frac{1}{2} \int_{\alpha}^{\beta} [f(\theta)]^2 \, d\theta$$

which brings us to the following theorem:

THEOREM 12.4 (*Area in Polar Coordinates*)	If f is continuous and nonnegative on the interval $[\alpha, \beta]$, where $0 < \beta - \alpha \leq 2\pi$, then the area bounded by $r = f(\theta), \theta = \alpha$, and $\theta = \beta$ is given by $$A = \frac{1}{2} \int_{\alpha}^{\beta} [f(\theta)]^2 \, d\theta$$

Example 3

Find the area of the region bounded by the graph of $r = 3 \cos 3\theta$.

Solution: From Figure 12.24 we see that the bounded region is divided into three equal petals. So we can determine the area of the specified region by calculating the area of one of the petals and multiplying by 3.

We choose the petal lying between $\theta = -\pi/6$ and $\theta = \pi/6$ since r is continuous and nonnegative there. Therefore, the total area is given by

$$A = 3 \left(\frac{1}{2}\right) \int_{-\pi/6}^{\pi/6} (3 \cos 3\theta)^2 \, d\theta = \frac{27}{2} \int_{-\pi/6}^{\pi/6} \frac{1 + \cos 6\theta}{2} \, d\theta$$

$$= \frac{27}{4} \int_{-\pi/6}^{\pi/6} (1 + \cos 6\theta) \, d\theta = \frac{27}{4} \left[\theta + \frac{\sin 6\theta}{6}\right]_{-\pi/6}^{\pi/6}$$

$$= \frac{27}{4} \left(\frac{\pi}{6} + \frac{\pi}{6}\right) = \frac{9\pi}{4} \qquad \blacksquare$$

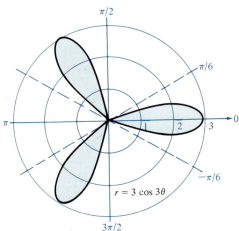

$r = 3 \cos 3\theta$

FIGURE 12.24

Two observations should be made about Example 3. First, even though the area of the petal between $\theta = -\pi/6$ and $\theta = \pi/6$ is partially below the "*x*-axis," its area is registered as positive by the integral $\int_{-\pi/6}^{\pi/6} (3 \cos 3\theta)^2 \, d\theta$. Second, we cannot indiscriminantly find areas in polar coordinates by simply integrating between 0 and 2π. For instance, if we had integrated between 0 and 2π in Example 3, we would have obtained an area of $9\pi/2$, which is twice the actual area. This problem occurs because the rose curve is traversed *twice* as θ increases from 0 to 2π. We can avoid this problem by making sure that we choose an interval $[\alpha, \beta]$ on which r is nonnegative, as required in Theorem 12.4.

Example 4

Find the area of the region lying inside the circle $r = -6 \cos \theta$ and outside the cardioid $r = 2 - 2 \cos \theta$.

Solution: First, we sketch the two graphs and find their points of intersection, as shown in Figure 12.25.

Since the region lies inside the graph of $r = -6 \cos \theta$ and outside the graph of $r = 2 - 2 \cos \theta$, and since both functions are nonnegative on the interval $[2\pi/3, 4\pi/3]$, the area of the specified region is given by

$A = \text{(area in circle)} - \text{(area in cardioid)}$

$$= \frac{1}{2} \int_{2\pi/3}^{4\pi/3} (-6 \cos \theta)^2 \, d\theta - \frac{1}{2} \int_{2\pi/3}^{4\pi/3} (2 - 2 \cos \theta)^2 \, d\theta$$

$$= \frac{1}{2} \int_{2\pi/3}^{4\pi/3} (36 \cos^2 \theta - 4 + 8 \cos \theta - 4 \cos^2 \theta) \, d\theta$$

$$= 2 \int_{2\pi/3}^{4\pi/3} (-1 + 2 \cos \theta + 8 \cos^2 \theta) \, d\theta$$

$$= 2 \int_{2\pi/3}^{4\pi/3} (3 + 2 \cos \theta + 4 \cos 2\theta) \, d\theta$$

$$= 2 \left[3\theta + 2 \sin \theta + 2 \sin 2\theta \right]_{2\pi/3}^{4\pi/3}$$

$$= 2[(4\pi - \sqrt{3} + \sqrt{3}) - (2\pi + \sqrt{3} - \sqrt{3})] = 4\pi \quad \blacksquare$$

In Example 4 we used the same limits of integration for the area of the region inside the circle as for the area of the region inside the cardioid. For the region under consideration, this was possible because the points of intersection have the same θ-values for each function. If the points of intersection have *different* θ-values for the two functions, then we must choose separately appropriate limits of integration for each function.

For instance, in Figure 12.25 if we wanted to find the area of the upper half of the region lying inside the cardioid and outside the circle, we would set up the subtraction as follows:

$A = \text{(area in cardioid)} - \text{(area in circle)}$

$$= \frac{1}{2} \int_{0}^{2\pi/3} (2 - 2 \cos \theta)^2 \, d\theta - \frac{1}{2} \int_{\pi/2}^{2\pi/3} (-6 \cos \theta)^2 \, d\theta$$

Note the difference in the limits. This difference stems from the fact that the point $(3, 2\pi/3)$ has the same coordinates on both the cardioid and the circle, whereas the pole has coordinates $(0, 0)$ on the cardioid and coordinates $(0, \pi/2)$ on the circle.

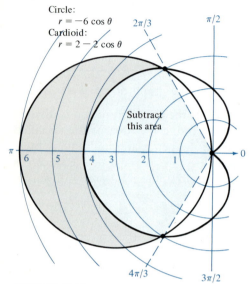

Circle:
$r = -6 \cos \theta$
Cardioid:
$r = 2 - 2 \cos \theta$

Subtract this area

FIGURE 12.25

Example 5

Find the area of the region common to the circle $r = -6 \cos \theta$ and the cardioid $r = 2 - 2 \cos \theta$ (Figure 12.26).

Solution: To simplify the situation we will consider only the top half of the designated region. For this subregion the two points of intersection are $(3, 2\pi/3)$ and the pole. Furthermore, from $\theta = \pi/2$ to $\theta = 2\pi/3$ the region is bounded by the circle $r = -6 \cos \theta$, whereas from $\theta = 2\pi/3$ to $\theta = \pi$ the region is bounded by the cardioid $r = 2 - 2 \cos \theta$. Since both functions are nonnegative on the interval $[\pi/2, \pi]$, the area is

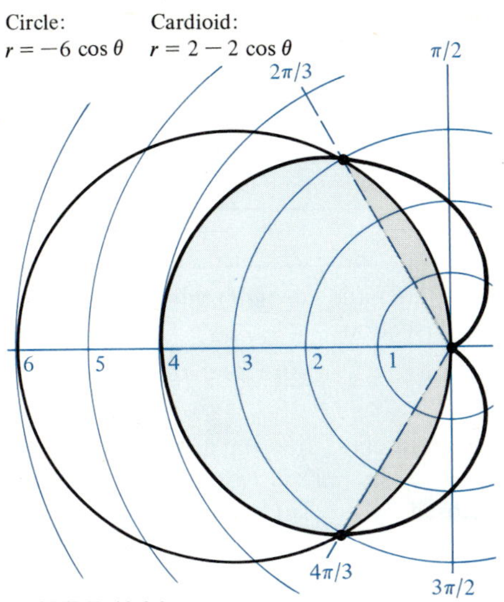

Circle: Cardioid:
$r = -6 \cos \theta$ $r = 2 - 2 \cos \theta$

$$A = 2 \left[\frac{1}{2} \int_{\pi/2}^{2\pi/3} (-6 \cos \theta)^2 \, d\theta + \frac{1}{2} \int_{2\pi/3}^{\pi} (2 - 2 \cos \theta)^2 \, d\theta \right]$$

$$= \int_{\pi/2}^{2\pi/3} 36 \left(\frac{1 + \cos 2\theta}{2} \right) d\theta$$

$$+ \int_{2\pi/3}^{\pi} \left[4 - 8 \cos \theta + 4 \left(\frac{1 + \cos 2\theta}{2} \right) \right] d\theta$$

$$= 18 \int_{\pi/2}^{2\pi/3} (1 + \cos 2\theta) \, d\theta + \int_{2\pi/3}^{\pi} (6 - 8 \cos \theta + 2 \cos 2\theta) \, d\theta$$

$$= 18 \left[\theta + \frac{\sin 2\theta}{2} \right]_{\pi/2}^{2\pi/3} + \left[6\theta - 8 \sin \theta + \sin 2\theta \right]_{2\pi/3}^{\pi} = 5\pi$$

FIGURE 12.26

Section Exercises (12.3)

In Exercises 1–12, find the points of intersection of the graphs of the given equations.

1. $r = 1 + \cos \theta$; $r = 1 - \cos \theta$
2. $r = 3(1 + \sin \theta)$; $r = 3(1 - \sin \theta)$
3. $r = 1 + \cos \theta$; $r = 1 - \sin \theta$
4. $r = 2 - 3 \cos \theta$; $r = \cos \theta$
[S] **5.** $r = 4 - 5 \sin \theta$; $r = 3 \sin \theta$
6. $r = 4 \sin 2\theta$; $r = 2$
[C][S] **7.** $r = 4 \sin 2\theta$; $r = 2 \sin \theta$

8. $\theta = \frac{\pi}{4}$; $r = 2$

9. $r = \frac{\theta}{2}$; $r = 2$

[C] **10.** $r = 3 + \sin \theta$; $r = 2 \csc \theta$

[C] **11.** $r = 2 + 3 \cos \theta$; $r = \frac{\sec \theta}{2}$

[C] **12.** $r = 3(1 - \cos \theta)$; $r = \frac{6}{1 - \cos \theta}$

In Exercises 13–30, find the area of the indicated region.
13. the region enclosed by one loop of $r = 2 \cos 3\theta$
14. the region enclosed by one loop of $r = 4 \sin 2\theta$
[S] **15.** the region enclosed by one loop of $r = \cos (3\theta/2)$
16. the region enclosed by one loop of $r = \cos 5\theta$
17. the region enclosed by the inner loop of $r = 1 + 2 \cos \theta$
18. the region enclosed by the graph of $r = 1 - \sin \theta$
[S] **19.** the region between the loops of $r = 1 + 2 \cos \theta$

20. the interior of the ellipse $r = \dfrac{3}{2 - \cos \theta}$

21. the interior of the ellipse $r = \dfrac{2}{3 - 2 \sin \theta}$

22. the region inside the cardioid $r = 1 - \sin \theta$ and above the polar axis
23. the region outside the inner loop and inside the outer loop of $r = 1 + 2 \sin \theta$ (see Figure 12.27)
24. the region common to the interiors of $r = 3(1 + \sin \theta)$ and $r = 3(1 - \sin \theta)$ (see Figure 12.28)
[S] **25.** the region common to the interiors of $r = 4 \sin 2\theta$ and $r = 2$
[C] **26.** the region enclosed by the inner loop of $r = 3 + 4 \sin \theta$

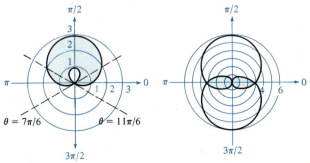

FIGURE 12.27   **FIGURE 12.28**

27. the region inside the circle $r = 3 \sin \theta$ but outside the limaçon $r = 2 - \sin \theta$

28. the region common to the interiors of $r = 4 \sin \theta$ and $r = 2$

29. the region common to the interiors of $r = 3 - 2 \sin \theta$ and $r = -3 + 2 \sin \theta$

30. the region common to the interiors of $r = 3 - 2 \sin \theta$ and $r = 3 - 2 \cos \theta$

For Review and Class Discussion

True or False

1. ____ The area of the region enclosed by the circle $r = \sin \theta$ is given by the definite integral $\frac{1}{2} \int_0^{2\pi} \sin^2 \theta \, d\theta$.

2. ____ If $f(\theta) > 0$ for all θ and $g(\theta) < 0$ for all θ, then the graphs of $r = f(\theta)$ and $r = g(\theta)$ do not intersect.

3. ____ If $f(\theta) = g(\theta)$ for $\theta = 0, \pi/2, \pi,$ and $3\pi/2$, then the graphs of $r = f(\theta)$ and $r = g(\theta)$ have at least four points of intersection.

4. ____ The graphs of $r = 2 + \sin \theta$ and $r = -2 + \sin \theta$ coincide.

5. ____ If n is an even integer, then the area of the region enclosed by $r = \sin (n\theta)$ is twice the area of the region enclosed by $r = \sin [(n + 1)\theta]$.

Miscellaneous Exercises (Ch. 12)

In Exercises 1–18, sketch the graph of the given equation.

1. $r = -2(1 + \cos \theta)$
2. $r = 3 - 4 \cos \theta$
3. $r = 4 - 3 \cos \theta$
4. $r = \cos 5\theta$
5. $r = -3 \cos 2\theta$
6. $r^2 = \cos 2\theta$
7. $r = 4$
8. $r = 2\theta$

S 9. $r = \dfrac{3}{\cos (\theta - \pi/4)}$
10. $r = \dfrac{4}{\cos (\theta + \pi/3)}$

11. $r = -\sec \theta$
12. $r = 3 \csc \theta$

13. $r^2 = 4 \sin^2 2\theta$
14. $r = \dfrac{4}{5 - 3 \cos \theta}$

15. $r = \dfrac{2}{1 - \sin \theta}$
C 16. $r = 2 \sin \theta \cos^2 \theta$

C S 17. $r = 4(\sec \theta - \cos \theta)$
18. $r = 4 \cos 2\theta \sec \theta$

In Exercises 19–23, find a rectangular equation that has the same graph as the given polar equation.

19. $r = 3 \cos \theta$
20. $r = 4 \sec \left(\theta - \dfrac{\pi}{3} \right)$

21. $r = -2(1 + \cos \theta)$
22. $r = 1 + \tan \theta$
S 23. $r = 4 \cos 2\theta \sec \theta$

24. Find a polar equation of the curve whose rectangular equation is $(x^2 + y^2)^2 = ax^2 y$.

25. Find a polar equation of the circle $x^2 + y^2 - 4x = 0$.

26. Find a polar equation of the parabola with focus at the pole and vertex at $(2, \pi)$.

27. Find a polar equation of the ellipse with a focus at the pole and vertices at $(5, 0)$ and $(1, \pi)$.

28. Find a polar equation of the hyperbola with a focus at the pole and vertices at $(1, 0)$ and $(7, 0)$.

29. Find a polar equation of the line with intercepts $(3, 0)$ and $(0, 4)$.

30. Find the tangent lines at the pole and all the points of horizontal and vertical tangency for the graph of $r^2 = 4 \sin 2\theta$. Sketch the graph of the equation.

C S 31. Find the tangent lines at the pole and all points of vertical and horizontal tangency for the graph of $r = 1 - 2 \cos \theta$. Sketch the graph of the equation.

32. Find the tangent lines at the pole and all points of vertical and horizontal tangency for the graph of $r = 4 \cos 3\theta$. Sketch the graph of the equation.

33. Verify that if the curve whose polar equation is $r = f(\theta)$ is rotated about the pole through an angle ϕ, then an equation for the rotated curve is $r = f(\theta - \phi)$.

34. If the polar form of an equation for a curve is $r = f(\sin \theta)$, then show that the form becomes the following:
    (a) $r = f(-\cos \theta)$ if the curve is rotated counterclockwise $\pi/2$ radians about the pole
    (b) $r = f(-\sin \theta)$ if the curve is rotated counterclockwise π radians about the pole
    (c) $r = f(\cos \theta)$ if the curve is rotated counterclockwise $3\pi/2$ radians about the pole

35. Use the results of Exercises 33 and 34 to write the equation of the limaçon $r = 2 - \sin \theta$ after it has been rotated counterclockwise through the following:

    (a) $\dfrac{\pi}{4}$ radians
    (b) $\dfrac{\pi}{2}$ radians

(c) π radians   (d) $\dfrac{3\pi}{2}$ radians

36. Use the result of Exercise 33 to write an equation of the parabola with focus at the pole and vertex at $(2, 7\pi/6)$.

In Exercises 37–43, find the area of the indicated region.
37. the region enclosed by the graph of $r = 2 + \cos\theta$
38. the region bounded by the polar axis and the graph of $r = e^\theta$, $0 \le \theta \le \pi$.
39. the region enclosed by the graph of $r = \sin\theta \cos^2\theta$
40. the region inside the rose curve $r = 4\sin 3\theta$
S **41.** the region common to the interiors of $r = 4\cos\theta$ and $r = 2$
42. the region enclosed by the graph of $r^2 = a^2 \sin 2\theta$
43. the region common to the interiors of $r = a$ and $r^2 = 2a^2 \sin 2\theta$

44. Find the angle between the circle $r = 3\sin\theta$ and the limaçon $r = 4 - 5\sin\theta$ at the point of intersection $(\frac{3}{2}, \pi/6)$.
S **45.** Show that the cardioids $r = 1 + \cos\theta$ and $r = 1 - \cos\theta$ are orthogonal at two of their points of intersection.
46. Show that

$$\int_{\theta_1}^{\theta_2} \sqrt{r^2 + \left(\frac{dr}{d\theta}\right)^2}\, d\theta$$

gives the arc length in polar coordinates for the curve $r = f(\theta)$ from $\theta = \theta_1$ to $\theta = \theta_2$.
S **47.** Use the result of Exercise 46 to find the perimeter of a cardioid $r = a(1 - \cos\theta)$.
48. Use the result of Exercise 46 to find the length of the spiral $r = a\theta$ on $[0, \pi]$.

49. Two of Kepler's laws concerning planetary motion are (1) each planet moves in an elliptical orbit with the sun at one focus and (2) the line joining the sun and any planet sweeps out equal areas in equal times. These laws are also true for satellites orbiting the earth. For these satellites the earth is at one focus of the elliptical orbit. If the path of a satellite is described by $r = b^2/(a + c\cos\theta)$, show that the area swept out by the line from the center of the earth to the satellite as it moves from $\theta = \theta_1$ to $\theta = \theta_2$ is

$$A = \frac{b^2}{2}\left[\frac{-c\sin\theta}{a + c\cos\theta} + \frac{2a}{b}\operatorname{Arctan}\left(\frac{b\tan(\theta/2)}{a + c}\right)\right]_{\theta_1}^{\theta_2}$$

C **50.** *Explorer 1*, orbited by the United States in 1958, had an apogee and perigee of 1580 mi and 220 mi, respectively. (For definitions, see Example 3 of Section 11.2.) Using 4000 mi as the radius of the earth and the result in Exercise 49, find the following:
(a) The area the line from the earth to the satellite sweeps out as θ varies from 0 radians to $\pi/12$ radians. (Make a sketch of the ellipse.)
(b) The angle θ_1 such that the area swept out by the line from the earth to the satellite, as θ ranges from θ_1 to π, is equal to that of part (a).
(c) Approximate the distance traveled by the satellite in part (a).
(d) Approximate the distance traveled by the satellite in part (b).
(e) Recalling that the distances of parts (c) and (d) were traveled in equal times, find the ratio of the average speed of the satellite in part (c) to that of part (d).

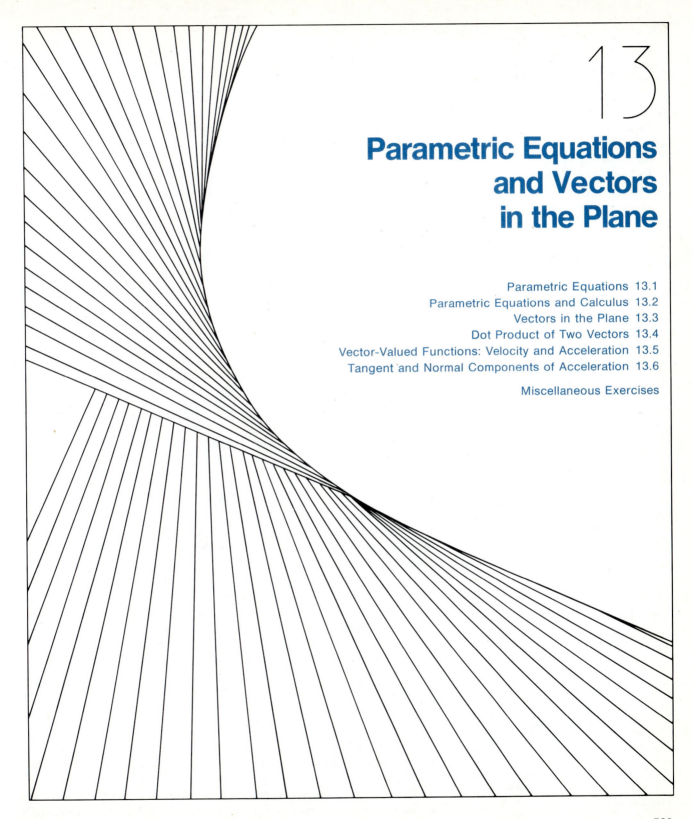

13

Parametric Equations and Vectors in the Plane

13.1 **Parametric Equations**

Purpose

- To demonstrate the use of parametric equations to describe curves in the plane.
- To investigate procedures for converting equations from parametric to rectangular form.

The nature of a curved line is generally given in such a way that the coordinates x and y corresponding to the different points of the curve are given in the form of functions of a single variable, which we shall call t.
— *Carl Gauss (1777–1855)*

Up to this point we have been representing a curve by a single equation involving only two variables. In this chapter we look at situations in which it is useful to introduce a third variable to represent a curve in the plane.

We begin by reviewing an example of linear motion discussed in Chapter 3. Recall that when an object is thrown vertically from ground level with an initial velocity of 48 feet per second, its height at any time t is given by the equation

$$s = -16t^2 + 48t$$

For instance, the object is at ground level ($s = 0$) when $t = 0$; $s = 32$ when $t = 1$; and s is again zero when $t = 3$. (See Figure 13.1.) Since the path of the moving object is linear, we need only two variables to completely describe the motion: one to denote the *height* of the object and the second to denote the *time* the object is at a particular height.

Now suppose this same object is thrown into the air at an angle of $45°$; then it travels along the parabolic path given by

$$y = -\frac{x^2}{72} + x$$

Such nonlinear motion in the plane is referred to as **curvilinear motion.** Curvilinear motion in the plane cannot be described completely by a single equation in only two variables. This is true because we need two variables to denote the *position* of the object and a third variable to denote the *time* the object is at a particular position. Thus to represent curvilinear motion in the plane, we use three variables: the usual position coordinates x and y and a third variable called a **parameter.** With this scheme a curve in the plane is represented by two *parametric equations,* one expressing the x-coordinate and the other expressing the y-coordinate in terms of the parameter. For example, the parabola $y = -(x^2/72) + x$ can be parametrically represented by the equations

$$x = 24\sqrt{2}t \qquad \text{and} \qquad y = -16t^2 + 24\sqrt{2}t$$

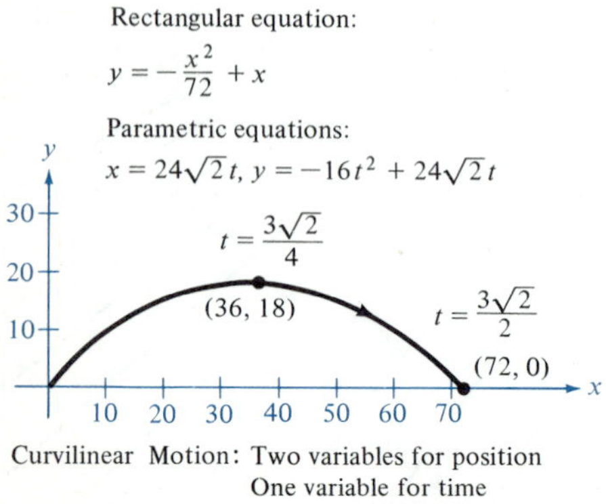

From these two equations we can determine that $(x, y) = (0, 0)$ when $t = 0$; $(x, y) = (36, 18)$ when $t = 3\sqrt{2}/4$; and $(x, y) = (72, 0)$ when $t = 3\sqrt{2}/2$. (See Figure 13.2.)

We will devote the remainder of this section to the study of parametric representations of curves in a plane. Although we have been informally

One variable for position
One variable for time
Linear Motion
FIGURE 13.1

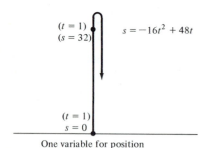

Rectangular equation:

$y = -\frac{x^2}{72} + x$

Parametric equations:

$x = 24\sqrt{2}t, \ y = -16t^2 + 24\sqrt{2}t$

Curvilinear Motion: Two variables for position
One variable for time

FIGURE 13.2

using the terms "curve" and "graph" interchangeably, a curve is actually a special type of graph, as we indicate in the following definition.

Definition: Plane Curves and Parametric Equations	A **plane curve** C is the set of all points $(f(t), g(t))$ satisfying the **parametric equations** $$x = f(t) \quad \text{and} \quad y = g(t)$$ where f and g are continuous functions of the **parameter** t on an interval I.

We should point out that when sketching the curve represented by a pair of parametric equations, we still plot points (x, y) as before, with each coordinate being determined from the value chosen for the parameter t. Once the points are plotted, we connect them by a smooth curve in the order of increasing (or decreasing) values of t.

Example 1

FIGURE 13.3

Sketch the curve described by the parametric equations $x = t/2$ and $y = t^2 - 4$.

Solution: From the table of values, Table 13.1, we make the sketch shown in Figure 13.3. Note that we have labeled several points with their corresponding value for t. From this we see how a point P traces out the curve as we increase the value of t.

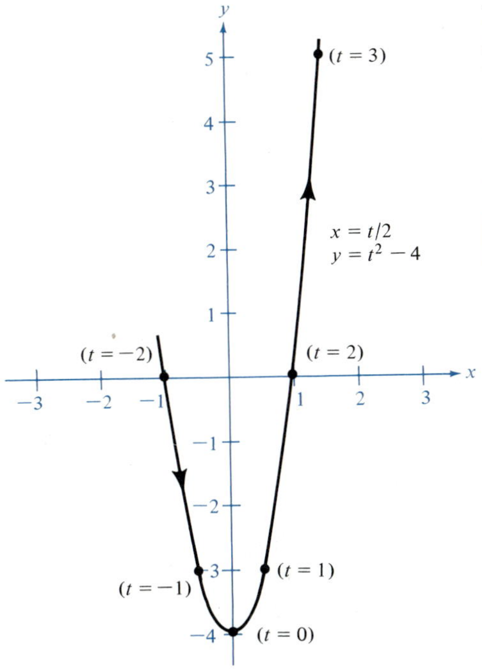

TABLE 13.1

t	-2	-1	0	1	2	3
x	-1	$-\frac{1}{2}$	0	$\frac{1}{2}$	1	$\frac{3}{2}$
y	0	-3	-4	-3	0	5

From Figure 13.3 it appears that the curve represented by the parametric equations $x = t/2$ and $y = t^2 - 4$ is a parabola. To verify this we can find the rectangular equation of the curve by a process called **eliminating the parameter**. In this particular instance we can eliminate the parameter by solving for t in the equation $x = t/2$ and substituting this value for t into the equation $y = t^2 - 4$. We suggest the following scheme for eliminating a parameter and obtaining the rectangular equation of a curve:

$$\text{parametric equations} \longrightarrow \text{solve for } t \text{ in one equation} \longrightarrow \text{substitute into second equation} \longrightarrow \text{rectangular equation}$$

$$x = \frac{t}{2} \qquad\qquad t = 2x \qquad\qquad y = (2x)^2 - 4 \qquad\qquad y = 4x^2 - 4$$

$$y = t^2 - 4$$

To eliminate the parameter in equations involving trigonometric functions, we suggest a slightly different approach, one that is based on familiar trigonometric identities. This approach is demonstrated in the next example.

Example 2

Given the parametric equations

$$x = 3 \cos t \quad \text{and} \quad y = 4 \sin t$$

(a) Eliminate the parameter and determine the corresponding rectangular equation.
(b) Sketch a graph of the curve represented by the given parametric equations.

Solution:

(a) Since the parametric equations involve the sine and cosine, we first think of a trigonometric identity that involves these two functions. In this case we consider the identity $\sin^2 t + \cos^2 t = 1$. From the parametric equations we have

$$\frac{y}{4} = \sin t \quad \text{and} \quad \frac{x}{3} = \cos t$$

Therefore, using this familiar trigonometric identity, we obtain

$$\sin^2 t + \cos^2 t = \left(\frac{y}{4}\right)^2 + \left(\frac{x}{3}\right)^2 = 1$$

and the rectangular equation is

$$\frac{y^2}{16} + \frac{x^2}{9} = 1$$

(b) Now from this rectangular equation, we recognize the curve to be an ellipse with a vertical major axis, where $a = 4$ and $b = 3$. (See Figure 13.4.) ∎

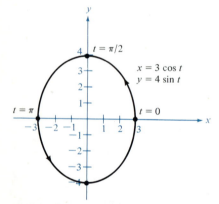

FIGURE 13.4

In addition to the methods illustrated in Examples 1 and 2, there are many other ingenious devices for eliminating the parameter from two parametric equations. For instance, we can sometimes eliminate the parameter by adding or subtracting multiples of the two equations. However, you should be aware that in some instances it may be impossible to eliminate the parameter. In such cases the curve must be plotted point by point directly from the parametric equations.

One word of caution is in order when converting equations from parametric to rectangular form: the range of x and y implied by the parametric equations may be altered by the change to rectangular form. In such instances it is necessary to alter the domain of the rectangular equation so that its graph matches the graph of the parametric equations. Such a situation is demonstrated in the next example.

Example 3

Given the parametric equations

$$x = \frac{1}{\sqrt{t+1}} \quad \text{and} \quad y = \frac{t}{t+1}$$

(a)   Find the corresponding rectangular equation.
(b)   Sketch the graph of the curve.

Solution:

(a)   Solving for t in the equation

$$x = \frac{1}{\sqrt{t+1}}, \qquad -1 < t < \infty$$

yields $\quad x^2 = \dfrac{1}{t+1} \quad$ or $\quad t = \dfrac{1-x^2}{x^2}$

By substituting this value for t into the given equation for y, we obtain

$$y = \frac{t}{t+1} = \frac{(1-x^2)/x^2}{[(1-x^2)/x^2]+1} = 1 - x^2$$

(b)   The rectangular equation, $y = 1 - x^2$, is defined for *all* values of x, but from the parametric equations we see that x is defined only when $-1 < t < \infty$. Thus we restrict the domain of x to

$$x = \frac{1}{\sqrt{t+1}} > 0$$

as shown in Figure 13.5.   ■

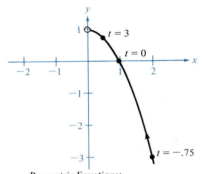

Parametric Equations:
$$x = \frac{1}{\sqrt{t+1}}, \ y = \frac{t}{t+1}$$

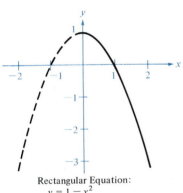

Rectangular Equation:
$$y = 1 - x^2$$

FIGURE 13.5

In Examples 1 through 3 it is important to realize that eliminating the parameter is primarily an aid to curve sketching. In the case of moving objects, once we know the path of the object, we still need the parametric equations to tell us the position, direction, and speed at a given time. For instance, in Example 3 the parametric equations tell us that the object is moving up its parabolic path toward the limiting position of $(0, 1)$ with an ever-decreasing speed. If this same path were traveled in a different manner, the parametric equations describing its position in time would be different.

Of course, it is not necessary that the parameter in a set of parametric equations represent time. The next example describes the use of a parameter other than time.

Example 4

Determine the path traced out by a point P on the circumference of a circle of radius a as the circle rolls along a straight line in a plane. Such a path is called a **cycloid**.

Solution: As our parameter, we let θ be the measure of the circle's rotation, and we let the point P begin at the origin. Thus when $\theta = 0$, P is at the origin; when $\theta = \pi$, P is at a maximum point $(\pi a, 2a)$; and when $\theta = 2\pi$, P is back on the x-axis at $(2\pi a, 0)$.

In Figure 13.6, $\angle APC = 180° - \theta$. Hence

$$\sin \theta = \sin(180° - \theta) = \sin(\angle APC) = \frac{|AC|}{a} = \frac{|BD|}{a}$$

$$\cos \theta = -\cos(180° - \theta) = -\cos(\angle APC) = \frac{|AP|}{-a}$$

which implies that

$$|AP| = -a \cos \theta \qquad \text{and} \qquad |BD| = a \sin \theta$$

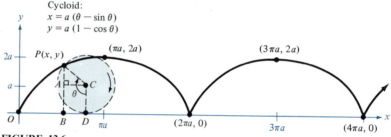

FIGURE 13.6

Now since the circle rolls along the x-axis, we know that

$$|OD| = |\overparen{PD}| = a\theta$$

Furthermore, since $\qquad |BA| = |DC| = a$

we have

$$x = |OD| - |BD| = a\theta - a \sin \theta$$
$$y = |BA| + |AP| = a - a \cos \theta$$

Therefore, the parametric equations are

$$x = a(\theta - \sin \theta) \qquad \text{and} \qquad y = a(1 - \cos \theta) \qquad \blacksquare$$

We should point out that in Example 4 it is *convenient* but not *necessary* to choose as the parameter a measure of the rotation of the circle. Had we chosen a different parameter, we would have obtained a different set of parametric equations to represent the cycloid.

Section Exercises (13.1)

In Exercises 1–20, sketch the curve represented by the parametric equations and write the corresponding rectangular equation by eliminating the parameter.

1. $x = 3t - 1$, $y = 2t + 1$

2. $x = 3 - 2t$, $y = 2 + 3t$

3. $x = t + 1$, $y = t^2$

4. $x = t + 1$, $y = t^3$

$\boxed{\text{S}}$ **5.** $x = t^3$, $y = \dfrac{t^2}{2}$

6. $x = 1 + \dfrac{1}{t}$, $y = t - 1$

7. $x = t - 1$, $y = \dfrac{t}{t - 1}$

$\boxed{\text{S}}$ **8.** $x = t^2 + t$, $y = t^2 - t$

9. $x = 3 \cos \theta$, $y = 3 \sin \theta$

10. $x = \cos \theta$, $y = 3 \sin \theta$

11. $x = 4 \sin 2\theta$, $y = 2 \cos 2\theta$

12. $x = \cos \theta$, $y = 2 \sin 2\theta$

$\boxed{\text{S}}$ **13.** $x = \cos \theta$, $y = 2 \sin^2 \theta$

14. $x = 4 \cos^2 \theta$, $y = 2 \sin \theta$

15. $x = 4 + 2 \cos \theta$, $y = -1 + \sin \theta$

16. $x = 4 + 2 \cos \theta$, $y = -1 + 2 \sin \theta$

17. $x = 4 + 2 \cos \theta$, $y = -1 + 4 \sin \theta$

18. $x = \sec \theta$, $y = \tan \theta$

$\boxed{\text{S}}$ **19.** $x = 4 \sec \theta$, $y = 3 \tan \theta$

20. $x = \cos^3 \theta$, $y = \sin^3 \theta$

21. Show that the parametric equations $x = x_1 + t(x_2 - x_1)$ and $y = y_1 + t(y_2 - y_1)$ represent the equation of the line passing through the points (x_1, y_1) and (x_2, y_2).

22. Find parametric equations for the line passing through the points $(-1, 1)$ and $(2, 3)$.

$\boxed{\text{S}}$ **23.** Find parametric equations for the line passing through the points $(0, 0)$ and $(5, -2)$.

24. Eliminate the parameter from the equations $x = h + a \cos \theta$ and $y = k + b \sin \theta$ to obtain the standard form of the equation of an ellipse.

$\boxed{\text{S}}$ **25.** Eliminate the parameter from the equations $x = h + a \sec \theta$ and $y = k + b \tan \theta$ to obtain the standard form of the equation of a hyperbola.

26. Eliminate the parameter from the equations $x = h + a\sqrt{t + 1}$ and $y = k + b\sqrt{t}$, and compare the result with that of Exercise 25.

27. Find two different pairs of parametric equations for the rectangular equation $y = x^3$.

In Exercises 28–32, sketch the curve represented by the given equations.

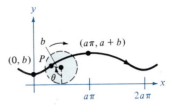

FIGURE 13.7

$\boxed{\text{C}}$ **28.** $x = \theta + \sin \theta$, $y = 1 - \cos \theta$ (cycloid)

$\boxed{\text{C}}$ **29.** $x = 2 \cot \theta$, $y = 2 \sin^2 \theta$ (witch of Agnesi)

$\boxed{\text{C}}$ **30.** $x = 2\theta - \sin \theta$, $y = 2 - \cos \theta$ (curtate cycloid)

$\boxed{\text{C}}\boxed{\text{S}}$ **31.** $x = \theta - \frac{3}{2} \sin \theta$, $y = 1 - \frac{3}{2} \cos \theta$ (prolate cycloid)

$\boxed{\text{C}}$ **32.** $x = \dfrac{3t}{1 + t^3}$, $y = \dfrac{3t^2}{1 + t^3}$ (folium of Descartes)

33. A wheel of radius a rolls along a straight line without slipping. Find the parametric equation for the curve described by a point P that is b units from the center of the wheel. This curve is called a *curtate cycloid* when $b < a$. (See Figure 13.7.)

34. A circle of radius 1 rolls around the outside of a circle of radius 2 without slipping. Show that the parametric equations for the curve described by a point on the circumference of the rolling wheel are $x = 3 \cos \theta - \cos 3\theta$, $y = 3 \sin \theta - \sin 3\theta$. This curve is called an *epicycloid*. (See Figure 13.8.)

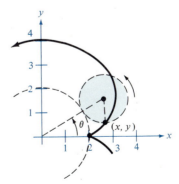

FIGURE 13.8

For Review and Class Discussion

True or False

1. _____ The two sets of parametric equations $x = t$, $y = t^2 + 1$ and $x = 3t, y = 9t^2 + 1$ correspond to the same rectangular equation.

2. _____ The graph of the parmetric equations $x = t^2$ and $y = t^2$ is the line $y = x$.

3. _____ If y is a function of t and x is a function of t, then y is a function of x.

4. _____ If $f(t_1) = 0$ and $g(t_2) = 0$, then the curve represented by the parametric equations $x = f(t)$ and $y = g(t)$ passes through the origin.

5. _____ The curve represented by the parametric equations $x = e^t + t$ and $y = t^3 - t$ passes through the origin.

13.2 Parametric Equations and Calculus

Purpose

- To use the methods of calculus to determine slope, arc length, and area parametrically.

During the last two centuries and a half, physical knowledge has been gradually made to rest upon a basis which it had not before. It has become mathematical. —Augustus DeMorgan (1806–1871)

In this section we investigate the problem of finding the slope dy/dx of a curve defined parametrically. If we can eliminate the parameter, then dy/dx can be found directly from the rectangular equation for the curve. If, however, we cannot find a rectangular equation for a given curve, then we must determine dy/dx indirectly from the parametric equations. The following theorem shows how this can be accomplished:

THEOREM 13.1 $\left(\text{Parametric Form of } \dfrac{dy}{dx}\right)$	If $x = f(t)$, $y = g(t)$, and $dx/dt \neq 0$, then $$\frac{dy}{dx} = \frac{dy/dt}{dx/dt}$$

Proof: In Figure 13.9 consider $\Delta t > 0$ and let

$$\Delta y = g(t + \Delta t) - g(t) \quad \text{and} \quad \Delta x = f(t + \Delta t) - f(t)$$

Now, by definition,

$$\frac{dy}{dx} = \lim_{\Delta x \to 0} \frac{\Delta y}{\Delta x}$$

and since $\Delta x \to 0$ as $\Delta t \to 0$, we can write

$$\frac{dy}{dx} = \lim_{\Delta t \to 0} \frac{g(t + \Delta t) - g(t)}{f(t + \Delta t) - f(t)}$$

Finally, if we divide both numerator and denominator by Δt, then we have

$$\frac{dy}{dx} = \lim_{\Delta t \to 0} \frac{[g(t + \Delta t) - g(t)]/\Delta t}{[f(t + \Delta t) - f(t)]/\Delta t}$$

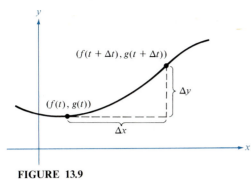

FIGURE 13.9

$$= \frac{\lim_{\Delta t \to 0} [g(t + \Delta t) - g(t)]/\Delta t}{\lim_{\Delta t \to 0} [f(t + \Delta t) - f(t)]/\Delta t} = \frac{g'(t)}{f'(t)} = \frac{dy/dt}{dx/dt} \qquad \blacksquare$$

Since dy/dx is a function of t, for curves described parametrically, we can use Theorem 13.1 to find higher-order derivatives. For instance,

$$\frac{d^2y}{dx^2} = \frac{d}{dx}\left[\frac{dy}{dx}\right] = \frac{\frac{d}{dt}\left[\frac{dy}{dx}\right]}{dx/dt}$$

Similarly,
$$\frac{d^3y}{dx^3} = \frac{d}{dx}\left[\frac{d^2y}{dx^2}\right] = \frac{\frac{d}{dt}\left[\frac{d^2y}{dx^2}\right]}{dx/dt}$$

Example 1

Find all points of horizontal tangency on the ellipse given by

$$x = 2 \sin\left(t + \frac{\pi}{3}\right) \qquad \text{and} \qquad y = 2 \cos\left(t - \frac{\pi}{3}\right)$$

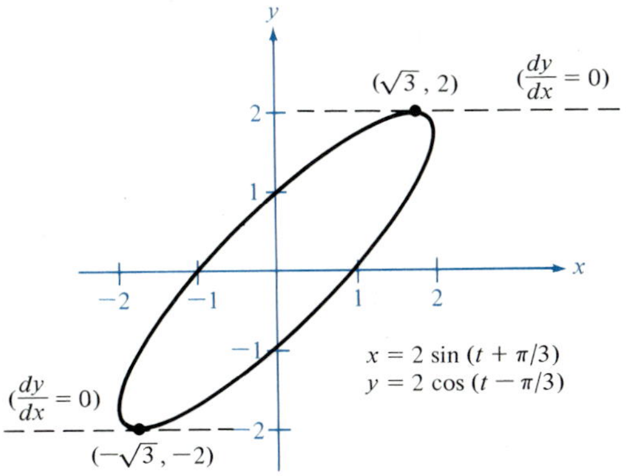

Solution: First, $\dfrac{dy}{dx} = \dfrac{dy/dt}{dx/dt} = \dfrac{-2 \sin[t - (\pi/3)]}{2 \cos[t + (\pi/3)]}$

$$= \frac{-\sin[t - (\pi/3)]}{\cos[t + \pi/3)]}$$

Now $dy/dx = 0$ when $t = \pi/3$ or when $t = 4\pi/3$. Thus to find the points at which the tangent line is horizontal, we substitute $t = \pi/3$ and $t = 4\pi/3$ into the original equations to obtain the points $(\sqrt{3}, 2)$ and $(-\sqrt{3}, -2)$, respectively. (See Figure 13.10.) $\qquad \blacksquare$

$(\sqrt{3}, 2)$ $\quad (\frac{dy}{dx} = 0)$

$x = 2 \sin(t + \pi/3)$
$y = 2 \cos(t - \pi/3)$

$(\frac{dy}{dx} = 0)$
$(-\sqrt{3}, -2)$

FIGURE 13.10

When working with the parametric equations $x = f(t)$ and $y = g(t)$, it is good to keep in mind that these equations need not define y as a function of x even though x and y are both functions of t. Graphically, this means that a curve defined parametrically can loop and cross itself in the plane. The next example presents such a case.

Example 2

The *prolate cycloid* given by

$$x = 2t - \pi \sin t \qquad \text{and} \qquad y = 2 - \pi \cos t$$

crosses itself at the point $(0, 2)$ (Figure 13.11). Find the equation of each tangent line at this point.

Solution: Since $x = 0$ and $y = 2$ when $t = \pm\pi/2$, and since

$$\frac{dy}{dx} = \frac{dy/dt}{dx/dt} = \frac{\pi \sin t}{2 - \pi \cos t}$$

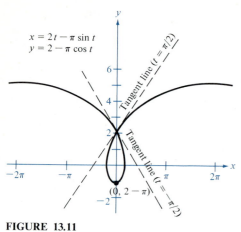

$x = 2t - \pi \sin t$
$y = 2 - \pi \cos t$

Tangent line ($t = \pi/2$)

Tangent line ($t = -\pi/2$)

$(0, 2 - \pi)$

FIGURE 13.11

we have
$$\frac{dy}{dx} = -\frac{\pi}{2} \qquad \text{when} \qquad t = -\frac{\pi}{2}$$

and
$$\frac{dy}{dx} = \frac{\pi}{2} \qquad \text{when} \qquad t = \frac{\pi}{2}$$

Therefore, the two tangent lines at $(0, 2)$ are

$$y - 2 = -\left(\frac{\pi}{2}\right)x \qquad \text{when} \qquad t = -\frac{\pi}{2}$$

and
$$y - 2 = \left(\frac{\pi}{2}\right)x \qquad \text{when} \qquad t = \frac{\pi}{2} \qquad \blacksquare$$

We have previously identified the usefulness of parametric equations for describing the path of a particle moving in the plane. We now want to derive a formula for determining the *distance* (length of arc) a particle travels along its path. Such a formula is given in the following theorem.

THEOREM 13.2
(Arc Length in Parametric Form)

If a curve is given by $x = f(t)$ and $y = g(t)$, where f' and g' are continuous on the interval $[a, b]$, then the arc length of the curve over this interval is given by

$$s = \int_a^b \sqrt{\left(\frac{dx}{dt}\right)^2 + \left(\frac{dy}{dt}\right)^2} \, dt$$

Proof: We will omit a rigorous development of this formula and merely show that it is consistent with the rectangular formula for arc length. Recall from Section 7.8 that the differential of arc length in terms of variable x is

$$ds = \sqrt{1 + \left(\frac{dy}{dx}\right)^2} \, dx$$

In variable t this differential has the form

$$ds = \sqrt{1 + \left(\frac{dy/dt}{dx/dt}\right)^2} \, dx = \frac{\sqrt{(dx/dt)^2 + (dy/dt)^2}}{dx/dt}\left(\frac{dx}{dt}\right) dt$$

$$= \sqrt{\left(\frac{dx}{dt}\right)^2 + \left(\frac{dy}{dt}\right)^2} \, dt$$

Now since
$$ds = \sqrt{\left(\frac{dx}{dt}\right)^2 + \left(\frac{dy}{dt}\right)^2} \, dt$$

then it follows that the length of the curve from $t = a$ to $t = b$ is given by

$$s = \int_a^b \sqrt{\left(\frac{dx}{dt}\right)^2 + \left(\frac{dy}{dt}\right)^2} \, dt \qquad \blacksquare$$

In the previous section we saw that if a circle rolls along a line, then a point on its circumference will trace out a path that is called a cycloid. If this circle rolls around the circumference of another circle, then the path of the point is called an **epicycloid** (see Figure 13.12).

Example 3

A circle of radius 1 is rolling around the circumference of a circle of radius 4 (Figure 13.12). The path of a point on the circumference of the smaller circle is given by

$$x = 5 \cos t - \cos 5t \quad \text{and} \quad y = 5 \sin t - \sin 5t$$

Find the distance traveled by the point in one complete trip about the larger circle.

Solution: Applying Theorem 13.2 we have

$$s = \int_0^{2\pi} \sqrt{\left(\frac{dx}{dt}\right)^2 + \left(\frac{dy}{dt}\right)^2} \, dt$$

$$= \int_0^{2\pi} \sqrt{(-5 \sin t + 5 \sin 5t)^2 + (5 \cos t - 5 \cos 5t)^2} \, dt$$

$$= 5 \int_0^{2\pi} \sqrt{2 - 2 \sin t \sin 5t - 2 \cos t \cos 5t} \, dt$$

$$= 5 \int_0^{2\pi} \sqrt{2 - 2 \cos 4t} \, dt = 5 \int_0^{2\pi} \sqrt{4(\sin^2 2t)} \, dt$$

$$= 10 \int_0^{2\pi} |\sin 2t| \, dt = 40 \int_0^{\pi/2} \sin 2t \, dt$$

$$= -20 \cos 2t \Big]_0^{\pi/2} = 40 \qquad \blacksquare$$

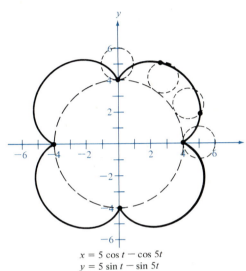

$x = 5 \cos t - \cos 5t$
$y = 5 \sin t - \sin 5t$

FIGURE 13.12

We have pointed out in this section that the parametric equations $x = f(t)$ and $y = g(t)$ need not define y as a function of x. Nevertheless, for those cases in which y is a function of x, we can evaluate definite integrals according to the following theorem:

THEOREM 13.3
(Definite Integrals in Parametric Form)

If y is a continuous function of x on the interval $a \leq x \leq b$, where

$$x = f(t) \quad \text{and} \quad y = g(t)$$

then

$$\int_a^b y \, dx = \int_{t_1}^{t_2} g(t) f'(t) \, dt$$

provided $f(t_1) = a$, $f(t_2) = b$, and both g and f' are continuous on $[t_1, t_2]$.

The proof of this theorem is left as an exercise (see Exercise 42).

Example 4

Find the area of the region bounded by the x-axis and one arch of the cycloid given by

$$x = a(t - \sin t) \qquad \text{and} \qquad y = a(1 - \cos t)$$

(See Figure 13.13.)

$x = a(t - \sin t)$
$y = a(1 - \cos t)$

FIGURE 13.13

Solution: Using Theorem 13.3 the desired area is determined by the integral

$$\int_0^{2a\pi} y \, dx = \int_0^{2\pi} a(1 - \cos t)(a)(1 - \cos t) \, dt$$

$$= a^2 \int_0^{2\pi} (1 - 2\cos t + \cos^2 t) \, dt$$

$$= a^2 \int_0^{2\pi} \left(1 - 2\cos t + \frac{1 + \cos 2t}{2}\right) dt$$

$$= \frac{a^2}{2} \int_0^{2\pi} (3 - 4\cos t + \cos 2t) \, dt$$

$$= \frac{a^2}{2} \left[3t - 4\sin t + \frac{\sin 2t}{2}\right]_0^{2\pi}$$

Therefore, the area is $\qquad A = 3a^2\pi$ ∎

Section Exercises (13.2)

In Exercises 1–10, find dy/dx and d^2y/dx^2 and evaluate at the specified value of the parameter.
1. $x = 2t, y = 3t - 1; t = 3$
2. $x = \sqrt{t}, y = 3t - 1; t = 1$
3. $x = t + 1, y = t^2 + 3t; t = -1$
S 4. $x = t^2 + 3t, y = t + 1; t = 0$
5. $x = 2\cos\theta, y = 2\sin\theta; \theta = \pi/4$
6. $x = \cos\theta, y = 3\sin\theta; \theta = 0$
S 7. $x = 2 + \sec\theta, y = 1 + 2\tan\theta; \theta = \pi/6$
8. $x = \sqrt{t}, y = \sqrt{t} - 1; t = 2$
9. $x = a\cos^3\theta, y = a\sin^3\theta; \theta = \pi/4$
10. $x = a(\theta - \sin\theta), y = a(1 - \cos\theta); \theta = \pi$

In Exercises 11–15, find equations for the lines tangent to the curve at the specified value of the parameter.

S 11. $x = 4\cos\theta, y = 3\sin\theta; \theta = 3\pi/4$
12. $x = 2 - 3\cos\theta, y = 3 + 2\sin\theta; \theta = 5\pi/3$
13. $x = t^2 - t + 2, y = t^3 - 3t; t = -1$ and $t = 2$
14. $x = t - 1, y = \dfrac{1}{t} + 1; t = 1$
15. $x = 2\cot\theta, y = 2\sin^2\theta; \theta = \pi/4$

In Exercises 16–23, find all points of horizontal tangency on the indicated curve.
16. $x = t + 1, y = t^2 + 3t$
17. $x = 1 - t, y = t^2$
18. $x = t^2 - t + 2, y = t^3 - 3t$
S 19. $x = 1 - t, y = t^3 - 3t$

20. $x = \cos\theta + \theta\sin\theta$, $y = \sin\theta - \theta\cos\theta$ (involute of a circle)

21. $x = \cot\theta$, $y = 2\sin\theta\cos\theta$

22. $x = \dfrac{3s}{1+s^3}$, $y = \dfrac{3s^2}{1+s^3}$ (folium of Descartes)

23. $x = 2\theta$, $y = 2(1 - \cos\theta)$

In Exercises 24–27, find the length of the given arc.

24. $x = t^2$, $y = 4t^3$; $-1 \le t \le 1$

S 25. $x = e^{-t}\cos t$, $y = e^{-t}\sin t$; $0 \le t \le \pi/2$

26. $x = \text{Arcsin } t$, $y = \ln\sqrt{1 - t^2}$; $0 \le t \le \frac{1}{2}$

27. $x = t^2$, $y = 2t$; $0 \le t \le 2$

28. Find the circumference of the circle $x = a\cos\theta$, $y = a\sin\theta$ by integration.

S 29. Find the perimeter of the hypocycloid $x = a\cos^3\theta$, $y = a\sin^3\theta$.

30. Find the length of one arch of the cycloid $x = a(\theta - \sin\theta)$, $y = a(1 - \cos\theta)$.

31. Find the length of the arc from $t = 0$ to $t = 1$ on the curve described by $x = \sqrt{t}$, $y = 3t - 1$.

32. Find the area of the region enclosed by the ellipse $x = a\sin\theta$, $y = b\cos\theta$.

33. Find the area of the region bounded by the curve $x = 2\cot\theta$, $y = 2\sin^2\theta$, and the x-axis.

34. Find the area of the region enclosed by the hypocycloid $x = \cos^3\theta$, $y = \sin^3\theta$.

S 35. Sketch the graph of the cissoid $x = 2\sin^2\theta$, $y = 2\sin^2\theta\tan\theta$ and find the area of the region enclosed between the curve and its asymptote in the first quadrant.

36. Let R be the region bounded by $x = f(t)$ and $y = g(t)$ on $[t_1, t_2]$. Show that the surface area of the solid of revolution is given by:

(a) $S = 2\pi\int_{t_1}^{t_2} g(t)\sqrt{(dx/dt)^2 + (dy/dt)^2}\,dt$, if R is revolved about the x-axis

(b) $S = 2\pi\int_{t_1}^{t_2} f(t)\sqrt{(dx/dt)^2 + (dy/dt)^2}\,dt$, if R is revolved about the y-axis

37. Find the surface area of the solid generated by revolving the hypocycloid $x = a\cos^3\theta$, $y = a\sin^3\theta$ about the x-axis.

38. Find the surface area of the solid generated by revolving the ellipse $x = a\cos\theta$, $y = b\sin\theta$ about its:
(a) major axis (prolate spheroid)
(b) minor axis (oblate spheroid)

S 39. A portion of a sphere is removed by a circular cone with its vertex at the center of the sphere. Find the surface area removed from the sphere if the vertex of the cone forms an angle of 2θ.

40. Show that the parametric form for curvature is

$$K = \frac{\left|\dfrac{dx}{dt}\dfrac{d^2y}{dt^2} - \dfrac{dy}{dt}\dfrac{d^2x}{dt^2}\right|}{\left[\left(\dfrac{dx}{dt}\right)^2 + \left(\dfrac{dy}{dt}\right)^2\right]^{3/2}}$$

by showing that this form is equivalent to

$$K = \frac{\left|\dfrac{d^2y}{dx^2}\right|}{\left[1 + \left(\dfrac{dy}{dx}\right)^2\right]^{3/2}}$$

41. Find the curvature of the ellipse given by $x = 3\cos t$, $y = -4\sin t$ when $t = \pi/2$. (See Exercise 40.)

42. Prove Theorem 13.3.

For Review and Class Discussion

True or False

1. ____ If $x = f(t)$ and $y = g(t)$, then $d^2y/dx^2 = g''(t)/f''(t)$.

2. ____ The curve given by $x = \sqrt[3]{t}$, $y = t$ has no horizontal tangents since dy/dt is never zero.

3. ____ The curve given by $x = t^3$, $y = t^2$ has a horizontal tangent at the origin since $dy/dt = 0$ when $t = 0$.

4. ____ A curve may have more than one tangent line at a given point.

5. ____ If $f'(t)$ and $g'(t)$ are positive for all real values of t, then the curve given by $x = f(t)$, $y = g(t)$ does not cross itself.

13.3 Vectors in the Plane

Purpose
- To introduce the geometry and algebra of vectors in the plane.

If new notation be advisable, permanently or temporarily, it should carry with it some mark of distinction from that which is already in use.
—*Augustus DeMorgan (1806–1871)*

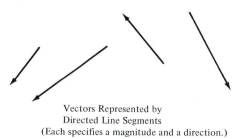

Vectors Represented by
Directed Line Segments
(Each specifies a magnitude and a direction.)

FIGURE 13.14

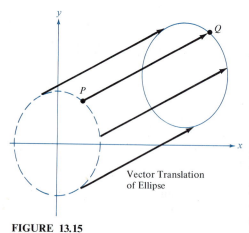

Vector Translation
of Ellipse

FIGURE 13.15

We are familiar with quantities in geometry and physics that can be characterized by a single real number scaled to an appropriate unit of measure. Some examples are circumference, area, volume, temperature, mass, voltage, and time. Such quantities are referred to as **scalar quantities** and the real number associated with each is called a **scalar.**

In contrast to scalar quantities, there are other physical and geometrical quantities involving direction in the plane, which cannot be completely characterized by a single real number. Some examples are force, velocity, acceleration, and translation in the plane. These quantities are called **vector quantities** and the mathematical object used to describe each is called a **vector.**

The two fundamental characteristics of a vector are its *magnitude* and its *direction.* Thus it is natural to represent a vector geometrically by a directed line segment, where the length and direction of the segment indicate the magnitude and direction, respectively, of the vector (Figure 13.14).

As a geometric illustration of the use of directed line segments, consider the translation of an ellipse in the plane, as shown in Figure 13.15. Since point P is translated to point Q, we say that the directed line segment $\overrightarrow{PQ}$ has its **initial point** at P and its **terminal point** at Q. Considering the translation of each point of the ellipse, we have a family of directed line segments, *each having the same length and direction as* $\overrightarrow{PQ}$. Since each directed line segment in the family has the same length and direction, we can geometrically represent the entire family by any one of its members. This suggests the following definition of a vector:

Definition of a Vector in the Plane	A **vector V** is the collection of all directed line segments in the plane having a given length and a given direction.

We will generally use the boldface letters **U, V, W,** and **Z** to denote vectors.

Note in this definition that a vector is actually an infinite collection of directed line segments. Though this definition is theoretically useful, in practice we work primarily with a few "representatives" of the entire collection of directed line segments. Therefore, to avoid the inconvenience of distinguishing between the entire class of directed line segments and one of its representatives, we will use the term "vector" in both cases.

Before we can make any meaningful applications, we must define two

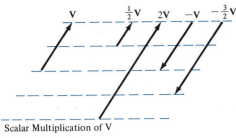

Scalar Multiplication of V

FIGURE 13.16

basic vector operations: **scalar multiplication** and **vector addition.** Each of these operations can be viewed either geometrically or algebraically and both views are essential to a good understanding of vectors and their applications. We begin with the geometrical point of view.

Scalar multiplication consists of multiplying a vector by a scalar to produce another vector. Specifically, if **V** is a vector, then the scalar multiple $k\mathbf{V}$ is a vector having the same (if $k > 0$) or opposite (if $k < 0$) direction as **V** but with a length k times that of **V.** Some scalar multiples of a given vector **V** are shown in Figure 13.16.

From Figure 13.16 we can see that scalar multiplication is one way to produce new vectors from a given vector. We can also generate new vectors by vector addition. To add two vectors geometrically, we position them (without changing their magnitude or direction) so that the initial point of one coincides with the terminal point of the other. The resulting sum is the vector formed by joining the initial point of the first vector to the terminal point of the second, as shown in Figure 13.17.

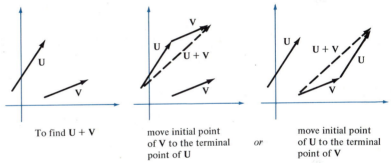

| To find **U** + **V** | move initial point of **V** to the terminal point of **U** | *or* | move initial point of **U** to the terminal point of **V** |

FIGURE 13.17

Having defined these two operations geometrically, we now want to discuss them from an algebraic point of view. We begin by defining two **unit vectors, i** and **j,** which are the vectors from the origin to the points $(1, 0)$ and $(0, 1)$, respectively (Figure 13.18).

Suppose a vector **V** has its initial point at the origin and its terminal point at (a, b). Then by scalar multiplication and vector addition, we can represent **V** as the sum

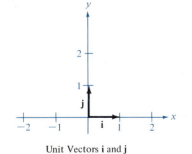

Unit Vectors **i** and **j**

FIGURE 13.18

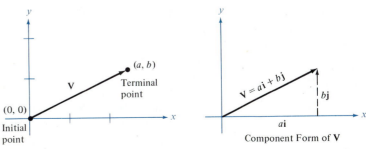

FIGURE 13.19

$$V = a\mathbf{i} + b\mathbf{j}$$

(See Figure 13.19, p. 583. This linear combination of vectors $\mathbf{i}$ and $\mathbf{j}$ is the **component form** of $\mathbf{V}$, and the numbers a and b are referred to as the respective horizontal and vertical *scalar components* (or simply *components*) of $\mathbf{V}$.

Definition of Component Form of a Vector	If a vector $\mathbf{V}$ is represented by the directed line segment from the point (x_1, y_1) to the point (x_2, y_2), then the **component form** of $\mathbf{V}$ is $$\mathbf{V} = a\mathbf{i} + b\mathbf{j}$$ where $a = x_2 - x_1$ is the **horizontal component** of $\mathbf{V}$ and $b = y_2 - y_1$ is the **vertical component** of $\mathbf{V}$.

If $(x_1, y_1) = (x_2, y_2)$, then we call $\mathbf{V}$ the **zero vector,** which is denoted by $\mathbf{0}$. The component representation of the zero vector is

$$\mathbf{0} = 0\mathbf{i} + 0\mathbf{j}$$

In component form we define the operations of scalar multiplication, vector addition, and vector subtraction as follows:

Operations with Vectors	If $\qquad \mathbf{U} = a_1\mathbf{i} + b_1\mathbf{j} \qquad$ and $\qquad \mathbf{V} = a_2\mathbf{i} + b_2\mathbf{j}$
	then $\qquad k\mathbf{U} = ka_1\mathbf{i} + kb_1\mathbf{j} \qquad\qquad$ (scalar multiplication)
	$\mathbf{U} + \mathbf{V} = (a_1 + a_2)\mathbf{i} + (b_1 + b_2)\mathbf{j} \qquad$ (vector addition)
	and $\qquad \mathbf{U} - \mathbf{V} = (a_1 - a_2)\mathbf{i} + (b_1 - b_2)\mathbf{j} \qquad$ (vector subtraction)

Example 1

Given the vectors

$$\mathbf{U} = 3\mathbf{i} - \mathbf{j} \qquad \text{and} \qquad \mathbf{V} = -2\mathbf{i} + 3\mathbf{j}$$

find $\mathbf{U} - \mathbf{V}$, $2\mathbf{V}$, and $\mathbf{U} + 2\mathbf{V}$ in terms of $\mathbf{i}$ and $\mathbf{j}$. Draw a figure to represent $\mathbf{U} + 2\mathbf{V}$.

Solution: By definition

$$\mathbf{U} - \mathbf{V} = (3 + 2)\mathbf{i} + (-1 - 3)\mathbf{j} = 5\mathbf{i} - 4\mathbf{j}$$

Since

$$2\mathbf{V} = 2(-2)\mathbf{i} + 2(3)\mathbf{j} = -4\mathbf{i} + 6\mathbf{j}$$

we have

$$\mathbf{U} + 2\mathbf{V} = -\mathbf{i} + 5\mathbf{j}$$

The sum $\mathbf{U} + 2\mathbf{V}$ is obtained geometrically in Figure 13.20 by moving $2\mathbf{V}$ so that its initial point coincides with the terminal point of $\mathbf{U}$ (or by moving $\mathbf{U}$ so that its initial point corresponds with the terminal point of $2\mathbf{V}$).

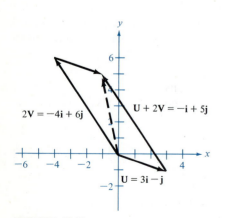

FIGURE 13.20

The **length** (or **magnitude**) of a vector **V** is denoted by |**V**|. Using the Distance Formula in the plane, we define the length of the vector **V** = $a\mathbf{i} + b\mathbf{j}$ to be

$$|\mathbf{V}| = \sqrt{a^2 + b^2}$$

Example 2

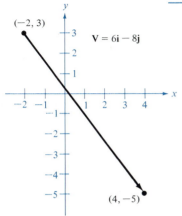

FIGURE 13.21

Write the component form of a vector **V** having initial point $(-2, 3)$ and terminal point $(4, -5)$. What is its length?

Solution: By definition

$$\mathbf{V} = (x_2 - x_1)\mathbf{i} + (y_2 - y_1)\mathbf{j}$$
$$= (4 + 2)\mathbf{i} + (-5 - 3)\mathbf{j} = 6\mathbf{i} - 8\mathbf{j}$$

Therefore, its length is

$$|\mathbf{V}| = \sqrt{6^2 + (-8)^2} = \sqrt{36 + 64} = \sqrt{100} = 10$$

(See Figure 13.21.) ∎

Using the component form of vectors, we can verify the following theorem, which identifies the properties of vector addition and scalar multiplication:

THEOREM 13.4
(*Properties of Vector Addition and Scalar Multiplication*)

For arbitrary vectors **U**, **V**, and **W** and scalars a and b, the following properties hold:

i. $\mathbf{U} + \mathbf{V} = \mathbf{V} + \mathbf{U}$

ii. $\mathbf{U} + (\mathbf{V} + \mathbf{W}) = (\mathbf{U} + \mathbf{V}) + \mathbf{W}$

iii. $(ab)\mathbf{V} = a(b\mathbf{V})$

iv. $a(\mathbf{U} + \mathbf{V}) = a\mathbf{U} + a\mathbf{V}$

v. $\mathbf{V} + \mathbf{0} = \mathbf{V}$

vi. $\mathbf{V} + (-\mathbf{V}) = \mathbf{0}$

vii. $|a\mathbf{V}| = |a|\,|\mathbf{V}|$

Note that the same symbol is used for absolute value and vector length. However, since absolute value applies to scalars and vector length applies to vectors, this double usage should never be a cause for confusion.

We call the vectors **i** and **j** *unit* vectors because their lengths are 1. These particular unit vectors have directions corresponding to the positive *x*- and *y*-axes, respectively. Our next theorem shows how to construct a unit vector in the direction of an arbitrary nonzero vector **V**.

THEOREM 13.5
(*Unit Vectors*)

If **V** is a nonzero vector, then the vector **V**/|**V**| is a **unit vector** having the same direction as **V**.

Proof: From the definition of scalar multiplication, we know that $(1/|\mathbf{V}|)\mathbf{V}$ has the same direction as **V** since $1/|\mathbf{V}| > 0$. Furthermore, **V**/|**V**|

is a unit vector since

$$\left|\frac{\mathbf{V}}{|\mathbf{V}|}\right| = \frac{|\mathbf{V}|}{|\mathbf{V}|} = 1$$

∎

Example 3

If $\mathbf{U} = 2\mathbf{j}$, $\mathbf{V} = 4\mathbf{i} - 7\mathbf{j}$, and $\mathbf{W} = -3\mathbf{i} + 5\mathbf{j}$, express $2\mathbf{V} + 2\mathbf{W}$ and $\mathbf{U} + (\mathbf{V} + \mathbf{W})$ in terms of $\mathbf{i}$ and $\mathbf{j}$. Write a unit vector in the direction of $\mathbf{W}$ and in the direction opposite to $\mathbf{W}$.

Solution: First, we determine that

$$\mathbf{V} + \mathbf{W} = (4 - 3)\mathbf{i} + (-7 + 5)\mathbf{j} = \mathbf{i} - 2\mathbf{j}$$

Now by Theorem 13.4

$$2\mathbf{V} + 2\mathbf{W} = 2(\mathbf{V} + \mathbf{W}) = 2\mathbf{i} - 4\mathbf{j}$$

Furthermore, $\mathbf{U} + (\mathbf{V} + \mathbf{W}) = 2\mathbf{j} + (\mathbf{i} - 2\mathbf{j}) = \mathbf{i}$

A unit vector in the direction of $\mathbf{W}$ is given by

$$\frac{\mathbf{W}}{|\mathbf{W}|} = \frac{-3\mathbf{i} + 5\mathbf{j}}{\sqrt{(-3)^2 + 5^2}} = \frac{-3\mathbf{i} + 5\mathbf{j}}{\sqrt{34}} = \left(\frac{-3}{\sqrt{34}}\right)\mathbf{i} + \left(\frac{5}{\sqrt{34}}\right)\mathbf{j}$$

A unit vector in the opposite direction is given by

$$\frac{-\mathbf{W}}{|\mathbf{W}|} = \left(\frac{3}{\sqrt{34}}\right)\mathbf{i} - \left(\frac{5}{\sqrt{34}}\right)\mathbf{j}$$

∎

Section Exercises (13.3)

In Exercises 1–4, find the component form and length of the vector **V**.

1.

2.

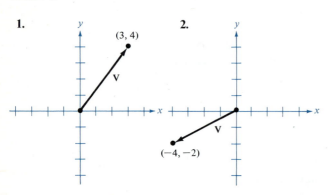

S 3.

4.

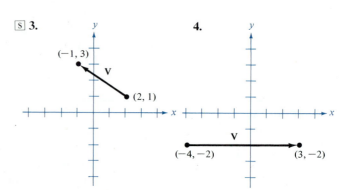

In Exercises 5–14, sketch **V** and find its component form.

5. **V** is a horizontal vector (pointing to the right) of length 3.

6. **V** is a unit vector making an angle of 45° with the positive x-axis.

[S] **7.** **V** is a unit vector making an angle of 150° with the positive x-axis.

8. **V** is a unit vector in the direction of the vector $\mathbf{W} = \mathbf{i} + 3\mathbf{j}$.

9. **V** is the sum of the vectors $\mathbf{V}_1 = 2\mathbf{i} + \mathbf{j}$ and $\mathbf{V}_2 = 3\mathbf{i} + 5\mathbf{j}$.

10. **V** is $-5\mathbf{W}$, where $\mathbf{W} = -\mathbf{i} + 3\mathbf{j}$.

11. **V** is $3\mathbf{V}_1 - 2\mathbf{V}_2$, where $\mathbf{V}_1 = 2\mathbf{i} + \mathbf{j}$ and $\mathbf{V}_2 = 3\mathbf{i} + 5\mathbf{j}$.

12. **V** is a unit vector parallel to the line tangent to $y = x^3$ at $(1, 1)$.

[S] **13.** **V** is a unit vector normal to the line tangent to $y = x^3$ at $(1, 1)$.

14. **V** has a magnitude of 3 units and is parallel to the line tangent to $y = x^3$ at $(1, 1)$.

In Exercises 15–20, find the component form of **V** and illustrate the indicated vector operations geometrically, where $\mathbf{U} = 2\mathbf{i} - \mathbf{j}$ and $\mathbf{W} = \mathbf{i} + 2\mathbf{j}$.

15. $\mathbf{V} = \frac{3}{2}\mathbf{U}$ **16.** $\mathbf{V} = \mathbf{U} + \mathbf{W}$

[S] **17.** $\mathbf{V} = \mathbf{U} + 2\mathbf{W}$ **18.** $\mathbf{V} = -\mathbf{U} + \mathbf{W}$

19. $\mathbf{V} = \frac{1}{2}(3\mathbf{U} + \mathbf{W})$ **20.** $\mathbf{V} = \mathbf{U} - 2\mathbf{W}$

In Exercises 21–24, find a unit vector in the direction of the given vector.

[S] **21.** $\mathbf{V} = 4\mathbf{i} - 3\mathbf{j}$ **22.** $\mathbf{V} = \mathbf{i} + \mathbf{j}$

23. $\mathbf{V} = 2\mathbf{j}$ **24.** $\mathbf{V} = \mathbf{i} - 2\mathbf{j}$

In Exercises 25–30, find a and b such that $\mathbf{V} = a\mathbf{U} + b\mathbf{W}$, where $\mathbf{U} = \mathbf{i} + 2\mathbf{j}$ and $\mathbf{W} = \mathbf{i} - \mathbf{j}$.

[S] **25.** $\mathbf{V} = 2\mathbf{i} + \mathbf{j}$ **26.** $\mathbf{V} = 3\mathbf{j}$

27. $\mathbf{V} = 3\mathbf{i}$ **28.** $\mathbf{V} = 3\mathbf{i} + 3\mathbf{j}$

29. $\mathbf{V} = \mathbf{i} + \mathbf{j}$ **30.** $\mathbf{V} = -\mathbf{i} + 7\mathbf{j}$

31. Three vertices of a parallelogram are (1, 2), (3, 1), and (8, 4). Find the three possible fourth vertices. (See Figure 13.22.)

32. Using vectors, prove that the diagonals of a parallelogram bisect each other.

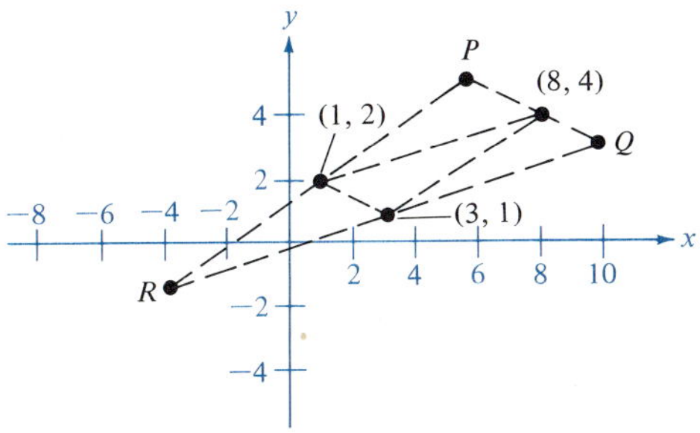

FIGURE 13.22

For Review and Class Discussion

True or False

1. ____ If **U** and **V** have the same magnitude and direction, then $\mathbf{U} = \mathbf{V}$.

2. ____ If $\mathbf{V} = \mathbf{U} + \mathbf{W}$, then $\mathbf{U} = \mathbf{V} - \mathbf{W}$.

3. ____ If $|\mathbf{U}| = |\mathbf{V}| = |\mathbf{U} + \mathbf{V}|$, then $\mathbf{U} = \mathbf{V} = \mathbf{0}$.

4. ____ If $\mathbf{V} = \mathbf{U} + \mathbf{W}$ and $\mathbf{U} = \mathbf{V} + \mathbf{W}$, then $\mathbf{W} = \mathbf{0}$.

5. ____ If **U** is a unit vector in the direction of **V**, then $\mathbf{V} = |\mathbf{V}|\,\mathbf{U}$.

6. ____ If $\mathbf{U} = a\mathbf{i} + b\mathbf{j}$ and $\mathbf{V} = b\mathbf{i} - a\mathbf{j}$, then $|\mathbf{U}| = |\mathbf{V}|$.

7. ____ If $\mathbf{U} = a\mathbf{i} + b\mathbf{j}$ is a unit vector, then $a^2 + b^2 = 1$.

8. ____ If $\mathbf{V} = a\mathbf{i} + b\mathbf{j} = \mathbf{0}$, then $a = b = 0$.

9. ____ If $a = b$, then $|a\mathbf{i} + b\mathbf{j}| = \sqrt{2}a$.

10. ____ If **U** and **V** have the same magnitude but opposite directions, then $\mathbf{U} + \mathbf{V} = \mathbf{0}$.

13.4 **Dot Product of Two Vectors**

Purpose
- To define the dot product of two vectors.
- To identify some uses of the dot product.

What goes beyond geometry goes beyond man.   —*Blaise Pascal (1623–1662)*

We have seen that adding two vectors results in a vector and that the scalar multiple of a vector is again a vector. We now define a multiplication operation on two vectors that results in a scalar value rather than another vector. This product is referred to as the "dot," "inner," or "scalar" product.

Definition of Dot Product	If $U = a_1i + b_1j$ and $V = a_2i + b_2j$, then the **dot product** of vectors U and V, denoted by $U \cdot V$, is given by $$U \cdot V = a_1a_2 + b_1b_2$$

Keep in mind that the dot product of two vectors is a scalar and *not* another vector.

Example 1

Find the dot product of $U = 5i - 2j$ and $V = \frac{3}{2}i + 3j$.

Solution: By definition

$$U \cdot V = a_1a_2 + b_1b_2 = 5(\tfrac{3}{2}) + (-2)(3) = \tfrac{15}{2} - 6 = \tfrac{3}{2} \qquad \blacksquare$$

The properties listed in Theorem 13.6 are direct results of the definition of dot product, and their verifications are left as an exercise. (See Exercise 44.)

THEOREM 13.6 *(Properties of Dot Product)*	If U, V, and W are vectors and c is any scalar, then:

   i. $U \cdot V = V \cdot U$ iv. $0 \cdot V = 0$
   ii. $U \cdot (V + W) = U \cdot V + U \cdot W$ v. $V \cdot V = |V|^2$
   iii. $c(U \cdot V) = (cU) \cdot V$

The dot product of two vectors proves to be a very useful concept, and we devote the remainder of this section to some of its applications. First, we show how the dot product can be used to determine the angle between two vectors. We identify the *angle between two nonzero vectors* U and V as the angle θ between representatives of U and V that have a common initial point, where $0 \le \theta \le \pi$. This means that if U and V have the same direction, then $\theta = 0$, and if U and V have opposite directions, then $\theta = \pi$. Note that θ is not defined if either U or V is zero.

THEOREM 13.7
(*Angle Between Two Vectors*)

If θ is the angle between two nonzero vectors **U** and **V**, then

$$\cos \theta = \frac{\mathbf{U} \cdot \mathbf{V}}{|\mathbf{U}| \, |\mathbf{V}|}$$

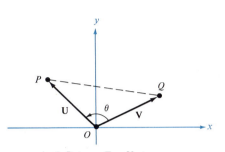

Angle Between Two Vectors

FIGURE 13.23

Proof: In Figure 13.23 if we apply the Law of Cosines to triangle OPQ, we obtain

$$|PQ|^2 = |\mathbf{U}|^2 + |\mathbf{V}|^2 - 2|\mathbf{U}| \, |\mathbf{V}| \cos \theta$$

Now if we let

$$\mathbf{U} = a_1\mathbf{i} + b_1\mathbf{j} \quad \text{and} \quad \mathbf{V} = a_2\mathbf{i} + b_2\mathbf{j}$$

and use the Distance Formula for $|PQ|^2$, then it follows that

$$
\begin{aligned}
(a_2 - a_1)^2 + (b_2 - b_1)^2 &= |\mathbf{U}|^2 + |\mathbf{V}|^2 - 2|\mathbf{U}| \, |\mathbf{V}| \cos \theta \\
&= (a_1^2 + b_1^2) + (a_2^2 + b_2^2) - 2|\mathbf{U}| \, |\mathbf{V}| \cos \theta
\end{aligned}
$$

or $\quad 2|\mathbf{U}| \, |\mathbf{V}| \cos \theta = 2a_1a_2 + 2b_1b_2 = 2(\mathbf{U} \cdot \mathbf{V})$

Consequently, $\qquad \cos \theta = \dfrac{\mathbf{U} \cdot \mathbf{V}}{|\mathbf{U}| \, |\mathbf{V}|}$ ∎

Example 2

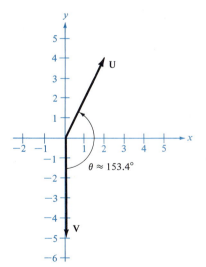

FIGURE 13.24

Determine the angle θ between the vectors $\mathbf{U} = 2\mathbf{i} + 4\mathbf{j}$ and $\mathbf{V} = -5\mathbf{j}$.

Solution: By Theorem 13.7

$$\cos \theta = \frac{\mathbf{U} \cdot \mathbf{V}}{|\mathbf{U}| \, |\mathbf{V}|} = \frac{0 - 20}{\sqrt{4 + 16} \sqrt{25}} = \frac{-20}{10\sqrt{5}} = \frac{-2}{\sqrt{5}}$$

$$\theta = \text{Arccos}\left(\frac{-2}{\sqrt{5}}\right) \approx 2.68 \text{ radians}$$

(See Figure 13.24.) ∎

Observe that Theorem 13.7 provides an alternative way to calculate the dot product of two vectors. Since

$$\cos \theta = \frac{\mathbf{U} \cdot \mathbf{V}}{|\mathbf{U}| \, |\mathbf{V}|}$$

it follows that $\qquad \mathbf{U} \cdot \mathbf{V} = |\mathbf{U}| \, |\mathbf{V}| \cos \theta$

The next theorem follows directly from this equation, since we consider two vectors to be *orthogonal* (perpendicular) if $\theta = \pi/2$. (We consider the zero vector to be orthogonal to every vector.)

THEOREM 13.8 *(Orthogonal Vectors)*	**U** and **V** are orthogonal if and only if $$\mathbf{U} \cdot \mathbf{V} = 0$$

Informally, we say two vectors are *parallel* if they have the same or opposite direction. It is easy to determine whether or not two vectors are parallel from their coordinate representation. Specifically, **U** and **V** have:

1. the *same direction* if and only if $\mathbf{U} = k\mathbf{V}$, $k > 0$
2. *opposite directions* if and only if $\mathbf{U} = k\mathbf{V}$, $k < 0$

Example 3

Given the vectors $\mathbf{U} = 3\mathbf{i} - \mathbf{j}$ and $\mathbf{V} = a\mathbf{i} + 4\mathbf{j}$.
(a)   Determine a such that **U** and **V** are orthogonal.
(b)   Determine a such that **U** and **V** have opposite directions.

Solution:

(a)   By Theorem 13.8 **U** and **V** are orthogonal if and only if $\mathbf{U} \cdot \mathbf{V} = 0$. Thus we have

$$3a - 4 = 0$$

and therefore, $$a = \tfrac{4}{3}$$

(b)   **U** and **V** will have opposite directions if $\mathbf{U} = k\mathbf{V}$, where k is negative. Thus we have

$$3\mathbf{i} - \mathbf{j} = k(a\mathbf{i} + 4\mathbf{j}) = ak\mathbf{i} + 4k\mathbf{j}$$

Equating components we have

$$3 = ak \quad \text{and} \quad -1 = 4k$$

Therefore, $k = -\tfrac{1}{4}$ and we conclude that

$$a = \frac{3}{k} = \frac{3}{-\frac{1}{4}} = -12$$ ∎

We conclude this section with a geometrical and a physical interpretation of the dot product. A geometrical interpretation of the dot product can be obtained by considering the *projection* of one vector onto another. Suppose **U** and **V** are vectors having the same initial point P and θ is the angle between the vectors. Then the **projection of U onto V** is a vector parallel to **V**, having P as its initial point and a magnitude of $|\mathbf{U}|\,|\cos\theta|$ (see Figure 13.25). The scalar value $|\mathbf{U}|\cos\theta$ is referred to as the **component of U in the direction of V**. To determine vector **W**, the projection of **U** onto **V**, we use Theorem 13.9.

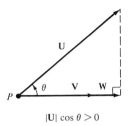
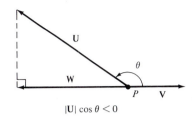

$|\mathbf{U}| \cos \theta > 0$ $|\mathbf{U}| \cos \theta < 0$

$|\mathbf{U}| \cos \theta$: Component of **U** in the direction of **V**
W: Projection of **U** onto **V**

FIGURE 13.25

THEOREM 13.9
(*Projection of **U** onto **V***)

If **W** is the projection of **U** onto **V** and θ is the angle between **U** and **V**, then

$$\mathbf{W} = \left(\frac{\mathbf{U} \cdot \mathbf{V}}{|\mathbf{V}|}\right)\frac{\mathbf{V}}{|\mathbf{V}|} = |\mathbf{U}| \cos \theta \, \frac{\mathbf{V}}{|\mathbf{V}|}$$

and

$$|\mathbf{W}| = \frac{|\mathbf{U} \cdot \mathbf{V}|}{|\mathbf{V}|} = |\mathbf{U}| \, |\cos \theta|$$

Proof: If **W** is the projection of **U** onto **V**, then **W** is simply a scalar multiple of **V**. From Figure 13.25 we can see that

$$|\mathbf{W}| = |\mathbf{U}| \, |\cos \theta|$$

Therefore, if we let $\mathbf{V}/|\mathbf{V}|$ be a unit vector in the direction of **V**, then we can write **W** as

$$\mathbf{W} = |\mathbf{U}| \cos \theta \, \frac{\mathbf{V}}{|\mathbf{V}|}$$

or

$$\mathbf{W} = \left(\frac{|\mathbf{U}| \, |\mathbf{V}| \cos \theta}{|\mathbf{V}|}\right)\frac{\mathbf{V}}{|\mathbf{V}|}$$

By Theorem 13.7 it follows that

$$\mathbf{W} = \left(\frac{\mathbf{U} \cdot \mathbf{V}}{|\mathbf{V}|}\right)\frac{\mathbf{V}}{|\mathbf{V}|}$$

From this equation we can see that the length of **W** is

$$|\mathbf{W}| = \frac{|\mathbf{U} \cdot \mathbf{V}|}{|\mathbf{V}|}$$

since $\mathbf{V}/|\mathbf{V}|$ is a unit vector. ∎

Example 4

Given the vectors $\mathbf{U} = 3\mathbf{i} - 5\mathbf{j}$ and $\mathbf{V} = 7\mathbf{i} + \mathbf{j}$, determine the component of **U** in the direction of **V** and the vector **W** that is the projection of **U** onto **V**.

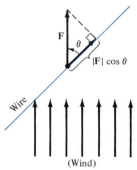

U = 3i − 5j
V = 7i + j
W = (projection of U onto V)
$$= \frac{56}{25}\mathbf{i} + \frac{8}{25}\mathbf{j}$$

FIGURE 13.26

Solution: Since $\mathbf{U} \cdot \mathbf{V} = |\mathbf{U}|\,|\mathbf{V}|\cos\theta$

then the component of **U** in the direction of **V** is

$$|\mathbf{U}|\cos\theta = \frac{\mathbf{U} \cdot \mathbf{V}}{|\mathbf{V}|} = \frac{21 - 5}{\sqrt{50}} = \frac{16}{5\sqrt{2}} = \frac{8\sqrt{2}}{5}$$

A unit vector in the direction of **V** is

$$\frac{\mathbf{V}}{|\mathbf{V}|} = \frac{7\mathbf{i} + \mathbf{j}}{5\sqrt{2}} = \frac{7}{5\sqrt{2}}\mathbf{i} + \frac{1}{5\sqrt{2}}\mathbf{j}$$

Therefore, the projection of **U** onto **V** is given by

$$\mathbf{W} = \left(\frac{\mathbf{U} \cdot \mathbf{V}}{|\mathbf{V}|}\right)\frac{\mathbf{V}}{|\mathbf{V}|} = \frac{8\sqrt{2}}{5}\left(\frac{7}{5\sqrt{2}}\mathbf{i} + \frac{1}{5\sqrt{2}}\mathbf{j}\right)$$

or $$\mathbf{W} = \frac{56}{25}\mathbf{i} + \frac{8}{25}\mathbf{j}$$

(See Figure 13.26.)   ■

Note in Theorem 13.9 that if **V** is a *unit* vector, then the projection of **U** onto **V** is simply

$$\mathbf{W} = (\mathbf{U} \cdot \mathbf{V})\mathbf{V}$$

This also means that if **U** is projected onto the unit vector **V**, then

$$\mathbf{U} \cdot \mathbf{V} = |\mathbf{U}|\cos\theta = \text{component of } \mathbf{U} \text{ in the direction of } \mathbf{V}$$

We will make special use of this result in Section 13.6.

The component interpretation of the dot product is useful in physics and mechanics. Recall from Section 7.5 that the work W done by a constant force F causing a displacement of d is given by $W = Fd$. However, we can use this formula only if the force is applied along the line of motion. If the constant force is not directed along the line of motion, then work is defined as the product of the *component of force* along the line of motion times the distance moved. To determine work under such circumstances, we can use the dot product, as indicated in the following definition:

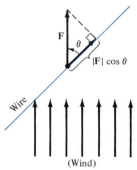

(Wind)

FIGURE 13.27

Definition of Work Done by a Force	The work W done by a constant force **F** that produces a displacement **V** is given by

$$W = (|\mathbf{F}|\cos\theta)|\mathbf{V}| = |\mathbf{F}|\,|\mathbf{V}|\cos\theta = \mathbf{F} \cdot \mathbf{V}$$

where θ is the angle between **F** and **V**.

To get a visual picture of this definition of work, we can imagine a bead strung on a taut wire. If a wind **F** is blowing on the bead, its displacement will depend upon the direction of the wind relative to the direction of the wire. Specifically, if θ is the angle between **F** and the wire, then the component of force acting on the bead is $|\mathbf{F}| \cos \theta$. (See Figure 13.27.) Note that the full force acts on the bead only if $\theta = 0$.

Example 5

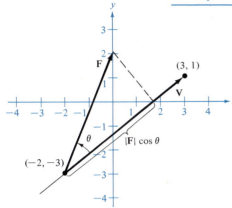

If a force represented by $\mathbf{F} = 2\mathbf{i} + 5\mathbf{j}$ moves a particle along a straight line from the point $(-2, -3)$ to $(3, 1)$, find the work done by the force.

Solution: The displacement can be represented by the vector

$$\mathbf{V} = (3 + 2)\mathbf{i} + (1 + 3)\mathbf{j} = 5\mathbf{i} + 4\mathbf{j}$$

Therefore, if the force is measured in pounds and the displacement in feet, then the work done is

$$W = \mathbf{F} \cdot \mathbf{V} = (2\mathbf{i} + 5\mathbf{j}) \cdot (5\mathbf{i} + 4\mathbf{j}) = 10 + 20 = 30 \text{ ft} \cdot \text{lb} \quad \blacksquare$$

FIGURE 13.28

Section Exercises (13.4)

In Exercises 1–10, find the angle θ between the given vectors.
1. $\mathbf{V} = \mathbf{i} + \mathbf{j}$, $\mathbf{W} = 2(\mathbf{i} - \mathbf{j})$
2. $\mathbf{V} = 3\mathbf{i} + \mathbf{j}$, $\mathbf{W} = 2\mathbf{i} - \mathbf{j}$
3. $\mathbf{V} = \mathbf{i} + \mathbf{j}$, $\mathbf{W} = 3\mathbf{i} - \mathbf{j}$
4. $\mathbf{V} = \mathbf{i} + 2\mathbf{j}$, $\mathbf{W} = 2\mathbf{i} - \mathbf{j}$
5. $\mathbf{V} = 2\mathbf{i} - 3\mathbf{j}$, $\mathbf{W} = -9\mathbf{i} - 6\mathbf{j}$
6. $\mathbf{V} = -\mathbf{i} + 2\mathbf{j}$, $\mathbf{W} = 4\mathbf{i} + 6\mathbf{j}$
7. $\mathbf{V} = 3\mathbf{i} + \mathbf{j}$, $\mathbf{W} = -2\mathbf{i} + 4\mathbf{j}$
8. $\mathbf{V} = \mathbf{i} - 5\mathbf{j}$, $\mathbf{W} = 4\mathbf{i} - 2\mathbf{j}$
9. $\mathbf{V} = \mathbf{i} \cos\left(\dfrac{\pi}{6}\right) + \mathbf{j} \sin\left(\dfrac{\pi}{6}\right)$, $\mathbf{W} = \mathbf{i} \cos\left(\dfrac{3\pi}{4}\right) + \mathbf{j} \sin\left(\dfrac{3\pi}{4}\right)$
10. $\mathbf{V} = 4\mathbf{i} - 3\mathbf{j}$, $\mathbf{W} = -2\mathbf{i} + 5\mathbf{j}$

In Exercises 11–16, find the angle θ between the given vector and the positive x-axis.
11. $\mathbf{V} = \mathbf{i} - \mathbf{j}$
12. $\mathbf{V} = \sqrt{3}\mathbf{i} + \mathbf{j}$
13. $\mathbf{V} = 3\mathbf{i} + 2\mathbf{j}$
14. $\mathbf{V} = 12\mathbf{i} + 5\mathbf{j}$
15. $\mathbf{V} = -3(\mathbf{i} + \mathbf{j})$
16. $\mathbf{V} = 240\mathbf{i} - 180\mathbf{j}$

In Exercises 17–19, let θ_U and θ_V represent the angles between the positive x-axis and the vectors **U** and **V**, respectively. In each case find $\mathbf{U} + \mathbf{V}$, assuming that the vertical components of **U** and **V** are positive.

17. $|\mathbf{U}| = 13$, $\theta_U = \dfrac{\pi}{6}$; $|\mathbf{V}| = 5$, $\theta_V = \dfrac{\pi}{3}$
18. $|\mathbf{U}| = 4.5$, $\theta_U = 130°$; $|\mathbf{V}| = 4.5$, $\theta_V = 75°$
19. $|\mathbf{U}| = 15$, $\theta_U = 170°$; $|\mathbf{V}| = 20$, $\theta_V = 25°$

In Exercises 20–27, determine whether **V** and **W** are orthogonal, parallel, or neither.
20. $\mathbf{V} = 4\mathbf{i}$, $\mathbf{W} = \mathbf{i} + \mathbf{j}$
21. $\mathbf{V} = 4\mathbf{j}$, $\mathbf{W} = \sqrt{3}\mathbf{i} + \mathbf{j}$
22. $\mathbf{V} = 2\mathbf{i} - 4\mathbf{j}$, $\mathbf{W} = 2\mathbf{i} + \mathbf{j}$
23. $\mathbf{V} = \mathbf{i} - 3\mathbf{j}$, $\mathbf{W} = 3\mathbf{i} - \mathbf{j}$
24. $\mathbf{V} = 6\mathbf{i} - 4\mathbf{j}$, $\mathbf{W} = -3\mathbf{i} + 2\mathbf{j}$
25. $\mathbf{V} = -\frac{1}{3}(\mathbf{i} - 2\mathbf{j})$, $\mathbf{W} = 2\mathbf{i} - 4\mathbf{j}$
26. $\mathbf{V} = 2\mathbf{i} + 18\mathbf{j}$, $\mathbf{W} = \frac{3}{2}\mathbf{i} - \frac{1}{6}\mathbf{j}$
27. $\mathbf{V} = 4\mathbf{i} + 3\mathbf{j}$, $\mathbf{W} = \frac{1}{2}\mathbf{i} - \frac{2}{3}\mathbf{j}$

In Exercises 28–32, (a) find the component of **V** in the direction of **W** and (b) find the projection of **V** onto **W**.
28. $\mathbf{V} = 2\mathbf{i} + 3\mathbf{j}$, $\mathbf{W} = 5\mathbf{i} + \mathbf{j}$
29. $\mathbf{V} = \mathbf{i} - 2\mathbf{j}$, $\mathbf{W} = \mathbf{i} + 3\mathbf{j}$
30. $\mathbf{V} = \mathbf{i} + \mathbf{j}$, $\mathbf{W} = 5\mathbf{i}$
31. $\mathbf{V} = 2\mathbf{i} - 3\mathbf{j}$, $\mathbf{W} = 5\mathbf{i} - \mathbf{j}$
32. $\mathbf{V} = 2\mathbf{i} - 3\mathbf{j}$, $\mathbf{W} = 3\mathbf{i} + 2\mathbf{j}$

33. Determine the projection of **W** onto **V** in Exercise 28.
34. Determine the projection of **W** onto **V** in Exercise 29.
35. Determine the projection of **W** onto **V** in Exercise 30.
36. Show that the unit vector making an angle θ with the positive x-axis is $\mathbf{U} = \mathbf{i} \cos \theta + \mathbf{j} \sin \theta$.
37. A force of 150 lb in a direction 30° above the horizontal is applied to a bolt. Find the horizontal and vertical components of the force. (See Figure 13.29.)
[c][s] 38. To carry a 100-lb cylindrical weight, two men lift on the ends of short ropes that are tied to an eyelet on the top center of the cylinder. If one of the ropes makes a 20° angle away from the vertical and the other a 30° angle, find the following:
    (a) the tension in each rope if the resultant force is vertical
    (b) the vertical component of each man's force (see Figure 13.30)
39. A man drags a heavy implement 10 ft across the floor, using a force of 85 lb. Find the work done if the direc-

tion of the force is 60° above the horizontal. (See Figure 13.31.)
[c] 40. An airplane's velocity with respect to the air is 580 mi/h and it is headed 32° north of west. The wind, at the altitude of the plane, is from the southwest and has a velocity of 60 mi/h. What is the true direction of the plane and what is its speed with respect to the ground? (See Figure 13.32.)
[c] 41. A ball is thrown into the air with an initial velocity of 80 ft/s and at an angle of 50° with the horizontal. Find the vertical and horizontal components of the velocity.
42. Using vectors, prove that the line joining the midpoints of two sides of a triangle is parallel to and one-half the length of the third side.
[s] 43. Show that the only two unit vectors that are orthogonal to the unit vector $\mathbf{U} = a\mathbf{i} + b\mathbf{j}$ are $\mathbf{V}_1 = b\mathbf{i} - a\mathbf{j}$ and $\mathbf{V}_2 = -b\mathbf{i} + a\mathbf{j}$.
44. Prove Theorem 13.6.

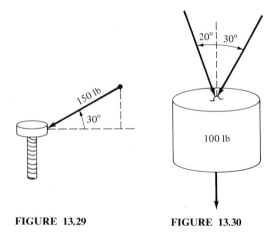

FIGURE 13.29   **FIGURE 13.30**

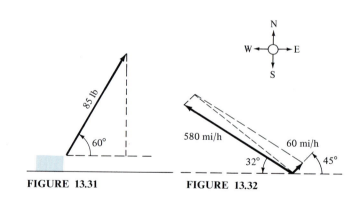

FIGURE 13.31   **FIGURE 13.32**

For Review and Class Discussion

True or False

1. _____ $\mathbf{U} \cdot \mathbf{0} = \mathbf{0}$.
2. _____ If $\mathbf{V} \cdot \mathbf{V} = 1$, then **V** is a unit vector.
3. _____ If $|\mathbf{V}| \neq 0$, then there are precisely two unit vectors that are orthogonal to **V**.
4. _____ If the projection of **U** onto **V** has the same magnitude as the projection of **V** onto **U**, then $|\mathbf{U}| = |\mathbf{V}|$.
5. _____ If $\mathbf{U} \cdot \mathbf{V} > 0$, then the angle between **U** and **V** is less than 90°.

13.5 **Vector-Valued Functions: Velocity and Acceleration**

Purpose

- To introduce vector-valued functions to describe motion in the plane.
- To summarize the calculus of vector-valued functions.

If a leaden ball, projected from the top of a mountain by the force of gunpowder, with a given velocity, and in a direction parallel to the horizon, is carried in a curved line to the distance of two miles before it falls to the ground; the same, if the resistance of the air were taken away, with a double velocity, would fly twice as far. And by increasing the velocity, we may at pleasure increase the distance till it even might go quite round the whole earth before it falls; or lastly, so that it might never fall to the earth, but go forwards into the celestial spaces, and proceed in its motion in infinitum.

—Isaac Newton (1642–1727)

In Section 13.1 we introduced parametric equations to study curvilinear motion in the plane. We did this because, from a pair of parametric equations, we can identify the exact *position* of an object at any specific time *t*. In this section we extend our study of curvilinear motion in the plane by describing a means for determining the velocity and acceleration of an object as it moves along a curved path. Since velocity and acceleration are vector quantities, we introduce the notion of a **vector-valued function.**

Definition of Vector-Valued Function	If f and g are real-valued functions of t, then the function $\mathbf{R}$ defined by $$\mathbf{R}(t) = f(t)\mathbf{i} + g(t)\mathbf{j}$$ is called a **vector-valued function of** t.

Note that the domain of a vector-valued function $\mathbf{R}$ consists of *scalars* (real numbers, t) while the range consists of *vectors,* $\mathbf{R}(t)$. For convenience we will refer to vector-valued functions as simply *vector functions.*

We can use vector functions to describe curves in the plane. Suppose a curve C is described by the parametric equations

$$x = f(t) \qquad \text{and} \qquad y = g(t)$$

Now consider the **position vector**

$$\mathbf{R}(t) = f(t)\mathbf{i} + g(t)\mathbf{j}$$

which has its initial point at the origin and its terminal point at (x, y). As t varies over the domain of $\mathbf{R}$, then the terminal point of $\mathbf{R}(t)$ will trace out the same curve C as that described by the parametric equations. (See Figure 13.33.)

We consider the positive direction along curve C to be the direction in which the terminal point of $\mathbf{R}(t)$ moves as the parameter t increases. Furthermore, if $\mathbf{R}(t)$ represents the position of a moving object, we define the **velocity vector** at time t to be the vector whose direction and magni-

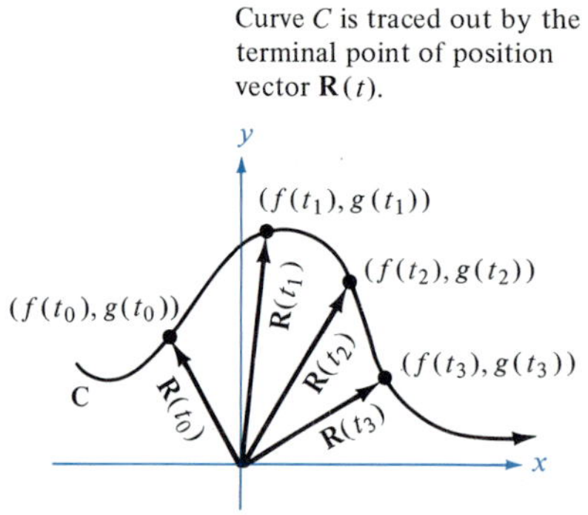

Curve C is traced out by the terminal point of position vector $\mathbf{R}(t)$.

FIGURE 13.33

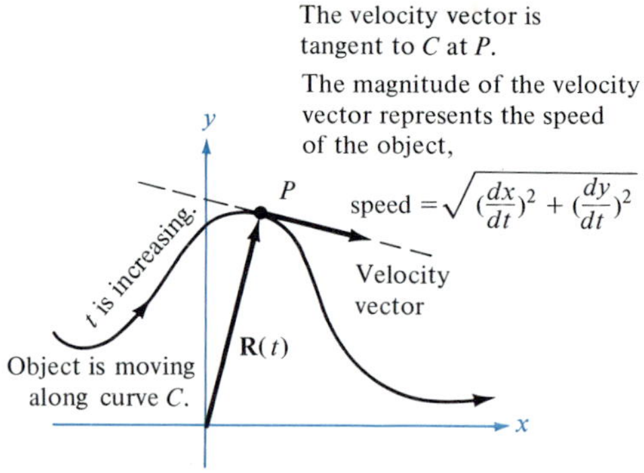

The velocity vector is tangent to C at P.

The magnitude of the velocity vector represents the speed of the object,

$$\text{speed} = \sqrt{(\tfrac{dx}{dt})^2 + (\tfrac{dy}{dt})^2}$$

FIGURE 13.34

tude correspond to the direction and speed of the object at time t (Figure 13.34). Speed has previously been defined as the distance (in this case arc length) traversed per unit of time. In other words, the speed of an object moving along C at time t is given by the derivative of the arc length s with respect to time. From Section 13.2 we know that this derivative is given by

$$\frac{ds}{dt} = \sqrt{\left(\frac{dx}{dt}\right)^2 + \left(\frac{dy}{dt}\right)^2}$$

Recall that for straight-line motion, the velocity function $v = v(t)$ was obtained by differentiating the position function $s = s(t)$ with respect to time. For curvilinear motion in the plane, we can find the velocity vector in a similar manner. We begin our discussion of the velocity vector with the definition of the limit of a vector function.

Definition of the Limit of a Vector Function

Let $\mathbf{R}$ be a vector function defined by

$$\mathbf{R}(t) = f(t)\mathbf{i} + g(t)\mathbf{j}$$

Then

$$\lim_{t \to t_0} \mathbf{R}(t) = \left[\lim_{t \to t_0} f(t)\right]\mathbf{i} + \left[\lim_{t \to t_0} g(t)\right]\mathbf{j}$$

provided $\lim_{t \to t_0} f(t)$ and $\lim_{t \to t_0} g(t)$ both exist.

Suppose x and y are differentiable functions of t and $\mathbf{R}(t) = x\mathbf{i} + y\mathbf{j}$. Then we have

$$\Delta \mathbf{R} = \mathbf{R}(t + \Delta t) - \mathbf{R}(t) = \Delta x\mathbf{i} + \Delta y\mathbf{j}$$

$$\frac{\Delta \mathbf{R}}{\Delta t} = \frac{\Delta x}{\Delta t}\mathbf{i} + \frac{\Delta y}{\Delta t}\mathbf{j}$$

Therefore,

$$\lim_{\Delta t \to 0} \frac{\Delta \mathbf{R}}{\Delta t} = \left[\lim_{\Delta t \to 0} \frac{\Delta x}{\Delta t}\right]\mathbf{i} + \left[\lim_{\Delta t \to 0} \frac{\Delta y}{\Delta t}\right]\mathbf{j}$$

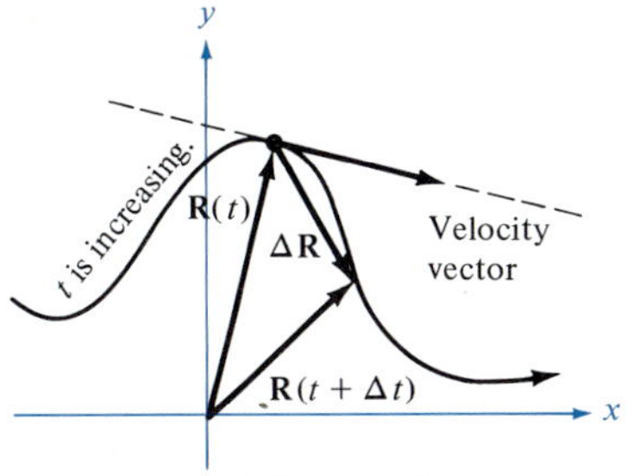

and it follows that
$$\frac{d\mathbf{R}}{dt} = \frac{dx}{dt}\mathbf{i} + \frac{dy}{dt}\mathbf{j}$$

From Figure 13.35 we can see that as $\Delta t \to 0$, the direction of $\Delta \mathbf{R}$ approaches the direction of the velocity vector. Hence the direction of $d\mathbf{R}/dt$ must be that of the velocity vector. Furthermore, the magnitude of $d\mathbf{R}/dt$ is given by

$$\left| \frac{d\mathbf{R}}{dt} \right| = \sqrt{\left(\frac{dx}{dt}\right)^2 + \left(\frac{dy}{dt}\right)^2}$$

As $\Delta t \to 0$, $\Delta \mathbf{R} \to$ velocity vector.

FIGURE 13.35

which is precisely the magnitude of the velocity vector. Hence $d\mathbf{R}/dt$ *is the velocity vector* since it possesses the appropriate magnitude and direction.

In a similar manner it can be shown that the second derivative of $\mathbf{R}(t)$ gives us the acceleration vector. We summarize these results as follows:

Velocity and Acceleration Vectors

If x and y are twice differentiable functions of t and $\mathbf{R}$ is a vector function defined by

$$\mathbf{R}(t) = x\mathbf{i} + y\mathbf{j}$$

then the **velocity vector** $\mathbf{V}(t)$ is given by

$$\mathbf{V}(t) = \frac{d\mathbf{R}}{dt} = \frac{dx}{dt}\mathbf{i} + \frac{dy}{dt}\mathbf{j}$$

The **acceleration vector** $\mathbf{A}(t)$ is given by

$$\mathbf{A}(t) = \frac{d\mathbf{V}}{dt} = \frac{d^2x}{dt^2}\mathbf{i} + \frac{d^2y}{dt^2}\mathbf{j}$$

Example 1

A particle moves along a curve C described by

$$\mathbf{R}(t) = \left(2\sin\frac{t}{2}\right)\mathbf{i} + \left(2\cos\frac{t}{2}\right)\mathbf{j}$$

For any time t find the following:
(a) the velocity and speed
(b) the acceleration and its magnitude
(c) the angle θ between the velocity and acceleration vectors

Solution:

(a)
$$\mathbf{V}(t) = \frac{d\mathbf{R}}{dt} = \left(\cos\frac{t}{2}\right)\mathbf{i} - \left(\sin\frac{t}{2}\right)\mathbf{j}$$

and, therefore,

$$\text{speed} = \left|\frac{d\mathbf{R}}{dt}\right| = \sqrt{\cos^2\frac{t}{2} + \sin^2\frac{t}{2}} = 1$$

(b)
$$\mathbf{A}(t) = \frac{d\mathbf{V}}{dt} = \left(-\frac{1}{2}\sin\frac{t}{2}\right)\mathbf{i} - \left(\frac{1}{2}\cos\frac{t}{2}\right)\mathbf{j}$$

and, therefore,

$$\left|\frac{d\mathbf{V}}{dt}\right| = \sqrt{\frac{1}{4}\left(\sin^2\frac{t}{2} + \cos^2\frac{t}{2}\right)} = \frac{1}{2}$$

(c)
$$\cos\theta = \frac{\mathbf{V}(t)\cdot\mathbf{A}(t)}{(1)(\frac{1}{2})}$$

$$= \frac{-\frac{1}{2}\sin(t/2)\cos(t/2) + \frac{1}{2}\sin(t/2)\cos(t/2)}{\frac{1}{2}} = 0$$

and therefore the angle of intersection of $\mathbf{A}(t)$ and $\mathbf{V}(t)$ is $\theta = \pi/2$. ∎

If we eliminate the parameter in the vector representation of curve C in Example 1, we have, from

$$x = 2\sin\frac{t}{2} \qquad \text{and} \qquad y = 2\cos\frac{t}{2}$$

the result
$$\frac{x^2}{4} + \frac{y^2}{4} = \sin^2\frac{t}{2} + \cos^2\frac{t}{2} = 1$$

or
$$x^2 + y^2 = 4$$

This is the equation of a circle of radius 2 centered at the origin. (See Figure 13.36.) Since the velocity vector $\mathbf{V}(t) = \cos(t/2)\mathbf{i} - \sin(t/2)\mathbf{j}$ has a constant magnitude but a changing direction as t increases, the particle moves around this circle at a constant speed. Note further that for this *circular path* the acceleration vector is always directed toward the center of the circle because it is always perpendicular to the velocity vector.

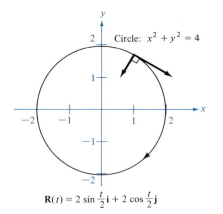

Circle: $x^2 + y^2 = 4$

$$\mathbf{R}(t) = 2\sin\frac{t}{2}\mathbf{i} + 2\cos\frac{t}{2}\mathbf{j}$$

FIGURE 13.36

Example 2

A particle moves along a curve C described by

$$\mathbf{R}(t) = e^{2t}\mathbf{i} + 3e^t\mathbf{j}$$

(a) Find $\mathbf{V}(t)$, $|\mathbf{V}(t)|$, $\mathbf{A}(t)$, and $|\mathbf{A}(t)|$.

(b) Sketch the path of the particle, and sketch the velocity and acceleration vectors having initial points at $t = 0$.

Solution:

(a)
$$\mathbf{V}(t) = 2e^{2t}\mathbf{i} + 3e^t\mathbf{j}$$

and
$$|\mathbf{V}(t)| = \sqrt{4e^{4t} + 9e^{2t}} = e^t\sqrt{4e^{2t} + 9}$$

$$\mathbf{A}(t) = 4e^{2t}\mathbf{i} + 3e^t\mathbf{j}$$

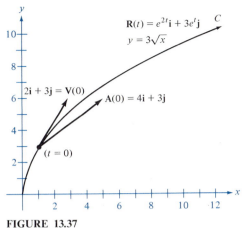

FIGURE 13.37

and $\quad |\mathbf{A}(t)| = \sqrt{16e^{4t} + 9e^{2t}} = e^t \sqrt{16e^{2t} + 9}$

(b) $\mathbf{V}(0) = 2\mathbf{i} + 3\mathbf{j}$ and $\mathbf{A}(0) = 4\mathbf{i} + 3\mathbf{j}$. Since $x = e^{2t} = (e^t)^2$ and $y = 3e^t$, we have

$$\sqrt{x} = e^t = \frac{y}{3} \quad \text{or} \quad y = 3\sqrt{x}$$

A sketch of $y = 3\sqrt{x}$ is given in Figure 13.37, along with the vectors $\mathbf{V}(0)$ and $\mathbf{A}(0)$. ∎

In the following theorem concerning the derivative of a vector function, keep in mind the two types of functions involved. A vector function $\mathbf{R}$ assigns to a number t a vector $\mathbf{R}(t)$, whereas a scalar function f assigns to a number t another number $f(t)$. Note also that the product of a scalar and a vector function is a vector function.

THEOREM 13.10
(*Properties of the Derivative of a Vector Function*)

Let $\mathbf{U}$ and $\mathbf{V}$ be vectors and f a scalar function of t; then

i. $\dfrac{d}{dt}[\mathbf{U}(t) + \mathbf{V}(t)] = \dfrac{d}{dt}[\mathbf{U}(t)] + \dfrac{d}{dt}[\mathbf{V}(t)]$

ii. $\dfrac{d}{dt}[f(t)\,\mathbf{U}(t)] = f(t)\dfrac{d}{dt}[\mathbf{U}(t)] + \mathbf{U}(t)\dfrac{d}{dt}[f(t)]$

iii. $\dfrac{d}{dt}[\mathbf{U}(t) \cdot \mathbf{V}(t)] = \left\{ \mathbf{U}(t) \cdot \dfrac{d}{dt}[\mathbf{V}(t)] \right\} + \left\{ \mathbf{V}(t) \cdot \dfrac{d}{dt}[\mathbf{U}(t)] \right\}$

iv. $\dfrac{d}{dt}[\mathbf{U}(s)] = \dfrac{d}{ds}[\mathbf{U}(s)]\dfrac{ds}{dt}$, where $s = f(t)$ (Chain Rule)

Proof: For each rule we consider

$$\mathbf{U}(t) = g_1(t)\mathbf{i} + g_2(t)\mathbf{j} \quad \text{and} \quad \mathbf{V}(t) = h_1(t)\mathbf{i} + h_2(t)\mathbf{j}$$

and, where convenient, we will use the prime notation, $\mathbf{U}'(t)$, to denote derivatives with respect to t. We will prove rules ii and iv and leave the others as exercises (see Exercises 40 and 41).

ii. Considering

$$f(t)\,\mathbf{U}(t) = f(t)[g_1(t)\mathbf{i} + g_2(t)\mathbf{j}] = f(t)\,g_1(t)\mathbf{i} + f(t)\,g_2(t)\mathbf{j}$$

we have

$$\frac{d}{dt}[f(t)\,\mathbf{U}(t)] = [f(t)\,g_1'(t) + g_1(t)f'(t)]\mathbf{i} + [f(t)\,g_2'(t) + g_2(t)f'(t)]\mathbf{j}$$

$$= f(t)[g_1'(t)\mathbf{i} + g_2'(t)\mathbf{j}] + f'(t)[g_1(t)\mathbf{i} + g_2(t)\mathbf{j}]$$

$$= f(t)\,\mathbf{U}'(t) + \mathbf{U}(t)f'(t)$$

iv. Let $\mathbf{U}(s) = g_1(s)\mathbf{i} + g_2(s)\mathbf{j}$, where $s = f(t)$. Then by the Chain Rule for scalar functions, we have

$$\frac{d}{dt}[\mathbf{U}(s)] = \frac{d}{dt}[g_1(s)]\mathbf{i} + \frac{d}{dt}[g_2(s)]\mathbf{j}$$

$$= \frac{dg_1}{ds}\frac{ds}{dt}\mathbf{i} + \frac{dg_2}{ds}\frac{ds}{dt}\mathbf{j} = \left[\frac{dg_1}{ds}\mathbf{i} + \frac{dg_2}{ds}\mathbf{j}\right]\frac{ds}{dt}$$

$$= \frac{d}{ds}[\mathbf{U}(s)]\frac{ds}{dt} \qquad \blacksquare$$

Thus far in this section, we have seen how the position function can be used to determine the velocity and acceleration of a moving object. In actual practice we are often asked to solve the reverse problem; that is, we are asked to determine the position function when we are given the velocity or the acceleration. To do this we use integration of vector functions, which can be handled on a component-by-component basis.

Integration of Vector Functions	If $\qquad\qquad \mathbf{R}(t) = f(t)\mathbf{i} + g(t)\mathbf{j}$

where f and g are integrable functions of t, then

$$\int \mathbf{R}(t)\,dt = \left[\int f(t)\,dt\right]\mathbf{i} + \left[\int g(t)\,dt\right]\mathbf{j} + \mathbf{C}$$

for some constant vector $\mathbf{C} = c_1\mathbf{i} + c_2\mathbf{j}$.

Example 3

Find $\mathbf{R}(t)$ given that

$$\mathbf{R}'(t) = e^{-2t}\mathbf{i} - e^{-t}\mathbf{j} \qquad \text{and} \qquad \mathbf{R}(0) = 2\mathbf{i} - \mathbf{j}$$

Solution:

$$\mathbf{R}(t) = \mathbf{i}\int e^{-2t}\,dt - \mathbf{j}\int e^{-t}\,dt = \mathbf{i}[-\tfrac{1}{2}e^{-2t} + c_1] - \mathbf{j}[-e^{-t} + c_2]$$

Since $\mathbf{R}(0) = 2\mathbf{i} - \mathbf{j}$, we have

$$2\mathbf{i} - \mathbf{j} = \mathbf{i}[-\tfrac{1}{2} + c_1] - \mathbf{j}[-1 + c_2]$$

Thus $\qquad\qquad 2 = -\tfrac{1}{2} + c_1 \qquad$ and $\qquad 1 = -1 + c_2$

$$c_1 = \tfrac{5}{2} \qquad\qquad\qquad\qquad c_2 = 2$$

Therefore, $\qquad\qquad \mathbf{R}(t) = (\tfrac{5}{2} - \tfrac{1}{2}e^{-2t})\mathbf{i} + (e^{-t} - 2)\mathbf{j} \qquad\qquad \blacksquare$

In our next example we use the integration of vector functions to determine the parametric equations that represent the path of a projectile traveling in a vertical plane. From the results of this example, we can verify the parametric equations for the path of the projectile discussed in Section 13.1.

Example 4 (*Path of a Projectile*)

Suppose a projectile is fired with an angle of elevation of θ radians and with an initial speed of v_0 feet per second. Assuming no air resistance and that the only force acting on the projectile is that due to gravity, find the velocity vector $\mathbf{V}(t)$ and position vector $\mathbf{R}(t)$ at any time t (Figure 13.38).

Solution: Since the force of gravity acts in a downward direction, the acceleration of the object is given by

$$\mathbf{A} = -g\mathbf{j}$$

where $g \approx 32$ ft/s². Now since

$$\mathbf{A}(t) = \mathbf{V}'(t) = \mathbf{R}''(t) = -g\mathbf{j}$$

it follows that
$$\mathbf{V}(t) = \mathbf{R}'(t) = -gt\mathbf{j} + \mathbf{C}$$

where $\mathbf{C}$ is a constant vector. Letting $t = 0$ we conclude that

$$\mathbf{V}(0) = 0\mathbf{j} + \mathbf{C} = \mathbf{C}$$

and write
$$\mathbf{V}(t) = \mathbf{R}'(t) = -gt\mathbf{j} + \mathbf{V}(0)$$

Integrating both sides of this equation gives us

$$\mathbf{R}(t) = -\frac{gt^2}{2}\mathbf{j} + t\mathbf{V}(0) + \mathbf{D}$$

for some vector $\mathbf{D}$. Since $\mathbf{R}(0) = \mathbf{0}$, we conclude that $\mathbf{D} = \mathbf{0}$, and therefore,

$$\mathbf{R}(t) = -\frac{gt^2}{2}\mathbf{j} + t\mathbf{V}(0)$$

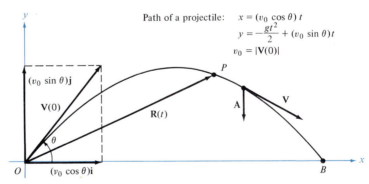

Path of a projectile:
$$x = (v_0 \cos \theta)\, t$$
$$y = -\frac{gt^2}{2} + (v_0 \sin \theta)t$$
$$v_0 = |\mathbf{V}(0)|$$

FIGURE 13.38

From Figure 13.38 we can see that

$$\mathbf{V}(0) = (v_0 \cos \theta)\mathbf{i} + (v_0 \sin \theta)\mathbf{j}$$

Consequently, we have

$$\mathbf{R}(t) = -\frac{gt^2}{2}\mathbf{j} + t[(v_0 \cos \theta)\mathbf{i} + (v_0 \sin \theta)\mathbf{j}]$$

or, equivalently,

$$\mathbf{R}(t) = t(v_0 \cos \theta)\mathbf{i} + \left(tv_0 \sin \theta - \frac{gt^2}{2}\right)\mathbf{j}$$

From this representation of $\mathbf{R}(t)$, we obtain the parametric equations

$$x = (v_0 \cos \theta)t \qquad \text{and} \qquad y = -\tfrac{1}{2}gt^2 + (v_0 \sin \theta)t \qquad \blacksquare$$

Example 5

Consider the projectile described in Section 13.1.
(a) Use the results of Example 4 to verify that its path is represented by the parametric equations

$$x = 24\sqrt{2}\,t \qquad \text{and} \qquad y = 24\sqrt{2}\,t - 16t^2$$

(b) Determine the total flight time of the projectile.
(c) Determine the range (distance $|OB|$, Figure 13.38).
(d) Determine the maximum height reached by the projectile.
(e) Determine the velocity vector at impact.

Solution:

(a) For the projectile under consideration, we have

$$v_0 = 48 \text{ ft/s} \qquad \text{and} \qquad \theta = \frac{\pi}{4}$$

Therefore, the position vector is given by

$$\mathbf{R}(t) = t\left(48 \cos \frac{\pi}{4}\right)\mathbf{i} + \left(48t \sin \frac{\pi}{4} - 16t^2\right)\mathbf{j}$$

$$= 24\sqrt{2}t\mathbf{i} + (24\sqrt{2}t - 16t^2)\mathbf{j}$$

from which we obtain the parametric equations

$$x = 24\sqrt{2}t \qquad \text{and} \qquad y = 24\sqrt{2}t - 16t^2$$

(b) To find the total flight time, we find t when $y = 0$. Thus we solve the equation

$$0 = 24\sqrt{2}t - 16t^2$$
$$0 = t(24\sqrt{2} - 16t)$$

and obtain the values

$$t = 0 \qquad \text{and} \qquad t = \frac{3\sqrt{2}}{2}$$

Since $t = 0$ corresponds to firing time, the time of flight is $3\sqrt{2}/2$ s.

(c) The range corresponds to the value of x at the time of impact, $t = 3\sqrt{2}/2$. Therefore, the range is

$$x = 24\sqrt{2}\left(\frac{3\sqrt{2}}{2}\right) = 72 \text{ ft}$$

(d) The maximum height is reached when the vertical component of the velocity vector $\mathbf{V}(t)$ is zero, that is, when $dy/dt = 0$. Since

$$\frac{dy}{dt} = 24\sqrt{2} - 32t = 0$$

has the solution $t = 3\sqrt{2}/4$, the maximum height is

$$y = 24\sqrt{2}\left(\frac{3\sqrt{2}}{4}\right) - 16\left(\frac{3\sqrt{2}}{4}\right)^2 = 36 - 18 = 18 \text{ ft}$$

(e) The velocity vector is given by

$$\mathbf{V}(t) = \mathbf{R}'(t) = 24\sqrt{2}\mathbf{i} + (24\sqrt{2} - 32t)\mathbf{j}$$

Therefore, at impact $t = 3\sqrt{2}/2$, the velocity vector is

$$\mathbf{V}\left(\frac{3\sqrt{2}}{2}\right) = 24\sqrt{2}\mathbf{i} + (24\sqrt{2} - 48\sqrt{2})\mathbf{j} = 24\sqrt{2}\mathbf{i} - 24\sqrt{2}\mathbf{j} \quad \blacksquare$$

Section Exercises (13.5)

In Exercises 1–6, find $\mathbf{R}'$ and $\mathbf{R}''$.

1. $\mathbf{R}(t) = 3t\mathbf{i} + (t - 1)\mathbf{j}$

2. $\mathbf{R}(t) = \left(\frac{1}{t}\right)\mathbf{i} + \left(\frac{t+1}{t-1}\right)\mathbf{j}$

3. $\mathbf{R}(t) = (a\cos t)\mathbf{i} + (b\sin t)\mathbf{j}$

4. $\mathbf{R}(\theta) = (\sec\theta)\mathbf{i} + (\tan\theta)\mathbf{j}$

[S] 5. $\mathbf{R}(\theta) = (\theta - \sin\theta)\mathbf{i} + (1 - \cos\theta)\mathbf{j}$

6. $\mathbf{R}(\theta) = (\cot\theta)\mathbf{i} + (2\sin\theta\cos\theta)\mathbf{j}$

In Exercises 7–14, find the velocity and acceleration vectors and their magnitudes.

7. $\mathbf{R}(t) = (t - \frac{3}{2})\mathbf{i} + 2t\mathbf{j}$

8. $\mathbf{R}(t) = (6 - t)\mathbf{i} + t\mathbf{j}$

9. $\mathbf{R}(t) = e^t\mathbf{i} + e^{t/2}\mathbf{j}$

10. $\mathbf{R}(t) = (2\cos t)\mathbf{i} + (3\sin t)\mathbf{j}$

[S] 11. $\mathbf{R}(t) = (\sin t)\mathbf{i} + (\cos 2t)\mathbf{j}$

12. $\mathbf{R}(t) = e^{-t}\mathbf{i} + e^t\mathbf{j}$

13. $\mathbf{R}(t) = t^2\mathbf{i} + t\mathbf{j}$

14. $\mathbf{R}(t) = t^3\mathbf{i} + t^2\mathbf{j}$

In Exercises 15–18, find the position vector for the given conditions.

[S] 15. $\mathbf{R}'(t) = 4e^{2t}\mathbf{i} + 3e^t\mathbf{j};\ \mathbf{R}(0) = 2\mathbf{i}$

16. $\mathbf{R}'(t) = 2t\mathbf{i} + \sqrt{t}\mathbf{j};\ \mathbf{R}(0) = \mathbf{i} + \mathbf{j}$

17. $\mathbf{A}(t) = -32\mathbf{j};\ \mathbf{V}(0) = 600\sqrt{3}\mathbf{i} + 600\mathbf{j},\ \mathbf{R}(0) = \mathbf{0}$

18. $\mathbf{A}(t) = (-4\cos t)\mathbf{i} + (-3\sin t)\mathbf{j};\ \mathbf{V}(0) = 3\mathbf{j},\ \mathbf{R}(0) = 4\mathbf{i}$

[S] 19. Find $\dfrac{d}{dt}[\mathbf{R}_1(t) \cdot \mathbf{R}_2(t)]$ if $\mathbf{R}_1(t) = 2t\mathbf{i} + [2/(t + 2)]\mathbf{j}$ and $\mathbf{R}_2(t) = t\mathbf{i} - (t + 2)\mathbf{j}$.

20. Find $\dfrac{d}{dt}[\|\mathbf{R}(t)\|]$ if $\mathbf{R}(t) = (3\cos t)\mathbf{i} + (3\sin t)\mathbf{j}$.

21. Find $\dfrac{d}{dt}[\|\mathbf{R}(t)\|]$ if $\mathbf{R}(t) = \frac{1}{2}t\mathbf{i} + t^2\mathbf{j}$.

22. Find $\|\mathbf{R}'(t)\|$ if $\mathbf{R}(t) = (3\cos t)\mathbf{i} + (3\sin t)\mathbf{j}$.

23. Find $\mathbf{R}(t)$ given that $\mathbf{R}'(t) = te^{-t^2}\mathbf{i} - e^{-t}\mathbf{j}$ and $\mathbf{R}(0) = \frac{1}{2}\mathbf{i} - \mathbf{j}$.

24. Find $\mathbf{R}(t)$ given that $\mathbf{R}'(t) = [1/(1 + t^2)]\mathbf{i} + (1/t^2)\mathbf{j}$ and $\mathbf{R}(1) = 2\mathbf{i}$.

25. Find the velocity and acceleration vectors for the hyperbola described by $\mathbf{R}(t) = (b \sec t)\mathbf{i} + (b \tan t)\mathbf{j}$.

26. Prove that if $\mathbf{V}(t)$ is the velocity vector at any point P on a cycloid, then $\mathbf{V}(t)$ is parallel to the line through P and the highest point on the generating circle.

S 27. A batted baseball leaves the bat at an angle of $45°$ and is caught by an outfielder 300 ft from the plate. What was the initial velocity of the ball and how high did it rise?

C 28. A bomber flying at 30,000 ft altitude and 540 mi/h has a destroyer as its target. Because of the speed of the plane, the bombs must be dropped before reaching the target. Determine the angle of depression from the plane to the target when the bomb should be dropped (neglect air resistance). Find the velocity of the bomb at the time of impact.

29. Find the angle at which a gun should be fired to obtain (a) the maximum range and (b) the maximum height.

30. Eliminate the parameter t from the parametric equations for the path of a projectile (see Example 4) to show that the Cartesian equation is $y = -[16(\sec^2 \theta)/v_0^2]x^2 + (\tan \theta)x$.

S 31. In Section 3.2 the equation $y = x - 0.005x^2$ was given as the path of a ball.
    (a) Use Exercise 30 to show that the initial velocity of the ball was 80 ft/s and that the angle at which it was thrown was $45°$ with the horizontal.
    (b) Write the vector-valued function for the path of the ball.
    (c) Find the speed and direction of the ball at the point when it has traveled 60 ft horizontally.

C 32. A gun with a muzzle velocity of 1200 ft/s is fixed at a target 1000 yd away. Neglecting air resistance, what should be the minimum angle of elevation of the gun?

33. A boy standing 20 ft from the base of a 40-ft silo is trying to throw a ball into an opening at the top. Find the minimum initial angle and the corresponding velocity for the ball to be thrown.

S 34. A particle moves at a constant rate along the circle of radius b described by $\mathbf{R}(t) = [b \cos (wt)]\mathbf{i} + [b \sin (wt)]\mathbf{j}$, where $w = d\theta/dt$ is the angular velocity.
    (a) Find the velocity vector and show that it is orthogonal to $\mathbf{R}(t)$.

(b) Show that the speed of the particle is bw.
(c) Find the acceleration vector and show that its direction is always toward the center of the circle.
(d) Show that the magnitude of the acceleration vector is $w^2 b$.

35. Find the velocity and acceleration vectors and their magnitudes for the cycloid described by $\mathbf{R}(t) = b(wt - \sin wt)\mathbf{i} + b(1 - \cos wt)\mathbf{j}$. The constant angular velocity of the generating circle is w.

36. Use the result of Exercise 35 to determine the times when the velocity of a particle moving on the circumference of a circle rolling at a constant rate will be (a) zero and (b) maximum.

37. A stone weighing 1 lb is attached to a 2-ft string and is whirled horizontally with one end of the string held fixed. If the string will break under a force of 10 lb, find the maximum velocity the stone can attain without breaking the string. (Hint: Use the results in Exercise 34 together with Newton's Second Law of Motion, $F = mA$, where the mass of the 1-lb stone is $m = \frac{1}{32}$.)

C 38. A 3000-lb automobile is negotiating a circular interchange of radius 300 ft at 30 mi/h.
    (a) If the roadway is level, find the force between the tires and the road so that the car stays on the circular path and does not skid. (Hint: Use $F = mA$, where the mass of the car is $m = 3000/32$.)
    (b) At what angle should the roadbed be banked?

S 39. Find the maximum velocity of a point on the circumference of the tire of an automobile when the automobile is traveling at 55 mi/h and the radius of the wheel is 1 ft. (See Exercise 35.) Compare this velocity with the velocity of the automobile.

40. In Theorem 13.10 prove that

$$\frac{d}{dt}[\mathbf{U}(t) + \mathbf{V}(t)] = \mathbf{U}'(t) + \mathbf{V}'(t)$$

41. In Theorem 13.10 prove that

$$\frac{d}{dt}[\mathbf{U}(t) \cdot \mathbf{V}(t)] = \mathbf{U}(t) \cdot \mathbf{V}'(t) + \mathbf{V}(t) \cdot \mathbf{U}'(t)$$

13.6 Tangent and Normal Components of Acceleration

Purpose

- To define the unit tangent vector **T** and the unit normal vector **N**.
- To discuss acceleration in terms of its tangential and normal components.

I shall have given a sufficient introduction to the study of curves when I have given a general method of drawing a straight line making right angles with a curve at an arbitrarily chosen point upon the curve. And I dare say that this is not only the most useful and most general problem in geometry that I know, but even that I have ever desired to know.

—René Descartes (1596–1650)

In curvilinear motion we have seen that if the position of a moving object at any time *t* is given by the vector

$$\mathbf{R}(t) = x(t)\mathbf{i} + y(t)\mathbf{j}$$

then the velocity and acceleration of the object at any time *t* are given by the respective vectors

$$\mathbf{V}(t) = \frac{dx}{dt}\mathbf{i} + \frac{dy}{dt}\mathbf{j} \quad \text{and} \quad \mathbf{A}(t) = \frac{d^2x}{dt^2}\mathbf{i} + \frac{d^2y}{dt^2}\mathbf{j}$$

We also know that $\mathbf{V}(t)$ represents the direction of motion at time *t* because $\mathbf{V}(t)$ is tangent to the path of motion at *t*. It is useful to think of the acceleration vector $\mathbf{A}(t)$ as representing the effect of a force on the moving object. Note, in curvilinear motion, there must be some force acting on the object to continually change its direction; otherwise, the object would move in a straight line.

Now given the existence of a nonzero acceleration vector in curvilinear motion, we want to determine how much of the acceleration acts along the tangent vector and how much does not. Specifically, we want to determine how much of the acceleration at any time *t* acts tangent to the path and how much acts *normal* (perpendicular) to the path of motion. To do this we introduce two unit vectors **T** and **N**, which will serve in much the same way as do **i** and **j** in the study of acceleration in the horizontal and vertical directions.

Since $\mathbf{V}(t) = \mathbf{R}'(t)$ is tangent to the path of motion at *t*, we define the **unit tangent vector T** to be the unit vector in the direction of **V**. To find a unit vector that is perpendicular to **T**, we consider the fact that

$$\mathbf{T} \cdot \mathbf{T} = |\mathbf{T}|^2 = 1$$

By differentiating both sides of this equation with respect to *t*, we have

$$2(\mathbf{T} \cdot \mathbf{T}') = 0$$

which by Theorem 13.8 implies that **T** and **T**′ are perpendicular. Hence $\mathbf{N} = \mathbf{T}'/|\mathbf{T}'|$ is a unit vector that is perpendicular to **T**. Since $-\mathbf{N}$ is also a unit vector perpendicular to **T**, we call **N** the **principal unit normal vector,** as indicated in the following definition:

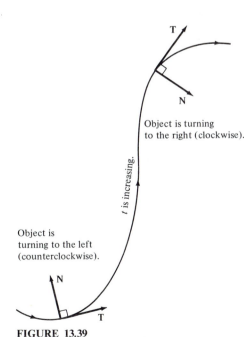

Object is turning to the right (clockwise).

t is increasing.

Object is turning to the left (counterclockwise).

FIGURE 13.39

Definition of the Unit Tangent and Unit Normal Vectors T and N	Let $\mathbf{R}(t)$ represent the position vector of a moving object at any time t, and let $\mathbf{V}(t) = \mathbf{R}'(t)$ be the velocity vector at t. Then the **unit tangent vector T** is defined to be

$$\mathbf{T} = \mathbf{T}(t) = \frac{\mathbf{V}(t)}{|\mathbf{V}(t)|}$$

whenever $\mathbf{V} \neq \mathbf{0}$. The **principal unit normal vector N** is defined to be

$$\mathbf{N} = \mathbf{N}(t) = \frac{\mathbf{T}'(t)}{|\mathbf{T}'(t)|}$$

whenever $\mathbf{T}' \neq \mathbf{0}$.

Geometrically, if C is the curve traversed by a moving object, then at a point P on the curve, $\mathbf{T}$ points in the direction in which the object is *moving*, while $\mathbf{N}$ points in the direction in which the object is *turning* (see Figure 13.39). The direction of $\mathbf{N}$ can also be interpreted as the direction toward the *concave* side of the curve. At a point of inflection of C, $\mathbf{N}$ is not defined because $\mathbf{T}' = \mathbf{0}$ at such a point. (See Exercise 20.)

Example 1

If $\mathbf{R} = (a \cos bt)\mathbf{i} + (a \sin bt)\mathbf{j}$, find $\mathbf{T}$ and $\mathbf{N}$ ($a > 0, b > 0$).

Solution:
$$\mathbf{R} = (a \cos bt)\mathbf{i} + (a \sin bt)\mathbf{j}$$
$$\mathbf{R}' = (-ab \sin bt)\mathbf{i} + (ab \cos bt)\mathbf{j}$$
$$|\mathbf{R}'| = \sqrt{(ab)^2 \sin^2 bt + (ab)^2 \cos^2 bt} = ab$$

Moreover,
$$\mathbf{T} = \frac{\mathbf{R}'}{|\mathbf{R}'|} = (-\sin bt)\mathbf{i} + (\cos bt)\mathbf{j}$$

$$\mathbf{T}' = (-b \cos bt)\mathbf{i} - (b \sin bt)\mathbf{j}$$
$$|\mathbf{T}'| = \sqrt{b^2(\cos^2 bt + \sin^2 bt)} = b$$

Finally,

$$\mathbf{N} = \frac{\mathbf{T}'}{|\mathbf{T}'|} = \frac{1}{b}[(-b \cos bt)\mathbf{i} - (b \sin bt)\mathbf{j}]$$

$$\mathbf{N} = (-\cos bt)\mathbf{i} - (\sin bt)\mathbf{j} = -\left(\frac{1}{a}\right)\mathbf{R}$$

Note that $\mathbf{N}$ has the opposite direction of $\mathbf{R}$, which is what we would expect, since the graph of $\mathbf{R} = (a \cos bt)\mathbf{i} + (a \sin bt)\mathbf{j}$ is a circle of radius a, centered at the origin (see Figure 13.40). ∎

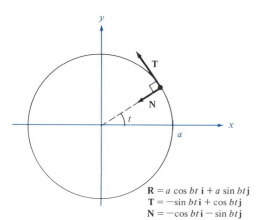

$\mathbf{R} = a \cos bt\,\mathbf{i} + a \sin bt\,\mathbf{j}$
$\mathbf{T} = -\sin bt\,\mathbf{i} + \cos bt\,\mathbf{j}$
$\mathbf{N} = -\cos bt\,\mathbf{i} - \sin bt\,\mathbf{j}$

FIGURE 13.40

In Example 1 note the similarity between $\mathbf{T}$ and $\mathbf{N}$:

$$\mathbf{T} = (-\sin bt)\mathbf{i} + (\cos bt)\mathbf{j}$$
$$\mathbf{N} = (-\cos bt)\mathbf{i} + (-\sin bt)\mathbf{j}$$

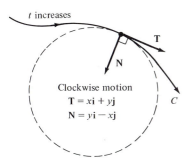

Clockwise motion

$T = x\mathbf{i} + y\mathbf{j}$
$N = y\mathbf{i} - x\mathbf{j}$

t increases

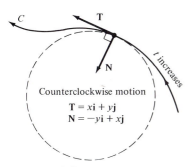

Counterclockwise motion

$T = x\mathbf{i} + y\mathbf{j}$
$N = -y\mathbf{i} + x\mathbf{j}$

t increases

FIGURE 13.41

This similarity is not merely a coincidence, since for any unit vector,

$$\mathbf{U} = a\mathbf{i} + b\mathbf{j} \qquad (a^2 + b^2 = 1)$$

there are precisely two unit vectors that are normal to **U**. They are

$$\mathbf{V}_1 = b\mathbf{i} - a\mathbf{j} \qquad \text{and} \qquad \mathbf{V}_2 = -b\mathbf{i} + a\mathbf{j}$$

(See Section 13.4, Exercise 43.) This result can be used as a convenient *alternative way of finding the principal unit normal vector* **N**. For if **T** = $x\mathbf{i} + y\mathbf{j}$ is a unit vector, then the unit vectors $y\mathbf{i} - x\mathbf{j}$ and $-y\mathbf{i} + x\mathbf{j}$ are both perpendicular to **T**. To determine which vector to choose as the principal unit normal vector **N**, we consider the fact that the principal unit normal vector must point toward the concave side of curve C (assuming that t is increasing). With respect to the closest circular approximation at some point on the curve, if the motion of the object is *counterclockwise*, then we choose $\mathbf{N} = -y\mathbf{i} + x\mathbf{j}$. If the motion is *clockwise*, we choose $\mathbf{N} = y\mathbf{i} - x\mathbf{j}$ (Figure 13.41).

For instance, in Example 1 the motion was counterclockwise, so we could have determined **N** in the following manner. Since

$$\mathbf{T} = x\mathbf{i} + y\mathbf{j} = (-\sin bt)\mathbf{i} + (\cos bt)\mathbf{j}$$

then

$$\mathbf{N} = -y\mathbf{i} + x\mathbf{j} = -(\cos bt)\mathbf{i} + (-\sin bt)\mathbf{j}$$

Example 2 *(Path of a Projectile)*

Using the position function

$$\mathbf{R} = (v_0 \cos \theta)t\mathbf{i} + [(v_0 \sin \theta)t - 16t^2]\mathbf{j}$$

where $v_0 = 20$ and $\theta = \text{Arcsin } \tfrac{4}{5}$

find **T** and **N** at $t = \tfrac{1}{2}$ and $t = \tfrac{3}{4}$.

Solution:

$$\mathbf{R} = 12t\mathbf{i} + (16t - 16t^2)\mathbf{j}$$
$$\mathbf{V} = 12\mathbf{i} + (16 - 32t)\mathbf{j} = 4[3\mathbf{i} + 4(1 - 2t)\mathbf{j}]$$
$$|\mathbf{V}| = 4\sqrt{9 + 16(1 - 2t)^2}$$

Therefore,

$$\mathbf{T} = \frac{\mathbf{V}}{|\mathbf{V}|} = \frac{3\mathbf{i} + 4(1 - 2t)\mathbf{j}}{\sqrt{9 + 16(1 - 2t)^2}}$$

From Figure 13.42 we can see that for this position function the motion is clockwise. Hence we know that

$$\mathbf{N} = \frac{4(1 - 2t)\mathbf{i} - 3\mathbf{j}}{\sqrt{9 + 16(1 - 2t)^2}}$$

Now when $t = \tfrac{1}{2}$,

$$\mathbf{T} = \frac{3\mathbf{i}}{\sqrt{9}} = \mathbf{i} \qquad \text{and} \qquad \mathbf{N} = \frac{-3\mathbf{j}}{\sqrt{9}} = -\mathbf{j}$$

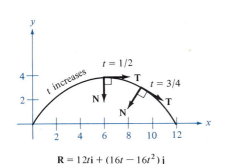

t increases

$t = 1/2$

$t = 3/4$

$\mathbf{R} = 12t\mathbf{i} + (16t - 16t^2)\mathbf{j}$

FIGURE 13.42

and when $t = \frac{3}{4}$,

$$\mathbf{T} = \frac{3\mathbf{i} - 2\mathbf{j}}{\sqrt{13}} \quad \text{and} \quad \mathbf{N} = \frac{-2\mathbf{i} - 3\mathbf{j}}{\sqrt{13}}$$ ∎

For an object moving along a curved path C in a plane, we are now prepared to determine the tangent and normal components of the acceleration. Recall from Theorem 13.9 that the projection of a vector $\mathbf{U}$ onto a unit vector $\mathbf{V}$ is given by $(\mathbf{U} \cdot \mathbf{V})\mathbf{V}$. Thus at any time t, the projection of the acceleration vector onto the unit vectors $\mathbf{T}$ and $\mathbf{N}$ is given by

$$(\mathbf{A} \cdot \mathbf{T})\mathbf{T} \quad \text{and} \quad (\mathbf{A} \cdot \mathbf{N})\mathbf{N}$$

respectively (Figure 13.43).

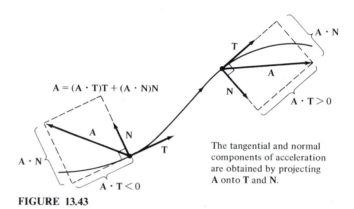

The tangential and normal components of acceleration are obtained by projecting $\mathbf{A}$ onto $\mathbf{T}$ and $\mathbf{N}$.

FIGURE 13.43

We call the two dot products $\mathbf{A} \cdot \mathbf{T}$ and $\mathbf{A} \cdot \mathbf{N}$ the **tangential** and **normal** components of acceleration. (The normal component of acceleration is also referred to as the **centripetal** component of acceleration.)

Definition of Tangential and Normal Components of Acceleration	If $\mathbf{A}$ is the acceleration of an object moving along a plane curve C at time t, then $$\mathbf{A} = (\mathbf{A} \cdot \mathbf{T})\mathbf{T} + (\mathbf{A} \cdot \mathbf{N})\mathbf{N}$$ and we call $\mathbf{A} \cdot \mathbf{T}$ and $\mathbf{A} \cdot \mathbf{N}$ the **tangential** and **normal components of acceleration,** respectively.

Example 3

Given $\mathbf{R} = 4t\mathbf{i} + (2 \ln t - t^2)\mathbf{j}$, $t > 0$, find the tangential and normal components of acceleration when $t = 2$ and sketch the result.

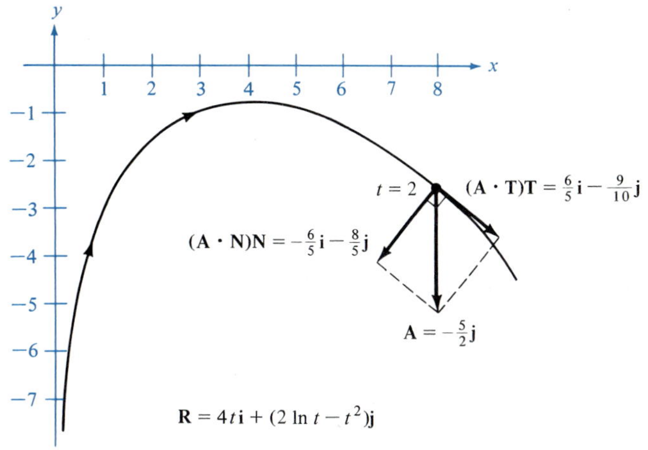

FIGURE 13.44

Solution:

$$\mathbf{V} = \mathbf{R}' = 4\mathbf{i} + \left(\frac{2 - 2t^2}{t}\right)\mathbf{j}$$

$$|\mathbf{V}| = \sqrt{16 + \frac{4 - 8t^2 + 4t^4}{t^2}} = \frac{2(1 + t^2)}{t}$$

$$\mathbf{T} = \frac{\mathbf{V}}{|\mathbf{V}|} = \frac{2t}{1 + t^2}\mathbf{i} + \frac{1 - t^2}{1 + t^2}\mathbf{j}$$

From Figure 13.44 we can see that as t increases, the motion is clockwise, and hence

$$\mathbf{N} = \frac{1 - t^2}{1 + t^2}\mathbf{i} - \frac{2t}{1 + t^2}\mathbf{j}$$

Furthermore, since $\quad \mathbf{A} = \mathbf{V}' = \dfrac{-2(1 + t^2)}{t^2}\mathbf{j}$

then at $t = 2$ we have

$$\mathbf{A} = \tfrac{-5}{2}\mathbf{j}, \qquad \mathbf{T} = \tfrac{4}{5}\mathbf{i} - \tfrac{3}{5}\mathbf{j}, \qquad \mathbf{N} = \tfrac{-3}{5}\mathbf{i} - \tfrac{4}{5}\mathbf{j}$$

Therefore, at $t = 2$ the tangential and normal components of acceleration are given by

$$\mathbf{A} \cdot \mathbf{T} = \tfrac{3}{2} \quad \text{and} \quad \mathbf{A} \cdot \mathbf{N} = 2 \qquad \blacksquare$$

We defined the tangential and normal components of acceleration geometrically by projecting the acceleration vector **A** onto **T** and **N**, respectively. In the following theorem we give a physical interpretation of these two components in terms of the speed, the rate of change of the speed, and the curvature.

THEOREM 13.11
(*Alternative Form of Tangential and Normal Components of Acceleration*)

The acceleration of an object whose position function is $\mathbf{R} = x\mathbf{i} + y\mathbf{j}$ is given by

$$\mathbf{A} = \frac{d^2s}{dt^2}\mathbf{T} + K\left(\frac{ds}{dt}\right)^2\mathbf{N}$$

where

$$\frac{ds}{dt} = |\mathbf{V}| = \sqrt{\left(\frac{dx}{dt}\right)^2 + \left(\frac{dy}{dt}\right)^2}$$

and K is the curvature of the path at time t.

Proof: The velocity of the object is given by

$$\frac{d\mathbf{R}}{dt} = \mathbf{V} = |\mathbf{V}|\mathbf{T} = \frac{ds}{dt}\mathbf{T}$$

By differentiating we have

$$\frac{d\mathbf{V}}{dt} = \mathbf{A} = \frac{d^2s}{dt^2}\mathbf{T} + \frac{ds}{dt}\mathbf{T'} = \frac{d^2s}{dt^2}\mathbf{T} + \frac{ds}{dt}|\mathbf{T'}|\mathbf{N}$$

Now since

$$\mathbf{T} = \frac{\frac{dx}{dt}\mathbf{i} + \frac{dy}{dt}\mathbf{j}}{\sqrt{\left(\frac{dx}{dt}\right)^2 + \left(\frac{dy}{dt}\right)^2}}$$

we have (after differentiating and simplifying)

$$\mathbf{T'} = \frac{\left(\frac{dx}{dt}\frac{d^2y}{dt^2} - \frac{dy}{dt}\frac{d^2x}{dt^2}\right)\left(-\frac{dy}{dt}\mathbf{i} + \frac{dx}{dt}\mathbf{j}\right)}{\left[\left(\frac{dx}{dt}\right)^2 + \left(\frac{dy}{dt}\right)^2\right]^{3/2}}$$

and

$$|\mathbf{T'}| = \left|\frac{\frac{dx}{dt}\frac{d^2y}{dt^2} - \frac{dy}{dt}\frac{d^2x}{dt^2}}{\left[\left(\frac{dx}{dt}\right)^2 + \left(\frac{dy}{dt}\right)^2\right]^{3/2}}\right|\sqrt{\left(\frac{dx}{dt}\right)^2 + \left(\frac{dy}{dt}\right)^2}$$

However, from Exercise 40 of Section 13.2, we note that the first factor is the curvature K, and the second factor is the speed ds/dt. Consequently, we have

$$|\mathbf{T'}| = K\frac{ds}{dt}$$

and we conclude that

$$\mathbf{A} = \frac{d^2s}{dt^2}\mathbf{T} + \frac{ds}{dt}|\mathbf{T'}|\mathbf{N} = \frac{d^2s}{dt^2}\mathbf{T} + \frac{ds}{dt}K\left(\frac{ds}{dt}\right)\mathbf{N}$$

$$= \frac{d^2s}{dt^2}\mathbf{T} + K\left(\frac{ds}{dt}\right)^2\mathbf{N}$$

■

Note that in the proof of Theorem 13.11, we obtained the equation

$$|\mathbf{T}'| = K\frac{ds}{dt}$$

From this equation we obtain the following formulation of curvature, which compares the rate of change in **T** to the rate of change in s (arc length) with respect to time:

Curvature

The **curvature** K at a point on the curve described by the position vector

$$\mathbf{R}(t) = x(t)\mathbf{i} + y(t)\mathbf{j}$$

is given by

$$K = \frac{|\mathbf{T}'(t)|}{ds/dt}$$

where **T** is the unit tangent vector.

This is reasonable if we consider that for Δt near (but not equal to) zero, we have

$$\frac{\mathbf{T}'(t)}{ds/dt} \approx \frac{[\mathbf{T}(t + \Delta t) - \mathbf{T}(t)]/\Delta t}{[s(t + \Delta t) - s(t)]/\Delta t} = \frac{\mathbf{T}(t + \Delta t) - \mathbf{T}(t)}{s(t + \Delta t) - s(t)} = \frac{\Delta \mathbf{T}}{\Delta s}$$

Geometrically, this means that for a given Δs, the greater the length of $\Delta \mathbf{T}$, the more bending there is to the curve at t (see Figure 13.45).

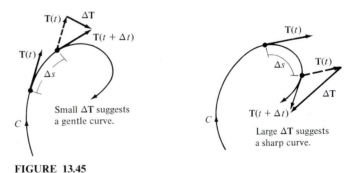

FIGURE 13.45

Now since the tangential and normal components of acceleration are given by

$$\mathbf{A} \cdot \mathbf{T} = \frac{d^2s}{dt^2} \quad \text{and} \quad \mathbf{A} \cdot \mathbf{N} = K\left(\frac{ds}{dt}\right)^2$$

we can make the following observations. First, the tangential component of acceleration is the rate of change of the speed; therefore, the tangential component is *negative* if the object is slowing down and *positive* if it is speeding up. Figure 13.46 shows two common examples of this.

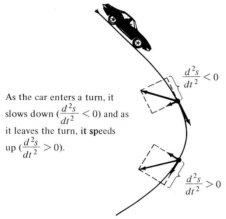

As the car enters a turn, it slows down ($\frac{d^2s}{dt^2} < 0$) and as it leaves the turn, it speeds up ($\frac{d^2s}{dt^2} > 0$).

As the satellite moves closer to the Earth, it speeds up ($\frac{d^2s}{dt^2} > 0$), and as it moves away from the Earth, it slows down ($\frac{d^2s}{dt^2} < 0$).

FIGURE 13.46

Secondly, since

$$\mathbf{A} \cdot \mathbf{N} = K\left(\frac{ds}{dt}\right)^2$$

we can see that the normal component of acceleration is always positive. Furthermore, if the speed is constant, then the normal component increases as the curvature increases. This explains why a car rounding a sharp curve at a moderate speed or a gentle curve at high speed requires a normal force (tire friction) of large magnitude to stay on the road.

The important uses of vector functions are reviewed in the following summary.

Summary of Velocity and Acceleration Vectors

Given the vector function $\mathbf{R}(t) = x(t)\mathbf{i} + y(t)\mathbf{j}$, we have:

$$\mathbf{V} = \mathbf{R}' = \text{velocity} \qquad \frac{ds}{dt} = |\mathbf{V}| = |\mathbf{R}'| = \text{speed}$$

$$\mathbf{A} = \mathbf{R}'' = \text{acceleration} \qquad K = \frac{|\mathbf{T}'|}{ds/dt} = \frac{|\mathbf{T}'|}{|\mathbf{R}'|} = \text{curvature}$$

$$\mathbf{T} = \frac{\mathbf{V}}{|\mathbf{V}|} = \text{unit tangent vector}$$

$$\mathbf{N} = \frac{\mathbf{T}'}{|\mathbf{T}'|} = \text{unit normal vector}$$

$$\mathbf{A} \cdot \mathbf{T} = \frac{d^2s}{dt^2} = \text{tangential component of acceleration}$$

$$\mathbf{A} \cdot \mathbf{N} = K\left(\frac{ds}{dt}\right)^2 = \text{normal component of acceleration}$$

$$\mathbf{A} = (\mathbf{A} \cdot \mathbf{T})\mathbf{T} + (\mathbf{A} \cdot \mathbf{N})\mathbf{N} = \frac{d^2s}{dt^2}\mathbf{T} + K\left(\frac{ds}{dt}\right)^2\mathbf{N}$$

Section Exercises (13.6)

In Exercises 1–10, find **T**, **N**, **A** · **T**, **A** · **N**, and K at the given time t.

S **1.** $\mathbf{R} = t\mathbf{i} + (1/t)\mathbf{j}$, $t = 1$

2. $\mathbf{R} = t\mathbf{i} + t^2\mathbf{j}$, $t = 1$

3. $\mathbf{R} = 4 \cos (2\pi t)\mathbf{i} + 4 \sin (2\pi t)\mathbf{j}$, $t = \frac{1}{8}$

4. $\mathbf{R} = a \cos (wt)\mathbf{i} + a \sin (wt)\mathbf{j}$, $t = t_0$

5. $\mathbf{R} = 2 \cos (\pi t)\mathbf{i} + \sin (\pi t)\mathbf{j}$, $t = \frac{1}{3}$

6. $\mathbf{R} = a \cos (wt)\mathbf{i} + b \sin (wt)\mathbf{j}$, $t = 0$

C S **7.** $\mathbf{R} = (e^t \cos t)\mathbf{i} + (e^t \sin t)\mathbf{j}$, $t = \pi/2$

8. $\mathbf{R} = a(wt - \sin wt)\mathbf{i} + a(1 - \cos wt)\mathbf{j}$, $t = t_0$

9. $\mathbf{R} = (t^3/3)\mathbf{i} + (t^2/2)\mathbf{j}$, $t = \sqrt{3}$

10. $\mathbf{R} = [\cos (wt) + wt \sin (wt)]\mathbf{i} + [\sin (wt) - wt \cos (wt)]\mathbf{j}$, $t = t_0$

S **11.** The equation $\mathbf{R}(t) = (v_0 t \cos \theta)\mathbf{i} + (v_0 t \sin \theta - 16t^2)\mathbf{j}$ describes the motion of a projectile fired at an angle of θ with the horizontal and with an initial speed of v_0. Find the following:
   (a) the tangential and normal components of acceleration
   (b) the tangential and normal acceleration at the maximum height

C **12.** A plane flying at an altitude of 30,000 ft and with a speed of 540 mi/h releases a bomb. If we neglect air resistance, find the following:
   (a) a vector equation for the motion of the bomb
   (b) the tangential and normal components of acceleration

13. Consider a weight spinning at the end of a string. (See Exercise 34 in Section 13.5.)
   (a) If the angular velocity is doubled, how is the centripetal (normal) component of acceleration changed?
   (b) If the angular velocity is unchanged but the length of the string is halved, how is the centripetal component of acceleration changed?

S **14.** Consider an object of mass m moving in a circular path of radius r with a velocity of magnitude v.
   (a) Show that $F = mv^2/r$ is the centripetal force.

   (b) Newton's Law of Universal Gravitation is $F = GMm/d^2$, where d is the distance between the centers of the two bodies of mass M and m. Use this and the result of part (a) to show that the speed required for circular motion is $v = \sqrt{GM/r}$.

15. Use Exercise 14 to find the speed v necessary to maintain an object in a circular orbit 100 mi above the surface of the earth. (Use 4000 mi as the radius of the earth and $GM = 9.56 \times 10^4$ mi³/s².)

16. Use Newton's Law of Universal Gravitation (Exercise 14) to show that $g_h = [r/(r + h)]^2 g$, where r is the radius of the earth, h is the height of a body above the earth's surface, g_h is the acceleration due to gravity at height h, and g is the acceleration due to gravity on the earth's surface.

17. Use the result of Exercise 14 to find the velocity of a satellite in circular orbit around the earth 1000 mi above the surface of the earth. *Echo I,* an experimental communications satellite launched on 12 August 1960, had such an orbit. Find the time required for one orbit by *Echo I.* (Disregard the rotation of the earth.)

C S **18.** The syncom satellites have synchronous orbits, meaning that if they are placed in a west-to-east orbit over the equator, they revolve in their circular orbits at the same rate as the earth. Thus they appear to be stationary over the earth. Use Exercise 14 to find the altitude and speed of such satellites.

19. Given the parametric equations $x = a \cos t$, $y = a \sin t$, show how to calculate **T**, using

$$\mathbf{T} = \frac{d\mathbf{R}}{ds} = \left(\frac{dx}{ds}\right)\mathbf{i} + \left(\frac{dy}{ds}\right)\mathbf{j}$$

(Hint: Calculate dx and dy and use the equation $ds^2 = dx^2 + dy^2$; i.e., $ds = \sqrt{dx^2 + dy^2}$.)

20. If C has a point of inflection at time t, show that $\mathbf{T}' = \mathbf{0}$ at this point.

Miscellaneous Exercises (Ch. 13)

In Exercises 1–10, find dy/dx and all points of horizontal tangency. Where possible, eliminate the parameter.

1. $x = 1 + 4t$, $y = 2 - 3t$

2. $x = e^t$, $y = e^{-t}$

S **3.** $x = 3 + 2 \cos \theta$, $y = 2 + 5 \sin \theta$

4. $x = t^2 - 3t + 2$, $y = t^3 - 3t^2 + 2$

5. $x = \dfrac{1}{t}$, $y = 2t + 3$

6. $x = 2t - 1$, $y = \dfrac{1}{t^2 - 2t}$

7. $x = \dfrac{1}{2t + 1}, \ y = \dfrac{2t(t + 1)}{2t + 1}$

8. $x = \cot \theta, \ y = \sin 2\theta$

9. $x = \cos^3 \theta, \ y = 4 \sin^3 \theta$

10. $x = 2\theta - \sin \theta, \ y = 2 - \cos \theta$

11. Show that the Cartesian equation of a cycloid is $x = a \operatorname{Arccos}[(a - y)/a] \pm \sqrt{2ay - y^2}$.

12. Find a parametric representation of the ellipse with center at $(-3, 4)$, major axis horizontal and 8 units in length, and minor axis 6 units in length.

⟨S⟩ 13. Find a parametric representation of the hyperbola with vertices at $(0, \pm 4)$ and foci at $(0, \pm 5)$.

14. The involute of a circle is described by the endpoint P of a string that is held taut as it is unwound from a spool (the spool does not rotate; see Figure 13.47). Show that a parametric representation of the involute of a circle is given by $x = r(\cos \theta + \theta \sin \theta), \ y = r(\sin \theta - \theta \cos \theta)$.

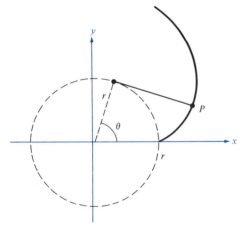

FIGURE 13.47

⟨S⟩ 15. Find the length of the involute of Exercise 14 when $\theta = \pi$.

16. Find the horizontal and vertical components of a vector if its magnitude is 100 and it makes an angle of $\pi/6$ radians with the positive x-axis.

⟨C⟩ 17. The angle between two vectors of magnitude 50 and 75 is $50°$. Find the magnitude of the sum of these two vectors.

⟨C⟩⟨S⟩ 18. Find the largest weight w that can be supported by the structure in Figure 13.48 if the strut can withstand a maximum compression of 500 lb.

19. Find the vector projection $\mathbf{W}$ of $\mathbf{U} = 3\mathbf{i} + 2\mathbf{j}$ onto $\mathbf{V} = \frac{1}{2}\mathbf{i} - \frac{3}{2}\mathbf{j}$.

20. Using vectors prove that the midpoints of the sides of *any* quadrilateral are the vertices of a parallelogram.

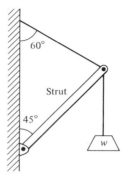

FIGURE 13.48

21. Using vectors prove that the medians of a triangle pass through a point that is $\frac{2}{3}$ of the distance from any vertex to the midpoint of the opposite side.

22. Prove that $\mathbf{W} = (\mathbf{W} \cdot \mathbf{U})\mathbf{U} + (\mathbf{W} \cdot \mathbf{V})\mathbf{V}$ for any vector $\mathbf{W}$ if $\mathbf{U}$ and $\mathbf{V}$ are orthogonal unit vectors.

23. If $\mathbf{U} \neq \mathbf{0}$ and $\mathbf{U} \cdot \mathbf{V} = \mathbf{U} \cdot \mathbf{W}$, does $\mathbf{V} = \mathbf{W}$? Explain.

24. Give a geometrical argument to show that the reflection of the vector $\mathbf{U}$ through the vector $\mathbf{V}$ is given by

$$\mathbf{W} = \left(\frac{2\mathbf{U} \cdot \mathbf{V}}{\mathbf{V} \cdot \mathbf{V}} \right) \mathbf{V} - \mathbf{U}$$

⟨S⟩ 25. Use the result of Exercise 24 to reflect the vector $\mathbf{U} = \mathbf{i} + 3\mathbf{j}$ through these vectors:
(a) $\mathbf{V} = \mathbf{i}$ (b) $\mathbf{V} = \mathbf{j}$ (c) $\mathbf{V} = \mathbf{i} + \mathbf{j}$
In each case show the result geometrically.

26. Find the work done in moving an object along the vector $\mathbf{V} = 3\mathbf{i} + 2\mathbf{j}$ if the applied force is $\mathbf{F} = 2\mathbf{i} - \mathbf{j}$.

27. For vectors $\mathbf{U}$ and $\mathbf{V}$, prove the triangle inequality, $|\mathbf{U} + \mathbf{V}| \leq |\mathbf{U}| + |\mathbf{V}|$. (Hint: Square both sides and use the dot product.)

28. Let $\mathbf{U} = (\cos \alpha)\mathbf{i} + (\sin \alpha)\mathbf{j}$ and $\mathbf{V} = (\cos \beta)\mathbf{i} + (\sin \beta)\mathbf{j}$. Draw these vectors and by interpreting the dot product $\mathbf{U} \cdot \mathbf{V}$ geometrically, prove that $\cos(\alpha - \beta) = \cos \alpha \cos \beta + \sin \alpha \sin \beta$.

29. Prove that $c_1 = c_2 = 0$ if $\mathbf{U}$ and $\mathbf{V}$ are noncollinear, nonzero vectors such that $c_1\mathbf{U} + c_2\mathbf{V} = \mathbf{0}$.

In Exercises 30–33, find $\mathbf{T}, \mathbf{N}, \mathbf{A} \cdot \mathbf{T},$ and $\mathbf{A} \cdot \mathbf{N}$.

30. $\mathbf{R}(t) = (1 + 4t)\mathbf{i} + (2 - 3t)\mathbf{j}$

⟨S⟩ 31. $\mathbf{R}(t) = (t \cos t)\mathbf{i} + (t \sin t)\mathbf{j}$

32. $\mathbf{R}(t) = e^t\mathbf{i} + e^{-t}\mathbf{j}$

33. $\mathbf{R}(t) = t\mathbf{i} + \sqrt{t}\mathbf{j}$

34. An automobile is in a circular traffic interchange and its speed is twice that which is posted. By what factor has the centripetal force increased from that which would occur at the proper speed?

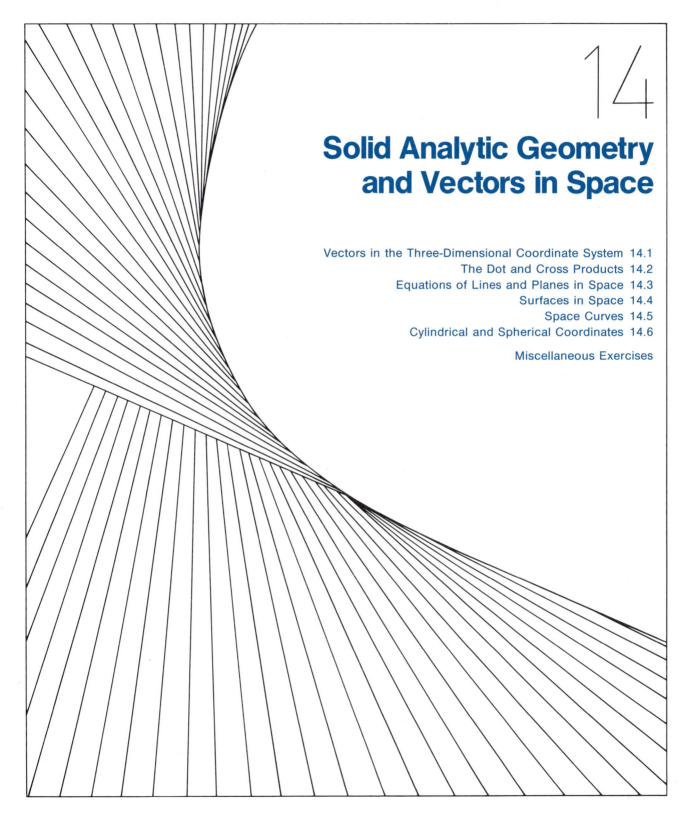

14

Solid Analytic Geometry and Vectors in Space

14.1 Vectors in the Three-Dimensional Coordinate System

Purpose
- To introduce the three-dimensional coordinate system and the formula for the distance between two points in space.
- To discuss the component form of vectors in three-dimensional space.

I hope that hereafter those who are clever enough to make use of the geometric methods herein suggested will find no great difficulty in applying them to plane or solid problems. I therefore think it proper to suggest to such a more extended line of investigation which will furnish abundant opportunities for practice. —René Descartes (1596–1650)

In Chapter 1 we described the Cartesian plane as the plane determined by two mutually perpendicular number lines called the *x*- and *y*-axes. These axes together with their point of intersection (the origin) allowed us to develop a two-dimensional coordinate system for identifying points in a plane and for discussing topics in plane analytic geometry. To identify a point in space, we need to introduce a third dimension to our model. The geometry of this three-dimensional model is referred to as **solid analytic** geometry.

To construct a three-dimensional coordinate system, we begin with the two-dimensional system (the *xy*-plane in a horizontal position) and through the origin, 0, we pass a vertical *z*-axis that is perpendicular to both the *x*- and *y*-axes (Figure 14.1).

This particular orientation of the *x*-, *y*-, and *z*-axes is called a **right-handed** system. As an aid to remembering the relationship between the three axes in a right-handed system, imagine that you are standing at the origin with your arms in the direction of the positive *x*- and *y*-axes. The system is right-handed if your *right* hand points in the direction of the *x*-axis, and it is left-handed if your *left* hand points in the direction of the *x*-axis (Figure 14.2). In this text we will work exclusively with the right-handed system.

Pairwise, the three axes of the three-dimensional system form three **coordinate planes.** The *xy-plane* is determined by the *x*- and *y*-axes, the *xz-plane* by the *x*- and *z*-axes, and the *yz-plane* by the *y*- and *z*-axes. The three coordinate planes separate space into eight **octants,** and we refer to the *first octant* as the one in which all three coordinates are positive.

A point *P* in three-dimensional space (3-space) is determined by an ordered triple (x, y, z), where the *x*-coordinate denotes the directed distance from the *yz*-plane to *P*, the *y*-coordinate the directed distance from the *xz*-plane to *P*, and the *z*-coordinate the directed distance from the *xy*-plane to *P*. Several points are shown in Figure 14.3.

Many of the geometric properties of 3-space are simple extensions of those established for the plane in Chapter 1. For instance, by using the Pythagorean Theorem twice, we can establish that the *distance d between two points* (x_1, y_1, z_1) *and* (x_2, y_2, z_2) *in 3-space is*

$$d = \sqrt{(x_2 - x_1)^2 + (y_2 - y_1)^2 + (z_2 - z_1)^2}$$

(See Figure 14.4.)

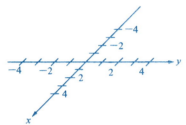

xy–plane is horizontal

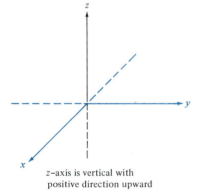

z–axis is vertical with positive direction upward

FIGURE 14.1

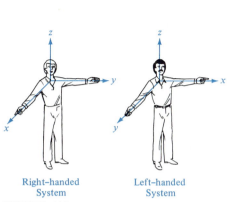

Right-handed
System

Left-handed
System

FIGURE 14.2

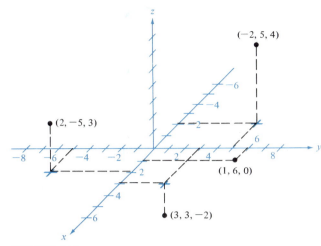

FIGURE 14.3

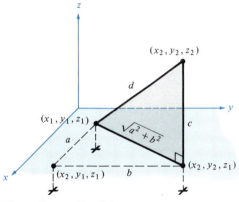

Distance Between Two Points:

$$d = \sqrt{a^2 + b^2 + c^2}$$

$$= \sqrt{(x_2 - x_1)^2 + (y_2 - y_1)^2 + (z_2 - z_1)^2}$$

FIGURE 14.4

The **Midpoint Rule** is also an obvious extension of the one for plane geometry. If (x_1, y_1, z_1) and (x_2, y_2, z_2) are two points in space, then the midpoint of the line segment connecting the two points is

$$\left(\frac{x_1 + x_2}{2}, \frac{y_1 + y_2}{2}, \frac{z_1 + z_2}{2} \right)$$

Since a *sphere* is considered to be the set of all points in space that lie at a fixed distance from a given point, we can use the Distance Formula to obtain the following *standard equation of a sphere of radius r with center at* (h, k, l):

$$(x - h)^2 + (y - k)^2 + (z - l)^2 = r^2$$

Example 1

Find the distance between $(5, -2, 3)$ and $(0, 4, -3)$. What is the midpoint of the line segment connecting these two points?

Solution: By the Distance Formula

$$d = \sqrt{(0 - 5)^2 + (4 + 2)^2 + (-3 - 3)^2} = \sqrt{25 + 36 + 36} = \sqrt{97}$$

The midpoint is

$$\left(\frac{5 + 0}{2}, \frac{-2 + 4}{2}, \frac{3 - 3}{2} \right) = \left(\frac{5}{2}, 1, 0 \right)$$

Example 2

Find the center and radius of the sphere $x^2 + y^2 + z^2 - 2x + 4y - 6z + 8 = 0$.

Solution: We can obtain the standard equation of this sphere by completing the square with each variable, as follows:

$$(x^2 - 2x \quad) + (y^2 + 4y \quad) + (z^2 - 6z \quad) = -8$$
$$(x^2 - 2x + 1) + (y^2 + 4y + 4) + (z^2 - 6z + 9) = -8 + 1 + 4 + 9$$
$$(x - 1)^2 + (y + 2)^2 + (z - 3)^2 = 6$$

Therefore, the center of the sphere is at $(1, -2, 3)$ and its radius is $\sqrt{6}$ (see Figure 14.5). ∎

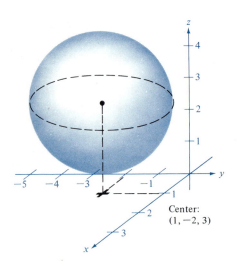

Center:
$(1, -2, 3)$

FIGURE 14.5

To denote vectors in 3-space, we use the previously discussed unit vectors **i** and **j**, along with a new unit vector **k**. Vector **k** points in the direction of the positive z-axis and has its initial point at the origin. The three mutually perpendicular unit vectors **i**, **j**, and **k** are shown in Figure 14.6.

Every vector **V** in 3-space having initial point (x_1, y_1, z_1) and terminal point (x_2, y_2, z_2) can be written in the *component form*

$$\mathbf{V} = a\mathbf{i} + b\mathbf{j} + c\mathbf{k}$$

where $a = x_2 - x_1$, $b = y_2 - y_1$, and $c = z_2 - z_1$ (see Figure 14.7). The *length* of vector **V** is given by

$$|\mathbf{V}| = \sqrt{a^2 + b^2 + c^2}$$

A *unit vector in the direction of* **V** is the vector

$$\frac{\mathbf{V}}{|\mathbf{V}|} = \frac{a\mathbf{i} + b\mathbf{j} + c\mathbf{k}}{\sqrt{a^2 + b^2 + c^2}}$$

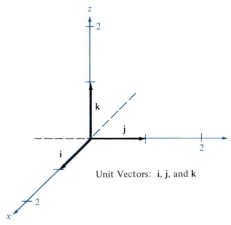

Unit Vectors: **i**, **j**, and **k**

FIGURE 14.6

The basic operations with three-dimensional vectors are defined as follows: If $\mathbf{U} = a_1\mathbf{i} + b_1\mathbf{j} + c_1\mathbf{k}$ and $\mathbf{V} = a_2\mathbf{i} + b_2\mathbf{j} + c_2\mathbf{k}$, then

$$\mathbf{U} + \mathbf{V} = (a_1 + a_2)\mathbf{i} + (b_1 + b_2)\mathbf{j} + (c_1 + c_2)\mathbf{k} \qquad \text{(vector addition)}$$
$$\mathbf{U} - \mathbf{V} = (a_1 - a_2)\mathbf{i} + (b_1 - b_2)\mathbf{j} + (c_1 - c_2)\mathbf{k} \qquad \text{(vector subtraction)}$$
$$c\mathbf{U} = ca_1\mathbf{i} + cb_1\mathbf{j} + cc_1\mathbf{k} \qquad \text{(scalar multiplication)}$$

where c is any real number. These operations have the same properties as those identified in Theorem 13.4 for vectors in the plane.

Example 3

If $\mathbf{U} = 3\mathbf{i} - 4\mathbf{j} + 2\mathbf{k}$ and **V** is the vector having $(-1, 5, -2)$ as its initial point and $(3, 0, 4)$ as its terminal point, find the following:

(a) $-2\mathbf{U} + \mathbf{V}$

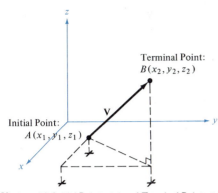

Terminal Point:
$B(x_2, y_2, z_2)$

Initial Point:
$A(x_1, y_1, z_1)$

Vector with Initial Point at A and Terminal Point at B:
$\mathbf{V} = \mathbf{AB} = (x_2 - x_1)\mathbf{i} + (y_2 - y_1)\mathbf{j} + (z_2 - z_1)\mathbf{k}$

FIGURE 14.7

(b) $|\mathbf{V}|$

(c) a unit vector in the direction of $\mathbf{U}$

Solution:

(a) The component form of vector $\mathbf{V}$ is

$$\mathbf{V} = (3 + 1)\mathbf{i} + (0 - 5)\mathbf{j} + (4 + 2)\mathbf{k} = 4\mathbf{i} - 5\mathbf{j} + 6\mathbf{k}$$

We also have $\qquad -2\mathbf{U} = -6\mathbf{i} + 8\mathbf{j} - 4\mathbf{k}$

and therefore, $\qquad -2\mathbf{U} + \mathbf{V} = -2\mathbf{i} + 3\mathbf{j} + 2\mathbf{k}$

(b) $\qquad\qquad |\mathbf{V}| = \sqrt{4^2 + (-5)^2 + (6)^2} = \sqrt{77}$

(c) A unit vector in the direction of $\mathbf{U}$ is

$$\frac{\mathbf{U}}{|\mathbf{U}|} = \frac{3\mathbf{i} - 4\mathbf{j} + 2\mathbf{k}}{\sqrt{9 + 16 + 4}} = \frac{3}{\sqrt{29}}\mathbf{i} - \frac{4}{\sqrt{29}}\mathbf{j} + \frac{2}{\sqrt{29}}\mathbf{k} \qquad \blacksquare$$

Summary of Formulas (Three-Dimensional Space)

If $A = (x_1, y_1, z_1)$ and $B = (x_2, y_2, z_2)$, then we have the following:

Distance between A and B:

$$d = \sqrt{(x_2 - x_1)^2 + (y_2 - y_1)^2 + (z_2 - z_1)^2}$$

Midpoint between A and B:

$$\left(\frac{x_1 + x_2}{2}, \frac{y_1 + y_2}{2}, \frac{z_1 + z_2}{2} \right)$$

Vector from A to B:

$$\mathbf{V} = \overrightarrow{AB} = (x_2 - x_1)\mathbf{i} + (y_2 - y_1)\mathbf{j} + (z_2 - z_1)\mathbf{k}$$

If $\mathbf{V} = a\mathbf{i} + b\mathbf{j} + c\mathbf{k}$, then:

Length of $\mathbf{V}$:

$$|\mathbf{V}| = \sqrt{a^2 + b^2 + c^2}$$

Unit vector in direction of $\mathbf{V}$:

$$\frac{\mathbf{V}}{|\mathbf{V}|} = \frac{a\mathbf{i} + b\mathbf{j} + c\mathbf{k}}{\sqrt{a^2 + b^2 + c^2}}$$

The standard equation of a **sphere** of radius r with center at (h, k, l) is

$$(x - h)^2 + (y - k)^2 + (z - l)^2 = r^2$$

Section Exercises (14.1)

1. Plot the points $(2, 1, 3)$, $(-1, 2, 1)$, and $(3, -2, 5)$.

2. Plot the points $(\frac{3}{2}, 4, -2)$, $(5, -2, 2)$, and $(5, -2, -2)$.

In Exercises 3–5, find the lengths of the sides of the triangle with the indicated vertices and determine whether the triangle is a right triangle, an isosceles triangle, or neither of these.

3. (0, 0, 0), (2, 2, 1), (2, −4, 4)
4. (5, 3, 4), (7, 1, 3), (3, 5, 3)

[S] 5. (1, −3, −2), (5, −1, 2), (−1, 1, 2)

6. Find the midpoint of the line segment joining the points (5, −9, 7) and (−2, 3, 3).
7. The endpoint of a line segment is (−2, 1, 1) and its midpoint is (0, 2, 5). Find the other endpoint.
8. Find the lengths of the medians of the triangle with vertices (1, −3, −2), (5, −1, 2), and (−1, 1, 2).
9. Find the equation of the sphere with center at (0, 2, 5) and radius 2.
10. Find the equation of the sphere with center at (−2, 1, 1) and tangent to the xy-coordinate plane.

[S] 11. Find the center and radius of the sphere $x^2 + y^2 + z^2 - 2x + 6y + 8z + 1 = 0$.

12. Find the center and radius of the sphere $4x^2 + 4y^2 + 4z^2 - 4x - 32y + 8z + 33 = 0$.

In Exercises 13–18, write the component form of vector **V** and sketch it.

13. **V** is the vector from (−1, 2, 3) to (3, 3, 4).
14. **V** is the vector from (2, −1, −2) to (−4, 3, 7).

[S] 15. **V** lies in the yz-plane, has magnitude 2, and makes an angle of 30° with the positive y-axis.

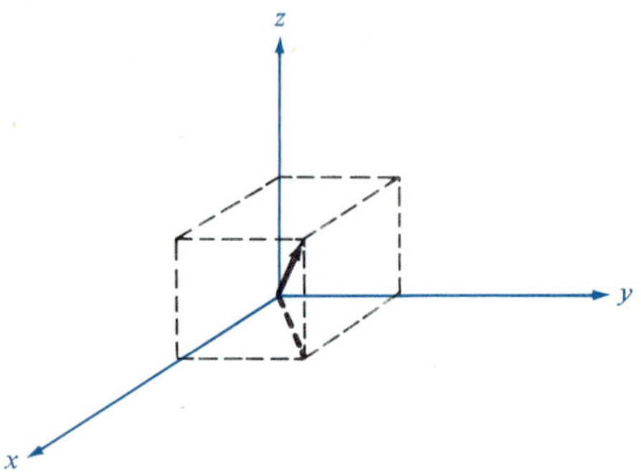

FIGURE 14.8

16. **V** lies in the xz-plane, has magnitude 5, and makes an angle of 45° with the positive z-axis.
17. **V** is a unit vector representing the diagonal of a cube, as shown in Figure 14.8.
18. **V** is a vector of magnitude 5, which forms the diagonal of a cube. (See Figure 14.8.)

In Exercises 19–26, let $\mathbf{U} = \mathbf{i} + 2\mathbf{j} + 3\mathbf{k}$, $\mathbf{V} = 2\mathbf{i} + 2\mathbf{j} - \mathbf{k}$, and $\mathbf{W} = 4\mathbf{i} - 4\mathbf{k}$.

19. Find $\mathbf{U} - \mathbf{V}$ and $\mathbf{V} - \mathbf{U}$.
20. Find $\mathbf{U} - \mathbf{V} + 2\mathbf{W}$.
21. Find $2\mathbf{U} + 4\mathbf{V} - \mathbf{W}$.
22. Find **Z**, where $2\mathbf{Z} - 3\mathbf{U} = \mathbf{W}$.

[S] 23. Find **Z**, where $2\mathbf{U} + \mathbf{V} - \mathbf{W} + 3\mathbf{Z} = \mathbf{0}$.

24. For what values of c is $|c\mathbf{U}| = 1$?
25. For what values of c is $|c\mathbf{V}| = 1$?
26. For what values of c is $|c\mathbf{V}| = 5$?

27. Sketch the vector $\mathbf{V} = 8\mathbf{i} + 8\mathbf{j} + 6\mathbf{k}$ and find a vector **U** such that:
    (a) **U** has the same direction as **V** and $\frac{1}{2}$ its magnitude
    (b) **U** has the opposite direction as **V** and $\frac{1}{4}$ its magnitude
28. Show that the points (0, −2, −5), (3, 4, 4), and (2, 2, 1) lie on a straight line.

[S] 29. Show that the points (1, −1, 5), (0, −1, 6), and (3, −1, 3) lie on a straight line.

30. Show that the points (2, 9, 1), (3, 11, 4), (0, 10, 2), and (1, 12, 5) are vertices of a parallelogram.

[S] 31. Let $\mathbf{V} = \mathbf{i} + \mathbf{j}$, $\mathbf{U} = \mathbf{j} + \mathbf{k}$, and $\mathbf{W} = a\mathbf{V} + b\mathbf{U}$.
    (a) Sketch **V** and **U**.
    (b) If $\mathbf{W} = \mathbf{0}$, show that a and b are both zero.
    (c) Find a and b such that $\mathbf{W} = \mathbf{i} + 2\mathbf{j} + \mathbf{k}$.
    (d) Prove that no choice of a and b yields $\mathbf{W} = \mathbf{i} + 2\mathbf{j} + 3\mathbf{k}$.

32. Using vectors prove that the line segment joining the midpoints of two sides of a triangle is parallel to and one-half the length of the third side.
33. Prove that the vectors $\mathbf{U} = (\cos\theta)\mathbf{i} - (\sin\theta)\mathbf{j}$ and $\mathbf{V} = (\sin\theta)\mathbf{i} + (\cos\theta)\mathbf{j}$ are unit vectors.

For Review and Class Discussion

True or False

1. ____ If a, b, and c are positive, then $\mathbf{V} = a\mathbf{i} + b\mathbf{j} + c\mathbf{k}$ lies in the first octant.
2. ____ $|\mathbf{V} + \mathbf{U}| \leq |\mathbf{V}| + |\mathbf{U}|$.
3. ____ If $|\mathbf{V} + \mathbf{U}| = |\mathbf{V}| + |\mathbf{U}|$, then **V** is a scalar multiple of **U**.
4. ____ If **U**, **V**, and **W** are nonzero vectors lying in the same plane, then there exist scalars a and b such that $\mathbf{W} = a\mathbf{U} + b\mathbf{V}$.
5. ____ If $\mathbf{U} = a\mathbf{i} + \mathbf{j}$, $\mathbf{V} = \mathbf{i} + b\mathbf{k}$, and $\mathbf{W} = c\mathbf{j} + \mathbf{k}$, where $\mathbf{U} + \mathbf{V} + \mathbf{W} = \mathbf{0}$, then **U**, **V**, and **W** have the same magnitude.

14.2 The Dot and Cross Products

Purpose

- To summarize the properties of the dot product for vectors in space.
- To define and discuss the properties of the cross product of two vectors.

Many results must be given of which the details are suppressed. These must not be taken on trust by the student, but must be worked out by his own pen, which must never be out of his own hand while engaged in any mathematical process. —Augustus DeMorgan (1806–1871)

In this section we will investigate two types of multiplication with vectors in space, the dot product and the cross product. Since we have already discussed the dot product of vectors in the plane (Section 13.4), we will limit our discussion of the dot product of vectors in space to a summary of the important definitions and properties.

As was true for vectors in the plane, the dot product of two vectors in 3-space is a *scalar* value, as specified in the following definition.

Definition of Dot Product	If $\mathbf{U} = a_1\mathbf{i} + b_1\mathbf{j} + c_1\mathbf{k}$ and $\mathbf{V} = a_2\mathbf{i} + b_2\mathbf{j} + c_2\mathbf{k}$, then the **dot product** $\mathbf{U} \cdot \mathbf{V}$ is given by $$\mathbf{U} \cdot \mathbf{V} = a_1a_2 + b_1b_2 + c_1c_2$$

As in the plane, we identify the angle between two nonzero vectors $\mathbf{U}$ and $\mathbf{V}$ in space as the angle θ between representatives of $\mathbf{U}$ and $\mathbf{V}$ that have a common initial point, where $0 \le \theta \le \pi$. (See Figure 14.9.) With this interpretation of θ in mind, we give the following summary of the properties of the dot product of vectors in space. The corresponding theorems from Section 13.4 are given for comparison.

Summary of Properties of the Dot Product in 3-Space	For nonzero vectors $\mathbf{U}$, $\mathbf{V}$, and $\mathbf{W}$ and scalar c:

1. **(Theorem 13.6)**

   i. $\mathbf{U} \cdot \mathbf{V} = \mathbf{V} \cdot \mathbf{U}$ iv. $\mathbf{0} \cdot \mathbf{V} = 0$

   ii. $\mathbf{U} \cdot (\mathbf{V} + \mathbf{W}) = \mathbf{U} \cdot \mathbf{V} + \mathbf{U} \cdot \mathbf{W}$ v. $\mathbf{V} \cdot \mathbf{V} = |\mathbf{V}|^2$

   iii. $c(\mathbf{U} \cdot \mathbf{V}) = (c\mathbf{U}) \cdot \mathbf{V}$

2. **(Theorem 13.7)** $\mathbf{U} \cdot \mathbf{V} = |\mathbf{U}|\,|\mathbf{V}| \cos \theta$

   where θ is the angle between $\mathbf{U}$ and $\mathbf{V}$.

3. **(Theorem 13.8)** $\mathbf{U} \cdot \mathbf{V} = 0$ if and only if $\mathbf{U}$ and $\mathbf{V}$ are orthogonal.

4. **(Theorem 13.9)** $\mathbf{W} = \left(\dfrac{\mathbf{U} \cdot \mathbf{V}}{|\mathbf{V}|}\right)\dfrac{\mathbf{V}}{|\mathbf{V}|}$

   is the projection of $\mathbf{U}$ onto $\mathbf{V}$.

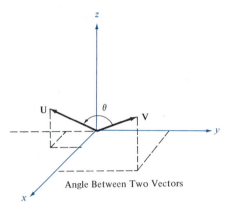

Angle Between Two Vectors

FIGURE 14.9

Example 1

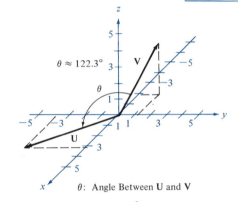

$\theta:$ Angle Between **U** and **V**

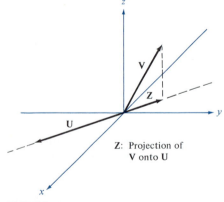

Z: Projection of **V** onto **U**

FIGURE 14.10

Given $\mathbf{U} = 3\mathbf{i} - 4\mathbf{j}$ and $\mathbf{V} = -2\mathbf{i} + \mathbf{j} + 3\mathbf{k}$, find the following:
(a) $\mathbf{V} \cdot \mathbf{U}$
(b)   the angle between $\mathbf{U}$ and $\mathbf{V}$
(c)   the projection of $\mathbf{V}$ onto $\mathbf{U}$

Solution:

(a)   By definition

$$\mathbf{V} \cdot \mathbf{U} = -2(3) + 1(-4) + 3(0) = -10$$

(b)   Since

$$|\mathbf{U}| = \sqrt{9 + 16 + 0} = 5 \qquad \text{and} \qquad |\mathbf{V}| = \sqrt{4 + 1 + 9} = \sqrt{14}$$

we have

$$\cos \theta = \frac{\mathbf{U} \cdot \mathbf{V}}{|\mathbf{U}|\,|\mathbf{V}|} = \frac{-10}{5\sqrt{14}} = -\frac{2}{\sqrt{14}} \approx -0.535$$

and
$$\theta \approx 122.3°$$

(See Figure 14.10.)

(c)   The projection of $\mathbf{V}$ onto $\mathbf{U}$ is

$$\mathbf{Z} = \left(\frac{\mathbf{U} \cdot \mathbf{V}}{|\mathbf{U}|}\right)\frac{\mathbf{U}}{|\mathbf{U}|} = \left(\frac{-10}{5}\right)\frac{3\mathbf{i} - 4\mathbf{j}}{5} = \frac{-6}{5}\mathbf{i} + \frac{8}{5}\mathbf{j}$$

(See Figure 14.10.)   ■

Example 2

Find the angles between the vector $\mathbf{V} = 2\mathbf{i} + 3\mathbf{j} + 4\mathbf{k}$ and the unit vectors $\mathbf{i}, \mathbf{j},$ and $\mathbf{k}$.

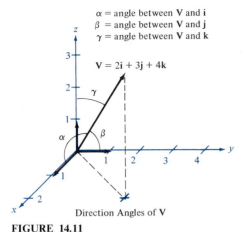

α = angle between **V** and **i**
β = angle between **V** and **j**
γ = angle between **V** and **k**

V = 2**i** + 3**j** + 4**k**

Direction Angles of **V**

FIGURE 14.11

Solution: Let α, β, and γ be the angles between **V** and **i, j,** and **k,** respectively (Figure 14.11). Then

$$\cos \alpha = \frac{\mathbf{V} \cdot \mathbf{i}}{|\mathbf{V}| \, |\mathbf{i}|} = \frac{2}{\sqrt{9 + 4 + 16}} = \frac{2}{\sqrt{29}}$$

$$\cos \beta = \frac{\mathbf{V} \cdot \mathbf{j}}{|\mathbf{V}| \, |\mathbf{j}|} = \frac{3}{\sqrt{29}}$$

$$\cos \gamma = \frac{\mathbf{V} \cdot \mathbf{k}}{|\mathbf{V}| \, |\mathbf{k}|} = \frac{4}{\sqrt{29}}$$

Thus

$$\alpha = \text{Arccos} \frac{2}{\sqrt{29}} \approx 68.20°$$

$$\beta = \text{Arccos} \frac{3}{\sqrt{29}} \approx 56.15°$$

$$\gamma = \text{Arccos} \frac{4}{\sqrt{29}} \approx 42.03°$$

∎

We call the three angles α, β, and γ the **direction angles** of **V.** The cosines of these angles are called the **direction cosines** of **V.** Thus if $\mathbf{V} = a\mathbf{i} + b\mathbf{j} + c\mathbf{k}$, then the direction cosines of **V** are

$$\cos \alpha = \frac{a}{|\mathbf{V}|}, \qquad \cos \beta = \frac{b}{|\mathbf{V}|}, \qquad \cos \gamma = \frac{c}{|\mathbf{V}|}$$

One interesting property of the direction cosines is that

$$\cos^2 \alpha + \cos^2 \beta + \cos^2 \gamma = 1$$

(See Exercise 10.)

One important use of the dot product is in finding vectors that are orthogonal (perpendicular) to a given vector. For example, to find a vector $\mathbf{V} = x\mathbf{i} + y\mathbf{j}$ that is orthogonal to $\mathbf{U} = a\mathbf{i} + b\mathbf{j}$, we find a solution to the equation

$$\mathbf{U} \cdot \mathbf{V} = xa + yb = 0$$

One solution is $x = b$ and $y = -a$, giving us $\mathbf{V} = b\mathbf{i} - a\mathbf{j}.$

The extension of this problem to 3-space is slightly more complicated and it requires us to solve a pair of equations simultaneously. For example, let

$$\mathbf{U} = a_1\mathbf{i} + b_1\mathbf{j} + c_1\mathbf{k} \qquad \text{and} \qquad \mathbf{V} = a_2\mathbf{i} + b_2\mathbf{j} + c_2\mathbf{k}$$

be two nonzero vectors, neither of which is a scalar multiple of the other. To find a nonzero vector, $\mathbf{W} = x\mathbf{i} + y\mathbf{j} + z\mathbf{k}$, that is orthogonal to both **U** and **V,** we must find x, y, and z such that

$$\mathbf{U} \cdot \mathbf{W} = a_1 x + b_1 y + c_1 z = 0 \qquad and \qquad \mathbf{V} \cdot \mathbf{W} = a_2 x + b_2 y + c_2 z = 0$$

To solve this system of equations, we first write

$$a_1x + b_1y = -c_1z$$
$$a_2x + b_2y = -c_2z$$

Then by multiplying the first equation by b_2 and the second by b_1, we obtain

$$a_1b_2x + b_1b_2y = -c_1b_2z$$
$$b_1a_2x + b_1b_2y = -b_1c_2z$$

Now subtracting the second equation from the first, we have

$$(a_1b_2 - b_1a_2)x = (b_1c_2 - c_1b_2)z$$

Since **U** is not a multiple of **V**, we know that $a_1b_2 - b_1a_2 \neq 0$; hence

$$x = \left(\frac{b_1c_2 - c_1b_2}{a_1b_2 - b_1a_2}\right)z$$

In a similar manner we can obtain the equation

$$y = -\left(\frac{a_1c_2 - c_1a_2}{a_1b_2 - b_1a_2}\right)z$$

Finally, since z is arbitrary, we can choose $z = a_1b_2 - b_1a_2$, and we conclude that the vector

$$\mathbf{W} = x\mathbf{i} + y\mathbf{j} + z\mathbf{k} = (b_1c_2 - c_1b_2)\mathbf{i} - (a_1c_2 - c_1a_2)\mathbf{j} + (a_1b_2 - b_1a_2)\mathbf{k}$$

is orthogonal to both **U** and **V**. (See Figure 14.12.) We call this particular choice of **W** the **cross product** of **U** and **V**.

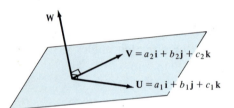

W is orthogonal to both **U** and **V**.

$\mathbf{W} = \mathbf{U} \times \mathbf{V}$

$= (b_1c_2 - c_1b_2)\mathbf{i} - (a_1c_2 - a_2c_1)\mathbf{j} + (a_1b_2 - a_2b_1)\mathbf{k}$

FIGURE 14.12

Definition of Cross Product	For vectors $\mathbf{U} = a_1\mathbf{i} + b_1\mathbf{j} + c_1\mathbf{k}$ and $\mathbf{V} = a_2\mathbf{i} + b_2\mathbf{j} + c_2\mathbf{k}$, the **cross product** $\mathbf{U} \times \mathbf{V}$ is given by $$\mathbf{U} \times \mathbf{V} = (b_1c_2 - b_2c_1)\mathbf{i} - (a_1c_2 - a_2c_1)\mathbf{j} + (a_1b_2 - a_2b_1)\mathbf{k}$$ (Notice the negative sign before the **j**-component.)

Since this formula is rather complicated, we like to write the cross product of **U** and **V** in the determinant form

$$\mathbf{U} \times \mathbf{V} = \begin{vmatrix} \mathbf{i} & \mathbf{j} & \mathbf{k} \\ a_1 & b_1 & c_1 \\ a_2 & b_2 & c_2 \end{vmatrix}$$

Each component of $\mathbf{U} \times \mathbf{V}$ is then determined from the 2×2 determinant formed by deleting the row and column containing **i**, **j**, and **k**, respectively. Thus

$$\mathbf{U} \times \mathbf{V} = \begin{vmatrix} \mathbf{i} & \mathbf{j} & \mathbf{k} \\ a_1 & b_1 & c_1 \\ a_2 & b_2 & c_2 \end{vmatrix}$$

$$= \begin{vmatrix} \mathbf{i} & \mathbf{j} & \mathbf{k} \\ a_1 & b_1 & c_1 \\ a_2 & b_2 & c_2 \end{vmatrix} \mathbf{i} - \begin{vmatrix} \mathbf{i} & \mathbf{j} & \mathbf{k} \\ a_1 & b_1 & c_1 \\ a_2 & b_2 & c_2 \end{vmatrix} \mathbf{j} + \begin{vmatrix} \mathbf{i} & \mathbf{j} & \mathbf{k} \\ a_1 & b_1 & c_1 \\ a_2 & b_2 & c_2 \end{vmatrix} \mathbf{k}$$

$$= \begin{vmatrix} b_1 & c_1 \\ b_2 & c_2 \end{vmatrix} \mathbf{i} - \begin{vmatrix} a_1 & c_1 \\ a_2 & c_2 \end{vmatrix} \mathbf{j} + \begin{vmatrix} a_1 & b_1 \\ a_2 & b_2 \end{vmatrix} \mathbf{k}$$

where $\begin{vmatrix} b_1 & c_1 \\ b_2 & c_2 \end{vmatrix} = (b_1 c_2 - b_2 c_1),$ $\begin{vmatrix} a_1 & c_1 \\ a_2 & c_2 \end{vmatrix} = (a_1 c_2 - a_2 c_1)$

and $\begin{vmatrix} a_1 & b_1 \\ a_2 & b_2 \end{vmatrix} = (a_1 b_2 - a_2 b_1)$

Example 3

Given $\mathbf{U} = \mathbf{i} - 2\mathbf{j} + \mathbf{k}$ and $\mathbf{V} = 3\mathbf{i} + \mathbf{j} - 2\mathbf{k}$, find the following:
(a) $\mathbf{U} \times \mathbf{V}$ (b) $\mathbf{V} \times \mathbf{U}$ (c) $\mathbf{U} \times \mathbf{U}$

Solution:

(a)
$$\mathbf{U} \times \mathbf{V} = \begin{vmatrix} \mathbf{i} & \mathbf{j} & \mathbf{k} \\ 1 & -2 & 1 \\ 3 & 1 & -2 \end{vmatrix} = \begin{vmatrix} -2 & 1 \\ 1 & -2 \end{vmatrix} \mathbf{i} - \begin{vmatrix} 1 & 1 \\ 3 & -2 \end{vmatrix} \mathbf{j} + \begin{vmatrix} 1 & -2 \\ 3 & 1 \end{vmatrix} \mathbf{k}$$

$$= (4 - 1)\mathbf{i} - (-2 - 3)\mathbf{j} + (1 + 6)\mathbf{k} = 3\mathbf{i} + 5\mathbf{j} + 7\mathbf{k}$$

(b)
$$\mathbf{V} \times \mathbf{U} = \begin{vmatrix} \mathbf{i} & \mathbf{j} & \mathbf{k} \\ 3 & 1 & -2 \\ 1 & -2 & 1 \end{vmatrix} = \begin{vmatrix} 1 & -2 \\ -2 & 1 \end{vmatrix} \mathbf{i} - \begin{vmatrix} 3 & -2 \\ 1 & 1 \end{vmatrix} \mathbf{j} + \begin{vmatrix} 3 & 1 \\ 1 & -2 \end{vmatrix} \mathbf{k}$$

$$= (1 - 4)\mathbf{i} - (3 + 2)\mathbf{j} + (-6 - 1)\mathbf{k} = -3\mathbf{i} - 5\mathbf{j} - 7\mathbf{k}$$

(c)
$$\mathbf{U} \times \mathbf{U} = \begin{vmatrix} \mathbf{i} & \mathbf{j} & \mathbf{k} \\ 1 & -2 & 1 \\ 1 & -2 & 1 \end{vmatrix} = (-2 + 2)\mathbf{i} - (1 - 1)\mathbf{j} + (-2 + 2)\mathbf{k} = \mathbf{0}$$

The results obtained in Example 3 suggest some interesting properties of the cross product. For instance,

$$\mathbf{U} \times \mathbf{V} = -(\mathbf{V} \times \mathbf{U}) \quad \text{and} \quad \mathbf{U} \times \mathbf{U} = \mathbf{0}$$

These properties along with several others are given in the following theorem:

THEOREM 14.1 (*Algebraic Properties of the Cross Product*)	If **U**, **V**, and **W** are vectors in space, and p and q are scalars, then: i. $\mathbf{U} \times \mathbf{V} = -(\mathbf{V} \times \mathbf{U})$ $\qquad$ iv. $\mathbf{U} \times \mathbf{0} = \mathbf{0} \times \mathbf{U} = \mathbf{0}$ ii. $\mathbf{U} \times \mathbf{U} = \mathbf{0}$ $\qquad\qquad\quad$ v. $\mathbf{U} \times (\mathbf{V} + \mathbf{W}) = (\mathbf{U} \times \mathbf{V})$ $\qquad\qquad\qquad\qquad\qquad\qquad\qquad\qquad\qquad\qquad + (\mathbf{U} \times \mathbf{W})$ iii. $p\mathbf{U} \times q\mathbf{V} = pq(\mathbf{U} \times \mathbf{V})$ $\quad$ vi. $\mathbf{U} \cdot (\mathbf{V} \times \mathbf{W}) = (\mathbf{U} \times \mathbf{V}) \cdot \mathbf{W}$

Proof: The verification of each of these six properties is straightforward, and we present the details for part vi only.

$$\text{Let} \qquad\qquad \mathbf{U} = a_1\mathbf{i} + b_1\mathbf{j} + c_1\mathbf{k}$$
$$\mathbf{V} = a_2\mathbf{i} + b_2\mathbf{j} + c_2\mathbf{k}$$
$$\mathbf{W} = a_3\mathbf{i} + b_3\mathbf{j} + c_3\mathbf{k}$$

Then

$$
\begin{aligned}
\mathbf{U} \cdot (\mathbf{V} \times \mathbf{W}) &= a_1(b_2c_3 - c_2b_3) - b_1(a_2c_3 - c_2a_3) + c_1(a_2b_3 - b_2a_3) \\
&= a_1b_2c_3 - a_1c_2b_3 - b_2a_2c_3 + b_1c_2a_3 + c_1a_2b_3 - c_1b_2a_3 \\
&= b_1c_2a_3 - c_1b_2a_3 - a_1c_2b_3 + c_1a_2b_3 + a_1b_2c_3 - b_1a_2c_3 \\
&= (b_1c_2 - c_1b_2)a_3 - (a_1c_2 - c_1a_2)b_3 + (a_1b_2 - b_1a_2)c_3 \\
&= (\mathbf{U} \times \mathbf{V}) \cdot \mathbf{W} \qquad\qquad\qquad\qquad\qquad\qquad\blacksquare
\end{aligned}
$$

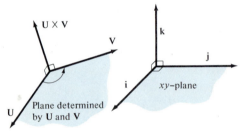

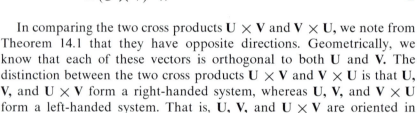

Right–handed Systems

FIGURE 14.13

In comparing the two cross products $\mathbf{U} \times \mathbf{V}$ and $\mathbf{V} \times \mathbf{U}$, we note from Theorem 14.1 that they have opposite directions. Geometrically, we know that each of these vectors is orthogonal to both **U** and **V**. The distinction between the two cross products $\mathbf{U} \times \mathbf{V}$ and $\mathbf{V} \times \mathbf{U}$ is that **U**, **V**, and $\mathbf{U} \times \mathbf{V}$ form a right-handed system, whereas **U**, **V**, and $\mathbf{V} \times \mathbf{U}$ form a left-handed system. That is, **U**, **V**, and $\mathbf{U} \times \mathbf{V}$ are oriented in roughly the same way as the vectors **i**, **j**, and **k** (Figure 14.13).

The following theorem summarizes some other geometric properties of the cross product.

THEOREM 14.2 (*Geometric Properties of the Cross Product*)	If **U** and **V** are nonzero vectors in space and θ is the angle between **U** and **V**, then: i. $\mathbf{U} \times \mathbf{V}$ is orthogonal to both **U** and **V**. ii. $\|\mathbf{U} \times \mathbf{V}\| = \|\mathbf{U}\| \,\|\mathbf{V}\| \sin \theta$. iii. $\mathbf{U} \times \mathbf{V} = \mathbf{0}$ if and only if **U** and **V** are parallel. iv. $|\mathbf{U} \times \mathbf{V}| = $ area of parallelogram having **U** and **V** as adjacent sides. v. $\dfrac{\|\mathbf{U} \times \mathbf{V}\|}{\mathbf{U} \cdot \mathbf{V}} = \tan \theta$.

Proof:

ii.   Let $\mathbf{U} = a_1\mathbf{i} + b_1\mathbf{j} + c_1\mathbf{k}$ and $\mathbf{V} = a_2\mathbf{i} + b_2\mathbf{j} + c_2\mathbf{k}$. Then

$$\|\mathbf{U}\| \,\|\mathbf{V}\| \sin \theta = \|\mathbf{U}\| \,\|\mathbf{V}\| \sqrt{1 - \cos^2 \theta} \qquad \text{for} \qquad 0 \le \theta \le \pi$$

Now since $\cos \theta = (\mathbf{U} \cdot \mathbf{V})/(|\mathbf{U}| \, |\mathbf{V}|)$, we have

$$|\mathbf{U}| \, |\mathbf{V}| \sin \theta = |\mathbf{U}| \, |\mathbf{V}| \sqrt{1 - \left(\frac{\mathbf{U} \cdot \mathbf{V}}{|\mathbf{U}| \, |\mathbf{V}|}\right)^2} = \sqrt{|\mathbf{U}|^2 \, |\mathbf{V}|^2 - (\mathbf{U} \cdot \mathbf{V})^2}$$

$$= \sqrt{(a_1{}^2 + b_1{}^2 + c_1{}^2)(a_2{}^2 + b_2{}^2 + c_2{}^2) - (a_1 a_2 + b_1 b_2 + c_1 c_2)^2}$$

By expanding, simplifying, and then regrouping, we obtain

$$|\mathbf{U}| \, |\mathbf{V}| \sin \theta = \sqrt{(b_1 c_2 - b_2 c_1)^2 + (a_1 c_2 - a_2 c_1)^2 + (a_1 b_2 - a_2 b_1)^2}$$
$$= |\mathbf{U} \times \mathbf{V}|$$

iii.   This property follows directly from part ii by noting that $|\mathbf{U} \times \mathbf{V}| = 0$ if and only if $\sin \theta = 0$.

iv.   From Figure 14.14, we can see that the area of the parallelogram is

$$A = (\text{base})(\text{height}) = |\mathbf{U}|(|\mathbf{V}| \sin \theta) = |\mathbf{U} \times \mathbf{V}|$$

v. $$\frac{|\mathbf{U} \times \mathbf{V}|}{\mathbf{U} \cdot \mathbf{V}} = \frac{|\mathbf{U}| \, |\mathbf{V}| \sin \theta}{|\mathbf{U}| \, |\mathbf{V}| \cos \theta} = \tan \theta \qquad \blacksquare$$

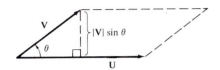

$|\mathbf{U} \times \mathbf{V}|$ = area of parallelogram having $\mathbf{U}$ and $\mathbf{V}$ as adjacent sides.

FIGURE 14.14

Example 4

Given $\mathbf{U} = 3\mathbf{i} + 2\mathbf{k}$ and $\mathbf{V} = \mathbf{i} - 4\mathbf{j} + \mathbf{k}$.
(a)   Verify that $\mathbf{U} \times \mathbf{V}$ is orthogonal to both $\mathbf{U}$ and $\mathbf{V}$.
(b)   Determine the area of the parallelogram having $\mathbf{U}$ and $\mathbf{V}$ as adjacent sides.

Solution:

(a)   Since

$$\mathbf{U} \times \mathbf{V} = \begin{vmatrix} \mathbf{i} & \mathbf{j} & \mathbf{k} \\ 3 & 0 & 2 \\ 1 & -4 & 1 \end{vmatrix} = (0 + 8)\mathbf{i} - (3 - 2)\mathbf{j} + (-12 - 0)\mathbf{k}$$
$$= 8\mathbf{i} - \mathbf{j} - 12\mathbf{k}$$

we obtain $\mathbf{U} \cdot (\mathbf{U} \times \mathbf{V}) = 24 - 0 - 24 = 0$
$\mathbf{V} \cdot (\mathbf{U} \times \mathbf{V}) = 8 + 4 - 12 = 0$

Therefore, $\mathbf{U} \times \mathbf{V}$ is orthogonal to both $\mathbf{U}$ and $\mathbf{V}$.

(b)   By part iv of Theorem 14.2, the area of the parallelogram having $\mathbf{U} = 3\mathbf{i} + 2\mathbf{k}$ and $\mathbf{V} = \mathbf{i} - 4\mathbf{j} + \mathbf{k}$ as adjacent sides is

$$|\mathbf{U} \times \mathbf{V}| = |8\mathbf{i} - \mathbf{j} - 12\mathbf{k}| = \sqrt{64 + 1 + 144} = \sqrt{209} \qquad \blacksquare$$

Example 5

Show that the quadrilateral with vertices at $A(5, 2, 0)$, $B(2, 6, 1)$, $C(2, 4, 7)$, and $D(5, 0, 6)$ is a parallelogram and find its area.

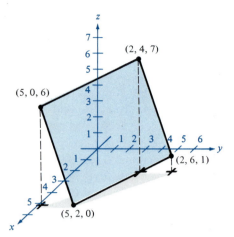

FIGURE 14.15

Solution: From Figure 14.15 we see that the sides of the quadrilateral correspond to the four vectors

$$\overrightarrow{AB} = -3\mathbf{i} + 4\mathbf{j} + \mathbf{k} \qquad \overrightarrow{CD} = 3\mathbf{i} - 4\mathbf{j} - \mathbf{k}$$
$$\overrightarrow{AD} = 0\mathbf{i} - 2\mathbf{j} + 6\mathbf{k} \qquad \overrightarrow{CB} = 0\mathbf{i} + 2\mathbf{j} - 6\mathbf{k}$$

and we see that $\overrightarrow{AB}$ is parallel to $\overrightarrow{CD}$ and $\overrightarrow{AD}$ is parallel to $\overrightarrow{CB}$. Thus $ABCD$ is a parallelogram with $\overrightarrow{AB}$ and $\overrightarrow{AD}$ as adjacent sides. Moreover, since

$$\overrightarrow{AB} \times \overrightarrow{AD} = \begin{vmatrix} \mathbf{i} & \mathbf{j} & \mathbf{k} \\ -3 & 4 & 1 \\ 0 & -2 & 6 \end{vmatrix} = (24 + 2)\mathbf{i} - (-18 - 0)\mathbf{j} + (6 - 0)\mathbf{k}$$

$$= 26\mathbf{i} + 18\mathbf{j} + 6\mathbf{k}$$

the area of the parallelogram is

$$|\overrightarrow{AB} \times \overrightarrow{AD}| = \sqrt{1036} \approx 32.19 \qquad \blacksquare$$

Example 6 *(Volume Using Cross Product)*

Find the volume of the parallelepiped having $\mathbf{U} = 3\mathbf{i} - 5\mathbf{j} + \mathbf{k}$, $\mathbf{V} = 2\mathbf{j} - 2\mathbf{k}$, and $\mathbf{W} = 3\mathbf{i} + \mathbf{j} + \mathbf{k}$ as edges.

Solution: From Figure 14.16 we see that the volume of a parallelepiped is given by

$$\text{volume} = (\text{height})(\text{area of base})$$

$$= \frac{\mathbf{U} \cdot (\mathbf{V} \times \mathbf{W})}{|\mathbf{V} \times \mathbf{W}|} |\mathbf{V} \times \mathbf{W}| = \mathbf{U} \cdot (\mathbf{V} \times \mathbf{W})$$

Therefore,

$$\text{volume} = (3\mathbf{i} - 5\mathbf{j} + \mathbf{k}) \cdot \begin{vmatrix} \mathbf{i} & \mathbf{j} & \mathbf{k} \\ 0 & 2 & -2 \\ 3 & 1 & 1 \end{vmatrix}$$

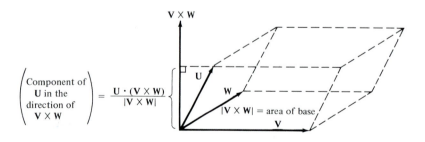

$$\begin{pmatrix} \text{Component of} \\ \mathbf{U} \text{ in the} \\ \text{direction of} \\ \mathbf{V} \times \mathbf{W} \end{pmatrix} = \frac{\mathbf{U} \cdot (\mathbf{V} \times \mathbf{W})}{|\mathbf{V} \times \mathbf{W}|}$$

Volume of Parallelepiped

FIGURE 14.16

$$= (3\mathbf{i} - 5\mathbf{j} + \mathbf{k}) \cdot \left(\begin{vmatrix} 2 & -2 \\ 1 & 1 \end{vmatrix} \mathbf{i} - \begin{vmatrix} 0 & -2 \\ 3 & 1 \end{vmatrix} \mathbf{j} + \begin{vmatrix} 0 & 2 \\ 3 & 1 \end{vmatrix} \mathbf{k} \right)$$

$$= 3 \begin{vmatrix} 2 & -2 \\ 1 & 1 \end{vmatrix} + 5 \begin{vmatrix} 0 & -2 \\ 3 & 1 \end{vmatrix} + \begin{vmatrix} 0 & 2 \\ 3 & 1 \end{vmatrix}$$

$$= 3(2 + 2) + 5(0 + 6) + (0 - 6) = 36 \text{ cubic units} \quad \blacksquare$$

The product $\mathbf{U} \cdot (\mathbf{V} \times \mathbf{W})$ used in Example 6 is referred to as the **triple scalar product** of **U**, **V**, and **W**. It is of interest to note that a triple scalar product such as the one in Example 6 can be evaluated as a single determinant. That is,

$$\mathbf{U} \cdot (\mathbf{V} \times \mathbf{W}) = \begin{vmatrix} 3 & -5 & 1 \\ 0 & 2 & -2 \\ 3 & 1 & 1 \end{vmatrix} = 36$$

where $\mathbf{U} = 3\mathbf{i} - 5\mathbf{j} + \mathbf{k}$, $\mathbf{V} = 2\mathbf{j} - 2\mathbf{k}$, and $\mathbf{W} = 3\mathbf{i} + \mathbf{j} + \mathbf{k}$.

Section Exercises (14.2)

In Exercises 1–6, find $\mathbf{U} \cdot \mathbf{V}$.

1. $\mathbf{U} = \mathbf{i}$, $\mathbf{V} = \mathbf{i}$
2. $\mathbf{U} = \mathbf{i}$, $\mathbf{V} = \mathbf{j}$
3. $\mathbf{U} = 2\mathbf{i} - \mathbf{j} + \mathbf{k}$, $\mathbf{V} = \mathbf{i} - \mathbf{k}$
4. $\mathbf{U} = 2\mathbf{i} + \mathbf{j} - 2\mathbf{k}$, $\mathbf{V} = \mathbf{i} - 3\mathbf{j} + 2\mathbf{k}$
5. $\mathbf{U} = 4\mathbf{i} - 5\mathbf{j} - \mathbf{k}$, $\mathbf{V} = 3\mathbf{i} - 2\mathbf{j} + 7\mathbf{k}$
6. $\mathbf{U} = 3\mathbf{i} + 4\mathbf{j}$, $\mathbf{V} = -2\mathbf{j} + 3\mathbf{k}$

7. Find the cosine of the angle between $\mathbf{U} = 5\mathbf{i}$ and $\mathbf{V} = 3\mathbf{i} + 4\mathbf{j}$.
8. Find the cosine of the angle between $\mathbf{U} = 2\mathbf{i} + 3\mathbf{j} + \mathbf{k}$ and $\mathbf{V} = -3\mathbf{i} + 2\mathbf{j}$.
9. Find the cosine of the angle between $\mathbf{U} = 3\mathbf{i} + 4\mathbf{j}$ and $\mathbf{V} = -2\mathbf{j} + 3\mathbf{k}$.
10. (a) Find the cosine of the angle between $\mathbf{U} = \mathbf{i} - 2\mathbf{j} + 2\mathbf{k}$ and each of the unit vectors $\mathbf{i}$, $\mathbf{j}$, and $\mathbf{k}$.
    (b) Show that the sum of the squares of the results of part (a) is equal to 1.
11. What is known about θ, the angle between vectors **U** and **V**, if:
    (a) $\mathbf{U} \cdot \mathbf{V} = 0$?   (b) $\mathbf{U} \cdot \mathbf{V} > 0$?
    (c) $\mathbf{U} \cdot \mathbf{V} < 0$?
12. Find the projection **W** of the vector $\mathbf{U} = 2\mathbf{i} + \mathbf{j} + 2\mathbf{k}$ on $\mathbf{V} = 3\mathbf{j} + 4\mathbf{k}$.
13. Find the projection **W** of the vector $\mathbf{U} = 4\mathbf{j} + \mathbf{k}$ on $\mathbf{V} = 2\mathbf{j} + 3\mathbf{k}$.

14. Find the projection of the vector $\mathbf{U} = \mathbf{i} + \mathbf{j} + \mathbf{k}$ on $\mathbf{V} = -2\mathbf{i} - \mathbf{j} + \mathbf{k}$.
15. Find the component of $\mathbf{U} = \mathbf{i} + \mathbf{j}$ in the direction of $\mathbf{V} = 2\mathbf{i} + 3\mathbf{j} + \mathbf{k}$.
16. Find the component of $\mathbf{U} = 2\mathbf{i} + \mathbf{j} + 2\mathbf{k}$ in the direction of $\mathbf{V} = 4\mathbf{i} + 3\mathbf{j}$.
17. Find the work done by a force $\mathbf{F} = 2\mathbf{i} + \mathbf{j} + 2\mathbf{k}$, where the displacement is described by $\mathbf{U} = \mathbf{i} - 3\mathbf{j} + \mathbf{k}$.
18. If $\mathbf{U} \cdot \mathbf{V} = \mathbf{U} \cdot \mathbf{W}$ and $\mathbf{U} \neq 0$, then is $\mathbf{V} = \mathbf{W}$? Explain.
19. Find c so that $\mathbf{U} - c\mathbf{V}$ is orthogonal to **U** if $\mathbf{U} = 3\mathbf{i} - 2\mathbf{j} + \mathbf{k}$ and $\mathbf{V} = -\mathbf{i} + \mathbf{j} + 2\mathbf{k}$.
20. Use vectors to prove that the diagonals of a rhombus are perpendicular.
21. Evaluate the following cross products:
    (a) $\mathbf{i} \times \mathbf{i}$ (b) $\mathbf{i} \times \mathbf{j}$
    (c) $\mathbf{j} \times \mathbf{k}$ (d) $\mathbf{k} \times \mathbf{i}$
    (e) $\mathbf{i} \times \mathbf{k}$ (f) $\mathbf{j} \times \mathbf{i}$

In Exercises 22–24, find $\mathbf{U} \times \mathbf{V}$ and show that it is orthogonal to both **U** and **V**.

22. $\mathbf{U} = \mathbf{i} + \mathbf{j} + \mathbf{k}$, $\mathbf{V} = 2\mathbf{i} + \mathbf{j} - \mathbf{k}$
23. $\mathbf{U} = \mathbf{j} + 6\mathbf{k}$, $\mathbf{V} = \mathbf{i} - 2\mathbf{j} - \mathbf{k}$
24. $\mathbf{U} = 2\mathbf{i} - 3\mathbf{j} + \mathbf{k}$, $\mathbf{V} = \mathbf{i} - 2\mathbf{j} + \mathbf{k}$

In Exercises 25–27, find the area of the triangle with the given vertices.

25. (0, 0, 0), (1, 2, 3), (−3, 0, 0)

26. (2, −3, 4), (0, 1, 2), (−1, 2, 0)

[S] **27.** (1, 3, 5), (3, 3, 0), (−2, 0, 5)

In Exercises 28–30, find the area of the parallelogram that has the given vectors as adjacent sides.

28. $\mathbf{U} = \mathbf{j}$, $\mathbf{V} = \mathbf{j} + \mathbf{k}$

[S] **29.** $\mathbf{U} = \mathbf{i} + \mathbf{j} + \mathbf{k}$, $\mathbf{V} = \mathbf{j} + \mathbf{k}$

30. $\mathbf{U} = 3\mathbf{i} + 2\mathbf{j} - \mathbf{k}$, $\mathbf{V} = \mathbf{i} + 2\mathbf{j} + 3\mathbf{k}$

31. Let $\mathbf{U} = (\cos\theta)\mathbf{i} + (\sin\theta)\mathbf{j}$ and $\mathbf{V} = (\cos\phi)\mathbf{i} + (\sin\phi)\mathbf{j}$.
(a) Find $|\mathbf{U}|$ and $|\mathbf{V}|$.
(b) Use $\mathbf{U}\cdot\mathbf{V}$ to show that $\theta - \phi$ is the angle between $\mathbf{U}$ and $\mathbf{V}$.

In Exercises 32–34, find the triple scalar product $\mathbf{U}\cdot(\mathbf{V}\times\mathbf{W})$.

32. $\mathbf{U} = \mathbf{i}$, $\mathbf{V} = \mathbf{j}$, $\mathbf{W} = \mathbf{k}$

[S] **33.** $\mathbf{U} = \mathbf{i} + \mathbf{j} + \mathbf{k}$, $\mathbf{V} = 2\mathbf{i} + \mathbf{j}$, $\mathbf{W} = \mathbf{k}$

34. $\mathbf{U} = 2\mathbf{i} + \mathbf{k}$, $\mathbf{V} = 3\mathbf{j}$, $\mathbf{W} = \mathbf{k}$

[S] **35.** Find the volume of the parallelepiped with $\mathbf{U} = \mathbf{i} + \mathbf{j}$, $\mathbf{V} = \mathbf{j} + \mathbf{k}$, and $\mathbf{W} = \mathbf{i} + \mathbf{k}$ as adjacent edges.

36. Find the volume of the parallelepiped whose vertices are (0, 0, 0), (3, 0, 0), (0, 5, 1), (3, 5, 1), (2, 0, 5), (5, 0, 5), (2, 5, 6), (5, 5, 6).

37. Find the altitude of the parallelepiped with $\mathbf{U} = \mathbf{i} + \mathbf{j}$, $\mathbf{V} = \mathbf{j} + \mathbf{k}$, and $\mathbf{W} = \mathbf{i} + \mathbf{k}$ if the base is the parallelogram determined by $\mathbf{U}$ and $\mathbf{V}$.

38. If $\mathbf{U}\times\mathbf{V} = \mathbf{0}$ and $\mathbf{U}\cdot\mathbf{V} = 0$, what conclusion can be drawn concerning $\mathbf{U}$ or $\mathbf{V}$?

39. Determine all vectors $\mathbf{U}$ such that $a\mathbf{i}\times\mathbf{U} = \mathbf{k}$.

40. Use the cross product to find the distance from the point (1, 1, 0) to the line through the points (−1, −2, 2) and (5, 5, 5).

41. Prove that $\mathbf{U}\times(\mathbf{V}+\mathbf{W}) = (\mathbf{U}\times\mathbf{V}) + (\mathbf{U}\times\mathbf{W})$.

42. Prove that $(c\mathbf{U})\times\mathbf{V} = c(\mathbf{U}\times\mathbf{V})$.

43. Prove that $\mathbf{U}\times(\mathbf{V}\times\mathbf{W}) = (\mathbf{U}\cdot\mathbf{W})\mathbf{V} - (\mathbf{U}\cdot\mathbf{V})\mathbf{W}$.

44. Prove that $\mathbf{U}\times\mathbf{V} = -(\mathbf{V}\times\mathbf{U})$.

For Review and Class Discussion

True or False

1. ____ The angle between two nonzero vectors in space is less than or equal to 180°.

2. ____ The dot product of two vectors is a vector.

3. ____ If $\mathbf{U}\times\mathbf{V} = \mathbf{0}$, then $\mathbf{U} = \mathbf{0}$ or $\mathbf{V} = \mathbf{0}$.

4. ____ For nonzero vectors $\mathbf{U}$ and $\mathbf{V}$, if $\mathbf{U}\cdot\mathbf{V} = 0$, then $\mathbf{V}\times(\mathbf{U}\times\mathbf{V}) = \mathbf{U}$.

5. ____ $\mathbf{V}\times(\mathbf{U}\times\mathbf{W}) = (\mathbf{V}\times\mathbf{U})\times\mathbf{W}$.

14.3 Equations of Lines and Planes in Space

Purpose

- To develop the parametric and symmetric equations of a line in space.
- To develop the equation of a plane in space.
- To demonstrate a procedure for finding the distance between a point and a plane.

The simplest surface is the plane. The simplest curves are the plane curves, and of these the simplest is the straight line. —*David Hilbert (1862–1943)*

In 2-space to determine the equation of a line, we find it convenient to use the *slope* of the line. However, in 3-space we will see that it is more convenient to use *vectors* to determine the equation of a line.

We begin by considering the line L through the distinct points (x_1, y_1, z_1) and (x_2, y_2, z_2). Then the vector

$$\mathbf{V} = (x_2 - x_1)\mathbf{i} + (y_2 - y_1)\mathbf{j} + (z_2 - z_1)\mathbf{k}$$

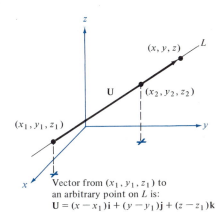

Vector from (x_1, y_1, z_1)
to (x_2, y_2, z_2) is:
$V = (x_2 - x_1)\mathbf{i} + (y_2 - y_1)\mathbf{j} + (z_2 - z_1)\mathbf{k}$

Vector from (x_1, y_1, z_1) to
an arbitrary point on L is:
$U = (x - x_1)\mathbf{i} + (y - y_1)\mathbf{j} + (z - z_1)\mathbf{k}$

FIGURE 14.17

from (x_1, y_1, z_1) to (x_2, y_2, z_2) lies on L. (See Figure 14.17.) If (x, y, z) is an arbitrary point on the line L, then the vector

$$\mathbf{U} = (x - x_1)\mathbf{i} + (y - y_1)\mathbf{j} + (z - z_1)\mathbf{k}$$

also lies on L. Since $\mathbf{U}$ and $\mathbf{V}$ both lie on L, they must be scalar multiples of each other, and we can write

$$\mathbf{U} = t\mathbf{V}$$

or

$$(x - x_1)\mathbf{i} + (y - y_1)\mathbf{j} + (z - z_1)\mathbf{k}$$
$$= t(x_2 - x_1)\mathbf{i} + t(y_2 - y_1)\mathbf{j} + t(z_2 - z_1)\mathbf{k}$$

Now if two vectors are equal, then their respective components must be equal, and it follows that

$$x - x_1 = t(x_2 - x_1)$$
$$y - y_1 = t(y_2 - y_1)$$
$$z - z_1 = t(z_2 - z_1)$$

This leads to the following description of parametric equations for a line in space:

Parametric Equations of a Line in Space	A set of **parametric equations** for the line L passing through (x_1, y_1, z_1) and (x_2, y_2, z_2) is given by $x = x_1 + (x_2 - x_1)t, \qquad y = y_1 + (y_2 - y_1)t, \qquad z = z_1 + (z_2 - z_1)t$

The vector

$$\mathbf{V} = (x_2 - x_1)\mathbf{i} + (y_2 - y_1)\mathbf{j} + (z_2 - z_1)\mathbf{k} = a\mathbf{i} + b\mathbf{j} + c\mathbf{k}$$

(see Figure 14.17) is called a **direction vector** for line L and its components are called **direction numbers** for L. Consequently, the parametric equations for a line L parallel to $\mathbf{V} = a\mathbf{i} + b\mathbf{j} + c\mathbf{k}$ and containing the point (x_1, y_1, z_1) have the form

$$x = x_1 + at, \qquad y = y_1 + bt, \qquad z = z_1 + ct$$

Furthermore, if the direction numbers a, b, and c are nonzero, we can eliminate the parameter t to obtain the equations

$$\frac{x - x_1}{a} = \frac{y - y_1}{b} = \frac{z - z_1}{c}$$

which are called **symmetric equations** for line L.

Example 1

Find a set of parametric equations and a set of symmetric equations for the line passing through $(1, -2, 4)$ and $(3, 2, 0)$.

Solution: Using the form

$$x = x_1 + (x_2 - x_1)t, \qquad y = y_1 + (y_2 - y_1)t, \qquad z = z_1 + (z_2 - z_1)t$$

and letting

$$(x_1, y_1, z_1) = (1, -2, 4) \qquad \text{and} \qquad (x_2, y_2, z_2) = (3, 2, 0)$$

we have the parametric equations

$$x = 1 + (3 - 1)t, \qquad y = -2 + [2 - (-2)]t, \qquad z = 4 + (0 - 4)t$$
$$x = 1 + 2t, \qquad\qquad y = -2 + 4t, \qquad\qquad z = 4 - 4t$$

Solving for t in each equation gives us

$$t = \frac{x - 1}{2} = \frac{y + 2}{4} = \frac{z - 4}{-4}$$

Consequently, we have the symmetric form

$$\frac{x - 1}{2} = \frac{y + 2}{4} = \frac{z - 4}{-4}$$

(See Figure 14.18.)   ∎

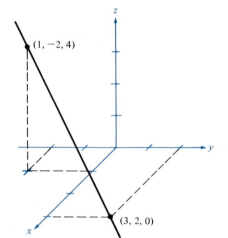

Equation for a line in space:
 parametric form,
  $x = 1 + 2t$, $y = -2 + 4t$, $z = 4 - 4t$
 symmetric form,
  $\dfrac{x - 1}{2} = \dfrac{y + 2}{4} = \dfrac{z - 4}{-4}$

FIGURE 14.18

Note in Example 1 that had we let $(x_1, y_1, z_1) = (3, 2, 0)$ and $(x_2, y_2, z_2) = (1, -2, 4)$, we would have obtained the parametric equations

$$x = 3 + (1 - 3)s, \qquad y = 2 + (-2 - 2)s, \qquad z = 0 + (4 - 0)s$$
$$x = 3 - 2s, \qquad\qquad y = 2 - 4s, \qquad\qquad z = 4s$$

The corresponding symmetric equations are

$$\frac{x-3}{-2} = \frac{y-2}{-4} = \frac{z}{4}$$

This shows that neither the parametric equations nor the symmetric equations for a line are unique. However, there is a similarity between the two sets of direction numbers 2, 4, −4 and −2, −4, 4. In general, it can be shown that any two sets of direction numbers for a line must be multiples of each other (see Exercise 46). For example, 1, 2, −2 is another set of direction numbers for the line in Example 1.

Example 2

Show that the two lines

$$L_1: x = 4 + 2t, \qquad y = 3 - 2t, \qquad z = -7 - 4t$$

and

$$L_2: x = s \qquad \qquad y = 3 + 3s, \qquad z = 1 - 2s$$

intersect, and find their angle of intersection.

Solution: To find possible points of intersection, we equate the respective coordinates to obtain the three equations

$$4 + 2t = s, \qquad 3 - 2t = 3 + 3s, \qquad -7 - 4t = 1 - 2s$$

or, equivalently,

$$2t - s = -4$$
$$-2t - 3s = 0$$
$$-4t + 2s = 8$$

Since the third equation is a multiple of the first, we can solve the first two simultaneously and obtain $-4s = -4$, or $s = 1$, which then yields $t = -\frac{3}{2}$. Substituting $s = 1$ into the equation for L_2 or $t = -\frac{3}{2}$ into the equation for L_1, both yield $(1, 6, -1)$ as the point of intersection.

The angle θ between the two lines is either equal to or supplementary to the angle between the direction vectors

$$\mathbf{V}_1 = 2\mathbf{i} - 2\mathbf{j} - 4\mathbf{k} \qquad \text{and} \qquad \mathbf{V}_2 = \mathbf{i} + 3\mathbf{j} - 2\mathbf{k}$$

Thus

$$\cos\theta = \frac{|\mathbf{V}_1 \cdot \mathbf{V}_2|}{|\mathbf{V}_1| \, |\mathbf{V}_2|} = \frac{|2 - 6 + 8|}{\sqrt{24}\sqrt{14}} = \frac{4}{4\sqrt{21}} = \frac{1}{\sqrt{21}}$$

and

$$\theta \approx \text{Arccos}\,\frac{1}{\sqrt{21}} \approx 77.4°$$

∎

We have seen that the set of parametric equations for a line in space can be obtained from a point on the line and a vector *parallel* to it. We will now show that an equation for a plane in space can be obtained from a point on the plane and a vector *normal* (orthogonal) to it.

To derive the equation for a plane, we assume that (x_1, y_1, z_1) is a point in the plane and

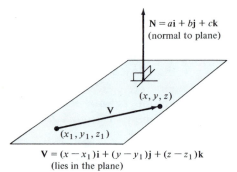

$\mathbf{N} = a\mathbf{i} + b\mathbf{j} + c\mathbf{k}$
(normal to plane)

(x, y, z)

$\mathbf{V}$

(x_1, y_1, z_1)

$\mathbf{V} = (x - x_1)\mathbf{i} + (y - y_1)\mathbf{j} + (z - z_1)\mathbf{k}$
(lies in the plane)

FIGURE 14.19

$$\mathbf{N} = a\mathbf{i} + b\mathbf{j} + c\mathbf{k}$$

is normal to the plane. (That is, $\mathbf{N}$ is orthogonal to every vector in the given plane. See Figure 14.19.) If (x, y, z) represents an arbitrary point in the plane, then the vector

$$\mathbf{V} = (x - x_1)\mathbf{i} + (y - y_1)\mathbf{j} + (z - z_1)\mathbf{k}$$

lies in the plane. Now by definition $\mathbf{N}$ is orthogonal to $\mathbf{V}$ and it follows that

$$\mathbf{N} \cdot \mathbf{V} = a(x - x_1) + b(y - y_1) + c(z - z_1) = 0$$

From this result we obtain the following *standard equation of a plane*:

Standard Equation of a Plane in Space

If the nonzero vector

$$\mathbf{N} = a\mathbf{i} + b\mathbf{j} + c\mathbf{k}$$

is normal to the plane containing the point (x_1, y_1, z_1), then the **standard equation of the plane** is

$$a(x - x_1) + b(y - y_1) + c(z - z_1) = 0$$

By regrouping terms, we obtain the **general equation** of a plane in space,

$$ax + by + cz + d = 0$$

where a, b, and c are the direction numbers for a line normal to the plane. Note the similarity of this equation to the general equation of a *line* in the *plane*. Moreover, keep in mind the similarities:

1. To write the equation of a line in the plane, we need a point on the line and the slope of the line.
2. To write the equation of a plane in 3-space, we need a point in the plane and a vector normal to the plane.

Example 3

Find the equation of the plane containing the points $(2, 1, 1)$, $(0, 4, 1)$, and $(-2, 1, 4)$.

Solution: To use the form

$$a(x - x_1) + b(y - y_1) + c(z - z_1) = 0$$

we need to know a point in the plane and a vector $\mathbf{N}$ that is normal to the plane. We already have three choices for the point. To obtain a normal vector, we use the cross product of the vectors $\mathbf{V}_1$ and $\mathbf{V}_2$ from the point $(2, 1, 1)$ to $(0, 4, 1)$ and to $(-2, 1, 4)$, respectively. We have

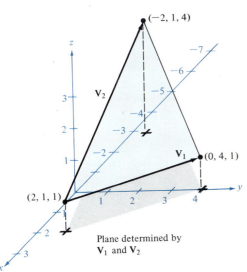

$$V_1 = (0 - 2)i + (4 - 1)j + (1 - 1)k = -2i + 3j$$
$$V_2 = (-2 - 2)i + (1 - 1)j + (4 - 1)k = -4i + 3k$$

as shown in Figure 14.20. Thus the vector

$$N = V_1 \times V_2 = \begin{vmatrix} i & j & k \\ -2 & 3 & 0 \\ -4 & 0 & 3 \end{vmatrix} = 9i + 6j + 12k$$

is normal to the given plane. Using the direction numbers for **N** and the point $(2, 1, 1)$, we determine the equation of the plane to be

$$9(x - 2) + 6(y - 1) + 12(z - 1) = 0$$
$$9x + 6y + 12z - 36 = 0$$
$$3x + 2y + 4z = 12$$

Plane determined by
V_1 and V_2

FIGURE 14.20

If two planes intersect, we can determine the angle between them from the angle between their normal lines. Specifically, if vectors N_1 and N_2 are normal to two intersecting planes, then the angle θ between the two planes is determined from the equation

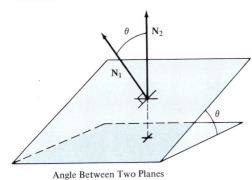

$$\cos \theta = \frac{|N_1 \cdot N_2|}{|N_1| \, |N_2|}$$

(See Figure 14.21.) Consequently, two planes with normal vectors N_1 and N_2 are:

1. *orthogonal* if $N_1 \cdot N_2 = 0$
2. *parallel* if N_1 is a scalar multiple of N_2

Angle Between Two Planes

FIGURE 14.21

Example 4

Determine the angle between the two planes $2x - 4y - 5z = 2$ and $x + 2y + z = 1$.

Solution: Using the coefficients in the equations of the given planes, we have the normal vectors

$$N_1 = 2i - 4j - 5k \quad \text{and} \quad N_2 = i + 2j + k$$

Thus
$$\cos \theta = \frac{|N_1 \cdot N_2|}{|N_1| \, |N_2|} = \frac{|2 - 8 - 5|}{\sqrt{4 + 16 + 25} \, \sqrt{1 + 4 + 1}}$$

$$= \frac{|-11|}{\sqrt{45} \, \sqrt{6}} = \frac{11}{3\sqrt{30}} \approx 0.6694$$

and $\theta \approx 47.98°$ is the angle between the two planes. (See Figure 14.22.)

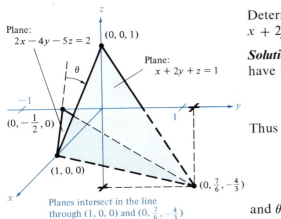

Plane:
$2x - 4y - 5z = 2$

Plane:
$x + 2y + z = 1$

Planes intersect in the line
through $(1, 0, 0)$ and $(0, \frac{7}{6}, -\frac{4}{3})$

FIGURE 14.22

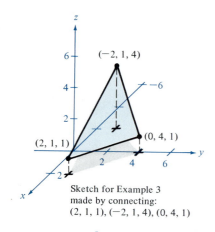

Sketch for Example 3
made by connecting:
$(2, 1, 1), (-2, 1, 4), (0, 4, 1)$

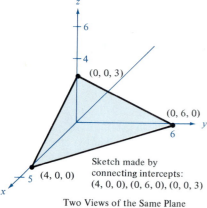

Sketch made by
connecting intercepts:
$(4, 0, 0), (0, 6, 0), (0, 0, 3)$

Two Views of the Same Plane

FIGURE 14.23

One way to sketch the plane having an equation of the form

$$ax + by + cz + d = 0$$

is to determine three points that satisfy the equation and then connect these points to form a triangular region in the plane. Three points that can be readily found are those where the plane intersects the coordinate axes. For instance, if we let $y = z = 0$ in the equation

$$3x + 2y + 4z = 12$$

we obtain $3x = 12$, or $x = 4$, and the point $(4, 0, 0)$ is called the *x-intercept*. Similarly, the *y-intercept* is $(0, 6, 0)$ and the *z-intercept* is $(0, 0, 3)$. This plane is sketched in Figure 14.23. Note that the sketch made by connecting the three intercepts gives a better picture of the plane than that obtained in Figure 14.20 by connecting three arbitrary points in the same plane.

Of course, some planes have less than three intercepts. In particular, this occurs when one or more of the coefficients in

$$ax + by + cz + d = 0$$

is zero. For instance, the plane

$$cz + d = 0$$

is parallel to the *xy*-plane and thus has no *x*- or *y*-intercepts. The plane

$$ax + by + d = 0$$

is parallel to the *z*-axis and it has no *z*-intercept, while the plane

$$ax + by + cz = 0$$

intersects all three axes at the origin. Figures 14.24 and 14.25 identify additional examples of this sort.

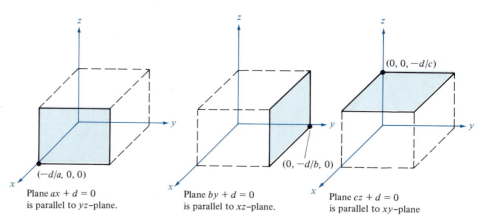

Plane $ax + d = 0$
is parallel to *yz*-plane.

Plane $by + d = 0$
is parallel to *xz*-plane.

Plane $cz + d = 0$
is parallel to *xy*-plane

Planes Parallel to Coordinate Planes

FIGURE 14.24

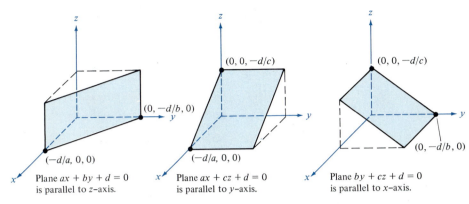

Planes Parallel to Coordinate Axes

FIGURE 14.25

Example 5

Find an equation for the plane containing the two lines in Example 2,

$$L_1: x = 4 + 2t, \qquad y = 3 - 2t, \qquad z = -7 - 4t$$

and

$$L_2: x = s, \qquad y = 3 + 3s, \qquad z = 1 - 2s$$

Solution: Two lines do not determine a plane unless they are parallel or intersecting lines. From Example 2 we know that these two lines do intersect at the point $(1, 6, -1)$. Using the direction numbers for lines L_1 and L_2, we obtain their respective direction vectors

$$\mathbf{V}_1 = 2\mathbf{i} - 2\mathbf{j} - 4\mathbf{k}$$

and

$$\mathbf{V}_2 = \mathbf{i} + 3\mathbf{j} - 2\mathbf{k}$$

We then determine the vector

$$\mathbf{N} = \mathbf{V}_1 \times \mathbf{V}_2 = \begin{vmatrix} \mathbf{i} & \mathbf{j} & \mathbf{k} \\ 2 & -2 & -4 \\ 1 & 3 & -2 \end{vmatrix} = 16\mathbf{i} + 8\mathbf{k}$$

which is normal to the plane containing $\mathbf{V}_1$ and $\mathbf{V}_2$. Finally, using the point $(4, 3, -7)$ and the direction numbers for $\mathbf{N}$, we determine the equation of the plane to be

$$16(x - 4) + 8(z + 7) = 0$$

or, in general form,

$$2x + z - 1 = 0 \qquad \blacksquare$$

Note that if the lines in Example 5 were parallel, we could not obtain a normal vector from the cross product of the direction vectors $\mathbf{V}_1$ and $\mathbf{V}_2$, because the cross product of parallel vectors is zero. In this case it is simplest to find two points on one line and one point on the other line and proceed as in Example 3.

Example 6

Find an equation of the plane S, which contains the line

$$\frac{x-5}{3} = \frac{y+1}{-2} = \frac{z-3}{6}$$

and is perpendicular to the plane $y - 3z = 6$.

Solution: The vector

$$V = 3i - 2j + 6k$$

is a direction vector for the given line, and

$$N_1 = j - 3k$$

is normal to the plane $y - 3z = 6$. N_1 in turn is parallel to plane S. Now because the vector

$$N = V \times N_1 = \begin{vmatrix} i & j & k \\ 3 & -2 & 6 \\ 0 & 1 & -3 \end{vmatrix} = 0i - (-9)j + 3k = 9j + 3k$$

is normal to both V and N_1, it is also normal to plane S. Therefore, using the direction numbers for N and the point $(5, -1, 3)$, we obtain

$$0(x-5) + 9(y+1) + 3(z-3) = 0$$
$$9y + 3z = 0$$

or $$3y + z = 0$$

as the equation of plane S.   ∎

Example 7   (*Distance Between Point and Plane*)

Find the distance p between the point $(-3, 2, 5)$ and the plane $x - 2y + 3z = 6$.

Solution: First, we choose a convenient point in the plane, say the z-intercept $(0, 0, 2)$. Then we let V be the vector from $(0, 0, 2)$ to the given point $(-3, 2, 5)$; that is

$$V = (-3-0)i + (2-0)j + (5-2)k = -3i + 2j + 3k$$

Now the vector $$N = i - 2j + 3k$$

is normal to the plane, and from Figure 14.26 we can see that the length of the projection of V onto N will give us the desired distance p. That is,

$$p = \frac{|N \cdot V|}{|N|} = \frac{|-3 - 4 + 9|}{\sqrt{1 + 4 + 9}} = \frac{2}{\sqrt{14}}$$

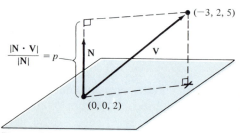

Distance Between a Point and a Plane
FIGURE 14.26

We can generalize the procedure used in Example 7 to derive a formula for the distance between a point (x_1, y_1, z_1) and a plane $ax + by + cz + d = 0$. If (x_2, y_2, z_2) lies in the plane and $\mathbf{N} = a\mathbf{i} + b\mathbf{j} + c\mathbf{k}$ is a normal vector, then

$$\text{distance} = \frac{|a(x_1 - x_2) + b(y_1 - y_2) + c(z_1 - z_2)|}{\sqrt{a^2 + b^2 + c^2}}$$

Now since (x_2, y_2, z_2) is in the plane, we have $ax_2 + by_2 + cz_2 = -d$, so that the previous equation can be written as

$$\text{distance} = \frac{|ax_1 + by_1 + cz_1 + d|}{\sqrt{a^2 + b^2 + c^2}}$$

which is the general formula for the **distance between a point and a plane.**

Section Exercises (14.3)

In Exercises 1–10, find a set of (a) parametric equations and (b) symmetric equations of the specified line. Express the direction numbers for each line in terms of integers.

1. the line through the points $(1, 3, 2)$ and $(0, 0, 0)$
2. the line through the points $(0, 0, 0)$ and $(-2, \frac{5}{2}, 1)$
3. the line through the points $(-2, 0, 3)$ and $(0, 4, 1)$
4. the line through the points $(-2, 0, 3)$ and $(4, 3, 3)$
5. the line through the points $(5, -3, -2)$ and $(-\frac{2}{3}, \frac{2}{3}, 1)$
6. the line through the points $(1, 0, 1)$ and $(1, 3, -2)$
7. the line through the point $(1, 0, 1)$ and parallel to the vector $\mathbf{V} = 3\mathbf{i} - 2\mathbf{j} + \mathbf{k}$
8. the line through the point $(-3, 5, 4)$ and parallel to the line $\dfrac{x - 1}{3} = \dfrac{y + 1}{-2} = \dfrac{z - 3}{1}$
9. the line through the point $(2, 3, 4)$ and parallel to the xz-coordinate plane and to the yz-coordinate plane.
10. the line through the point $(2, 3, 4)$ and perpendicular to the plane $3x + 2y - z = 6$

11. A line L passes through the point $(-2, 3, 1)$ and is parallel to the vector $\mathbf{V} = 4\mathbf{i} - \mathbf{k}$. Determine which of the following points lie on the line:
    (a) $(2, 3, 0)$ (b) $(-6, 3, 2)$
    (c) $(2, 1, 0)$ (d) $(10, 3, -2)$
    (e) $(6, 3, -2)$
12. A line L passes through the points $(2, 0, -3)$ and $(4, 2, -2)$. Determine which of the following points are on L:
    (a) $(4, 1, -2)$ (b) $(3, 1, -\frac{5}{2})$

(c) $(\frac{5}{2}, \frac{1}{2}, -\frac{11}{4})$ (d) $(-1, -3, -4)$
(e) $(0, -2, -4)$

13. (a) Find the point of intersection of the lines

$$x = 4t + 2 \qquad\qquad x = 2s + 2$$
$$y = 3 \qquad \text{and} \qquad y = 2s + 3$$
$$z = -t + 1 \qquad\qquad z = s + 1$$

(b) Find the cosine of the angle of intersection.
14. Determine if these lines intersect:

$$\frac{x}{3} = \frac{y - 2}{-1} = \frac{z + 1}{1}$$

and

$$\frac{x - 1}{4} = \frac{y + 2}{1} = \frac{z + 3}{-3}$$

In Exercises 15–28, find an equation of the specified plane.
15. the plane through the point $(3, 2, 2)$ with normal vector $\mathbf{V} = 2\mathbf{i} + 3\mathbf{j} - \mathbf{k}$
16. the plane through the point $(3, 2, 2)$ and perpendicular to the line

$$\frac{x - 1}{4} = \frac{y + 2}{1} = \frac{z + 3}{-3}$$

17. the plane through the points $(0, 0, 0)$, $(1, 2, 3)$, and $(-2, 3, 3)$
18. the plane through the points $(1, 2, -3)$, $(2, 3, 1)$, and $(0, -2, -1)$

19. the plane through the points $(1, 2, 3)$, $(3, 2, 1)$, and $(-1, -2, 2)$

20. the plane through the point $(1, 2, 3)$ and parallel to the yz-coordinate plane

21. the plane through the point $(1, 2, 3)$ and parallel to the xy-coordinate plane

22. the plane containing the y-axis and making an angle of $\pi/6$ radians with the positive x-axis

[S] 23. the plane determined by the two intersecting lines

$$\frac{x-1}{-2} = \frac{y-4}{1} = \frac{z}{1} \quad \text{and} \quad \frac{x-2}{-3} = \frac{y-1}{4} = \frac{z-2}{-1}$$

24. the plane determined by the line $\dfrac{x}{2} = \dfrac{y-4}{-1} = \dfrac{z}{1}$ and the point $(2, 2, 1)$

[S] 25. the plane through the points $(2, 2, 1)$ and $(-1, 1, -1)$ and perpendicular to the plane $2x - 3y + z = 3$

26. the plane through the points $(3, 2, 1)$ and $(3, 1, -5)$ and perpendicular to the plane $6x + 7y + 2z = 10$

27. the plane parallel to the x-axis and passing through the points $(1, -2, -1)$ and $(2, 5, 6)$

28. the plane parallel to the z-axis and passing through the points $(4, 2, 1)$ and $(-3, 5, 7)$

[S] 29. Find a set of parametric equations for the line of intersection of the planes $3x + 2y - z = 7$ and $x - 4y + 2z = 0$.

30. Find a set of parametric equations for the line of intersection of the planes $x - 3y + 6z = 4$ and $5x + y - z = 4$.

31. Find the point of intersection (if any) of the line

$$\frac{x-1}{3} = \frac{y+1}{-2} = \frac{z-3}{1}$$

and the plane $2x + 3y = 10$.

32. Find the point of intersection (if any) of the line

$$\frac{x-1}{4} = \frac{y}{2} = \frac{z-3}{6}$$

and the plane $2x + 3y = -5$.

[S] 33. Find the point of intersection (if any) of the line

$$\frac{x-\frac{1}{2}}{1} = \frac{y+\frac{3}{2}}{-1} = \frac{z+1}{2}$$

and the plane $2x - 2y + z = 12$.

34. Find the angle between the planes $3x + 2y - z = 7$ and $x - 4y + 2z = 0$.

35. Find the angle between the planes $x - 3y + 6z = 4$ and $5x + y - z = 4$.

In Exercises 36–42, mark the intercepts and sketch the plane.

36. $3x + 6y + 2z = 6$ 37. $4x + 2y + 6z = 12$

38. $2x - y + z = 4$ 39. $2x - y + 3z = 4$

40. $x + 2y = 4$ 41. $y + z = 5$

42. $x - 3z = 3$

43. Find the distance from the point $(1, 2, 3)$ to the plane $2x - y + z = 4$.

44. Find the distance from the origin to the plane $2x + 3y + z = 12$.

[S] 45. Find the distance between the two lines

$$\frac{x}{1} = \frac{y}{2} = \frac{z}{3} \quad \text{and} \quad \frac{x-1}{-1} = \frac{y-4}{1} = \frac{z+1}{1}$$

46. If a_1, b_1, c_1 and a_2, b_2, c_2 are two sets of direction numbers for the same line, show that there exists a scalar d such that

$$a_1 = a_2 d, \qquad b_1 = b_2 d, \qquad c_1 = c_2 d$$

For Review and Class Discussion

True or False

1. _____ The line given by $x = a_1 + a_2 t$, $y = b_1 + b_2 t$, $z = c_1 + c_2 t$ is parallel to the vector $\mathbf{V} = a_2\mathbf{i} + b_2\mathbf{j} + c_2\mathbf{k}$.

2. _____ Every plane in space can be represented by an equation of the form $ax + by + cz + d = 0$.

3. _____ If $\mathbf{V}_1$ and $\mathbf{V}_2$ lie in plane P and $\mathbf{V}_1 \times \mathbf{V}_2 = a\mathbf{i} + b\mathbf{j} + c\mathbf{k}$, then P is represented by the equation $ax + by + cz = 0$.

4. _____ If $\mathbf{V} = a_1\mathbf{i} + b_1\mathbf{j} + c_1\mathbf{k}$ is any vector in the plane given by $a_2 x + b_2 y + c_2 z + d_2 = 0$, then $a_1 a_2 + b_1 b_2 + c_1 c_2 = 0$.

5. _____ Every pair of lines in space are either intersecting or parallel.

14.4 Surfaces in Space

Purpose
- To identify the equations of cylindrical surfaces and surfaces of revolution.
- To identify the basic quadric surfaces: paraboloids, ellipsoids, and hyperboloids.

By generalization of the circular cylinder we obtain the elliptic cylinder; it is generated by a straight line moved along an ellipse in such a way as to be always perpendicular to the plane of the curve. We get the parabolic and the hyperbolic cylinder from the parabola and the hyperbola by the same procedure. —David Hilbert (1862–1943)

So far we have studied three types of graphs in space:

Sphere: $(x - h)^2 + (y - k)^2 + (z - l)^2 = r^2$ (Section 14.1)
Plane: $ax + by + cz + d = 0$ (Section 14.3)
Line: $x = x_1 + at, y = y_1 + bt, z = z_1 + ct$ (Section 14.3)

Notice that both a sphere and a plane can be represented by a single equation involving x, y, and z, whereas a line requires more than one equation for its representation. In general, we call the graph of an equation involving x, y, and z a **surface in space,** and we call the graph of the set of parametric equations

$$x = f(t), \qquad y = g(t), \qquad z = h(t)$$

a **curve in space,** or a **space curve.** Since the representation of surfaces in space is simpler than that of curves in space, we will first look at some general surfaces and postpone our study of space curves until the next section.

One of the simpler types of surface in space is called a **cylindrical surface,** or, more simply, a **cylinder.**

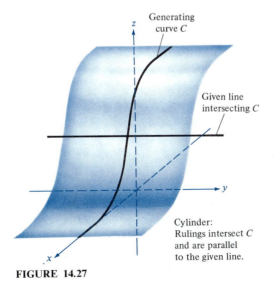

Generating curve C

Given line intersecting C

Cylinder: Rulings intersect C and are parallel to the given line.

FIGURE 14.27

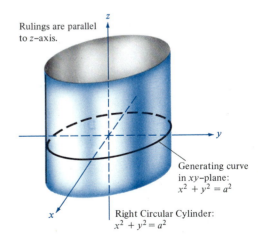

Rulings are parallel to z–axis.

Generating curve in xy–plane: $x^2 + y^2 = a^2$

Right Circular Cylinder: $x^2 + y^2 = a^2$

FIGURE 14.28

To generate a cylinder we begin with a plane curve and a line intersecting this curve; then the cylinder consists of all lines that intersect the curve and are parallel to the given line (Figure 14.27).

| **Definition of Cylinder** | A **cylinder** is a surface consisting of all straight lines parallel to a given line and intersecting a given curve C. |

The given curve is called the **generating curve** of the cylinder and the parallel lines are called **rulings.** For our purposes we will assume that curve C lies in one of the three coordinate planes and that the rulings are perpendicular to that plane. For example, the generating curve of the right circular cylinder shown in Figure 14.28 is $x^2 + y^2 = a^2$, a circle in the xy-plane. The rulings are perpendicular to the xy-plane and are therefore parallel to the z-axis.

Notice in Figure 14.28 that we can generate any *one* of the rulings by fixing the values of x and y and then allowing z to take on all real values. In this sense the value of z is arbitrary and it is not necessary to include z in the equation for this cylinder. Therefore, the equation of the cylinder in Figure 14.28 is simply $x^2 + y^2 = a^2$, the equation of its generating curve.

| **Equations of Cylinders** | An equation in only two of the three variables x, y, and z represents, in space, a cylinder whose rulings are parallel to the axis of the missing variable. |

Example 1

Sketch the surface represented by the equation $z = y^2$.

Solution: The graph is a cylinder whose generating curve, $z = y^2$, is a parabola in the yz-plane. The rulings of the cylinder are parallel to the x-axis (Figure 14.29).

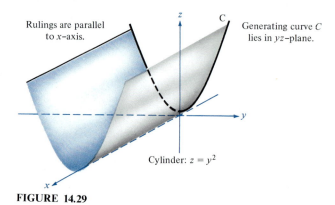

Rulings are parallel to x–axis.

Generating curve C lies in yz–plane.

Cylinder: $z = y^2$

FIGURE 14.29

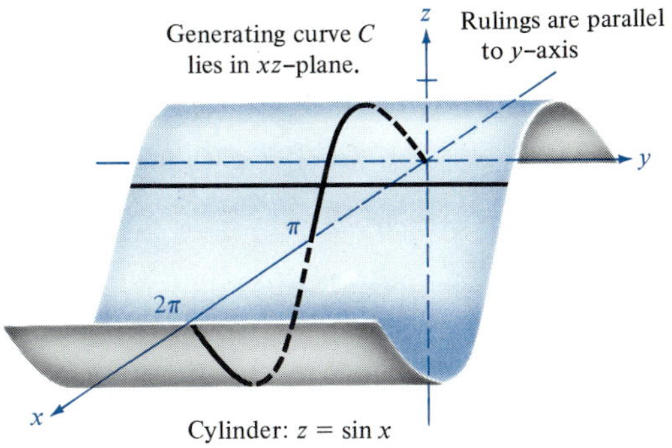

Generating curve C
lies in xz-plane.

Rulings are parallel
to y-axis

π

2π

Cylinder: $z = \sin x$

FIGURE 14.30

Example 2

Sketch the surface represented by the equation $z = \sin x$ $(0 \leq x \leq 2\pi)$.

Solution: The surface is a cylinder with rulings parallel to the y-axis. The generating curve is a sine curve along the x-axis. (See Figure 14.30.) ∎

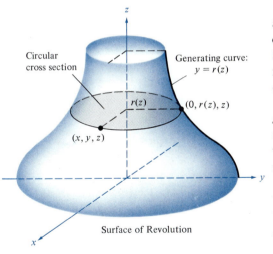

Circular
cross section

Generating curve:
$y = r(z)$

$r(z)$

$(0, r(z), z)$

(x, y, z)

Surface of Revolution

FIGURE 14.31

Another common type of surface is the surface generated by revolving a plane curve about a line in the plane. Such a surface is called a **surface of revolution** and the specified line is called the **axis of revolution.** In Section 7.9 we studied techniques for calculating the area of a surface of revolution, and now we will show how to represent such surfaces by equations involving x, y, and z.

For a surface of revolution, it is important to note that each *cross section* (the intersection of the surface with a plane perpendicular to the axis of revolution) is a circle. Figure 14.31 shows a surface of revolution whose generating curve C lies in the yz-plane with the z-axis as the axis of revolution. Two circular cross sections are also shown.

To determine an equation for a surface of revolution, let us assume that we have formed the surface by revolving the curve $y = r(z)$ about the z-axis, as in Figure 14.31. Now since $r(z)$ is the distance between the z-axis and an arbitrary point (x, y, z) on the surface, the variables x, y, and z must be related by the equation

$$\sqrt{(x - 0)^2 + (y - 0)^2 + (z - z)^2} = r(z)$$
$$\sqrt{x^2 + y^2} = r(z)$$

By squaring both sides we obtain

$$x^2 + y^2 = [r(z)]^2$$

as the equation for a surface of revolution having the z-axis as its axis of revolution. Informally, the equation of a surface of revolution is simply the equation for an arbitrary circular cross section whose radius is expressed as a function of the variable corresponding to the axis of revolution.

Equations of Surfaces of Revolution

If r is the generating curve (radius function) of a surface of revolution and if the axis of revolution is:

1. the *x-axis*, then the equation of the surface is

$$y^2 + z^2 = [r(x)]^2$$

2. the *y-axis*, then the equation of the surface is

$$x^2 + z^2 = [r(y)]^2$$

3. the *z-axis*, then the equation of the surface is

$$x^2 + y^2 = [r(z)]^2$$

In each case r is determined from the equation of the generating curve. For example, the equation of the surface of revolution formed by revolving the graph of $y = 1/z$ about the *z*-axis is

$$x^2 + y^2 = [r(z)]^2 = \left(\frac{1}{z}\right)^2$$

Similarly, the equation of the surface formed by revolving the graph of $x = y^3$ about the *y*-axis is

$$x^2 + z^2 = [r(y)]^2 = (y^3)^2 = y^6$$

Example 3

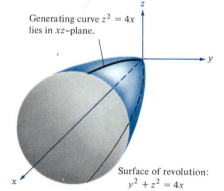

Generating curve $z^2 = 4x$ lies in *xz*–plane.

Surface of revolution:
$y^2 + z^2 = 4x$

FIGURE 14.32

Find an equation for the surface of revolution generated by revolving the graph of $z^2 = 4x$ about the *x*-axis. Sketch the surface.

Solution: Since the axis of revolution is the *x*-axis, the surface has an equation of the form

$$y^2 + z^2 = [r(x)]^2$$

where $r(x) = z = \pm\sqrt{4x}$. Therefore, the equation of the surface is

$$y^2 + z^2 = 4x$$

and its graph is shown in Figure 14.32. ■

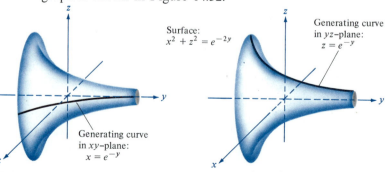

Surface:
$x^2 + z^2 = e^{-2y}$

Generating curve in *xy*–plane:
$x = e^{-y}$

Generating curve in *yz*–plane:
$z = e^{-y}$

FIGURE 14.33

Note that the generating curve for a surface of revolution is not unique. For instance, the surface $x^2 + z^2 = e^{-2y}$ can be formed by revolving either the plane curve $x = e^{-y}$ about the y-axis or the plane curve $z = e^{-y}$ about the y-axis. (See Figure 14.33.)

Example 4

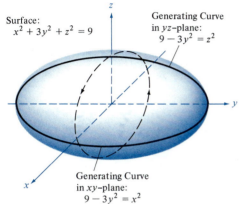

Surface:
$x^2 + 3y^2 + z^2 = 9$

Generating Curve
in yz-plane:
$9 - 3y^2 = z^2$

Generating Curve
in xy-plane:
$9 - 3y^2 = x^2$

FIGURE 14.34

Find a generating curve and the axis for the surface of revolution represented by the equation $x^2 + 3y^2 + z^2 = 9$.

Solution: We assume that the equation has one of the following forms:

$$x^2 + y^2 = [r(z)]^2, \qquad y^2 + z^2 = [r(x)]^2, \qquad x^2 + z^2 = [r(y)]^2$$

Since the coefficients of x^2 and z^2 are equal, we choose the third form and write

$$x^2 + z^2 = 9 - 3y^2$$

and conclude that the y-axis is the axis of revolution. Finally, we can choose the generating curve to be either

$$9 - 3y^2 = [r(y)]^2 = x^2$$

in the xy-plane, or

$$9 - 3y^2 = [r(y)]^2 = z^2$$

in the yz-plane. (See Figure 14.34.) ∎

A third common type of surface in space is one whose equation is of the form

$$Ax^2 + By^2 + Cz^2 + Dx + Ey + Fz + G = 0$$

We call the graph of such an equation a **quadric surface.** (Note that the surfaces in Examples 3 and 4 may be classified as quadric surfaces in addition to being surfaces of revolution.) We confine our discussion of quadric surfaces to those having their center at the origin and axes along the coordinate axes.

The six basic quadric surfaces are:

1. the ellipsoid
2. the hyperboloid of one sheet
3. the hyperboloid of two sheets
4. the elliptic paraboloid
5. the hyperbolic paraboloid
6. the cone

The simplest way to sketch the graph of a quadric surface is to find its intercepts and then determine its *trace* in each of the coordinate planes or in planes parallel to the coordinate planes. We identify some traces for each quadric surface in Table 14.1.

TABLE 14.1   Quadric Surfaces

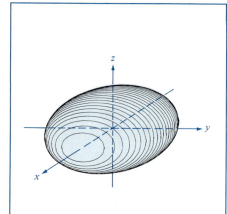

Ellipsoid

$$\frac{x^2}{a^2} + \frac{y^2}{b^2} + \frac{z^2}{c^2} = 1$$

Traces parallel to xy-plane:
Ellipses

Traces parallel to xz-plane:
Ellipses

Traces parallel to yz-plane:
Ellipses

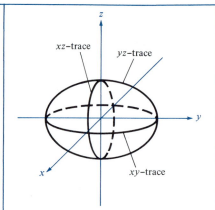

FIGURE 14.35

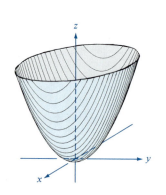

Elliptic Paraboloid

$$\frac{x^2}{a^2} + \frac{y^2}{b^2} = \frac{z}{c}$$

Traces parallel to xy-plane:
Ellipses

Traces parallel to xz-plane:
Parabolas

Traces parallel to yz-plane:
Parabolas

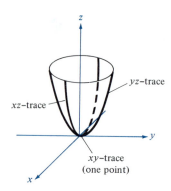

FIGURE 14.36

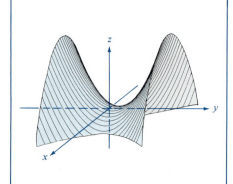

Hyperbolic Paraboloid

$$\frac{x^2}{a^2} - \frac{y^2}{b^2} = \frac{z}{c}$$

Traces parallel to xy-plane:
Hyperbolas

Traces parallel to xz-plane:
Parabolas

Traces parallel to yz-plane:
Parabolas

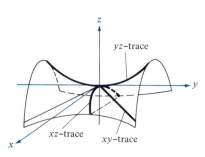

FIGURE 14.37

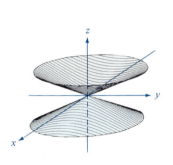

Hyperboloid of One Sheet

$$\frac{x^2}{a^2} + \frac{y^2}{b^2} - \frac{z^2}{c^2} = 1$$

Traces parallel to xy-plane:
Ellipses

Traces parallel to xz-plane:
Hyperbolas

Traces parallel to yz-plane:
Hyperbolas

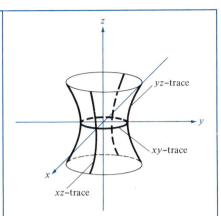

FIGURE 14.38

Elliptic Cone

$$\frac{x^2}{a^2} + \frac{y^2}{b^2} - \frac{z^2}{c^2} = 0$$

Traces parallel to xy-plane:
Ellipses

Traces parallel to xz-plane:
Hyperbolas

Traces parallel to yz-plane:
Hyperbolas

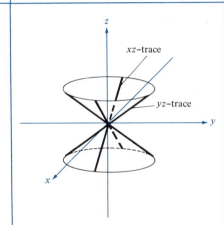

FIGURE 14.39

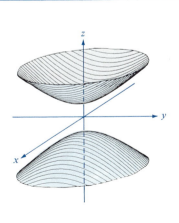

Hyperboloid of Two Sheets

$$\frac{x^2}{a^2} + \frac{y^2}{b^2} - \frac{z^2}{c^2} = -1$$

Traces parallel to xy-plane:
Ellipses

Traces parallel to xz-plane:
Hyperbolas

Traces parallel to yz-plane:
Hyperbolas

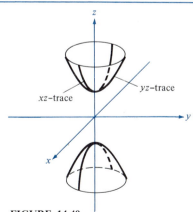

FIGURE 14.40

An **ellipsoid** is shown in Figure 14.35. If $a = c$, then the xz trace is a circle. Similarly, if $a = b$ or if $b = c$, the corresponding trace is a circle. Note further that if $a = b = c$, then the surface is a *sphere*. If any *one* of the traces is an ellipse, the surface is called an ellipsoid.

An **elliptic paraboloid** is shown in Figure 14.36. The axis of the paraboloid is the variable of the *first*-degree term. If $c > 0$ (as is the case in Figure 14.36), the paraboloid opens upward, and if $c < 0$, it opens downward. Its traces in the xz- and yz-planes are both parabolas, while its xy trace is a single point, the origin. If $a = b$, the surface is a **paraboloid of revolution,** generated by rotating a parabola about its axis.

A **hyperbolic paraboloid** is shown in Figure 14.37. The xy trace consists of two intersecting lines. The traces in planes parallel $(z = z_1)$ to the xy-plane are hyperbolas. Both the xz and yz traces are parabolas, one opens upward and the other downward. The hyperbolic paraboloid is sometimes referred to as being *saddle-shaped*.

A **hyperboloid of one sheet** is shown in Figure 14.38. The axis of the hyperboloid corresponds to the negative variable in its equation. The trace perpendicular to the axis (in this case the xy trace) is an ellipse if $a \neq b$. The traces in the other coordinate planes are hyperbolas.

A **cone** is shown in Figure 14.39. The axis of the cone corresponds to the negative variable in its equation. The xy trace is a single point at the origin, whereas both the xy and yz traces are pairs of lines intersecting at the origin. If $a = b$, the surface is a *circular* cone.

A **hyperboloid of two sheets** is shown in Figure 14.40. In this case two terms of the equation are negative and the axis of the hyperboloid corresponds to the positive term. There is no trace in the coordinate plane perpendicular to the axis, while the traces in the other two coordinate planes are both hyperbolas. Note further that the surface consists of two parts, one for $z \geq c$ and the other for $z \leq -c$. The traces in planes parallel to the xy-plane $(|z| > c)$ are ellipses.

In Table 14.1 only one of several orientations of the quadric surface is shown. If the surface is oriented along a different axis, then its equation will change accordingly. Some illustrations of these changes are shown in the following examples.

Example 5

Describe and sketch the surface given by $4x^2 - 3y^2 + 12z^2 + 12 = 0$.

Solution: To express the given equation in standard form, we first divide by -12. Thus we have the equation

$$\frac{x^2}{-3} + \frac{y^2}{4} - z^2 - 1 = 0$$

or
$$\frac{-x^2}{3} + \frac{y^2}{4} - z^2 = 1$$

which represents a hyperboloid of two sheets. The axis of the hyperboloid is the y-axis. The yz trace (when $x = 0$) is the hyperbola

$$\frac{y^2}{4} - \frac{z^2}{1} = 1$$

and the xy trace (when $z = 0$) is the hyperbola

$$\frac{y^2}{4} - \frac{x^2}{3} = 1$$

There is no xz trace. (See Figure 14.41.)

Hyperboloid of Two Sheets:
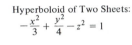

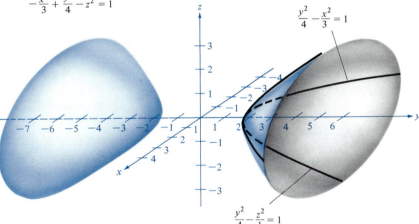

FIGURE 14.41

Example 6

Describe and sketch the surface given by $x - y^2 - 4z^2 = 0$.

Solution: Since one of the terms is not quadratic, the surface will be a paraboloid and in this case its axis is the x-axis. Dividing the equation by -4, we obtain

$$\frac{y^2}{4} + \frac{z^2}{1} = \frac{x}{4}$$

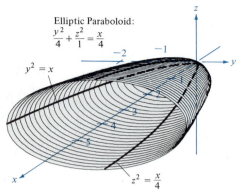

Elliptic Paraboloid:
$$\frac{y^2}{4} + \frac{z^2}{1} = \frac{x}{4}$$

$y^2 = x$

$z^2 = \dfrac{x}{4}$

FIGURE 14.42

The yz trace (when $x = 0$) is the single point at the origin; the xy trace is the parabola

$$\frac{y^2}{4} = \frac{x}{4} \qquad \text{or} \qquad y^2 = x$$

and the xz trace (when $y = 0$) is the parabola

$$z^2 = \frac{x}{4}$$

A trace parallel to the yz-plane, say at $x = 4$, is the ellipse

$$\frac{y^2}{4} + \frac{z^2}{1} = 1$$

Thus we have the elliptic paraboloid shown in Figure 14.42. ∎

Section Exercises (14.4)

In Exercises 1–10, describe and sketch each surface.

1. $z = 3$
2. $x = 4$
3. $y^2 + z^2 = 9$
4. $x^2 + z^2 = 16$
$\boxed{\text{S}}$ **5.** $x^2 - y = 0$
6. $y^2 + z = 4$
7. $4x^2 + y^2 = 4$
8. $z - \sin y = 0$
9. $y^2 - z^2 = 4$
10. $z - e^y = 0$

In Exercises 11–24, identify and sketch the given quadric surface.

11. $\dfrac{x^2}{9} + \dfrac{y^2}{16} + \dfrac{z^2}{9} = 1$
12. $\dfrac{x^2}{9} + \dfrac{y^2}{16} + \dfrac{z^2}{16} = 1$

$\boxed{\text{S}}$ **13.** $4x^2 - y^2 + 4z^2 = 4$
14. $9x^2 + 4y^2 - 8z^2 = 72$
15. $4x^2 - 4y + z^2 = 0$
16. $z = 4x^2 + y^2$

$\boxed{\text{S}}$ **17.** $4x^2 - y^2 + 4z = 0$
18. $z^2 - x^2 - \dfrac{y^2}{4} = 1$

$\boxed{\text{S}}$ **19.** $15x^2 - 4y^2 + 15z^2 = -4$
20. $z^2 = x^2 + \dfrac{y^2}{4}$

21. $y^2 = 4x^2 + 9z^2$
22. $4y = x^2 + z^2$
23. $12z = -3y^2 + 4x^2$
24. $z^2 = 2x^2 + 2y^2$

In Exercises 25–30, find an equation for the surface of revolution generated by revolving the given curve about the specified axis.

25. $z^2 = 4y$ in the yz-plane about the y-axis.
26. $z = 2y$ in the yz-plane about the y-axis
$\boxed{\text{S}}$ **27.** $z = 2y$ in the yz-plane about the z-axis
28. $2z = \sqrt{4 - x^2}$ in the xz-plane about the x-axis
29. $xy = 2$ in the xy-plane about the x-axis
30. $z = \ln y$ in the yz-plane about the z-axis

31. Find an equation of a generating curve if the equation of its surface of revolution is $x^2 + y^2 - 2z = 0$.

$\boxed{\text{S}}$ **32.** Find an equation of a generating curve if the equation of its surface of revolution is $x^2 + z^2 = \sin^2 y$.

33. Find the length of the major and minor axes and the coordinates of the foci of the ellipse generated when the surface $z = (x^2/2) + (y^2/4)$ is intersected by the planes (a) $z = 2$ and (b) $z = 8$.

34. Find the coordinates of the focus of the parabola formed when the surface $z = (x^2/2) + (y^2/4)$ is intersected by the planes (a) $y = 4$ and (b) $x = 2$.

14.5 Space Curves

Purpose
- To represent curves in space by parametric equations and as the intersection of two surfaces.
- To use vector functions to describe motion along a space curve.

In all this discussion I have considered only curves that can be described upon a plane surface, but my remarks can easily be made to apply to all those curves which can be conceived of as generated by the regular movement of the points of a body in three dimensional space.

—René Descartes (1596–1650)

Having discussed ways to represent lines, planes, and surfaces in space, we now want to discuss ways to represent *curves* in space. As is true for curves in the plane (Sections 13.1 and 13.5), curves in space can be represented by parametric equations or by vector functions. Furthermore, space curves can be represented in yet a third way—as the intersection of two surfaces. We consider, first, curves represented by parametric equations.

Definition of a Curve in Space	A **space curve** C is the set of all points $(f(t), g(t), h(t))$ satisfying the parametric equations $$x = f(t), \qquad y = g(t), \qquad z = h(t)$$ where f, g, and h are continuous functions of t on an interval I.

Example 1

Sketch the curve C represented by $x = 2 \sin t$, $y = 3 \cos t$, $z = t$.

Solution: To eliminate the parameter in the first two equations, we write

$$\frac{x}{2} = \sin t \qquad \text{and} \qquad \frac{y}{3} = \cos t$$

which, by squaring and adding, yields

$$\frac{x^2}{4} + \frac{y^2}{9} = 1$$

Therefore, curve C lies entirely on the elliptic cylinder represented by

$$\frac{x^2}{4} + \frac{y^2}{9} = 1$$

(See Figure 14.43.)

The curve in Figure 14.43 is called a **helix**. A table of values is shown in Table 14.2.

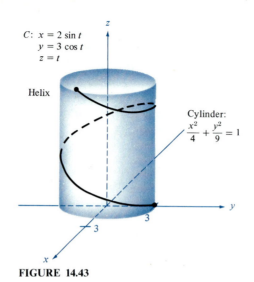

C: $x = 2 \sin t$
$y = 3 \cos t$
$z = t$

Helix

Cylinder:
$\dfrac{x^2}{4} + \dfrac{y^2}{9} = 1$

FIGURE 14.43

TABLE 14.2

t	x	y	z
0	0	3	0
$\dfrac{\pi}{2}$	2	0	$\dfrac{\pi}{2}$
π	0	-3	π
$\dfrac{3\pi}{2}$	-2	0	$\dfrac{3\pi}{2}$
2π	0	3	2π
3π	0	-3	3π

As indicated earlier, a space curve may also be determined by the intersection of two surfaces. We illustrate this case in the next example.

Example 2

Sketch the curve C represented by the intersection of the semiellipsoid

$$\frac{x^2}{12} + \frac{y^2}{24} + \frac{z^2}{4} = 1, \qquad z \geq 0$$

and the parabolic cylinder $y = x^2$. Find parametric equations for curve C.

Solution: A sketch of the ellipsoid and the parabolic cylinder is shown in Figure 14.44.

If we let $x = t$, then from the equation $y = x^2$ we obtain $y = t^2$, from which it follows that

$$\frac{z^2}{4} = 1 - \frac{x^2}{12} - \frac{y^2}{24} = 1 - \frac{t^2}{12} - \frac{t^4}{24}$$

or

$$z = 2\sqrt{\frac{24 - 2t^2 - t^4}{24}} = \frac{1}{\sqrt{6}} \sqrt{(6 + t^2)(4 - t^2)}$$

Therefore, for this choice of parameter, the parametric equations for curve C are

$$x = t, \qquad y = t^2, \qquad z = \sqrt{\frac{(6 + t^2)(4 - t^2)}{6}}$$

From these equations we can see that the three points shown on curve C in Figure 14.44 correspond to the values $t = -2$, $t = 0$, and $t = 2$. ■

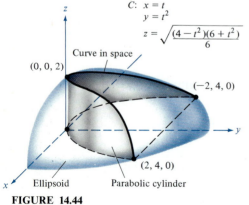

C: $x = t$
$y = t^2$
$z = \sqrt{\dfrac{(4 - t^2)(6 + t^2)}{6}}$

Curve in space

$(0, 0, 2)$

$(-2, 4, 0)$

$(2, 4, 0)$

Ellipsoid   Parabolic cylinder

FIGURE 14.44

The length of a curve in space is obtained by extending the formula in Theorem 13.2,

$$s = \int_a^b \sqrt{\left(\frac{dx}{dt}\right)^2 + \left(\frac{dy}{dt}\right)^2}\, dt$$

to include the third dimension $z = h(t)$. Thus if C is a curve represented by the equations

$$x = f(t), \qquad y = g(t), \qquad z = h(t)$$

where f', g', and h' are continuous on interval $[a, b]$ then the **arc length** of C over this integral is given by

$$s = \int_a^b \sqrt{\left(\frac{dx}{dt}\right)^2 + \left(\frac{dy}{dt}\right)^2 + \left(\frac{dz}{dt}\right)^2}\, dt$$

Example 3

Find the length of the space curve represented by the equations

$$x = t, \qquad y = \frac{4}{3}t^{3/2}, \qquad z = \frac{t^2}{2}$$

from $(0, 0, 0)$ to $(2, 8\sqrt{2}/3, 2)$. (See Figure 14.45.)

Solution: Since $x = t$, we consider t on the interval $0 \le t \le 2$. Thus the arc length on the given interval is

$$s = \int_0^2 \sqrt{\left(\frac{dx}{dt}\right)^2 + \left(\frac{dy}{dt}\right)^2 + \left(\frac{dz}{dt}\right)^2}\, dt$$

$$= \int_0^2 \sqrt{(1)^2 + (2\sqrt{t})^2 + (t)^2}\, dt$$

$$= \int_0^2 \sqrt{1 + 4t + t^2}\, dt = \int_0^2 \sqrt{(t + 2)^2 - 3}\, dt$$

$$= \left[\frac{t + 2}{2}\sqrt{(t + 2)^2 - 3} - \frac{3}{2}\ln |(t + 2) + \sqrt{(t + 2)^2 - 3}|\right]_0^2$$

$$= 2\sqrt{13} - \tfrac{3}{2}\ln(4 + \sqrt{13}) - 1 + \tfrac{3}{2}\ln 3 \approx 4.816 \qquad \blacksquare$$

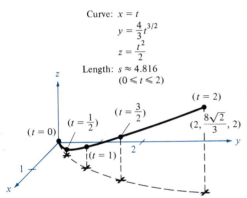

Curve: $x = t$
$y = \frac{4}{3}t^{3/2}$
$z = \frac{t^2}{2}$

Length: $s \approx 4.816$
$(0 \le t \le 2)$

$(t = 0)$ $(t = \frac{1}{2})$ $(t = \frac{3}{2})$ $(t = 2)$ $(2, \frac{8\sqrt{2}}{3}, 2)$ $(t = 1)$

FIGURE 14.45

To discuss the motion of a particle along a curve in space, it is convenient to represent the curve by a vector function denoted by the **position vector**

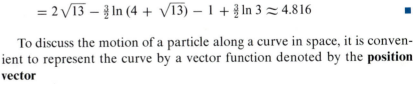

$$\mathbf{R}(t) = x(t)\mathbf{i} + y(t)\mathbf{j} + z(t)\mathbf{k}$$

The properties of vector functions having *three* components are similar to those for vector functions with two components. Thus for all vector functions in this section, we will assume the properties established in Sections 13.5 and 13.6.

If the vector

$$\mathbf{R}(t) = x(t)\mathbf{i} + y(t)\mathbf{j} + z(t)\mathbf{k}$$

represents the *position* of an object at time t, then the vectors

$$\mathbf{V}(t) = \mathbf{R}'(t) = x'(t)\mathbf{i} + y'(t)\mathbf{j} + z'(t)\mathbf{k}$$

and $$\mathbf{A}(t) = \mathbf{R}''(t) = x''(t)\mathbf{i} + y''(t)\mathbf{j} + z''(t)\mathbf{k}$$

represent the **velocity** and **acceleration,** respectively, at time t. Furthermore, the **speed** of the object at time t is

$$\frac{ds}{dt} = |\mathbf{R}'(t)| = |\mathbf{V}(t)| = \sqrt{\left(\frac{dx}{dt}\right)^2 + \left(\frac{dy}{dt}\right)^2 + \left(\frac{dz}{dt}\right)^2}$$

and the **rate of change of speed** is

$$\frac{d^2s}{dt^2} = \frac{d}{dt}[|\mathbf{R}'(t)|] = \frac{d}{dt}[|\mathbf{V}(t)|]$$

Example 4

An object moves along a curve C having parametric equations $x = t$, $y = t^3$, and $z = 3t$. Find the velocity and acceleration vectors, and the speed of the object at $t = 1$. Sketch the curve, and show the velocity and acceleration vectors at $t = 1$.

Solution: Since the position vector for the curve is

$$\mathbf{R}(t) = t\mathbf{i} + t^3\mathbf{j} + 3t\mathbf{k}$$

we have

$$\mathbf{V}(t) = \mathbf{i} + 3t^2\mathbf{j} + 3\mathbf{k}$$

$$\mathbf{A}(t) = 6t\mathbf{j}$$

$$|\mathbf{V}(t)| = \sqrt{10 + 9t^4}$$

Thus when $t = 1$, it follows that

$$\mathbf{V} = \mathbf{i} + 3\mathbf{j} + 3\mathbf{k} \quad \text{and} \quad \mathbf{A} = 6\mathbf{j}$$

And the speed of the object is

$$\frac{ds}{dt} = |\mathbf{V}(1)| = \sqrt{19}$$

A sketch is shown in Figure 14.46. ∎

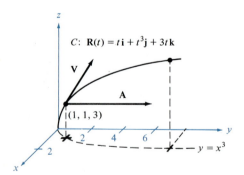

$C: \mathbf{R}(t) = t\mathbf{i} + t^3\mathbf{j} + 3t\mathbf{k}$

$(1, 1, 3)$

$y = x^3$

FIGURE 14.46

For space curves represented by

$$\mathbf{R}(t) = x(t)\mathbf{i} + y(t)\mathbf{j} + z(t)\mathbf{k}$$

the **unit tangent vector T** and the **unit normal vector N** are defined as

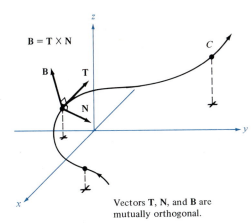

B = T X N

Vectors **T**, **N**, and **B** are
mutually orthogonal.

FIGURE 14.47

$$T = \frac{R'(t)}{|R'(t)|} = \frac{V(t)}{|V(t)|}, \qquad N = \frac{T'(t)}{|T'(t)|}$$

whenever $V(t)$ and $T'(t)$ are nonzero vectors.

In space the vectors **T** and **N** are interpreted in much the same way as they are in the plane. For an object moving along a curve C in space, the vector **T** points in the direction the object is moving, whereas the vector **N** is orthogonal to **T** and points in the direction the object is turning (the concave side of C). Figure 14.47 shows vectors **T**, **N**, and **B**, where $B = T \times N$ is orthogonal to both **T** and **N**. Vector **B** is called the **unit binormal vector** to the curve C.

Example 5

Determine the unit vectors **T** and **N** for the curve in Example 4. Find **T** and **N** when $t = 2$.

Solution: Since $\qquad R(t) = t\mathbf{i} + t^3\mathbf{j} + 3t\mathbf{k}$

we have

$$T = \frac{R'(t)}{|R'(t)|} = \frac{\mathbf{i} + 3t^2\mathbf{j} + 3\mathbf{k}}{\sqrt{10 + 9t^4}}$$

$$T' = \left[\frac{1}{(10 + 9t^4)^{1/2}}\right](6t\mathbf{j}) + (\mathbf{i} + 3t^2\mathbf{j} + 3\mathbf{k})\left[\frac{-18t^3}{(10 + 9t^4)^{3/2}}\right]$$

$$= \frac{6(-3t^3\mathbf{i} + 10t\mathbf{j} - 9t^3\mathbf{k})}{(10 + 9t^4)^{3/2}}$$

$$|T'| = 6\sqrt{\frac{9t^6 + 100t^2 + 81t^6}{(10 + 9t^4)^3}} = 6\sqrt{\frac{10t^2(10 + 9t^4)}{(10 + 9t^4)^3}} = \frac{6\sqrt{10}t}{10 + 9t^4}$$

Therefore, it follows that

$$N = \frac{T'}{|T'|} = (-3t^3\mathbf{i} + 10t\mathbf{j} - 9t^3\mathbf{k})\left[\frac{6}{(10 + 9t^4)^{3/2}}\right]\left(\frac{10 + 9t^4}{6\sqrt{10}t}\right)$$

$$= \frac{-3t^2\mathbf{i} + 10\mathbf{j} - 9t^2\mathbf{k}}{\sqrt{100 + 90t^4}}$$

When $t = 2$, we have

$$T = \frac{\mathbf{i} + 12\mathbf{j} + 3\mathbf{k}}{\sqrt{154}}$$

$$N = \frac{-12\mathbf{i} + 10\mathbf{j} - 36\mathbf{k}}{\sqrt{1540}} = \frac{-6\mathbf{i} + 5\mathbf{j} - 18\mathbf{k}}{\sqrt{385}}$$

■

For curves in space the formulas for the tangential and normal components of acceleration and for curvature follow those obtained in Section 13.6:

$$\text{tangential component of acceleration} = \mathbf{A} \cdot \mathbf{T} = \frac{d^2s}{dt^2}$$

$$\text{normal component of acceleration} = \mathbf{A} \cdot \mathbf{N} = K \left(\frac{ds}{dt}\right)^2$$

$$\text{curvature} = K = \frac{|\mathbf{T}'|}{ds/dt}$$

In the following theorem we show how the *dot* and *cross product* can be used to calculate these three quantities.

THEOREM 14.3
(*Alternative Formulas for* $\mathbf{A} \cdot \mathbf{T}$, $\mathbf{A} \cdot \mathbf{N}$, *and* K)

If $\mathbf{R}(t) = x(t)\mathbf{i} + y(t)\mathbf{j} + z(t)\mathbf{k}$ is the position vector for an object moving along a curve C in space, then at time t we have:

i. $\mathbf{A} \cdot \mathbf{T} = \dfrac{\mathbf{R}' \cdot \mathbf{R}''}{|\mathbf{R}'|}$ = tangential component of acceleration

ii. $\mathbf{A} \cdot \mathbf{N} = \dfrac{|\mathbf{R}' \times \mathbf{R}''|}{|\mathbf{R}'|}$ = normal component of acceleration

iii. $K = \dfrac{|\mathbf{R}' \times \mathbf{R}''|}{|\mathbf{R}'|^3}$ = curvature

---

Proof:

i. Since $\mathbf{A} = \mathbf{R}''$ and $\mathbf{T} = \mathbf{R}'/|\mathbf{R}'|$, it follows that

$$\mathbf{A} \cdot \mathbf{T} = \mathbf{T} \cdot \mathbf{A} = \frac{\mathbf{R}'}{|\mathbf{R}'|} \cdot \mathbf{R}'' = \frac{\mathbf{R}' \cdot \mathbf{R}''}{|\mathbf{R}'|}$$

ii. Using the analogy to Theorem 13.11, where

$$\mathbf{R}'' = \frac{d^2s}{dt^2}\mathbf{T} + K\left(\frac{ds}{dt}\right)^2 \mathbf{N}$$

and the fact that

$$\mathbf{R}' = |\mathbf{R}'|\,\mathbf{T} = \frac{ds}{dt}\mathbf{T}$$

it follows that

$$\mathbf{R}' \times \mathbf{R}'' = \left(\frac{ds}{dt}\right)\left(\frac{d^2s}{dt^2}\right)\mathbf{T} \times \mathbf{T} + K\left(\frac{ds}{dt}\right)^3 \mathbf{T} \times \mathbf{N}$$

$$= \left(\frac{ds}{dt}\right)\left(\frac{d^2s}{dt^2}\right)\mathbf{0} + K\left(\frac{ds}{dt}\right)^3 \mathbf{T} \times \mathbf{N}$$

Furthermore, since $|\mathbf{T} \times \mathbf{N}| = 1$, we conclude that

$$\frac{|\mathbf{R}' \times \mathbf{R}''|}{ds/dt} = K\left(\frac{ds}{dt}\right)^2 = \mathbf{A} \cdot \mathbf{N}$$

iii. From the last equation in part ii, it follows that

$$\frac{|\mathbf{R}' \times \mathbf{R}''|}{(ds/dt)^3} = K$$

■

Example 6

A particle moves along a space curve according to the vector function

$$\mathbf{R}(t) = 3t\mathbf{i} - t\mathbf{j} + t^2\mathbf{k}$$

Find (a) the velocity, speed, and acceleration, (b) the acceleration components $\mathbf{A} \cdot \mathbf{T}$ and $\mathbf{A} \cdot \mathbf{N}$, and (c) the curvature.

Solution:

(a) Since $\mathbf{R} = 3t\mathbf{i} - t\mathbf{j} + t^2\mathbf{k}$, we have

$$\mathbf{V} = \mathbf{R}' = 3\mathbf{i} - \mathbf{j} + 2t\mathbf{k} \qquad \text{(velocity)}$$
$$|\mathbf{V}| = |\mathbf{R}'| = \sqrt{9 + 1 + 4t^2} = \sqrt{10 + 4t^2} \qquad \text{(speed)}$$
$$\mathbf{A} = \mathbf{R}'' = 2\mathbf{k} \qquad \text{(acceleration)}$$

(b) By Theorem 14.3 we have

$$\mathbf{A} \cdot \mathbf{T} = \frac{\mathbf{R}' \cdot \mathbf{R}''}{|\mathbf{R}'|} = \frac{4t}{\sqrt{10 + 4t^2}} \qquad \begin{array}{l}\text{(tangential}\\\text{component)}\end{array}$$

Since
$$\mathbf{R}' \times \mathbf{R}'' = \begin{vmatrix} \mathbf{i} & \mathbf{j} & \mathbf{k} \\ 3 & -1 & 2t \\ 0 & 0 & 2 \end{vmatrix} = -2\mathbf{i} - 6\mathbf{j}$$

it follows from Theorem 14.3 that

$$\mathbf{A} \cdot \mathbf{N} = \frac{|\mathbf{R}' \times \mathbf{R}''|}{|\mathbf{R}'|} = \frac{\sqrt{40}}{\sqrt{10 + 4t^2}} = \frac{2\sqrt{5}}{\sqrt{5 + 2t^2}} \qquad \begin{array}{l}\text{(normal}\\\text{component)}\end{array}$$

(c) Again, by Theorem 14.3 we obtain

$$K = \frac{|\mathbf{R}' \times \mathbf{R}''|}{|\mathbf{R}'|^3} = \frac{\sqrt{40}}{(10 + 4t^2)^{3/2}} = \frac{\sqrt{5}}{(5 + 2t^2)^{3/2}} \qquad \text{(curvature)}$$

■

Section Exercises (14.5)

In Exercises 1–8, sketch the space curve represented by the parametric equations.

[S] **1.** $x = 2\cos t,\ y = 2\sin t,\ z = t$
2. $x = 3\cos t,\ y = 4\sin t,\ z = t/2$
3. $x = t,\ y = 2t - 5,\ z = 3t$
4. $x = -t + 1,\ y = 4t + 2,\ z = 2t + 3$
5. $x = 2\cos t,\ y = 2\sin t,\ z = e^{-t}$
[S] **6.** $x = t,\ y = t^2,\ z = 3t/2$
7. $x = t,\ y = t^2,\ z = 2t^3/3$
8. $x = \cos t + t\sin t,\ y = \sin t - t\cos t,\ z = t$

In Exercises 9–15, sketch the curve represented by the intersection of the two specified surfaces. Find a set of parametric equations for each curve.

9. $z = x^2 + y^2,\ x + y = 0$
10. $z = x^2 + y^2,\ z = 4$
[S] **11.** $x^2 + z^2 = 4,\ y^2 + z^2 = 4$ (first octant)
12. $x^2 + y^2 = 4,\ z = x^2$
13. $\frac{x^2}{9} + \frac{y^2}{16} + \frac{z^2}{4} = 1,\ x = y^2\ (z > 0)$
[S] **14.** $x^2 + y^2 + z^2 = 4,\ x + z = 2$
15. $x^2 + y^2 + z^2 = 16,\ xy = 4$ (first octant)

In Exercises 16–20, find the length of each curve.

16. $\mathbf{R}(t) = t\mathbf{i} + 3t\mathbf{j}$, from $t = 0$ to $t = 4$
[S] **17.** $\mathbf{R}(t) = a(\cos t)\mathbf{i} + a(\sin t)\mathbf{j} + bt\mathbf{k}$, from $t = 0$ to $t = 2\pi$
18. $\mathbf{R}(t) = a(\cos^3 t)\mathbf{i} + a(\sin^3 t)\mathbf{j}$, from $t = 0$ to $t = 2\pi$
19. $\mathbf{R}(t) = (\sin t - t\cos t)\mathbf{i} + (\cos t + t\sin t)\mathbf{j} + t^2\mathbf{k}$, from $t = 0$ to $t = \pi/2$
20. $\mathbf{R}(t) = 4t\mathbf{i} + 3(\cos t)\mathbf{j} + 3(\sin t)\mathbf{k}$, from $t = 0$ to $t = \pi/2$

In Exercises 21–28, a particle moves according to the given formula. Find (a) the velocity, speed, and acceleration of the particle; (b) the acceleration components $\mathbf{A}\cdot\mathbf{T}$ and $\mathbf{A}\cdot\mathbf{N}$; and (c) the curvature of the path.

21. $\mathbf{R}(t) = 4t\mathbf{i}$
22. $\mathbf{R}(t) = 4t\mathbf{i} - 4t\mathbf{j} + 2t\mathbf{k}$
23. $\mathbf{R}(t) = 4t^2\mathbf{i}$
24. $\mathbf{R}(t) = a\cos(wt)\mathbf{i} + a\sin(wt)\mathbf{j}$
[S] **25.** $\mathbf{R}(t) = 4t\mathbf{i} + (3\cos t)\mathbf{j} + (3\sin t)\mathbf{k}$
26. $\mathbf{R}(t) = t\mathbf{i} + t^2\mathbf{j} + \left(\frac{t^2}{2}\right)\mathbf{k}$
27. $\mathbf{R}(t) = (e^t\cos t)\mathbf{i} + (e^t\sin t)\mathbf{j} + e^t\mathbf{k}$
28. $\mathbf{R}(t) = t\mathbf{i} + 3t\mathbf{j} + \left(\frac{t^2}{2}\right)\mathbf{k}$

[S] **29.** Find a parametric representation of the tangent line to the helix $\mathbf{R}(t) = (2\cos t)\mathbf{i} + (2\sin t)\mathbf{j} + t\mathbf{k}$ at the point $(-\sqrt{2}, \sqrt{2}, 3\pi/4)$.
30. Find a parametric representation of the tangent line to the curve $\mathbf{R}(t) = t\mathbf{i} + t^2\mathbf{j} + \frac{2}{3}t^3\mathbf{k}$ at the point $(3, 9, 18)$.

For Review and Class Discussion

True or False

1. ____ If f, g, and h are first-degree polynomial functions, then the curve given by $x = f(t), y = g(t), z = h(t)$ is a line.

2. ____ If the curve given by $x = f(t), y = g(t), z = h(t)$ is a line, then f, g, and h are first-degree polynomial functions of t.

3. ____ $\frac{d}{dt}[|\mathbf{R}(t)|] = |\mathbf{R}'(t)|$.

4. ____ The curve determined by the intersection of the sphere $x^2 + y^2 + z^2 = 1$ and the plane $ax + by + cz = 0$ has a constant curvature of 1.

5. ____ The curve given by $x = f(t), y = g(t), z = t$ lies in a plane.

14.6 **Cylindrical and Spherical Coordinates**

Purpose
- To introduce two additional coordinate systems for representing points in space.

We want to take the reader on a leisurely walk, as it were, in the big garden that is geometry, so that each may pick for himself a bouquet to his liking.   —David Hilbert (1862–1943)

In plane analytic geometry we studied two coordinate systems, the rectangular and the polar systems. We saw that under certain circumstances one system was more convenient than the other. A similar situation exists in three-dimensional geometry, and in this section we introduce two new space coordinate systems. The first of these, called the **cylindrical coordinate system,** is an extension of the polar coordinate system to three dimensions. In this cylindrical system a point P in space is represented by an ordered triple (r, θ, z) where (r, θ) is a polar representation of the projection of P onto the xy-plane and z is the directed distance from the xy-plane to the point P. (See Figure 14.48.)

If we superimpose rectangular coordinates upon Figure 14.48, we can readily see how to convert coordinates from one system to the other (see Figure 14.49). The transformation *from cylindrical to rectangular* coordinates is given by the equations

$$x = r \cos \theta, \qquad y = r \sin \theta, \qquad z = z$$

and the transformation *from rectangular to cylindrical* coordinates is given by the equations

$$r^2 = x^2 + y^2, \qquad \tan \theta = \frac{y}{x}, \qquad z = z$$

A cylindrical coordinate system is useful primarily for problems involving an *axis* of symmetry. For instance, the graph of the equation $r = c$, where c is positive, is a right circular cylinder of radius c with the z-axis as its axis of symmetry (Figure 14.50). In fact, the simplicity of the equation $r = c$ for a cylinder having the z-axis as its axis of symmetry is

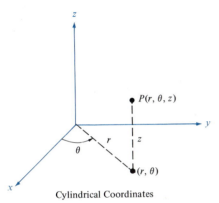

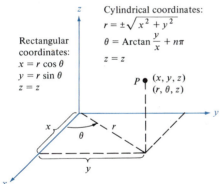

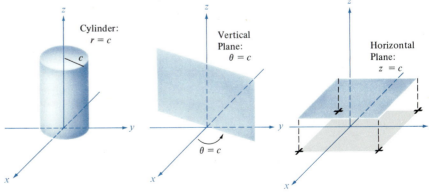

FIGURE 14.48

Cylindrical Coordinates

FIGURE 14.49

Rectangular coordinates:
$x = r \cos \theta$
$y = r \sin \theta$
$z = z$

Cylindrical coordinates:
$r = \pm\sqrt{x^2 + y^2}$
$\theta = \text{Arctan} \dfrac{y}{x} + n\pi$
$z = z$

FIGURE 14.50

the reason this system of coordinates is called the "cylindrical" coordinate system. Note that the graph of $\theta = c$ is a plane containing the z-axis and the graph of $z = c$ is a plane parallel to the xy-plane (see Figure 14.50).

Example 1

(a) Express the point $(r, \theta, z) = (4, 5\pi/6, 3)$ in rectangular coordinates.
(b) Express the point $(x, y, z) = (2, \sqrt{2}, -1)$ in cylindrical coordinates.

Solution:

(a) From the equations given with Figure 14.49, we have

$$x = 4 \cos \frac{5\pi}{6} = 4 \left(-\frac{\sqrt{3}}{2} \right) = -2\sqrt{3}$$

$$y = 4 \sin \frac{5\pi}{6} = 4 \left(\frac{1}{2} \right) = 2$$

$$z = 3$$

Thus the point is $(-2\sqrt{3}, 2, 3)$ in rectangular coordinates.

(b) Since

$$r = \pm\sqrt{4 + 2} = \pm\sqrt{6}$$

$$\theta = \text{Arctan} \frac{\sqrt{2}}{2} + n\pi \approx \text{Arctan} \,(0.707) + n\pi \approx 0.6155 + n\pi$$

$$z = -1$$

we have two choices for r and infinitely many choices for θ. If we choose $\theta = 0.6155$ (quadrant I), then we must choose $r = +\sqrt{6}$, or if we choose $\theta = 0.6155 + \pi$ (quadrant III), then $r = -\sqrt{6}$. Therefore, two possible representations of the given point are

$$(\sqrt{6}, 0.6155, -1) \quad \text{or} \quad (-\sqrt{6}, 0.6155 + \pi, -1)$$

(In general, we will choose a value of θ that will allow us to use a positive value for r.) ∎

Example 2 *(Rectangular to Cylindrical)*

Find equations in cylindrical coordinates for these surfaces:
(a) $x^2 + y^2 = 4z^2$
(b) $y^2 = x$
Identify the graph of each.

Solution:

(a) From Figure 14.39 we can see that the graph of $x^2 + y^2 = 4z^2$ is a cone with its axis along the z-axis. If we replace $x^2 + y^2$ by r^2, its equation in cylindrical coordinates is

$$r^2 = 4z^2 \qquad \text{or} \qquad r = 2z$$

(b) The graph of the surface $y^2 = x$ is a parabolic cylinder with rulings parallel to the z-axis. If we replace y^2 by $r^2 \sin \theta$ and x by $r \cos \theta$, the equation in cylindrical coordinates is

$$r^2 \sin^2 \theta = r \cos \theta$$

and for $r \neq 0$, $\qquad r \sin^2 \theta = \cos \theta$

$$r = \csc^2 \theta \cos \theta$$

or, equivalently, $\qquad r = \csc \theta \cot \theta$ ∎

Example 3 (*Cylindrical to Rectangular*)

Find a rectangular equation for the graph of the equation $r^2 \cos 2\theta - z^2 = 1$.

Solution: If we replace $\cos 2\theta$ by $\cos^2 \theta - \sin^2 \theta$, we obtain

$$r^2(\cos^2 \theta - \sin^2 \theta) - z^2 = 1$$
$$r^2 \cos^2 \theta - r^2 \sin^2 \theta - z^2 = 1$$

Now replacing $r \cos \theta$ by x and $r \sin \theta$ by y yields

$$x^2 - y^2 - z^2 = 1$$

which is an elliptic hyperboloid of two sheets having its transverse axis along the x-axis. ∎

A **spherical coordinate system** can also be used to represent points in space. In this system a point P is represented by an ordered triple (ρ, θ, ϕ) where ρ (Greek lowercase letter rho) is the distance between the point P and the origin, θ is the same angle as in cylindrical coordinates, and ϕ measures the nondirected angle *between* the positive z-axis and the line segment OP (see Figure 14.51).

From Figure 14.51 we can see that the transformation from *spherical to rectangular* coordinates is given by the equations

$$x = \rho \sin \phi \cos \theta$$
$$y = \rho \sin \phi \sin \theta$$
$$z = \rho \cos \phi$$

(note: $\sqrt{x^2 + y^2} = r = \rho \sin \phi$), and the transformation from *rectangular to spherical* coordinates is given by the equations

$$\rho^2 = x^2 + y^2 + z^2$$

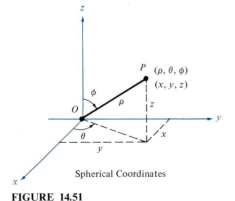

Spherical Coordinates

FIGURE 14.51

$$\tan \theta = \frac{y}{x}$$

$$\cos \phi = \frac{z}{\sqrt{x^2 + y^2 + z^2}}$$

We consider ρ to be positive and ϕ to be undirected. Thus for any point (ρ, θ, ϕ) we consider $\rho \geq 0$ and $0 \leq \phi \leq \pi$.

A spherical coordinate system is useful primarily for three-dimensional problems having a *point* or *center* of symmetry. In fact, the term "spherical" is used because the graph of the simple equation $\rho = c$, where $c > 0$, is a sphere of radius c centered at the origin. The graph of $\theta = c$ is a plane containing the z-axis (as was true in cylindrical coordinates), and the graph of $\phi = c$ is a half cone with its vertex at the origin and its axis along the z-axis (see Figure 14.52).

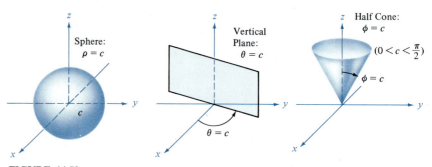

FIGURE 14.52

Example 4 (*Rectangular to Spherical*)

Find an equation in spherical coordinates for these surfaces:
(a) cone, $x^2 + y^2 = z^2$
(b) sphere, $x^2 + y^2 + z^2 - 4z = 0$

Solution:

(a) Making the appropriate replacements for x, y, and z in the equation $x^2 + y^2 = z^2$ gives us

$$\rho^2 \sin^2 \phi \cos^2 \theta + \rho^2 \sin^2 \phi \sin^2 \theta = \rho^2 \cos^2 \phi$$
$$\rho^2 \sin^2 \phi (\cos^2 \theta + \sin^2 \theta) = \rho^2 \cos^2 \phi$$
$$\rho^2 \sin^2 \phi = \rho^2 \cos^2 \phi$$

which, for $\rho > 0$, yields the equation

$$\frac{\sin^2 \phi}{\cos^2 \phi} = \tan^2 \phi = 1$$

$$\tan \phi = \pm 1$$

Therefore, $$\phi = \frac{\pi}{4} \quad \text{or} \quad \phi = \frac{3\pi}{4}$$

The equation $\phi = \pi/4$ represents the *upper* half cone and $\phi = 3\pi/4$ represents the *lower* half cone.

(b) Since $\rho^2 = x^2 + y^2 + z^2$ and $z = \rho \cos \phi$, the given equation has the spherical form

$$\rho^2 - 4\rho \cos \phi = 0$$
$$\rho(\rho - 4 \cos \phi) = 0$$

If we exclude the case of $\rho = 0$ (the origin), then the equation for the sphere is

$$\rho - 4 \cos \phi = 0 \quad \text{or} \quad \rho = 4 \cos \phi$$

Note that the graph of this equation includes a point where $\rho = 0$, so nothing is lost by discarding the factor ρ. ∎

Section Exercises (14.6)

In Exercises 1–18, convert the given point from the coordinate system in which it is given to the coordinate system specified.

1. $(1, \sqrt{3}, 4)$, rectangular to cylindrical
2. $(\sqrt{3}, -1, 2)$, rectangular to cylindrical
S **3.** $(2, -2, -4)$, rectangular to cylindrical
4. $(-3, 2, -1)$, rectangular to cylindrical
5. $(2, \pi/3, 2)$, cylindrical to rectangular
6. $(3, -\pi/4, 1)$, cylindrical to rectangular
S **7.** $(4, 7\pi/6, 3)$, cylindrical to rectangular
8. $(1, 1, 1)$, rectangular to spherical
S **9.** $(-2, 2\sqrt{3}, 4)$, rectangular to spherical
10. $(2, 2, 4\sqrt{2})$, rectangular to spherical
S **11.** $(4, \pi/6, \pi/4)$, spherical to rectangular
C **12.** $(12, 3\pi/4, \pi/9)$, spherical to rectangular
13. $(12, -\pi/4, 0)$, spherical to rectangular
14. $(4, \pi/18, \pi/2)$, spherical to cylindrical
15. $(6, -\pi/6, \pi/3)$, spherical to cylindrical
16. $(2, 2\pi/3, -2)$, cylindrical to spherical
S **17.** $(4, -\pi/6, 6)$, cylindrical to spherical
18. $(-4, \pi/3, 4)$, cylindrical to spherical

In Exercises 19–24, find an equation in rectangular coordinates for each equation in cylindrical coordinates.
19. $r = 2$
20. $z = 2$

21. $\theta = \frac{\pi}{6}$
S **22.** $r = \frac{z}{2}$
23. $r = 2 \sin \theta$
24. $r^2 = z$

In Exercises 25–30, find an equation in rectangular coordinates for each equation in spherical coordinates.

25. $\rho = 5$
26. $\theta = \frac{3\pi}{4}$

27. $\phi = \frac{\pi}{6}$
28. $\rho = 2 \sec \phi$

S **29.** $\rho = 4 \cos \phi$
30. $\phi = \frac{\pi}{2}$

In Exercises 31–36, find an equation of the given surface in (a) cylindrical coordinates and (b) spherical coordinates.
31. $x^2 + y^2 + z^2 = 16$
32. $x^2 + y^2 = z^2$
S **33.** $x^2 + y^2 + z^2 - 2z = 0$
34. $x^2 + y^2 = z$
S **35.** $x^2 + y^2 = 4y$
36. $x^2 + y^2 = 16$

37. Sketch the solid that has the following description in cylindrical coordinates: $0 \le \theta \le 2\pi$, $0 \le r \le a$, $r \le z \le a$.

38. Sketch the solid that has the following description in spherical coordinates: $0 \le \theta \le 2\pi$, $\pi/4 \le \phi \le \pi/2$, $0 \le \rho \le 1$.

⟦S⟧ **39.** Sketch the solid that has the following description in spherical coordinates: $0 \le \theta \le 2\pi$, $0 \le \phi \le \pi/6$, $0 \le \rho \le a \sec \phi$.

40. Sketch the solid that has the following description in cylindrical coordinates: $0 \le \theta \le 2\pi$, $2 \le r \le 4$, $z^2 \le -r^2 + 6r - 8$.

Miscellaneous Exercises (Ch. 14)

1. Given the two points $(2, 1, 3)$ and $(-4, 2, -1)$.
(a) Plot the points.
(b) Find the coordinates of the midpoint of the line segment joining the points.
(c) Give a parametric representation of the line through the points.
(d) Find an equation of the plane containing the points and parallel to the y-axis.

2. Given the two points $(2, 1, 0)$ and $(2, 2, 2)$.
(a) Plot the points.
(b) Find the coordinates of the midpoint of the line segment joining the points.
(c) Give a parametric representation of the line through the points.
(d) Find an equation of the plane containing the points and parallel to the yz-coordinate plane.
(e) Find an equation of the plane containing the point $(2, 1, 0)$ and orthogonal to the line through the given points.

3. Sketch the graph of the equation $x = 1$ in the following:
(a) a one-dimensional system
(b) a two-dimensional system
(c) a three-dimensional system

⟦S⟧ **4.** Sketch the graph of the equation $y = x^2$ in the following:
(a) a two-dimensional system
(b) a three-dimensional system

In Exercises 5–14, let $\mathbf{V}_1 = 3\mathbf{i} - 2\mathbf{j} + \mathbf{k}$, $\mathbf{V}_2 = 2\mathbf{i} - 4\mathbf{j} - 3\mathbf{k}$, and $\mathbf{V}_3 = -\mathbf{i} + 2\mathbf{j} + 2\mathbf{k}$.

5. Find the magnitude of $\mathbf{V}_1$.
6. Find $\mathbf{V}_1 \cdot \mathbf{V}_2$.
7. Show that $\mathbf{V}_1 \cdot \mathbf{V}_1 = |\mathbf{V}_1|^2$.
⟦C⟧ **8.** Find the angle between $\mathbf{V}_1$ and $\mathbf{V}_2$.
9. Find $\mathbf{V}_1 \times \mathbf{V}_2$ and show that it is orthogonal to $\mathbf{V}_1$ and to $\mathbf{V}_2$.
10. Determine a unit vector perpendicular to the plane containing $\mathbf{V}_2$ and $\mathbf{V}_3$.
11. Show that $\mathbf{V}_1 \cdot (\mathbf{V}_2 + \mathbf{V}_3) = \mathbf{V}_1 \cdot \mathbf{V}_2 + \mathbf{V}_1 \cdot \mathbf{V}_3$.
12. Show that $\mathbf{V}_1 \times (\mathbf{V}_2 + \mathbf{V}_3) = (\mathbf{V}_1 \times \mathbf{V}_2) + (\mathbf{V}_1 \times \mathbf{V}_3)$.
13. Find the volume of the solid whose edges are $\mathbf{V}_1$, $\mathbf{V}_2$, and $\mathbf{V}_3$.

14. Find the work done in moving an object along the vector $\mathbf{V}_1$ if the applied force is $\mathbf{V}_3$.

15. Prove that $ax + by + cz = |\mathbf{V}|d$ is the equation of a plane orthogonal to $\mathbf{V} = a\mathbf{i} + b\mathbf{j} + c\mathbf{k}$ and d units from the origin.

16. Find the angle between the yz-plane and the vector $\mathbf{V} = 2\mathbf{i} + \mathbf{j} + \mathbf{k}$.

⟦S⟧ **17.** Find a set of parametric equations for the line perpendicular to the xz-coordinate plane passing through the point $(1, 2, 3)$.

18. Find a set of parametric equations for the line through the point $(1, 2, 3)$ and parallel to the line $x = y = z$.

19. Find an equation of the plane through the point $(1, 2, 3)$ and orthogonal to the line $x = y = z$.

20. Find an equation of the plane containing the lines $(x - 1)/(-2) = y = z + 1$ and $(x + 1)/(-2) = (y - 1)/1 = (z - 2)/1$.

21. Find a set of parametric equations for the line of intersection of the planes $3x - 3y - 7z = -4$ and $x - y + 2z = 3$.

22. Find the point of intersection (if any) of the lines

$$\frac{x - 7}{5} = \frac{y - 6}{4} = \frac{z - 8}{5}$$

and

$$\frac{x - 8}{6} = \frac{y - 6}{4} = \frac{z - 9}{6}$$

⟦C⟧ **23.** Find the angle between the plane $x + y + z = 4$ and the line $(x + 1)/(-1) = (y - 2)/2 = (z + 3)/2$.

⟦S⟧ **24.** Find the distance between the planes $x - 2y + 2z = 5$ and $x - 2y + 2z = 10$.

25. Find the distance between the skew lines

$$\frac{x + 1}{1} = \frac{y}{2} = \frac{z - 1}{3}$$

and

$$\frac{x - 1}{-1} = \frac{y - 3}{1} = \frac{z + 2}{-2}$$

26. Find the distance from the origin to the plane $x - 2y + 2z = 10$.

In Exercises 27–38, sketch the graph of the specified surface.

27. $x + 2y + 3z = 6$ **28.** $y = z^2$

29. $x^2 + z^2 = 4$

30. $x^2 + y^2 + z^2 - 2x + 4y - 6z + 5 = 0$

[S] **31.** $16x^2 + 16y^2 - 9z^2 = 0$ **32.** $\dfrac{x^2}{16} + \dfrac{y^2}{9} + z^2 = 1$

33. $y = \frac{1}{2}z$ **34.** $\dfrac{x^2}{16} + \dfrac{y^2}{9} - z^2 = 1$

35. $\dfrac{x^2}{16} - \dfrac{y^2}{9} + z^2 = -1$ **36.** $\dfrac{x^2}{25} + \dfrac{y^2}{4} - \dfrac{z^2}{100} = 1$

37. $y^2 - 4x^2 = z$ [S] **38.** $y = \cos z$

39. Find the speed, the normal and tangential components of acceleration, and the curvature of the path

$$\mathbf{R}(t) = (\sin wt - wt \cos wt)\mathbf{i} + (\cos wt + wt \sin wt)\mathbf{j}$$

40. Write the equation $x^2 + y^2 + z^2 = 16$ in cylindrical and spherical coordinates.

[S] **41.** Write the equation $r^2(\cos^2 \theta - \sin^2 \theta) + z^2 = 1$ in rectangular coordinates.

42. Write the equation $\rho = \csc \phi$ in rectangular coordinates.

43. Write the equation $x^2 - y^2 = 2z$ in cylindrical and spherical coordinates.

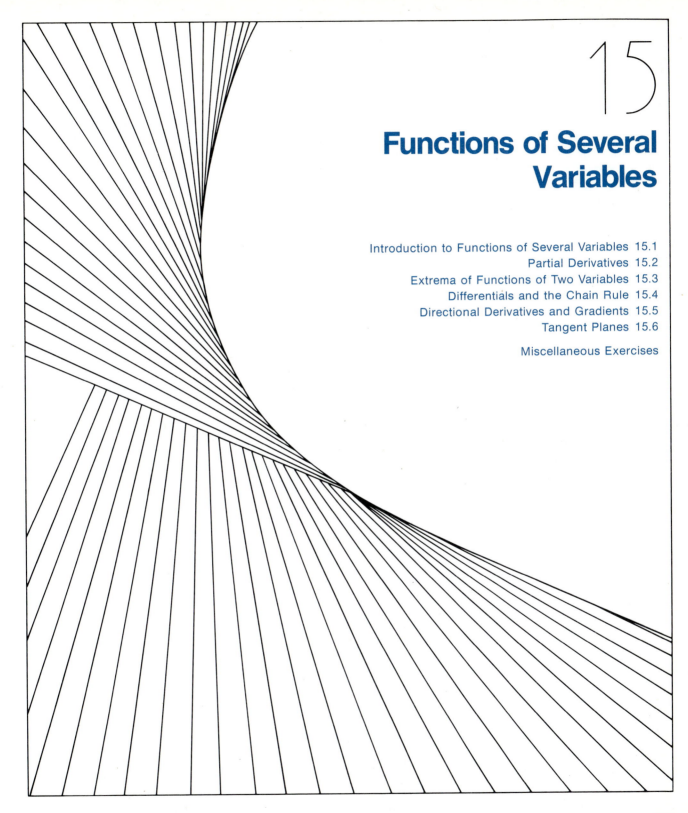

15

Functions of Several Variables

15.1 **Introduction to Functions of Several Variables**

Purpose

- To introduce the concept of a function of several variables.
- To discuss the graphs of functions of two variables.
- To define limits and continuity for functions of two variables.

The nature of a curved surface is defined by an equation between the coordinates of its points, which we represent by $f(x, y, z) = 0$. Generally speaking, on passing through the surface the value of f changes its sign, so that, as long as the continuity is not interrupted, the values are positive on one side and negative on the other. —*Carl Gauss (1777–1855)*

So far we have dealt only with functions of a single independent variable. These functions have forms like $y = f(x)$, $r = f(\theta)$, or $\mathbf{R} = x(t)\mathbf{i} + y(t)\mathbf{j}$, where unique values for y, r, or $\mathbf{R}$ are determined for each value assigned to x, θ, or t, respectively. Many familiar quantities are functions of not one but two or more variables. For instance, the area A of a rectangle ($A = lw$) or the volume V of a right circular cylinder ($V = \pi r^2 h$) are both functions of *two* variables. The volume of a rectangular solid is a function of *three* variables ($V = lwh$).

We denote a function of two or more variables by notation similar to that for functions of a single variable. For instance, the volume of a right circular cylinder can be denoted by $f(r, h) = \pi r^2 h$, or, similarly, the volume of a rectangular solid by $g(l, w, h) = lwh$.

We give the following definition of a function of *two* variables. Similar definitions can be given for functions of three, four, or n variables.

Definition of a Function of Two Variables

If to each ordered pair (x, y) in some set D, there corresponds a real number $f(x, y)$, then f is called a **function of x and y.** The set D is the **domain** of f and the corresponding set of values for $f(x, y)$ is the **range** of f.

Geometrically we can use the three-dimensional coordinate system to represent a function of two variables x and y by considering the domain

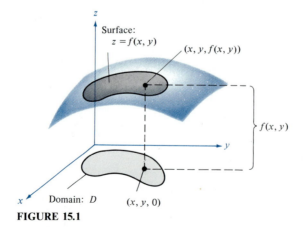

FIGURE 15.1

D to be a region in the xy-plane and by letting $z = f(x, y)$. Thus the set of all points (x, y, z) whose coordinates satisfy the equation $z = f(x, y)$ is called the **graph** of the function f. Usually this graph will take the form of a surface in space, where $f(x, y)$ gives the directed distance from $(x, y, 0)$ to $(x, y, f(x, y))$. (See Figure 15.1.)

When we refer to the function given by the equation $z = f(x, y)$, it is assumed (unless specifically restricted, as in Figure 15.1) that the domain is the set of all (x, y) for which the equation has meaning.

Example 1

Determine the domain and range of the function defined by

$$f(x, y) = \sqrt{64 - 4x^2 - y^2}$$

Sketch the graph of f.

Solution: Since D is not otherwise specified, we assume the domain of f to be the set of all points (x, y) such that

$$64 - 4x^2 - y^2 \geq 0 \qquad \text{or} \qquad \frac{x^2}{16} + \frac{y^2}{64} \leq 1$$

In other words, D is the set of all points lying on or inside the ellipse

$$\frac{x^2}{16} + \frac{y^2}{64} = 1$$

The range of f is all values $z = f(x, y)$ such that

$$0 \leq z \leq \sqrt{64} = 8$$

The graph of f is a semiellipsoid having elliptical xy and xz traces and a circular yz trace (see Figure 15.2).

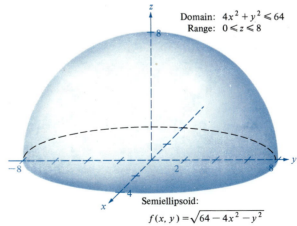

Domain: $4x^2 + y^2 \leq 64$
Range: $0 \leq z \leq 8$

Semiellipsoid:
$$f(x, y) = \sqrt{64 - 4x^2 - y^2}$$

FIGURE 15.2

An alternative way of graphing a function of two variables is one commonly used in topographical maps (Figure 15.3). Each of the curves in Figure 15.3 represents the intersection of the surface $z = f(x, y)$ with a plane $z = c$. (In Figure 15.3 c has the values 0, 10, 20, . . . , 70.) We call these curves **level curves,** and the collection of all such curves, $c = c_1, c_2, c_3, . . . , c_n$, is called a **contour map** of f. For example, to make a contour map of the semiellipsoid $f(x, y) = \sqrt{64 - 4x^2 - y^2}$, we can choose $c_1 = 0, c_2 = 1, c_3 = 2, . . . , c_9 = 8$, as shown in Figure 15.4.

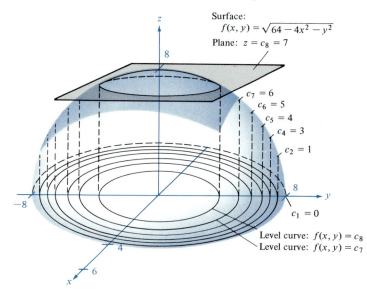

Surface: $f(x, y) = \sqrt{64 - 4x^2 - y^2}$
Plane: $z = c_8 = 7$

$c_7 = 6$
$c_6 = 5$
$c_5 = 4$
$c_4 = 3$
$c_2 = 1$
$c_1 = 0$

Level curve: $f(x, y) = c_8$
Level curve: $f(x, y) = c_7$

FIGURE 15.4

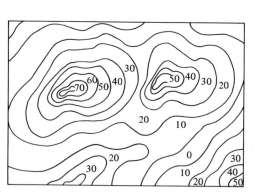

Topographical Map of Two Peaks.
One is over 70 units high and
the other is over 50 units high.

FIGURE 15.3

Note that a contour map depicts the variation of z with respect to x and y by the change in the spacing of the level curves. Much space between level curves indicates z is changing slowly, while little space between level curves indicates a rapid change in z.

Example 2

Sketch a contour map for the function $f(x, y) = 6 - x^2 - y$ by using $c = 0, 2, 4, 6$.

Solution: We consider curves of the form $f(x, y) = c$. For $c = 0$ we have the parabola

$$0 = 6 - x^2 - y \quad \text{or} \quad y = 6 - x^2$$

For $c = 2, 4$, and 6, we have, respectively, the parabolas

$$y = 4 - x^2, \quad y = 2 - x^2, \quad y = -x^2$$

(See Figure 15.5.) ∎

Example 3

Sketch a contour map for the function $f(x, y) = \sin xy$ by using $c = -1, 0, 1$.

Solution: For $c = 0$ we have the family of hyperbolas represented by

$$\sin xy = 0$$
$$xy = n\pi, \qquad n = \dots, -2, -1, 0, 1, 2, \dots$$

Similarly, for $c = 1$ we have the family of hyperbolas

$$\sin xy = 1$$
$$xy = \frac{(4n + 1)\pi}{2}$$

and for $c = -1$ we have the family of hyperbolas

$$\sin xy = -1$$
$$xy = \frac{(4n + 3)\pi}{2}$$

The surface is pictured in Figure 15.6 and the level curves are shown in Figure 15.7. ∎

In a contour map we assume that the level curves are chosen to be representative of the behavior of the function everywhere and that the value of $f(x, y)$ changes continuously from one level curve to the next. Thus the most appropriate use of a contour map is in representing *continuous* functions such as height above sea level, atmospheric temperature, or barometric pressure.

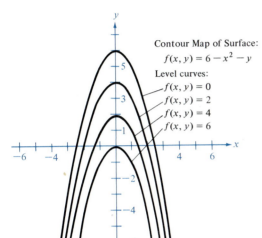

Contour Map of Surface:
$$f(x, y) = 6 - x^2 - y$$
Level curves:
$f(x, y) = 0$
$f(x, y) = 2$
$f(x, y) = 4$
$f(x, y) = 6$

FIGURE 15.5

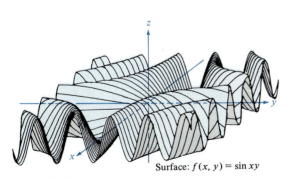

Surface: $f(x, y) = \sin xy$

FIGURE 15.6

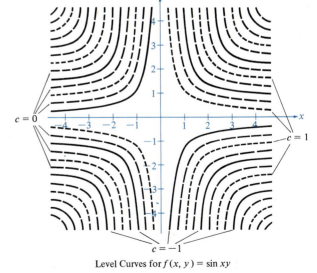

Level Curves for $f(x, y) = \sin xy$

FIGURE 15.7

Before defining continuity for functions of two variables, we need to discuss the limit of such functions. We denote the limit of a function of two variables by

$$\lim_{(x,y)\to(a,b)} f(x,y) = L$$

and define it as follows:

Definition of the Limit of a Function of Two Variables	Let f be a function of two variables that is defined [except possibly at (a, b)] in the interior of a circle centered at (a, b). Then $$\lim_{(x,y)\to(a,b)} f(x,y) = L$$ if and only if for each $\epsilon > 0$ there corresponds a $\delta > 0$ such that $$|f(x,y) - L| < \epsilon \quad \text{whenever} \quad 0 < \sqrt{(x-a)^2 + (y-b)^2} < \delta$$

Geometrically this limit means that for any point (x, y) within the circle of radius δ, the value $f(x, y)$ lies between $L + \epsilon$ and $L - \epsilon$ (see Figure 15.8).

Although this definition closely parallels the limit definition (Section 2.4) for a function of a single variable, there is one basic difference. When we write

$$(x, y) \to (a, b)$$

we mean that the point (x, y) is allowed to approach (a, b) from *any* direction. If the value of

$$\lim_{(x,y)\to(a,b)} f(x,y)$$

is not the same for all possible approaches to (a, b), then the limit does not exist. For example, if $(x, y) \to (0, 0)$ along the path $x = 0$, then

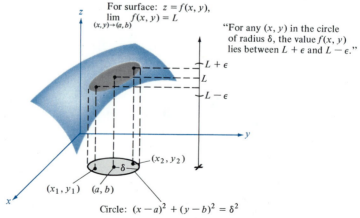

FIGURE 15.8

$$\lim_{(0,y)\to(0,0)} \frac{x^2 - y^2}{x^2 + y^2} = -1$$

whereas along the path $y = 0$,

$$\lim_{(x,0)\to(0,0)} \frac{x^2 - y^2}{x^2 + y^2} = +1$$

Hence
$$\lim_{(x,y)\to(0,0)} \frac{x^2 - y^2}{x^2 + y^2}$$

does not exist.

To evaluate the limit

$$\lim_{(x,y)\to(a,b)} f(x, y)$$

we can use direct substitution as we did with functions of a single variable, provided the function is in reduced form. However, if direct substitution yields an *indeterminate* form, we should investigate the limit along different paths, which may show that the limit does not exist, or we may find it convenient to use another coordinate system to evaluate the limit (see Example 5).

Example 4

Determine the following limits, if they exist:

(a) $\displaystyle\lim_{(x,y)\to(0,1)} (x^2y - 3y)$

(b) $\displaystyle\lim_{(x,y)\to(2,-1)} \frac{3xy}{x^2 + y^2}$

(c) $\displaystyle\lim_{(x,y)\to(0,0)} \left(\frac{x^2 - y^2}{x^2 + y^2}\right)^2$

Solution:

(a)   By direct substitution

$$\lim_{(x,y)\to(0,1)} (x^2y - 3y) = 0 - 3 = -3$$

(b)   Direct substitution yields

$$\lim_{(x,y)\to(2,-1)} \frac{3xy}{x^2 + y^2} = \frac{-6}{4 + 1} = \frac{-6}{5}$$

(c)   In this case direct substitution yields the indeterminate form

$$\lim_{(x,y)\to(0,0)} \left(\frac{x^2 - y^2}{x^2 + y^2}\right)^2 = \frac{0}{0}$$

By letting (x, y) approach the origin along either axis, we obtain a limit of 1. However, if we let (x, y) approach the origin along the line $y = x$, we have

$$\lim_{(x,x)\to(0,0)} \left(\frac{x^2 - x^2}{x^2 + x^2}\right)^2 = \lim_{(x,x)\to(0,0)} \left(\frac{0}{2x^2}\right)^2 = 0$$

Thus the limit does not exist. (See Figure 15.9.)

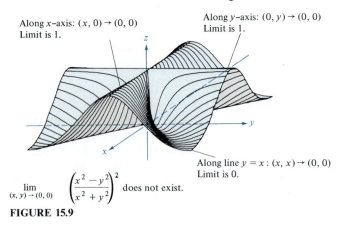

Along x-axis: $(x, 0) \to (0, 0)$
Limit is 1.

Along y-axis: $(0, y) \to (0, 0)$
Limit is 1.

Along line $y = x$: $(x, x) \to (0, 0)$
Limit is 0.

$$\lim_{(x, y) \to (0, 0)} \left(\frac{x^2 - y^2}{x^2 + y^2}\right)^2 \text{ does not exist.}$$

FIGURE 15.9 ∎

Example 5

Use polar coordinates to show that

$$\lim_{(x,y)\to(0,0)} \frac{5x^2 y}{x^2 + y^2} = 0$$

Solution: Since $r \to 0$ as $(x, y) \to (0, 0)$, it follows that

$$\lim_{(x,y)\to(0,0)} \frac{5x^2 y}{x^2 + y^2} = \lim_{r\to 0} \frac{5(r^2 \cos^2 \theta)(r \sin \theta)}{r^2}$$

$$= \lim_{r\to 0} (5r \cos^2 \theta \sin \theta)$$

$$= (5 \cos^2 \theta \sin \theta) \lim_{r\to 0} r = 0 \quad ∎$$

Having defined the limit of a function of two variables, we give the following definition of continuity:

Definition of Continuity of a Function of Two Variables	A function f of two variables **is continuous at** (a, b) provided $$\lim_{(x,y)\to(a,b)} f(x, y) = f(a, b)$$ Furthermore, if D is a subset of the domain of f, then f **is continuous on** D if f is continuous at each point (a, b) in D.

The definitions of limits and continuity can be extended to functions of three variables by considering points (x, y, z) within the *sphere* $(x - a)^2 + (y - b)^2 + (z - c)^2 = \delta^2$.

Example 6

Discuss the continuity of the following functions:

(a) $f(x, y) = \dfrac{x - 2y}{x^2 + y^2}$

(b) $g(x, y) = \dfrac{2}{y - x^2}$

(c) $h(x, y, z) = \dfrac{1}{x^2 + y^2 - z}$

Solution:

(a) Since rational functions are continuous except where the denominator is zero (see Theorem 2.3), we conclude that

$$f(x, y) = \frac{x - 2y}{x^2 + y^2}$$

is continuous at each point in the xy-plane except $(0, 0)$. (See Figure 15.10.)

(b) The function given by

$$g(x, y) = \frac{2}{y - x^2}$$

is continuous except at points on the parabola $y = x^2$. (See Figure 15.11.)

(c) In this case $h(x, y, z) = \dfrac{1}{x^2 + y^2 - z}$

is continuous at each point in space, except at points on the paraboloid $z = x^2 + y^2$. ∎

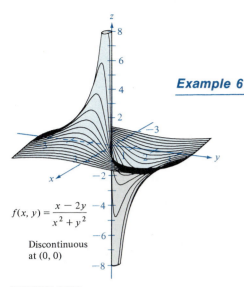

$f(x, y) = \dfrac{x - 2y}{x^2 + y^2}$

Discontinuous at $(0, 0)$

FIGURE 15.10

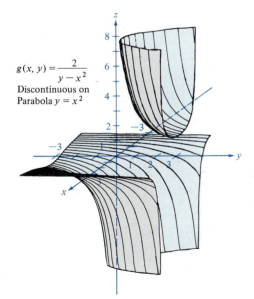

$g(x, y) = \dfrac{2}{y - x^2}$

Discontinuous on Parabola $y = x^2$

FIGURE 15.11

Section Exercises (15.1)

1. If $f(x, y) = x/y$, find the following:
(a) $f(3, 2)$ (b) $f(-1, 4)$
(c) $f(\frac{1}{2}, y)$ (d) $f(x, \frac{2}{3})$
(e) $f(x + \Delta x, y)$ (f) $f(x, y + \Delta y)$

2. If $f(x, y) = 4 - x^2 - 4y^2$, find the following:
(a) $f(0, 0)$ (b) $f(0, 1)$
(c) $f(x, \frac{1}{2})$ (d) $f(2, y)$
(e) $f(x + \Delta x, y)$ (f) $f(x, y + \Delta y)$

In Exercises 3–12, describe the region R, in the xy-coordinate plane, that corresponds to the domain of the given function.

[S] **3.** $f(x, y) = \sqrt{4 - x^2 - y^2}$
4. $f(x, y) = \sqrt{4 - x^2 - 4y^2}$
5. $f(x, y) = \sqrt{x^2 + y^2 - 1}$
[S] **6.** $f(x, y) = \text{Arcsin } (x + y)$

7. $z = \dfrac{x + y}{xy}$

8. $z = \dfrac{xy}{x - y}$

9. $f(x, y) = \ln (4 - x - y)$
10. $f(x, y) = \ln (4 - xy)$
11. $f(x, y) = e^{x/y}$
12. $f(x, y) = 4 - x^2 - y^2$

In Exercises 13–20, sketch the graph of the surface specified by each function.
13. $z = 4 - x^2 - y^2$
14. $z = \sqrt{x^2 + y^2}$

15. $f(x, y) = y^2$
16. $f(x, y) = \frac{1}{12} \sqrt{144 - 16x^2 - 9y^2}$
17. $f(x, y) = 6 - 2x - 3y$
18. $f(x, y) = y^2 - x^2 + 1$
[S] **19.** $f(x, y) = e^{-x}$
20. $f(x, y) = \begin{cases} xy, x \geq 0, y \geq 0 \\ 0, \text{ elsewhere} \end{cases}$

In Exercises 21–30, describe the level curves for each function. In each case sketch several level curves.
21. $f(x, y) = \sqrt{25 - x^2 - y^2}$
22. $f(x, y) = x^2 + y^2$ **23.** $z = 6 - 2x - 3y$

24. $z = xy$ [S] **25.** $f(x, y) = \dfrac{x}{x^2 + y^2}$

26. $f(x, y) = \text{Arctan}\dfrac{y}{x}$ **27.** $f(x, y) = \ln (x - y)$

28. $f(x, y) = \dfrac{x + y}{x - y}$ [S] **29.** $f(x, y) = e^{xy}$

30. $f(x, y) = \cos (x + y)$

In Exercises 31–40, discuss the continuity of each function and evaluate the indicated limit, if it exists.

31. $\lim\limits_{(x, y) \to (1, 1)} \dfrac{xy}{x^2 + y^2}$

32. $\lim\limits_{(x, y) \to (0, 0)} 1 - \dfrac{\cos (x^2 + y^2)}{x^2 + y^2}$

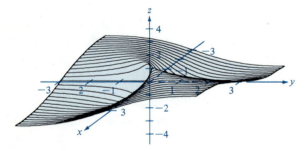

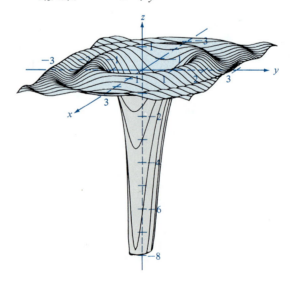

S 33. $\displaystyle\lim_{(x,y)\to(0,0)} \frac{xy}{x^3 + y^3}$

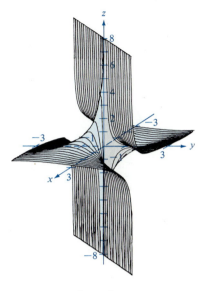

34. $\displaystyle\lim_{(x,y)\to(0,0)} \frac{y}{x^2 + y^2}$

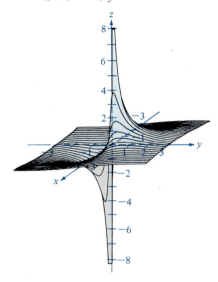

35. $\displaystyle\lim_{(x,y)\to(1,3)} \frac{6x - 2y}{9x^2 - y^2}$

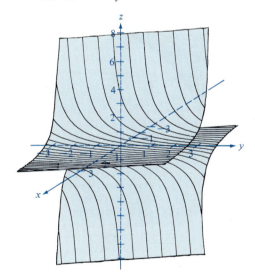

S 36. $\displaystyle\lim_{(x,y)\to(0,0)} \frac{2x - y^2}{2x^2 + y}$

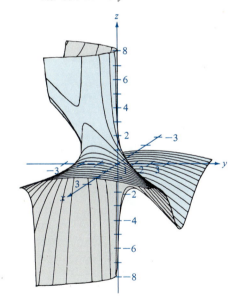

37. $\lim_{(x,y)\to(0,0)} \dfrac{\sin xy}{xy}$

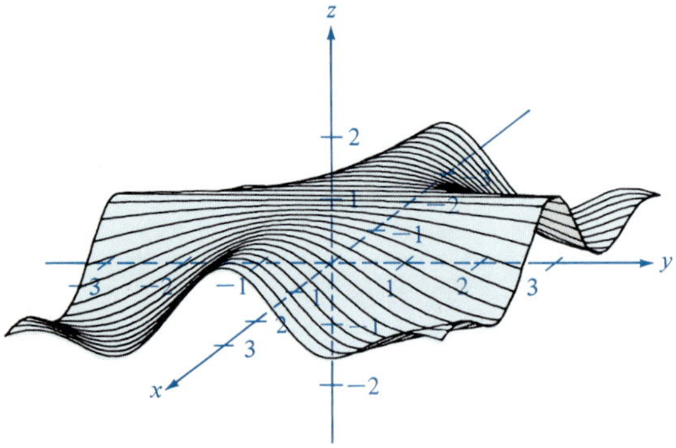

38. $\lim_{(x,y)\to(0,0)} e^{xy}$

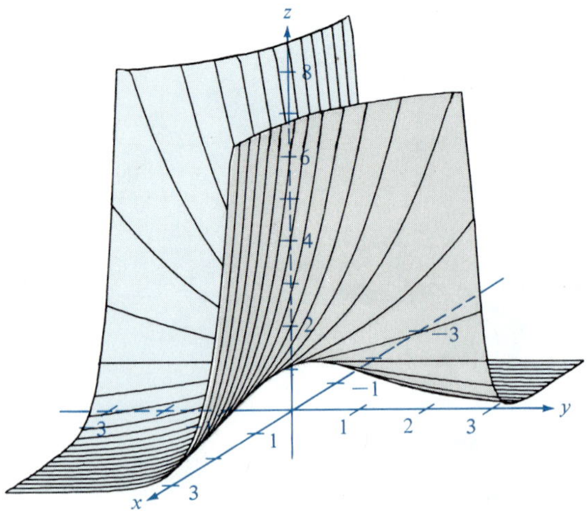

39. $\lim_{(x,y)\to(0,1)} \dfrac{1 - x^2 - y^2}{\sqrt{|1 - x^2 - y^2|}}$

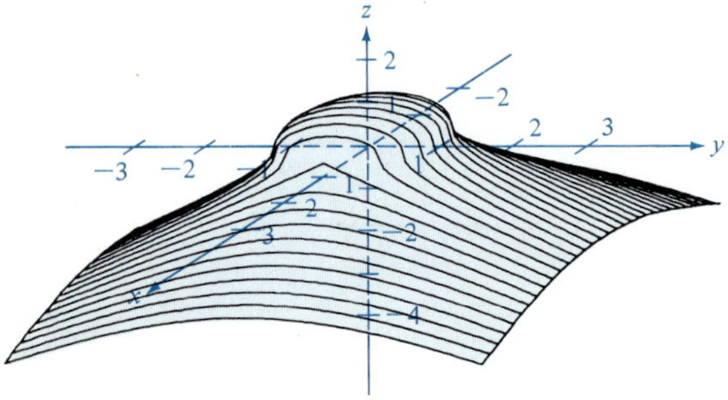

40. $\lim_{(x,y)\to(0,0)} \ln(x^2 + y^2)$

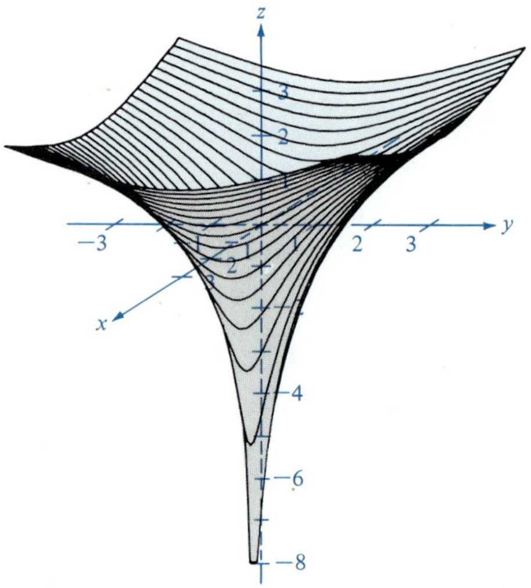

41. Show that $\lim_{(x,y)\to(0,0)} [xy/(x^2 + y^2)]$ does not exist. (See sketch in Exercise 31.)

42. Show that $\lim_{(x,y)\to(0,0)} [\sin(x + y)/(x + y)] = 1$. (Hint: Use polar coordinates.)

[S] **43.** Show that $\lim_{(x,y)\to(0,0)} [xy^2/(x^2 + y^2)] = 0$. (Hint: Use polar coordinates.)

44. Determine (if possible) how $f(0, 0)$ should be defined so that $f(x, y) = (x^3 + y^3)/(x^2 + y^2)$ is a continuous function.

45. Determine (if possible) how $f(0, 0)$ should be defined so that $f(x, y) = (x + y)/(x - y)$ is continuous.

46. Discuss the continuity of $f(x, y, z) = 1/\sqrt{x^2 + y^2 + z^2}$.

[S] **47.** Discuss the continuity of $f(x, y, z) = z/(x^2 + y^2 - 4)$.

48. Discuss the continuity of $f(x, y, z) = \sin z/(e^x + e^y)$.

49. Discuss the continuity of $w = xy \sin z$.

50. (a) Sketch the graph of the surface $f(x, y) = x^2 + y^2$.

(b) On the surface of part (a), sketch the graph of $f(1, y)$.

(c) On the surface of part (a), sketch the graph of $f(x, 1)$.

51. (a) Sketch the graph of the surface $f(x, y) = xy$ in the first octant.

(b) On the surface of part (a), sketch the graph of $f(x, x)$.

For Review and Class Discussion

True or False

1. _____ If $\lim_{(x,y)\to(0,0)} f(x, y) = 0$, then $\lim_{x\to 0} f(x, 0) = 0$.

2. _____ If $\lim_{x\to 0} f(x, 0) = 0$ and $\lim_{y\to 0} f(0, y) = 0$, then $\lim_{(x,y)\to(0,0)} f(x, y) = 0$.

3. _____ If f is continuous for all nonzero x and y and $f(0, 0) = 0$, then $\lim_{(x,y)\to(0,0)} f(x, y) = 0$.

4. _____ The level curves of $f(x, y) = \sqrt{9 - x^2 - y^2}$ are circles.

5. _____ If g and h are continuous functions of x and y, respectively, and $f(x, y) = g(x) + h(y)$, then f is continuous.

15.2 Partial Derivatives

Purpose

- To define the partial derivatives of a function of two variables.
- To provide a geometric interpretation of partial derivatives.
- To demonstrate how to find partial derivatives of order two and higher.

How thoroughly it is ingrained in mathematical science that every real advance goes hand in hand with the invention of sharper tools and simpler methods which at the same time assist in understanding earlier theories and in casting aside some more complicated developments.

—David Hilbert (1862–1943)

In the applications of functions of several variables, the question often arises, "How will the function be affected if I change one or some or all of its independent variables?" We can answer this question, at least in part, by considering the independent variables one at a time. For instance, the gross national product is a function of many variables such as tax rates, unemployment, and wars. Thus an economist who wants to determine the effect of a tax increase would hold all other variables constant while raising or lowering taxes. Or to determine the effect of a certain catalyst in an experiment, a chemist could conduct the experiment several times by using varying amounts of the catalyst while keeping constant other variables such as temperature and pressure.

Mathematically we follow a similar procedure to determine the rate of change of a function f with respect to one of its several independent variables. In this procedure we find the derivative of f with respect to one independent variable at a time while holding the others constant. This process is called **partial differentiation** and the result is referred to as the **partial derivative** of f with respect to the chosen independent variable. For functions of two variables, we give the following definition:

Definition of Partial Derivatives	If $z = f(x, y)$, then the **first-partial derivatives of f with respect to x and to y** are the functions $\partial f/\partial x$ and $\partial f/\partial y$ defined as

$$\frac{\partial f}{\partial x} = \lim_{\Delta x \to 0} \frac{f(x + \Delta x, y) - f(x, y)}{\Delta x}$$

$$\frac{\partial f}{\partial y} = \lim_{\Delta y \to 0} \frac{f(x, y + \Delta y) - f(x, y)}{\Delta y}$$

Note that the definition indicates that partial derivatives are determined by temporarily treating a function of two variables as a function of one variable by considering the other variable to be fixed. For instance, if $z = f(x, y)$, then to find $\partial f/\partial x$ we *consider y constant* and differentiate with respect to x. To find $\partial f/\partial y$ we *consider x constant* and differentiate with respect to y.

Example 1

Find $\partial f/\partial x$ and $\partial f/\partial y$ where $f(x, y) = 3x - x^2 y^2 + 2x^3 y$.

Solution: Considering y to be constant, we have

$$\frac{\partial f}{\partial x} = 3 - 2xy^2 + 6x^2 y$$

Now considering x to be constant and differentiating with respect to y gives us

$$\frac{\partial f}{\partial y} = -2x^2 y + 2x^3$$

∎

There are several notations for first partial derivatives. We list the common ones here along with the notation for a partial derivative evaluated at some point (a, b).

Notation for First Partial Derivatives	If $z = f(x, y)$, then

$$\frac{\partial f}{\partial x} = f_x = z_x = \frac{\partial z}{\partial x} = \frac{\partial}{\partial x} f(x, y) = f_x(x, y)$$

and
$$\frac{\partial f}{\partial y} = f_y = z_y = \frac{\partial z}{\partial y} = \frac{\partial}{\partial y} f(x, y) = f_y(x, y)$$

The first partials evaluated at the point (a, b) are denoted by

$$\frac{\partial f}{\partial x}\bigg|_{(a,b)} \qquad \text{or} \qquad f_x(a, b)$$

and
$$\frac{\partial f}{\partial y}\bigg|_{(a,b)} \qquad \text{or} \qquad f_y(a, b)$$

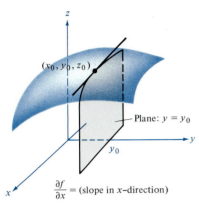

$\dfrac{\partial f}{\partial x}$ = (slope in x–direction)

FIGURE 15.12

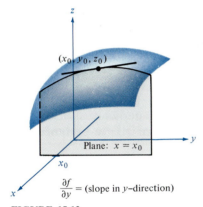

$\dfrac{\partial f}{\partial y}$ = (slope in y–direction)

FIGURE 15.13

The partial derivatives of a function of two variables, $z = f(x, y)$, have a simple geometric interpretation. If y is fixed, say, $y = y_0$, then $z = f(x, y_0)$ represents a curve that is the intersection of the plane $y = y_0$ and the surface $z = f(x, y)$. (See Figure 15.12.) Thus

$$\frac{\partial f}{\partial x}\bigg|_{(x_0, y_0)} = f_x(x_0, y_0) = \lim_{\Delta x \to 0} \frac{f(x_0 + \Delta x, y_0) - f(x_0, y_0)}{\Delta x}$$

represents the slope of this curve at the point (x_0, y_0, z_0). (Note that both the curve and the tangent line lie in the plane $y = y_0$.) Similarly,

$$\frac{\partial f}{\partial y}\bigg|_{(x_0, y_0)} = f_y(x_0, y_0) = \lim_{\Delta y \to 0} \frac{f(x_0, y_0 + \Delta y) - f(x_0, y_0)}{\Delta y}$$

represents the slope of the intersection of $z = f(x, y)$ and the plane $x = x_0$ at (x_0, y_0, z_0). (See Figure 15.13.)

Informally we say that the values of $\partial f/\partial x$ and $\partial f/\partial y$ at some point (x_0, y_0, z_0) on the surface $z = f(x, y)$ denote the **slope of the surface in the x and y directions,** respectively.

All the definitions and notations of this section can easily be extended to functions of three or more variables

$$u = f(x, y, z) \qquad \text{or} \qquad w = f(x_1, x_2, \ldots, x_n)$$

even though there is no simple geometric interpretation for partial derivatives of functions of three or more variables. However, we can interpret the partial derivatives of functions of several variables as *rates of change* no matter how many variables are involved.

Example 2

The volume of a frustum of a cone (Figure 15.14) is given by $V = \frac{1}{3}\pi h(R^2 + Rr + r^2)$. When $R = 10$, $r = 4$, and $h = 6$, what is the rate of change in volume with respect to the upper radius r? With respect to the height h?

Solution: Since
$$\frac{\partial V}{\partial r} = \frac{1}{3}\pi h(0 + R + 2r)$$

we have, for $R = 10$, $r = 4$, and $h = 6$,

$$\frac{\partial V}{\partial r} = \frac{1}{3}\pi(6)(18) = 36\pi$$

Furthermore,

$$\frac{\partial V}{\partial h} = \frac{1}{3}\pi(R^2 + Rr + r^2) = \frac{1}{3}\pi(100 + 40 + 16) = 52\pi$$

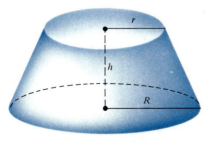

Frustum of a Cone

FIGURE 15.14

As is true for ordinary derivatives, it is possible to take a second, third, and so forth, partial derivative of a function of several variables. For instance, if $z = f(x, y)$, we can take the following second partial derivatives:

1. both with respect to x,

$$\frac{\partial}{\partial x}\left(\frac{\partial f}{\partial x}\right) = \frac{\partial^2 f}{\partial x^2} = \frac{\partial^2 z}{\partial x^2} = f_{xx} = z_{xx}$$

2. both with respect to y,

$$\frac{\partial}{\partial y}\left(\frac{\partial f}{\partial y}\right) = \frac{\partial^2 f}{\partial y^2} = \frac{\partial^2 z}{\partial y^2} = f_{yy} = z_{yy}$$

3. first with respect to x, then y,

$$\frac{\partial}{\partial y}\left(\frac{\partial f}{\partial x}\right) = \frac{\partial^2 f}{\partial y\,\partial x} = \frac{\partial^2 z}{\partial y\,\partial x} = f_{xy} = z_{xy}$$

4. first with respect to y, then x,

$$\frac{\partial}{\partial x}\left(\frac{\partial f}{\partial y}\right) = \frac{\partial^2 f}{\partial x\,\partial y} = \frac{\partial^2 z}{\partial x\,\partial y} = f_{yx} = z_{yx}$$

Partials 3 and 4 are referred to as "mixed" partial derivatives. Notice that the symbol $\partial^2 f/\partial y\,\partial x$ means the partial derivative with respect to x first, then with respect to y.

Observe that for a function of two variables, there are two first partials and four second partials. For a function of three variables, there are three first partials, f_x, f_y, f_z, and nine second partials,

$$f_{xx}, f_{xy}, f_{xz}, f_{yx}, f_{yy}, f_{yz}, f_{zx}, f_{zy}, f_{zz}$$

six of which are mixed partials. We will see in the next two examples that some of these mixed partials are equal. In fact, it can be shown that *if f has continuous second partials,* then the order in which the partial derivatives are taken is immaterial. That is

$$f_{xy} = f_{yx}, \qquad f_{xz} = f_{zx}$$

and so on.

To take partial derivatives of order three and higher, we follow this same pattern. For instance, if $z = f(x, y)$, we have

$$z_{xxx} = \frac{\partial}{\partial x}\left(\frac{\partial^2 f}{\partial x^2}\right) = \frac{\partial^3 f}{\partial x^3} \qquad \text{and} \qquad z_{xxy} = \frac{\partial}{\partial y}\left(\frac{\partial^2 f}{\partial x^2}\right) = \frac{\partial^3 f}{\partial y\,\partial x^2}$$

Example 3

Find the second partial derivatives of $z = 3xy^2 - 2y + 5x^2 y^2$. Determine the value of $z_{xy}(-1, 2)$.

Solution:

Since

$$\frac{\partial z}{\partial x} = 3y^2 + 10xy^2 \quad \text{and} \quad \frac{\partial z}{\partial y} = 6xy - 2 + 10x^2y$$

we have

$$\frac{\partial^2 z}{\partial y \, \partial x} = 6y + 20xy \quad \text{and} \quad \frac{\partial^2 z}{\partial x \, \partial y} = 6y + 20xy$$

Furthermore,

$$\frac{\partial^2 z}{\partial x^2} = 10y^2 \quad \text{and} \quad \frac{\partial^2 z}{\partial y^2} = 6x + 10x^2$$

Finally,

$$z_{xy}(-1, 2) = \frac{\partial^2 z}{\partial y \, \partial x}\bigg|_{(-1, 2)} = 12 - 40 = -28 \qquad \blacksquare$$

Example 4

Find the second partial derivatives of $f(x, y, z) = ye^x + x \sin 2z$.

Solution: Since

$$f_x = ye^x + \sin 2z$$
$$f_y = e^x$$
$$f_z = 2x \cos 2z$$

it follows that

$$
\begin{array}{lll}
f_{xx} = ye^x, & f_{xy} = e^x, & f_{xz} = 2 \cos 2z \\
f_{yx} = e^x, & f_{yy} = 0, & f_{yz} = 0 \\
f_{zx} = 2 \cos 2z, & f_{zy} = 0, & f_{zz} = -4x \sin 2z
\end{array}
$$

$\blacksquare$

Section Exercises (15.2)

In Exercises 1–16, find all the first partial derivatives and evaluate them at the indicated points.

1. $z = x^2 - 3xy + y^2$, $(-1, 2)$

$\boxed{s}$ **2.** $z = xe^{x/y}$, $(0, 1)$

3. $f(x, y) = e^{-(x^2+y^2)}$

4. $f(x, y) = \dfrac{x^2}{2y} + \dfrac{4y^2}{x}$

5. $z = \sin(2x - y)$

6. $z = \sin 3x \cos 3y$

$\boxed{s}$ **7.** $z = \text{Arctan} \dfrac{y}{x}$, $(2, -2)$

8. $z = \sqrt{x^2 + y^2}$, $(-3, 4)$

9. $f(x, y) = e^y \sin xy$

10. $f(x, y) = \cos(x^2 + y^2)$

11. $z = \ln \sqrt{x^2 + y^2}$, $(-2, 0)$

12. $z = \dfrac{xy}{x - y}$, $(2, -2)$

13. $f(x, y, z) = \sqrt{x^2 + y^2 + z^2}$, $(2, -1, 2)$

14. $f(x, y, z) = \dfrac{xy}{x + y + z}$, $(1, 2, 0)$

15. $w = \sin(x + 2y + 3z)$

$\boxed{s}$ **16.** $w = \ln \sqrt{x^2 + y^2 + z^2}$

In Exercises 17–24, find $\partial^2 z/\partial x^2$, $\partial^2 z/\partial y^2$, $\partial^2 z/\partial x \, \partial y$, and $\partial^2 z/\partial y \, \partial x$.

$\boxed{s}$ **17.** $z = x^2 - 2xy + 3y^2$

18. $z = 2e^{xy^2}$

19. $z = e^x \tan y$

20. $z = x^4 - 3x^2y^2 + y^4$

$\boxed{s}$ **21.** $z = \dfrac{xy}{x - y}$

22. $z = \sin(x - 2y)$

23. $z = \text{Arctan} \dfrac{y}{x}$

24. $z = \sqrt{x^2 + y^2}$

In Exercises 25–28, show that each function is a solution of $(\partial^2 z/\partial x^2) + (\partial^2 z/\partial y^2) = 0$.

25. $z = kxy$, where k is a constant

26. $z = \left(\frac{e^y - e^{-y}}{2}\right) \sin x$

27. $z = e^x \sin y$

28. $z = \text{Arctan}\, \frac{y}{x}$

In Exercises 29–30, show that each function is a solution of $\partial^2 z/\partial t^2 = c^2(\partial^2 z/\partial x^2)$.

[S] **29.** $z = \sin(x - ct)$

30. $z = \sin wct \sin wx$, c is a constant

In Exercises 31–32, show that each function is a solution of $\partial z/\partial t = c^2(\partial^2 z/\partial x^2)$.

31. $z = e^{-t} \cos \frac{x}{c}$

32. $z = e^{-t} \sin \frac{x}{c}$

In Exercises 33–36, use the limit definition of partial derivatives to find $\partial f/\partial x$ and $\partial f/\partial y$.

33. $f(x, y) = 2x + 3y$

34. $f(x, y) = \frac{1}{x + y}$

[S] **35.** $f(x, y) = x^2 - 2xy + y^2$

36. $f(x, y) = \sqrt{x + y}$

37. Find the slope, in the x direction, of the sphere $x^2 + y^2 + z^2 = 49$ at the point $(2, 3, 6)$.

38. Find the slope, in the y direction, of the paraboloid $z = x^2 + 4y^2$ at the point $(2, 1, 8)$.

[S] **39.** At the point $(1, 3, 0)$ on the paraboloid $z = 9x^2 - y^2$, find the slope of the curve cut by the plane $x = 1$.

40. At the point $(2, 1, 2\sqrt{2})$ on the ellipsoid $x^2 + 4y^2 + z^2 = 16$, find the slope of the curve cut by the plane $y = 1$.

41. A company manufactures two types of wood-burning stoves: a free-standing model and a fireplace-insert model. The cost function for producing x free-standing stoves and y fireplace-insert stoves is

$$C = 32\sqrt{xy} + 175x + 205y + 1050$$

Find the marginal costs ($\partial C/\partial x$ and $\partial C/\partial y$) when $x = 80$ and $y = 20$.

42. Let N be the number of applicants to a university, p the charge for food and housing at the university, and t the tuition. If $N = f(p, t)$, explain why $\partial N/\partial p < 0$ and $\partial N/\partial t < 0$.

43. The range of a projectile fired at an angle θ above the horizontal with a velocity v_0 is $R = (v_0^2 \sin 2\theta)/32$. Evaluate $\partial R/\partial v_0$ and $\partial R/\partial \theta$ when $v_0 = 2000$ ft/s and $\theta = 5°$.

44. According to the *ideal gas law*, $PV = kT$, where P is pressure, V is volume, T is temperature, and k is a constant of proportionality. A tank contains 2600 in.^{3} of nitrogen at a pressure of 20 lb/in.^{2} and a temperature of 40°F.

(a) Determine k.
(b) Evaluate $\partial P/\partial T$ for the given values.
(c) Evaluate $\partial V/\partial P$ for the given values.

45. The temperature at any point (x, y) in a steel plate is given by

$$T = 500 - 0.6x^2 - 1.5y^2$$

where x and y are measured in feet. At the point $(2, 3)$, find the rate of change of the temperature with respect to the distance moved along the plate in the directions of the x- and y-axes, respectively.

For Review and Class Discussion

True or False

1. ____ If $z = xy^2$, then $\partial z/\partial x = (2xy)(\partial y/\partial x) + y^2$.

2. ____ If $z = f(x, y)$ and $\partial z/\partial x = \partial z/\partial y$, then $z = c(x + y)$.

3. ____ If $z = f(x)\, g(y)$, then $(\partial z/\partial x) + (\partial z/\partial y) = f'(x)\, g(y) + f(x)\, g'(y)$.

4. ____ If $z = e^{xy}$, then $\partial^2 z/\partial y\, \partial x = (xy + 1)e^{xy}$.

5. ____ If a cylindrical surface, $z = f(x, y)$, has rulings parallel to the y-axis, then $\partial z/\partial y = 0$.

15.3 **Extrema of Functions of Two Variables**

Purpose
- To define the relative extrema of a function of two variables.
- To define the critical points of a function of two variables.
- To discuss a second-partial-derivative test for relative extrema.

For since the fabric of the universe is most perfect and the work of a most wise Creator, nothing at all takes place in the universe in which some rule of maximum or minimum does not appear.   —*Leonhard Euler (1707–1783)*

A considerable portion of our study of the derivative of a function of one variable dealt with the finding and testing of the extreme values of a function (Chapters 4 and 5). In this section we take up this subject for functions of several variables and we will find the ideas to be much the same. Although most theorems and definitions will be given in terms of functions of two variables, analogous ones may be stated for functions of three or more variables.

THEOREM 15.1
(*Extreme Value Theorem*)

Let f be a function of two variables x and y, defined and continuous on a closed region R in the xy-plane. Then:

i.   There is at least one point in R where f takes on a minimum value.

ii. There is at least one point in R where f takes on a maximum value.

(See Figure 15.15.)

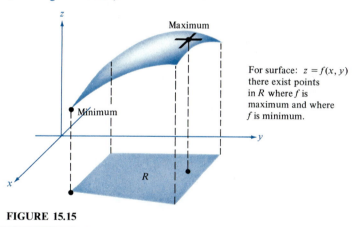

For surface: $z = f(x, y)$ there exist points in R where f is maximum and where f is minimum.

FIGURE 15.15

Note: By a *closed region R* in the xy-plane we mean all points within R as well as all points on the boundary of R. This is comparable to a closed interval $[a, b]$ of real numbers, which includes the endpoints a and b. Similarly, an *open region* is comparable to an open interval and does *not* include the points on the boundary.

We define first the relative (local) extrema of a function of two variables.

Definition of Local Extrema	If f is a function defined at (x_0, y_0), then $f(x_0, y_0)$ is called:

1. a **relative maximum** of f if there is a circular region R containing (x_0, y_0) such that $f(x, y) \leq f(x_0, y_0)$ for all (x, y) in R

2. a **relative minimum** of f if there is a circular region R containing (x_0, y_0) such that $f(x, y) \geq f(x_0, y_0)$ for all (x, y) in R

To say that $z_0 = f(x_0, y_0)$ is a relative maximum of f means geometrically that the point (x_0, y_0, z_0) is at least as high as all "nearby" points on the graph of $z = f(x, y)$. Similarly, $z_0 = f(x_0, y_0)$ is a relative minimum if (x_0, y_0, z_0) is at least as low as all "nearby" points on the graph of $z = f(x, y)$. In Figure 15.16 several relative maximum and minimum points are pictured.

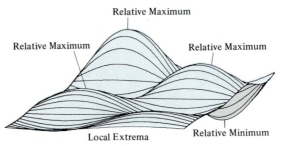

FIGURE 15.16

To locate the relative extrema of a function of two variables, we look to its partial derivatives. As was pointed out in Section 15.2, if surface S is the graph of $z = f(x, y)$ and if f_x exists, then at (x_0, y_0, z_0), $f_x(x_0, y_0)$ is the slope of the surface in the x direction. From Figure 15.16 it would appear that if $f(x_0, y_0)$ is a local extremum, then $f_x(x_0, y_0) = 0$ and, similarly, $f_y(x_0, y_0) = 0$. This result is verified in the following theorem:

THEOREM 15.2 (*First Partials Are Zero at Local Extrema*)	If $f(x_0, y_0)$ is a local extremum of f on an open region R in the xy-plane, and the first partial derivatives of f exist in R, then $$f_x(x_0, y_0) = 0 = f_y(x_0, y_0)$$

Proof: Consider the function of a single variable,

$$g(x) = f(x, y_0)$$

By hypothesis it has a local extremum at x_0 and it is differentiable there. Hence

$$g'(x_0) = \lim_{\Delta x \to 0} \frac{f(x_0 + \Delta x, y_0) - f(x_0, y_0)}{\Delta x} = f_x(x_0, y_0) = 0$$

Similarly, $h(y) = f(x_0, y)$ has a local extremum at y_0 and being differentiable there, it satisfies the equation

$$h'(y_0) = \lim_{\Delta y \to 0} \frac{f(x_0, y_0 + \Delta y) - f(x_0, y_0)}{\Delta y} = f_y(x_0, y_0) = 0$$

Thus it follows that

$$f_x(x_0, y_0) = f_y(x_0, y_0) = 0$$

if $f(x_0, y_0)$ is a local extremum.   ■

A point (x_0, y_0) for which both $f_x(x_0, y_0) = 0$ and $f_y(x_0, y_0) = 0$ is called a **critical point** of f. The contrapositive of Theorem 15.2 suggests that if the first partial derivatives exist, then to find local extrema we need only examine values of $f(x, y)$ at critical points. However, as is true for a function of one variable, the critical points of a function of two variables do not always yield relative maxima or minima. For example, some critical points yield "saddle points" (see Figure 15.17), which are neither relative maxima nor relative minima.

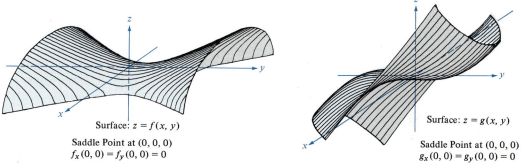

Surface: $z = f(x, y)$

Saddle Point at $(0, 0, 0)$
$f_x(0, 0) = f_y(0, 0) = 0$

Surface: $z = g(x, y)$

Saddle Point at $(0, 0, 0)$
$g_x(0, 0) = g_y(0, 0) = 0$

FIGURE 15.17

Example 1

Determine the local extrema of $f(x, y) = 2x^2 + y^2 + 8x - 6y + 20$.

Solution: Since

$$f_x = 4x + 8 \qquad \text{and} \qquad f_y = 2y - 6$$

we solve the system of equations

$$4x + 8 = 0 \qquad \text{and} \qquad 2y - 6 = 0$$

to obtain the solution $(-2, 3)$. Now for all $(x, y) \neq (-2, 3)$, completing the square shows us that

$$f(x, y) = 2(x + 2)^2 + (y - 3)^2 + 3 > 3$$

Hence a relative *minimum* of f occurs at $(-2, 3)$ and its value is $f(-2, 3) = 3$. (See Figure 15.18.)

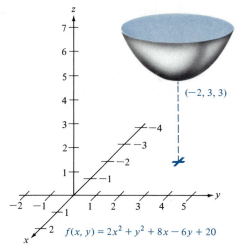

$f(x, y) = 2x^2 + y^2 + 8x - 6y + 20$

FIGURE 15.18   ∎

For a function such as the one in Example 1 it is relatively easy to determine the *type* of local extremum at the critical points. This can be done by algebraic arguments or by sketching a graph of the function. For more complicated functions such procedures are generally not so fruitful, and hence we seek a test that gives conditions under which a critical point will yield a relative maximum, a relative minimum, or neither.

The basic test for determining the relative maxima and minima of a function of two variables is the "second-partials test," which is the counterpart of the Second-Derivative Test for functions of one variable. We give the following test, omitting its rather complicated proof:

THEOREM 15.3
(*Second-Partials Test*)

Let f be a function of two variables that has continuous first and second partial derivatives on an open region R. Suppose further that $f_x(a, b) = f_y(a, b) = 0$ at some point (a, b) in R. Then:

i. $f(a, b)$ is a local **minimum** of f if

$$f_{xx}(a, b) f_{yy}(a, b) - [f_{xy}(a, b)]^2 > 0 \qquad \text{and} \qquad f_{xx}(a, b) > 0$$

ii. $f(a, b)$ is a local **maximum** of f if

$$f_{xx}(a, b) f_{yy}(a, b) - [f_{xy}(a, b)]^2 > 0 \qquad \text{and} \qquad f_{xx}(a, b) < 0$$

iii. $(a, b, f(a, b))$ is a **saddle point** if

$$f_{xx}(a, b) f_{yy}(a, b) - [f_{xy}(a, b)]^2 < 0$$

Note that in parts i and ii, $f_{xx}(a, b)$ and $f_{yy}(a, b)$ must agree in sign in order for $f_{xx}(a, b) f_{yy}(a, b) - [f_{xy}(a, b)]^2$ to be positive. Hence we may replace $f_{xx}(a, b) > 0$ by $f_{yy}(a, b) > 0$ in these two parts. Also, note that if

$$f_{xx}(a, b) f_{yy}(a, b) - [f_{xy}(a, b)]^2 = 0$$

the test gives no information and it is then necessary to rely on sketches or methods similar to the one used in Example 1.

Example 2

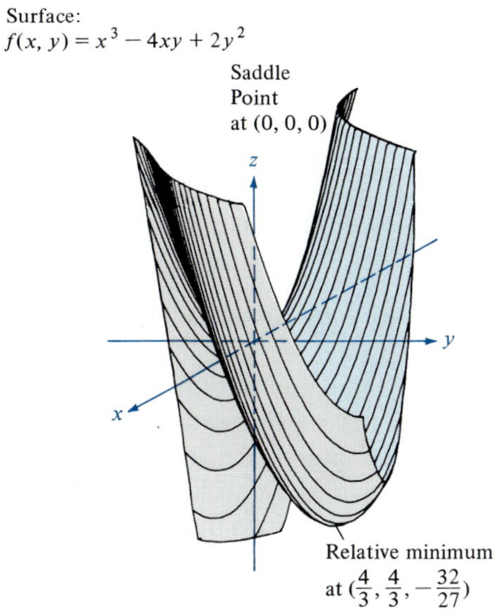

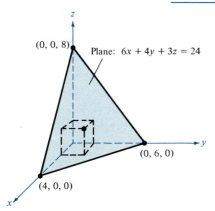

Surface:
$f(x, y) = x^3 - 4xy + 2y^2$

Saddle
Point
at $(0, 0, 0)$

Relative minimum
at $(\frac{4}{3}, \frac{4}{3}, -\frac{32}{27})$

FIGURE 15.19

Find the local extrema of $f(x, y) = x^3 - 4xy + 2y^2$, if any exist.

Solution: Since

$$f_x(x, y) = 3x^2 - 4y \quad \text{and} \quad f_y(x, y) = -4x + 4y$$

we obtain the system

$$3x^2 - 4y = 0$$
$$-4x + 4y = 0 \quad \text{or} \quad x = y$$

Thus by substitution
$$3(y)^2 - 4y = 0$$
$$y(3y - 4) = 0$$

and
$$y = 0 \quad \text{or} \quad y = \tfrac{4}{3}$$

Therefore, the critical points are $(0, 0)$ and $(\frac{4}{3}, \frac{4}{3})$. Since

$$f_{xx}(x, y) = 6x, \qquad f_{yy}(x, y) = 4, \qquad f_{xy}(x, y) = -4$$

it follows that

$$f_{xx}(0, 0)f_{yy}(0, 0) - [f_{xy}(0, 0)]^2 = 0 - 16 < 0$$

and by part iii of Theorem 15.3, we conclude that $(0, 0, 0)$ is a saddle point of f. Furthermore, since $f_{xx}(\frac{4}{3}, \frac{4}{3}) = 8 > 0$ and

$$f_{xx}(\tfrac{4}{3}, \tfrac{4}{3})f_{yy}(\tfrac{4}{3}, \tfrac{4}{3}) - [f_{xy}(\tfrac{4}{3}, \tfrac{4}{3})]^2 = 8(4) - 16 = 16 > 0$$

we conclude by part i of Theorem 15.3 that $f(\frac{4}{3}, \frac{4}{3}) = -\frac{32}{27}$ is a relative *minimum* of f. (See Figure 15.19.) ∎

Example 3

$(0, 0, 8)$

Plane: $6x + 4y + 3z = 24$

$(0, 6, 0)$

$(4, 0, 0)$

FIGURE 15.20

A rectangular box is resting on the xy-plane with one vertex at the origin. Find the maximum volume of the box if its vertex opposite the origin lies in the plane $6x + 4y + 3z = 24$ (see Figure 15.20).

Solution: Since one vertex of the box lies in the plane

$$6x + 4y + 3z = 24 \quad \text{or} \quad z = \tfrac{1}{3}(24 - 6x - 4y)$$

we can denote the volume of the box by

$$V = xyz = \tfrac{1}{3}xy(24 - 6x - 4y) = \tfrac{1}{3}(24xy - 6x^2y - 4xy^2)$$

Now from the system

$$V_x = \frac{1}{3}(24y - 12xy - 4y^2) = \frac{y}{4}(24 - 12x - 4y) = 0$$

$$V_y = \frac{1}{3}(24x - 6x^2 - 8xy) = \frac{x}{3}(24 - 6x - 8y) = 0$$

we obtain the solutions $x = 0, y = 0$ and $x = \frac{4}{3}, y = 2$. Since the volume is zero for $x = 0$ and $y = 0$, we test the values $x = \frac{4}{3}$ and $y = 2$ to see if they yield a maximum volume. From

$$V_{xx}(x, y) = -4y, \qquad V_{yy}(x, y) = \frac{-8x}{3}, \qquad V_{xy}(x, y) = \frac{1}{3}(24 - 12x - 8y)$$

and

$$V_{xx}(\tfrac{4}{3}, 2) \, V_{yy}(\tfrac{4}{3}, 2) - [V_{xy}(\tfrac{4}{3}, 2)]^2 = (-8)(\tfrac{-32}{9}) - [\tfrac{-8}{3}]^2 = \tfrac{64}{3} > 0$$

we conclude by Theorem 15.3 that V is maximum for $x = \frac{4}{3}$ and $y = 2$, and this maximum volume is

$$V = xyz = \tfrac{4}{3}(2)(\tfrac{1}{3})(24 - 8 - 8) = \tfrac{64}{9} \text{ cubic units} \qquad \blacksquare$$

In Example 3 we were able to convert a function of three variables to a function of two variables to obtain a solution. In general, this is not necessary since the definitions of local extrema and critical points can be extended to functions of three or more variables. Specifically, if all first partials of $w = f(x, y, z, \ldots)$ exist, then it can be shown that a local maximum or minimum can occur at $(x, y, z, \ldots)$ only if

$$f_x = 0, \qquad f_y = 0, \qquad f_z = 0, \qquad \ldots$$

(compare to Theorem 15.2), which means the critical points are obtained by solving this system of equations. The extension of Theorem 15.3 to three or more variables is also possible, though we will not consider such an extension in this text.

Example 3 also illustrates a very common type of application where the variables are subject to a *constraint* (or *side condition*). For instance, in that example we maximized the function $V = xyz$ subject to the condition that x, y, and z satisfy the equation $6x + 4y + 3z = 24$. A solution was obtained from the equation $V = xyz$ by replacing z by its value in the equation $3z = 24 - 6x - 4y$. Another method that can be used to solve this type of problem makes use of variables called **Lagrange multipliers.** We first outline this procedure and then demonstrate its use in the remaining examples in this section.

Method of Lagrange Multipliers	To find the local extrema of $f(x, y, z)$ subject to the constraint $g(x, y, z) = 0$, find the critical numbers for the new function F defined by

$$F(x, y, z, \lambda) = f(x, y, z) + \lambda g(x, y, z)$$

The variable λ (the Greek lowercase letter lambda) is called a **Lagrange multiplier.**

Example 4

Solve Example 3 by using a Lagrange multiplier.

Solution: First, we let

$$f(x, y, z) = xyz \quad \text{and} \quad g(x, y, z) = 6x + 4y + 3z - 24$$

Then we define a new function F by

$$\begin{aligned} F(x, y, z, \lambda) &= f(x, y, z) + \lambda g(x, y, z) \\ &= xyz + \lambda(6x + 4y + 3z - 24) \end{aligned}$$

Now we have the system

$$\begin{aligned} F_x &= yz + 6\lambda = 0 \\ F_y &= xz + 4\lambda = 0 \\ F_z &= xy + 3\lambda = 0 \\ F_\lambda &= 6x + 4y + 3z - 24 = 0 \end{aligned}$$

Multiplying the first equation by x, the second by y, and the third by z gives us the system

$$\begin{aligned} xyz + 6x\lambda &= 0 \\ xyz + 4y\lambda &= 0 \\ xyz + 3z\lambda &= 0 \end{aligned}$$

from which it follows that $6x = 4y = 3z$, or

$$y = \tfrac{3}{2}x \quad \text{and} \quad z = 2x$$

Substituting these values for y and z into $F_\lambda = 0$, we obtain

$$\begin{aligned} F_\lambda = 6x + 4(\tfrac{3}{2}x) + 3(2x) - 24 &= 0 \\ 18x &= 24 \\ x &= \tfrac{4}{3} \end{aligned}$$

Finally, the critical values are

$$x = \tfrac{4}{3}, \quad y = 2, \quad z = \tfrac{8}{3}$$

Thus the maximum volume is

$$V = xyz = (\tfrac{4}{3})(2)(\tfrac{8}{3}) = \tfrac{64}{9} \qquad \blacksquare$$

The real power of Lagrange multipliers becomes more apparent in problems that have more than one constraint. In the next example we demonstrate their use when two constraints are placed on a function of three variables.

Example 5

If $T(x, y, z) = 20 + 2x + 2y + z^2$ represents the temperature at each point on the hemisphere $x^2 + y^2 + z^2 = 11$ ($z > 0$), find the maximum temperature on the curve of intersection of the plane $x + y + z - 3 = 0$ and the hemisphere.

Solution: In this case we have two constraints,

$$g(x, y, z) = x^2 + y^2 + z^2 - 11 = 0 \qquad \text{(hemisphere)}$$
$$h(x, y, z) = x + y + z - 3 = 0 \qquad\qquad \text{(plane)}$$

Let

$$F(x, y, z, \lambda, \mu) = 20 + 2x + 2y + z^2$$
$$+ \lambda(x^2 + y^2 + z^2 - 11) + \mu(x + y + z - 3)$$

(μ is the Greek lowercase letter mu). Then we have the system

$$F_x = 2 + 2x\lambda + \mu = 0$$
$$F_y = 2 + 2y\lambda + \mu = 0$$
$$F_z = 2z + 2z\lambda + \mu = 0$$
$$F_\lambda = x^2 + y^2 + z^2 - 11 = 0$$
$$F_\mu = x + y + z - 3 = 0$$

Subtracting F_y from F_x yields

$$2\lambda(x - y) = 0$$

which means that $\lambda = 0$ or $x = y$.

1.   If $\lambda = 0$, then the first equation yields $\mu = -2$ and, subsequently, the third equation yields $z = 1$. The fourth and fifth equations now become

$$x^2 + y^2 - 10 = 0 \qquad \text{and} \qquad x + y - 2 = 0$$

and by substitution we obtain

$$x^2 + (2 - x)^2 - 10 = 0$$
$$2x^2 - 4x - 6 = 0$$
$$x^2 - 2x - 3 = 0$$

with solutions $x = 3$ and $x = -1$. The corresponding y-values are $y = -1$ and $y = 3$, respectively. Thus for $\lambda = 0$ the critical points are $(3, -1, 1)$ and $(-1, 3, 1)$.

2.   If $x = y$, then the fourth and fifth equations become

$$2x^2 + z^2 = 11 \qquad \text{and} \qquad 2x + z = 3$$

and by substitution we obtain

$$2x^2 + (3 - 2x)^2 = 11$$

$$6x^2 - 12x - 2 = 0$$
$$3x^2 - 6x - 1 = 0$$

The solutions are $x = (3 \pm 2\sqrt{3})/3$, and we discard $x = (3 + 2\sqrt{3})/3$ since $z = 3 - 2x$ would then be negative. Thus for $x = y$ the critical point is $((3 - 2\sqrt{3})/3, (3 - 2\sqrt{3})/3, (3 + 4\sqrt{3})/3)$.

Finally, since

$$T(3, -1, 1) = T(-1, 3, 1) = 25$$

and $\quad T\left(\dfrac{3 - 2\sqrt{3}}{3}, \dfrac{3 - 2\sqrt{3}}{3}, \dfrac{3 + 4\sqrt{3}}{3}\right) = \dfrac{91}{3} \approx 30.33$

we conclude that $T = 30.33$ is the maximum temperature on the curve of intersection. ∎

Note in Examples 4 and 5 that the system of equations that arises in the method of Lagrange multipliers is not, in general, a linear system. Because of this nonlinearity, the solution of the system often requires some ingenuity and we encourage you to tailor your method of solution to each individual system. Of course, if the system happens to be linear, we can resort to familiar methods, as demonstrated in the next example.

Example 6

Find the minimum of the function

$$f(x, y, z, w) = x^2 + y^2 + z^2 + w^2$$

subject to the constraint $3x + 2y - 4z + w = 3$.

Solution: Let

$$F(x, y, z, w, \lambda) = x^2 + y^2 + z^2 + w^2 + \lambda(3x + 2y - 4z + w - 3)$$

then we obtain the system

$F_x = 2x + 3\lambda = 0$	$x = -\tfrac{3}{2}\lambda$
$F_y = 2y + 2\lambda = 0$	$y = -\lambda$
$F_z = 2z - 4\lambda = 0$	$z = 2\lambda$
$F_w = 2w + \lambda = 0$	$w = -\tfrac{1}{2}\lambda$
$F_\lambda = 3x + 2y - 4z + w - 3 = 0$	$3x + 2y - 4z + w = 3$

Substituting for x, y, z, and w into the last equation yields

$$3(-\tfrac{3}{2}\lambda) + 2(-\lambda) - 4(2\lambda) + (-\tfrac{1}{2}\lambda) = 3$$
$$-15\lambda = 3$$
$$\lambda = -\tfrac{1}{5}$$

Therefore, the critical numbers are

$$x = \tfrac{3}{10}, \qquad y = \tfrac{1}{5}, \qquad z = -\tfrac{2}{5}, \qquad w = \tfrac{1}{10}$$

and the minimum value of f subject to the given constraint is

$$f(\tfrac{3}{10}, \tfrac{1}{5}, -\tfrac{2}{5}, \tfrac{1}{10}) = \tfrac{9}{100} + \tfrac{1}{25} + \tfrac{4}{25} + \tfrac{1}{100} = \tfrac{3}{10} \qquad \blacksquare$$

Section Exercises (15.3)

In Exercises 1–12, examine each function for relative extrema and saddle points.

1. $f(x, y) = 2x^2 + 3y^2 - 4x - 12y + 13$

2. $f(x, y) = 5 + 3x - 4y - 3x^2 - 2y^2$

S **3.** $f(x, y) = x^2 - y^2 - 2x - 4y - 4$

4. $f(x, y) = x^2 - 3xy - y^2$

5. $f(x, y) = xy$

6. $f(x, y) = 120x + 120y - xy - x^2 - y^2$

S **7.** $f(x, y) = x^3 - 3xy + y^3$

8. $f(x, y) = xy^2 - x^2y + x - y$

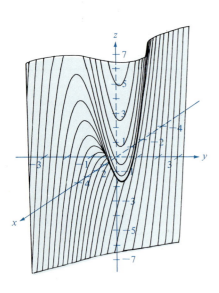

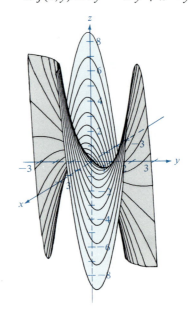

9. $f(x, y) = xy + \dfrac{a^3}{x} + \dfrac{a^3}{y}$

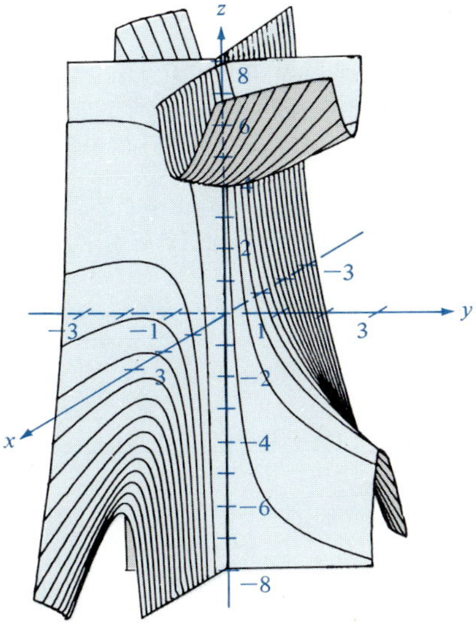

10. $f(x, y) = y^3 - 3yx^2 - 3y^2 - 3x^2 + 1$

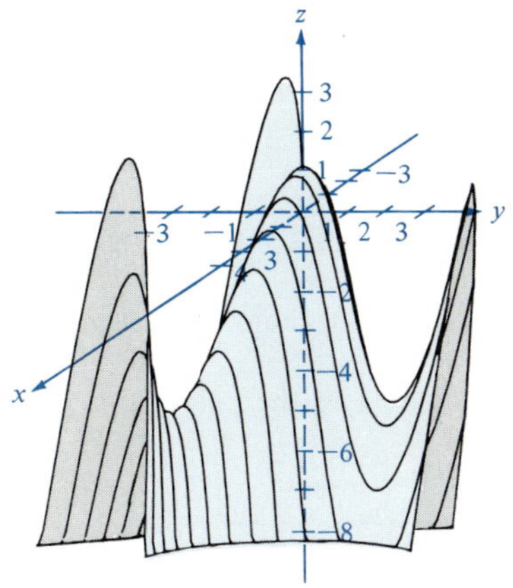

11. $f(x, y) = e^{-x} \sin y$

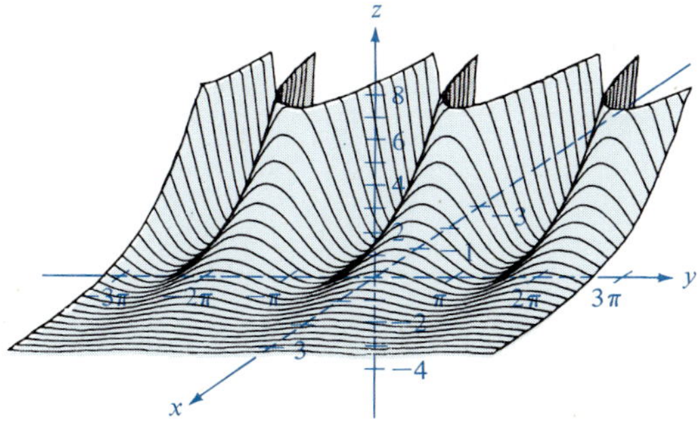

12. $f(x, y) = \dfrac{-4x}{x^2 + y^2 + 1}$

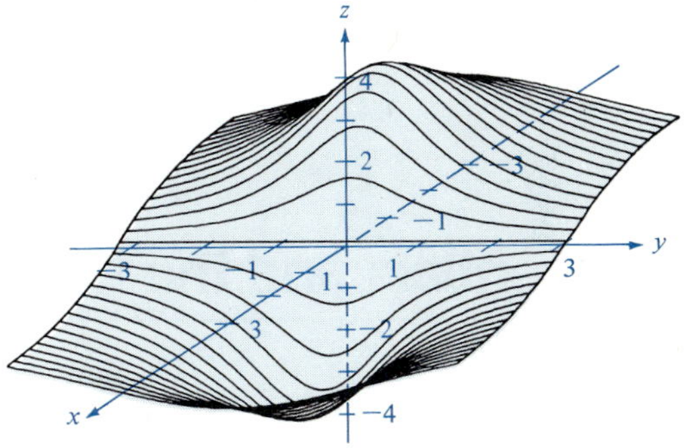

[S] **13.** Find the dimensions of a rectangular package of largest volume that may be sent by parcel post if the sum of the length and the girth (perimeter of a cross section) cannot exceed 108 inches.

14. Use partial derivatives to find the minimum distance from the origin to the plane $2x + 3y + z = 12$.

15. Use partial derivatives to find the minimum distance from the point (1, 2, 3) to the plane $2x + 3y + z = 12$.

16. The material for constructing the base of an open box costs 1.5 times as much as the material for the sides. For a fixed amount of money C_0, find the dimensions of the box of largest volume that can be made in this manner.

17. The volume of an ellipsoid $(x^2/a^2) + (y^2/b^2) + (z^2/c^2) = 1$ is given by $4\pi abc/3$. If $a + b + c$ is a fixed constant, show that the ellipsoid of maximum volume is a sphere.

18. Find three positive numbers, x, y, and z, whose sum is 32 and for which $P = xy^2z$ is a maximum.

19. Find three positive numbers whose sum is 30 and whose product is maximum.

[S] **20.** Find three positive numbers whose sum is 30 and whose sum of squares is minimum.

21. Show that the rectangular box of given volume and minimum surface area is a cube.

22. Show that the rectangular box of maximum volume inscribed in a sphere of radius r is a cube.

In Exercises 23–32, solve each problem by using Lagrange multipliers.

23. Minimize $f(x, y) = x^2 + y^2$, subject to the constraint $2x - 4y + 5 = 0$.

24. Maximize $f(x, y) = x^2 - y^2$, subject to the constraint $y - x^2 = 0$.

S 25. Maximize $f(x, y, z) = xyz$, subject to the constraints $x + y + z = 32$ and $x - y + z = 0$.

26. Minimize $f(x, y, z) = x^2 + y^2 + z^2$, subject to the constraints $x + 2z = 4$ and $x + y = 8$.

27. Find the minimum distance from the origin to the line $2x + 3y = -1$.

S 28. Find the minimum distance from the point $(2, 1, 1)$ to the plane $x + y + z = 1$.

S 29. Find the dimensions of the rectangular box (edges parallel to the coordinate axes) of maximum volume that can be inscribed in the ellipsoid $(x^2/a^2) + (y^2/b^2) + (z^2/c^2) = 1$.

30. Find the values of x and y that minimize $x^2 - 8x + y^2 - 12y + 48$, subject to the constraint $x + y = 8$.

S 31. Find any extrema to $f(x, y, z) = xyz$ subject to the constraints $x^2 + z^2 = 5$ and $x - 2y = 0$.

C 32. Find the highest point on the curve of intersection of the sphere $x^2 + y^2 + z^2 = 36$ and the plane $2x + y - z = 2$ (maximize z).

For Review and Class Discussion

True or False

1. _____ The function $f(x, y) = \sqrt{1 - x^2 - y^2}$ has one relative maximum and infinitely many relative minima.

2. _____ If f has a relative maximum at (x, y, z), then $f_x(x, y) = f_y(x, y) = 0$.

3. _____ The function given by $f(x, y) = \sqrt[3]{x^2 + y^2}$ has a relative minimum at the origin.

4. _____ Of all parallelepipeds having a fixed surface area, the cube has the largest volume.

5. _____ If f is continuous for all x and y and has two relative minima, then f must have at least one relative maximum.

15.4 Differentials and the Chain Rule

Purpose
- To define the total differential.
- To use the total differential as an approximation for Δz.
- To describe the Chain Rule for a function of two or more variables.
- To demonstrate implicit differentiation with functions of two or more variables.

It is not easy to demonstrate all the axioms, and to reduce demonstrations wholly to intuitive knowledge. And if we had chosen to wait for that, perhaps we should not yet have the science of geometry.

—*Gottfried Leibniz (1646–1716)*

For a function of a single variable, $y = f(x)$, we defined (Section 5.4) the differential to be $dy = f'(x) \, dx$. In this section we generalize the notion of a differential to a function of two or more variables and then show some applications of such differentials. We begin with the definition of the total differential of a function of two variables.

Definition of Total Differential

If $z = f(x, y)$ and both first partial derivatives of z exist, then the **total differential** dz is defined to be

$$dz = \frac{\partial f}{\partial x} \, dx + \frac{\partial f}{\partial y} \, dy$$

This definition also extends readily to a function of three or more variables. For instance, if $w = f(x, y, z, u)$, then

$$dw = \frac{\partial f}{\partial x} dx + \frac{\partial f}{\partial y} dy + \frac{\partial f}{\partial z} dz + \frac{\partial f}{\partial u} du$$

Example 1

Given $z = 2x \sin y - 3x^2 y^2$, find dz.

Solution: By definition

$$dz = \frac{\partial z}{\partial x} dx + \frac{\partial z}{\partial y} dy$$

$$= (2 \sin y - 6xy^2)\, dx + (2x \cos y - 6x^2 y)\, dy \qquad \blacksquare$$

For a function given by $y = f(x)$, we can use the differential $dy = f'(x)\, dx$ as a convenient approximation (for small Δx) to the value $\Delta y = f(x + \Delta x) - f(x)$ (Section 5.4). This is also possible for a function of two variables, as the following theorem shows:

THEOREM 15.4
($\Delta z \approx dz$)

Let $z = f(x, y)$ define a function of two variables, where $f, f_x,$ and f_y are continuous in an open region containing (x, y). Then

$$\Delta z - dz = \epsilon_1 \Delta x + \epsilon_2 \Delta y$$

where ϵ_1 and ϵ_2 both approach zero as Δx and Δy approach zero.

Proof: Let S be a surface defined by $z = f(x, y)$, where $f, f_x,$ and f_y are continuous at (x, y) in the domain of f. Let A, B, and C be points on surface S, as shown in Figure 15.21.

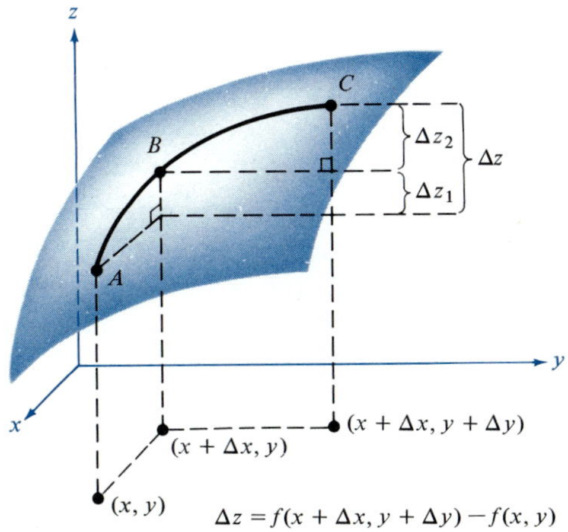

From Figure 15.21 we can see that the change in f from point A to point C is given by

$$\Delta z = f(x + \Delta x, y + \Delta y) - f(x, y)$$

or

$$\Delta z = \Delta z_1 + \Delta z_2$$
$$= [f(x + \Delta x, y) - f(x, y)] + [f(x + \Delta x, y + \Delta y) - f(x + \Delta x, y)]$$

Now from A to B, y is fixed while x changes; hence by the Mean Value Theorem, there is a value x_1 between x and $x + \Delta x$ such that

$$\Delta z_1 = f(x + \Delta x, y) - f(x, y) = f_x(x_1, y) \Delta x$$

Similarly, from B to C, x is fixed and y changes; thus there is a y_1 between y and $y + \Delta y$ such that

$$\Delta z_2 = f(x + \Delta x, y + \Delta y) - f(x + \Delta x, y) = f_y(x + \Delta x, y_1) \Delta y$$

FIGURE 15.21

$\Delta z = f(x + \Delta x, y + \Delta y) - f(x, y)$

Therefore, it follows that

$$\Delta z = \Delta z_1 + \Delta z_2 = f_x(x_1, y) \, \Delta x + f_y(x + \Delta x, y_1) \, \Delta y$$

Now let us define ϵ_1 and ϵ_2 by

$$\epsilon_1 = f_x(x_1, y) - f_x(x, y)$$
$$\epsilon_2 = f_y(x + \Delta x, y_1) - f_y(x, y)$$

Then by the continuity of f_x and f_y, and the fact that $x \le x_1 \le x + \Delta x$ and $y \le y_1 \le y + \Delta y$, it follows that $\epsilon_1 \to 0$ and $\epsilon_2 \to 0$ as Δx and $\Delta y \to 0$. Thus we can write

$$\begin{aligned}
\Delta z = \Delta z_1 + \Delta z_2 &= [\epsilon_1 + f_x(x, y)] \, \Delta x + [\epsilon_2 + f_y(x, y)] \, \Delta y \\
&= [f_x(x, y) \, \Delta x + f_y(x, y) \, \Delta y] + \epsilon_1 \, \Delta x + \epsilon_2 \, \Delta y \\
&= [dz] + \epsilon_1 \, \Delta x + \epsilon_2 \, \Delta y
\end{aligned}$$

from which it follows that

$$\Delta z - dz = \epsilon_1 \, \Delta x + \epsilon_2 \, \Delta y \qquad \blacksquare$$

Example 2

Let f be given by $f(x, y) = x^2 y^3 - 2xy$. Calculate Δz and dz as (x, y) changes from $(3, -1)$ to $(2.99, -0.95)$.

Solution: First, note that $x = 3$, $y = -1$, $\Delta x = -0.01$, and $\Delta y = 0.05$. Thus

$$dz = \frac{\partial f}{\partial x} \Delta x + \frac{\partial f}{\partial y} \Delta y = (2xy^3 - 2y) \, \Delta x + (3x^2 y^2 - 2x) \, \Delta y$$

$$= (-6 + 2)(-0.01) + (27 - 6)(0.05) = 0.04 + 1.05 = 1.09$$

Furthermore,

$$\begin{aligned}
\Delta z &= f(x + \Delta x, y + \Delta y) - f(x, y) \\
&= f(2.99, -0.95) - f(3, -1) = -1.984 + 3 = 1.016
\end{aligned}$$

Hence the error involved in using dz in place of Δz is 0.074. $\qquad \blacksquare$

Example 3

The possible error involved in measuring each dimension of a rectangular box is ± 0.1 millimeters (mm). If the dimensions of the box are measured to be $x = 50$ cm, $y = 20$ cm, and $z = 15$ cm, use dV to estimate the relative error in the calculated volume of the box.

Solution: The volume of the box is given by $V = xyz$ and thus

$$dV = \frac{\partial V}{\partial x} dx + \frac{\partial V}{\partial y} dy + \frac{\partial V}{\partial z} dz = yz \, dx + xz \, dy + xy \, dz$$

Now since 1 mm equals 0.1 cm, we have $dx = dy = dz = \pm 0.1$ mm $= \pm 0.01$ cm. Therefore,

$$dV = 300(\pm 0.01) + 750(\pm 0.01) + 1000(\pm 0.01)$$
$$= 2050(\pm 0.01) = \pm 20.5 \text{ cm}^3$$

Since the total volume is

$$V = (50)(20)(15) = 15{,}000 \text{ cm}^3$$

the relative error, $\Delta V/V$, is approximately

$$\frac{\Delta V}{V} \approx \frac{dV}{V} = \frac{20.5}{15{,}000} \approx 0.14\%$$

∎

We have used Theorem 15.4 to support our use of dV as a reasonable estimate for ΔV when Δx and Δy are small. This theorem has other uses, one of which is that it guarantees the differentiability of f.

Definition of Differentiability of $z = f(x, y)$	A function f given by $z = f(x, y)$ is **differentiable** at (x, y) provided Δz can be expressed in the form $$\Delta z = dz + \epsilon_1 \Delta x + \epsilon_2 \Delta y$$ $$= f_x(x, y)\, \Delta x + f_y(x, y)\, \Delta y + \epsilon_1 \Delta x + \epsilon_2 \Delta y$$ where both ϵ_1 and $\epsilon_2 \to 0$ as Δx and $\Delta y \to 0$.

Note that by Theorem 15.4 this definition means that f is *differentiable* at (x, y) provided $f, f_x,$ and f_y are continuous in an open region containing (x, y).

As is true for a function of a single variable, if f is differentiable at some point in its domain, then it is also continuous there. It is worth noting that the differentiability of f at (x, y) assumes the existence of f_x and f_y there. However, the existence of f_x and f_y is not sufficient to guarantee differentiability, as the next example illustrates.

Example 4

Let

$$f(x, y) = \begin{cases} \dfrac{3xy}{x^2 + y^2}, & \text{if } (x, y) \neq (0, 0) \\ 0, & \text{if } (x, y) = (0, 0) \end{cases}$$

Show that $f_x(0, 0)$ and $f_y(0, 0)$ both exist but that f is not differentiable at $(0, 0)$.

Solution: First, we have

$$f_x(0, 0) = \lim_{x \to 0} \frac{f(x, 0) - f(0, 0)}{x} = \lim_{x \to 0} \frac{0 - 0}{x} = 0$$

$$f_y(0, 0) = \lim_{y \to 0} \frac{f(0, y) - f(0, 0)}{y} = \lim_{y \to 0} \frac{0 - 0}{y} = 0$$

However, along $y = x$ we have

$$\lim_{(x, x) \to (0, 0)} f(x, y) = \lim_{(x, x) \to (0, 0)} \frac{3x^2}{2x^2} = \frac{3}{2}$$

while along $y = -x$ we have

$$\lim_{(x, -x) \to (0, 0)} f(x, y) = \lim_{(x, -x) \to (0, 0)} \frac{-3x^2}{2x^2} = \frac{-3}{2}$$

From this we conclude that

$$\lim_{(x, y) \to (0, 0)} f(x, y) \neq f(0, 0)$$

which means that f is not continuous at $(0, 0)$. This in turn means that f is not differentiable at $(0, 0)$. ∎

As with differentials, the definition of differentiability can be extended to three or more variables. Thus if $w = f(x, y, z)$, then w is **differentiable at** (x, y, z) provided

$$\Delta w = f(x + \Delta x, y + \Delta y, z + \Delta z) - f(x, y, z)$$

can be written in the form

$$\Delta w = f_x \Delta x + f_y \Delta y + f_z \Delta z + \epsilon_1 \Delta x + \epsilon_2 \Delta y + \epsilon_3 \Delta z$$

where ϵ_1, ϵ_2, and $\epsilon_3 \to 0$ as Δx, Δy, and $\Delta z \to 0$.

We now consider another very important application of Theorem 15.4. It provides the basis for the Chain Rules for differentiating composite functions of more than one variable. We give two versions of the Chain Rule: The first case involves w as a function of x and y where x and y are functions of one variable; the second case involves w as a function of x and y where x and y are in turn functions of two variables.

THEOREM 15.5
(Chain Rule: One Independent Variable)

Let $w = f(x, y)$ be a differentiable function of x and y, where $x = g(t)$ and $y = h(t)$ are differentiable functions of t. Then w is a function of t, and

$$\frac{dw}{dt} = \frac{\partial w}{\partial x} \frac{dx}{dt} + \frac{\partial w}{\partial y} \frac{dy}{dt}$$

Proof: Since x and y are functions of t, a change of Δt in t produces changes in x and y of Δx and Δy, respectively. By Theorem 15.4 we have

$$\Delta w = \frac{\partial w}{\partial x} \Delta x + \frac{\partial w}{\partial y} \Delta y + \epsilon_1 \Delta x + \epsilon_2 \Delta y$$

where ϵ_1 and $\epsilon_2 \to 0$ as Δx and $\Delta y \to 0$ (i.e., as $\Delta t \to 0$). Thus for $\Delta t \neq 0$ we have

$$\frac{\Delta w}{\Delta t} = \frac{\partial w}{\partial x} \frac{\Delta x}{\Delta t} + \frac{\partial w}{\partial y} \frac{\Delta y}{\Delta t} + \epsilon_1 \frac{\Delta x}{\Delta t} + \epsilon_2 \frac{\Delta y}{\Delta t}$$

from which it follows that

$$\frac{dw}{dt} = \lim_{\Delta t \to 0} \frac{\Delta w}{\Delta t} = \frac{\partial w}{\partial x} \frac{dx}{dt} + \frac{\partial w}{\partial y} \frac{dy}{dt} + 0 \left(\frac{dx}{dt} \right) + 0 \left(\frac{dy}{dt} \right)$$

$$= \frac{\partial w}{\partial x} \frac{dx}{dt} + \frac{\partial w}{\partial y} \frac{dy}{dt} \qquad \blacksquare$$

Example 5

Let $w = x^2 y - y^2$ where $x = \sin t$ and $y = e^t$. Find dw/dt when $t = 0$.

Solution: By the Chain Rule for one independent variable, we have

$$\frac{dw}{dt} = \frac{\partial w}{\partial x} \frac{dx}{dt} + \frac{\partial w}{\partial y} \frac{dy}{dt} = 2xy(\cos t) + (x^2 - 2y)e^t$$

When $t = 0$, $x = 0$ and $y = 1$, so it follows that

$$\frac{dw}{dt} = 0 - 2 = -2 \qquad \blacksquare$$

THEOREM 15.6
(*Chain Rule: Two Independent Variables*)

Let $w = f(x, y)$ be a differentiable function of x and y, where $x = g(s, t)$ and $y = h(s, t)$ are differentiable functions of s and t. Then w is indirectly a function of s and t, and

$$\frac{\partial w}{\partial s} = \frac{\partial w}{\partial x} \frac{\partial x}{\partial s} + \frac{\partial w}{\partial y} \frac{\partial y}{\partial s}$$

$$\frac{\partial w}{\partial t} = \frac{\partial w}{\partial x} \frac{\partial x}{\partial t} + \frac{\partial w}{\partial y} \frac{\partial y}{\partial t}$$

Proof: The proof is similar to that for Theorem 15.5, where for $\partial w/\partial s$ we hold t constant and let s change by an amount Δs. Similarly, for $\partial w/\partial t$ we hold s constant and let t change by an amount Δt. In both instances the desired result follows from Theorem 15.4 in the manner shown in the proof of Theorem 15.5. We leave the details as an exercise. $\blacksquare$

Example 6

If $w = 2xy$, $x = s^2 + t^2$, and $y = s/t$, find $\partial w/\partial s$ and $\partial w/\partial t$.

Solution: If we hold t fixed, then by Theorem 15.6 we have

$$\frac{\partial w}{\partial s} = \frac{\partial w}{\partial x}\frac{\partial x}{\partial s} + \frac{\partial w}{\partial y}\frac{\partial y}{\partial s} = 2y(2s) + 2x\left(\frac{1}{t}\right)$$

$$= 4\left(\frac{s^2}{t}\right) + \frac{2s^2 + 2t^2}{t} = \frac{6s^2 + 2t^2}{t}$$

Similarly, holding s fixed gives us

$$\frac{\partial w}{\partial t} = \frac{\partial w}{\partial x}\frac{\partial x}{\partial t} + \frac{\partial w}{\partial y}\frac{\partial y}{\partial t} = 2y(2t) + 2x\left(\frac{-s}{t^2}\right)$$

$$= 4s - \frac{2s^3 + 2st^2}{t^2} = \frac{4st^2 - 2s^3 - 2st^2}{t^2} = \frac{2st^2 - 2s^3}{t^2} \quad\blacksquare$$

The Chain Rules (Theorems 15.5 and 15.6) can be extended to any number of variables. Thus if w is a differentiable function of the n variables $x_1, x_2, \ldots, x_n$ where each x_i is a differentiable function of the m variables $t_1, t_2, \ldots, t_m$, then for $w = f(x_1, x_2, \ldots, x_n)$ we have

$$\frac{\partial w}{\partial t_1} = \frac{\partial w}{\partial x_1}\frac{\partial x_1}{\partial t_1} + \frac{\partial w}{\partial x_2}\frac{\partial x_2}{\partial t_1} + \cdots + \frac{\partial w}{\partial x_n}\frac{\partial x_n}{\partial t_1}$$

$$\frac{\partial w}{\partial t_2} = \frac{\partial w}{\partial x_1}\frac{\partial x_1}{\partial t_2} + \frac{\partial w}{\partial x_2}\frac{\partial x_2}{\partial t_2} + \cdots + \frac{\partial w}{\partial x_n}\frac{\partial x_n}{\partial t_2}$$

$$\vdots$$

$$\frac{\partial w}{\partial t_m} = \frac{\partial w}{\partial x_1}\frac{\partial x_1}{\partial t_m} + \frac{\partial w}{\partial x_2}\frac{\partial x_2}{\partial t_m} + \cdots + \frac{\partial w}{\partial x_n}\frac{\partial x_n}{\partial t_m}$$

If each x_i is a differentiable function of a single variable t, then for $w = f(x_1, x_2, \ldots, x_n)$ we have

$$\frac{dw}{dt} = \frac{\partial w}{\partial x_1}\frac{dx_1}{dt} + \frac{\partial w}{\partial x_2}\frac{dx_2}{dt} + \cdots + \frac{\partial w}{\partial x_n}\frac{dx_n}{dt}$$

Example 7

If $w = xy + yz + xz$, $x = s\cos t$, $y = s\sin t$, and $z = t$, find $\partial w/\partial s$ and $\partial w/\partial t$ when $s = 1$ and $t = 2\pi$.

Solution: By an extension of Theorem 15.6, we have

$$\frac{\partial w}{\partial s} = \frac{\partial w}{\partial x}\frac{\partial x}{\partial s} + \frac{\partial w}{\partial y}\frac{\partial y}{\partial s} + \frac{\partial w}{\partial z}\frac{\partial z}{\partial s}$$

$$= (y + z)(\cos t) + (x + z)(\sin t) + (y + x)(0)$$

When $s = 1$ and $t = 2\pi$, we have $x = 1$, $y = 0$, and $z = 2\pi$ and thus

$$\frac{\partial w}{\partial s} = 2\pi(1) + (1 + 2\pi)(0) + 0 = 2\pi$$

Furthermore,

$$\frac{\partial w}{\partial t} = \frac{\partial w}{\partial x}\frac{\partial x}{\partial t} + \frac{\partial w}{\partial y}\frac{\partial y}{\partial t} + \frac{\partial w}{\partial z}\frac{\partial z}{\partial t}$$

$$= (y + z)(-s \sin t) + (x + z)(s \cos t) + (y + x)(1)$$

and for $s = 1$ and $t = 2\pi$, it follows that

$$\frac{\partial w}{\partial t} = (0 + 2\pi)(0) + (1 + 2\pi)(1) + (0 + 1)(1) = 2 + 2\pi \qquad \blacksquare$$

We conclude this section with an application of the total differential that is useful for determining the derivative of a function defined *implicitly*. Suppose that x and y are related by the equation $F(x, y) = 0$, where it is assumed that y is a function of x. We could use the methods of Section 3.8 to determine dy/dx. However, the total differential provides an alternative procedure. Since $F(x, y) = 0$, we have

$$\frac{\partial F}{\partial x} dx + \frac{\partial F}{\partial y} dy = 0$$

from which it follows that

$$\frac{dy}{dx} = \frac{-\partial F/\partial x}{\partial F/\partial y} = \frac{-F_x}{F_y} \qquad (\text{if } F_y \neq 0)$$

Suppose we use this formula to find dy/dx in Example 3 of Section 3.8. In that example we have $F(x, y) = x^3 + xy + y^2 + x - 4 = 0$, hence

$$F_x = 3x^2 + y + 1 \qquad \text{and} \qquad F_y = x + 2y$$

Therefore, $$\frac{dy}{dx} = \frac{-F_x}{F_y} = \frac{-3x^2 - y - 1}{x + 2y}$$

as was previously obtained.

The total differential can be used to obtain formulas for partial derivatives in much the same way as we obtained the formula for dy/dx above. Suppose x, y, and z are related by the equation $F(x, y, z) = 0$, where it is assumed that z is a differentiable function of x and y. Under this assumption we know that

$$dz = \frac{\partial z}{\partial x} dx + \frac{\partial z}{\partial y} dy$$

Furthermore, we have

$$\frac{\partial F}{\partial x} dx + \frac{\partial F}{\partial y} dy + \frac{\partial F}{\partial z} dz = 0$$

or, equivalently, $$F_x \, dx + F_y \, dy + F_z \, dz = 0$$

Solving for dz in this last equation gives us

$$dz = \left(\frac{-F_x}{F_z}\right)dx + \left(\frac{-F_y}{F_z}\right)dy$$

By comparing this representation of dz to the definition of dz, we obtain the following formulas for the partials of z:

$$\frac{\partial z}{\partial x} = \frac{-F_x}{F_z} \quad \text{and} \quad \frac{\partial z}{\partial y} = \frac{-F_y}{F_z}$$

We summarize this discussion in the following theorem.

THEOREM 15.7
(*Finding Derivatives and Partial Derivatives Implicitly*)

i. If the equation $F(x, y) = 0$ defines y implicitly as a differentiable function of x, then

$$\frac{dy}{dx} = \frac{-F_x}{F_y} \quad (F_y \neq 0)$$

ii. If the equation $F(x, y, z) = 0$ defines z implicitly as a differentiable function of x and y, then

$$\frac{\partial z}{\partial x} = \frac{-F_x}{F_z} \quad \text{and} \quad \frac{\partial z}{\partial y} = \frac{-F_y}{F_z} \quad (F_z \neq 0)$$

This theorem can be extended to differentiable functions defined implicitly with any number of variables.

Example 8

Use Theorem 15.7 to find $\partial z/\partial x$ and $\partial z/\partial y$, if $3x^2 z - x^2 y^2 + 2z^3 + 3yz - 5 = 0$.

Solution: Let

$$F(x, y, z) = 3x^2 z - x^2 y^2 + 2z^3 + 3yz - 5$$

Then

$$F_x = 6xz - 2xy^2$$
$$F_y = -2x^2 y + 3z$$
$$F_z = 3x^2 + 6z^2 + 3y$$

Therefore,

$$\frac{\partial z}{\partial x} = \frac{-F_x}{F_z} = \frac{2xy^2 - 6xz}{3x^2 + 6z^2 + 3y}$$

$$\frac{\partial z}{\partial y} = \frac{-F_y}{F_z} = \frac{2x^2 y - 3z}{3x^2 + 6z^2 + 3y}$$

Note that we could have obtained $\partial z/\partial x$ less conveniently by holding y constant and differentiating F implicitly with respect to x. Thus

$$\left[3x^2 \frac{\partial z}{\partial x} + 6xz\right] - [2xy^2 + 0] + 6z^2 \frac{\partial z}{\partial x} + 3y \frac{\partial z}{\partial x} = 0$$

Solving for $\partial z/\partial x$ we obtain

$$\frac{\partial z}{\partial x} = \frac{2xy^2 - 6xz}{3x^2 + 6z^2 + 3y}$$

Similarly, if we hold x constant and differentiate F implicitly with respect to y, we obtain

$$3x^2 \frac{\partial z}{\partial y} - 2x^2 y + 6z^2 \frac{\partial z}{\partial y} + \left[3y \frac{\partial z}{\partial y} + 3z \right] = 0$$

Solving for $\partial z/\partial y$ then gives us

$$\frac{\partial z}{\partial y} = \frac{2x^2 y - 3z}{3x^2 + 6z^2 + 3y} \qquad \blacksquare$$

Section Exercises (15.4)

In Exercises 1–10, find the total differential.

1. $z = 3x^2 y^3$

2. $z = x^2/y$

$\boxed{s}$ **3.** $z = x \cos y - y \cos x$

4. $z = e^x \sin y$

5. $z = -1/(x^2 + y^2)$

6. $u = \frac{1}{2}(e^{x^2+z^2} - e^{-x^2-z^2})$

7. $w = 2z^3 y \sin x$

8. $w = x^2 yz^2 + \sin yz$

$\boxed{s}$ **9.** $u = \dfrac{x + y}{z - 2y}$

10. $w = e^x \cos y + z$

11. Let $f(x, y) = 9 - x^2 - y^2$.
  (a) Evaluate $f(1, 2)$ and $f(1.05, 2.1)$ and calculate Δz.
  (b) Use the total differential dz to approximate Δz of part a.

$\boxed{c}$ **12.** Repeat Exercise 11 using the function $f(x, y) = \sqrt{x^2 + y^2}$.

$\boxed{c}$ **13.** The volume of a cone is $V = \frac{1}{3}\pi r^2 h$. A cone of height $h = 6$ and radius $r = 3$ is constructed and in the process errors of Δr and Δh are made in the radius and height, respectively. Complete the accompanying table to show the relationship between ΔV and dV for the given errors.

Δr	Δh	dV	ΔV	$\Delta V - dV$
0.1	0.1			
0.1	-0.1			
0.001	0.002			
-0.001	0.002			

$\boxed{s}$ **14.** The volume of a cylinder is $V = \pi r^2 h$. If r and h are measured with a possible error of 4% and 2%, respectively, approximate the maximum possible percentage error in measuring the volume.

15. The centripetal acceleration of a particle moving in a circle is $a = V^2/r$, where V is the velocity and r is the radius of the circle. Approximate the maximum percentage error in measuring the acceleration due to possible errors of 2% in V and 1% in r.

16. A triangle is measured and two adjacent sides are found to be 3 and 4 inches in length with an included angle of $\pi/4$ radians. If the possible errors in measurement are $\frac{1}{16}$ inch in the sides and 0.02 radians in the angle, approximate the maximum possible error in computing the area.

In Exercises 17–20, find dw/dt by using the Chain Rule, and evaluate it at the specified value of t.

17. $w = x^2 + y^2$, $x = e^t$, $y = e^{-t}$; $t = 0$

$\boxed{s}$ **18.** $w = \sqrt{x^2 + y^2}$, $x = \sin t$, $y = e^t$; $t = 0$

19. $w = x^2 + y^2 + z^2$, $x = e^t \cos t$, $y = e^t \sin t$, $z = e^t$; $t = 0$

20. $w = \ln \dfrac{y}{x}$, $x = \cos t$, $y = \sin t$; $t = \dfrac{\pi}{4}$

In Exercises 21–24, find $\partial w/\partial s$ and $\partial w/\partial t$ by using the Chain Rule, and evaluate each partial derivative at the indicated values.

21. $w = x^2 + y^2$, $x = s + t$, $y = s - t$; $s = 2$, $t = -1$

22. $w = y^3 - 3x^2y$, $x = e^s$, $y = e^t$; $s = 0$, $t = 1$

S **23.** $w = x^2 - y^2$, $x = s \cos t$, $y = s \sin t$

24. $w = \sin(2x + 3y)$, $x = s + t$, $y = s - t$

In Exercises 25–30, find the specified derivatives (a) by the Chain Rule and (b) by converting w to a function of the independent variables *before* differentiating. Compare the results of parts (a) and (b).

S **25.** $w = xy$, $x = 2 \sin t$, $y = \cos t$; dw/dt

26. $w = \cos(x - y)$, $x = t^2$, $y = 1$; dw/dt

27. $w = xy + xz + yz$, $x = t - 1$, $y = t^2 - 1$, $z = t$; dw/dt

28. $w = \sqrt{4 - 2x^2 - 2y^2}$, $x = r \cos \theta$, $y = r \sin \theta$; $\partial w/\partial r$, $\partial w/\partial \theta$

S **29.** $w = \text{Arctan}(y/x)$, $x = r \cos \theta$, $y = r \sin \theta$; $\partial w/\partial r$, $\partial w/\partial \theta$

30. $w = yx/z$, $x = u + v$, $y = u - v$, $z = v^2$; $\partial w/\partial u$, $\partial w/\partial v$

In Exercises 31–37, use Theorem 15.7 to find the indicated partials.

31. Find $\partial z/\partial x$ and $\partial z/\partial y$ for $x^2 + y^2 + z^2 = 25$.

32. Find $\partial z/\partial x$ and $\partial z/\partial y$ for $xz + yz + xy = 0$.

33. Find $\partial w/\partial x$, $\partial w/\partial y$, and $\partial w/\partial z$ for $xyz + xzw - yzw + w^2 = 5$.

S **34.** Find $\partial w/\partial x$, $\partial w/\partial y$, and $\partial w/\partial z$ for $x^2 + y^2 + z^2 + 6xw - 8w^2 = 5$.

35. Find $\partial z/\partial x$ and $\partial z/\partial y$ for $\tan(x + y) + \tan(y + z) = 1$.

36. Find $\partial z/\partial x$ and $\partial z/\partial y$ for $z = e^x \sin(y + z)$.

37. Find all first and second partial derivatives of z for $x^2 + 2yz + z^2 = 1$.

38. If $w = f(x, y)$, $x = r \cos \theta$, and $y = r \sin \theta$, show that
$$\left(\frac{\partial w}{\partial x}\right)^2 + \left(\frac{\partial w}{\partial y}\right)^2 = \left(\frac{\partial w}{\partial r}\right)^2 + \left(\frac{1}{r^2}\right)\left(\frac{\partial w}{\partial \theta}\right)^2$$

39. If $w = f(x, y)$, $x = u - v$, and $y = v - u$, show that
$$\frac{\partial w}{\partial u} + \frac{\partial w}{\partial v} = 0$$

40. Demonstrate the result of Exercise 39 for the function $w = (x - y)\sin(y - x)$.

S **41.** The length, width, and depth of a rectangular chamber are increasing at the rate of 3 ft/min, 2 ft/min, and $\frac{1}{2}$ ft/min, respectively. Find the rate at which the volume is changing the instant the length, width, and depth are 10 ft, 6 ft, and 4 ft, respectively.

42. The radius and height of a right circular cylinder are decreasing at 6 in./min and 4 in./min, respectively. Find the rate at which the volume is changing at the instant the radius and height are 12 in. and 36 in., respectively.

43. Find the rate at which the surface area is changing in Exercise 42.

44. The ideal gas law is $pV = RT$, where R is a constant. If p and V are functions of time, find dT/dt, the rate at which the temperature changes with respect to time.

15.5 **Directional Derivatives and Gradients**

Purpose

- To define the concept of a directional derivative.
- To give a geometric interpretation of the directional derivative.
- To define the gradient and show its relationship to the directional derivative.

Although geometers have given much attention to general investigations of curved surfaces and their results cover a significant portion of the domain of higher geometry, this subject is still so far from being exhausted, that it can well be said that, up to this time, but a small portion of an exceedingly fruitful field has been cultivated. —Carl Gauss (1777–1855)

For the surface $z = f(x, y)$ we interpreted the partial derivatives f_x and f_y to be the slope of the surface in the x and y directions, respectively. We now want to show how to determine the slope of $z = f(x, y)$ in *any* direction by the use of a **directional derivative.** In general terms, this corresponds to determining the rate of change in $f(x, y)$ with respect to simultaneous changes in both x and y.

We develop our definition of the directional derivative geometrically. Let $z = f(x, y)$ be a surface, and let $P = (x, y, z)$ and $Q = (x + \Delta x,$

$y + \Delta y, z + \Delta z)$ be two points on the surface. If we begin to move from P to Q, does the surface slope up or down, and how steep is this slope? The answer is provided by the directional derivative, which denotes the slope of the tangent line at P in the direction of Q. The direction from P to Q is determined by the projection of the vector $\overrightarrow{PQ}$ onto the xy-plane, as shown in Figure 15.22.

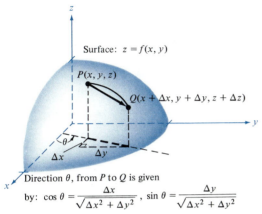

Direction θ, from P to Q is given

by: $\cos\theta = \dfrac{\Delta x}{\sqrt{\Delta x^2 + \Delta y^2}}$, $\sin\theta = \dfrac{\Delta y}{\sqrt{\Delta x^2 + \Delta y^2}}$

FIGURE 15.22

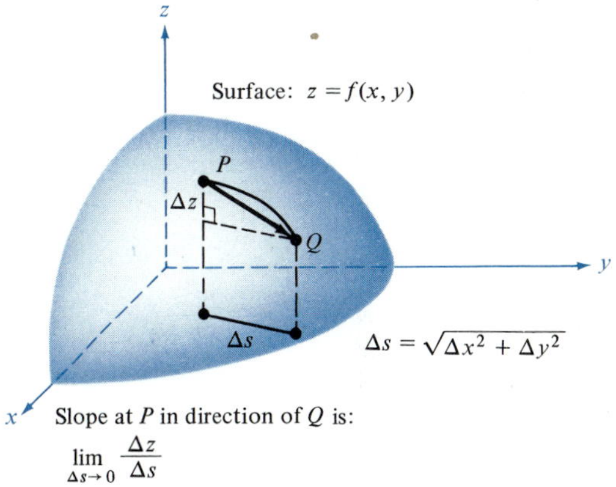

Surface: $z = f(x, y)$

$\Delta s = \sqrt{\Delta x^2 + \Delta y^2}$

Slope at P in direction of Q is:

$\lim\limits_{\Delta s \to 0} \dfrac{\Delta z}{\Delta s}$

FIGURE 15.23

To find the slope of the surface at P in the direction of Q, we let $\Delta s = \sqrt{\Delta x^2 + \Delta y^2}$ and take the limit of the ratio $\Delta z/\Delta s$, as $\Delta s \to 0$ (Figure 15.23). Now, since

$$\Delta z = f(x + \Delta x, y + \Delta y) - f(x, y)$$
$$= f(x + \Delta s \cos\theta, y + \Delta s \sin\theta) - f(x, y)$$

we have the following definition:

Definition of Directional Derivative

Let f be a function of two variables x and y; then the **directional derivative of f in the direction θ,** denoted by $D_\theta f$, is

$$D_\theta f(x, y) = \lim_{\Delta s \to 0} \frac{f(x + \Delta s \cos\theta, y + \Delta s \sin\theta) - f(x, y)}{\Delta s}$$

To calculate directional derivatives by this definition is comparable to finding the derivative of a function of one variable by the four-step process. A simpler "working" formula for finding directional derivatives involves the partial derivatives f_x and f_y.

THEOREM 15.8

(*Formula for the Directional Derivative*)

If f is a differentiable function of x and y, then the directional derivative of f in the direction of θ is

$$D_\theta f(x, y) = f_x(x, y)\cos\theta + f_y(x, y)\sin\theta$$

Proof: We begin by defining a new function

$$g(h) = f(x + h\cos\theta, y + h\sin\theta) = f(u, v)$$

in which we keep x, y, and θ fixed. Then by the Chain Rule,

$$g'(h) = f_u(u, v)\frac{\partial u}{\partial h} + f_v(u, v)\frac{\partial v}{\partial h}$$

$$= f_u(u, v)\cos\theta + f_v(u, v)\sin\theta$$

Now if $h = 0$, then

$$u = x + (0)\cos\theta = x$$

and

$$v = y + (0)\sin\theta = y$$

Thus it follows that

$$g'(0) = f_x(x, y)\cos\theta + f_y(x, y)\sin\theta$$

Furthermore, it is *also* true that

$$g'(0) = \lim_{h\to 0}\frac{g(h) - g(0)}{h}$$

$$= \lim_{h\to 0}\frac{f(x + h\cos\theta, y + h\sin\theta) - f(x, y)}{h}$$

$$= D_\theta f(x, y)$$

Hence we conclude that

$$D_\theta f = f_x(x, y)\cos\theta + f_y(x, y)\sin\theta \qquad\blacksquare$$

Note that we can verify two special cases of Theorem 15.8. That is, if $\theta = 0$ or $\pi/2$, we know from our previous discussion that we should obtain the partial derivatives f_x and f_y. Applying Theorem 15.8 for these two values of θ yields

$$D_0 f(x, y) = f_x(x, y)\cos(0) + f_y(x, y)\sin(0) = f_x(x, y)$$

and

$$D_{\pi/2} f(x, y) = f_x(x, y)\cos\left(\frac{\pi}{2}\right) + f_y(x, y)\sin\left(\frac{\pi}{2}\right) = f_y(x, y)$$

as expected.

Example 1

If $f(x, y) = 3x^2 - 3xy - y^2$, then at $(1, 2)$ find the directional derivative of f in the direction $\theta = \pi/6$.

Solution: By Theorem 15.8 the directional derivative is

$$D_\theta f = f_x(x, y)\cos\theta + f_y(x, y)\sin\theta$$

$$= (6x - 3y)\cos\theta + (-3x - 2y)\sin\theta$$

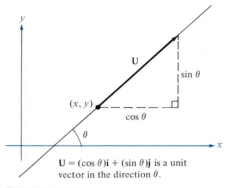

$U = (\cos \theta)i + (\sin \theta)j$ is a unit vector in the direction θ.

FIGURE 15.24

and for $\theta = \pi/6$, $x = 1$, and $y = 2$, we have

$$D_{\pi/6} f(1, 2) = (6 - 6) \left(\frac{\sqrt{3}}{2} \right) + (-3 - 4) \left(\frac{1}{2} \right) = -\frac{7}{2} \qquad \blacksquare$$

Suppose we let $\mathbf{U} = (\cos \theta)\mathbf{i} + (\sin \theta)\mathbf{j}$ be the unit vector in the direction θ, as shown in Figure 15.24. With this vector in mind, the formula for $D_\theta f$ in Theorem 15.8 looks very much like the dot product of two vectors—namely, the dot product of the vectors

$$(\cos \theta)\mathbf{i} + (\sin \theta)\mathbf{j} \qquad \text{and} \qquad f_x(x, y)\mathbf{i} + f_y(x, y)\mathbf{j}$$

This latter vector is a very important one and it has a variety of uses. We give it a special name in the following definition:

Definition of Gradient	If $z = f(x, y)$, then the **gradient** of f, denoted by ∇f, is the vector $$\nabla f = f_x(x, y)\mathbf{i} + f_y(x, y)\mathbf{j}$$ (The symbol "∇f" is read "del f.")

The ∇ symbol is in boldface to emphasize the fact that the gradient of f is a vector. The symbol **grad f** is also used to denote the gradient. Using the del notation, the formula in Theorem 15.8, for the directional derivative, becomes

$$D_\theta f = f_x(x, y) \cos \theta + f_y(x, y) \sin \theta$$
$$= [f_x(x, y)\mathbf{i} + f_y(x, y)\mathbf{j}] \cdot [(\cos \theta)\mathbf{i} + (\sin \theta)\mathbf{j}] = \nabla f \cdot \mathbf{U}$$

In the remaining portion of this text, we will use the following convenient form of the directional derivative.

THEOREM 15.9 (*Directional Derivative in Terms of the Gradient*)	If f is a function of x and y and ∇f is the gradient of f, then the **directional derivative of f in the direction of U** is given by $$D_{\mathbf{U}} f = \nabla f \cdot \mathbf{U}$$ where $\mathbf{U}$ is a unit vector.

Example 2

If $f(x, y) = 3x^2 - 2y^2$, find the directional derivative at $(-1, 3)$ in the direction from $(-1, 3)$ toward $(1, -2)$.

Solution: First, a vector in the specified direction is

$$\mathbf{V} = (1 + 1)\mathbf{i} + (-2 - 3)\mathbf{j} = 2\mathbf{i} - 5\mathbf{j}$$

Hence a unit vector in this direction is

$$U = \frac{V}{|V|} = \frac{2}{\sqrt{29}}i - \frac{5}{\sqrt{29}}j$$

Now
$$\nabla f = f_x i + f_y j = (6x)i + (-4y)j$$

and at $(-1, 3)$ the gradient is

$$\nabla f = -6i - 12j$$

Therefore,
$$D_U f = \nabla f \cdot U = \frac{-12}{\sqrt{29}} + \frac{60}{\sqrt{29}} = \frac{48}{\sqrt{29}}$$

is the rate of change in f in the direction from $(-1, 3)$ to $(1, -2)$. ∎

Two important properties of the gradient are given in the following theorem:

THEOREM 15.10
(*Uses of* ∇f)

If f has continuous partials f_x and f_y, then at (x, y) the gradient of f has these properties:

i. ∇f has the direction of the maximum rate of change in f.
ii. $|\nabla f|$ is the maximum value of the directional derivative at (x, y).

Proof: By the definition of dot product, we have

$$\nabla f \cdot U = |\nabla f| \, |U| \cos \phi$$

where ϕ denotes the angle between ∇f and U. Since $|U| = 1$, we have

$$D_U f = \nabla f \cdot U = |\nabla f| \cos \phi$$

and it follows that the largest value for $D_U f$ will occur when $\cos \phi = 1$, that is, when $\phi = 0$. Hence the largest value for the directional derivative occurs when ∇f and U have the same direction, and furthermore, this largest value for $D_U f$ is precisely

$$D_U f = |\nabla f| \cos \phi = |\nabla f|$$

(when $\phi = 0$). ∎

To illustrate the property that the gradient has the direction of maximum change in f, consider a skier coming down a mountainside. If the value of $f(x, y)$ denotes the altitude of the skier, then at (x, y), $-\nabla f$ indicates the *compass direction* the skier should take to ski the path of steepest descent. ($+\nabla f$ indicates the compass direction of steepest ascent.) Note that the vector $\nabla f = f_x(x, y)i + f_y(x, y)j$ indicates direction in a horizontal plane (north, south, east, etc.) and does not itself point up or down the mountainside.

As another illustration of the gradient, consider the temperature $T(x, y)$ at any point (x, y) on a hot plate. In this case, at point (x, y), ∇T gives the direction of greatest temperature increase, and $-\nabla T$ indicates the direction in which heat would flow—toward the coldest spot.

Example 3

The temperature distribution on a metal plate is given by the function $T(x, y) = 20 - 4x^2 - 9y^2$.

(a) In what direction from $(2, -3)$ does the temperature increase most rapidly? What is this rate of increase?

(b) In what direction from $(2, -3)$ does the temperature decrease most rapidly?

Solution: From $T(x, y) = 20 - 4x^2 - y^2$ we have

$$\nabla T = T_x(x, y)\mathbf{i} + T_y(x, y)\mathbf{j} = -8x\mathbf{i} - 2y\mathbf{j}$$

(a) At $(2, -3)$ the temperature increases most rapidly in the direction of

$$\nabla T = -16\mathbf{i} + 6\mathbf{j}$$

and the rate of increase is

$$|\nabla T| = \sqrt{256 + 36} = \sqrt{292} \approx 17.09$$

(b) The temperature decreases most rapidly in the direction of

$$-\nabla T = 16\mathbf{i} - 6\mathbf{j} \qquad \blacksquare$$

The notions of the directional derivative and the gradient can be readily generalized to functions of three or more variables. However, the geometrical interpretation of the directional derivative as the "slope" of a tangent line no longer applies.

We list here, without formal verification, the definitions and properties of the directional derivative and the gradient of a function of three variables.

Directional Derivative and Gradient (**Summary of Three-Variable Case**)

If f is a function of three variables, x, y, and z, then:

1. The directional derivative of f in the direction of $\mathbf{U}$ is given by

$$D_{\mathbf{U}}f = f_x(x, y, z)\cos\alpha + f_y(x, y, z)\cos\beta + f_z(x, y, z)\cos\gamma$$

where $\mathbf{U} = \cos\alpha\,\mathbf{i} + \cos\beta\,\mathbf{j} + \cos\gamma\,\mathbf{k}$ is a unit vector.

2. The gradient of f is defined to be

$$\nabla f = f_x(x, y, z)\mathbf{i} + f_y(x, y, z)\mathbf{j} + f_z(x, y, z)\mathbf{k}$$

where ∇f has the direction of greatest increase in f, and $|\nabla f|$ is the maximum value of $D_{\mathbf{U}}f$ at (x, y, z).

3. The directional derivative of f in the direction of $\mathbf{U}$ is most readily calculated by using the formula

$$D_{\mathbf{U}}f = \nabla f \cdot \mathbf{U}$$

Example 4

Find the directional derivative of $f(x, y, z) = x^2 e^{2yz}$ at the point $(-1, 0, 2)$ and in the direction from $(-1, 0, 2)$ to $(1, 3, -2)$. Find the maximum value of $D_U f$ at $(-1, 0, 2)$.

Solution: First, a vector in the specified direction is

$$\mathbf{V} = (1 + 1)\mathbf{i} + (3 - 0)\mathbf{j} + (-2 - 2)\mathbf{k} = 2\mathbf{i} + 3\mathbf{j} - 4\mathbf{k}$$

A unit vector in the direction of $\mathbf{V}$ is

$$\mathbf{U} = \frac{\mathbf{V}}{|\mathbf{V}|} = \frac{2}{\sqrt{29}}\mathbf{i} + \frac{3}{\sqrt{29}}\mathbf{j} - \frac{4}{\sqrt{29}}\mathbf{k}$$

Since

$$\nabla f = f_x \mathbf{i} + f_y \mathbf{j} + f_z \mathbf{k} = 2xe^{2yz}\mathbf{i} + 2zx^2 e^{2yz}\mathbf{j} + 2yx^2 e^{2yz}\mathbf{k}$$

then at $(-1, 0, 2)$ we have

$$\nabla f = -2\mathbf{i} + 4\mathbf{j}$$

Therefore, in the direction of $\mathbf{U}$, the value of the directional derivative is

$$D_U f = \nabla f \cdot \mathbf{U} = \frac{-4}{\sqrt{29}} + \frac{12}{\sqrt{29}} = \frac{8}{\sqrt{29}}$$

The maximum value of $D_U f$ at $(-1, 0, 2)$ is

$$|\nabla f| = \sqrt{4 + 16} = 2\sqrt{5}$$ ∎

Section Exercises (15.5)

In Exercises 1–10, answer the questions about the function $f(x, y) = 3 - (x/3) - (y/2)$.

1. Sketch the graph of f in the first octant and plot the point $(3, 2, 1)$ on the surface.

2. Find $D_\theta f(3, 2)$ when $\theta = \pi/4$.

S **3.** Find $D_\theta f(3, 2)$ when $\theta = 2\pi/3$.

4. Find $D_\theta f(3, 2)$ when $\theta = 4\pi/3$.

5. Find ∇f.

6. Find $D_U f(3, 2)$ if $\mathbf{U} = (1/\sqrt{2})(\mathbf{i} + \mathbf{j})$. Compare this result with that of Exercise 2.

7. Find $D_U f(3, 2)$ where $\mathbf{U}$ is the unit vector in the direction of $\mathbf{V} = -3\mathbf{i} - 4\mathbf{j}$.

S **8.** Find $D_U f(3, 2)$ where $\mathbf{U}$ is in the direction of the vector from $(1, 2)$ to $(-2, 6)$.

S **9.** Find the maximum value of the directional derivative at $(3, 2)$.

10. Find a unit vector $\mathbf{U}$ orthogonal to ∇f and calculate $D_U f(3, 2)$. Discuss the meaning of this result.

In Exercises 11–16, answer questions about the function $f(x, y) = 9 - x^2 - y^2$.

11. Sketch the graph of f in the first octant and plot the point $(1, 2, 4)$ on the surface.

12. Find $D_\theta f(1, 2)$ when $\theta = -\pi/4$.

13. Find $D_\theta f(1, 2)$ when $\theta = \pi/3$.

14. Find $D_U f(1, 2)$ when $\mathbf{U}$ is the unit vector in the direction of $\mathbf{V} = -(4\mathbf{i} + 3\mathbf{j})$.

15. Find $\nabla f(1, 2)$ and $|\nabla f(1, 2)|$.

S **16.** Find a unit vector $\mathbf{U}$ orthogonal to ∇f and calculate $D_U f(1, 2)$.

In Exercises 17–22, find the directional derivative of f at the given point in the direction of $\mathbf{V}$. Also, find the maximum value of the directional derivative at the point.

17. $f(x, y) = xy$; $(2, 3)$; $\mathbf{V} = \mathbf{i} + \mathbf{j}$

18. $f(x, y, z) = x^2 + y^2 + z^2$; $(1, 2, -1)$; $\mathbf{V} = \mathbf{i} - 2\mathbf{j} + 3\mathbf{k}$

S **19.** $f(x, y) = \sqrt{x^2 + y^2}$; $(3, 4)$; $\mathbf{V} = 3\mathbf{i} - 4\mathbf{j}$

20. $f(x, y, z) = xyz$; $(2, 1, 1)$; $\mathbf{V} = 2\mathbf{i} + \mathbf{j} + 2\mathbf{k}$

21. $f(x, y) = e^x \sin y$; $(1, \pi/2)$; $\mathbf{V} = -\mathbf{i}$

22. $f(x, y) = e^{-(x^2+y^2)}$; $(0, 0)$; $\mathbf{V} = \mathbf{i} + \mathbf{j}$

⌐S⌐ **23.** The temperature field at any point in a plate is given by $T = x/(x^2 + y^2)$. Find the direction of greatest increase in heat at the point $(3, 4)$.

24. The surface of a mountain is described by the equation $h(x, y) = 4000 - 0.001x^2 - 0.004y^2$. Suppose that a mountain climber is at the point $(500, 300, 3390)$. In what direction should the mountain climber move in order to ascend at the greatest rate?

25. Assuming the functions f and g are differentiable, show the following:

(a) $\nabla (fg) = f \nabla g + g \nabla f$

(b) $\nabla \left(\dfrac{f}{g} \right) = \dfrac{g \nabla f - f \nabla g}{g^2}$

For Review and Class Discussion

True or False

1. ___ If $f(x, y) = \sqrt{1 - x^2 - y^2}$, then $D_\theta f(0, 0) = 0$ for any angle θ.

2. ___ If $f(x, y) = x + y$, then $-1 \le D_\theta f(x, y) \le 1$.

3. ___ If $D_\theta f(x, y)$ exists, then $D_\theta f(x, y) = -D_{\theta + \pi} f(x, y)$.

4. ___ If $D_\theta f(x_0, y_0) = c$ for any angle θ, then $c = 0$.

5. ___ If $F(x, y, z) = f(x, y) - z$, then $\nabla F = f_x(x, y)\mathbf{i} + f_y(x, y)\mathbf{j} - \mathbf{k}$.

15.6 Tangent Planes

Purpose

■ To demonstrate the use of the gradient in finding the tangent plane to a surface.

Consider a point P of the surface together with all the curves through P that lie on the surface. It is a remarkable fact that in general the tangents to these curves at the point P all lie in a common plane, which is therefore called the tangent plane to the surface at P. —David Hilbert (1862–1943)

In Section 15.5 we saw how the directional derivative, $D_\mathbf{U} f$, can be used to measure the slope of tangent lines to the surface $z = f(x, y)$. Furthermore, in Theorem 15.8 we saw that the directional derivative of f in *any* direction θ can be determined from the formula

$$D_\theta f(x, y) = f_x(x, y) \cos \theta + f_y(x, y) \sin \theta$$

Geometrically this means that every tangent line to a point on the surface $z = f(x, y)$ lies in the plane determined by the tangents in the x and y directions. We call this plane the **tangent plane** to f at (x_0, y_0, z_0). (See Figure 15.25.)

To find the equation of the tangent plane, we consider the cross product of the two tangent vectors

$$\mathbf{T}_x = \mathbf{i} + f_x(x_0, y_0)\mathbf{k} \quad \text{and} \quad \mathbf{T}_y = \mathbf{j} + f_y(x_0, y_0)\mathbf{k}$$

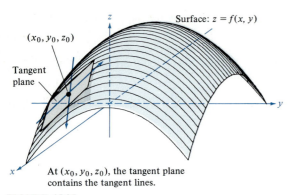

At (x_0, y_0, z_0), the tangent plane contains the tangent lines.

FIGURE 15.25

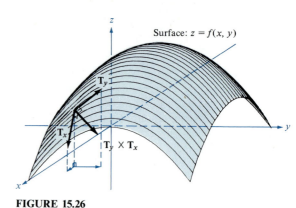

FIGURE 15.26

(See Figure 15.26.) Taking this cross product we have

$$\mathbf{T}_y \times \mathbf{T}_x = \begin{vmatrix} \mathbf{i} & \mathbf{j} & \mathbf{k} \\ 0 & 1 & f_y(x_0, y_0) \\ 1 & 0 & f_x(x_0, y_0) \end{vmatrix} = f_x(x_0, y_0)\mathbf{i} + f_y(x_0, y_0)\mathbf{j} - \mathbf{k}$$

which is precisely the formula for the gradient ∇F, where $F(x, y, z) = f(x, y) - z$. Now since the cross product of two vectors is orthogonal to both vectors, it follows that the gradient ∇F at (x_0, y_0, z_0) is normal to the tangent plane at (x_0, y_0, z_0). Thus the equation for the tangent plane to the surface $z = f(x, y)$ at the point (x_0, y_0, z_0) is given by

$$f_x(x_0, y_0)(x - x_0) + f_y(x_0, y_0)(y - y_0) - (z - z_0) = 0$$

Note: It is important that you see the distinction between ∇f and ∇F. When a surface is given by $z = f(x, y)$, ∇f is a vector in the xy-plane pointing in the compass direction of the maximum change in f (Section 15.5). However, when the surface is given by $F(x, y, z) = 0$, ∇F is a three-dimensional vector that is normal to the surface.

THEOREM 15.11
[*Tangent Plane to $z = f(x, y)$*]

If $z = f(x, y)$ is differentiable at (x_0, y_0, z_0), then the tangent plane to f at (x_0, y_0, z_0) is given by

$$f_x(x_0, y_0)(x - x_0) + f_y(x_0, y_0)(y - y_0) - (z - z_0) = 0$$

Example 1

Find the tangent plane to $f(x, y) = \sin xy$ at the point $(0, 1, 0)$.

Solution: Since

$$f_x(x, y) = y \cos xy \qquad \text{and} \qquad f_y(x, y) = x \cos xy$$

we have $f_x(0, 1) = 1$ $\qquad$ and $\qquad$ $f_y(0, 1) = 0$

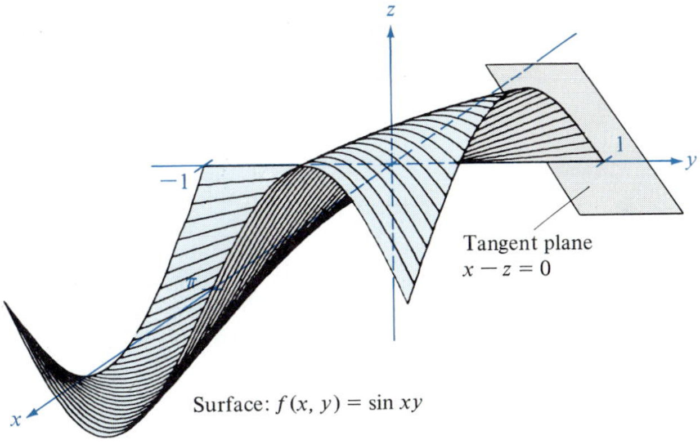

Surface: $f(x, y) = \sin xy$

Tangent plane
$x - z = 0$

FIGURE 15.27

Now applying Theorem 15.11, the tangent plane at $(0, 1, 0)$ is given by

$$1(x - 0) + (0)(y - 1) - (z - 0) = 0$$

or

$$x - z = 0$$

(See Figure 15.27.)   ∎

THEOREM 15.12 [*Tangent Plane to* $F(x, y, z) = 0$]	If F is differentiable at (x_0, y_0, z_0), then the tangent plane to F at (x_0, y_0, z_0) is given by $F_x(x_0, y_0, z_0)(x - x_0) + F_y(x_0, y_0, z_0)(y - y_0) + F_z(x_0, y_0, z_0)(z - z_0) = 0$

Example 2

Find the angle θ between the *xy*-plane and the tangent plane to the hemisphere $f(x, y) = \sqrt{9 - x^2 - y^2}$ at the point $(2, 2, 1)$.

Solution: If we let

$$F(x, y, z) = f(x, y) - z = \sqrt{9 - x^2 - y^2} - z$$

then $\nabla F(2, 2, 1)$ is normal to the tangent plane at $(2, 2, 1)$, and since the unit vector $\mathbf{k}$ is normal to the *xy*-plane, we conclude that θ is the angle between $\nabla F(2, 2, 1)$ and $\mathbf{k}$. Thus we have

$$\nabla F = F_x(x, y, z)\mathbf{i} + F_y(x, y, z)\mathbf{j} + F_z(x, y, z)\mathbf{k}$$

$$= \frac{-x}{\sqrt{9 - x^2 - y^2}}\mathbf{i} + \frac{-y}{\sqrt{9 - x^2 - y^2}}\mathbf{j} - \mathbf{k}$$

$$\nabla F(2, 2, 1) = -2\mathbf{i} - 2\mathbf{j} - \mathbf{k}$$

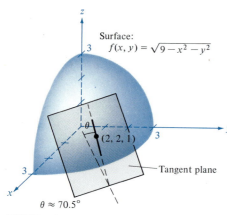

FIGURE 15.28

Therefore,

$$\cos \theta = \frac{|\nabla F(2, 2, 1) \cdot \mathbf{k}|}{|\nabla F(2, 2, 1)| \, |\mathbf{k}|} = \frac{|-1|}{\sqrt{2^2 + 2^2 + 1^2}(1)} = \frac{1}{3}$$

$$\theta = \text{Arccos} \tfrac{1}{3} \approx 70.5°$$

(See Figure 15.28.) ∎

We can generalize the procedure shown in Example 2 to conclude that the *angle between the xy-plane and the tangent plane to the surface*, $z = f(x, y)$, at (x_0, y_0, z_0) is given by

$$\cos \theta = \frac{1}{\sqrt{[f_x(x_0, y_0)]^2 + [f_y(x_0, y_0)]^2 + 1}}$$

Similarly, the *angle between the xy-plane and the tangent plane to the surface*, $F(x, y, z) = 0$, at (x_0, y_0, z_0) is given by

$$\cos \theta = \frac{1}{\sqrt{[F_x(x_0, y_0, z_0)]^2 + [F_y(x_0, y_0, z_0)]^2 + [F_z(x_0, y_0, z_0)]^2}}$$

Example 3

Find an equation for the tangent plane to the elliptic paraboloid $x^2 + 4y^2 - 10z = 0$ at the point $(2, -2, 2)$.

Solution: Let $F(x, y, z) = x^2 + 4y^2 - 10z$; then

$$F_x = 2x, \qquad F_y = 8y, \qquad F_z = -10$$

Therefore, at $(2, -2, 2)$ the equation of the tangent plane is

$$(4)(x - 2) + (-16)(y + 2) + (-10)(z - 2) = 0$$

which simplifies to

$$2x - 8y - 5z - 10 = 0$$ ∎

Example 4

Find an equation of the tangent plane and the equations for the normal line to the graph of $yx^2 + zy^2 - xz^2 = 18$ at the point $(-2, 0, 3)$.

Solution: Considering $F(x, y, z) = yx^2 + zy^2 - xz^2 - 18$, we get

$$F_x = 2xy - z^2, \qquad F_y = x^2 + 2yz, \qquad F_z = y^2 - 2xz$$

and at $(-2, 0, 3)$

$$F_x = -9, \qquad F_y = 4, \qquad F_z = 12$$

The equation of the tangent plane is, therefore,

$$(-9)(x + 2) + (4)(y - 0) + (12)(z - 3) = 0$$
$$-9x + 4y + 12z - 54 = 0$$

At $(-2, 0, 3)$ the gradient

$$\nabla F = -9\mathbf{i} + 4\mathbf{j} + 12\mathbf{k}$$

is normal to the surface and has direction numbers -9, 4, and 12. Therefore, symmetric equations for the normal line at $(-2, 0, 3)$ are

$$\frac{x + 2}{-9} = \frac{y - 0}{4} = \frac{z - 3}{12}$$

■

Example 5

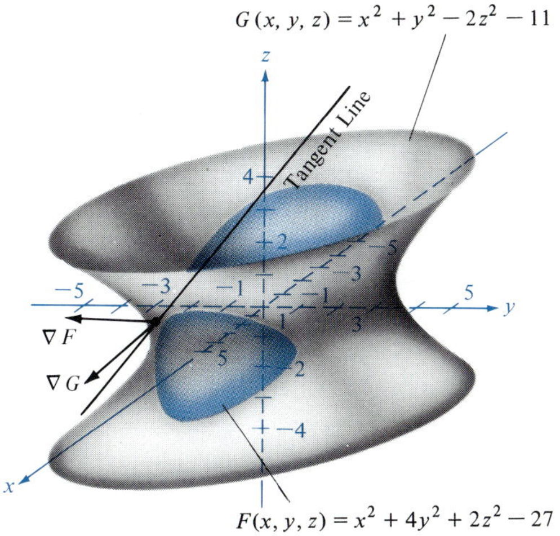

$G(x, y, z) = x^2 + y^2 - 2z^2 - 11$

Find symmetric equations for the tangent line to the curve of intersection of the surfaces

$$x^2 + 4y^2 + 2z^2 = 27 \quad \text{and} \quad x^2 + y^2 - 2z^2 = 11$$

at the point $(3, -2, 1)$. (See Figure 15.29.)

Solution: Let

$$F(x, y, z) = x^2 + 4y^2 + 2z^2 - 27$$

and

$$G(x, y, z) = x^2 + y^2 - 2z^2 - 11$$

Then

$$\nabla F = 2x\mathbf{i} + 8y\mathbf{j} + 4z\mathbf{k} \quad \text{and} \quad \nabla G = 2x\mathbf{i} + 2y\mathbf{j} - 4z\mathbf{k}$$

and at $(3, -2, 1)$

$$\nabla F = 6\mathbf{i} - 16\mathbf{j} + 4\mathbf{k} \quad \text{and} \quad \nabla G = 6\mathbf{i} - 4\mathbf{j} - 4\mathbf{k}$$

Now the vector

$$\nabla F \times \nabla G = \begin{vmatrix} \mathbf{i} & \mathbf{j} & \mathbf{k} \\ 6 & -16 & 4 \\ 6 & -4 & -4 \end{vmatrix} = 8(10\mathbf{i} + 6\mathbf{j} + 9\mathbf{k})$$

with direction numbers 10, 6, and 9, has the direction of the required tangent line. Therefore, symmetric equations for the tangent line at $(3, -2, 1)$ are

$$\frac{x - 3}{10} = \frac{y + 2}{6} = \frac{z - 1}{9}$$

■

FIGURE 15.29

Tangent Line

$F(x, y, z) = x^2 + 4y^2 + 2z^2 - 27$

∇F

∇G

Two surfaces are said to be *tangent* at a point (x_0, y_0, z_0) if they have a common tangent plane at that point.

Section Exercises (15.6)

In Exercises 1–14, find the equation of the tangent plane at the indicated point.

1. $f(x, y) = 25 - x^2 - y^2$; $(3, 1, 15)$
2. $f(x, y) = \sqrt{x^2 + y^2}$; $(3, 4, 5)$
 ⒮ 3. $f(x, y) = \dfrac{y}{x}$; $(1, 2, 2)$

4. $f(x, y) = 2 - \dfrac{2x}{3} - y$; $(3, -1, 1)$

5. $f(x, y) = x^2 - y^2$; $(5, 4, 9)$

6. $f(x, y) = \text{Arctan} \dfrac{y}{x}$; $(1, 0, 0)$

7. $f(x, y) = e^x(\sin y + 1)$; $(0, \pi/2, 2)$
8. $f(x, y) = x^3 - 3xy + y^3$; $(1, 2, 3)$
9. $f(x, y) = \ln \sqrt{x^2 + y^2}$; $(3, 4, \ln 5)$
10. $f(x, y) = \cos y$; $(5, \pi/4, \sqrt{2}/2)$
11. $x^2 + 4y^2 + z^2 = 36$; $(2, -2, 4)$
12. $x^2 + 2z^2 = y^2$; $(1, 3, -2)$
 ⒮ 13. $xy^2 + 3x - z^2 = 4$; $(2, 1, -2)$
14. $y = x(2z - 1)$; $(4, 4, 1)$

In Exercises 15–24, find an equation for the tangent plane and find symmetric equations of the normal line at the specified point.
 ⒮ 15. $x^2 + y^2 + z = 9$; $(1, 2, 4)$

16. $x^2 + y^2 + z^2 = 9$; $(1, 2, 2)$
17. $xy - z = 0$; $(-2, -3, 6)$
18. $x^2 + y^2 - z^2 = 0$; $(5, 12, 13)$
 ⒮ 19. $z = \text{Arctan}\,(y/x)$; $(1, 1, \pi/4)$
20. $xyz = 10$; $(1, 2, 5)$
21. $3x^2 + 2y^2 - z = 15$; $(2, 2, 5)$
22. $xy - z^2 = 0$; $(2, 2, 2)$
23. $x^2 - y^2 + z = 0$; $(1, 2, 3)$
24. $x^2 + y^2 = 5$; $(2, 1, 3)$

25. Show that at the point (x_0, y_0, z_0) the equation of the tangent plane to the ellipsoid

$$\frac{x^2}{a^2} + \frac{y^2}{b^2} + \frac{z^2}{c^2} = 1 \qquad \text{is} \qquad \frac{x_0 x}{a^2} + \frac{y_0 y}{b^2} + \frac{z_0 z}{c^2} = 1$$

26. Show that the tangent plane to the cone $z^2 = a^2 x^2 + b^2 y^2$ passes through the origin.

In Exercises 27–30, (a) find symmetric equations of the tangent line to the curve of intersection of the given surfaces at the given point; (b) find the cosine of the angle between the gradients of the surfaces at the given point.
 ⒮ 27. $x^2 + y^2 = 5$, $z = x$; $(2, 1, 2)$
28. $z = x^2 + y^2$, $z = 4 - y$; $(2, -1, 5)$
29. $x^2 + z^2 = 25$, $y^2 + z^2 = 25$; $(3, 3, 4)$
30. $z = \sqrt{x^2 + y^2}$, $2x + y + 2z = 20$; $(3, 4, 5)$

For Review and Class Discussion

True or False

1. _____ If $f_x(x_0, y_0) = 0$ and $f_y(x_0, y_0) = 0$, then at the point (x_0, y_0, z_0), the tangent plane to the surface given by $z = f(x, y)$ is horizontal.

2. _____ The gradient $\nabla f(x_0, y_0)$ is normal to the surface given by $z = f(x, y)$ at the point (x_0, y_0, z_0).

3. _____ The plane $z = 1$ is tangent to the surface given by $z = \sin(xy)$ at infinitely many points.

4. _____ The plane $x_0 x + y_0 y + z_0 z = c^2$ is tangent to the sphere $x^2 + y^2 + z^2 = c^2$ at the point (x_0, y_0, z_0).

5. _____ If the plane given by $a(x - x_0) + b(y - y_0) + c(z - z_0) = 0$ is tangent to the surface given by $f(x, y, z) = 0$ at (x_0, y_0, z_0), then $f_x(x_0, y_0, z_0) = a$, $f_y(x_0, y_0, z_0) = b$, and $f_z(x_0, y_0, z_0) = c$.

Miscellaneous Exercises (Ch. 15)

In Exercises 1–10, find all first partial derivatives.

1. $f(x, y) = e^x \cos y$

2. $f(x, y) = \dfrac{xy}{x + y}$

3. $f(x, y) = xe^y + ye^x$
4. $f(x, y) = \ln(x^2 + y^2 + 1)$

5. $f(x, y) = \dfrac{xy}{x^2 + y^2}$

6. $f(x, y, z) = \sqrt{x^2 + y^2 + z^2}$

7. $f(x, y, z) = z \, \text{Arctan} \, \dfrac{y}{x}$

S 8. $f(x, y, z) = \dfrac{1}{\sqrt{1 - x^2 - y^2 - z^2}}$

9. $u(x, t) = ce^{-n^2 t} \sin(nx)$

10. $u(x, t) = c \sin(akx) \cos(kt)$

11. Find $\partial z / \partial x$ if $x^2 y - 2xyz - xz - z^2 = 0$.

12. Find $\partial z / \partial y$ if $xz^2 - y \sin z = 0$.

In Exercises 13–16, find all second partial derivatives and verify that the second mixed partials are equal.

13. $f(x, y) = 3x^2 - xy + 2y^3$

14. $f(x, y) = \dfrac{x}{x + y}$

S 15. $f(x, y) = x \sin y + y \cos x$

16. $f(x, y) = \cos(x - 2y)$

A function is said to be *harmonic* if it satisfies Laplace's equation, $(\partial^2 z / \partial x^2) + (\partial^2 z / \partial y^2) = 0$. In Exercises 17–20, show that z is harmonic.

17. $z = x^2 - y^2$

18. $z = x^3 - 3xy^2$

S 19. $z = \dfrac{y}{x^2 + y^2}$

20. $z = e^x \sin y$

S 21. Find $\partial u / \partial r$ and $\partial u / \partial t$ if $u = x^2 + y^2 + z^2$ and $x = r \cos t$, $y = r \sin t$, and $z = r$. Use the Chain Rule and check the result by substituting for x, y, and z before differentiating.

22. Find du/dt if $u = y^2 - x$ and $x = \cos t$ and $y = \sin t$. Use the Chain Rule and check the result by substitution before differentiating.

23. A function $f(x, y)$ is said to be **homogeneous of degree** n if $f(tx, ty) = t^n f(x, y)$. Determine whether or not the following functions are homogeneous and if so determine the degree:

(a) $f(x, y) = x^3 - 3xy^2 + y^3$

S (b) $f(x, y) = \dfrac{xy}{\sqrt{x^2 + y^2}}$

S (c) $f(x, y) = e^{xy}$

(d) $f(x, y) = x \sin \dfrac{y}{x}$

24. Show that if $f(x, y)$ is homogeneous of degree n, then

$$x \frac{\partial f}{\partial x} + y \frac{\partial f}{\partial y} = nf(x, y)$$

[Hint: Let $g(t) = f(tx, ty) = t^n f(x, y)$. Find $g'(t)$ and then let $t = 1$.]

25. Demonstrate the result of Exercise 24 for the homogeneous function $f(x, y) = 2x^3 - 3xy^2$.

26. Demonstrate the result of Exercise 24 for the homogeneous function $f(x, y) = xy / \sqrt{x^2 + y^2}$.

27. Find the directional derivative of $f(x, y) = x^2 y$ at the point $(2, 1)$ in the direction of the vector $\mathbf{V} = \mathbf{i} - \mathbf{j}$.

28. Find the directional derivative of $f(x, y, z) = y^2 + xz$ at the point $(1, 2, 2)$ in the direction of the vector $\mathbf{V} = 2\mathbf{i} - \mathbf{j} + 2\mathbf{k}$.

In Exercises 29–32, find an equation of the tangent plane and the symmetric equations for a normal line to the surface at the specified point.

29. $f(x, y) = x^2 y$; $(2, 1, 4)$

30. $f(x, y) = \sqrt{25 - y^2}$; $(2, 3, 4)$

S 31. $f(x, y) = 9 + 4x - 6y - x^2 - y^2$; $(2, -3, 4)$

32. $f(x, y) = \sqrt{9 - x^2 - y^2}$; $(1, 2, 2)$

33. Find symmetric equations of the tangent line to the curve of intersection of $z = x^2 - y^2$ and $z = 3$ at the point $(2, 1, 3)$.

34. Find symmetric equations of the tangent line to the curve of intersection of $z = 25 - y^2$ and $y = x$ at the point $(4, 4, 9)$.

35. The legs of a right triangle are measured and found to be 5 in. and 12 in., with a possible error of $\frac{1}{16}$ in. Approximate the maximum possible error in computing the length of the hypotenuse. Approximate the maximum percentage error.

36. To determine the height of a tower, the angle of elevation to the top of the tower was measured from a point $100 \pm \frac{1}{2}$ ft from the base. The angle is measured at $33°$ with a possible error of $1°$. Assuming the ground to be horizontal, approximate the maximum error in determining the height of the tower.

37. Locate and classify any extrema to the function $f(x, y) = x^3 - 3xy + y^2$.

38. Locate and classify any extrema to the function $z = x^2 y$ subject to the condition $x + 2y = 2$.

S 39. Given n points $(x_1, y_1), (x_2, y_2), \ldots, (x_n, y_n)$, where the x_i's are not all alike, then the **least squares regression line** is the line $y = mx + b$ that minimizes the sum of the squares of the vertical distances from the points to the line. Use calculus to show that

$$E(m, b) = \sum_{i=1}^{n} (mx_i + b - y_i)^2$$

is minimum when m and b are the unique solutions to the system of equations

$$nb + \left(\sum_{i=1}^{n} x_i\right) m = \sum_{i=1}^{n} y_i$$

$$\left(\sum_{i=1}^{n} x_i\right) b + \left(\sum_{i=1}^{n} x_i^2\right) m = \sum_{i=1}^{n} x_i y_i$$

In Exercises 40–42, use the result of Exercise 39 to find the equation of the least squares regression line to each set of points. Plot the points and sketch the regression line.

40. $(-2, 0)$, $(0, 1)$, $(2, 1)$

$\boxed{S}$ **41.** $(-1, 0)$, $(0, 1)$, $(0, 2)$, $(1, 2)$

$\boxed{C}$ **42.** $(0, 1)$, $(1, 0)$, $(1, 2)$, $(1, 3)$, $(2, 2)$, $(3, 3)$, $(3, 4)$, $(5, 5)$

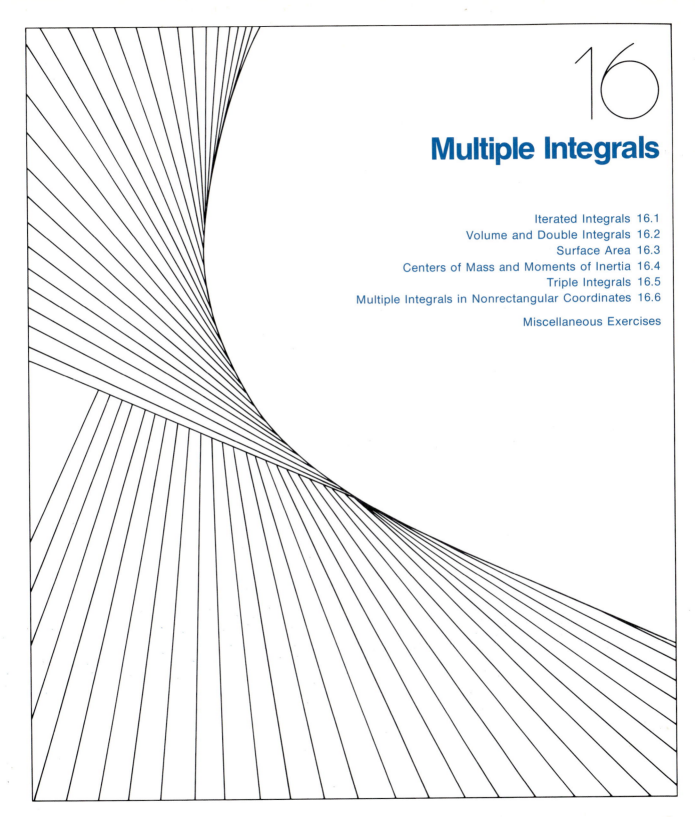

16

Multiple Integrals

16.1 Iterated Integrals

Purpose
- To introduce iterated integrals and demonstrate their evaluation.
- To represent areas of a plane region by iterated integrals.
- To describe the region having an area corresponding to a given iterated integral.

Thus the whole thing will immediately reduce to finding the area of some plane figure. —Gottfried Leibniz (1646–1716)

In Chapter 15 we showed that it is meaningful to differentiate functions of more than one variable by differentiating with respect to one variable at a time while holding the other variables constant. It should not be surprising to learn that we can also *integrate* functions of two or more variables by a similar procedure. For example, if we are given the partial derivative $f_x(x, y) = 2xy$, then by holding y constant, we can integrate with respect to x to obtain

$$f(x, y) = x^2 y + C(y)$$

This procedure is sometimes referred to as **partial integration with respect to** x. Note that the "constant of integration," $C(y)$, is assumed to be a function of y since y is fixed while integrating with respect to x.

To evaluate the definite integral of a function of two or more variables, we can use the Fundamental Theorem of Calculus (Section 6.4) with one variable while holding the others constant. For instance,

$$\int_1^{2y} 2xy \; \boxed{dx} = x^2 y \Big]_1^{2y} = (2y)^2 y - (1)^2 y = \underbrace{4y^3 - y}$$

x is the variable of integration and y is fixed

replace x by the limits of integration

the result is a function of y

Notice that we omit the constant of integration just as we do for the definite integral of a function of one variable.

A partial integral with respect to y can be evaluated in a similar manner, and we generalize these procedures as follows:

$$\int_{h_1(y)}^{h_2(y)} f_x(x, y) \, dx = f(x, y) \Big]_{h_1(y)}^{h_2(y)} = f(h_2(y), y) - f(h_1(y), y)$$

and

$$\int_{g_1(x)}^{g_2(x)} f_y(x, y) \, dy = f(x, y) \Big]_{g_1(x)}^{g_2(x)} = f(x, g_2(x)) - f(x, g_1(x))$$

Example 1

Evaluate the integrals

$$\int_1^x (2x^2 y^{-2} + 2y) \, dy \qquad \text{and} \qquad \int_y^{5y} \sqrt{x - y} \, dx$$

Solution: Considering x to be constant while integrating with respect to y, we obtain

$$\int_1^x (2x^2y^{-2} + 2y)\, dy = \left[\frac{-2x^2}{y} + y^2\right]_1^x$$

$$= \left(\frac{-2x^2}{x} + x^2\right) - \left(\frac{-2x^2}{1} + 1\right) = 3x^2 - 2x - 1$$

Similarly, considering y to be constant, we have

$$\int_y^{5y} \sqrt{x - y}\, dx = \frac{2}{3}(x - y)^{3/2}\Big]_y^{5y} = \frac{2}{3}(5y - y)^{3/2} - (y - y)^{3/2}$$

$$= \frac{2}{3}(4y)^{3/2} = \frac{16}{3}y^{3/2} \qquad\blacksquare$$

Notice in Example 1 that the integral

$$\int_1^x (2x^2y^{-2} + 2y)\, dy = 3x^2 - 2x - 1$$

defines a function of one variable x and, as such, can *itself* be integrated, as shown in the next example.

Example 2

Evaluate the integral

$$\int_0^1 \left[\int_1^x (2x^2y^{-2} + 2y)\, dy\right] dx$$

Solution: Using the results from Example 1, we have

$$\int_0^1 \left[\int_1^x (2x^2y^{-2} + 2y)\, dy\right] dx = \int_0^1 (3x^2 - 2x - 1)\, dx$$

$$= \left[x^3 - x^2 - x\right]_0^1 = -1 \qquad\blacksquare$$

We call integrals of the form

$$\int_a^b \left[\int_{g_1(x)}^{g_2(x)} f(x, y)\, dy\right] dx \qquad \text{or} \qquad \int_c^d \left[\int_{h_1(y)}^{h_2(y)} f(x, y)\, dx\right] dy$$

iterated integrals, and we normally shorten the notation by omitting the brackets. Thus by definition

$$\int_a^b \int_{g_1(x)}^{g_2(x)} f(x, y)\, dy\, dx = \int_a^b \left[\int_{g_1(x)}^{g_2(x)} f(x, y)\, dy\right] dx$$

One of the simplest applications of an iterated integral is in finding the area of a plane region. For instance, if R is the region bounded by $a \leq x \leq b$ and $g_1(x) \leq y \leq g_2(x)$, then by the methods of Section 7.1, we know that the area of R is given by

$$\int_a^b [g_2(x) - g_1(x)]\, dx$$

(See Figure 16.1.) This same area is also given by the iterated integral

$$\int_a^b \int_{g_1(x)}^{g_2(x)} dy\, dx$$

because

$$\int_a^b \int_{g_1(x)}^{g_2(x)} dy\, dx = \int_a^b y \Big]_{g_1(x)}^{g_2(x)} dx = \int_a^b [g_2(x) - g_1(x)]\, dx$$

Figure 16.1 shows the two basic types of plane regions whose area can be determined by an iterated integral.

Area in the Plane by Iterated Integrals

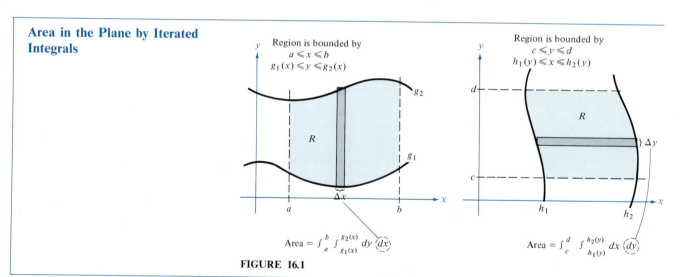

Area = $\int_a^b \int_{g_1(x)}^{g_2(x)} dy\, \widetilde{dx}$ Area = $\int_c^d \int_{h_1(y)}^{h_2(y)} dx\, \widetilde{dy}$

FIGURE 16.1

In Figure 16.1 note that the position (vertical or horizontal) of the narrow rectangle indicates the order of integration. The "outer" variable of integration always corresponds to the width (thickness) of the rectangle. (Compare this with the procedures of Section 7.1 for the area between two curves.) Note further that the outer limits of integration for an iterated integral are constant, whereas the inner limits may be functions of the outer variable.

Example 3

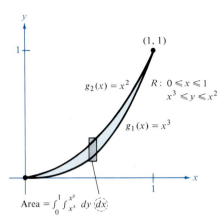

$g_2(x) = x^2$

$R: 0 \leqslant x \leqslant 1$
$x^3 \leqslant y \leqslant x^2$

$g_1(x) = x^3$

(1, 1)

Area $= \int_0^1 \int_{x^3}^{x^2} dy \, (dx)$

FIGURE 16.2

Set up an iterated integral for the area of the region bounded by $0 \leq x \leq 1$ and $x^3 \leq y \leq x^2$. (See Figure 16.2.)

Solution: Since the limits for x are constant, we let x be the outer variable and write

$$\text{area} = \int_0^1 \int_{x^3}^{x^2} dy \, dx = \int_0^1 y \Big]_{x^3}^{x^2} dx = \int_0^1 (x^2 - x^3) \, dx$$

$$= \left[\frac{x^3}{3} - \frac{x^4}{4} \right]_0^1 = \frac{1}{3} - \frac{1}{4} = \frac{1}{12} \qquad \blacksquare$$

In setting up iterated integrals the most difficult task is likely to be the determination of the correct limits for the specified order of integration. This task can often be simplified by making a sketch of the region R and identifying the appropriate bounds for x and y. The next example illustrates this procedure.

Example 4

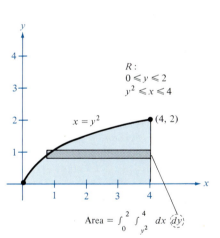

$R:$
$0 \leqslant y \leqslant 2$
$y^2 \leqslant x \leqslant 4$

$x = y^2$

(4, 2)

Area $= \int_0^2 \int_{y^2}^4 dx \, (dy)$

FIGURE 16.3

Given the iterated integral

$$\int_0^2 \int_{y^2}^4 dx \, dy$$

(a) Sketch the region R whose area is given by this integral.
(b) Rewrite the integral so that x is the outer variable.
(c) Show that both orders of integration yield the same value.

Solution:

(a) From the limits of integration, we know that

$$y^2 \leq x \leq 4$$

which means that the region R is bounded on the left by the parabola $y^2 = x$ and on the right by the line $x = 4$. Furthermore, since

$$0 \leq y \leq 2$$

we have the region shown in Figure 16.3.

(b) If we interchange the order of integration so that x is the outer variable, we see that x has the constant bounds $0 \leq x \leq 4$, and by solving for y in the equation $y^2 = x$, we conclude that the bounds for y are $0 \leq y \leq \sqrt{x}$ (see Figure 16.4). Therefore, with x as the outer variable, we obtain the integral

$$\int_0^4 \int_0^{\sqrt{x}} dy \, dx$$

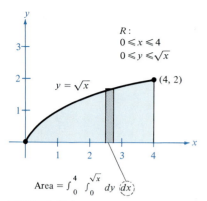

$$\text{Area} = \int_0^4 \int_0^{\sqrt{x}} dy \; \widehat{(dx)}$$

FIGURE 16.4

(c)   Evaluating both integrals, we obtain

$$\int_0^2 \int_{y^2}^4 dx \, dy = \int_0^2 x \Big]_{y^2}^4 dy = \int_0^2 (4 - y^2) \, dy = \left[4y - \frac{y^3}{3}\right]_0^2 = \frac{16}{3}$$

and

$$\int_0^4 \int_0^{\sqrt{x}} dy \, dx = \int_0^4 y \Big]_0^{\sqrt{x}} dx = \int_0^4 \sqrt{x} \, dx = \frac{2}{3} x^{3/2} \Big]_0^4 = \frac{16}{3} \quad \blacksquare$$

To designate an iterated integral or an area of a region without specifying a particular order of integration, we use the **double integral** symbol

$$\iint_R dA$$

where $dA = dx \, dy$ or $dA = dy \, dx$.

Example 5

Use an iterated integral to calculate the area denoted by

$$\iint_R dA$$

where R is the region bounded by $y = x$ and $y = x^2 - x$.

Solution: From a sketch (Figure 16.5) of region R, we can see that vertical rectangles of width dx are more convenient than horizontal ones. (Check Section 7.1 to see why this is true.) Therefore, x is the outer variable of integration and its constant bounds are $0 \le x \le 2$. Thus we have

$$\iint_R dA = \int_0^2 \int_{x^2-x}^x dy \, dx = \int_0^2 y \Big]_{x^2-x}^x dx$$

$$= \int_0^2 [x - (x^2 - x)] \, dx = \int_0^2 (2x - x^2) \, dx$$

$$= \left[x^2 - \frac{x^3}{3}\right]_0^2 = 4 - \frac{8}{3} = \frac{4}{3} \quad \blacksquare$$

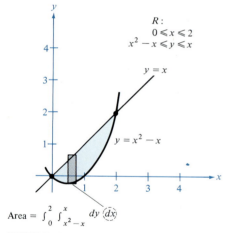

$$\text{Area} = \int_0^2 \int_{x^2-x}^x dy \; \widehat{(dx)}$$

FIGURE 16.5

Sometimes it is not possible to calculate the area of a region by one iterated integral. Under such circumstances we divide the region into subregions such that the area of each subregion can be calculated by an iterated integral. The total area is then the sum of the iterated integrals. We illustrate this case in the next example.

Example 6

Find the area of the region R that lies below the parabola $y = 4x - x^2$, above the x-axis, and above the line $y = -3x + 6$. (See Figure 16.6.)

Solution: Considering the graph of this region, we divide R into the two subregions R_1 and R_2 shown in Figure 16.6. In both regions it is convenient to use x as the outer variable. Thus we have

$$\iint\limits_{R} dA = \iint\limits_{R_1} dA + \iint\limits_{R_2} dA = \int_{1}^{2} \int_{-3x+6}^{4x-x^2} dy \, dx + \int_{2}^{4} \int_{0}^{4x-x^2} dy \, dx$$

$$= \int_{1}^{2} (4x - x^2 + 3x - 6) \, dx + \int_{2}^{4} (4x - x^2) \, dx$$

$$= \left[\frac{7x^2}{2} - \frac{x^3}{3} - 6x \right]_{1}^{2} + \left[2x^2 - \frac{x^3}{3} \right]_{2}^{4}$$

$$= (14 - \tfrac{8}{3} - 12 - \tfrac{7}{2} + \tfrac{1}{3} + 6) + (32 - \tfrac{64}{3} - 8 + \tfrac{8}{3}) = \tfrac{15}{2} \quad \blacksquare$$

At this point you may be wondering why we need double (iterated) integrals since we can already find the area between curves in the plane by single integrals. The real need for double integrals will be demonstrated in Sections 16.2 through 16.4, where they are used to find volume, centers of mass, moments, and surface area.

In this section we have chosen to introduce double integrals by way of a simple application so that we could give primary attention to the procedures for finding the limits of integration. As the examples of this section show, this task is aided greatly by a sketch of the region under consideration.

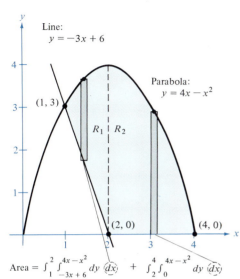

Area $= \int_{1}^{2} \int_{-3x+6}^{4x-x^2} dy \, \widehat{dx} + \int_{2}^{4} \int_{0}^{4x-x^2} dy \, \widehat{dx}$

FIGURE 16.6

Section Exercises (16.1)

In Exercises 1–10, evaluate the specified integral.

1. $\displaystyle\int_{0}^{x} (2x - y) \, dy$

2. $\displaystyle\int_{x}^{x^2} \frac{y}{x} \, dy$

3. $\displaystyle\int_{1}^{2y} \frac{y}{x} \, dx$

4. $\displaystyle\int_{0}^{\cos y} y \, dx$

5. $\displaystyle\int_{0}^{\sqrt{4-x^2}} x^2 y \, dy$

6. $\displaystyle\int_{x^2}^{\sqrt{x}} (x^2 + y^2) \, dy$

S 7. $\displaystyle\int_{e^y}^{y} \frac{y \ln x}{x} \, dx$

8. $\displaystyle\int_{-\sqrt{1-y^2}}^{\sqrt{1-y^2}} (x^2 + y^2) \, dx$

9. $\displaystyle\int_{0}^{x^3} y e^{-y/x} \, dy$

10. $\displaystyle\int_{y}^{\pi/2} \sin^3 x \cos y \, dx$

In Exercises 11–20, evaluate the specified double integral.

11. $\displaystyle\int_{0}^{1} \int_{0}^{2} (x + y) \, dy \, dx$

12. $\displaystyle\int_{0}^{1} \int_{0}^{x} \sqrt{1 - x^2} \, dy \, dx$

S 13. $\displaystyle\int_{1}^{2} \int_{0}^{4} (x^2 - 2y^2 + 1) \, dx \, dy$

14. $\displaystyle\int_{0}^{1} \int_{y}^{2y} (1 + 2x^2 + 2y^2) \, dx \, dy$

15. $\displaystyle\int_{0}^{1} \int_{0}^{\sqrt{1-y^2}} (x + y) \, dx \, dy$

16. $\int_0^2 \int_{3y^2-6y}^{2y-y^2} 3y \, dx \, dy$

17. $\int_0^2 \int_0^{\sqrt{4-y^2}} \frac{2}{\sqrt{4-y^2}} \, dx \, dy$

18. $\int_0^{\pi/2} \int_0^{2\cos\theta} r \, dr \, d\theta$

S 19. $\int_0^{\pi/2} \int_0^{\sin\theta} \theta r \, dr \, d\theta$

20. $\int_0^\infty \int_0^\infty xye^{-(x^2+y^2)} \, dx \, dy$

In Exercises 21–27, sketch the region R whose area is given by the iterated integral, switch the order of integration and show that both orders yield the same area.

21. $\int_0^1 \int_0^2 dy \, dx$

22. $\int_1^2 \int_2^4 dx \, dy$

23. $\int_0^1 \int_{-\sqrt{1-y^2}}^{\sqrt{1-y^2}} dx \, dy$

S 24. $\int_0^2 \int_0^x dy \, dx + \int_2^4 \int_0^{4-x} dy \, dx$

25. $\int_0^2 \int_{x/2}^1 dy \, dx$

26. $\int_0^4 \int_{\sqrt{x}}^2 dy \, dx$

S 27. $\int_0^1 \int_{y^2}^{\sqrt[3]{y}} dx \, dy$

In Exercises 28–37, use a double integral to find the area of the specified region.

28.
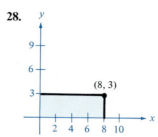

29.
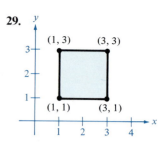

30.
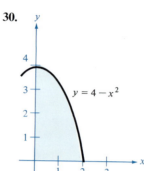

31.
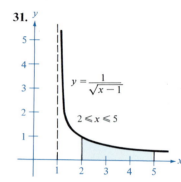

32.
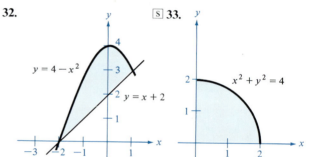

S 33.

34. R is bounded by $y = x^{3/2}$ and $y = x$.

35. R is the triangle with vertices $(0,0)$, $(5,0)$, and $(3,2)$.

36. R is bounded by $xy = 9$, $y = x$, $y = 0$, and $x = 9$.

37. R is bounded by $(x^2/a^2) + (y^2/b^2) = 1$.

For Review and Class Discussion

True or False

1. —— $\int_a^b \int_c^d f(x, y) \, dy \, dx = \int_c^d \int_a^b f(x, y) \, dx \, dy$.

2. —— $\int_a^b \int_c^d f(x) \, g(y) \, dy \, dx$

$$= \left[\int_a^b f(x) \, dx\right] \left[\int_c^d g(y) \, dy\right].$$

3. —— $\int_0^1 \int_0^x f(x, y) \, dy \, dx = \int_0^1 \int_0^y f(x, y) \, dx \, dy$.

4. —— The area of the ellipse $(x^2/a^2) + (y^2/b^2) = 1$ is given by the iterated integral $\int_{-a}^a \int_{-b}^b dy \, dx$.

5. —— $\int_0^1 f_x(x, y) \, dx = f(1, y) - f(0, y)$.

16.2 **Volume and Double Integrals**

Purpose

- To calculate the volume of solid regions by double integrals.

What is it indeed that gives us the feeling of elegance in a solution, in a demonstration? It is the harmony of the diverse parts, their symmetry, their happy balance; in a word it is all that introduces order, all that gives unity, that permits us to see clearly and to comprehend at once both the ensemble and the details.   —*Henri Poincaré (1854–1912)*

In the previous section we demonstrated the use of an iterated integral as an alternative way to find the area of a plane region. In this section we discuss a more imperative use of iterated integrals, namely, to find the volume of a solid region.

In Chapter 7 we used integration to find the volume of two special types of solid regions: solids of revolution (Sections 7.2 and 7.3) and solids with known cross sections (Section 7.4). For volumes of solids with known cross sections, we used single integrals of the form

$$\int_a^b A(x)\, dx \qquad \text{or} \qquad \int_c^d A(y)\, dy$$

where $A(x)$ and $A(y)$ represented the area of a cross section. In all cases the cross sections were common geometric shapes for which formulas are available to denote their area.

For instance, consider the solid region bounded by the plane $z = f(x, y) = 2 - x - 2y$ and the three coordinate planes (Figure 16.7). From Figure 16.7 we can see that each cross section is a triangular region perpendicular to the xy-plane. Its base in the xy-plane ($z = 0$) is given by

$$y = \frac{2 - x}{2}$$

and in the xz-plane ($y = 0$), its height is

$$z = 2 - x$$

Therefore, the area of each triangular cross section is given by

$$A(x) = \frac{1}{2}(\text{base})(\text{height}) = \frac{1}{2}\left(\frac{2 - x}{2}\right)(2 - x) = \frac{(2 - x)^2}{4}$$

and the volume of the solid is

$$\int_0^2 \frac{(2 - x)^2}{4}\, dx = -\frac{(2 - x)^3}{12}\bigg]_0^2 = \frac{8}{12} = \frac{2}{3}$$

Of course, this procedure will work just as well no matter how one finds the formula for $A(x)$. In particular, we can find $A(x)$ by integration, as follows. Let S be the solid region between the surface $z = f(x, y)$ and the plane region R, where R is bounded by $a \le x \le b$ and $g_1(x) \le y \le g_2(x)$ (Figure 16.8). We assume that f is continuous and nonnegative in R.

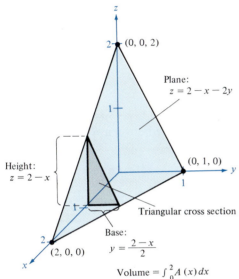

Plane:
$z = 2 - x - 2y$

$(0, 0, 2)$

$(0, 1, 0)$

Height:
$z = 2 - x$

Triangular cross section

$(2, 0, 0)$

Base:
$y = \dfrac{2 - x}{2}$

Volume $= \int_0^2 A(x)\, dx$

FIGURE 16.7

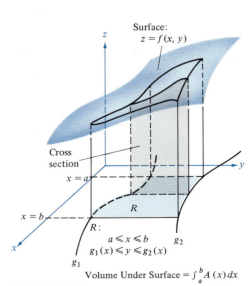

Surface:
$z = f(x, y)$

Cross section

$x = a$

$x = b$

R

R:
$a \le x \le b$
$g_1(x) \le y \le g_2(x)$

g_1

g_2

Volume Under Surface $= \int_a^b A(x)\, dx$

FIGURE 16.8

In this case, for each fixed value of x, there is no simple formula for the area $A(x)$. Instead, we must represent $A(x)$ by an integral. With x fixed we can integrate with respect to y and find the area of a cross section to be

$$A(x) = \int_{g_1(x)}^{g_2(x)} f(x, y)\, dy$$

where $f(x, y)$ represents the height and dy the width of an arbitrary vertical rectangle in the cross section.

Now since $A(x)$ is itself an integral, then the volume of the solid is given by an iterated integral

$$\int_a^b \int_{g_1(x)}^{g_2(x)} f(x, y)\, dy\, dx$$

If the region R has constant y bounds $(c \leq y \leq d)$ and variable x bounds $[h_1(y) \leq x \leq h_2(y)]$, then the volume between R and $z = f(x, y)$ is given by

$$\int_c^d \int_{h_1(y)}^{h_2(y)} f(x, y)\, dx\, dy$$

In general, if f is continuous and nonnegative over a region R, then the volume between $z = f(x, y)$ and the region R in the xy-plane is given by the *double integral*

$$\iint\limits_R f(x, y)\, dA$$

where $dA = dy\, dx$ or $dA = dx\, dy$.

Actually, a double integral can be (and, in fact, usually is) defined independently of iterated integrals. What follows is a brief development of such a definition.

Consider the solid region lying between the surface $z = f(x, y)$ and the plane region R, as shown in Figure 16.9. To determine the volume of this solid, we first subdivide the region R into a grid of n rectangles, where the dimensions of the ith rectangle are Δx_i by Δy_i. Then choosing a point (x_i, y_i) in the ith rectangle, we construct a box of height $f(x_i, y_i)$ with the ith rectangle as its base. Now we can approximate the volume of the solid by summing up the volumes of the n boxes. That is,

$$\text{volume of solid} \approx \sum_{i=1}^{n} f(x_i, y_i)\, \Delta x_i\, \Delta y_i$$

Finally, if we let $\|\Delta\|$ be the area of the largest rectangle in the grid and then take the limit as $\|\Delta\| \to 0$, it follows that

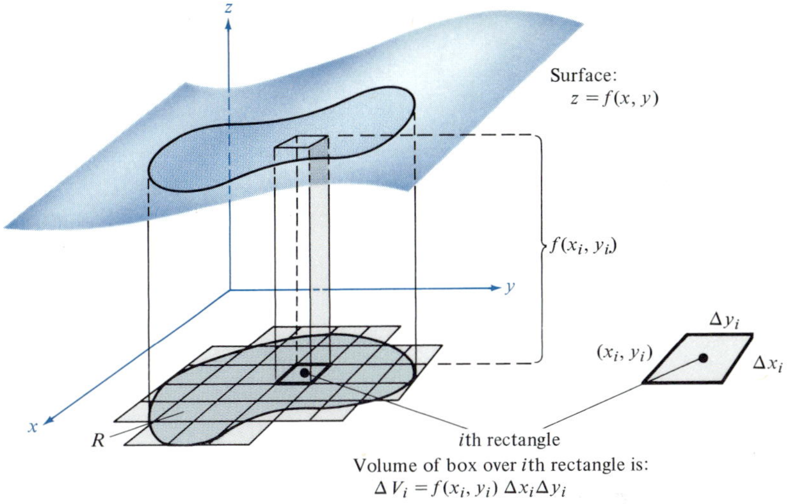

Surface:
$z = f(x, y)$

$f(x_i, y_i)$

(x_i, y_i)

Δy_i

Δx_i

*i*th rectangle

Volume of box over *i*th rectangle is:
$\Delta V_i = f(x_i, y_i)\,\Delta x_i \Delta y_i$

FIGURE 16.9

$$\text{volume of solid} = \lim_{\|\Delta\| \to 0} \sum_{i=1}^{n} f(x_i, y_i)\,\Delta x_i\,\Delta y_i$$

which we denote by the double integral

$$\text{volume of solid} = \iint\limits_{R} f(x, y)\,dA$$

Definition of Double Integral

If f is defined on a bounded region R in the xy-plane, then the **double integral of f over R** is given by

$$\iint\limits_{R} f(x, y)\,dA = \lim_{\|\Delta\| \to 0} \sum_{i=1}^{n} f(x_i, y_i)\,\Delta x_i\,\Delta y_i$$

provided the limit exists.

Although this definition of a double integral uses volume as the geometric model, we should not limit the interpretation of a double integral to volume applications. Like single definite integrals, double integrals have a variety of other applications.

We should also point out that the preceding limit definition of a double integral is primarily of theoretical value. The *evaluation* of double integrals involves the use of one of these iterated forms:

$$\int_{a}^{b} \int_{g_1(x)}^{g_2(x)} f(x, y)\,dy\,dx \qquad \text{or} \qquad \int_{c}^{d} \int_{h_1(y)}^{h_2(y)} f(x, y)\,dx\,dy$$

Before looking at some examples of double integrals, we list their basic properties.

Properties of Double Integrals

If f is continuous over the bounded plane region R, then:

1. $\displaystyle\iint\limits_{R} cf(x, y)\, dA = c \iint\limits_{R} f(x, y)\, dA$

2. $\displaystyle\iint\limits_{R} [f(x, y) + g(x, y)]\, dA = \iint\limits_{R} f(x, y)\, dA + \iint\limits_{R} g(x, y)\, dA$

3. $\displaystyle\iint\limits_{R} f(x, y)\, dA = \iint\limits_{R_1} f(x, y)\, dA + \iint\limits_{R_2} f(x, y)\, dA$

where R is composed of two subregions R_1 and R_2.

We began this section by using the formula $\int_a^b A(x)\, dx$ to find the volume of a solid bounded in the first octant by a plane. We now show how this same volume can be determined by an iterated integral.

Example 1

Find the volume of the solid bounded in the first octant by the plane $z = 2 - x - 2y$.

Solution: To set up the iterated integral for the volume, it is helpful to sketch both the solid and the region R in the xy-plane. From Figure 16.10 we can see that if $z = 0$, then R is bounded in the xy-plane by the lines $x = 0$, $y = 0$, and $y = (2 - x)/2$. For a rectangle of width dx, we obtain

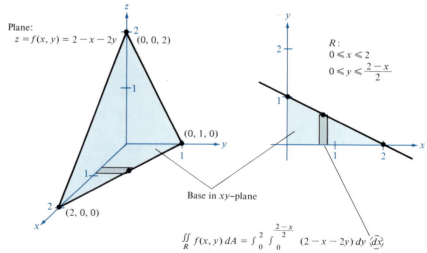

$$\iint\limits_{R} f(x, y)\, dA = \int_0^2 \int_0^{\frac{2-x}{2}} (2 - x - 2y)\, dy\ \widehat{dx}$$

FIGURE 16.10

constant bounds for x: $0 \leq x \leq 2$

variable bounds for y: $0 \leq y \leq \dfrac{2 - x}{2}$

Thus the volume of the solid region is

$$\int_0^2 \int_0^{(2-x)/2} (2 - x - 2y)\, dy\, dx = \int_0^2 \left[(2 - x)y - y^2 \right]_0^{(2-x)/2} dx$$

$$= \int_0^2 \left[(2 - x)\left(\frac{2 - x}{2}\right) - \left(\frac{2 - x}{2}\right)^2 \right] dx$$

$$= \frac{1}{4} \int_0^2 (2 - x)^2\, dx$$

$$= \frac{-1}{12}(2 - x)^3 \Big]_0^2 = \frac{8}{12} = \frac{2}{3} \qquad \blacksquare$$

Following the pattern of Example 1, we outline a procedure for determining the volume of a solid region:

1. Write the equation of the surface in the form $z = f(x, y)$ and sketch the solid region.
2. Sketch region R in the xy-plane and determine the order and limits of integration.
3. Evaluate the double integral

$$\iint_R f(x, y)\, dA$$

using the order and limits determined in the second step.

Example 2

Find the volume of the solid bounded by the paraboloid $x^2 + 4y^2 + z = 4$ and the xy-plane.

Solution: Writing z as a function of x and y, we have

$$z = 4 - x^2 - 4y^2$$

If we let $z = 0$, then R is bounded in the xy-plane by the ellipse $x^2 + 4y^2 = 4$. (See Figure 16.11.) For a rectangle of width dx, we obtain

variable bounds for y: $-\frac{1}{2}\sqrt{4 - x^2} \leq y \leq \frac{1}{2}\sqrt{4 - x^2}$

constant bounds for x: $-2 \leq x \leq 2$

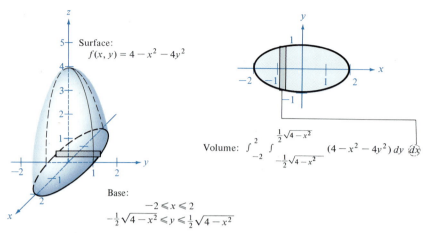

FIGURE 16.11

Therefore, we have

$$V = \int_{-2}^{2} \int_{(-1/2)\sqrt{4-x^2}}^{(1/2)\sqrt{4-x^2}} (4 - x^2 - 4y^2) \, dy \, dx$$

$$= \int_{-2}^{2} \left[(4 - x^2)y - \frac{4y^3}{3} \right]_{(-1/2)\sqrt{4-x^2}}^{(1/2)\sqrt{4-x^2}} dx = \frac{2}{3} \int_{-2}^{2} (4 - x^2)^{3/2} \, dx$$

Using trigonometric substitution, we let $x = 2 \sin \theta$, $dx = 2 \cos \theta \, d\theta$, and we have

$$V = \frac{2}{3} \int_{-2}^{2} (4 - x^2)^{3/2} \, dx = \frac{2}{3} \int_{-\pi/2}^{\pi/2} 16 \cos^4 \theta \, d\theta$$

$$= \left(\frac{32}{3} \right)\left(\frac{1}{4} \right)\left[\frac{3\theta}{2} + \sin (2\theta) + \frac{\sin (4\theta)}{8} \right]_{-\pi/2}^{\pi/2} = \frac{8}{3}\left[\frac{3\pi}{4} + \frac{3\pi}{4} \right] = 4\pi$$

∎

Notice in Example 2 that in the evaluation of the integral

$$\int_{-\pi/2}^{\pi/2} \cos^4 \theta \, d\theta = 2 \int_{0}^{\pi/2} \cos^4 \theta \, d\theta$$

$$= 2 \left(\frac{1}{4} \right)\left[\frac{3\theta}{2} + \sin (2\theta) + \frac{\sin (4\theta)}{8} \right]_{0}^{\pi/2}$$

$$= 2 \left(\frac{1}{4} \right)\left[\frac{3}{2}\left(\frac{\pi}{2} \right) + 0 + 0 - (0 + 0 + 0) \right]$$

the only nonzero term is $2(3\pi/16)$. Such integrals can readily be evaluated by Wallis's Formulas, as introduced in Section 10.5, Exercise 51. We summarize these formulas as follows:

Wallis's Formulas

$$\int_0^{\pi/2} \sin^n \theta \, d\theta = \int_0^{\pi/2} \cos^n \theta \, d\theta$$

$$= \begin{cases} \left[\dfrac{(1)(3)(5) \cdots (n-1)}{(2)(4)(6) \cdots (n)}\right]\left(\dfrac{\pi}{2}\right), & n \text{ even}, \ 0 < n \\[2ex] \dfrac{(2)(4)(6) \cdots (n-1)}{(3)(5)(7) \cdots (n)}, & n \text{ odd}, \ 1 < n \end{cases}$$

Thus

$$\int_0^{\pi/2} \cos^2 \theta \, d\theta = \frac{\pi}{4}, \qquad \int_0^{\pi/2} \cos^4 \theta \, d\theta = \frac{3\pi}{16}, \qquad \int_0^{\pi/2} \cos^6 \theta \, d\theta = \frac{5\pi}{32}$$

In Examples 1 and 2 the order of integration was arbitrary. In both instances we could have used y as the outer variable, as shown in Figure 16.12. There are, however, some occasions in which one order of integration is much more convenient than the other. Example 3 shows such a case.

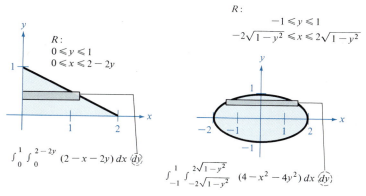

$$R: \quad 0 \le y \le 1 \quad 0 \le x \le 2 - 2y$$

$$\int_0^1 \int_0^{2-2y} (2 - x - 2y) \, dx \, dy$$

$$R: \quad -1 \le y \le 1 \quad -2\sqrt{1-y^2} \le x \le 2\sqrt{1-y^2}$$

$$\int_{-1}^1 \int_{-2\sqrt{1-y^2}}^{2\sqrt{1-y^2}} (4 - x^2 - 4y^2) \, dx \, dy$$

Reversing the Order of Integration
(Compare with Figures 16.10 and 16.11.)

FIGURE 16.12

Example 3

Find the volume under the cylindrical surface $f(x, y) = e^{-x^2}$ bounded by the xz-plane and the planes $y = x$ and $x = 1$.

Solution: In the xy-plane the bounds of region R are the lines $y = 0$, $x = 1$, and $y = x$ (see Figure 16.13). The two possible orders of integration are given in Figure 16.14. In attempting to evaluate these two iterated integrals, we discover that the one on the right requires the antiderivative $\int e^{-x^2} \, dx$, which we know is not an elementary function. On the other hand, we can evaluate the integral on the left in the following manner:

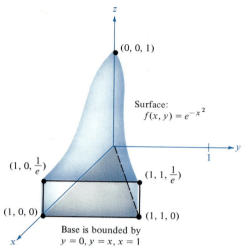

Surface:
$$f(x, y) = e^{-x^2}$$

$(0, 0, 1)$

$(1, 0, \frac{1}{e})$

$(1, 1, \frac{1}{e})$

$(1, 0, 0)$

$(1, 1, 0)$

Base is bounded by
$y = 0, y = x, x = 1$

FIGURE 16.13

$R:$
$0 \leqslant x \leqslant 1$
$0 \leqslant y \leqslant x$

$(1, 1)$

$(1, 0)$

$$\int_0^1 \int_0^x e^{-x^2} \, dy \, dx$$

FIGURE 16.14

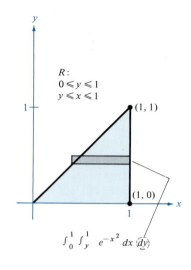

$R:$
$0 \leqslant y \leqslant 1$
$y \leqslant x \leqslant 1$

$(1, 1)$

$(1, 0)$

$$\int_0^1 \int_y^1 e^{-x^2} \, dx \, dy$$

$$\int_0^1 \int_0^x e^{-x^2} \, dy \, dx = \int_0^1 e^{-x^2} y \Big]_0^x \, dx = \int_0^1 x e^{-x^2} \, dx$$

$$= -\frac{1}{2} e^{-x^2} \Big]_0^1 = -\frac{1}{2} \left(\frac{1}{e} - 1 \right) = \frac{e - 1}{2e} \qquad \blacksquare$$

In the plane we extended the technique for finding area *under* a curve to finding area *between* two curves. A similar extension can be used to find the volume between two surfaces.

Example 4

Find the volume of the solid bounded above by the paraboloid $z = 1 - x^2 - y^2$ and below by the plane $z = 1 - y$. (See Figure 16.15.)

Solution: By equating z-values we find that the curve of intersection (see Figure 16.15) determines a circular region R given by

$$1 - y = 1 - x^2 - y^2$$
$$x^2 = y - y^2$$
$$x^2 + (y - \tfrac{1}{2})^2 = (\tfrac{1}{2})^2$$

Now the desired volume is the difference between the volume under the paraboloid and the volume under the plane. Thus we have

$$\text{volume} = \int_0^1 \int_{-\sqrt{y-y^2}}^{\sqrt{y-y^2}} (1 - x^2 - y^2) \, dx \, dy - \int_0^1 \int_{-\sqrt{y-y^2}}^{\sqrt{y-y^2}} (1 - y) \, dx \, dy$$

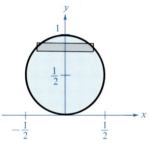

Paraboloid:
$z = 1 - x^2 - y^2$

Plane:
$z = 1 - y$

$R:$
$0 \leqslant y \leqslant 1$
$-\sqrt{y - y^2} \leqslant x \leqslant \sqrt{y - y^2}$

FIGURE 16.15

$$= \int_0^1 \int_{-\sqrt{y-y^2}}^{\sqrt{y-y^2}} (y - y^2 - x^2) \, dx \, dy$$

$$= \int_0^1 \left[(y - y^2)x - \frac{x^3}{3} \right]_{-\sqrt{y-y^2}}^{\sqrt{y-y^2}} dy$$

$$= \frac{4}{3} \int_0^1 (y - y^2)^{3/2} \, dy = \left(\frac{4}{3}\right)\left(\frac{1}{8}\right) \int_0^1 [1 - (2y - 1)^2]^{3/2} \, dy$$

Letting $2y - 1 = \sin \theta$ and $dy = [(\cos \theta)/2] \, d\theta$, we have

$$\left(\frac{4}{3}\right)\left(\frac{1}{8}\right) \int_0^1 [1 - (2y - 1)^2]^{3/2} = \left(\frac{4}{3}\right)\left(\frac{1}{8}\right) \int_{-\pi/2}^{\pi/2} \frac{\cos^4 \theta}{2} \, d\theta$$

$$= \frac{1}{6} \int_0^{\pi/2} \cos^4 \theta \, d\theta$$

and by Wallis's Formula we have

$$\frac{1}{6} \int_0^{\pi/2} \cos^4 \theta \, d\theta = \left(\frac{1}{6}\right)\left(\frac{3\pi}{16}\right) = \frac{\pi}{32} \quad \blacksquare$$

Example 5

Sketch R, the region of integration, and evaluate the integral

$$\int_0^2 \int_x^{3x-x^2} (3x^2 - 2xy) \, dy \, dx$$

Solution: Since y has the limits $x \leq y \leq 3x - x^2$, region R is bounded above by the parabola $y = 3x - x^2$ and below by the line $y = x$ (see Figure 16.16). The value of the integral is

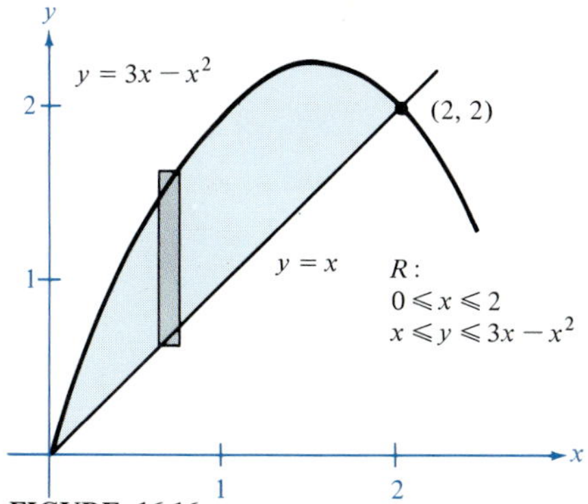

$$\int_0^2 \int_x^{3x-x^2} (3x^2 - 2xy) \, dy \, dx$$

$$= \int_0^2 \left[3x^2 y - xy^2 \right]_x^{3x-x^2} dx$$

$$= \int_0^2 [(9x^3 - 3x^4 - 9x^3 + 6x^4 - x^5) - (3x^3 - x^3)] \, dx$$

$$= \int_0^2 (-2x^3 + 3x^4 - x^5) \, dx$$

$$= \left[-\frac{x^4}{2} + \frac{3x^5}{5} - \frac{x^6}{6} \right]_0^2 = \frac{8}{15} \quad \blacksquare$$

$y = 3x - x^2$

$(2, 2)$

$y = x$

$R:$
$0 \leqslant x \leqslant 2$
$x \leqslant y \leqslant 3x - x^2$

FIGURE 16.16

Section Exercises (16.2)

In Exercises 1–6, sketch the region R and evaluate the integral.

1. $\int_0^2 \int_0^1 (1 + 2x + 2y)\, dy\, dx$

2. $\int_0^\pi \int_0^{\pi/2} \sin^2 x \cos^2 y\, dy\, dx$

3. $\int_0^6 \int_{y/2}^3 (x + y)\, dx\, dy$

$\boxed{\text{s}}$ 4. $\int_0^1 \int_y^{\sqrt{y}} x^2 y^2\, dx\, dy$

5. $\int_{-a}^a \int_{-\sqrt{a^2-x^2}}^{\sqrt{a^2-x^2}} (x + y)\, dy\, dx$

6. $\int_0^1 \int_{y-1}^0 e^{x+y}\, dx\, dy + \int_0^1 \int_0^{1-y} e^{x+y}\, dx\, dy$

In Exercises 7–12, set up the integrals for both $dx\, dy$ and $dy\, dx$ and use the most convenient order to evaluate the given double integral over the region R.

7. $\iint\limits_R xy\, dA$; R: rectangle   with   vertices $(0,0)$, $(0,5)$, $(3,5)$, $(3,0)$

8. $\iint\limits_R \sin x \sin y\, dA$; R: rectangle   with   vertices $(-\pi, 0)$, $(\pi, 0)$, $(\pi, \pi/2)$, $(-\pi, \pi/2)$

$\boxed{\text{s}}$ 9. $\iint\limits_R \dfrac{y}{x^2 + y^2}\, dA$; R: triangle   bounded   by $y = x$, $y = 2x$, $x = 2$

10. $\iint\limits_R \dfrac{y}{1 + x^2}\, dA$; R: region   bounded   by $y = 0$, $y = \sqrt{x}$, $x = 4$

11. $\iint\limits_R x\, dA$; R: sector of a circle in the first quadrant bounded by $y = \sqrt{25 - x^2}$, $3x - 4y = 0$, and $y = 0$

12. $\iint\limits_R (x^2 + y^2)\, dA$: R: semicircle   bounded   by $y = \sqrt{4 - x^2}$ and $y = 0$

In Exercises 13–24, use a double integral to find the volume of the specified solid.

13. the solid bounded by the coordinate planes and the plane $2x + 3y + 4z = 12$

14. the solid bounded by the coordinate planes and the plane $x + y + z = 1$

15. the solid in the first octant bounded by the surfaces $z = xy$, $y = x$, and $x = 1$

16. the solid bounded by the surfaces $y = 0$, $z = 0$, $y = x$, $z = x$, $x = 0$, and $x = 5$

17. the   solid   bounded   by $z = 0$ and $z = x^2$,   where $0 \le x \le 2$ and $0 \le y \le 4$

18. the sphere of radius r

19. the solid in the first octant bounded by $y = 0$, $z = 0$, $y = x$, and the cylinder $x^2 + z^2 = 1$

20. the solid in the first octant bounded by the surfaces $y = 1 - x^2$ and $z = 1 - x^2$

$\boxed{\text{s}}$ 21. the solid of intersection of the cylinders $x^2 + z^2 = 1$ and $y^2 + z^2 = 1$. Show that the volume is a multiple of the result of Exercise 19.

22. the solid in the first octant bounded by the plane $z = x + y$ and the cylinder $x^2 + y^2 = 4$

23. the solid in the first octant bounded by the surface $z = e^{-(x+y)/2}$ (improper integral)

24. the solid bounded by $z = 0$ and $z = 1/(1 + y^2)$, where $0 \le x \le 2$ and $0 \le y$

$\boxed{\text{s}}$ 25. Approximate the double integral $\int_0^4 \int_0^2 (x + y)\, dy\, dx$ by dividing the region of integration R into 8 equal squares and finding the sum $\sum_{i=1}^8 f(x_i, y_i)\, \Delta x_i\, \Delta y_i$. Let (x_i, y_i) be the *center* of the ith square. Evaluate the double integral and compare it to the approximation.

26. Follow the instructions for Exercise 25 using the integral $\int_0^4 \int_0^2 xy\, dy\, dx$.

27. Follow the instructions for Exercise 25 using the integral $\int_0^4 \int_0^2 (x^2 + y^2)\, dy\, dx$.

In Exercises 28–31, use Wallis's Formula as an aid in finding the volume of the specified solid.

28. the solid bounded by the surfaces $x^2 = 9 - y$ and $z^2 = 9 - y$ in the first octant

$\boxed{\text{s}}$ 29. the solid bounded by the paraboloid $z = 4 - x^2 - y^2$ and the xy-plane

30. the   solid   bounded   by $z = \sin^2 x$ and $z = 0$,   where $0 \le x \le \pi$ and $0 \le y \le 5$

31. the solid above the xy-plane bounded by the paraboloid $z = x^2 + y^2$ and the cylinder $x^2 + y^2 = 4$

For Review and Class Discussion

True or False

1. ____ The volume of the sphere $x^2 + y^2 + z^2 = 1$ is given by the integral

$$V = 8 \int_0^1 \int_0^1 \sqrt{1 - x^2 - y^2} \, dx \, dy$$

2. ____ If f is continuous over R_1 and R_2 and

$$\iint_{R_1} dA = \iint_{R_2} dA$$

then

$$\iint_{R_1} f(x, y) \, dA = \iint_{R_2} f(x, y) \, dA$$

3. ____ If $f(x, y) \le g(x, y)$ for all (x, y) in R, and both f and g are continuous over R, then

$$\iint_R f(x, y) \, dA \le \iint_R g(x, y) \, dA$$

4. ____ $\int_{-1}^1 \int_{-1}^1 \cos(x^2 + y^2) \, dx \, dy$

$$= 4 \int_0^1 \int_0^1 \cos(x^2 + y^2) \, dx \, dy.$$

5. ____ $\int_0^1 \int_0^1 \dfrac{1}{1 + x^2 + y^2} \, dx \, dy < \dfrac{\pi}{4}.$

16.3 Surface Area

Purpose

- To use a surface integral to calculate the area of a surface over a region.
- To use a line integral to calculate the area of the lateral surface of a cylindrical solid.
- To identify some physical applications of surface and line integrals.

The effects of heat are subject to constant laws which cannot be discovered without the aid of mathematical analysis. The object of the theory is to demonstrate these laws; it reduces all physical researches on the propagation of heat to problems of integral calculus. —*Joseph Fourier (1768–1830)*

Consider the solid region between the surface $z = f(x, y)$ and the region R in the xy-plane (Figure 16.17). At this point in the text, we know quite a bit about this solid region. For instance, we can calculate the area of its base (Section 7.1), its volume (Section 16.2), the perimeter of its base (Section 7.8), the area of various cross sections, its maximum height (Section 15.3), and the angle between the upper surface and the lateral surface at a given point (Section 15.6). Two rather obvious omissions from this list are the following:

1. the area of the upper surface
2. the lateral surface area

In this section we demonstrate the use of integration to find each of these two areas, beginning with the area of the upper surface.

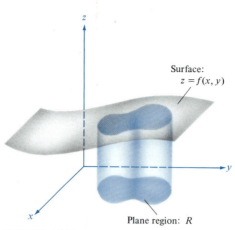

Surface:
$z = f(x, y)$

Plane region: R

FIGURE 16.17

Area of a Surface

If f is differentiable over a region R in the xy-plane, then the area of the surface $z = f(x, y)$, which lies over R, is given by

$$\iint\limits_R dS = \iint\limits_R \sqrt{1 + [f_x(x, y)]^2 + [f_y(x, y)]^2}\, dA$$

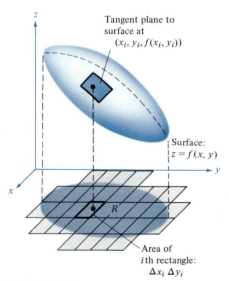

Tangent plane to surface at $(x_i, y_i, f(x_i, y_i))$

Surface:
$z = f(x, y)$

R

Area of ith rectangle: $\Delta x_i\, \Delta y_i$

FIGURE 16.18

Proof: We begin by covering R with a grid of n rectangles, where the area of the ith rectangle is $\Delta x_i\, \Delta y_i$. We then let (x_i, y_i) be a point in the ith rectangle and consider the tangent plane to the surface at the point $(x_i, y_i, f(x_i, y_i))$ (Figure 16.18).

If θ_i is the angle between the xy-plane and the tangent plane at $(x_i, y_i, f(x_i, y_i))$, then from Section 15.6 we know that

$$\cos \theta_i = \frac{1}{\sqrt{1 + [f_x(x_i, y_i)]^2 + [f_y(x_i, y_i)]^2}}$$

and hence

$$\sec \theta_i = \sqrt{1 + [f_x(x_i, y_i)]^2 + [f_y(x_i, y_i)]^2}$$

Finally, from geometry we know that the area of that portion of the tangent plane that lies over the ith rectangle is given by

$$\sec \theta_i\, (\Delta x_i\, \Delta y_i)$$

(see Figure 16.19).

Now if we let $\|\Delta\|$ represent the area of the largest rectangle in R, then summing all such areas and taking the limit as $\|\Delta\| \to 0$, we have

$$\text{surface area} = \lim_{\|\Delta\| \to 0} \sum_{i=1}^{n} \sqrt{1 + [f_x(x_i, y_i)]^2 + [f_y(x_i, y_i)]^2}\, \Delta x_i\, \Delta y_i$$

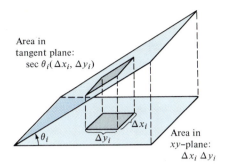

Area in tangent plane: $\sec \theta_i (\Delta x_i, \Delta y_i)$

Area in xy-plane: $\Delta x_i \, \Delta y_i$

FIGURE 16.19

$$= \iint\limits_{R} \sqrt{1 + [f_x(x, y)]^2 + [f_y(x, y)]^2} \, dA$$

As an aid to remembering the formula for surface area, it is helpful to note its similarity to the arc length formula. This is shown in Table 16.1.

TABLE 16.1

Length on the x-axis: $\int dx$	Arc length in xy-plane: $\int ds = \int \sqrt{1 + \left(\dfrac{dy}{dx}\right)^2} \, dx$
Area in xy-plane: $\int dA$	Surface area in space: $\iint dS = \iint \sqrt{1 + \left(\dfrac{\partial z}{\partial x}\right)^2 + \left(\dfrac{\partial z}{\partial y}\right)^2} \, dA$

■

Example 1

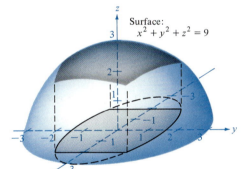

Surface: $x^2 + y^2 + z^2 = 9$

Find the area S of that portion of the hemisphere $x^2 + y^2 + z^2 = 9$ that lies above the region R bounded by

$$-2 \leq x \leq 2 \quad \text{and} \quad \frac{-\sqrt{9 - x^2}}{2} \leq y \leq \frac{\sqrt{9 - x^2}}{2}$$

Solution: Solving for z we have

$$z = \sqrt{9 - x^2 - y^2}$$

from which it follows that

$$f_x(x, y) = \frac{\partial z}{\partial x} = \frac{-x}{\sqrt{9 - x^2 - y^2}}$$

and

$$f_y(x, y) = \frac{\partial z}{\partial y} = \frac{-y}{\sqrt{9 - x^2 - y^2}}$$

The portion of the surface that lies above R is shown in Figure 16.20, and we conclude that

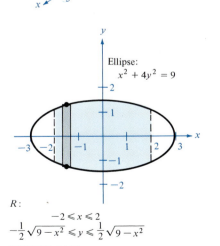

Ellipse: $x^2 + 4y^2 = 9$

R:
$$-2 \leq x \leq 2$$
$$-\frac{1}{2}\sqrt{9 - x^2} \leq y \leq \frac{1}{2}\sqrt{9 - x^2}$$

FIGURE 16.20

$$S = \int_{-2}^{2} \int_{-\sqrt{9-x^2}/2}^{\sqrt{9-x^2}/2} \sqrt{1 + \left(\frac{-x}{\sqrt{9 - x^2 - y^2}}\right)^2 + \left(\frac{-y}{\sqrt{9 - x^2 - y^2}}\right)^2} \, dy \, dx$$

$$= \int_{-2}^{2} \int_{-\sqrt{9-x^2}/2}^{\sqrt{9-x^2}/2} \frac{3}{\sqrt{(9 - x^2) - y^2}} \, dy \, dx$$

$$= 3 \int_{-2}^{2} \left. \text{Arcsin} \frac{y}{\sqrt{9 - x^2}} \right]_{-\sqrt{9-x^2}/2}^{\sqrt{9-x^2}/2} dx$$

$$= 3 \int_{-2}^{2} \frac{\pi}{3} \, dx = \pi x \Big]_{-2}^{2} = 4\pi$$

■

Example 2

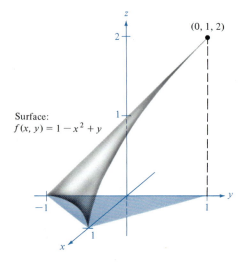

Surface:
$f(x, y) = 1 - x^2 + y$

$(0, 1, 2)$

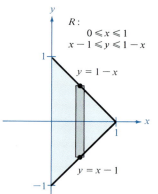

R:
$0 \leqslant x \leqslant 1$
$x - 1 \leqslant y \leqslant 1 - x$

$y = 1 - x$

$y = x - 1$

FIGURE 16.21

Find the area of that portion of the surface $f(x, y) = 1 - x^2 + y$ that lies above the triangular region $0 \leq x \leq 1$, $x - 1 \leq y \leq 1 - x$.

Solution: The partial derivatives of f are given by

$$f_x(x, y) = -2x \qquad \text{and} \qquad f_y(x, y) = 1$$

From Figure 16.21 we see that the bounds for R are $0 \leq x \leq 1$ and $x - 1 \leq y \leq 1 - x$. Thus by Theorem 16.1 the area of the given surface is

$$\int_0^1 \int_{x-1}^{1-x} \sqrt{1 + 4x^2 + 1} \, dy \, dx$$

$$= \int_0^1 y \sqrt{2 + 4x^2} \Big]_{x-1}^{1-x} dx$$

$$= \int_0^1 (2 \sqrt{2 + 4x^2} - 2x \sqrt{2 + 4x^2}) \, dx$$

$$= \frac{1}{2} \left[2x \sqrt{2 + 4x^2} + 2 \ln (2x + \sqrt{2 + 4x^2}) \right]_0^1 - \left[\frac{1}{6} (2 + 4x^2)^{3/2} \right]_0^1$$

$$= \sqrt{6} + \ln (2 + \sqrt{6}) - \ln \sqrt{2} - \tfrac{1}{6}(6 \sqrt{6} - 2 \sqrt{2})$$

$$= \ln \left(\frac{2 + \sqrt{6}}{\sqrt{2}} \right) + \frac{\sqrt{2}}{3} \approx 1.146 + 0.471 = 1.617 \qquad \blacksquare$$

The second type of area we wish to find is the area of the *lateral surface* of a cylindrical region (Figure 16.22). To obtain the area of a lateral surface such as that shown in Figure 16.22, we can integrate the height, $f(x, y)$, of the lateral surface along the curve C between P and Q, as indicated next.

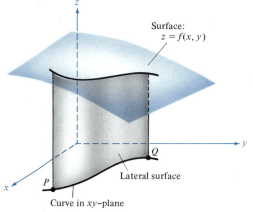

Surface:
$z = f(x, y)$

Q

Lateral surface

P

Curve in xy–plane

FIGURE 16.22

Lateral Surface Area

If s represents the arc length measured along a curve C, and if f is continuous and nonnegative along C, then the area of the lateral surface between f and the xy-plane is given by

$$\text{lateral surface area} = \int_C f(x, y)\, ds$$

where

$$ds = \sqrt{1 + \left(\frac{dy}{dx}\right)^2}\, dx,\ \sqrt{1 + \left(\frac{dx}{dy}\right)^2}\, dy,\ \text{or}\ \sqrt{\left(\frac{dx}{dt}\right)^2 + \left(\frac{dy}{dt}\right)^2}\, dt$$

and integration along C is in the direction of increasing values of x, y, or t, respectively.

Example 3

Find the lateral surface area L of that portion of the circular cylinder $x^2 + y^2 = 1$ that lies between the xy-plane and the parabolic cylinder $f(x, y) = 1 - y^2$.

Solution: The lateral surface is shown in Figure 16.23. Since dy/dx and dx/dy do not exist at some points on curve C, we choose to represent C parametrically as

$$x = \sin t \quad \text{and} \quad y = \cos t$$

where $0 \le t \le 2\pi$. Then we have

$$f(x, y) = 1 - y^2 = 1 - \cos^2 t = \sin^2 t$$

$$ds = \sqrt{\left(\frac{dx}{dt}\right)^2 + \left(\frac{dy}{dt}\right)^2}\, dt = \sqrt{(\cos t)^2 + (-\sin t)^2}\, dt = dt$$

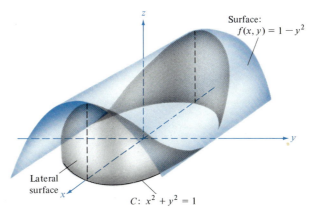

Surface:
$f(x, y) = 1 - y^2$

Lateral
surface

$C: x^2 + y^2 = 1$

FIGURE 16.23

Therefore, it follows that

$$L = \int_C f(x, y)\, ds = \int_0^{2\pi} \sin^2 t\, dt = \frac{1}{2} \int_0^{2\pi} (1 - \cos 2t)\, dt$$

$$= \left[\frac{1}{2} \left(t - \frac{\sin 2t}{2} \right) \right]_0^{2\pi} = \pi$$

∎

Example 4

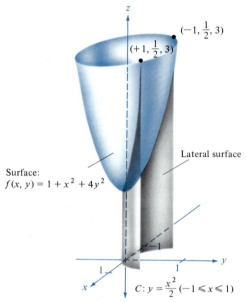

Surface:
$f(x, y) = 1 + x^2 + 4y^2$

$(-1, \frac{1}{2}, 3)$

$(+1, \frac{1}{2}, 3)$

Lateral surface

$C: y = \frac{x^2}{2} \ (-1 \leqslant x \leqslant 1)$

FIGURE 16.24

Find the lateral surface area L of that portion of the cylinder $y = x^2/2$, $-1 \leq x \leq 1$, that lies between the xy-plane and the hyperboloid $f(x, y) = 1 + x^2 + 4y^2$.

Solution: Figure 16.24 shows the graph of the surface. Since $dy/dx = x$ exists along the entire curve C, we can use the form

$$ds = \sqrt{1 + \left(\frac{dy}{dx} \right)^2}\, dx = \sqrt{1 + x^2}\, dx$$

Since x is increasing in the interval $[-1, 1]$, we have

$$L = \int_{-1}^1 (1 + x^2 + 4y^2) \sqrt{1 + x^2}\, dx = \int_{-1}^1 (1 + x^2 + x^4) \sqrt{1 + x^2}\, dx$$

Using the trigonometric substitution $x = \tan \theta$, we have

$$L = \int_{-\pi/4}^{\pi/4} (1 + \tan^2 \theta + \tan^4 \theta) \sec^3 \theta\, d\theta$$

$$= 2 \int_0^{\pi/4} (\sec^3 \theta - \sec^5 \theta + \sec^7 \theta)\, d\theta$$

Thus, by Formula 69, Section 10.9, we have

$$L = \left[\frac{\tan \theta \sec^5 \theta}{3} - \frac{\tan \theta \sec^3 \theta}{12} + \frac{7 \tan \theta \sec \theta}{8} + \frac{7}{8} \ln |\tan \theta + \sec \theta| \right]_0^{\pi/4}$$

$$= \frac{49 \sqrt{2}}{24} + \frac{7}{8} \ln (1 + \sqrt{2}) \approx 3.656.$$

∎

The two types of integrals introduced in this section are special cases of **surface integrals** and **line integrals**:

Surface integral for surface area: $\iint\limits_R dS$

Line integral for arc length: $\int_C ds$

Line integral for lateral surface area: $\int_C f(x, y)\, ds$

There are many physical applications of these two types of integrals. For example, in the next section (Example 2) we will see that a surface integral of the form

$$\iint\limits_{R} \rho\,(x, y)\,dS$$

can be used to find the mass of a nonplanar lamina, where ρ is a density function. Line integrals and their applications will be discussed in Chapter 17.

Section Exercises (16.3)

In Exercises 1–4, find the area of the specified surface.

1. the part of the surface $z = 2x + 2y$ lying above the triangular region whose vertices are $(0, 0, 0)$, $(2, 0, 0)$, and $(0, 2, 0)$

2. the part of the surface $z = x^2$ lying above the triangular region whose vertices are $(0, 0, 0)$, $(2, 0, 0)$, and $(0, 2, 0)$

[S] 3. the part of the cone $z = \sqrt{x^2 + y^2}$ inside the cylinder $x^2 + y^2 = 1$

4. the part of the surface $z = 2 + x^{3/2}$ lying above the quadrangular region whose vertices are $(0, 0, 0)$, $(0, 4, 0)$, $(3, 4, 0)$, and $(3, 0, 0)$

5. Show that the surface area of the cone $z = k\sqrt{x^2 + y^2}$ $(k > 0)$ over a region R in the xy-plane is $A\sqrt{k^2 + 1}$, where A is the area of region R.

[S] 9. Find the lateral surface area of the cylinder $y = 1 - x^2$ above the xy-plane and below the surface $z = xy$ in the first octant.

6. Find the lateral surface area of the cylinder $y = 1 - x^2$ between the planes $z = 0$ and $z = h$ in the first octant.

7. Find the lateral surface area of the cylinder $y = 1 - x^2$ between the planes $z = 0$ and $z = y + 1$ in the first octant.

8. The formula for surface area can be extended to cover some situations in which the function $z = f(x, y)$ is differentiable in the *interior* of a region R. For example, the hemisphere $z = \sqrt{r^2 - x^2 - y^2}$ is differentiable in the interior of the circle $x^2 + y^2 = r^2$, but it is not differentiable on the boundary of the circle. Show that the integral for the surface area of this hemisphere is an improper integral, which, nevertheless, yields the correct surface area.

10. Find the lateral surface of the cylinder $x^2 + y^2 = 4$ above the xy-plane and below the surface $z = x^2 - y^2 + 4$.

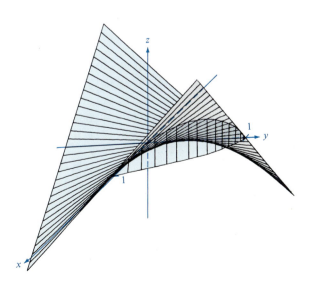

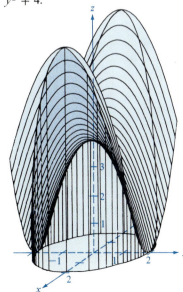

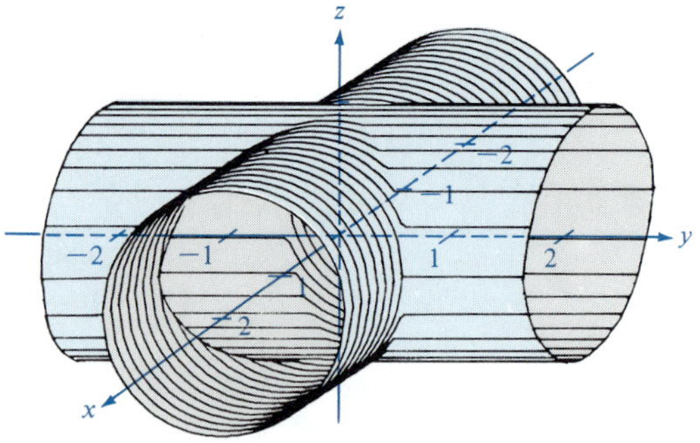

[S] **11.** Find the surface area of the solid of intersection of the cylinders $x^2 + z^2 = a^2$ and $y^2 + z^2 = a^2$.

16.4 Centers of Mass and Moments of Inertia

Purpose

- To use double integrals to find the mass, center of mass, and moments of inertia of a lamina of variable density.

Let us consider a paraboloid of revolution. Let us find its center of gravity by using the same method which we applied for maxima and minima and for the tangents of curves; let us illustrate, with new and brilliant applications of this method, how wrong those are who believe that it may fail.

—*Pierre de Fermat (1601–1665)*

In Section 7.7 we discussed several applications of integration to a lamina of *constant* density ρ. For example, if lamina R, bounded by $x = a$, $x = b$, $y = f(x)$, and $y = g(x)$, has constant density ρ, then by a single integral, the mass of the lamina is given by

$$\text{mass} = \rho(\text{area of } R) = \rho \int_a^b [f(x) - g(x)]\, dx$$

Using double integrals this mass is given by

$$\text{mass} = \rho(\text{area of } R) = \rho \iint_R dA = \iint_R \rho\, dA$$

The use of double integrals to determine mass allows for the natural extension of the formula to find the mass of a lamina of *variable* density.

Mass of a Planar Lamina of Variable Density

If $\rho(x, y)$ is the density of a lamina corresponding to a plane region R, then the mass m of the lamina is given by

$$m = \iint_R \rho(x, y)\, dA$$

Note that density is normally expressed as mass per unit volume. However, for a planar lamina we consider density as mass per unit surface area.

Example 1

Find the mass of the lamina corresponding to the circular region $x^2 + y^2 = 1$ if the density at a point is proportional to the square of its distance from the center.

Solution: Since the density at (x, y) is proportional to the square of the distance from the center, we have

$$\rho(x, y) = k(x^2 + y^2)$$

provided the lamina is centered at the origin. From Figure 16.25 we can see that the mass of the lamina is given by

$$\text{mass} = \int_{-1}^{1} \int_{-\sqrt{1-x^2}}^{\sqrt{1-x^2}} k(x^2 + y^2)\, dy\, dx = k \int_{-1}^{1} \left[x^2 y + \frac{y^3}{3} \right]_{-\sqrt{1-x^2}}^{\sqrt{1-x^2}} dx$$

$$= k \int_{-1}^{1} \left[2x^2 \sqrt{1 - x^2} + \frac{2}{3}(1 - x^2)\sqrt{1 - x^2} \right] dx$$

$$= 2k \int_{0}^{1} \left[\left(\frac{4}{3}\right) x^2 \sqrt{1 - x^2} + \frac{2}{3} \sqrt{1 - x^2} \right] dx$$

$$= \frac{4k}{3} \int_{0}^{1} [2x^2 \sqrt{1 - x^2} + \sqrt{1 - x^2}]\, dx$$

Letting $x = \sin \theta$ and $dx = \cos \theta\, d\theta$, we have

$$\text{mass} = \frac{4k}{3} \int_{0}^{\pi/2} (2 \sin^2 \theta \cos^2 \theta + \cos^2 \theta)\, d\theta$$

$$= \frac{4k}{3} \int_{0}^{\pi/2} [2(1 - \cos^2 \theta) \cos^2 \theta + \cos^2 \theta]\, d\theta$$

$$= \frac{4k}{3} \left[3 \int_{0}^{\pi/2} \cos^2 \theta\, d\theta - 2 \int_{0}^{\pi/2} \cos^4 \theta\, d\theta \right]$$

$$= \frac{4k}{3} \left[3 \left(\frac{\pi}{4}\right) - 2 \left(\frac{3\pi}{16}\right) \right] = \frac{k\pi}{2}$$

by Wallis's Formula.   ∎

A second extension of the formula

$$\text{mass} = (\text{density}) \times (\text{area})$$

involves a nonplanar lamina in which the surface area is given by

$$\text{surface area} = \iint_{R} \sqrt{1 + [f_x(x, y)]^2 + [f_y(x, y)]^2}\, dA$$

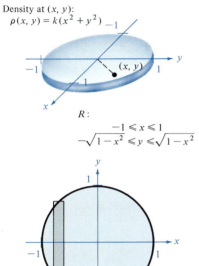

Lamina of variable density

Density at (x, y):
$\rho(x, y) = k(x^2 + y^2)$

(x, y)

R:
$-1 \leqslant x \leqslant 1$
$-\sqrt{1 - x^2} \leqslant y \leqslant \sqrt{1 - x^2}$

FIGURE 16.25

If the density of such a lamina is $\rho(x, y)$, then the mass is given by the *surface integral*

$$\text{mass} = \iint\limits_{R} \rho(x, y) \sqrt{1 + [f_x(x, y)]^2 + [f_y(x, y)]^2} \, dA$$

Example 2

Find the mass of the lamina corresponding to that portion of the paraboloid $f(x, y) = 1 - x^2 - y^2$ lying above the *xy*-plane. Assume that the density at a point is proportional to the distance between the point and the *xz*-plane.

Solution: Since y denotes the distance from any point on the paraboloid to the *xz*-plane, we can express the density as $\rho(x, y) = ky$. Because of the symmetry of the paraboloid, we consider only the first octant portion shown in Figure 16.26. Thus we obtain the formula

$$\text{mass} = 4 \int_0^1 \int_0^{\sqrt{1-x^2}} ky \sqrt{1 + [f_x(x, y)]^2 + [f_y(x, y)]^2} \, dy \, dx$$

Now substituting $f_x(x, y) = -2x$ and $f_y(x, y) = -2y$, we have

$$\text{mass} = 4k \int_0^1 \int_0^{\sqrt{1-x^2}} y \sqrt{1 + 4x^2 + 4y^2} \, dy \, dx$$

$$= \frac{k}{3} \int_0^1 (1 + 4x^2 + 4y^2)^{3/2} \Big]_0^{\sqrt{1-x^2}} dx$$

$$= \frac{k}{3} \int_0^1 [5\sqrt{5} - (1 + 4x^2)^{3/2}] \, dx$$

$$= \frac{k}{3} \int_0^1 [5\sqrt{5} - \sqrt{1 + 4x^2} - 4x^2 \sqrt{1 + 4x^2}] \, dx$$

Then, using Formulas 26 and 27, Section 10.9, we have

$$\text{mass} = \frac{k}{3} \left[5\sqrt{5}x - \frac{1}{4}\{2x\sqrt{1 + 4x^2} + \ln(2x + \sqrt{1 + 4x^2})\} \right.$$

$$\left. - \frac{1}{16}\{2x(8x^2 + 1)\sqrt{1 + 4x^2} - \ln(2x + \sqrt{1 + 4x^2})\} \right]_0^1$$

$$= \frac{k}{3} \left[5\sqrt{5} - \frac{1}{4}\{2\sqrt{5} + \ln(2 + \sqrt{5})\} \right.$$

$$\left. - \frac{1}{16}\{18\sqrt{5} - \ln(2 + \sqrt{5})\} \right]$$

$$= \frac{k}{3} \left[\frac{27}{8}\sqrt{5} - \frac{3}{16} \ln(2 + \sqrt{5}) \right] \approx 2.425k \qquad \blacksquare$$

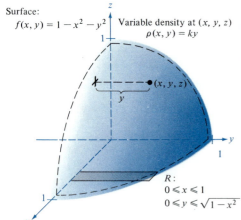

Surface:
$f(x, y) = 1 - x^2 - y^2$

Variable density at (x, y, z)
$\rho(x, y) = ky$

(x, y, z)

R:
$0 \leqslant x \leqslant 1$
$0 \leqslant y \leqslant \sqrt{1 - x^2}$

FIGURE 16.26

Recall from Section 7.7 that the center of mass of a lamina was defined to be the point $(\bar{x}, \bar{y})$, where

$$\bar{x} = \frac{\text{moment about } y\text{-axis}}{\text{total mass}} = \frac{M_y}{m}$$

$$\bar{y} = \frac{\text{moment about } x\text{-axis}}{\text{total mass}} = \frac{M_x}{m}$$

and where the moments M_y and M_x were determined by integrating ρx and ρy over the appropriate region. These formulas are the basis for the following rules for finding the center of mass of a region with variable density.

Moments and Center of Mass of a Planar Lamina	If $\rho(x, y)$ gives the density of a lamina corresponding to the plane region R, then the moments of the lamina about the x- and y-axes are given by $$M_x = \iint_R y\rho(x, y)\, dA \quad \text{and} \quad M_y = \iint_R x\rho(x, y)\, dA$$ Furthermore, if m is the total mass of the lamina, then the center of mass is given by $$(\bar{x}, \bar{y}) = \left(\frac{M_y}{m}, \frac{M_x}{m}\right)$$

Note that if ρ is constant and the region R is given by $a \leq x \leq b$, $0 \leq y \leq f(x)$, then

$$M_x = \int_a^b \int_0^{f(x)} \rho y\, dy\, dx = \rho \int_a^b \frac{y^2}{2}\bigg]_0^{f(x)} dx = \frac{\rho}{2} \int_a^b f(x)^2\, dx$$

and

$$M_y = \int_a^b \int_0^{f(x)} \rho x\, dy\, dx = \rho \int_a^b xy\bigg]_0^{f(x)} dx = \rho \int_a^b xf(x)\, dx$$

which agree with our previous definitions of moments in Section 7.7.

Example 3

Find the center of mass of the lamina corresponding to the parabolic region $0 \leq y \leq 4 - x^2$ if the density at the point (x, y) is proportional to the distance between (x, y) and the x-axis. (See Figure 16.27.)

Solution: Since the lamina is symmetric with respect to the y-axis, we know that $\bar{x} = 0$. The mass of the region is given by

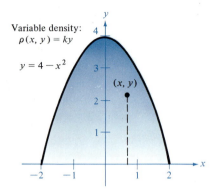

Variable density:
$\rho(x, y) = ky$

$y = 4 - x^2$

(x, y)

FIGURE 16.27

$$\text{mass} = \int_{-2}^{2} \int_{0}^{4-x^2} ky \, dy \, dx = \frac{k}{2} \int_{-2}^{2} y^2 \Big]_{0}^{4-x^2} dx$$

$$= \frac{k}{2} \int_{-2}^{2} (16 - 8x^2 + x^4) \, dx$$

$$= \frac{k}{2} \left[16x - \frac{8x^3}{3} + \frac{x^5}{5} \right]_{-2}^{2} = k \left(32 - \frac{64}{3} + \frac{32}{5} \right) = \frac{256k}{15}$$

The moment about the x-axis is

$$M_x = \int_{-2}^{2} \int_{0}^{4-x^2} (y)(ky) \, dy \, dx = \frac{k}{3} \int_{-2}^{2} y^3 \Big]_{0}^{4-x^2} dx$$

$$= \frac{k}{3} \int_{-2}^{2} (64 - 48x^2 + 12x^4 - x^6) \, dx$$

$$= \frac{k}{3} \left[64x - 16x^3 + \frac{12x^5}{5} - \frac{x^7}{7} \right]_{-2}^{2} = \frac{4096k}{105}$$

Thus
$$\bar{y} = \frac{M_x}{m} = \frac{4096k/105}{256k/15} = \frac{16}{7}$$

and the center of mass is $(0, \frac{16}{7})$. ∎

The moments M_x and M_y used in determining the center of mass of a region are sometimes referred to as the **first moments** about the x- and y-axes, respectively. In each case the moment is the product of a mass times a distance:

$$M_x = \iint_{R} (\text{distance})\text{mass} = \iint_{R} (y)\rho(x, y) \, dA$$

$$M_y = \iint_{R} (\text{distance})\text{mass} = \iint_{R} (x)\rho(x, y) \, dA$$

We now look at another type of moment referred to as the **second moment**, or **moment of inertia** about a line. In the same way that mass is a measure of the tendency of matter to resist a change in straight line motion, the moment of inertia about a line is a measure of the tendency of matter to resist a change in rotational motion. We denote the second moments about the x- and y-axes by I_x and I_y, respectively. In each of these cases, the second moment is the product of a mass times the square of a distance:

$$I_x = \iint_{R} (\text{distance})^2\text{mass} = \iint_{R} (y)^2\rho(x, y) \, dA$$

$$I_y = \iint_{R} (\text{distance})^2\text{mass} = \iint_{R} (x)^2\rho(x, y) \, dA$$

(Compare these formulas with those for M_x and M_y.)

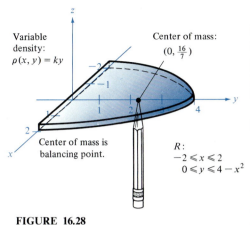

Variable
density:
$\rho(x, y) = ky$

Center of mass:
$(0, \frac{16}{7})$

Center of mass is
balancing point.

$R:$
$-2 \leqslant x \leqslant 2$
$0 \leqslant y \leqslant 4 - x^2$

FIGURE 16.28

We use the first moments M_x and M_y to determine an idealized point $(\bar{x}, \bar{y})$, called the *center of mass* of a region R. This allows us (in certain applications) to treat a region as if its mass were concentrated at just one point. For instance, the center of mass of the parabolic lamina in Example 3 was calculated to be the point $(0, \frac{16}{7})$. This means that the lamina should balance on the point of a pencil placed at $(0, \frac{16}{7})$ (see Figure 16.28).

For the second moments I_x and I_y, there is a similar idealized point $(\bar{\bar{x}}, \bar{\bar{y}})$, determined by the formulas

$$\bar{\bar{x}} = \sqrt{\frac{I_y}{m}} \quad \text{and} \quad \bar{\bar{y}} = \sqrt{\frac{I_x}{m}}$$

where m is the mass of the lamina R being considered. We call $\bar{\bar{x}}$ the *radius of gyration of R about the y-axis* and $\bar{\bar{y}}$ the *radius of gyration of R about the x-axis*. Note that the formulas

$$m\bar{\bar{x}}^2 = I_y \quad \text{and} \quad m\bar{\bar{y}}^2 = I_x$$

imply that $(\bar{\bar{x}}, \bar{\bar{y}})$ is the point at which all the mass of lamina R can be concentrated without changing the moments of inertia of R with respect to the coordinate axes.

Example 4

Find the radius of gyration about the x-axis of the lamina in Example 3.

Solution: We already know that the mass of this lamina is $256k/15$. To find $\bar{\bar{y}}$, we use the formula

$$I_x = \int_{-2}^{2} \int_{0}^{4-x^2} y^2(ky)\, dy\, dx = \frac{k}{4} \int_{-2}^{2} y^4 \Big]_{0}^{4-x^2} dx$$

$$= \frac{k}{4} \int_{-2}^{2} (256 - 256x^2 + 96x^4 - 16x^6 + x^8)\, dx$$

$$= \frac{k}{4} \left[256x - \frac{256x^3}{3} + \frac{96x^5}{5} - \frac{16x^7}{7} + \frac{x^9}{9} \right]_{-2}^{2} = \frac{32{,}768k}{315}$$

Therefore,

$$\bar{\bar{y}} = \sqrt{\frac{I_x}{m}} = \sqrt{\frac{32{,}768k/315}{256k/15}} = \sqrt{\frac{128}{21}} \approx 2.47 \qquad \blacksquare$$

Example 5

Find the point $(\bar{\bar{x}}, \bar{\bar{y}})$ for the lamina corresponding to the region $0 \leq y \leq \sin x, 0 \leq x \leq \pi$, where the density at (x, y) is given by $\rho(x, y) = x$.

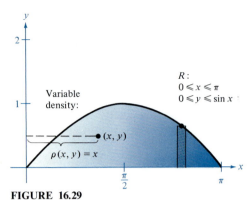

Variable density:

R:
$0 \leq x \leq \pi$
$0 \leq y \leq \sin x$

(x, y)

$\rho(x, y) = x$

FIGURE 16.29

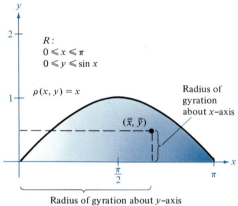

R:
$0 \leq x \leq \pi$
$0 \leq y \leq \sin x$

$\rho(x, y) = x$

Radius of gyration about x–axis

$(\bar{\bar{x}}, \bar{\bar{y}})$

Radius of gyration about y–axis

FIGURE 16.30

Solution: The region R is shown in Figure 16.29. The mass is given by

$$m = \int_0^\pi \int_0^{\sin x} x \, dy \, dx = \int_0^\pi xy \Big]_0^{\sin x} dx$$

$$= \int_0^\pi x \sin x \, dx = \Big[-x \cos x + \sin x \Big]_0^\pi = \pi$$

The moment of inertia about the x-axis is

$$I_x = \int_0^\pi \int_0^{\sin x} xy^2 \, dy \, dx = \frac{1}{3} \int_0^\pi xy^3 \Big]_0^{\sin x} dx$$

$$= \frac{1}{3} \int_0^\pi x \sin^3 x \, dx = \frac{1}{3} \int_0^\pi \frac{x}{2} \sin x \, (1 - \cos 2x) \, dx$$

$$= \frac{1}{6} \int_0^\pi x[\sin x - (\sin x \cos 2x)] \, dx$$

$$= \frac{1}{6} \int_0^\pi x\left[\sin x - \frac{1}{2}(\sin 3x - \sin x)\right] dx$$

$$= \frac{1}{12} \int_0^\pi (3x \sin x - x \sin 3x) \, dx$$

$$= \frac{1}{12}\left[3 \sin x - 3x \cos x - \frac{1}{9} \sin 3x + \frac{1}{3} x \cos 3x \right]_0^\pi = \frac{2\pi}{9}$$

The moment of inertia about the y-axis is

$$I_y = \int_0^\pi \int_0^{\sin x} x^3 \, dy \, dx = \int_0^\pi x^3 y \Big]_0^{\sin x} dx = \int_0^\pi x^3 \sin x \, dx$$

$$= \Big[(3x^2 - 6)(\sin x) - (x^3 - 6x)(\cos x) \Big]_0^\pi = \pi^3 - 6\pi$$

Thus

$$\bar{\bar{x}} = \sqrt{\frac{I_y}{m}} = \sqrt{\frac{\pi^3 - 6\pi}{\pi}} = \sqrt{\pi^2 - 6} \approx 1.97$$

$$\bar{\bar{y}} = \sqrt{\frac{I_x}{m}} = \sqrt{\frac{2\pi/9}{\pi}} = \sqrt{\frac{2}{9}} \approx 0.47$$

The point $(\bar{\bar{x}}, \bar{\bar{y}})$ is shown in Figure 16.30.   ∎

In Figure 16.30 note that if the entire mass of the lamina were located at the point $(\bar{\bar{x}}, \bar{\bar{y}})$, the moments of inertia about the axes would be unchanged.

Section Exercises (16.4)

1. Find the center of mass of the rectangular lamina with vertices $(0, 0)$, $(a, 0)$, $(0, b)$, and (a, b) when the density is the following:

(a) $\rho = k$ (b) $\rho = ky$
(c) $\rho = kxy$ (d) $\rho = k(x^2 + y^2)$

In Exercises 2–7, find the mass and center of mass of the given lamina of specified density.

2. lamina: triangle with vertices $(0, 0)$, $(b/2, h)$, $(b, 0)$; density: $\rho = k$

S 3. lamina: triangle with vertices $(0, 0)$, $(0, a)$, $(a, 0)$; density: $\rho = x^2 + y^2$

4. lamina: semicircle bounded by $y = 0$, $y = \sqrt{a^2 - x^2}$; density: ρ is proportional to the distance from the x-axis

5. lamina: region in first quadrant bounded by $x^2 + y^2 = a^2$; density: $\rho = k$

6. lamina: region bounded by $y = x^2$ and $y^2 = x$; density: ρ is proportional to the distance from the y-axis

7. lamina: region bounded by $y = x^2$ and $y^2 = x$; density: ρ is proportional to the square of the distance from the origin

8. Find the mass of the triangular region of the plane $2x + 3y + 6z = 12$ lying in the first octant, where $\rho = x^2 + y^2$.

S 9. Find the mass of the hemisphere $z = \sqrt{r^2 - x^2 - y^2}$, where ρ is proportional to the distance from the xy-plane.

In Exercises 10–14, find the moments of inertia I_x and I_y and the point $(\bar{\bar{x}}, \bar{\bar{y}})$ for the given lamina of specified density.

10. lamina: triangular region with vertices $(0, 0)$, $(b, 0)$, (b, h); density: $\rho = k$

11. lamina: square region with vertices $(0, 0)$, $(a, 0)$, $(0, b)$, (a, b); density: $\rho = ky$

S 12. lamina: region bounded by $y = \sqrt{a^2 - x^2}$ and $y = 0$; density: ρ is proportional to the distance from the x-axis

13. lamina: region bounded by $y = 4 - x^2$, $y = 0$; density: ρ is proportional to the distance from the y-axis

14. lamina: region bounded by $y = x$ and $y = x^2$; density: $\rho = kxy$

15. The polar moment of inertia I_o is defined to be

$$I_o = I_x + I_y = \iint\limits_R (x^2 + y^2) \, \rho(x, y) \, dA$$

Find the polar moment of inertia of the lamina of Exercise 10.

16. Find the polar moment of inertia I_o (see Exercise 15) of the lamina of Exercise 2.

The moment of inertia of a mass (lamina) distributed over a region R with respect to a given line L is defined to be

$$I_L = \iint\limits_R D^2 \, \rho \, dA$$

where $\rho(x, y)$ is the density and $D(x, y)$ is the distance from the point (x, y) to the line L. Use this definition in Exercises 17–19.

17. Show that the moment of inertia about the line $x = a$ is

$$\iint\limits_R (x - a)^2 \, \rho(x, y) \, dA$$

18. Show that the moment of inertia about the line $y = b$ is

$$\iint\limits_R (y - b)^2 \, \rho(x, y) \, dA$$

S 19. Find the moment of inertia of a uniform circular disk $x^2 + y^2 = r^2$ about the line $x = a \, (a > r)$.

16.5 **Triple Integrals**

Purpose
- To demonstrate the use of triple integrals to find volumes, centers of mass, and moments of inertia.
- To define a triple integral and demonstrate its evaluation as an iterated integral.

Mathematics is often considered a difficult and mysterious science, because of the numerous symbols which it employs. Of course, nothing is more incomprehensible than a symbolism which we do not understand. Also a symbolism, which we only partially understand and are unaccustomed to use, is difficult to follow. In exactly the same way the technical terms of any profession or trade are incomprehensible to those who have never been trained to use them. But this is not because they are difficult in themselves. On the contrary they have invariably been introduced to make things easy.

—Alfred Whitehead (1861–1947)

Let us review our limit definitions of single and double integrals:

$$\int_a^b f(x)\,dx = \lim_{\|\Delta\|\to 0} \sum_{i=1}^n f(x_i)\,\Delta x_i$$

and

$$\iint_R f(x,y)\,dA = \lim_{\|\Delta\|\to 0} \sum_{i=1}^n f(x_i, y_i)\,\Delta x_i\,\Delta y_i$$

In this section we want to develop a similar definition for a triple integral over some solid region S.

Let S be a bounded solid region and suppose we approximate S by a network of n adjacent boxes, where the volume of the ith box is $\Delta x_i\,\Delta y_i\,\Delta z_i = \Delta V_i$ (Figure 16.31). Now by letting $\|\Delta\|$ be the volume of the largest box in the network, we can take the limit as $\|\Delta\| \to 0$, to obtain

$$\text{volume of } S = \lim_{\|\Delta\|\to 0} \sum_{i=1}^n \Delta x_i\,\Delta y_i\,\Delta z_i = \lim_{\|\Delta\|\to 0} \sum_{i=1}^n \Delta V_i$$

Or, if (x_i, y_i, z_i) is a point in the ith box and $\rho(x_i, y_i, z_i)$ is the density of S at (x_i, y_i, z_i), then we have

$$\text{mass of } S = \lim_{\|\Delta\|\to 0} \sum_{i=1}^n \rho(x_i, y_i, z_i)\,\Delta V_i$$

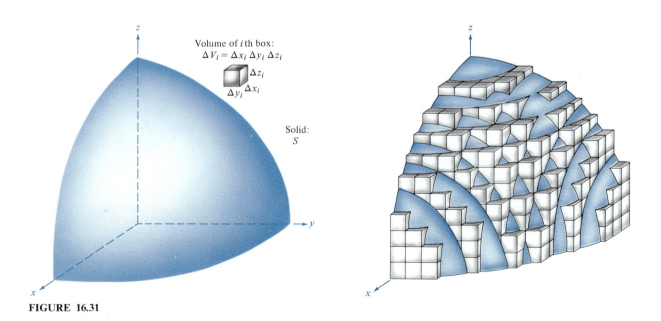

Volume of ith box:
$\Delta V_i = \Delta x_i\,\Delta y_i\,\Delta z_i$

Δz_i
$\Delta y_i \quad \Delta x_i$

Solid: S

FIGURE 16.31

| **Definition of Triple Integral** | If f is continuous over a solid region S, then the **triple integral of f over S** is defined to be |

$$\iiint_S f(x, y, z)\, dV = \lim_{\|\Delta\|\to 0} \sum_{i=1}^n f(x_i, y_i, z_i)\, \Delta x_i\, \Delta y_i\, \Delta z_i$$

where the differential dV represents any one of the six combinations

$$dx\,dy\,dz, \qquad dx\,dz\,dy, \qquad dy\,dx\,dz,$$
$$dy\,dz\,dx, \qquad dz\,dx\,dy, \qquad dz\,dy\,dx$$

As we would expect, a triple integral is most easily evaluated as an iterated integral, using one of the six possible permutations of dx, dy, and dz. For instance, the order $dz\,dy\,dx$ corresponds to the iterated integral

$$\int_a^b \int_{g_1(x)}^{g_2(x)} \int_{h_1(x,y)}^{h_2(x,y)} f(x, y, z)\, dz\, dy\, dx$$

where *both* x and y are held constant for the innermost integration and x is held constant for the second integration.

Example 1

Evaluate the iterated integral

$$\int_0^2 \int_0^x \int_0^{x+y} e^x(y + 2z)\, dz\, dy\, dx$$

Solution: Holding x and y constant, we have

$$\int_0^2 \int_0^x \int_0^{x+y} e^x(y + 2z)\, dz\, dy\, dx = \int_0^2 \int_0^x e^x(yz + z^2)\Big]_0^{x+y} dy\, dx$$

$$= \int_0^2 \int_0^x e^x(x^2 + 3xy + 2y^2)\, dy\, dx$$

Now holding x constant we have

$$\int_0^2 \int_0^x e^x(x^2 + 3xy + 2y^2)\, dy\, dx = \int_0^2 \left[e^x \left(x^2 y + \frac{3xy^2}{2} + \frac{2y^3}{3} \right) \right]_0^x$$

$$= \frac{19}{6} \int_0^2 x^3 e^x\, dx$$

$$= \frac{19}{6} \left[e^x(x^3 - 3x^2 + 6x - 6) \right]_0^2$$

$$= 19 \left(\frac{e^2}{3} + 1 \right)$$

To find the limits for a particular order of integration, it is generally advisable to first determine the innermost limits, which may be functions of the outer two variables. Then if we project the solid S onto the coordinate plane of the outer two variables, we can determine their limits of integration by the methods used for double integrals. For instance, to evaluate

$$\iiint\limits_S f(x, y, z)\, dx\, dy\, dz$$

we first determine the limits for x, and then the integral has the form

$$\iint\limits_{R_{yz}} \left[\int_{g_1(y,z)}^{g_2(y,z)} f(x, y, z)\, dx \right] dy\, dz$$

Now by projecting solid S onto the yz-plane, we can determine the limits for y and z as we did for double integrals.

Example 2
_____

Evaluate the triple integral

$$\iiint\limits_S dV$$

where S is the solid region inside the ellipsoid $4x^2 + y^2 + 16z^2 = 16$.

Solution: To write this triple integral as an iterated integral, we observe that the roles of x, y, and z in the equation $4x^2 + y^2 + 16z^2 = 16$ are similar, and the order of integration is probably immaterial. Arbitrarily choosing the order $dz\, dy\, dx$, we first determine the limits for z to be

$$16z^2 = 16 - 4x^2 - y^2$$
$$z^2 = \tfrac{1}{16}(16 - 4x^2 - y^2)$$
$$z = \pm\tfrac{1}{4}\sqrt{16 - 4x^2 - y^2}$$

The projection of S onto the xy-plane is the ellipse $4x^2 + y^2 = 16$ (see Figure 16.32). Since x must have constant limits, we conclude that

$$-2 \le x \le 2 \quad \text{and} \quad -2\sqrt{4 - x^2} \le y \le 2\sqrt{4 - x^2}$$

Considering only that portion of the ellipsoid lying in the first octant, we have

$$\iiint\limits_S dV$$

$$= 8 \int_0^2 \int_0^{2\sqrt{4-x^2}} \int_0^{\sqrt{16-4x^2-y^2}/4} dz\, dy\, dx$$

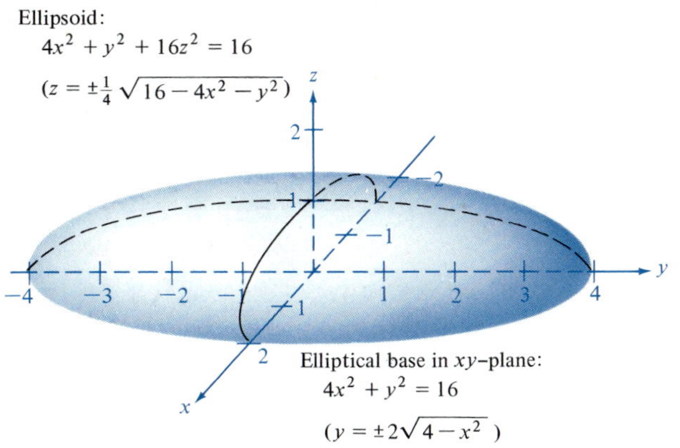

Ellipsoid:
$$4x^2 + y^2 + 16z^2 = 16$$
$$(z = \pm \tfrac{1}{4} \sqrt{16 - 4x^2 - y^2})$$

Elliptical base in xy-plane:
$$4x^2 + y^2 = 16$$
$$(y = \pm 2\sqrt{4 - x^2})$$

FIGURE 16.32

$$= 8 \int_0^2 \int_0^{2\sqrt{4-x^2}} z \,\Big]_0^{\sqrt{16-4x^2-y^2}/4} \, dy \, dx$$

$$= 2 \int_0^2 \int_0^{2\sqrt{4-x^2}} \sqrt{16 - 4x^2 - y^2} \, dy \, dx$$

$$= \int_0^2 \left[y \sqrt{16 - 4x^2 - y^2} + (16 - 4x^2) \, \text{Arcsin} \, \frac{y}{\sqrt{16 - 4x^2}} \right]_0^{2\sqrt{4-x^2}} dx$$

$$= \int_0^2 \left[0 + (16 - 4x^2) \left(\frac{\pi}{2} \right) - 0 - 0 \right] dx$$

$$= \frac{\pi}{2} \int_0^2 (16 - 4x^2) \, dx = \frac{\pi}{2} \left[16x - \frac{4x^3}{3} \right]_0^2 = \frac{32\pi}{2} \quad \blacksquare$$

Example 3

For the ellipsoid given in Example 2, set up the iterated integrals for these integration orders: $dy \, dx \, dz$ and $dx \, dy \, dz$.

Solution: For the order $dy \, dx \, dz$, we first determine the limits for y to be

$$y^2 = 16 - 16z^2 - 4x^2$$
$$y = \pm 2\sqrt{4 - 4z^2 - x^2}$$

The projection of S in the xz-plane is the ellipse $4x^2 + 16z^2 = 16$. Since z is the outer variable, its limits are the constants $16z^2 = 16$, or $z = \pm 1$. The limits for x are

$$4x^2 = 16 - 16z^2$$
$$x^2 = 4 - 4z^2$$
$$x = \pm 2\sqrt{1 - z^2}$$

Therefore,

$$\iiint_S dV = 8 \int_0^1 \int_0^{2\sqrt{1-z^2}} \int_0^{2\sqrt{4-4z^2-x^2}} dy\, dx\, dz$$

Similarly, for $dx\, dy\, dz$, we determine the limits for x to be

$$4x^2 = 16 - 16z^2 - y^2 \qquad \text{or} \qquad x = \pm \frac{\sqrt{16 - 16z^2 - y^2}}{2}$$

In the yz-plane we have the ellipse $y^2 + 16z^2 = 16$, from which we conclude that

$$z = \pm 1 \qquad \text{and} \qquad y = \pm 4\sqrt{1 - z^2}$$

Therefore,

$$\iiint_S dV = 8 \int_0^1 \int_0^{4\sqrt{1-z^2}} \int_0^{\sqrt{16-16z^2-y^2}/2} dx\, dy\, dz \qquad \blacksquare$$

Example 4

Evaluate the iterated integral

$$\int_0^{\sqrt{\pi/2}} \int_x^{\sqrt{\pi/2}} \int_1^3 \sin y^2\, dz\, dy\, dx$$

by changing the order of integration.

Solution: Note that after one integration in the given order, we encounter the integral $2 \int \sin (y^2)\, dy$, which is not an elementary function. To avoid this problem we change the order of integration to $dz\, dx\, dy$ so that y is the outer variable.

Since the solid region S is given by

$$0 \le x \le \sqrt{\frac{\pi}{2}}, \qquad x \le y \le \sqrt{\frac{\pi}{2}}, \qquad 1 \le z \le 3$$

(Figure 16.33), the projection of S in the xy-plane yields the bounds $0 \le y \le \sqrt{\pi}/2$, and $0 \le x \le y$. Therefore, we have

$$\int_0^{\sqrt{\pi/2}} \int_0^y \int_1^3 \sin (y^2)\, dz\, dx\, dy = \int_0^{\sqrt{\pi/2}} \int_0^y z \sin (y^2) \Big]_1^3 dx\, dy$$

$$= 2 \int_0^{\sqrt{\pi/2}} \int_0^y \sin (y^2)\, dx\, dy$$

$$= 2 \int_0^{\sqrt{\pi/2}} x \sin (y^2) \Big]_0^y dy$$

$$= 2 \int_0^{\sqrt{\pi/2}} y \sin (y^2)\, dy$$

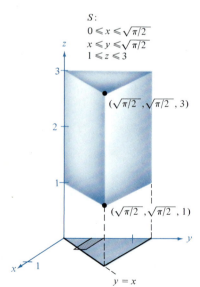

S:
$0 \le x \le \sqrt{\pi/2}$
$x \le y \le \sqrt{\pi/2}$
$1 \le z \le 3$

$(\sqrt{\pi/2}, \sqrt{\pi/2}, 3)$

$(\sqrt{\pi/2}, \sqrt{\pi/2}, 1)$

$y = x$

FIGURE 16.33

$$= -\cos(y^2)\Big]_0^{\sqrt{\pi/2}} = 1 \qquad \blacksquare$$

The **center of mass** of a solid region S is given by $(\bar{x}, \bar{y}, \bar{z})$, where

$$m = \iiint_S \rho(x, y, z)\, dV \qquad M_{xz} = \iiint_S y\rho(x, y, z)\, dV$$

$$M_{yz} = \iiint_S x\rho(x, y, z)\, dV \qquad M_{xy} = \iiint_S z\rho(x, y, z)\, dV$$

and
$$\bar{x} = \frac{M_{yz}}{m}, \qquad \bar{y} = \frac{M_{xz}}{m}, \qquad \bar{z} = \frac{M_{xy}}{m}$$

The quantities M_{yz}, M_{xz}, and M_{xy} are called the **first moments** of region S about the yz-, xz-, and xy-planes, respectively.

Example 5

Find the center of mass of the unit cube $0 \le x \le 1$, $0 \le y \le 1$, $0 \le z \le 1$, if the density at a point (x, y, z) is proportional to the square of its distance from the origin. (See Figure 16.34.)

Solution: We begin by finding the mass

$$m = \iiint_S \rho(x, y, z)\, dV$$

of the cube. Since the density at (x, y, z) is proportional to the square of the distance between $(0, 0, 0)$ and (x, y, z), we have

$$\rho(x, y, z) = k(x^2 + y^2 + z^2)$$

Because of the symmetry of the region, the order of integration is immaterial and we have

$$m = \int_0^1 \int_0^1 \int_0^1 k(x^2 + y^2 + z^2)\, dz\, dy\, dx$$

$$= k \int_0^1 \int_0^1 \left[(x^2 + y^2)z + \frac{z^3}{3}\right]_0^1 dy\, dx$$

$$= k \int_0^1 \int_0^1 \left(x^2 + y^2 + \frac{1}{3}\right) dy\, dx = k \int_0^1 \left[\left(x^2 + \frac{1}{3}\right)y + \frac{y^3}{3}\right]_0^1 dx$$

$$= k \int_0^1 \left(x^2 + \frac{2}{3}\right) dx = k \left[\frac{x^3}{3} + \frac{2x}{3}\right]_0^1 = k$$

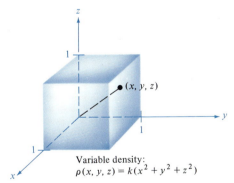

Variable density:
$\rho(x, y, z) = k(x^2 + y^2 + z^2)$

FIGURE 16.34

The first moment about the *yz*-plane is

$$M_{yz} = k \int_0^1 \int_0^1 \int_0^1 x(x^2 + y^2 + z^2) \, dz \, dy \, dx$$

and since *x* is considered constant for the first two partial integrations, we write

$$M_{yz} = k \int_0^1 x \left[\int_0^1 \int_0^1 (x^2 + y^2 + z^2) \, dz \, dy \right] dx$$

Now in this form the two inner integrals are the same as for the mass *m*; hence we have

$$M_{yz} = k \int_0^1 x \left(x^2 + \frac{2}{3} \right) dx = k \int_0^1 \left(x^3 + \frac{2x}{3} \right) dx$$

$$= k \left[\frac{x^4}{4} + \frac{x^2}{3} \right]_0^1 = \frac{7k}{12}$$

Therefore,

$$\bar{x} = \frac{M_{yx}}{m} = \frac{7k/12}{k} = \frac{7}{12}$$

Finally, from the symmetry of *x*, *y*, and *z* in this region, we have $\bar{x} = \bar{y} = \bar{z}$ and the center of mass is $(\frac{7}{12}, \frac{7}{12}, \frac{7}{12})$. ■

It is of interest to note that the first moments for solid regions are taken about a plane, whereas second moments for solids are taken about a line. The moments of a solid region *S*, given by

$$I_x = \iiint_S (y^2 + z^2) \rho(x, y, z) \, dV$$

$$I_y = \iiint_S (x^2 + z^2) \rho(x, y, z) \, dV$$

$$I_z = \iiint_S (x^2 + y^2) \rho(x, y, z) \, dV$$

are called the **second moments** (or **moments of inertia**) about the *x*-, *y*-, and *z*-axes, respectively. For problems involving all three of these moments of inertia, sometimes considerable effort can be saved by applying the additive property of integrals and writing

$$I_x = I_{xz} + I_{xy}, \qquad I_y = I_{yz} + I_{xy}, \qquad I_z = I_{yz} + I_{xz}$$

where

$$I_{xy} = \iiint_S z^2 \rho(x, y, z) \, dV, \qquad I_{xz} = \iiint_S y^2 \rho(x, y, z) \, dV$$

$$I_{yz} = \iiint_S x^2 \rho(x, y, z) \, dV$$

Example 6

Find the moments of inertia about the x- and y-axes for the solid region bounded above by the hemisphere $z = \sqrt{4 - x^2 - y^2}$ and below by the xy-plane. Let the density at (x, y, z) be proportional to the distance between (x, y, z) and the xy-plane.

Solution: The density of the region is given by $\rho(x, y, z) = kz$. Considering the symmetry of this problem, we know that $I_x = I_y$. Thus we need to compute only one moment, say,

$$I_x = \iiint_S (y^2 + z^2) \rho(x, y, z) \, dV$$

From Figure 16.35 we choose the order $dz \, dy \, dx$ and write

$$I_x = \int_{-2}^{2} \int_{-\sqrt{4-x^2}}^{\sqrt{4-x^2}} \int_{0}^{\sqrt{4-x^2-y^2}} (y^2 + z^2)(kz) \, dz \, dy \, dx$$

$$= k \int_{-2}^{2} \int_{-\sqrt{4-x^2}}^{\sqrt{4-x^2}} \left[\frac{y^2 z^2}{2} + \frac{z^4}{4} \right]_{0}^{\sqrt{4-x^2-y^2}} dy \, dx$$

$$= k \int_{-2}^{2} \int_{-\sqrt{4-x^2}}^{\sqrt{4-x^2}} \left[\frac{y^2(4 - x^2 - y^2)}{2} + \frac{(4 - x^2 - y^2)^2}{4} \right] dy \, dx$$

$$= \frac{k}{4} \int_{-2}^{2} \int_{-\sqrt{4-x^2}}^{\sqrt{4-x^2}} [(4 - x^2)^2 - y^4] \, dy \, dx$$

$$= \frac{k}{4} \int_{-2}^{2} \left[(4 - x^2)^2 y - \frac{y^5}{5} \right]_{-\sqrt{4-x^2}}^{\sqrt{4-x^2}} dx$$

$$= \frac{k}{4} \int_{-2}^{2} \frac{8}{5} (4 - x^2)^{5/2} \, dx = \frac{4k}{5} \int_{0}^{2} (4 - x^2)^{5/2} \, dx$$

Letting $x = 2 \sin \theta$ and $dx = 2 \cos \theta \, d\theta$, we have

$$\frac{4k}{5} \int_{0}^{2} (4 - x^2)^{5/2} \, dx = \frac{4k}{5} \int_{0}^{\pi/2} 64 \cos^6 \theta \, d\theta = \left(\frac{256k}{5} \right) \left(\frac{5\pi}{32} \right) = 8k\pi$$

Thus $I_x = 8k\pi = I_y$. ■

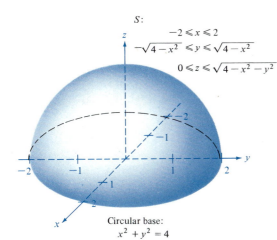

S:
$$-2 \leqslant x \leqslant 2$$
$$-\sqrt{4 - x^2} \leqslant y \leqslant \sqrt{4 - x^2}$$
$$0 \leqslant z \leqslant \sqrt{4 - x^2 - y^2}$$

Circular base:
$$x^2 + y^2 = 4$$

FIGURE 16.35

Section Exercises (16.5)

In Exercises 1–6, evaluate each triple integral.

1. $\int_{0}^{3} \int_{0}^{2} \int_{0}^{1} (x + y + z) \, dx \, dy \, dz$

2. $\int_{-1}^{1} \int_{-1}^{1} \int_{-1}^{1} x^2 y^2 z^2 \, dx \, dy \, dz$

3. $\int_{0}^{9} \int_{0}^{y/3} \int_{0}^{\sqrt{y^2 - 9x^2}} z \, dz \, dx \, dy$

4. $\int_{0}^{\sqrt{2}} \int_{0}^{\sqrt{2-x^2}} \int_{2x^2+y^2}^{4-y^2} y \, dz \, dy \, dx$

[S] **5.** $\int_0^2 \int_{-\sqrt{4-x^2}}^{\sqrt{4-x^2}} \int_0^{x^2} x \, dz \, dy \, dx$

6. $\int_0^{\pi/2} \int_0^{y/2} \int_0^{1/y} \sin y \, dz \, dx \, dy$

In Exercises 7–12, use a triple integral to find the volume of the specified solid.

7. solid bounded by $x = 4 - y^2, z = 0, z = x$

8. solid bounded by $z = xy, z = 0, x = 0, x = 1, y = 0,$ $y = 1$

9. solid bounded by the sphere $x^2 + y^2 + z^2 = r^2$

10. solid bounded by $z = 9 - x^2 - y^2, z = 0$

[S] **11.** solid bounded by the cylinders $z = 4 - x^2$ and $y = 4 - x^2$ in the first octant

12. solid bounded by the cylinder $z = 9 - x^2$ and the planes $y = -x + 2, y = 0, z = 0, x \geq 0$

In Exercises 13–16, sketch the solid region whose volume is given by the triple integral.

13. $\int_0^4 \int_0^{(4-x)/2} \int_0^{(12-3x-6y)/4} dx \, dy \, dz$

14. $\int_0^4 \int_0^{\sqrt{16-x^2}} \int_0^{10-x-y} dz \, dy \, dx$

[S] **15.** $\int_0^1 \int_y^1 \int_0^{\sqrt{1-y^2}} dz \, dx \, dy$

16. $\int_0^2 \int_{2x}^4 \int_0^{\sqrt{y^2-4x^2}} dz \, dy \, dx$

In Exercises 17–24, find the required coordinate(s) for the center of mass of the specified solid.

17. The solid bounded by $x = 0, y = 0, z = 0,$ and $2x + 3y + 6z = 12; \rho = k;$ find $\bar{x}$.

18. The solid bounded by $x = 0, y = 0, z = 0,$ and $(x/a) + (y/b) + (z/c) = 1$, where a, b, and c are positive; $\rho = k$; find $\bar{y}$.

[S] **19.** The cube bounded by $x = 0, x = b, y = 0, y = b, z = 0,$ $z = b; \rho = kxy;$ find $(\bar{x}, \bar{y}, \bar{z})$.

20. The solid bounded by $x = 0, x = a, y = 0, y = b,$ $z = 0, z = c; \rho = kz;$ find $(\bar{x}, \bar{y}, \bar{z})$.

21. The cone bounded by $z = (h/r)\sqrt{x^2 + y^2}$ and $z = h;$ $\rho = k;$ find $(\bar{x}, \bar{y}, \bar{z})$.

22. The circular wedge bounded by $y = \sqrt{4 - x^2}, y = 0,$ $z = y, z = 0; \rho = k;$ find $(\bar{x}, \bar{y}, \bar{z})$.

23. The hemisphere bounded by $z = \sqrt{r^2 - x^2 - y^2}$ and $z = 0; \rho = k;$ find $(\bar{x}, \bar{y}, \bar{z})$.

24. The solid bounded by $x = -2, x = 2, y = 1, y = 0,$ $z = 1/(y^2 + 1), z = 0; \rho = k;$ find $(\bar{x}, \bar{y}, \bar{z})$.

In Exercises 25–29, find the moment of inertia about the given axis of the solid having the specified density.

25. The cube whose base has the vertices $(0, 0), (0, b), (b, 0),$ $(b, b); \rho = k;$ find I_x.

26. The solid bounded by $x = -b/2, x = b/2, y = a,$ $y = a + b, z = 0, z = b; \rho = k;$ find I_z.

27. The cube whose base has the vertices $(b/2, b/2),$ $(b/2, -b/2), (-b/2, b/2), (-b/2, -b/2); \rho = k;$ find I_z.

[S] **28.** The solid bounded by $x^2 + y^2 = a^2, z = 0, z = h;$ $\rho = k(x^2 + y^2);$ find I_z.

29. The solid bounded by $x^2 + y^2 = a^2, z = 0, z = h;$ $\rho = k;$ find I_z.

30. Assume that the solid bounded by $x^2 + y^2 = r^2, z = h,$ and $z = -h$ has uniform density and mass m. Show that the moment of inertia about the y-axis is given by

$$I_y = m\left(\frac{r^2}{4} + \frac{h^2}{3}\right)$$

[S] **31.** Assume that the solid bounded by $x^2 + y^2 = r^2, z = 0,$ and $z = h$ has uniform density and mass m. Show that $I_z = mr^2/2.$

16.6 Multiple Integrals in Nonrectangular Coordinates

Purpose

- To define a double integral in polar coordinates.
- To define triple integrals in cylindrical and spherical coordinates.
- To demonstrate the use of multiple integrals in cylindrical and spherical coordinates.

The feeling of mathematical elegance is only the satisfaction due to any adaptation of the solution to the needs of our mind, and it is because of this very adaptation that this solution can be for us an instrument.

—Henri Poincaré (1854–1912)

In this chapter we have encountered some rather complicated multiple integrals in rectangular coordinates. Frequently we had to evaluate integrals involving radicals of sums or differences of squares, and often such integrals required sophisticated integration techniques. Sometimes an

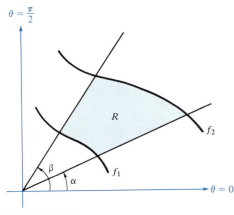

$\theta = \frac{\pi}{2}$

R

f_2

β

f_1

α

$\theta = 0$

FIGURE 16.36

equation for a graph can be written more simply in nonrectangular coordinates (Chapters 12 and 14). Thus some of the integrals of this chapter may have been simpler in a polar or a spherical system. For this reason we want to consider multiple integrals in nonrectangular systems and to suggest appropriate situations in which polar, cylindrical, or spherical coordinates may be more convenient than rectangular coordinates.

We begin with a description of a double integral in polar coordinates and recall that polar and rectangular coordinates satisfy the relationships

$$x = r \cos \theta, \qquad y = r \sin \theta, \qquad x^2 + y^2 = r^2$$

Suppose the polar region R shown in Figure 16.36 is bounded by the curves $r = f_1(\theta)$ and $r = f_2(\theta)$, and by the lines $\theta = \alpha$ and $\theta = \beta$, where α and β make positive angles with the polar axis. Now let us subdivide R by a grid of n sections, R_i, of the type shown in Figure 16.37. We will refer to these sections as "polar rectangles" even though they are actually portions of sectors of a circle.

The area of each of these polar rectangles is the difference of the areas of two circular sectors, and thus

$$\Delta A_i = \tfrac{1}{2}r_i^2(\Delta \theta_i) - \tfrac{1}{2}r_{i-1}^2(\Delta \theta_i) = \tfrac{1}{2}(r_i + r_{i-1})(r_i - r_{i-1})\,\Delta \theta_i$$
$$= \tfrac{1}{2}(r_i + r_{i-1})\,\Delta r_i \Delta \theta_i$$

If we denote the *average radius* $\tfrac{1}{2}(r_i + r_{i-1})$ by $\bar{r}_i$, then

$$\Delta A_i = \bar{r}_i\,\Delta r_i\,\Delta \theta_i$$

Now let us choose a point $(\bar{r}_i, \bar{\theta}_i)$ in R_i and let $\|\Delta\|$ be the length of the longest diagonal of the R_i. If f is continuous on region R, then the **double integral of f on R** is defined in polar coordinates to be

$$\lim_{\|\Delta\| \to 0} \sum_{i=1}^{n} f(\bar{r}_i, \bar{\theta}_i)\bar{r}_i\,\Delta r_i\,\Delta \theta_i = \iint\limits_{R} f(r, \theta)r\,dr\,d\theta$$

Note that in polar coordinates $dA = r\,dr\,d\theta$, and that we evaluate a double integral as an iterated integral just as we did in rectangular coordinates. For instance, if $\rho = f(r, \theta)$ is the density of region R given in Figure 16.36, then the mass m of the region is given by the iterated integral

$$m = \iint\limits_{R} \rho\, dA = \int_{\alpha}^{\beta} \int_{g_1(\theta)}^{g_2(\theta)} f(r, \theta)r\,dr\,d\theta$$

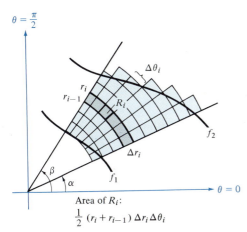

$\theta = \frac{\pi}{2}$

$\Delta \theta_i$

r_i

r_{i-1}

R_i

f_2

Δr_i

β

f_1

α

$\theta = 0$

Area of R_i:

$\frac{1}{2}(r_i + r_{i-1})\,\Delta r_i \Delta \theta_i$

FIGURE 16.37

Example 1

Find the volume of the solid inside both the sphere $x^2 + y^2 + z^2 = 16$ and the cylinder $x^2 + y^2 = 7$.

Solution: It is convenient to work with half of the solid and double the result. From Figure 16.38 we can see that the height of the upper half of

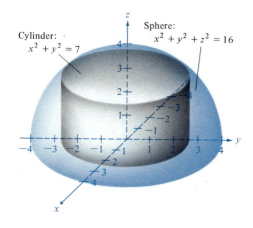

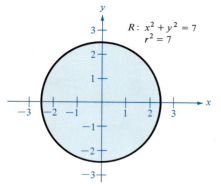

FIGURE 16.38

the solid is $z = \sqrt{16 - x^2 - y^2}$, and in the xy-plane we have

$$-\sqrt{7 - y^2} \le x \le \sqrt{7 - y^2} \quad \text{and} \quad -\sqrt{7} \le y \le \sqrt{7}$$

Thus in rectangular coordinates, the total volume is given by

$$V = 2 \iint_R f(x, y)\, dA = 2 \int_{-\sqrt{7}}^{\sqrt{7}} \int_{-\sqrt{7-y^2}}^{\sqrt{7-y^2}} \sqrt{16 - x^2 - y^2}\, dx\, dy$$

In polar coordinates we have the height as

$$z = \sqrt{16 - x^2 - y^2} = \sqrt{16 - r^2}$$

and in the polar plane,

$$0 \le r \le \sqrt{7} \quad \text{and} \quad 0 \le \theta \le 2\pi$$

Therefore, the volume is given by

$$V = 2 \iint_R g(r, \theta)\, dA = 2 \int_0^{2\pi} \int_0^{\sqrt{7}} \sqrt{16 - r^2}\, r\, dr\, d\theta$$

$$= -\int_0^{2\pi} \left[\frac{2}{3}(16 - r^2)^{3/2} \right]_0^{\sqrt{7}} d\theta = \int_0^{2\pi} -\frac{2}{3}(27 - 64)\, d\theta$$

$$= \int_0^{2\pi} \frac{74}{3}\, d\theta = \frac{148}{3}\pi$$

(You should attempt to evaluate the corresponding rectangular double integral to see the convenience of polar coordinates in this case.) ∎

Example 2

Determine the mass of the lamina corresponding to the circular region $x^2 + y^2 = 1$, where the density at each point is proportional to the square of the distance from the origin. (Compare to Example 1, Section 16.4.)

Solution: Since mass m is given by

$$m = \iint_R \rho(x, y)\, dA$$

and since $\rho(x, y) = k(x^2 + y^2)$, we have, in polar coordinates,

$$m = \iint_R kr^2\, dA = \int_0^{2\pi} \int_0^1 kr^2 r\, dr\, d\theta$$

$$= k \int_0^{2\pi} \frac{r^4}{4} \Big]_0^1 d\theta = k \left(\frac{\theta}{4} \right) \Big]_0^{2\pi} = \frac{k\pi}{2}$$ ∎

Again, the convenience of polar coordinates for evaluating integrals involving circular regions is evident. In some instances double integrals cannot be evaluated (except by approximation techniques) unless they are converted to polar coordinates. The next example shows such a case.

Example 3

Evaluate

$$\int_{0}^{3\sqrt{3}/2} \int_{x/\sqrt{3}}^{\sqrt{9-x^2}} e^{-(x^2+y^2)} \, dy \, dx$$

using polar coordinates.

Solution: First, region R lies between the graphs of $y = x/\sqrt{3}$ and $y = \sqrt{9 - x^2}$ from $x = 0$ to $x = 3\sqrt{3}/2$ (see Figure 16.39). In polar coordinates R lies inside the circle $r = 3$, and between the lines $\theta = \pi/6$ and $\theta = \pi/2$. Therefore,

$$\int_{0}^{3\sqrt{3}/2} \int_{x/\sqrt{3}}^{\sqrt{9-x^2}} e^{-(x^2+y^2)} \, dy \, dx = \int_{\pi/6}^{\pi/2} \int_{0}^{3} e^{-r^2} r \, dr \, d\theta$$

$$= \int_{\pi/6}^{\pi/2} -\frac{1}{2} e^{-r^2} \Big]_{0}^{3} \, d\theta$$

$$= \int_{\pi/6}^{\pi/2} \frac{1}{2}(1 - e^{-9}) \, d\theta = \frac{\pi}{6}(1 - e^{-9}) \quad \blacksquare$$

For triple integrals in cylindrical coordinates, the fundamental element of volume ΔV is the "cylindrical block" shown in Figure 16.40. If

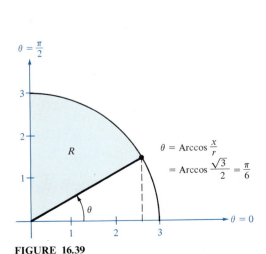

$\theta = \frac{\pi}{2}$

R

$\theta = \text{Arccos} \dfrac{x}{r}$

$= \text{Arccos} \dfrac{\sqrt{3}}{2} = \dfrac{\pi}{6}$

θ

$\theta = 0$

FIGURE 16.39

Volume of cylindrical block:
$$\Delta V_i = \bar{r}_i \, \Delta r_i \, \Delta \theta_i \, \Delta z_i$$

$\theta = 0$

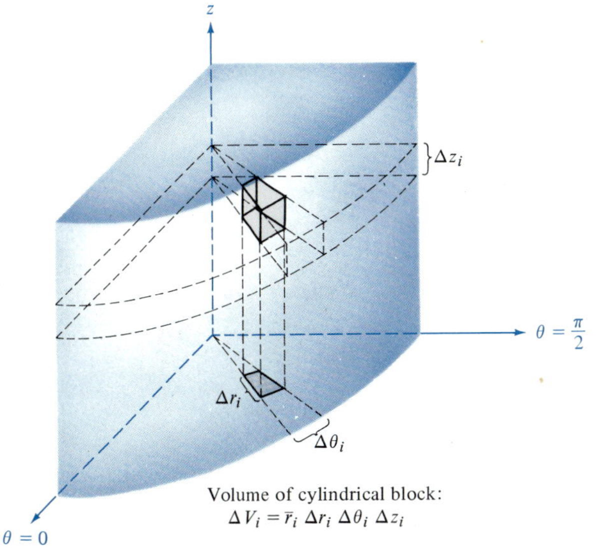

FIGURE 16.40

we let $\bar{r}_i$ be the average radius of the base of this wedge, then we have

$$\Delta V_i = (\text{area of base}) \times (\text{height}) = (\bar{r}_i \, \Delta \, r_i \, \Delta \, \theta_i) \, \Delta \, z_i$$

Choosing a point $(\bar{r}_i, \bar{\theta}_i, \bar{z}_i)$ in the ith block and letting $\|\Delta\|$ be the length of the longest diagonal of the blocks, we define, in cylindrical coordinates, the **triple integral** of f on S to be

$$\lim_{\|\Delta\| \to 0} \sum_{i=1}^{n} f(r_i, \theta_i, z_i) \, \Delta V_i = \iiint_S f(r, \theta, z) r \, dr \, d\theta \, dz$$

provided f is continuous on S. As with other multiple integrals, this triple integral is evaluated as an iterated integral. There are, of course, five other possible orders of integration. It is sometimes convenient to view the triple integral

$$\int_\alpha^\beta \int_{g_1(\theta)}^{g_2(\theta)} \int_{h_1(r, \theta)}^{h_2(r, \theta)} f(r, \theta, z) r \, dr \, d\theta \, dz$$

as a triple sum, where the first summation is along the r direction and it produces a solid sector, as shown in Figure 16.41(a). The second summation adds these sectors in the θ direction to form a layer of height Δz, Figure 16.41(b). And, finally, summing these layers along the z direction completes the coverage of the entire region S. (Try to visualize this order of summation for the block in Figure 16.40.)

Recall from Section 14.6 that cylindrical coordinates are especially useful when there is a line of symmetry and we can use the z-axis as this line.

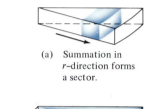

(a) Summation in r–direction forms a sector.

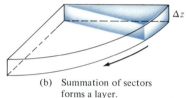

(b) Summation of sectors forms a layer.

FIGURE 16.41

Example 4

Find the moment of inertia about the axis of symmetry of the solid bounded by the paraboloid $z = x^2 + y^2$ and the plane $z = 4$. Let the density at each point be proportional to its distance from the axis. (See Figure 16.42.)

Solution: Since the z-axis is the axis of symmetry and since $\rho(x, y, z) = k\sqrt{x^2 + y^2}$, it follows that

$$I_z = \iiint_S k(x^2 + y^2) \sqrt{x^2 + y^2} \, dV$$

In cylindrical coordinates $0 \le r \le \sqrt{z} = \sqrt{x^2 + y^2}$, and, therefore,

$$I_z = k \int_0^4 \int_0^{2\pi} \int_0^{\sqrt{z}} r^2(r) r \, dr \, d\theta \, dz$$

Thus it follows that

$$I_z = k \int_0^4 \int_0^{2\pi} \frac{r^5}{5} \Big]_0^{\sqrt{z}} d\theta \, dz = k \int_0^4 \int_0^{2\pi} \frac{z^{5/2}}{5} d\theta \, dz$$

$$= k \int_0^4 \frac{z^{5/2}}{5}(2\pi) \, dz = k \left[\left(\frac{2\pi}{5} \right) \left(\frac{2}{7} \right) z^{7/2} \right]_0^4 = \frac{512 \, k\pi}{35} \qquad \blacksquare$$

Since cylindrical coordinates are a three-dimensional extension of polar coordinates, triple integrals in cylindrical coordinates often lend themselves to the following integration order:

$$\iiint\limits_S f(r, \theta, z) r \, dz \, dr \, d\theta = \iint\limits_R \left[\int_{g_1(r, \theta)}^{g_2(r, \theta)} f(r, \theta, z) \, dz \right] r \, dr \, d\theta$$

where R is the projection of the solid in the polar plane. From a sketch of R, the limits for r and θ can be obtained as though we were dealing with a double integral in polar coordinates. For instance, in Example 4, S is bounded below by $z = x^2 + y^2 = r^2$ and above by $z = 4$. Furthermore, the region R is given by $x^2 + y^2 = 4 = r^2$. Therefore, we have

$$\iiint\limits_S k(x^2 + y^2)^{3/2} \, dV = \iint\limits_R \left[\int_{x^2+y^2}^4 k(x^2 + y^2)^{3/2} \, dz \right] r \, dr \, d\theta$$

$$= \int_0^{2\pi} \int_0^2 \left[\int_{r^2}^4 k(r^2)^{3/2} \, dz \right] r \, dr \, d\theta$$

$$= \int_0^{2\pi} \int_0^2 \int_{r^2}^4 kr^4 \, dz \, dr \, d\theta$$

We continue our discussion of multiple integrals by considering triple integrals in spherical coordinates. Recall from Section 14.6 that

$$x = \rho \sin \phi \cos \theta, \qquad y = \rho \sin \phi \sin \theta, \qquad z = \rho \cos \phi$$

For solids in spherical coordinates, the fundamental element of volume, ΔV, is the "spherical block" shown in Figure 16.43. If we let $(\bar{\rho}_i, \bar{\theta}_i, \bar{\phi}_i)$ be *any point* in the block, then

$$\Delta V_i \approx \bar{\rho}_i^2 \sin \phi_i \, \Delta \rho_i \, \Delta \theta_i \, \Delta \phi_i$$

Taking the limit of a sum in the usual way, we write the triple integral of f on S as

$$\iiint\limits_S f(\rho, \theta, \phi) \, dV = \iiint\limits_S f(\rho, \theta, \phi) \rho^2 \sin \phi \, d\rho \, d\theta \, d\phi$$

provided f is continuous on S. There are five other possible orders of integration.

The symbol ρ in spherical coordinates is not to be confused with density. For problems involving spherical coordinates and a density function, we will use a different symbol to denote density.

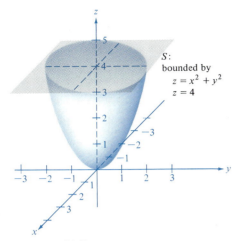

FIGURE 16.42

S:
bounded by
$z = x^2 + y^2$
$z = 4$

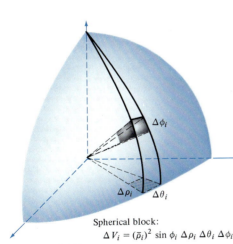

Spherical block:
$\Delta V_i = (\bar{\rho}_i)^2 \sin \phi_i \, \Delta \rho_i \, \Delta \theta_i \, \Delta \phi_i$

FIGURE 16.43

Example 5

Find the volume and the center of mass of the solid region bounded above by the sphere $x^2 + y^2 + z^2 = 9$ and below by the upper nappe of the cone $z^2 = x^2 + y^2$. Assume the solid is of uniform density. (See Figure 16.44.)

Solution: The equation of the sphere in spherical coordinates is

$$\rho^2 = x^2 + y^2 + z^2 = 9 \qquad \text{or} \qquad \rho = 3$$

Furthermore, the sphere and cone intersect when

$$(x^2 + y^2) + z^2 = (z^2) + z^2 = 9 \qquad \text{or} \qquad z = \frac{3}{\sqrt{2}}$$

Therefore, since $z = \rho \cos \phi$, it follows that

$$\left(\frac{3}{\sqrt{2}}\right)\left(\frac{1}{3}\right) = \cos \phi \qquad \text{or} \qquad \phi = \frac{\pi}{4}$$

Consequently, the volume is

$$V = \iiint_S dV = \int_0^{\pi/4} \int_0^{2\pi} \int_0^3 \rho^2 \sin \phi \, d\rho \, d\theta \, d\phi$$

$$= \int_0^{\pi/4} \int_0^{2\pi} 9 \sin \phi \, d\theta \, d\phi = \int_0^{\pi/4} 18\pi \sin \phi \, d\phi$$

$$= -18\pi \cos \phi \Big]_0^{\pi/4} = 18\pi - 9\sqrt{2}\pi \approx 16.56$$

By symmetry, the center of mass is on the z-axis and we need only calculate $\bar{z} = M_{xy}/V$. Since $z = \rho \cos \phi$, it follows that

$$M_{xy} = \iiint_S z \, dV = \int_0^{\pi/4} \int_0^{2\pi} \int_0^3 (\rho \cos \phi)\rho^2 \sin \phi \, d\rho \, d\theta \, d\phi$$

$$= \int_0^{\pi/4} \int_0^{2\pi} \frac{(3)^4}{4} \cos \phi \sin \phi \, d\theta \, d\phi$$

$$= \frac{81}{4} \int_0^{\pi/4} 2\pi \cos \phi \sin \phi \, d\phi = \left(\frac{81\pi}{2}\right) \frac{\sin^2 \phi}{2} \Big]_0^{\pi/4} = \frac{81\pi}{8}$$

Consequently,

$$\bar{z} = \frac{81\pi}{8\pi(18 - 9\sqrt{2})} = \frac{9}{8(2 - \sqrt{2})} = \frac{9(2 + \sqrt{2})}{16} \approx 1.92$$

from which it follows that the center of mass is at the point $(0, 0, 1.92)$ in rectangular coordinates.

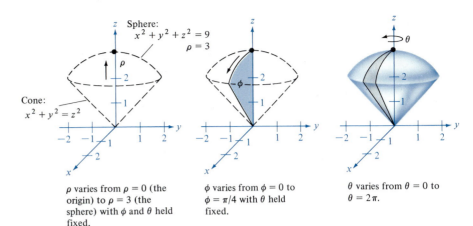

ρ varies from $\rho = 0$ (the origin) to $\rho = 3$ (the sphere) with ϕ and θ held fixed.

ϕ varies from $\phi = 0$ to $\phi = \pi/4$ with θ held fixed.

θ varies from $\theta = 0$ to $\theta = 2\pi$.

FIGURE 16.44 ∎

We conclude this section with a summary of multiple integrals in nonrectangular coordinates.

Summary of Multiple Integrals in Nonrectangular Coordinates

Polar coordinates: If f is continuous on the plane region R, then

$$\iint_R f(r, \theta)\, dA = \lim_{\|\Delta\|\to 0} \sum_{i=1}^{n} f(\bar{r}_i, \bar{\theta}_i)\bar{r}_i\, \Delta r_i\, \Delta \theta_i = \iint_R f(r, \theta)r\, dr\, d\theta$$

(or $dA = r\, d\theta\, dr$).

Cylindrical coordinates: If f is continuous on the solid region S, then

$$\iiint_S f(r, \theta, z)\, dV = \lim_{\|\Delta\|\to 0} \sum_{i=1}^{n} f(r_i, \theta_i, z_i)\, \Delta V_i = \iiint_S f(r, \theta, z)r\, dr\, d\theta\, dz$$

(or $dV = r\, dr\, dz\, d\theta$, $r\, dr\, d\theta\, dz$, etc.).

Spherical coordinates: If f is continuous on the solid region S, then

$$\iiint_S f(\rho, \theta, \phi)\, dV = \lim_{\|\Delta\|\to 0} \sum_{i=1}^{n} f(\rho_i, \theta_i, \phi_i)\, \Delta V_i = \iiint_S f(\rho, \theta, \phi)\rho^2 \sin\phi\, d\rho\, d\theta\, d\phi$$

(or $dV = \rho^2 \sin\phi\, d\theta\, d\phi\, d\rho$, etc.).

Section Exercises (16.6)

In Exercises 1–10, use polar coordinates to evaluate the integral $\iint_R f(x, y)\, dA$.

1. $f(x, y) = x + y$; R: $x^2 + y^2 \le 4$, $0 \le x$, $0 \le y$

2. $f(x, y) = e^{-(x^2+y^2)}$; R: $x^2 + y^2 \le 4$, $0 \le x$, $0 \le y$

⬜ **3.** $f(x, y) = \text{Arctan } \dfrac{y}{x}$; R: $x^2 + y^2 \le 1$, $0 \le x$, $0 \le y$

4. $f(x, y) = 9 - x^2 - y^2$; R: $x^2 + y^2 \le 9$, $0 \le x$, $0 \le y$

5. Find the volume under the surface $f(x, y) = xy$ and over the region bounded by $x^2 + y^2 \le 1$, $0 \le x$, $0 \le y$.

6. Find the volume under the paraboloid $f(x, y) = x^2 + y^2 + 1$ and inside the cylinder $x^2 + y^2 = 4$.

[S] 7. Find the surface area of the paraboloid $f(x, y) = x^2 + y^2$ for $0 \le z \le 1$.

8. Find the surface area of the cone $z^2 = x^2 + y^2$ for $0 \le z \le 1$.

9. Find the surface area of the hemisphere $f(x, y) = \sqrt{a^2 - x^2 - y^2}$ inside the cylinder $x^2 + y^2 = b^2 (b < a)$.

10. A hole of radius b inches is drilled through the center of a sphere of radius R inches. Find the volume of the remaining portion of the sphere.

In Exercises 11–16, sketch the solid region whose volume is given by the integral.

11. $8 \int_0^{\pi/2} \int_0^3 \sqrt{9 - r^2} \, r \, dr \, d\theta$

12. $\int_0^{\pi/2} \int_0^3 r e^{-r^2} \, dr \, d\theta$

[S] 13. $\int_0^{\pi/2} \int_0^3 \int_0^{e^{-r^2}} r \, dz \, dr \, d\theta$

14. $\int_0^{2\pi} \int_0^{\sqrt 3} \int_0^{3-r^2} r \, dz \, dr \, d\theta$

15. $4 \int_0^{\pi/2} \int_{\pi/6}^{\pi/2} \int_0^4 \rho^2 \sin \phi \, d\rho \, d\phi \, d\theta$

16. $\int_0^{2\pi} \int_0^{\pi} \int_2^5 \rho^2 \sin \phi \, d\rho \, d\phi \, d\theta$

In Exercises 17–23, use cylindrical coordinates to evaluate each triple integral.

[S] 17. Find the volume of a right circular cone having altitude h and a circular base of radius r_0.

18. Determine the center of mass of the cone of Exercise 17 if its density at any point is proportional to the distance of the point from the base.

19. Determine the center of mass of the cone of Exercise 17 if its density at any point is proportional to the distance of the point from the axis of the cone.

20. Assuming that the cone of Exercise 17 has constant density, show that the moment of inertia about its axis is $\frac{3}{10} m r_0^2$.

21. Show that the moment of inertia about the axis of a cylindrical shell of constant density bounded by $x^2 + y^2 = a^2$, $x^2 + y^2 = b^2 (b > a)$, $z = 0$, and $z = h$ is $\frac{1}{2}m(a^2 + b^2)$.

22. For the right circular cylinder $r = 2a \sin \theta$, of height h and constant density k, show that I_z is $\frac{3}{2}ma^2$.

[S] 23. Find the volume of the solid inside both the sphere

$x^2 + y^2 + z^2 = a^2$ and the cylinder $[x - (a/2)]^2 + y^2 = (a/2)^2$.

In Exercises 24–31, use spherical coordinates to evaluate each triple integral.

24. Find the mass of the solid inside the sphere $x^2 + y^2 + z^2 = a^2$ if its density at any point is proportional to the distance of the point from the origin.

25. Find the mass of the solid inside the sphere $x^2 + y^2 + z^2 = a^2$ if its density at a point is proportional to the distance of the point from the z-axis.

26. Find the center of mass of a solid of constant density lying between two concentric hemispheres of radius r and R, where $r < R$.

27. Find the center of mass of a hemispherical solid of radius r and constant density k.

28. Find the center of mass of the solid of constant density bounded by the hemisphere $\rho = \cos \phi$, $\pi/4 \le \phi \le \pi/2$, and the cone $\phi = \pi/4$.

29. Show that the moment of inertia about the diameter of a sphere of radius r, mass m, and constant density is $\frac{2}{5}mr^2$.

[S] 30. Find the volume of the solid inside the sphere $x^2 + y^2 + z^2 = 1$ and outside the cone $z^2 = x^2 + y^2$.

31. Find the moment of inertia about the z-axis of the solid in Exercise 28.

In Exercises 32–35, convert the integral from rectangular coordinates to (a) cylindrical coordinates and (b) spherical coordinates.

32. $\int_{-2}^{2} \int_{-\sqrt{4-x^2}}^{\sqrt{4-x^2}} \int_{x^2+y^2}^{4} x \, dz \, dy \, dx$

33. $\int_0^2 \int_0^{\sqrt{4-x^2}} \int_0^{\sqrt{16-x^2-y^2}} \sqrt{x^2 + y^2} \, dz \, dy \, dx$

34. $\int_0^1 \int_0^{\sqrt{1-x^2}} \int_0^{\sqrt{1-x^2-y^2}} \sqrt{x^2 + y^2 + z^2} \, dz \, dy \, dx$

[S] 35. $\int_{-a}^{a} \int_{-\sqrt{a^2-x^2}}^{\sqrt{a^2-x^2}} \int_a^{a+\sqrt{a^2-x^2-y^2}} x \, dz \, dy \, dx$

36. An important integral in probability is $\int_{-\infty}^{\infty} e^{-x^2/2} \, dx$. One way to evaluate this integral is to let $I = \int_0^{\infty} e^{-x^2/2} \, dx$, and then

$$I^2 = \left(\int_0^{\infty} e^{-x^2/2} \, dx \right)\left(\int_0^{\infty} e^{-y^2/2} \, dy \right)$$

$$= \int_0^{\infty} \int_0^{\infty} e^{-(x^2+y^2)/2} \, dA$$

Use polar coordinates to evaluate this double integral and thus determine I^2 and I.

Miscellaneous Exercises (Ch. 16)

In Exercises 1–6, evaluate each double integral.

S 1. $\displaystyle\int_0^1 \int_0^{1+x} (3x + 2y)\, dy\, dx$ **2.** $\displaystyle\int_0^2 \int_{x^2}^{2x} (x^2 + 2y)\, dy\, dx$

3. $\displaystyle\int_0^3 \int_0^{\sqrt{9-x^2}} 4x\, dy\, dx$ **4.** $\displaystyle\int_0^{\sqrt{3}} \int_{2-\sqrt{4-y^2}}^{2+\sqrt{4-y^2}} dx\, dy$

5. $\displaystyle\int_{-2}^4 \int_{y^2/4}^{(4+y)/2} (x - y)\, dx\, dy$

6. $\displaystyle\int_{-2}^2 \int_0^{4-y^2} (8x - 2y^2)\, dx\, dy$

In Exercises 7–10, write the limits to the double integral $\iint_R f(x, y)\, dA$ for both orders of integration. Compute the area by letting $f(x, y) = 1$.

7. R is a triangle with vertices $(0, 0)$, $(3, 0)$, and $(0, 1)$.

8. R is a triangle with vertices $(0, 0)$, $(3, 0)$, and $(2, 2)$.

S 9. R is the larger region between the circle $x^2 + y^2 = 25$ and the line $x = 3$.

10. R is the region bounded by the parabolas $y = 6x - x^2$ and $y = x^2 - 2x$.

11. Use a double integral to find the area of the region enclosed by $y^2 = x^2 - x^4$.

12. Find the volume of the solid bounded by the planes $z = x + y$, $z = 0$, $x = 0$, $x = 3$, and $y = x$.

13. Find the volume of the solid bounded by $z = x^2 - y + 4$, $z = 0$, $x = 0$, and $x = 4$.

14. Find the volume of the solid outside the cylinder $x^2 + y^2 = 1$ and inside the hyperboloid $x^2 + y^2 - z^2 = 1$ between the planes $z = 0$ and $z = h$. Use cylindrical coordinates.

S 15. Find the surface area of the solid bounded by $z = 16 - x^2 - y^2$ and $z = 0$. (Use polar coordinates.)

16. Find the center of mass of a wedge of constant density bounded by the cylinder $x^2 + y^2 = a^2$ and the plane $z = cy\ (0 < c)$, where $0 \leq y$ and $0 \leq z$.

17. Evaluate the triple integral

$$\int_0^a \int_0^b \int_0^c (x^2 + y^2 + z^2)\, dx\, dy\, dz$$

18. Evaluate the triple integral

$$\int_0^2 \int_0^{\sqrt{4-x^2}} \int_0^{\sqrt{4-x^2-y^2}} xyz\, dz\, dy\, dx$$

19. Calculate (using cylindrical coordinates) the volume of the solid within the cylinder $r = 2\cos\theta$ and the sphere $r^2 + z^2 = 4$.

20. Calculate (using cylindrical coordinates) the volume of the solid within the paraboloid $r^2 + z = 16$ and the cylinder $r = 2\sin\theta$ above the xy-plane.

21. Give a geometrical interpretation of

$$\int_0^{2\pi} \int_{-\pi/2}^{\pi/2} \int_0^{6\sin\phi} \rho^2 \sin\phi\, d\rho\, d\phi\, d\theta$$

In Exercises 22–28, evaluate each integral using the coordinate system that makes the integration easiest.

22. $\displaystyle\int_0^4 \int_0^{\sqrt{16-y^2}} (x^2 + y^2)\, dx\, dy$

S 23. $\displaystyle\int_0^h \int_0^x \sqrt{x^2 + y^2}\, dy\, dx$

24. $\displaystyle\int_{-2}^2 \int_{-\sqrt{4-x^2}}^{\sqrt{4-x^2}} \int_0^{(x^2+y^2)/2} (x^2 + y^2)\, dz\, dy\, dx$

25. $\displaystyle\int_{-3}^3 \int_{-\sqrt{9-x^2}}^{\sqrt{9-x^2}} \int_{x^2+y^2}^9 \sqrt{x^2 + y^2}\, dz\, dy\, dx$

26. $\displaystyle\int_0^5 \int_0^{\sqrt{25-x^2}} \int_0^{\sqrt{25-x^2-y^2}} \frac{1}{\sqrt{x^2 + y^2 + z^2}}\, dz\, dy\, dx$

S 27. Show that the center of mass of the portion of the solid of constant density in the first octant bounded by $x^2 + y^2 + z^2 = a^2$ is $(3a/8, 3a/8, 3a/8)$.

28. Show that the moment of inertia of a sphere of radius a with respect to a line passing through its center is $4\pi ka^6/9$, if the density is proportional to the distance from the center.

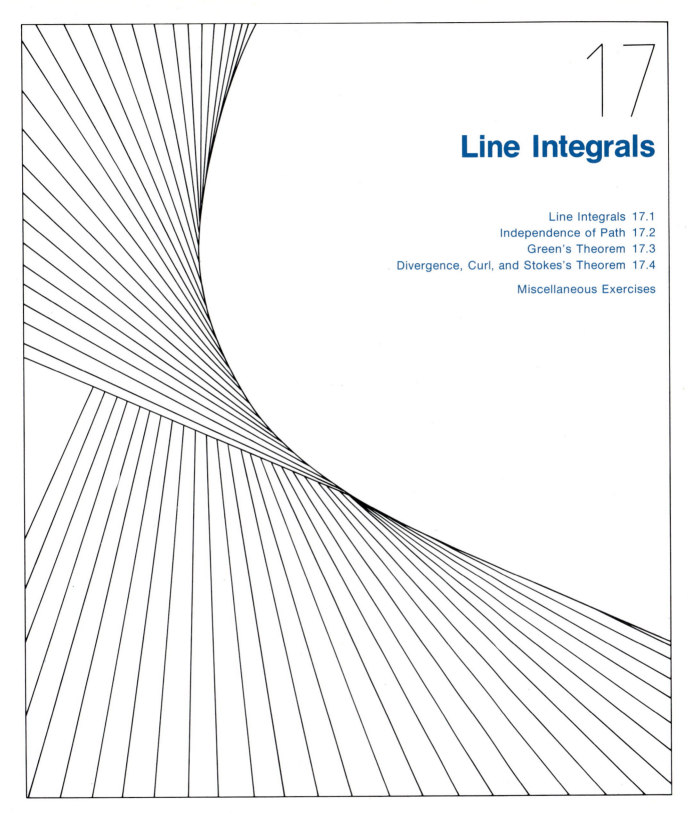

17

Line Integrals

17.1 **Line Integrals**

Purpose
- To define a line integral.
- To introduce the parametric, vector, and differential forms of a line integral.
- To use line integrals to calculate work done in a force field.

The truth is, the science of Nature has been already too long made only a work of the brain and the fancy: It is now high time that it should return to the plainness and soundness of observations on material and obvious things. —*Robert Hooke (1635–1703)*

Recall from our definition of the definite integral that we began with the geometric model of the area of a region under a curve. Later we encountered numerous other applications of definite integrals, which, though unrelated to area, nevertheless involved the limit of a sum *along a line* (usually the *x*-axis). In this chapter we follow this same pattern with line integrals *along a curve*. That is, we begin by using lateral surface area (see Section 16.3) as a model for defining line integrals. Then we extend the use of line integrals to *other* applications that also involve the limit of a sum along a curve. (Note that the term "line integral" is somewhat misleading in that the integration actually takes place along a curve rather than a line.)

Before proceeding we make note that throughout this chapter we will assume that all curves are **smooth.** That is, if a curve *C* is given by the parametric equations

$$x = f(t), \quad y = g(t) \qquad \text{with} \qquad a \le t \le b$$

then *C* is considered **smooth** if f' and g' are continuous on $[a, b]$.

Consider a (smooth) curve *C* and a function $z = f(x, y)$ that is continuous and nonnegative on a region containing *C*. (See Figure 17.1.) Suppose we partition the curve *C* from *P* to *Q* into *n* subarcs of length Δs_i. Then choosing a point (x_i, y_i) in each subarc, we form a rectangle of height $f(x_i, y_i)$ and width Δs_i. Now we approximate the lateral surface area *L* of the region between *f* and *C* by summing the areas of these rectangles

$$L \approx \sum_{i=1}^{n} f(x_i, y_i) \, \Delta s_i$$

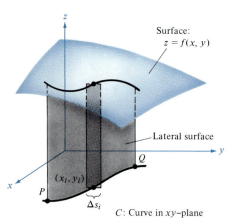

Surface:
$z = f(x, y)$

Lateral surface

(x_i, y_i)

Δs_i

C: Curve in *xy*–plane

FIGURE 17.1

Finally, if we let $\|\Delta\|$ represent the length of the largest subarc, we have

$$L = \lim_{\|\Delta\| \to 0} \sum_{i=1}^{n} f(x_i, y_i) \, \Delta s_i$$

We call this limit the **line integral** of *f* along *C* and denote it by

$$\int_C f(x, y) \, ds$$

Definition of Line Integral	Let f be a continuous function of x and y on a region containing a (smooth) curve C; then the **line integral of f along C from P to Q** is given by

$$\int_C f(x, y)\, ds = \lim_{\|\Delta\|\to 0} \sum_{i=1}^{n} f(x_i, y_i)\, \Delta s_i$$

As with other integrals, we do not usually evaluate line integrals by the limit definition. For curves represented parametrically, we use the following rule for converting line integrals to definite integrals.

Parametric Evaluation of a Line Integral	If the curve C is given by the parametric equations

$$x = g(t) \qquad \text{and} \qquad y = h(t)$$

where $a \leq t \leq b$, then

$$\int_C f(x, y)\, ds = \int_a^b f(g(t), h(t)) \sqrt{[g'(t)]^2 + [h'(t)]^2}\, dt$$

This definition and evaluation procedure can be directly extended to a space curve C defined by

$$x = g(t), \qquad y = h(t), \qquad z = k(t)$$

where

$$ds = \sqrt{[g'(t)]^2 + [h'(t)]^2 + [k'(t)]^2}\, dt$$

Example 1

Evaluate the line integral $\int_C xy\, ds$, where C is described by the parametric equations $x = 4 - t$ and $y = t$, with $0 \leq t \leq 2$.

Solution: Since $x'(t) = -1$ and $y'(t) = 1$, we have

$$\int_C xy\, ds = \int_0^2 (4 - t)t \sqrt{(-1)^2 + (1)^2}\, dt = \sqrt{2} \int_0^2 (4t - t^2)\, dt$$

$$= \sqrt{2} \left[2t^2 - \frac{t^3}{3} \right]_0^2 = \frac{16\sqrt{2}}{3} \qquad \blacksquare$$

Since line integrals are defined as the limit of a sum, we would expect them to share the basic properties of definite integrals given in Section 6.4. Indeed, this is the case, and we list these properties without verification.

Properties of Line Integrals

1. $\displaystyle\int_C kf(x,y)\,ds = k\int_C f(x,y)\,ds$, where k is a constant.

2. $\displaystyle\int_C [f(x,y) \pm g(x,y)]\,ds = \int_C f(x,y)\,ds \pm \int_C g(x,y)\,ds$

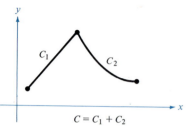

$C = C_1 + C_2$

Piecewise Smooth Curve

FIGURE 17.2

On occasion it is convenient to extend the use of line integrals from smooth curves to composite curves, which are called **paths** (or **piecewise smooth curves**). Specifically, if C_1 and C_2 are smooth curves such that the terminal point of C_1 coincides with the initial point of C_2, then the composite of C_1 and C_2 is a path, and we denote it by $C = C_1 + C_2$ (see Figure 17.2). This concept can, of course, be generalized to composites of three or more smooth curves.

Line Integrals Along Paths

If a path C is the composite of curves C_1 and C_2, then

$$\int_C f(x,y)\,ds = \int_{C_1} f(x,y)\,ds + \int_{C_2} f(x,y)\,ds$$

Example 2

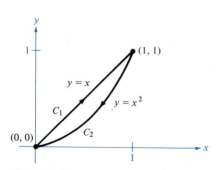

FIGURE 17.3

Evaluate the line integral $\int_C x\,ds$ along the path $C = C_1 + C_2$, where

$$C_1: x = t, \quad y = t \qquad C_2: x = 1 - t, \quad y = (1-t)^2$$
$$0 \le t \le 1 \qquad\qquad 0 \le t \le 1$$

Solution: The path is shown in Figure 17.3, where we first integrate up the line $y = x$, then back down the parabola $y = x^2$. For C_1 we have $ds = \sqrt{2}\,dt$, and for C_2 we have $ds = \sqrt{1 + 4(1-t)^2}\,dt$. Therefore,

$$\int_C x\,ds = \int_{C_1} x\,ds + \int_{C_2} x\,ds = \int_0^1 t\sqrt{2}\,dt + \int_0^1 (1-t)\sqrt{1 + 4(1-t)^2}\,dt$$

$$= \frac{\sqrt{2}}{2}t^2 \Big]_0^1 - \frac{1}{8}\left(\frac{2}{3}\right)[1 + 4(1-t)^2]^{3/2}\Big]_0^1 = \frac{\sqrt{2}}{2} + \frac{1}{12}(5^{3/2} - 1)$$

$$\approx 1.56 \qquad\qquad\blacksquare$$

In applications, line integrals often occur in the form

$$\int_C M(x,y)\,dx + \int_C N(x,y)\,dy$$

where the first integral is called the **line integral of M along C with respect** to *x* and the second is called the **line integral of *N* along *C* with respect to *y*.** This sum is usually written in the condensed form

$$\int_C M(x, y)\, dx + N(x, y)\, dy$$

Example 3

Evaluate $\int_C y\, dx + xy\, dy$ if *C* is the parabolic curve $x = y^2$ from (1, 1) to (4, 2).

Solution: In determining a set of parametric equations for *C*, we must preserve the given direction along the curve. (See Figure 17.4.) Thus for increasing *t*, a set of parametric equations is

$$x = t^2, \quad y = t \qquad \text{with} \qquad 1 \le t \le 2$$

Thus $dx = 2t\, dt$, $dy = dt$, and we have

$$\int_C y\, dx + xy\, dy = \int_1^2 t^2(2t)\, dt + t^3\, dt = \int_1^2 3t^3\, dt = \frac{3t^4}{4}\Big]_1^2 = \frac{45}{4} \quad \blacksquare$$

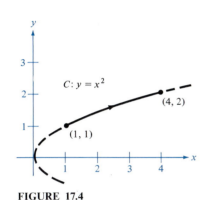

FIGURE 17.4

It is worth noting that line integrals in the form

$$\int_C M(x, y)\, dx + N(x, y)\, dy$$

often lend themselves to evaluation using a rectangular coordinate representation of *C* rather than a parametric one. For instance, in Example 3, since *x* is a function of *y*, we can readily convert the line integral to variable *y* alone.

$$x = y^2, \qquad dx = 2y\, dy, \qquad \text{and} \qquad 1 \le y \le 2$$

from which it follows that

$$\int_C y^2\, dx + xy\, dy = \int_1^2 y^2(2y\, dy) + y^3\, dy = \int_1^2 3y^3\, dy = \frac{45}{4}$$

Example 4

Evaluate $\int_C y\, dx + x^2\, dy$, where *C* is the parabola $y = 4x - x^2$ from (4, 0) to (1, 3). (See Figure 17.5.)

Solution: Since $y = 4x - x^2$, $dy = (4 - 2x)\, dx$, and since the specified direction means that *x* changes from $x = 4$ to $x = 1$, we write

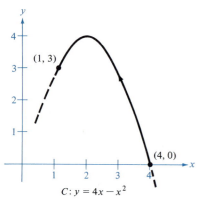

C: $y = 4x - x^2$

FIGURE 17.5

$$\int_C y \, dx + x^2 \, dy = \int_4^1 (4x - x^2) \, dx + x^2(4 - 2x) \, dx$$

$$= \int_4^1 (4x + 3x^2 - 2x^3) \, dx = \left[2x^2 + x^3 - \frac{x^4}{2} \right]_4^1 = 34\tfrac{1}{2}$$

Note that a possible parametric representation would be

$$x = 4 - t, \quad y = (4 - t)t \qquad \text{with} \qquad 0 \le t \le 3$$

You should evaluate the line integral in this parametric form and compare answers. ∎

If we had moved clockwise along C in Example 4 (Figure 17.5), the results would have changed signs. In general, if curve C is traced in one direction, then the curve traced in the opposite direction is denoted $-C$, and we have

$$\int_{-C} M(x, y) \, dx + N(x, y) \, dy = - \int_C M(x, y) \, dx + N(x, y) \, dy$$

One of the most important physical applications of line integrals is in calculating work done in a **force field.** Figure 17.6 shows an inverse square force field similar to the gravitational field of the sun. Note that for this particular force field the magnitude of the force along a circular path about the sun is constant, whereas the magnitude of the force along a parabolic path varies from point to point. Force fields can be represented as vector functions, as follows:

$$\mathbf{F}(x, y) = M(x, y)\mathbf{i} + N(x, y)\mathbf{j} \qquad \text{(in the plane)}$$

$$\mathbf{F}(x, y, z) = M(x, y, z)\mathbf{i} + N(x, y, z)\mathbf{j} + P(x, y, z)\mathbf{k} \quad \text{(in 3-space)}$$

To see how a line integral can be used to find work done in a force field, consider an object moving along a path in the field, as shown in

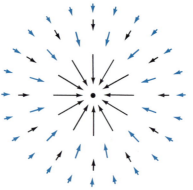

Inverse Square Force Field F

Vectors along a Parabolic Path in the Force Field F

FIGURE 17.6

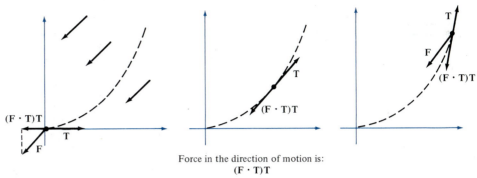

Force in the direction of motion is:
$$(\mathbf{F} \cdot \mathbf{T})\mathbf{T}$$

FIGURE 17.7

Figure 17.7. To determine the work done by the force, we need consider only that part of the force that is acting in the same (or opposite) direction as that in which the object is moving. In vector terms this means that at each point on the path, we consider the projection of the force vector onto the tangent vector. Specifically, if we let $\mathbf{T}$ be the unit tangent vector to the path C, then the component of the force $\mathbf{F}$ in the direction of motion is $\mathbf{F} \cdot \mathbf{T}$. Therefore, the total work done as the object moves along C is

$$W = \int_C \mathbf{F} \cdot \mathbf{T} \, ds$$

Example 5

Find the work done by the constant force $\mathbf{F} = -\mathbf{i} - \mathbf{j}$ acting on an object that moves along the path given by $\mathbf{R}(t) = t\mathbf{i} + t^2\mathbf{j}$ from $(0, 0)$ to $(2, 4)$.

Solution: Since
$$\mathbf{T} = \frac{\mathbf{R}'}{|\mathbf{R}'|}$$

we have
$$\mathbf{T} = \frac{\mathbf{i} + 2t\mathbf{j}}{\sqrt{1 + 4t^2}}$$

And since $x = t$ and $y = t^2$, we have

$$ds = \sqrt{\left(\frac{dx}{dt}\right)^2 + \left(\frac{dy}{dt}\right)^2} \, dt = \sqrt{1 + 4t^2} \, dt$$

Therefore,

$$W = \int_C \mathbf{F} \cdot \mathbf{T} \, ds = \int_0^2 (-\mathbf{i} - \mathbf{j}) \left(\frac{\mathbf{i} + 2t\mathbf{j}}{\sqrt{1 + 4t^2}}\right) \sqrt{1 + 4t^2} \, dt$$

$$= \int_0^2 (-1 - 2t) \, dt = \left[-t - t^2\right]_0^2 = -6$$

(The units for W depend on the units of distance and force.) ∎

Perhaps the negative result in Example 5 is a bit disturbing to you. Negative values for the work done by a force occur when the object is moving *against* the force, that is, when the projection $(\mathbf{F} \cdot \mathbf{T})\mathbf{T}$ has the opposite direction of $\mathbf{T}$, as shown in Figure 17.7.

From Example 5 we also observe that when a curve C is described by a vector function $\mathbf{R}(t)$, the line integral for work has the **vector form**

$$W = \int_C \mathbf{F} \cdot \mathbf{T} \, ds = \int_C \mathbf{F} \cdot \frac{\mathbf{R}'}{|\mathbf{R}'|} |\mathbf{R}'| \, dt = \int_C \mathbf{F} \cdot \mathbf{R}' \, dt = \int_C \mathbf{F} \cdot d\mathbf{R}$$

From this vector form of the line integral for work, we can obtain the following two alternative forms. For instance, if

$$\mathbf{R}(t) = x(t)\mathbf{i} + y(t)\mathbf{j} \qquad \text{and} \qquad \mathbf{F}(t) = f(t)\mathbf{i} + g(t)\mathbf{j}$$

then

$$d\mathbf{R} = \left(\frac{dx}{dt}\mathbf{i} + \frac{dy}{dt}\mathbf{j} \right) dt$$

and we obtain the form

$$\int_C \mathbf{F} \cdot \mathbf{T} \, ds = \int_C \mathbf{F} \cdot d\mathbf{R} = \int_C [f(t)\mathbf{i} + g(t)\mathbf{j}] \cdot \left(\frac{dx}{dt}\mathbf{i} + \frac{dy}{dt}\mathbf{j} \right) dt$$

$$= \int_C \left[f(t) \frac{dx}{dt} + g(t) \frac{dy}{dt} \right] dt$$

which we call the **parametric form**. If

$$\mathbf{R}(t) = x(t)\mathbf{i} + y(t)\mathbf{j} \qquad \text{and} \qquad \mathbf{F}(x, y) = M(x, y)\mathbf{i} + N(x, y)\mathbf{j}$$

then we have the **differential form**:

$$\int_C \mathbf{F} \cdot \mathbf{T} \, ds = \int_C \mathbf{F} \cdot d\mathbf{R} = \int_C [M(x, y)\mathbf{i} + N(x, y)\mathbf{j}] \cdot \left(\frac{dx}{dt}\mathbf{i} + \frac{dy}{dt}\mathbf{j} \right) dt$$

$$= \int_C M(x, y) \, dx + N(x, y) \, dy$$

Forms of the Line Integral
$\int_C \mathbf{F} \cdot \mathbf{T} \, ds$

Vector form: $\displaystyle\int_C \mathbf{F} \cdot d\mathbf{R}$

Parametric form: $\displaystyle\int_C \left[f(t) \frac{dx}{dt} + g(t) \frac{dy}{dt} \right] dt$

Differential form: $\displaystyle\int_C M(x, y) \, dx + N(x, y) \, dy$

Each of these forms is easily extended to three variables, as we demonstrate in the next example.

Example 6

Evaluate $\int_C \sqrt{y}\, dx - 2z\, dy + 3x\, dz$, where C is given by

$$x = t, \quad y = t^2, \quad z = t^3, \qquad \text{from} \qquad t = 1 \text{ to } t = 2$$

Solution: Since $dx = dt$, $dy = 2t\, dt$, and $dz = 3t^2\, dt$, we obtain

$$\int_C \sqrt{y}\, dx - 2z\, dy + 3x\, dz = \int_1^2 [t\, dt - 2t^3(2t\, dt) + 3t(3t^2\, dt)]$$

$$= \int_1^2 (t + 9t^3 - 4t^4)\, dt = \left[\frac{t^2}{2} + \frac{9t^4}{4} - \frac{4t^5}{5}\right]_1^2 = \frac{209}{20} = 10.45 \quad \blacksquare$$

Section Exercises (17.1)

In Exercises 1–6, evaluate $\int_C (x^2 + y^2)\, ds$ along the specified path.

S **1.** the line from $(0, 0)$ to $(1, 1)$

2. the line from $(0, 0)$ to $(3, 9)$

S **3.** counterclockwise around the triangle with vertices $(0, 0)$, $(1, 0)$, and $(0, 1)$

4. counterclockwise around the square with vertices $(0, 0)$, $(1, 0)$, $(1, 1)$, and $(0, 1)$

5. counterclockwise around the circle $x^2 + y^2 = 1$ from $(1, 0)$ to $(0, 1)$

6. counterclockwise around the circle $x^2 + y^2 = 4$ from $(2, 0)$ to $(-2, 0)$

In Exercises 7–12, evaluate the integral

$$\int_C (2x - y)\, dx + (x + 3y)\, dy$$

along the given path.

7. the x-axis from $x = 0$ to $x = 5$

8. the y-axis from $y = 0$ to $y = 2$

9. the line segments from $(0, 0)$ to $(3, 0)$ and from $(3, 0)$ to $(3, 3)$

10. the line segments from $(0, 0)$ to $(0, -3)$ and from $(0, -3)$ to $(2, -3)$

S **11.** the parabolic path $x = t$, $y = 2t^2$ from $(0, 0)$ to $(2, 8)$

12. the elliptic path $x = 4 \sin t$, $y = 3 \cos t$ from $(0, 3)$ to $(4, 0)$

In Exercises 13–18, evaluate $\int_C \mathbf{F} \cdot d\mathbf{R}$ for the given values of $\mathbf{F}$ and C.

13. $\mathbf{F} = xy\mathbf{i} + y\mathbf{j}$; C: $\mathbf{R} = 4t\mathbf{i} + t\mathbf{j}$, $0 \le t \le 1$

14. $\mathbf{F} = xy\mathbf{i} + y\mathbf{j}$; C: $\mathbf{R} = (4 \cos t)\mathbf{i} + (4 \sin t)\mathbf{j}$, $0 \le t \le \pi/2$

S **15.** $\mathbf{F} = 3x\mathbf{i} + 4y\mathbf{j}$; C: $\mathbf{R} = (2 \cos t)\mathbf{i} + (2 \sin t)\mathbf{j}$, $0 \le t \le \pi/2$

16. $\mathbf{F} = 3x\mathbf{i} + 4y\mathbf{j}$; C: $\mathbf{R} = t\mathbf{i} + \sqrt{4 - t^2}\,\mathbf{j}$, $-2 \le t \le 2$

17. $\mathbf{F} = x^2y\mathbf{i} + (x - z)\mathbf{j} + xyz\mathbf{k}$; C: $\mathbf{R} = t\mathbf{i} + t^2\mathbf{j} + 2\mathbf{k}$, $0 \le t \le 1$

18. $\mathbf{F} = x^2\mathbf{i} + y^2\mathbf{j} + z^2\mathbf{k}$; C: $\mathbf{R} = (\sin t)\mathbf{i} + (\cos t)\mathbf{j} + t^2\mathbf{k}$, $0 \le t \le \pi/2$

In Exercises 19–24, find the work done by the force field on an object moving along the given path.

S **19.** $\mathbf{F} = -x\mathbf{i} - 2y\mathbf{j}$, along $y = x^3$ from $(0, 0)$ to $(2, 8)$

20. $\mathbf{F} = x^2\mathbf{i} - xy\mathbf{j}$, along the hypocycloid $x = \cos^3 t$, $y = \sin^3 t$ from $(1, 0)$ to $(0, 1)$

21. $\mathbf{F} = 2x\mathbf{i} + y\mathbf{j}$, counterclockwise around the triangle whose vertices are $(0, 0)$, $(1, 0)$, and $(1, 1)$

22. $\mathbf{F} = -y\mathbf{i} - x\mathbf{j}$, along the semicircle passing through $(-2, 0)$, $(0, 2)$, and $(2, 0)$

23. $\mathbf{F} = x\mathbf{i} + y\mathbf{j} - 5z\mathbf{k}$; C: $\mathbf{R} = (2 \cos t)\mathbf{i} + (2 \sin t)\mathbf{j} + t\mathbf{k}$, $0 \le t \le 2\pi$

24. $\mathbf{F} = yz\mathbf{i} + xz\mathbf{j} + xy\mathbf{k}$, along the line from $(0, 0, 0)$ to $(5, 3, 2)$

If on the curve C given by $\mathbf{R} = x(t)\mathbf{i} + y(t)\mathbf{j}$, the tangent vector $\mathbf{T}$ to $\mathbf{R}$ is orthogonal to the force vector

$$\mathbf{F} = M(x, y)\mathbf{i} + N(x, y)\mathbf{j}$$

then $\int_C \mathbf{F} \cdot d\mathbf{R} = 0$ regardless of the initial and terminal points of C. Demonstrate this property in Exercises 25–28.

25. $\mathbf{F} = y\mathbf{i} - x\mathbf{j}$; C: $\mathbf{R} = t\mathbf{i} - 2t\mathbf{j}$

26. $\mathbf{F} = -3y\mathbf{i} + x\mathbf{j}$; C: $\mathbf{R} = t\mathbf{i} + t^3\mathbf{j}$

S **27.** $\mathbf{F} = (x^3 - 2x^2)\mathbf{i} + [x - (y/2)]\mathbf{j}$; C: $\mathbf{R} = t\mathbf{i} + t^2\mathbf{j}$

28. $\mathbf{F} = x\mathbf{i} + y\mathbf{j}$; C: $\mathbf{R} = (3 \sin t)\mathbf{i} + (3 \cos t)\mathbf{j}$

29. The density of a wire (lying along curve C) is described by f, where $f(x, y, z)$ is the mass per unit length at the point (x, y, z). The total mass of the wire is given by

$$\int_C f(x, y, z)\, ds$$

Compute the mass of two coils of a spring in the shape of a helix whose vector equation is

$$\mathbf{R} = (3 \cos t)\mathbf{i} + (3 \sin t)\mathbf{j} + 2t\mathbf{k}$$

and the density is

$$f(x, y, z) = \tfrac{1}{2}(x^2 + y^2 + z^2)$$

For Review and Class Discussion

True or False

1. _____ If C is given by $x(t) = t$, $y(t) = t$, $0 \le t \le 1$, then $\int_C xy\, ds = \int_0^1 t^2\, dt$.

2. _____ If $C_2 = -C_1$, then $\int_{C_1} f(x, y)\, ds + \int_{C_2} f(x, y)\, ds = 0$.

3. _____ The vector functions $\mathbf{R}_1 = t\mathbf{i} + t^2\mathbf{j}$, $0 \le t \le 1$, and $\mathbf{R}_2 = (1 - t)\mathbf{i} + (1 - t)^2\mathbf{j}$, $0 \le t \le 1$, define the same curve.

4. _____ If $\mathbf{F}$ and $\mathbf{T}$ are orthogonal, then $\int_C \mathbf{F} \cdot \mathbf{T}\, ds = 0$.

5. _____ If $\int_C \mathbf{F} \cdot \mathbf{T}\, ds = 0$, then $\mathbf{F}$ and $\mathbf{T}$ are orthogonal.

17.2 Independence of Path

Purpose
- To identify conditions under which the value of a line integral is independent of its parameter.
- To identify conditions under which the value of a line integral is independent of its path.

Mathematics has a triple end. It should furnish an instrument for the study of nature. Furthermore, it has a philosophic end, and, I venture to say, an end esthetic. —*Henri Poincaré (1854–1912)*

When we evaluate a line integral along a curve C, given parametrically, it might seem that the value of the line integral would change if we changed the parametric representation of C. For instance, suppose an object moves along C, first at a slow pace and then at a faster pace. Would the work required be different in these two cases? In Example 1 we will see that, at least under certain conditions, the amount of work is the same.

Example 1 (*Independence of Parameter*)

Suppose an object is moving through a force field given by $\mathbf{F} = 3xy\mathbf{i} + y\mathbf{j}$, where $|\mathbf{F}|$ is measured in pounds. Find the work done if the object moves along the following curves C, where the distance is measured in feet. (See Figure 17.8.)

(a) $\mathbf{R} = t\mathbf{i} + t^3\mathbf{j}$, $0 \le t \le 2$

(b) $\mathbf{R} = u^2\mathbf{i} + u^6\mathbf{j}, \; 0 \le u \le \sqrt{2}$
(c) $\mathbf{R} = (2 - w)\mathbf{i} + (2 - w)^3\mathbf{j}, \; 0 \le w \le 2$

Solution:

(a)   Since $x = t$, $y = t^3$, and $d\mathbf{R} = (\mathbf{i} + 3t^2\mathbf{j})\,dt$, we have

$$W = \int_C \mathbf{F} \cdot d\mathbf{R} = \int_C (3xy\mathbf{i} + y\mathbf{j}) \cdot d\mathbf{R} = \int_0^2 (3t^4\mathbf{i} + t^3\mathbf{j}) \cdot (\mathbf{i} + 3t^2\mathbf{j})\,dt$$

$$= \int_0^2 (3t^4 + 3t^5)\,dt = \left[\frac{3t^5}{5} + \frac{t^6}{2}\right]_0^2 = \frac{256}{5}\ \text{ft} \cdot \text{lb}$$

(b)   Since $x = u^2$, $y = u^6$, and $d\mathbf{R} = (2u\mathbf{i} + 6u^5\mathbf{j})\,du$, we have

$$W = \int_C \mathbf{F} \cdot d\mathbf{R} = \int_C (3xy\mathbf{i} + y\mathbf{j}) \cdot d\mathbf{R}$$

$$= \int_0^{\sqrt{2}} (3u^8\mathbf{i} + u^6\mathbf{j}) \cdot (2u\mathbf{i} + 6u^5\mathbf{j})\,du$$

$$= \int_0^{\sqrt{2}} (6u^9 + 6u^{11})\,du = \left[\frac{3u^{10}}{5} + \frac{u^{12}}{2}\right]_0^{\sqrt{2}} = \frac{256}{5}\ \text{ft} \cdot \text{lb}$$

(c)   Since $x = 2 - w$, $y = (2 - w)^3$, and $d\mathbf{R} = [-\mathbf{i} - 3(2 - w)^2\mathbf{j}]\,dw$, we have

$$W = \int_C \mathbf{F} \cdot d\mathbf{R} = \int_C (3xy\mathbf{i} + y\mathbf{j}) \cdot d\mathbf{R}$$

$$= \int_0^2 [3(2 - w)^4\mathbf{i} + (2 - w)^3\mathbf{j}] \cdot [-\mathbf{i} - 3(2 - w)^2\mathbf{j}]\,dw$$

$$= \int_0^2 [-3(2 - w)^4 - 3(2 - w)^5]\,dw$$

$$= \left[\frac{3(2 - w)^5}{5} + \frac{(2 - w)^6}{2}\right]_0^2 = -\frac{256}{5}\ \text{ft} \cdot \text{lb} \qquad \blacksquare$$

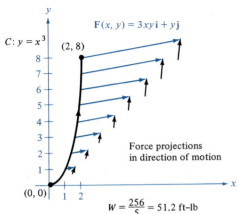

$C: y = x^3$

$(2, 8)$

$\mathbf{F}(x, y) = 3xy\mathbf{i} + y\mathbf{j}$

Force projections
in direction of motion

$(0, 0)$

$W = \frac{256}{5} = 51.2\ \text{ft–lb}$

FIGURE 17.8

Note in Example 1 that the work obtained with the third set of parametric equations has the opposite sign of that obtained with the first two sets. This occurred because on the interval $0 \le w \le 2$, $\mathbf{R} = (2 - w)\mathbf{i} + (2 - w)^3\mathbf{j}$ traces the curve in the opposite direction [from $(2, 8)$ to $(0, 0)$] to that in the first two parts.

We formalize the result suggested by Example 1 in the following theorem, which states that the value of the line integral along a curve C is *independent of the parameter* used, provided the direction along C is not changed.

Theorem 17.1	If the parametric representations $\mathbf{R}_1(t)$ and $\mathbf{R}_2(u)$ maintain the same di-
(Independence of Parameter)	rection along the curve C, then

$$\int_C \mathbf{F} \cdot d\mathbf{R}_1 = \int_C \mathbf{F} \cdot d\mathbf{R}_2$$

In Example 1 would the same amount of work be done if the object moved from $(0, 0)$ to $(2, 8)$ along a *different* curve? We examine this question in the next example.

Example 2

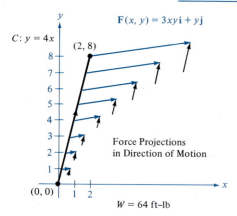

FIGURE 17.9

For the force field in Example 1, $\mathbf{F}(x, y) = 3xy\mathbf{i} + y\mathbf{j}$, find the work done on the object as it moves along the curve C given by

$$\mathbf{R}(t) = t\mathbf{i} + 4t\mathbf{j}, \qquad 0 \le t \le 2$$

(See Figure 17.9.)

Solution: In this case we have $x = t$, $y = 4t$, and $d\mathbf{R} = (\mathbf{i} + 4\mathbf{j})\, dt$, which yields

$$W = \int_C \mathbf{F} \cdot d\mathbf{R} = \int_C (3xy\mathbf{i} + y\mathbf{j}) \cdot d\mathbf{R}$$

$$= \int_0^2 (12t^2\mathbf{i} + 4t\mathbf{j}) \cdot (\mathbf{i} + 4\mathbf{j})\, dt$$

$$= \int_0^2 (12t^2 + 16t)\, dt = \left[4t^3 + 8t^2 \right]_0^2 = 64 \text{ ft} \cdot \text{lb} \quad \blacksquare$$

Based on the different results of Examples 1 and 2, it appears that the amount of work done on an object moving between two points in a force field can depend upon the path traveled. We should not find this result to be surprising. What may, in fact, be surprising is that there is a fairly large class of force fields for which the value of the line integral between two points in the field is *independent of the path* traveled. In such force fields the only restriction on the paths traveled from P to Q is that they all

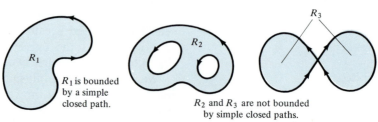

FIGURE 17.10

lie within a region whose boundary is a simple closed path. We say that a path is **simple** if it does not cross itself, and it is **closed** if it has the same initial and terminal point. Thus for practical purposes, a region whose boundary is a simple closed path is one that has no holes. (See Figure 17.10.)

Theorem 17.2 (*Independence of Path*)	In a region R bounded by a simple closed path, let C_1 and C_2 be two paths with the same initial and terminal points. If in region R, $\mathbf{F}(x, y) = M(x, y)\mathbf{i} + N(x, y)\mathbf{j}$ has the properties: i. $\dfrac{\partial M}{\partial y}$ and $\dfrac{\partial N}{\partial x}$ are continuous ii. $\dfrac{\partial M}{\partial y} = \dfrac{\partial N}{\partial x}$ then $$\int_{C_1} \mathbf{F} \cdot d\mathbf{R}_1 = \int_{C_2} \mathbf{F} \cdot d\mathbf{R}_2$$

Note that the result in Theorem 17.2 has the differential form

$$\int_{C_1} M(x, y)\, dx + N(x, y)\, dy = \int_{C_2} M(x, y)\, dx + N(x, y)\, dy$$

Also, keep in mind that this test for path independence involves the partials of *M with respect to y* and *N with respect to x;* not $\partial M/\partial x$ and $\partial N/\partial y$.

Example 3

Show

$$\int_{C_1} M\, dx + N\, dy = \int_{C_2} M\, dx + N\, dy$$

Let $\mathbf{F} = y\mathbf{i} + x\mathbf{j}$ and C_1 and C_2 be given by

$$C_1: x = t, \quad y = t, \qquad 0 \le t \le 1$$
$$C_2: x = t, \quad y = t^2, \qquad 0 \le t \le 1$$

Solution: From Figure 17.11 we see that C_1 and C_2 have the same initial and terminal points. Furthermore, since

$$\mathbf{F} = y\mathbf{i} + x\mathbf{j} = M(x, y)\mathbf{i} + N(x, y)\mathbf{j}$$

we see that

$$\frac{\partial M}{\partial y} = 1 \qquad \text{and} \qquad \frac{\partial N}{\partial x} = 1$$

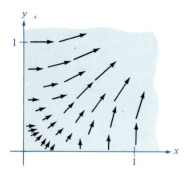

$\mathbf{F} = y\mathbf{i} + x\mathbf{j}$ is a path-independent force field.

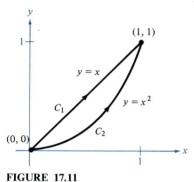

FIGURE 17.11

Thus from Theorem 17.2 we expect the line integrals of $\mathbf{F}$ along C_1 and C_2 to be equal. We verify this expectation as follows.

For C_1, $x = t$, $y = t$, $dx = dt$, and $dy = dt$. Thus we have

$$\int_{C_1} M\,dx + N\,dy = \int_{C_1} y\,dx + x\,dy = \int_0^1 [t\,dt + t\,dt]$$

$$= \int_0^1 2t\,dt = t^2 \Big]_0^1 = 1$$

For C_2, $x = t$, $y = t^2$, $dx = dt$, and $dy = 2t\,dt$. Thus we have

$$\int_{C_2} M\,dx + N\,dy = \int_{C_2} y\,dx + x\,dy = \int_0^1 [t^2\,dt + t(2t\,dt)]$$

$$= \int_0^1 3t^2\,dt = t^3 \Big]_0^1 = 1$$ ∎

Theorem 17.2 has the following important consequence. If $\int_C \mathbf{F} \cdot d\mathbf{R}$ is independent of path, then its value can be determined solely from the initial and terminal points of C. The next theorem gives us a method of doing just that. We call this theorem the *Fundamental Theorem of Line Integrals*.

Theorem 17.3
(Fundamental Theorem of Line Integrals)

If $\int_C \mathbf{F} \cdot d\mathbf{R}$ is path-independent in a region containing C, where (a_1, b_1) and (a_2, b_2) are the initial and terminal points of C, then there exists a function U such that

$$\mathbf{F} = \frac{\partial U}{\partial x}\mathbf{i} + \frac{\partial U}{\partial y}\mathbf{j} = \nabla U$$

and

$$\int_C \mathbf{F} \cdot d\mathbf{R} = U(a_2, b_2) - U(a_1, b_1)$$

Example 4

If $\mathbf{F} = (y + 1)\mathbf{i} + (x - 1)\mathbf{j}$, show that $\int_C \mathbf{F} \cdot d\mathbf{R}$ is path-independent and use the Fundamental Theorem of Line Integrals to evaluate this integral given that C has initial point $(2, -1)$ and terminal point $(3, 2)$.

Solution:

1. Since

$$\mathbf{F}(x, y) = (y + 1)\mathbf{i} + (x - 1)\mathbf{j} = M(x, y)\mathbf{i} + N(x, y)\mathbf{j}$$

we have

$$M = y + 1 \qquad N = x - 1$$

$$\frac{\partial M}{\partial y} = 1 \qquad \frac{\partial N}{\partial x} = 1$$

Therefore, $\int_C \mathbf{F} \cdot d\mathbf{R}$ is path-independent.

2. To apply the Fundamental Theorem, we must find $U(x, y)$ such that

$$\frac{\partial U}{\partial x} = M = y + 1 \qquad \text{and} \qquad \frac{\partial U}{\partial y} = N = x - 1$$

We begin with the *partial integration* (see Table 17.1) of M with respect to x and obtain

$$U = \int M \, dx + h(y) = \int (y + 1) \, dx + h(y) = xy + x + h(y)$$

$$\frac{\partial U}{\partial y} = x + h'(y)$$

Since we also know that

$$\frac{\partial U}{\partial y} = N = x - 1$$

we have

$$h'(y) = -1 \qquad \text{and} \qquad h(y) = -y + C$$

Finally, we conclude that

$$U(x, y) = xy + x + h(y) = xy + x - y + C$$

TABLE 17.1 Three Examples of Finding U such that $\partial U/\partial x = M$ and $\partial U/\partial y = N$

$\mathbf{F} = M\mathbf{i} + N\mathbf{j}$	$\mathbf{F} = 2xy\mathbf{i} + x^2\mathbf{j}$	$\mathbf{F} = y^3\mathbf{i} + (3xy^2 + 1)\mathbf{j}$	$\mathbf{F} = (2xe^y + y)\mathbf{i} + (x^2e^y + x + 2y)\mathbf{j}$
Test: $\dfrac{\partial M}{\partial y} = \dfrac{\partial N}{\partial x}$	$\dfrac{\partial M}{\partial y} = \dfrac{\partial N}{\partial x} = 2x$	$\dfrac{\partial M}{\partial y} = \dfrac{\partial N}{\partial x} = 3y^2$	$\dfrac{\partial M}{\partial y} = \dfrac{\partial N}{\partial x} = 2xe^y + 1$
$U = \int M \, dx + h(y)$	$U = x^2y + h(y)$	$U = xy^3 + h(y)$	$U = x^2e^y + xy + h(y)$
$\dfrac{\partial U}{\partial y} = N$	$x^2 + h'(y) = x^2$	$3xy^2 + h'(y) = 3xy^2 + 1$	$x^2e^y + x + h'(y) = x^2e^y + x + 2y$
$h'(y)$	$h'(y) = 0$	$h'(y) = 1$	$h'(y) = 2y$
$h(y)$	$h(y) = C$	$h(y) = y + C$	$h(y) = y^2 + C$
U	$U = x^2y + C$	$U = xy^3 + y + C$	$U = x^2e^y + xy + y^2 + C$

3.   By the Fundamental Theorem we have

$$\int_C \mathbf{F} \cdot d\mathbf{R} = U(2, -1) - U(3, 2)$$

$$= [2(-1) + 2 + 1 + C] - [3(2) + 3 - 2 + C] = -6$$ ■

Note from Example 4 that there are three basic steps in applying the Fundamental Theorem of Line Integrals:

1.   Test $\int_C \mathbf{F} \cdot d\mathbf{R}$ for path independence.
2.   Find U.
3.   Evaluate $U(a_2, b_2) - U(a_1, b_1)$.

Of these three steps the second is the most challenging, and we provide three additional examples of this step in Table 17.1.

Example 5

Let $\mathbf{F} = 2xy\mathbf{i} + x^2\mathbf{j}$, where C is any path from $(0, 0)$ to $(1, 4)$. Evaluate the line integral $\int_C \mathbf{F} \cdot d\mathbf{R}$.

Solution: From Table 17.1 we know that this integral is path-independent and $U = x^2y$. Thus by the Fundamental Theorem, we have

$$\int_C \mathbf{F} \cdot d\mathbf{R} = x^2y \Big]_{(0, 0)}^{(1, 4)} = 4$$ ■

We conclude this section with some observations about the Fundamental Theorem of Line Integrals. First, note that the vector function $\mathbf{F}$ is actually the *gradient* ∇U of the function U. In other words, as long as we restrict ourselves to regions bounded by simple closed paths, the force fields that are path-independent are precisely those that are the gradient of some function U. We call such force fields **conservative** and the function U is called the **potential function** for $\mathbf{F}$. Second, from the Fundamental Theorem we can conclude that if $\int_C \mathbf{F} \cdot d\mathbf{R}$ is path-independent and C is a *closed* path, then the value of the line integral is zero. This follows from the fact that if C starts and ends at (a, b), the value of the line integral is

$$\int_C \mathbf{F} \cdot d\mathbf{R} = U(a, b) - U(a, b) = 0$$

We summarize these observations as follows:

Properties of F and $\int_C \mathbf{F} \cdot d\mathbf{R}$

In a region R bounded by a simple closed path, we have the following:

i.   **F** is conservative if it is the gradient of some function U.

ii. $\int_C \mathbf{F} \cdot d\mathbf{R}$ is path-independent if and only if **F** is conservative.

iii. If C is a closed path and $\int_C \mathbf{F} \cdot d\mathbf{R}$ is path-independent, then $\int_C \mathbf{F} \cdot d\mathbf{R} = 0$.

Example 6

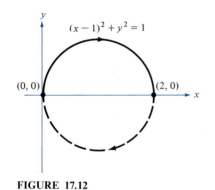

FIGURE 17.12

Let $\mathbf{F} = (y^3 + 1)\mathbf{i} + (3xy^2 + 1)\mathbf{j}$ represent a force field.

(a) Find the work done in moving an object from $(0, 0)$ to $(2, 0)$ along the semicircle $(x - 1)^2 + y^2 = 1$.

(b) Find the work done in moving the object completely around the circle. (See Figure 17.12.)

(c) Find the potential function of **F**.

Solution:

(a) Since $\mathbf{F} = M(x, y)\mathbf{i} + N(x, y)\mathbf{j} = (y^3 + 1)\mathbf{i} + (3xy^2 + 1)\mathbf{j}$, we have

$$\frac{\partial M}{\partial y} = 3y^2 \quad \text{and} \quad \frac{\partial N}{\partial x} = 3y^2$$

and we conclude that $\int_C \mathbf{F} \cdot d\mathbf{R}$ is path-independent. Therefore, it does not matter which path we follow from $(0, 0)$ to $(2, 0)$, and we choose the simpler straight-line path along the x-axis. Thus we have $y = 0$, $dy = 0$, $0 \le x \le 2$, and

$$W = \int_C \mathbf{F} \cdot d\mathbf{R} = \int_C M\,dx + N\,dy$$

$$= \int_C [(y^3 + 1)\,dx + (3xy^2 + 1)\,dy]$$

$$= \int_0^2 [(0 + 1)\,dx + 0] = \int_0^2 dx = x \Big]_0^2 = 2$$

(b) If we integrate around the entire circle, C is a closed path, and since $\int_C \mathbf{F} \cdot d\mathbf{R}$ is path-independent, we have

$$W = \int_C \mathbf{F} \cdot d\mathbf{R} = 0$$

(c) Following the pattern of Table 17.1, we determine U as follows:

$$U = \int M\,dx = \int (y^3 + 1)\,dx = xy^3 + x + h(y)$$

$$\frac{\partial U}{\partial y} = 3xy^2 + h'(y) = 3xy^2 + 1 = N$$

Thus

$$h'(y) = 1, \qquad h(y) = y + C$$

and it follows that the potential function is

$$U = xy^3 + x + y + C$$

Note that we can use this potential function to verify the result of part (a):

$$\int_C \mathbf{F} \cdot d\mathbf{R} = U(2, 0) - U(0, 0) = 2 - 0 = 2 \qquad \blacksquare$$

Section Exercises (17.2)

In Exercises 1–4, show that the value of the line integral $\int_C \mathbf{F} \cdot d\mathbf{R}$ is the same for the given parametric representations of C.

1. $\mathbf{F} = x^2\mathbf{i} + xy\mathbf{j}$
   (a) $\mathbf{R}_1 = t\mathbf{i} + t^2\mathbf{j}, \ 0 \le t \le 1$
   (b) $\mathbf{R}_2 = (\sin \theta)\mathbf{i} + (\sin^2 \theta)\mathbf{j}, \ 0 \le \theta \le \pi/2$
2. $\mathbf{F} = (x^2 + y^2)\mathbf{i} - x\mathbf{j}$
   (a) $\mathbf{R}_1 = t\mathbf{i} + \sqrt{t}\,\mathbf{j}, \ 0 \le t \le 4$
   (b) $\mathbf{R}_2 = w^2\mathbf{i} + w\mathbf{j}, \ 0 \le w \le 2$
3. $\mathbf{F} = y\mathbf{i} - x\mathbf{j}$ [S]
   (a) $\mathbf{R}_1 = (\sec \theta)\mathbf{i} + (\tan \theta)\mathbf{j}, \ 0 \le \theta \le \pi/3$
   (b) $\mathbf{R}_2 = \sqrt{t+1}\,\mathbf{i} + \sqrt{t}\,\mathbf{j}, \ 0 \le t \le 3$
4. $\mathbf{F} = y\mathbf{i} - x^2\mathbf{j}$
   (a) $\mathbf{R}_1 = (2 + t)\mathbf{i} + (3 - t)\mathbf{j}, \ 0 \le t \le 3$
   (b) $\mathbf{R}_2 = (2 + \ln w)\mathbf{i} + (3 - \ln w)\mathbf{j}, \ 1 \le w \le e^3$

In Exercises 5–12, find the value of the line integral $\int_C \mathbf{F} \cdot d\mathbf{R}$ over the given paths.

5. $\mathbf{F} = 2xy\mathbf{i} + x^2\mathbf{j}$
   (a) $\mathbf{R}_1 = t\mathbf{i} + t^2\mathbf{j}, \ 0 \le t \le 1$
   (b) $\mathbf{R}_2 = t\mathbf{i} + t^3\mathbf{j}, \ 0 \le t \le 1$
6. $\mathbf{F} = ye^{xy}\mathbf{i} + xe^{xy}\mathbf{j}$
   (a) $\mathbf{R}_1 = t\mathbf{i} - \frac{3}{2}(t - 2)\mathbf{j}, \ 0 \le t \le 2$
   (b) the line segments from $(0, 3)$ to $(0, 0)$ and then from $(0, 0)$ to $(2, 0)$
7. $\mathbf{F} = y\mathbf{i} - x\mathbf{j}$ [S]
   (a) $\mathbf{R}_1 = t\mathbf{i} + t\mathbf{j}, \ 0 \le t \le 1$
   (b) $\mathbf{R}_2 = t\mathbf{i} + t^2\mathbf{j}, \ 0 \le t \le 1$
   (c) $\mathbf{R}_3 = t\mathbf{i} + t^3\mathbf{j}, \ 0 \le t \le 1$
8. $\mathbf{F} = xy^2\mathbf{i} + 2x^2y\mathbf{j}$
   (a) $\mathbf{R}_1 = t\mathbf{i} + (1/t)\mathbf{j}, \ 1 \le t \le 3$
   (b) $\mathbf{R}_2 = (t + 1)\mathbf{i} - \frac{1}{3}(t - 3)\mathbf{j}, \ 0 \le t \le 2$

9. $\int_C y^2 \, dx + 2xy \, dy$

(a) C_1:

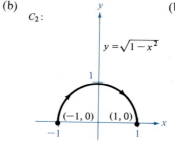

(b) C_2:

(c) C_3:

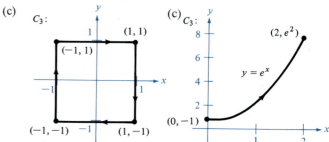

10. $\int_C (2x - 3y + 1) \, dx - (3x + y - 5) \, dy$

(a) C_1:

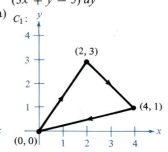

(b) C_2:

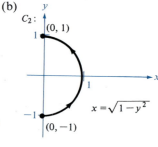

(c) C_3:

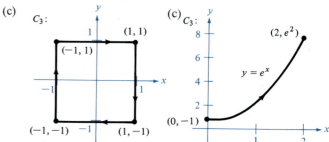

11. $\int_C 2xy\,dx + (x^2 + y^2)\,dy$
(a) along the ellipse $(x^2/25) + (y^2/16) = 1$ from $(5, 0)$ to $(0, 4)$
(b) along the parabola $y = 4 - x^2$ from $(2, 0)$ to $(0, 4)$

12. $\int_C (x^2 + y^2)\,dx + 2xy\,dy$
(a) $\mathbf{R}_1 = t^3\mathbf{i} + t^2\mathbf{j},\ 0 \le t \le 2$
(b) $\mathbf{R}_2 = (2\cos t)\mathbf{i} + (2\sin t)\mathbf{j},\ 0 \le t \le \pi/2$

In Exercises 13–22, determine if **F** is conservative and, if so, find the potential function U.

13. $\mathbf{F} = (2x - 3y)\mathbf{i} - 3(x - y^2)\mathbf{j}$
14. $\mathbf{F} = xe^y\mathbf{i} + ye^x\mathbf{j}$
[S] **15.** $\mathbf{F} = xe^{x^2y}(2y\mathbf{i} + x\mathbf{j})$
16. $\mathbf{F} = 2xy^3\mathbf{i} + 3y^2x^2\mathbf{j}$
17. $\mathbf{F} = \dfrac{x\mathbf{i} + y\mathbf{j}}{x^2 + y^2}$
18. $\mathbf{F} = \dfrac{2y}{x}\mathbf{i} - \dfrac{x^2}{y^2}\mathbf{j}$
19. $\mathbf{F} = (e^x\cos y)\mathbf{i} + (e^x\sin y)\mathbf{j}$
20. $\mathbf{F} = \dfrac{2x}{y}\mathbf{i} - \dfrac{x^2}{y^2}\mathbf{j}$
21. $\mathbf{F} = (e^x\sin y)\mathbf{i} + (e^x\cos y + 2)\mathbf{j}$
22. $\mathbf{F} = \dfrac{2x}{(x^2 + y^2)^2}\mathbf{i} + \dfrac{2y}{(x^2 + y^2)^2}\mathbf{j}$

In Exercises 23–26, evaluate the given line integral using the Fundamental Theorem.

23. $\int_C \mathbf{F}\cdot d\mathbf{R}$ where $\mathbf{F} = y\mathbf{i} + x\mathbf{j}$, from $(0, 0)$ to $(3, 8)$

24. $\int_C \mathbf{F}\cdot d\mathbf{R}$ where $\mathbf{F} = 2(x + y)\mathbf{i} + 2(x + y)\mathbf{j}$, from $(-1, 1)$ to $(3, 2)$

25. $\displaystyle\int_{(0,-\pi)}^{(3\pi/2,\,\pi/2)} \cos x\sin y\,dx + \sin x\cos y\,dy$

26. $\displaystyle\int_{(1,1)}^{(2\sqrt{3},\,2)} \frac{y\,dx - x\,dy}{x^2 + y^2}$

[S] **27.** Evaluate the integral

$$\int_C e^x\sin y\,dx + e^x\cos y\,dy$$

where C is one arch of the cycloid $x = \theta - \sin\theta$, $y = 1 - \cos\theta$ from $(0, 0)$ to $(2\pi, 0)$.

28. Evaluate the integral

$$\int_C \frac{2x}{(x^2 + y^2)^2}\,dx + \frac{2y}{(x^2 + y^2)^2}\,dy$$

where C is the circle centered at $(4, 5)$ with a radius of 3.

29. Find the work done in moving an object from $(0, 0)$ to $(5, 9)$, where the force field is given by

$$\mathbf{F} = (9x^2y^2)\mathbf{i} + (6x^3y - 1)\mathbf{j}$$

30. Find the work done in moving an object from $(-1, 1)$ to $(3, 2)$, where the force field is given by

$$\mathbf{F} = \frac{2x}{y}\mathbf{i} - \frac{x^2}{y^2}\mathbf{j}$$

The three-dimensional counterpart of Theorem 17.2 states that the line integral

$$\int_C M\,dx + N\,dy + P\,dz$$

is independent of path if

$$\frac{\partial M}{\partial y} = \frac{\partial N}{\partial x}, \qquad \frac{\partial M}{\partial z} = \frac{\partial P}{\partial x}, \qquad \frac{\partial N}{\partial z} = \frac{\partial P}{\partial y}$$

Use this result in Exercises 31–34.

[S] **31.** Evaluate

$$\int_C (z + 2y)\,dx + (2x - z)\,dy + (x - y)\,dz$$

along these lines:
(a) broken line from $(0, 0, 0)$ to $(1, 0, 0)$ to $(1, 1, 0)$ to $(1, 1, 1)$
(b) straight line from $(0, 0, 0)$ to $(1, 1, 1)$
(c) broken line from $(0, 0, 0)$ to $(0, 0, 1)$ to $(1, 1, 1)$

32. Repeat Exercise 31 using the integral

$$\int_C zy\,dx + xz\,dy + xy\,dz$$

33. Evaluate

$$\int_{(0,0,0)}^{(\pi/2,\,3,\,4)} -\sin x\,dx + z\,dy + y\,dz$$

34. Evaluate

$$\int_{(0,0,0)}^{(3,\,4,\,0)} 6x\,dx - 4z\,dy + (-4y + 20z)\,dz$$

For Review and Class Discussion

True or False

1. ____ If C_1 and C_2 are defined by $\mathbf{R}_1 = t\mathbf{i} + (1 - t)\mathbf{j}$, $0 \le t \le 1$, and $\mathbf{R}_2 = (1 - t)\mathbf{i} + t\mathbf{j}$, $0 \le t \le 1$, then $\int_{C_1} \mathbf{F} \cdot d\mathbf{R}_1 = \int_{C_2} \mathbf{F} \cdot d\mathbf{R}_2$.

2. ____ If C_1, C_2, and C_3 have the same initial and terminal points and $\int_{C_1} \mathbf{F} \cdot d\mathbf{R}_1 = \int_{C_2} \mathbf{F} \cdot d\mathbf{R}_2$, then $\int_{C_1} \mathbf{F} \cdot d\mathbf{R}_1 = \int_{C_3} \mathbf{F} \cdot d\mathbf{R}_3$.

3. ____ If $\mathbf{F} = y\mathbf{i} + x\mathbf{j}$ and C is given by $\mathbf{R} = (4 \sin t)\mathbf{i} + (3 \cos t)\mathbf{j}$, $0 \le t \le \pi$, then $\int_C \mathbf{F} \cdot d\mathbf{R} = 0$.

4. ____ If $\mathbf{F}$ is conservative in a region R bounded by a simple closed path and C lies within R, then $\int_C \mathbf{F} \cdot d\mathbf{R}$ is path-independent.

5. ____ If $\mathbf{F} = M\mathbf{i} + N\mathbf{j}$ and $\partial M/\partial x = \partial N/\partial y$, then $\mathbf{F}$ is conservative.

17.3 Green's Theorem

Purpose

■ To introduce Green's Theorem in the plane.

Such is the advantage of a well constructed language that its simplified notation often becomes the source of profound theories.

—Pierre-Simon Laplace (1749–1827)

In this section we investigate an important theorem that transforms a line integral around the boundary of a plane region R into a double integral over R and conversely. This theorem is known as **Green's Theorem,** and though the result applies to a wide variety of plane regions, we will restrict our initial statement of the theorem to a plane region bounded by a simple closed path. Furthermore, we will assume the *positive direction* along the path to be that direction which places the enclosed region R always to the left of the path, as shown in Figure 17.13.

Theorem 17.4
(Green's Theorem)

Let R be a region bounded by a simple closed path C. If M, N, $\partial M/\partial y$, and $\partial N/\partial x$ are all continuous on R, then

$$\int_C M\,dx + N\,dy = \iint_R \left(\frac{\partial N}{\partial x} - \frac{\partial M}{\partial y}\right) dA$$

Proof: We give a proof only for the special case in which the region R (see Figure 17.14) can be represented by both of the forms

$$R: a \le x \le b, \quad f_1(x) \le y \le f_2(x)$$
$$R: c \le y \le d, \quad g_1(y) \le x \le g_2(y)$$

The line integral $\int_C M\,dx$ can be written as

$$\int_C M\,dx = \int_{C_1} M\,dx + \int_{C_2} M\,dx = \int_a^b M(x, f_1(x))\,dx + \int_b^a M(x, f_2(x))\,dx$$

$$= \int_a^b [M(x, f_1(x)) - M(x, f_2(x))]\,dx$$

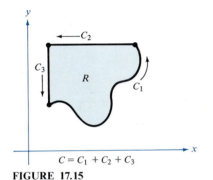

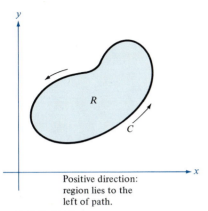

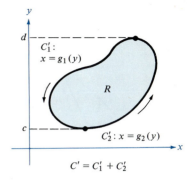

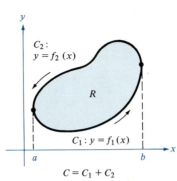

FIGURE 17.13

Positive direction: region lies to the left of path.

$C = C_1 + C_2$

FIGURE 17.14

$C = C_1 + C_2 + C_3$

FIGURE 17.15

$C = C_1 + C_2 + C_3 + C_4$

FIGURE 17.16

Furthermore, we have

$$\iint_R \frac{\partial M}{\partial y}\, dA = \int_a^b \int_{f_1(x)}^{f_2(x)} \frac{\partial M}{\partial y}\, dy\, dx = \int_a^b M(x,y)\Big]_{f_1(x)}^{f_2(x)} dx$$

$$= \int_a^b [M(x, f_2(x)) - M(x, f_1(x))]\, dx$$

Consequently,

$$\int_C M\, dx = -\iint_R \frac{\partial M}{\partial y}\, dA$$

In a similar manner we can use $g_1(y)$ and $g_2(y)$ to show that

$$\int_C N\, dy = \iint_R \frac{\partial N}{\partial x}\, dA$$

Thus

$$\int_C M\, dx + N\, dy = -\iint_R \frac{\partial M}{\partial y}\, dA + \iint_R \frac{\partial N}{\partial x}\, dA = \iint_R \left(\frac{\partial N}{\partial x} - \frac{\partial M}{\partial y}\right) dA \qquad \blacksquare$$

Although we presented a proof for only a special case of Green's Theorem, the theorem is valid for any region bounded by a simple closed path. Figures 17.15 and 17.16 show two more complicated regions for which Green's Theorem is valid.

Example 1

Use Green's Theorem to evaluate the line integral

$$\int_C y^3\, dx + (x^3 + 3xy^2)\, dy$$

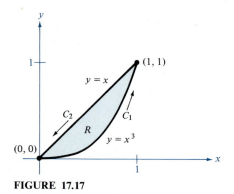

FIGURE 17.17

where C is the boundary of the region between $y = x^3$ and $y = x$ between $x = 0$ and $x = 1$. (See Figure 17.17.)

Solution: Since $M = y^3$ and $N = x^3 + 3xy^2$, we have

$$\int_C y^3 \, dx + (x^3 + 3xy^2) \, dy = \iint_R \left(\frac{\partial N}{\partial x} - \frac{\partial M}{\partial y} \right) dA$$

$$= \int_0^1 \int_{x^3}^x [(3x^2 + 3y^2) - 3y^2] \, dy \, dx$$

$$= \int_0^1 \int_{x^3}^x 3x^2 \, dy \, dx = \int_0^1 3x^2 y \Big]_{x^3}^x \, dx$$

$$= \int_0^1 (3x^3 - 3x^5) \, dx = \left[\frac{3x^4}{4} - \frac{x^6}{2} \right]_0^1 = \frac{1}{4}$$

∎

Example 2

Use Green's Theorem to evaluate the line integral in Example 1, where C is the circle of radius 3 centered at the origin.

Solution: We know from previous work that integration over circular regions is accomplished more simply in polar coordinates than in rectangular. Thus we choose to represent C as

$$r = 3, \qquad 0 \le \theta \le 2\pi$$

(See Figure 17.18.) Applying Green's Theorem we have

$$\int_C y^3 \, dx + (x^3 + 3xy^2) \, dy = \iint_R [(3x^2 + 3y^2) - 3y^2] \, dA = \iint_R 3x^2 \, dA$$

Now in polar coordinates we have $x = r \cos \theta$ and $dA = r \, dr \, d\theta$. Thus the line integral is given by

$$\int_C y^3 \, dx + (x^3 + 3xy^2) \, dy = \int_0^{2\pi} \int_0^3 3(r \cos \theta)^2 r \, dr \, d\theta$$

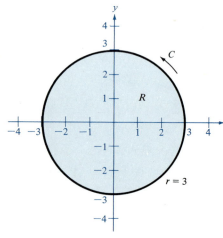

FIGURE 17.18

$$= 3 \int_0^{2\pi} \int_0^3 r^3 \cos^2 \theta \, dr \, d\theta = 3 \int_0^{2\pi} \frac{r^4}{4} \cos^2 \theta \Big]_0^3 \, d\theta$$

$$= 3 \int_0^{2\pi} \frac{81}{4} \cos^2 \theta \, d\theta = \frac{243}{8} \int_0^{2\pi} (1 + \cos 2\theta) \, d\theta$$

$$= \frac{243}{8} \left[\theta + \frac{\sin 2\theta}{2} \right]_0^{2\pi} = \frac{243\pi}{4}$$

∎

Green's Theorem can be used in reverse to obtain a useful line integral for the area of a region R bounded by a simple closed path C. If we choose M and N such that $[(\partial N/\partial x) - (\partial M/\partial y)] = 1$, then

$$A = \iint_R (1)\, dA = \int_C M\, dx + N\, dy$$

For instance, if we choose $M = -y/2$ and $N = x/2$, we obtain the equation

$$A = \iint_R \left(\frac{1}{2} + \frac{1}{2}\right) dA = \int_C \frac{x}{2}\, dy - \frac{y}{2}\, dx = \frac{1}{2} \int_C x\, dy - y\, dx$$

which we formalize in the following theorem.

Theorem 17.5
(*Line Integral for Area of a Region*)

If R is a region bounded by a simple closed path C, then the area of R is given by

$$A = \frac{1}{2} \int_C x\, dy - y\, dx$$

Example 3

Use a line integral to calculate the area of the ellipse

$$\frac{x^2}{a^2} + \frac{y^2}{b^2} = 1$$

Solution: In parametric form we can orient this elliptical path counterclockwise by letting

$$x = a \cos t \quad \text{and} \quad y = b \sin t, \qquad 0 \le t \le 2\pi$$

Therefore, we have

$$A = \frac{1}{2} \int_C x\, dy - y\, dx$$

$$= \frac{1}{2} \int_0^{2\pi} (a \cos t)(b \cos t)\, dt - (b \sin t)(-a \sin t)\, dt$$

$$= \frac{ab}{2} \int_0^{2\pi} (\cos^2 t + \sin^2 t)\, dt = \frac{ab}{2} t \Big]_0^{2\pi} = \pi ab \qquad \blacksquare$$

Green's Theorem can be extended to regions other than those bounded by simple closed paths. One such case is seen in the next example in which we consider a region containing a "hole." Note in Figure 17.19 that the line integral over the segment used to subdivide R is taken twice but in opposite directions, hence canceling itself.

Example 4

Let R be the region inside the ellipse $(x^2/9) + (y^2/4) = 1$ and outside the circle $x^2 + y^2 = 1$ and let C be the boundary of this region. Evaluate the line integral

$$\int_C 2xy\,dx + (x^2 + 2x)\,dy$$

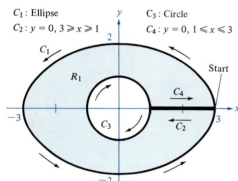

C_1: Ellipse
C_2: $y = 0,\ 3 \geqslant x \geqslant 1$
C_3: Circle
C_4: $y = 0,\ 1 \leqslant x \leqslant 3$

FIGURE 17.19

Solution: If we consider C as the composite path

$$C = C_1 + C_2 + C_3 + C_4$$

shown in Figure 17.19, then we can apply Green's Theorem and obtain

$$\int_C 2xy\,dx + (x^2 + 2x)\,dy = \iint_R \left(\frac{\partial N}{\partial x} - \frac{\partial M}{\partial y}\right) dA = \iint_R (2x + 2 - 2x)\,dA$$

$$= \iint_R 2\,dA = 2\iint_R dA = 2(\text{area of } R)$$

$$= 2(\text{area of ellipse} - \text{area of circle})$$

Therefore, using the result of Example 3 with $a = 3$ and $b = 2$, we have

$$\int_C 2xy\,dx + (x^2 + 2x)\,dy = 2(6\pi - \pi) = 10\pi$$

∎

Section Exercises (17.3)

In Exercises 1–4, verify Green's Theorem by evaluating *both* integrals

$$\int_C y^2\,dx + x^2\,dy = \iint_R \left(\frac{\partial N}{\partial x} - \frac{\partial M}{\partial y}\right) dA$$

for the given path.

1. C is the boundary of the square with vertices $(0, 0)$, $(4, 0)$, $(4, 4)$, and $(0, 4)$.

2. C is the boundary of the triangle with vertices $(0, 0)$, $(4, 0)$, and $(4, 4)$.

⎡S⎤ **3.** C is the boundary of the region lying between $y = x$ and $y = x^2/4$.

4. C is the boundary of the circle given by $x^2 + y^2 = 1$.

In Exercises 5–8, use Green's Theorem to evaluate the integral

$$\int_C (y - x)\,dx + (2x - y)\,dy$$

for the given path.

5. C is the boundary of the region lying between $y = x$ and $y = x^2 - x$.

6. C is the boundary of the ellipse given by $x = 2\cos\theta$ and $y = \sin\theta$.

⎡S⎤ **7.** C is the boundary of the region lying inside the rectangle with vertices $(5, 3)$, $(-5, 3)$, $(-5, -3)$, and $(5, -3)$ and outside the square with vertices $(1, 1)$, $(-1, 1)$, $(-1, -1)$, and $(1, -1)$.

8. C is the boundary of the region lying inside the circle $x^2 + y^2 = 16$ and outside the circle $x^2 + y^2 = 1$.

In Exercises 9–16, use Green's Theorem to evaluate the given line integral.

9. $\int_C 2xy\,dx + (x + y)\,dy$; C is the boundary of the region lying between $y = 0$ and $y = 4 - x^2$.

10. $\int_C y^2\,dx + xy\,dy$; C is the boundary of the region lying between $y = 0$, $y = \sqrt{x}$, and $x = 4$.

11. $\int_C (x^2 - y^2)\,dx + 2xy\,dy$; C is the boundary of the circle $x^2 + y^2 = a^2$.

12. $\int_C (x^2 - y^2)\, dx + 2xy\, dy$; C is the boundary of the cardioid $r = 1 + \cos\theta$

S 13. $\int_C 2 \text{Arctan}\,(y/x)\, dx + \ln(x^2 + y^2)\, dy$; C is the boundary of the ellipse $x = 4 + 2\cos\theta$, $y = 4 + \sin\theta$.

14. $\int_C e^x \sin 2y\, dx + 2e^x \cos 2y\, dy$; C is the boundary of the circle $x^2 + y^2 = a^2$.

15. $\int_C (\sin x)(\cos y)\, dx + [xy + (\cos x)(\sin y)]\, dy$; C is the boundary of the region lying between $y = x$ and $y = \sqrt{x}$.

16. $\int_C (e^{-x^2/2} - y)\, dx + (e^{-y^2/2} + x)\, dy$; C is the boundary of the region lying inside the circle $x = 5\cos\theta$, $y = 5\sin\theta$ and outside the ellipse $x = 2\cos\theta$, $y = \sin\theta$.

In Exercises 17–20, use a line integral to find the area of the given region.

17. R is the circular region bounded by $x^2 + y^2 = a^2$.

18. R is the triangular region bounded by $x = 0$, $2x - 3y = 0$, and $x + 3y = 9$.

S 19. R is the region bounded by $y = 2x + 1$ and $y = 4 - x^2$.

20. R is the region inside the loop of the folium of Descartes given by $x = 3t/(t^3 + 1)$, $y = 3t^2/(t^3 + 1)$.

21. Use Green's Theorem to show that the centroid of the region having area A and bounded by the simple closed path C is

$$\bar{x} = \frac{1}{2A} \int_C x^2\, dy \qquad \text{and} \qquad \bar{y} = \frac{-1}{2A} \int_C y^2\, dx$$

In Exercises 22–25, use the result of Exercise 21 to find the centroid of the given region.

22. R is the region bounded by $y = 0$ and $y = 4 - x^2$.

S 23. R is the region bounded by $y = \sqrt{a^2 - x^2}$ and $y = 0$.

24. R is the region bounded by $y = x^3$ and $y = x$, where $0 \le x \le 1$.

25. R is the triangular region with vertices $(-a, 0)$, $(a, 0)$, and (b, c). Assume that $-a \le b \le a$.

26. Show that the area of a plane region in polar coordinates is

$$A = \frac{1}{2} \int_C r^2\, d\theta$$

In Exercises 27–30, use the result of Exercise 26 to find the area of the given region.

27. R is the region bounded by the cardioid $r = a(1 - \cos\theta)$.

28. R is the region bounded by the rose curve $r = a\cos 3\theta$.

S 29. R is the region bounded by the inner loop of the limacon $r = 1 + 2\cos\theta$.

30. R is the elliptical region bounded by $r = 3/(2 - \cos\theta)$.

17.4 **Divergence, Curl, and Stokes's Theorem**

Purpose
- To introduce the concepts of divergence and curl.
- To discuss two extensions of Green's Theorem: Stokes's Theorem and the Divergence Theorem.

I have decided to furnish, not a rigorous demonstration, but a demonstration of such a nature that those who examine it will not remain in any doubt as to the possibility of a rigorous demonstration.

—*Christian Huygens (1629–1695)*

In this section we investigate two extensions of Green's Theorem: Stokes's Theorem and the Divergence Theorem. Although each of these two theorems can be stated in purely mathematical terms, their basic use lies in physical applications involving curl and divergence, respectively. Thus it is fitting that we begin with an informal discussion of one such application: the flow of a fluid through a region in space.

Suppose the fluid in the cylinder shown in Figure 17.20 is being stirred at a constant rate from the center so that the fluid is swirling around

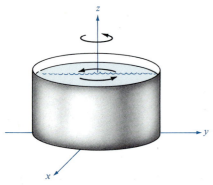

FIGURE 17.20

inside the cylinder. From experience we know that in this situation some portions of the fluid rotate more rapidly than others. For instance, the fluid near the outer edge of the cylinder rotates more slowly than the fluid near the center. This may be due in part to the stirring procedure and in part to the tendency of the fluid to adhere to the walls of the cylinder. In any event, the velocity of the fluid flow varies from point to point, thus forming a velocity field representable by a vector function **F**. We define the **curl** of **F** to be the tendency of the fluid to rotate about an axis. For example, in Figure 17.20 the curl of the velocity field would measure the tendency of the fluid to rotate about the *z*-axis.

Suppose that a grid is placed in the fluid in Figure 17.20 so that the rotating fluid passes through the grid surface. Now at any point on the grid surface, the **divergence** of **F**, denoted by *div* **F**, represents the rate of mass flow per unit volume at that point. That is, *div* **F**, indicates the rate of loss or gain of fluid per unit volume. For instance, suppose that at some point on the grid, fluid is jettisoned into the cylinder. At this point, called a **source**, we have *div* **F** > 0, since there is a gain in fluid flow at this point. Conversely, suppose at some other point that we have a vacuum jet through which fluid is being siphoned out of the cylinder. At this point, called a **sink**, we have a loss of fluid flow and *div* **F** < 0. If *div* **F** $= 0$ throughout a region, then there are no sources or sinks within the region. This condition is referred to as a **condition of incompressibility.** That is, a fluid whose velocity field has zero divergence throughout is said to be **incompressible.**

To define curl and divergence mathematically, we make use of the differential operator ∇ introduced in Section 15.5. Recall that for a *scalar* function f, the ∇ operator produced the gradient vector

$$\nabla f = \frac{\partial f}{\partial x}\mathbf{i} + \frac{\partial f}{\partial y}\mathbf{j} + \frac{\partial f}{\partial z}\mathbf{k}$$

Here we use ∇ as an operator on a *vector* function to define **curl F** and *div* **F**, where

$$\mathbf{F}(x, y, z) = M(x, y, z)\mathbf{i} + N(x, y, z)\mathbf{j} + P(x, y, z)\mathbf{k}$$

Definition of Curl

Let $\mathbf{F} = M\mathbf{i} + N\mathbf{j} + P\mathbf{k}$ be a vector field such that $\partial M/\partial x$, $\partial N/\partial y$, and $\partial P/\partial z$ exist. Then the **curl** of **F** is given by

$$\text{curl } \mathbf{F} = \nabla \times \mathbf{F} = \begin{vmatrix} \mathbf{i} & \mathbf{j} & \mathbf{k} \\ \dfrac{\partial}{\partial x} & \dfrac{\partial}{\partial y} & \dfrac{\partial}{\partial z} \\ M & N & P \end{vmatrix}$$

$$= \left(\frac{\partial P}{\partial y} - \frac{\partial N}{\partial z}\right)\mathbf{i} - \left(\frac{\partial P}{\partial x} - \frac{\partial M}{\partial z}\right)\mathbf{j} + \left(\frac{\partial N}{\partial x} - \frac{\partial M}{\partial y}\right)\mathbf{k}$$

Definition of Divergence

Let $\mathbf{F} = M\mathbf{i} + N\mathbf{j} + P\mathbf{k}$ be a vector field such that $\partial M/\partial x$, $\partial N/\partial y$, and $\partial P/\partial z$ exist. Then the **divergence** of $\mathbf{F}$ is given by

$$div\ \mathbf{F} = \nabla \cdot \mathbf{F} = \frac{\partial M}{\partial x} + \frac{\partial N}{\partial y} + \frac{\partial P}{\partial z}$$

Keep in mind that **curl F** is a vector function, whereas *div* **F** is a scalar function. Also, note that if **curl F** = **0**, then

$$\frac{\partial P}{\partial y} = \frac{\partial N}{\partial z}, \qquad \frac{\partial P}{\partial x} = \frac{\partial M}{\partial z}, \qquad \frac{\partial N}{\partial x} = \frac{\partial M}{\partial y}$$

which are the conditions for path independence given prior to Exercise 31 of Section 17.2. Thus **curl F** = **0**, is considered equivalent to the path independence of $\int_C \mathbf{F} \cdot d\mathbf{R}$.

Example 1

Find *div* **F** and **curl F** at $(2, 0, 1)$, where

$$\mathbf{F}(x, y, z) = x^3y^2z\mathbf{i} + x^2z\mathbf{j} + x^2y\mathbf{k}$$

Indicate which regions in space contain sources or sinks.

Solution:

$$\mathbf{curl\ F} = \begin{vmatrix} \mathbf{i} & \mathbf{j} & \mathbf{k} \\ \dfrac{\partial}{\partial x} & \dfrac{\partial}{\partial y} & \dfrac{\partial}{\partial z} \\ x^3y^2z & x^2z & x^2y \end{vmatrix}$$

$$= \begin{vmatrix} \dfrac{\partial}{\partial y} & \dfrac{\partial}{\partial z} \\ x^2z & x^2y \end{vmatrix}\mathbf{i} - \begin{vmatrix} \dfrac{\partial}{\partial x} & \dfrac{\partial}{\partial z} \\ x^3y^2z & x^2y \end{vmatrix}\mathbf{j} + \begin{vmatrix} \dfrac{\partial}{\partial x} & \dfrac{\partial}{\partial y} \\ x^3y^2z & x^2z \end{vmatrix}\mathbf{k}$$

$$= (x^2 - x^2)\mathbf{i} - (2xy - x^3y^2)\mathbf{j} + (2xz - 2x^3yz)\mathbf{k}$$

$$= (x^3y^2 - 2xy)\mathbf{j} + (2xz - 2x^3yz)\mathbf{k}$$

$$div\ \mathbf{F} = \frac{\partial}{\partial x}[x^3y^2z] + \frac{\partial}{\partial y}[x^2z] + \frac{\partial}{\partial z}[x^2y] = 3x^2y^2z$$

At the point $(2, 0, 1)$ we have

$$\mathbf{curl\ F} = 0\mathbf{j} + 4\mathbf{k} = 4\mathbf{k} \qquad \text{and} \qquad div\ \mathbf{F} = 0$$

We observe that for any point on the three coordinate planes $x = 0$, $y = 0$, and $z = 0$, we have *div* **F** = 0. Furthermore, when x and y are nonzero, we have *div* **F** > 0 for $z > 0$ and *div* **F** < 0 for $z < 0$. Therefore, the sources lie above the xy-plane and the sinks lie below the xy-plane. ∎

It is of interest to note that if **F** is a plane vector field, then we have **F** = $M\mathbf{i} + N\mathbf{j}$ and

$$\text{curl } \mathbf{F} = \left(\frac{\partial N}{\partial x} - \frac{\partial M}{\partial y}\right)\mathbf{k}$$

where the coefficient of **k** has the form of the integrand in Green's Theorem. Thus Green's Theorem can be written as

$$\int_C \mathbf{F} \cdot \mathbf{T} \, ds = \int_C \mathbf{F} \cdot d\mathbf{R} = \iint_R (\text{curl } \mathbf{F}) \cdot \mathbf{k} \, dA$$

This suggests that for a plane surface *the line integral of the tangential component of* **F** *taken along C is equal to the double integral over R of the normal component of* **curl F**. This result has a three-dimensional version, which is called **Stokes's Theorem.**

To extend this result to three dimensions, we replace the plane region *R* by a surface *S* that is oriented so that we can identify an outer normal vector **N** and a positive direction along *C*, the boundary of *S*. (See Figures 17.21 and 17.22.) To refresh your memory of surfaces and surface integrals, refer to Section 16.3.

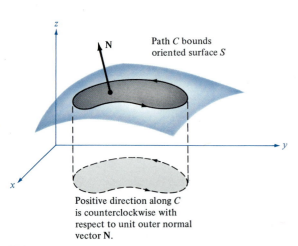

Positive direction along *C* is counterclockwise with respect to unit outer normal vector **N**.

FIGURE 17.21

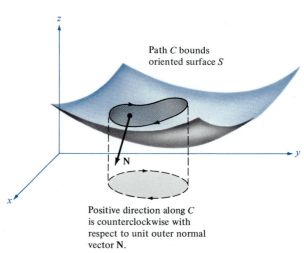

Positive direction along *C* is counterclockwise with respect to unit outer normal vector **N**.

FIGURE 17.22

Theorem 17.6
(*Stokes's Theorem*)

Let *S* be an oriented surface (see Figures 17.21 and 17.22) bounded by a simple closed path *C* and having unit outer normal vector **N**. Let **F** be a vector field with continuous first partial derivatives on *S*; then

$$\int_C \mathbf{F} \cdot \mathbf{T} \, ds = \int_C \mathbf{F} \cdot d\mathbf{R} = \iint_S (\text{curl } \mathbf{F}) \cdot \mathbf{N} \, dS$$

Example 2

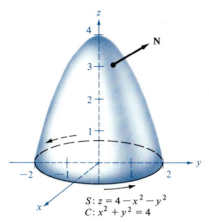

S: $z = 4 - x^2 - y^2$
C: $x^2 + y^2 = 4$

FIGURE 17.23

Verify Stokes's Theorem for $\mathbf{F} = 2z\mathbf{i} + x\mathbf{j} + y^2\mathbf{k}$, where S is the surface of the paraboloid $z = 4 - x^2 - y^2$ and C is the trace of S in the xy-plane. (See Figure 17.23.)

Solution:

(a) (*Double Integral Method*) Considering S to be given by

$$f(x, y, z) = x^2 + y^2 + z - 4 = 0$$

then the outer normal vector to S is

$$\mathbf{N} = \frac{\nabla f}{|\nabla f|} = \frac{2x\mathbf{i} + 2y\mathbf{j} + \mathbf{k}}{\sqrt{1 + 4x^2 + 4y^2}}$$

Furthermore,

$$dS = \sqrt{1 + (f_x)^2 + (f_y)^2} \, dA = \sqrt{1 + 4x^2 + 4y^2} \, dx \, dy$$

and

$$\mathbf{curl \ F} = \begin{vmatrix} \mathbf{i} & \mathbf{j} & \mathbf{k} \\ \dfrac{\partial}{\partial x} & \dfrac{\partial}{\partial y} & \dfrac{\partial}{\partial z} \\ 2z & x & y^2 \end{vmatrix} = 2y\mathbf{i} + 2\mathbf{j} + \mathbf{k}$$

Consequently, we have

$$\iint_S (\mathbf{curl \ F}) \cdot \mathbf{N} \, dS = \int_{-2}^{2} \int_{-\sqrt{4-y^2}}^{\sqrt{4-y^2}} (4xy + 4y + 1) \, dx \, dy$$

$$= \int_{-2}^{2} \left[2x^2 y + 4xy + x \right]_{-\sqrt{4-y^2}}^{\sqrt{4-y^2}} dy$$

$$= \int_{-2}^{2} (8y \sqrt{4 - y^2} + 2 \sqrt{4 - y^2}) \, dy$$

$$= \left[-\frac{8}{3}(4 - y^2)^{3/2} + y\sqrt{4 - y^2} + 4 \, \text{Arcsin} \, \frac{y}{2} \right]_{-2}^{2}$$

$$= \left[0 + 0 + 4\left(\frac{\pi}{2}\right) \right] - \left[0 + 0 - 4\left(\frac{\pi}{2}\right) \right] = 4\pi$$

(b) (*Line Integral Method*) Evaluating this integral by means of a line integral, we have

$$\int_C \mathbf{F} \cdot \mathbf{T} \, ds = \int_C M \, dx + N \, dy + P \, dz = \int_C 2z \, dx + x \, dy + y \, dz$$

Since $z = 0$ on C, we have $dz = 0$ and

$$\int_C \mathbf{F} \cdot \mathbf{T} \, ds = 0 + \int_C x \, dy + 0$$

Finally, using the parametric equations $x = 2 \cos t$ and $y = 2 \sin t$ for C, we have $dy = 2 \cos t \, dt$ and

$$\int_C \mathbf{F} \cdot \mathbf{T} \, ds = \int_0^{2\pi} 2 \cos t (2 \cos t) \, dt = 2 \int_0^{2\pi} (1 + \cos 2t) \, dt$$

$$= 2 \left[t + \frac{\sin 2t}{2} \right]_0^{2\pi} = 4\pi \qquad \blacksquare$$

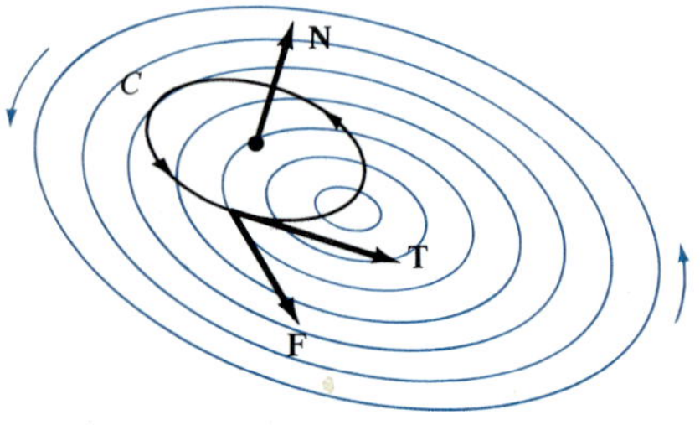

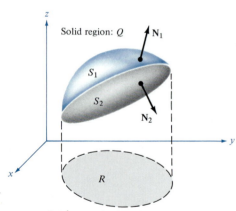

Circular path: C

Flow field: $\mathbf{F}$

FIGURE 17.24

Stokes's Theorem allows us to gain more insight into the informal interpretation of curl given earlier in this section. For instance, consider the path C to be a circle within the flow of a fluid and let $\mathbf{T}$ be the unit tangent vector as shown in Figure 17.24. Then the value of the integral $\int_C \mathbf{F} \cdot \mathbf{T} \, ds$ depends upon the value of $\mathbf{F} \cdot \mathbf{T}$. The closer $\mathbf{F}$ and $\mathbf{T}$ are in direction, the larger is their dot product. This means that the extent to which $\mathbf{F}$ keeps the direction of $\mathbf{T}$ is the extent to which the flow is a rotation along C. As a result, the integral $\int_C \mathbf{F} \cdot \mathbf{T} \, ds$ is referred to as the **circulation** around C. Moreover, the equation

$$\int_C \mathbf{F} \cdot \mathbf{T} \, ds = \iint_S (\mathbf{curl} \; \mathbf{F}) \cdot \mathbf{N} \, dS$$

indicates that when S is small, we can approximate the circulation around C by

$$(\text{circulation around } C) \approx [(\mathbf{curl} \; \mathbf{F}) \cdot \mathbf{N}](\text{area of } S)$$

and we can approximate the rate of circulation per unit of area by

$$(\mathbf{curl} \; \mathbf{F}) \cdot \mathbf{N} \approx \frac{\text{circulation}}{\text{area}} = \text{rate of circulation}$$

From this we note that the closer the directions of **curl F** and **N**, the greater is the rate of circulation.

A second three-dimensional extension of Green's Theorem is called the **Divergence Theorem**. This theorem deals with a surface S that forms the complete boundary of a solid region Q and a vector $\mathbf{N}$, which is the unit outer normal vector to S (see Figure 17.25).

Solid region: Q

$\mathbf{N}_1$

S_1

S_2

$\mathbf{N}_2$

R

Solid region Q is bounded by a closed surface S made up of a top surface S_1 and a bottom surface S_2.

FIGURE 17.25

Theorem 17.7
(*Divergence Theorem*)

Let Q be a solid region bounded by a closed surface S with outer normal vector $\mathbf{N}$, and let $\mathbf{F}$ be a vector field with continuous partial derivatives in Q. Then

$$\iint_S \mathbf{F} \cdot \mathbf{N} \, dS = \iiint_Q div \, \mathbf{F} \, dV$$

It should be pointed out that if the surface S is composed of two (or more) subsurfaces S_1 and S_2, as in Figure 17.25, then

$$\iint_S \mathbf{F} \cdot \mathbf{N}\, dS = \iint_{S_1} \mathbf{F} \cdot \mathbf{N}_1\, dS + \iint_{S_2} \mathbf{F} \cdot \mathbf{N}_2\, dS = \iiint_Q \operatorname{div} \mathbf{F}\, dV$$

This suggests that the divergence theorem may be used as a convenient way to evaluate a surface integral when the complete surface is composed of several subsurfaces that bound a solid region. Example 3 illustrates such a case.

Example 3

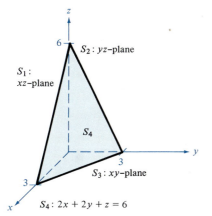

S_2: yz-plane

S_1: xz-plane

S_4

S_3: xy-plane

S_4: $2x + 2y + z = 6$

FIGURE 17.26

Let Q be the solid region bounded by the coordinate planes and the plane $2x + 2y + z = 6$ and let $\mathbf{F} = x\mathbf{i} + y^2\mathbf{j} + z\mathbf{k}$. Find

$$\iint_S \mathbf{F} \cdot \mathbf{N}\, dS$$

Solution: From Figure 17.26 we see that Q is bounded by four subsurfaces; hence we would need four surface integrals to evaluate $\iint_S \mathbf{F} \cdot \mathbf{N}\, dS$.

However, by the Divergence Theorem we need only one triple integral, as seen in the following development. Since

$$\operatorname{div} \mathbf{F} = \frac{\partial M}{\partial x} + \frac{\partial N}{\partial y} + \frac{\partial P}{\partial z} = 1 + 2y + 1$$

we have

$$\iint_S \mathbf{F} \cdot \mathbf{N}\, dS = \iiint_Q \operatorname{div} \mathbf{F}\, dV = \int_0^3 \int_0^{3-y} \int_0^{6-2x-2y} (1 + 2y + 1)\, dz\, dx\, dy$$

$$= \int_0^3 \int_0^{3-y} (2z + 2yz) \Big]_0^{6-2x-2y} dx\, dy$$

$$= \int_0^3 \int_0^{3-y} (12 - 4x + 8y - 4xy - 4y^2)\, dx\, dy$$

$$= \int_0^3 \left[12x - 2x^2 + 8xy - 2x^2y - 4xy^2 \right]_0^{3-y} dy$$

$$= \int_0^3 (18 + 6y - 10y^2 + 2y^3)\, dy$$

$$= \left[18y + 3y^2 - \frac{10y^3}{3} + \frac{y^4}{2} \right]_0^3 = \frac{63}{2}$$

∎

Example 4

Consider the region Q enclosed by the xy-plane and the surface S as shown in Figure 17.23 of Example 2. Verify the Divergence Theorem for this region.

Solution: Consider the surface S_1 to be the surface of the paraboloid and S_2 its circular base. Thus we have

$$S_1: z = 4 - x^2 - y^2 \quad \text{and} \quad S_2: x^2 + y^2 = 4$$

From Example 2 we know that $\mathbf{F} = 2z\mathbf{i} + x\mathbf{j} + y^2\mathbf{k}$, and the unit normal vectors to S_1 and S_2 are given by

$$\mathbf{N}_1 = \frac{2x\mathbf{i} + 2y\mathbf{j} + \mathbf{k}}{\sqrt{1 + 4x^2 + 4y^2}} \quad \text{and} \quad \mathbf{N}_2 = -\mathbf{k}$$

Thus we have

$$\iint_S \mathbf{F} \cdot \mathbf{N}\, dS = \iint_{S_1} \mathbf{F} \cdot \mathbf{N}_1\, dS + \iint_{S_2} \mathbf{F} \cdot \mathbf{N}_2\, dS$$

$$= \int_{-2}^{2} \int_{-\sqrt{4-y^2}}^{\sqrt{4-y^2}} (4xz + 2xy + y^2)\, dx\, dy - \int_{-2}^{2} \int_{-\sqrt{4-y^2}}^{\sqrt{4-y^2}} y^2\, dx\, dy$$

$$= \int_{-2}^{2} \int_{-\sqrt{4-y^2}}^{\sqrt{4-y^2}} [4x(4 - x^2 - y^2) + 2xy]\, dx\, dy$$

$$= \int_{-2}^{2} (8x^2 - x^4 - 2x^2y^2 + x^2y)\Big]_{-\sqrt{4-y^2}}^{\sqrt{4-y^2}} dy = \int_{-2}^{2} 0\, dy = 0$$

On the other hand, since

$$div\, \mathbf{F} = \frac{\partial}{\partial x}[z^2] + \frac{\partial}{\partial y}[x] + \frac{\partial}{\partial z}[y^2] = 0$$

we also obtain

$$\iiint_Q 0\, dV = 0$$

Recall that divergence can be interpreted as the *mass flow per unit of volume*. Therefore, in Example 3 if $\mathbf{F}$ represents the velocity of a fluid, then the fact that the triple integral was positive means that the flow out of Q exceeded the flow into Q. In Example 4 the integral was zero, and this means that there is no fluid gain or loss through the surface.

In this section we have used vector fields determined by the flow of a fluid as our model. We should point out that the results of this section also apply to other vector fields, such as gravitational or electric fields.

As a final comment, we point out that if $\mathbf{F}$ is a three-dimensional vector field, then **F** is conservative under conditions similar to those summarized in Section 17.2. That is, if **F** satisfies the conditions of Stokes's Theorem, then the following four statements are equivalent.

Equivalent Conditions for a Conservative Vector Field in Space	i. $\mathbf{F} = \nabla U$ for some $U(x, y, z)$. ii. $\int_C \mathbf{F} \cdot d\mathbf{R}$ is independent of path. iii. **curl F** $= \mathbf{0}$. iv. **F** is a conservative vector field.

Example 5 (*Three-Dimensional Force Field*)

Given $\mathbf{F}(x, y, z) = e^{-z}\mathbf{i} + 2y\mathbf{j} - xe^{-z}\mathbf{k}$.
(a)   Show that **F** is conservative.
(b)   Find the potential function U.

Solution:

(a)   Since

$$\text{curl } \mathbf{F} = \begin{vmatrix} \mathbf{i} & \mathbf{j} & \mathbf{k} \\ \dfrac{\partial}{\partial x} & \dfrac{\partial}{\partial y} & \dfrac{\partial}{\partial z} \\ e^{-z} & 2y & -xe^{-z} \end{vmatrix}$$

$$= (0 - 0)\mathbf{i} - (-e^{-z} + e^{-z})\mathbf{j} + (0 - 0)\mathbf{k} = \mathbf{0}$$

F is conservative.

(b)   Since **F** is conservative there exists a function U such that

$$\mathbf{F} = e^{-z}\mathbf{i} + 2y\mathbf{j} - xe^{-z}\mathbf{k} = \nabla U = \frac{\partial U}{\partial x}\mathbf{i} + \frac{\partial U}{\partial y}\mathbf{j} + \frac{\partial U}{\partial z}\mathbf{k}$$

It follows that

$$M = \frac{\partial U}{\partial x} = e^{-z}, \qquad N = \frac{\partial U}{\partial y} = 2y, \qquad P = \frac{\partial U}{\partial z} = -xe^{-z}$$

Partial integration with respect to x yields

$$U = \int e^{-z}\, dx = xe^{-z} + h(y, z)$$

and $\qquad\qquad \dfrac{\partial U}{\partial y} = 0 + h_y(y, z) = N = 2y$

$$h(y, z) = y^2 + k(z)$$

Now we have

$$U = xe^{-z} + y^2 + k(z)$$

and $\qquad \dfrac{\partial U}{\partial z} = -xe^{-z} + 0 + k'(z) = P = -xe^{-z}$

$$k'(z) = 0, \qquad k(z) = C$$

Finally, then, the potential function is

$$U = xe^{-z} + y^2 + C \qquad \blacksquare$$

Section Exercises (17.4)

In Exercises 1–8, find (a) the divergence and (b) the curl of the vector field **F**.

$\boxed{\text{S}}$ **1.** $\mathbf{F} = xyz\mathbf{i} + y\mathbf{j} + z\mathbf{k}$ at the point $(1, 2, 1)$

2. $\mathbf{F} = x^2 z\mathbf{i} - 2xz\mathbf{j} + yz\mathbf{k}$ at the point $(2, -1, 3)$

3. $\mathbf{F} = (e^x \sin y)\mathbf{i} - (e^x \cos y)\mathbf{j}$ at the point $(0, 0, 3)$

4. $\mathbf{F} = e^{-xyz}(\mathbf{i} + \mathbf{j} + \mathbf{k})$ at the point $(3, 2, 0)$

$\boxed{\text{S}}$ **5.** $\mathbf{F} = \text{Arctan}\,(x/y)\,\mathbf{i} + \ln\sqrt{x^2 + y^2}\,\mathbf{j} + \mathbf{k}$

6. $\mathbf{F} = \dfrac{yz}{y - z}\mathbf{i} + \dfrac{xz}{x - z}\mathbf{j} + \dfrac{xy}{x - y}\mathbf{k}$

7. $\mathbf{F} = \sin\,(x - y)\mathbf{i} + \sin\,(y - z)\mathbf{j} + \sin\,(z - x)\mathbf{k}$

8. $\mathbf{F} = \sqrt{x^2 + y^2 + z^2}\,(\mathbf{i} + \mathbf{j} + \mathbf{k})$

$\boxed{\text{S}}$ **9.** Given the vector field $\mathbf{F} = xyz\mathbf{i} + y\mathbf{j} + z\mathbf{k}$.
  (a) Find **curl** $(\mathbf{curl\ F}) = \nabla \times (\nabla \times \mathbf{F})$.
  (b) Find $div\,(\mathbf{curl\ F}) = \nabla \cdot (\nabla \times \mathbf{F})$.

10. Given the vector field $\mathbf{F} = x^2 z\mathbf{i} - 2xz\mathbf{j} + yz\mathbf{k}$.
  (a) Find **curl** $(\mathbf{curl\ F}) = \nabla \times (\nabla \times \mathbf{F})$.
  (b) Find $div\,(\mathbf{curl\ F}) = \nabla \cdot (\nabla \times \mathbf{F})$.

In Exercises 11–14, verify Stokes's Theorem by evaluating $\int_C \mathbf{F} \cdot \mathbf{T}\,ds$ as a line integral and as a double integral.

11. $\mathbf{F} = (-y + z)\mathbf{i} + (x - z)\mathbf{j} + (x - y)\mathbf{k}$, where S is the upper half of the sphere $x^2 + y^2 + z^2 = 1$

12. $\mathbf{F} = (-y + z)\mathbf{i} + (x - z)\mathbf{j} + (x - y)\mathbf{k}$, where S is the portion of the paraboloid $z = 4 - x^2 - y^2$ lying above the xy-plane

$\boxed{\text{S}}$ **13.** $\mathbf{F} = xyz\mathbf{i} + y\mathbf{j} + z\mathbf{k}$, where S is the portion of the plane $3x + 4y + 2z = 12$ lying in the first octant

14. $\mathbf{F} = z^2\mathbf{i} + x^2\mathbf{j} + y^2\mathbf{k}$, where S is the cylinder $z = x^2$, $0 \le x \le a$, $0 \le y \le a$

In Exercises 15–17, use Stokes's Theorem to evaluate $\int_C \mathbf{F} \cdot d\mathbf{R}$.

$\boxed{\text{S}}$ **15.** $\mathbf{F} = 2y\mathbf{i} + 3z\mathbf{j} - x\mathbf{k}$, where C is the triangle whose vertices are $(0, 0, 0)$, $(0, 2, 0)$, and $(1, 1, 1)$

16. $\mathbf{F} = \text{Arctan}\,(x/y)\mathbf{i} + \ln\sqrt{x^2 + y^2}\,\mathbf{j} + \mathbf{k}$, where C is the triangle whose vertices are $(0, 0, 0)$, $(1, 1, 1)$, and $(0, 0, 2)$

17. $\mathbf{F} = xyz\mathbf{i} + y\mathbf{j} + z\mathbf{k}$, where S is the surface given by $z = x^2$, $0 \le x \le a$, $0 \le y \le a$

In Exercises 18–20, verify the Divergence Theorem by evaluating $\int\int_S \mathbf{F} \cdot \mathbf{N}\,dS$ as a surface integral and as a triple integral.

18. $\mathbf{F} = 2x\mathbf{i} - 2y\mathbf{j} + z^2\mathbf{k}$, where Q is the cube $0 \le x \le a$, $0 \le y \le a$, $0 \le z \le a$

$\boxed{\text{S}}$ **19.** $\mathbf{F} = 2x\mathbf{i} - 2y\mathbf{j} + z^2\mathbf{k}$, where Q is the cylinder $x^2 + y^2 = 1$, $0 \le z \le h$

20. $\mathbf{F} = (2x - y)\mathbf{i} - (2y - z)\mathbf{j} + z\mathbf{k}$, where Q is the solid bounded by the three coordinate planes and the plane $2x + 4y + 2z = 12$

In Exercises 21–24, use the Divergence Theorem to evaluate $\int\int_S \mathbf{F} \cdot \mathbf{N}\,dS$.

21. $\mathbf{F} = x^2\mathbf{i} + y^2\mathbf{j} + z^2\mathbf{k}$, where S is the surface of the cube bounded by the three coordinate planes and $x = a$, $y = a$, $z = a$

22. $\mathbf{F} = xy\mathbf{i} + yz\mathbf{j} - yz\mathbf{k}$ and S is the surface of the hemisphere given by

$$z = \sqrt{a^2 - x^2 - y^2} \qquad \text{and} \qquad z = 0$$

$\boxed{\text{S}}$ **23.** $\mathbf{F} = x\mathbf{i} + y\mathbf{j} + z\mathbf{k}$, where S is the sphere given by $x^2 + y^2 + z^2 = 4$

24. $\mathbf{F} = xyz\mathbf{j}$, where z is the surface of the cylinder $x^2 + y^2 = 9$, $0 \le z \le 4$

25. If $\mathbf{F}_1$ and $\mathbf{F}_2$ are vector functions whose second partials are continuous, show that:
  (a) **curl** $(\mathbf{F}_1 + \mathbf{F}_2) = \mathbf{curl\ F}_1 + \mathbf{curl\ F}_2$
  (b) $div\,(\mathbf{F}_1 + \mathbf{F}_2) = div\,\mathbf{F}_1 + div\,\mathbf{F}_2$
  (c) $div\,(\mathbf{curl\ F}_1) = 0$

26. If f has continuous second partials, show that

$$\mathbf{curl}\,(\nabla f) = \nabla \times (\nabla f) = \mathbf{0}$$

27. Use the Divergence Theorem to show that the volume bounded by a surface S is

$$\int\int_S x\,dy\,dz = \int\int_S y\,dz\,dx = \int\int_S z\,dx\,dy$$

28. Verify the result of Exercise 27 for the cube $0 \le x \le a$, $0 \le y \le a$, $0 \le z \le a$.

In Exercises 29–32, show that **F** is conservative and find the potential function U.

29. $\mathbf{F}(x, y, z) = e^z(y\mathbf{i} + x\mathbf{j} + xy\mathbf{k})$

30. $\mathbf{F}(x, y, z) = 3x^2y^2z\mathbf{i} + 2x^3yz\mathbf{j} + x^3y^2\mathbf{k}$

[S] **31.** $\mathbf{F}(x, y, z) = \frac{1}{y}\mathbf{i} - \frac{x}{y^2}\mathbf{j} + (2z - 1)\mathbf{k}$

32. $\mathbf{F}(x, y, z) = \frac{x}{x^2 + y^2}\mathbf{i} + \frac{y}{x^2 + y^2}\mathbf{j} + \mathbf{k}$

Miscellaneous Exercises (Ch. 17)

1. Evaluate $\int_C (x^2 + y^2)\, ds$ along:

(a) the line segment from $(-1, -1)$ to $(2, 2)$

(b) the circle $x^2 + y^2 = 16$, counterclockwise starting at $(4, 0)$

2. Evaluate $\int_C xy\, ds$:

(a) along the line segment from $(0, 0)$ to $(5, 4)$

(b) counterclockwise around the triangle whose vertices are $(0, 0)$, $(4, 0)$, and $(0, 2)$

[S] **3.** Evaluate $\int_C (2x - y)\, dx + (x + 3y)\, dy$:

(a) along the line segment from $(0, 0)$ to $(2, -3)$

(b) counterclockwise around the circle $x = 3\cos t$ and $y = 3\sin t$

4. Evaluate $\int_C (2x - y)\, dx + (x + 2y)\, dy$ along the involute $x = \cos\theta + \theta\sin\theta$, $y = \sin\theta - \theta\cos\theta$ from $\theta = 0$ to $\theta = \pi/2$.

In Exercises 5–10, evaluate the integral $\int_C \mathbf{F} \cdot d\mathbf{R}$.

5. $\mathbf{F} = xy\mathbf{i} + x^2\mathbf{j}$; C: $\mathbf{R} = t^2\mathbf{i} + t^3\mathbf{j}$; $0 \le t \le 1$

6. $\mathbf{F} = (x - y)\mathbf{i} + (x + y)\mathbf{j}$; C: $\mathbf{R} = (4\cos\theta)\mathbf{i} + (3\sin\theta)\mathbf{j}$, $0 \le \theta \le 2\pi$

7. $\mathbf{F} = x\mathbf{i} + y\mathbf{j} + z\mathbf{k}$; C: $\mathbf{R} = (2\cos\theta)\mathbf{i} + (2\sin\theta)\mathbf{j} + \theta\mathbf{k}$, $0 \le \theta \le 2\pi$

8. $\mathbf{F} = (2y - z)\mathbf{i} + (z - x)\mathbf{j} + (x - y)\mathbf{k}$; C: the curve of intersection of the cylinders $x^2 + z^2 = 4$ and $y^2 + z^2 = 4$ from $(2, 2, 0)$ to $(0, 0, 2)$

[S] **9.** $\mathbf{F} = (y - z)\mathbf{i} + (z - x)\mathbf{j} + (x - y)\mathbf{k}$; C: the curve of intersection of the paraboloid $z = x^2 + y^2$ and the plane $x + y = 0$ from the point $(-2, 2, 8)$ to $(2, -2, 8)$

10. $\mathbf{F} = (x^2 - z)\mathbf{i} + (y^2 + z)\mathbf{j} + x\mathbf{k}$; C: the curve of intersection of the cylinders $x^2 + y^2 = 4$ and $z = x^2$ from $(0, -2, 0)$ to $(0, 2, 0)$

In Exercises 11–18, determine if **F** is conservative, and if it is, find the potential function U.

11. $\mathbf{F} = \frac{1}{y}\mathbf{i} - \frac{y}{x^2}\mathbf{j}$

12. $\mathbf{F} = -\frac{y}{x^2}\mathbf{i} + \frac{1}{x}\mathbf{j}$

13. $\mathbf{F} = (6xy^2 - 3x^2)\mathbf{i} + (6x^2y + 3y^2 - 7)\mathbf{j}$

14. $\mathbf{F} = (-2y^3 \sin 2x)\mathbf{i} + 3y^2(1 + \cos 2x)\mathbf{j}$

15. $\mathbf{F} = (4xy + z)\mathbf{i} + (2x^2 + 6y)\mathbf{j} + 2z\mathbf{k}$

16. $\mathbf{F} = (4xy + z^2)\mathbf{i} + (2x^2 + 6yz)\mathbf{j} + 2xz\mathbf{k}$

[S] **17.** $\mathbf{F} = (yz\mathbf{i} - xz\mathbf{j} - xy\mathbf{k})/y^2z^2$

18. $\mathbf{F} = \sin z(y\mathbf{i} + x\mathbf{j} + \mathbf{k})$

19. Consider the integral $\int_C \mathbf{F} \cdot d\mathbf{R}$, where

$$\mathbf{F} = \frac{-y}{x^2 + y^2}\mathbf{i} + \frac{x}{x^2 + y^2}\mathbf{j}$$

and

$$\mathbf{R} = (\cos t)\,\mathbf{i} + (\sin t)\,\mathbf{j}$$

Since **F** is conservative and C is a circle, it is expected that the line integral will have a value of zero. However, upon direct integration we obtain $\int_C \mathbf{F} \cdot d\mathbf{R} = 2\pi$. Which is correct and why?

In Exercises 20–24, use Green's Theorem to evaluate each line integral.

20. $\int_C y\, dx + \int_C 2x\, dy$; C is the boundary of the square with vertices $(0, 0)$, $(0, 2)$, $(2, 0)$, and $(2, 2)$

21. $\int_C xy\, dx + \int_C (x^2 + y^2)\, dy$; C is the boundary of the square with vertices $(0, 0)$, $(0, 2)$, $(2, 0)$, and $(2, 2)$

22. $\int_C (x^2 - y^2)\, dx + \int_C 2xy\, dy$; C is the circle $x^2 + y^2 = a^2$

23. $\int_C xy\, dx + \int_C x^2\, dy$; C is the boundary of the region lying between the graphs of $y = x^2$ and $y = x$

24. $\int_C y^2\, dx + x^{2/3}\, dy$; C is the hypocycloid $x^{2/3} + y^{2/3} = 1$

In Exercises 25–28, find (a) the divergence and (b) the curl of the vector field **F**.

25. $\mathbf{F} = x^2\mathbf{i} + y^2\mathbf{j} + z^2\mathbf{k}$

26. $\mathbf{F} = xy^2\mathbf{j} - zx^2\mathbf{k}$

[S] 27. $\mathbf{F} = (\cos y + y \cos x)\mathbf{i} + (\sin x - x \sin y)\mathbf{j} + xyz\mathbf{k}$

28. $\mathbf{F} = (3x - y)\mathbf{i} + (y - 2z)\mathbf{j} + (z - 3x)\mathbf{k}$

29. Verify Stokes's Theorem by evaluating $\int_C \mathbf{F} \cdot \mathbf{T}\, ds$ as a line integral *and* as a double integral for the case in which

$$\mathbf{F} = (\cos y + y \cos x)\mathbf{i} + (\sin x - x \sin y)\mathbf{j} + xyz\mathbf{k}$$

S is the cylinder given by $z = y^2$, and C is the square in the xy-plane with vertices $(0, 0)$, $(a, 0)$, (a, a), and $(0, a)$.

30. Verify the Divergence Theorem by evaluating $\iint_S \mathbf{F} \cdot \mathbf{N}\, dS$ as a surface integral *and* as a triple integral for the case in which

$$\mathbf{F} = x\mathbf{i} + y\mathbf{j} + z\mathbf{k}$$

and Q is the solid region bounded by the coordinate planes and the plane $2x + 3y + 4z = 12$.

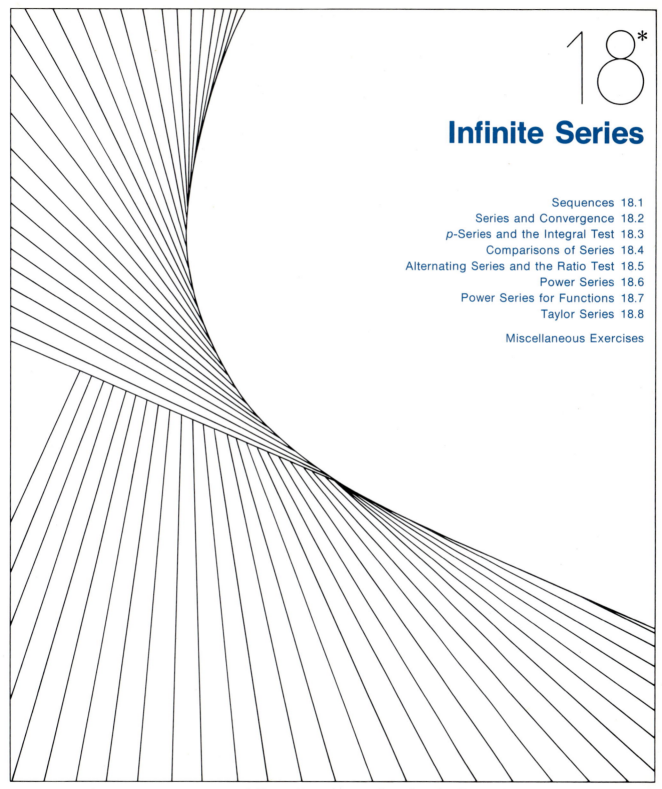

18*

Infinite Series

* Chapter 18 may be covered any time after Chapter 10.

18.1 **Sequences**

Purpose
- To define sequence and limit of a sequence.
- To discuss some techniques for determining the general term of a sequence.

God made integers, all else is the work of man. —*Leopold Kronecker* (1823–1891)

In mathematics the word "sequence" is used in much the same way as in ordinary English. When we say that a collection of objects or events is "in sequence" we usually mean that the collection is ordered so that it has an identified first member, second member, third member, and so on. We define a sequence mathematically as a function whose domain is the set of positive integers. For instance, the equation

$$a(n) = \frac{1}{2^n}$$

defines the sequence

$$\frac{1}{2}, \frac{1}{4}, \frac{1}{8}, \frac{1}{16}, \ldots, \frac{1}{2^n}, \ldots$$

the terms of which correspond respectively to

$$a(1), a(2), a(3), a(4), \ldots, a(n), \ldots$$

We will generally write a sequence such as this in the more convenient *subscript* form

$$a_1, a_2, a_3, a_4, \ldots, a_n, \ldots$$

or we will denote it by the symbol $\{a_n\}$, where a_n is the *n*th term of the sequence.

Definition of a Sequence	A **sequence** $\{a_n\}$ is a function whose domain is the set of positive integers. The functional values $a_1, a_2, a_3, \ldots, a_n, \ldots$ are called the **terms** of the sequence.

Example 1

List the first four terms of the sequences given by:

(a) $a_n = 3 + (-1)^n$ (b) $b_n = \frac{2n}{1+n}$ (c) $c_n = \frac{n^2}{2^n - 1}$

Solution:

(a) $\{a_n\} = \{3 + (-1)^n\}$
   $= \{3 + (-1)^1, 3 + (-1)^2, 3 + (-1)^3, 3 + (-1)^4, \ldots\}$
   $= \{2, 4, 2, 4, \ldots\}$

(b) $\{b_n\} = \left\{\dfrac{2n}{1+n}\right\} = \left\{\dfrac{2(1)}{1+1}, \dfrac{2(2)}{1+2}, \dfrac{2(3)}{1+3}, \dfrac{2(4)}{1+4}, \ldots\right\}$

$$= \left\{ \frac{2}{2}, \frac{4}{3}, \frac{6}{4}, \frac{8}{5}, \dots \right\}$$

(c) $\qquad \{c_n\} = \left\{ \frac{n^2}{2^n - 1} \right\} = \left\{ \frac{1^2}{2^1 - 1}, \frac{2^2}{2^2 - 1}, \frac{3^2}{2^3 - 1}, \frac{4^2}{2^4 - 1}, \dots \right\}$

$$= \left\{ \frac{1}{1}, \frac{4}{3}, \frac{9}{7}, \frac{16}{15}, \dots \right\} \qquad\qquad \blacksquare$$

In this chapter our primary interest is in sequences whose terms approach a limiting value. Such sequences are said to **converge.** For instance, the terms of the sequence

$$\left\{ \frac{1}{2}, \frac{1}{4}, \frac{1}{8}, \frac{1}{16}, \dots, \frac{1}{2^n}, \dots \right\}$$

approach 0 as n increases, and we write

$$\lim_{n \to \infty} a_n = \lim_{n \to \infty} \frac{1}{2^n} = 0$$

Similarly, for the sequence $\{b_n\}$ in Example 1, we have

$$\lim_{n \to \infty} \frac{2n}{1 + n} = 2 \lim_{n \to \infty} \frac{n}{1 + n} = 2(1) = 2$$

The precise statement of the limit of an infinite sequence is given in the following definition.

Definition of Limit of a Sequence

We write $\qquad\qquad \lim_{n \to \infty} a_n = L$

if for each $\epsilon > 0$ there exists an integer $M > 0$ such that $|a_n - L| < \epsilon$ whenever $n > M$. If $\lim_{n \to \infty} a_n$ is finite, we say the sequence $\{a_n\}$ **converges,** and if $\lim_{n \to \infty} a_n$ is infinite or does not exist, then the sequence $\{a_n\}$ **diverges.**

Geometrically, this definition means that eventually (for $n > M$) the terms of the sequence will lie within the band between the lines $y = L + \epsilon$ and $y = L - \epsilon$. Figure 18.1 shows one possible situation for which $\lim_{n \to \infty} a_n = L$.

This definition of the limit of a sequence is similar to the definition given in Section 2.4 for the limit

$$\lim_{x \to \infty} f(x) = L$$

Both cases require that for each $\epsilon > 0$, there exists a number $M > 0$ such that

$$|f(x) - L| < \epsilon \qquad (|a_n - L| < \epsilon)$$

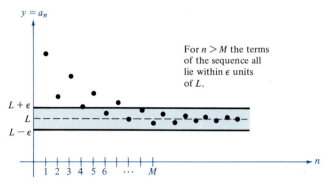

FIGURE 18.1

whenever $x > M (n > M)$. The difference, of course, is that n is restricted to positive integers while x may assume any positive values, including the integers. This latter distinction suggests that the existence of the limit as $x \to \infty$ will guarantee the existence of the limit as $n \to \infty$ since $x > M$ includes all integers $n > M$. Thus the following theorem:

THEOREM 18.1
(*Limit of a Sequence*)

If $\lim_{x \to \infty} f(x) = L$ and if $f(n) = a_n$, then $\lim_{n \to \infty} a_n = L$.

As a consequence of Theorem 18.1, we will use, for sequences, the various properties of limits of functions identified in Section 2.1.

We point out that the converse of Theorem 18.1 is not necessarily true; the limit as $n \to \infty$ may exist when the limit as $x \to \infty$ does not exist. For instance, consider

$$a_n = \frac{1}{n} + \cos 2\pi n \quad \text{and} \quad f(x) = \frac{1}{x} + \cos 2\pi x, \quad x > 1$$

From Figure 18.2 we see that $f(x)$ continuously oscillates between relative maximum values near 1 and relative minimum values near -1. However, the terms of the sequence

$$\{a_n\} = \left\{ \frac{1}{n} + \cos 2\pi n \right\} = \left\{ 1 + 1, \frac{1}{2} + 1, \frac{1}{3} + 1, \frac{1}{4} + 1, \ldots, \frac{1}{n} + 1, \ldots \right\}$$

approach 1 as $n \to \infty$. Therefore, we have

$$\lim_{n \to \infty} \left(\frac{1}{n} + \cos 2\pi n \right) = 1$$

whereas $\quad \lim_{x \to \infty} \left(\frac{1}{x} + \cos 2\pi x \right)$ does not exist

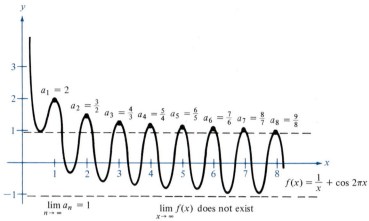

FIGURE 18.2

Theorem 18.1 opens up the possibility of using L'Hôpital's Rule to determine the limit of a sequence, provided we temporarily think of n as varying continuously over the real numbers rather than just the integers. We demonstrate the use of L'Hôpital's Rule in the next example.

Example 2

Determine the convergence or divergence of the sequences given by:

(a) $a_n = 3 + (-1)^n$ (b) $b_n = \dfrac{n}{1 - 2n}$ (c) $c_n = \dfrac{n^2}{2^n - 1}$

Solution:

(a) Since the sequence

$$\{a_n\} = \{3 + (-1)^n\} = \{2, 4, 2, 4, \ldots\}$$

oscillates between 2 and 4, we know that $\lim_{n \to \infty} a_n$ does not exist and we conclude that $\{a_n\}$ diverges.

(b) For $\qquad\qquad \{b_n\} = \left\{ \dfrac{n}{1 - 2n} \right\}$

we have $\qquad \lim_{n \to \infty} \dfrac{n}{1 - 2n} = \lim_{n \to \infty} \left[\dfrac{1}{(1/n) - 2} \right] = \dfrac{1}{-2}$

and $\{b_n\}$ converges to $-\frac{1}{2}$.

(c) Finally, for $\qquad\qquad \{c_n\} = \left\{ \dfrac{n^2}{2^n - 1} \right\}$

we consider Theorem 18.1 and apply L'Hôpital's Rule twice to obtain

$$\lim_{n\to\infty} \frac{n^2}{2^n - 1} = \lim_{n\to\infty} \frac{2n}{(\ln 2)2^n} = \lim_{n\to\infty} \frac{2}{(\ln 2)^2 2^n} = 0$$

which implies that $\{c_n\}$ converges to 0.   ■

Another useful theorem that can be rewritten for sequences is the Squeeze Theorem of Section 2.4.

THEOREM 18.2 (*Squeeze Theorem for Sequences*)	If $$\lim_{n\to\infty} a_n = \lim_{n\to\infty} b_n = L$$ and if there exists an N such that $a_n \leq c_n \leq b_n$ for all $n > N$, then $$\lim_{n\to\infty} c_n = L$$

The usefulness of this Squeeze Theorem will be seen in the next example which involves the symbol $n!$ (read "n factorial"). If n is a positive integer, then n **factorial** is defined as

$$n! = 1 \cdot 2 \cdot 3 \cdot 4 \cdots (n-1) \cdot n$$

Moreover, for $n = 0$ we let $0! = 1$. Thus we have

$$0! = 1$$
$$1! = 1$$
$$2! = 1 \cdot 2 = 2$$
$$3! = 1 \cdot 2 \cdot 3 = 6$$
$$4! = 1 \cdot 2 \cdot 3 \cdot 4 = 24$$

Factorials follow the same conventions for order of operation as exponents. That is, just as $2x^3$ and $(2x)^3$ imply a different order of operations, $2n!$ and $(2n)!$ imply the following orders:

$$2n! = 2(n!) = 2(1 \cdot 2 \cdot 3 \cdot 4 \cdots n)$$
$$(2n)! = 1 \cdot 2 \cdot 3 \cdot 4 \cdots n \cdot (n + 1) \cdots (2n)$$

Example 3

Show that the sequence

$$\left\{ \frac{(-1)^n}{n!} \right\}$$

converges to 0.

Solution: First, consider the sequences

$$\left\{ \frac{-1}{2^n} \right\} \qquad \text{and} \qquad \left\{ \frac{1}{2^n} \right\}$$

TABLE 18.1

n	1	2	3	4	5	6
$\left\{\dfrac{-1}{2^n}\right\}$	$-\dfrac{1}{2}$	$-\dfrac{1}{4}$	$-\dfrac{1}{8}$	$-\dfrac{1}{16}$	$-\dfrac{1}{32}$	$-\dfrac{1}{64}$
$\left\{\dfrac{(-1)^n}{n!}\right\}$	$\dfrac{-1}{1}$	$\dfrac{1}{2}$	$-\dfrac{1}{6}$	$\dfrac{1}{24}$	$-\dfrac{1}{120}$	$\dfrac{1}{720}$
$\left\{\dfrac{1}{2^n}\right\}$	$\dfrac{1}{2}$	$\dfrac{1}{4}$	$\dfrac{1}{8}$	$\dfrac{1}{16}$	$\dfrac{1}{32}$	$\dfrac{1}{64}$

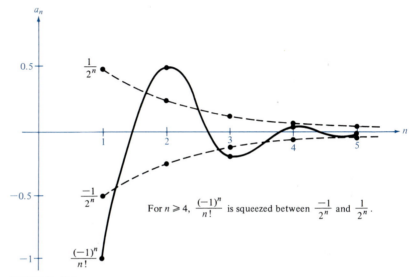

For $n \geqslant 4$, $\dfrac{(-1)^n}{n!}$ is squeezed between $\dfrac{-1}{2^n}$ and $\dfrac{1}{2^n}$.

FIGURE 18.3

both of which converge to zero. By comparing these sequences to $\{(-1)^n/n!\}$, as shown in Table 18.1, we see that for $n \geq 4$ we have

$$\frac{-1}{2^n} \leq \frac{(-1)^n}{n!} \leq \frac{1}{2^n}$$

(see Figure 18.3). Therefore, by the Squeeze Theorem it follows that

$$\lim_{n \to \infty} \frac{(-1)^n}{n!} = 0$$

∎

Example 3 is a special case of the next theorem.

THEOREM 18.3

For the sequence $\{a_n\}$, if

$$\lim_{n \to \infty} |a_n| = 0 \qquad \text{then} \qquad \lim_{n \to \infty} a_n = 0$$

Proof: Consider the two sequences $\{|a_n|\}$ and $\{-|a_n|\}$. Since both of these sequences converge to 0 and since

$$-|a_n| \le a_n \le |a_n|$$

we can apply the Squeeze Theorem to conclude that $\{a_n\}$ converges to 0. ∎

The result in Example 3 suggests something about the rate at which $n!$ increases as $n \to \infty$. From Figure 18.3 we can see that both $1/2^n$ and $1/n!$ approach zero as $n \to \infty$. Yet $1/n!$ approaches 0 *faster* than $1/2^n$ does. This suggests that

$$\frac{2^n}{n!} \to 0 \qquad \text{as} \qquad n \to \infty$$

and in fact we could show that for any fixed number k,

$$\lim_{n \to \infty} \frac{k^n}{n!} = 0$$

This means that *the factorial function grows faster than the exponential function.* We will find this to be a very useful result as we work with limits of sequences.

It is sometimes the case that the terms of a sequence are generated by some rule that does not explicitly identify the nth term of the sequence. Under such circumstances we are required to discover a *pattern* in the sequence and come up with a description of the nth term. Once the nth term is specified, then we can discuss the convergence or divergence of the sequence. This is demonstrated in our next example.

Example 4

Given $f(x) = e^{x/3}$, determine the convergence or divergence of the sequence given by

$$a_n = 1 - f^{(n-1)}(0)$$

where $f^{(0)} = f$ and $f^{(n)}$ is the nth derivative of f.

Solution: We determine the first several terms of the sequence as shown in Table 18.2. The pattern in the sequence $\{1 - f^{(n-1)}(0)\}$ suggests that the nth term is

$$a_n = 1 - \frac{1}{3^{n-1}}$$

and therefore, the sequence converges to 1 because

$$\lim_{n \to \infty} a_n = \lim_{n \to \infty} \left(1 - \frac{1}{3^{n-1}} \right) = 1 - 0 = 1$$

TABLE 18.2

n	1	2	3	4	5	$\cdots$	n
$f^{(n-1)}(x)$	$e^{x/3}$	$\dfrac{e^{x/3}}{3}$	$\dfrac{e^{x/3}}{3^2}$	$\dfrac{e^{x/3}}{3^3}$	$\dfrac{e^{x/3}}{3^4}$	$\cdots$	$\dfrac{e^{x/3}}{3^{n-1}}$
$1 - f^{(n-1)}(0)$	$1 - 1$	$1 - \dfrac{1}{3}$	$1 - \dfrac{1}{3^2}$	$1 - \dfrac{1}{3^3}$	$1 - \dfrac{1}{3^4}$	$\cdots$	$1 - \dfrac{1}{3^{n-1}}$

∎

Without a specific rule for generating the terms of a sequence or some knowledge of the context in which the terms of the sequence are obtained, it is not possible to determine the convergence or divergence of the sequence merely from its first several terms. For instance, the first three terms of the four sequences given next are identical, yet from the description of their individual nth terms, we see that the first two sequences converge to 0, the third sequence converges to $\frac{1}{9}$, and the fourth one diverges:

$$\{a_n\} = \left\{\frac{1}{2}, \frac{1}{4}, \frac{1}{8}, \frac{1}{16}, \ldots, \frac{1}{2^n}, \ldots\right\}$$

$$\{b_n\} = \left\{\frac{1}{2}, \frac{1}{4}, \frac{1}{8}, \frac{1}{15}, \ldots, \frac{6}{(n+1)(n^2-n+6)}, \ldots\right\}$$

$$\{c_n\} = \left\{\frac{1}{2}, \frac{1}{4}, \frac{1}{8}, \frac{7}{62}, \ldots, \frac{n^2-3n+3}{9n^2-25n+18}, \ldots\right\}$$

$$\{d_n\} = \left\{\frac{1}{2}, \frac{1}{4}, \frac{1}{8}, 0, \ldots, \frac{-n(n+1)(n-4)}{6(n^2+3n-2)}, \ldots\right\}$$

Thus if only the first several terms of a sequence are given, there are many forms for an nth term of the sequence. In such a situation we can only determine the convergence or divergence of the sequence on the basis of our choice for the nth term.

Example 5

Determine an nth term for the sequence

$$\left\{-\frac{1}{1}, \frac{3}{2}, -\frac{7}{6}, \frac{15}{24}, -\frac{31}{120}, \ldots\right\}$$

and then decide if the sequence converges or diverges.

Solution: Observe that the numerators are one less than 2^n for $n = 1, 2, 3, 4, 5, \ldots$. Hence we can generate the numerators by the rule $2^n - 1$. Now if we factor the denominators successively, we have

$$1, 1 \cdot 2, 1 \cdot 2 \cdot 3, 1 \cdot 2 \cdot 3 \cdot 4, 1 \cdot 2 \cdot 3 \cdot 4 \cdot 5, \ldots$$

This suggests that the denominators are represented by $n!$. Furthermore, since the signs alternate, we can write

$$a_n = (-1)^n \left(\frac{2^n - 1}{n!} \right)$$

as an nth term for the given sequence. From Theorem 18.3 and the discussion following Example 3 about the growth of $n!$, it follows that

$$\lim_{n \to \infty} |a_n| = \lim_{n \to \infty} \frac{2^n - 1}{n!} = 0 = \lim_{n \to \infty} a_n$$

and we conclude that $\{a_n\}$ converges to 0. ■

Before concluding this section we provide the following list of some common sequence patterns.

1. Changes in sign:

$$\{(-1)^n\} = \{-1, 1, -1, 1, -1, 1, \ldots\}$$
$$\{(-1)^{n+1}\} = \{1, -1, 1, -1, 1, -1, \ldots\}$$
$$\{(-1)^{n(n+1)/2}\} = \{-1, -1, 1, 1, -1, -1, \ldots\}$$

2. Arithmetic sequences (successive terms differ by a constant value):

$$\{2n\} = \{2, 4, 6, 8, 10, 12, \ldots\}$$
$$\{2n - 1\} = \{1, 3, 5, 7, 9, 11, \ldots\}$$
$$\{an + b\} = \{a + b, 2a + b, 3a + b, 4a + b, 5a + b, 6a + b, \ldots\}$$

3. Binary (powers of 2) and geometric (powers of r) sequences:

$$\{2^{n-1}\} = \{1, 2, 4, 8, 16, 32, \ldots\}$$
$$\{ar^{n-1}\} = \{a, ar, ar^2, ar^3, ar^4, ar^5, \ldots\}$$

4. Power sequences:

$$\{n^2\} = \{1, 4, 9, 16, 25, 36, \ldots\}$$
$$\{n^m\} = \{1, 2^m, 3^m, 4^m, 5^m, 6^m, \ldots\}$$

5. Sequences of products:

$$\{n!\} = \{1, 1 \cdot 2, 1 \cdot 2 \cdot 3, 1 \cdot 2 \cdot 3 \cdot 4, 1 \cdot 2 \cdot 3 \cdot 4 \cdot 5, \ldots\}$$
$$\{2^n n!\} = \{2, 2 \cdot 4, 2 \cdot 4 \cdot 6, 2 \cdot 4 \cdot 6 \cdot 8, 2 \cdot 4 \cdot 6 \cdot 8 \cdot 10, \ldots\}$$
$$\left\{ \frac{(2n)!}{2^n n!} \right\} = \{1, 1 \cdot 3, 1 \cdot 3 \cdot 5, 1 \cdot 3 \cdot 5 \cdot 7, 1 \cdot 3 \cdot 5 \cdot 7 \cdot 9, \ldots\}$$

Section Exercises (18.1)

In Exercises 1–8, write out the first five terms of the specified sequence.

1. $a_n = 2^n$

2. $a_n = \dfrac{n}{n+1}$

3. $a_n = \left(-\dfrac{1}{2}\right)^n$

4. $a_n = \sin\left(\dfrac{n\pi}{2}\right)$

S **5.** $a_n = \dfrac{3^n}{n!}$

6. $a_n = 5 - \dfrac{1}{n} + \dfrac{1}{n^2}$

7. $a_n = \dfrac{(-1)^{n(n+1)/2}}{n^2}$ **8.** $a_n = \dfrac{3n!}{(n-1)!}$

In Exercises 9–24, write an expression for the nth term of the sequence.

9. $\{1, 4, 7, 10, \ldots\}$
10. $\{3, 7, 11, 15, \ldots\}$
 S 11. $\{-1, 2, 7, 14, 23, \ldots\}$
12. $\{1, \frac{1}{4}, \frac{1}{9}, \frac{1}{16}, \ldots\}$
13. $\{\frac{2}{3}, \frac{3}{4}, \frac{4}{5}, \frac{5}{6}, \ldots\}$
14. $\{2, \frac{3}{3}, \frac{4}{5}, \frac{5}{7}, \frac{6}{9}, \ldots\}$
 S 15. $\{2, -1, \frac{1}{2}, -\frac{1}{4}, \frac{1}{8}, \ldots\}$
16. $\{\frac{1}{2}, \frac{1}{3}, \frac{2}{9}, \frac{4}{27}, \frac{8}{81}, \ldots\}$
17. $\{2, 1 + \frac{1}{2}, 1 + \frac{1}{3}, 1 + \frac{1}{4}, 1 + \frac{1}{5}, \ldots\}$
18. $\{1 + \frac{1}{2}, 1 + \frac{3}{4}, 1 + \frac{7}{8}, 1 + \frac{15}{16}, 1 + \frac{31}{32}, \ldots\}$
19. $\left\{\dfrac{1}{2 \cdot 3}, \dfrac{2}{3 \cdot 4}, \dfrac{3}{4 \cdot 5}, \dfrac{4}{5 \cdot 6}, \ldots\right\}$
20. $\left\{1, \dfrac{1}{2}, \dfrac{1}{6}, \dfrac{1}{24}, \dfrac{1}{120}, \ldots\right\}$
 S 21. $\left\{1, \dfrac{1}{1 \cdot 3}, \dfrac{1}{1 \cdot 3 \cdot 5}, \dfrac{1}{1 \cdot 3 \cdot 5 \cdot 7}, \ldots\right\}$
22. $\{2, -4, 6, -8, 10, \ldots\}$
23. $\{1, -1, -1, 1, 1, -1, -1, \ldots\}$
24. $\left\{1, x, \dfrac{x^2}{2}, \dfrac{x^3}{6}, \dfrac{x^4}{24}, \dfrac{x^5}{120}, \ldots\right\}$

In Exercises 25–45, determine the convergence or divergence of each sequence. If the sequence converges, find its limit.

25. $a_n = \dfrac{n+1}{n}$ **26.** $a_n = \dfrac{1}{n^{3/2}}$

 S 27. $a_n = (-1)^n \left(\dfrac{n}{n+1}\right)$ **28.** $a_n = \dfrac{n-1}{n} - \dfrac{n}{n-1}$, $n \geq 2$

29. $a_n = \dfrac{3n^2 - n + 4}{2n^2 + 1}$ **30.** $a_n = \dfrac{\sqrt{n}}{\sqrt{n}+1}$

31. $a_n = \dfrac{n^2 - 1}{n + 1}$ **32.** $a_n = 1 + (-1)^n$

33. $a_n = \dfrac{1 + (-1)^n}{n}$ **34.** $a_n = \dfrac{\ln(n^2)}{n}$

35. $a_n = \cos\left(\dfrac{n\pi}{2}\right)$ **36.** $a_n = \dfrac{n}{\sqrt{n^2 + 1}}$

37. $a_n = \dfrac{3^n}{4^n}$ **38.** $a_n = \dfrac{(n-2)!}{n!}$

 S 39. $a_n = f^{(n-1)}(2), f(x) = \ln x$

40. $a_n = \dfrac{n^2}{2n + 1} - \dfrac{n^2}{2n - 1}$

 S 41. $a_n = 3 - \dfrac{1}{2^n}$ **42.** $a_n = \dfrac{n!}{n^n}$

 S 43. $a_n = \left(1 + \dfrac{k}{n}\right)^n$ **44.** $a_n = 2^{1/n}$

45. $a_n = \dfrac{n^p}{e^n}$ $(p > 0)$

 C 46. Consider the sequence $\{x_n\}$ defined by

$$x_n = x_{n-1} - \dfrac{f(x_{n-1})}{f'(x_{n-1})}$$

If $f(x) = x^2 + x - 1$ and we let $x_1 = 0.5$, find the next three terms of the sequence.

47. Show that $\{ar^n/(1 - r)\}$ diverges if $|r| \geq 1$ and converges to 0 if $|r| < 1$.

48. Prove that if $\lim_{n\to\infty} s_n = L$ and $L > 0$, then there exists a number N such that $s_n > 0$ for every $n > N$.

49. If $\{s_n\}$ is a convergent sequence, show that $\lim_{n\to\infty} s_{n-1} = \lim_{n\to\infty} s_n$.

 C 50. Consider the sequence $\{A_n\}$, whose nth term is given by $A_n = P[1 + (i/n)]^n$, where P is the principal, A_n is the amount at compound interest after 1 year, and i is the yearly interest rate.
 (a) Is $\{A_n\}$ a convergent sequence?
 (b) Find the first ten terms of the sequence if $P = 9000$ and $i = 0.075$.

For Review and Class Discussion

True or False

1. ____ If $0 < a_n < b_n$ for all n and $\{b_n\}$ converges, then $\{a_n\}$ converges.

2. ____ If $0 < a_n < b_n$ for all n and $\{b_n\}$ converges to 0, then $\{a_n\}$ converges to 0.

3. ____ If n is a positive integer, then $n! \leq n^n \leq (2n)!$.

4. ____ If $\{a_n\}$ converges to 3 and $\{b_n\}$ converges to 2, then $\{a_n + b_n\}$ converges to 5.

5. ____ If $b_n = a_{n+1}$ and $\{a_n\}$ converges to L, then $\{b_n\}$ converges to L.

6. _____ If $\{a_n\}$ converges, then $\lim_{n \to \infty} (a_n - a_{n+1}) = 0$.

7. _____ If $n > 1$, then $n! = n(n - 1)!$.

8. _____ If both $\{a_n\}$ and $\{b_n\}$ diverge, then $\{a_n + b_n\}$ diverges.

9. _____ If $b_n = a_{2n}$ and $\{a_n\}$ converges, then $\{b_n\}$ converges.

10. _____ If $\{a_n\}$ converges, then $\{a_n/n\}$ converges to 0.

18.2 Series and Convergence

Purpose

- To define a series and the convergence of a series.
- To introduce the *n*th-Term Test for divergence.
- To introduce geometric and telescoping series and describe conditions for their convergence.

The sum of the harmonic series $\frac{1}{1} + \frac{1}{2} + \frac{1}{3} + \frac{1}{4} + \cdots$ surpasses any given number and hence is infinite. —Jacob Bernoulli (1654–1705)

In this section we will investigate an important application of infinite sequences, namely, their use in representing infinite summations. As a simple illustration, suppose we write the decimal representation of $\frac{1}{3}$ as

$$\frac{1}{3} = 0.33333\ldots = \frac{3}{10} + \frac{3}{10^2} + \frac{3}{10^3} + \frac{3}{10^4} + \frac{3}{10^5} + \cdots$$

We consider this representation to be an *infinite summation* whose value is $\frac{1}{3}$.

To obtain a better picture of what an infinite summation really is, let $\{a_n\}$ be a sequence from which we form another sequence $\{S_n\}$ in the manner

$$S_1 = a_1$$
$$S_2 = a_1 + a_2$$
$$S_3 = a_1 + a_2 + a_3$$
$$\vdots$$
$$S_n = a_1 + a_2 + a_3 + \cdots + a_n = \sum_{i=1}^{n} a_i$$

As $n \to \infty$, the infinite summation

$$a_1 + a_2 + a_3 + a_4 + \cdots$$

denoted by the expression

$$\sum_{n=1}^{\infty} a_n$$

is called an **infinite series** or simply a **series**. The sequence $\{S_n\}$ is called a **sequence of partial sums,** and we have the following:

1. If $\lim_{n \to \infty} S_n = S$ (S is finite), the series $\sum_{n=1}^{\infty} a_n$ **converges** and S is called the **sum of the series.**

2. If $\lim_{n \to \infty} S_n$ does not exist, the series $\sum_{n=1}^{\infty} a_n$ **diverges.**

Since a series is a special type of sequence, the following properties of infinite series are direct consequences of the corresponding properties of limits of sequences.

Properties of Infinite Series

If c, A, and B are real numbers such that

$$\sum_{n=1}^{\infty} a_n = A, \qquad \sum_{n=1}^{\infty} b_n = B, \qquad S_N = \sum_{n=1}^{N} a_n$$

then:

1. $\displaystyle\sum_{n=1}^{\infty} c a_n = c \sum_{n=1}^{\infty} a_n = cA$

2. $\displaystyle\sum_{n=1}^{\infty} (a_n \pm b_n) = \sum_{n=1}^{\infty} a_n \pm \sum_{n=1}^{\infty} b_n = A \pm B$

3. $\displaystyle\sum_{n=N+1}^{\infty} a_n = \sum_{n=1}^{\infty} a_n - \sum_{n=1}^{N} a_n = A - S_N$

Note that Property 3 suggests that *if the first N terms of a series are dropped, this will not destroy the convergence (or divergence) of the series.* That is, if a series converges, then dropping its first N terms will not change the fact that it converges though it may change the value to which it converges. Furthermore, this same property suggests that if the terms of two series are identical after the first N terms, then either both series converge or they both diverge.

In the remainder of this chapter, we shall seek answers to these two questions:

1. Does the series converge or does it diverge?
2. If the series converges, to what value does it converge?

These questions are not always easy to answer, especially the second one. We begin our pursuit for answers to these questions with a simple test for *divergence*.

THEOREM 18.4
(nth-Term Test for Divergence)

If the series given by $\sum_{n=1}^{\infty} a_n$ converges, then $\lim_{n\to\infty} a_n = 0$. Equivalently, if $\lim_{n\to\infty} a_n \neq 0$, then the series *diverges*.

Proof: Assume the given series converges and that

$$\sum_{n=1}^{\infty} a_n = \lim_{n\to\infty} S_n = L$$

Then since

$$S_n = S_{n-1} + a_n$$

and
$$\lim_{n\to\infty} S_n = \lim_{n\to\infty} S_{n-1} = L$$

(see Exercise 49, Section 18.1), we can write

$$L = \lim_{n\to\infty} S_n = \lim_{n\to\infty} (S_{n-1} + a_n)$$

$$= \lim_{n\to\infty} S_{n-1} + \lim_{n\to\infty} a_n = L + \lim_{n\to\infty} a_n$$

which requires that $\lim_{n\to\infty} a_n = 0$. This also establishes the equivalent contrapositive statement that the series *diverges* if $\lim_{n\to\infty} a_n \neq 0$. ∎

Note that this theorem does *not* state that a series converges if $\lim_{n\to\infty} a_n = 0$, but rather that a series diverges if $\lim_{n\to\infty} a_n \neq 0$. In other words, the condition that the nth term of a series approaches zero as $n \to \infty$ is *necessary* for convergence to occur, but it is *not sufficient* to guarantee convergence.

Example 1

Determine which, if any, of the following series can be said to diverge by the nth-Term Test.

(a) $\displaystyle\sum_{n=0}^{\infty} 2^n$ (b) $\displaystyle\sum_{n=0}^{\infty} \frac{1}{2^n}$ (c) $\displaystyle\sum_{n=1}^{\infty} \frac{n!}{2n! + 1}$ (d) $\displaystyle\sum_{n=1}^{\infty} \frac{10}{n}$

Solution: For the series in parts (b) and (d), we have, respectively,

$$\lim_{n\to\infty} \frac{1}{2^n} = 0 \quad \text{and} \quad \lim_{n\to\infty} \frac{10}{n} = 0$$

Hence the nth-Term Test does not apply and we can draw *no* conclusions about convergence or divergence. We will see later in this section that one of these two series diverges and the other converges.

For the series in parts (a) and (c), we have, respectively,

$$\lim_{n\to\infty} 2^n = \infty \quad \text{and} \quad \lim_{n\to\infty} \frac{n!}{2n! + 1} = \frac{1}{2}$$

which implies, by the nth-Term Test, that both of the series

$$\sum_{n=0}^{\infty} 2^n \quad \text{and} \quad \sum_{n=1}^{\infty} \frac{n!}{2n! + 1}$$

diverge. ∎

Since the sum of a series is actually defined as the limit of a sequence of partial sums, it is natural to attempt to determine the convergence or divergence of a series by examining $\lim_{n\to\infty} S_n$. If this limit exists, then the corresponding series converges to this limit value. Of course, the task of finding the nth term of the sequence $\{S_n\}$ often requires some cleverness. The next example demonstrates some important considerations for determining the convergence of a series from its sequence of partial sums.

Example 2

Use the sequence $\{S_n\}$ to determine if the series

$$\sum_{n=1}^{\infty} \frac{2}{4n^2 - 1}$$

converges or diverges. If it converges, determine its sum.

Solution: Let us write out the first several terms of $\{S_n\}$:

$$S_1 = \tfrac{2}{3} \approx 0.667$$
$$S_2 = \tfrac{2}{3} + \tfrac{2}{15} = 0.800$$
$$S_3 = \tfrac{2}{3} + \tfrac{2}{15} + \tfrac{2}{35} \approx 0.857$$
$$S_4 = \tfrac{2}{3} + \tfrac{2}{15} + \tfrac{2}{35} + \tfrac{2}{63} \approx 0.889$$
$$S_5 = \tfrac{2}{3} + \tfrac{2}{15} + \tfrac{2}{35} + \tfrac{2}{63} + \tfrac{2}{99} \approx 0.909$$

If we had sufficient patience, we could continue this process and determine that

$$S_{10} \approx 0.952, \qquad S_{50} \approx 0.990, \qquad S_{100} \approx 0.995, \qquad S_{500} \approx 0.999$$

By that time we might suspect that this series converges to 1. But how can we be sure? In this case we can develop a formula for S_n by observing that the nth term of the series is

$$a_n = \frac{2}{4n^2 - 1} = \frac{2}{(2n - 1)(2n + 1)}$$

which has a *partial fraction* representation of

$$a_n = \frac{1}{2n - 1} - \frac{1}{2n + 1}$$

Now the terms of $\{S_n\}$ can be written in the form

$$S_1 = (\tfrac{1}{1} - \tfrac{1}{3}) = 1 - \tfrac{1}{3}$$
$$S_2 = (\tfrac{1}{1} - \tfrac{1}{3}) + (\tfrac{1}{3} - \tfrac{1}{5}) = 1 - \tfrac{1}{5}$$
$$S_3 = (\tfrac{1}{1} - \tfrac{1}{3}) + (\tfrac{1}{3} - \tfrac{1}{5}) + (\tfrac{1}{5} - \tfrac{1}{7}) = 1 - \tfrac{1}{7}$$
$$\vdots$$
$$S_n = \left(\frac{1}{1} - \frac{1}{3}\right) + \left(\frac{1}{3} - \frac{1}{5}\right) + \cdots + \left(\frac{1}{2n - 1} - \frac{1}{2n + 1}\right)$$
$$= 1 - \frac{1}{2n + 1}$$

and we can verify that the series converges to 1, since

$$\sum_{n=1}^{\infty} \frac{2}{4n^2 - 1} = \lim_{n \to \infty} S_n = \lim_{n \to \infty} \left(1 - \frac{1}{2n + 1} \right) = 1$$

■

In Example 2 there is something very special about the new form for the series, which is obtained by replacing

$$\left(\frac{2}{4n^2 - 1} \right) \quad \text{by} \quad \left(\frac{1}{2n - 1} - \frac{1}{2n + 1} \right)$$

Notice that in the new form

$$\sum_{n=1}^{\infty} \left(\frac{1}{2n - 1} - \frac{1}{2n + 1} \right)$$

$$= \left(\frac{1}{1} - \frac{1}{3} \right) + \left(\frac{1}{3} - \frac{1}{5} \right) + \left(\frac{1}{5} - \frac{1}{7} \right) + \left(\frac{1}{7} - \frac{1}{9} \right) + \cdots$$

$$= 1 - \frac{1}{3} + \frac{1}{3} - \frac{1}{5} + \frac{1}{5} - \frac{1}{7} + \frac{1}{7} - \frac{1}{9} + \cdots$$

each term, after the first one, is canceled by its successor. A series written in such a form is called a **telescoping** series.

When possible, it is advantageous to write a series in telescoping form, for then it is usually a simple matter to derive a general formula for S_n from which we can then determine the sum of the series. As we saw in Example 2, one way to write a series in telescoping form is to decompose the nth term of the series into its partial fractions.

Example 3

Write the series

$$\sum_{n=0}^{\infty} \frac{3}{2^n}$$

in telescoping form and then determine its sum.

Solution: First, we wish to determine A such that

$$\frac{3}{2^n} = \frac{A}{2^n} - \frac{A}{2^{n+1}}$$

This requires that

$$\frac{3}{2^n} = \frac{2A - A}{2^{n+1}}$$

$$\frac{6}{2^{n+1}} = \frac{A}{2^{n+1}}$$

Thus we have $6 = A$, and we can write the series as

$$\sum_{n=0}^{\infty} \frac{3}{2^n} = \sum_{n=0}^{\infty} \left(\frac{6}{2^n} - \frac{6}{2^{n+1}} \right)$$

Now since n starts at 0, we have, as shown in Figure 18.4,

$$S_0 = \left(\frac{6}{1} - \frac{6}{2} \right) = 6 - \frac{6}{2}$$

$$S_1 = \left(\frac{6}{1} - \frac{6}{2} \right) + \left(\frac{6}{2} - \frac{6}{4} \right) = 6 - \frac{6}{4}$$

$$S_2 = \left(\frac{6}{1} - \frac{6}{2} \right) + \left(\frac{6}{2} - \frac{6}{4} \right) + \left(\frac{6}{4} - \frac{6}{8} \right) = 6 - \frac{6}{8}$$

$$S_3 = \left(\frac{6}{1} - \frac{6}{2} \right) + \left(\frac{6}{2} - \frac{6}{4} \right) + \left(\frac{6}{4} - \frac{6}{8} \right) + \left(\frac{6}{8} - \frac{6}{16} \right) = 6 - \frac{6}{16}$$

$$\vdots$$

$$S_n = \left(\frac{6}{1} - \frac{6}{2} \right) + \left(\frac{6}{2} - \frac{6}{4} \right) + \left(\frac{6}{4} - \frac{6}{8} \right) + \cdots + \left(\frac{6}{2^n} - \frac{6}{2^{n+1}} \right)$$

$$= 6 - \frac{6}{2^{n+1}}$$

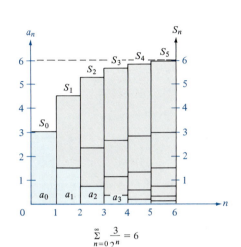

$$\sum_{n=0}^{\infty} \frac{3}{2^n} = 6$$

FIGURE 18.4

Therefore, the series converges to

$$\lim_{n \to \infty} S_n = \lim_{n \to \infty} \left(6 - \frac{6}{2^{n+1}} \right) = 6 - 0 = 6 \qquad \blacksquare$$

In the remaining portion of this section we will discuss a special type of series that has a simple arithmetic test for convergence. The series in Example 3 is of this type and has the form

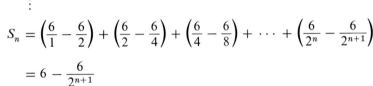

$$\sum_{n=0}^{\infty} \frac{3}{2^n} = \sum_{n=0}^{\infty} 3 \left(\frac{1}{2} \right)^n = \sum_{n=0}^{\infty} ar^n$$

which we call a **geometric series.**

Definition of Geometric Series	The series given by

$$\sum_{n=0}^{\infty} ar^n = a + ar + ar^2 + \cdots + ar^n + \cdots \qquad (a \ne 0)$$

is called a **geometric series** with ratio r.

The conditions for the convergence or divergence of a geometric series are given in the following theorem.

THEOREM 18.5
(*Convergence of a Geometric Series*)

The geometric series

$$\sum_{n=0}^{\infty} ar^n = a + ar + ar^2 + \cdots$$

has these properties:

i.   If $|r| \geq 1$, it *diverges*.
ii. If $|r| < 1$, it *converges* and has the sum

$$\sum_{n=0}^{\infty} ar^n = \frac{a}{1-r}$$

Proof: Let $S_n = a + ar + ar^2 + \cdots + ar^{n-1}$

Then multiplication by r yields

$$rS_n = ar + ar^2 + ar^3 + \cdots + ar^n$$

By subtracting these two equations, we obtain the telescoping sum

$$S_n = a + ar + ar^2 + \cdots + ar^{n-1}$$
$$-rS_n = \quad - ar - ar^2 - \cdots - ar^{n-1} - ar^n$$
$$S_n - rS_n = a - ar^n$$

Therefore, $S_n(1 - r) = a(1 - r^n)$

$$S_n = \frac{a}{1-r}(1 - r^n)$$

Consequently, if $|r| < 1$, then $\lim_{n\to\infty} r^n = 0$ and we obtain

$$\lim_{n\to\infty} S_n = \lim_{n\to\infty} \frac{a}{1-r}(1 - r^n) = \frac{a}{1-r} \lim_{n\to\infty} (1 - r^n) = \frac{a}{1-r}$$

which means the series *converges* to $a/(1 - r)$.
   If $|r| \geq 1$, then

$$\lim_{n\to\infty} ar^n \neq 0$$

and by Theorem 18.4 we conclude that the series diverges.   ∎

Example 4

Show that the series given in Example 3 is a geometric series and verify that its sum is 6.

Solution: The series in Example 3 can be written as

$$\sum_{n=0}^{\infty} \frac{3}{2^n} = \sum_{n=0}^{\infty} 3\left(\frac{1}{2}\right)^n = 3(1) + 3\left(\frac{1}{2}\right) + 3\left(\frac{1}{2}\right)^2 + \cdots$$

from which we conclude that the series is geometric with $a = 3$ and $r = \frac{1}{2}$. Since $r < 1$, it converges to the value

$$\frac{a}{1 - r} = \frac{3}{1 - \frac{1}{2}} = 6$$

∎

Example 5

Express the repeating decimal $0.080808\ldots$ as a ratio of two integers by finding the limit of the geometric series

$$\frac{8}{10^2} + \frac{8}{10^4} + \frac{8}{10^6} + \cdots$$

Solution: Since

$$a + ar + ar^2 + ar^3 + \cdots$$

$$= \frac{8}{10^2} + \frac{8}{10^4} + \frac{8}{10^6} + \frac{8}{10^8} + \cdots$$

$$= \left(\frac{8}{10^2}\right)\left(\frac{1}{10^2}\right)^0 + \left(\frac{8}{10^2}\right)\left(\frac{1}{10^2}\right)^1 + \left(\frac{8}{10^2}\right)\left(\frac{1}{10^2}\right)^2 + \left(\frac{8}{10^2}\right)\left(\frac{1}{10^2}\right)^3 + \cdots$$

we have $a = 8/10^2$ and $r = 1/10^2$. Thus

$$0.080808\ldots = \sum_{n=0}^{\infty} \left(\frac{8}{10^2}\right)\left(\frac{1}{10^2}\right)^n = \frac{8/10^2}{1 - (1/10^2)} = \frac{8}{99}$$

∎

Section Exercises (18.2)

In Exercises 1–6, find the first five terms of the sequence of partial sums.

C **1.** $1 + \frac{1}{4} + \frac{1}{9} + \frac{1}{16} + \frac{1}{25} + \cdots$

C **2.** $\frac{1}{2 \cdot 3} + \frac{2}{3 \cdot 4} + \frac{3}{4 \cdot 5} + \frac{4}{5 \cdot 6} + \frac{5}{6 \cdot 7} + \cdots$

C S **3.** $3 - \frac{9}{2} + \frac{27}{4} - \frac{81}{8} + \frac{243}{16} - \cdots$

C **4.** $\frac{1}{1} + \frac{1}{3} + \frac{1}{5} + \frac{1}{9} + \frac{1}{11} + \cdots$

C **5.** $\sum_{n=1}^{\infty} \frac{3}{2^{n-1}}$

6. $\sum_{n=1}^{\infty} \frac{(-1)^{n+1}}{n!}$

In Exercises 7–12, verify that the infinite series diverges.

7. $\frac{1}{2} + \frac{2}{3} + \frac{3}{4} + \frac{4}{5} + \cdots$

8. $3 - \frac{9}{2} + \frac{27}{4} - \frac{81}{8} + \frac{243}{16} - \cdots$

S **9.** $\frac{1}{4} + \frac{3}{8} + \frac{7}{16} + \frac{15}{32} + \frac{31}{64} + \cdots$

10. $\frac{1}{20} - \frac{3}{40} + \frac{5}{60} - \frac{7}{80} + \frac{9}{100} - \cdots$

11. $\frac{3}{4} - \frac{5}{8} + \frac{9}{16} - \frac{17}{32} + \frac{33}{64} - \cdots$

12. $\frac{1}{2} + \frac{2}{4} + \frac{6}{8} + \frac{24}{16} + \frac{120}{32} + \frac{720}{64} + \cdots$

In Exercises 13–18, verify that the infinite series converges.

S **13.** $2 + \frac{3}{2} + \frac{9}{8} + \frac{27}{32} + \frac{81}{128} + \cdots$

14. $2 - 1 + \frac{1}{2} - \frac{1}{4} + \frac{1}{8} - \cdots$

15. $1 + 0.9 + 0.81 + 0.729 + 0.6561 + \cdots$

16. $1 - 0.6 + 0.36 - 0.216 + 0.1296 - \cdots$

S **17.** $\frac{1}{1 \cdot 2} + \frac{1}{2 \cdot 3} + \frac{1}{3 \cdot 4} + \frac{1}{4 \cdot 5} + \frac{1}{5 \cdot 6} + \cdots$

$$\left[\text{Hint: } \frac{1}{n(n + 1)} = \frac{1}{n} - \frac{1}{n + 1}.\right]$$

18. $\dfrac{1}{1\cdot 3} + \dfrac{1}{2\cdot 4} + \dfrac{1}{3\cdot 5} + \dfrac{1}{4\cdot 6} + \dfrac{1}{5\cdot 7} + \cdots$

$\left[\text{Hint: } \dfrac{1}{n(n+2)} = \dfrac{1}{2n} - \dfrac{1}{2(n+2)}.\right]$

In Exercises 19–22, express the repeated decimal as a geometric series and write its sum as the ratio of two integers.

19. $0.666\overline{6}$ **20.** $0.23\overline{23}$

21. $0.075\overline{75}$ **22.** $0.215\overline{15}$

In Exercises 23–28, find the sum of the series.

S 23. $\displaystyle\sum_{n=1}^{\infty} 2\left(\dfrac{3}{4}\right)^{n-1}$ **24.** $\displaystyle\sum_{n=0}^{\infty} \dfrac{4}{2^n}$

25. $\displaystyle\sum_{n=2}^{\infty} \dfrac{1}{n^2 - 1}$ **26.** $\displaystyle\sum_{n=1}^{\infty} \dfrac{1}{n(n+1)}$

27. $\displaystyle\sum_{n=1}^{\infty} \dfrac{1}{(2n+1)(2n+3)}$ **28.** $\displaystyle\sum_{n=1}^{\infty} \dfrac{4}{n(n+2)}$

In Exercises 29–40, determine the convergence or divergence of the series.

29. $\displaystyle\sum_{n=1}^{\infty} \dfrac{n+10}{10n+1}$ **30.** $\displaystyle\sum_{n=0}^{\infty} \dfrac{4}{2^n}$

31. $\displaystyle\sum_{n=1}^{\infty} \left(\dfrac{1}{n} - \dfrac{1}{n+2}\right)$ **32.** $\displaystyle\sum_{n=1}^{\infty} \dfrac{n+1}{2n-1}$

S 33. $\displaystyle\sum_{n=1}^{\infty} \dfrac{3n-1}{2n+1}$ **34.** $\displaystyle\sum_{n=0}^{\infty} \dfrac{1}{4^n}$

35. $\displaystyle\sum_{n=0}^{\infty} (1.075)^n$ **36.** $\displaystyle\sum_{n=1}^{\infty} \dfrac{2^n}{100}$

37. $\displaystyle\sum_{n=2}^{\infty} \dfrac{n}{\ln n}$ **S 38.** $\displaystyle\sum_{n=1}^{\infty} \dfrac{2^n}{n^2}$

39. $\displaystyle\sum_{n=1}^{\infty} \left(1 + \dfrac{k}{n}\right)^n$ **40.** $\displaystyle\sum_{n=1}^{\infty} \dfrac{1}{n(n+3)}$

41. A ball is dropped from a height of 8 ft. Each time it drops h feet, it rebounds $0.7h$ feet. Find the total distance traveled by the ball.

42. Express $0.428571428571428571\ldots$ as the quotient of two integers.

43. Prove that the repeating decimal expansion of a number that ends in zeros is equal to a decimal number ending in nines if the last nonzero digit is decreased by one in the first representation (e.g., $\frac{3}{4} = 0.7500\ldots = 0.74999\ldots$).

44. Prove that every decimal with a repeating pattern of digits is a rational number.

45. Show that the series $\sum_{n=1}^{\infty} a_n$ can always be written in the telescopic form

$$\sum_{n=1}^{\infty} [(c - S_{n-1}) - (c - S_n)]$$

if $S_0 = 0$ and S_n is the nth partial sum.

46. Let $\sum_{n=1}^{\infty} a_n$ be a convergent series and let $R_N = a_{N+1} + a_{N+2} + \cdots$ be the remainder of the series after the first N terms. Prove that $\lim_{N\to\infty} R_N = 0$.

For Review and Class Discussion

True or False

1. _____ If $\lim_{n\to\infty} a_n = 0$, then $\sum_{n=1}^{\infty} a_n$ converges.

2. _____ If $\sum_{n=1}^{\infty} a_n = L$, then $\sum_{n=0}^{\infty} a_n = L + a_0$.

3. _____ If $a_n = b_n - b_{n+1}$ and $\lim_{n\to\infty} b_n = 0$, then $\sum_{n=1}^{\infty} a_n = b_1$.

4. _____ If $|r| < 1$, then $\sum_{n=1}^{\infty} ar^n = a/(1-r)$.

5. _____ The series $\sum_{n=1}^{\infty} n/1000(n+1)$ diverges.

18.3 *p*-Series and the Integral Test

Purpose
- To determine the convergence or divergence of a *p*-series as well as to approximate the limit of a convergent *p*-series.
- To introduce the Integral Test and Root Test for convergence (or divergence) of a series.
- To use the Integral Test to approximate the limit of a series.

Our minds are finite, and yet even in these circumstances of finitude we are surrounded by possibilities that are infinite, and the purpose of human life is to grasp as much as we can out of that infinitude.

—Alfred Whitehead (1861–1947)

In this section we investigate a second type of series that has a simple arithmetic test for convergence or divergence. Along with geometric series, this new type of series (called *p*-series) will serve as the basic series for the comparison tests of the next section.

We begin by comparing the following two series. In both series we assume $p > 0$.

$$\sum_{n=1}^{\infty} \frac{1}{p^n} = \frac{1}{p} + \frac{1}{p^2} + \frac{1}{p^3} + \frac{1}{p^4} + \cdots$$

$$\sum_{n=1}^{\infty} \frac{1}{n^p} = \frac{1}{1^p} + \frac{1}{2^p} + \frac{1}{3^p} + \frac{1}{4^p} + \cdots$$

The first of the two series is geometric with $a = 1$ and $r = 1/p$. Hence it converges for $p > 1$ and diverges for $0 < p \leq 1$. In the second series the roles of n and p have been interchanged, but, surprisingly, the values of p for which the new series converges remain unchanged. We call the second series a **p-series,** and we outline its conditions for convergence or divergence in the following theorem.

THEOREM 18.6
(Convergence of a p-Series)

The *p*-series

$$\sum_{n=1}^{\infty} \frac{1}{n^p} = \frac{1}{1^p} + \frac{1}{2^p} + \frac{1}{3^p} + \cdots$$

has these properties:

i. If $0 < p \leq 1$, it *diverges*.
ii. If $p > 1$, it *converges*.
iii. If $\sum_{n=1}^{\infty} 1/n^p = S$, the **remainder** $R_N = S - S_N$ is bounded by

$$0 < R_N < \frac{1}{N^{p-1}(p-1)}$$

Proof: Recall from Section 10.11 (Exercise 27) that the improper integral

$$\int_1^{\infty} \frac{1}{x^p}\, dx$$

converges if $p > 1$ and diverges if $0 < p \leq 1$. We can establish the con-

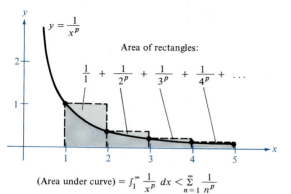

$$\text{(Area under curve)} = \int_1^\infty \frac{1}{x^p}\, dx < \sum_{n=1}^\infty \frac{1}{n^p}$$

FIGURE 18.5

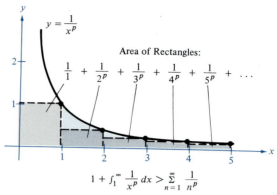

$$1 + \int_1^\infty \frac{1}{x^p}\, dx > \sum_{n=1}^\infty \frac{1}{n^p}$$

FIGURE 18.6

vergence or divergence of a p-series by a comparison to this improper integral, which represents the area of the shaded region in Figures 18.5 and 18.6.

By combining the inequalities given with Figures 18.5 and 18.6, we obtain

$$\int_1^\infty \frac{1}{x^p}\, dx < \sum_{n=1}^\infty \frac{1}{n^p} < 1 + \int_1^\infty \frac{1}{x^p}\, dx$$

Therefore, if $0 < p \le 1$, the divergence of $\int_1^\infty (1/x^p)\, dx$ implies the divergence of $\sum_{n=1}^\infty (1/n^p)$. And if $p > 1$, the convergence of $1 + \int_1^\infty (1/x^p)\, dx$ implies the convergence of $\sum_{n=1}^\infty (1/n^p)$. Finally, for $p > 1$, we have

$$\int_N^\infty \frac{1}{x^p}\, dx = \frac{x^{1-p}}{1-p}\Bigg]_N^\infty = \left(\frac{-N^{1-p}}{1-p}\right) = \frac{1}{N^{p-1}(p-1)}$$

and we obtain

$$S - S_N = R_N = \sum_{n=N+1}^\infty \frac{1}{n^p} < \int_N^\infty \frac{1}{x^p}\, dx = \frac{1}{N^{p-1}(p-1)} \qquad \blacksquare$$

Note the basic difference between Theorems 18.5 and 18.6. Both give conditions under which convergence or divergence occurs. However, Theorem 18.5 gives a *formula* for the sum of a geometric series, whereas Theorem 18.6 merely gives a *bound* for the remainder of a p-series. Except for a geometric series, it is usually quite difficult to determine the exact value to which a series converges. Thus we will be content (until Section 18.8) merely to specify reasonable approximations to the sums of convergent series.

Example 1

Show that the series $\sum_{n=1}^\infty (1/n^3)$ converges and estimate the remainder after five terms.

Solution: Since $p = 3$, we know the series converges, and by Theorem 18.6 we know that the remainder after five terms is bounded by

$$R_5 < \frac{1}{N^{p-1}(p-1)} = \frac{1}{5^2(2)} = \frac{1}{50} = 0.02$$

Using this remainder estimate, we know that the sum of this series can be approximated by

$$S_5 < S < S_5 + 0.02$$

Finally, since

$$S_5 = \frac{1}{1^3} + \frac{1}{2^3} + \frac{1}{3^3} + \frac{1}{4^3} + \frac{1}{5^3} \approx 1.186$$

we conclude that

$$1.186 < S < 1.206 \qquad \blacksquare$$

We make special note of the p-series

$$\sum_{n=1}^{\infty} \frac{1}{n}$$

which is called the **harmonic series.** This is a divergent series ($p = 1$) that we will use quite frequently in later sections as a basic series to which others are compared.

The proof of Theorem 18.6 suggests that perhaps we can use an improper integral to determine the divergence or convergence of a series provided the nth term of the series has an appropriate form. For instance, in that proof the function f was continuous, positive, and decreasing. The next theorem shows that if the terms of a series satisfy these properties, then we can test for convergence or divergence by the use of an improper integral.

THEOREM 18.7
(*Integral Test*)

If f is continuous, positive, and decreasing for $x \geq 1$, and if $a_n = f(n)$, then

$$\sum_{n=1}^{\infty} a_n \qquad \text{and} \qquad \int_1^{\infty} f(x)\, dx$$

either both converge or both diverge. Furthermore, if the series converges, then the remainder R_N is bounded by

$$0 < R_N < \int_N^{\infty} f(x)\, dx$$

The proof of this theorem is a generalization of the proof of Theorem 18.6 and is therefore omitted.

Example 2

Determine if the series

$$\sum_{n=1}^{\infty} \frac{1}{(n+1)\ln(n+1)}$$

converges or diverges.

Solution: Observe first that the terms of this series are smaller than the corresponding terms of the harmonic series $\sum_{n=1}^{\infty} 1/(n+1)$, which diverges. So the question is, does the factor $\ln(n+1)$ reduce the terms sufficiently to bring about convergence? Since

$$f(x) = \frac{1}{(x+1)\ln(x+1)}$$

is continuous, positive, and decreasing for $x \geq 1$, we can apply the Integral Test. Thus

$$\int_{1}^{\infty} \frac{1}{(x+1)\ln(x+1)}\,dx = \ln\ln(x+1)\Big]_{1}^{\infty} = \infty$$

implies that the series *diverges*.   ∎

Example 3

Determine if the series $\qquad \sum_{n=1}^{\infty} \frac{n}{e^n}$

converges or diverges. If it converges, estimate the sum with ten terms.

Solution: Since $\qquad f(x) = \frac{x}{e^x}$

satisfies the conditions for the Integral Test, we integrate by parts to obtain

$$\int_{1}^{\infty} \frac{x}{e^x}\,dx = \left[-xe^{-x} - e^{-x}\right]_{1}^{\infty} = 0 - \left(-\frac{2}{e}\right) = \frac{2}{e}$$

Thus by Theorem 18.7 the given series converges, and, furthermore, since

$$R_{10} < \int_{10}^{\infty} \frac{x}{e^x}\,dx = \frac{11}{e^{10}} \approx 0.0005$$

and $\qquad S_{10} = \dfrac{1}{e^1} + \dfrac{2}{e^2} + \dfrac{3}{e^3} + \cdots + \dfrac{10}{e^{10}} \approx 0.92037$

we conclude that

$$0.92037 < S < 0.92087$$ ∎

The next test for convergence or divergence of a series places fewer restrictions on a_n than the Integral Test does. It works especially well for series in which the nth term is raised to the nth power.

THEOREM 18.8
(*Root Test*)

The series $\sum_{n=1}^{\infty} a_n$:

i.   converges if $\lim_{n \to \infty} \sqrt[n]{|a_n|} < 1$

ii. diverges if $\lim_{n \to \infty} \sqrt[n]{|a_n|} > 1$

If $\lim_{n \to \infty} \sqrt[n]{|a_n|} = 1$, the Root Test fails to give conclusive information and another test must be tried.

Example 4

Determine the convergence or divergence of

$$\sum_{n=1}^{\infty} \frac{e^{2n}}{n^n}$$

Solution: Since

$$\lim_{n \to \infty} \sqrt[n]{|a_n|} = \lim_{n \to \infty} \sqrt[n]{\frac{e^{2n}}{n^n}} = \lim_{n \to \infty} \frac{e^{2n/n}}{n^{n/n}} = \lim_{n \to \infty} \frac{e^2}{n} = 0 < 1$$

we conclude (by the Root Test) that the series converges.   ∎

Example 5

Determine the convergence or divergence of

$$\sum_{n=1}^{\infty} \frac{n^3}{3^n}$$

Solution: First, we have

$$\lim_{n \to \infty} \sqrt[n]{|a_n|} = \lim_{n \to \infty} \sqrt[n]{\frac{n^3}{3^n}} = \lim_{n \to \infty} \frac{n^{3/n}}{3}$$

Since the limit in the numerator yields the indeterminate form ∞^0, we apply L'Hôpital's Rule as follows:

$$y = \lim_{x \to \infty} x^{3/x}$$

$$\ln y = \ln \left(\lim_{x \to \infty} x^{3/x} \right) = \lim_{x \to \infty} (\ln x^{3/x}) = \lim_{x \to \infty} \left(\frac{3}{x} \ln x \right)$$

$$= \lim_{x \to \infty} \frac{3 \ln x}{x} = \lim_{x \to \infty} \frac{3/x}{1} = 0$$

Since $\ln y = 0$, we conclude that

$$y = \lim_{x \to \infty} x^{3/x} = 1$$

and it follows that

$$\lim_{n \to \infty} \frac{n^{3/n}}{3} = \frac{1}{3} < 1$$

and by the Root Test the series converges.

∎

Section Exercises (18.3)

In Exercises 1–4, determine the convergence or divergence of the given *p*-series.

1. $1 + \dfrac{1}{\sqrt{2}} + \dfrac{1}{\sqrt{3}} + \dfrac{1}{\sqrt{4}} + \cdots$

2. $1 + \frac{1}{4} + \frac{1}{9} + \frac{1}{16} + \frac{1}{25} + \cdots$

ⓢ 3. $1 + \dfrac{1}{2\sqrt{2}} + \dfrac{1}{3\sqrt{3}} + \dfrac{1}{4\sqrt{4}} + \dfrac{1}{5\sqrt{5}} + \cdots$

4. $1 + \dfrac{1}{\sqrt[3]{4}} + \dfrac{1}{\sqrt[3]{9}} + \dfrac{1}{\sqrt[3]{16}} + \dfrac{1}{\sqrt[3]{25}} + \cdots$

In Exercises 5–8, determine the convergence or divergence of the given series by the Integral Test.

5. $\frac{1}{2} + \frac{1}{5} + \frac{1}{10} + \frac{1}{17} + \frac{1}{26} + \cdots$

6. $\frac{1}{3} + \frac{1}{5} + \frac{1}{7} + \frac{1}{9} + \frac{1}{11} + \cdots$

ⓢ 7. $\dfrac{\ln 2}{2} + \dfrac{\ln 3}{3} + \dfrac{\ln 4}{4} + \dfrac{\ln 5}{5} + \dfrac{\ln 6}{6} + \cdots$

8. $\dfrac{1}{4} + \dfrac{2}{7} + \dfrac{3}{12} + \dfrac{4}{19} + \dfrac{5}{28} + \cdots + \dfrac{n}{n^2 + 3}$

In Exercises 9–12, determine the convergence or divergence of the given series by the Root Test.

9. $\dfrac{1}{(\ln 3)^3} + \dfrac{1}{(\ln 4)^4} + \dfrac{1}{(\ln 5)^5} + \dfrac{1}{(\ln 6)^6} + \cdots$

10. $1 + \dfrac{2}{3} + \dfrac{3}{3^2} + \dfrac{4}{3^3} + \dfrac{5}{3^4} + \dfrac{6}{3^5} + \cdots$

ⓢ 11. $(\frac{1}{3})^1 + (\frac{2}{5})^2 + (\frac{3}{7})^3 + (\frac{4}{9})^4 + (\frac{5}{11})^5 + \cdots$

12. $\displaystyle\sum_{n=1}^{\infty} \left(\frac{2n}{n+1}\right)^n$

In Exercises 13–20, determine the convergence or divergence of the given series.

13. $\displaystyle\sum_{n=1}^{\infty} \frac{1}{2n - 1}$

14. $\displaystyle\sum_{n=1}^{\infty} \frac{n^2}{e^n}$

15. $\displaystyle\sum_{n=1}^{\infty} \frac{1}{n \sqrt[3]{n}}$

16. $3 \displaystyle\sum_{n=1}^{\infty} \frac{1}{n^{0.95}}$

ⓢ 17. $\displaystyle\sum_{n=1}^{\infty} (\sqrt[n]{n} - 1)^n$

18. $\displaystyle\sum_{n=1}^{\infty} (2\sqrt[n]{n} + 1)^n$

19. $\displaystyle\sum_{n=0}^{\infty} \left(\frac{2}{3}\right)^n$

20. $\displaystyle\sum_{n=1}^{\infty} \frac{n}{\sqrt{n^2 + 1}}$

ⓒⓈ 21. Estimate the sum $\sum_{n=1}^{\infty} 1/n^2$ using six terms (with remainder).

ⓒ 22. Estimate the sum $\sum_{n=1}^{\infty} 1/n^5$ using four terms (with remainder).

ⓒⓈ 23. How many terms are required in approximating the series $\sum_{n=1}^{\infty} 1/n^4$ so that $R_N < 0.001$?

ⓒ 24. How many terms are required in approximating the series $\sum_{n=1}^{\infty} 1/n^{3/2}$ so that $R_N < 0.001$?

ⓒ 25. Estimate the sum $\sum_{n=1}^{\infty} 1/(n^2 + 1)$ using ten terms (with remainder).

ⓒ 26. Estimate the sum $\sum_{n=1}^{\infty} 1/(n + 1)[\ln (n + 1)]^3$ using five terms (with remainder).

For Review and Class Discussion

True or False

1. _____ If $\lim_{n \to \infty} S_n = S$, then $\lim_{n \to \infty} (S - S_n) = \lim_{n \to \infty} R_n = 0$.

2. _____ If $\lim_{n \to \infty} S_n = S$, then there exists an N such that $R_N < 0.00001$.

3. _____ The series $\sum_{n=1}^{\infty} 1/(n - 3)$ satisfies the hypothesis of the Integral Test.

4. _____ $1.16 < \sum_{n=1}^{\infty} 1/n^3 < 1.26$.

5. _____ Since

$$\int_1^{\infty} e^{-x} \sin x \, dx = \frac{-e^{-x}(\sin x + \cos x)}{2} \Bigg]_1^{\infty}$$

$$= \frac{\sin(1) + \cos(1)}{2e} \approx 0.254$$

we can apply the Integral Test to conclude that $\sum_{n=1}^{\infty} e^{-n} \sin n$ converges.

18.4 Comparisons of Series

Purpose
- To introduce two comparison tests for convergence and divergence of series.
- To compare the rates of convergence (or divergence) of some common series.

The harmonic series, with as many terms, will surpass the geometric and if it be permitted to take a jump into geometry, then it also follows that the area between a hyperbola and its asymptote is infinite.

—*Jacob Bernoulli (1654–1705)*

In the previous section we developed several tests for convergence or divergence of special series. In each case the terms of the series were quite simple in form and they had to possess certain characteristics in order to apply the tests for convergence. The slightest deviation from these special characteristics could make a test nonapplicable. For instance:

1. The series $\sum_{n=1}^{\infty} \dfrac{1}{(2)^n}$ is geometric, but $\sum_{n=1}^{\infty} \dfrac{1}{n(2)^n}$ is not.

2. The series $\sum_{n=1}^{\infty} \dfrac{1}{n^3}$ is a *p*-series, but $\sum_{n=1}^{\infty} \dfrac{1}{n^3 + 1}$ is not.

3. The term $a_n = \dfrac{n}{(n^2 + 3)^2}$ is easily integrated, whereas $b_n = \dfrac{n^2}{(n^2 + 3)^2}$ is not.

In this section we discuss two new tests that permit us to *compare* a series, having similar but more complicated terms, to our previous series types that have direct tests for convergence or divergence. The proof of both of these tests depends on the completeness property of the real numbers. For our purposes we simply state a consequence of this completeness property, which is then assumed in the proofs: *If $\{a_n\}$ is nondecreasing and bounded (that is, $a_n \leq a_{n+1} \leq C$ for all n), then $\{a_n\}$ converges.* (See Figure 18.7.)

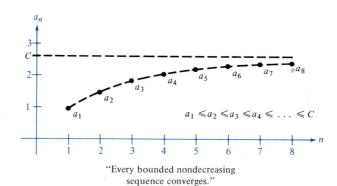

"Every bounded nondecreasing
sequence converges."

FIGURE 18.7

THEOREM 18.9 (*Comparison Test*)	Suppose N is a positive integer and $0 \leq a_n \leq b_n$ for all $n > N$; then: i. If $\sum_{n=1}^{\infty} b_n$ converges, then $\sum_{n=1}^{\infty} a_n$ converges. ii. If $\sum_{n=1}^{\infty} a_n$ diverges, then $\sum_{n=1}^{\infty} b_n$ diverges.

Proof:

i. Let $L = \sum_{n=1}^{\infty} b_n$ and $S_n = \sum_{i=1}^{n} a_i$. Since $0 \leq a_n \leq b_n$ for $n > N$, we know that the sequence

$$\{S_{N+n}\} = \{S_N, S_{N+1}, S_{N+2}, \ldots\}$$

is nondecreasing and bounded above by $S_N + L$; hence it must converge. And since

$$\lim_{n \to \infty} S_{N+n} = \lim_{n \to \infty} S_n = \sum_{n=1}^{\infty} a_n$$

it follows that $\sum_{n=1}^{\infty} a_n$ converges.

ii. This part of the theorem is logically equivalent to part i because it is the *contrapositive* of part i. ■

We need to be careful when using the Comparison Test. Students tend to misuse the test because they fail to keep in mind that $a_n \leq b_n$ for *both* parts of the theorem. Informally, the test says that:

i. A series that is term by term "smaller" than a convergent series must also converge.

ii. A series that is term by term "larger" than a divergent series must also diverge.

Example 1

Use the Comparison Test to show that the series

$$\sum_{n=1}^{\infty} \frac{1}{2 + 3^n}$$

converges.

Solution: Suspecting convergence, we compare

$$\sum_{n=1}^{\infty} \frac{1}{2 + 3^n} \qquad \text{to} \qquad \sum_{n=1}^{\infty} \frac{1}{3^n} \qquad \text{(convergent geometric series)}$$

Term-by-term comparison yields

$$a_n = \frac{1}{2 + 3^n} < \frac{1}{3^n} = b_n \qquad (n \geq 1)$$

Thus it follows by the Comparison Test that the series

$$\sum_{n=1}^{\infty} \frac{1}{2 + 3^n}$$

converges. ■

Example 2

Use the Comparison Test to see if the series

$$\sum_{n=1}^{\infty} \frac{1}{2 + \sqrt{n}}$$

converges or diverges.

Solution: Suspecting divergence, we compare

$$\sum_{n=1}^{\infty} \frac{1}{2 + \sqrt{n}} \qquad \text{to} \qquad \sum_{n=1}^{\infty} \frac{1}{n^{1/2}} \qquad \text{(divergent *p*-series)}$$

Term-by-term comparison yields

$$\frac{1}{2 + \sqrt{n}} \leq \frac{1}{\sqrt{n}}$$

which does not meet the requirements for divergence. (Remember that if term by term comparison reveals a series that is *smaller* than a divergent series, the Comparison Test tells us nothing.) Still expecting the series to diverge, we compare

$$\sum_{n=1}^{\infty} \frac{1}{2 + \sqrt{n}} \quad \text{to} \quad \sum_{n=1}^{\infty} \frac{1}{n} \quad \text{(divergent harmonic series)}$$

In this case term-by-term comparison yields

$$a_n = \frac{1}{n} \le \frac{1}{2 + \sqrt{n}} = b_n \quad (n \ge 4)$$

and by part ii of the Comparison Test, the series

$$\sum_{n=1}^{\infty} \frac{1}{2 + \sqrt{n}}$$

must diverge because it is term-by-term ($n \ge 4$) greater than a divergent series. ∎

Notice that in the Comparison Test we listed no estimate for the sum of a series. Nevertheless, the Comparison Test can be used to obtain an approximation for a given series provided the series to which it is compared converges *rapidly*. For instance, consider the two series

$$\sum_{n=0}^{\infty} \frac{1}{n!} \quad \text{and} \quad \sum_{n=0}^{\infty} \frac{1}{2^n} = 2$$

where

$$\frac{1}{n!} \le \frac{1}{2^n}$$

for $n > 1$. Since the factorial series is converging more rapidly than the geometric series (see Table 18.3), and since

$$\sum_{n=0}^{10} \frac{1}{2^n} \approx 1.999$$

TABLE 18.3

N	0	2	4	6	8	10	limit
$\sum_{n=0}^{N} \frac{1}{n!}$	1.000	2.500	2.708	2.718	2.718	2.718	L
$\sum_{n=0}^{N} \frac{1}{2^n}$	1.000	1.750	1.938	1.984	1.996	1.999	2

is within 0.001 units of its limit, we know that

$$\sum_{n=0}^{10} \frac{1}{n!} \approx 2.718$$

must be within 0.001 units of L. Thus by summing only *ten* terms of the series $\sum_{n=0}^{\infty} (1/n!)$, we are guaranteed to obtain a value within 0.001 of the actual sum.

However, if we make a comparison to a series that converges slowly, then we may have to sum an unreasonably large number of terms to obtain this same degree of accuracy. For example, if we made a comparison to the slowly converging series

$$\sum_{n=1}^{\infty} \frac{1}{n^2} = \frac{\pi^2}{6} \approx 1.645$$

we would need to sum 1000 terms to guarantee an accuracy of 0.001 units. (See Table 18.4.)

TABLE 18.4

N	1	10	50	100	150
$\sum_{n=1}^{N} \frac{1}{n^2}$	1.000	1.550	1.625	1.635	1.638

It is often the case that a given series closely resembles a *p*-series or a geometric series, yet we cannot establish the proper term-by-term comparison necessary to apply the Comparison Test (Theorem 18.9). Under these (and other) circumstances there is a second comparison test, called the **Limit Comparison Test,** that may be applicable.

THEOREM 18.10
(*Limit Comparison Test*)

Suppose a_n, $b_n > 0$ and

$$\lim_{n \to \infty} \left(\frac{a_n}{b_n} \right) = L > 0$$

then the two series

$$\sum_{n=1}^{\infty} a_n \quad \text{and} \quad \sum_{n=1}^{\infty} b_n$$

either both converge or both diverge.

Proof: Since a_n, $b_n > 0$ and $\lim_{n \to \infty} (a_n/b_n) = L$, there exists an $N > 0$ such that

$$0 < \frac{a_n}{b_n} < L + 1$$

for $n \geq N$. This means that

$$0 < a_n < (L + 1)b_n$$

Hence by the Comparison Test, the convergence of $\sum_{n=1}^{\infty} b_n$ implies the convergence of $\sum_{n=1}^{\infty} a_n$.

A similar argument, using $\lim_{n \to \infty} (b_n/a_n) = (1/L)$, shows that the convergence of $\sum_{n=1}^{\infty} a_n$ implies the convergence of $\sum_{n=1}^{\infty} b_n$. The divergence follows from the logical equivalence of the contrapositive, and hence the theorem is established. ■

Example 3

Determine if the series

$$\sum_{n=1}^{\infty} \frac{k}{cn + d}, \qquad k, c, d > 0$$

converges or diverges.

Solution: Suspecting divergence, we compare

$$\sum_{n=1}^{\infty} \frac{k}{cn + d} \qquad \text{to} \qquad \sum_{n=1}^{\infty} \frac{1}{n} \qquad \text{(divergent harmonic series)}$$

Without knowing the specific values of k, c, and d, term-by-term comparison is not feasible. Thus we consider the limit

$$\lim_{n \to \infty} \frac{a_n}{b_n} = \lim_{n \to \infty} \left(\frac{k}{cn + d}\right)\left(\frac{n}{1}\right) = \lim_{n \to \infty} \frac{kn}{cn + d} = \frac{k}{c} \neq 0$$

Since this limit is greater than 0, we conclude from the Limit Comparison Test that the series

$$\sum_{n=1}^{\infty} \frac{k}{cn + d}$$

also diverges. This latter series is referred to as the **general harmonic series.** ■

The Limit Comparison Test works well for comparing "messy" algebraic series to a p-series. In choosing an appropriate p-series, we must choose one with an nth term of the same magnitude as the nth term of the given series. For example:

1. Given $\sum \dfrac{1}{3n^2 - 4n + 5}$, choose $\sum \dfrac{1}{n^2}$.

2. Given $\sum \dfrac{1}{\sqrt{3n-2}}$, choose $\sum \dfrac{1}{\sqrt{n}}$.

3. Given $\sum \dfrac{n^2-10}{4n^5+n^3}$, choose $\sum \dfrac{1}{n^3}$.

4. Given $\sum \dfrac{\sqrt{n}}{\sqrt{n^3+1}}$, choose $\sum \dfrac{1}{n}$.

In other words, *when choosing a series to which we want to make a comparison, we disregard all but the highest powers of n in both the numerator and denominator.*

Example 4

Determine if the series $\displaystyle\sum_{n=1}^{\infty} \dfrac{4\sqrt{n}-1}{n^2+2\sqrt{n}}$

converges or diverges.

Solution: Disregarding all but the highest powers of n in the numerator and the denominator, we compare

$$\sum_{n=1}^{\infty} \dfrac{4\sqrt{n}-1}{n^2+2\sqrt{n}} \qquad \text{to} \qquad \sum_{n=1}^{\infty} \dfrac{\sqrt{n}}{n^2} = \sum_{n=1}^{\infty} \dfrac{1}{n^{3/2}} \qquad \text{(convergent } p\text{-series)}$$

Since

$$\lim_{n\to\infty} \dfrac{a_n}{b_n} = \lim_{n\to\infty} \left(\dfrac{4\sqrt{n}-1}{n^2+2\sqrt{n}}\right)\left(\dfrac{n^{3/2}}{1}\right)$$

$$= \lim_{n\to\infty} \dfrac{4n^2-n^{3/2}}{n^2+2\sqrt{n}} = 4$$

we conclude by the Limit Comparison Test that the given series,

$$\sum_{n=1}^{\infty} \dfrac{4\sqrt{n}-1}{n^2+2\sqrt{n}}$$

also converges. ■

Example 5

Determine if the series $\displaystyle\sum_{n=1}^{\infty} \dfrac{n2^n+5}{4n^3+3n}$

converges or diverges.

Solution: A reasonable comparison would be

$$\sum_{n=1}^{\infty} \frac{n2^n + 5}{4n^3 + 3n} \quad \text{to} \quad \sum_{n=1}^{\infty} \frac{2^n}{n^2}$$

Note that the second series diverges by the *n*th-Term Test since

$$\lim_{n \to \infty} \frac{2^n}{n^2} \neq 0$$

From the limit

$$\lim_{n \to \infty} \frac{a_n}{b_n} = \lim_{n \to \infty} \left(\frac{n2^n + 5}{4n^3 + 3n} \right) \left(\frac{n^2}{2^n} \right) = \lim_{n \to \infty} \frac{1 + (5/2^n n)}{4 + (3/n^2)} = \frac{1}{4}$$

we conclude that the given series diverges. ∎

Section Exercises (18.4)

In Exercises 1–20, use one of the comparison tests to show convergence or divergence.

1. $\displaystyle\sum_{n=0}^{\infty} \frac{1}{n^2 + 1}$

2. $\displaystyle\sum_{n=1}^{\infty} \frac{1}{4n - 1}$

S 3. $\displaystyle\sum_{n=0}^{\infty} \frac{n}{n^2 + 1}$

4. $\displaystyle\sum_{n=2}^{\infty} \frac{1}{\sqrt{n^2 - 1}}$

5. $\displaystyle\sum_{n=0}^{\infty} \frac{1}{\sqrt{n^2 + 1}}$

6. $\displaystyle\sum_{n=0}^{\infty} \frac{1}{\sqrt{n^3 + 1}}$

7. $\displaystyle\sum_{n=0}^{\infty} \frac{1}{3^n + 1}$

S 8. $\displaystyle\sum_{n=0}^{\infty} \frac{1}{2^n - 5}$

9. $\displaystyle\sum_{n=1}^{\infty} \frac{n + 3}{n(n + 2)}$

10. $\displaystyle\sum_{n=1}^{\infty} \frac{1}{n(n^2 + 1)}$

S 11. $\displaystyle\sum_{n=1}^{\infty} \left(\frac{1}{n^2} - \frac{1}{2^n} \right)$

12. $\displaystyle\sum_{n=1}^{\infty} \left(\frac{1}{n + 1} - \frac{1}{n + 2} \right)$

13. $\displaystyle\sum_{n=1}^{\infty} \left(\frac{1}{n^2} - \frac{1}{n} \right)$

14. $\displaystyle\sum_{n=1}^{\infty} \frac{\ln n}{n + 1}$

15. $\displaystyle\sum_{n=1}^{\infty} \frac{1}{n \sqrt{n^2 + 1}}$

16. $\displaystyle\sum_{n=1}^{\infty} \frac{n}{(n + 1)2^{n-1}}$

S 17. $\displaystyle\sum_{n=0}^{\infty} \frac{1}{n!}$

18. $\displaystyle\sum_{n=1}^{\infty} \frac{1}{n + \sqrt{n^2 + 1}}$

19. $\displaystyle\sum_{n=1}^{\infty} \sin \frac{1}{n}$

20. $\displaystyle\sum_{n=1}^{\infty} \tan \frac{1}{n}$

In Exercises 21–28, test for convergence or divergence without using any of the eight tests more than once.
(a) *n*th-Term Test   (b) Geometric Series Test
(c) telescoping series   (d) *p*-Series Test
(e) Integral Test   (f) Root Test
(g) Comparison Test   (h) Limit Comparison Test

21. $\displaystyle\sum_{n=1}^{\infty} \frac{\sqrt{n}}{n}$

22. $\displaystyle\sum_{n=0}^{\infty} 5 \left(-\frac{1}{5} \right)^n$

23. $\displaystyle\sum_{n=1}^{\infty} \frac{1}{3^n + 2}$

24. $\displaystyle\sum_{n=4}^{\infty} \frac{1}{3n^2 - 2n - 15}$

25. $\displaystyle\sum_{n=1}^{\infty} \frac{n}{2n + 3}$

26. $\displaystyle\sum_{n=2}^{\infty} \left(\frac{\ln n}{n} \right)^n$

27. $\displaystyle\sum_{n=1}^{\infty} \frac{n}{(n^2 + 1)^2}$

28. $\displaystyle\sum_{n=1}^{\infty} \frac{3}{n(n + 3)}$

S 29. Determine the values of *p* for which the series $\sum_{n=2}^{\infty} (\ln n)/n^p$ converges.

30. Find the error when $\sum_{n=0}^{\infty} 2(\frac{1}{3})^n$ is approximated by its first ten terms.

C 31. Estimate the error when $\sum_{n=1}^{\infty} (1/n^n)$ is approximated by the first five terms.

32. Prove that if $P(n)$ and $Q(n)$ are polynomials of degree j and k, respectively, then the series $\sum_{n=1}^{\infty} P(n)/Q(n)$ converges if $j < k - 1$ and diverges if $j \geq k - 1$.

In Exercises 33–36, use the *Polynomial Test* as given in Exercise 32 to determine if the following series converge or diverge.

33. $\frac{1}{2} + \frac{2}{5} + \frac{3}{10} + \frac{4}{17} + \frac{5}{26} + \cdots$

[S] **34.** $\frac{1}{3} + \frac{1}{8} + \frac{1}{15} + \frac{1}{24} + \frac{1}{35} + \cdots$

35. $\sum_{n=1}^{\infty} \frac{1}{n^3 + 1}$

36. $\sum_{n=1}^{\infty} \frac{n^2}{n^3 + 1}$

37. (Positive nth-Term Divergence Test) Use the Limit Comparison Test with the harmonic series to show that the series $\sum_{n=1}^{\infty} a_n$ $(a_n > 0)$ diverges if $\lim_{n \to \infty} n a_n \neq 0$.

For Review and Class Discussion

True or False

1. ____ If $0 \leq a_n \leq b_n$ and $\sum_{n=1}^{\infty} a_n$ converges, then $\sum_{n=1}^{\infty} b_n$ converges.

2. ____ If $0 \leq a_n \leq b_n$ and $\sum_{n=1}^{\infty} a_n$ converges, then $\sum_{n=1}^{\infty} b_n$ diverges.

3. ____ If $0 \leq a_{n+10} \leq b_n$ and $\sum_{n=1}^{\infty} b_n$ converges, then $\sum_{n=1}^{\infty} a_n$ converges.

4. ____ If $(1/n) \leq k a_n$, then $\sum_{n=1}^{\infty} a_n$ diverges.

5. ____ If $b_n \leq a_n \leq 0$ and $\sum_{n=1}^{\infty} b_n$ converges, then $\sum_{n=1}^{\infty} a_n$ converges.

18.5 Alternating Series and the Ratio Test

Purpose
- To introduce alternating series and the Ratio Test.
- To estimate the sum of a series by the use of a remainder test.
- To summarize the various tests for convergence or divergence.

For those ultimate ratios with which quantities vanish are not truly the ratios of ultimate quantities, but limits towards which the ratios of quantities decreasing without limit do always converge. —Isaac Newton (1642–1727)

The geometric series

$$\sum_{n=1}^{\infty} \frac{1}{(-2)^{n-1}} = \sum_{n=1}^{\infty} (-1)^{n-1} \left(\frac{1}{2^{n-1}} \right) = 1 - \frac{1}{2} + \frac{1}{4} - \frac{1}{8} + \frac{1}{16} - \cdots$$

is called an **alternating series** since the terms of the series alternate in sign.

An alternating series can occur in two ways:

$$\sum_{n=1}^{\infty} (-1)^n a_n = -a_1 + a_2 - a_3 + a_4 - \cdots$$

and

$$\sum_{n=1}^{\infty} (-1)^{n-1} a_n = a_1 - a_2 + a_3 - a_4 + \cdots$$

where $a_n > 0$. In one case the *odd* terms are negative and in the other case the *even* terms are negative. The conditions for the convergence of an alternating series are given in the following theorem.

| THEOREM 18.11 (*Alternating Series Test*) | The alternating series |

$$\sum_{n=1}^{\infty} (-1)^{n-1} a_n = a_1 - a_2 + a_3 - a_4 + \cdots$$

converges, provided:

i. $0 < a_{n+1} \le a_n$ for $n \ge 1$, *and*

ii. $\lim_{n \to \infty} a_n = 0$

Proof: First, we assume that $0 < a_{n+1} \le a_n$. Thus the even partial sum

$$S_{2n} = (a_1 - a_2) + (a_3 - a_4) + (a_5 - a_6) + \cdots + (a_{2n-1} - a_{2n})$$

has all nonnegative terms and therefore $\{S_{2n}\}$ is a nondecreasing sequence. But we can also write

$$S_{2n} = a_1 - (a_2 - a_3) - (a_4 - a_5) - \cdots - (a_{2n-2} - a_{2n-1}) - a_{2n}$$

which implies that each term after the first is negative (or zero) and therefore $S_{2n} \le a_1$ for every integer n. Thus $\{S_{2n}\}$ is a bounded, nondecreasing sequence that converges to some value L, with

$$\lim_{n \to \infty} S_{2n} = L \le a_1$$

Considering the fact that

$$S_{2n-1} = S_{2n} + a_{2n}$$

we then have

$$\lim_{n \to \infty} S_{2n-1} = \lim_{n \to \infty} S_{2n} + \lim_{n \to \infty} a_{2n} = L + \lim_{n \to \infty} a_{2n}$$

By assumption

$$\lim_{n \to \infty} a_{2n} = \lim_{n \to \infty} a_n = 0$$

Therefore, we have

$$\lim_{n \to \infty} S_{2n-1} = L + 0 = L$$

Since both the odd and even sequences of partial sums converge to the same limit L, it follows that

$$\lim_{n \to \infty} S_n = L$$

and consequently the given alternating series converges. ∎

Example 1

Determine whether the following series converge or diverge:

(a) $\displaystyle\sum_{n=1}^{\infty} \frac{n}{(-2)^{n-1}}$ (b) $\displaystyle\sum_{n=1}^{\infty} \frac{(-1)^n n}{\ln 2n}$ (c) $\displaystyle\sum_{n=1}^{\infty} (-1)^{n+1} \left(\frac{3n+2}{4n^2-3} \right)$

Solution:

(a) The given series is alternating, since

$$\sum_{n=1}^{\infty} \frac{n}{(-2)^{n-1}} = \sum_{n=1}^{\infty} (-1)^{n-1} \left(\frac{n}{2^{n-1}} \right) = \frac{1}{1} - \frac{2}{2} + \frac{3}{4} - \frac{4}{8} + \cdots$$

The inequality $\quad a_{n+1} = \dfrac{n+1}{2^n} \leq \dfrac{n}{2^{n-1}} = a_n$

is satisfied if $\quad (n+1)2^{n-1} \leq n2^n$

or if $\qquad\qquad \dfrac{1}{2} \leq \dfrac{n}{n+1}$

But this latter inequality is satisfied for $n \geq 1$; hence $a_{n+1} \leq a_n$ for all n. Furthermore, by L'Hôpital's Rule,

$$\lim_{n\to\infty} \frac{n}{2^{n-1}} = \lim_{n\to\infty} \frac{1}{2^{n-1}(\ln 2)} = 0$$

Therefore, by the Alternating Series Test, the series

$$\sum_{n=1}^{\infty} \frac{n}{(-2)^{n-1}}$$

converges.

(b) Since, by L'Hôpital's Rule, we have

$$\lim_{n\to\infty} a_n = \lim_{n\to\infty} \frac{n}{\ln 2n} = \lim_{n\to\infty} \frac{1}{1/n} = \lim_{n\to\infty} n = \infty \neq 0$$

the given series diverges. (Yes, the nth-Term Test for Divergence can be applied to an alternating series.)

(c) Sometimes it is convenient to use the derivative to establish that $a_{n+1} \leq a_n$. In this case let

$$f(x) = \frac{3x+2}{4x^2-3}$$

Then $\qquad\qquad f'(x) = \dfrac{-12x^2-16x-9}{(4x^2-3)^2}$

is always negative. Hence $f(n) = a_n$ is a decreasing function and it follows that $a_{n+1} \leq a_n$ for $n \geq 1$. Furthermore, since

$$\lim_{n \to \infty} \frac{3n + 2}{4n^2 - 3} = \lim_{n \to \infty} \frac{3}{8n} = 0$$

the series converges by the Alternating Series Test.   ∎

For a convergent alternating series, the partial sum S_N can be a useful approximation for the sum S of the series. Just how close S_N is to S is stated in the following theorem.

THEOREM 18.12
(*Alternating Series Remainder*)

If

$$\sum_{n=1}^{\infty} (-1)^{n-1} a_n$$

is a convergent alternating series with sum S, then the remainder R_N involved in approximating S by S_N is less in magnitude than the first nelgected term. That is

$$|S - S_N| = |R_N| \leq a_{N+1}$$

Proof: Let

$$\sum_{n=1}^{\infty} (-1)^n a_n$$

be a convergent alternating series with sum S. Then the series obtained by deleting the first N terms is a convergent alternating series whose sum is

$$R_N = S - S_N = \sum_{n=1}^{\infty} (-1)^{n-1} a_n - \sum_{n=1}^{N} (-1)^n a_n$$

$$= (-1)^N a_{N+1} + (-1)^{N+1} a_{N+2} + (-1)^{N+2} a_{N+3} + \cdots$$

$$= (-1)^N (a_{N+1} - a_{N+2} + a_{N+3} - \cdots)$$

Now $|R_N| = a_{N+1} - a_{N+2} + a_{N+3} - a_{N+4} + a_{N+5} - \cdots$

$$= a_{N+1} - (a_{N+2} - a_{N+3}) - (a_{N+4} - a_{N+5}) - \cdots$$

As in the proof of Theorem 18.11, it then follows that $|R_N| \leq a_{N+1}$. Consequently,

$$|S - S_N| = |R_N| \leq a_{N+1}$$

which establishes the theorem.   ∎

Example 2

Show that the series

$$\sum_{n=1}^{\infty} (-1)^{n-1} \left(\frac{1}{n!} \right) = \frac{1}{1!} - \frac{1}{2!} + \frac{1}{3!} - \frac{1}{4!} + \frac{1}{5!} - \frac{1}{6!} + \cdots$$

converges. Find S_6 and determine how close this estimate is to the actual sum of the series.

Solution: The series converges by the Alternating Series Test because

$$\frac{1}{(n + 1)!} < \frac{1}{n!}$$

for $n \geq 1$ and

$$\lim_{n \to \infty} \frac{1}{n!} = 0$$

Now

$$S_6 = 1 - \frac{1}{2} + \frac{1}{6} - \frac{1}{24} + \frac{1}{120} - \frac{1}{720} \approx 0.63194$$

and by Theorem 18.12

$$|S - S_6| = |R_6| \leq a_7 = \frac{1}{5040} \approx 0.0002$$

Therefore, the sum S is such that

$$S_6 - 0.0002 \leq S \leq S_6 + 0.0002$$

or

$$0.63174 \leq \sum_{n=1}^{\infty} (-1)^{n-1} \left(\frac{1}{n!}\right) \leq 0.63214$$

[The actual sum of this series is $(e - 1)/e \approx 0.63212$. See Section 18.8, Exercise 34.] ∎

On occasion a series may have both positive and negative terms and not be an alternating series. For instance, the series

$$\sum_{n=1}^{\infty} \frac{\sin n}{n^2} = \frac{\sin 1}{1} + \frac{\sin 2}{4} + \frac{\sin 3}{9} + \cdots$$

has both positive and negative terms, yet it is not an alternating series. One way to obtain some information about the convergence of this series might be to investigate the convergence of the series

$$\sum_{n=1}^{\infty} \frac{|\sin n|}{n^2}$$

The next theorem indicates some possible information to be gained by such an investigation.

THEOREM 18.13 (*Absolute Convergence*)	If the series $\sum_{n=1}^{\infty} |a_n|$ converges, then the series $\sum_{n=1}^{\infty} a_n$ also converges.

Proof: Since some terms of the series may be negative, we have $0 \leq a_n + |a_n| \leq 2|a_n|$ for all n. Therefore, the series

$$\sum_{n=1}^{\infty} (a_n + |a_n|)$$

converges by comparison to the convergent series

$$\sum_{n=1}^{\infty} 2|a_n|$$

Furthermore, since $a_n = (a_n + |a_n|) - |a_n|$

we can write

$$\sum_{n=1}^{\infty} a_n = \sum_{n=1}^{\infty} (a_n + |a_n|) - \sum_{n=1}^{\infty} |a_n|$$

where both series on the right converge. Hence it follows that $\sum_{n=1}^{\infty} a_n$ converges. ∎

A series $\sum_{n=1}^{\infty} a_n$, for which $\sum_{n=1}^{\infty} |a_n|$ converges, is called an **absolutely convergent** series. It should be pointed out that the converse of Theorem 18.13 is not true since there are convergent series that are *not* absolutely convergent. The alternating harmonic series is a case in point. The series

$$\sum_{n=1}^{\infty} \frac{(-1)^{n+1}}{n} = \frac{1}{1} - \frac{1}{2} + \frac{1}{3} - \frac{1}{4} + \cdots$$

converges by the Alternating Series Test, whereas the series

$$\sum_{n=1}^{\infty} \left| \frac{(-1)^{n+1}}{n} \right| = \sum_{n=1}^{\infty} \frac{1}{n} = \frac{1}{1} + \frac{1}{2} + \frac{1}{3} + \frac{1}{4} + \cdots$$

is the divergent harmonic series.

Example 3
***

Determine whether the series

$$\sum_{n=1}^{\infty} \frac{\sin n}{n^2}$$

converges or diverges.

Solution: Since $|\sin n| \leq 1$ for all n, the series

$$\sum_{n=1}^{\infty} \left| \frac{\sin n}{n^2} \right|$$

converges by comparison to the convergent *p*-series

$$\sum_{n=1}^{\infty} \frac{1}{n^2}$$

Therefore, by Theorem 18.13 the series

$$\sum_{n=1}^{\infty} \frac{\sin n}{n^2}$$

converges, since it was shown to be absolutely convergent. ∎

The difficulty of choosing convergence tests appropriate for alternating series, for series of positive terms, or for series with mixed terms suggests the need of a more general test to use with series having both positive and negative terms. Such a test is available, and it is called the **Ratio Test.**

THEOREM 18.14
(*Ratio Test*)

The series $\sum_{n=1}^{\infty} a_n$:

i. converges if $\displaystyle\lim_{n\to\infty} \left| \frac{a_{n+1}}{a_n} \right| < 1$

(in fact, it converges *absolutely*)

ii. diverges if $\displaystyle\lim_{n\to\infty} \left| \frac{a_{n+1}}{a_n} \right| > 1$

Proof:

i. Let $\displaystyle\lim_{n\to\infty} \left| \frac{a_{n+1}}{a_n} \right| = r < 1$

and choose R such that $0 \le r < R < 1$. By the definition of the limit of a sequence, there exists some $N > 0$ such that

$$\left| \frac{a_{n+1}}{a_n} \right| < R$$

for all $n > N$, and thus

$$|a_{n+1}| < |a_n|R$$

Furthermore, $\quad |a_{N+1}| < |a_N|R$

$|a_{N+2}| < |a_{N+1}|R < |a_N|R^2$

$|a_{N+3}| < |a_{N+2}|R < |a_{N+1}|R^2 < |a_N|R^3$

$\vdots$

Now since the geometric series

$$\sum_{n=1}^{\infty} a_N R^n = a_N R + a_N R^2 + \cdots + a_N R^n + \cdots$$

converges, then by the Comparison Test the series

$$\sum_{n=1}^{\infty} |a_{N+n}| = |a_{N+1}| + |a_{N+2}| + \cdots + |a_{N+n}| + \cdots$$

also converges. This in turn implies that the series $\sum_{n=1}^{\infty} |a_n|$ converges, since discarding a finite number of terms ($n = N - 1$) does not affect convergence. Consequently, by Theorem 18.13 the series $\sum_{n=1}^{\infty} a_n$ converges.

ii.   The proof of this part is similar, except that we choose R such that

$$\lim_{n \to \infty} \left| \frac{a_{n+1}}{a_n} \right| = r > R > 1$$

and show that there exists some $M > 0$ such that $|a_{M+n}| > |a_M| R^n$. ∎

   The Ratio Test should not be viewed as a cure for all ills related to tests for convergence, for the Ratio Test fails to give us any useful information when

$$\lim_{n \to \infty} \left| \frac{a_{n+1}}{a_n} \right| = 1$$

For instance, we know that the first of the two series

$$\sum_{n=1}^{\infty} \frac{1}{n} \qquad \text{and} \qquad \sum_{n=1}^{\infty} \frac{1}{n^2}$$

diverges and the second one converges, while in both cases

$$\lim_{n \to \infty} \left| \frac{a_{n+1}}{a_n} \right| = 1$$

Nevertheless, the Ratio Test is widely applicable. It is particularly useful for series that *converge rapidly* or that *diverge rapidly*. Series involving factorials or exponentials are frequently of this type.

Example 4

Determine whether the following series converge or diverge:

(a) $\displaystyle \sum_{n=0}^{\infty} \frac{2^n}{n!}$ (b) $\displaystyle \sum_{n=1}^{\infty} (-1)^n \frac{\sqrt{n}}{n + 1}$

Solution:

(a) We use the Ratio Test and obtain

$$\lim_{n \to \infty} \left| \frac{a_{n+1}}{a_n} \right| = \lim_{n \to \infty} \frac{2^{n+1}}{(n+1)!} \div \frac{2^n}{n!} = \lim_{n \to \infty} \frac{2^{n+1}}{(n+1)!} \cdot \frac{n!}{2^n}$$

$$= \lim_{n \to \infty} \frac{2}{n+1} = 0$$

Therefore, the series converges.

(b) Since

$$\lim_{n \to \infty} \left| \frac{a_{n+1}}{a_n} \right| = \lim_{n \to \infty} \left[\left(\frac{\sqrt{n+1}}{n+2} \right) \left(\frac{n+1}{\sqrt{n}} \right) \right]$$

$$= \lim_{n \to \infty} \left[\sqrt{\frac{n+1}{n}} \left(\frac{n+1}{n+2} \right) \right] = \sqrt{1}(1) = 1$$

the Ratio Test gives us no useful information, so we use the Alternating Series Test. Now to see that

$$a_{n+1} = \frac{\sqrt{n+1}}{n+2} < \frac{\sqrt{n}}{n+1} = a_n$$

we multiply both sides of the inequality by $(n+2)(n+1)$, to obtain

$$(\sqrt{n+1})^3 < \sqrt{n}(n+2) \qquad \text{or} \qquad \sqrt{(n+1)^3} < \sqrt{n(n+2)^2}$$

Since we know that

$$(n+1)^3 = n^3 + 3n^2 + 3n + 1 < n^3 + 4n^2 + 4n = n(n+2)^2$$

it follows that $a_{n+1} \le a_n$ for $n \ge 1$. Furthermore, by L'Hôpital's Rule,

$$\lim_{n \to \infty} \frac{\sqrt{n}}{n+1} = \lim_{n \to \infty} \frac{1/2 \sqrt{n}}{1} = \lim_{n \to \infty} \frac{1}{2\sqrt{n}} = 0$$

Therefore, the series

$$\sum_{n=1}^{\infty} (-1)^n \frac{\sqrt{n}}{n+1}$$

converges. ∎

Example 5

Determine whether the following series converge or diverge:

(a) $\displaystyle\sum_{n=0}^{\infty} \frac{n^2 2^{n+1}}{3^n}$ (b) $\displaystyle\sum_{n=0}^{\infty} \frac{n^n}{n!}$

TABLE 18.5 Summary of Tests for Series

Test	Series	Converges	Sum or Remainder	Diverges
nth-Term	$\sum_{n=1}^{\infty} a_n$			$\lim_{n\to\infty} a_n \neq 0$
Geometric Series	$\sum_{n=0}^{\infty} ar^n$	$|r| < 1$	$S = \dfrac{a}{1-r}$	$|r| \geq 1$
Telescoping Series	$\sum_{n=1}^{\infty}(b_n - b_{n+1})$	$\lim_{n\to\infty} b_n = 0$	$S = b_1$	
p-Series	$\sum_{n=1}^{\infty} \dfrac{1}{n^p}$	$p > 1$	$0 < R_N < \dfrac{1}{N^{p-1}(p-1)}$	$p \leq 1$
Alternating Series	$\sum_{n=1}^{\infty}(-1)^{n-1} a_n$	$0 \leq a_{n+1} \leq a_n$ and $\lim_{n\to\infty} a_n = 0$	$|R_N| \leq a_{N+1}$	
Integral (f is continuous, positive, and decreasing)	$\sum_{n=1}^{\infty} a_n,\ a_n = f(n)$	$\displaystyle\int_1^{\infty} f(x)\,dx$ converges	$0 < R_N < \displaystyle\int_N^{\infty} f(x)\,dx$	$\displaystyle\int_1^{\infty} f(x)\,dx$ diverges
Root	$\sum_{n=1}^{\infty} a_n$	$\lim_{n\to\infty} \sqrt[n]{|a_n|} < 1$		$\lim_{n\to\infty} \sqrt[n]{|a_n|} > 1$
Ratio	$\sum_{n=1}^{\infty} a_n$	$\lim_{n\to\infty} \left|\dfrac{a_{n+1}}{a_n}\right| < 1$		$\lim_{n\to\infty} \left|\dfrac{a_{n+1}}{a_n}\right| > 1$
Comparison	$\sum_{n=1}^{\infty} a_n$	$0 \leq a_n \leq b_n$ and $\sum_{n=1}^{\infty} b_n$ converges	$0 \leq R_n \leq \sum_{n=N}^{\infty} b_n$	$0 \leq b_n \leq a_n$ and $\sum_{n=1}^{\infty} b_n$ diverges
Limit Comparison $(a_n, b_n > 0)$	$\sum_{n=1}^{\infty} a_n$	$\lim_{n\to\infty}\dfrac{a_n}{b_n} = L > 0$ and $\sum_{n=1}^{\infty} b_n$ converges		$\lim_{n\to\infty}\dfrac{a_n}{b_n} = L > 0$ and $\sum_{n=1}^{\infty} b_n$ diverges

Solution:

(a) Since
$$\lim_{n\to\infty}\left|\frac{a_{n+1}}{a_n}\right| = \lim_{n\to\infty}\left[(n+1)^2\left(\frac{2^{n+2}}{3^{n+1}}\right)\left(\frac{3^n}{n^2 2^{n+1}}\right)\right]$$
$$= \lim_{n\to\infty}\frac{2(n+1)^2}{3n^2} = \frac{2}{3} < 1$$

the series converges by the Ratio Test.

(b) Since
$$\lim_{n\to\infty}\left|\frac{a_{n+1}}{a_n}\right| = \lim_{n\to\infty}\left[\frac{(n+1)^{n+1}}{(n+1)!}\left(\frac{n!}{n^n}\right)\right]$$
$$= \lim_{n\to\infty}\left[\frac{(n+1)^{n+1}}{(n+1)}\left(\frac{1}{n^n}\right)\right] = \lim_{n\to\infty}\frac{(n+1)^n}{n^n}$$
$$= \lim_{n\to\infty}\left(1 + \frac{1}{n}\right)^n = e > 1$$

the series diverges by the Ratio Test.   ∎

In the last four sections we have discussed ten different tests for determining the convergence or divergence of an infinite series. The choice of an appropriate way to test a given series is sometimes difficult. Skill in choosing and applying the various tests will come only with practice. As a general aid to choosing appropriate tests for convergence, we summarize in Table 18.5 the nature of the various tests we have studied.

With this outline in mind, a useful pattern for choosing an appropriate test is the following:

1.   Check to see if the *n*th term goes to zero; if not, the series diverges.
2.   Check to see if the series is one of the special types.
3.   Check to see if the Integral, Root, or Ratio Tests can be readily applied.
4.   Check to see if the series can be compared favorably to one of the special types.

Keep in mind that in some instances more than one test is applicable, but our objective is to learn to choose the most efficient way to test a series for convergence or divergence.

Section Exercises (18.5)

In Exercises 1–22, use the Alternating Series Test or the Ratio Test to test for convergence or divergence of each series.

1. $\displaystyle\sum_{n=1}^{\infty}\frac{(-1)^{n+1}}{n}$

2. $\displaystyle\sum_{n=1}^{\infty}\frac{(-1)^{n-1}n}{2n-1}$

[S] 3. $\displaystyle\sum_{n=1}^{\infty}\frac{(-1)^{n+1}}{2n-1}$

4. $\displaystyle\sum_{n=2}^{\infty}\frac{(-1)^n}{\ln(n)}$

5. $\displaystyle\sum_{n=1}^{\infty}\frac{(-1)^n n^2}{n^2+1}$

6. $\displaystyle\sum_{n=1}^{\infty}\frac{(-1)^{n+1}n}{n^2+1}$

7. $\sum_{n=0}^{\infty} \dfrac{n!}{3^n}$

8. $\sum_{n=1}^{\infty} \dfrac{1}{n2^n}$

$\boxed{\text{C}}\boxed{\text{S}}$ 25. $\sum_{n=0}^{\infty} \dfrac{(-1)^n}{n!}$ $\left(\text{the sum is } \dfrac{1}{e}\right)$

$\boxed{\text{S}}$ 9. $\sum_{n=0}^{\infty} \dfrac{3^n}{n!}$

$\boxed{\text{S}}$ 10. $\sum_{n=1}^{\infty} n\left(\dfrac{2}{3}\right)^n$

$\boxed{\text{C}}$ 26. $\sum_{n=0}^{\infty} \dfrac{(-1)^n}{(2n)!}$ (the sum is cos 1)

11. $\sum_{n=1}^{\infty} \dfrac{n}{2^n}$

12. $\sum_{n=1}^{\infty} \dfrac{n^2}{2^n}$

$\boxed{\text{C}}$ 27. $\sum_{n=1}^{\infty} \dfrac{(-1)^{n+1}}{n2^n}$ $\left[\text{the sum is } \ln\left(\dfrac{3}{2}\right)\right]$

$\boxed{\text{S}}$ 13. $\sum_{n=0}^{\infty} \dfrac{(-1)^{n+1}n!}{1 \cdot 3 \cdot 5 \cdots (2n+1)}$

14. $\sum_{n=1}^{\infty} \dfrac{(-1)^{n+1}(n+2)}{n(n+1)}$

28. Prove that the alternating p-series $\sum_{n=1}^{\infty} [(-1)^{n-1}/n^p]$ converges if $p > 0$.

15. $\sum_{n=0}^{\infty} \dfrac{(-1)^n 2^n}{n!}$

16. $\sum_{n=1}^{\infty} \dfrac{(-1)^{n-1}(\frac{3}{2})^n}{n^2}$

29. Prove that if $\sum_{n=1}^{\infty} |a_n|$ converges, then $\sum_{n=1}^{\infty} a_n^2$ converges. Is the converse true?

$\boxed{\text{S}}$ 17. $\sum_{n=1}^{\infty} \dfrac{(-1)^n 2 \cdot 4 \cdot 6 \cdots (2n)}{2 \cdot 5 \cdot 8 \cdots (3n-1)}$

18. $\sum_{n=1}^{\infty} \dfrac{(-1)^n \sin(n)}{n^2}$

30. Show that the Ratio Test gives no information for a p-series.

19. $\sum_{n=1}^{\infty} \dfrac{(-1)^n}{\sqrt{n}}$

20. $\sum_{n=1}^{\infty} \dfrac{(2n)!}{n^5}$

In Exercises 31–42, test for convergence or divergence using any appropriate test from this chapter. Identify the test used.

21. $\sum_{n=1}^{\infty} \dfrac{(-1)^n \ln(n)}{n}$

22. $\sum_{n=1}^{\infty} \dfrac{n^n}{n!}$

$\boxed{\text{S}}$ 31. $\sum_{n=1}^{\infty} \dfrac{(-1)^n 3^{n-2}}{2^n}$

32. $\sum_{n=1}^{\infty} \dfrac{10}{3\sqrt{n^3}}$

23. Examine the following series for convergence or absolute convergence:

(a) $\sum_{n=1}^{\infty} \dfrac{(-1)^{n+1}}{(n+1)^2}$

(b) $\sum_{n=1}^{\infty} \dfrac{(-1)^{n+1}}{n+1}$

$\boxed{\text{S}}$ 33. $\sum_{n=1}^{\infty} \dfrac{10n+3}{n2^n}$

34. $\sum_{n=1}^{\infty} \dfrac{2^n}{4n^2-1}$

(c) $\sum_{n=2}^{\infty} \dfrac{(-1)^n}{\ln n}$

35. $\sum_{n=1}^{\infty} (-1)^n \ln\left(\dfrac{n+2}{n}\right)$

36. $\sum_{n=1}^{\infty} \dfrac{1}{\sqrt{n}+2}$

In Exercises 24–27, approximate the sum of each series with an error less than 0.001.

$\boxed{\text{S}}$ 37. $\sum_{n=1}^{\infty} \dfrac{\cos(n)}{2^n}$

38. $\sum_{n=2}^{\infty} \dfrac{(-1)^n}{n \ln(n)}$

$\boxed{\text{C}}$ 24. $\sum_{n=1}^{\infty} \dfrac{(-1)^{n+1}}{2n^3-1}$

39. $\sum_{n=1}^{\infty} \dfrac{n7^n}{n!}$

40. $\sum_{n=1}^{\infty} \dfrac{\ln(n)}{n^2}$

41. $\sum_{n=1}^{\infty} \dfrac{(-1)^n 3^{n-1}}{n!}$

42. $\sum_{n=1}^{\infty} \dfrac{(-1)^n 3^n}{n2^n}$

For Review and Class Discussion

True or False

1. ____ $\sum_{n=1}^{\infty} [(-1)^{n(n-1)/2}/n]$ is an alternating series.

2. ____ If $0 < a_{n+1} < a_n$ for $n > 100$ and $\lim_{n \to \infty} a_n = 0$, then $\sum_{n=1}^{\infty} (-1)^n a_n$ converges.

3. ____ If $\sum_{n=1}^{\infty} (-1)^n a_n$ converges, then $\sum_{n=1}^{\infty} (-1)^{n-1} a_n$ converges.

4. ____ If $\sum_{n=1}^{\infty} a_n$ converges, then $\sum_{n=1}^{\infty} (-1)^n a_n$ converges.

5. ____ If both $\sum_{n=1}^{\infty} a_n$ and $\sum_{n=1}^{\infty} - a_n$ converge, then $\sum_{n=1}^{\infty} |a_n|$ converges.

18.6 **Power Series**

Purpose
- To define a power series and determine its interval of convergence.
- To identify the properties of a function represented by a power series.

Moreover, it often happens that when integration is complicated, that the development in series presents no difficulty; and the formulas thus derived open a fruitful source for the solution of many important problems.
—Carl Gauss (1777–1855)

Up to this point we have been dealing with series whose terms are constants. Now we want to consider series whose terms are variable. In particular, if x is a variable, then a series of the form

$$\sum_{n=0}^{\infty} a_n x^n = a_0 + a_1 x + a_2 x^2 + a_3 x^3 + \cdots + a_n x^n + \cdots$$

is called a **power series,** and, more generally, we call a series of the form

$$\sum_{n=0}^{\infty} a_n (x - c)^n = a_0 + a_1(x - c) + a_2(x - c)^2 + \cdots$$
$$+ a_n(x - c)^n + \cdots$$

a **power series centered at** c. Note that we begin a power series at $n = 0$, and to simplify the nth term, we agree that $(x - c)^0 = 1$ even if $x = c$.

Since a power series has variable terms, it can be viewed as a function of x,

$$f(x) = \sum_{n=0}^{\infty} a_n (x - c)^n$$

where the *domain of f* is the set of all x for which the power series converges. The determination of this "domain of convergence" is one of the primary problems associated with power series.

Quite obviously, every power series converges at its center c, since for $x = c$

$$f(x) = \sum_{n=0}^{\infty} a_n (c - c)^n = a_0(1) + 0 + 0 + \cdots + 0 + \cdots = a_0$$

Thus $x = c$ always lies in the domain of f. It may happen that the domain of f consists only of the value $x = c$, as the next example shows. If there are other values for which a power series converges, we will see that the Ratio Test often provides a convenient way to find them.

Example 1

Show that the domain of

$$f(x) = \sum_{n=0}^{\infty} n! x^n$$

consists only of $x = 0$.

Solution: When $x = 0$,

$$f(0) = \sum_{n=0}^{\infty} n!0^n = 0!0^0 + 1!0^1 + 2!0^2 + 3!0^3 + \cdots$$

$$= 1 + 0 + 0 + \cdots = 1$$

For $|x| > 0$, let $u_n = n!x^n$; then

$$\lim_{n \to \infty} \left| \frac{u_{n+1}}{u_n} \right| = \lim_{n \to \infty} \left| \frac{(n+1)!x^{n+1}}{n!x^n} \right| = |x| \lim_{n \to \infty} (n+1) = \infty$$

Therefore, by the Ratio Test we conclude that the series diverges for $|x| > 0$, and converges only when $x = 0$. ∎

Example 2

Find the domain of

$$f(x) = \sum_{n=0}^{\infty} 3(x-2)^n$$

Solution: Since this series becomes a geometric series for each value assigned to x, we know that it converges only when

$$|r| = |x-2| < 1$$

The domain of f is therefore all x such that

$$-1 < x - 2 < 1 \qquad \text{or} \qquad 1 < x < 3$$

Thus the given series converges for all x in the interval $(1, 3)$, which we note is centered at 2. ∎

Example 3

Show that the power series

$$\sum_{n=0}^{\infty} \frac{(-1)^n x^{2n+1}}{(2n+1)!}$$

converges for all x.

Solution: If we let $u_n = (-1)^n x^{2n+1}/(2n+1)!$, then

$$\lim_{n \to \infty} \left| \frac{u_{n+1}}{u_n} \right| = \lim_{n \to \infty} \left| \frac{[(-1)^{n+1} x^{2n+3}]/(2n+3)!}{[(-1)^n x^{2n+1}]/(2n+1)!} \right|$$

$$= \lim_{n \to \infty} \frac{x^2}{(2n+3)(2n+2)}$$

Now for any assigned value for x, we have

$$\lim_{n \to \infty} \left| \frac{u_{n+1}}{u_n} \right| = \lim_{n \to \infty} \frac{x^2}{(2n + 3)(2n + 2)} = 0$$

Therefore, by the Ratio Test the series converges for all x.   ■

Examples 1, 2, and 3 demonstrate the three types of *domains* for power series. We outline the general situation in Theorem 18.15, which we state without proof.

THEOREM 18.15
(*Convergence of a Power Series*)

For a power series centered at c, precisely one of the following is true:

i.   The series converges only for $x = c$ ($|x - c| = 0$).
ii.   The series converges for all x ($|x - c| < \infty$).
iii. There exists an $R > 0$ such that the series converges for $|x - c| < R$, and diverges for $|x - c| > R$.

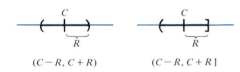

$(C - R, C + R)$ $(C - R, C + R]$

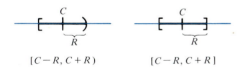

$[C - R, C + R)$ $[C - R, C + R]$

Intervals of Convergence

FIGURE 18.8

In Theorem 18.15, part iii, we call R the **radius of convergence** of the power series, and in parts i and ii, we consider the radius of convergence to be 0 and ∞, respectively. Furthermore, since Theorem 18.15 states that the domain of convergence of a power series is always an interval, we call this domain the **interval of convergence.**

Note that Theorem 18.15 says nothing about the convergence of the series at the *endpoints* of its interval of convergence. Each endpoint must be tested separately for convergence or divergence. As a result, an interval of convergence can be any one of the four types shown in Figure 18.8.

We demonstrated in Examples 1 and 3 the usefulness of the Ratio Test in determining the interval of convergence for a power series. For convenience we describe in the following theorem a Ratio Test for determining the *radius of convergence* of a power series. From the radius of convergence, we can specify the exact interval of convergence after testing for convergence at the endpoints.

THEOREM 18.16
(*Power Series Ratio Test*)

For the power series $\sum_{n=0}^{\infty} a_n (x - c)^n$, if

$$\lim_{n \to \infty} \left| \frac{a_{n+1}}{a_n} \right| = L \qquad (0 \le L \le \infty)$$

then R, the *radius of convergence* of this series, is given by

$$R = \frac{1}{L}$$

where $R = 0$ if $L = \infty$, and $R = \infty$ if $L = 0$.

Proof: We prove only the case where $0 < L < \infty$. Suppose that for the given series we have

$$\lim_{n \to \infty} \left| \frac{a_{n+1}}{a_n} \right| = L$$

This means that

$$\lim_{n \to \infty} \left| \frac{a_{n+1}(x - c)^{n+1}}{a_n(x - c)^n} \right| = |x - c| \lim_{n \to \infty} \left| \frac{a_{n+1}}{a_n} \right| = |x - c|L$$

Therefore, by the Ratio Test (Theorem 18.14), the series converges if $|x - c|L < 1$ and diverges if $|x - c|L > 1$. Dividing both sides of these inequalities by L yields the result that the series converges if $|x - c| < 1/L$ and diverges if $|x - c| > 1/L$. Hence by Theorem 18.15 the radius of convergence is $R = 1/L$. ∎

Example 4

Find the interval of convergence for the series

$$\sum_{n=1}^{\infty} \frac{x^n}{n}$$

Solution: Since $a_n = 1/n$, and

$$\lim_{n \to \infty} \left| \frac{a_{n+1}}{a_n} \right| = \lim_{n \to \infty} \left| \frac{1/(n + 1)}{1/n} \right| = \lim_{n \to \infty} \left| \frac{n}{n + 1} \right| = 1$$

the radius of convergence is $R = 1$. Now since the series is centered at $x = 0$, it will converge in the interval $(-1, 1)$ and we proceed to test for convergence at the endpoints. When $x = 1$, we obtain the series

$$\sum_{n=1}^{\infty} \frac{1}{n} = \frac{1}{1} + \frac{1}{2} + \frac{1}{3} + \cdots$$

which is the *divergent* harmonic series. When $x = -1$, we obtain the series

$$\sum_{n=1}^{\infty} \frac{(-1)^n}{n} = -1 + \frac{1}{2} - \frac{1}{3} + \frac{1}{4} - \cdots$$

which *converges* by the Alternating Series Test. Therefore, the interval of convergence for the series

$$\sum_{n=1}^{\infty} \frac{x^n}{n}$$

is $[-1, 1)$. ∎

Example 5

Find the interval of convergence for the series

$$\sum_{n=0}^{\infty} \frac{(-1)^n(x+1)^n}{2^n}$$

Solution: Since

$$\lim_{n \to \infty} \left| \frac{a_{n+1}}{a_n} \right| = \lim_{n \to \infty} \left| \frac{(-1)^{n+1}/2^{n+1}}{(-1)^n/2^n} \right| = \lim_{n \to \infty} \left| \frac{2^n}{2^{n+1}} \right| = \frac{1}{2}$$

the radius of convergence is $R = 1/L = 2$. Since the series is centered at $x = -1$, it will converge in the interval $(-3, 1)$. Furthermore, at the endpoints we have the series

$$\sum_{n=0}^{\infty} \frac{(-1)^n(-2)^n}{2^n} = \sum_{n=0}^{\infty} \frac{2^n}{2^n} = \sum_{n=0}^{\infty} 1 \qquad (x = -3)$$

and

$$\sum_{n=0}^{\infty} \frac{(-1)^n(2)^n}{2^n} = \sum_{n=0}^{\infty} (-1)^n \qquad (x = 1)$$

both of which diverge. Thus the interval of convergence is $(-3, 1)$.   ■

When a function is represented by a power series

$$f(x) = \sum_{n=0}^{\infty} a_n(x - c)^n$$

it is natural to ask if its derivative or antiderivative can also be represented by a power series. Theorem 18.17 (p. 860) which we state without proof, answers this question affirmatively.

Theorem 18.17 indicates simply that a function defined by a power series behaves like a polynomial. It is continuous in the interval of convergence, and both its derivative and its antiderivative can be determined by differentiating and integrating, respectively, each term of the given power series. Furthermore, the *radius* of convergence for the two series obtained by differentiating and integrating is the same as for the original power series.

Note, however, that Theorem 18.17 does *not* imply that the *intervals* of convergence are identical for the series that represent $f(x)$, $f'(x)$, and $\int f(x)\, dx$. In fact, the convergence of these series often differs at the endpoints, as shown in the next example.

THEOREM 18.17
(*Properties of a Function Defined by a Power Series*)

If

$$f(x) = \sum_{n=0}^{\infty} a_n(x - c)^n$$

$$= a_0 + a_1(x - c) + a_2(x - c)^2 + a_3(x - c)^3 + \cdots$$

then on the interval $(c - R, c + R)$, where $R > 0$ is the radius of convergence:

i. f is differentiable, and

$$f'(x) = \sum_{n=0}^{\infty} na_n(x - c)^{n-1}$$

$$= a_1 + 2a_2(x - c) + 3a_3(x - c)^2 + \cdots$$

ii. f is integrable on each closed subinterval, and

$$\int f(x)\, dx = \left[\sum_{n=0}^{\infty} \frac{a_n(x - c)^{n+1}}{n + 1} \right] + C$$

$$= C + a_0(x - c) + a_1\frac{(x - c)^2}{2} + a_2\frac{(x - c)^3}{3} + \cdots$$

iii. f and $f^{(n)}$ are continuous for every n.

Example 6

Find the intervals of convergence of $\int f(x)\, dx$, $f(x)$, and $f'(x)$, where

$$f(x) = \sum_{n=1}^{\infty} \frac{x^n}{n} = x + \frac{x^2}{2} + \frac{x^3}{3} + \cdots$$

Solution: By Theorem 18.17 we have

$$f'(x) = \sum_{n=1}^{\infty} x^{n-1} = 1 + x + x^2 + x^3 + \cdots$$

and

$$\int f(x)\, dx = C + \left[\sum_{n=1}^{\infty} \frac{x^{n+1}}{n(n + 1)} \right] = C + \frac{x^2}{1 \cdot 2} + \frac{x^3}{2 \cdot 3} + \frac{x^4}{3 \cdot 4} + \cdots$$

By the Ratio Test we know that each series has a radius of convergence of $R = 1$. Now considering the interval $(-1, 1)$, we have the following:

1. The series for $f(x)$,

$$\sum_{n=1}^{\infty} \frac{x^n}{n}$$

converges for $x = -1$, diverges for $x = 1$, and its interval of convergence is $[-1, 1)$.

2. The series for $f'(x)$,

$$\sum_{n=1}^{\infty} x^{n-1}$$

diverges for $x = \pm 1$, and its interval of convergence is $(-1, 1)$.

3. The series for $\int f(x)\, dx$,

$$\sum_{n=1}^{\infty} \frac{x^{n+1}}{n(n+1)}$$

converges for $x = \pm 1$, and its interval of convergence is $[-1, 1]$.

∎

This example suggests that of the three series, the one for the derivative, $f'(x)$, may be the least likely to converge at the endpoints. This is indeed the case, and, in general, if the series for $f'(x)$ converges at the endpoints $x = c \pm R$, then the series for $f(x)$ will also converge there (see Exercise 31).

The next example shows how to approximate the value of the definite integral of a function represented by a power series. This technique is useful for a function such as e^{x^2}, which has no elementary antiderivative but can be readily represented by a power series.

Example 7
_____

Using $R_N \leq 0.01$, approximate the value of the definite integral $\int_0^1 f(x)\, dx$, where

$$f(x) = \sum_{n=0}^{\infty} \frac{(-1)^n x^{2n}}{(2n)!}$$

Solution: First of all, the Ratio Test shows that the series converges for all x ($R = \infty$) and so it follows that f is integrable on $[0, 1]$. By Theorem 18.17 we have

$$\int_0^1 f(x)\, dx = \int_0^1 \left[\sum_{n=0}^{\infty} \frac{(-1)^n x^{2n}}{(2n)!} \right] dx = \left[\sum_{n=0}^{\infty} \frac{(-1)^n x^{2n+1}}{(2n+1)(2n)!} \right]_0^1$$

$$= \left[\sum_{n=0}^{\infty} \frac{(-1)^n x^{2n+1}}{(2n+1)!} \right]_0^1$$

$$= \frac{1}{1!} - \frac{1}{3!} + \frac{1}{5!} - \frac{1}{7!} + \frac{1}{9!} - \cdots$$

$$= \sum_{n=0}^{\infty} (-1)^n \frac{1}{(2n+1)!}$$

By the Alternating Series Remainder, we want to choose N such that

$$|R_N| \le a_{N+1} = \frac{1}{(2N+3)!} < 0.01$$

If $N = 1$, then

$$a_{N+1} = \frac{1}{5!} = \frac{1}{120} < 0.01$$

and we have

$$|R_1| \le \frac{1}{120} \le 0.01$$

Therefore,

$$\int_0^1 f(x)\, dx \approx S_1 = 1 - \frac{1}{3!} \approx 0.8333$$

(The actual value of this integral is $\sin 1 \approx 0.8415$. See Exercise 35 of Section 18.8.)   ■

Section Exercises (18.6)

In Exercises 1–24, find the interval of convergence for each power series. Include a check for convergence at the endpoints of the interval.

1. $\sum_{n=0}^{\infty} \left(\frac{x}{2}\right)^n$

2. $\sum_{n=0}^{\infty} \left(\frac{x}{k}\right)^n$

S 3. $\sum_{n=1}^{\infty} \frac{(-1)^n x^n}{n}$

4. $\sum_{n=0}^{\infty} (-1)^{n+1} n x^n$

5. $\sum_{n=0}^{\infty} \frac{x^n}{n!}$

S 6. $\sum_{n=0}^{\infty} \frac{(3x)^n}{(2n)!}$

S 7. $\sum_{n=0}^{\infty} (2n)! \left(\frac{x}{2}\right)^n$

8. $\sum_{n=0}^{\infty} \frac{(-1)^n x^n}{(n+1)(n+2)}$

9. $\sum_{n=1}^{\infty} \frac{(-1)^{n+1} x^n}{4^n}$

10. $\sum_{n=0}^{\infty} \frac{(-1)^n n!(x-4)^n}{3^n}$

S 11. $\sum_{n=1}^{\infty} \frac{(-1)^{n+1}(x-5)^n}{n5^n}$

12. $\sum_{n=0}^{\infty} \frac{(x-2)^{n+1}}{(n+1)3^{n+1}}$

13. $\sum_{n=0}^{\infty} \frac{(-1)^{n+1}(x-1)^{n+1}}{n+1}$

14. $\sum_{n=1}^{\infty} \frac{(-1)^{n+1}(x-c)^n}{nc^n}$

15. $\sum_{n=1}^{\infty} \frac{(x-c)^{n-1}}{c^{n-1}}, \ 0 < c$

16. $\sum_{n=1}^{\infty} \left(\frac{1 \cdot 3 \cdot 5 \cdots (2n-1)}{2 \cdot 4 \cdot 6 \cdots 2n}\right) \left(\frac{x^{2n+1}}{2n+1}\right)$

S 17. $\sum_{n=1}^{\infty} \frac{(-1)^{n+1} x^{2n-1}}{2n-1}$

18. $\sum_{n=1}^{\infty} \frac{n!(x-c)^n}{1 \cdot 3 \cdot 5 \cdots (2n-1)}$

19. $\sum_{n=1}^{\infty} \frac{n}{n+1}(-2x)^{n-1}$

20. $\sum_{n=0}^{\infty} \frac{(-1)^n x^{2n}}{n!}$

21. $\sum_{n=0}^{\infty} \frac{x^{2n+1}}{(2n+1)!}$

22. $\sum_{n=1}^{\infty} \frac{n! x^n}{(2n)!}$

S 23. $\sum_{n=1}^{\infty} \frac{k(k+1)(k+2) \cdots (k+n-1)x^n}{n!}, \ k \ge 1$

24. $\sum_{n=1}^{\infty} \frac{(-1)^n 2^{2n-1} x^{2n}}{(2n)!}$

S 25. If $f(x) = \sum_{n=0}^{\infty} (x/2)^n$, find the interval of convergence of the following:
   (a) $f(x)$ (see Exercise 1)
   (b) $f'(x)$
   (c) $\int f(x)\, dx$

26. If

$$f(x) = \sum_{n=1}^{\infty} \frac{(-1)^{n+1}(x-5)^n}{n5^n}$$

find the interval of convergence of the following:
(a) $f(x)$ (see Exercise 11)
(b) $f'(x)$
(c) $\int f(x)\, dx$

27. If
$$f(x) = \sum_{n=0}^{\infty} \frac{(-1)^{n+1}(x-1)^{n+1}}{n+1}$$

find the interval of convergence of the following:
(a) $f(x)$ (see Exercise 13)
(b) $f'(x)$
(c) $f''(x)$
(d) $\int f(x)\, dx$

28. If
$$f(x) = \sum_{n=1}^{\infty} \frac{(-1)^{n+1}(x-1)^n}{n}$$

find the following:
(a) the interval of convergence
(b) a power series for $f'(x)$
(c) a power series for $f''(x)$
(d) a power series for $\int f(x)\, dx$

$\boxed{\text{S}}$ 29. Let
$$f(x) = \sum_{n=0}^{\infty} \frac{(-1)^n x^{2n}}{(2n+1)!}$$

(a) Find the interval of convergence.
(b) Find the power series for $\int f(x)\, dx$.
(c) Approximate $\int_0^1 f(x)\, dx$, using $R_N \le 0.01$.

30. Let
$$f(x) = \sum_{n=0}^{\infty} \frac{(-1)^n x^{2n+1}}{(2n+1)!}$$

and
$$g(x) = \sum_{n=0}^{\infty} \frac{(-1)^n x^{2n}}{(2n)!}$$

(a) Find the interval of convergence of $f(x)$ and $g(x)$.
(b) Show that $f'(x) = g(x)$.
(c) Show that $g'(x) = -f(x)$.
(d) From the results of parts (b) and (c), identify the functions f and g.

31. If the series for $f'(x)$ converges at the endpoints of its interval of convergence, show that the series for $f(x)$ does also.

For Review and Class Discussion

True or False

1. _____ If a power series converges for both $x = 1$ and $x = 3$, then it converges for $x = 2$ also.

2. _____ If the power series $\sum_{n=0}^{\infty} a_n x^n$ converges for $x = 2$, then it converges for $x = -2$ also.

3. _____ If the power series $\sum_{n=0}^{\infty} a_n x^n$ converges for $x = 2$, then it converges for $x = -1$ also.

4. _____ If the interval of convergence for $\sum_{n=0}^{\infty} a_n x^n$ is $(-1, 1)$, then the interval of convergence for $\sum_{n=0}^{\infty} a_n (x-1)^n$ is $(0, 2)$.

5. _____ If $f(x) = \sum_{n=0}^{\infty} a_n x^n$ converges for $|x| < 2$, then $\int_0^1 f(x)\, dx = \sum_{n=0}^{\infty} [a_n/(n+1)]$.

18.7 **Power Series for Functions**

Purpose
- To develop a geometric power series for functions of the form $f(x) = a/(x - b)$.
- To use these geometric series to find a power series for various elementary functions.

Now here is a difficulty that really is great. It may happen that a finite equation may also be expressed as an infinite one, so that the equation obtained may really be the same as the given equation although it does not appear to be. For example,

$$\frac{x}{1+x} = x - x^2 + x^3 - x^4 + x^5 - x^6 + \cdots$$

—Gottfried Leibniz (1646–1716)

In the previous section we established some properties of power series and of functions represented by power series. Informally you could say that we were seeking answers to the question: Given a power series, what function does it represent? In this and the next section, we want to turn this question around and ask: Given a function, what is its power series representation? We accomplish this objective in this section by deriving power series for some basic elementary functions. Then, using properties of series and operations with series, we produce power series for other functions that are combinations (sums, products, quotients, or compositions) of the basic functions. We begin our development with a simple function that resembles the sum of a geometric series.

Consider the function given by

$$f(x) = \frac{1}{1 - x}$$

The form of f closely resembles the sum of the convergent geometric series

$$\sum_{n=0}^{\infty} ar^n = \frac{a}{1 - r}$$

where $|r| < 1$. Therefore, if we let $a = 1$ and $r = x$, then a power series representation for $1/(1 - x)$, centered at $x = 0$, is simply

$$\frac{1}{1 - x} = \sum_{n=0}^{\infty} x^n = 1 + x + x^2 + x^3 + \cdots, \qquad |x| < 1$$

Of course, we should be aware that this series represents $f(x) = 1/(1 - x)$ only on the interval $(-1, 1)$, whereas f is defined for all $x \neq 1$ (see Figure 18.9).

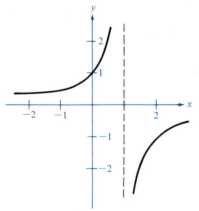

$f(x) = \dfrac{1}{1 - x}$, Domain: all $x \neq 1$

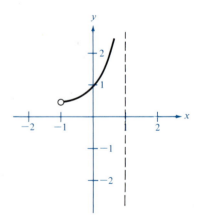

$f(x) = \sum_{n=0}^{\infty} x^n$, Domain: $-1 < x < 1$

FIGURE 18.9

The fact that the series $\sum_{n=0}^{\infty} x^n$ represents $1/(1-x)$ just on the interval $(-1, 1)$ should not be distressing because it is possible to represent $1/(1-x)$ by a power series centered at *any* point other than $x = 1$ (its vertical asymptote). For instance, to obtain the power series centered at $x = -1$, we write

$$\frac{1}{1-x} = \frac{1}{2-(x+1)} = \frac{\frac{1}{2}}{1-[(x+1)/2]} = \frac{a}{1-r}$$

which suggests that we let $a = \frac{1}{2}$ and $r = (x+1)/2$. We then have

$$\frac{1}{1-x} = \sum_{n=0}^{\infty} \left(\frac{1}{2}\right)\left(\frac{x+1}{2}\right)^n, \qquad |x+1| < 2$$

from which it follows that, on the interval $(-3, 1)$, $1/(1-x)$ is represented by the power series

$$\frac{1}{1-x} = \frac{1}{2} \sum_{n=0}^{\infty} \left(\frac{x+1}{2}\right)^n$$

$$= \frac{1}{2}\left[1 + \frac{(x+1)}{2} + \frac{(x+1)^2}{4} + \frac{(x+1)^3}{8} + \cdots\right]$$

centered at $x = -1$.

This procedure for writing a geometric power series centered at an arbitrary point is outlined in the following theorem.

THEOREM 18.18
(*Geometric Power Series*)

Given the function

$$f(x) = \frac{a}{x-b}$$

i. Its power series representation, *centered at $x = 0$*, is

$$\frac{a}{x-b} = -\frac{a}{b} \sum_{n=0}^{\infty} \left(\frac{x}{b}\right)^n, \qquad |x| < |b| \neq 0$$

ii. Its power series representation, *centered at $x = c$*, is

$$\frac{a}{x-b} = -\frac{a}{b-c} \sum_{n=0}^{\infty} \left(\frac{x-c}{b-c}\right)^n, \qquad |x-c| < |b-c| \neq 0$$

Proof:
i. This part follows directly from Theorem 18.5, since for $b \neq 0$

$$\frac{a}{x-b} = \frac{-a}{b-x} = \frac{-a/b}{1-(x/b)} = -\frac{a}{b} \sum_{n=0}^{\infty} \left(\frac{x}{b}\right)^n, \qquad \left|\frac{x}{b}\right| < 1$$

ii.   Since $b \neq c$, we write

$$\frac{a}{x-b} = \frac{-a}{b-x} = \frac{-a}{(b-c)-(x-c)} = \frac{-a/(b-c)}{1-[(x-c)/(b-c)]}$$

and it follows that

$$\frac{a}{x-b} = \left(\frac{-a}{b-c}\right) \sum_{n=0}^{\infty} \left(\frac{x-c}{b-c}\right)^n, \qquad \left|\frac{x-c}{b-c}\right| < 1$$

∎

Example 1

Find a power series centered at $x = 0$ for $f(x) = 3/(x + 2)$.

Solution: Writing $f(x)$ in the form $a/(x - b)$, we have

$$\frac{3}{x+2} = \frac{3}{x-(-2)} = \frac{a}{x-b}$$

which suggests that $a = 3$ and $b = -2$. Therefore, by Theorem 18.18 we have

$$\frac{3}{x+2} = \frac{-3}{-2} \sum_{n=0}^{\infty} \left(\frac{x}{-2}\right)^n, \qquad |x| < 2$$

$$= \frac{3}{2}\left[1 - \frac{x}{2} + \frac{x^2}{4} - \frac{x^3}{8} + \cdots\right]$$

∎

Example 2

Find a power series centered at $x = 1$ for $f(x) = 1/x$.

Solution: First, we note that $c = 1$ and $f(x)$ has the form

$$\frac{1}{x} = \frac{1}{x-0} = \frac{a}{x-b}$$

which implies that $a = 1$ and $b = 0$. Therefore, by part ii of Theorem 18.18, we have

$$\frac{1}{x} = \frac{-1}{0-1} \sum_{n=0}^{\infty} \left(\frac{x-1}{0-1}\right)^n$$

$$= \sum_{n=0}^{\infty} (-1)^n (x-1)^n, \qquad |x-1| < 1$$

$$= 1 - (x-1) + (x-1)^2 - (x-1)^3 + \cdots$$

∎

The true versatility of Theorem 18.18 will be seen later in this section following a discussion of some operations (addition, multiplication, and composition) with power series. These operations, used in conjunction with Theorem 18.18, with the derivative and the integral, will give us a means for developing power series for a wide variety of elementary functions. For instance, we will find power series for functions as diverse as

$$\frac{5x - 1}{x^2 - x - 2} \qquad \ln x, \qquad x \text{ Arctan } x$$

When manipulating power series, we must be aware of possible changes in the interval of convergence. For example, if we add the two convergent series

$$\sum_{n=0}^{\infty} x^n, \quad |x| < 1 \qquad \text{and} \qquad \sum_{n=0}^{\infty} \left(\frac{x}{2}\right)^n, \quad |x| < 2$$

then the resulting series $\qquad \displaystyle\sum_{n=0}^{\infty} \left(1 + \frac{1}{2^n}\right) x^n$

converges in the *intersection* ($|x| < 1$) of the intervals of convergence for the two original series.

With this observation in mind, we summarize the basic operations with power series. For simplicity, we state the results for series centered at $x = 0$.

Operations with Power Series

Given

$$f(x) = \sum_{n=0}^{\infty} a_n x^n \qquad \text{and} \qquad g(x) = \sum_{n=0}^{\infty} b_n x^n$$

we have:

1. $f(kx) = \displaystyle\sum_{n=0}^{\infty} a_n k^n x^n$ 3. $f(x) \pm g(x) = \displaystyle\sum_{n=0}^{\infty} (a_n \pm b_n) x^n$

2. $f(x^N) = \displaystyle\sum_{n=0}^{\infty} a_n x^{nN}$ 4. $f(x)\, g(x) = \displaystyle\sum_{n=0}^{\infty} c_n x^n,$

$$\text{where } c_n = \sum_{i=0}^{n} a_i b_{n-i}$$

Proof: Parts 1, 2, and 3 are the obvious properties of a sum. Part 4 can be verified by multiplication. Thus the product of

$$\sum_{n=0}^{\infty} a_n x^n = a_0 + a_1 x + a_2 x^2 + a_3 x^3 + \cdots$$

and
$$\sum_{n=0}^{\infty} b_n x_n = b_0 + b_1 x + b_2 x^2 + b_3 x^3 + \cdots$$

is

$$\left(\sum_{n=0}^{\infty} a_n x^n\right)\left(\sum_{n=0}^{\infty} b_n x^n\right)$$

$$= a_0 b_0 + (a_0 b_1 + a_1 b_0)x + (a_0 b_2 + a_1 b_1 + a_2 b_0)x^2 + \cdots$$

$$= \sum_{n=0}^{\infty} c_n x^n$$

This implies that
$$c_0 = a_0 b_0$$
$$c_1 = a_0 b_1 + a_1 b_0$$
$$c_2 = a_0 b_2 + a_1 b_1 + a_2 b_0$$
$$\vdots$$
$$c_n = \sum_{i=0}^{n} a_i b_{n-i}$$

■

Example 3

Find a power series centered at $x = 0$ for
$$f(x) = \frac{5x - 1}{x^2 - x - 2}$$

Solution: Using partial fractions we can rewrite $f(x)$ as
$$\frac{5x - 1}{x^2 - x - 2} = \frac{2}{x + 1} + \frac{3}{x - 2}$$

Now applying Theorem 18.18 we obtain
$$\frac{2}{x + 1} = \frac{2}{x - (-1)} = \frac{-2}{-1} \sum_{n=0}^{\infty} \left(\frac{x}{-1}\right)^n = \sum_{n=0}^{\infty} 2(-1)^n x^n, \qquad |x| < 1$$

and
$$\frac{3}{x - 2} = \frac{-3}{2} \sum_{n=0}^{\infty} \left(\frac{x}{2}\right)^n = \sum_{n=0}^{\infty} \frac{-3x^n}{2^{n+1}}, \qquad |x| < 2$$

Finally, by adding these two series, we have
$$\frac{5x - 1}{x^2 - x - 2} = \sum_{n=0}^{\infty} \left[2(-1)^n - \frac{3}{2^{n+1}}\right]x^n$$

$$= \sum_{n=0}^{\infty} \left[\frac{(-1)^n 2^{n+2} - 3}{2^{n+1}} \right] x^n$$

$$= \frac{1}{2} - \frac{11}{4}x + \frac{13}{8}x^2 - \frac{35}{16}x^3 + \cdots, \qquad |x| < 1$$

■

It is possible to determine the power series representation for a rational function such as the one in Example 3 by long division. For instance,

$$
\begin{array}{r}
\frac{1}{2} - \frac{11}{4}x + \frac{13}{8}x^2 - \frac{35}{16}x^3 + \cdots \\
-2 - x + x^2 \overline{\smash{\big)}\,-1 + 5x} \\
-1 - \frac{x}{2} + \frac{x^2}{2} \\[4pt]
\hline
\frac{11x}{2} - \frac{x^2}{2} \\[4pt]
\frac{11x}{2} + \frac{11x^2}{4} - \frac{11x^3}{4} \\[4pt]
\hline
-\frac{13x^2}{4} + \frac{11x^2}{4} \\[4pt]
-\frac{13x^2}{4} - \frac{13x^3}{8} + \frac{13x^4}{8} \\[4pt]
\hline
\frac{35x^3}{8} - \frac{13x^4}{8} \\[4pt]
\frac{35x^3}{8} + \frac{35x^4}{16} - \frac{35x^5}{16} \\[4pt]
\hline
\end{array}
$$

One disadvantage of this alternative procedure is that the general term of the series is not given and we may be hard pressed to determine it. And without knowing the general term of the series, we cannot readily determine the interval of convergence.

Example 4

Find a power series centered at $x = 1$ for $f(x) = \ln x$.

Solution: From Example 2 we know that

$$\frac{1}{x} = \sum_{n=0}^{\infty} (-1)^n (x - 1)^n, \qquad |x - 1| < 1$$

By integration we have

$$\ln x = \int \frac{1}{x} dx + C = C + \sum_{n=0}^{\infty} (-1)^n \frac{(x - 1)^{n+1}}{n + 1}$$

If we let $x = 1$, then $C = 0$ and therefore,

$$\ln x = \sum_{n=0}^{\infty} (-1)^n \left[\frac{(x-1)^{n+1}}{n+1} \right]$$

$$= \frac{(x-1)}{1} - \frac{(x-1)^2}{2} + \frac{(x-1)^3}{3} - \frac{(x-1)^4}{4} + \cdots, \quad 0 < x \le 2$$

Note that the series converges at $x = 2$. ■

This last example suggests that power series can be used to obtain tables of values for functions such as $\ln x$, e^x, $\sin x$, and so forth. For instance, to obtain a value for $\ln (1.1)$ correct to four decimal places, we can use the alternating series remainder R_N and determine N such that $\ln (1.1) \approx S_N$. This is accomplished in the following manner. We need N such that

$$|R_N| < a_{n+1} \le 0.00005$$

If we choose $N = 3$, we can see from the series for $\ln x$ that

$$a_{N+1} = \frac{(1.1 - 1)^4}{4} = \frac{(0.1)^4}{4} = \frac{0.0001}{4} = 0.000025 < 0.00005$$

Therefore,

$$\ln (1.1) \approx S_3 = (0.1) - \frac{(0.01)}{2} + \frac{(0.001)}{3} \approx 0.09533$$

is correct to four decimal places.

Example 5

Find a power series centered at $x = 0$ for $f(x) = x \operatorname{Arctan} x$.

Solution: Since

$$\frac{d}{dx}[\operatorname{Arctan} x] = \frac{1}{1 + x^2}$$

we use the series

$$f(x) = \frac{1}{1 + x} = \sum_{n=0}^{\infty} (-1)^n x^n, \qquad |x| < 1$$

Therefore, $$f(x^2) = \frac{1}{1 + x^2} = \sum_{n=0}^{\infty} (-1)^n x^{2n}$$

Now by integrating we have

$$\operatorname{Arctan} x = \int \frac{1}{1 + x^2} dx + C = C + \sum_{n=0}^{\infty} \frac{(-1)^n x^{2n+1}}{2n + 1}$$

If we let $x = 0$, then $C = 0$ and therefore,

$$\text{Arctan } x = \sum_{n=0}^{\infty} \frac{(-1)^n x^{2n+1}}{2n+1}$$

Finally, multiplying by x yields

$$x \text{ Arctan } x = \sum_{n=0}^{\infty} (-1)^n \frac{x^{2n+2}}{2n+1}$$

$$= \frac{x^2}{1} - \frac{x^4}{3} + \frac{x^6}{5} - \frac{x^8}{7} + \cdots, \qquad |x| \leq 1$$

Again, note that the integrand series converges at the endpoints $x = \pm 1$.

◾

Example 6

Let S_n be the nth partial sum of the harmonic series

$$1 + \tfrac{1}{2} + \tfrac{1}{3} + \tfrac{1}{4} + \tfrac{1}{5} + \cdots$$

Show that

$$\frac{\ln x}{x} = \sum_{n=0}^{\infty} (-1)^n S_{n+1}(x-1)^n$$

Solution: From Examples 2 and 4 we have the two series

$$\frac{1}{x} = \sum_{n=0}^{\infty} (-1)^n (x-1)^n$$

and

$$\ln x = \sum_{n=0}^{\infty} (-1)^n \frac{(x-1)^{n+1}}{n+1}$$

Now by multiplying these two series together, we obtain

$$1 - (x-1) + (x-1)^2 - (x-1)^3 + (x-1)^4 - \cdots$$

$$(x-1) - \frac{(x-1)^2}{2} + \frac{(x-1)^3}{3} - \frac{(x-1)^4}{4} + \cdots$$

$$\overline{\qquad\qquad\qquad\qquad\qquad\qquad\qquad\qquad\qquad}$$

$$(x-1) - (x-1)^2 + (x-1)^3 - (x-1)^4 + (x-1)^5 - \cdots$$

$$-\frac{(x-1)^2}{2} + \frac{(x-1)^3}{2} - \frac{(x-1)^4}{2} + \frac{(x-1)^5}{2} - \cdots$$

$$+\frac{(x-1)^3}{3} - \frac{(x-1)^4}{3} + \frac{(x-1)^5}{3} - \cdots$$

$$-\frac{(x-1)^4}{4} + \frac{(x-1)^5}{4} - \cdots$$

$$+\frac{(x-1)^5}{5} - \cdots$$

$$\overline{\qquad\qquad\qquad\qquad\qquad\qquad\qquad\qquad\qquad}$$

$$S_1(x-1) - S_2(x-1)^2 + S_3(x-1)^3 - S_4(x-1)^4 + S_5(x-1)^5 - \cdots$$

$$= \sum_{n=0}^{\infty} (-1)^{n+1} S_n (x - 1)^n$$

where

$$S_1 = 1$$
$$S_2 = 1 + \tfrac{1}{2}$$
$$S_3 = 1 + \tfrac{1}{2} + \tfrac{1}{3}$$

and so on.   ∎

Section Exercises (18.7)

In Exercises 1–12, use the methods of this section to find geometric series centered at c for the specified functions, and find the interval of convergence.

1. $f(x) = \dfrac{1}{2 - x}; c = 0$

2. $f(x) = \dfrac{3}{4 - x}; c = 0$

[S] **3.** $f(x) = \dfrac{1}{2 - x}; c = 5$

4. $f(x) = \dfrac{3}{4 - x}; c = -2$

5. $f(x) = \dfrac{3}{2x - 1}; c = 0$

6. $f(x) = \dfrac{3}{2x - 1}; c = 2$

7. $f(x) = \dfrac{1}{2x - 5}; c = -3$

8. $f(x) = \dfrac{1}{2x - 5}; c = 0$

9. $f(x) = \dfrac{3}{x + 2}; c = 0$

10. $f(x) = \dfrac{4}{3x + 2}; c = 2$

[S] **11.** $f(x) = \dfrac{3x}{x^2 + x - 2}; c = 0$

12. $f(x) = \dfrac{4x - 7}{2x^2 + 3x - 2}; c = 0$

[S] **13.** Find a power series about $c = 0$ for $f(x) = 2/(1 - x^2)$ by:
(a) using partial fractions

(b) finding a series representation for $2/(1 - u)$, where $u = x^2$
(c) using long division

14. Find a power series representation for $f(x) = 4/(4 + x^2)$ about $c = 0$, and find the interval of convergence.

In Exercises 15–22, use the power series

$$\frac{1}{x + 1} = \sum_{n=0}^{\infty} (-1)^n x^n$$

to determine a power series representation about $c = 0$ for the given function. Specify the interval of convergence.

15. $f(x) = \dfrac{-1}{(x + 1)^2}$ (Hint: Use differentiation.)

16. $f(x) = \dfrac{2}{(x + 1)^3}$

[S] **17.** $f(x) = \ln (x + 1)$
18. $f(x) = \ln (x^2 + 1)$

19. $f(x) = \dfrac{1}{4x^2 + 1}$

20. $f(x) = \text{Arctan } 2x$ (Hint: Use the result of Exercise 19.)

21. $f(x) = \ln \sqrt{\dfrac{1 + x}{1 - x}}$

22. $f(x) = \dfrac{1}{x^2 - x + 1}$ $\left(\text{Hint: } \dfrac{1}{x^2 - x + 1} = \dfrac{x + 1}{x^3 + 1}.\right)$

[C] **23.** Complete the accompanying table to demonstrate the inequalities

$$x - \frac{x^2}{2} \leq \ln(x+1) \leq x - \frac{x^2}{2} + \frac{x^3}{3}$$

in $0 \leq x \leq 1$.

c **24.** Complete a table for the same values of x as in Exercise 23 to demonstrate the inequalities

$$x - \frac{x^2}{2} + \frac{x^3}{3} - \frac{x^4}{4} \leq \ln(x+1)$$

$$\leq x - \frac{x^2}{2} + \frac{x^3}{3} - \frac{x^4}{4} + \frac{x^5}{5}$$

x	$x - \dfrac{x^2}{2}$	$\ln(x+1)$	$x - \dfrac{x^2}{2} + \dfrac{x^3}{3}$
0.0			
0.2			
0.4			
0.6			
0.8			
1.0			

18.8 Taylor Series

Purpose
- To derive and use Taylor's Theorem for generating the power series for a given function.
- To use the methods of the previous section to obtain power series representations of elementary functions.

Thus, provided that we have a method of computing the first derivative of any function, we can obtain, by merely repeating the same operation, all the derived functions, and consequently all the terms of the series.

—Joseph Lagrange (1736–1813)

In Section 18.7 we showed how to derive the power series for many functions by using the basic geometric series

$$\frac{a}{x-b} = \frac{-a}{b-c} \sum_{n=0}^{\infty} \left(\frac{x-c}{b-c}\right)^n$$

We complete this chapter with a discussion of a *general* procedure for deriving the power series for *any* differentiable elementary function.

Suppose that the function f is represented by a power series, centered at $x = c$,

$$f(x) = \sum_{n=0}^{\infty} a_n(x-c)^n, \qquad |x-c| < R$$

Then by Theorem 18.17 we know that the nth derivative of f exists for $|x-c| < R$. Thus by successive differentiation we have

$$f^{(0)}(x) = a_0 + a_1(x-c) + a_2(x-c)^2 + a_3(x-c)^3 + \cdots$$
$$f^{(1)}(x) = a_1 + 2a_2(x-c) + 3a_3(x-c)^2 + 4a_4(x-c)^3 + \cdots$$
$$f^{(2)}(x) = 2a_2 + 3!a_3(x-c) + 4 \cdot 3a_4(x-c)^2 + 5 \cdot 4a_5(x-c)^3 + \cdots$$

$$f^{(3)}(x) = 3!a_3 + 4!a_4(x - c) + 5 \cdot 4 \cdot 3a_5(x - c)^2$$
$$+ 6 \cdot 5 \cdot 4(x - c)^3 + \cdots$$

$$\vdots$$

$$f^{(n)}(x) = n!a_n + (n + 1)!a_{n+1}(x - c) + \cdots$$

Now evaluating each of these derivatives at $x = c$ yields

$$f^{(0)}(c) = a_0 = 0!a_0$$
$$f^{(1)}(c) = a_1 = 1!a_1$$
$$f^{(2)}(c) = 2a_2 = 2!a_2$$
$$f^{(3)}(c) = 3!a_3$$
$$\vdots$$
$$f^{(n)}(c) = n!a_n$$

By solving for a_n in the last equation, we find that the nth coefficient of the power series representation of $f(x)$ is

$$a_n = \frac{f^{(n)}(c)}{n!}$$

We call this result **Taylor's Theorem.**

THEOREM 18.19
(Taylor's Theorem)

If f is represented by a power series

$$f(x) = a_0 + a_1(x - c) + a_2(x - c)^2 + a_3(x - c)^3 + \cdots$$

then the coefficients are given by

$$a_n = \frac{f^{(n)}(c)}{n!}$$

and

$$f(x) = f(c) + f'(c)(x - c) + \frac{f''(c)}{2!}(x - c)^2 + \frac{f'''(c)}{3!}(x - c)^3 + \cdots$$

$$= \sum_{n=0}^{\infty} \frac{f^{(n)}(c)}{n!}(x - c)^n$$

The series obtained in Theorem 18.19 is often referred to as a **Taylor series** for $f(x)$ at $x = c$. If the series is centered at $x = 0$, it is called a **Maclaurin series.**

Example 1

Find the power series centered at $x = 0$ for $f(x) = \sin x$.

Solution: Successive differentiation of $f(x)$ yields

$$f(x) = \sin x \qquad f(0) = 0$$
$$f'(x) = \cos x \qquad f'(0) = 1$$
$$f''(x) = -\sin x \qquad f''(0) = 0$$
$$f'''(x) = -\cos x \qquad f'''(0) = -1$$
$$f^{(4)}(x) = \sin x \qquad f^{(4)}(0) = 0$$
$$f^{(5)}(x) = \cos x \qquad f^{(5)}(0) = 1$$
$$f^{(6)}(x) = -\sin x \qquad f^{(6)}(0) = 0$$
$$f^{(7)}(x) = -\cos x \qquad f^{(7)}(0) = -1$$

and so on. The pattern repeats after the third derivative. Therefore, by Taylor's Theorem we have

$$\sin x = f(0) + f'(0)x + \frac{f''(0)x^2}{2} + \frac{f'''(0)x^3}{3!} + \frac{f^{(4)}(0)x^4}{4!} + \cdots$$

$$= x - \frac{x^3}{3!} + \frac{x^5}{5!} - \frac{x^7}{7!} + \cdots = \sum_{n=0}^{\infty} \frac{(-1)^n x^{2n+1}}{(2n+1)!} \qquad \blacksquare$$

To obtain a better understanding of just how a Taylor series such as

$$\sum_{n=0}^{\infty} \frac{f^{(n)}(0)}{n!} x^n$$

(centered at $x = 0$) represents a function $f(x)$, consider the following sequence of partial sums:

$$S_0(x) = f(0)$$
$$S_1(x) = f(0) + f'(0)x$$

$$S_2(x) = f(0) + f'(0)x + \frac{f''(0)}{2!} x^2$$

$$S_3(x) = f(0) + f'(0)x + \frac{f''(0)}{2!} x^2 + \frac{f'''(0)}{3!} x^3$$

$$\vdots$$

$$S_n(x) = f(0) + f'(0)x + \frac{f''(0)}{2!} x^2 + \cdots + \frac{f^{(n)}(0)}{n!} x_n$$

The members of this sequence are called the **Taylor polynomials** for $f(x)$. On the interval of convergence, the graphs of these Taylor polynominals become closer and closer approximations to the graph of $f(x)$, as $n \to \infty$. For example, the graphs of a few Taylor polynomials for

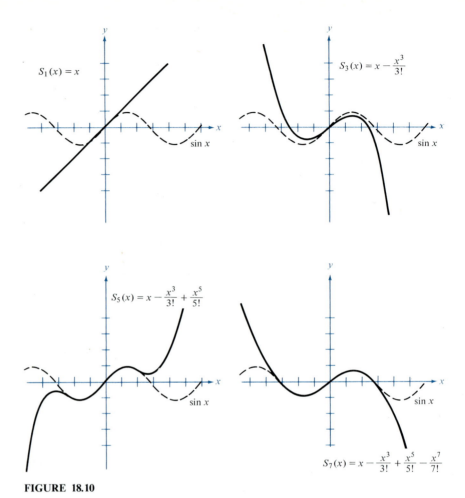

FIGURE 18.10

$$\sin x = \sum_{n=0}^{\infty} \frac{(-1)^n x^{2n+1}}{(2n + 1)!}$$

are shown in Figure 18.10. On the interval $(-\pi, \pi)$, notice that the graphs of $S_1(x)$, $S_3(x)$, $S_5(x)$, and $S_7(x)$ become successively better approximations to the graph of $\sin x$. Furthermore, from Figure 18.10 it appears that the closer x is to 0, the better the polynomial $S_n(x)$ approximates $\sin x$.

To reinforce this conclusion, consider the values shown in Table 18.6. Using the Ratio Test we see that the power series for $\sin x$ converges for *all* x. However, from Table 18.6 and Figure 18.10, we see that the further x is from 0, the more terms we need to obtain a good approximation.

A formal description of how closely a Taylor polynomial approximates a function f is given in the following theorem.

TABLE 18.6

x		0	$\pi/4$	$\pi/2$	$3\pi/4$	π
$S_1 = x$		0	0.7854	1.5708	2.3562	3.1416
$S_3 = x - \dfrac{x^3}{3!}$		0	0.7047	0.9248	0.1761	-2.0261
$S_5 = x - \dfrac{x^3}{3!} + \dfrac{x^5}{5!}$		0	0.7071	1.0045	0.7812	0.5240
$S_7 = x - \dfrac{x^3}{3!} + \dfrac{x^5}{5!} - \dfrac{x^7}{7!}$		0	0.7071	0.9998	0.7012	-0.0752
$S_9 = x - \dfrac{x^3}{3!} + \dfrac{x^5}{5!} - \dfrac{x^7}{7!} + \dfrac{x^9}{9!}$		0	0.7071	1.0000	0.7074	0.0069
$\sin x$		0	0.7071	1.0000	0.7071	0

THEOREM 18.20
(Taylor's Formula with Remainder)

Let f have derivatives up through order $n + 1$ for every x in an interval I containing c. Then for each x in I, there exists z between x and c such that

$$f(x) = f(c) + f'(c)(x - c) + \frac{f''(c)}{2!}(x - c)^2 + \cdots + \frac{f^{(n)}(c)}{n!}(x - c)^n + R_n$$

where
$$R_n = \frac{f^{(n+1)}(z)}{(n + 1)!}(x - c)^{n+1}$$

Proof: To find R_n we fix x in I, $x \neq c$, and write

$$R_n = f(x) - S_n(x)$$

where $S_n(x)$ is the nth partial sum of the series for $f(x)$. Then we let g be a function of t defined by

$$g(t) = f(x) - f(t) - f'(t)(x - t) - \frac{f''(t)}{2!}(x - t)^2 - \cdots$$

$$- \frac{f^{(n)}(t)}{n!}(x - t)^n - R_n \frac{(x - t)^{n+1}}{(x - c)^{n+1}}$$

Now differentiation with respect to t yields

$$g'(t) = -f'(t) + [f'(t) - f''(t)(x - t)]$$

$$+ \left[f''(t)(x - t) - \frac{f'''(t)}{2!}(x - t)^2 \right] + \cdots$$

$$+ \left[\frac{f^{(n)}(t)}{(n-1)!}(x-t)^{n-1} - \frac{f^{(n+1)}(t)}{n!}(x-t)^n \right] + (n+1)R_n \frac{(x-t)^{n+1}}{(x-c)^{n+1}}$$

which simplifies to

$$g'(t) = -\frac{f^{(n+1)}(t)}{n!}(x-t)^n + (n+1)R_n \frac{(x-t)^{n+1}}{(x-c)^{n+1}}$$

for all t between c and x. Now g has the properties that

$$g(c) = f(x) - [S_n(x) + R_n] = f(x) - f(x) = 0$$

and $\qquad g(x) = f(x) - f(x) - 0 - 0 \cdots = f(x) - f(x) = 0$

Therefore, g satisfies the conditions of Rolle's Theorem, and it follows that there is a number z between c and x such that $g'(z) = 0$. Substituting z for t in the equation for $g'(t)$ and then solving for R_n, we obtain

$$0 = -\frac{f^{(n+1)}(z)}{n!}(x-z)^n + (n+1)R_n \frac{(x-t)^{n+1}}{(x-c)^{n+1}}$$

or $\qquad R_n = \frac{f^{(n+1)}(z)}{(n+1)!}(x-c)^{n+1}$

Finally, using this value for R_n we have

$$0 = g(c) = f(x) - f(c) - f'(c)(x-c) - \frac{f''(c)}{2!}(x-c)^2 - \cdots$$

$$- \frac{f^{(n)}(c)}{n!}(x-c)^n - \frac{f^{(n+1)}(z)}{(n+1)!}(x-c)^{n+1}$$

and it follows that

$$f(x) = f(c) + f'(c)(x-c) + \frac{f''(c)}{2!}(x-c)^2 + \frac{f'''(c)}{3!}(x-c)^3 + \cdots$$

$$+ \frac{f^{(n)}(c)}{n!}(x-c)^n + \frac{f^{(n+1)}(z)}{(n+1)!}(x-c)^{n+1}$$

$$= S_n(x) + R_n$$

■

Example 2

If $f(x) = \sin x$, use Theorem 18.20 to approximate $\sin(0.1)$ by $S_3(0.1)$, and determine the accuracy.

Solution: From Example 1 we know that

$$\sin x = f(0) + f'(0)x + \frac{f''(0)}{2!}x^2 + \frac{f'''(0)}{3!}x^3 + R_n$$

Therefore,

$$\sin x = 0 + x + 0 - \frac{x^3}{3!} + \frac{f^{(4)}(z)}{4!}x^4$$

and

$$\sin (0.1) = 0.1 - \frac{(0.1)^3}{3!} + \frac{f^{(4)}(z)}{4!}(0.1)^4$$

where $0 < z < 0.1$. Summing the first two terms, we have

$$\sin (0.1) \approx 0.1 - 0.000167 = 0.099833$$

Furthermore,

$$|R_n| = \frac{\sin z}{4!}(0.1)^4 < \frac{0.0001}{4!} = 0.000004$$

Consequently, the approximation $\sin (0.1) \approx 0.099833$ is accurate to five decimal places. ∎

Example 3

Find a power series, centered at $x = 0$, for $f(x) = e^{x^2}$.

Solution: To use Taylor's Theorem we must calculate successive derivatives of $f(x) = e^{x^2}$. By calculating just the first two,

$$f'(x) = 2xe^{x^2} \quad \text{and} \quad f''(x) = (4x^2 + 2)e^{x^2}$$

we recognize this to be a rather cumbersome task. Fortunately there is an alternative. Suppose we first consider the power series for $g(x) = e^x$. Then we have

$$\begin{aligned} g(x) &= e^x & g(0) &= 1 \\ g'(x) &= e^x & g'(0) &= 1 \\ g''(x) &= e^x & g''(0) &= 1 \\ &\vdots & &\vdots \\ g^{(n)}(x) &= e^x & g^{(n)}(0) &= 1 \end{aligned}$$

Therefore,

$$g(x) = e^x = \sum_{n=0}^{\infty} \frac{g^{(n)}(0)}{n!}x^n = \sum_{n=0}^{\infty} \frac{1}{n!}x^n$$

Now since $e^{x^2} = g(x^2)$, we have

$$e^{x^2} = g(x^2) = \sum_{n=0}^{\infty} \frac{1}{n!}(x^2)^n$$

$$= 1 + x^2 + \frac{x^4}{2!} + \frac{x^6}{3!} + \frac{x^8}{4!} + \cdots$$

as the power series for e^{x^2}. ∎

Example 3 illustrates an important point in determining power series representations of functions. Though Taylor's Theorem is applicable to a wide variety of functions, it is frequently tedious to use because of the complexity of the derivatives. Therefore, the most practical use of Taylor's Theorem is in developing power series for a *basic list* of elementary functions. Then from this basic list, we can determine power series for other functions by the operations of addition, subtraction, multiplication, division, differentiation, integration, or composition with known power series.

We provide the following list of power series for some basic elementary functions.

Power Series for Elementary Functions

$$\frac{1}{x} = 1 - (x - 1) + (x - 1)^2 - (x - 1)^3 + (x - 1)^4 - \cdots + (-1)^n(x - 1)^n + \cdots, \qquad 0 < x < 2$$

$$\frac{1}{1 + x} = 1 - x + x^2 - x^3 + x^4 - x^5 + \cdots + (-1)^n x^n + \cdots, \qquad -1 < x < 1$$

$$\ln x = (x - 1) - \frac{(x - 1)^2}{2} + \frac{(x - 1)^3}{3} - \frac{(x - 1)^4}{4} + \frac{(x - 1)^5}{5} - \cdots + \frac{(-1)^{n-1}(x - 1)^n}{n} + \cdots,$$
$$0 < x \leq 2$$

$$e^x = 1 + x + \frac{x^2}{2!} + \frac{x^3}{3!} + \frac{x^4}{4!} + \frac{x^5}{5!} + \cdots + \frac{x^n}{n!} + \cdots, \qquad -\infty < x < \infty$$

$$\sin x = x - \frac{x^3}{3!} + \frac{x^5}{5!} - \frac{x^7}{7!} + \frac{x^9}{9!} - \cdots + \frac{(-1)^{n+1} x^{2n+1}}{(2n + 1)!} + \cdots, \qquad -\infty < x < \infty$$

$$\cos x = 1 - \frac{x^2}{2!} + \frac{x^4}{4!} - \frac{x^6}{6!} + \frac{x^8}{8!} - \cdots + \frac{(-1)^n x^{2n}}{(2n)!} + \cdots, \qquad -\infty < x < \infty$$

$$\text{Arctan } x = x - \frac{x^3}{3} + \frac{x^5}{5} - \frac{x^7}{7} + \frac{x^9}{9} - \cdots + \frac{(-1)^{n+1} x^{2n+1}}{2n + 1} + \cdots, \qquad -1 \leq x \leq 1$$

$$\text{Arcsin } x = x + \frac{x^3}{2 \cdot 3} + \frac{1 \cdot 3 x^5}{2 \cdot 4 \cdot 5} + \frac{1 \cdot 3 \cdot 5 x^7}{2 \cdot 4 \cdot 6 \cdot 7} + \frac{1 \cdot 3 \cdot 5 \cdot 7 x^9}{2 \cdot 4 \cdot 6 \cdot 8 \cdot 9} + \cdots + \frac{(2n)! \, x^{2n+1}}{(2^n n!)^2 (2n + 1)} + \cdots, \quad -1 \leq x \leq 1$$

$$(1 + x)^k = 1 + kx + \frac{k(k - 1)x^2}{2!} + \frac{k(k - 1)(k - 2)x^3}{3!} + \frac{k(k - 1)(k - 2)(k - 3)x^4}{4!} + \cdots, \quad -1 < x < 1^*$$

$$(1 + x)^{-k} = 1 - kx + \frac{k(k + 1)x^2}{2!} - \frac{k(k + 1)(k + 2)x^3}{3!} + \frac{k(k + 1)(k + 2)(k + 3)x^4}{4!} - \cdots, \quad -1 < x < 1^*$$

*The convergence at $x = \pm 1$ depends on the value k.

Proof: Some of the preceding series representations have been previously established, and we prove only two of the remaining ones; the binomial series for $(1 + x)^k$ and Arcsin x.

Using Taylor's Theorem for $f(x) = (1 + x)^k$, we obtain

$$f(x) = (1 + x)^k \qquad\qquad f(0) = 1$$
$$f'(x) = k(1 + x)^{k-1} \qquad\qquad f'(0) = k$$
$$f''(x) = k(k - 1)(1 + x)^{k-2} \qquad f''(0) = k(k - 1)$$
$$f'''(x) = k(k - 1)(k - 2)(1 + x)^{k-3} \qquad f'''(0) = k(k - 1)(k - 2)$$
$$\vdots \qquad\qquad\qquad\qquad\qquad \vdots$$
$$f^{(n)}(x) = k(k - 1)(k - 2) \cdots \qquad f^{(n)}(0) = k(k - 1)(k - 2) \cdots$$
$$(k - n + 1)(1 + x)^{k-n} \qquad\qquad (k - n + 1)$$

Therefore,

$$(1 + x)^k = 1 + kx + \frac{k(k - 1)x^2}{2} + \left[\frac{k(k - 1)(k - 2)}{3!}x^3\right] + \cdots$$

And by the Ratio Test, we have $R = 1$.

Now to obtain the series for $f(x) = \text{Arcsin } x$, we consider the binomial series

$$(1 + x)^{-1/2} = \frac{1}{\sqrt{1 + x}}$$

$$= 1 - \frac{1}{2}x + \frac{1 \cdot 3}{2!2^2}x^2 - \frac{1 \cdot 3 \cdot 5}{3!2^3}x^3 + \frac{1 \cdot 3 \cdot 5 \cdot 7}{4!2^4}x^4 - \cdots$$

Therefore, replacing x by $(-x^2)$, we obtain the series

$$\frac{1}{\sqrt{1 - x^2}} = 1 + \frac{1}{2}x^2 + \frac{1 \cdot 3}{2!2^2}x^4 + \frac{1 \cdot 3 \cdot 5}{3!2^3}x^6 + \frac{1 \cdot 3 \cdot 5 \cdot 7}{4!2^4}x^8 + \cdots$$

Now by integrating each term, it follows that

$$\text{Arcsin } x = x + \frac{x^3}{2 \cdot 3} + \frac{1 \cdot 3}{2!2^2 \cdot 5}x^5 + \frac{1 \cdot 3 \cdot 5}{3!2^3 \cdot 7}x^7 + \frac{1 \cdot 3 \cdot 5 \cdot 7}{4!2^4 \cdot 9}x^9 + \cdots$$

$$= x + \frac{x^3}{2 \cdot 3} + \frac{1 \cdot 3}{2 \cdot 4 \cdot 5}x^5 + \frac{1 \cdot 3 \cdot 5}{2 \cdot 4 \cdot 6 \cdot 7}x^7$$

$$+ \frac{1 \cdot 3 \cdot 5 \cdot 7}{2 \cdot 4 \cdot 6 \cdot 8 \cdot 9}x^9 + \cdots$$

In the next four examples in this section, we demonstrate how to make use of the basic list of power series to obtain series for functions not on the list.

Example 4

Determine the power series for $g(x) = e^{2x+1}$.

Solution: First, we consider $f(x)$ to be $e^{2x+1} = e^{2x}e$. Then using the power series

$$f(x) = e^x = 1 + x + \frac{x^2}{2!} + \frac{x^3}{3!} + \frac{x^4}{4!} + \cdots$$

we write

$$ee^{2x} = ef(2x) = e\left[1 + 2x + \frac{(2x)^2}{2!} + \frac{(2x)^3}{3!} + \frac{(2x)^4}{4!} + \cdots\right]$$

$$= e\sum_{n=0}^{\infty} \frac{(2x)^n}{n!} = e\sum_{n=0}^{\infty} \frac{2^n}{n!}x^n, \qquad -\infty < x < \infty$$ ∎

Example 5

Find the power series for $f(x) = \cos\sqrt{x}$.

Solution: Using the power series

$$\cos x = 1 - \frac{x^2}{2!} + \frac{x^4}{4!} - \frac{x^6}{6!} + \frac{x^8}{8!} - \cdots$$

we can replace x by $\sqrt{x}$ to obtain the series

$$\cos\sqrt{x} = 1 - \frac{x}{2!} + \frac{x^2}{4!} - \frac{x^3}{6!} + \frac{x^4}{8!} - \cdots$$

which converges for all x in the domain of $\cos\sqrt{x}$. ∎

Example 6

Find the power series for $g(x) = \sqrt[3]{1 + x}$.

Solution: Using the binomial series

$$(1 + x)^k = 1 + kx + \frac{k(k-1)x^2}{2!} + \frac{k(k-1)(k-2)x^3}{3!} + \cdots$$

we let $k = \frac{1}{3}$ and write

$$(1 + x)^{1/3} = 1 + \frac{x}{3} - \frac{2x^2}{3^2 2!} + \frac{2 \cdot 5x^3}{3^3 3!} - \frac{2 \cdot 5 \cdot 8x^4}{3^4 4!} + \cdots$$

which converges for $-1 \leq x \leq 1$. ∎

Example 7

Find the power series for $f(x) = \sin^2 x$.

Solution: In this case instead of squaring the power series for sin x, we write

$$\sin^2 x = \frac{1 - \cos 2x}{2} = \frac{1}{2} - \frac{\cos 2x}{2}$$

Now using the series

$$\cos x = 1 - \frac{x^2}{2!} + \frac{x^4}{4!} - \frac{x^6}{6!} + \frac{x^8}{8!} - \cdots$$

we replace x by $2x$ to obtain the series

$$\cos 2x = 1 - \frac{(2x)^2}{2!} + \frac{(2x)^4}{4!} - \frac{(2x)^6}{6!} + \frac{(2x)^8}{8!} - \cdots$$

$$= 1 - \frac{2^2}{2!}x^2 + \frac{2^4}{4!}x^4 - \frac{2^6}{6!}x^6 + \frac{2^8}{8!}x^8 - \cdots$$

Therefore,

$$-\frac{\cos 2x}{2} = -\frac{1}{2} + \frac{2}{2!}x^2 - \frac{2^3}{4!}x^4 + \frac{2^5}{6!}x^6 - \frac{2^7}{8!}x^8 + \cdots$$

and finally,

$$\sin^2 x = \frac{1}{2} - \frac{\cos 2x}{2}$$

$$= \frac{1}{2} - \frac{1}{2} + \frac{2x^2}{2!} - \frac{2^3}{4!}x^4 + \frac{2^5}{6!}x^6 - \frac{2^7}{8!}x^8 + \cdots$$

This series converges for $-\infty < x < \infty$.   ■

As mentioned in the previous section, power series can be used to obtain tables of values for transcendental functions. They are also useful for estimating the value of definite integrals for which antiderivatives cannot be found. The next example demonstrates this use.

Example 8

Estimate $\int_0^1 e^{-x^2}\, dx$ to two decimal places. (See Example 3, Section 10.10.)

Solution: By replacing x with $-x^2$ in the series for e^x, we have

$$e^{-x^2} = 1 - x^2 + \frac{x^4}{2!} - \frac{x^6}{3!} + \frac{x^8}{4!} - \cdots$$

Therefore, by integration we obtain

$$\int_0^1 e^{-x^2} dx = \left[x - \frac{x^3}{3} + \frac{x^5}{5 \cdot 2!} - \frac{x^7}{7 \cdot 3!} + \frac{x^9}{9 \cdot 4!} - \cdots \right]_0^1$$

$$= 1 - \frac{1}{3} + \frac{1}{10} - \frac{1}{42} + \frac{1}{216} - \cdots$$

Summing the first four terms, we have

$$\int_0^1 e^{-x^2} dx \approx 0.74$$

where the error is less than $(1/216) \approx 0.005$. ∎

Section Exercises (18.8)

In Exercises 1–8, find the Taylor series (centered at c) for the given function.

1. $f(x) = e^{2x}$, $c = 0$ **2.** $f(x) = e^x$, $c = 1$
3. $f(x) = \cos(x^2)$, $c = 0$ **4.** $f(x) = \cos x$, $c = \pi/4$
ⓢ **5.** $f(x) = \sin(2x)$, $c = 0$
6. $f(x) = \tan x$, $c = 0$ (first three terms)
7. $f(x) = \sec x$, $c = 0$ (first three terms)
8. $f(x) = \ln(x^2 + 1)$, $c = 0$

(In the remaining problems, assume all series are centered at $c = 0$.)

In Exercises 9–12, use the binomial series to find the power series for the given function.

9. $f(x) = \dfrac{1}{(1 + x)^2}$ **10.** $f(x) = \dfrac{1}{\sqrt{1 - x}}$

ⓢ **11.** $f(x) = \dfrac{1}{\sqrt{4 + x^2}}$ **12.** $f(x) = \sqrt{1 + x}$

13. Find the power series for $f(x) = \sin x$ by differentiating the power series for the cosine function.
14. Find the power series for $f(x) = \cos^2 x$ by using the identity $\cos^2 x = \frac{1}{2}(1 + \cos 2x)$.
ⓢ **15.** Find the power series for $f(x) = e^{-x^2/2}$ by using the series for e^x.
16. Find the first three terms of the power series for $f(x) = \sec^2 x$ by differentiating the series for $\tan x$ (see Exercise 6).
17. Find the power series for $f(x) = \sinh x = (e^x - e^{-x})/2$ by using the series for e^x.
18. Find the power series for $f(x) = \cosh x = (e^x + e^{-x})/2$ by differentiating the series found in Exercise 17.
19. Use the power series for the functions $\sin x$, $\cos x$, and e^x to prove the following:

(a) $\sin x = \dfrac{e^{ix} - e^{-ix}}{2i}$

(b) $\cos x = \dfrac{e^{ix} + e^{-ix}}{2}$ $(i^2 = -1)$

20. Integrate the power series for $f(x) = 1/\sqrt{x^2 + 1}$ to find the series representation of $\sinh^{-1} x = \ln(x + \sqrt{x^2 + 1})$.
ⓢ **21.** Find the power series for $f(x) = (\sin x)/x$.
22. Use the series from Exercise 21 to show that

$$\lim_{x \to 0} \frac{\sin x}{x} = 1$$

In Exercises 23–28, use power series to approximate the given integral accurately to four decimal places.

ⓒⓢ **23.** $\displaystyle\int_0^{\pi/2} \frac{\sin x}{x} dx$ ⓒ **24.** $\displaystyle\int_0^1 \cos x^2 \, dx$

ⓒ **25.** $\displaystyle\int_0^{\pi/2} \sqrt{x} \cos x \, dx$ ⓒ **26.** $\displaystyle\int_0^{1/2} \frac{\ln(x + 1)}{x} dx$

ⓒ **27.** $\displaystyle\int_0^{1/2} \frac{\text{Arctan } x}{x} dx$ ⓒ **28.** $\displaystyle\int_1^2 e^{-x^2} dx$

29. The standard normal probability density function is $f(x) = (1/\sqrt{2\pi})e^{-x^2/2}$, and the probability that x falls between a and b is

$$P(a \le x \le b) = \frac{1}{\sqrt{2\pi}} \int_a^b e^{-x^2/2} dx$$

Find $P(0 \le x \le 1)$, accurate to four decimal places.
ⓒ **30.** For what values of x can $\sin x$ be replaced by the Taylor polynomial $x - (x^3/3!)$ if the error cannot exceed 0.001?
ⓢ **31.** For what values of $x < 0$ can e^x be replaced by the polynomial $1 + x + (x^2/2!) + (x^3/3!)$ if the error cannot exceed 0.001?

32. How many terms in the expansion of $\ln(x + 1)$ must be used to guarantee finding $\ln 1.5$ if the error cannot exceed 0.0001?

33. Estimate the error in approximating $e^{0.6}$ by the fifth-degree polynomial

$$1 + x + \frac{x^2}{2!} + \frac{x^3}{3!} + \frac{x^4}{4!} + \frac{x^5}{5!}$$

34. Show that

$$\sum_{n=1}^{\infty} (-1)^{n-1}\left(\frac{1}{n!}\right) = \frac{e-1}{e}$$

35. Show that

$$\sum_{n=0}^{\infty} (-1)^{n}\left[\frac{1}{(2n+1)!}\right] = \sin 1$$

Miscellaneous Exercises (Ch. 18)

1. Find the general term of the sequence $\{1, \frac{1}{2}, \frac{1}{6}, \frac{1}{24}, \frac{1}{120}, \ldots\}$.

2. Find the general term of the sequence $\{\frac{1}{2}, \frac{2}{5}, \frac{3}{10}, \frac{4}{17}, \ldots\}$.

In Exercises 3–10, determine the convergence or divergence of $\{a^n\}$.

3. $a_n = \dfrac{n+1}{n^2}$

4. $a_n = \dfrac{1}{\sqrt{n}}$

5. $a_n = \dfrac{n^3}{n^2+1}$

$\boxed{\text{S}}$ **6.** $a_n = \dfrac{n}{\ln(n)}$

$\boxed{\text{S}}$ **7.** $a_n = \sqrt{n+1} - \sqrt{n}$

8. $a_n = \left(1 + \dfrac{1}{2n}\right)^n$

9. $a_n = \dfrac{\sin\sqrt{n}}{\sqrt{n}}$

10. $a_n = (b^n + c^n)^{1/n}$ (b and c are positive real numbers)

In Exercises 11–14, for each series find the first five terms of the sequence of partial sums $\{S_n\}$.

11. $\displaystyle\sum_{n=0}^{\infty} \left(\frac{3}{2}\right)^n$

12. $\displaystyle\sum_{n=1}^{\infty} \frac{(-1)^{n+1}}{2n}$

$\boxed{\text{S}}$ **13.** $\displaystyle\sum_{n=1}^{\infty} \frac{(-1)^{n+1}}{(2n)!}$

14. $\displaystyle\sum_{n=1}^{\infty} \frac{1}{n(n+1)}$

In Exercises 15–20, find the sum of the infinite series. (Express each repeating decimal as an infinite series.)

15. $\displaystyle\sum_{n=0}^{\infty} \left(\frac{2}{3}\right)^n$

16. $\displaystyle\sum_{n=0}^{\infty} \frac{2^{n+2}}{3^n}$

$\boxed{\text{S}}$ **17.** $\displaystyle\sum_{n=0}^{\infty} \left(\frac{1}{2^n} - \frac{1}{3^n}\right)$

18. $\displaystyle\sum_{n=0}^{\infty} \left[\left(\frac{2}{3}\right)^n - \frac{1}{(n+1)(n+2)}\right]$

19. $0.090909\ldots$

20. $0.923076923076\ldots$

In Exercises 21–30, determine the convergence or divergence of each series.

21. $\displaystyle\sum_{n=1}^{\infty} \frac{2^n}{n^3}$

22. $\displaystyle\sum_{n=1}^{\infty} \frac{1}{\sqrt{n^2+2n}}$

23. $\displaystyle\sum_{n=1}^{\infty} \frac{1}{\sqrt{n^3+2n}}$

24. $\displaystyle\sum_{n=1}^{\infty} \frac{n+1}{n(n+2)}$

$\boxed{\text{S}}$ **25.** $\displaystyle\sum_{n=1}^{\infty} \frac{1}{(n^3+2n)^{1/3}}$

26. $\displaystyle\sum_{n=1}^{\infty} \frac{n!}{e^n}$

27. $\displaystyle\sum_{n=1}^{\infty} \frac{(-1)^n n}{\ln(n)}$

28. $\displaystyle\sum_{n=1}^{\infty} \frac{(-1)^n \sqrt{n}}{n+1}$

29. $\displaystyle\sum_{n=1}^{\infty} \frac{(-1)^n 1 \cdot 3 \cdot 5 \cdots (2n-1)}{2 \cdot 4 \cdot 6 \cdots (2n)(2n+1)}$

30. $\displaystyle\sum_{n=1}^{\infty} \frac{1 \cdot 3 \cdot 5 \cdots (2n-1)}{2 \cdot 5 \cdot 8 \cdots (3n-1)}$

In Exercises 31–34, find the interval of convergence of the given power series.

31. $\displaystyle\sum_{n=0}^{\infty} \frac{(-1)^n (x-2)^n}{(n+1)^2}$

32. $\displaystyle\sum_{n=0}^{\infty} (2x)^n$

$\boxed{\text{S}}$ **33.** $\displaystyle\sum_{n=0}^{\infty} n!(x-2)^n$

34. $\displaystyle\sum_{n=0}^{\infty} \frac{(x-2)^n}{2^n}$

35. Find the power series for $f(x) = \sin x$ centered at $c = 3\pi/4$.

36. Find the power series for $f(x) = \sqrt{x}$ centered at $c = 4$ and find the interval of convergence.

[S] **37.** Find the power series for $f(x) = 3^x$ centered at $c = 0$.

38. Find the first three terms of the power series for $f(x) = \csc x$ centered at $c = \pi/2$.

[S] **39.** (a) Find the Taylor polynomial $P(x)$ of degree three for $f(x) = \text{Arcsin } x$.

[C] (b) Complete the accompanying table for $f(x)$ and $P(x)$ of part (a).

(c) Use the table of part (b) to sketch the graphs of $f(x)$ and $P(x)$ on the same axes.

[C] **40.** Repeat Exercise 39 using the function $f(x) = \text{Arctan } x$.

[C] **41.** Sketch the graphs of $f(x) = e^x$ and $P_5(x)$ on the same axis, where $P_5(x)$ is the fifth-degree Taylor polynomial for $f(x)$.

[C] **42.** Use a Taylor polynomial to approximate sin 95° with an error less than 0.001.

x	-1	-0.75	-0.50	-0.25	0	0.25	0.50	0.75	1
$f(x)$									
$P(x)$									

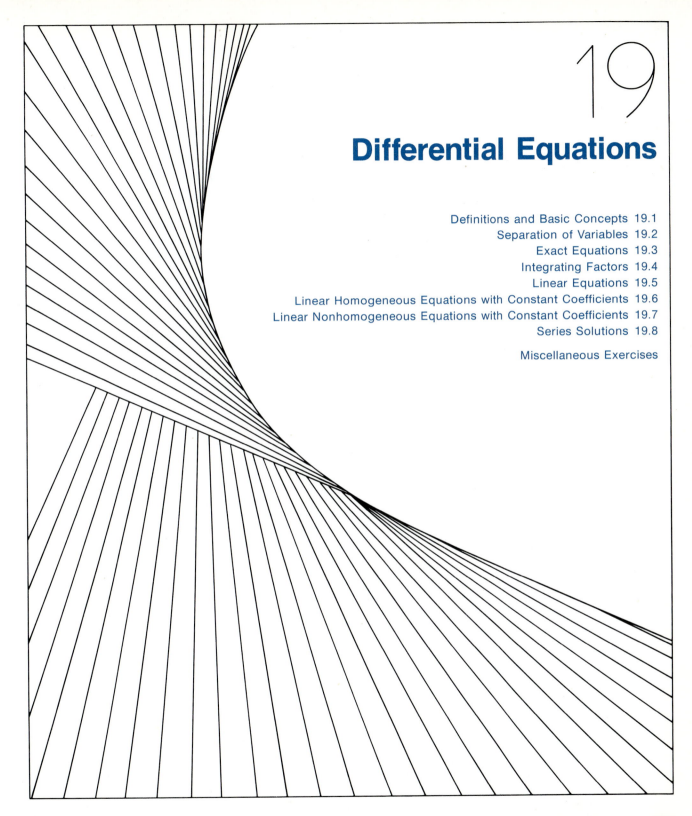

Differential Equations

19.1 **Definitions and Basic Concepts**

Purpose
- To classify differential equations according to type, order, and degree.
- To introduce the concepts of the general solution and particular solutions of a differential equation.
- To find the particular solution that satisfies a set of initial conditions.

The secret of being a bore is to tell everything.

—François Voltaire (1694–1778)

In Section 8.3 we identified several physical phenomena whose behavior can be described by simple *differential equations*. In particular, we saw that problems involving radioactive decay, population growth, chemical reactions, Newton's Law of Cooling, and Newton's Second Law of Motion can all be formulated in terms of differential equations. In this chapter we will see that these are only a few of the many types of problems of physics, chemistry, biology, and engineering that can be solved through the use of such equations.

A **differential equation** is an equation involving an unknown function (dependent variable) and one or more of its derivatives. If this function has only one independent variable, then the derivatives are ordinary and the equation is called an **ordinary differential equation.** For instance,

$$\frac{d^2y}{dx^2} + 3\frac{dy}{dx} - 2y = 0$$

is an ordinary differential equation in which $y = f(x)$ is a differentiable function of x. If the function depends on two or more variables, the derivatives are partial and the equation is called a **partial differential equation.** In this chapter we restrict our discussion of differential equations to the ordinary type.

In addition to **type** (ordinary or partial), differential equations can be classified by *order* and *degree*. The **order** of a differential equation is the order of the highest derivative appearing in the equation. Note that the preceding differential equation is of order two. The **degree** of a differential equation is the highest power to which the highest-order derivative is raised (provided the equation is written in polynomial form in terms of the derivatives and the dependent variable). These classifications are useful in deciding which procedures to use to solve a given differential equation.

Example 1

Classify each of the following differential equations according to type, order, and degree (when applicable).

(a) $(y''')^2 + 4y = 2$

(b) $\frac{d^2s}{dt^2} = -32$

(c) $\frac{\partial^2 u}{\partial x^2} + \frac{\partial^2 u}{\partial y^2} = 0$

(d) $y' - 2y(y')^2 + x = 0$

(e) $y' - 3y = e^x$

(f) $y'' - \sin y = 0$

Solution:

	Equation	Type	Order	Degree
(a)	$(y''')^2 + 4y = 2$	ordinary	3	2
(b)	$\dfrac{d^2s}{dt^2} = -32$	ordinary	2	1
(c)	$\dfrac{\partial^2 u}{\partial x^2} + \dfrac{\partial^2 u}{\partial y^2} = 0$	partial	2	1
(d)	$y' - 2y(y')^2 + x = 0$	ordinary	1	2
(e)	$y' - 3y = e^x$	ordinary	1	1
(f)	$y'' - \sin y = 0$	ordinary	2	none

Note that we do not specify a degree for the last equation since the term involving y is not of polynomial form. ∎

A function $y = f(x)$ is called a **solution** of a given differential equation if the equation is satisfied when y and its derivatives are replaced by $f(x)$ and its derivatives, respectively. For example,

$$y = e^{-2x}$$

is a solution to the differential equation

$$y' + 2y = 0$$

because $y' = -2e^{-2x}$ and by substitution we have

$$-2e^{-2x} + 2(e^{-2x}) = 0$$

Furthermore, we can readily see that

$$y = 2e^{-2x}, \qquad y = 3e^{-2x}, \qquad \text{and} \qquad y = \tfrac{1}{2}e^{-2x}$$

are also solutions to this same differential equation. In fact, the functions

$$y = Ce^{-2x}$$

where C is any real number, all are solutions. Such a solution containing one or more arbitrary constants is referred to as the **general solution** of a given differential equation.

Recall from Section 6.1, Example 6, that the second-order differential equation

$$s''(t) = -32$$

has the general solution

$$s(t) = -16t^2 + C_1 t + C_2$$

which contains two arbitrary constants. It is usually the case that a differential equation of order n has a general solution involving n arbitrary constants.

A **particular solution** of a differential equation is any solution that is obtained by assigning specific values to the constants in the general solution. Occasionally a differential equation has other solutions, not obtainable from the general solution by assigning values to the arbitrary constants; such solutions are called **singular solutions.** (Additional information on singular solutions can be found in more advanced texts on differential equations.)

Geometrically, the general solution of a given first-order differential equation represents a family of curves **(solution curves),** one for each value assigned to the arbitrary constant. For instance, we can easily verify that

$$y = \frac{C}{x}$$

is the general solution to the differential equation

$$xy' + y = 0$$

Figure 19.1 shows some of the solution curves corresponding to different values of C.

In practice, particular solutions of a differential equation are usually obtained from **initial or boundary conditions** placed upon the unknown function and its derivatives. For instance, in Figure 19.1 if we want to obtain the particular solution whose curve passes through the point $(-1, 1)$, we have the initial condition of

$$f(-1) = 1 \quad \text{or} \quad y = 1 \quad \text{when} \quad x = -1$$

Substituting this condition into the general solution yields

$$y = \frac{C}{x}$$

$$1 = \frac{C}{-1}$$

$$-1 = C$$

and the particular solution is

$$y = -\frac{1}{x}$$

Similarly, the second-order differential equation

$$s''(t) = -32$$

with initial conditions $s'(0) = 64$ and $s(0) = 80$ (see Example 6, Section 6.1) has by successive antidifferentiation and substitution

$$s'(t) = \int -32 \, dt = -32t + C_1 = -32t + 64$$

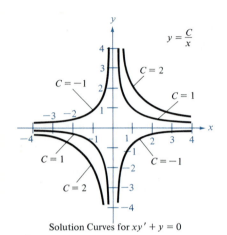

$y = \dfrac{C}{x}$

Solution Curves for $xy' + y = 0$

FIGURE 19.1

$$s(t) = \int (-32t + 64)\, dt = -16t^2 + 64t + C_2$$

which yields the particular solution

$$s(t) = -16t^2 + 64t + 80 \qquad \blacksquare$$

Example 2

Determine whether or not the given functions are solutions to the differential equation $y'' - y = 0$.

(a) $y = \sin x$ (b) $y = e^x$

(c) $y = e^{2x}$ (d) $y = 4e^{-x}$

(e) $y = 0$ (f) $y = Ce^x$

Solution:

(a)   Since $y = \sin x$, $y' = \cos x$, and $y'' = -\sin x$, we have

$$y'' - y = -\sin x - \sin x = -2 \sin x \neq 0$$

Hence $y = \sin x$ is *not* a solution.

(b)   Since $y = e^x$, $y' = e^x$, and $y'' = e^x$, we have

$$y'' - y = e^x - e^x = 0$$

Hence $y = e^x$ *is* a solution.

(c)   Since $y = e^{2x}$, $y' = 2e^{2x}$, and $y'' = 4e^{2x}$, we have

$$y'' - y = 4e^{2x} - e^{2x} = 3e^{2x} \neq 0$$

Hence $y = e^{2x}$ is *not* a solution.

(d)   Since $y = 4e^{-x}$, $y' = -4e^{-x}$, and $y'' = 4e^{-x}$, we have

$$y'' - y = 4e^{-x} - 4e^{-x} = 0$$

Hence $y = 4e^{-x}$ *is* a solution.

(e)   Since $y = 0$, $y' = 0$, and $y'' = 0$, we have

$$y'' - y = 0 - 0 = 0$$

Hence $y = 0$ *is* a solution.

(f)   Since $y = Ce^x$, $y' = Ce^x$, and $y'' = Ce^x$, we have

$$y'' - y = Ce^x - Ce^x = 0$$

Hence $y = Ce^x$ *is* a solution for any value of C. $\qquad \blacksquare$

Example 3

Verify that $y = Cx^3$ is a solution of the differential equation

$$xy' - 3y = 0$$

Find the particular solution determined by the initial condition

$$y = 2 \qquad \text{when} \qquad x = -3$$

Solution: Since $y = Cx^3$ and $y' = 3Cx^2$, we have

$$xy' - 3y = x(3Cx^2) - 3(Cx^3) = 0$$

Hence $y = Cx^3$ is a solution. Furthermore, from the initial condition we determine that

$$\begin{aligned} y &= Cx^3 \\ 2 &= C(-3)^3 \\ -\tfrac{2}{27} &= C \end{aligned}$$

Therefore, the particular solution is $y = -2x^3/27$. ■

In the remainder of this chapter, we will discuss procedures for finding the general solution to several classes of differential equations. We will concentrate mainly on first- and second-order equations and linear equations with constant coefficients. By a linear differential equation with constant coefficients

$$a_0, a_1, a_2, \ldots, a_n$$

we mean any equation of the form

$$a_n \frac{d^n y}{dx^n} + a_{n-1} \frac{d^{n-1} y}{dx^{n-1}} + \cdots + a_1 \frac{dy}{dx} + a_0 y = F(x)$$

which is linear (of first degree) in y and its derivatives.

Section Exercises (19.1)

In Exercises 1–12, classify each differential equation according to type, order, and degree (when applicable).

1. $\dfrac{dy}{dx} + 3xy = x^2$
2. $y'' + 2y' + y = 1$
3. $\dfrac{d^2x}{dt^2} + 2\dfrac{dx}{dt} - 4x = e^t$
4. $\dfrac{d^2u}{dt^2} + \dfrac{du}{dt} = \sec t$
5. $y^{(4)} + 3(y')^2 - 4y = 0$
6. $x^2y'' + 3xy' = 0$

[S] 7. $(y'')^2 + 3y' - 4y = 0$

8. $\dfrac{\partial u}{\partial t} = C^2 \dfrac{\partial^2 u}{\partial x^2}$
9. $\dfrac{\partial u}{\partial x} + \dfrac{\partial u}{\partial y} = 2u$
10. $\dfrac{\partial^2 u}{\partial x\, \partial y} = \dfrac{\partial u}{\partial y}$
11. $\dfrac{d^2y}{dx^2} = \sqrt{1 + \left(\dfrac{dy}{dx}\right)^2}$
12. $\sqrt{\dfrac{d^2y}{dx^2}} = \dfrac{dy}{dx}$

In Exercises 13–18, verify that each equation represents a solution of the given differential equation.

13. $y = Ce^{4x}$; $\dfrac{dy}{dx} = 4y$

14. $x^2 + y^2 = Cy$; $y' = \dfrac{2xy}{x^2 - y^2}$

15. $y = C_1 \cos x + C_2 \sin x$; $y'' + y = 0$

S **16.** $y = C_1 e^{-x} \cos x + C_2 e^{-x} \sin x$; $y'' + 2y' + 2y = 0$

17. $u = e^{-t} \sin bx$; $b^2 \dfrac{\partial u}{\partial t} = \dfrac{\partial^2 u}{\partial x^2}$

18. $u = \dfrac{y}{x^2 + y^2}$; $\dfrac{\partial^2 u}{\partial x^2} + \dfrac{\partial^2 u}{\partial y^2} = 0$

19. Determine whether or not the given functions are solutions to the differential equation $y^{(4)} - 16y = 0$.
(a) $y = 3 \cos x$
(b) $y = 3 \cos 2x$
(c) $y = e^{-2x}$
(d) $y = 5 \ln x$
(e) $y = C_1 e^{2x} + C_2 e^{-2x} + C_3 \sin 2x + C_4 \cos 2x$

20. Determine whether or not the given functions are solutions to the differential equation

$$x \frac{\partial u}{\partial x} - y \frac{\partial u}{\partial y} = 0$$

(a) $u = e^{x+y}$
(b) $u = 5$

(c) $u = x^2 y^2$
(d) $u = \sin xy$
(e) $u = (xy)^n$

In Exercises 21–25, verify that the general solutions satisfy the given differential equations. Then find the particular solution satisfying the given initial conditions

S **21.** $y = Ce^{-2x}$, $y' + 2y = 0$; $y = 3$ when $x = 0$

22. $2x^2 + 3y^2 = C$, $2x + 3yy' = 0$; $y = 2$ when $x = 1$

23. $y = C_1 \sin 3x + C_2 \cos 3x$, $y'' + 9y = 0$; $y = 2$ and $y' = 1$ when $x = \pi/6$

24. $y = C_1 + C_2 \ln x$, $xy'' + y' = 0$; $y = 5$ and $y' = \frac{1}{2}$ when $x = 1$

25. $y = C_1 x + C_2 x^3$, $x^2 y'' - 3xy' + 3y = 0$; $y = 0$ and $y' = 4$ when $x = 2$

26. The general solution to the differential equation $2yy' - x = 0$ is given by $4y^2 - x^2 = C$. Sketch the particular solution curves given by $C = 0$, $C = \pm 1$, and $C = \pm 4$.

S **27.** The general solution to the differential equation $yy' + x = 0$ is given by $x^2 + y^2 = C$. Sketch the solution curves given by $C = 0$, $C = 1$, and $C = 4$.

For Review and Class Discussion

True or False

1. ____ The general solution of the differential equation $y' = f(x)$ is $y = \int f(x)\,dx + C$.

2. ____ For any value of C, $y = (x - C)^3$ is a solution of $y' = 3y^{2/3}$.

3. ____ Exactly one member of the family of curves $y = Ce^x$ passes through the point $(0, 1)$.

4. ____ If $y = f(x)$ is a solution to a first-order differential equation, then $y = f(x) + C$ is also a solution.

5. ____ Exactly one member of the family of curves $y = C_1 x + C_2 x^3$ satisfies the boundary conditions $y(1) = 1$ and $y(-1) = 1$.

19.2 **Separation of Variables**

Purpose

- To introduce the technique of separation of variables to solve first-order differential equations of the form $f(y)\,dy = g(x)\,dx$.
- To expand the technique of separation of variables to solve first-order differential equations involving homogeneous functions.

Now here is a method that demands only a very simple application of the principles of the differential and integral calculus.

—*Joseph Lagrange (1736–1813)*

With this section we begin our discussion of special techniques for solving specific kinds of ordinary differential equations. We start with procedures for solving first-order differential equations whose variables can be separated. This method is called **separation of variables** and is outlined as follows:

Separation of Variables

If the first-order differential equation

$$\frac{dy}{dx} = F(x, y)$$

can be written with *variables separated* in the differential form

$$f(y)\, dy = g(x)\, dx$$

where f and g are continuous, then the general solution has the form

$$\int f(y)\, dy = \int g(x)\, dx + C$$

where C is an arbitrary constant.

Example 1

Find the general solution of the equation

$$(x^2 + 4)\frac{dy}{dx} = xy$$

by separation of variables.

Solution: The given equation has the differential form

$$(x^2 + 4)\, dy = xy\, dx$$

We separate the variables by dividing by $(x^2 + 4)y$ to obtain

$$\frac{1}{y}\, dy = \frac{x}{x^2 + 4}\, dx$$

By integration, we have

$$\int \frac{1}{y}\, dy = \int \frac{x}{x^2 + 4}\, dx$$

$$\ln |y| = \tfrac{1}{2} \ln (x^2 + 4) + C_1$$

For convenience, we let $C_1 = \ln |C|$ and write

$$\ln |y| = \ln \sqrt{x^2 + 4} + \ln |C| = \ln |C\sqrt{x^2 + 4}|$$

Therefore, the general solution is

$$y = C\sqrt{x^2 + 4}$$ ∎

We can readily verify that the general solution obtained in Example 1 satisfies the given differential equation. We encourage you to verify each solution obtained in this chapter. In some cases it is not feasible to write a solution in the explicit form $y = G(x)$. In such cases implicit differentiation can be used to verify the solution. The next example is a case in point.

Example 2

Solve the equation

$$ye^{x^2}\, dx + \frac{y^2 - 1}{x}\, dy = 0$$

given the initial condition $y(0) = 1$.

Solution: Note that to separate the variables, we need to rid the first term of y and the second term of $1/x$. Thus we multiply by x/y and obtain

$$xe^{x^2}\, dx + \frac{y^2 - 1}{y}\, dy = 0$$

Integrating, we get

$$\int xe^{x^2}\, dx + \int \frac{y^2 - 1}{y}\, dy = 0$$

$$\frac{1}{2}\int 2xe^{x^2}\, dx + \int \left(y - \frac{1}{y}\right) dy = 0$$

$$\tfrac{1}{2}e^{x^2} + \tfrac{1}{2}y^2 - \ln|y| = C_1$$

$$e^{x^2} + y^2 - \ln y^2 = 2C_1 = C$$

Since $y = 1$ when $x = 0$, we find C to be

$$1 + 1 - \ln 1 = C$$

$$2 = C$$

In implicit form the particular solution is

$$e^{x^2} + y^2 - \ln y^2 = 2$$ ∎

Example 3

In the preservation of food, cane sugar is inverted into a mixture of two simpler sugars: glucose and fructose. In dilute solution the inversion rate is proportional to the concentration $y(t)$ of unaltered sugar. If the concentration is $\frac{1}{50}$ when $t = 0$ and $\frac{1}{200}$ after 3 h, find the concentration of unaltered sugar after 6 h and after 12 h.

Solution: Since the rate of inversion is proportional to $y(t)$, we have the differential equation

$$\frac{dy}{dt} = ky$$

Separating the variables and integrating, we get

$$\frac{1}{y}\, dy = k\, dt$$

$$\int \frac{1}{y} \, dy = \int k \, dt$$

$$\ln |y| = kt + C_1$$

$$|y| = e^{kt+C_1} = e^{kt} e^{C_1}$$

We can simplify this equation by letting $e^{C_1} = C$. Thus we have

$$y = Ce^{kt}$$

Now since $y = \frac{1}{50}$ when $t = 0$, we have

$$\frac{1}{50} = C$$

Furthermore, since $y = \frac{1}{200}$ when $t = 3$, we have

$$\frac{1}{200} = \frac{1}{50} e^{3k}$$

$$\frac{1}{4} = e^{3k}$$

$$-\ln 4 = 3k$$

$$\frac{-\ln 4}{3} = k$$

Therefore, the concentration is given by

$$y(t) = \frac{1}{50} e^{-\ln(4)t/3}$$

After 6 h the concentration is

$$y(6) = \frac{1}{50} e^{-2\ln 4} = \frac{1}{50}(4^{-2}) = \frac{1}{800}$$

and after 12 h the concentration is

$$y(12) = \frac{1}{50} e^{-4\ln 4} = \frac{1}{50}(4^{-4}) = \frac{1}{2800} \qquad \blacksquare$$

Example 4

Find the equation of the curve that passes through the point $(1, 3)$ and whose slope is y/x^2.

Solution: Since the slope of the curve is given by y/x^2, we have the differential equation

$$\frac{dy}{dx} = \frac{y}{x^2}$$

with the initial condition $y(1) = 3$. Separating the variables and integrating, we get

$$\frac{1}{y} \, dy = \frac{1}{x^2} \, dx$$

$$\int \frac{1}{y} \, dy = \int \frac{1}{x^2} \, dx$$

$$\ln |y| = -\frac{1}{x} + C_1$$

$$|y| = e^{-(1/x)+C_1} = e^{-1/x}e^{C_1}$$

$$y = Ce^{-1/x}$$

Now since $y = 3$ when $x = 1$, it follows that

$$3 = Ce^{-1}$$

$$3e = C$$

Therefore, the equation of the specified curve is

$$y = (3e)e^{-1/x} = 3e^{(x-1)/x}$$

(See Figure 19.2.) ■

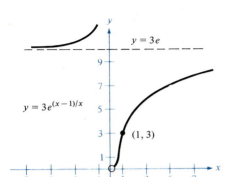

FIGURE 19.2

Not all first-order differential equations are separable. However, some of those that are not can be made separable by a change of variables. This is true for differential equations of the form $y' = f(x, y)$, where f is a **homogeneous function.**

Definition of a Homogeneous Function	The function given by $z = f(x, y)$ is said to be **homogeneous of degree** n if $$f(tx, ty) = t^n f(x, y)$$

Example 5

Determine which of the following functions are homogeneous. For those that are, specify the degree.

(a) $f(x, y) = x^2 y - 4x^3 + 3xy^2$ (b) $f(x, y) = xy - 3x^2 + 2y$

(c) $f(x, y) = x + y \sin \dfrac{y}{x}$ (d) $f(x, y) = \dfrac{x + 3y}{2\sqrt{xy}}$

(e) $f(x, y) = 3e^{x/y}$

Solution:

(a) $f(tx, ty) = (tx)^2(ty) - 4(tx)^3 + 3(tx)(ty)^2$
$$= t^3(x^2 y) - t^3(4x^3) + t^3(3xy^2)$$
$$= t^3(x^2 y - 4x^3 + 3xy^2) = t^3 f(x, y)$$

Thus f is homogeneous of degree 3.

(b) $f(tx, ty) = (tx)(ty) - 3(tx)^2 + 2(ty) = t^2(xy) - t^2(3x^2) + t(2y)$
$$\neq t^n(xy - 3x^2 + 2y)$$

Thus f is not homogeneous.

(c) $f(tx, ty) = tx + ty \sin \dfrac{ty}{tx} = t\left(x + y \sin \dfrac{y}{x}\right) = t^1 f(x, y)$

Thus f is homogeneous of degree 1.

(d) $f(tx, ty) = \dfrac{tx + 3ty}{2\sqrt{txty}} = \dfrac{t(x + 3y)}{2t\sqrt{xy}} = \dfrac{x + 3y}{2\sqrt{xy}} = t^0 f(x, y)$

Thus f is homogeneous of degree 0.

(e) $f(tx, ty) = 3e^{tx/ty} = 3e^{x/y} = t^0 f(x, y)$

Thus f is homogeneous of degree 0. ∎

Note in part (a) of Example 5 that each term of the homogeneous function has the same total degree in x and y; whereas the nonhomogeneous function in part (b) has terms that differ in total degree. This observation suggests a quick test for algebraic functions: *An algebraic function is homogeneous if all its terms have the same total degree.* For instance, in part (d) of Example 5, the term $2\sqrt{xy}$ is of first degree, as are both terms in the numerator; hence

$$f(x, y) = \frac{x + 3y}{2\sqrt{xy}}$$

is homogeneous. In order for a transcendental function to be homogeneous, it must be possible to write it in the form

$$f(x, y) = F\left(\frac{y}{x}\right) \quad \text{or} \quad G\left(\frac{x}{y}\right)$$

Note that the sine and natural logarithm functions in parts (c) and (e) of Example 5 involve either y/x or x/y. These observations suggest the change of variables in Theorem 19.1 that can be used to transform "homogeneous" differential equations into equations whose variables are separable.

Example 6

Find the general solution to the differential equation

$$\frac{dy}{dx} = \frac{3x + 2y}{x}$$

Solution: Since $(3x + 2y)/x$ is homogeneous, we can apply Theorem 19.1 and let

$$v = \frac{y}{x}, \quad y = vx, \quad \text{and} \quad \frac{dy}{dx} = v + x\frac{dv}{dx}$$

Then we have

$$\frac{dy}{dx} = \frac{3x + 2y}{x}$$

$$v + x\frac{dv}{dx} = \frac{3x + 2(vx)}{x} = \frac{x(3 + 2v)}{x} = 3 + 2v$$

$$x\frac{dv}{dx} = 3 + 2v - v = 3 + v$$

$$x\, dv = (3 + v)\, dx$$

$$\frac{1}{3 + v}\, dv = \frac{1}{x}\, dx$$

Now by integration we obtain

$$\int \frac{1}{3 + v}\, dv = \int \frac{1}{x}\, dx$$

$$\ln|3 + v| = \ln|x| + C_1 = \ln|x| + \ln|C| = \ln|Cx|$$

Thus

$$3 + v = Cx$$

and substituting for v, we have

$$3 + \frac{y}{x} = Cx$$

$$3x + y = Cx^2$$

and the general solution can be written as

$$y = Cx^2 - 3x \qquad \blacksquare$$

THEOREM 19.1
(Changing a Homogeneous Differential Equation to Separable Variables Form)

i. If f is homogeneous, the differential equation

$$\frac{dy}{dx} = f(x, y)$$

can be transformed into an equation whose variables are separable by letting $v = y/x$. Then

$$y = vx \qquad \text{and} \qquad \frac{dy}{dx} = v + x\frac{dv}{dx}$$

ii. If both M and N are homogeneous of the same degree, the differential equation

$$M(x, y)\, dx + N(x, y)\, dy = 0$$

can be transformed into an equation whose variables are separable by letting $v = y/x$. Then

$$y = vx \qquad \text{and} \qquad dy = v\, dx + x\, dv$$

Example 7

Find the general solution to the differential equation

$$(x^2 - y^2)\,dx + 3xy\,dy = 0$$

Solution: This differential equation is of the form

$$M(x, y)\,dx + N(x, y)\,dy = 0$$

where both M and N are homogeneous of degree 2. Applying Theorem 19.1 we let

$$v = \frac{y}{x}, \qquad y = vx, \qquad \text{and} \qquad dy = v\,dx + x\,dv$$

and obtain

$$(x^2 - y^2)\,dx + 3xy\,dy = 0$$
$$(x^2 - v^2x^2)\,dx + 3x(vx)(x\,dv + v\,dx) = 0$$
$$(x^2 - v^2x^2 + 3v^2x^2)\,dx + (3x^3v)\,dv = 0$$
$$x^2(1 + 2v^2)\,dx + x^2(3vx)\,dv = 0$$

Now dividing by x^2, we can separate variables:

$$(1 + 2v^2)\,dx = -3vx\,dv$$

$$\frac{1}{x}\,dx = \frac{-3v}{1 + 2v^2}\,dv$$

Finally, we integrate to obtain

$$\int \frac{1}{x}\,dx = \int \frac{-3v}{1 + 2v^2}\,dv = -\frac{3}{4}\int \frac{4v}{1 + 2v^2}\,dv$$

$$\ln|x| = -\tfrac{3}{4}\ln(1 + 2v^2) + C_1$$
$$4\ln|x| = -3\ln(1 + 2v^2) + \ln|C|$$

$$\ln(x^4) = \ln\left|\frac{C}{(1 + 2v^2)^3}\right|$$

$$x^4 = \frac{C}{(1 + 2v^2)^3}$$

Substituting for v we have

$$x^4 = \frac{C}{[1 + (2y^2/x^2)]^3}$$

$$x^4\left(1 + \frac{2y^2}{x^2}\right)^3 = C$$

$$x^4(x^2 + 2y^2)^3 = Cx^6$$

$$(x^2 + 2y^2)^3 = Cx^2$$

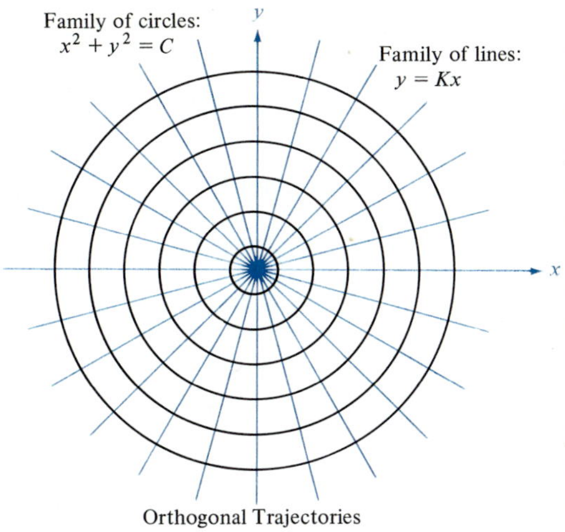

Family of circles:
$x^2 + y^2 = C$

Family of lines:
$y = Kx$

Orthogonal Trajectories

FIGURE 19.3

As previously mentioned, the general solution $U(x, y) = C$ represents a family of solution curves, one for each value assigned to the parameter C. A problem that often arises in electrostatics, thermodynamics, and hydrodynamics is this: Given a one-parameter family of curves $f(x, y, C) = 0$, find a second family of curves $g(x, y, K) = 0$ with the property that each member of one family of curves is orthogonal (perpendicular) to each member of the other family. Two such families of curves are said to be **mutually orthogonal** and each family is called the **orthogonal trajectory** of the other. For example, the family of all straight lines through the origin is an orthogonal trajectory of the family of all concentric circles centered at the origin (see Figure 19.3).

In electrostatics one of the orthogonal families of curves is called *equipotential curves* and the other *lines of force;* in thermodynamics one family is called *isothermal lines* and the other *heat flow lines;* and in hydrodynamics one family is called *velocity potential curves* and the other *flow lines* or *stream lines*.

In the next example we demonstrate the use of a differential equation to find a family of curves orthogonal to a given family.

Example 8

Find the orthogonal trajectory of the family of hyperbolas

$$y = \frac{C}{x} \qquad C \neq 0$$

and sketch several members of each family.

Solution: First, we solve the given equation for the constant:

$$xy = C$$

Then we find a differential equation for the given family of curves by differentiating implicitly with respect to x to obtain

$$xy' + y = 0 \qquad \text{or} \qquad y' = -\frac{y}{x}$$

Since $y' = -y/x$ represents the slope of the given family of curves at (x, y), it follows that the orthogonal family has the negative reciprocal slope, x/y. Thus the differential equation for the orthogonal family is

$$\frac{dy}{dx} = \frac{x}{y}$$

Now by separating variables and integrating, we get

$$y \, dy = x \, dx$$

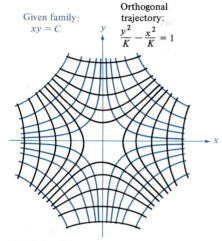

Given family: $xy = C$

Orthogonal trajectory: $\dfrac{y^2}{K} - \dfrac{x^2}{K} = 1$

FIGURE 19.4

$$\int y \, dy = \int x \, dx$$

$$\frac{y^2}{2} = \frac{x^2}{2} + C_1$$

Therefore, the family of orthogonal trajectories is given by

$$\frac{y^2}{K} - \frac{x^2}{K} = 1, \qquad 2C_1 = K \neq 0$$

which is a family of hyperbolas centered at the origin with the y-axis as the transverse axis if $K > 0$ and the x-axis as the transverse axis if $K < 0$. Several members of each family are shown in Figure 19.4. ∎

We summarize the steps for finding a family of curves orthogonal to a given family as follows:

1. Solve for the arbitrary constant in the equation for the given family of curves.
2. Find the slope of the given family by implicit differentiation and solve for y'.
3. Let $dy/dx = -1/y'$ be the slope of the orthogonal family.
4. The general solution to the differential equation in the third step represents the orthogonal trajectories of the given family.

When working through the exercise set for this section, keep in mind that, at this point, we can solve only those differential equations whose variables can be *separated*, either directly by algebraic manipulation or indirectly by a change of variables.

Section Exercises (19.2)

In Exercises 1–14, solve the given equation by separation of variables.

1. $\dfrac{dy}{dx} = \dfrac{x}{y}$

2. $\dfrac{dy}{dx} = \dfrac{x^2 + 2}{3y^2}$

3. $(2 + x)y' = 2y$

4. $xy' = y$

S 5. $yy' = \sin x$

6. $\sqrt{1 - 4x^2}\, y' = 1$

7. $yy' - e^x = 0,\ y(0) = 4$

8. $\sqrt{x} + \sqrt{y}\, y' = 0,\ y(1) = 4$

S 9. $y(x + 1) + y' = 0,\ y(-2) = 1$

10. $xyy' - \ln x = 0,\ y(1) = 0$

11. $(1 + x^2)y' - (1 + y^2) = 0,\ y(0) = \sqrt{3}$

12. $\sqrt{1 - x^2}\, y' - \sqrt{1 - y^2} = 0,\ y(0) = 1$

13. $dP - kP\, dt = 0,\ P(0) = P_0$

14. $dT + k(T - 70)\, dt = 0,\ T(0) = 140$

In Exercises 15–20, determine whether or not the given functions are homogeneous, and, if so, determine the degree.

15. $f(x, y) = x^3 - 4xy^2 + y^3$

S 16. $f(x, y) = \dfrac{xy}{\sqrt{x^2 + y^2}}$

17. $f(x, y) = 2 \ln xy$

18. $f(x, y) = \tan(x + y)$

19. $f(x, y) = 2 \ln \dfrac{x}{y}$

20. $f(x, y) = \tan \dfrac{y}{x}$

In Exercises 21–30, solve the homogeneous differential equation.

21. $y' = \dfrac{x + y}{2x}$

22. $y' = \dfrac{2x + y}{y}$

23. $y' = \dfrac{x - y}{x + y}$

24. $y' = \dfrac{x^2 + y^2}{2xy}$

[S] **25.** $y' = \dfrac{xy}{x^2 - y^2}$

26. $x\,dy - (3x + 2y)\,dx = 0$, $y(1) = 2$

27. $x\,dy - (2xe^{-y/x} + y)\,dx = 0$

28. $-y^2\,dx + x(x + y)\,dy = 0$

[S] **29.** $\left[x \sec\left(\dfrac{y}{x}\right) + y\right] dx - x\,dy = 0$, $y(1) = 0$

30. $(y - \sqrt{x^2 - y^2})\,dx - x\,dy = 0$, $y(1) = 0$

31. Find an equation for the curve that passes through the point $(1, 1)$ and has a slope of $-9x/16y$.

32. Find an equation for the curve that passes through the point $(8, 2)$ and has a slope of $2y/3x$.

[c] **33.** The amount A of an investment P increases at a rate proportional to A at any instant of time t.
   (a) Find an equation for the amount A as a function of t.
   (b) If the initial investment is \$1000.00 and the rate is 11%, find the amount after 10 years.
   (c) If the rate is 11%, find the time necessary to double the investment.

[c] **34.** The rate of growth of a population of fruit flies is proportional to the size of the population at any instant. If there were 180 flies after the second day of the experiment and 300 flies after the fourth day, how many flies were there in the original population?

[c][s] **35.** The rate of decomposition of radioactive radium is proportional to the amount present at a given instant. Find the percentage that remains of a present amount after 25 years if the half-life of radioactive radium is 1600 years.

[c] **36.** In a chemical reaction a certain compound changes into another compound at a rate proportional to the unchanged amount. If initially there was 20 g of the original compound and 16 g after 1 h, when will 75% of the compound be changed?

[c] **37.** Newton's Law of Cooling states that the rate of change in the temperature of an object is proportional to the difference between its temperature and the temperature of the surrounding air. Suppose a room is kept at a constant temperature of 70° and an object cooled from 350° to 150° in 45 min. At what time will the object cool to a temperature of 80°?

[c] **38.** If s gives the position of an object moving in a straight line, the velocity and acceleration are given by

$$v = \dfrac{ds}{dt}, \qquad a = \dfrac{dv}{dt} = \dfrac{ds}{dt}\dfrac{dv}{ds} = v\dfrac{dv}{ds}$$

Suppose that a boat weighing 640 lb has a motor that exerts a force of 60 lb. If the resistance (in pounds) to the motion of the boat is 3 times the velocity (in feet per second) and if the boat starts from rest, find its maximum speed. Use the equation

$$\text{force} = F = ma = mv\dfrac{dv}{ds} = \dfrac{w}{32}v\dfrac{dv}{ds}$$

where w is the weight of the boat.

In Exercises 39–44, find the orthogonal trajectories of the given family and sketch several members of each family.

39. $x^2 + y^2 = C$

[S] **40.** $2x^2 - y^2 = C$

41. $x^2 = Cy$

42. $y^2 = 2Cx$

43. $y^2 = Cx^3$

44. $y = Ce^x$

For Review and Class Discussion

True or False

1. ____ The differential equation $y' + y = x$ is written in separated variables form.

2. ____ The differential equation $y' = xy - 2y + x - 2$ can be written in separated variables form.

3. ____ The differential equation $y' = \sin(x + y) + \sin(x - y)$ can be written in separated variables form.

4. ____ The function $f(x, y) = x^2 + xy + 2$ is homogeneous.

5. ____ The families $x^2 + y^2 = 2Cy$ and $x^2 + y^2 = 2kx$ are mutually orthogonal.

19.3 **Exact Equations**

Purpose
- To recognize first-order differential equations that are exact.
- To solve first-order exact differential equations.

Many matters which, when I was young, baffled me by the vagueness of all that had been said about them, are now amenable to an exact technique, which makes possible the kind of progress that is customary in science.
— *Bertrand Russell (1872–1970)*

In this section we introduce a method of solving the first-order differential equation

$$M(x, y)\, dx + N(x, y)\, dy = 0$$

for the special case in which this equation represents the exact differential of a function.

Definition of an Exact Differential Equation

The differential equation

$$M(x, y)\, dx + N(x, y)\, dy = 0$$

is **exact** if there exists a function $U(x, y)$ such that the total differential of U is given by

$$dU = \frac{\partial U}{\partial x}\, dx + \frac{\partial U}{\partial y}\, dy = M(x, y)\, dx + N(x, y)\, dy$$

From this definition we can see that the general solution to the exact differential equation

$$dU = M(x, y)\, dx + N(x, y)\, dy = 0$$

is given by

$$U(x, y) = C$$

To determine whether or not a differential equation is exact, we observe that if

$$M(x, y)\, dx + N(x, y)\, dy = 0$$

is exact, we have

$$\frac{\partial U}{\partial x} = M(x, y) \qquad \text{and} \qquad \frac{\partial U}{\partial y} = N(x, y)$$

From Section 15.2, we know that if U has continuous second partial derivatives, then we have

$$\frac{\partial M}{\partial y} = \frac{\partial^2 U}{\partial y\, \partial x} = \frac{\partial^2 U}{\partial x\, \partial y} = \frac{\partial N}{\partial x}$$

which leads us to the following test for exactness.

THEOREM 19.2
(*Test for Exactness*)

The differential equation

$$M(x, y)\, dx + N(x, y)\, dy = 0$$

is exact if

$$\frac{\partial M}{\partial y} = \frac{\partial N}{\partial x}$$

Be sure you see the distinction between the *definition* of exactness, which requires that

$$\frac{\partial U}{\partial x} = M(x, y) \quad \text{and} \quad \frac{\partial U}{\partial y} = N(x, y)$$

and the *test* for exactness, which requires that

$$\frac{\partial M}{\partial y} = \frac{\partial N}{\partial x}$$

Example 1

Test each of the following differential equations for exactness.

(a) $3x^2\, dx - 2y\, dy = 0$ (b) $\dfrac{dy}{dx} = \dfrac{2xy}{x^2 + y^2}$

(c) $ye^x\, dx + (2y + e^x)\, dy = 0$ (d) $\dfrac{dy}{dx} = \dfrac{-\cos y}{y^2 - x \sin y}$

Solution:

(a)   Since $M(x, y) = 3x^2$ and $N(x, y) = -2y$, we have

$$\frac{\partial M}{\partial y} = 0 \quad \text{and} \quad \frac{\partial N}{\partial x} = 0$$

Thus the equation is exact. [Note that every equation $f(x)\, dx + g(y)\, dy = 0$, with variables separated, is also exact.]

(b)   In differential form the equation is

$$-2xy\, dx + (x^2 + y^2)\, dy = 0$$

Hence $M(x, y) = -2xy$, $N(x, y) = x^2 + y^2$, and we have

$$\frac{\partial M}{\partial y} = -2x \quad \text{and} \quad \frac{\partial N}{\partial x} = 2x$$

Thus the equation is not exact.

(c)   Since $M(x, y) = ye^x$ and $N(x, y) = 2y + e^x$, we have

$$\frac{\partial M}{\partial y} = e^x \quad \text{and} \quad \frac{\partial N}{\partial x} = e^x$$

Thus the equation is exact.

(d)   In differential form the equation is

$$\cos y \, dx + (y^2 - x \sin y) \, dy = 0$$

Hence $M(x, y) = \cos y$ and $N(x, y) = 2y - x \sin y$, and we have

$$\frac{\partial M}{\partial y} = -\sin y \qquad \text{and} \qquad \frac{\partial N}{\partial x} = -\sin y$$

Thus the equation is exact.   ∎

Once we have determined that a differential equation is exact, we know that there exists a function U such that

$$dU = M(x, y) \, dx + N(x, y) \, dy = 0$$

Furthermore, we know that the general solution is given by

$$U(x, y) = C$$

A method for finding U was demonstrated in Example 4 of Section 17.2 and we summarize this procedure in the following theorem.

THEOREM 19.3
(*General Solution for Exact Equations*)

If the differential equation

$$M(x, y) \, dx + N(x, y) \, dy = 0$$

is exact, then the general solution is given by

$$U(x, y) = C$$

where

$$U(x, y) = \int M(x, y) \, dx + f(y)$$

and

$$N(x, y) - \frac{\partial}{\partial y} \left[\int M(x, y) \, dx \right] = f'(y)$$

[Note that $\int M(x, y) \, dx$ denotes a partial integration with respect to x.]

Proof: By taking the partial derivative of $U(x, y)$ with respect to x, we have

$$\frac{\partial U}{\partial x} = \frac{\partial}{\partial x} \left[\int M(x, y) \, dx + f(y) \right] = \frac{\partial}{\partial x} \left[\int M(x, y) \, dx \right] + \frac{\partial}{\partial x} [f(y)]$$

$$= M(x, y)$$

and by taking the partial derivative of $U(x, y)$ with respect to y, we have

$$\frac{\partial U}{\partial y} = \frac{\partial}{\partial y} \left[\int M(x, y) \, dx + f(y) \right] = \frac{\partial}{\partial y} \left[\int M(x, y) \, dx \right] + \frac{\partial}{\partial y} [f(y)]$$

$$= \frac{\partial}{\partial y}\left[\int M(x,y)\,dx\right] + f'(y)$$

$$= \frac{\partial}{\partial y}\left[\int M(x,y)\,dx\right] + N(x,y) - \frac{\partial}{\partial y}\left[\int M(x,y)\,dx\right] = N(x,y)$$

Therefore, $U(x,y)$ satisfies the requirements

$$\frac{\partial U}{\partial x} = M(x,y) \qquad \text{and} \qquad \frac{\partial U}{\partial y} = N(x,y)$$

and we may conclude that $U(x,y) = C$ is the general solution of the exact equation

$$M(x,y)\,dx + N(x,y)\,dy = 0 \qquad\blacksquare$$

Example 2

Show that the differential equation

$$(2xy - 3x^2)\,dx + (x^2 - 2y)\,dy = 0$$

is exact and find its general solution.

Solution: Since $M(x,y) = 2xy - 3x^2$ and $N(x,y) = x^2 - 2y$, we have

$$\frac{\partial M}{\partial y} = 2x \qquad \text{and} \qquad \frac{\partial N}{\partial x} = 2x$$

Thus the equation is exact. Applying Theorem 19.3 we get

$$U(x,y) = \int M(x,y)\,dx + f(y)$$

$$= \int (2xy - 3x^2)\,dx + f(y) = x^2 y - x^3 + f(y)$$

where

$$f'(y) = N(x,y) - \frac{\partial}{\partial y}\left[\int M(x,y)\,dx\right] = x^2 - 2y - \frac{\partial}{\partial y}[x^2 y - x^3]$$

$$= x^2 - 2y - x^2 = -2y$$

Now by integrating with respect to y, we have

$$f(y) = -\int 2y\,dy = -y^2 + C_1$$

Therefore,

$$U(x,y) = x^2 y - x^3 - y^2 + C_1$$

and the general solution is

$$x^2 y - x^3 - y^2 + C_1 = C_2$$

or simply

$$x^2y - x^3 - y^2 = C$$

■

It is worth mentioning that Theorem 19.3 has an alternative form that basically involves switching the roles of $M(x, y)$ and $N(x, y)$. Under this formulation of the theorem, we would have

$$U(x, y) = \int N(x, y)\, dy + g(x)$$

where

$$M(x, y) - \frac{\partial}{\partial x}\left[\int N(x, y)\, dy\right] = g'(x)$$

We demonstrate the use of this option in the following example.

Example 3

Show that the differential equation

$$\frac{y}{x^2}\, dx - \frac{1}{x}\, dy = 0$$

is exact and find its general solution.

Solution: Since $M(x, y) = y/x^2$ and $N(x, y) = -1/x$, we have

$$\frac{\partial M}{\partial y} = \frac{1}{x^2} \qquad \text{and} \qquad \frac{\partial N}{\partial x} = \frac{1}{x^2}$$

Thus the equation is exact. By the alternative form of Theorem 19.3, we have

$$U(x, y) = \int N(x, y)\, dy + g(x) = -\int \frac{1}{x}\, dy + g(x) = -\frac{y}{x} + g(x)$$

where

$$g'(x) = M(x, y) - \frac{\partial}{\partial x}\left[\int N(x, y)\, dy\right] = \frac{y}{x^2} - \frac{\partial}{\partial x}\left[-\frac{y}{x}\right]$$

$$= \frac{y}{x^2} - \frac{y}{x^2} = 0$$

By integrating we have

$$g(x) = C_1$$

Therefore,

$$U(x, y) = -\frac{y}{x} + C_1$$

and the general solution is

$$-\frac{y}{x} = C$$ ∎

In determining whether to apply Theorem 19.3 in its given form or its alternative form, you should decide which of the following integrals is simpler:

$$\int M(x, y)\, dx \qquad \text{or} \qquad \int N(x, y)\, dy$$

If the first is simpler, use the given form. If the second is simpler, use the alternative form.

Example 4

Solve the differential equation

$$ye^{xy}\, dx + (3 + xe^{xy})\, dy = 0$$

subject to the initial condition $y(0) = 0$.

Solution: Since $M(x, y) = ye^{xy}$ and $N(x, y) = 3 + xe^{xy}$, we have

$$\frac{\partial M}{\partial y} = xye^{xy} + e^{xy} \qquad \text{and} \qquad \frac{\partial N}{\partial x} = xye^{xy} + e^{xy}$$

Thus the equation is exact. By Theorem 19.3 we have

$$U(x, y) = \int M(x, y)\, dx + f(y) = \int ye^{xy}\, dx + f(y) = e^{xy} + f(y)$$

where

$$f'(y) = N(x, y) - \frac{\partial}{\partial y}\left[\int M(x, y)\, dx\right] = 3 + xe^{xy} - \frac{\partial}{\partial y}[e^{xy}]$$

$$= 3 + xe^{xy} - xe^{xy} = 3$$

By integrating we have

$$f(y) = \int 3\, dy = 3y + C_1$$

Therefore,

$$U(x, y) = e^{xy} + 3y + C_1$$

and the general solution is

$$e^{xy} + 3y = C$$

Substituting the initial condition $y = 0$ when $x = 0$, we have

$$1 + 0 = C$$

and thus the particular solution is

$$e^{xy} + 3y = 1$$ ∎

Section Exercises (19.3)

In Exercises 1–15, test the given differential equation for exactness and solve the equations that are exact.

1. $(2x - 3y) dx + (2y - 3x) dy = 0$
2. $ye^x dx + e^x dy = 0$, $y(0) = 5$
S 3. $(3y^2 + 10xy^2) dx + (6xy - 2 + 10x^2y) dy = 0$
4. $2 \cos (2x - y) dx - \cos (2x - y) dy = 0$, $y(\pi/4) = 0$
5. $(4x^3 - 6xy^2) dx + (4y^3 - 6xy) dy = 0$
6. $2y^2 e^{xy^2} dx + 2xye^{xy^2} dy = 0$

S 7. $\dfrac{-y}{x^2 + y^2} dx + \dfrac{x}{x^2 + y^2} dy = 0$, $y(1) = 1$
8. $xe^{-(x^2+y^2)} dx + ye^{-(x^2+y^2)} dy = 0$
9. $\left(\dfrac{y}{x - y}\right)^2 dx + \left(\dfrac{x}{x - y}\right)^2 dy = 0$
10. $(ye^y \cos xy) dx + e^y(x \cos xy + \sin xy) dy = 0$

11. $\dfrac{y}{x - 1} dx + [\ln (x - 1) + 2y] dy = 0$, $y(2) = 4$
12. $\dfrac{x}{\sqrt{x^2 + y^2}} dx + \dfrac{y}{\sqrt{x^2 + y^2}} dy = 0$, $y(4) = 3$
S 13. $\dfrac{x}{x^2 + y^2} dx - \dfrac{y}{x^2 + y^2} dy = 0$
14. $(e^{2x} \sin 3y) dx + (e^{2x} \cos 3y) dy = 0$
15. $(2x \tan y + 5) dx + (x^2 \sec^2 y) dy = 0$

16. Find an equation for the curve passing through the point $(2, 1)$ with a slope of $y' = (y - x)/(3y - x)$.
17. Find an equation for the curve passing through the point $(0, 2)$ with a slope of

$$\frac{dy}{dx} = \frac{-2xy}{x^2 + y^2}$$

For Review and Class Discussion

True or False

1. _____ The differential equation $f(x) dx + g(y) dy = 0$ is exact.
2. _____ The differential equation $2xy dx + (y^2 - x^2) dy = 0$ is exact.
3. _____ $U = C$ is the general solution of the equation $(\partial U/\partial x) dx + (\partial U/\partial y) dy = 0$.
4. _____ If $M dx + N dy = 0$ is exact, then $xM dx + xN dy = 0$ is also exact.
5. _____ If $M dx + N dy = 0$ is exact, then $[f(x) + M] dx + [g(y) + N] dy = 0$ is also exact.

19.4 **Integrating Factors**

Purpose
- To introduce the concept of an integrating factor.
- To find integrating factors that are functions of x or y alone.
- To recognize some common differentials that can be used in finding integrating factors by inspection.

The finding of an integrating factor is not in theory any simpler than the original problem. Nevertheless, in many cases such a factor is easily found by trial and error. —Richard Courant (1888–1972)

If the differential equation

$$M(x, y)\, dx + N(x, y)\, dy = 0$$

is not exact, it may be possible to make it exact by multiplying by an appropriate function $F(x, y)$. Such a function is called an **integrating factor** for the differential equation. For instance, if the differential equation

$$2y\, dx + x\, dy = 0 \qquad \text{(not exact)}$$

is multiplied by the integrating factor $F(x, y) = x$, the resulting equation

$$2xy\, dx + x^2\, dy = 0$$

is exact, and the left side is the total differential of $x^2 y$. That is,

$$d[x^2 y] = 2xy\, dx + x^2\, dy = 0$$

Similarly, if the equation

$$y\, dx - x\, dy = 0$$

is multiplied by the integrating factor $F(x, y) = 1/y^2$, the left side becomes the exact differential of x/y. That is,

$$d\left[\frac{x}{y}\right] = \frac{y\, dx - x\, dy}{y^2} = \frac{1}{y}\, dx - \frac{x}{y^2}\, dy = 0$$

Finding an integrating factor can be a very difficult problem. However, there are two classes of differential equations whose integrating factors can be found in a routine way, namely, those differential equations that possess integrating factors that are functions of either x alone or y alone.

To see this, suppose that $F(x, y)$ is an integrating factor for

$$M(x, y)\, dx + N(x, y)\, dy = 0$$

Then

$$F(x, y)\, M(x, y)\, dx + F(x, y)\, N(x, y)\, dy = 0$$

is exact and we have

$$\frac{\partial}{\partial x}[F(x, y)\, N(x, y)] = \frac{\partial}{\partial y}[F(x, y)\, M(x, y)]$$

$$F(x, y)\,\frac{\partial N}{\partial x} + N(x, y)\,\frac{\partial F}{\partial x} = F(x, y)\,\frac{\partial M}{\partial y} + M(x, y)\,\frac{\partial F}{\partial y}$$

Now if F is a function of x alone, then

$$\frac{\partial F}{\partial x} = \frac{dF}{dx} \quad \text{and} \quad \frac{\partial F}{\partial y} = 0$$

and we have

$$N(x, y) \frac{dF}{dx} = F(x) \frac{\partial M}{\partial y} - F(x) \frac{\partial N}{\partial x}$$

$$\frac{1}{F(x)} \frac{dF}{dx} = \frac{1}{N(x, y)} \left[\frac{\partial M}{\partial y} - \frac{\partial N}{\partial x} \right]$$

Since the left-hand side of this equation is a function of x alone, the right-hand side must be also. Thus by letting

$$h(x) = \frac{1}{N(x, y)} \left[\frac{\partial M}{\partial y} - \frac{\partial N}{\partial x} \right]$$

and integrating, we have

$$\ln |F(x)| = \int h(x) \, dx$$

which implies that

$$F(x) = e^{\int h(x) \, dx}$$

is an integrating factor for the original equation. For the case in which F is a function of y alone, the result is similar, and we summarize both cases in the following theorem.

THEOREM 19.4
(Integrating Factors That Are Functions of x or y Alone)

For the differential equation

$$M(x, y) \, dx + N(x, y) \, dy = 0$$

i.　If

$$\frac{\dfrac{\partial M}{\partial y} - \dfrac{\partial N}{\partial x}}{N} = h(x)$$

then $e^{\int h(x) \, dx}$ is an integrating factor.

ii.　If

$$\frac{\dfrac{\partial N}{\partial x} - \dfrac{\partial M}{\partial y}}{M} = k(y)$$

then $e^{\int k(y) \, dy}$ is an integrating factor.

Note that if either $h(x)$ or $k(y)$ is a constant, the theorem still applies. Furthermore, as an aid to remembering the formulas, note that the subtracted partial identifies both the denominator and the variable for the integrating factor.

Example 1

Find the general solution to the differential equation

$$2y\,dx + (x + y)\,dy = 0, \qquad y > 0$$

Solution: Since $M(x, y) = 2y$ and $N(x, y) = x + y$, we have

$$\frac{\partial M}{\partial y} = 2 \qquad \text{and} \qquad \frac{\partial N}{\partial x} = 1$$

Thus the equation is not exact. Applying Theorem 19.4 we observe that

$$\frac{\dfrac{\partial N}{\partial y} - \dfrac{\partial M}{\partial x}}{M} = \frac{1 - 2}{2y} = \frac{-1}{2y} = k(y)$$

is a function of y alone. Therefore, $e^{\int k(y)\,dy}$ is an integrating factor. Since

$$\int k(y)\,dy = -\int \frac{1}{2y}\,dy = -\frac{1}{2}\ln|y| = \ln\left(\frac{1}{\sqrt{y}}\right)$$

the integrating factor is

$$e^{\int k(y)\,dy} = e^{\ln(1/\sqrt{y})} = \frac{1}{\sqrt{y}}$$

Thus

$$\left(\frac{1}{\sqrt{y}}\right)2y\,dx + \left(\frac{1}{\sqrt{y}}\right)(x + y)\,dy = 2\sqrt{y}\,dx + \left(\frac{x}{\sqrt{y}} + \sqrt{y}\right)dy = 0$$

is exact, and by Theorem 19.3 with $M = 2\sqrt{y}$ and $N = (x/\sqrt{y}) + \sqrt{y}$, we have

$$U(x, y) = 2\int \sqrt{y}\,dx + f(y) = 2x\sqrt{y} + f(y)$$

where

$$f'(y) = \frac{x}{\sqrt{y}} + \sqrt{y} - \frac{\partial}{\partial y}[2x\sqrt{y}] = \frac{x}{\sqrt{y}} + \sqrt{y} - \frac{x}{\sqrt{y}} = \sqrt{y}$$

$$f(y) = \tfrac{2}{3}y^{3/2} + C_1$$

Therefore,

$$U(x, y) = 2x\sqrt{y} + \tfrac{2}{3}y^{3/2} + C_1$$

and the general solution is

$$2x\sqrt{y} + \tfrac{2}{3}y^{3/2} = C_2$$

or

$$3x + y = \frac{C}{\sqrt{y}}$$

Example 2

Find the general solution to the differential equation

$$(y^2 - x)\,dx + 2y\,dy = 0$$

Solution: Since $M(x, y) = y^2 - x$ and $N(x, y) = 2y$, we have

$$\frac{\partial M}{\partial y} = 2y \quad \text{and} \quad \frac{\partial N}{\partial x} = 0$$

Thus the equation is not exact. However, since

$$\frac{\frac{\partial M}{\partial y} - \frac{\partial N}{\partial x}}{N} = \frac{2y - 0}{2y} = 1 = h(x)$$

we can apply Theorem 19.4 to conclude that

$$e^{\int h(x)\,dx} = e^{\int dx} = e^x$$

is an integrating factor. Thus

$$(y^2 e^x - xe^x)\,dx + 2ye^x\,dy = 0$$

is exact, and by Theorem 19.3 we have

$$U(x, y) = \int (y^2 e^x - xe^x)\,dx + f(y) = y^2 e^x - (xe^x - e^x) + f(y)$$

$$= y^2 e^x - xe^x + e^x + f(y)$$

where

$$f'(y) = 2ye^x - \frac{\partial}{\partial y}[y^2 e^x - xe^x + e^x] = 2ye^x - 2ye^x = 0$$

$$f(y) = C_1$$

Therefore,

$$U(x, y) = y^2 e^x - xe^x + e^x + C_1$$

and the general solution is

$$y^2 e^x - xe^x + e^x = C$$

or

$$y^2 - x + 1 = Ce^{-x} \qquad \blacksquare$$

There are many techniques for finding integrating factors. Of these, Theorem 19.4 is one of the most straightforward and widely applicable. In the remainder of this section, we will demonstrate ways to find integrating factors by *inspection*. Finding integrating factors in this manner depends largely upon recognition of certain exact differentials and a bit of ingenuity. Six of the most common exact differentials are listed on the following page. Learn to recognize them.

Six Common Exact Differentials	1. $d[xy] = y\,dx + x\,dy$

1. $d[xy] = y\,dx + x\,dy$

2. $d[x^m y^n] = x^{m-1} y^{n-1}(my\,dx + nx\,dy)$

3. $d\left[\dfrac{x}{y}\right] = \dfrac{y\,dx - x\,dy}{y^2}$

4. $d\left[\dfrac{y}{x}\right] = \dfrac{x\,dy - y\,dx}{x^2}$

5. $d[\ln \sqrt{x^2 + y^2}] = \dfrac{x\,dx + y\,dy}{x^2 + y^2}$

6. $d\left[\text{Arctan } \dfrac{y}{x}\right] = \dfrac{x\,dy - y\,dx}{x^2 + y^2}$

When finding integrating factors by inspection, you may find it helpful to regroup the terms of a differential equation in order to recognize familiar differentials. The next example is a case in point.

Example 3

Find the general solution of the equation

$$(x^2 y + y^2)\,dx - x^3\,dy = 0$$

Solution: Observe that this equation is not exact and that Theorem 19.4 is not applicable. However, by regrouping according to recognizable differentials, we have

$$(x^2 y\,dx - x^3\,dy) + y^2\,dx = 0$$
$$x^2(y\,dx - x\,dy) + y^2\,dx = 0$$

Now multiplication by the integrating factor $1/x^2 y^2$ yields

$$\frac{y\,dx - x\,dy}{y^2} + \frac{1}{x^2}\,dx = 0$$

$$d\left[\frac{x}{y}\right] + x^{-2}\,dx = 0$$

By integrating we obtain the general solution

$$\frac{x}{y} - \frac{1}{x} = C$$

From the differential

$$d[x^m y^n] = x^{m-1} y^{n-1}(my\,dx + nx\,dy)$$

we observe that $F(x, y) = x^{m-1}y^{n-1}$ can be used as an integrating factor for any differential of the form

$$my \, dx + nx \, dy = 0$$

Our next example capitalizes on this observation.

Example 4

Find the general solution of the equation

$$(3y^4 + 4xy) \, dx + (5xy^3 + 2x^2) \, dy = 0$$

Solution: This equation is not exact, nor does Theorem 19.4 apply. However, *by grouping terms of equal degree and factoring,* we obtain

$$(3y^4 \, dx + 5xy^3 \, dy) + (4xy \, dx + 2x^2 \, dy) = 0$$
$$y^3(3y \, dx + 5x \, dy) + x(4y \, dx + 2x \, dy) = 0$$

For the left grouping $m = 3$, $n = 5$, and $x^{m-1}y^{n-1} = x^2y^4$. For the right grouping $m = 4$, $n = 2$, and $x^{m-1}y^{n-1} = x^3y$. However, the first group already has y^3 as a factor, while the second group has x as a factor; hence the single integrating factor is

$$\frac{x^2y^4}{y^3} = x^2y = \frac{x^3y}{x}$$

Multiplication by this integrating factor yields

$$x^2y^4(3y \, dx + 5x \, dy) + x^3y(4y \, dx + 2x \, dy) = 0$$
$$d[x^3y^5] + d[x^4y^2] = 0$$

Finally, by integrating we have the general solution

$$x^3y^5 + x^4y^2 = C \qquad \blacksquare$$

Example 5

Find the general solution of

$$(x + x^2 + y^2) \, dy - y \, dx = 0$$

Solution: By regrouping and multiplying by $1/(x^2 + y^2)$, we have

$$(x^2 + y^2) \, dy + (x \, dy - y \, dx) = 0$$
$$dy + \frac{x \, dy - y \, dx}{x^2 + y^2} = 0$$

and by integrating we have the general solution

$$y + \text{Arctan} \, \frac{y}{x} = C \qquad \blacksquare$$

In the next example we show how a differential equation can serve as an aid to sketching a force field

$$\mathbf{F}(x, y) = M(x, y)\mathbf{i} + N(x, y)\mathbf{j}$$

Example 6

Sketch the force field given by

$$\mathbf{F}(x, y) = \frac{2y}{\sqrt{x^2 + y^2}}\mathbf{i} - \frac{y^2 - x}{\sqrt{x^2 + y^2}}\mathbf{j}$$

by finding and sketching the family of tangent curves to **F**.

Solution: At the point (x, y) in the plane, the vector

$$\mathbf{F}(x, y) = \frac{2y}{\sqrt{x^2 + y^2}}\mathbf{i} - \frac{y^2 - x}{\sqrt{x^2 + y^2}}\mathbf{j}$$

has a slope of

$$\frac{dy}{dx} = \frac{-(y^2 - x)/\sqrt{x^2 + y^2}}{2y/\sqrt{x^2 + y^2}} = \frac{-(y^2 - x)}{2y}$$

which in differential form is

$$2y\,dy = -(y^2 - x)\,dx$$
$$(y^2 - x)\,dx + 2y\,dy = 0$$

From Example 2 we know that the general solution to this differential equation is

$$y^2 - x + 1 = Ce^{-x}$$

or
$$y^2 = x - 1 + Ce^{-x}$$

Figure 19.5 shows several representative curves from this family. Note that the force vector at (x, y) is tangent to the curve passing through (x, y). ∎

Force field:

$$\mathbf{F}(x, y) = \frac{2y}{\sqrt{x^2 + y^2}}\mathbf{i} - \frac{y^2 - x}{\sqrt{x^2 + y^2}}\mathbf{j}$$

Family of tangent curves to **F**:
$$y^2 = x - 1 + Ce^{-x}$$

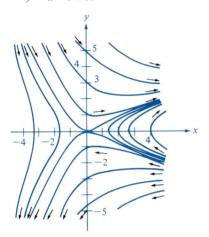

FIGURE 19.5

Section Exercises (19.4)

In Exercises 1–10, show that the given equation has an integrating factor that is a function of x or y alone. Then using that integrating factor, find the general solution of the equation.

1. $y\,dx - (x + 6y^2)\,dy = 0$

2. $(2x^3 + y)\,dx - x\,dy = 0$

3. $(5x^2 - y)\,dx + x\,dy = 0$

4. $(5x^2 - y^2)\,dx + 2y\,dy = 0$

S 5. $(x + y)\,dx + (\tan x)\,dy = 0$

6. $(2x^2y - 1)\,dx + x^3\,dy = 0$

7. $y^2\,dx + (xy - 1)\,dy = 0$

8. $(x^2 + 2x + y)\,dx + 2\,dy = 0$

9. $2y\,dx + (x - \sin\sqrt{y})\,dy = 0$

10. $(-2y^3 + 1)\,dx + (3xy^2 + x^3)\,dy = 0$

In Exercises 11–14, find an integrating factor of the form $x^m y^n$ and use this factor to find the general solution of the equation.

11. $(4x^2 y + 2y^2)\,dx + (3x^3 + 4xy)\,dy = 0$

12. $(3y^2 + 5x^2 y)\,dx + (3xy + 2x^3)\,dy = 0$

S 13. $(-y^5 + x^2 y)\,dx + (2xy^4 - 2x^3)\,dy = 0$

14. $-y^3\,dx + (xy^2 - x^2)\,dy = 0$

In Exercises 15–19, rewrite the equation as the sum or difference of exact differentials. Then integrate to find the general solution.

15. $(2yx^2 + 2y^3 + x)\,dy - y\,dx = 0$

16. $(x^2 y + y^3 + y)\,dx + (x^3 + xy^2 - x)\,dy = 0$

S 17. $(x^3 - x^2 - y^2)\,dx + x^2 y\,dy = 0$

18. $(x^3 + xy^2 + y)\,dx + (x^2 y + y^3 - x)\,dy = 0$

19. $\left(3ye^x - \dfrac{1}{x^2 y^3}\right)dx + 4xe^x\,dy = 0$

20. Find the exact differential of $\sqrt{x^2 + y^2}$ and use this differential to solve the equation

$$(y + y\sqrt{x^2 + y^2})\,dy + x\,dx = 0$$

S 21. Find the exact differential of $\sqrt{x^2 - y^2}$ and use this differential to solve the equation

$$(x + y\sqrt{x^2 - y^2})\,dx - (y - x\sqrt{x^2 - y^2})\,dy = 0$$

22. Show that $1/x^2$, $1/y^2$, $1/xy$, and $1/(x^2 + y^2)$ are each integrating factors for the equation

$$y\,dx - x\,dy = 0$$

23. Show that the differential equation

$$(axy^2 + by)\,dx + (bx^2 y + ax)\,dy = 0$$

is exact only if $a = b$. If $a \neq b$, show that $x^m y^n$ is an integrating factor, where $m = -(2b + a)/(a + b)$ and $n = -(2a + b)/(a + b)$.

In Exercises 24–27, sketch the family of tangent curves to the given force field.

24. $\mathbf{F}(x, y) = \dfrac{x}{\sqrt{x^2 + y^2}}\mathbf{i} - \dfrac{y}{\sqrt{x^2 + y^2}}\mathbf{j}$

25. $\mathbf{F}(x, y) = \dfrac{y}{\sqrt{x^2 + y^2}}\mathbf{i} - \dfrac{x}{\sqrt{x^2 + y^2}}\mathbf{j}$

26. $\mathbf{F}(x, y) = (1 + x^2)\mathbf{i} - 2xy\mathbf{j}$

27. $\mathbf{F}(x, y) = 4x^2 y\mathbf{i} - \left(2xy^2 + \dfrac{x}{y^2}\right)\mathbf{j}$

For Review and Class Discussion

True or False

1. _____ If $\partial M/\partial y = nN/x$ and $\partial N/\partial x = mM/y$, then $x^n y^m$ is an integrating factor for $M\,dx + N\,dy = 0$.

2. _____ Each of the functions $-1/x^2$, $1/y^2$, $-1/yx$, and $-1/(x^2 + y^2)$ is an integrating factor for $y\,dx - x\,dy = 0$.

3. _____ If $M = yf(xy)$ and $N = xg(xy)$, then $1/(xM - yN)$ is an integrating factor for $M\,dx + N\,dy = 0$.

4. _____ If

$$\left[\dfrac{\partial M}{\partial y} - \dfrac{\partial N}{\partial x}\right]\Big/N = h(x)$$

then $h(x)$ is an integrating factor for $M\,dx + N\,dy = 0$.

5. _____ If $M/N = P/Q$ and $M\,dx + N\,dy = 0$ is exact, then $P\,dx + Q\,dy = 0$ is exact.

19.5 **Linear Equations**

Purpose
- To recognize and solve first-order linear differential equations.
- To solve a nonlinear Bernoulli equation by transforming it into a linear equation.
- To summarize the solution techniques for first-order differential equations.

Concern for man himself and his fate must always form the chief interest of all technical endeavors, . . . in order that the creations of our mind shall be a blessing and not a curse to mankind. Never forget this in the midst of your diagrams and equations.   —Albert Einstein (1879–1955)

In the previous section we saw that many first-order differential equations can be made exact by the use of an integrating factor. In this section we will see how integrating factors can be used to solve one of the most important classes of first-order differential equations, namely, first-order *linear* differential equations.

First-Order Linear Differential Equations

An equation of the form

$$A(x)\frac{dy}{dx} + B(x)y = C(x)$$

is called a **first-order linear differential equation.** The **standard form** of this equation is

$$\frac{dy}{dx} + P(x)y = Q(x)$$

In differential form the linear equation

$$\frac{dy}{dx} + P(x)y = Q(x)$$

is

$$[P(x)y - Q(x)]\,dx + dy = 0$$

This equation possesses an integrating factor that is a function of x alone. To see this, we let $M(x, y) = P(x)y - Q(x)$ and $N(x, y) = 1$. Then

$$\frac{\frac{\partial M}{\partial y} - \frac{\partial N}{\partial x}}{N} = \frac{P(x) - 0}{1} = P(x)$$

Thus by Theorem 19.4 the integrating factor is

$$e^{\int P(x)\,dx}$$

Now since

$$e^{\int P(x)\,dx}[P(x)y - Q(x)]\,dx + e^{\int P(x)\,dx}\,dy = 0$$

is exact, we can apply Theorem 19.3 to conclude that the general solution is given by $U(x, y) = C$, where

$$U(x, y) = \int [yP(x)e^{\int P(x)\,dx} - Q(x)e^{\int P(x)\,dx}]\,dx + f(y)$$

$$= ye^{\int P(x)\,dx} - \int Q(x)e^{\int P(x)\,dx}\,dx + f(y)$$

It can be shown that $f(y)$ is constant, and the resulting general solution is

$$ye^{\int P(x)\,dx} = \int Q(x)e^{\int P(x)\,dx}\,dx + C$$

THEOREM 19.5
(*General Solution of a First-Order Linear Equation*)

The general solution of the first-order linear differential equation

$$\frac{dy}{dx} + P(x)y = Q(x)$$

is given by

$$ye^{\int P(x)\,dx} = \int Q(x)e^{\int P(x)\,dx}\,dx + C$$

Example 1

Find the general solution to the linear equation

$$xy' - 2y = x^2$$

Solution: The standard form of this linear equation is

$$\frac{dy}{dx} - \left(\frac{2}{x}\right)y = x$$

Thus $P(x) = -2/x$, and we have

$$\int P(x)\,dx = -\int \frac{2}{x}\,dx = -\ln x^2$$

and

$$e^{\int P(x)\,dx} = e^{-\ln x^2} = \frac{1}{x^2}$$

Finally, since $Q(x) = x$, the general solution is

$$y\left(\frac{1}{x^2}\right) = \int x\left(\frac{1}{x^2}\right)\,dx + C = \int \frac{1}{x}\,dx + C = \ln |x| + C$$

$$y = x^2 \ln |x| + Cx^2$$

∎

Example 2

Find the general solution to the linear equation

$$y' - y \tan t = 1$$

Solution: Since $P(t) = -\tan t$, we have

$$\int P(t)\, dt = -\int \tan t\, dt = \ln |\cos t|$$

and

$$e^{\int P(t)\, dt} = e^{\ln |\cos t|} = |\cos t|$$

A quick check will show that $\cos t$ is an integrating factor, as well as $|\cos t|$. Thus since $Q(t) = 1$, the general solution is

$$y \cos t = \int \cos t\, dt + C$$

$$y \cos t = \sin t + C$$

$$y = \tan t + C \sec t \qquad \blacksquare$$

Our next example illustrates the use of a first-order linear differential equation to solve a problem involving chemical mixtures.

Example 3

A tank contains 50 gal of a solution composed of 90% water and 10% alcohol. A second solution containing 50% water and 50% alcohol is added to the tank at the rate of 4 gal/min. As the second solution is being added, the tank is being drained of 5 gal/min. Assuming the solution in the tank is stirred constantly, how much alcohol is in the tank after 10 min?

Solution: Let a be the number of gallons of alcohol in the tank at any time t. We know that $a = 5$ when $t = 0$. Since the number of gallons of solution in the tank at any time is $50 - t$ and since the tank loses 5 gal of solution per minute, it must lose

$$\left(\frac{5}{50 - t}\right)a$$

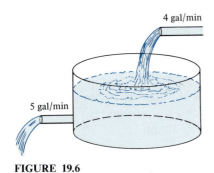

4 gal/min

5 gal/min

FIGURE 19.6

gallons of alcohol per minute (see Figure 19.6). Furthermore, since the tank is gaining 2 gal of alcohol per minute, the rate of change of alcohol in the tank is given by

$$\frac{da}{dt} = 2 - \left(\frac{5}{50 - t}\right)a \qquad \text{or} \qquad \frac{da}{dt} + \left(\frac{5}{50 - t}\right)a = 2$$

To solve this linear equation, we let $P(t) = 5/(50 - t)$ and obtain

$$\int P(t)\, dt = \int \frac{5}{50 - t}\, dt = -5 \ln |50 - t|$$

Since $t < 50$, we can drop the absolute value signs and conclude that

$$e^{\int P(t)\,dt} = e^{-5\ln(50-t)} = \frac{1}{(50-t)^5}$$

Thus the general solution is

$$a\left[\frac{1}{(50-t)^5}\right] = \int \frac{2}{(50-t)^5}\,dt + C$$

$$\frac{a}{(50-t)^5} = \left(\frac{1}{2}\right)\left[\frac{1}{(50-t)^4}\right] + C$$

$$a = \frac{50-t}{2} + C(50-t)^5$$

Since $a = 5$ when $t = 0$, we have

$$5 = \frac{50}{2} + C(50)^5$$

$$\frac{-20}{50^5} = C$$

which means that the particular solution is

$$a = \frac{50-t}{2} - 20\left(\frac{50-t}{50}\right)^5$$

Finally, when $t = 10$ the amount of alcohol in the tank is

$$a = \frac{50-10}{2} - 20\left(\frac{50-10}{50}\right)^5 = 13.45 \text{ gal}$$

which represents a solution containing 33.6% alcohol. ∎

A well-known nonlinear equation that reduces to a linear one with an appropriate substitution is the **Bernoulli equation:**

$$y' + P(x)y = Q(x)y^n$$

Note that this equation is linear if $n = 0$ and has separable variables if $n = 1$. Thus in the following development, we assume that $n \neq 0$ and $n \neq 1$. We begin by multiplying by y^{-n} to obtain

$$y^{-n}y' + P(x)y^{1-n} = Q(x)$$

Further multiplication by $(1 - n)$ yields

$$(1 - n)y^{-n}y' + (1 - n)P(x)y^{1-n} = (1 - n)Q(x)$$

or

$$\frac{d}{dx}[y^{1-n}] + (1 - n)P(x)y^{1-n} = (1 - n)Q(x)$$

which is a linear equation in the variable y^{1-n}. Thus if we let $z = y^{1-n}$, we obtain the linear equation

$$\frac{dz}{dx} + (1 - n)P(x)z = (1 - n)Q(x)$$

Finally, by Theorem 19.5 the *general solution to the Bernoulli equation* is

$$ze^{\int(1-n)P(x)\,dx} = \int (1 - n)Q(x)e^{\int(1-n)P(x)\,dx}\,dx + C$$

or

$$y^{1-n}e^{\int(1-n)P(x)\,dx} = \int (1 - n)Q(x)e^{\int(1-n)P(x)\,dx}\,dx + C$$

Example 4 (*Bernoulli*)

Find the general solution to the Bernoulli equation

$$y' + xy = xy^{-3}e^{-x^2}$$

Solution: For this Bernoulli equation we have $n = -3$, $1 - n = 4$, and $P(x) = x$. Thus we have

$$\int (1 - n)P(x)\,dx = \int 4x\,dx = 2x^2$$

and

$$e^{\int(1-n)P(x)\,dx} = e^{2x^2}$$

Since $Q(x) = xe^{-x^2}$, the general solution is

$$y^4 e^{2x^2} = \int 4xe^{-x^2}e^{2x^2}\,dx + C$$

$$y^4 e^{2x^2} = 4\int xe^{x^2}\,dx + C$$

$$y^4 e^{2x^2} = 2e^{x^2} + C$$

or

$$y^4 = 2e^{-x^2} + Ce^{-2x^2} \qquad \blacksquare$$

At this point we have introduced several methods for solving first-order differential equations. Of these, the separable variables case is usually the simplest, while solution by an integrating factor is usually left as a last resort. The following summary lists, in their normal order of difficulty, the various methods we have studied.

Summary of First-Order Differential Equations	Method	Form of Equation
	Separable Variables	$f(y)\,dy = g(x)\,dx$
	Homogeneous	$y' = f(x, y)$, where $f(tx, ty) = t^n f(x, y)$
	Linear	$y' + P(x)y = Q(x)$
	Bernoulli	$y' + P(x)y = Q(x)y^n$
	Exact	$M(x, y)\,dx + N(x, y)\,dy = 0$, where $\dfrac{\partial M}{\partial y} = \dfrac{\partial N}{\partial x}$
	Integrating Factor	$F(x, y)M(x, y)\,dx + F(x, y)N(x, y)\,dy = 0$ is exact

Example 5 (*Linear*)

Find the general solution to the equation

$$(x^4 + 2y)\,dx - x\,dy = 0$$

Solution: We first note that the equation is not likely to lend itself to separation of variables because the factor $(x^4 + 2y)$ is a sum that includes *both* variables. Furthermore, we can see by inspection that the equation is not homogeneous. In checking for linearity we obtain

$$-x\,dy + (x^4 + 2y)\,dx = 0$$

$$-x\frac{dy}{dx} + x^4 + 2y = 0$$

$$\frac{dy}{dx} - x^3 - \frac{2y}{x} = 0$$

$$y' - \left(\frac{2}{x}\right)y = x^3$$

Thus the equation is linear and we have

$$\int P(x)\,dx = -\int \frac{2}{x}\,dx = -2\ln|x| = \ln\frac{1}{x^2}$$

and

$$e^{\int P(x)\,dx} = e^{\ln(1/x^2)} = \frac{1}{x^2}$$

Therefore, the general solution is

$$y\left(\frac{1}{x^2}\right) = \int x^3\left(\frac{1}{x^2}\right)dx + C$$

$$\frac{y}{x^2} = \frac{1}{2}x^2 + C$$

$$y = \tfrac{1}{2}x^4 + Cx^2$$

■

Example 6 (*Exact*)

Find the general solution to the equation

$$ye^x \, dx + (2y + e^x) \, dy = 0$$

Solution: We begin by noting that the variables cannot be separated, the equation is not homogeneous, and it is not linear. In checking for exactness we have $M(x, y) = ye^x$ and $N(x, y) = 2y + e^x$. Thus

$$\frac{\partial M}{\partial y} = e^x \quad \text{and} \quad \frac{\partial N}{\partial x} = e^x$$

and the equation is exact. By Theorem 19.3 we have

$$U(x, y) = \int ye^x \, dx + f(y) = ye^x + f(y)$$

where

$$f'(y) = 2y + e^x - \frac{\partial}{\partial y}[ye^x] = 2y + e^x - e^x = 2y$$

$$f(y) = y^2 + C_1$$

Therefore, we have

$$U(x, y) = ye^x + y^2 + C_1$$

and the general solution is

$$ye^x + y^2 = C \qquad \blacksquare$$

Keep in mind that many first-order differential equations are solvable by more than one method. Thus after you have spent a reasonable amount of time trying to solve a particular equation by one method, it is wise to check the other methods to see if the solution can be obtained more simply.

Section Exercises (19.5)

In Exercises 1–15, solve the first-order linear differential equation.

1. $\dfrac{dy}{dx} + \left(\dfrac{1}{x}\right)y = 3x + 4$

2. $\dfrac{dy}{dx} + \left(\dfrac{2}{x}\right)y = 3x + 1$

3. $\dfrac{dy}{dx} = e^x - y,\ y(0) = \frac{3}{2}$

4. $y' + 2y = \sin x,\ y(\pi/2) = \frac{2}{5}$

$\boxed{\text{S}}$ **5.** $y' - y = \cos x,\ y(0) = \frac{1}{2}$

6. $y' + 2xy = 2x,\ y(0) = -1$

7. $(3y + \sin 2x) \, dx - dy = 0$

8. $[(y - 1) \sin x] \, dx - dy = 0$

$\boxed{\text{S}}$ **9.** $(x - 1)y' + y = x^2 - 1$

10. $y' + 5y = e^{5x}$

11. $y' \cos^2 x + y - 1 = 0$

12. $x^3 y' + 2y = e^{1/x^2}$

13. $y' + y \tan x = \sec x + \cos x,\ y(0) = 1$

14. $y' + \left(\frac{1}{x}\right)y = 0$ (Solve with two methods.)

15. $y' + (2x - 1)y = 0, y(1) = 2$ (Solve with two methods.)

In Exercises 16–20, solve the Bernoulli differential equation.

16. $y' + 2xy = xy^2$

[S] **17.** $y' + \left(\frac{1}{x}\right)y = xy^2$

18. $y' + \left(\frac{1}{x}\right)y = x\sqrt{y}$

19. $y' - y = x^3\sqrt[3]{y}$

20. $yy' - 2y^2 = e^x$

21. The differential equation

$$L\frac{dI}{dt} + RI = E$$

arises in the study of electrical circuits. I is the current (in amperes, A), R is the resistance (in ohms, Ω), L is the inductance (in henrys, H), E is the electromotive force (in volts, V), and t is the time (in seconds). Solve this differential equation for the case in which the voltage is a constant E_0.

22. At time $t = 0$ a tank contains v_0 gallons of a solution of which, by weight, q_0 pounds is a soluble concentrate. Another solution containing q_1 pounds of the concentrate per gallon is running into the tank at the rate of r_1 gallons per minute. The solution in the tank is kept well stirred and is withdrawn at the rate of r_2 gallons per minute. If Q is the amount of concentrate in the solution at any time t, show that

$$\frac{dQ}{dt} + \frac{r_2 Q}{v_0 + (r_1 - r_2)t} = q_1 r_1, \qquad Q(0) = q_0$$

In Exercises 23–25, use the result of Exercise 22.

[S] **23.** A 200-gal tank is full of a solution containing 25 lb of concentrate. Starting at time $t = 0$, distilled water is admitted to the tank at the rate of 10 gal/min, and the well-stirred solution is withdrawn at the same rate.

(a) Find the amount of the concentrate in the solution as a function of t.

(b) Find the time at which concentrate must be added if it is done when the amount of concentrate in the tank reaches 15 lb.

24. Repeat Exercise 23 if solution entering the tank contains 0.05 lb of concentrate per gallon.

25. A 200-gal tank is half full of distilled water. At time $t = 0$, a solution containing 0.5 lb of concentrate per gallon enters the tank at the rate of 5 gal/min, and the well-stirred mixture is withdrawn at the rate of 3 gal/min.

(a) At what time will the tank be full?

(b) At the time the tank is full, how many pounds of concentrate will it contain?

In Exercises 26–41, solve the first-order differential equation by *any* appropriate method studied so far.

26. $(x + 1)\,dx - (y^2 + 2y)\,dy = 0$

[S] **27.** $e^{2x+y}\,dx - e^{x-y}\,dy = 0$

28. $(1 + 2e^{2x+y})\,dx + e^{2x+y}\,dy = 0$

29. $(1 + y^2)\,dx + (2xy + y + 2)\,dy = 0$

30. $(x + 1)\,dy + (y - e^x)\,dx = 0$

31. $(y\cos x - \cos x)\,dx + dy = 0$

32. $y' = 2x\sqrt{1 - y^2}$

[S] **33.** $2xy\,dx + (x^2 + \cos y)\,dy = 0$

34. $(x + y)\,dx - x\,dy = 0$

35. $(3y^2 + 4xy)\,dx + (2xy + x^2)\,dy = 0$

36. $(y^2 + xy)\,dx - x^2\,dy = 0$

37. $(2y - e^x)\,dx + x\,dy = 0$

38. $y\,dx + (3x + 4y)\,dy = 0$

[S] **39.** $(x^2y^4 - 1)\,dx + x^3y^3\,dy = 0$

40. $x\,dx + (y + e^y)(x^2 + 1)\,dy = 0$

41. $3y\,dx - (x^2 + 3x + y^2)\,dy = 0$

For Review and Class Discussion

True or False

1. ____ $y' + x\sqrt{y} = x^2$ is a first-order linear differential equation.

2. ____ $y' + xy = e^x y$ is a first-order linear differential equation.

3. ____ $(dx/dy) + y^2x = e^y$ is a first-order linear differential equation in x.

4. ____ $\int P(x)e^{\int P(x)\,dx}\,dx = e^{\int P(x)\,dx} + C$.

5. ____ $y'' + P(x)y' = Q(x)$ is a first-order linear differential equation in y'.

19.6 Linear Homogeneous Equations with Constant Coefficients

Purpose

- To demonstrate the relationship between linear independence and the general solution of a linear differential equation with constant coefficients.
- To solve higher-order linear homogeneous differential equations with constant coefficients by means of characteristic equations.

A truly realistic mathematics should be conceived, in line with physics, as a branch of the theoretical construction of the one real world.

—*Hermann Weyl* (1885–1955)

Up to this point we have restricted our discussion of solution methods to first-order differential equations. In this and the following section, we discuss the solution of one class of higher-order differential equations, namely, those that are linear with *constant coefficients*.

Linear Differential Equation with Constant Coefficients

The differential equation

$$a_n y^{(n)} + a_{n-1} y^{(n-1)} + \cdots + a_1 y' + a_0 y = F(x)$$

is called a **linear differential equation with constant coefficients**. If $F(x) = 0$, the equation is **homogeneous**; otherwise, it is **nonhomogeneous**.

Before looking at the general solution of a linear differential equation with constant coefficients, we introduce the concept of linear independence.

Definition of Linear Independence

The functions $y_1, y_2, \ldots, y_n$ are said to be **linearly independent** if the *only* solution to the equation

$$C_1 y_1 + C_2 y_2 + \cdots + C_n y_n = 0$$

is

$$C_1 = C_2 = \cdots = C_n = 0$$

A collection of functions that are not linearly independent is said to be **linearly dependent.**

From this definition it is easy to see that two functions are linearly dependent if and only if one of the functions is a constant multiple of the other. For example, $y_1 = x$ and $y_2 = 3x$ are linearly dependent since

$$C_1 x + C_2(3x) = 0$$

has the nonzero solution $C_1 = -3$ and $C_2 = 1$.

A quick test for linear independence of a collection of more than two functions can be obtained by the use of the Wronskian of the collection of functions. For *n* functions the **Wronskian** $w(y_1, y_2, \ldots, y_n)$ is defined as the determinant

$$w(y_1, y_2, \ldots, y_n) = \begin{vmatrix} y_1 & y_2 & \cdots & y_n \\ y_1' & y_2' & \cdots & y_n' \\ \vdots & \vdots & \vdots & \\ y_1^{(n-1)} & y_2^{(n-1)} & \cdots & y_n^{(n-1)} \end{vmatrix}$$

Wronskian Test for Linear Independence

If $y_1, y_2, \ldots, y_n$ are solutions to an *n*th-order linear homogeneous equation, then the following test is valid.

1. If $w(y_1, y_2, \ldots, y_n) \neq 0$, the functions $y_1, y_2, \ldots, y_n$ are *linearly independent*.

2. If $w(y_1, y_2, \ldots, y_n) = 0$, the functions $y_1, y_2, \ldots, y_n$ are *linearly dependent*.

Example 1

Test the following sets of solutions for linear independence.

(a) $\{1, x, x^2\}$　　(b) $\{e^x, xe^x\}$　　(c) $\{1, x, 2x - 3\}$

Solution:

(a) Since

$$w(1, x, x^2) = \begin{vmatrix} 1 & x & x^2 \\ 0 & 1 & 2x \\ 0 & 0 & 2 \end{vmatrix} = (1) \begin{vmatrix} 1 & 2x \\ 0 & 2 \end{vmatrix} = 2 \neq 0$$

the collection is linearly independent.

(b) Since xe^x is not a constant multiple of e^x, the two solutions form a linearly independent collection.

(c) Since

$$w(1, x, 2x - 3) = \begin{vmatrix} 1 & x & 2x - 3 \\ 0 & 1 & 2 \\ 0 & 0 & 0 \end{vmatrix} = (1) \begin{vmatrix} 1 & 2 \\ 0 & 0 \end{vmatrix} = 0$$

the collection is linearly dependent. ∎

The following theorem points out the importance of linear independence in constructing the general solution of a linear homogeneous differential equation with constant coefficients.

THEOREM 19.6
(*General Solution of a*
Homogeneous Linear Equation)

If $y_1, y_2, \ldots, y_n$ are linearly independent solutions of the differential equation

$$a_n y^{(n)} + a_{n-1} y^{(n-1)} + \cdots + a_1 y' + a_0 y = 0$$

then the general solution is given by

$$y = C_1 y_1 + C_2 y_2 + \cdots + C_n y_n$$

where $C_1, C_2, \ldots, C_n$ are arbitrary constants.

From Theorem 19.6 we can see that the problem of finding the general solution of an nth-order linear homogeneous differential equation with constant coefficients can be reduced to the problem of finding a collection of n linearly independent particular solutions

$$y_1, y_2, \ldots, y_n$$

To gain some insight into the solution of this problem, we consider the second-order equation

$$y'' + ay' + by = 0$$

The nature of this equation suggests that it may have solutions of the form $y = e^{mx}$. By substituting into the original equation, we have

$$\frac{d^2}{dx^2}[e^{mx}] + a\frac{d}{dx}[e^{mx}] + be^{mx} = 0$$

$$m^2 e^{mx} + ame^{mx} + be^{mx} = 0$$

$$e^{mx}(m^2 + am + b) = 0$$

Now since e^{mx} is never zero, we conclude that $y = e^{mx}$ *is a solution to the given differential equation provided m is a solution to the polynomial equation*

$$m^2 + am + b = 0$$

This equation is called the **characteristic equation** of the differential equation

$$y'' + ay' + by = 0$$

Note that the characteristic equation can be determined from its differential equation by simply replacing y'' by m^2, y' by m, and y by 1.

Example 2

Find the general solution of the differential equation $y'' - 4y = 0$.

Solution: In this case the characteristic equation is

$$m^2 - 4 = 0 \quad \text{so} \quad m^2 = 4 \quad \text{or} \quad m = \pm 2$$

Thus

$$y_1 = e^{m_1 x} = e^{2x} \quad \text{and} \quad y_2 = e^{m_2 x} = e^{-2x}$$

are each solutions of the given differential equation. Furthermore, since these two solutions are linearly independent, we can apply Theorem 19.6 to conclude that the general solution is

$$y = C_1 e^{2x} + C_2 e^{-2x} \qquad \blacksquare$$

The characteristic equation in Example 2 has two distinct real roots. From algebra we know that this is only one of *three* possibilities for quadratic equations. In general, the quadratic equation

$$m^2 + am + b = 0$$

has roots

$$m_1 = \frac{-a + \sqrt{a^2 - 4b}}{2} \quad \text{and} \quad m_2 = \frac{-a - \sqrt{a^2 - 4b}}{2}$$

which fall into one of three cases.

Case I: two distinct real roots, $m_1 \neq m_2$
Case II: two equal real roots, $m_1 = m_2$
Case III: two complex conjugate roots, $m_1 = \alpha + \beta i$ and $m_2 = \alpha - \beta i$

In terms of the differential equation

$$y'' + ay' + by = 0$$

these three cases correspond to three different types of general solutions.

General Solutions to
$y'' + ay' + by = 0$

Case I (Distinct Real Roots): If $m_1 \neq m_2$ are distinct roots of the characteristic equation, then the general solution is

$$y = C_1 e^{m_1 x} + C_2 e^{m_2 x}$$

Case II (Equal Real Roots): If $m_1 = m_2$ are equal real roots of the characteristic equation, then the general solution is

$$y = C_1 e^{m_1 x} + C_2 x e^{m_1 x} = (C_1 + C_2 x) e^{m_1 x}$$

Case III (Complex Roots): If $m_1 = \alpha + \beta i$ and $m_2 = \alpha - \beta i$ are complex roots of the characteristic equation, then the general solution is

$$y = C_1 e^{\alpha x} \cos(\beta x) + C_2 e^{\alpha x} \sin(\beta x)$$

Example 3

Find the general solution of the differential equation $y'' + 6y' + 12y = 0$.

Solution: The characteristic equation

$$m^2 + 6m + 12 = 0$$

has two complex roots:

$$m = \frac{-6 \pm \sqrt{36 - 48}}{2} = \frac{-6 \pm \sqrt{-12}}{2} = -3 \pm \sqrt{-3} = -3 \pm \sqrt{3}\,i$$

Thus $\alpha = -3$ and $\beta = \sqrt{3}$ and the general solution is

$$y = C_1 e^{-3x} \cos \sqrt{3}\,x + C_2 e^{-3x} \sin \sqrt{3}\,x$$ ∎

Example 4

Solve the differential equation $y'' + 4y' + 4y = 0$, subject to the initial conditions $y(0) = 2$ and $y'(0) = 1$.

Solution: The characteristic equation

$$m^2 + 4m + 4 = 0$$
$$(m + 2)^2 = 0$$
$$m = -2$$

has two equal real roots. Therefore, the general solution is

$$y = C_1 e^{-2x} + C_2 x e^{-2x}$$

Now since $y = 2$ when $x = 0$, we have

$$2 = C_1(1) + C_2(0)(1) = C_1$$

Furthermore, since $y' = 1$ when $x = 0$, we have

$$y' = -2C_1 e^{-2x} + C_2(-2xe^{-2x} + e^{-2x})$$
$$1 = -2(2)(1) + C_2[-2(0)(1) + 1]$$
$$1 = -4 + C_2$$
$$5 = C_2$$

Therefore, the solution is

$$y = 2e^{-2x} + 5xe^{-2x}$$ ∎

For higher-order homogeneous differential equations, we find the general solution using much the same procedure used for second-order equations. That is, we begin by determining the n roots of the characteristic equation and then, based on these *n* roots, we form a linearly independent collection of *n* solutions. The major difference between the two procedures is that with equations of order three or more, roots of the characteristic equation may occur more than twice. When this happens, we form additional (linearly independent) solutions by multiplying by increased powers of *x*.

For example, the differential equation

$$y''' + 3y'' + 3y' + y = 0$$

has the characteristic equation

$$m^3 + 3m^2 + 3m + 1 = 0$$
$$(m + 1)^3 = 0$$
$$m = -1$$

Since the root $m = -1$ occurs three times, the general solution is

$$y = C_1 e^{-x} + C_2 x e^{-x} + C_3 x^2 e^{-x}$$

Similarly, the differential equation

$$y^{(4)} + 2y'' + y = 0$$

has the characteristic equation

$$m^4 + 2m^2 + 1 = 0$$
$$(m^2 + 1)^2 = 0$$
$$m = \pm i$$

Since the roots $m_1 = \alpha + \beta i = 0 + i$ and $m_2 = \alpha - \beta i = 0 - i$ each occur twice, the general solution is

$$y = C_1 \cos x + C_2 \sin x + C_3 x \cos x + C_4 x \sin x \qquad \blacksquare$$

Example 5

Find the general solution of the differential equation

$$y^{(5)} - 2y^{(4)} + 5y''' - 8y'' + 4y' = 0$$

Solution: The characteristic equation is

$$m^5 - 2m^4 + 5m^3 - 8m^2 + 4m = 0$$
$$m(m^4 - 2m^3 + 5m^2 - 8m + 4) = 0$$

Using synthetic division (or long division), we can verify that $m = 1$ is a double root and conclude that the equation factors as

$$m(m - 1)^2(m^2 + 4) = 0$$

Hence we have the five roots $m = 0, 1, 1, 2i, -2i$, which implies that the general solution is

$$y = C_1 + C_2 e^x + C_3 x e^x + C_4 \cos 2x + C_5 \sin 2x$$ ■

Example 5 points out that the major problem in solving higher-order linear differential equations is factoring the characteristic equation. Once the roots of the characteristic equation are known, the general solution follows easily.

Section Exercises (19.6)

In Exercises 1–26, find the general solution of the linear differential equation.

1. $y'' - y' = 0$

2. $y'' + 2y' = 0$

[S] **3.** $y'' - y' - 6y = 0$

4. $y'' + 6y' + 5y = 0$

5. $2y'' + 3y' - 2y = 0$

6. $16y'' - 16y' + 3y = 0$

[S] **7.** $y'' + 6y' + 9y = 0$

8. $y'' - 10y' + 25y = 0$

9. $16y'' - 8y' + y = 0$

10. $9y'' - 12y' + 4y = 0$

11. $y'' + y = 0$

12. $y'' + 4y = 0$

13. $y'' - 9y = 0$

14. $y'' - 2y = 0$

15. $y'' - 2y' + 4y = 0$

16. $y'' - 4y' + 21y = 0$

[S] **17.** $y'' - 3y' + y = 0$

18. $3y'' + 4y' - y = 0$

[S] **19.** $9y'' - 12y' + 11y = 0$

20. $2y'' - 6y' + 7y = 0$

21. $y^{(4)} - y = 0$

22. $y^{(4)} - y'' = 0$

23. $y''' - 6y'' + 11y' - 6y = 0$

24. $y''' - y'' - y' + y = 0$

25. $y''' - 3y'' + 7y' - 5y = 0$

26. $y''' - 3y'' + 3y' - y = 0$

27. Solve the differential equation $y'' - y' - 30y = 0$, subject to the initial conditions $y(0) = 1$ and $y'(0) = -4$.

28. Solve the differential equation $y'' + 2y' + 3y = 0$, subject to the initial conditions $y(0) = 2$ and $y'(0) = 1$.

29. A spring with spring modulus k supports a mass m, as shown in Figure 19.7. Let y denote the displacement of the mass from its equilibrium at time t. At $t = 0$ the

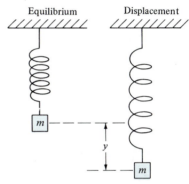

FIGURE 19.7

mass is pulled down a distance y_0 from the equilibrium position and released. Assuming that the resulting oscillations are free and undamped, we can use Newton's Second Law together with Hooke's Law to obtain

$$F = ma$$
$$-ky = my''$$
$$my'' + ky = 0$$

Solve this differential equation.

30. A 32-lb weight is suspended on a spring as in Exercise 29. The weight stretches the spring $\frac{2}{3}$ ft from its natural position. If the weight is pulled $\frac{1}{2}$ ft below the equilibrium position and released, find the equation describing the resulting motion and find the frequency of the oscillations (frequency = 1/period).

31. For the case in which the characteristic equation

$$m^2 + am + b = 0$$

has two equal real roots given by $m = -a/2$, show that

$$y = C_1 e^{mx} + C_2 x e^{mx}$$

is the general solution of

$$y'' + ay' + by = 0$$

32. For the case in which the characteristic equation

$$m^2 + am + b = 0$$

has two complex roots given by $m_1 = \alpha + \beta i$ and $m_2 = \alpha - \beta i$, show that

$$y = C_1 e^{\alpha x} \cos \beta x + C_2 e^{\alpha x} \sin \beta x$$

is the general solution of

$$y'' + ay' + by = 0$$

For Review and Class Discussion

True or False

1. _____ $y = C_1 e^{3x} + C_2 e^{-3x}$ is the general solution to $y'' - 6y' + 9 = 0$.

2. _____ $y = (C_1 + C_2 x) \sin x + (C_3 + C_4 x) \cos x$ is the general solution of $y^{(4)} + 2y'' + y = 0$.

3. _____ $y = e^{\alpha x} \sin \beta x$ is a solution of $a_n y^{(n)} + a_{n-1} y^{(n-1)} + \cdots + a_1 y' + a_0 y = 0$ if and only if $y = e^{\alpha x} \cos \beta x$ is also a solution.

4. _____ $y = x$ is a solution of $a_n y^{(n)} + a_{n-1} y^{(n-1)} + \cdots + a_1 y' + a_0 y = 0$ if and only if $a_1 = a_0 = 0$.

5. _____ It is possible to choose a and b so that $y = x^2 e^x$ is a solution to $y'' + ay' + by = 0$.

19.7 Linear Nonhomogeneous Equations with Constant Coefficients

Purpose

- To demonstrate the undetermined coefficients method of finding a particular solution to a linear nonhomogeneous differential equation.
- To demonstrate the variation of parameter method of finding a particular solution to a linear nonhomogeneous differential equation.

I have given this equation only as a particular solution of the formula which in general contains all the curves that a string in motion can assume.

—Leonhard Euler (1707–1783)

In the preceding section we learned how to solve a *homogeneous* linear equation. In this section we extend this solution procedure to cover *nonhomogeneous* linear equations.

The first step in finding the general solution of the nonhomogeneous equation

$$a_n y^{(n)} + a_{n-1} y^{(n-1)} + \cdots + a_1 y' + a_0 y = F(x)$$

is finding the general solution of the corresponding homogeneous equation [where $F(x) = 0$]. We denote this solution by

$$y_h = C_1 y_1 + C_2 y_2 + \cdots + C_n y_n$$

Next we attempt to find a particular solution, y_p, of the nonhomogeneous equation. If we are successful in finding y_p, we may conclude that the general solution of the nonhomogeneous equation is

$$y = y_h + y_p$$

THEOREM 19.7 (*General Solution of a Nonhomogeneous Linear Equation*)	Let $$a_n y^{(n)} + a_{n-1} y^{(n-1)} + \cdots + a_1 y' + a_0 y = F(x)$$ be a linear nonhomogeneous differential equation with constant coefficients $a_n, a_{n-1}, \ldots, a_1, a_0$. If y_p is a particular solution to this nonhomogeneous equation and y_h is the general solution to the corresponding homogeneous equation, then $$y = y_p + y_h$$ is the **general solution** to the nonhomogeneous equation.

Proof: We provide a proof for only the second-order case. Assume that y_h is the general solution to $a_2 y'' + a_1 y' + a_0 y = 0$ and y_p is a particular solution to $a_2 y'' + a_1 y' + a_0 y = F(x)$. Then the sum $y = y_p + y_h$ is a solution to the nonhomogeneous equation because

$$a_2 y'' + a_1 y' + a_0 y = a_2(y_p'' + y_h'') + a_1(y_p' + y_h') + a_0(y_p + y_h)$$
$$= (a_2 y_p'' + a_1 y_p' + a_0 y_p) + (a_2 y_h'' + a_1 y_h' + a_0 y_h)$$
$$= 0 + F(x) = F(x)$$

Conversely, if y is any solution to the nonhomogeneous equation, then $y - y_p$ is a solution to the homogeneous equation because

$$a_2(y'' - y_p'') + a_1(y' - y_p') + a_0(y - y_p)$$
$$= (a_2 y'' + a_1 y' + a_0 y) - (a_2 y_p'' + a_1 y_p' + a_0 y_p)$$
$$= F(x) - F(x) = 0$$

Hence $y - y_p$ is of the form $y - y_p = y_h$, which in turn implies that any solution to the nonhomogeneous equation is of the form

$$y = y_p + y_h$$

As you might expect, the problem of finding a particular solution of a nonhomogeneous linear equation is not trivial. However, for the special case in which $F(x)$ consists only of sums or products of

$$x^n, \quad e^{mx}, \quad \cos \beta x, \quad \sin \beta x$$

the difficulty in finding a particular solution is greatly reduced. In such cases we can find a particular solution y_p by the method of **undetermined coefficients.** In this method we assume that y_p is an expression, similar to $F(x)$, that contains unknown coefficients. We then determine these coefficients by substituting y_p and its derivatives into the nonhomogeneous equation.

In selecting y_p *similar to* $F(x)$, we choose a generalized form of $F(x)$. That is, y_p has the form of $F(x)$, as given, plus all its general-derivative forms. For instance:

If $F(x) = 7x^3$, choose $y_p = Ax^3 + (Bx^2 + Cx + D)$.

If $F(x) = 4xe^x$, choose $y_p = Axe^x + (Be^x)$.

If $F(x) = \sin 2x$, choose $y_p = A \sin 2x + (B \cos 2x)$.

If $F(x) = x^2 + \cos 2x$, choose

$$y_p = Ax^2 + (Bx + C) + D \cos 2x + (E \sin 2x)$$

Example 1

Find the general solution of the nonhomogeneous equation $y'' - 2y' - 3y = 2 \sin x$.

Solution: First, we find the general solution of the homogeneous equation, using the characteristic equation

$$m^2 - 2m - 3 = 0$$
$$(m + 1)(m - 3) = 0$$
$$m = -1, 3$$

Thus we have

$$y_h = C_1 e^{-x} + C_2 e^{3x}$$

Next we choose y_p to be similar to $F(x) = 2 \sin x$ and obtain

$$y_p = A \cos x + B \sin x \qquad y_p' = -A \sin x + B \cos x$$
$$y_p'' = -A \cos x - B \sin x$$

Now substituting y_p and its derivatives into the nonhomogeneous equation, we have

$$(-A \cos x - B \sin x) - 2(-A \sin x + B \cos x)$$
$$- 3(A \cos x + B \sin x) = 2 \sin x$$
$$(-4A - 2B) \cos x + (2A - 4B) \sin x = 2 \sin x$$

Equating coefficients of $\cos x$ and $\sin x$ on both sides of this equation yields

$$-4A - 2B = 0 \qquad \text{and} \qquad 2A - 4B = 2$$

Solving these two equations simultaneously, we have $A = \frac{1}{5}$ and $B = -\frac{2}{5}$. Therefore,

$$y_p = \tfrac{1}{5} \cos x - \tfrac{2}{5} \sin x$$

and the general solution of the nonhomogeneous equation is

$$y = y_h + y_p = C_1 e^{-x} + C_2 e^{3x} + \tfrac{1}{5} \cos x - \tfrac{2}{5} \sin x \qquad \blacksquare$$

We summarize the method of undetermined coefficients as follows:

Method of Undetermined Coefficients: $a_n y^{(n)} + a_{n-1} y^{(n-1)} + \cdots + a_1 y' + a_0 y = F(x)$

1. Find y_h, the general solution of the homogeneous equation formed by letting $F(x) = 0$.
2. Use $F(x)$ to determine the form of the particular solution y_p. See the accompanying table.

Term in $F(x)$	Corresponding terms in y_p
Cx^k	$A_0 + A_1 x + \cdots + A_{k-1} x^{k-1} + A_k x^k$
Ce^{mx}	Ae^{mx}
$C \cos \beta x$ or $C \sin \beta x$	$A \cos \beta x + B \sin \beta x$

3. If a term of y_p coincides with a term in y_h, multiply the term in y_p by the smallest power of x that removes the duplication.
4. Substitute y_p and its derivatives into the nonhomogeneous equation and solve for the coefficients.
5. Form the general solution: $y = y_h + y_p$.

Example 2

Determine a suitable choice for y_p for the following.

	$F(x)$	y_h
(a)	x^2	$C_1 + C_2 x$
(b)	$4 \sin 3x$	$C_1 e^x \cos 3x + C_2 e^x \sin 3x$
(c)	e^{2x}	$C_1 e^x + C_2 e^{2x} + C_3 x e^{2x}$

Solution:

(a) Since $F(x) = x^2$, the normal choice for y_p would be

$$A + Bx + Cx^2$$

However, since

$$y_h = C_1 + C_2 x$$

already contains a linear term, we multiply by x^2 to obtain

$$y_p = Ax^2 + Bx^3 + Cx^4$$

(b) Since $F(x) = 4 \sin 3x$ and since each term in y_h contains a factor of e^x, we simply let

$$y_p = A \cos 3x + B \sin 3x$$

(c) Since $F(x) = e^{2x}$, the normal choice for y_p would be Ae^{2x}. However, since

$$y_h = C_1 e^x + C_2 e^{2x} + C_3 x e^{2x}$$

already contains an e^{2x} term, we multiply by x^2 to obtain

$$y_p = Ax^2e^{2x}$$

■

Example 3

Find the general solution to $y'' - 2y' = x + 2e^x$.

Solution: The characteristic equation is

$$m^2 - 2m = 0$$
$$m(m - 2) = 0$$
$$m = 0, 2$$

and thus we have

$$y_h = C_1 + C_2e^{2x}$$

Since $F(x) = x + 2e^x$, the normal choice for y_p would be

$$(A + Bx) + Ce^x$$

However, since y_h already contains the constant term C_1, we multiply the *polynomial* $(A + Bx)$ by x to obtain

$$y_p = Ax + Bx^2 + Ce^x$$
$$y_p' = A + 2Bx + Ce^x \qquad y_p'' = 2B + Ce^x$$

Substituting into the differential equation, we have

$$(2B + Ce^x) - 2(A + 2Bx + Ce^x) = x + 2e^x$$
$$(2B - 2A) - 4Bx - Ce^x = x + 2e^x$$

Equating coefficients of like terms yields

$$2B - 2A = 0, \qquad -4B = 1, \qquad -C = 2$$

Solving for A, B, and C gives us $A = B = -\frac{1}{4}$ and $C = -2$. Therefore,

$$y_p = -\tfrac{1}{4}x - \tfrac{1}{4}x^2 - 2e^x$$

and the general solution is

$$y = C_1 + C_2e^{2x} - \tfrac{1}{4}x - \tfrac{1}{4}x^2 - 2e^x$$

■

Example 4

Find the general solution to $y''' + 4y' = 5xe^x$

Solution: The characteristic equation is

$$m^3 + 4m = 0$$
$$m(m^2 + 4) = 0$$
$$m = 0, 2i, -2i$$

and thus we have

$$y_h = C_1 + C_2 \cos 2x + C_3 \sin 2x$$

Since $F(x) = 5xe^x$, we have

$$y_p = (A + Bx)e^x \qquad\qquad y_p'' = (A + Bx)e^x + 2Be^x$$
$$y_p' = (A + Bx)e^x + Be^x \qquad y_p''' = (A + Bx)e^x + 3Be^x$$

Substituting into the differential equation, we have

$$(A + Bx)e^x + 3Be^x + 4(A + Bx)e^x + 4Be^x = 5xe^x$$
$$(5A + 7B)e^x + 5Bxe^x = 5xe^x$$

By equating coefficients we have

$$5A + 7B = 0 \qquad \text{and} \qquad 5B = 5$$

which implies that $A = -\frac{7}{5}$ and $B = 1$. Therefore,

$$y_p = (-\tfrac{7}{5} + x)e^x = -\tfrac{7}{5}e^x + xe^x$$

and the general solution is

$$y = C_1 + C_2 \cos 2x + C_3 \sin 2x - \tfrac{7}{5}e^x + xe^x \qquad\blacksquare$$

We have seen that the undetermined coefficients method for finding a particular solution to a nonhomogeneous linear equation works nicely if $F(x)$ consists of sums or products of x^n, e^{mx}, $\cos \beta x$, and $\sin \beta x$. When $F(x)$ is not of this form, we can attempt to find a particular solution with a more general method called the method of **variation of parameters.** The most common use of the variation of parameters method is for the case where $F(x)$ contains a term whose successive derivatives do not eventually become zero or repetitive. The functions $1/x$ and $\tan x$ are two cases in point. In this method we assume that *the particular solution has the form of y_h* except that the constants in y_h are replaced by variables.

Method of Variation of Parameters: $a_n y^{(n)} + a_{n-1} y^{(n-1)} + \cdots + a_1 y' + a_0 y = F(x)$

1. Find the general solution y_h of the homogeneous equation:

$$y_h = C_1 y_1 + C_2 y_2 + \cdots + C_n y_n$$

2. Replace the constants in y_h by variables to form y_p:

$$y_p = v_1 y_1 + v_2 y_2 + \cdots + v_n y_n$$

3. Solve the following system for $v_1', v_2', \ldots, v_n'$:

$$
\begin{aligned}
v_1' y_1 + v_2' y_2 + \cdots + v_n' y_n &= 0 \\
v_1' y_1' + v_2' y_2' + \cdots + v_n' y_n' &= 0 \\
v_1' y_1'' + v_2' y_2'' + \cdots + v_n' y_n'' &= 0 \\
\vdots \qquad\qquad \vdots \qquad\qquad \vdots \quad &\ \ \vdots \\
v_1' y_1^{(n-1)} + v_2' y_2^{(n-1)} + \cdots + v_n' y_n^{(n-1)} &= F(x)
\end{aligned}
$$

4. Integrate to find $v_1, v_2, \ldots, v_n$.
5. Form the general solution: $y = y_h + y_p$.

In solving the system of equations that arises in the variation of parameters method, we can appeal to a well-known technique from linear algebra called **Cramer's Rule.** For a second-order differential equation, the system

$$v_1'y_1 + v_2'y_2 = 0$$
$$v_1'y_1' + v_2'y_2' = F(x)$$

has the following solution:

$$v_1' = \frac{\begin{vmatrix} 0 & y_2 \\ F(x) & y_2' \end{vmatrix}}{\begin{vmatrix} y_1 & y_2 \\ y_1' & y_2' \end{vmatrix}} = \frac{-F(x)y_2}{w(y_1, y_2)}, \qquad v_2' = \frac{\begin{vmatrix} y_1 & 0 \\ y_1' & F(x) \end{vmatrix}}{\begin{vmatrix} y_1 & y_2 \\ y_1' & y_2' \end{vmatrix}} = \frac{F(x)y_1}{w(y_1, y_2)}$$

Similarly, a third-order differential equation yields the system

$$v_1'y_1 + v_2'y_2 + v_3'y_3 = 0$$
$$v_1'y_1' + v_2'y_2' + v_3'y_3' = 0$$
$$v_1'y_1'' + v_2'y_2'' + v_3'y_3'' = F(x)$$

which by Cramer's Rule has the following solution:

$$v_1' = \frac{\begin{vmatrix} 0 & y_2 & y_3 \\ 0 & y_2' & y_3' \\ F(x) & y_2'' & y_3'' \end{vmatrix}}{\begin{vmatrix} y_1 & y_2 & y_3 \\ y_1' & y_2' & y_3' \\ y_1'' & y_2'' & y_3'' \end{vmatrix}} = \frac{F(x)\begin{vmatrix} y_2 & y_3 \\ y_2' & y_3' \end{vmatrix}}{w(y_1, y_2, y_3)} = F(x)\left[\frac{w(y_2, y_3)}{w(y_1, y_2, y_3)}\right]$$

$$v_2' = -F(x)\left[\frac{w(y_1, y_3)}{w(y_1, y_2, y_3)}\right], \qquad v_3' = F(x)\left[\frac{w(y_1, y_2)}{w(y_1, y_2, y_3)}\right]$$

[Recall that the Wronskian, $w(y_1, y_2, \ldots, y_n)$ was introduced in Section 19.6.]

Example 5

Solve the nonhomogeneous equation

$$y'' - 2y' + y = \frac{e^x}{2x}$$

Solution: The characteristic equation is

$$m^2 - 2m + 1 = 0$$
$$(m - 1)^2 = 0$$
$$m = 1$$

and thus the general solution of the homogeneous equation is

$$y_h = C_1 y_1 + C_2 y_2 = C_1 e^x + C_2 x e^x$$

Replacing C_1 and C_2 by v_1 and v_2, we have

$$y_p = v_1 y_1 + v_2 y_2 = v_1 e^x + v_2 x e^x$$

The resulting system of equations is

$$v_1' e^x + v_2' x e^x = 0$$

$$v_1' e^x + v_2'(x e^x + e^x) = \frac{e^x}{2x}$$

Now since

$$w(y_1, y_2) = \begin{vmatrix} e^x & x e^x \\ e^x & x e^x + e^x \end{vmatrix} = x e^{2x} + e^{2x} - x e^{2x} = e^{2x}$$

we can apply Cramer's Rule to obtain

$$v_1' = \frac{-F(x) y_2}{w(y_1, y_2)} = \frac{-(e^x/2x) x e^x}{e^{2x}} = \frac{-e^{2x}/2}{e^{2x}} = -\frac{1}{2}$$

$$v_2' = \frac{F(x) y_1}{w(y_1, y_2)} = \frac{(e^x/2x) e^x}{e^{2x}} = \frac{e^{2x}/2x}{e^{2x}} = \frac{1}{2x}$$

Therefore, by integration we have

$$v_1 = -\int \frac{1}{2} \, dx = -\frac{x}{2}$$

$$v_2 = \frac{1}{2} \int \frac{1}{x} \, dx = \frac{1}{2} \ln x = \ln \sqrt{x}$$

from which it follows that a particular solution is

$$y_p = -\tfrac{1}{2} x e^x + (\ln \sqrt{x}) x e^x$$

Finally, the general solution is

$$y = C_1 e^x + c_2 x e^x - \tfrac{1}{2} x e^x + x e^x \ln \sqrt{x} \qquad \blacksquare$$

Example 6

Solve the nonhomogeneous equation $y''' + y' = \tan x$.

Solution: The characteristic equation is

$$m^3 + m = 0$$
$$m(m^2 + 1) = 0$$
$$m = 0, \, i, \, -i$$

and thus the general solution of the homogeneous equation is

$$y_h = C_1(1) + C_2 \cos x + C_3 \sin x$$

Replacing C_1, C_2, and C_3 by v_1, v_2, and v_3, we have

$$y_p = v_1(1) + v_2 \cos x + v_3 \sin x$$

and the resulting system of equations is

$$v_1{}'(1) + v_2{}' \cos x + v_3{}' \sin x = 0$$
$$v_1{}'(0) - v_2{}' \sin x + v_3{}' \cos x = 0$$
$$v_1{}'(0) - v_2{}' \cos x - v_3{}' \sin x = \tan x$$

Since

$$w(y_1, y_2, y_3) = \begin{vmatrix} 1 & \cos x & \sin x \\ 0 & -\sin x & \cos x \\ 0 & -\cos x & -\sin x \end{vmatrix} = 1$$

we have

$$v_1{}' = \tan x \, w(y_2, y_3) = \tan x \begin{vmatrix} \cos x & \sin x \\ -\sin x & \cos x \end{vmatrix} = \tan x$$

$$v_2{}' = -\tan x \, w(y_1, y_3) = -\tan x \begin{vmatrix} 1 & \sin x \\ 0 & \cos x \end{vmatrix} = -\sin x$$

$$v_3{}' = \tan x \, w(y_1, y_2) = \tan x \begin{vmatrix} 1 & \cos x \\ 0 & -\sin x \end{vmatrix} = -\frac{\sin^2 x}{\cos x}$$

By integrating we have

$$v_1 = \int \tan x \, dx = -\ln |\cos x|$$

$$v_2 = -\int \sin x \, dx = \cos x$$

$$v_3 = -\int \frac{\sin^2 x}{\cos x} \, dx = \int \frac{\cos^2 x - 1}{\cos x} \, dx$$

$$= \sin x - \ln |\sec x + \tan x|$$

Finally, a particular solution is

$$y_p = -\ln |\cos x| + \cos^2 x + \sin^2 x - \sin x \ln |\sec x + \tan x|$$
$$= \ln |\sec x| + 1 - \sin x \ln |\sec x + \tan x|$$

and the general solution of the nonhomogeneous equation is

$$y = C_1 + C_2 \cos x + C_3 \sin x + \ln |\sec x| - \sin x \ln |\sec x + \tan x| \quad \blacksquare$$

Section Exercises (19.7)

In Exercises 1–15, solve the differential equation by the method of undetermined coefficients.

1. $y'' - 3y' + 2y = 2x$
2. $y'' - 2y' - 3y = x^2 - 1$
⟨S⟩ **3.** $y'' + y = x^3$, $y(0) = 1$, $y'(0) = 0$
4. $y'' + 4y = 4$, $y(0) = 1$, $y'(0) = 6$
5. $y'' + 2y' = 2e^x$
6. $y'' - 9y = 5e^{3x}$
7. $y'' - 10y' + 25y = 5 + 6e^x$
8. $16y'' - 8y' + y = 4(x + e^x)$
9. $y'' + y' = 2 \sin x$, $y(0) = 0$, $y'(0) = -3$
10. $y'' + y' - 2y = 3 \cos 2x$, $y(0) = -1$, $y'(0) = 2$
⟨S⟩ **11.** $y'' + 9y = \sin 3x$
12. $y'' + 4y' + 5y = \sin x + \cos x$
13. $y''' - 3y' + 2y = 2e^{-2x}$
14. $y''' - y'' = 4x^2$, $y(0) = 1$, $y'(0) = 1$, $y''(0) = 1$
15. $y' - 4y = xe^x - xe^{4x}$, $y(0) = \frac{1}{3}$

In Exercises 16–20, solve the differential equation by the method of variation of parameters.

16. $y'' + y = \sec x$
⟨S⟩ **17.** $y'' + 4y = \csc 2x$
18. $y'' - 4y' + 4y = x^2 e^{2x}$
19. $y'' - 2y' + y = e^x \ln x$
20. $y'' - 4y' + 4y = e^{2x}/x$

In Exercises 29 and 30 of Section 19.6, we used a second-order *homogeneous* linear equation to describe the motion of a spring undergoing free, undamped motion. When an ex-

ternal force is applied, we say the resulting motion is forced, not free. In Exercises 21–22, solve the *nonhomogeneous* linear equation arising from the given forced motion.

⟨S⟩ **21.** A 5-kg mass is attached to a spring. The spring constant is 140 N/m and the external force is $F(t) = 5 \sin 2t$. Furthermore, the force due to air resistance is $-80\, dy/dt$ newtons and the initial velocity in the upward direction is 2 m/s. Find the resulting motion using the differential equation

$$5y'' + 80y' + 140y = 5 \sin 2t$$

and the initial conditions $y(0) = 0$ and $y'(0) = 2$.

22. Rework Exercise 21 for the case in which the forced motion is $F(t) = 10 \sin t$.

In Exercises 23–24, use the simple electric circuit equation

$$\frac{d^2q}{dt^2} + \left(\frac{R}{L}\right)\frac{dq}{dt} + \left(\frac{1}{LC}\right)q = \left(\frac{1}{L}\right)E(t)$$

where R is the resistance (in ohms), C is the capacitance (in farads, F), L is the inductance (in henrys), $E(t)$ is the electromotive force (in volts), and q is the charge on the capacitor (in coulombs, C).

23. Find the charge q as a function of time for the electrical circuit in which $R = 20\ \Omega$, $C = .02$ F, $L = 2$ H, and $E(t) = 12 \sin 5t$. Assume that $q(0) = 0$ and $q'(0) = 0$.

24. Find the charge q as a function of time for the electrical circuit in which $R = 15\ \Omega$, $C = .02$ F, $L = 1$ H, and $E(t) = 10 \sin 5t$. Assume that $q(0) = 0$ and $q'(0) = 0$.

19.8 Series Solutions

Purpose
- To introduce the power series method of finding the general solution of a differential equation.
- To introduce the Taylor polynomial method of finding a particular solution of a differential equation satisfying an initial condition.

The mathematician's patterns, like the painter's or the poet's, must be beautiful; the ideas, like the colors or the words, must fit together in a harmonious way. Beauty is the first test: there is no permanent place in the world for ugly mathematics.   —Godfrey Hardy (1877–1947)

In all the methods (for solving differential equations) considered so far, we have been looking for explicit solutions such as $y = x + C$ or $y = C_1 e^x + C_2 e^{-x}$. For example, in Sections 19.1–19.5 we discussed solution methods that apply to a wide variety of *first-order* differential equations.

Then in Sections 19.6–19.7 we looked at methods of solving *higher-order linear equations with constant coefficients.* We conclude Chapter 19 by looking at methods for approximating the solution of a differential equation. These approximation methods work especially well with *linear equations whose coefficients are either constant or powers of x.*

For the sake of brevity, we will limit our discussion to the statement and demonstration of these approximation methods and exclude any theoretical development.

We begin with the general **power series solution** method. Recall from Chapter 18 that a power series represents a function f on an interval of convergence and that we can successively differentiate the power series to obtain series for f', f'', and so on. For instance, if

$$f(x) = a_0 + a_1 x + a_2 x^2 + a_3 x^3 + \cdots = \sum_{n=0}^{\infty} a_n x^n$$

then

$$f'(x) = a_1 + 2a_2 x + 3a_3 x^2 + \cdots = \sum_{n=0}^{\infty} na_n x^{n-1}$$

$$f''(x) = 2a_2 + 6a_3 x + \cdots = \sum_{n=0}^{\infty} n(n-1)a_n x^{n-2}$$

We use these properties in the *power series* method of solution demonstrated in the first two examples. In our first example we use a simple differential equation that can be solved by other methods.

Example 1

Use a power series to find the general solution to the differential equation $y' - 2y = 0$.

Solution:

1. Assume $y = \sum_{n=0}^{\infty} a_n x^n$ is a solution; then $y' = \sum_{n=0}^{\infty} na_n x^{n-1}$. Substituting for y' and $-2y$, we obtain the following series form for the given differential equation:

$$y' - 2y = \sum_{n=0}^{\infty} na_n x^{n-1} - 2 \sum_{n=0}^{\infty} a_n x^n = 0$$

$$\sum_{n=0}^{\infty} na_n x^{n-1} = 2 \sum_{n=0}^{\infty} a_n x^n$$

2. Next we adjust the summation indices so that x^n appears in each series. In this case we can accomplish this by replacing n by $n + 1$ in

the left-hand series to obtain

$$\sum_{n=-1}^{\infty}(n+1)a_{n+1}x^n = 2\sum_{n=0}^{\infty}a_n x^n$$

3. Now by equating coefficients of like terms, we obtain the **recursion formula:**

$$(n+1)a_{n+1} = 2a_n$$

$$a_{n+1} = \frac{2a_n}{n+1} \qquad n \geq 0$$

which generates the following results in terms of a_0:

$$a_1 = 2a_0$$

$$a_2 = \frac{2a_1}{2} = \frac{2^2 a_0}{2}$$

$$a_3 = \frac{2a_2}{3} = \frac{2^3 a_0}{2\cdot 3} = \frac{2^3 a_0}{3!}$$

$$a_4 = \frac{2a_3}{4} = \frac{2^4 a_0}{2\cdot 3\cdot 4} = \frac{2^4 a_0}{4!}$$

$$\vdots$$

$$a_n = \frac{2^n a_0}{n!}$$

4. Using these values as the coefficients for the solution series, we have

$$y = \sum_{n=0}^{\infty}\frac{2^n a_0}{n!}x^n = a_0 \sum_{n=0}^{\infty}\frac{2^n}{n!}x^n = a_0 e^{2x}$$

(See Section 18.8 for the series representation of e^{2x}.) ■

The next example cannot be solved by any of the methods discussed in Sections 19.1–19.7.

Example 2

Use a power series to solve the differential equation $y'' + xy' + y = 0$.

Solution:

1. Assume $\sum_{n=0}^{\infty}a_n x^n$ is a solution; then

$$y' = \sum_{n=0}^{\infty}na_n x^{n-1}, \qquad xy' = \sum_{n=0}^{\infty}na_n x^n, \qquad y'' = \sum_{n=0}^{\infty}n(n-1)a_n x^{n-2}$$

Substituting for y'', xy', and y, we obtain the following series form for the given differential equation:

$$\sum_{n=0}^{\infty} n(n-1)a_n x^{n-2} + \sum_{n=0}^{\infty} na_n x^n + \sum_{n=0}^{\infty} a_n x^n = 0$$

or, equivalently,

$$\sum_{n=0}^{\infty} n(n-1)a_n x^{n-2} = -\sum_{n=0}^{\infty} na_n x^n - \sum_{n=0}^{\infty} a_n x^n = -\sum_{n=0}^{\infty} (n+1)a_n x^n$$

2. Next, to obtain equal powers on x, we adjust the summation indices by replacing n by $n + 2$ in the left-hand sum to obtain

$$\sum_{n=-2}^{\infty} (n+2)(n+1)a_{n+2} x^n = -\sum_{n=0}^{\infty} (n+1)a_n x^n$$

3. Now by equating coefficients we have

$$(n+2)(n+1)a_{n+2} = -(n+1)a_n$$

from which we get the recursion formula

$$a_{n+2} = \frac{-(n+1)}{(n+2)(n+1)} a_n = -\frac{a_n}{n+2} \qquad n \geq 0$$

Thus the coefficients of the solution series are

$$a_2 = -\frac{a_0}{2} \qquad\qquad\qquad a_3 = -\frac{a_1}{3}$$

$$a_4 = -\frac{a_2}{4} = \frac{a_0}{2 \cdot 4} \qquad\qquad a_5 = -\frac{a_3}{5} = \frac{a_1}{3 \cdot 5}$$

$$a_6 = -\frac{a_4}{6} = -\frac{a_0}{2 \cdot 4 \cdot 6} \qquad a_7 = -\frac{a_5}{7} = \frac{-a_1}{3 \cdot 5 \cdot 7}$$

$$\vdots \qquad\qquad\qquad\qquad \vdots$$

$$a_{2n} = \frac{(-1)^n a_0}{2 \cdot 4 \cdot 6 \cdots (2n)} = \frac{(-1)^n a_0}{2^n(n!)} \qquad a_{2n+1} = \frac{(-1)^n a_1}{3 \cdot 5 \cdot 7 \cdots (2n+1)}$$

4. In this case we represent the general solution as the sum of two series, one for the even-powered terms with coefficients in terms of a_0 and one for the odd-powered terms with coefficients in terms of a_1.

$$y = a_0 \left(1 - \frac{x^2}{2} + \frac{x^4}{2 \cdot 4} - \cdots\right) + a_1 \left(x - \frac{x^3}{3} + \frac{x^5}{3 \cdot 5} - \cdots\right)$$

$$= a_0 \sum_{n=0}^{\infty} \frac{(-1)^n x^{2n}}{2^n(n!)} + a_1 \sum_{n=0}^{\infty} \frac{(-1)^n x^{2n+1}}{3 \cdot 5 \cdot 7 \cdots (2n+1)}$$

Note that the solution has two arbitrary constants a_0 and a_1, as we would expect in the general solution of a second-order differential equation.

■

A second type of series solution method involves a differential equation with initial conditions and makes use of Taylor's Theorem (see Section 18.8).

Example 3 (*Approximation by Taylor's Theorem*)

Use Taylor's Theorem to find the series solution of

$$y' = y^2 - x \qquad y = 1 \quad \text{when} \quad x = 0$$

Then use the first six terms of this series solution to approximate values of y for $0 \le x \le 1$.

Solution: Recall from Taylor's Theorem that for $c = 0$

$$y = y(0) + y'(0)x + \frac{y''(0)}{2!}x^2 + \frac{y'''(0)}{3!}x^3 + \cdots$$

Since $y(0) = 1$ and $y' = y^2 - x$, it follows that

$$y(0) = 1$$
$$y' = y^2 - x \qquad\qquad y'(0) = 1$$
$$y'' = 2yy' - 1 \qquad\qquad y''(0) = 2 - 1 = 1$$
$$y''' = 2yy'' + 2(y')^2 \qquad\qquad y'''(0) = 2 + 2 = 4$$
$$y^{(4)} = 2yy''' + 6y'y'' \qquad\qquad y^{(4)}(0) = 8 + 6 = 14$$
$$y^{(5)} = 2yy^{(4)} + 8y'y''' + 6(y'')^2 \qquad y^{(5)}(0) = 28 + 32 + 6 = 66$$

Therefore, we can approximate the values of the solution from the series

$$y = y(0) + y'(0)x + \frac{y''(0)}{2!}x^2 + \frac{y'''(0)}{3!}x^3 + \frac{y^{(4)}(0)}{4!}x^4 + \frac{y^{(5)}(0)}{5!}x^5 + \cdots$$

$$= 1 + x + \frac{1}{2}x^2 + \frac{4}{3!}x^3 + \frac{14}{4!}x^4 + \frac{66}{5!}x^5 + \cdots$$

Using the first six terms of this series, we compute several values for y in the interval $0 \le x \le 1$, as shown in Table 19.1.

TABLE 19.1

x	0.0	0.1	0.2	0.3	0.4	0.5	0.6	0.7	0.8	0.9	1.0
y	1.0000	1.1057	1.2264	1.3691	1.5432	1.7620	2.0424	2.4062	2.8805	3.4985	4.3000

■

TABLE 19.2

x	0.0	0.1	0.2	0.3	0.4	0.5	0.6	0.7	0.8	0.9	1.0
y	1.0000	1.1057	1.2265	1.3696	1.5463	1.7750	2.0855	2.5261	3.1746	4.1529	5.6475

From our experience with Taylor polynomials (see Section 18.8), we know that, for a fixed number of terms, the further we move from the center of expansion (in this case $c = 0$), the less accurate our estimate will be. This observation is supported by comparing the y-values in Table 19.1 with those in Table 19.2, for which we used the first *nine* terms in the series rather than the first *six* terms. Specifically, by continuing the procedure of Example 3, we can compute the first nine terms of the solution to be

$$y = 1 + x + \frac{1}{2}x^2 + \frac{4}{3!}x^3 + \frac{14}{4!}x^4 + \frac{66}{5!}x^5 + \frac{352}{6!}x^6 + \frac{2{,}636}{7!}x^7 + \frac{13{,}532}{8!}x^8 + \cdots$$

Section Exercises (19.8)

In Exercises 1–12, use power series to solve the differential equation.

1. $y' - y = 0$

2. $y' - ky = 0$

[S] **3.** $y'' - 9y = 0$

4. $y'' - k^2y = 0$

5. $y'' + 4y = 0$

6. $y'' + k^2y = 0$

7. $xy' - 2y = 0$

8. $y' - 2xy = 0$

[S] **9.** $xy'' - y = 0$

10. $y'' - xy' - y = 0$

11. $(x^2 + 4)y'' + y = 0$

12. $x^2y'' + xy' - y = 0$

13. Use Taylor's Theorem to find the series solution of $y' + (2x - 1)y = 0$ given that $y = 2$ when $x = 0$. Use the first five terms of the series to approximate y when $x = \frac{1}{2}$.

14. Use Taylor's Theorem to find the series solution of $y' - 2xy = 0$ given that $y = 1$ when $x = 0$. Use the first four terms of the series to approximate y when $x = 1$.

[S] **15.** Use Taylor's Theorem to find the series solution of $y'' - 2xy = 0$ given that $y = 1$ when $x = 0$ and $y' = -3$ when $x = 0$. Use the first six terms of the series to approximate y when $x = \frac{1}{4}$.

16. Use Taylor's Theorem to find the series solution of $y'' - 2xy' + y = 0$ given that $y = 1$ when $x = 0$ and $y' = 2$ when $x = 0$. Use the first eight terms of the series to approximate y when $x = \frac{1}{2}$.

Miscellaneous Exercises (Ch. 19)

In Exercises 1–30, find the general solution of the given first-order differential equation.

1. $\dfrac{dy}{dx} - \dfrac{y}{x} = 2 + \sqrt{x}$

2. $y' + xy = 2y$

3. $\dfrac{dy}{dx} - \dfrac{2y}{x} = \dfrac{dy}{x\,dx}$

4. $\dfrac{dy}{dx} - 3x^2y = e^{x^3}$

[S] **5.** $\dfrac{dy}{dx} - \dfrac{y}{x} = \dfrac{x}{y}$

6. $y' - \dfrac{3y}{x^2} = \dfrac{1}{x^2}$

7. $(10x + 8y + 2)\,dx + (8x + 5y + 2)\,dy = 0$

8. $y' + \dfrac{2y}{x} = -x^9 y^5$

9. $y' - 2y = e^x$

10. $(y + x^3 + xy^2)\,dx - x\,dy = 0$

S **11.** $(2x - 2y^3 + y)\,dx + (x - 6xy^2)\,dy = 0$

12. $3x^2 y^2\,dx + (2x^3 y + x^3 y^4)\,dy = 0$

13. $x\,dy = (x + y + 2)\,dx$

14. $ye^{xy}\,dx + xe^{xy}\,dy = 0$

15. $\ln(1 + y)\,dx + \left(\dfrac{1}{1 + y}\right)dy = 0$

16. $(2x + y - 3)\,dx + (x - 3y + 1)\,dy = 0$

17. $dy = (y \tan x + 2e^x)\,dx$

18. $y\,dx - (x + \sqrt{xy})\,dy = 0$

19. $(x - y - 5)\,dx - (x + 3y - 2)\,dy = 0$

20. $y' = x^2 y^2 - 9x^2$

21. $2xy' - y = x^3 - x$

22. $\dfrac{dy}{dx} = 2x\sqrt{1 - y^2}$

23. $x + yy' = \sqrt{x^2 + y^2}$

24. $xy' + y = \sin x$

25. $yy' + y^2 = 1 + x^2$

26. $2x\,dx + 2y\,dy = (x^2 + y^2)\,dx$

S **27.** $(1 + x^2)\,dy = (1 + y^2)\,dx$

28. $y' = \dfrac{x^4 + 3x^2 y^2 + y^4}{x^3 y}$

29. $y' - \dfrac{ay}{x} = bx^3$

30. $y' = y + 2x(y - e^x)$

In Exercises 31–36, find the general solution of the given second-order differential equation.

31. $y'' + y = x^3 + x$

32. $y'' + 2y = e^{2x} + x$

33. $y'' + y = 2\cos x$

34. $y'' + 5y' + 4y = x^2 + \sin 2x$

S **35.** $y'' - 2y' + y = 2xe^x$

36. $y'' + 2y' + y = \dfrac{1}{e^x x^2}$

37. Find the orthogonal trajectories of the family of circles $(x - C)^2 + y^2 = C^2$.

38. Find the orthogonal trajectories of the family of lines $y - 2x = C$.

39. Use power series to solve the differential equation $(x - 4)y' + y = 0$.

40. Use power series to find the first six terms of the series solution to $x^2 y'' + y' - 3y = 0$.

TABLE OF SQUARE ROOTS AND CUBE ROOTS

n	$\sqrt{n}$	$\sqrt[3]{n}$	n	$\sqrt{n}$	$\sqrt[3]{n}$	n	$\sqrt{n}$	$\sqrt[3]{n}$
1	1.00000	1.00000	31	5.56776	3.14138	61	7.81025	3.93650
2	1.41421	1.25992	32	5.65685	3.17480	62	7.87401	3.95789
3	1.73205	1.44225	33	5.74456	3.20753	63	7.93725	3.97906
4	2.00000	1.58740	34	5.83095	3.23961	64	8.00000	4.00000
5	2.23607	1.70998	35	5.91608	3.27107	65	8.06226	4.02073
6	2.44949	1.81712	36	6.00000	3.30193	66	8.12404	4.04124
7	2.64575	1.91293	37	6.08276	3.33222	67	8.18535	4.06155
8	2.82843	2.00000	38	6.16441	3.36198	68	8.24621	4.08166
9	3.00000	2.08008	39	6.24500	3.39121	69	8.30662	4.10157
10	3.16228	2.15443	40	6.32456	3.41995	70	8.36660	4.12129
11	3.31662	2.22398	41	6.40312	3.44822	71	8.42615	4.14082
12	3.46410	2.28943	42	6.48074	3.47603	72	8.48528	4.16017
13	3.60555	2.35133	43	6.55744	3.50340	73	8.54400	4.17934
14	3.74166	2.41014	44	6.63325	3.53035	74	8.60233	4.19834
15	3.87298	2.46621	45	6.70820	3.55689	75	8.66025	4.21716
16	4.00000	2.51984	46	6.78233	3.58305	76	8.71780	4.23582
17	4.12311	2.57128	47	6.85565	3.60883	77	8.77496	4.25432
18	4.24264	2.62074	48	6.92820	3.63424	78	8.83176	4.27266
19	4.35890	2.66840	49	7.00000	3.65931	79	8.88819	4.29084
20	4.47214	2.71442	50	7.07107	3.68403	80	8.94427	4.30887
21	4.58258	2.75892	51	7.14143	3.70843	81	9.00000	4.32675
22	4.69042	2.80204	52	7.21110	3.73251	82	9.05539	4.34448
23	4.79583	2.84387	53	7.28011	3.75629	83	9.11043	4.36207
24	4.89898	2.88450	54	7.34847	3.77976	84	9.16515	4.37952
25	5.00000	2.92402	55	7.41620	3.80295	85	9.21954	4.39683
26	5.09902	2.96250	56	7.48331	3.82586	86	9.27362	4.41400
27	5.19615	3.00000	57	7.54983	3.84850	87	9.32738	4.43105
28	5.29150	3.03659	58	7.61577	3.87088	88	9.38083	4.44796
29	5.38516	3.07232	59	7.68115	3.89300	89	9.43398	4.46475
30	5.47723	3.10723	60	7.74597	3.91487	90	9.48683	4.48140

TABLE OF SQUARE ROOTS AND CUBE ROOTS (Continued)

n	$\sqrt{n}$	$\sqrt[3]{n}$	n	$\sqrt{n}$	$\sqrt[3]{n}$	n	$\sqrt{n}$	$\sqrt[3]{n}$
91	9.53939	4.49794	128	11.3137	5.03968	165	12.8452	5.48481
92	9.59166	4.51436	129	11.3578	5.05277	166	12.8841	5.49586
93	9.64365	4.53065	130	11.4018	5.06580	167	12.9228	5.50688
94	9.69536	4.54684	131	11.4455	5.07875	168	12.9615	5.51785
95	9.74679	4.56290	132	11.4891	5.09164	169	13.0000	5.52877
96	9.79796	4.57886	133	11.5326	5.10447	170	13.0384	5.53966
97	9.84886	4.59470	134	11.5758	5.11723	171	13.0767	5.55050
98	9.89949	4.61044	135	11.6190	5.12993	172	13.1149	5.56130
99	9.94987	4.62606	136	11.6619	5.14256	173	13.1529	5.57205
100	10.0000	4.64159	137	11.7047	5.15514	174	13.1909	5.58277
101	10.0499	4.65701	138	11.7473	5.16765	175	13.2288	5.59344
102	10.0995	4.67233	139	11.7898	5.18010	176	13.2665	5.60408
103	10.1489	4.68755	140	11.8322	5.19249	177	13.3041	5.61467
104	10.1980	4.70267	141	11.8743	5.20483	178	13.3417	5.62523
105	10.2470	4.71769	142	11.9164	5.21710	179	13.3791	5.63574
106	10.2956	4.73262	143	11.9583	5.22932	180	13.4164	5.64622
107	10.3441	4.74746	144	12.0000	5.24148	181	13.4536	5.65665
108	10.3923	4.76220	145	12.0416	5.25359	182	13.4907	5.66705
109	10.4403	4.77686	146	12.0830	5.26564	183	13.5277	5.67741
110	10.4881	4.79142	147	12.1244	5.27763	184	13.5647	5.68773
111	10.5357	4.80590	148	12.1655	5.28957	185	13.6015	5.69802
112	10.5830	4.82028	149	12.2066	5.30146	186	13.6382	5.70827
113	10.6301	4.83459	150	12.2474	5.31329	187	13.6748	5.71848
114	10.6771	4.84881	151	12.2882	5.32507	188	13.7113	5.72865
115	10.7238	4.86294	152	12.3288	5.33680	189	13.7477	5.73879
116	10.7703	4.87700	153	12.3693	5.34848	190	13.7840	5.74890
117	10.8167	4.89097	154	12.4097	5.36011	191	13.8203	5.75897
118	10.8628	4.90487	155	12.4499	5.37169	192	13.8564	5.76900
119	10.9087	4.91868	156	12.4900	5.38321	193	13.8924	5.77900
120	10.9545	4.93242	157	12.5300	5.39469	194	13.9284	5.78896
121	11.0000	4.94609	158	12.5698	5.40612	195	13.9642	5.79889
122	11.0454	4.95968	159	12.6095	5.41750	196	14.0000	5.80879
123	11.0905	4.97319	160	12.6491	5.42884	197	14.0357	5.81865
124	11.1355	4.98663	161	12.6886	5.44012	198	14.0712	5.82848
125	11.1803	5.00000	162	12.7279	5.45136	199	14.1067	5.83827
126	11.2250	5.01330	163	12.7671	5.46256	200	14.1421	5.84804
127	11.2694	5.02653	164	12.8062	5.47370			

Appendix B

TABLE OF EXPONENTIAL AND LOGARITHMIC FUNCTIONS

x	e^x	e^{-x}	$\ln x$	x	e^x	e^{-x}	$\ln x$
0.0	1.0000	1.0000	——	2.5	12.182	0.0821	0.9163
0.1	1.1052	0.9048	−2.3026	2.6	13.464	0.0743	0.9555
0.2	1.2214	0.8187	−1.6094	2.7	14.880	0.0672	0.9933
0.3	1.3499	0.7408	−1.2040	2.8	16.445	0.0608	1.0296
0.4	1.4918	0.6703	−0.9163	2.9	18.174	0.0550	1.0647
0.5	1.6487	0.6065	−0.6932	3.0	20.086	0.0498	1.0986
0.6	1.8221	0.5488	−0.5108	3.1	22.198	0.0450	1.1314
0.7	2.0138	0.4966	−0.3567	3.2	24.533	0.0408	1.1632
0.8	2.2255	0.4493	−0.2231	3.3	27.113	0.0369	1.1939
0.9	2.4596	0.4066	−0.1054	3.4	29.964	0.0334	1.2238
1.0	2.7183	0.3679	0.0000	3.5	33.115	0.0302	1.2528
1.1	3.0042	0.3329	0.0953	3.6	36.598	0.0273	1.2809
1.2	3.3201	0.3012	0.1823	3.7	40.447	0.0247	1.3083
1.3	3.6693	0.2725	0.2624	3.8	44.701	0.0224	1.3350
1.4	4.0552	0.2466	0.3365	3.9	49.402	0.0202	1.3610
1.5	4.4817	0.2231	0.4055	4.0	54.598	0.0183	1.3863
1.6	4.9530	0.2019	0.4700	4.1	60.340	0.0166	1.4110
1.7	5.4739	0.1827	0.5306	4.2	66.686	0.0150	1.4351
1.8	6.0496	0.1653	0.5878	4.3	73.700	0.0136	1.4586
1.9	6.6859	0.1496	0.6419	4.4	81.451	0.0123	1.4816
2.0	7.3891	0.1353	0.6931	4.5	90.017	0.0111	1.5041
2.1	8.1662	0.1225	0.7419	4.6	99.484	0.0101	1.5261
2.2	9.0250	0.1108	0.7885	4.7	109.95	0.0091	1.5476
2.3	9.9742	0.1003	0.8329	4.8	121.51	0.0082	1.5686
2.4	11.023	0.0907	0.8755	4.9	134.29	0.0074	1.5892

TABLE OF EXPONENTIAL AND LOGARITHMIC FUNCTIONS (Continued)

x	e^x	e^{-x}	$\ln x$	x	e^x	e^{-x}	$\ln x$
5.0	148.41	0.0067	1.6094	7.5	1808.04	0.0006	2.0149
5.1	164.02	0.0061	1.6292	7.6	1998.20	0.0005	2.0282
5.2	181.27	0.0055	1.6487	7.7	2208.35	0.0005	2.0412
5.3	200.34	0.0050	1.6677	7.8	2440.60	0.0004	2.0541
5.4	221.41	0.0045	1.6864	7.9	2697.28	0.0004	2.0669
5.5	244.69	0.0041	1.7048	8.0	2980.96	0.0003	2.0794
5.6	270.43	0.0037	1.7228	8.1	3294.47	0.0003	2.0919
5.7	298.87	0.0033	1.7405	8.2	3640.95	0.0003	2.1041
5.8	330.30	0.0030	1.7579	8.3	4023.87	0.0002	2.1163
5.9	365.04	0.0027	1.7750	8.4	4447.07	0.0002	2.1282
6.0	403.43	0.0025	1.7918	8.5	4914.77	0.0002	2.1401
6.1	445.86	0.0022	1.8083	8.6	5431.66	0.0002	2.1518
6.2	492.75	0.0020	1.8246	8.7	6002.91	0.0002	2.1633
6.3	544.57	0.0018	1.8406	8.8	6634.24	0.0002	2.1748
6.4	601.85	0.0017	1.8563	8.9	7331.97	0.0001	2.1861
6.5	665.14	0.0015	1.8718	9.0	8103.08	0.0001	2.1972
6.6	735.10	0.0014	1.8871	9.1	8955.29	0.0001	2.2083
6.7	812.41	0.0012	1.9021	9.2	9897.13	0.0001	2.2192
6.8	897.85	0.0011	1.9169	9.3	10938.02	0.0001	2.2300
6.9	992.27	0.0010	1.9315	9.4	12088.38	0.0001	2.2407
7.0	1096.63	0.0009	1.9459	9.5	13359.73	0.0001	2.2513
7.1	1211.97	0.0008	1.9601	9.6	14764.78	0.0001	2.2618
7.2	1339.43	0.0007	1.9741	9.7	16317.61	0.0001	2.2721
7.3	1480.30	0.0007	1.9879	9.8	18033.74	0.0001	2.2824
7.4	1635.98	0.0006	2.0015	9.9	19930.37	0.0001	2.2925
				10.0	22026.47	0.0000	2.3026

Appendix C

TRIGONOMETRIC TABLES

1 degree ≈ 0.01745 radians
1 radian ≈ 57.29578 degrees

For $0 \leqslant \theta \leqslant 45$, read from upper left.
For $45 \leqslant \theta \leqslant 90$, read from lower right.
For $90 \leqslant \theta \leqslant 360$, use the identities:

θ	Quadrant II	Quadrant III	Quadrant IV
$\sin \theta$	$\sin(180-\theta)$	$-\sin(\theta-180)$	$-\sin(360-\theta)$
$\cos \theta$	$-\cos(180-\theta)$	$-\cos(\theta-180)$	$\cos(360-\theta)$
$\tan \theta$	$-\tan(180-\theta)$	$\tan(\theta-180)$	$-\tan(360-\theta)$
$\cot \theta$	$-\cot(180-\theta)$	$\cot(\theta-180)$	$-\cot(360-\theta)$

Degrees	Radians	sin	cos	tan	cot		
0° 00′	.0000	.0000	1.0000	.0000	—	1.5708	90° 00′
10	.0029	.0029	1.0000	.0029	343.774	1.5679	50
20	.0058	.0058	1.0000	.0058	171.885	1.5650	40
30	.0087	.0087	1.0000	.0087	114.589	1.5621	30
40	.0116	.0116	.9999	.0116	85.940	1.5592	20
50	.0145	.0145	.9999	.0145	68.750	1.5563	10
1° 00′	.0175	.0175	.9998	.0175	57.290	1.5533	89° 00′
10	.0204	.0204	.9998	.0204	49.104	1.5504	50
20	.0233	.0233	.9997	.0233	42.964	1.5475	40
30	.0262	.0262	.9997	.0262	38.188	1.5446	30
40	.0291	.0291	.9996	.0291	34.368	1.5417	20
50	.0320	.0320	.9995	.0320	31.242	1.5388	10
2° 00′	.0349	.0349	.9994	.0349	28.636	1.5359	88° 00′
10	.0378	.0378	.9993	.0378	26.432	1.5330	50
20	.0407	.0407	.9992	.0407	24.542	1.5301	40
30	.0436	.0436	.9990	.0437	22.904	1.5272	30
40	.0465	.0465	.9989	.0466	21.470	1.5243	20
50	.0495	.0494	.9988	.0495	20.206	1.5213	10
3° 00′	.0524	.0523	.9986	.0524	19.081	1.5184	87° 00′
10	.0553	.0552	.9985	.0553	18.075	1.5155	50
20	.0582	.0581	.9983	.0582	17.169	1.5126	40
30	.0611	.0610	.9981	.0612	16.350	1.5097	30
40	.0640	.0640	.9980	.0641	15.605	1.5068	20
50	.0669	.0669	.9978	.0670	14.924	1.5039	10
		cos	sin	cot	tan	Radians	Degrees

Degrees	Radians	sin	cos	tan	cot		
4° 00′	.0698	.0698	.9976	.0699	14.301	1.5010	86° 00′
10	.0727	.0727	.9974	.0729	13.727	1.4981	50
20	.0756	.0756	.9971	.0758	13.197	1.4952	40
30	.0785	.0785	.9969	.0787	12.706	1.4923	30
40	.0814	.0814	.9967	.0816	12.251	1.4893	20
50	.0844	.0843	.9964	.0846	11.826	1.4864	10
5° 00′	.0873	.0872	.9962	.0875	11.430	1.4835	85° 00′
10	.0902	.0901	.9959	.0904	11.059	1.4806	50
20	.0931	.0929	.9957	.0934	10.712	1.4777	40
30	.0960	.0958	.9954	.0963	10.385	1.4748	30
40	.0989	.0987	.9951	.0992	10.078	1.4719	20
50	.1018	.1016	.9948	.1022	9.788	1.4690	10
6° 00′	.1047	.1045	.9945	.1051	9.514	1.4661	84° 00′
10	.1076	.1074	.9942	.1080	9.255	1.4632	50
20	.1105	.1103	.9939	.1110	9.010	1.4603	40
30	.1134	.1132	.9936	.1139	8.777	1.4573	30
40	.1164	.1161	.9932	.1169	8.556	1.4544	20
50	.1193	.1190	.9929	.1198	8.345	1.4515	10
7° 00′	.1222	.1219	.9925	.1228	8.144	1.4486	83° 00′
10	.1251	.1248	.9922	.1257	7.953	1.4457	50
20	.1280	.1276	.9918	.1287	7.770	1.4428	40
30	.1309	.1305	.9914	.1317	7.596	1.4399	30
40	.1338	.1334	.9911	.1346	7.429	1.4370	20
50	.1367	.1363	.9907	.1376	7.269	1.4341	10
		cos	sin	cot	tan	Radians	Degrees

TRIGONOMETRIC TABLES (Continued)

Degrees	Radians	sin	cos	tan	cot	Radians	Degrees
8°00'	.1396	.1392	.9903	.1405	7.115	1.4312	82°00'
10	.1425	.1421	.9899	.1435	6.968	1.4283	50
20	.1454	.1449	.9894	.1465	6.827	1.4254	40
30	.1484	.1478	.9890	.1495	6.691	1.4224	30
40	.1513	.1507	.9886	.1524	6.561	1.4195	20
50	.1542	.1536	.9881	.1554	6.435	1.4166	10
9°00'	.1571	.1564	.9877	.1584	6.314	1.4137	81°00'
10	.1600	.1593	.9872	.1614	6.197	1.4108	50
20	.1629	.1622	.9868	.1644	6.084	1.4079	40
30	.1658	.1650	.9863	.1673	5.976	1.4050	30
40	.1687	.1679	.9858	.1703	5.871	1.4021	20
50	.1716	.1708	.9853	.1733	5.769	1.3992	10
10°00'	.1745	.1736	.9848	.1763	5.671	1.3963	80°00'
10	.1774	.1765	.9843	.1793	5.576	1.3934	50
20	.1804	.1794	.9838	.1823	5.485	1.3904	40
30	.1833	.1822	.9833	.1853	5.396	1.3875	30
40	.1862	.1851	.9827	.1883	5.309	1.3846	20
50	.1891	.1880	.9822	.1914	5.226	1.3817	10
11°00'	.1920	.1908	.9816	.1944	5.145	1.3788	79°00'
10	.1949	.1937	.9811	.1974	5.066	1.3759	50
20	.1978	.1965	.9805	.2004	4.989	1.3730	40
30	.2007	.1994	.9799	.2035	4.915	1.3701	30
40	.2036	.2022	.9793	.2065	4.843	1.3672	20
50	.2065	.2051	.9787	.2095	4.773	1.3643	10
12°00'	.2094	.2079	.9781	.2126	4.705	1.3614	78°00'
10	.2123	.2108	.9775	.2156	4.638	1.3584	50
20	.2153	.2136	.9769	.2186	4.574	1.3555	40
30	.2182	.2164	.9763	.2217	4.511	1.3526	30
40	.2211	.2193	.9757	.2247	4.449	1.3497	20
50	.2240	.2221	.9750	.2278	4.390	1.3468	10
13°00'	.2269	.2250	.9744	.2309	4.331	1.3439	77°00'
10	.2298	.2278	.9737	.2339	4.275	1.3410	50
20	.2327	.2306	.9730	.2370	4.219	1.3381	40
30	.2356	.2334	.9724	.2401	4.165	1.3352	30
40	.2385	.2363	.9717	.2432	4.113	1.3323	20
50	.2414	.2391	.9710	.2462	4.061	1.3294	10
14°00'	.2443	.2419	.9703	.2493	4.011	1.3265	76°00'
10	.2473	.2447	.9696	.2524	3.962	1.3235	50
20	.2502	.2476	.9689	.2555	3.914	1.3206	40
30	.2531	.2504	.9681	.2586	3.867	1.3177	30
40	.2560	.2532	.9674	.2617	3.821	1.3148	20
50	.2589	.2560	.9667	.2648	3.776	1.3119	10
15°00'	.2618	.2588	.9659	.2679	3.732	1.3090	75°00'
10	.2647	.2616	9652	.2711	3.689	1.3061	50
20	.2676	.2644	.9644	.2742	3.647	1.3032	40
30	.2705	.2672	.9636	.2773	3.606	1.3003	30
40	.2734	.2700	.9628	.2805	3.566	1.2974	20
50	.2763	.2728	.9621	.2836	3.526	1.2945	10
16°00'	.2793	.2756	.9613	.2867	3.487	1.2915	74°00'
10	.2822	.2784	.9605	.2899	3.450	1.2886	50
20	.2851	.2812	.9596	.2931	3.412	1.2857	40
30	.2880	.2840	.9588	.2962	3.376	1.2828	30
40	.2909	.2868	.9580	.2994	3.340	1.2799	20
50	.2938	.2896	.9572	.3026	3.305	1.2770	10
17°00'	.2967	.2924	.9563	.3057	3.271	1.2741	73°00'
10	.2996	.2952	.9555	.3089	3.237	1.2712	50
20	.3025	.2979	.9546	.3121	3.204	1.2683	40
30	.3054	.3007	.9537	.3153	3.172	1.2654	30
40	.3083	.3035	.9528	.3185	3.140	1.2625	20
50	.3113	.3062	.9520	.3217	3.108	1.2595	10
		cos	sin	cot	tan	Radians	Degrees

Degrees	Radians	sin	cos	tan	cot	Radians	Degrees
18°00'	.3142	.3090	.9511	.3249	3.078	1.2566	72°00'
10	.3171	.3118	.9502	.3281	3.047	1.2537	50
20	.3200	.3145	.9492	.3314	3.018	1.2508	40
30	.3229	.3173	.9483	.3346	2.989	1.2479	30
40	.3258	.3201	.9474	.3378	2.960	1.2450	20
50	.3287	.3228	.9465	.3411	2.932	1.2421	10
19°00'	.3316	.3256	.9455	.3443	2.904	1.2392	71°00'
10	.3345	.3283	.9446	.3476	2.877	1.2363	50
20	.3374	.3311	.9436	.3508	2.850	1.2334	40
30	.3403	.3338	.9426	.3541	2.824	1.2305	30
40	.3432	.3365	.9417	.3574	2.798	1.2275	20
50	.3462	.3393	.9407	.3607	2.773	1.2246	10
20°00'	.3491	.3420	.9397	.3640	2.747	1.2217	70°00'
10	.3520	.3448	.9387	.3673	2.723	1.2188	50
20	.3549	.3475	.9377	.3706	2.699	1.2159	40
30	.3578	.3502	.9367	.3739	2.675	1.2130	30
40	.3607	.3529	.9356	.3772	2.651	1.2101	20
50	.3636	.3557	.9346	.3805	2.628	1.2072	10
21°00'	.3665	.3584	.9336	.3839	2.605	1.2043	69°00'
10	.3694	.3611	.9325	.3872	2.583	1.2014	50
20	.3723	.3638	.9315	.3906	2.560	1.1985	40
30	.3752	.3665	.9304	.3939	2.539	1.1956	30
40	.3782	.3692	.9293	.3973	2.517	1.1926	20
50	.3811	.3719	.9283	.4006	2.496	1.1897	10
22°00'	.3840	.3746	.9272	.4040	2.475	1.1868	68°00'
10	.3869	.3773	.9261	.4074	2.455	1.1839	50
20	.3898	.3800	.9250	.4108	2.434	1.1810	40
30	.3927	.3827	.9239	.4142	2.414	1.1781	30
40	.3956	.3854	.9228	.4176	2.394	1.1752	20
50	.3985	.3881	.9216	.4210	2.375	1.1723	10
23°00'	.4014	.3907	.9205	.4245	2.356	1.1694	67°00'
10	.4043	.3934	.9194	.4279	2.337	1.1665	50
20	.4072	.3961	.9182	.4314	2.318	1.1636	40
30	.4102	.3987	.9171	.4348	2.300	1.1606	30
40	.4131	.4014	.9159	.4383	2.282	1.1577	20
50	.4160	.4041	.9147	.4417	2.264	1.1548	10
24°00'	.4189	.4067	.9135	.4452	2.246	1.1519	66°00'
10	.4218	.4094	.9124	.4487	2.229	1.1490	50
20	.4247	.4120	.9112	.4522	2.211	1.1461	40
30	.4276	.4147	.9100	.4557	2.194	1.1432	30
40	.4305	.4173	.9088	.4592	2.177	1.1403	20
50	.4334	.4200	.9075	.4628	2.161	1.1374	10
25°00'	.4363	.4226	.9063	.4663	2.145	1.1345	65°00'
10	.4392	.4253	.9051	.4699	2.128	1.1316	50
20	.4422	.4279	.9038	.4734	2.112	1.1286	40
30	.4451	.4305	.9026	.4770	2.097	1.1257	30
40	.4480	.4331	.9013	.4806	2.081	1.1228	20
50	.4509	.4358	.9001	.4841	2.066	1.1199	10
26°00'	.4538	.4384	.8988	.4877	2.050	1.1170	64°00'
10	.4567	.4410	.8975	.4913	2.035	1.1141	50
20	.4596	.4436	.8962	.4950	2.020	1.1112	40
30	.4625	.4462	.8949	.4986	2.006	1.1083	30
40	.4654	.4488	.8936	.5022	1.991	1.1054	20
50	.4683	.4514	.8923	.5059	1.977	1.1025	10
27°00'	.4712	.4540	.8910	.5095	1.963	1.0996	63°00'
10	.4741	.4566	.8897	.5132	1.949	1.0966	50
20	.4771	.4592	.8884	.5169	1.935	1.0937	40
30	.4800	.4617	.8870	.5206	1.921	1.0908	30
40	.4829	.4643	.8857	.5243	1.907	1.0879	20
50	.4858	.4669	.8843	.5280	1.894	1.0850	10
		cos	sin	cot	tan	Radians	Degrees

TRIGONOMETRIC TABLES (Continued)

Degrees	Radians	sin	cos	tan	cot		
28°00′	.4887	.4695	.8829	.5317	1.881	1.0821	62°00′
10	.4916	.4720	.8816	.5354	1.868	1.0792	50
20	.4945	.4746	.8802	.5392	1.855	1.0763	40
30	.4974	.4772	.8788	.5430	1.842	1.0734	30
40	.5003	.4797	.8774	.5467	1.829	1.0705	20
50	.5032	.4823	.8760	.5505	1.816	1.0676	10
29°00′	.5061	.4848	.8746	.5543	1.804	1.0647	61°00′
10	.5091	.4874	.8732	.5581	1.792	1.0617	50
20	.5120	.4899	.8718	.5619	1.780	1.0588	40
30	.5149	.4924	.8704	.5658	1.767	1.0559	30
40	.5178	.4950	.8689	.5696	1.756	1.0530	20
50	.5207	.4975	.8675	.5735	1.744	1.0501	10
30°00′	.5236	.5000	.8660	.5774	1.732	1.0472	60°00′
10	.5265	.5025	.8646	.5812	1.720	1.0443	50
20	.5294	.5050	.8631	.5851	1.709	1.0414	40
30	.5323	.5075	.8616	.5890	1.698	1.0385	30
40	.5325	.5100	.8601	.5930	1.686	1.0356	20
50	.5381	.5125	.8587	.5969	1.675	1.0327	10
31°00′	.5411	.5150	.8572	.6009	1.664	1.0297	59°00′
10	.5440	.5175	.8557	.6048	1.653	1.0268	50
20	.5469	.5200	.8542	.6088	1.643	1.0239	40
30	.5498	.5225	.8526	.6128	1.632	1.0210	30
40	.5527	.5250	.8511	.6168	1.621	1.0181	20
50	.5556	.5275	.8496	.6208	1.611	1.0152	10
32°00′	.5585	.5299	.8480	.6249	1.600	1.0123	58°00′
10	.5614	.5324	.8465	.6289	1.590	1.0094	50
20	.5643	.5348	.8450	.6330	1.580	1.0065	40
30	.5672	.5373	.8434	.6371	1.570	1.0036	30
40	.5701	.5398	.8418	.6412	1.560	1.0007	20
50	.5730	.5422	.8403	.6453	1.550	.0977	10
33°00′	.5760	.5446	.8387	.6494	1.540	.9948	57°00′
10	.5789	.5471	.8371	.6536	1.530	.9919	50
20	.5818	.5495	.8355	.6577	1.520	.9890	40
30	.5847	.5519	.8339	.6619	1.511	.9861	30
40	.5876	.5544	.8323	.6661	1.501	.9832	20
50	.5905	.5568	.8307	.6703	1.492	.9803	10
34°00′	.5934	.5592	.8290	.6745	1.483	.9774	56°00′
10	.5963	.5616	.8274	.6787	1.473	.9745	50
20	.5992	.5640	.8258	.6830	1.464	.9716	40
30	.6021	.5664	.8241	.6873	1.455	.9687	30
40	.6050	.5688	.8225	.6916	1.446	.9657	20
50	.6080	.5712	.8208	.6959	1.437	.9628	10
35°00′	.6109	.5736	.8192	.7002	1.428	.9599	55°00′
10	.6138	.5760	.8175	.7046	1.419	.9570	50
20	.6167	.5783	.8158	.7089	1.411	.9541	40
30	.6196	.5807	.8141	.7133	1.402	.9512	30
40	.6225	.5831	.8124	.7177	1.393	.9483	20
50	.6254	.5854	.8107	.7221	1.385	.9454	10
36°00′	.6283	.5878	.8090	.7265	1.376	.9425	54°00′
10	.6312	.5901	.8073	.7310	1.368	.9396	50
20	.6341	.5925	.8056	.7355	1.360	.9367	40
30	.6370	.5948	.8039	.7400	1.351	.9338	30
40	.6400	.5972	.8021	.7445	1.343	.9308	20
50	.6429	.5995	.8004	.7490	1.335	.9279	10
		cos	sin	cot	tan	Radians	Degrees

Degrees	Radians	sin	cos	tan	cot		
37°00′	.6458	.6018	.7986	.7536	1.327	.9250	53°00′
10	.6487	.6041	.7969	.7581	1.319	.9221	50
20	.6516	.6065	.7951	.7627	1.311	.9192	40
30	.6545	.6088	.7934	.7673	1.303	.9163	30
40	.6574	.6111	.7916	.7720	1.295	.9134	20
50	.6603	.6134	.7898	.7766	1.288	.9105	10
38°00′	.6632	.6157	.7880	.7813	1.280	.9076	52°00′
10	.6661	.6180	.7862	.7860	1.272	.9047	50
20	.6690	.6202	.7844	.7907	1.265	.9018	40
30	.6720	.6225	.7826	.7954	1.257	.8988	30
40	.6749	.6248	.7808	.8002	1.250	.8959	20
50	.6778	.6271	.7790	.8050	1.242	.8930	10
39°00′	.6807	.6293	.7771	.8098	1.235	.8901	51°00′
10	.6836	.6316	.7753	.8146	1.228	.8872	50
20	.6865	.6338	.7735	.8195	1.220	.8843	40
30	.6894	.6361	.7716	.8243	1.213	.8814	30
40	.6923	.6383	.7698	.8292	1.206	.8785	20
50	.6952	.6406	.7679	.8342	1.199	.8756	10
40°00′	.6981	.6428	.7660	.8391	1.192	.8727	50°00′
10	.7010	.6450	.7642	.8441	1.185	.8698	50
20	.7039	.6472	.7623	.8491	1.178	.8668	40
30	.7069	.6494	.7604	.8541	1.171	.8639	30
40	.7098	.6517	.7585	.8591	1.164	.8610	20
50	.7127	.6539	.7566	.8642	1.157	.8581	10
41°00′	.7156	.6561	.7547	.8693	1.150	.8552	49°00′
10	.7185	.6583	.7528	.8744	1.144	.8523	50
20	.7214	.6604	.7509	.8796	1.137	.8494	40
30	.7243	.6626	.7490	.8847	1.130	.8465	30
40	.7272	.6648	.7470	.8899	1.124	.8436	20
50	.7301	.6670	.7451	.8952	1.117	.8407	10
42°00′	.7330	.6691	.7431	.9004	1.111	.8378	48°00′
10	.7359	.6713	.7412	.9057	1.104	.8348	50
20	.7389	.6734	.7392	.9110	1.098	.8319	40
30	.7418	.6756	.7373	.9163	1.091	.8290	30
40	.7447	.6777	.7353	.9217	1.085	.8261	20
50	.7476	.6799	.7333	.9271	1.079	.8232	10
43°00′	.7505	.6820	.7314	.9325	1.072	.8203	47°00′
10	.7534	.6841	.7294	.9380	1.066	.8174	50
20	.7563	.6862	.7274	.9435	1.060	.8145	40
30	.7592	.6884	.7254	.9490	1.054	.8116	30
40	.7621	.6905	.7234	.9545	1.048	.8087	20
50	.7650	.6926	.7214	.9601	1.042	.8058	10
44°00′	.7679	.6947	.7193	.9657	1.036	.8029	46°00′
10	.7709	.6967	.7173	.9713	1.030	.7999	50
20	.7738	.6988	.7153	.9770	1.024	.7970	40
30	.7767	.7009	.7133	.9827	1.018	.7941	30
40	.7796	.7030	.7112	.9884	1.012	.7912	20
50	.7825	.7050	.7092	.9942	1.006	.7883	10
45°00′	.7854	.7071	.7071	1.0000	1.000	.7854	45°00′
		cos	sin	cot	tan	Radians	Degrees

Answers to Odd-Numbered Exercises

CHAPTER 1

SECTION 1.1

1. $[-2,0)$ $-2 \leqslant x < 0$

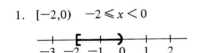

$(-\infty,-4]$ $x \leqslant -4$

$[3,11/2]$ $3 \leqslant x \leqslant 11/2$

$(-1,7)$ $-1 < x < 7$

$[100,\infty)$ $100 \leqslant x$

$(10,\infty)$ $10 < x$

$(\sqrt{2},8]$ $\sqrt{2} < x \leqslant 8$

$(1/3,22/7]$ $1/3 < x \leqslant 22/7$

3. $x \geqslant 12$

5. $x < -1/2$

7. $1/2 \leqslant x$

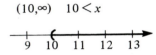

9. $1 < x$

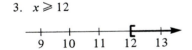

11. $-1/2 < x < 7/2$

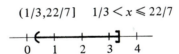

13. $x < -4$

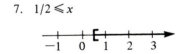

15. $6 < x$

17. $-1 < x < 1$

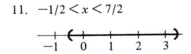

19. $x \leqslant -7$ or $13 \leqslant x$

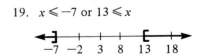

21. $a - b < x < a + b$

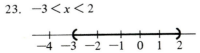

23. $-3 < x < 2$

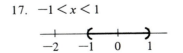

25. $0 < x < 3$

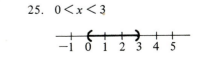

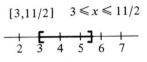
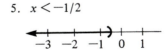
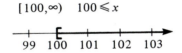
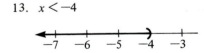

A9

27. a) 31/3
    b) −31/3
    c) 31/3
    d) −29/6

29. a) 3/2
    b) −3/2
    c) 3/2
    d) −9/4

31. 4/3, 20/3

True or False

1. False, it is possible that $a = b$
2. False, 355/113 is rational and π is irrational
3. False, let $a = 0$ and $b = 1$
4. True
5. False, if the intervals (a,b) and (c,d) intersect, then they must have infinitely many points in common
6. True
7. False, $|0| = 0$
8. True
9. True
10. False, let $x = 1$

SECTION 1.2

1.
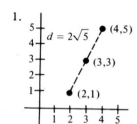
$d = 2\sqrt{5}$

3.
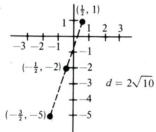
$d = 2\sqrt{10}$

5.
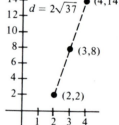
$d = 2\sqrt{37}$

7.
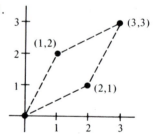
$d = \sqrt{8 - 2\sqrt{3}}$

9.

$d_1^2 + d_2^2 = d_3^2$
$45 + 5 = 50$

11.

The length of each side is $\sqrt{5}$

13.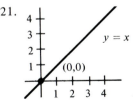

Collinear since $d_1 + d_2 = d_3$

$$2\sqrt{5} + \sqrt{5} = 3\sqrt{5}$$

15.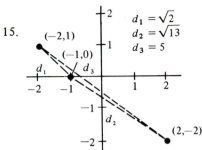

$d_1 = \sqrt{2}$
$d_2 = \sqrt{13}$
$d_3 = 5$

Not on a line since $d_1 + d_2 > d_3$

17. $x = \pm 3$

19. $2y = 3x - 1$

21. $\left(\dfrac{3x_1 + x_2}{4}, \dfrac{3y_1 + y_2}{4}\right), \left(\dfrac{x_1 + x_2}{2}, \dfrac{y_1 + y_2}{2}\right), \left(\dfrac{x_1 + 3x_2}{4}, \dfrac{y_1 + 3y_2}{4}\right)$

True or False

1. True
2. False
3. True
4. True

5. False, (0,0) is equidistant from (0,1) and (1,0)
6. False, the distance is $|2b|$
7. True

8. True
9. True
10. True

SECTION 1.3

1. Symmetric with respect to the y-axis

3. Symmetric with respect to the y-axis

5. Symmetric with respect to the x-axis

7. Symmetric with respect to the origin

9. Symmetric with respect to the origin

11. $(0,-3), (3/2,0)$

13. $(0,-2), (-2,0), (1,0)$

15. $(0,0), (-3,0), (3,0)$

17. $(0,1/2), (1,0)$

19. $(0,0)$

21.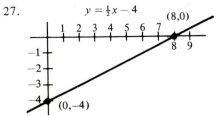

23.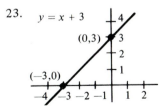

$y = x + 3$

25.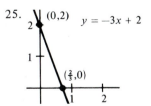

$y = -3x + 2$

27.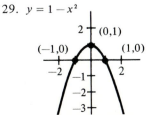

$y = \tfrac{1}{2}x - 4$

29. $y = 1 - x^2$

31. $y = -2x^2 + x + 1$

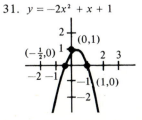

33. $y = x^3 + 2$

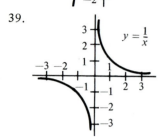

35. $x^2 + 4y^2 = 4$

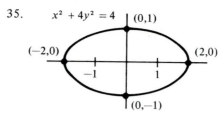

37. $y = (x + 2)^2$

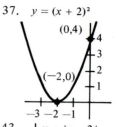

39.

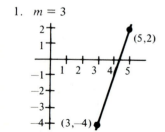

41.

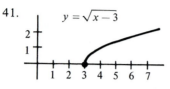

43.

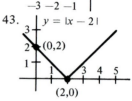

45. $y = x^3 - 3x$

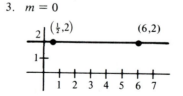

47. $(1,1)$

49. $(5,2)$

51. $(2,1), (-1,-2)$

53. $(-1,-1), (0,0), (1,1)$

55. $(-1,-5), (0,-1), (2,1)$

57. $(1,2)$ is not on the graph
$(1,-1)$ is on the graph
$(4,5)$ is on the graph

59. $(1,1/5)$ is on the graph
$(2,1/2)$ is on the graph
$(-1,-2)$ is not on the graph

61. a) $k = 4$, b) $k = -1/8$
c) $k =$ any real number
d) $k = 1$

True or False

1. True
2. False, $x^2 + y^2 = 0$ has only one solution point
3. True

4. False, if $a < 0$, the x-intercepts are $(\sqrt{-a},0)$ and $(-\sqrt{-a},0)$
5. True
6. False

7. False
8. True
9. True
10. True

SECTION 1.4

1. $m = 3$

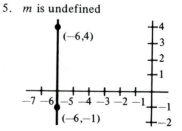

3. $m = 0$

5. m is undefined

7. $m = 4/3$

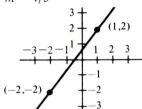

9. $2x - y - 3 = 0$

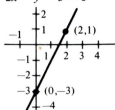

11. $3x + y = 0$

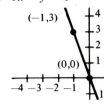

13. $x - 2 = 0$

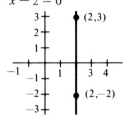

15. $y + 2 = 0$

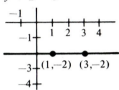

17. $3x - 4y + 12 = 0$

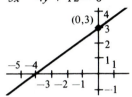

19. $2x - 3y = 0$

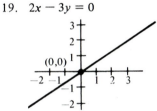

21. $2x + y - 5 = 0$

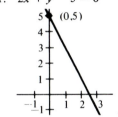

23. $4x - y + 2 = 0$

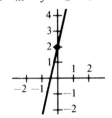

25. $9x - 12y + 8 = 0$

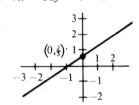

27. $x - 3 = 0$

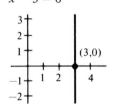

29. $3x + 2y = 6$

31. $12x + 3y = -2$

33. $x + y = 3,\ 2x - y = 0$

35. $2x + 4y = 5,\ x - 3y = 0$

37. a) $2x - y - 3 = 0$
    b) $x + 2y - 4 = 0$

39. $40x + 24y - 53 = 0$
    $24x - 40y + 9 = 0$

41. a) $x = 2,$
    b) $y = 5$

43. Collinear

45. Noncollinear

47. $\left(0, \dfrac{-a^2 + b^2 + c^2}{2c}\right)$

53.

C	$-17.7\overline{7}$	-10	10	20	$32.2\overline{2}$	177
F	0	14	50	68	90	350.6

55. $7/5$

57. 7

59. $9/5$

61. 2

True or False

1. True
2. True
3. False, the slope of a vertical line is undefined
4. True
5. True
6. True
7. False
8. False, let $a = 1/2$ and $b = 2$
9. False, $y = x + 1$ has intercepts $(-1,0)$ and $(0,1)$
10. True

SECTION 1.5

1. $x^2 + y^2 - 9 = 0$

3. $x^2 + y^2 - 4x + 2y - 11 = 0$

5. $x^2 + y^2 + 2x - 4y = 0$

7. $x^2 + y^2 - 6x - 4y + 3 = 0$

9. $x^2 + y^2 - 50 = 0$

11. $(x - 1)^2 + (y + 3)^2 = 4$

13. $(x - 1)^2 + (y + 3)^2 = 0$

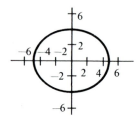

15. $\left(x - \frac{1}{2}\right)^2 + \left(y - \frac{1}{2}\right)^2 = 2$

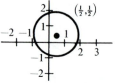

17. $\left(x + \frac{1}{2}\right)^2 + \left(y + \frac{5}{4}\right)^2 = \frac{9}{4}$

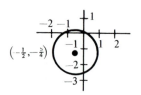

19. $\frac{x^2}{25} + \frac{y^2}{16} = 1$

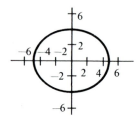

21. $\frac{x^2}{16} + \frac{y^2}{25} = 1$

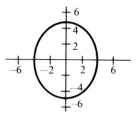

23. $\frac{x^2}{9} + \frac{y^2}{16} = 1$

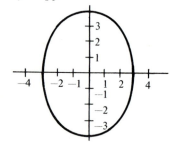

25. $4x^2 + y^2 = 1$

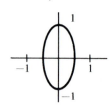

27. $\frac{(x - 1)^2}{25} + \frac{(y + 2)^2}{16} = 1$

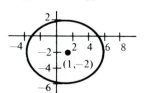

29. $\dfrac{16(x-1)^2}{9} + 4\left(y+\dfrac{1}{2}\right)^2 = 1$

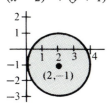

31. $4\left(x-\dfrac{1}{2}\right)^2 + (y+1)^2 = 1$

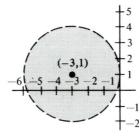

33. $\dfrac{(x-3)^2}{4} + \dfrac{(y+1)^2}{3} = 1$

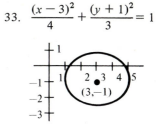

35. $(x-1)^2 + (y+1)^2 = 25$

37. $\dfrac{(x-4)^2}{9} + \dfrac{(y-5)^2}{16} = 1$

39. $(x-2)^2 + (y+1)^2 \leqslant 4$

41. $(x+3)^2 + (y-1)^2 < 9$

True or False

1. False, since $c = 2\pi r$, the circumference will be irrational if the radius is rational
2. True
3. False, the points could be collinear
4. False, $x^2 + 2y^2 + 1 = 0$ has no solution points
5. True

SECTION 1.6

1. a) -1 d) $2b-3$
   b) -3 e) $2x-5$
   c) -9 f) $-5/2$

3. a) 0 d) 3
   b) 1 e) $\sqrt{x+\Delta x+3}$
   c) $\sqrt{3}$ f) $\sqrt{c+3}$

5. a) 1 d) 1
   b) -1 e) 1
   c) -1 f) $|x-1|/(x-1)$

7. $3+\Delta x$

9. $3x^2 + 3x\,\Delta x + (\Delta x)^2$

11. $\dfrac{-1}{\sqrt{x-1}\,(1+\sqrt{x-1})}$

13. a) -1 d) 17
   b) 2 e) $-7/4$
   c) 56 f) $9x^2 - 9x + 2$

15. a) $1/3$ d) 1
   b) $-3/4$ e) $(1-x^2)/x^2$
   c) -2 f) $1/(x^2-1)$

17. ± 3

19. $10/7$

21. Domain: $[1,\infty)$
    Range: $[0,\infty)$
    $f(x) = \sqrt{x-1}$

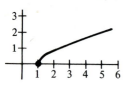

23. Domain: $(-\infty,\infty)$
    Range: $[0,\infty)$
    $f(x) = x^2$

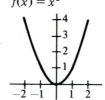

25. Domain: $[-3,3]$
    Range: $[0,3]$
    $f(x) = \sqrt{9-x^2}$

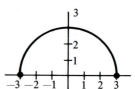

27. Domain: $(-\infty,0)$ and $(0,\infty)$
    Range: $(0,\infty)$
    $f(x) = 1/|x|$

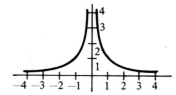

29. Domain: $(-\infty,0)$ and $(0,\infty)$
    Range: -1 and 1
    $f(x) = |x|/x$

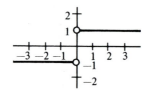

31. y is not a function of x

33. y is a function of x

35. y is a function of x

37. y is not a function of x

39. y is a function of x

41. $f(x) = 2x - 3$, $f^{-1}(x) = (x+3)/2$

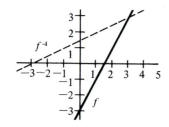

43. $f(x) = x^3$, $f^{-1}(x) = \sqrt[3]{x}$

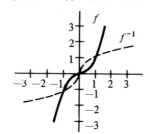

45. $f(x) = \sqrt{x}$, $f^{-1}(x) = x^2$, $0 \leqslant x$

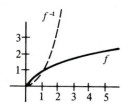

47. $f(x) = 1/x$, $f^{-1}(x) = 1/x$

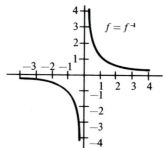

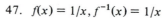

49. One-to-one as long as $a \neq 0$ and $f^{-1}(x) = (x-b)/a$

51. Not one-to-one

53. Not one-to-one

True or False

1. True
2. False, let $f(x) = 3$
3. False, let $f(x) = x^2$, then $f(-1) = f(1)$
4. True

5. True
6. True
7. False, let $f(x) = x^2$, then $f(x + \Delta x) - f(x) = 2x \Delta x + (\Delta x)^2$

8. False, let $f(x) = x^2$, then $f(2x) = 4x^2 \neq 2f(x)$
9. False, let $f(x) = -x$, then $f(f(x)) = x$ and yet $f(1) = -1$
10. True

MISCELLANEOUS EXERCISES

1. $v = 850a + 300,000$

3. $s = 6x^2$

5. $d = 45t$

7. 27/16

9. (3,4)

11. $m = 2$, y-intercept: $(0,-3)$

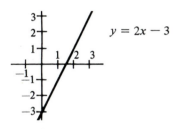

$y = 2x - 3$

13. $m = \frac{2}{5}$, y-intercept: $\left(0, \frac{6}{5}\right)$

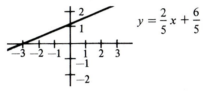

$y = \frac{2}{5}x + \frac{6}{5}$

15. (4,1)

17. $45°$

19. $120.96°$

21. $45°$

23. $71.57°$

25. a) $7x - 16y + 78 = 0$ d) $y = 4$
     b) $5x - 3y + 22 = 0$ e) $x = -2$
     c) $3x + 5y - 14 = 0$ f) $2x + y = 0$

27. Radius: 2
    Center: $\left(\frac{1}{2}, -1\right)$
    $\left(x - \frac{1}{2}\right)^2 + (y + 1)^2 = 4$

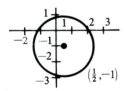

$(\frac{1}{2}, -1)$

29. -21

31. $\dfrac{(x-1)^2}{16} + \dfrac{(y-4)^2}{9} = 1$

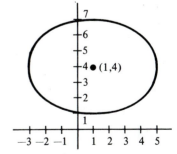

(1,4)

33.

-2 -1 0 1 2 3 4 5 6

35.

-7 -6 -5 -4 -3 -2 -1 0 1

37. $y = 1 + \dfrac{1}{x}$

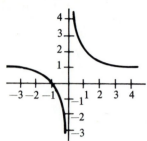

39. $y = 7 - 6x - x^2$

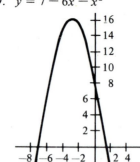

41. $(3,-2), (1,4), (-1,0)$

43. $x^2 + y^2 - 2x - 4y = 4$

    a) on the circle   e) outside
    b) inside   f) outside
    c) outside   g) outside
    d) inside   h) outside

45. y is a function of x

47. y is not a function of x

49. y is a function of x

51. a) $-x^2 + 2x + 2$ d) $(1 - x^2)/(2x + 1)$
    b) $-x^2 - 2x$ e) $-4x^2 - 4x$
    c) $-2x^3 - x^2 + 2x + 1$ f) $-2x^2 + 3$

CHAPTER 2

SECTION 2.1

1. 4

3. 4

5. -2

7. $1/6$

9. Limit doesn't exist

11. 2

13. 12

15. Limit doesn't exist

17. 2

19. $2x - 2$

21. $1/10$

23. $3/2$

25. Limit doesn't exist

27. $\sqrt{3}/6$

29. $-1/4$

31. a) -1 b) 6   c) $2/3$

33.

x	1.5	1.9	1.99	2	2.01	2.1	2.5
$f(x)$	11.5	13.5	13.95	14	14.05	14.5	16.5

$$\lim_{x \to 2} (5x + 4) = 14$$

35.

x	1.9	1.99	1.999	2	2.001	2.01	2.1
$f(x)$	0.256	0.251	0.250	undefined	0.250	0.249	0.244

$$\lim_{x \to 2} \frac{x - 2}{x^2 - 4} = \frac{1}{4}$$

37.

x	-0.1	-0.01	-0.001	0	0.001	0.01	0.1
$f(x)$	0.358	0.354	0.354	undefined	0.354	0.353	0.349

$$\lim_{x \to 0} \frac{\sqrt{x+2} - \sqrt{2}}{x} = \frac{1}{2\sqrt{2}} \approx .354$$

39.

x	-0.1	-0.01	-0.001	0	0.001	0.01	0.1
$f(x)$	-0.263	-0.251	-0.250	undefined	-0.250	-0.249	-0.238

$$\lim_{x \to 0} \frac{\frac{1}{2+x} - \frac{1}{2}}{x} = -\frac{1}{4}$$

41.
a) 1
b) 1
c) 1

43.
a) 0
b) 0
c) 0

45.
a) 3
b) -3
c) Does not exist

True or False

1. False
2. True

3. True
4. True

5. False

SECTION 2.2

1. Continuous

3. Nonremovable discontinuity at $x = 1$

5. Continuous

7. Removable discontinuity at $x = -2$
Nonremovable discontinuity at $x = 5$

9. Continuous

11. Nonremovable discontinuity at $x = 2$

13. Nonremovable discontinuity at $x = -2$

15. Continuous

17. Nonremovable discontinuity at $x = n$, where n is any integer

19. Nonremovable discontinuity at $x = 1$

21.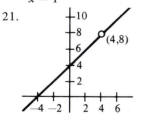

$y = \dfrac{x^2 - 16}{x - 4}$

Discontinuous at $x = 4$

23.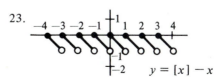

$y = [x] - x$

Discontinuous at every integer

25. The Intermediate Value Theorem

27. f is a polynomial form and therefore continuous; $f(2) = 0$

29. f has a nonremovable discontinuity only at $x = 1$; $f(3) = 6$

31. $a = -1, b = 1$

33. f has neither a maximum nor a minimum on $(0,1)$. The Extreme Value Theorem does not apply since the interval $(0,1)$ is open.

True or False

1. True
2. True
3. True

4. False, let $f(x) = x$ on the interval $(0,1]$

5. False, the rational function $f(x) = p(x)/q(x)$ has at most n discontinuities, where n is the degree of $q(x)$

SECTION 2.3

1. $\lim\limits_{x \to 0^+} \dfrac{1}{x^2} = \infty = \lim\limits_{x \to 0^-} \dfrac{1}{x^2}$

   Therefore, $x = 0$ is an even vertical asymptote.

3. $\lim\limits_{x \to 1^+} \dfrac{2+x}{1-x} = -\infty, \quad \lim\limits_{x \to 1^-} \dfrac{2+x}{1-x} = \infty$

   Therefore, $x = 1$ is an odd vertical asymptote.

5. $\lim\limits_{x \to 0^+} \left(1 - \dfrac{4}{x^2}\right) = -\infty = \lim\limits_{x \to 0^-} \left(1 - \dfrac{4}{x^2}\right)$

   Therefore, $x = 0$ is an even vertical asymptote.

7. $\lim\limits_{x \to -3^+} \dfrac{1}{(x+3)^4} = \infty = \lim\limits_{x \to -3^-} \dfrac{1}{(x+3)^4}$

   Therefore, $x = -3$ is an even vertical asymptote.

9. $\displaystyle\lim_{x \to -2^+} \frac{x}{(x+2)(x-1)} = \infty, \quad \lim_{x \to -2^-} \frac{x}{(x+2)(x-1)} = -\infty$

Therefore, $x = -2$ is an odd vertical asymptote.

$\displaystyle\lim_{x \to 1^+} \frac{x}{(x+2)(x-1)} = \infty, \quad \lim_{x \to 1^-} \frac{x}{(x+2)(x-1)} = -\infty$

Therefore, $x = 1$ is an odd vertical asymptote.

11. 2/3

13. 0   15. $-\infty$ 17. ∞

19. 5   21. 0   23. $-1/2$

25. -1 27. 2   29. 1

31. $y = \dfrac{2 + x}{1 - x}$

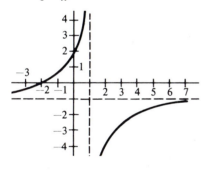

33. $y = \dfrac{x^2}{x^2 + 9}$

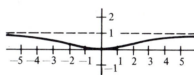

35. $xy^2 = 4$

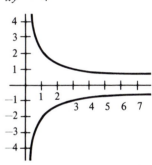

37. $y = \dfrac{2x}{1 - x}$

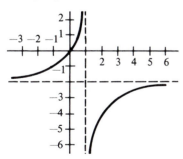

39. $y = 2 - \dfrac{3}{x^2}$

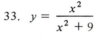

41. $y = \dfrac{x^3}{\sqrt{x^2 - 4}}$

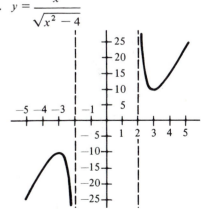

43.

x	1	10	10^2	10^4	10^6
$f(x)$	1	.513	.501	.500	.500

$\displaystyle\lim_{x \to \infty} [x - \sqrt{x(x-1)}] = 1/2$

45. $y = \dfrac{x^2 + 1}{x}$

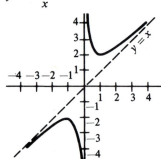

47. $y = \dfrac{x^3}{2x^2 - 8}$

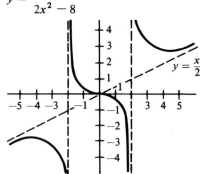

49. 0.50

51. Approaches 100%

53. $32\pi/3$

True or False

1. False, see Example 8, Section 2.3
2. True
3. False, let $p(x) = x^2 - 1$
4. False, see Exercise 23, Section 2.3

5   True
6. True
7. False, let $f(x) = \begin{cases} 1/x, & x \neq 0 \\ 0, & x = 0 \end{cases}$

8. True
9. True
10. True

SECTION 2.4

1. Let $\delta = \dfrac{1}{300} = .003\overline{3}$

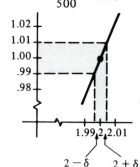

$$2 - \delta \quad 2 + \delta$$

3. Let $\delta = \dfrac{1}{500} = .002$

$$2 - \delta \quad 2 + \delta$$

5. Let $\delta = .01$

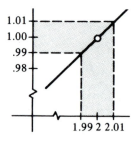

7. Let $\delta = \epsilon$

9. Any δ will work

11. Let $\delta = \epsilon^3$

13. Assume $1 < x < 3$ and let $\delta = \epsilon/5$

15. Let $\delta = \epsilon$

17. Assume $1 < x < 3$ and let $\delta = 2\epsilon$

19. Assume $1 < x < 3$ and let $\delta = \epsilon/4$

21. Let $\delta = \epsilon^2$

23. Let $\delta = 1/M$

25. Let $\delta = 1/\sqrt{M}$

27. Let $M = \dfrac{2}{\epsilon} + 1$

MISCELLANEOUS EXERCISES

1. 7

3. 77

5. 10/3

7. −1/4

9. −1

11. 3

13. 1

15. 2

17. 0

19. ∞

21.

x	1	10^2	10^4	10^6
$f(x)$	.732	.071	.007	.001

$\lim_{x \to \infty} (\sqrt{2x + 1} - \sqrt{2x - 1}) = 0$

23.

x	1	10^2	10^4	10^6
$f(x)$	.414	.499	.499	.500

$\lim_{x \to \infty} (\sqrt{x^2 + x} - x) = 1/2$

25. False

27. True

29. False

31. Discontinuity at $x = 1$

33. Discontinuity at $x = 2$

35. Discontinuity at $x = 0$, (undefined in the interval $(-2, 0]$)

37. $c = -1/2$

39. $y = \dfrac{x^2 - 1}{x^2 + 1}$

41. $y^2 = \dfrac{x + 2}{x(x - 2)}$

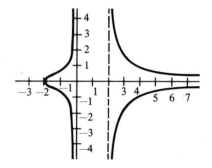

CHAPTER 3

SECTION 3.1

1.
$$f(x) = 3$$
i) $f(x + \Delta x) = 3$

ii) $f(x + \Delta x) - f(x) = 0$

iii) $\dfrac{f(x + \Delta x) - f(x)}{\Delta x} = 0$

iv) $\lim\limits_{\Delta x \to 0} \dfrac{f(x + \Delta x) - f(x)}{\Delta x} = 0$

3.
$$f(x) = -5x$$
i) $f(x + \Delta x) = -5x - 5\,\Delta x$

ii) $f(x + \Delta x) - f(x) = -5\,\Delta x$

iii) $\dfrac{f(x + \Delta x) - f(x)}{\Delta x} = -5$

iv) $\lim\limits_{\Delta x \to 0} \dfrac{f(x + \Delta x) - f(x)}{\Delta x} = -5$

5.
$$f(x) = 2x^2 + x - 1$$

i)
$$f(x + \Delta x) = 2x^2 + 4x\,\Delta x + 2(\Delta x)^2 + x + \Delta x - 1$$

ii)
$$f(x + \Delta x) - f(x) = 4x\,\Delta x + 2(\Delta x)^2 + \Delta x$$

iii)
$$\frac{f(x + \Delta x) - f(x)}{\Delta x} = 4x + 2\,\Delta x + 1$$

iv)
$$\lim_{\Delta x \to 0} \frac{f(x + \Delta x) - f(x)}{\Delta x} = 4x + 1$$

7.
$$f(x) = \frac{1}{x - 1}$$

i)
$$f(x + \Delta x) = \frac{1}{x + \Delta x - 1}$$

ii)
$$f(x + \Delta x) - f(x) = \frac{-\Delta x}{(x + \Delta x - 1)(x - 1)}$$

iii)
$$\frac{f(x + \Delta x) - f(x)}{\Delta x} = \frac{-1}{(x + \Delta x - 1)(x - 1)}$$

iv)
$$\lim_{\Delta x \to 0} \frac{f(x + \Delta x) - f(x)}{\Delta x} = \frac{-1}{(x - 1)^2}$$

9.
$$f(t) = t^3 - 12t$$

i)
$$f(t + \Delta t) = t^3 + 3t^2\,\Delta t + 3t(\Delta t)^2 + (\Delta t)^3 - 12t - 12\,\Delta t$$

ii)
$$f(t + \Delta t) - f(t) = 3t^2\,\Delta t + 3t(\Delta t)^2 + (\Delta t)^3 - 12\,\Delta t$$

iii)
$$\frac{f(t + \Delta t) - f(t)}{\Delta t} = 3t^2 + 3t\,\Delta t + (\Delta t)^2 - 12$$

iv)
$$\lim_{\Delta t \to 0} \frac{f(t + \Delta t) - f(t)}{\Delta t} = 3t^2 - 12$$

11. $f'(x) = 2x$

Tangent line: $y = 4x - 3$

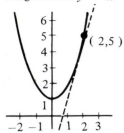

13. $f'(x) = 3x^2$

Tangent line: $y = 12x - 16$

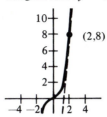

15. $f'(x) = \dfrac{1}{2\sqrt{x+1}}$

Tangent line: $4y = x + 5$

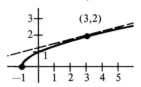

17. $f'(x) = -1/x^2$

Tangent line: $y = -x + 2$

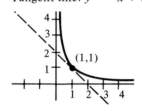

19. $y = 3x - 2$ or $y = 3x + 2$

21. $y = 2x + 1$ and $y = -2x + 9$

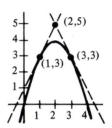

True or False

1. True
2. True
3. False, at the points $(1,1)$ and $(-1,-1)$, the slope is 3
4. False, let $y = x^3$, then the tangent line at $(1,1)$ also crosses the curve at $(-2,-8)$

5. True
6. False, the equation should be $y - 1 = -2(x + 1)$
7. True
8. True

9. True
10. False

SECTION 3.2

1. Average rate: 2
   Instantaneous rates: $f'(1) = 2$
   $\qquad\qquad\qquad\quad f'(2) = 2$

3. Average rate: $-1/4$
   Instantaneous rates: $f'(0) = -1$
   $\qquad\qquad\qquad\quad f'(3) = -1/16$

5. Average rate: -4
   Instantaneous rates: $h'(-1) = -8$
   $\qquad\qquad\qquad\quad h'(3) = 0$

7. Average rate: 3
   Instantaneous rates: $h'(4)$ is undefined
   $\qquad\qquad\qquad\quad h'(5) = 5/3$

9. a)

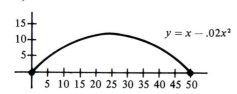

$y = x - .02x^2$

11. a)

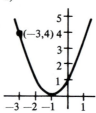

$(-3,4)$

b) 50 ft

c) $x = 25$ ft

d) $f'(x) = 1 - .04x, f'(0) = 1, f'(10) = .6,$
$f'(25) = 0, f'(30) = -.2, f'(50) = -1$

e) 0

b) $m_{sec} = 2x + 2 + \Delta x$

c) i) -2 v) -6
    ii) -3 vi) -5
    iii) -3.75 vii) -4.25
    iv) -3.99 viii) -4.01

d) $f'(-3) = -4$

True or False

1. True
2. True

3. False, let $f(x) = x - x^2$, $x = 0$,
   and $\Delta x = .1$, then $f'(0) = 1$ and
   $\Delta y/\Delta x = .9$

4. True
5. True

SECTION 3.3

1. $y' = 0$

3. $f'(x) = 1$

5. $g'(x) = 2x$

7. $f'(t) = -4t + 3$

9. $s'(t) = 3t^2 - 2$

11. $f'(x) = -1/x^2; -1$

13. $f'(t) = 3/5t^2; 5/3$

15. $y' = -1/x^4; -1$

17. $y' = 8x + 4; 4$

19. $f'(x) = 32x - 128x^3$

21. $f'(x) = 2x + \dfrac{4}{x^2}$

23. $f'(x) = 3x^2 - 3 + \dfrac{8}{x^5}$

25. $f'(x) = 1 - \dfrac{8}{x^3}$

27. $f'(x) = -2\pi/9x^3$

29. $f'(x) = 3x^2 - 1$

31. $f'(x) = \dfrac{4}{5x^{1/5}}$

33. $f'(x) = \dfrac{1}{3x^{2/3}} + \dfrac{1}{5x^{4/5}}$

35. $f'(x) = \dfrac{-1}{2x^{3/2}} - \dfrac{2}{x^3} - \dfrac{4}{x^5}$

37. $y = -2x + 2$

39. $(0, 2), (\sqrt{3/2}, -1/4), (-\sqrt{3/2}, -1/4)$

41. No horizontal tangents

43. 8

45. a) 10,000
    b) 4,000
    c) 0
    d) $-8,000$

47. a) 4/9
    b) 1/3
    c) 0
    d) $-5/9$

True or False

1. True
2. False, let $f(x) = x$ and $g(x) = x + 1$
3. True
4. False, $\dfrac{dy}{dx} = -1/3x^2$

5. False, $\dfrac{d}{dx}[\sqrt{cx}] = \sqrt{c}\dfrac{d}{dx}[\sqrt{x}]$
6. False, $dy/dx = 0$
7. True

8. True
9. True
10. False, $f'(x) = \dfrac{\sqrt{a}}{2\sqrt{x}}$

SECTION 3.4

1. $f''(x) = 2$

3. $f''(x) = 0$

5. $f'''(x) = 24x - 12$

7. $f^{(4)}(x) = -24/x^5$

9. $f''(t) = 4/9t^{7/3}$

11. $f'''(x) = -9/2x^5$

13. $f'''(x) = 12$

15. a) $v(t) = 2t,\ a(t) = 2$
    b) Moving to the right when $0 < t$

17. a) $v(t) = 3t^2 - 6t,\ a(t) = 6t - 6$

    b) Moving to the left when $0 < t < 2$
       Moving to the right when $2 < t$

19. a) $v(t) = 12t^3 - 84t^2 + 120t$
       $a(t) = 36t^2 - 168t + 120$

    b) Moving to the right when $0 < t < 2$
       Moving to the left when $2 < t < 5$
       Moving to the right when $5 < t$

21. a)

t	0	.5	1	1.5	2	2.5	3	3.5	4
$s(t)$	4	3.375	2	.625	0	.875	4	10.125	20
$v(t)$	0	-2.25	-3	-2.25	0	3.75	9	15.75	24
$a(t)$	-6	-3		0	3	6	9	12	15	18

b)

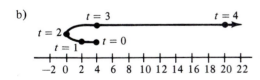

c) The speed is increasing

d) The speed is decreasing

23. a) $s(t) = -16t^2 + 384t$

    b) 144 ft/sec

    c) 224 ft/sec when $t = 5$,
       64 ft/sec when $t = 10$

    d) 2304 ft

25. $-80\sqrt{6}$ ft/sec (-195.96 ft /sec)

27. $v_0 = 8\sqrt{15}$ ft/sec (30.98 ft/sec)

29. 739.84 ft

31. The one dropped from a height of 100 ft

True or False

1. False, let $f(x) = x^2$, then $f''(x) = 2$
2. True
3. True
4. False, the speed could be zero

5. True
6. True
7. True

8. True
9. False, 55 mph $\approx$ 80.67 ft/sec
10. True

SECTION 3.5

1. a) $x = -3$
    b) Node at $(-3,0)$

3. a) $x = -2, x = 0, x = 2$
    b) Nodes at these three points

5. a) $x = -1$
    b) Discontinuous

7. a) $x = 0$
    b) Cusp at $(0,0)$

9. Differentiable everywhere

11. a) $x = 0$
    b) Discontinuous

13. a) $x = 1$
    b) $f(x)$ is undefined for $x < 1$

15.

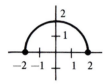

17. $\lim\limits_{x \to 0^-} \dfrac{f(x) - f(0)}{x} = \lim\limits_{x \to 0^-} \dfrac{x}{x} = 1$

$\lim\limits_{x \to 0^+} \dfrac{f(x) - f(0)}{x} = \lim\limits_{x \to 0^+} \dfrac{x^2}{x} = 0$

$$\lim_{x \to -2^+} \frac{f(x) - f(-2)}{x - (-2)} = \lim_{x \to -2^+} \frac{\sqrt{4 - x^2}}{x + 2}$$

$$= \lim_{x \to -2^+} \frac{\sqrt{2 - x}}{\sqrt{x + 2}}$$

$$= \frac{2}{0} \text{(undefined)}$$

$$\lim_{x \to 2^-} \frac{f(x) - f(2)}{x - 2} = \lim_{x \to 2^-} \frac{\sqrt{4 - x^2}}{x - 2}$$

$$= \lim_{x \to 2^-} \frac{-\sqrt{2 + x}}{\sqrt{2 - x}}$$

$$= \frac{-2}{0} \text{ (undefined)}$$

True or False

1. False
2. True
3. False, let $f(x) = |x|/x$

4. False, let $f(x) = x^{4/3}$, then f is differentiable at $x = 0$, but f' is not

5. False, let $f(x) = \sqrt[3]{x}$, then the graph of f has a (vertical) tangent line at $(0,0)$, but f is not differentiable there

SECTION 3.6

1. $2x^2, 0$

3. $-7/x^4, -7$

5. $(x - 1)^2 (5x^2 + 2x + 2), 0$

7. $(3x - 5)(x - 1)$, 5

9. $3\left(2x^5 + x^2 - 2x + \dfrac{1}{x^2}\right)$, 6

11. $-5/(2x - 3)^2$

13. $2/(x + 1)^2$

15. $6x/(4 - 3x^2)^2$

17. $\dfrac{5}{6x^{1/6}} + \dfrac{1}{x^{2/3}}$

19. $\dfrac{-t^2 - 2t}{(t^2 + 2t + 2)^2}$

21. $6s^2(s^3 - 2)$

23. $\dfrac{2x^2 + 8x - 1}{(x + 2)^2}$

25. $15x^4 - 48x^3 - 33x^2 - 32x - 20$

27. $-4c^2x/(c^2 + x^2)^2$

29. $y = -x + 4$

31. $y = -x - 2$

33. $(0,0)$ and $(2,4)$

35. 31.55

37. a) -0.48
    b) 0.12
    c) 0.015

39. $T_0 = 75°$; $\lim\limits_{t \to \infty} T = 40$

    a) -9.33
    b) -3.64
    c) -1.62
    d) -0.07

41. $v(t) = -2/(t + 1)^3$
    $a(t) = 6/(t + 1)^4$

True or False

1. False, $dy/dx = f(x)g'(x) + g(x)f'(x)$

2. True

3. True

4. False, $\dfrac{d^2y}{dx^2} = f(x)g''(x) + 2g'(x)f'(x) + g(x)f''(x)$

5. True

6. True

7. False, $\dfrac{dy}{dx} = \dfrac{-f'(x)}{[f(x)]^2}$

8. True

9. True

10. False, let $f(x) = x$, $g(x) = x - 1$, and $c = 1$

SECTION 3.7

1. $6(2x - 7)^2$

3. $12x(x^2 - 1)^2$

5. $-1/(x - 2)^2$

7. $-2/(t - 3)^3$

9. $-9x^2/(x^3 - 4)^2$

11. $2x(x - 2)^3(3x - 2)$

13. $(x + 3)^2(4x - 3)$

15. $\dfrac{1}{2\sqrt{t + 1}}$

17. $\dfrac{t + 1}{\sqrt{t^2 + 2t - 1}}$

19. $6x/(9x^2 + 4)^{2/3}$

21. $\dfrac{2x}{\sqrt{x^2 + 4}}$

23. $\dfrac{4x}{3(x^2 - 9)^{1/3}}$

25. $\dfrac{-1}{2(x + 2)^{3/2}}$

27. $-3x^2/(x^3 - 1)^{4/3}$

29. $\dfrac{1}{2\sqrt{x}(\sqrt{x} + 1)^2}$

31. $x/|x|$

33. $\dfrac{2x(x^2 - 1)}{|x^2 - 1|}$

35. $\dfrac{-(3x^{1/6} + 2)}{(x^{1/2} + x^{1/3})^2 6x^{2/3}}$

37. $\dfrac{x + 2}{(x + 1)^{3/2}}$

39. $\dfrac{1 - 3x^2 - 4x^{3/2}}{2\sqrt{x}\,(x^2 + 1)^2}$

41. $-5/(t - 1)^2$

43. $\dfrac{3t(t^2 + 3t - 2)}{(t^2 + 2t - 1)^{3/2}}$

45. $\dfrac{5}{6(t + 1)^{1/6}}$

47. $\dfrac{-1}{2x^{3/2}\sqrt{x + 1}}$

49. $\dfrac{t}{\sqrt{1 + t}}$

51. $\dfrac{t^2 + 4}{4t^{3/2}\sqrt{t^2 - 4}}$

53. $\dfrac{-3x}{2(x^2 + x + 1)^{3/2}}$

55. $9x - 5y - 2 = 0$

57. $y = 0$

59. $12(5x^2 - 1)(x^2 - 1)$

61. $\dfrac{2(x^2 - 1)}{|x^2 - 1|}$

63. $\dfrac{3t^2 + 2}{t^3(t^2 + 1)^{3/2}}$

65. a) 4/3
    b) 2

67. $\dfrac{dy}{dx} = \dfrac{2(x^2 - 1)(x^2 - x + 1)}{x^3}$

True or False

1. False, $y' = (1/2)(1 - x)^{-1/2}(-1)$
2. False, y' is a 19th degree polynomial

3. True
4. True

5. True

SECTION 3.8

1. $-x/y, -3/\sqrt{7}$

3. $-y/x, -1/4$

5. $-\sqrt{y/x}, -5/4$

7. $\dfrac{y - 3x^2}{2y - x}, 1/2$

9. $\dfrac{36x}{2y(x^2 + 9)^2}$, undefined at (3,0)

11. $\dfrac{1 - 3x^2y^3}{3x^3y^2 - 1}, -1$

13. $-\sqrt[3]{\dfrac{y}{x}}, -1/2$

15. $\dfrac{4xy - 3x^2 - 3y^2}{2x(3y - x)}, -15/28$

17. a) $y = \pm\sqrt{16 - x^2}, \quad -4 \le x \le 4$

    b) $y' = \dfrac{\mp x}{\sqrt{16 - x^2}}$

    c) $y' = -x/y$

    d) $y = \sqrt{16 - x^2}$ (upper semicircle)

       $y = -\sqrt{16 - x^2}$ (lower semicircle)

    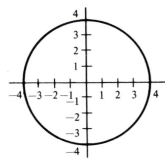

19. a) $y = \pm\dfrac{3}{4}\sqrt{16 - x^2}, \quad -4 \le x \le 4$

    b) $y' = \mp\dfrac{3}{4}\dfrac{x}{\sqrt{16 - x^2}}$

    c) $y' = -9x/16y$

    d) $y = \dfrac{3}{4}\sqrt{16 - x^2}$ (upper semi-ellipse)

       $y = -\dfrac{3}{4}\sqrt{16 - x^2}$ (lower semi-ellipse)

    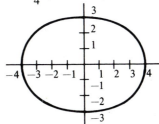

21. $\dfrac{2(x + y)}{x^2} = \dfrac{10}{x^3}$

25. $3x/4y$

31. At $(1,2)$, the slope of the ellipse is -1 and the slope of the parabola is 1

    At $(1,-2)$, the slope of the ellipse is 1 and the slope of the parabola is -1

    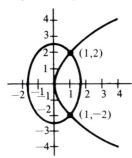

23. $-16/y^3$

29. a) Tangent line: $y = 3$
       Normal line: $x = 0$

    b) Tangent line: $2x + \sqrt{5}y = 9$
       Normal line: $\sqrt{5}x - 2y = 0$

33. The curves intersect at the point $(3,10)$, with slopes of 9 and $-1/9$

MISCELLANEOUS EXERCISES

---

1. $3x(x-2)$

3. $\dfrac{x+1}{2x^{3/2}}$

5. $\dfrac{-4}{3t^3}$

7. $\dfrac{3x^2}{2\sqrt{x^3+1}}$

9. $2(6x^3 - 9x^2 + 16x - 7)$

11. $5s(s^2-1)^{3/2}(s^3+5)^{2/3}(2s^3-s+5)$

13. $\dfrac{3x^2+1}{2\sqrt{x^3+x}}$

15. $\dfrac{-(x^2+1)}{(x^2-1)^2}$

17. $3x^3/|x|$

19. $9/(x^2+9)^{3/2}$

21. $\dfrac{2(t+2)}{(1-t)^4}$

23. $\dfrac{2(6x^3-15x^2-18x+5)}{(x^2+1)^3}$

25. $20|x^3|$

27. $\dfrac{-(2x+3y)}{3(x+y^2)}$

29. $\dfrac{2y\sqrt{x}-y\sqrt{y}}{2x\sqrt{y}-x\sqrt{x}}$

31. x/y

33. Tangent line: $3x-y+7=0$
     Normal line: $x+3y-1=0$

35. Tangent line: $x+2y-10=0$
     Normal line: $2x-y=0$

37. Tangent line: $2x-3y-3=0$
     Normal line: $3x+2y-11=0$

39. 56 ft/sec

41. a) $(0,-1)$ and $(-2,7/3)$
     b) $(-3,2)$ and $(1,-2/3)$
     c) $(-1+\sqrt{2}, 2(1-2\sqrt{2})/3)$
         and $(-1-\sqrt{2}, 2(1+2\sqrt{2})/3)$

43.
$$f(x)=\frac{x+1}{x-1}$$

   i)
$$f(x+\Delta x)=\frac{x+\Delta x+1}{x+\Delta x-1}$$

   ii)
$$f(x+\Delta x)-f(x)=\frac{-2\,\Delta x}{(x-1)(x+\Delta x-1)}$$

   iii)
$$\frac{f(x+\Delta x)-f(x)}{\Delta x}=\frac{-2}{(x-1)(x+\Delta x-1)}$$

   iv)
$$\lim_{\Delta x \to 0}\frac{f(x+\Delta x)-f(x)}{\Delta x}=\frac{-2}{(x-1)^2}$$

45.
$$f(x)=\frac{1}{\sqrt{x}}$$

   i)
$$f(x+\Delta x)=\frac{1}{\sqrt{x+\Delta x}}$$

   ii)
$$f(x+\Delta x)-f(x)=\frac{-\Delta x}{\sqrt{x}\sqrt{x+\Delta x}(\sqrt{x}+\sqrt{x+\Delta x})}$$

iii) $\dfrac{f(x + \Delta x) - f(x)}{\Delta x} = \dfrac{-1}{\sqrt{x}\sqrt{x + \Delta x}\,(\sqrt{x} + \sqrt{x + \Delta x})}$

iv) $\displaystyle\lim_{\Delta x \to 0} \dfrac{f(x + \Delta x) - f(x)}{\Delta x} = \dfrac{-1}{2x^{3/2}}$

47. a)

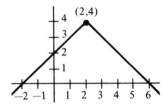

b) Continuous, but not differentiable since the derivative from the left and right are not equal

CHAPTER 4

SECTION 4.1

1. $[0,2], f'(1) = 0$

3. $[1,2], f'\left(\dfrac{6 - \sqrt{3}}{3}\right) = 0$

$[2,3], f'\left(\dfrac{6 + \sqrt{3}}{3}\right) = 0$

5. f is not differentiable on $[-1,1]$

7. f is not differentiable on $[-1,1]$

9. $[-1,3], f'(-2 + \sqrt{5}) = 0$

11. $f(x) \neq 0$

13. $f'(-1/2) = -1$

15. $f'(8/27) = 1$

17. $f'\left(\dfrac{-2 + \sqrt{6}}{2}\right) = 2/3$

19. $f'(\sqrt{3}/3) = 1$

True or False

1. False, f is not continuous on $[-1,1]$

2. False, let $f(x) = (x^3 - 4x)/(x^2 - 1)$

3. True

4. True

5. False, let $f(x) = \begin{cases} 1, & x \neq 0,1 \\ 0, & x = 0,1 \end{cases}$

SECTION 4.2

1. No critical numbers
   Increasing on $(-\infty,\infty)$

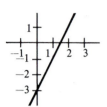

3. Critical number: $x = 1$
   Decreasing on $(-\infty,1)$
   Increasing on $(1,\infty)$

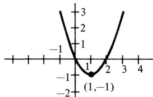

5. Critical numbers: $x = 0, 2$
   Increasing on $(-\infty,0)$
   Decreasing on $(0,2)$
   Increasing on $(2,\infty)$

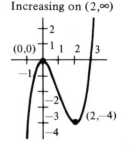

7. Domain: $(-\infty,-2), (-2,2), (2,\infty)$
   Critical number: $x = 0$
   Increasing on $(-\infty,-2), (-2,0)$
   Decreasing on $(0,2), (2,\infty)$

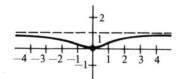

9. Critical number: $x = 0$
   Decreasing on $(-\infty,0)$
   Increasing on $(0,\infty)$

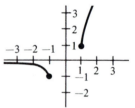

11. Domain: $(-\infty,-1), (1,\infty)$
    Critical numbers: $x = \pm 1$
    Decreasing on $(-\infty,-1)$
    Increasing on $(1,\infty)$

13. Critical number: $x = 0$
    Decreasing on $(-\infty,0)$
    Increasing on $(0,\infty)$

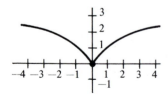

15. Domain: $(-\infty,0), (0,\infty)$
    Critical number: $x = 4$
    Increasing on $(-\infty,0)$
    Decreasing on $(0,4)$
    Increasing on $(4,\infty)$

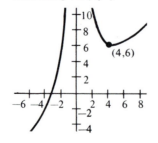

17. Critical number: $x = 0$
    Increasing on $(-\infty,0)$
    Decreasing on $(0,\infty)$

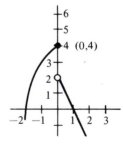

True or False

1. True
2. False, let $f(x) = x$, $g(x) = x$ on the interval $(-\infty, 0)$
3. True
4. False, let $f(x) = x^2$ on $(-1, 0)$
5. True
6. False, let $f(x) = x^3$
7. True
8. False, let $f(x) = x^3$ on $(-1, 1)$
9. False, let $f(x) = \sqrt[3]{x}$ on $(-1, 1)$
10. False, c must be in the domain of f

SECTION 4.3

1. $(1, 5)$, relative maximum

3. $(3, -9)$, relative minimum

5. $(-2, 20)$, relative maximum
   $(1, -7)$, relative minimum

7. $(0, 15)$, relative maximum
   $(4, -17)$, relative minimum

9. $(3/2, -27/16)$, relative minimum

11. no relative extrema

13. $(1, 2)$, relative minimum
    $(-1, -2)$, relative maximum

15. $(0, 0)$, relative maximum

17. $(2, -44)$, relative minimum

19. $(-3, -8)$, relative maximum
    $(1, 0)$, relative minimum

21. $(2, 2)$, minimum
    $(-1, 8)$, maximum

23. $(0, 0)$, minimum
    $(2, 4)$, maximum

25. $(-1, -4)$ and $(2, -4)$, minima
    $(0, 0)$ and $(3, 0)$, maxima

27. $(0, 0)$, minimum
    $(-1, 5)$, maximum

29. $(1, -1)$, minimum
    $(0, -1/2)$, maximum

31. $a = -1/2$, $b = 3/2$, $c = 0$, $d = 0$

33. Yes

35. Yes

37. No

39. No

41. a) 96 ft/sec   b) 144 ft
    c) Downward   d) 80 ft

43. $r = 2R/3$

45. $T = 10°C$

47. $|f''(0)| = 2$ max

49. $\left| f''\left(\sqrt[3]{\dfrac{-20 + \sqrt{432}}{2}} \right) \right| \cong 1.96$ max

51. $|f^{(4)}(\tfrac{1}{2})| = 360$ max

53. $|f^{(4)}(0)| = 56/81$ max

55. a) $h'(0) < 0$
    b) $g'(-6) < 0$ and $g'(0) > 0$
    c) $j'(0) > 0$ and $j'(8) < 0$

True or False

1. True
2. False, let $f(x) = x^3$ and $c = 0$
3. False, let $f(x) = -\sqrt{x}$ and $c = 0$
4. True
5. True

SECTION 4.4

1. (3,9), relative maximum

3. (5,0), relative minimum

5. (0,3), relative maximum
   (2,−1), relative minimum

7. (3,−25), relative minimum

9. (0,−3), relative minimum

11. (−2,−4), relative maximum
    (2,4) relative minimum

13. (2,0), point of inflection

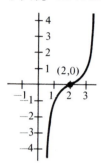

15. (2,−16), relative minimum
    (−2,16), relative maximum
    (0,0), point of inflection

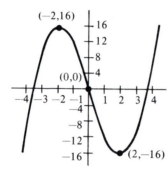

17. (−2,−4), relative minimum
    (0,0), relative maximum
    (2, − 4), relative minimum
    $(2/\sqrt{3},-20/9)$ point of inflection
    $(-2/\sqrt{3},-20/9)$ point of inflection

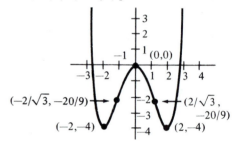

19. (1,2), relative minimum
    (−1,2), relative minimum

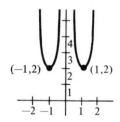

21. (−2,−2), relative minimum

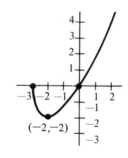

23. (0,0), point of inflection

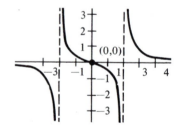

25. (2,0), point of inflection

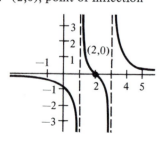

27. $a = 1/2, b = -6, c = 45/2, d = -24$

29. (4,16)

31. $x = 0$

33. $x = 100\sqrt{15} \approx 387$ units

True or False

1. False, let $f(x) = x^4$
2. True
3. True
4. True
5. False, 0 is not in the domain of f

SECTION 4.5

1.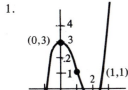

3.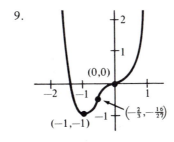

5.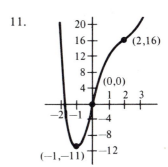

7.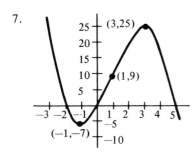

9.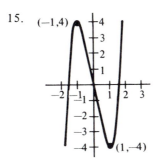

11.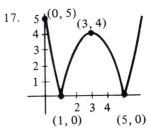

13.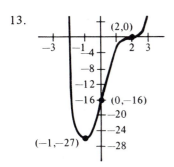

15.

17.

19.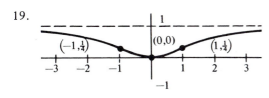

21.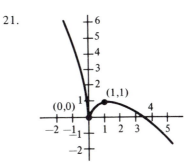

23. Domain: $(-\infty,2)(2,\infty)$

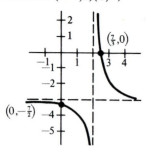

25. Domain: $(-\infty,-1),(-1,1),$
    $(1,\infty)$

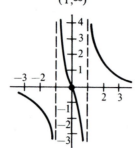

27. Domain: $(-\infty,4]$

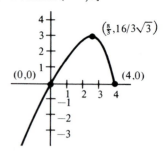

29. Domain: $(-\infty,0),(0,\infty)$

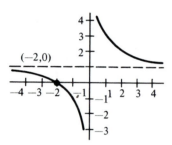

31. $a<0$ and $b^2<3ac$

33. $a<0$ and $b^2=3ac$

35. $a<0$ and $b^2>3ac$

SECTION 4.6

1. $K=0$

3. $K=4/17^{3/2}\approx.057, r=17^{3/2}/4\approx17.523$

5. $K=1/a, r=a$

7. a)

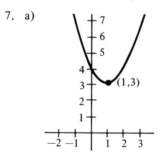

9. $(x-1)^2+\left(y-\dfrac{5}{2}\right)^2=\dfrac{1}{4}$

11. $K=\dfrac{16}{(16-3x^2)^{3/2}}$. In the inter-

    val $[-2,2]$, K is minimum when
    $x=0$ and maximum when $x=\pm2$

13. $a=1/4, b=2$

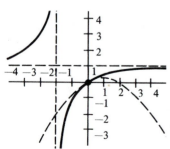

b) $K=\dfrac{2}{[1+4(x-1)^2]^{3/2}}$

c) $(1,3)$

d) $\displaystyle\lim_{x\to\pm\infty}K=0$

True and False

1. False
2. True

3. True
4. False, let $f(x) = x^2/2$, then the maximum curvature is 1

5. True

MISCELLANEOUS EXERCISES

1.

3.

5.

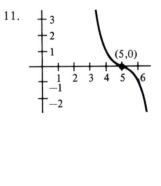

7.

9.

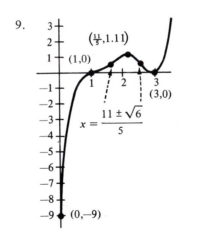

11.

13.

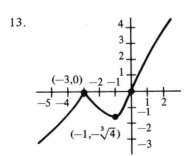

15.

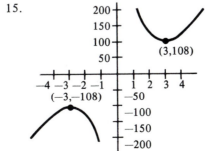

17.

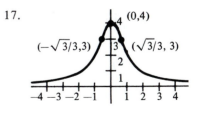

19.

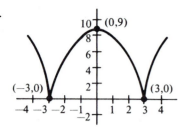

21.

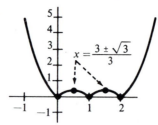

23.

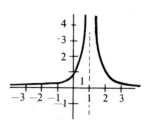

25. $(1,3)$, maximum
$(1,1)$, minimum

27. $c = \dfrac{-2 + \sqrt{85}}{3}$

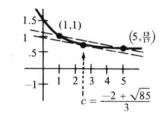

29. $c = (14/9)^3$

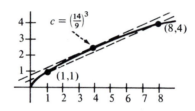

31. $c = 1$

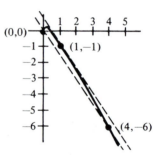

33. f is not differentiable in $(1,7)$

35. $-.347$

CHAPTER 5

SECTION 5.1

1. $l = w = 25$ ft

3. 12 and 6

5. 55 and 55

7. $\sqrt{192}$ and $\sqrt{192}$

9. $x = 25, y = 100/3$

11. $(7/2, \sqrt{7/2})$

13. $V = 128$ when $x = 2$

15. 0.392 ft $\times$ 1.215 ft $\times$ 2.215 ft

17. $50m \times 100/\pi$ m

19. $r \approx 1.51$ in., $h = 2r \approx 3.02$ in.

21. $x = 3, y = 3/2$

23. $x = y = 5\sqrt{2}/2 \approx 3.54$

25. $50/3$ in. $\times$ $50/3$ in. $\times$ $100/3$ in.

27. Radius: $8/(4 + \pi)$

29. 1 mi down coast from nearest point

31. $(4,0)$ and $(0,6)$

33. $w = 8\sqrt{3}, h = 8\sqrt{6}$

37. $4\pi r^3 / 3\sqrt{3}$

39. $I = 12$ amps

SECTION 5.2

1. a) $-7/12$ ft/sec
    b) $-3/2$ ft/sec
    c) $-48/7$ ft/sec

3. a) -750 mi/hr
    b) 20 min

5. $-\dfrac{14\sqrt{10}}{5} \approx -8.85$ ft/sec

7. $\dfrac{dA}{dt} = 2\pi r\,\dfrac{dr}{dt}$, if $\dfrac{dr}{dt}$ is constant, $\dfrac{dA}{dt}$ is not

9. 72π in.^{3}/min

11. $8/405\pi$ ft/min

13. a) 24.6%
     b) $1/32$ ft/min

15. a) $25/3$ ft/sec
     b) $10/3$ ft/sec

17. a) 9 cm^3/sec
     b) 900 cm^3/sec

19. a) 0 cm/min
     b) 12 cm/min

21. a) $8/25$ cm/min
     b) 0 cm/min
     c) $-8/25$ cm/min
     d) -0.0039 cm/min

23. 300 mi/hr

SECTION 5.3

1. 4500

3. 300

5. 200

7. 200

9. $x = 30$, $p = \$60$

11. $x = 1500$, $p = \$35.00$

13. 20

15. 200

17. Line should run from the power station to a point across the river $\dfrac{3}{2\sqrt{7}}$ mi downstream.

19. 8%

21. $x = 3$

23. a) 6.5%
     b) -1.3
     c) $-4/3$
     d) $x = 40/3$, $p = \sqrt{10/3}$

25. $p = 4\sqrt{3}/3$, $x = 32/3$

27. a) \$80.00
     b) \$99.29

29. \$92.50

SECTION 5.4

1. $6x\,dx$

3. $\dfrac{-3}{(2x-1)^2}\,dx$

5. $\dfrac{1-2x^2}{\sqrt{1-x^2}}\,dx$

7. a) $dy = 1, \Delta y = 1.0625$
   b) $dy = -1/2, \Delta y = -.4375$
   c) $dy = -1, \Delta y = -.7500$

9. $dy = -.0625, \Delta y = -.0525$

11. a) $dA = 2x\,\Delta x, \Delta A = 2x\,\Delta x + (\Delta x)^2$

13. Using $x = 1, \Delta x = 2,$
    $\sqrt[4]{83} \approx 3.01852$

15. Using $x = 27, \Delta x = -2,$
    $1/\sqrt[3]{25} \approx .34156$

17. $\sqrt[3]{63} \approx 3.97906$
    $\sqrt[4]{83} \approx 3.01835$
    $1/\sqrt{101} \approx 0.0995$
    $1/\sqrt[3]{25} \approx .34200$
    $\sqrt{50} \approx 7.07107$

19. a) 2.88π in.^{3}
    b) $.96\pi$ in.^{2}
    c) $1\%, 2/3\%$

21. a) $1/4\%$
    b) 216 sec

23. $d(uv) = u\,dv + v\,du$

$$d\left(\frac{u}{v}\right) = \frac{v\,du - u\,dv}{v^2}$$

True or False

1. True
2. False, if $dx < 0$, then $dy < 0$
3. True
4. True
5. False, let $f(x) = \sqrt{x}, x = 1,$ and $\Delta x = 3,$ then $dy = 3/2$ and $\Delta y = 1$

SECTION 5.5

1.

n	x_n	$f(x_n)$	$f'(x_n)$	$f(x_n)/f'(x_n)$	$x_n - f(x_n)/f'(x_n)$
1	1.700	−.110	3.400	−.032	1.732

3. .682

5. 1.146

7. 3.317

9. .57

11. Newton's method fails since $f'(1) = 0$

13. $x_{i+1} = \dfrac{(n-1)\,x_i^n + a}{nx_i^{n-1}}$

15. $x_{n+1} = \dfrac{3x_n^4 + 6}{4x_n^3}$

n	x_n
1	1.5
2	1.569
3	1.565
4	1.565

True or False

1. True
2. False, let $f(x) = (x^2 - 1)/(x - 1)$
3. True
4. True
5. True

MISCELLANEOUS EXERCISES

1. Approximately 4:55 p.m.
   64 miles

3. $48.00

5. $(0,0), (5,0), (0,10)$

7. 14.05 ft

11. $50\sqrt[3]{2} \approx 63$ mi/hr

13. a) $\sqrt{2}$ units/sec
    b) 1 unit/sec
    c) 1/2 unit/sec

15. 2/25 ft/min

17. 30,250 ft lbs

19. 5.133

21. $\Delta p = dp = -1/4$

23. a) 2%
    b) 3%

25. $-1.164, 1.453$

27. a) $x = 89, P = \$25,681.70$
    b) $x = 88, P = \$25,236.80$
    c) $x = 87, P = \$24,360.30$
    (*Note:* The producer should not pass on to the buyer the total tax.)

CHAPTER 6

SECTION 6.1

1. $\dfrac{x^4}{4} + 2x + C$

3. $\dfrac{2x^{5/2}}{5} + x^2 + x + C$

5. $\dfrac{3x^{5/3}}{5} + C$

7. $\dfrac{-1}{2x^2} + C$

9. $\dfrac{-1}{4x} + C$

11. $\dfrac{2\sqrt{x}}{15}(3x^2 + 5x + 15) + C$

13. $x^3 + \dfrac{x^2}{2} - 2x + C$

15. $t - \dfrac{2}{t} + C$

17. $\dfrac{2y^{7/2}}{7} + C$

19. $x + C$

21. $\dfrac{(1 + 2x)^5}{5} + C$

23. $\dfrac{(x^3 - 1)^5}{15} + C$

25. $\dfrac{(x^2 - 1)^8}{16} + C$

27. $4\sqrt{1 + x^2} + C$

29. $\dfrac{15(1 + x^2)^{4/3}}{8} + C$

31. $-3\sqrt{2x + 3} + C$

33. $\dfrac{-1}{2(x^2 + 2x - 3)} + C$

35. $\dfrac{-2}{1 + \sqrt{x}} + C$

37. $\dfrac{-1}{3(1+x^3)} + C$

39. $\dfrac{1}{2}\sqrt{1+x^4} + C$

41. $\sqrt{x} + C$

43. $\sqrt{2x} + C$

45. $-6\sqrt{36+x^2} + C$

47. $\dfrac{t^5}{5} - \dfrac{4t^3}{3} + 4t + C$

49. $\dfrac{y^{3/2}}{5}(30 - 2y) + C$

51. $\dfrac{1}{6}(2x-1)^3 + C_1$ or $\dfrac{4x^3}{3} - 2x^2 + x + C_2$, where $C_2 = C_1 - \dfrac{1}{6}$

53. $y = x^2 - x + 1$

55. $y = \dfrac{5 - (1-x^2)^{3/2}}{3}$

57. $f(x) = x^2 + x + 4$

59. $f(x) = -4\sqrt{x} + 3x$

61. $s = -16t^2 + 1600$, 10 sec

63. $40\sqrt{22} \approx 187.617$ ft/sec

65. a) $\dfrac{1+\sqrt{17}}{2} \approx 2.562$ sec

    b) $-16\sqrt{17} \approx -65.970$ ft/sec

67. a) $154/39 \approx 3.949$ ft/sec^2

    b) $1859/3 \approx 619.667$ ft

69. a) 300 ft

    b) 60 ft/sec ≈ 40.909 m/hr

73. $C = x^2 - 12x + 50$, $C/x = x - 12 + \dfrac{50}{x}$

75. $R = 10x - 3x^2 - \dfrac{2}{3}x^3$, $p = 10 - 3x - \dfrac{2}{3}x^2$

True or False

1. False

2. False, $\int(2x+1)^2\,dx = \dfrac{1}{6}(2x+1)^3 + C$

3. False, $\int x(x^2+1)\,dx = \dfrac{1}{4}(x^2+1)^2 + C$

4. False, $\dfrac{d}{dx}\left[\dfrac{-1}{x^2}\right] \neq \dfrac{1}{x}$

5. True
6. False, $1/2x$ cannot be moved outside the integral
7. True
8. True
9. True
10. True

SECTION 6.2

1. 35

3. 158/85

5. $4c$

7. 238

9. $\displaystyle\sum_{i=1}^{9} 1/3i$

11. $\displaystyle\sum_{j=1}^{8} [2(j/8) + 3]$

13. $\displaystyle\sum_{k=1}^{6} [(k/6)^2 + 2](1/6)$

15. $\displaystyle\sum_{i=1}^{n}\left[\left(\dfrac{2i}{n}\right)^3 - \left(\dfrac{2i}{n}\right)\right]\left(\dfrac{2}{n}\right)$

17. $\displaystyle\sum_{i=1}^{n}\left[2\left(1 + \dfrac{3i}{n}\right)^2\right]\left(\dfrac{3}{n}\right)$

19. 420

21. 2470

23. $1015/n^3$

25. 26/3   27. 4   29. Yes

31. Yes   33. No

SECTION 6.3

1. $s \approx .518, S \approx .768$ 3. $s \approx .646, S \approx .746$ 5. $s \approx .659, S \approx .859$

7. $A = 2$ 9. $A = 7/3$ 11. $A = 52/3$

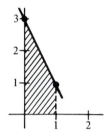   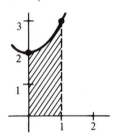   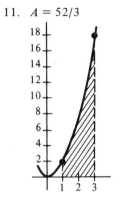

13. $A = 3/4$ 15. $A = 7/12$ 17. 36

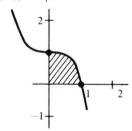   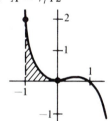   19. 0

21. 10/3

23. $A = 12$ 25. $A = 14$ 27. $A = \frac{1}{2} \pi r^2$

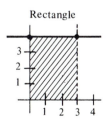
Rectangle

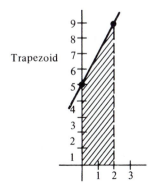
Trapezoid

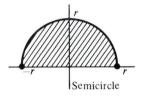
Semicircle

True or False

1. True
2. True
3. True
4. False, the right end points are $c_1 = 1 + \dfrac{3}{n}$,

$c_2 = 1 + \dfrac{6}{n}, \ldots, c_n = 4$

5. False, see Example 5, Section 6.3
6. True
7. True
8. False, $\displaystyle\int_0^1 x^2 \, dx = \lim_{x \to \infty} \sum_{i=1}^{n} \dfrac{i^2}{n^3}$
9. True
10. True

SECTION 6.4

1. 1
3. $-5/2$
5. $-10/3$
7. $1/3$
9. $1/2$
11. 36
13. -4
15. $2/3$
17. $-1/18$
19. $-27/20$
21. 2
23. 0
25. $6 - \dfrac{3}{2}\sqrt[3]{4} \approx 3.619$
27. 1
29. 4
31. $A = 6$

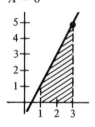

33. $A = 10/3$

33.

35. $A = 1/4$

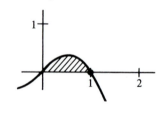

37. 10
39. 6
41. $32/3$
43. $8/5$
45. $12\sqrt{3}/5 \approx 4.157$

47. Average $= 8/3$

$x = \pm\, 2\sqrt{3}/3 \approx \pm 1.155$

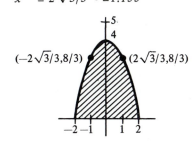

49. Average $= 4/3$

$x = \sqrt{2 + \dfrac{2\sqrt{5}}{3}} \approx 1.868$

$x = \sqrt{2 - \dfrac{2\sqrt{5}}{3}} \approx .714$

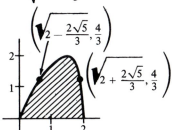

51. Average $= -2/3$

$x = \dfrac{4 + 2\sqrt{3}}{3} \approx 2.488$

$x = \dfrac{4 - 2\sqrt{3}}{3} \approx .179$

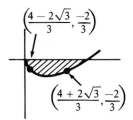

53. 0

55. a) 8/3   b) 16/3
    c) $-8/3$ d) 14/3
    e) 8

True or False

1. True
2. True
3. False, $\displaystyle\int_{1}^{2} (2x - 3)\, dx = \left[x^2 - 3x\right]_{1}^{2} =$

$(4 - 6) - (1 - 3) = 0$

4. False, $\displaystyle\int_{0}^{b} f(x)\, dx = F(b) - F(0)$

5. True

MISCELLANEOUS EXERCISES

1. $x^{2/3} + C$

3. $\dfrac{2x^3}{3} + \dfrac{x^2}{2} - x + C$

5. $\dfrac{2\sqrt{x}}{15}(15 + 10x + 3x^2) + C$

7. $\dfrac{2}{3}\sqrt{x^3 + 3} + C$

9. $\dfrac{x^7}{7} + \dfrac{3x^5}{5} + x^3 + x + C$

11. $y = 2 - x^2$

13. 240 ft/sec

15. a) 3 sec   c) $t = 3/2$ sec
     b) 144 ft   d) 108 ft

17. 37/210

19. 5

21. a) $\displaystyle\sum_{i=1}^{10} (2i-1)$ b) $\displaystyle\sum_{i=1}^{n} i^3$ 23. a) $S = \dfrac{5mb^2}{8}, s = \dfrac{3mb^2}{8}$ c) $\dfrac{mb^2}{2}$

   c) $\displaystyle\sum_{i=1}^{10} (4i+2)$ b) $S = \dfrac{mb^2(n+1)}{2n}, s = \dfrac{mb^2(n-1)}{2n}$ d) $\dfrac{mb^2}{2}$

25. 16   27. 0   29. 2

31. 422/5   33. $s \approx .734, S \approx .834$

$$\text{Average} = \frac{S+s}{2} \approx .784$$

35. Average $= 2/5, x = 29/4$ 37. Average $= 2, x = 2$ 39. a) \$20,650/M
                                                                       b) \$43,150/M

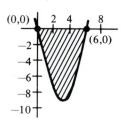

CHAPTER 7

SECTION 7.1

1. $A = 36$ 3. $A = 9/2$ 5. $A = 9$

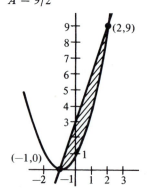

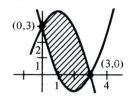

7. $A = \dfrac{8\sqrt{2}}{3}$

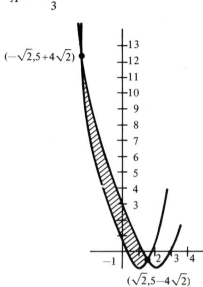

$(-\sqrt{2}, 5+4\sqrt{2})$

$(\sqrt{2}, 5-4\sqrt{2})$

9. $A = 1/12$

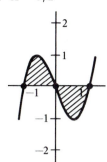

$(1,1)$

$(0,0)$

11. $A = 3/2$

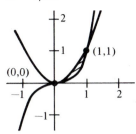

13. $A = \dfrac{75 - 17\sqrt{17}}{6} \approx .818$

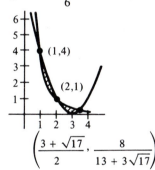

$(1,4)$

$(2,1)$

$\left(\dfrac{3 + \sqrt{17}}{2}, \dfrac{8}{13 + 3\sqrt{17}} \right)$

15. $A = 3/2$

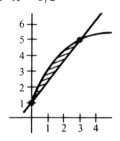

17. $A = 9/2$

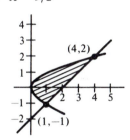

$(4,2)$

$(1,-1)$

19. $A = 6$

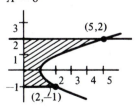

$(5,2)$

$(2,-1)$

21. $A = 8$

23. $A = \dfrac{ac}{2}$

25. $A = \displaystyle\int_{-1}^{1} (x^2 - x^4)\, dx = \dfrac{4}{15}$

27. 41.23 billion lb

True or False

1. True
2. True

3. False, $\displaystyle\int_{0}^{1} (x - x^2)\, dx$ represents the area of the given region

4. True
5. False

SECTION 7.2

1. a) 144π
    b) 96π

3. $4\pi/3$

5. a) $\pi/3$
    b) $\pi/3$
    c) $4\pi/3$

7. a) 8π
    b) $128\pi/5$
    c) $256\pi/15$
    d) $192\pi/5$

9. a) $32\pi/3$
    b) $64\pi/3$

11. $512\pi/15$

13. 18π

17. 100π

19. 30 ft

21. d, c, a, e, b

SECTION 7.3

1. $16\pi/3$

3. $8\pi/3$

5. $128\pi/5$

7. 8π

9. $16\pi/3$

11. 16π

13. 64π

15. $8\pi/3$

17. $192\pi/5$

19. a) $128\pi/7$
     b) $64\pi/5$
     c) $96\pi/5$
     d) $320\pi/7$

21. a) $\pi a^3/15$
     b) $\pi a^3/15$
     c) $4\pi a^3/15$

23. Diameter $= 2\sqrt{4 - 2\sqrt{3}} \approx 1.464$

SECTION 7.4

1. a) $128/3$ c) $16\pi/3$
    b) $32\sqrt{3}/3$ d) $32/3$

3. a) $\pi/80$ d) $3/80$
    b) $1/10$ e) $\pi/20$
    c) $\sqrt{3}/40$

5. $2r^3/3$

7. $45\pi/2$ ft$^3 \approx 528.7$ gal

SECTION 7.5

1. 30.625 in. lb ≈ 2.55 ft lb

3. 360 in. lb $= 30$ ft lb

5. 551,116.8π ft-lb ≈ 865.7 ft-ton

7. a) 53,913.6π ft lb ≈ 84.7 ft ton
    b) 20,217.6π ft lb ≈ 31.8 ft ton

9. 2,995.2π ft lb ≈ 4.7 ft ton

11. 95.65 mile ton $\approx 505,043.5$ ft ton

13. 337.5 ft lb

15. 168.75 ft lb

17. 7841.25 ft lb

SECTION 7.6

1. 1123.2 lb
3. 748.8 lb
5. 1064.96 lb
7. 748.8 lb
9. 15,163.2 lb
11. 112.5 lb
13. 967.2 lb

SECTION 7.7

1. $\bar{x} = -\dfrac{6}{7}$
3. $\bar{x} = 12$
5. a) $\bar{x} = 17$
   b) $\bar{x} = 12 + h$

7. (0,0)
9. (62/27, 64/27)

11. If the y-axis passes through the centers of both figures and the x-axis lies at the base of the square, then
$$(\bar{x},\bar{y}) = \left(0, \frac{4 + 3\pi}{4 + \pi}\right)$$

13. $(12/5, 3/4), M_x = 4, M_y = 64/5$
15. $(8/5, 0), M_x = 0, M_y = 256/15$
17. $(0, 4a/3\pi)$
19. $(1/2, 2/5)$

SECTION 7.8

1. $\dfrac{2}{3}(\sqrt{8} - 1) \approx 1.219$
3. $33/16 = 2.0625$
5. $779/240 \approx 3.246$

7. $\displaystyle\int_1^3 \frac{\sqrt{x^4 + 1}}{x^2}\, dx$
9. $\displaystyle\int_{-2}^1 \sqrt{2 + 4x + 4x^2}\, dx$
11. 4/3

SECTION 7.9

1. $\dfrac{\pi}{6}(17\sqrt{17} - 1) \approx 36.177$
3. $\dfrac{47\pi}{16} \approx 9.228$
5. $\dfrac{\pi}{27}(145\sqrt{145} - 10\sqrt{10})$
   ≈ 199.48

7. $2\pi r (r - \sqrt{r^2 - a^2}) = 2\pi rh$, where h is the height of the segment of the sphere

MISCELLANEOUS EXERCISES

1. $A = 4/5$

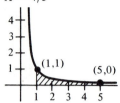

3. $A = 1$

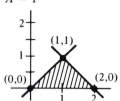

5. $A = 4/3$

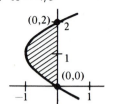

7. $A = 1/2$

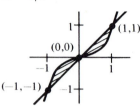

9. $A = 64/3$

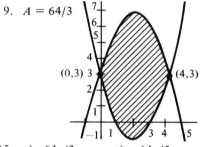

11. $A = 14/3$

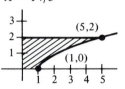

13. $A = 1/6$

15. a) $64\pi/3$ c) $64\pi/3$
    b) $128\pi/3$ d) $160\pi/3$

17. 15,600 lb on wall at shallow end
    62,400 lb on wall at deep end
    72,800 lb on side walls

19. a) 64π
    b) 48π

21. 50 in lb $\approx$ 4.167 ft lb

23. $104,000\pi$ ft lb $\approx$ 163.4 ft ton

25. 250 ft lb

27. 70 ft ton

29. $\pi r \sqrt{r^2 + h^2}$

31. $(a/5, a/5)$

CHAPTER 8

SECTION 8.1

1. $\log_2 8 = 3$

3. $\log_{27} 9 = 2/3$

5. $10^{-2} = .01$

7. $\ln 1 = 0$

9. $e^{.6931} = 2$

11. $e^{-.6931} = 1/2$

13. $x = 3$

15. $x = -3$

17. $x = 1/3$

19. $b = 3$

21. $x = 1/9$

23. $x = 4$

25. $x = 3$

27. $x = -1$ or 2

29. $x = 8$

31.
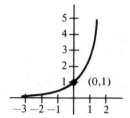

33.
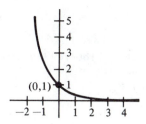

35.
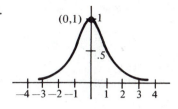

37.
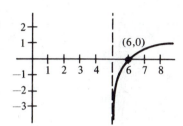

39.
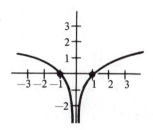

41.
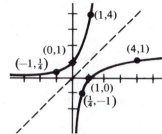

43.
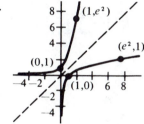

45.
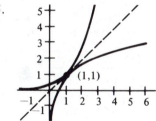

47. $\log_2 x + \log_2 y + \log_2 z$

49. $(1/2) \ln (a - 1)$

51. $-\log_2 5$

53. $2 + \ln 3$

55. -1

57. $\ln \left[\dfrac{x^3 y^2}{z^4} \right]$

59. $\ln \left[\dfrac{x}{x^2 - 1} \right]^2$

61. a) $4 = e^{\ln 4} \approx e^{1.386}$ d) $1/3 = e^{-\ln 3} \approx e^{-1.099}$

b) $\pi = e^{\ln \pi} \approx e^{1.145}$ e) $2e = e^{(1 + \ln 2)} \approx e^{1.693}$

c) $\sqrt{5} = e^{(\ln 5)/2} \approx e^{.805}$

63. a) 1.7917   d) .5493

b) $-.4055$ e) -1.3862

c) 4.3944   f) 3.1779

67. $e = 2.718281828\underline{4}\ldots$

$\dfrac{271801}{99990} = 2.718281828\underline{1}\ldots$

69. $2.7182\underline{5}\ldots$

71.

r	2%	4%	6%	8%	10%	12%
t	34.7	17.3	11.6	8.7	6.9	5.8

73. a) \$2061.03   d) \$2112.06

b) \$2088.15   e) \$2116.84

c) \$2102.35   f) \$2117.00

75. 397,635 years

True or False

1. False, $2^{3^2} = 2^9 = 512$
2. False, $2(1)^3 = 2$
3. True
4. False, $\log_a x + \log_a y = \log_a (xy)$

5. True
6. False, 271801/99990 is rational and e is irrational
7. True

8. True
9. True
10. False, $\dfrac{1}{2}(\ln x) = \ln (x^{1/2})$

SECTION 8.2

1. $\dfrac{2}{x}$

3. $\dfrac{2(x^3 - 1)}{x(x^3 - 4)}$

5. $\dfrac{4(\ln x)^3}{x}$

7. $\dfrac{2x^2 - 1}{x(x^2 - 1)}$

9. $\dfrac{1 - x^2}{x(x^2 + 1)}$

11. $\dfrac{1 - 2\ln x}{x^3}$

13. $\dfrac{2}{x \ln (x^2)}$

15. $\dfrac{1}{1 - x^2}$

17. $\dfrac{-4}{x(x^2 + 4)}$

19. $\dfrac{\sqrt{x^2 + 1}}{x^2}$

21. 1

23. $\log_3 e \left[\dfrac{3x - 2}{2x(x - 1)} \right]$

25. $\log_{10} e \left[\dfrac{x^2 + 1}{x(x^2 - 1)} \right]$

27. $\dfrac{dy}{dx} = \dfrac{3x^2 - 12x + 11}{2y} = \dfrac{3x^2 - 12x + 11}{2\sqrt{(x - 1)(x - 2)(x - 3)}}$

29. $\dfrac{dy}{dx} = \dfrac{-4xy}{3(x^4 - 1)} = \dfrac{-4x}{3(x^2 - 1)^{4/3}(x^2 + 1)^{2/3}}$

31. $\dfrac{dy}{dx} = \dfrac{6y(2 - x^2)}{(x^2 - 1)(x^2 - 4)} = \dfrac{6(2 - x^2)}{(x - 1)^2(x - 2)^2}$

33. $\dfrac{dy}{dx} = \dfrac{2y}{x^2}(1 - \ln x) = 2x^{2(1-x)/x}(1 - \ln x)$

35. $\dfrac{dy}{dx} = y\left[\dfrac{1}{x(1 + x)} - \dfrac{\ln (1 + x)}{x^2} \right] = \dfrac{1}{x}\left[(1 + x)^{(1-x)/x} - \dfrac{\ln (1 + x)}{x} \right]$

39.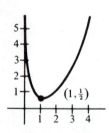

41.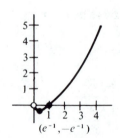

43. .567

47. a) 12.4
    b) 6.7
    c) 4.6

True or False

1. True
2. True

3. False, let $f(x) = \ln x$
4. True

5. False, $y' = 0$

SECTION 8.3

1. $\ln |x + 1| + C$

3. $-\dfrac{1}{2}\ln |3 - 2x| + C$

5. $\ln \sqrt{x^2 + 1} + C$

7. $4\ln 2 - \dfrac{3}{2} \approx 1.273$

9. $1/4$

11. $7/3$

13. $-\ln 3$

15. $2\sqrt{x + 1} + C$

17. $\dfrac{1}{3}\ln |x^3 + 3x^2 + 9x + 1| + C$

19. $3\ln |1 + x^{1/3}| + C$

21. $-\dfrac{2}{3}\ln |1 - x\sqrt{x}| + C$

23. $\ln |x - 1| + \dfrac{1}{2(x - 1)^2} + C$

25. $3 - 2\ln 4 \approx .227$

27. $4 + 5\ln 5 \approx 12.047$

29. a) $21\pi/64 \approx 1.031$
    b) $2\pi \ln 4 \approx 8.701$
    c) $2\pi(3 - \ln 4) \approx 10.139$

31. 900

33. 6015

35. 95.76%

37. $22.35°$

39. $11.75°$

41. $287,462$ ft-lb

True or False

1. False
2. True
3. True

4. False, the integrand, $1/x$, is not continuous throughout the entire interval $[-1, e]$

5. True

SECTION 8.4

1. $2e^{2x}$

3. $2(x - 1)e^{1 - 2x + x^2}$

5. $\dfrac{e^{\sqrt{x}}}{2\sqrt{x}}$

7. $3(e^x + e^{-x})^2(e^x - e^{-x})$

9. $2x$

11. $\dfrac{2e^{2x}}{1 + e^{2x}}$

13. $\dfrac{e^x - e^{-x}}{e^x + e^{-x}}$

15. $x^2 e^x$

17. $\ln 5(5^{x-2})$

19. $e^{-x}\left(\dfrac{1}{x} - \ln x\right)$

21. $\dfrac{1}{2}(1 - e^{-2}) \approx .432$

23. $\dfrac{e}{3}(e^2 - 1) \approx 5.789$

25. $x - \ln(1 + e^x) + C$

27. $\dfrac{1}{2a}e^{ax^2} + C$

29. $\dfrac{e}{3}(e^2 - 1) \approx 5.789$

31. $\dfrac{7}{\ln 4} \approx 5.044$

33. $-\dfrac{2}{3}(1 - e^x)^{3/2} + C$

35. $\ln |e^x - e^{-x}| + C$

37. $-\dfrac{5}{2} e^{-2x} + e^{-x} + C$

39. 4

41.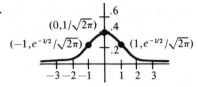

43.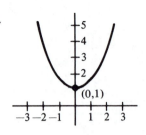

45.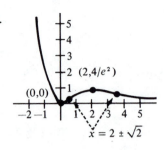

47. $y = x + 1$

49.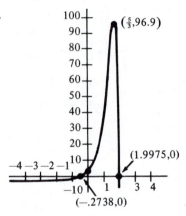

51. $e^5 - 1 \approx 147.413$

53. $2\left(1 - \dfrac{1}{\ln 3}\right) \approx .180$

55. $\sqrt{2\pi}\,(1 - e^{-1/2}) \approx .986$

59.

x	e^x	$1 + x + \dfrac{x^2}{2} + \dfrac{x^3}{6} + \dfrac{x^4}{24}$
2.0	7.3891	7.0000
1.0	2.7183	2.7083
0.5	1.6487	1.6484
0.1	1.1051709	1.1051708
0.0	1.0	1.0

61. $t = 5 \ln 19 \approx 14.72$

True or False

1. True

2. True

3. False, $f'(x) = ex^{e-1}$

4. True

5. True

SECTION 8.5

1. 3

3. 0

5. 2

7. $0, n < 2$
$1/2, n = 2$
$\infty, n > 2$

9. 0

11. 3/2

13. ∞

15. 0

17. $-3/2$

19. 1

21. 0

23. 1

25. 1

27. 0

29. 0

31.

x	10	10^2	10^4	10^6	10^8	10^{10}
$\dfrac{(\ln x)^4}{x}$	2.811	4.498	.720	.036	.001	.000

35. ∞

37. $v_0 + 32t$

True or False

1. False, L'Hopital's Rule doesn't apply since
$\lim\limits_{x \to 0} (x^2 + x + 1) \neq 0$

2. False, $y' = \dfrac{e^x(x-2)}{x^3}$

3. True

4. False, let $f(x) = x$ and $g(x) = x + 1$

5. False, let $f(x) = x$ and $g(x) = x + 1$

MISCELLANEOUS EXERCISES

1. .30105

3. .0792

5. 1.4313

7. 3

9. $5 \pm \sqrt{7}$

11. $1/2x$

13. $-2x$

15. $\dfrac{1 + 2 \ln x}{2 \sqrt{\ln x}}$

17. $\dfrac{-y}{x(2y + \ln x)}$

19. $\dfrac{x}{(a + bx)^2}$

21. $\dfrac{1}{x(a + bx)}$

23. $xe^x(x + 2)$

25. $\dfrac{e^{2x} - e^{-2x}}{\sqrt{e^{2x} + e^{-2x}}}$

27. $(\ln 3)3^{x-1}$

29. $\dfrac{-e^y - ye^x + y}{e^x + xe^y - x}$

31. $-\dfrac{1}{6} e^{-3x^2} + C$

33. $\dfrac{1}{7} \ln |7x - 2| + C$

35. $\ln |\ln (3x)| + C$

37. $\dfrac{x^2}{2} + 3 \ln |x| + C$

39. $\dfrac{5}{3} + \dfrac{e^3}{3} - e - \dfrac{1}{e} \approx 5.276$

41. $\ln |e^x - 1| + C$

43. $1 - e^{-1/2} \approx .393$

45. $\dfrac{1}{3} \ln |x^3 - 1| + C$

47. $-\dfrac{1}{4} + \ln\left(\dfrac{4}{3}\right) \approx .038$

49. $2 \sqrt{\ln x} + C$

51. a) .39   c) .25
    b) .78   d) .95

53. a) \$525.64
    b) \$824.36
    c) \$74,206.58

55. \$3,499.38

57. a) 7.53 lb
    b) 12.77 min

61. $\dfrac{1}{K}\left[32 - (32 - Kv_0)e^{-Kt}\right]$

63. $s(t) = \dfrac{32t}{K} - \dfrac{32}{K^2}(1 - e^{-Kt})$

65. a) $y = 28\,e^{0.6-0.012s}$

b)

speed	50	55	60	65	70
mileage	28.0	26.4	24.8	23.4	22.0

67. 0

69. ∞

71. 1

73. $1000\,e^{.09} \approx 1094.17$

CHAPTER 9

SECTION 9.1

1. a)

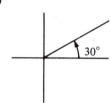

c)

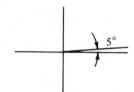

e)

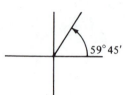

b)

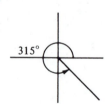

d)

3. a) $\pi/2$
   b) $2\pi/3$
   c) $-7\pi/4$
   d) $4\pi/3$
   e) $\pi/9$

5. a) $210°$ f) $157.56°$
   b) $225°$ g) $27.85°$
   c) $72°$ h) $171.89°$
   d) $20°$ i) $415.39°$
   e) $8°$ j) $57.30°$

7. a) $\sqrt{3}/3$
   b) 2
   c) 4/3
   d) 5/12
   e) 17/8

9. a) $-.1736$
   b) -1.4142
   c) $.7071$
   d) 1.1383
   e) -4.3864

11. $\pi/4, 3\pi/4, 5\pi/4, 7\pi/4$

13. $\pi/4, \pi, 5\pi/4, 0$

15. $0, 2\pi/3, 4\pi/3$

17. $\pi, 0, 3\pi/4, 5\pi/4$

19. $\pi/2, 3\pi/2$

21.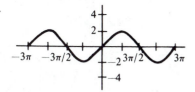

23.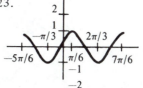

25.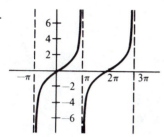

27.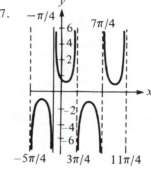

29.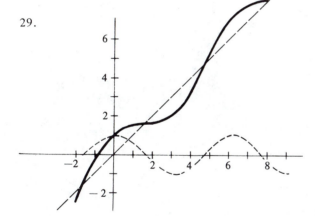

31.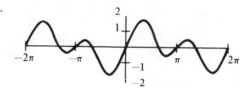

33. $\dfrac{100\sqrt{3}}{\sqrt{3}}$

35. $\dfrac{25\sqrt{3}}{\sqrt{3}}$

37. 15.56

39. 9.19

41. 1

43. 1

45. 2

47.

x	1	.1	.01	.001
$\dfrac{\sin 3x}{\sin 4x}$	$-.1864$	.7589	.7501	.7500

True or False

1. False
2. False

3. True
4. True

5. True

SECTION 9.2

1. $2\cos 2x$

3. $-\sin(x/2)$

5. $2x\sec(x^2)\tan(x^2)$

7. $x\cos x + \sin x$

9. $-\pi\sin\pi x$

11. $\cos 2x$

13. $\dfrac{1}{2}\cot x\ \sqrt{\sin x}$

15. $\pi\sec^2\left(\pi x - \dfrac{\pi}{2}\right)$

17. $\sin\left(\dfrac{1}{x}\right) - \dfrac{1}{x}\cos\left(\dfrac{1}{x}\right)$

19. $6\sec^3 2x\tan 2x$

21. $-\csc 2x\cot 2x$

23. $\tan^2 x$

25. $\sin^2 x$

27. $2e^x\cos x$

29. $\csc x$

31. $2e^x\tan(e^x)\sec^2(e^x)$

33. $3\cos 3x(1 - 3\sin^2 3x)$

35. $-\sec x\csc x$

37. $-\sin x\cos(\cos x)$

39. $\dfrac{-x^2}{x^2 + 1}$

45. $2/3$

47. -2

49. $1/2$

51. 1

53.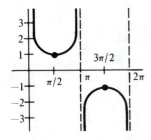

55.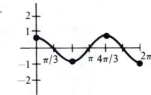

57.

59.

61. 2.029

63. $1/20$ rad/sec

65. a) $\dfrac{11}{1500}$ rad/sec

    b) $\dfrac{33}{1500}$ rad/sec

    c) $.027$ rad/sec

69. a) 0

    b) 10π ft/sec

    c) $10\sqrt{3}\pi$ ft/sec

71. a) $(0.222548, -0.217234)$

True or False

1. True
2. False, at $(\pi/4, \pi/4), f'(\pi/4) = \sqrt{2}/2$ and $g'(\pi/4) = -\sqrt{2}/2$

3. True
4. False

5. False, $f'(x) = 4\sin 2x\cos 2x$

SECTION 9.3

1. $-\dfrac{1}{2}\cos 2x + C$

3. $\dfrac{1}{2}\sin x^2 + C$

5. $2(\sqrt{3}-1) \approx 1.464$

7. $\dfrac{1}{3}\ln 2 \approx .231$

9. $\dfrac{1}{2}\ln|\sin 2x| + C$

11. $-1 + \sec(1) \approx .851$

13. $\dfrac{1}{2}\tan^2 x + C$

15. $-x - \cot x + C$

17. $\sin e^x + C$

19. $\dfrac{1}{5}\tan^5 x + C$

21. $3/4$

23. $e - 1$

25. $\ln|\sec x - 1| + C$

27. $\ln|\csc x - \cot x| + \cos x + C$

29. $-\ln|1 + \cos x| + C$

31. $-\dfrac{2}{3}(1 + \cos\theta)^{3/2} + C$

33. $\pi/4$

35. $-\theta + \csc 2\theta - \cot 2\theta + C$

37. 2

39. 2π

47. $2/\pi$

True or False

1. True
2. True
3. True

4. False, $\int \tan x\, dx = -\ln|\cos x| + C$

5. False, $\int \sin^2 2x \cos 2x\, dx = \dfrac{1}{6}\sin^3 2x + C$

SECTION 9.4

1. $\pi/6$
7. $\pi/4$
13. 1

3. $\pi/3$
9. $.663$
15. $y = x$

5. $\pi/6$
11. $1/2$
17. $y = \sqrt{1 - 4x^2}$

19. $y = \dfrac{\sqrt{x^2 - 1}}{x}$

21. $(\sin(1/2) + \pi)/3$

23. $1/3$

25. $\pi/2$

27. 0

SECTION 9.5

1. $\dfrac{2}{\sqrt{1 - 4x^2}}$

3. $\dfrac{2}{\sqrt{2x - x^2}}$

5. $\dfrac{-3}{\sqrt{4 - x^2}}$

7. $\dfrac{5}{1 + 25x^2}$

9. $\dfrac{1}{|x|\sqrt{x^2 - 1}}$

11. 0

13. $\dfrac{-t}{\sqrt{1-t^2}}$

15. $\dfrac{1}{2+3t^2}$

17. $\dfrac{1}{1-x^4}$

19. $\text{Arcsin } x$

21. $\pi/18$

23. $\pi/6$

25. $\text{Arcsec } |2x| + C$

27. $\dfrac{1}{2}\left[x^2 - \ln(x^2 + 1)\right] + C$

29. $\text{Arcsin }(x + 1) + C$

31. $\dfrac{1}{2}\text{Arcsin }(t^2) + C$

33. $\dfrac{1}{2}\text{Arcsec }\left|\dfrac{x-1}{2}\right| + C$

35. $\pi^2/32 \approx .308$

37. $(\sqrt{3} - 2)/2 \approx -.134$

39. $\text{Arcsin }(e^x) + C$

41. $\dfrac{1}{3}\text{Arctan }\left(\dfrac{x-3}{3}\right) + C$

43. $2\,\text{Arctan }\sqrt{x} + C$

45. $\pi/4$

47. $(0, \pi/2)$

49. $\left(\dfrac{\sqrt{3}}{2}, \dfrac{\pi}{3} - \sqrt{3}\right)$, minimum

51. $1/2$

53. $\pi/6$

   $\left(\dfrac{-\sqrt{3}}{2}, -\dfrac{\pi}{3} + \sqrt{3}\right)$, maximum

55. $\pi^2/4 \approx 2.467$

57. $(\sqrt{2}/2, \pi/4)$

59. $3/200$ rad/sec

61. $v = \sqrt{\dfrac{32}{k}}\,\tan\left[\text{Arctan }500\sqrt{\dfrac{k}{32}} - \sqrt{32\,k}\,t\right]$

True or False

1. True

2. True

3. False, $\displaystyle\int \dfrac{x}{1-x^2}\,dx = -\sqrt{1-x^2} + C$

4. True

5. False, dy/dx is undefined when $x = \pm 1$

SECTION 9.6

1. $-2x \cosh(1 - x^2)$

3. $\coth x$

5. $\operatorname{csch} x$

7. $\sinh^2 x$

9. $\operatorname{sech} x$

11. $\dfrac{y}{x}\left[\cosh x + x(\sinh x)\ln x\right]$

13. $-2e^{-2x}$

15. $-\dfrac{1}{2}\cosh(1 - 2x) + C$

17. $\dfrac{1}{3}\cosh^3(x - 1) + C$

19. $\ln|\sinh x| + C$

21. $-\coth\left(\dfrac{x^2}{2}\right) + C$

23. $\operatorname{csch}\left(\dfrac{1}{x}\right) + C$

25. $\dfrac{3}{\sqrt{9x^2 - 1}}$

27. $|\sec x|$

29. $2 \sec 2x$

31. $2\sinh^{-1}(2x)$

33. $\dfrac{1}{10}\ln 9 \approx .220$

35. $\pi/4$

37. $\dfrac{1}{2}$ Arctan $(x^2) + C$

39. $-\operatorname{csch}^{-1}(e^x) + C$

41. $2 \sinh^{-1}\sqrt{x} + C$

43. $(\ln(1 + \sqrt{2}), 1/\sqrt{2})$ and $(-\ln(1 + \sqrt{2}), 1/\sqrt{2})$

45.

47. $\dfrac{-\sqrt{a^2 - x^2}}{x}$

MISCELLANEOUS EXERCISES

1. $-9\sin(3x + 1)$

3. $\dfrac{-\sec^2\sqrt{1-x}}{2\sqrt{1-x}}$

5. $2x^2\cos(x^2) + \sin(x^2)$

7. $(x\cos x - 2\sin x)/x^3$

9. $(1 - x^2)^{-3/2}$

11. $2\cot x$

13. $x\sec^2 x + \tan x$

15. $\dfrac{x}{|x|\sqrt{x^2 - 1}} + \operatorname{Arcsec} x$

17. $1 + \cos 4x$

19. $\dfrac{1}{2} + x\operatorname{Arctan} x$

21. $\dfrac{1}{\sqrt{1 - (x - 2)^2}}$

23. $\dfrac{-(e^y + 2x\sin(x^2))}{xe^y}$

25. $(\operatorname{Arcsin} x)^2$

27. $\operatorname{Arctan}(\sin x) + C$

29. $\dfrac{1}{4}(\operatorname{Arctan} 2x)^2 + C$

31. $\dfrac{1}{n+1}\tan^{n+1} x + C$

33. $-\dfrac{1}{2}\ln(1 + e^{-2x}) + C$

35. $\dfrac{1}{2}\operatorname{Arctan}(e^{2x}) + C$

37. $\dfrac{1}{2}\operatorname{Arcsin}(x^2) + C$

39. $\dfrac{1}{2}\ln(16 + x^2) + C$

41. $\dfrac{1}{2}\sec 2x + C$

43. $\dfrac{1}{4}\left(\operatorname{Arctan}\dfrac{x}{2}\right)^2 + C$

45. $\dfrac{1}{2}\ln|\sec(x^2)| + C$

47. $K = \dfrac{|2\sec^2 x\tan x|}{(1 + \sec^4 x)^{3/2}}$, $\displaystyle\lim_{x \to \pi/2^-} K = 0$

49. 1.122

53. 1.895

55. $\sqrt{\dfrac{32}{L}}\sqrt{A^2 + B^2}$

57. $(a^{2/3} + b^{2/3})^{3/2}$

CHAPTER 10

SECTION 10.1

1. $\dfrac{1}{15}(3x - 2)^5 + C$

3. $-\dfrac{1}{5}(-2x + 5)^{5/2} + C$

5. $\dfrac{v^2}{2} - \dfrac{1}{6(3v - 1)^2} + C$

7. $-\dfrac{1}{3}\ln|-t^3 + 9t + 1| + C$

9. $\dfrac{x^2}{2} + x + \ln|x - 1| + C$

11. $\dfrac{1}{3}\ln\left|\dfrac{3x - 1}{3x + 1}\right| + C$

13. $-\dfrac{1}{2}\cos(t^2) + C$

15. $e^{\sin x} + C$

17. $-e^{-t} + 2t + e^t + C$

19. $\dfrac{1}{3}\sec 3x + C$

21. $2\ln|1 + e^x| + C$

23. $-\cot x - \csc x + C$

25. $\ln(t^2 + 4) - \dfrac{1}{2}\text{Arctan}\left(\dfrac{t}{2}\right) + C$

27. $3\ln|t + \sqrt{t^2 - 1}| + C$

29. $\dfrac{1}{2}\text{Arcsec}\,\dfrac{|x|}{2} + C$

31. $-\dfrac{1}{2}\text{Arcsin}(2t - 1) + C$

33. $\dfrac{1}{24}\ln\left|\dfrac{3x + 4}{3x - 4}\right| + C$

35. $\dfrac{1}{2}\text{Arcsin}(t^2) + C$

37. $\dfrac{1}{2}\text{Arctan}\left(\dfrac{\tan x}{2}\right) + C$

39. $\dfrac{1}{2}\ln\left|\cos\left(\dfrac{2}{t}\right)\right| + C$

41. $\dfrac{1}{2}\left(1 - \dfrac{1}{e}\right)$

43. $1/2$

45. 4

47. $\dfrac{1}{2}\ln 2$

49. $\dfrac{1}{2}\left(\text{Arcsec}\,4 - \dfrac{\pi}{3}\right) \approx .135$

51. $4/3$

53. $1/2$

55. $b = \sqrt{\ln\left(\dfrac{3\pi}{3\pi - 4}\right)} \approx .743$

True or False

1. True
2. True
3. False
4. True

5. False
6. True
7. False
8. True

9. True
10. False

SECTION 10.2

1. a) $(x + 3)^2 - 9$

   b) $\left(x - \dfrac{11}{2}\right)^2 - \dfrac{117}{4}$

   c) $\left(x^2 + \dfrac{3}{2}\right)^2 - \dfrac{17}{4}$

3. $\pi/2$

5. $\ln|x^2 + 6x + 13| - 3\,\text{Arctan}$
   $\left(\dfrac{x + 3}{2}\right) + C$

7. $-\dfrac{1}{8} \ln 3$

9. $\text{Arcsin} \left(\dfrac{x+2}{2} \right) + C$

11. $-\ln \left(\dfrac{1 + \sqrt{x^2 - 2x + 2}}{|x - 1|} \right) + C$

13. $\text{Arcsin}\,(x - 1) + C$

15. $4 - 2\sqrt{3} + \dfrac{\pi}{6} \approx 1.059$

17. $\dfrac{1}{2\sqrt{6}} \ln \left| \dfrac{\sqrt{2}(x+1) + \sqrt{3}}{\sqrt{2}(x+1) - \sqrt{3}} \right| + C$

19. $\dfrac{1}{2} \text{Arctan}\,(x^2 + 1) + C$

21. $\dfrac{1}{4} \text{Arcsin}\,(4x - 2) + C$

23. $\dfrac{1}{4} \text{Arcsin} \left(\dfrac{4x - 1}{9} \right) + C$

25. $-\dfrac{x^2}{2} - 4x - \dfrac{10}{3} \ln \left| \dfrac{x - 5}{x + 1} \right| + C$

27. $2\sqrt{9 - 8x - x^2} + 9\,\text{Arcsin} \left(\dfrac{x + 4}{5} \right) + C$

29. $\pi/3$

31. $2\pi^2/3$

33. $\dfrac{52}{31} \approx 1.677$

True or False

1. True
2. False, if $ax^2 + bx + c$ is a perfect square, the power rule is required

3. True
4. False

5. True

SECTION 10.3

1. $\dfrac{2}{5}(x - 3)^{3/2}(x + 2) + C$

3. $-\dfrac{2}{105}(1 - x)^{3/2}(15x^2 + 12x + 8) + C$

5. $\dfrac{\sqrt{2x - 1}}{15}(3x^2 + 2x - 13) + C$

7. $2(\sqrt{x - 1} - \text{Arctan}\,\sqrt{x - 1}) + C$

9. $2(2 - \sqrt{3})\sqrt{t} + C$

11. $2\left[\sqrt{x} - \ln\,(1 + \sqrt{x}) \right] + C$

13. $-(x + 2\sqrt{x + 1}) + C$

15. $-\dfrac{2}{3}\left[x^{3/2} + (x + 1)^{3/2} \right] + C$

17. $\dfrac{1}{9}\left[3\sqrt{2x} - \ln\,(3\sqrt{2x} + 1) \right] + C$

19. $2(\sqrt{2} - 1)\sqrt{x} + C$

21. $4/15$

23. $144/5$

25. $1209/28$

27. $7088/105$

29. $2 - \ln 3$

31. $4/15$

33. $32\pi/105$

35. $\dfrac{8}{15}(1 + 6\sqrt{3}) \approx 6.076$

37. a) 0.353
    b) 0.586

True or False

1. False, $\displaystyle\int_1^5 x\sqrt{x-1}\,dx =$

   $\displaystyle\int_0^2 (u^2+1)(u)(2u)\,du$

2. True
3. False

4. True
5. False, $\sqrt{x^2-x^3} = |x|\sqrt{1-x}$

SECTION 10.4

1. $\dfrac{1}{2}\ln\left|\dfrac{x-1}{x+1}\right| + C$

3. $\ln\left|\dfrac{x-1}{x+2}\right| + C$

5. $\dfrac{3}{2}\ln|2x-1| - 2\ln|x+1| + C$

7. $5\ln|x-2| - \ln|x+2| - 3\ln|x| + C$

9. $x^2 + \dfrac{3}{2}\ln|x-4| - \dfrac{1}{2}\ln|x+2| + C$

11. $\dfrac{1}{x} + \ln|x^4 + x^3| + C$

13. $\dfrac{x^2}{2} + 3x + 6\ln|x-1| - \dfrac{4}{x-1} - \dfrac{1}{2(x-1)^2} + C$

15. $3\ln|x-3| - \dfrac{9}{x-3} + C$

17. $\ln\left|\dfrac{x^2+1}{x}\right| + C$

19. $\dfrac{1}{6}\left[\ln\left|\dfrac{x-2}{x+2}\right| + \sqrt{2}\,\text{Arctan}\left(\dfrac{x}{\sqrt{2}}\right)\right] + C$

21. $\dfrac{1}{16}\ln\left|\dfrac{4x^2-1}{4x^2+1}\right| + C$

23. $\dfrac{\sqrt{2}}{2}\,\text{Arctan}\left(\dfrac{x}{\sqrt{2}}\right) - \dfrac{1}{2(x^2+2)} + C$

25. $\ln|x+1| + \sqrt{2}\,\text{Arctan}\left(\dfrac{x-1}{\sqrt{2}}\right) + C$

27. $\ln 2$

29. $\dfrac{1}{2}\ln\left(\dfrac{8}{5}\right) - \dfrac{\pi}{4} + \text{Arctan}(2) \approx .557$

31. $\ln\left|\dfrac{\cos x}{\cos x - 1}\right| + C$

33. $\dfrac{1}{5}\ln\left|\dfrac{e^x-1}{e^x+4}\right| + C$

35. $\ln\left|\dfrac{-1+\sin x}{2+\sin x}\right| + C$

37. $6 - \dfrac{7}{4}\ln 7 \approx 2.595$

39. $x = \dfrac{n[e^{k(n+1)t} - 1]}{e^{k(n+1)t} + n}$

True or False

1. False, $\dfrac{x^2-1}{(x-2)(x-3)} = 1 - \dfrac{3}{x-2} + \dfrac{8}{x-3}$

2. False, $\dfrac{1}{(x-1)(2x-2)(x-3)} = \dfrac{1/2}{(x-1)^2(x-3)} = \dfrac{1}{8}\left[\dfrac{-1}{x-1} - \dfrac{2}{(x-1)^2} + \dfrac{1}{x-3}\right]$

3. False, if $a = 0$, then $x^3 - x^2 = x^2(x-1)$

4. True

5. True

SECTION 10.5

1. $-\dfrac{1}{4}\cos^4 x + C$

3. $\dfrac{1}{12}\sin^6(2x) + C$

5. $-\dfrac{1}{3}\cos^3 x + \dfrac{2}{5}\cos^5 x - \dfrac{1}{7}\cos^7 x + C$

7. $\dfrac{1}{4}[2x + \sin(2x)] + C$

9. $\dfrac{x}{16} - \dfrac{\sin 4x}{64} + \dfrac{\sin^3 2x}{48} + C$

11. $\dfrac{1}{16}[2x - \sin(2x)] + C$

13. $\dfrac{1}{2}\tan^2 x + C$

15. $\dfrac{1}{3}\sec^3 x + C$

17. $\dfrac{1}{9}\sec^3(3x) - \dfrac{1}{3}\sec(3x) + C$

19. $\dfrac{1}{4}[\ln|\csc^2(2x)| - \cot^2(2x)] + C$

21. $-\cot\theta - \dfrac{1}{3}\cot^3\theta + C$

23. $\ln|\csc(t) - \cot(t)| + \cos(t) + C$

25. $-\dfrac{1}{10}[\cos(5x) + 5\cos x] + C$

27. $\dfrac{1}{8}[2\sin(2\theta) - \sin(4\theta)] + C$

29. $\ln|\csc x - \cot x| + \cos x + C$

31. π

33. $\dfrac{1}{2}(1 - \ln 2)$

35. $\ln 2$

37. $3\sqrt{2}/10$

39. $4/3$

45. $\pi^2/2$

47. π

49. 0

SECTION 10.6

1. $\dfrac{x}{25\sqrt{25-x^2}} + C$

3. $5\ln\left|\dfrac{5-\sqrt{25-x^2}}{x}\right| + \sqrt{25-x^2} + C$

5. $\sqrt{x^2+9} + C$

7. 2π

9. $\ln|x + \sqrt{x^2-9}| + C$

11. $\sqrt{3} - \dfrac{\pi}{3}$

13. $-\dfrac{(1-x^2)^{3/2}}{3x^3} + C$

15. $-\dfrac{1}{3} \ln \left| \dfrac{3 + \sqrt{4x^2 + 9}}{2x} \right| + C$

17. $\dfrac{-1}{\sqrt{x^2 + 3}} + C$

19. $\dfrac{1}{15}(x^2 - 4)^{3/2}(3x^2 + 8) + C$

21. $\dfrac{1}{3}(1 + e^{2x})^{3/2} + C$

23. $\dfrac{1}{20}\left[5 \text{ Arctan} \left(\dfrac{1}{2} \right) - 2 \right] \approx 0.0159$

25. $\dfrac{1}{3}(x^2 + 2x + 2)^{3/2} + C$

27. $\dfrac{1}{2}[e^x \sqrt{1 - e^{2x}} + \text{Arcsin} (e^x)] + C$

29. $\dfrac{1}{4}\left[\dfrac{x}{x^2 + 2} + \dfrac{1}{\sqrt{2}} \text{ Arctan} \left(\dfrac{x}{\sqrt{2}} \right) \right] + C$

31. $\dfrac{1}{2}\left[3 \ln (x^2 + 1) + \dfrac{x}{x^2 + 1} + \text{Arctan } x \right] + C$

33. $1 - \dfrac{\sqrt{3}\pi}{6}$

35. $\dfrac{25\pi}{12}$

37. 187.2π lb

39. $6\pi^2$

True or False

1. True

4. True

2. False, $\displaystyle\int \dfrac{\sqrt{x^2 - 1}}{x} dx = \int \tan^2 \theta \, d\theta$

5. False, $\displaystyle\int \dfrac{\sqrt{1 - x^2}}{x^2} dx = \int (\csc^2 \theta - 1) \, d\theta$

3. False, $\displaystyle\int_0^3 \dfrac{dx}{(1 + x^2)^{3/2}} = \int_0^{\pi/3} \cos \theta \, d\theta$

SECTION 10.7

1. $x \sin x + \cos x + C$

3. $-e^{-x}(x + 1) + C$

5. $x \ln x - x + C$

7. $x \tan x + \ln |\cos x| + C$

9. $x \text{ Arcsec } (2x) + \dfrac{1}{2} \ln |2x - \sqrt{4x^2 - 1}| + C$

11. $x \text{ Arctan } x - \dfrac{1}{2} \ln (1 + x^2) + C$

13. $\dfrac{-1}{2(\ln x)^2} + C$

15. $\dfrac{2}{15}(x - 1)^{3/2}(3x + 2) + C$

17. $\dfrac{2}{27}\sqrt{2 + 3x}(3x - 4) + C$

19. $\dfrac{1}{5}e^{2x}(2 \sin x - \cos x) + C$

21. $\dfrac{1}{8}(2 \sec^3 x \tan x + 3 \sec x \tan x + 3 \ln |\sec x + \tan x|) + C$

23. $\dfrac{1}{8}(\sin 2x - 2x \cos 2x) + C$

25. $\dfrac{1}{8}[2x^2 - 2x \sin 2x - \cos 2x] + C$

27. $\dfrac{1}{6}(3x\sqrt{4 + 9x^2} + 4 \ln |3x + \sqrt{4 + 9x^2}|) + C$

29. $\dfrac{1}{2}(x\sqrt{x^2 - 1} + \ln |x + \sqrt{x^2 - 1}|) + C$

31. $\dfrac{e[\sin(1) - \cos(1)] + 1}{2} \approx .909$ 33. π

43. $\sqrt{26} - \sqrt{2} + \ln(1 + \sqrt{2}) - \ln\left(\dfrac{1 + \sqrt{26}}{5}\right) \approx 4.367$ 45. $\dfrac{\pi}{32}[102\sqrt{2} - \ln(2\sqrt{2} + 3)] \approx 13.989$

47. $50[2\sqrt{2} + \ln(2\sqrt{2} + 3)] \approx 229.559$ ft   49. 2π

51. $18[2\sqrt{2} + \ln(2\sqrt{2} + 3)] \approx 82.641$ meters   55. $b_n = \dfrac{8h}{(n\pi)^2} \sin\dfrac{n\pi}{2}$

SECTION 10.8

1. $\ln\left|\dfrac{1 - \sqrt{1 - x}}{1 + \sqrt{1 - x}}\right| + C$ 3. $\pi/6$

7. $\sqrt{x^2 - 4} - 2\,\text{Arcsec}\left|\dfrac{x}{2}\right| + C$

5. $2\sqrt{x} - 3\sqrt[3]{x} + 6\sqrt[6]{x} - 6\ln(\sqrt[6]{x} + 1) + C$

9. $9/16$ 11. $2[\sqrt{x - 1} - \ln(1 + \sqrt{x - 1})] + C$

13. $\dfrac{-\sqrt{x^2 + 4x}}{2x} + C$ 15. $4\left[\dfrac{1}{1 + \sqrt[4]{x}} + \ln\left(\dfrac{\sqrt[4]{x}}{1 + \sqrt[4]{x}}\right)\right] + C$

17. $-\ln\left|\dfrac{2\sqrt{x^2 + x + 1}}{x} + \dfrac{x + 2}{x}\right| + C$ 19. $\dfrac{-\sqrt{4 - x^2}}{2x^2} + \dfrac{1}{4}\ln\left|\dfrac{2 + \sqrt{4 - x^2}}{x}\right| + C$

21. $\ln 2$ 23. $\dfrac{1}{2}\ln(3 - 2\cos\theta) + C$

25. $2\sin\sqrt{\theta} + C$ 27. $-\csc\theta + C$

31. a) $141/10$ d) $\bar{x} = 2517/564 \approx 4.463$
    b) $489\pi/14 \approx 109.731$ $\bar{y} = 815/658 \approx 1.239$
    c) $2517\pi/20 \approx 395.369$

SECTION 10.9

1. $-\dfrac{x}{2}(2 - x) + \ln|1 + x| + C$ 3. $\dfrac{-\sqrt{1 - x^2}}{x} + C$

5. $\dfrac{x^3}{9}(-1 + 3\ln x) + C$ 7. $\dfrac{\sqrt{x^2 - 4}}{4x} + C$

9. $\dfrac{1}{2}e^{x^2} + C$

11. $\dfrac{2}{9}\left(\ln|1 - 3x| + \dfrac{1}{1 - 3x}\right) + C$

13. $e^x \operatorname{Arccos} e^x - \sqrt{1 - e^{2x}} + C$

15. $\dfrac{1}{16}x^4(4 \ln|x| - 1) + C$

17. $\dfrac{1}{27}\left[3x - \dfrac{25}{3x - 5} + 10 \ln|3x - 5|\right] + C$

19. $\dfrac{1}{2}[x^2 + \cot x^2 + \csc x^2] + C$

21. $\operatorname{Arctan}(\sin x) + C$

23. $x - \dfrac{1}{2}\ln(1 + e^{2x}) + C$

25. $\dfrac{\sqrt{2}}{2}\operatorname{Arctan}\dfrac{1 + \sin\theta}{\sqrt{2}} + C$

27. $\dfrac{-\sqrt{2 + 9x^2}}{2x} + C$

29. $\dfrac{1}{2}\left[e^x\sqrt{1 + e^{2x}} + \ln(e^x + \sqrt{1 + e^{2x}})\right] + C$

31. $\dfrac{1}{16}[6x - 3\sin(2x)\cos(2x) - 2\sin^3(2x)\cos(2x)] + C$

33. $-2(\cot\sqrt{x} + \csc\sqrt{x}) + C$

35. $(t^4 - 12t^2 + 24)\sin t + (4t^3 - 24t)\cos t + C$

37. $\dfrac{1}{2}[(x^2 + 1)\operatorname{Arcsec}(x^2 + 1) - \ln(x^2 + 1 + \sqrt{x^4 + 2x^2})] + C$

39. $\dfrac{1}{4}[2\ln|x| - 3\ln|3 + 2\ln|x|\,|] + C$

41. $\sqrt{2 - 2x - x^2} - \sqrt{3}\ln\left|\dfrac{\sqrt{3} + \sqrt{2 - 2x - x^2}}{x + 1}\right| + C$

43. $\dfrac{1}{2}\operatorname{Arctan}(x^2 - 3) + C$

45. $\dfrac{1}{2}\ln|x^2 - 3 + \sqrt{x^4 - 6x^2 + 5}| + C$

47. $-\dfrac{1}{3}\sqrt{4 - x^2}\,(x^2 + 8) + C$

49. $-2\sqrt{\dfrac{1 - x}{x}} + C$

51. $\dfrac{2}{1 + e^x} - \dfrac{1}{2(1 + e^x)^2} + \ln|1 + e^x| + C$

SECTION 10.10

	Exact	Trapezoidal	Simpson's		Trapezoidal	Simpson's
1.	2.6667	2.7500	2.6667	21.	1/2	0
3.	4.0000	4.2500	4.0000	23.	$5e/64 \approx .212$	$13e/1024 \approx .035$
5.	4.0000	4.0625	4.0000	25.	.05	.01
7.	.5000	.5090	.5004	27.	130	
9.	.7854	.7828	.7854	29. a)	.3414	
11.		.957	.978	b)	.4772	

13.   3.41   3.22

15.   .342   .372

17.   .334   .305

19.   .194   .186

SECTION 10.11

1. 4   3. 6   5. Diverges

7. $-1/4$ 9. Diverges   11. $\ln(2+\sqrt{3})$

13. Diverges   15. 0   17. Diverges

19. 1   21. 1/2   23. π

25. $\pi/4$ 27. $p > 1$ 29. 6

33. Converges   35. $8\pi^2$

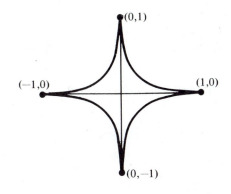

True or False

1. True   3. False, let $f(x) = \dfrac{1}{x+1}$ 4. True

2. True   5. True

MISCELLANEOUS EXERCISES

1. $x + \dfrac{9}{8}\ln|x-3| - \dfrac{25}{8}\ln|x+5| + C$ 3. $\tan\theta + \sec\theta + C$

5. $\dfrac{e^{2x}}{13}\,[2\sin(3x) - 3\cos(3x)] + C$ 7. $-\dfrac{1}{x}\,[1 + \ln(2x)] + C$

9. $\dfrac{1}{2}\left[4\,\text{Arcsin}\left(\dfrac{x}{2}\right) + x\,\sqrt{4-x^2}\right] + C$ 11. $\dfrac{3\,\sqrt{4-x^2}}{x} + C$

13. $\dfrac{2}{3}\left[\tan^3\left(\dfrac{x}{2}\right) + 3\tan\left(\dfrac{x}{2}\right)\right] + C$

15. $\dfrac{3}{2}\ln\left|\dfrac{x-3}{x+3}\right| + C$

17. $\dfrac{1}{4}\left[6\ln|x-1| - \ln(x^2+1) + 6\,\text{Arctan}\,(x)\right] + C$

19. $\dfrac{1}{2}\left[3\ln(x^2+1) - \dfrac{1}{x^2+1}\right] + C$

21. $16\,\text{Arcsin}\left(\dfrac{x}{4}\right) + C$

23. $\dfrac{1}{2}\,\text{Arctan}\left(\dfrac{e^x}{2}\right) + C$

25. $\dfrac{1}{2}\ln(x^2+4x+8) - \text{Arctan}\left(\dfrac{x+2}{2}\right) + C$

27. $\dfrac{1}{8}(\sin 2\theta - 2\theta\cos 2\theta) + C$

29. $\dfrac{1}{2}(2\theta - \cos 2\theta) + C$

31. $\dfrac{4}{3}\left[x^{3/4} - 3x^{1/4} + 3\,\text{Arctan}\,(x^{1/4})\right] + C$

33. $2\sqrt{1-\cos x} + C$

35. $x\ln|x^2+x| + \ln|x+1| - 2x + C$

37. $\sin x\,\ln(\sin x) - \sin x + C$

39. $x - \dfrac{1}{2(x^2+1)} + C$

41. $\dfrac{2\pi}{3\sqrt{3}} \approx 1.21$

43. 1.40 (Using Simpson's Rule, $n = 8$)

45. $.74$ (Using Simpson's Rule, $n = 4$)

47. $\dfrac{1}{2}\ln\left|\dfrac{2-\sqrt{x^2+4}}{x}\right| + C$

49. $\dfrac{2}{15}(4+x)^{3/2}(3x-8) + C$

51. a) 1.00 c) $.85$
    b) $.78$ d) 1.01

53. a) $\Gamma(1) = 1$
    b) $\Gamma(2) = 1$
    c) $\Gamma(3) = 2$

55. $0.015846 < \displaystyle\int_2^8 \dfrac{1}{x^5-1}\,dx < 0.015851$

57. $(17/5, 0)$

59. 3.82

CHAPTER 11

SECTION 11.1

1. Vertex: $(0,0)$
   Focus: $(0,1/16)$
   Directrix: $y = -1/16$

3. Vertex: $(0,0)$
   Focus: $(-3/2,0)$
   Directrix: $x = 3/2$

5. Vertex: $(0,0)$
   Focus: $(0,-2)$
   Directrix: $y = 2$

7. Vertex: $(1,-2)$
   Focus: $(1,-4)$
   Directrix: $y = 0$

9. Vertex: $(5,-1/2)$
   Focus: $(11/2,-1/2)$
   Directrix: $x = 9/2$

11. Vertex: $(1,1)$
    Focus: $(1,2)$
    Directrix: $y = 0$

13. Vertex: (8,−1)
    Focus: (9,−1)
    Directrix: $x = 7$

15. Vertex: (−2,−3)
    Focus: (−4,−3)
    Directrix: $x = 0$

17. Vertex: (−1,2)
    Focus: (0,2)
    Directrix: $x = −2$

19. Vertex: (−2,2)
    Focus: (−2,1)
    Directrix: $y = 3$

21. $x^2 + 6y = 0$

23. $y^2 − 4y + 8x − 20 = 0$

25. $x^2 − 8y + 32 = 0$

27. $x^2 + 8y − 16 = 0$

29. $5x^2 − 14x − 3y + 9 = 0$

31. $x^2 + y − 4 = 0$

33. For vertex at $(h,3)$: $(x − h)^2 = 8(y − 3)$
    For vertex at $(h,−1)$: $(x − h)^2 = −8(y + 1)$

35. a) $x^2 − 800y = 0$

    b) $100\left[\sqrt{5} + 4\ln\left(\dfrac{1 + \sqrt{5}}{2}\right)\right] \approx 416.1 \text{ ft}$

37. a) $10\sqrt{3} \approx 17.32$ feet from the point directly
       below the end of the pipe.

    b) $4x^2 + 25y − 1200 = 0$

True or False

1. False

2. True, $y = 4 − x^2$

3. False, the parabola given by
   $y = 3 − x^2$ is the only parabola
   passing through $(−1,2),(0,3)$
   and $(1,2)$

4. False

5. True

SECTION 11.2

1. Center: (0,0)
   Foci: (±3,0)
   Vertices: (±5,0)
   $e = 3/5$

3. Center: (0,0)
   Foci: (0,±3)
   Vertices: (0,±5)
   $e = 3/5$

5. Center: (0,0)
   Foci: (±2,0)
   Vertices: (±3,0)
   $e = 2/3$

7. Center: (0,0)
   Foci: $(\pm\sqrt{3},0)$
   Vertices: (±2,0)
   $e = \sqrt{3}/2$

9. Center: (0,0)
   Foci: (0,±1)
   Vertices: $(0,\pm\sqrt{3})$
   $e = \sqrt{3}/3$

11. Center: (0,0)
    Foci: $(0,\pm\sqrt{3}/2)$
    Vertices: (0,±1)
    $e = \sqrt{3}/2$

13. Center: (1,5)
    Foci: (1,9),(1,1)
    Vertices: (1,10),(1,0)
    $e = 4/5$

15. Center: (−2,3)
    Foci: $(-2,3\pm\sqrt{5})$
    Vertices: (−2,6),(−2,0)
    $e = \sqrt{5}/3$

17. Center: (1,−1)
    Foci: $(1\pm 3\sqrt{10}/20,-1)$
    Vertices: $(1\pm\sqrt{10}/4,-1)$
    $e = 3/5$

19. Center: $(1/2, -1)$

    Foci: $\left(\dfrac{1}{2} \pm \sqrt{2}, -1\right)$

    Vertices: $\left(\dfrac{1}{2} \pm \sqrt{5}, -1\right)$

    $e = \sqrt{10}/5$

21. $\dfrac{x^2}{9} + \dfrac{y^2}{5} = 1$

23. $\dfrac{x^2}{25} + \dfrac{y^2}{16} = 1$

25. $\dfrac{(x-2)^2}{4} + \dfrac{(y-2)^2}{1} = 1$

27. $\dfrac{(x-3)^2}{9} + \dfrac{(y-5)^2}{16} = 1$

29. $\dfrac{x^2}{24} + \dfrac{y^2}{49} = 1$

33. At $(3, 5\sqrt{7}/4)$: $15x + 4\sqrt{7}y - 80 = 0$

    At $(3, -5\sqrt{7}/4)$: $15x - 4\sqrt{7}y - 80 = 0$

35. $(0, 25/3)$

37. a) 2π

    b) $8\pi/3$

    c) $\dfrac{2\pi}{9}(9 + 4\sqrt{3}\pi) \approx 21.48$

    d) $16\pi/3$

    e) $\dfrac{4\pi}{3}[6 + \sqrt{3}\ln(2+\sqrt{3})] \approx 34.69$

    f) 9.69

39. $\dfrac{x^2}{36} + \dfrac{y^2}{9} = 1$

43. a) $.937$

    b) $\dfrac{(x+65{,}059.5)^2}{(65{,}059.5)^2} + \dfrac{y^2}{(22{,}727.2)^2} = 1$

True or False

1. True
2. True
3. False
4. True
5. False

SECTION 11.3

1. Center: $(0,0)$
   Vertices: $(\pm 1, 0)$
   Foci: $(\pm\sqrt{2}, 0)$

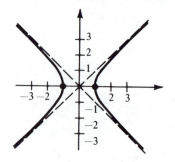

3. Center: $(0,0)$
   Vertices: $(0, \pm 1)$
   Foci: $(0, \pm\sqrt{5})$

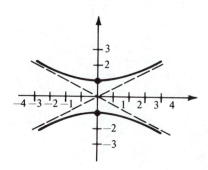

5. Center: $(0,0)$
   Vertices: $(0, \pm 5)$
   Foci: $(0, \pm 13)$

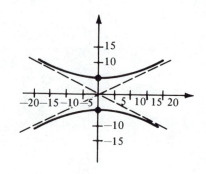

7. Center: $(0,0)$
   Vertices: $(\pm\sqrt{3},0)$
   Foci: $(\pm\sqrt{5},0)$

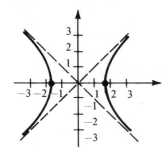

9. Center: $(0,0)$
   Vertices: $(0,\pm2)$
   Foci: $(0,\pm3)$

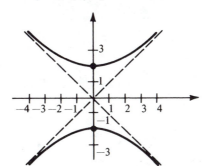

11. Center: $(1,-2)$
    Vertices: $(-1,-2),(3,-2)$
    Foci: $(1\pm\sqrt{5},-2)$

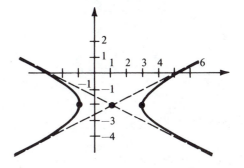

13. Center: $(2,-6)$
    Vertices: $(2,-5),(2,-7)$
    Foci: $(2,-6\pm\sqrt{2})$

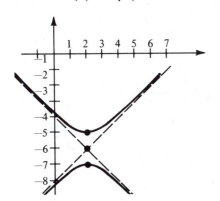

15. Center: $(2,-3)$
    Vertices: $(1,-3),(3,-3)$
    Foci: $(2\pm\sqrt{10},-3)$

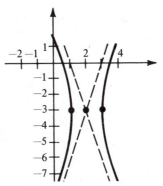

17. Center: $(1,-3)$
    Vertices: $(1,-3\pm\sqrt{2})$
    Foci: $(1,-3\pm2\sqrt{5})$

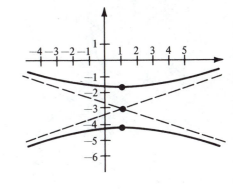

19. Degenerate hyperbola: graph is two intersecting lines with center at $(-1,-3)$

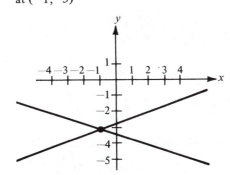

21. $\dfrac{y^2}{4} - \dfrac{x^2}{12} = 1$

23. $\dfrac{x^2}{1} - \dfrac{y^2}{9} = 1$

25. $\dfrac{(x-3)^2}{9} - \dfrac{(y-2)^2}{4} = 1$

27. $\dfrac{y^2}{9} - \dfrac{(x-2)^2}{9/4} = 1$

29. $\dfrac{(x-6)^2}{9} - \dfrac{(y-2)^2}{7} = 1$

31. At $(6,\sqrt{3})$: $\quad 2x - 3\sqrt{3}y - 3 = 0$
    At $(6,-\sqrt{3})$: $\quad 2x + 3\sqrt{3}y - 3 = 0$

33. At $(4,6)$: $\quad 3x + 4y - 36 = 0$
    At $(4,-6)$: $\quad 3x - 4y - 36 = 0$

35. a) $\dfrac{3}{8}(15 - 16 \ln 2) \approx 1.466$

    b) $39\pi/16$ c) $27\pi/2$

41. Ellipse

43. Parabola

45. Hyperbola

47. Circle

True or False

1. True   2. True   3. True   4. False   5. True

SECTION 11.4

1. $\dfrac{(y')^2}{2} - \dfrac{(x')^2}{2} = 1$

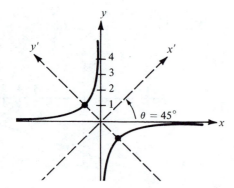

3. $y' = \dfrac{(x')^2}{6} - \dfrac{x'}{3}$

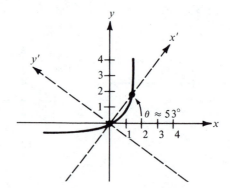

5. $\dfrac{(x')^2}{1/4} - \dfrac{(y')^2}{1/6} = 1$

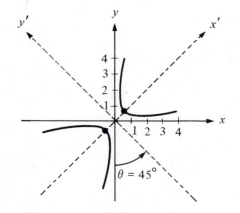

7. $\dfrac{(x' - 3\sqrt{2})^2}{16} - \dfrac{(y' - \sqrt{2})^2}{16} = 1$

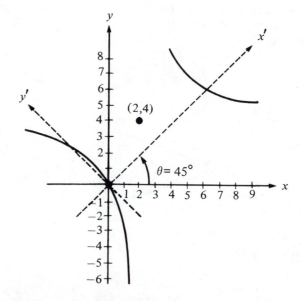

9. $\dfrac{(x')^2}{3} + \dfrac{(y')^2}{2} = 1$

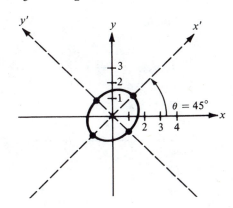

11. $x' = -(y')^2$

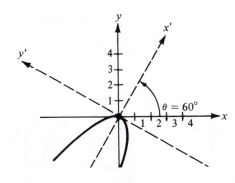

13. $\dfrac{(x')^2}{3} - \dfrac{(y')^2}{5} = 1$

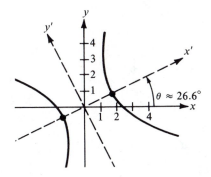

15. $\dfrac{(x')^2}{1.096} - \dfrac{(y')^2}{6.153} = 1$

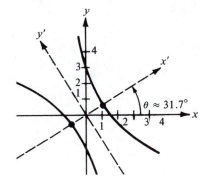

17. a) Parabola
    b) Hyperbola
    c) Ellipse

MISCELLANEOUS EXERCISES

1. Circle
   Center: $(1/2, -3/4)$

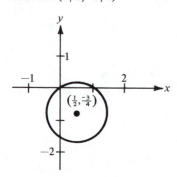

3. Hyperbola
   Center: $(-4, 3)$
   Vertices: $(-4 \pm \sqrt{2}, 3)$

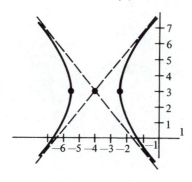

5. Ellipse
   Center: $(2, -3)$
   Vertices: $\left(2, -3 \pm \dfrac{\sqrt{2}}{2}\right)$

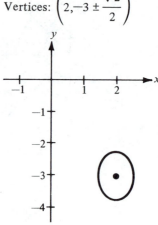

7. Parabola
   Vertex: $(3, 0)$

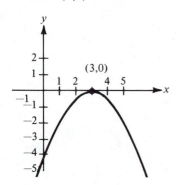

9. Parabola
   Vertex: $\left(-\dfrac{1}{\sqrt{8}}, \dfrac{1}{\sqrt{8}}\right)$

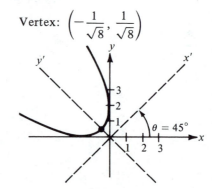

11. $\dfrac{y^2}{1} - \dfrac{x^2}{8} = 1$

13. $\dfrac{(x - 2)^2}{25} + \dfrac{y^2}{21} = 1$

15. $x^2 - 2xy + y^2 - 8x - 8y = 0$

17. $\dfrac{x^2}{4} + \dfrac{y^2}{16/3} = 1$

19. $\dfrac{x^2}{4} - \dfrac{y^2}{12} = 1$

21. $4x + 4y - 7 = 0$

23. $a = \sqrt{5}$

27. Prolate spheroid: $V = 16\pi$

$$S = 8\pi + \dfrac{36\pi}{\sqrt{5}} \text{ Arcsin } \dfrac{\sqrt{5}}{3} \approx 67.673$$

31. 192π ft^3

33.   4.212 ft

Oblate spheroid: $V = 24\pi$

$$S = 18\pi + \frac{12\pi}{\sqrt{5}} \ln \left| \frac{3 + \sqrt{5}}{3 - \sqrt{5}} \right| \approx 89.001$$

CHAPTER 12

SECTION 12.1

1.

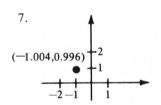

3.

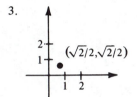

5.

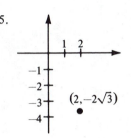

7.

(−1.004, 0.996)

9. $(\sqrt{2}, \pi/4), (-\sqrt{2}, 5\pi/4)$

11. $(5, 2.214), (-5, 5.356)$

13. $(\sqrt{6}, 5\pi/4). (-\sqrt{6}, \pi/4)$

15. $(2\sqrt{13}, .983), (-2\sqrt{13}, 4.124)$

17. $r = a$

19. $r^2 = 9 \cos(2\theta)$

21. $r = 2a \cos \theta$

23. $r = \dfrac{2a}{1 - \sin \theta}$ or $r = \dfrac{-2a}{1 + \sin \theta}$

25. $x^2 + y^2 - 4y = 0$

27. $(x^2 + y^2 + 2y)^2 = x^2 + y^2$

29. $(x^2 + y^2)^2 = 2y(3x^2 - y^2)$

31. $2x - 3y = 6$

37. $r = \dfrac{2}{1 - \sin \theta}$ or $r = \dfrac{-2}{1 + \sin \theta}$

39. $r^2 = \dfrac{225}{25 - 16 \cos^2 \theta}$

41. $r = \dfrac{-9}{4 + 5 \sin \theta}$ or $r = \dfrac{9}{4 - 5 \sin \theta}$

43. $y^2 - 4x - 4 = 0$

45. $3x^2 + 4y^2 - 4x - 4 = 0$

47. $32y^2 - 4x^2 + 36y + 9 = 0$

49.   11,002 miles and 1432 miles

True or False

1. True
2. True
3. True

4. False, 2 radians ≈ 114.6°, which places the point in the second quadrant

5. True

SECTION 12.2

1. Cardioid

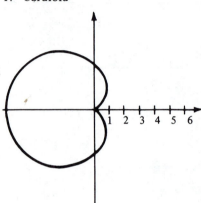

3. Limaçon

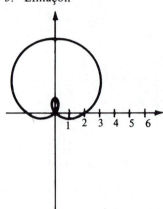

5. Limaçon

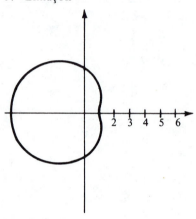

7. Circle

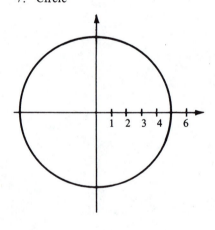

9. Spiral

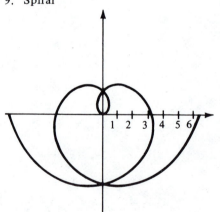

11. Circle

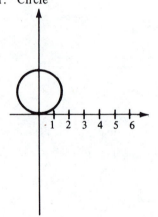

13. Rose curve

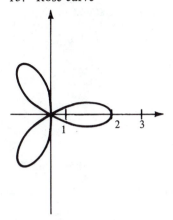

15. Rose curve

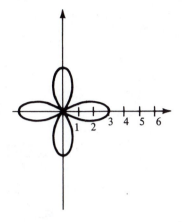

17. Lemniscate

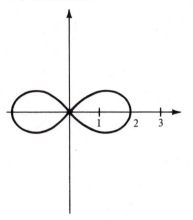

19. Rose curve

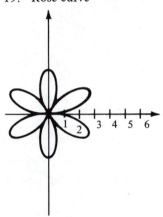

21. Line

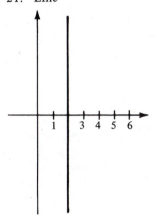

23. Ellipse

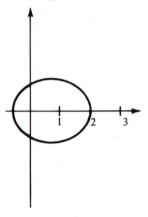

25. −1

27. 0

29. π

31.

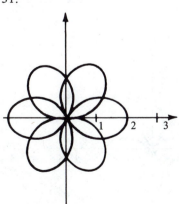

33.

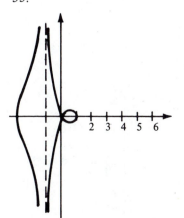

37. π/2

39. 0

41. π/3

True or False

1. True
2. False, the graph of $r = \cos\theta$ has a vertical tangent at $(0,1)$
3. False
4. False, the diameter is 1
5. False, the slope at $(2,\pi)$ is 2

SECTION 12.3

1. $(1,\pi/2),(1,3\pi/2),(0,0)$

3. $\left(\dfrac{2-\sqrt{2}}{2},\dfrac{3\pi}{4}\right),\left(\dfrac{2+\sqrt{2}}{2},\dfrac{7\pi}{4}\right),(0,0)$

5. $(3/2,\pi/6),(3/2,5\pi/6),(0,0)$

7. $(1.936,1.318),(1.936,1.824),(0,0)$

9. $(2,4),(-2,-4)$

11. $(.581,\pm.535),(2.581,\pm1.376)$

13. $\pi/3$

15. $\pi/6$

17. $\dfrac{2\pi - 3\sqrt{3}}{2}$

19. $\pi + 3\sqrt{3}$

21. $\dfrac{12\sqrt{5}\pi}{25}$

23. $\pi + 3\sqrt{3}$

25. $\dfrac{4}{3}(4\pi - 3\sqrt{3})$

27. $3\sqrt{3}$

29. $11\pi - 24$

True or False

1. False, the area is given by $\dfrac{1}{2}\displaystyle\int_0^\pi \sin^2\theta\,d\theta$

2. False, the graphs of $f(\theta) = 1$ and $g(\theta) = -1$ coincide

3. False, if $f(\theta) = 0$ and $g(\theta) = \sin 2\theta$, then there is only one point of intersection

4. True

5. True, the area enclosed by the first is $\pi/2$ and the area enclosed by the second is $\pi/4$

MISCELLANEOUS EXERCISES

1. Cardioid

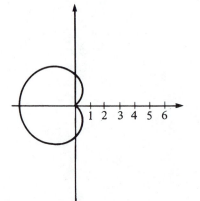

3. Limaçon

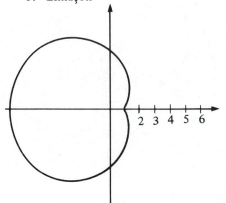

5. Rose curve

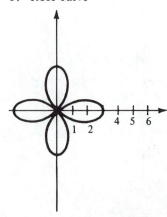

7. Circle

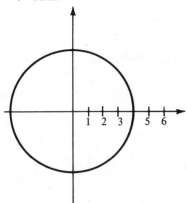

9. Line

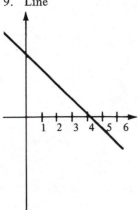

11. Line

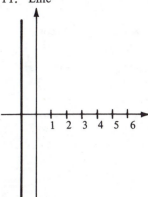

13. Rose curve

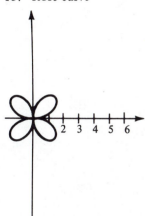

15. Parabola

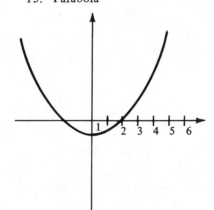

17. Semicubical parabola

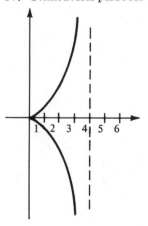

19. $x^2 + y^2 - 3x = 0$

21. $(x^2 + y^2 + 2x)^2 = 4(x^2 + y^2)$

23. $y^2 = x^2 \left(\dfrac{4-x}{4+x} \right)$

25. $r = 4 \cos \theta$

27. $r = \dfrac{5}{3 - 2 \cos \theta}$

29. $r = \dfrac{12}{3 \sin \theta + 4 \cos \theta}$

31. Tangents at the pole: $\theta = \pm\pi/3$

Points of vertical tangency:
$(-1,0),(3,\pi),(1/2,\pm1.318)$

Points of horizontal tangency:
$(-.686,\pm.568),(2.186,\pm2.206)$

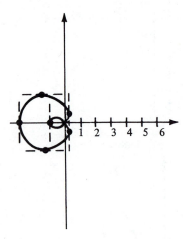

35. a) $r = 2 - \sin\left(\theta - \dfrac{\pi}{4}\right)$

b) $r = 2 + \cos\theta$

c) $r = 2 + \sin\theta$

d) $r = 2 - \cos\theta$

37. $9\pi/2$

39. $\pi/32$

41. $\dfrac{8\pi - 6\sqrt{3}}{3}$

43. $\dfrac{a^2}{3}(6 - 3\sqrt{3} + \pi)$

47. $8a$

CHAPTER 13

SECTION 13.1

1. $2x - 3y + 5 = 0$

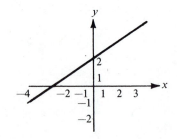

3. $y = (x - 1)^2$

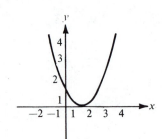

5. $y = \dfrac{1}{2}x^{2/3}$

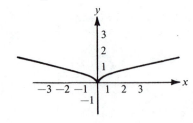

7. $y = \dfrac{x + 1}{x}$

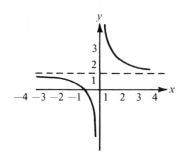

9. $x^2 + y^2 = 9$

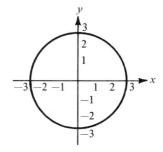

11. $\dfrac{x^2}{16} + \dfrac{y^2}{4} = 1$

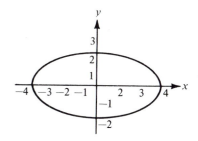

13. $y = 2 - 2x^2, \; -1 \leqslant x \leqslant 1$

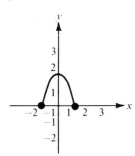

15. $\dfrac{(x - 4)^2}{4} + \dfrac{(y + 1)^2}{1} = 1$

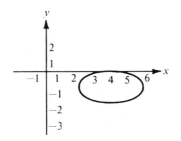

17. $\dfrac{(x - 4)^2}{4} + \dfrac{(y + 1)^2}{16} = 1$

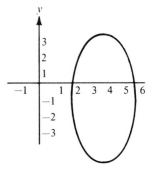

19. $\dfrac{x^2}{16} - \dfrac{y^2}{9} = 1$

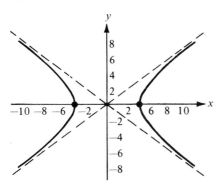

23. $x = 5t, \; y = -2t$ (Solution is not unique.)

25. $\dfrac{(x - h)^2}{a^2} - \dfrac{(y - k)^2}{b^2} = 1$

27. $x = t, \; y = t^3$

$x = \sqrt[3]{s}, \; y = s$ (Solution is not unique.)

29.

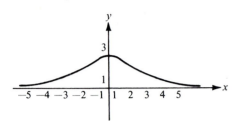

31.

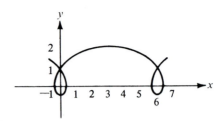

33. $x = a\theta - b \sin\theta$
   $y = a - b \cos\theta$

True or False

1. True
2. False, since neither x nor y can be negative, the graph consists of only that portion of the line $y = x$ which lies in the first quadrant
3. False, let $x = t^2$ and $y = t$
4. False, let $x = \sin t$ and $y = \cos t$
5. False

SECTION 13.2

1. $\dfrac{dy}{dx} = \dfrac{3}{2}, \dfrac{d^2 y}{dx^2} = 0$

3. $\dfrac{dy}{dx} = 2t + 3, \dfrac{d^2 y}{dx^2} = 2$

$\dfrac{dy}{dx} = 1, \dfrac{d^2 y}{dx^2} = 2$

5. $\dfrac{dy}{dx} = -\cot\theta, \dfrac{d^2 y}{dx^2} = -\dfrac{\csc^3\theta}{2}$

$\dfrac{dy}{dx} = -1, \dfrac{d^2 y}{dx^2} = -\sqrt{2}$

7. $\dfrac{dy}{dx} = 4, \dfrac{d^2 y}{dx^2} = -6\sqrt{3}$

$\dfrac{dy}{dx} = 2 \csc\theta, \dfrac{d^2 x}{dx^2} = -2 \cot^3\theta$

9. $\dfrac{dy}{dx} = -\tan\theta, \dfrac{d^2 y}{dx^2} = \dfrac{\sec^4\theta \, \csc\theta}{3a}$

$\dfrac{dy}{dx} = -1, \dfrac{d^2 y}{dx^2} = \dfrac{4\sqrt{2}}{3a}$

11. $3x - 4y + 12\sqrt{2} = 0$

13. $y = 2, 3x - y - 10 = 0$

15. $x + 2y - 4 = 0$

17. $(1,0)$

19. $(0,-2), (2,2)$

21. $(-1,-1), (1,1)$

23. $(4n\pi, 0), (2(2n - 1)\pi, 4)$

25. $\sqrt{2}(1 - e^{-\pi/2}) \approx 1.12$

27. $2\sqrt{5} + \ln|2 + \sqrt{5}|$

29. $6a$

31. $\dfrac{1}{12}(6\sqrt{37} + \ln(6 + \sqrt{37}))$ 33. 4π

35. $3\pi/2$

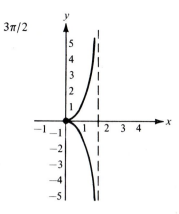

37. $12\pi a^2/5$

39. $2\pi a^2(1 - \cos\theta)$

41. $4/9$

True or False

1. False, $\dfrac{d^2 y}{dx^2} = \dfrac{\dfrac{d}{dt}\left[\dfrac{g'(t)}{f'(t)}\right]}{f'(t)} = \dfrac{f'(t)g''(t) - g'(t)f''(t)}{f'(t)^3}$

2. False, the graph of $y = x^3$ has a horizontal tangent at the origin

3. False, the graph of $y = x^{2/3}$ does not have a horizontal tangent at the origin

4. True

5. True

SECTION 13.3

1. $\mathbf{V} = 3\mathbf{i} + 4\mathbf{j}$

3. $\mathbf{V} = -3\mathbf{i} + 2\mathbf{j}$

5. $\mathbf{V} = 3\mathbf{i}$

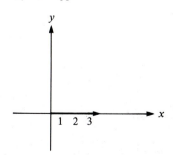

7. $\mathbf{V} = \dfrac{-\sqrt{3}\mathbf{i} \pm \mathbf{j}}{2}$

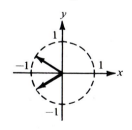

9. $\mathbf{V} = 5\mathbf{i} + 6\mathbf{j}$

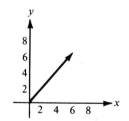

11. $\mathbf{V} = -7\mathbf{j}$

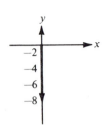

13. $\mathbf{V} = \dfrac{\pm(3\mathbf{i} - \mathbf{j})}{\sqrt{10}}$

15. $\mathbf{V} = 3\mathbf{i} - \dfrac{3}{2}\mathbf{j}$

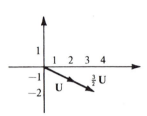

17. $\mathbf{V} = 4\mathbf{i} + 3\mathbf{j}$

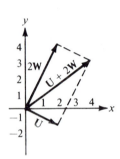

19. $\mathbf{V} = \dfrac{7\mathbf{i} - \mathbf{j}}{2}$

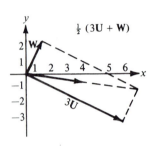

21. $\dfrac{4\mathbf{i} - 3\mathbf{j}}{5}$

23. $\mathbf{j}$

25. $a = 1, b = 1$

27. $a = 1, b = 2$

29. $a = 2/3, b = 1/3$

31. $(-4,-1), (6,5), (10,3)$

True or False

1. True

2. True

3. True

4. True

5. True

6. True

7. True

8. True

9. False, $|a\mathbf{i} + b\mathbf{j}| = \sqrt{2}|a|$

10. True

SECTION 13.4

1. $\pi/2$

3. $\text{Arccos}\left(\dfrac{1}{\sqrt{5}}\right) \approx 63.4°$

5. $\pi/2$

7. $\text{Arccos}\left(\dfrac{-1}{5\sqrt{2}}\right) \approx 98.1°$

9. $7\pi/12$

11. $\pi/4$

13. $\text{Arccos}\left(\dfrac{3}{\sqrt{13}}\right) \approx 33.7°$

15. $3\pi/4$

17. $\dfrac{1}{2}[(5 + 13\sqrt{3})\mathbf{i} + (13 + 5\sqrt{3})\mathbf{j}]$

19. $[-15\cos(10°) + 20\cos(25°)]\mathbf{i} +$
    $[15\sin(10°) + 20\sin(25°)]\mathbf{j}$

21. Neither

23. Neither

25. Parallel

27. Orthogonal

29. a) $-\sqrt{10}/2$ b) $-\dfrac{1}{2}(i + 3j)$

31. a) $\sqrt{26}/2$ b) $\dfrac{1}{2}(5i - j)$

33. $2i + 3j$

35. $\dfrac{5}{2}(i + j)$

37. Horizontal component $= 75\sqrt{3}$ lb
   vertical component $= 75$ lb

39. 425 ft lb

41. Horizontal component $= 80\cos(50°)$ ft/sec ≈ 51.423 ft/sec
   vertical component $= 80\sin(50°)$ ft/sec ≈ 61.284 ft/sec

True or False

1. False, $U \cdot 0$ is the *scalar*, zero

3. True

5. True

2. True

4. True

SECTION 13.5

1. $R'(t) = 3i + j$
   $R''(t) = 0$

3. $R'(t) = -(a \sin t)i + (b \cos t)j$
   $R''(t) = -(a \cos t)i - (b \sin t)j$

5. $R'(\theta) = (1 - \cos \theta)i + \sin \theta\, j$
   $R''(\theta) = \sin \theta\, i + \cos \theta\, j$

7. $V(t) = i + 2j$, $|V(t)| = \sqrt{5}$
   $A(t) = 0$ $|A(t)| = 0$

9. $V(t) = e^t i + \dfrac{1}{2} e^{t/2} j$, $|V(t)| = \dfrac{1}{2} \sqrt{4e^{2t} + e^t}$

   $A(t) = e^t i + \dfrac{1}{4} e^{t/2} j$, $|A(t)| = \dfrac{1}{4} \sqrt{16e^{2t} + e^t}$

11. $V(t) = \cos t\, i - 2 \sin 2t\, j$, $|V(t)| = \sqrt{\cos^2 t + 4 \sin^2 2t}$
    $A(t) = -\sin t\, i - 4 \cos 2t\, j$, $|A(t)| = \sqrt{\sin^2 t + 16 \cos^2 2t}$

13. $V(t) = 2ti + j$, $|V(t)| = \sqrt{4t^2 + 1}$
    $A(t) = 2i$ $|A(t)| = 2$

15. $R(t) = 2e^{2t}i + 3(e^t - 1)j$

17. $R(t) = (600\sqrt{3}t)i + (600t - 16t^2)j$

19. $4t$

21. $\dfrac{1 + 8t^2}{2\sqrt{1 + 4t^2}}$

23. $R(t) = \left(-\dfrac{1}{2} e^{-t^2} + 1\right)i + (e^{-t} - 2)j$

25. $V(t) = (b \sec t \tan t)i + (b \sec^2 t)j$
    $A(t) = b \sec t(\sec^2 t + \tan^2 t)i + (2b \sec^2 t \tan t)j$

27. $40\sqrt{6}$ ft/sec
    75 ft

29. a) $\pi/4$
    b) $\pi/2$

31. b) $R(t) = 40\sqrt{2}ti + (40\sqrt{2}t - 16t^2)j$
    c) Speed: $8\sqrt{58}$ ft/sec
       Direction: $8\sqrt{2}(5i + 2j)$

33. $v_0 = 4\sqrt{170}$ ft/sec
    $\theta = \text{Arctan}(4) \approx 76.0°$

35. $V(t) = bw[(1 - \cos(wt))i + \sin(wt)j]$,
    $|V(t)| = \sqrt{2}bw\sqrt{1 - \cos(wt)}$
    $A(t) = bw^2[\sin(wt)i + \cos(wt)j]$, $|A(t)| = bw^2$

37. $8\sqrt{10}$ ft/sec

39. 110 mi/hr

SECTION 13.6

1. $\mathbf{T} = \dfrac{1}{\sqrt{2}}(\mathbf{i} - \mathbf{j})$

   $\mathbf{N} = \dfrac{1}{\sqrt{2}}(\mathbf{i} + \mathbf{j})$

   $\mathbf{A} \cdot \mathbf{T} = -\sqrt{2}$

   $\mathbf{A} \cdot \mathbf{N} = \sqrt{2}$

   $K = 1/\sqrt{2}$

3. $\mathbf{T} = \dfrac{1}{\sqrt{2}}(-\mathbf{i} + \mathbf{j})$

   $\mathbf{N} = \dfrac{-1}{\sqrt{2}}(\mathbf{i} + \mathbf{j})$

   $\mathbf{A} \cdot \mathbf{T} = 0$

   $\mathbf{A} \cdot \mathbf{N} = 16\pi^2$

   $K = 1/4$

5. $\mathbf{T} = \dfrac{1}{\sqrt{13}}(-2\sqrt{3}\mathbf{i} + \mathbf{j})$

   $\mathbf{N} = -\dfrac{1}{\sqrt{13}}(\mathbf{i} + 2\sqrt{3}\mathbf{j})$

   $\mathbf{A} \cdot \mathbf{T} = \dfrac{3\pi^2\sqrt{3}}{2\sqrt{13}}$

   $\mathbf{A} \cdot \mathbf{N} = \dfrac{4\pi^2}{\sqrt{13}}$

   $K = \dfrac{16}{13^{3/2}}$

7. $\mathbf{T} = \dfrac{1}{\sqrt{2}}(-\mathbf{i} + \mathbf{j})$

   $\mathbf{N} = -\dfrac{1}{\sqrt{2}}(\mathbf{i} + \mathbf{j})$

   $\mathbf{A} \cdot \mathbf{T} = \sqrt{2}e^{\pi/2}$

   $\mathbf{A} \cdot \mathbf{N} = \sqrt{2}e^{\pi/2}$

   $K = \dfrac{1}{\sqrt{2}}e^{-\pi/2}$

9. $\mathbf{T} = \dfrac{1}{2}(\sqrt{3}\mathbf{i} + \mathbf{j})$

   $\mathbf{N} = \dfrac{1}{2}(\mathbf{i} - \sqrt{3}\mathbf{j})$

   $\mathbf{A} \cdot \mathbf{T} = 7/2$

   $\mathbf{A} \cdot \mathbf{N} = \sqrt{3}/2$

   $K = \sqrt{3}/24$

11. a) $\mathbf{A} \cdot \mathbf{T} = \dfrac{-32(v_0 \sin\theta - 32t)}{\sqrt{v_0^2 - 64v_0\, t \sin\theta + 1024t^2}}$, $\mathbf{A} \cdot \mathbf{N} = \dfrac{32v_0 \cos\theta}{\sqrt{v_0^2 - 64v_0\, t \sin\theta + 1024t^2}}$

    b) Tangential component: 0
       normal component: 32

13. a) The normal component is quadrupled
    b) The centripetal component is halved

15. $\sqrt{\dfrac{956}{41}} \approx 4.83$ mi/sec

17. $v = \sqrt{19.2} \approx 4.37$ mi/sec
    $t \approx 7189$ sec ≈ 2 hr

MISCELLANEOUS EXERCISES

1. $\dfrac{dy}{dx} = -\dfrac{3}{4}$, no horizontal tangents

   $y = \dfrac{-3x + 11}{4}$

3. $\dfrac{dy}{dx} = -\dfrac{5}{2}\cot\theta$, (3,7), (3,−3)

   $\dfrac{(x-3)^2}{4} + \dfrac{(y-2)^2}{25} = 1$

5. $\dfrac{dy}{dx} = -2t^2$, no horizontal tangents

$y = 3 + \dfrac{2}{x}$

7. $\dfrac{dy}{dx} = -(2t^2 + 2t + 1)$, no horizontal tangents

$y = \dfrac{1 - x^2}{2x}$

9. $\dfrac{dy}{dx} = -4 \tan \theta, (\pm 1, 0)$

$x^{2/3} + \left(\dfrac{y}{4}\right)^{2/3} = 1$

13. $x = 3 \tan \theta$
$y = 4 \sec \theta$

15. $\dfrac{1}{2} \pi^2 r$

17. $\sqrt{(50 + 75 \cos (50°))^2 + (75 \sin (50°))^2} \approx 113.78$

19. $-\dfrac{3}{10}(\mathbf{i} - 3\mathbf{j})$

23. No, **V** and **W** could both be orthogonal to **U** but have opposite directions

25. a) $\mathbf{i} - 3\mathbf{j}$
b) $-\mathbf{i} + 3\mathbf{j}$
c) $3\mathbf{i} + \mathbf{j}$

31. $\mathbf{T} = \dfrac{(-t \sin t + \cos t)\mathbf{i} + (t \cos t + \sin t)\mathbf{j}}{\sqrt{t^2 + 1}}$

$\mathbf{N} = \dfrac{-(t \cos t + \sin t)\mathbf{i} + (-t \sin t + \cos t)\mathbf{j}}{\sqrt{t^2 + 1}}$

$\mathbf{A} \cdot \mathbf{T} = \dfrac{t}{\sqrt{t^2 + 1}}$

$\mathbf{A} \cdot \mathbf{N} = \dfrac{t^2 + 2}{\sqrt{t^2 + 1}}$

33. $\mathbf{T} = \dfrac{2\sqrt{t}\,\mathbf{i} + \mathbf{j}}{\sqrt{4t + 1}}$

$\mathbf{N} = \dfrac{\mathbf{i} - 2\sqrt{t}\,\mathbf{j}}{\sqrt{4t + 1}}$

$\mathbf{A} \cdot \mathbf{T} = \dfrac{-1}{4t^{3/2} \sqrt{4t + 1}}$

$\mathbf{A} \cdot \mathbf{N} = \dfrac{1}{2t \sqrt{4t + 1}}$

CHAPTER 14

SECTION 14.1

1.

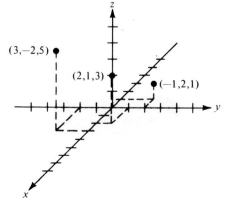

3. $3, 6, 3\sqrt{5}$, right triangle

5. $6, 6, 2\sqrt{10}$, isosceles triangle

7. $(2, 3, 9)$

9. $x^2 + y^2 + z^2 - 4y - 10z + 25 = 0$

11. Center: $(1, -3, -4)$
Radius: 5

13. $\mathbf{V} = 4\mathbf{i} + \mathbf{j} + \mathbf{k}$

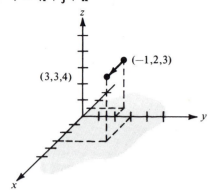

15. $\mathbf{V} = \sqrt{3}\mathbf{j} + \mathbf{k}$ or $\mathbf{V} = \sqrt{3}\mathbf{j} - \mathbf{k}$

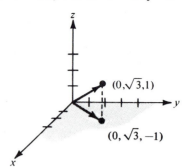

17. $\mathbf{V} = \dfrac{\sqrt{3}}{3}(\mathbf{i} + \mathbf{j} + \mathbf{k})$

19. $\mathbf{U} - \mathbf{V} = -\mathbf{i} + 4\mathbf{k}$
$\mathbf{V} - \mathbf{U} = \mathbf{i} - 4\mathbf{k}$

21. $6(\mathbf{i} + 2\mathbf{j} + \mathbf{k})$

23. $-2\mathbf{j} - 3\mathbf{k}$

25. $\pm 1/3$

27. a) $4\mathbf{i} + 4\mathbf{j} + 3\mathbf{k}$

b) $-\left(2\mathbf{i} + 2\mathbf{j} + \dfrac{3}{2}\mathbf{k}\right)$

31. c) $a = 1, b = 1$

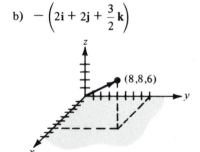

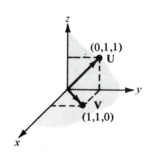

True or False

1. True

2. True

3. True

4. True

5. True

SECTION 14.2

1. 1

3. 1

5. 15

7. 3/5

9. $-8\sqrt{13}/65$

11. a) $\theta = \pi/2$
b) $0 < \theta < \pi/2$
c) $\pi/2 < \theta < \pi$

13. $\dfrac{11}{13}(2\mathbf{j} + 3\mathbf{k})$

15. $5\sqrt{14}/14$

17. 1

19. $-14/3$

21. a) **0**   d) **j**
    b) **k**   e) $-\mathbf{j}$
    c) **i**   f) $-\mathbf{k}$

23. $11\mathbf{i} + 6\mathbf{j} - \mathbf{k}$

25. $3\sqrt{13}/2$

27. $9\sqrt{6}/2$

29. $\sqrt{2}$

31. a) $|\mathbf{U}| = 1, |\mathbf{V}| = 1$

33. -1

35. 2

37. $2/\sqrt{3}$

39. $\mathbf{U} = b\mathbf{i} + \dfrac{1}{a}\mathbf{j}$, where b is any scalar

True or False

1. True
2. False, the dot product of two vectors is a scalar

3. False, let $\mathbf{U} = \mathbf{i}$ and $\mathbf{V} = \mathbf{i}$
4. False, let $\mathbf{U} = \mathbf{i}$, $\mathbf{V} = 2\mathbf{j}$, then $\mathbf{V} \times (\mathbf{U} \times \mathbf{V}) = 2\mathbf{j} \times 2\mathbf{k} = 4\mathbf{i}$

5. False, let $\mathbf{V} = \mathbf{i}$, $\mathbf{U} = \mathbf{j}$, $\mathbf{W} = \mathbf{i} + \mathbf{j}$, then $\mathbf{V} \times (\mathbf{U} \times \mathbf{W}) = \mathbf{j}$, but $(\mathbf{V} \times \mathbf{U}) \times \mathbf{W} = -\mathbf{i}$

SECTION 14.3

Parametric Equations	*Symmetric Equations*	*Direction Numbers*
1. $x = t, y = 3t, z = 2t$	$x = \dfrac{y}{3} = \dfrac{z}{2}$	1,3,2
3. $x = -2 + 2t, y = 4t, z = 3 - 2t$	$\dfrac{x+2}{2} = \dfrac{y}{4} = \dfrac{z-3}{-2}$	2,4,-2
5. $x = 5 + 17t, y = -3 - 11t, z = -2 - 9t$	$\dfrac{x-5}{17} = \dfrac{y+3}{-11} = \dfrac{z+2}{-9}$	17,-11,-9
7. $x = 1 + 3t, y = -2t, z = 1 + t$	$\dfrac{x-1}{3} = \dfrac{y}{-2} = \dfrac{z-1}{1}$	3,-2,1
9. $x = 2, y = 3, z = 4 + t$	(none)	0,0,1

11. a, b, d

13. a) $(2,3,1)$
    b) $7\sqrt{17}/51$

15. $2x + 3y - z = 10$

17. $3x + 9y - 7z = 0$

19. $4x - 3y + 4z = 10$

21. $z = 3$

23. $x + y + z = 5$

25. $7x + y - 11z = 5$

27. $y - z = -1$

29. $x = 2, y = 1 + t, z = 1 + 2t$

31. No points of intersection

33. $(2,-3,2)$

35. $\cos\theta = |-2\sqrt{138}/207|, \theta \approx 83.5°$

37.

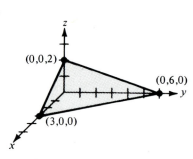

39.

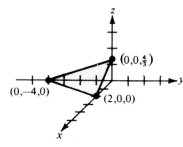

41.

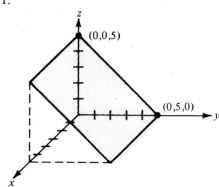

43. $\sqrt{6}/6$

45. $10\sqrt{26}/13$

True or False

1. True
2. True

3. False
4. True

5. False

SECTION 14.4

1. Plane

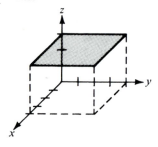

3. Right circular cylinder

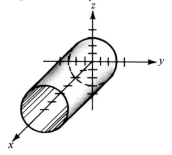

5. Parabolic cylinder

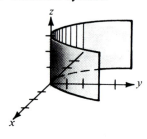

7. Elliptical cylinder

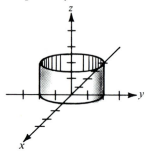

9. Hyperbolic cylinder

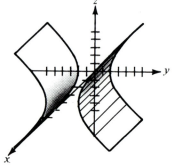

11. Ellipsoid

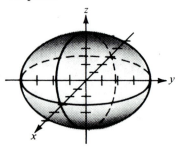

13. Hyperboloid of one sheet

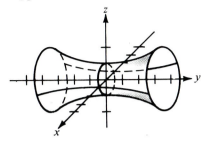

15. Elliptical paraboloid

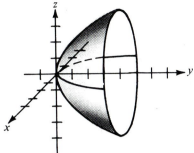

17. Hyperbolic paraboloid

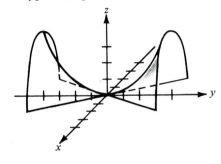

19. Hyperboloid of two sheets

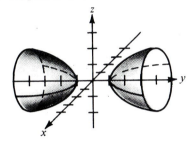

21. Elliptic cone

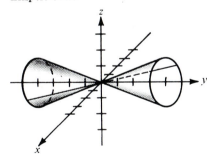

23. Hyperbolic paraboloid

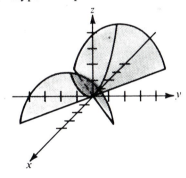

25. $x^2 + z^2 = 4y$

27. $4x^2 + 4y^2 = z^2$

29. $y^2 + z^2 = 4/x^2$

31. $y = \sqrt{2z}$

33. a) Major axis: $4\sqrt{2}$ b) Major axis: $8\sqrt{2}$
      Minor axis: 4 Minor axis: 8
      Foci: $(0,\pm2,2)$ Foci: $(0,\pm4,8)$

SECTION 14.5

1.

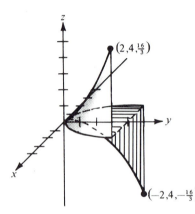

3.

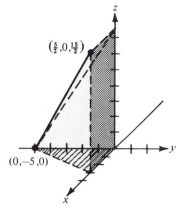

$\left(\frac{5}{2}, 0, \frac{15}{2}\right)$

$(0, -5, 0)$

5.

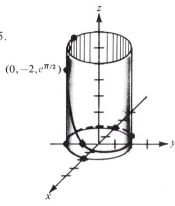

$(0, -2, e^{\pi/2})$

7.

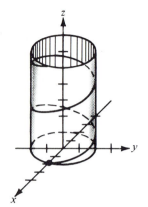

$\left(2, 4, \frac{16}{3}\right)$

$\left(-2, 4, -\frac{16}{3}\right)$

9. $x = t, y = -t, z = 2t^2$

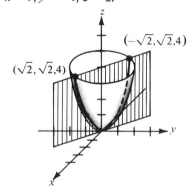

$(-\sqrt{2}, \sqrt{2}, 4)$

$(\sqrt{2}, \sqrt{2}, 4)$

11. $x = t, y = t, z = \sqrt{4 - t^2}$

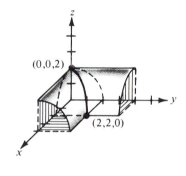

$(0, 0, 2)$

$(2, 2, 0)$

13. $x = t^2, y = t$

$z = \dfrac{1}{6}\sqrt{144 - 9t^2 - 16t^4}$

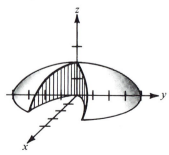

15. $x = t, y = 4/t,$

$z = \dfrac{1}{t}\sqrt{-t^4 + 16t^2 - 16}$

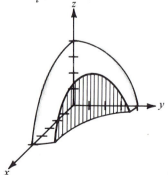

17. $2\pi\sqrt{a^2 + b^2}$

19. $\sqrt{5}\pi^2/8$

21. a) $\mathbf{V} = 4\mathbf{i}, \ |\mathbf{V}| = 4$
         $\mathbf{A} = \mathbf{0}$
     b) $\mathbf{T} = \mathbf{i}$
         $\mathbf{N} = \mathbf{0}$
     c) $K = 0$

23. a) $\mathbf{V} = 8t\mathbf{i}, \ |\mathbf{V}| = 8t$
         $\mathbf{A} = 8\mathbf{i}$
     b) $\mathbf{T} = \mathbf{i}$
         $\mathbf{N} = \mathbf{0}$
     c) $K = 0$

25. a) $\mathbf{V} = 4\mathbf{i} - (3 \sin t)\mathbf{j} + (3 \cos t)\mathbf{k}$
$|\mathbf{V}| = 5$
$\mathbf{A} = -(3 \cos t)\mathbf{j} - (3 \sin t)\mathbf{k}$

b) $\mathbf{A} \cdot \mathbf{T} = 0$
$\mathbf{A} \cdot \mathbf{N} = 3$

c) $K = 3/25$

27. a) $\mathbf{V} = (-e^t \sin t + e^t \cos t)\mathbf{i} + (e^t \cos t + e^t \sin t)\mathbf{j} + e^t\mathbf{k}$
$|\mathbf{V}| = \sqrt{3}\, e^t$
$\mathbf{A} = (-2e^t \sin t)\mathbf{i} + (2e^t \cos t)\mathbf{j} + e^t\mathbf{k}$

b) $\mathbf{A} \cdot \mathbf{T} = 8$
$\mathbf{A} \cdot \mathbf{N} = 0$

c) $K = \sqrt{2}/3e^t$

29. $x = -\sqrt{2}(1 + t),\ y = \sqrt{2}(1 - t),\ z = \dfrac{3\pi}{4} + t$

True or False

1. True
2. False, the graph of $x = t^3$, $y = t^3$, $z = t^3$ is a line
3. False, let $\mathbf{R}(t) = \cos t\, \mathbf{i} + \sin t\, \mathbf{j} + \mathbf{k}$, then $\dfrac{d}{dt}\, |\mathbf{R}(t)| = 0$, but $|\mathbf{R}'(t)| = 1$
4. True
5. False, see Exercise 1, Section 14.5

SECTION 14.6

1. $(2, \pi/3, 4)$
7. $(-2\sqrt{3}, -2, 3)$
13. $(0, 0, 12)$
19. $x^2 + y^2 = 4$
25. $x^2 + y^2 + z^2 = 25$
31. a) $r^2 + z^2 = 16$
    b) $\rho = 4$

3. $(2\sqrt{2}, -\pi/4, -4)$
9. $(4\sqrt{2}, 2\pi/3, \pi/4)$
15. $(3\sqrt{3}, -\pi/6, 3)$
21. $\sqrt{3}y = x$
27. $3x^2 + 3y^2 - z^2 = 0$
33. a) $r^2 + (z - 1)^2 = 1$
    b) $\rho = 2 \cos \phi$

5. $(1, \sqrt{3}, 2)$
11. $(\sqrt{6}, \sqrt{2}, 2\sqrt{2})$
17. $\left(2\sqrt{13}, -\dfrac{\pi}{6}, \text{Arccos } \dfrac{3}{\sqrt{13}}\right)$
23. $x^2 + y^2 - 2y = 0$
29. $x^2 + y^2 + (z - 2)^2 = 4$
35. a) $r = 4 \sin \theta$
    b) $\rho = \dfrac{4 \sin \theta}{\sin \phi}$

37.

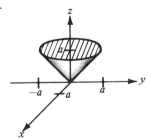

39.

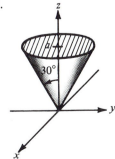

MISCELLANEOUS EXERCISES

1. a)

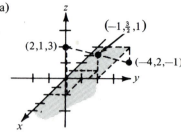

   b) $(-1, 3/2, 1)$

   c) $x = 2 + 6t$
   $y = 1 - t$
   $z = 3 + 4t$

   d) $2x - 3z + 5 = 0$

5. $\sqrt{14}$

9. $10\mathbf{i} + 11\mathbf{j} - 8\mathbf{k}$

13. 4

17. $x = 1, y = 2 + t, z = 3$

19. $x + y + z = 6$

21. $x = t, \; y = -1 + t, z = 1$

25. $26\sqrt{59}/59 \approx 3.385$

3. a) Point

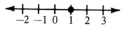

   b) Line

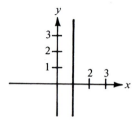

   c) Plane

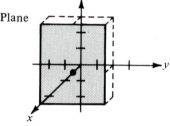

23. .615 radians $\approx 35.26°$

27.

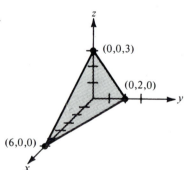

29.

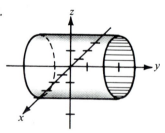

31.

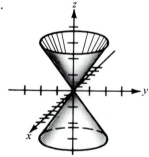

33.

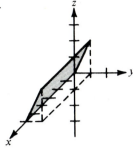

35.

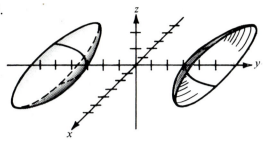

37.

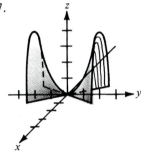

39. $|\mathbf{V}| = w^2 t$
$\mathbf{A} \cdot \mathbf{N} = w^3 t$
$\mathbf{A} \cdot \mathbf{T} = w^2$
$K = 1/wt$

41. $x^2 - y^2 + z^2 = 1$

43. $r^2 \cos 2\theta = 2z$
$\rho = 2 \sec 2\theta \cos \phi \csc^2 \phi$

CHAPTER 15

SECTION 15.1

1. a) $3/2$ d) $3x/2$

   b) $-1/4$ e) $\dfrac{x + \Delta x}{2}$

   c) $1/2y$ f) $\dfrac{x}{y + \Delta y}$

9. The half plane below the line $y = -x + 4$

3. All points inside and on the boundary of the circle $x^2 + y^2 = 4$

5. All points outside and on the boundary of the circle $x^2 + y^2 = 1$

7. All points in the xy-plane except those on the x and y axes

11. All points in the xy-plane except those on the x-axis

13.

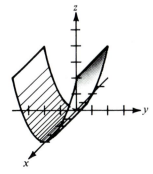

15.

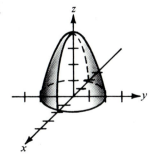

17.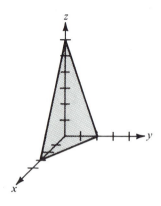

19.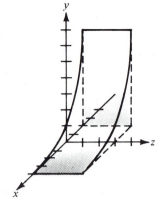

21. The level curves are circles of radius 5 or less centered at the origin

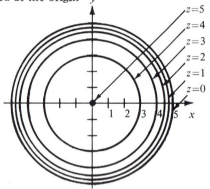

23. The level curves are lines with a slope of $-2/3$

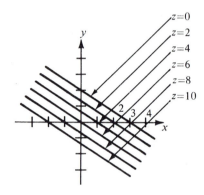

25. The level curves are circles passing through the origin and centered on the x-axis

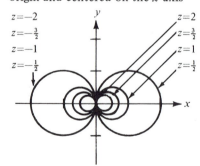

27. The level curves are lines of slope 1 passing through the fourth quadrant

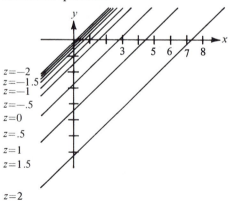

29. The level curves are hyperbolas centered at the origin with the x and y axes as asymptotes. (If $z = 1$, the level curve is the x and y axes.)

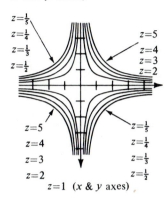

31. 1/2, continuous

33. Does not exist, discontinuous when $y = -x$

35. 1/3, discontinuous when $y = \pm 3x$

37. 1, discontinuous when $x = 0$ or $y = 0$

39. 0, discontinuous when $x^2 + y^2 = 1$

45. Not possible since $\lim\limits_{(x,y) \to (0,0)} f(x,y)$ does not exist

47. Discontinuous at points on the cylinder $x^2 + y^2 = 4$

49. Continuous

51.

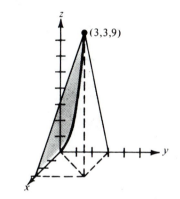

True or False

1. True

2. False

3. False

4. True, when $z = 3$, the level curve is a circle of radius zero.

5. True

SECTION 15.2

1. $\dfrac{\partial z}{\partial x} = 2x - 3y,\ -8$

   $\dfrac{\partial z}{\partial y} = -3x + 2y,\ 7$

7. $\dfrac{\partial z}{\partial x} = \dfrac{-y}{x^2 + y^2},\ 1/4$

   $\dfrac{\partial z}{\partial y} = \dfrac{x}{x^2 + y^2},\ 1/4$

13. $\dfrac{\partial f}{\partial x} = \dfrac{x}{\sqrt{x^2 + y^2 + z^2}},\ 2/3$

3. $\dfrac{\partial f}{\partial x} = -2xe^{-(x^2 + y^2)}$

   $\dfrac{\partial f}{\partial y} = -2ye^{-(x^2 + y^2)}$

9. $\dfrac{\partial f}{\partial x} = ye^y \cos(xy)$

   $\dfrac{\partial f}{\partial y} = e^y[x \cos(xy) + \sin(xy)]$

15. $\dfrac{\partial w}{\partial x} = \cos(x + 2y + 3z)$

5. $\dfrac{\partial z}{\partial x} = 2\cos(2x - y)$

   $\dfrac{\partial z}{\partial y} = -\cos(2x - y)$

11. $\dfrac{\partial z}{\partial x} = \dfrac{x}{x^2 + y^2},\ -1/2$

    $\dfrac{\partial z}{\partial y} = \dfrac{y}{x^2 + y^2},\ 0$

17. $\dfrac{\partial^2 z}{\partial x^2} = 2$

$$\frac{\partial w}{\partial y} = \frac{y}{\sqrt{x^2 + y^2 + z^2}} \,, -1/3$$

$$\frac{\partial w}{\partial y} = 2\cos(x + 2y + 3z)$$

$$\frac{\partial^2 z}{\partial y^2} = 6$$

$$\frac{\partial w}{\partial z} = \frac{z}{\sqrt{x^2 + y^2 + z^2}} \,, 2/3$$

$$\frac{\partial w}{\partial z} = 3\cos(x + 2y + 3z)$$

$$\frac{\partial^2 z}{\partial y\,\partial x} = \frac{\partial^2 z}{\partial x\,\partial y} = -2$$

19. $\dfrac{\partial^2 z}{\partial x^2} = e^x \tan y$

$$\frac{\partial^2 z}{\partial y^2} = 2e^x \sec^2 y \tan y$$

$$\frac{\partial^2 z}{\partial y\,\partial x} = \frac{\partial^2 z}{\partial x\,\partial y} = e^x \sec^2 y$$

21. $\dfrac{\partial^2 z}{\partial x^2} = \dfrac{2y^2}{(x-y)^3}$

$$\frac{\partial^2 z}{\partial y^2} = \frac{2x^2}{(x-y)^3}$$

$$\frac{\partial^2 z}{\partial y\,\partial x} = \frac{\partial^2 z}{\partial x\,\partial y} = \frac{-2xy}{(x-y)^3}$$

23. $\dfrac{\partial^2 z}{\partial x^2} = \dfrac{2xy}{(x^2 + y^2)^2}$

$$\frac{\partial^2 z}{\partial y^2} = \frac{-2xy}{(x^2 + y^2)^2}$$

$$\frac{\partial^2 z}{\partial y\,\partial x} = \frac{\partial^2 z}{\partial x\,\partial y} = \frac{y^2 - x^2}{(x^2 + y^2)^2}$$

33. $\dfrac{\partial f}{\partial x} = 2, \ \dfrac{\partial f}{\partial y} = 3$

35. $\dfrac{\partial f}{\partial x} = 2(x - y), \ \dfrac{\partial f}{\partial y} = 2(y - x)$ 37. $-1/3$

39. -6

41. At $(80,20)\,\dfrac{\partial c}{\partial x} = 183$ and $\dfrac{\partial c}{\partial y} = 237$

43. At $v_0 = 2000$ and $\theta = 5°$, $\dfrac{\partial R}{\partial v_0} \approx 21.7$ and $\dfrac{\partial R}{\partial \theta} \approx 246{,}202$ ft/rad ≈ 4297 ft/deg

b) Increase in θ

45. At $(2,3)$, $\dfrac{\partial T}{\partial x} = -2.4°$, $\dfrac{\partial T}{\partial y} = -9°$

True or False

1. False, $\partial z/\partial x = y^2$
2. False, let $z = x + y + 1$

3. True
4. True

5. True

SECTION 15.3

1. Relative minimum: $(1,2,-1)$
7. Relative minimum: $(1,1,-1)$
    Saddle point: $(0,0,0)$

3. Saddle point: $(1,-2,-1)$
9. Relative minimum: $(a, a, 3a^2)$

5. Saddle point: $(0,0,0)$
11. No Relative Extrema

13. 18 by 18 by 36

15. $1/\sqrt{14}$

19. 10,10,10

23. $x = -1/2, y = 1$

25. $x = 8, y = 16, z = 8$

27. $1/\sqrt{13}$

29. $2a/\sqrt{3}$ by $2b/\sqrt{3}$ by $2c/\sqrt{3}$

31. Relative maximum: $\left(\pm 2\sqrt{\dfrac{5}{6}}, \pm\sqrt{\dfrac{5}{6}}, \sqrt{\dfrac{5}{3}} \right)$

Relative minimum: $\left(\pm 2\sqrt{\dfrac{5}{6}}, \pm\sqrt{\dfrac{5}{6}}, -\sqrt{\dfrac{5}{3}} \right)$

True or False

1. False, the points on the circle $x^2 + y^2 = 1$ are minimum points but not relative minima

2. False, let $f(x,y) = |1 - x - y|$
3. True

4. True
5. False, let $f(x,y) = x^4 - 2x^2 + y^2$

SECTION 15.4

1. $dz = 6xy^3\, dx + 9x^2 y^2\, dy$

3. $dz = (\cos y + y \sin x)\, dx - (x \sin y + \cos x)\, dy$

5. $dz = \dfrac{2}{(x^2 + y^2)^2}\, (x\, dx + y\, dy)$

7. $dw = (2z^3 y \cos x)\, dx + (2z^3 \sin x)\, dy + (6z^2 \, y \sin x)\, dz$

9. $du = \dfrac{1}{z - 2y}\, dx + \dfrac{z + 2x}{(z - 2y)^2}\, dy - \dfrac{x + y}{(z - 2y)^2}\, dz$

11. a) $-.5125$
    b) $-.5$

13.

dV	ΔV	$\Delta V - dV$
4.7124	4.8391	.1267
2.8274	2.8264	$-.0010$
.0566	.0566	.0000
$-.0189$	$-.0189$	.0000

15. 5%

17. $\dfrac{dw}{dt} = 2(e^{2t} - e^{-2t}),\ 0$

19. $\dfrac{dw}{dt} = 4e^{2t},\ 4$

21. $\dfrac{\partial w}{\partial s} = 4s,\ 8$

$\dfrac{\partial w}{\partial t} = 4t,\ -4$

23. $\dfrac{\partial w}{\partial s} = 2s \cos (2t)$

$\dfrac{\partial w}{\partial t} = -2s^2 \sin (2t)$

25. $\dfrac{dw}{dt} = 2 \cos (2t)$

27. $\dfrac{dw}{dt} = 3(2t^2 - 1)$

29. $\dfrac{\partial w}{\partial r} = 0,\ \dfrac{\partial w}{\partial \theta} = 1$

31. $\dfrac{\partial z}{\partial x} = -x/z,\ \dfrac{\partial z}{\partial y} = -y/z$

33. $\dfrac{\partial w}{\partial x} = -\dfrac{z(y + w)}{xz - yz + 2w},\ \dfrac{\partial w}{\partial y} = -\dfrac{z(x - w)}{xz - yz + 2w},\ \dfrac{\partial w}{\partial z} = -\dfrac{xy + xw - yw}{xz - yz + 2w}$

35. $\dfrac{\partial z}{\partial x} = -\dfrac{\sec^2 (x + y)}{\sec^2 (y + z)},\ \dfrac{\partial z}{\partial y} = -\left[\dfrac{\sec^2 (x + y)}{\sec^2 (y + z)} + 1\right]$

37. $\dfrac{\partial z}{\partial x} = -\dfrac{x}{y + z},\ \dfrac{\partial z}{\partial y} = -\dfrac{z}{y + z}$

$\dfrac{\partial^2 z}{\partial x^2} = -\dfrac{(y + z)^2 + x^2}{(y + z)^3},\ \dfrac{\partial^2 z}{\partial y^2} = \dfrac{z(2y + z)}{(y + z)^3},\ \dfrac{\partial^2 z}{\partial x\, \partial y} = \dfrac{\partial^2 z}{\partial y\, \partial x} = \dfrac{xy}{(x + y)^3}$

41. 182 ft^3/min

43. -816π in.^{2}/min

SECTION 15.5

1.

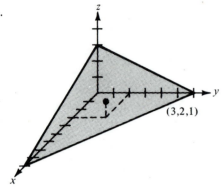

$(3,2,1)$

3. $\dfrac{2-3\sqrt{3}}{12}$

5. $-\dfrac{1}{3}\mathbf{i}-\dfrac{1}{2}\mathbf{j}$

7. $3/5$

9. $\sqrt{13}/6$

11.

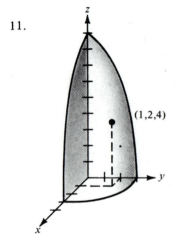

$(1,2,4)$

13. $-(1+2\sqrt{3})$

15. $\nabla f(1,2)=-2\mathbf{i}-4\mathbf{j},\ 2\sqrt{5}$

17. $D_{\mathbf{U}}f=5\sqrt{2}/2,\ |\nabla f|=\sqrt{13}$

19. $D_{\mathbf{U}}f=-7/25,\ |\nabla f|=1$

21. $D_{\mathbf{U}}f=-e,\ |\nabla f|=e$

23. $\dfrac{1}{625}\,(7\mathbf{i}-24\mathbf{j})$

True or False

1. True
2. False, $D_{\pi/4}\,f(x,y)=\sqrt{2}>1$
3. True
4. True
5. True

SECTION 15.6

1. $6x+2y+z=35$

5. $10x-8y-z=9$

9. $3x+4y-25z=25(1-\ln 5)$

3. $2x-y+z=2$

7. $2x-z=-2$

11. $x-4y+2z=18$

13. $x + y + z = 1$

15. $2x + 4y + z = 14, \dfrac{x-1}{2} = \dfrac{y-2}{4} = \dfrac{z-4}{1}$

17. $3x + 2y + z = -6, \dfrac{x+2}{3} = \dfrac{y+3}{2} = \dfrac{z-6}{1}$

19. $x - y + 2z = \pi/2, \dfrac{x-1}{1} = \dfrac{y-1}{-1} = \dfrac{z-(\pi/4)}{2}$

21. $12x + 8y - z = 35, \dfrac{x-2}{12} = \dfrac{y-2}{8} = \dfrac{z-5}{-1}$

23. $2x - 4y + z = -3, \dfrac{x-1}{2} = \dfrac{y-2}{-4} = \dfrac{z-3}{1}$

27. a) $\dfrac{x-2}{1} = \dfrac{y-1}{-2} = \dfrac{z-2}{1},$ b) $\sqrt{10}/5$

29. a) $\dfrac{x-3}{4} = \dfrac{y-3}{4} = \dfrac{z-4}{-3},$ b) $16/25$

True or False

1. True
2. False, $\nabla F(x_0, y_0, z_0)$ is normal to the surface
3. True
4. True
5. False, it is possible that the partials have the values ka, kb and kc

MISCELLANEOUS EXERCISES

1. $f_x = e^x \cos y$

   $f_y = -e^x \sin y$

3. $f_x = e^y + ye^x$

   $f_y = xe^y + e^x$

5. $f_x = \dfrac{y(y^2 - x^2)}{(x^2 + y^2)^2}$

   $f_y = \dfrac{x(x^2 - y^2)}{(x^2 + y^2)^2}$

7. $f_x = \dfrac{-yz}{x^2 + y^2}$

   $f_y = \dfrac{xz}{x^2 + y^2}$

   $f_z = \operatorname{Arctan} \dfrac{y}{x}$

9. $u_x = cne^{-n^2 t} \cos nx$

   $u_t = -cn^2 e^{-n^2 t} \sin nx$

11. $-\dfrac{2yz + z - 2xy}{2xy + x + 2z}$

13. $f_{xx} = 6, f_{yy} = 12y$

    $f_{xy} = f_{yx} = -1$

15. $f_{xx} = -y \cos x, f_{yy} = -x \sin y$

    $f_{xy} = f_{yx} = \cos y - \sin x$

21. $\dfrac{\partial u}{\partial r} = 4r, \dfrac{\partial u}{\partial t} = 0$

23. a) Homogeneous of degree 3
    b) Homogeneous of degree 1
    c) Not homogeneous
    d) Homogeneous of degree 1

27. 0

29. $4x + 4y - z = 8, \dfrac{x-2}{4} = \dfrac{y-1}{4} = \dfrac{z-4}{-1}$

31. $z = 4, x = 2,$ and $y = -3$

33. $\dfrac{x-2}{1} = \dfrac{y-1}{2}, z = 3$

35. .082 in., .6%

37. Saddle point: $(0,0,0)$

Relative minimum: $\left(\dfrac{3}{2}, \dfrac{9}{4}, -\dfrac{27}{16}\right)$

41. $y = x + \dfrac{5}{4}$

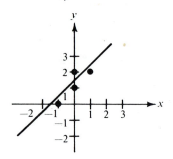

CHAPTER 16

SECTION 16.1

1. $3x^2/2$

3. $y \ln (2y)$

5. $\dfrac{4x^2 - x^4}{2}$

7. $\dfrac{y}{2}(\ln^2 y - y^2)$

9. $x^2\left(1 - e^{-x^2} - x^2 e^{-x^2}\right)$

11. 3

13. $20/3$

15. $2/3$

17. 4

19. $\dfrac{\pi^2}{32} + \dfrac{1}{8}$

21. $\displaystyle\int_0^1 \int_0^2 dy\,dx = \int_0^2 \int_0^1 dx\,dy = 2$

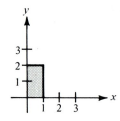

23. $\displaystyle\int_0^1 \int_{-\sqrt{1-y^2}}^{\sqrt{1-y^2}} dx\,dy = \int_{-1}^1 \int_0^{\sqrt{1-x^2}} dy\,dx = \pi$

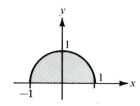

25. $\displaystyle\int_0^2\int_{x/2}^1 dy\,dx = \int_0^1\int_0^{2y} dx\,dy = 1$

27. $\displaystyle\int_0^1\int_{y^2}^{\sqrt[3]{y}} dx\,dy = \int_0^1\int_{x^3}^{\sqrt{x}} dy\,dx = 5/12$

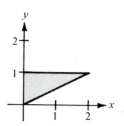

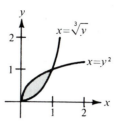

29. 4

31. 2

33. π

35. 5

37. πab

True or False

1. True

2. True

3. False, let $f(x,y) = x$

4. False, see Exercise 37, Section 16.1

5. True

SECTION 16.2

1. 8

3. 36

5. 0

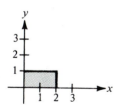

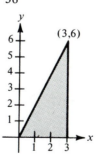

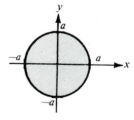

7. $\displaystyle\int_0^3\int_0^5 xy\,dy\,dx, \;\; \int_0^5\int_0^3 xy\,dx\,dy, \; 225/4$

9. $\displaystyle\int_0^2\int_x^{2x} \frac{y}{x^2+y^2}\,dy\,dx, \;\; \int_0^2\int_{y/2}^{y} \frac{y}{x^2+y^2}\,dx\,dy + \int_2^4\int_{y/2}^{2} \frac{y}{x^2+y^2}\,dx\,dy, \; \ln\frac{5}{2}$

11. $\displaystyle\int_0^3\int_{4y/3}^{\sqrt{25-y^2}} x\,dx\,dy, \;\; \int_0^4\int_0^{3x/4} x\,dy\,dx + \int_4^5\int_0^{\sqrt{25-x^2}} x\,dy\,dx, \; 25$

13. 12

15. 1/8

17. 32/3

19. 1/3   21. 16/3   23. 4

25. 24 (approximation is exact)   27. 52 (approximate), 53.33 (exact)

29. 8π 31. 6π

True or False

1. False, $V = 8 \int_0^1 \int_0^{\sqrt{1-y^2}} \sqrt{1 - x^2 - y^2}\ dx\ dy$

2. False, $\int_0^1 \int_0^1 x\ dx\ dy \neq \int_1^2 \int_1^2 x\ dx\ dy$

3. True

4. True

5. True, $\int_0^1 \int_0^1 \dfrac{1}{1 + x^2 + y^2}\ dx\ dy < \int_0^1 \int_0^1 \dfrac{1}{1 + x^2}\ dx\ dy = \pi/4$

SECTION 16.3

1. 6   3. $\sqrt{2}\,\pi$ 7. $[33 \ln |2 + \sqrt{5}| + 46\sqrt{5}]/64$

9. $(25\sqrt{5} - 11)/120$ 11. $16a^2$

SECTION 16.4

1. a) $(a/2, b/2)$ c) $(2a/3, 2b/3)$ 3. $a^4/6, (2a/5, 2a/5)$

    b) $(a/2, 2b/3)$ d) $\left(\dfrac{3a^3 + 2ab^2}{4a^2 + 4b^2}, \dfrac{2a^2 b + 3b^3}{4a^2 + 4b^2} \right)$

5. $k\pi a^2/4, (4a/3\pi, 4a/3\pi)$ 7. $6k/35, (275/432, 275/432)$

9. $k\pi r^3$ 11. $I_x = kab^4/4, I_y = ka^3 b^2/6$

                                           $(\bar{\bar{x}}, \bar{\bar{y}}) = (a^2/3, b^2/2)$

13. $I_x = 64k/3, I_y = 32k/3$ 15. $\dfrac{kbh}{12}\ (h^2 + 3b^2)$

     $(\bar{\bar{x}}, \bar{\bar{y}}) = (4/3, 8/3)$

                                          19. $\dfrac{k\pi r^2}{4}\ (r^2 + 4a^2)$

SECTION 16.5

1. 18

3. 729/4

5. 128/15

7. 256/15

9. $4\pi r^3/3$

11. 256/15

13.

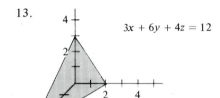

$3x + 6y + 4z = 12$

15.

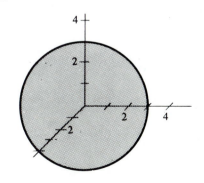

$x^2 + z^2 = 1$

17. 3/2

19. $(2b/3, 2b/3, b/2)$

21. $(0,0, 3h/4)$

23. $(0,0, 3r/8)$

25. $2kb^5/3$

27. $kb^5/6$

29. $k\pi ha^4/2$

SECTION 16.6

1. 16/3

3. $\pi^2/16$

5. 1/8

7. $(\pi/6)(5^{3/2} - 1)$

9. $2\pi a(a - \sqrt{a^2 - b^2})$

11.

13.

15.

17. $\pi r_0^2 h/3$

19. $h/5$

23. $\dfrac{4a^3}{3}\left(\dfrac{\pi}{2}-\dfrac{2}{3}\right)$

25. $k\pi^2 a^4/4$

27. $(0,0,3r/8)$

31. $k\pi/192$

33. a) $\displaystyle\int_0^{\pi/2}\int_0^4\int_0^{\sqrt{16-r^2}} r^2\, dz\, dr\, d\theta$

b) $\displaystyle\int_0^{\pi/2}\int_0^{\pi/6}\int_0^4 \rho^3\,\sin^2\phi\, d\rho\, d\phi\, d\theta + \int_0^{\pi/2}\int_{\pi/6}^{\pi/2}\int_4^{2\csc\phi} \rho^3\,\sin^2\phi\, d\rho\, d\phi\, d\theta$

35. a) $\displaystyle\int_0^{2x}\int_0^a\int_0^{a+\sqrt{a^2-r^2}} r^2\,\cos\theta\, dz\, dr\, d\theta$

b) $\displaystyle\int_0^{\pi/4}\int_0^{2\pi}\int_0^{2a\csc\phi} \rho^3\,\sin^2\phi\,\cos\theta\, d\rho\, d\theta\, d\phi$

MISCELLANEOUS EXERCISES

1. $29/6$

3. 36

5. $27/5$

7. $\displaystyle\int_0^3\int_0^{(3-x)/3} f(x,y)\, dy\, dx,\ \int_0^1\int_0^{3-3y} f(x,y)\, dx\, dy,\ 3/2$

9. $\displaystyle\int_{-5}^3\int_{-\sqrt{25-x^2}}^{\sqrt{25-x^2}} f(x,y)\, dy\, dx,$

$\displaystyle\int_{-5}^{-4}\int_{-\sqrt{25-y^2}}^{\sqrt{25-y^2}} f(x,y)\, dx\, dy + \int_{-4}^4\int_{-\sqrt{25-y^2}}^3 f(x,y)\, dx\, dy + \int_4^5\int_{-\sqrt{25-y^2}}^{\sqrt{25-y^2}} f(x,y)\, dx\, dy,$

$\dfrac{25\pi}{2} + 12 + 25\,\text{Arcsin}\,\dfrac{3}{5}$

11. $4/3$

13. $3296/15$

15. $\dfrac{\pi}{6}(65^{3/2}-1)$

17. $abc\,(a^2+b^2+c^2)/3$

19. $\dfrac{16}{3}\left(\dfrac{\pi}{2}-\dfrac{2}{3}\right)$

21. The integral represents the volume of a torus formed by revolving a circle of radius 3 and center at $(0,3,0)$ about the z-axis.

23. $\dfrac{h^3}{6}\,[\ln|\sqrt{2}+1|+\sqrt{2}\,]$

25. $324\pi/5$

CHAPTER 17

SECTION 17.1

1. $\dfrac{2\sqrt{2}}{3}$

3. $\dfrac{2(1 + \sqrt{2})}{3}$

5. $\dfrac{\pi}{2}$

7. 25

9. $\dfrac{63}{2}$

11. $\dfrac{316}{3}$

13. $\dfrac{35}{6}$

15. 2

17. $-\dfrac{17}{15}$

19. -66

21. 0

23. $-10\pi^2$

25. $\displaystyle\int_C \mathbf{F} \cdot d\mathbf{R} = \int_C (-2t\mathbf{i} - t\mathbf{j}) \cdot (\mathbf{i} - 2\mathbf{j})\, dt = 0$

27. $\displaystyle\int_C \mathbf{F} \cdot d\mathbf{R} = \int_C \left[(t^3 - 2t^2)\mathbf{i} + \left(t - \dfrac{t^2}{2} \right)\mathbf{j} \right] \cdot (\mathbf{i} + 2t\mathbf{j})\, dt = 0$

29. $\dfrac{2\sqrt{13}\,\pi}{3}(27 + 64\pi^2) \approx 4973.8$

True or False

1. False, $\displaystyle\int_C xy\, ds = \sqrt{2}\int_0^1 t^2\, dt$

2. True

3. False, the directions are different
4. True

5. False

SECTION 17.2

1. $\dfrac{11}{15}$

3. $-\ln(2 + \sqrt{3})$

5. 1

7. $\displaystyle\int_{C_1} \mathbf{F} \cdot d\mathbf{R} = 0, \int_{C_2} \mathbf{F} \cdot d\mathbf{R} = -\dfrac{1}{3}, \int_{C_3} \mathbf{F} \cdot d\mathbf{R} = -\dfrac{1}{2}$

9. $\displaystyle\int_{C_1} \mathbf{F} \cdot d\mathbf{R} = 64, \int_{C_2} \mathbf{F} \cdot d\mathbf{R} = 0, \int_{C_3} \mathbf{F} \cdot d\mathbf{R} = 0$

11. $\displaystyle\int_{C_1} \mathbf{F} \cdot d\mathbf{R} = \dfrac{64}{3}, \int_{C_2} \mathbf{F} \cdot d\mathbf{R} = \dfrac{64}{3}$

13. $U = x^2 - 3xy + y^3 + C$

15. $U = e^{x^2 y} + C$

17. $U = \ln\sqrt{x^2 + y^2} + C$

19. Not conservative

21. $U = e^x \sin y + 2y + C$

23. 24

25. -1

27. 0

29. 30, 366

31. 2

33. 11

True or False

1. False, C_1 and C_2 have opposite directions
2. False
3. True, $U = xy$, $(a_1, b_1) = (0, 3)$, $(a_2, b_2) = (0, -3)$, $\displaystyle\int_C \mathbf{F} \cdot d\mathbf{R} = U(0, 3) - U(0, -3) = 0$

4. True
5. False, the requirement is $\dfrac{\partial M}{\partial y} = \dfrac{\partial N}{\partial x}$

SECTION 17.3

1. 0

3. 32/15

5. 4/3

7. 56

9. 32/3

11. $\dfrac{16a^3}{3}$

13. 0

15. 1/12

17. πa^2

19. 32/3

23. $(0, 4a/3\pi)$

25. $(b/3, c/3)$

27. $3\pi a^2/2$

29. $\pi - (3\sqrt{3}/2)$

SECTION 17.4

1. a) 4
   b) $2\mathbf{j} - \mathbf{k}$

3. a) 0
   b) $-2\mathbf{k}$

5. a) $\dfrac{2y}{x^2 + y^2}$

   b) $\dfrac{2x}{x^2 + y^2}\mathbf{k}$

7. a) $\cos(x - y) + \cos(y - z) + \cos(z - x)$
   b) $\cos(y - z)\mathbf{i} + \cos(z - x)\mathbf{j} + \cos(x - y)\mathbf{k}$

9. a) $z\mathbf{j} + y\mathbf{k}$
   b) 0

11. 2π

13. 0

15. 1

17. $a^5/4$

19. πh^2

21. $3a^4$

23. 32π

29. $U = xye^z + C$

31. $U = \dfrac{x}{y} + z^2 - z + C$

MISCELLANEOUS EXERCISES

1. a) $6\sqrt{2}$
   b) 128π

3. a) 35/2
   b) 18π

5. 5/7

7. $2\pi^2$

9. 64/3

11. Not conservative

13. $U = 3x^2 y^2 - x^3 + y^3 - 7y + C$

15. Not conservative

17. $U = \dfrac{x}{yz} + C$

19. $\displaystyle\int_C \mathbf{F}\cdot d\mathbf{R} = 2\pi$, even though $\mathbf{F}$ is conservative, $\mathbf{F}$ is undefined at a point within the region enclosed by C.

21. 4

23. 1/12

25. a) $div\ \mathbf{F} = 2x + 2y + 2z$
    b) $\mathbf{curl\ F} = 0$

27. a) $div\ \mathbf{F} = -y\sin x - x\cos y + xy$
    b) $\mathbf{curl\ F} = xz\mathbf{i} - yz\mathbf{j}$

29. $-2a^6/5$

CHAPTER 18

SECTION 18.1

1. $2, 4, 8, 16, 32$

3. $-\dfrac{1}{2}, \dfrac{1}{4}, -\dfrac{1}{8}, \dfrac{1}{16}, -\dfrac{1}{32}$

5. $3, \dfrac{9}{2}, \dfrac{27}{6}, \dfrac{81}{24}, \dfrac{243}{120}$

7. $-1, -\dfrac{1}{4}, \dfrac{1}{9}, \dfrac{1}{16}, -\dfrac{1}{25}$

9. $3n - 2$

11. $n^2 - 2$

13. $\dfrac{n+1}{n+2}$

15. $\dfrac{(-1)^{n-1}}{2^{n-2}}$

17. $\dfrac{n+1}{n}$

19. $\dfrac{n}{(n+1)(n+2)}$

21. $\dfrac{2^n n!}{(2n)!}$

23. $(-1)^{n(n-1)/2}$

25. Converges to 1

27. Diverges

29. Converges to 3/2

31. Diverges

33. Converges to 0

35. Diverges

37. Converges to 0

39. Diverges

41. Converges to 3

43. Converges to e^k

45. Converges to 0

True or False

1. False, let $b_n = \dfrac{4n+1}{n}$ and $a_n = 2 + (-1)^n$

2. True

3. True

4. True

5. True

6. True

7. True

8. False, let $a_n = (-1)^n$ and $b_n = (-1)^{n+1}$

9. True

10. True

SECTION 18.2

1. 1, 1.25, 1.361, 1.424, 1.464

3. 3, -1.5, 5.25, -4.875, 10.3125

5. 3, 4.5, 5.25, 5.625, 5.8125

7. Use Theorem 18.4. $\lim\limits_{n \to \infty} a_n = \lim\limits_{n \to \infty} \dfrac{n}{n+1} = 1 \neq 0$

9. Use Theorem 18.4. $\lim\limits_{n \to \infty} a_n = \lim\limits_{n \to \infty} \dfrac{2^n - 1}{2^{n+1}} = \dfrac{1}{2} \neq 0$

11. Use Theorem 18.4. $\lim\limits_{n \to \infty} a_n = \lim\limits_{n \to \infty} \dfrac{2^n + 1}{2^{n+1}} = \dfrac{1}{2} \neq 0$

13. Use Theorem 18.5. $a_n = 2\left(\dfrac{3}{4}\right)^n$, $a = 2$, $r = \dfrac{3}{4} < 1$

15. Use Theorem 18.5. $a_n = \left(\dfrac{9}{10}\right)^n$, $a = 1$, $r = \dfrac{9}{10} < 1$

17. Use telescoping series. $a_n = \dfrac{1}{n} - \dfrac{1}{n+1}$

19. $.66\overline{6} = \sum\limits_{n=0}^{\infty} \dfrac{3}{5}(.1)^n = \dfrac{2}{3}$

21. $.075\overline{75} = \sum\limits_{n=0}^{\infty} \dfrac{3}{40}(.01)^n = \dfrac{5}{66}$

23. 8

25. 3/4

27. 1/6

29. Converges

31. Converges

33. Diverges

35. Diverges

37. Diverges

39. Diverges

41. 136/3 ft

True or False

1. False

2. True

3. True

4. False, the series must begin with $n = 0$ in order for the limit to be $\dfrac{a}{1 - r}$

5. True

SECTION 18.3

1. Diverges

3. Converges

5. Converges

7. Diverges

9. Converges

11. Converges

13. Diverges

15. Converges

17. Converges

19. Converges

21. $R_6 = 0.167$, $1.491 < \sum\limits_{n=1}^{\infty} \dfrac{1}{n^2} < 1.658$

23. $N = 7$

25. $R_{10} = 0.100$, $0.982 < \sum\limits_{n=1}^{\infty} \dfrac{1}{n^2 + 1} < 1.082$

True or False

1. True
2. True
3. False; $f(x) = \dfrac{1}{x-3}$ is discontinuous at $x = 3$

4. True
5. False; $f(x) = e^{-x} \sin x$ is not decreasing

SECTION 18.4

1. Converges
7. Converges
13. Diverges

3. Diverges
9. Diverges
15. Converges

5. Diverges
11. Converges
17. Converges

19. Diverges

21. Diverges, p-series, $p = 1/2$

23. Converges, compare with $\displaystyle\sum_{n=1}^{\infty}\left(\frac{1}{3}\right)^n$

25. Diverges, $\displaystyle\lim_{n\to\infty} a_n = 1/2 \neq 0$

27. Converges, Integral Test

29. $p > 1$

31. 0.000023

33. Diverges

35. Converges

True or False

1. False
2. False

3. True
4. True

5. True

SECTION 18.5

1. Converges
7. Diverges
13. Converges
19. Converges

3. Converges
9. Converges
15. Converges
21. Converges

5. Diverges
11. Converges
17. Converges

23. a) Converges and converges absolutely
    b) Converges, but does not converge absolutely
    c) Converges, but does not converge absolutely

25. .368

27. .405

31. Diverges (Ratio Test)

33. Converges (Limit Comparison Test with $b_n = 1/2^n$)

35. Converges (Alternating Series Test)

37. Converges (Converges absolutely by comparison with $b_n = 1/2^n$)

39. Converges (Ratio Test)

41. Converges (Ratio Test)

True or False

1. False
2. True

3. True
4. False, let $a_n = (-1)^n/n$

5. False, let $a_n = (-1)^n/n$

SECTION 18.6

1. $-2 < x < 2$

3. $-1 < x \leqslant 1$

5. $-\infty < x < \infty$

7. $x = 0$

9. $-4 < x < 4$

11. $0 < x \leqslant 10$

13. $0 < x \leqslant 2$

15. $0 < x < 2c$

17. $-1 \leqslant x \leqslant 1$

19. $-\dfrac{1}{2} < x < \dfrac{1}{2}$

21. $-\infty < x < \infty$

23. $-1 < x < 1$

25. a) $-2 < x < 2$
    b) $-2 < x < 2$
    c) $-2 \leqslant x < 2$

27. a) $0 < x \leqslant 2$
    b) $0 < x < 2$
    c) $0 < x < 2$
    d) $0 \leqslant x \leqslant 2$

29. a) $-\infty < x < \infty$
    b) $\displaystyle\sum_{n=0}^{\infty} \frac{(-1)^n x^{2n+1}}{(2n+1)(2n+1)!}$
    c) $.944$

True or False

1. True
2. False, let $a_n = (-1)^n/n2^n$

3. True
4. True

5. True

SECTION 18.7

1. $\displaystyle\sum_{n=0}^{\infty} \frac{x^n}{2^{n+1}}, \quad -2 < x < 2$

3. $\displaystyle\sum_{n=0}^{\infty} \frac{(x-5)^n}{(-3)^{n+1}}, \quad 2 < x < 8$

5. $-3 \displaystyle\sum_{n=0}^{\infty} (2x)^n, \quad -\dfrac{1}{2} < x < \dfrac{1}{2}$

7. $-\dfrac{1}{11} \displaystyle\sum_{n=0}^{\infty} \left[\dfrac{2}{11}(x+3)\right]^n, \quad -\dfrac{17}{2} < x < \dfrac{5}{2}$

9. $\dfrac{3}{2} \displaystyle\sum_{n=0}^{\infty} \left(\dfrac{x}{-2}\right)^n, \quad -2 < x < 2$

11. $\displaystyle\sum_{n=0}^{\infty} \left[\dfrac{1}{(-2)^n} - 1\right] x^n, \quad -1 < x < 1$

13. $2 \displaystyle\sum_{n=0}^{\infty} x^{2n}$

15. $\displaystyle\sum_{n=1}^{\infty} n(-1)^n x^{n-1}, \quad -1 < x < 1$

17. $\displaystyle\sum_{n=0}^{\infty} \frac{(-1)^n x^{n+1}}{n+1}, \quad -1 < x \leqslant 1$

19. $\displaystyle\sum_{n=0}^{\infty} (-1)^n (2x)^{2n}, \quad -\dfrac{1}{2} < x < \dfrac{1}{2}$

21. $\displaystyle\sum_{n=0}^{\infty} \frac{x^{2n+1}}{2n+1}$, $-1 < x < 1$

23.

x	$x - \dfrac{x^2}{2}$	$\ln(x+1)$	$x - \dfrac{x^2}{2} + \dfrac{x^3}{3}$
0.0	.000	.000	.000
0.2	.180	.182	.183
0.4	.320	.336	.341
0.6	.420	.470	.492
0.8	.480	.588	.651
1.0	.500	.693	.833

SECTION 18.8

1. $\displaystyle\sum_{n=0}^{\infty} \frac{(2x)^n}{n!}$

3. $\displaystyle\sum_{n=0}^{\infty} \frac{(-1)^n x^{4n}}{(2n)!}$

5. $\displaystyle\sum_{n=0}^{\infty} \frac{(-1)^n (2x)^{2n+1}}{(2n+1)!}$

7. $1 + \dfrac{x^2}{2!} + \dfrac{5x^4}{4!} + \left(\dfrac{61x^6}{6!} + \cdots \right)$

9. $\displaystyle\sum_{n=0}^{\infty} (-1)^n (n+1) x^n$

11. $\dfrac{1}{2}\left[1 + \displaystyle\sum_{n=1}^{\infty} \frac{(-1)^n 1\cdot 3\cdot 5 \cdots (2n-1) x^{2n}}{2^{3n} n!} \right]$

13. $\displaystyle\sum_{n=0}^{\infty} \frac{(-1)^n x^{2n+1}}{(2n+1)!}$

15. $\displaystyle\sum_{n=0}^{\infty} \frac{(-1)^n x^{2n}}{2^n n!}$

17. $\displaystyle\sum_{n=0}^{\infty} \frac{x^{2n+1}}{(2n+1)!}$

21. $\displaystyle\sum_{n=0}^{\infty} \frac{(-1)^n x^{2n}}{(2n+1)!}$

23. 1.3708

25. 0.7040

27. 0.4872

29. 0.3413

31. $-0.394 < x < 0$

33. 0.00012

MISCELLANEOUS EXERCISES

1. $a_n = \dfrac{1}{n!}$

3. Converges to 0

5. Diverges

7. Converges to 0

9. Converges to 0

11. 1, 2.5, 4.75, 8.125, 13.3875

13. .5, .45833, .45972, .45970, .45970

15. 3

17. 1/2

19. 1/11

21. Diverges

23. Converges

25. Diverges

27. Diverges

29. Converges

31. $1 \leqslant x \leqslant 3$

33. $x = 2$

35. $\dfrac{\sqrt{2}}{2} \displaystyle\sum_{n=0}^{\infty} \dfrac{(-1)^{n(n+1)/2} \left(x - \dfrac{3\pi}{4}\right)^n}{n!}$

37. $\displaystyle\sum_{n=0}^{\infty} \dfrac{(x \ln 3)^n}{n!}$

39. a) $P(x) = x + \dfrac{x^3}{6}$

b)

x	-1	$-.75$	$-.5$	$-.25$	0	.25	.5	.75	1
$f(x)$	-1.571	$-.848$	$-.524$	$-.253$	0	.253	.524	.848	1.571
$P(x)$	-1.167	$-.820$	$-.521$	$-.253$	0	.253	.521	.820	1.167

c)

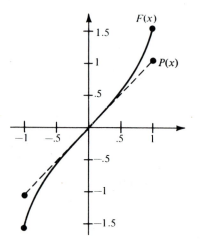

41.

CHAPTER 19

SECTION 19.1

Type	Order	Degree
1. Ordinary	1	1
3. Ordinary	2	1
5. Ordinary	4	1
7. Ordinary	2	2
9. Partial	1	1
11. Ordinary	2	2

13. $y = Ce^{4x}, y' = 4Ce^{4x} = 4y$

15. $y = C_1 \cos x + C_2 \sin x$
$y' = -C_1 \sin x + C_2 \cos x$
$y'' = -C_1 \cos x - C_2 \sin x$
$y'' + y = 0$

17. $u = e^{-t} \sin bx, \dfrac{\partial u}{\partial t} = -e^{-t} \sin bx,$

$\dfrac{\partial u}{\partial x} = be^{-t} \cos bx,$

$\dfrac{\partial^2 u}{\partial x^2} = -b^2 e^{-t} \sin bx,$

$b^2 \dfrac{\partial u}{\partial t} = -b^2 e^{-t} \sin bx$

19. a) is not a solution
b) is a solution
c) is a solution
d) is not a solution
e) is a solution

21. $y = 3e^{-2x}$

23. $y = 2 \sin 3x - \dfrac{1}{3} \cos 3x$

25. $y = -2x + \dfrac{1}{2}x^3$

27. $C = 0$: Point $(0, 0)$
$C = 1$: Circle of radius 1
$C = 4$: Circle of radius 2

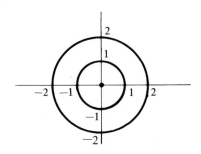

True or False

1. True
2. True
3. True

4. False, $y = e^x$ is a solution to $y' = y$ but $y = e^x + C$ is not a solution

5. False, the system $1 = C_1 + C_2$ and $1 = -C_1 - C_2$ has no solution

SECTION 19.2

1. $y^2 - x^2 = c$

3. $y = C(x^2 + 4x + 4)$

5. $y^2 = C - 2 \cos x$

7. $y^2 = 2e^x + 14$

9. $y = e^{-(x^2 + 2x)/2}$

11. $\dfrac{y - x}{1 + xy} = \sqrt{3}$

13. $P = P_0 e^{kt}$

15. Homogeneous of degree 3

17. Not homogeneous

19. Homogeneous of degree 0

21. $x = C(x - y)^2$

23. $y^2 + 2xy - x^2 = C$

25. $y = Ce^{-x^2/2y^2}$

27. $e^{y/x} = C + 2 \ln x$

29. $x = e^{\sin(y/x)}$

31. $9x^2 + 16y^2 = 25$

33. a) $A = Pe^{kt}$
    b) $A = \$3,004.17$
    c) $t = 6.3$ years

35. 98.9% of the original amount

37. $t = 119.7$ min

39. $y = Kx$

41. $x^2 + 2y^2 = K$

43. $2x^2 + 3y^2 = K$

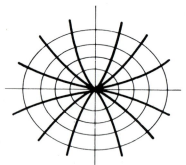

Curves: $y^2 = Cx^3$
Ellipses: $2x^2 + 3y^2 = K$

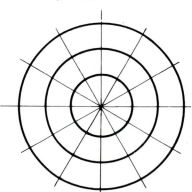

Circles: $x^2 + y^2 = C$
Lines: $y = Kx$

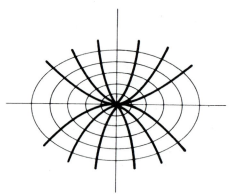

Parabolas: $x^2 = Cy$
Ellipses: $x^2 + 2y^2 = K$

True or False

1. False, the differentials dy and dx have not been separated
2. True, $y' = xy - 2y + x - 2 = (x - 2)(y + 1)$
3. True, $y' = \sin(x + y) + \sin(x - y) = 2 \sin x \cos y$
4. False
5. True, for $x^2 + y^2 = 2Cy$, $y' = -2xy/(y^2 - x^2)$ and for $x^2 + y^2 = 2Kx$, $y' = (y^2 - x^2)/2xy$

SECTION 19.3

1. $x^2 - 3xy + y^2 = C$

3. $3xy^2 + 5x^2y^2 - 2y = C$

5. Not exact

7. $\text{Arctan}\dfrac{x}{y} = \dfrac{\pi}{4}$

9. Not exact

11. $y \ln(x - 1) + y^2 = 16$

13. Not exact

15. $x^2 \tan y + 5x = C$

17. $3x^2y + y^3 = 8$

True or False

1. True, $\dfrac{\partial M}{\partial y} = \dfrac{\partial N}{\partial x} = 0$

2. False, $\dfrac{\partial M}{\partial y} = 2x, \dfrac{\partial N}{\partial x} = -2x$

3. True

4. False, $y \, dx + x \, dy = 0$ is exact but $xy \, dx + x^2 dy = 0$ is not exact

5. True

SECTION 19.4

1. Integrating factor: $1/y^2$
    General solution: $\dfrac{x}{y} - 6y = C$

3. Integrating factor: $1/x^2$
    General solution: $\dfrac{y}{x} + 5x = C$

5. Integrating factor: $\cos x$
    General solution:
    $y \sin x + x \sin x + \cos x = C$

7. Integrating factor: $1/y$
    General solution: $xy - \ln y = C$

9. Integrating factor: $1/\sqrt{y}$
    General solution:
    $x\sqrt{y} + \cos\sqrt{y} = C$

11. Integrating factor: xy^2
    General solution:
    $x^4y^3 + x^2y^4 = C$

13. Integrating factor: $1/x^2y^3$
    General solution: $\dfrac{y^2}{x} + \dfrac{x}{y^2} = C$

    (or any integrating factor of the
    form $x^m y^n$, where $2m + n = -7$,
    may be used)

15. $2y \, dy + \dfrac{x \, dy - y \, dx}{x^2 + y^2} = C$

    $y^2 + \text{Arctan}\dfrac{y}{x} = C$

17. $\dfrac{x \, dx + y \, dy}{x^2 + y^2} - \dfrac{1}{x^2} \, dx = 0$

    $\ln\sqrt{x^2 + y^2} + \dfrac{1}{x} = C$

19. $(3x^2y^4 \, dx + 4x^3y^3 \, dy) - e^{-x} \, dx = 0$
    $x^3y^4 + e^{-x} = C$

21. $d[\sqrt{x^2 - y^2}] = \dfrac{x \, dx - y \, dy}{\sqrt{x^2 - y^2}}$

    $\sqrt{x^2 - y^2} + xy = C$

25. $x^2 + y^2 = C$

27. $2x^2y^4 + x^2 = C$

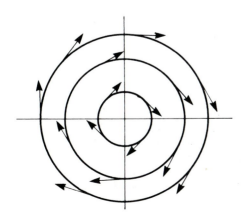

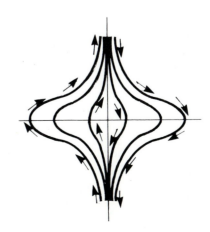

True or False

1. True
2. True
3. True

4. False, $e^{\int h(x)dx}$ is an integrating factor
5. False, let $\dfrac{M}{N} = \dfrac{y^3}{3xy^2}$ and $\dfrac{P}{Q} = \dfrac{y}{3x}$

SECTION 19.5

1. $y = x^2 + 2x + \dfrac{C}{x}$

3. $y = \dfrac{1}{2}e^x + e^{-x}$

5. $y = \dfrac{1}{2}(\sin x - \cos x) + e^x$

7. $y = -\dfrac{1}{13}(3\sin 2x + 2\cos 2x) + Ce^{3x}$

9. $y = \dfrac{x^3 - 3x + C}{3(x - 1)}$

11. $y = 1 + Ce^{-\tan x}$

13. $y = \sin x + (x + 1)\cos x$

15. $y = 2e^{x - x^2}$

17. $y = \dfrac{1}{-x^2 + Cx}$

19. $y^{2/3} = -(x^3 + \dfrac{9}{2}x^2 + \dfrac{27}{2}x + \dfrac{81}{4} + Ce^{2x/3})$

21. $I = \dfrac{E_0}{R} + Ce^{-Rt/L}$

23. a) $Q = 25e^{-t/20}$
    b) 10.2 min

25. a) 50 min
    b) 82.32 lb

27. $2e^x + e^{-2y} = C$

29. $x + xy^2 + \dfrac{1}{2}y^2 + 2y = C$

31. $y = Ce^{-\sin x} + 1$

33. $x^2y + \sin y = C$

35. $x^3y^2 + x^4y = C$

37. $y = \dfrac{e^x(x - 1) + C}{x^2}$

39. $x^4y^4 - 2x^2 = C$

41. $3\,\text{Arctan}\dfrac{x}{y} - y = C$

True or False

1. False
2. True
3. True

4. True
5. True

SECTION 19.6

1. $y = C_1 + C_2 e^x$

3. $y = C_1 e^{3x} + C_2 e^{-2x}$

5. $y = C_1 e^{x/2} + C_2 e^{-2x}$

7. $y = C_1 e^{-3x} + C_2 x e^{-3x}$

9. $y = C_1 e^{x/4} + C_2 x e^{x/4}$

11. $y = C_1 \sin x + C_2 \cos x$

13. $y = C_1 e^{3x} + C_2 e^{-3x}$

15. $y = e^x (C_1 \cos \sqrt{3} x + C_2 \sin \sqrt{3} x)$

17. $y = C_1 e^{(3 + \sqrt{5})x/2} + C_2 e^{(3 - \sqrt{5})x/2}$

19. $y = e^{2x/3} \left(C_1 \sin \dfrac{\sqrt{7} x}{3} + C_2 \cos \dfrac{\sqrt{7} x}{3} \right)$

21. $y = C_1 e^x + C_2 e^{-x} + C_3 \sin x + C_4 \cos x$

23. $y = C_1 e^x + C_2 e^{2x} + C_3 e^{3x}$

25. $y = C_1 e^x + e^x (C_2 \sin 2x + C_3 \cos 2x)$

27. $y = \dfrac{1}{11}(e^{6x} + 10 e^{-5x})$

29. $y = y_0 \cos \left(\sqrt{\dfrac{k}{m}}\, t \right)$

True or False

1. False, the general solution is $y = C_1 e^{3x} + C_2 x e^{3x}$
2. True
3. True, assuming that the coefficients $a_n, \ldots, a_1, a_0$ are real

4. True
5. False

SECTION 19.7

1. $y = C_1 e^x + C_2 e^{2x} + x + \dfrac{3}{2}$

3. $y = 6 \sin x + \cos x - 6x + x^3$

5. $y = C_1 + C_2 e^{-2x} + \dfrac{2}{3} e^x$

7. $y = (C_1 + C_2 x) e^{5x} + \dfrac{3}{8} e^x + \dfrac{1}{5}$

9. $y = -1 + 2 e^{-x} - \sin x - \cos x$

11. $y = C_1 \sin 3x + \left(C_2 - \dfrac{x}{6} \right) \cos 3x$

13. $y = C_1 e^x + C_2 x e^x + \left(C_3 + \dfrac{2x}{9} \right) e^{-2x}$

15. $y = \left(\dfrac{4}{9} - \dfrac{x^2}{2} \right) e^{4x} - \dfrac{1}{9}(1 + 3x) e^x$

17. $y = \left[C_1 + \dfrac{1}{4} \ln |\sin 2x| \right] \sin 2x + \left[C_2 - \dfrac{x}{2} \right] \cos 2x$

19. $y = C_1 e^x + C_2 x e^x + \dfrac{1}{4} x^2 e^x [\ln x^2 - 3]$

21. $y = \dfrac{1}{1200} [225 e^{-2t} - 201 e^{-14t}] + \dfrac{1}{200} [3 \sin 2t - 4 \cos 2t]$

23. $q = \dfrac{3}{25} [e^{-5t} + 5t e^{-5t} - \cos 5t]$

SECTION 19.8

1. $y = a_0 \displaystyle\sum_{n=0}^{\infty} \frac{x^n}{n!} = a_0 e^x$

3. $y = a_0 \displaystyle\sum_{n=0}^{\infty} \frac{(3x)^{2n}}{(2n)!} + \frac{a_1}{3} \displaystyle\sum_{n=0}^{\infty} \frac{(3x)^{2n+1}}{(2n+1)!} = C_0 \displaystyle\sum_{n=0}^{\infty} \frac{(3x)^n}{n!} + C_1 \displaystyle\sum_{n=0}^{\infty} \frac{(-3x)^n}{n!}$

$= C_0 e^{3x} + C_1 e^{-3x}$　Note: $C_0 + C_1 = a_0$ and $C_0 - C_1 = \dfrac{a_1}{3}$

5. $y = a_0 \displaystyle\sum_{n=0}^{\infty} \frac{(-1)^n (2x)^{2n}}{(2n)!} + \frac{a_1}{4} \displaystyle\sum_{n=0}^{\infty} \frac{(-1)^n (2x)^{2n+1}}{(2n+1)!} = C_0 \cos 2x + C_1 \sin 2x$

7. $y = a_2 x^2$

9. $y = a_0 + a_1 \displaystyle\sum_{n=1}^{\infty} \frac{x^n}{n!(n-1)!}$

11. $y = a_0 \left(1 - \dfrac{x^2}{8} + \dfrac{x^4}{128} - \ldots \right) + a_1 \left(x - \dfrac{x^3}{24} + \dfrac{7x^5}{1920} - \ldots \right)$

13. $y = 2 + \dfrac{2}{1!} x - \dfrac{2}{2!} x^2 - \dfrac{10}{3!} x^3 + \dfrac{2}{4!} x^4 + \ldots$　When $x = \dfrac{1}{2}, y \approx 2 + 1 - \dfrac{1}{4} - \dfrac{5}{24} + \dfrac{1}{96} = \dfrac{245}{96} \approx 2.552$

15. $y = 1 - \dfrac{3}{1!} x + \dfrac{2}{3!} x^3 - \dfrac{12}{4!} x^4 + \dfrac{16}{6!} x^6 - \dfrac{120}{7!} x^7$　When $x = \dfrac{1}{4}, y \approx 1 - \dfrac{3}{4} + \dfrac{1}{3(4^3)} - \dfrac{1}{2(4^4)} + \dfrac{1}{45(4^6)} - \dfrac{1}{42(4^7)} \approx .253$

MISCELLANEOUS EXERCISES

1. $y = x \ln x^2 + 2x^{3/2} + Cx$

3. $y = C(1-x)^2$

5. $y^2 = x^2 \ln x^2 + Cx^2$

7. $5x^2 + 8xy + 2x + \dfrac{5}{2} y^2 + 2y = C$

9. $y = Ce^{2x} - e^x$

11. $xy - 2xy^3 + x^2 = C$

13. $y = x \ln |x| - 2 + Cx$

15. $x + \ln |\ln|1 + y|| = C$

17. $y = e^x(1 + \tan x) + C \sec x$

19. $x^2 - 2xy - 10x - 3y^2 + 4y = C$

21. $y = \dfrac{1}{5} x^3 - x + C\sqrt{x}$

23. $y^2 = 2Cx + C^2$

25. $y^2 = x^2 - x + \dfrac{3}{2} + Ce^{-2x}$

27. $\dfrac{y - x}{1 + xy} = C$

29. $y = \dfrac{bx^4}{4 - a} + Cx^a$

31. $y = C_1 \sin x + C_2 \cos x - 5x + x^3$

33. $y = (C_1 + x)\sin x + C_2 \cos x$

35. $y = \left(C_1 + C_2 x + \dfrac{1}{3} x^3 \right) e^x$

37. Family of circles: $x^2 + (y - K)^2 = K^2$

39. $y = a_0 \displaystyle\sum_{n=0}^{\infty} \frac{x^n}{4^n}$

Index

FORMULAS FROM GEOMETRY

Triangle:

$h = a \sin\theta$

$\text{Area} = \frac{1}{2} bh$

(Law of Cosines)
$c^2 = a^2 + b^2 - 2ab \cos\theta$

Right Triangle:

(Pythagorean Theorem)
$c^2 = a^2 + b^2$

Equilateral Triangle:

$h = \frac{\sqrt{3}\,s}{2}$

$\text{Area} = \frac{\sqrt{3}\,s^2}{4}$

Parallelogram:

$\text{Area} = bh$

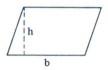

Trapezoid:

$\text{Area} = \frac{h}{2}(a+b)$

Circle:

$\text{Area} = \pi r^2$
$\text{Circumference} = 2\pi r$

Sector of Circle:

(θ in radians)

$\text{Area} = \frac{\theta r^2}{2}$

$s = r\theta$

Circular Ring:

(p = average radius,
w = width of ring)
$\text{Area} = \pi(R^2 - r^2)$
$\quad = 2\pi p w$

Sector of Circular Ring:

(p = average radius,
w = width of ring,
θ in radians)
$\text{Area} = \theta p w$

Ellipse:

$\text{Area} = \pi ab$

$\text{Circumference} \approx 2\pi \sqrt{\dfrac{a^2+b^2}{2}}$

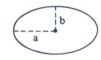

Cone:

(A = area of base)

$\text{Volume} = \frac{Ah}{3}$

Right Circular Cone:

$\text{Volume} = \frac{\pi r^2 h}{3}$

$\text{Lateral Surface Area} = \pi r\sqrt{r^2 + h^2}$

Frustum of Right Circular Cone:

$\text{Volume} = \frac{\pi(r^2 + rR + R^2)h}{3}$

$\text{Lateral Surface Area} = \pi s(R+r)$

Right Circular Cylinder:

$\text{Volume} = \pi r^2 h$
$\text{Lateral Surface Area} = 2\pi rh$

Sphere:

$\text{Volume} = \frac{4}{3}\pi r^3$

$\text{Surface Area} = 4\pi r^2$

Wedge:

(A = area of upper face,
 B = area of base)

$A = B \sec\theta$

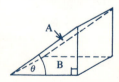